HANDBOOK of MICROBIOLOGY

Volume I
Organismic Microbiology

EDITORS

Allen I. Laskin, Ph.D.
Esso Research and Engineering Company
Linden, New Jersey

Hubert A. Lechevalier, Ph.D.
Institute of Microbiology
Rutgers University
New Brunswick, New Jersey

Published by

A DIVISION OF
THE CHEMICAL RUBBER CO.
18901 Cranwood Parkway • Cleveland, Ohio 44128

HANDBOOK OF MICROBIOLOGY

Volume I: Organismic Microbiology

International Standard Book Number (ISBN)

Complete Set 0-87819-580-7
Volume I 0-87819-581-5

Library of Congress Catalog Card Number 72-88766

HANDBOOK SERIES

Handbook of Chemistry and Physics, 53rd edition
Standard Mathematical Tables, 20th edition
Handbook of tables for Mathematics, 4th edition
Handbook of tables for Organic Compound Identification, 3rd edition
Handbook of Biochemistry, selected data for Molecular Biology, 2nd edition
Handbook of Clinical Laboratory Data, 2nd edition
Manual for Clinical Laboratory Procedures, 2nd edition
Manual of Nuclear Medicine Procedures, 1st edition
Handbook of Laboratory Safety, 2nd edition
Handbook of tables for Probability and Statistics, 2nd edition
Fenaroli's Handbook of Flavor Ingredients, 1st edition
Handbook of Analytical Toxicology, 1st edition
Manual of Laboratory Procedures in Toxicology, 1st edition
Handbook of tables for Applied Engineering Science, 2nd edition
Handbook of Chromatography, 1st edition
Handbook of Environmental Control, 1st edition
Handbook of Food Additives, 2nd edition
Handbook of Radioactive Nuclides, 1st edition
Handbook of Lasers, 1st edition
Handbook of Microbiology, 1st edition
Handbook of Engineering in Medicine and Biology, 1st edition
Atlas of Spectral Data and Physical Constants for Organic Compounds, 1st edition
*Handbook of Marine Sciences, 1st edition
*Handbook of Material Science, 1st edition
*Handbook of Spectroscopy, 1st edition

*Currently in preparation.

CRC PRESS

PREFACE

Microbiology is a tree with many branches. The work of a Microbiologist may touch faraway fields and the diversity of information needed is great. It has thus been impossible to bind the pages of the *CRC Handbook of Microbiology* within a single cover. This first volume of the *Handbook of Microbiology* contains information dealing with microorganisms themselves. With the dedicated assistance of our Advisory Board and Contributors, the Editors have assembled information on various groups of microorganisms: bacteria, fungi, algae, protozoa, and viruses. In addition, information of a general nature is included, and also information that is most likely to be valuable to those interested in the organismic aspects of microbiology.

The guidelines given the authors by the Editors were very few. They were asked only to be as brief as possible, consistent with giving meaningful information, and to present their data, as much as practicable, in the form of tables. As a result, there is a lack of uniformity from one presentation to another. The Editors feel that there is virtue in this diversity, especially for the first edition of such a handbook. Experience will show us which type of presentation is most useful, and an attempt at more uniformity might be desirable in future editions. For this, the Editors must depend on the users of the Handbook, and all constructive comments, suggestions, and criticisms will be highly appreciated.

The Editors thank the Advisory Board and all the authors for their unselfish labors and express their gratitude to Mrs. Lisbeth Hammer for her excellent editorial work and to Mrs. Verna Lepping for the accuracy and intelligence of her secretarial assistance.

A. I. Laskin
H. A. Lechevalier
New Jersey, 1972

ADVISORY BOARD

Thomas B. Platt, Ph.D.
Bioanalytical Section
The Squibb Institute of Medical Research
New Brunswick, New Jersey

Otto J. Plescia, Ph.D.
Institute of Microbiology
Rutgers University
New Brunswick, New Jersey

G. Pontecorvo, Ph.D.
Department of Cell Genetics
Imperial Cancer Research Fund
London, England

Chase Van Baalen, Ph.D.
Marine Science Institute
University of Texas
Port Aransas, Texas

Claude Vezina, Ph.D.
Microbiology Department
Ayerst Laboratories
St. Laurent, P.Q., Canada

L. C. Vining, Ph.D.
National Research Council
Atlantic Regional Laboratory
Halifax, N.S., Canada

E. D. Weinberg, Ph.D.
Department of Microbiology
Indiana University
Bloomington, Indiana

Burton I. Wilner, Ph.D.
Orinda, California

CONTRIBUTORS

Sheldon Aaronson, Ph. D.
Department of Biology
Queens College
City University of New York
Flushing, New York

Hans-Wolfgang Ackermann, M.D.
Department of Microbiology
Faculty of Medicine
Laval University
Quebec, P. Q., Canada

Roger A. Anderson, Ph.D.
Department of Biological Sciences
University of Denver
Denver, Colorado

H. L. Barnett, Ph.D.
Department of Plant Pathology and Bacteriology
Division of Plant Sciences
West Virginia University
Morgantown, West Virginia

Everett S. Beneke, Ph.D.
Department of Botany and Plant Pathology
Michigan State University
East Lansing, Michigan

Paul M. Borick, Ph.D.
Research Division
Ethicon, Inc.
Somerville, New Jersey

J. Michael Bowes, Ph.D.
Department of Bacteriology
University of California
Davis, California

G. Brochu, M.D.
Department of Microbiology
Faculty of Medicine
Laval University
Quebec, P. Q., Canada

Carl F. Clancy, Ph.D.
Jefferson Medical College
Thomas Jefferson University
Philadelphia, Pennsylvania

Rita R. Colwell, Ph.D.
Department of Microbiology
University of Maryland
College Park, Maryland

Dennis P. Cummings, Ph.D.
Research Division
Miles Laboratories, Inc.
West Haven, Connecticut

A. Keith Dunker, Ph.D.
Department of Molecular Biophysics and Biochemistry
Yale University
New Haven, Connecticut

Martin Dworkin, Ph.D.
Department of Microbiology
University of Minnesota
Minneapolis, Minnesota

M. S. Finstein, Ph.D.
Department of Environmental Sciences
Rutgers University
New Brunswick, New Jersey

Paul Fiset, M.D., Ph. D.
School of Medicine
University of Maryland
Baltimore, Maryland

Eugene R. L. Gaughran, Ph.D.
Research Center
Johnson & Johnson
New Brunswick, New Jersey

John J. Gavin, Ph.D.
Molecular Biology Department
Research Division
Miles Laboratories, Inc.
Elkhart, Indiana

Emma Gergely, M. L. S.
Cooper Laboratories, Inc.
Cedar Knolls, New Jersey

Ruth E. Gordon, Ph.D.
Institute of Microbiology
Rutgers University
New Brunswick, New Jersey

Albert Goze, M.D.
Laboratoire de Bacteriologie
Universitye Paul Sabatier
Toulouse, France

Michael L. Higgins, Ph.D.
Department of Microbiology
School of Medicine
Temple University
Philadelphia, Pennsylvania

Barry B. Hunter, Ph.D.
Department of Biology
California State College
California, Pennsylvania

Seymour H. Hutner, Ph.D.
Haskins Laboratories
Pace College
New York, New York

Corinne Johnson, Ph.D.
Department of Biochemistry
College of Dentistry
Brookdale Dental Center of
New York University
New York, New York

S. S. Kasatiya, D.V.M., D.Sc.
Department of Social Affairs
Government of Quebec
Laval, P. Q., Canada

S. P. Lapage, M.B., F.R.C.P., Dip. Bact.
National Collection of Type Cultures
Central Public Health Laboratory
London, England

Hubert A. Lechevalier, Ph.D.
Institute of Microbiology
Rutgers University
New Brunswick, New Jersey

Mary P. Lechevalier, M.S.
Institute of Microbiology
Rutgers University
New Brunswick, New Jersey

James D. Macmillan, Ph.D.
Department of Biochemistry and
Microbiology
College of Agriculture and Environmental Science
Rutgers University
New Brunswick, New Jersey

Karl Maramorosch, Ph.D.
Insect Physiology and Virology Program
Boyce Thompson Institute
Yonkers, New York

Daniel W. McNeil, B.Sc.
Research Center
Colgate-Palmolive Company
Piscataway, New Jersey

William F. Myers, Ph.D.
Department of Microbiology
School of Medicine
University of Maryland
Baltimore, Maryland

Pierre Nicolle, M.D. (retired)
Bacteriophage Department
Institut Pasteur
Paris, France

Leslie A. Page, Ph.D.
National Animal Disease Laboratory
U.S. Department of Agriculture
Ames, Iowa

Norbert Pfennig, Ph.D.
Institut für Microbiologie
Gesellschaft für Strahlen- und
Umweltforschung m. b. H.
Göttingen, Germany

Herman J. Phaff, Ph.D.
Department of Food Technology
College of Agriculture and Environmental Science
University of California
Davis, California

Leo Pine, Ph.D.
Department of Health, Education and Welfare
Center for Disease Control
Research and Development Unit
Atlanta, Georgia

Shmuel Razin, Ph.D.
Department of Clinical Microbiology
Hadassah Medical School
Hebrew University
Jerusalem, Israel

K. F. Redway, B.Sc.
National Collection of Type Cultures
Central Public Health Laboratory
London, England

Fred J. Roisen, Ph.D.
Department of Anatomy
College of Medicine and Dentistry
Rutgers Medical School
New Brunswick, New Jersey

Antonio H. Romano, Ph.D.
Microbiology Section
University of Connecticut
Storrs, Connecticut

V. B. D. Skerman, D.Sc.
Department of Microbiology
University of Queensland
St. Lucia, Queensland, Australia

Robert M. Smibert, Ph.D.
Anaerobe Laboratory
Virginia Polytechnic Institute and State University
Blacksburg, Virginia

Louis DS. Smith, Ph.D.
Anaerobe Laboratory
Virginia Polytechnic Institute and State University
Blacksburg, Virginia

James T. Staley, Ph.D.
Department of Microbiology
School of Medicine
University of Washington
Seattle, Washington

Ruth Ann Taber, M.S.
Department of Plant Sciences
Texas A & M University
College Station, Texas

Willard A. Taber, Ph.D.
Department of Biology
Texas A & M University
College Station, Texas

Virginia L. Thomas, Assoc. Gov't. Eng.
Department of Microbiology
College of Medicine and Dentistry
Rutgers Medical School
New Brunswick, New Jersey

Hans G. Trüper, Ph.D.
Department of Microbiology
University of Bonn
Bonn, Germany

Wolf V. Vishniac, Ph.D.
Department of Biology
University of Rochester
Rochester, New York

Burton I. Wilner, Ph.D.
Orinda, California

Robert L. Wiseman, Ph.D.
Department of Molecular Biophysics and Biochemistry
Yale University
New Haven, Connecticut

Charles L. Wisseman, Jr., M.D.
Department of Microbiology
School of Medicine
University of Maryland
Baltimore, Maryland

TABLE OF CONTENTS

BACTERIA

INTRODUCTION TO THE BACTERIA

DR. HUBERT LECHEVALIER

Bacteria are procaryotic organisms that, if photosynthetic, do not produce oxygen. Most bacteria are quite small, being rods, cocci or filaments that range from 0.5 to 1 μm in diameter. Since the resolution of the light microscope is of the order of 0.2 to 0.3 μm, it is easily understandable that no great progress was made in the cytology of bacteria before the introduction of the electron microscope and the development of allied methods of shadowing, thin-sectioning and staining.[1-3]

The following diagram is an attempt at schematically illustrating the various cytological features that can be recognized in bacteria. Of course, no single bacterium harbors all the features illustrated. Actual micrographs of most morphological features illustrated in the diagram can be found in the *Pictorial Atlas of Pathogenic Microorganisms,* edited by G. Henneberg.[4]

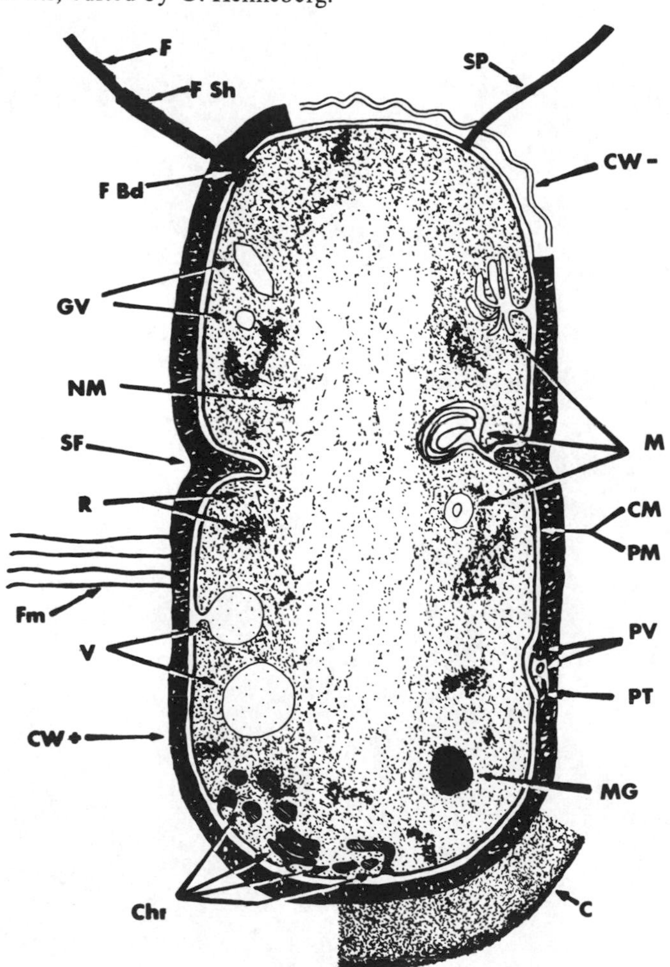

C	=	capsule
Chr	=	chromatophores
CM	=	cytoplasmic membrane
CW+	=	cell wall of Gram-positive bacteria
CW–	=	cell wall as in some Gram-negative bacteria
F	=	flagellum
FBd	=	flagellar basal body
Fm	=	fimbriae
FSh	=	flagellar sheath
GV	=	gas vacuoles
M	=	mesosome
MG	=	metachromatic granule
NM	=	nuclear material
PM	=	plasma membrane = CM
PT	=	periplasmic tubules
PV	=	periplasmic vesicles
R	=	ribosomes and polysomes
SF	=	septum formation
SP	=	sex pilus
V	=	vacuoles, probably containing reserve material

There is no rational order for presenting bacteria. Thus, the order that we have chosen is entirely empirical; the section starts with autotrophic bacteria and ends with heterotrophic rods and cocci not covered under other subdivisions.

REFERENCES

1. Hayat, M. A., *Principles and Techniques of Electron Microscopy: Biological Applications,* Vol. 1. Van Nostrand Reinhold Co., New York (1970).
2. Dawes, C. J., *Biological Techniques in Electron Microscopy.* Barnes & Noble, Inc., New York (1971).
3. Kay, D. H., *Techniques for Electron Microscopy,* 2nd ed. F. A. Davis Co., Philadelphia, Pennsylvania (1965).
4. Henneberg, G., *Pictorial Atlas of Pathogenic Microorganisms,* Vol. 3. Gustav Fischer Verlag, Stuttgart, Germany (1969).

CHEMOAUTOTROPHIC BACTERIA

DR. CORINNE JOHNSON AND DR. WOLF VISHNIAC

Chemoautotrophic bacteria possess the ability to grow by oxidation of inorganic compounds, fixing carbon dioxide as a sole source of carbon. A precise definition of this group has eluded microbiologists; an attempt to classify autotrophs by physiological criteria has been presented by Kelly.[27]

Chemoautotrophic bacteria may be conveniently divided into three groups: the hydrogen bacteria, the sulfur-oxidizing bacteria, and the nitrifying bacteria. The majority of the members of these three groups are aerobes; some species are able to reduce nitrate instead of O_2 and are facultative anaerobes; a few are obligate anaerobes, reducing CO_2. Most of the organisms are Gram-negative, but the occurrence of Gram-positive hydrogen bacteria and of at least one Gram-positive sulfur bacterium suggests the occurrence of autotrophy beyond the limits of easily definable taxonomic boundaries. The autotrophs, to the extent that they have been studied in detail, fix CO_2 chiefly by the reactions of the Calvin cycle.

The autotrophs interact with other microorganisms that participate in the sulfur, nitrogen, and iron cycles in nature. These ecological interactions have been discussed by Alexander[6] and by Brock.[8]

THE HYDROGEN BACTERIA

All of the hydrogen-oxidizing bacteria are facultative heterotrophs and are versatile in utilizing carbon compounds for growth. It is thought that the universal facultative heterotrophy occurs because in hydrogen oxidation NAD^+ is reduced spontaneously, as in the oxidation of organic substrates by heterotrophs. Such reduction of NAD^+ is carried out either by a single enzyme (*Pseudomonas ruhlandii* and *P. saccharophila*) or, more commonly, by a two-step reaction. Specialized respiratory chains for the oxidation of inorganic compounds coupled to reverse electron transport phosphorylation, as found in the thiobacilli and nitrifying bacteria, are not required. A recent classification of the hydrogen bacteria[10,11] emphasizes the capacities of these organisms to grow on organic substrates and their similarities to heterotrophic organisms that possess similar nutritional capabilities.

There is no unity of cell type in the hydrogen bacteria. Included among them are Gram-negative rods with peritrichous flagellation, Gram-negative rods with polar flagellation, and Gram-negative cocci. Also included in this group are the methane bacteria, which are obligate anaerobes and for which CO_2 is the terminal electron acceptor. As suggested by the survey of Belyayeva,[7] one of the hydrogen bacteria was identified as an *Achromobacter* species.[38] While sulfate-reducing bacteria can utilize hydrogen as an electron donor, they do not couple this process to CO_2 fixation; hence they are excluded from this group because of their inability to grow autotrophically. A number of actinomycetes[15,44] have shown themselves to be facultative autotrophs. In addition, Eberhardt[12] has described an unnamed Gram-positive short rod capable of autotrophic growth with hydrogen, oxygen and carbon dioxide. This obligately aerobic, non-sporulating, non-motile yellow organism is of unknown taxonomic affiliation.

As the ability to oxidize molecular hydrogen is sometimes lost in cultures that are grown heterotrophically, it has been suggested[10,11] that hydrogen bacteria be classified with the heterotrophic species they resemble when grown on carbon substrates. One of the most interesting physiological problems of the hydrogen bacteria is the existence of mixotrophy, i.e., the simultaneous utilization of hydrogen and an organic substrate for growth (for example, fructose or lactate) with concomitant fixation of carbon dioxide. The control of the synthesis of hydrogenase and ribulose diphosphate carboxylase under varying conditions of mixotrophy and conversion from growth on hydrogen to growth on an organic substrate is a complex problem in metabolic control, which has been explored only in general terms. The subject has been discussed by Rittenberg;[37] a more general treatment of the enzymology of hydrogen oxidation has been presented by Peck.[35]

Characteristics of Hydrogen Bacteria[10,11,15,38]

Species	DNA Base Composition, moles % G+C	Inorganic Substrates	Terminal Electron Acceptor	Facultative Heterotrophy	Cell Type
Alcaligenes					
eutrophus	66.3–66.8	H_2	O_2, NO_3^-	+	Gram-negative rods, peritrichous flagella
paradoxus	68–70	H_2	O_2	+	Gram-negative rods, long, fragile, peritrichous flagella
Paracoccus					
denitrificans	66.3–66.8	H_2	O_2, NO_3^-	+	Gram-negative cocci, non-motile
Pseudomonas					
facilis	61.7–63.8	H_2	O_2	+	Gram-negative rods, polar flagella
flava	67.3	H_2	O_2	+	Gram-negative rods, polar flagella
palleronii	66.8	H_2	O_2	+	Gram-negative rods, polar flagella
ruhlandii		H_2	O_2	+	Gram-negative rods, polar flagella
saccharophila	68.9	H_2	O_2	+	Gram-negative rods, polar flagella
Methanobacterium					
soehngenii		H_2	CO_2	*a*	Gram-variable rods, non-motile
Methanococcus sp.		H_2	CO_2	*a*	Gram-variable cocci, non-motile
Methanosarcina sp.		H_2	CO_2	*a*	Gram-variable cocci in regular cubical packages, non-motile
Mycobacterium					
phlei 134[b]		H_2	O_2	+	Gram-positive rods, non-motile, rarely branching or filamentous, acid-fast
Nocardia					
saturnea 71[b]		H_2	O_2	+	Gram-positive rods, cocci, or branching mycelia, acid-fast

Species	DNA Base Composition, moles % G+C	Inorganic Substrates	Terminal Electron Acceptor	Facultative Heterotrophy	Cell Type
Nocardia (continued)					
saturnea 99[b]		H_2	O_2	+	Gram-positive rods, cocci, or branching mycelia, acid-fast
petreophila 102[b]		H_2	O_2	+	Gram-positive rods, cocci, or branching mycelia, acid-fast
autotrophica 394[b]		H_2	O_2	+	Gram-positive rods, cocci, or branching mycelia, acid-fast
Streptomyces sp. 418[b]		H_2	O_2	+	Mycelia
Streptosporangium sp. 242[b]		H_2	O_2	+	Mycelia
Unidentified		H_2	O_2	+	Gram-positive short rods, non-motile, non-sporulating[12]

[a] Bryant *et al.*[9] have separated *Methanobacillus omelianskii* into two species of bacteria, one of which was capable of producing methane from H_2 and CO_2. It is now suggested that all the methane bacteria are capable of this reaction and thus grow autotrophically. Some species may, in addition, be able to decompose organic substrates; others may be obligate utilizers of 1-carbon compounds, namely CO_2, formate, and methanol.

[b] The strain numbers are given as listed in Reference 15; also see Reference 34.

THE THIOBACILLI

The Gram-negative thiobacilli are a group of organisms with physiological and morphological similarity and grow by oxidizing reduced sulfur compounds. They are similar to other Gram-negative rod-shaped bacteria in ultrastructure.[28,32,36,39] In addition, several species show polyhedral inclusion bodies (Figure 1), whose function is not understood. Most of the species are distinguished by physiological characters, which have been examined by numerical taxonomy,[17] and also by the measurement of DNA base ratios.[18] By these criteria, two species – *Thiobacillus neapolitanus* and *T. thioparus,* which resemble each other in appearance and behavior in culture – are shown to be of diverse evolutionary origin. Facultatively heterotrophic thiobacilli show complex relations between energy metabolism and carbon metabolism. This group includes *T. novellus, T. intermedius,* and two more recently described organisms: *Thiobacillus* A2,[45] and an as yet uncharacterized thiobacillus.[55] *Thiobacillus perometabolis* requires both organic matter and thiosulfate.[30]

Thermophilic thiobacilli have been described several times in the past, as reviewed by Zavarzin and Zhilina.[56] To this group should be added the organism described by Williams and Hoare;[55] the affiliation of still another thermophilic sulfur bacterium is uncertain. Egorova and Deryugina[13] have described a thermophilic, Gram-positive, spore-forming organism with polar flagellation, which they classified as *Thiobacillus thermophilica,* although its morphology would exclude it from this genus as it is presently described. Nevertheless, this finding suggests the occurrence of sulfur-oxidizing autotrophic bacteria in other taxa.

One species of thiobacillus, *T. ferrooxidans,* in addition to thiosulfate, also oxidizes ferrous iron according to the equation

$$4\,Fe^{++} + O_2 + 4\,H^+ \rightarrow 4\,Fe^{+++} + 2\,H_2O$$

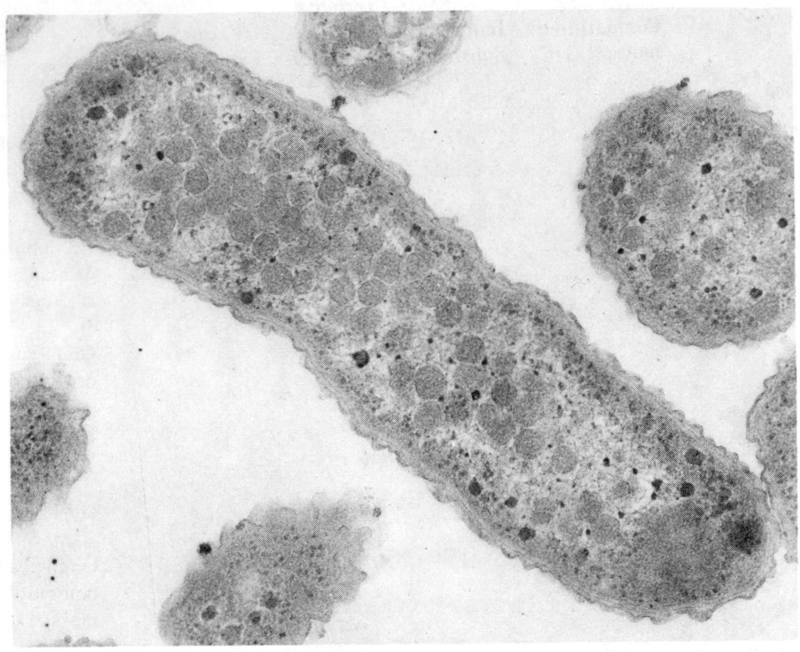

FIGURE 1. Electron Micrograph of a Thin Section of *Thiobacillus thioparus*. Cell structure typical of the thiobacilli with polyhedral inclusion bodies characteristic of several species. From Shively, J. M., Decker, G. L., and Greenawalt, J. W., *J. Bacteriol., 101,* 620 (1970). Reproduced by permission of the American Society for Microbiology, Washington, D.C.

and fixes CO_2 at the expense of this reaction. According to a personal communication (A. A. Imshenetsky), there also exists an antimony-oxidizing bacterium that may fall into this taxonomic group.

A variety of sulfur compounds are oxidized by the thiobacilli in nature: sulfide, sulfur, thiosulfate, and sulfite. Thiosulfate oxidation has been more extensively studied than that of the other sulfur compounds. During the oxidation of thiosulfate, both substrate-level phosphorylation and oxidative phosphorylation occur; the substrate-level phosphorylation occurs during the oxidation of the sulfite moiety.[14] One general scheme for the oxidation of sulfur compounds is presented below.[35] Some of the features of this scheme are subject to controversy, and reviews by Peck[35] and by Trudinger[46] should be consulted. A discussion of the comparative enzymology of sulfur oxidation in *T. thioparus* and other thiobacilli has recently been presented.[31]

$$SO_3^{=} \xrightarrow{\frac{1}{2}\,O_2;\ nADP} SO_4^{=} + nATP$$

$$S^{=} \rightarrow [RSSH] \rightarrow SO_3^{=} \xrightarrow[nADP]{\frac{1}{2}\,O_2;\ AMP} SO_4^{=} + nATP + ADP$$

$$S^{\circ} \rightarrow [RSSH] \qquad S_2O_3^{=} \rightleftarrows [RSSH] + SO_3^{=} \qquad S_4O_6^{=} + nATP \rightarrow SO_3^{=}$$

Characteristics of Thiobacilli[17,18]

Species	DNA Base Composition, moles % G+C	Inorganic Substrates[a]	Terminal Electron Acceptor	Facultative Heterotrophy	Cell Type
Thiobacillus					
denitrificans	64	$S_2O_3^=$	O_2, NO_3^-	–	Gram-negative rods, polar flagella
ferrooxidans	56–57	$S_2O_3^=$, Fe^{++}	O_2	+	Gram-negative rods, polar flagella
intermedius		$S_2O_3^=$	O_2	+	Gram-negative rods, polar flagella
neapolitanus	56–57	$S_2O_3^=$	O_2	–	Gram-negative rods, polar flagella
novellus	66–68	$S_2O_3^=$	O_2	+	Gram-negative rods, non-motile[45]
thiooxidans	51–52	S^0, $S_2O_3^=$	O_2	+	Gram-negative rods, polar flagella
thioparus	62–66	$S_2O_3^=$	O_2	–	Gram-negative rods, polar flagella
Thiobacillus sp.[55]	66.2	$S_2O_3^=$	O_2	+	Gram-negative rods, non-motile, thermophilic
Thiobacillus A2[45]		$S_2O_3^=$	O_2, NO_3^- on organic substrates	+	Gram-negative rods, non-motile
Thiobacillus					
perometabolis[30 b]		$S_2O_3^=$	O_2	+	Gram-negative rods, motility not reported
thermophilica imshenetskii[13 c]		$S_2O_3^=$	O_2	–	Gram-positive rods, motile, spore-forming, thermophilic

[a] Although $S_2O_3^=$ is listed as a substrate throughout, it is only an example of reduced sulfur compounds that may be utilized. Many–if not most–of the thiobacilli also utilize $S^=$, S^0, $SO_3^=$, and polythionates.

[b] *Thiobacillus perometabolis* requires thiosulfate and organic matter simultaneously for optimum growth. It grows slowly on organic matter alone, and not at all in a purely mineral medium. It could be considered to be an obligate mixotroph.

[c] Since this organism is Gram-positive, it presumably is not a *Thiobacillus* species *sensu strictu*, although it was so named by the authors (see text).

THE NITRIFYING BACTERIA (NITROBACTERIACEAE)

The nitrifying bacteria are of critical importance in the nitrogen cycle in nature. They possess complex membrane systems and cell types that are distinctive in ultrastructure, forming the criteria for distinguishing species (see Figures 2, 3, and 4 for several examples).

Functionally there are two major groups: organisms that oxidize ammonia and organisms that oxidize nitrite. The nitrifying bacteria grow slowly, with mean generation times of about ten to twelve hours;[48] *Nitrobacter agilis*, a facultative autotroph, grows even more slowly heterotrophically.[40]

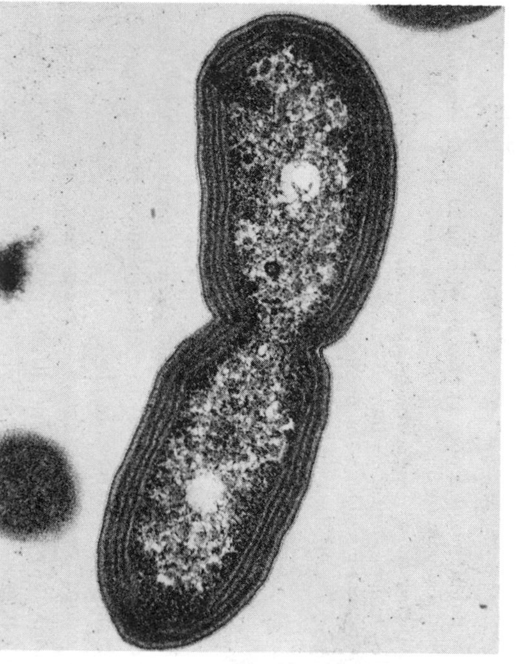

FIGURE 2. Electron Micrograph of a Thin Section of a Marine *Nitrosomonas* Showing Peripheral Cytomembranes.

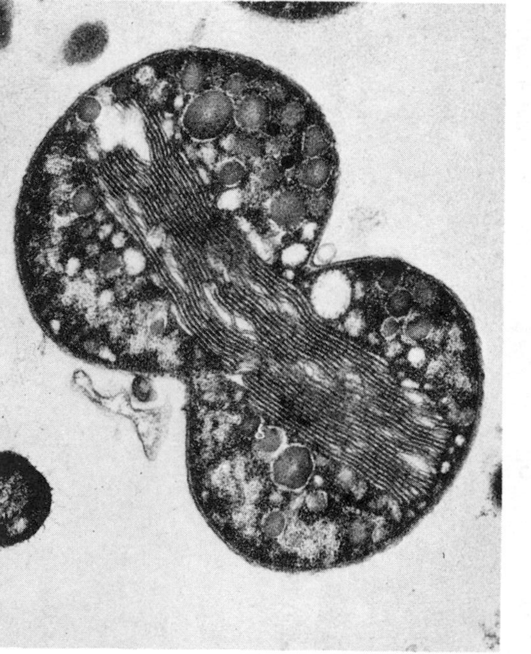

FIGURE 3. Electron Micrograph of a Thin Section of *Nitrosocystis oceanus* Showing Central Membranes in Cells.

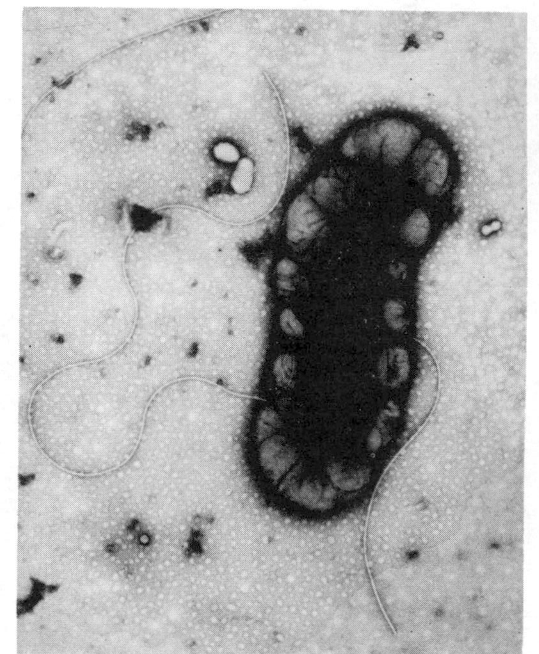

FIGURE 4. Electron Micrograph of Negatively Stained *Nitrospira briensis* Showing a Spiral Cell.

From Watson, S. W., and Mandel, M., *J. Bacteriol.*, *107*, 567-568 (1971). Reproduced by permission of the American Society for Microbiology, Washington, D.C.

Ammonia oxidation has been studied in *Nitrosomonas* and *Nitrosocystis* and may be outlined as follows:[35]

$$NH \rightarrow$$

$$NH_4 \rightarrow NH_2OH \rightarrow [X] \rightarrow NO_2^-$$

A detailed discussion of the oxidation of hydroxylamine is included in the review by Wallace and Nicholas;[48] the cell-free oxidation of ammonia to nitrite has been reported more recently.[42,49] In *Nitrobacter* and related organisms, nitrite oxidation to nitrate is catalyzed by a particulate complex, which also oxidizes formate. The properties of nitrite oxidase are reviewed by Wallace and Nicholas.[48]

Characteristics of Nitrobacteriaceae[52]

Species	DNA Base Composition, moles % G+C	Inorganic Substrates	Terminal Electron Acceptor	Facultative Heterotrophy	Cell Type
Nitrosomonas sp.	47.4–51.0	NH_3	O_2	–	Gram-negative rods, peripheral membranes in lamellae
Nitrosocystis					
oceanus[34]	50.5–51.0	NH_3	O_2	–	Gram-negative cocci, membranes in lamellae at center of cells
Nitrosospira					
briensis[50]	54.1	NH_3	O_2	–	Gram-negative spirals, peritrichous flagella, no cytomembranes
Nitrosolobus					
multiformis[51]	54.6	NH_3	O_2	–	Gram-negative lobular cells, partially compartmentalized by membranes
Nitrobacter sp.	60.2–61.7	NO_2^-	O_2	+	Gram-negative pear-shaped cells, polar caps of membranes, reproduced by budding
Nitrococcus					
mobilis[53]	61.2	NO_2^-	O_2	–	Gram-negative cocci, peritrichous flagella, internal tubular network of membranes
Nitrospina					
gracilis[53]	57.7	NO_2^-	O_2	–	Gram-negative long, slender rods, non-motile, no cytomembranes

NAD^+ REDUCTION

One of the problems facing thiobacilli and nitrifying autotrophs is that the oxidation of sulfur compounds, ammonia, and nitrite occurs at an oxidation/reduction potential higher than that of the NAD^+/NADH couple; thus the oxidation of these inorganic substrates cannot be directly coupled to the reduction of NAD^+. An ATP-dependent generation of NADH is observed in cell-free extracts of the thiobacilli and nitrifying autotrophs, and a mechanism of reversed electron flow dependent on ATP has been suggested to account for the pyridine nucleotide reduction.[1-5]

The nature of obligate autotrophy may well be a function of the mechanism of energy conservation in the organisms so characterized. It appears, in at least some experiments, that cell-free extracts of an obligately autotrophic thiobacillus are capable of oxidizing NADH, but that ATP is not synthesized during this process.[14] Yet these same extracts are able to link ATP synthesis to the oxidation of thiosulfate. More detailed work[41] has supported these findings in principle, although the authors were unable to detect the oxidation of NADH. Several other studies, however, have substantiated the occurrence of NADH oxidation in cell-free extracts of autotrophic bacteria. Smith, London and Stanier[41] have extended the studies of obligately autotrophic bacteria to the blue-green algae and investigated their carbon metabolism. In agreement with the conclusion that NADH oxidation does not lead to the formation of ATP, it appears that the enzymes ordinarily participating in the Krebs cycle appeared to function not catalytically but as a source of metabolites from which amino acids are derived. In particular, the enzyme system oxidizing α-ketoglutaric acid is absent. Therefore, the scheme shown in Figure 5 has been proposed.[41] This scheme is supported by enzymic studies and by the restricted use of acetate in amino acid biosynthesis. ^{14}C-acetate contributes radioactivity only to those amino acids for which either acetate or α-ketoglutarate are direct precursors.[16,22,41]

Additional studies of the thiobacilli show other features that contribute to the inability of the strict autotrophs to oxidize organic compounds. In *T. neapolitanus* and *T. thioparus* the enzymes of the Embden-Meyerhof-Parnas pathway are present, with the exception of phosphofructokinase;[21] this is also true in *Nitrosocystis oceanus.*[54] In these organisms the glycolytic pathway may be used primarily for

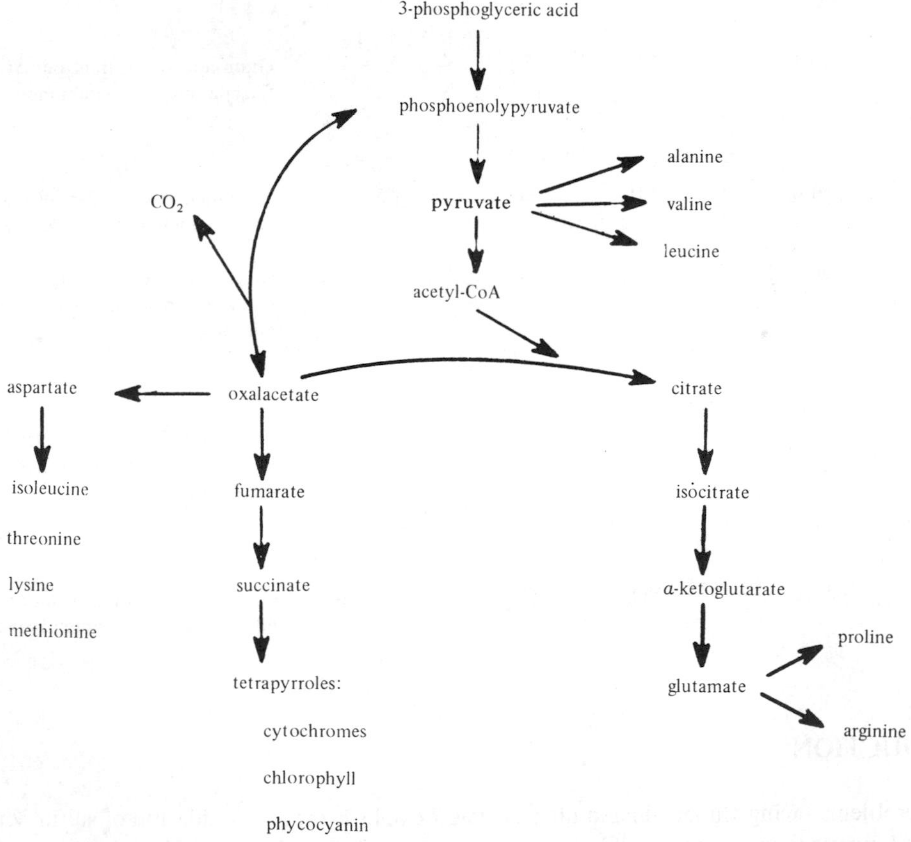

FIGURE 5. Proposed Scheme of Interrupted Krebs Cycle in Strictly Autotrophic Thiobacilli and Blue-Green Algae.

gluconeogenesis. The enzymes of the Entner-Doudoroff pathway, present in the facultative heterotroph *T. intermedius,* in the glucose utilizing strain of *T. ferrooxidans,* and in *T. perometabolis,* are not present in *T. neapolitanus* and *T. thioparus;* neither is the dissimilatory pentose phosphate pathway, which effects glucose catabolism in *T. novellus.*[33] Characteristics of the strict autotrophs among the thiobacilli are thus complex, and it is not possible to point to a single enzymic reaction as the cause of strict autotrophy.

In other aspects of the biosynthesis of amino acids, the thiobacilli resemble heterotrophic organisms.[19,23-26] Differences between *T. thioparus* and *T. neapolitanus* and common heterotrophs appear, however, in the biosynthetic utilization of pyruvate.[20]

When ^{14}C-acetate is incorporated into amino acids in *Nitrosomonas* and *Nitrobacter,* it is not restricted to the glutamate family but is found generally distributed.[47] In *Nitrosocystis oceanus* the enzymes of the Krebs cycle are all present, but succinic dehydrogenase activity is low. Since *N. oceanus* possesses an intact pentose phosphate pathway, but has not been grown heterotrophically, it has been suggested that the low succinic dehydrogenase activity might drastically slow growth.[54]

An interesting observation of "catabolite repression" has been made in *T. novellus,*[29] in which growth on glucose depresses synthesis of enzymes of sulfur oxidation.

Problems of organic metabolism in autotrophs have been reviewed by Rittenberg[37] and by Kelly.[27]

ACKNOWLEDGMENT

The authors thank Dr. Stanley W. Watson for making available material in advance of publication.

REFERENCES

1. Aleem, M. I. H., Path of Carbon and Assimilatory Power in Chemosynthetic Bacteria, I, *Nitrobacter agilis, Biochim. Biophys. Acta, 107,* 14 (1965).
2. Aleem, M. I. H., Generation of Reducing Power in Chemosynthesis, II, Energy-Linked Reduction of Pyridine Nucleotides in the Chemoautotroph *Nitrosomonas europaea, Biochim. Biophys. Acta, 113,* 216 (1966).
3. Aleem, M. I. H., Generation of Reducing Power in Chemosynthesis, III, Energy-Linked Reduction of Pyridine Nucleotides in *Thiobacillus novellus, J. Bacteriol., 91,* 729 (1966).
4. Aleem, M. I. H., Generation of Reducing Power in Chemosynthesis, IV, Energy-Linked Reduction of Pyridine Nucleotides by Succinate in *Thiobacillus novellus, Biochim. Biophys. Acta, 128,* 1 (1966).
5. Aleem, M. I. H., Generation of Reducing Power in Chemosynthesis, VI, Energy-Linked Reactions in the Chemoautotroph *Thiobacillus neapolitanus, Antonie van Leeuwenhoek J. Microbiol. Serol., 35,* 379 (1969).
6. Alexander, M., Biochemical Ecology of Microorganisms, *Annu. Rev. Microbiol., 25,* 361 (1971).
7. Belyayeva, M. J., The Distribution of Hydrogen Bacteria in Nature (in Russian), *Proc. Univ. Kazan, 114,* 229 (1954).
8. Brock, T. D., *Principles of Microbial Ecology.* Prentice-Hall, Inc., Englewood Cliffs, New Jersey (1966).
9. Bryant, M. P., Wolin, E. A., Wolin, M. J., and Wolfe, R. S., *Methanobacillus omelianskii,* a Symbiotic Association of Two Species of Bacteria, *Arch. Mikrobiol., 59,* 20 (1967).
10. Davis, D. H., Doudoroff, M., Stanier, R. Y., and Mandel, M., Proposal to Reject the Genus *Hydrogenomonas:* Taxonomic Implications, *Int. J. Syst. Bacteriol., 19,* 375 (1969).
11. Davis, D. H., Stanier, R. Y., Doudoroff, M., and Mandel, M., Taxonomic Studies on Some Gram-Negative Polarly Flagellated Hydrogen Bacteria and Related Species, *Arch. Mikrobiol., 70,* 1 (1970).
12. Eberhardt, U., On Chemolithotrophy and Hydrogenase of a Gram-Positive Knallgas Bacterium, *Arch. Mikrobiol., 66,* 91 (1969).
13. Egorova, A. A., and Deryugina, Z. P., The Spore-Forming Thermophilic Thiobacterium *Thiobacillus thermophilica imschenetskii, nov. sp.* (in Russian), *Mikrobiologiya, 32,* 376 (1963).
14. Hempfling, W. P., and Vishniac, W., Oxidative Phosphorylation in Extracts of *Thiobacillus* X, *Biochem. Z., 342,* 272 (1965).
15. Hirsch, P., Wasserstoffaktivierung und Chemoautotrophie bei Actinomyzeten, *Arch. Mikrobiol., 39,* 360 (1961).
16. Hoare, D. S., Hoare, S. L., and Moore, R. B., The Photoassimilation of Organic Compounds by Autotrophic Blue-Green Algae, *J. Gen. Microbiol., 49,* 351 (1967).
17. Hutchinson, M., Johnstone, K. I., and White, D., Taxonomy of the Genus *Thiobacillus:* the Outcome of Numerical Taxonomy Applied to the Group as a Whole, *J. Gen. Microbiol., 57,* 397 (1969).
18. Jackson, J. T., Moriarty, D. J. H., and Nicholas, D. J. D., Deoxyribonucleic Acid Base Composition and Taxonomy of the Thiobacilli and Some Nitrifying Bacteria, *J. Gen. Microbiol., 53,* 53 (1968).

19. Johnson, C. L., and Vishniac, W., Growth Inhibition in *Thiobacillus neapolitanus* by Histidine, Methionine, Phenylalanine and Threonine, *J. Bacteriol., 104,* 1145 (1970).
20. Johnson, E. J., and Abraham, S., Assimilation and Metabolism of Exogenous Organic Compounds by the Strict Autotrophs *Thiobacillus thioparus* and *Thiobacillus neapolitanus, J. Bacteriol., 97,* 1198 (1969).
21. Johnson, E. J., and Abraham, S., Enzymes of Intermediary Carbohydrate Metabolism in the Obligate Autotrophs *Thiobacillus thioparus* and *Thiobacillus neapolitanus, J. Bacteriol., 100,* 962 (1969).
22. Kelly, D. P., The Incorporation of Acetate by the Chemoautotroph *Thiobacillus neapolitanus* Strain C, *Arch. Mikrobiol., 58,* 99 (1967).
23. Kelly, D. P., Influence of Amino Acids and Organic Antimetabolites on Growth and Biosynthesis of the Chemoautotroph *Thiobacillus neapolitanus* Strain C, *Arch. Mikrobiol., 56,* 91 (1967).
24. Kelly, D. P., Regulation of Chemoautotrophic Metabolism, I, Toxicity of Phenylalanine to *Thiobacilli, Arch. Mikrobiol., 69,* 39 (1969).
25. Kelly, D. P., Regulation of Chemoautotrophic Metabolism, II, Competition between Amino Acids for Incorporation into *Thiobacillus, Arch. Mikrobiol., 69,* 343 (1969).
26. Kelly, D. P., Regulation of Chemoautotrophic Metabolism, III, DAHP Synthetase in *Thiobacillus neapolitanus, Arch. Mikrobiol., 69,* 360 (1969).
27. Kelly, D. P., Autotrophy: Concepts of Lithotrophic Bacteria and Their Organic Metabolism, *Annu. Rev. Microbiol., 25,* 177 (1971).
28. Kocur, M., Martinec, T., and Mazanec, K., Fine Structure of *Thiobacillus novellus, J. Gen. Microbiol., 52,* 343 (1968).
29. LeJohn, H. B., van Caseele, L., and Lees, H., Catabolite Repression in the Facultative Chemoautotroph *Thiobacillus novellus, J. Bacteriol., 94,* 1484 (1967).
30. London, J., and Rittenberg, S. C., *Thiobacillus perometabolis nov. sp.*, a Non-autotrophic Thiobacillus, *Arch. Mikrobiol., 59,* 218 (1967).
31. Lyric, R. M., and Suzuki, I., Properties of Thiosulfate-Oxidizing Enzyme and Proposed Pathway of Thiosulfate Oxidation, *Can. J. Biochem., 48,* 355 (1970).
32. Mahoney, R. P., and Edwards, M. R., Fine Structure of *Thiobacillus thiooxidans, J. Bacteriol., 92,* 487 (1966).
33. Matin, A., and Rittenberg, S. C., Enzymes of Carbohydrate Metabolism in *Thiobacillus* species, *J. Bacteriol., 107,* 179 (1971).
34. Murray, R. G. C., and Watson, S. W., Structure of *Nitrosocystis oceanus* and Comparison with *Nitrosomonas* and *Nitrobacter, J. Bacteriol., 89,* 1594 (1965).
35. Peck, H. D., Jr., Energy-Coupling Mechanisms in Chemolithotrophic Bacteria, *Annu. Rev. Microbiol., 22,* 489 (1968).
36. Remsen, C., and Lundgren, D. G., Electron Microscopy of the Cell Envelope of *T. ferrooxidans* Prepared by Freeze-Etching and Chemical Fixation Techniques, *J. Bacteriol., 92,* 1765 (1966).
37. Rittenberg, S. C., The Roles of Exogenous Organic Matter in the Physiology of Chemolithotrophic Bacteria, *Advan. Microbial Physiol., 3,* 159 (1969).
38. Savelyeva, N. D., and Zhilina, T. N., Taxonomy of Hydrogen Bacteria (in Russian), *Mikrobiologiya, 37,* 84 (1968).
39. Shively, J. M., Decker, G. L., and Greenawalt, J. W., Comparative Ultrastructure of the Thiobacilli, *J. Bacteriol., 101,* 618 (1970).
40. Smith, A. J., and Hoare, D. S., Acetate Assimilation by *Nitrobacter agilis* in Relation to Its Obligate Autotrophy, *J. Bacteriol., 95,* 844 (1968).
41. Smith, A. J., London, J., and Stanier, R. Y., Biochemical Basis of Obligate Autotrophy in Blue-Green Algae and Thiobacilli, *J. Bacteriol., 94,* 972 (1967).
42. Suzuki, I., and Kwok, S. C., Cell-Free Ammonia Oxidation by *Nitrosomonas europaea* Extracts: Effects of Polyamines, Mg^{++} and Albumin, *Biochem. Biophys. Res. Commun., 39,* 950 (1970).
43. Tabita, T. R., and Lundgren, D. G., *Bacteriol. Proc.*, p. 125 (1970).
44. Takamiya, A., and Tubaki, K., A New Form of *Streptomyces* Capable of Growing Autotrophically, *Arch. Mikrobiol., 25,* 58 (1956).
45. Taylor, B. T., and Hoare, D. S., New Facultative *Thiobacillus* and a Reevaluation of the Heterotrophic Potential of *Thiobacillus novellus, J. Bacteriol., 100,* 487 (1969).
46. Trudinger, P. A., Assimilatory and Dissimilatory Metabolism of Inorganic Sulphur Compounds by Microorganisms, *Advan. Microbial Physiol., 3,* 135 (1969).
47. Wallace, W., Knowles, S. E., and Nicholas, D. J. D., Intermediary Metabolism of Carbon Compounds by Nitrifying Bacteria, *Arch. Mikrobiol., 70,* 26 (1970).
48. Wallace, W., and Nicholas, D. J. D., The Biochemistry of Nitrifying Organisms, *Biol. Rev., 44,* 359 (1969).
49. Watson, S. W., Asbell, M. A., and Valois, T. W., Ammonia Oxidation by Cell-Free Extracts of *Nitrosocystis oceanus, Biochem. Biophys. Res. Commun., 38,* 1113 (1970).
50. Watson, S. W., Reisolation of *Nitrosospira briensis, Arch. Mikrobiol., 75,* 179 (1971).
51. Watson, S. W., Graham, L. B., Remsen, C. C., and Valois, T. W., A Lobular Ammonia-Oxidizing Bacterium, *Nitrosolobus multiformis, nov. gen., nov. sp., Arch. Mikrobiol., 76,* 183 (1971).

52. Watson, S. W., and Mandel, M., Comparison of the Morphology and Deoxyribonucleic Acid Composition of 27 Strains of Nitrifying Bacteria, *J. Bacteriol., 107,* 563 (1971).
53. Watson, S. W., and Waterbury, J. B., Characteristics of Two Marine Nitrite-Oxidizing Bacteria, *Nitrospina gracilis, nov. gen., nov. sp.* and *Nitrococcus mobilis, nov. gen., nov. sp., Arch. Mikrobiol., 77,* 203 (1971).
54. Williams, P. J. L. B., and Watson, S. W., Autotrophy in *Nitrosocystis oceanus, J. Bacteriol., 96,* 640 (1968).
55. Williams, R. A. D., and Hoare, D. S., Physiology of a New Facultatively Autotrophic Thermophilic Thiobacillus, *J. Gen. Microbiol., 70,* 555 (1972).
56. Zavarzin, G. A., and Zhilina, T. N., Thionic Acid Bacteria from Hot Springs (in Russian), *Mikrobiologiya, 33,* 753 (1964).

THE RHODOSPIRILLALES (PHOTOTROPHIC OR PHOTOSYNTHETIC BACTERIA)

Dr. NORBERT PFENNIG AND DR. HANS G. TRÜPER

The phototrophic bacteria are a physiological group of different kinds of Gram-negative aquatic bacteria. They are considered to be evolutionary rather old, representing the first phototrophic organisms on our planet.[1] The common characteristic of the Rhodospirillales is their ability to perform an anaerobic type of photosynthesis, which – unlike the photosynthesis of Cyanophyceae and green plants – proceeds without the production of oxygen. All Rhodospirillales contain photosynthetic pigments of the bacteriochlorophyll type (see Table 1), different from plant chlorophylls, and typical carotenoid pigments (see Table 2).

TABLE 1

NOMENCLATURE OF BACTERIAL CHLOROPHYLLS

Designation of Jensen *et al.*[2]	Former Designations	Characteristic Absorption Maxima of Living Cells, nm
Bacteriochlorophyll *a*	Bacteriochlorophyll	375, 590, 805, 830–890
Bacteriochlorophyll *b*	Bacteriochlorophyll *b*	400, 605, 850, 1020–1040
Bacteriochlorophyll *c*	Chlorobium chlorophyll 660	Long wavelength absorption maximum 745–755
Bacteriochlorophyll *d*	Chlorobium chlorophyll 650	Long wavelength absorption maximum 705–740

TABLE 2

CAROTENOID GROUPS OF PHOTOTROPHIC BACTERIA[3,4]

Group	Name	Major Components
1	Normal spirilloxanthin series	Lycopene, rhodopin, spirilloxanthin
2	Alternative spirilloxanthin series and keto-carotenoids of spheroidenone type	Spheroidene, hydroxy-spheroidene, spheroidenone, hydroxy-spheroidenone, spirilloxanthin
3	Okenone series	Okenone
4	Rhodopinal series (former warmingone series)	Lycopenal, lycopenol, rhodopin, rhodopinal, rhodopinol
5	Chlorobactene series	Chlorobactene, hydroxy-chlorobactene, β-isorenieratene, isorenieratene

Photosynthesis of the Rhodospirillales depends on the presence of oxidizable external electron donors, such as reduced sulfur compounds, molecular hydrogen or organic carbon compounds. As far as studied, all strains contain cytochromes, ubiquinones, and non-heme iron proteins (ferredoxins) as constituents of their photosynthetic electron transport systems.[5] The fixation of molecular nitrogen is apparently a common property of the group.[6] Carbon dioxide is photoassimilated through the reductive pentose phosphate cycle, with the key enzymes ribulose 1.5-diphosphate carboxylase and ribulose 5-phosphate kinase, as well as by further carbon dioxide fixation reactions involving ferredoxin as an electron donor. Under anaerobic conditions in the dark, all phototrophic bacteria carry out a fermentative maintenance metabolism, which proceeds with the excretion of carbon dioxide, organic acids (predominantly acetate), and, in the case of sulfur bacteria, sulfide originating from elemental sulfur.[4]

The present classification,[7] which is based on pure-culture studies, divides the Rhodospirillales into two suborders, the Rhodospirillineae and the Chlorobiineae. Type or neotype strains of all species presently in pure culture have been proposed and are deposited in recognized culture collections.[7]

The Rhodospirillineae comprise those phototrophic bacteria that contain bacteriochlorophyll *a* or *b* as the major bacteriochlorophyll and that carry the photopigments in intracytoplasmic membrane systems continuous to the cytoplasmic membrane. The different types of these membrane systems,[4,8] and the respective organisms are given in Table 3. The suborder is further divided into the families Rhodospirillaceae (the former Athiorhodaceae) and Chromatiaceae (the former Thiorhodaceae).

The Chlorobiineae, at present represented by one family, the Chlorobiaceae, comprise those phototrophic bacteria whose major bacteriochlorophyll is *c* or *d*, although small amounts of bacteriochlorophyll *a* are present in the photosynthetic reaction centers. The suborder is further characterized by the chlorobium vesicles (see Table 3), special organelles encoated in a non-unit single membrane and underlying the cytoplasmic membrane. These vesicles contain the photosynthetic apparatus.

PHOTOTAXIS

In all motile, flagellated purple sulfur and non-sulfur bacteria, the individual cells show a tactic response following irritation by a sudden change in environmental conditions (light in phototaxis, air and nutrients in aero- and chemotaxis). The reaction of the cells is a reversal of the moving direction of the flagellum, followed by the reversal of the direction of movement of the cell. This phenomenon became first known for light as "Schreckbewegung" (shock movement) by the work of Engelmann around 1882 and has since been studied in detail and reviewed by Clayton.[4,9] Since in procaryotic microbes no oriented movement of the cells occurs (e.g., toward light), the bacterial type of the taxis is known as phobophototaxis. A slight decrease – 1 to 3% is sufficient – in the intensity of illumination in space or time causes a reversal of the swimming direction of the cell. With this type of response, cells initially distributed at random must finally accumulate in areas of optimal light intensity (scotophobia). The action spectra for phobophototaxis in the purple bacteria coincide with the absorption spectra of the cells, showing that in the bacteria, in contrast to the eucaryotic algae, both carotenoids and bacteriochlorophylls are active. Clayton confirmed and extended Manten's conclusion that it is the abrupt decrease in the rate of photosynthesis following a decrease in light intensity that induces the tactic response. It has also been established that the saturating light intensities for photosynthesis and phototaxis are equal.

IDENTIFICATION

Rhodospirillaceae

Rhodospirillaceae (syn. Athiorhodaceae) are purple non-sulfur bacteria. The family is characterized by its inability to grow well with elemental sulfur and sulfide as photosynthetic electron donors. Instead, simple organic carbon compounds serve this function and are photometabolized. Many strains can grow photolithotrophically with molecular hydrogen as the electron donor. Most strains require vitamins. The photosynthetic membrane systems of Rhodospirillaceae are shown in Table 3. The basic properties[10-13] of the accepted species are listed in Table 4.

In general, all species are microaerophilic. Many representatives may grow at full atmospheric oxygen tension in the light or dark. In strains able to grow under microaerophilic to aerobic conditions, the photopigment content and the intracytoplasmic membrane system decrease as the concentration of dissolved oxygen increases. The formation of photopigments becomes derepressed below certain oxygen concentrations. Under strictly anaerobic conditions, good growth is only possible in light; under anaerobic conditions in the dark, very slow reproduction occurs, due to fermentative energy-providing processes.

Cell division occurs by binary fission, except for the genus *Rhodomicrobium* and the species *Rhodopseudomonas palustris, Rps. viridis,* and *Rps. acidophila*; these exceptions divide by budding. The

TABLE 3

INTRACYTOPLASMIC MEMBRANES AND CHLOROBIUM VESICLES IN PHOTOTROPHIC BACTERIA

Type of Membrane System	Rhodospirillaceae	Chromatiaceae	Chlorobiaceae
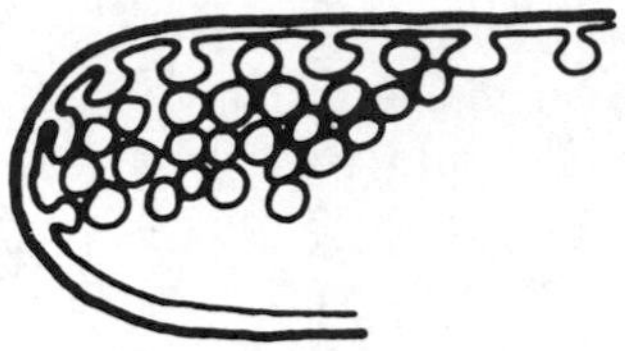	*Rhodopseudomonas* *capsulata* *globiformis* *sphaeroides* *Rhodospirillum* *rubrum*	*Amoebobacter* *Chromatium* *Lamprocystis* *Thiocapsa* *Thiocystis* *Thiodictyon* *Thiospirillum*	
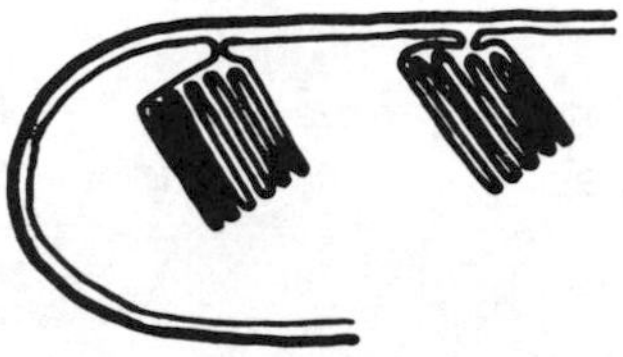	*Rhodospirillum* *fulvum*[a] *molischianum*[a] *photometricum*[a]	*Ectothiorhodospira*	
	*Rhodomicrobium*[b] *Rhodopseudomonas* *acidophila*[b] *palustris*[b] *viridis*[b]		
	*Rhodocyclus* *purpureus*[c] *Rhodopseudomonas* *gelatinosa*[c] *Rhodospirillum* *tenue*[c]		
		*Thiocapsa* *pfennigii*	
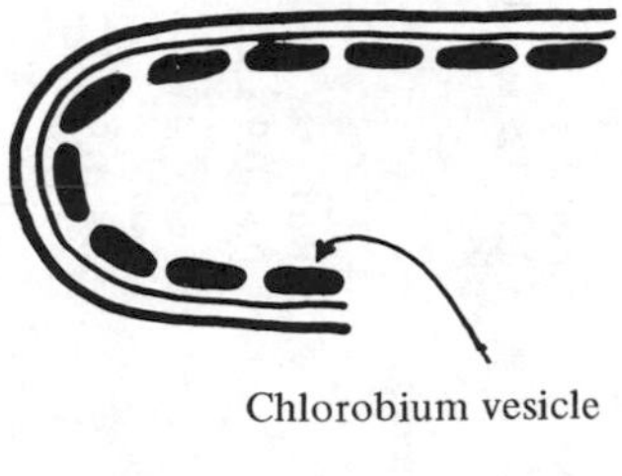 Chlorobium vesicle			*Chlorobium*[d] *Chloropseudomonas*[d] *Pelodictyon*[d] *Prosthecochloris*[d]

[a]Non-facultative aerobic.
[b]Multiply by budding.
[c]Smallest forms.
[d]Green sulfur bacteria.

TABLE 4

GENERAL PROPERTIES OF THE RHODOSPIRILLACEAE

Organism	Cell Shape and Dimensions, Diameter (Width) x Length, μm	Slime Formed	Motile by Flagella	Mode of Multiplication[a]	Type of Intracytoplasmic Membrane System	Color of Cell Suspension Under Anaerobic Growth Conditions	Bacteriochlorophyll	Predominant Carotenoids[b]	DNA Base Ratio,[14] moles % G + C	Growth Factors Required for the Majority of Strains	Ability to Grow Aerobically or Microaerophilically in the Dark[c]
Rhodospirillum											
rubrum	Spiral 0.8–1.0 x 7–10	–	+	bf	Vesicles	Red	*a*	sp	63.8–65.8	Biotin	ae
tenue	Spiral 0.3–0.5 x 3–6	–	+	bf	Tubes	Purple-violet or brown-orange	*a*	rl, ly, rh	64.8	None	ae
fulvum	Spiral 0.5–0.7 x 3.5	–	+	bf	Stacks	Brown	*a*	ly, rh	64.3–65.3	pAB	m
molischianum	Spiral 0.7–1.0 x 5–8	–	+	bf	Stacks	Brown	*a*	ly, rh	61.7–64.8	Amino acids	m
photometricum	Spiral 1.2–1.5 x 7–10	–	+	bf	Stacks	Brown	*a*	ly, rh	65.8	Yeast extract	m
Rhodopseudomonas											
palustris	Rod 0.6–0.9 x 1.2–2	–	+	bud	Polar stack	Red-brown	*a*	sp, ly, rh	64.8–66.3	pAB ± biotin	ae
viridis	Rod 0.6–0.9 x 1.2–2	–	+	bud	Polar stack	Green	*b*	ts	66.3–71.4	pAB + biotin	m
acidophila	Rod 1.0–1.3 x 2–5	–	+	bud	Polar stack	Purple-red or orange-brown	*a*	rh, rg	62.2–66.8	None	m

gelatinosa	Rod 0.4–0.5 x 1–2	+	+	bf	Tubes	Yellow-brown to pinkish	*a*	sp, sn	70.5–72.4	Biotin + thiamine	ae
capsulata	Rod/sphere 0.5–1.2 x 2–2.5	+	+	bf	Vesicles	Yellow to brown	*a*	sp, sn	65.5–66.8	Thiamine	ae
sphaeroides	Sphere/ovoid 0.7 x 2–2.5	–	+	bf	Vesicles	Green-brown to brown	*a*	sp, sn	68.4–69.9	Biotin + thiamine + nicotinic acid	ae
globiformis	Sphere 1.6–1.8	–	+	bf	Vesicles	Purple-red	*a*	?	66.3	Biotin + pAB + reduced S source	m
Rhodomicrobium											
vannielii	Ovoid + stalk 1.0–1.2 x 2–2.8	–	+	bud	Polar stack	Orange-brown	*a*	sp, βc	61.8–63.8	None	m
Rhodocyclus purpureus	Half-circle/circle 0.6–0.7 x 2.7–5	–	–	bf	Tubes	Purple-violet	*a*	rl, rh	65.3	Vitamin B_{12}	m

a bf = binary fission; bud = budding.
b βc = carotene; ly = lycopene; rg = rhodopin glucoside; rh = rhodopin; rl = rhodopinal; sn = spheroidenone; sp = spirilloxanthin; ts = tetrahydroxyspirilloxanthin.
c ae = aerobic; m = microaerophilic.

motile cells of *Rhodomicrobium* are peritrichously flagellated, whereas the species of *Rhodospirillum* and *Rhodopseudomonas* have polar flagella. Exospore-like heat-resistant bodies have been observed in *Rhodomicrobium*. None of the known species of the Rhodospirillaceae contain gas vacuoles.

The photosynthetic hydrogen donors (and carbon sources) utilized by the Rhodospirillaceae[10-12] are listed in Table 5. Of taxonomical importance is the ability of *Rhodopseudomonas gelatinosa* to liquefy gelatin. Storage products of the Rhodospirillaceae are polysaccharides, poly-β-hydroxybutyrate, and polyphosphates.

In nature, the Rhodospirillaceae occur preferably in all aquatic environments.[4] Their number usually increases with the amount of organic material present (e.g., pollution by organic waste products). They are not found in the open ocean. Members of the Rhodospirillaceae also occur in small numbers in most soils, increasing with the moisture content of the soil. *Rhodopseudomonas acidophila* and *Rhodomicrobium vannielii* are specifically adapted to grow in acid natural habitats, such as peat bogs, swamps, etc.[15] Special marine forms have so far not been encountered. Optimal growth temperatures lie between 20°C and 40°C.

TABLE 5

PHOTOSYNTHETIC ELECTRON DONORS (AND CARBON SOURCES) UTILIZED BY THE RHODOSPIRILLACEAE

Substance	*R. rubrum*	*R. tenue*	***R. fulvum***	*R. molischianum*	*R. photometricum*	*Rps. palustris*	*Rps. viridis*	*Rps. acidophila*	*Rps. gelatinosa*	***Rps. capsulata***	*Rps. sphaeroides*	*Rps. globiformis*	*Rm. vannielii*	*Rc. purpureus*
Thiosulfate	–	–	–	–	–	+	–	–	–	–	–	–	–	–
Formate	–	–			–	+	–	–	±			–	–	–
Acetate	+	+	+	+	+	+	+	+	+	+	+	–	+	+
Propionate	+	+	+	+	±	+	–	+	±	+	±	–	+	–
Butyrate	+	+	+	+	+	+	–	+	±	+	±	–	+	–
Valerate	+	+	+	+	+	+	–	+		+	±	–	+	–
Caproate	+	+	+	+	–	+	–	±		+	±	–	+	+
Caprylate		+	+	+	–	+		–		+	±	–	+	–
Pelargonate		+	+	+	±	–		–		+	±	–	–	–
Glycolate		–		–	+	+		±				–	–	–
Pyruvate	+	+	+	+	+	+	+	+	+	+	+	±	+	+
Lactate	+	+	–	±	+	+	–	+	+			–	+	–
Malonate		–		–		+		±				–	+	–
Malate	+	+	+	+	+	+	+	+	+	+	+	+	+	+
Succinate	+	+	+	+	+	+	+	+	+	+	+	+	+	–
Fumarate	+	+	+	+	+	+		+	+	+	+	+	+	+
Tartrate	–	–		–	–	–		±	±	–	+	+	–	–
Citrate	–	–	–	–	–	–	–	+	+	–	+	–	–	–
Gluconate	–									–	+	+		
Benzoate	–	–	+	–	–	+		–	–		–	–	–	+
Methanol	–	–		–	–	–		±	–			–	–	–
Ethanol	+	–	+	+	+	+	+	+	+	–	+	+	+	–
Glycerol	–	–	–	–	+	+	–	±	–	–	+	–	–	–
Glucose	–	–	±	–	+	–	+	±	+	+	+	+	–	–
Fructose	±	–	–	–	+	–		–		+	+	+	–	–
Mannitol	–	–	–	–	+	–		–	±	–	+	+	–	–
Aspartate	+	–	±	±	±	–		–		+		–	–	–
Glutamate	+				–	+	+	–		+		–	–	–
Arginine	+	+	–	–	–	–		–				–	–	–
Casamino acids	+	+	–	–	+	+	+	+				–	+	–
Yeast extract	+	+	+	+	+	+		+	+			+	+	–
Molecular hydrogen	+	+	+			+	–	+	±	+	+		+	+

Chromatiaceae

Chromatiaceae (syn. Thiorhodaceae) are purple sulfur bacteria. All species multiply by binary fission, and the motile species are polarly flagellated. Depending on the culture conditions or environmental conditions in nature, all species may develop either as single cells or form mostly non-motile aggregates or families of variable size and shape embedded in slime. Correct identification, therefore, is only possible by pure-culture studies. The photosynthetic membrane systems of Chromatiaceae are shown in Table 3. The general properties[4,11-13] of the species contained in this family[7] are listed in Table 6.

Most members of the Chromatiaceae are strictly anaerobic and therefore obligately phototrophic. Only *Thiocapsa roseopersicina* is able to grow in the dark under microaerophilic conditions. All species are capable of photolithotrophic carbon dioxide fixation in the presence of sulfide. Under these conditions, elemental sulfur is accumulated in the form of globules inside or outside the cells. The ultimate oxidation product of sulfide is sulfate. Many strains are able to utilize molecular hydrogen as electron donor. All strains photoassimilate a number of simple organic carbon compounds, of which acetate and pyruvate are most widely used. Strains lacking assimilatory sulfate reduction utilize organic substrates only in the presence of sulfide or other reduced sulfur compounds as a source of cell sulfur. Storage materials are polysaccharides, poly-β-hydroxybutyrate, and polyphosphate.

The large-celled species *Chromatium okenii, C. weissei, C. warmingii, C. buderi*, and *Thiospirillum jenense*, and probably *T. sanguineum,* differ from the rest of the species of the family not only in size but in their inability to utilize organic carbon sources other than acetate or pyruvate, their vitamin B_{12} requirement, their relatively low DNA base ratios,[14] and their lack of assimilatory sulfate reduction.[4] The genus *Ectothiorhodospira* comprises the smallest forms of the family. Its species do not store intracellular sulfur globules, but deposit them extracellularly.

In nature, the Chromatiaceae occur in the anaerobic and sulfide-containing parts of aquatic environments, from moist and muddy soils to ditches, ponds, lakes, rivers, sulfur springs, salt lakes, estuaries (especially salt marshes) and other marine habitats.[4] They do not occur in the open ocean. Definite marine forms with salt requirements are *Chromatium buderi, C. gracile,* and *Ectothiorhodospira mobilis,* whereas *Ectothiorhodospira halophila* is an extreme halophile. In general, the optimal growth temperature lies between 20°C and 30°C. Species containing gas vacuoles become buoyant at low temperatures.

Chlorobiaceae

Chlorobiaceae (syn. Chlorobacteriaceae) are green sulfur bacteria. The family is especially characterized by the possession of unique organelles, the chlorobium vesicles, which carry the photosynthetic apparatus (see Table 3). Multiplication proceeds by binary fission. In *Pelodictyon clathratiforme* ternary fission also occurs, resulting in the formation of three-dimensional nets. Only one genus, *Chloropseudomonas,* is motile by polar flagella. The general properties[4,11-13] of the accepted species[17] are listed in Table 7.

All forms are strictly anaerobic and obligately phototrophic. All species are capable of photolithotrophic carbon dioxide fixation in the presence of sulfide. Under these conditions, elemental sulfur is accumulated in the form of globules outside – never inside – the cells and further oxidized to sulfate. Many strains are able to use molecular hydrogen as an electron donor for growth, provided that hydrogen sulfide is present in addition as a source of cell sulfur, since assimilatory sulfate reduction is lacking. A number of simple organic substrates are photoassimilated in the presence of both sulfide and carbon dioxide; acetate is the most widely used compound. Only polyphosphates generally occur as storage materials. Polysaccharides were identified only in *Chloropseudomonas ethylica,* the only potentially photoorganotrophic green sulfur bacterium. Many strains require vitamin B_{12} for growth.[4]

In nature, the Chlorobiaceae live in the same habitats as the Chromatiaceae. The brown-colored species occur only in deeper layers of ponds and lakes and in the hypolimnion of meromictic lakes.[4] A species with obligate salt requirement, which has been isolated only from marine estuaries, is *Prosthecochloris aestuarii.* The Chlorobiaceae generally prefer pH values slightly more acid than the Chromatiaceae. The optimal pH range of the Chlorobiaceae is between 6.5 and 7.0, i.e., generally lower than that of the Chromatiaceae. The

TABLE 6

GENERAL PROPERTIES OF THE CHROMATIACEAE

Organism	Cell Shape and Dimensions, Diameter (Width) x Length, μm	Motile by Polar Flagella	Aggregates Formed in Pure Culture	Slime Formed in Pure Culture	Gas Vacuoles	Type of Intracytoplasmic Membrane System	Cells May Contain Sulfur Globules	Color of Cell Suspension	Bacteriochlorophyll	Predominant Carotenoids[a]	DNA Base Ratio,[14] moles % G + C	Vitamin B_{12} Required	NaCl Required	Hydrogenase Activity
Chromatium														
okenii	Rod 4.5-6.0 x 8-15	+	None	–	–	Vesicles	+	Purple-red	*a*	ok	48.0-50.0	+	–	–
weissei	Rod 3.5-4.5 x 7-9	+	None	–	–	Vesicles	+	Purple-red	*a*	ok	48.0-50.0	+	–	–
warmingii	Rod 3.5-4.0 x 5-11	+	None	–	–	Vesicles	+	Purple-violet	*a*	rl	55.1-60.2	+	–	+
buderi	Rod 3.5-4.5 x 4.5-9	+	None	–	–	Vesicles	+	Purple-violet	*a*	rl	62.2-62.8	+	+	–
minus	Rod 2.0 x 2.5-6	+	None	–	–	Vesicles	+	Purple-red	*a*	ok	52.0-62.2	–	–	+
vinosum	Rod 2.0 x 2.5-6	+	None	–	–	Vesicles	+	Brown-red	*a*	sp, ly, rh	61.3-66.3	–	–	+
violascens	Rod 2.0 x 2.5-6	+	None	–	–	Vesicles	+	Purple-violet	*a*	rl	61.8-64.3	–	–	+
gracile	Rod 1.0-1.3 x 2-6	+	None	+	–	Vesicles	+	Brown-red	*a*	sp, ly, rh	68.9-70.4	–	+	+
minutissimum	Rod 1.0-1.2 x 2.0	+	None	–	–	Vesicles	+	Brown-red	*a*	sp, ly, rh	63.7	–	–	+
Thiocystis														
violacea	Sphere 2.5-3.0	+	(Clumps)	+	–	Vesicles	+	Purple-violet	*a*	rl	62.8-67.9	–	–	+
gelatinosa	Sphere 3.0	+	None	+	–	Vesicles	+	Purple-red	*a*	ok	61.3	–	–	+
Thiosarcina														
rosea	Sphere 2.0-3.0	±	Packets	?	–	?	+	Pink-red	*a*	?	?	?	?	?

Thiospirillum jenense	Spiral 2.5-4.5 x 30-40	+	None	–	–	Vesicles	+	Orange-brown	*a*	ly, rh	45.5	+	–	–
sanguineum	Spiral 2.5-4.0 x 40	+	None	–	–	?	+	Purple-red	*a*	ok?	?	?	?	?
rosenbergii	Spiral 1.5-2.5 x 4-12	+	None	–	–	?	+	Red	*a*	?	?	?	?	?
Thiocapsa roseopersicina	Sphere 1.2-3.0	–	None	+	–	Vesicles	+	Pink-red	*a*	sp	63.3-66.3	–	–	+
pfennigii	Sphere 1.2-1.5	–	None	–	–	Tubes	+	Orange-brown	*b*	?	69.4-69.9	–	–	+
Lamprocystis roseopersicina	Sphere 3.0-3.5	+	None	–	+	Vesicles	+	Purple	*a*	la, lo	63.8	–	–	?
Thiodictyon elegans	Rod 1.5-2.0 x 3-8	–	Nets	–	+	Vesicles	+	Purple-violet	*a*	rl, rh	65.3	–	–	?
bacillosum	Rod 1.5-2.0 x 3-6	–	Clumps	–	+	Vesicles	+	Purple-violet	*a*	rl, rh	66.3	–	–	?
Thiopedia rosea	Ovoid 1.0-2.0x1.2-2.5	–	Platelets	–	+	Vesicles	+	Purple-violet	*a*	rl	?	?	–	?
Amoebobacter roseus	Sphere 2.0-3.0	–	None	+	+	Vesicles	+	Pink-red	*a*	sp	64.3	+	–	+
pendens	Sphere 1.5-2.5	–	None	+	+	Vesicles	+	Pink-red	*a*	sp	65.3	–	–	+
Ectothiorhodospira mobilis	Spiral 0.7-1.0 x 2-2.6	+	None	–	–	Stacks	–	Brown-red	*a*	sp, rh	67.3-69.9	+	+	+
shaposhnikovii	Spiral 0.8-0.9 x 1.5-2.5	+	None	–	–	Stacks	–	Brown-red	*a*	sp, rh	62.3	–	–	+
halophila	Spiral 0.8x5.0	+	None	–	–	Stacks	–	Red	*a*	sp	68.4	–	+	?

[a] la = lycopenal; lo = lycopenol; ly = lycopene; ok = okenone; rh = rhodopin; rl = rhodopinal; sp = spirilloxanthin.

optimal growth temperature, in most cases, lies between 20°C and 30°C. Species containing gas vacuoles become buoyant at lower temperatures.

TABLE 7

GENERAL PROPERTIES OF THE CHLOROBIACEAE

Organism	Cell Shape and Dimensions, Diameter (Width) x Length, μm	Motile by Flagella	Aggregates Formed in Pure Culture	Gas Vacuoles	Color of Cell Suspension	Predominant Bacterio-chlorophyll	Predominant Carotenoid[b]	DNA Base Ratio,[14] moles % G + C
Chlorobium								
limicola[a]	Rod 0.7-1.1 x 0.9-1.5	–	None	–	Green	*c, d*	chl	51.0-58.1
vibrioforme[a]	Vibrio 0.5-0.7 x 1.0-2.0	–	None	–	Green	*d, c*	chl	52.0-57.1
phaeobacteroides	Rod 0.6-0.8 x 1.3-2.7	–	None	–	Brown	*d*	irt	49.0-50.0
phaeovibrioides	Vibrio 0.3-0.4 x 0.7-1.4	–	None	–	Brown	*d*	irt	52.0-53.0
Prosthecochloris								
aestuarii	Sphere with prosthecae 0.5-0.7 x 1.0-1.2	–	None	–	Green	*c*	chl	50.0-56.0
Chloropseudomonas								
ethylica	Rod 0.7-0.9 x 1.0-1.5	+	None	–	Green	*c*	chl	55.6
Pelodictyon								
luteolum	Ovoid 0.6-0.9 x 1.2-2.0	–	Clumps, spheres	+	Green	*c, d*	chl	53.5-58.1
clathratiforme	Rod 0.7-1.2 x 1.5-2.5	–	Nets	+	Green	*c, d*	chl	48.5
Clathrochloris								
sulfurica	Sphere 0.5-1.5	–	Strings?	+	Green	?	?	?

[a] The thiosulfate utilizing strains of these two spieces are assembled as formae: *C. limicola* f. *thiosulfatophilum* and *C. vibrioforme* f. *thiosulfatophilum*.
[b] chl = chlorobactene, irt = isorenieratene.

REFERENCES

1. Stanier, R. Y., in *20th Symposium of the Society for General Microbiology*, p. 1. Cambridge University Press, Cambridge, England (1970).
2. Jensen, A., Asmundrud, O., and Eimhjellen, K. E., *Biochim. Biophys. Acta*, *88*, 466 (1964).
3. Liaaen-Jensen, S., *Annu. Rev. Microbiol.*, *19*, 163 (1965).
4. Pfennig, N., *Annu. Rev. Microbiol.*, *21*, 285 (1967).
5. San Pietro, A. (Ed.), *Methods in Enzymology*, Vol. 23, Part A, S. P. Colowick and N. O. Kaplan, Chief Eds. Academic Press, New York (1971).
6. Gest, H., and Kamen, M. D., in *Encyclopedia of Plant Physiology*, Vol. 5, Part 2, p. 568, W. Ruhland, Ed. Springer Verlag, Berlin, Germany (1960).
7. Pfennig, N., and Trüper, H. G., *Int. J. Syst. Bacteriol.*, *21*, 11-24 (1971).
8. Cohen-Bazire, G., and Sistrom, R. W., in *The Chlorophylls*, p. 313, L. P. Vernon and G. R. Seely, Eds. Academic Press, New York (1966).

9. Clayton, R. K., in *Encyclopedia of Plant Physiology,* Vol. 17, Part 1, p. 371, W. Ruhland, Ed. Springer Verlag, Berlin, Germany (1957).
10. Van Niel, C. B., *Bacteriol. Rev., 8,* 1-118 (1944).
11. Van Niel, C. B., in *Bergey's Manual of Determinative Bacteriology,* 7th ed., p. 35, R. S. Breed, E. G. D. Murray and N. R. Smith, Eds. Williams and Wilkins, Baltimore, Maryland (1957).
12. Pfennig, N., and Trüper, H. G., in *Bergey's Manual of Determinative Bacteriology,* 8th ed. (In preparation).
13. Kondratieva, E. N., *Photosynthetic Bacteria.* Russian: Izdadelstvo Akademiya Nauk SSSR, Moskva (1963); English translation: Israel Program for Scientific Translations, Jerusalem (1965).
14. Mandel, M., Leadbetter, E. R., Pfennig, N., and Trüper, H. G., *Int. J. Syst. Bacteriol., 21,* 222 (1971).
15. Pfennig, N., *J. Bacteriol., 99,* 597 (1969).
16. Winogradsky, S., *Zur Morphologie und Physiologie der Schwefelbakterien.* Arthur Felix, Leipzig, Germany (1888).
17. Trüper, H. G., and Pfennig, N., *Int. J. Syst. Bacteriol., 21,* 8 (1971).

BUDDING AND PROSTHECATE BACTERIA

DR. JAMES T. STALEY

The budding and prosthecate bacteria are a remarkable group of microorganisms that differ from typical eubacteria in their mode of reproduction, distinctive shapes, and complex life cycles. It is a heterogeneous assemblage, including some that reproduce by budding, others that produce cellular appendages (prosthecae), and still others that are both budding and prosthecate. Nonetheless, they are considered eubacteria, because motile strains have flagella and all have rigid cell walls. With rare exceptions, they are free-living bacteria that are widely distributed in aquatic and soil environments.

Binary fission versus budding. Most bacteria reproduce by binary transverse fission, as illustrated in Figure 1a. In this process a newly formed cell elongates along its longitudinal axis as it grows until its length approximately doubles (little, if any, change occurs in cell diameter). Then a transverse septum (cell wall and membrane) is synthesized near the center of the cell, partitioning the cell into two equal daughter or, more precisely, sister cells. When division is complete, the cells separate. During the entire process the cell remains symmetrical with regard to both longitudinal and transverse axes. Each sister cell receives nearly equal numbers of cell-wall molecular subunits from the mother cell, so that the mother cell loses its identity upon division.

Reproduction by budding is considerably different (Figure 1b). In this process growth occurs until the cell matures sufficiently for reproduction. Then the new cell is synthesized at a specific location on the surface of the mother cell. The most distinctive feature of bud formation is that the cell envelope of the bud is synthesized *de novo*, so that virtually none of the parental wall material is incorporated into the bud. This is most clearly evident in certain of the prosthecate budding forms, as illustrated for *Ancalomicrobium adetum* (Figure 2).

Another characteristic feature of budding is that the cells do not maintain symmetry with respect to the transverse axis throughout the division cycle, although symmetry may occur at and immediately following the time of division. Cells retain their own identity upon division, so that the genealogy of a clone is established as multiplication proceeds.

Most bacteriologists regard budding as a relatively rare process restricted to a small group of bacteria. Such is hardly the case, for representatives of this group are found in three of the four nutritional groups: photoheterotrophs, chemoheterotrophs, and chemoautotrophs. Furthermore, a number of bacteria divide by processes that are intermediate between budding and binary transverse fission, such as the genus *Caulobacter*, whose reproduction will be discussed in more detail later.

Prosthecae. Prosthecae are one of several types of bacterial appendages. They are structurally different

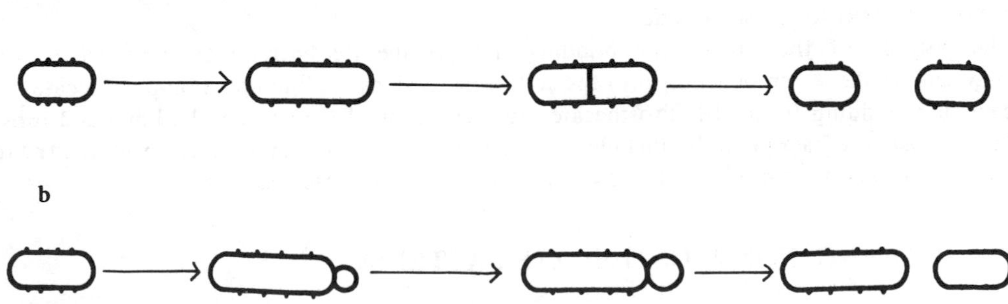

FIGURE 1. Major Differences Between Reproduction by Binary Transverse Fission and Budding. Markings on the cell surface represent cell wall subunits. In binary transverse fission (a) the cell wall material is distributed equally to the two daughter cells, although perhaps not in the manner illustrated above. In budding (b) the bud is synthesized from newly formed wall material, so that the mother cell retains the original cell wall subunits.

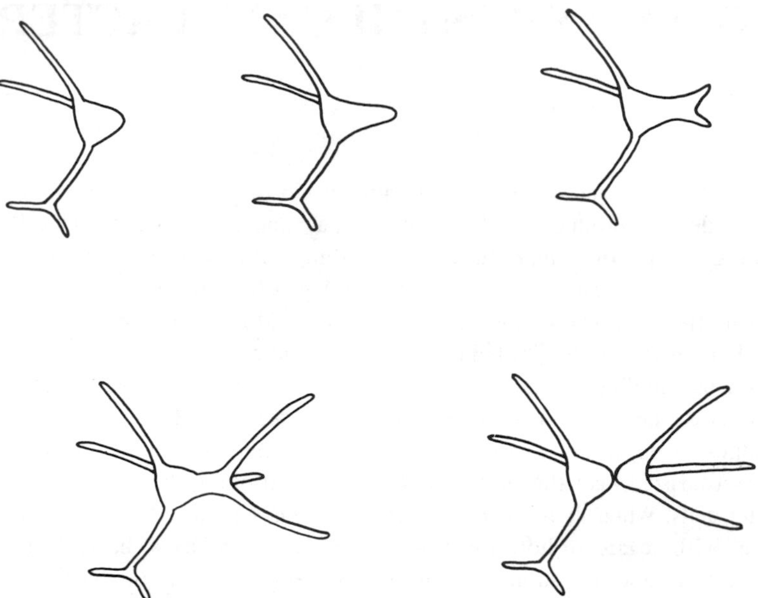

FIGURE 2. Bud Formation in *Ancalomicrobium adetum*. Growth occurs at only one location on the cell (upper left). When a certain stage is reached, the newly forming bud begins to differentiate and to form prosthecae (upper right). Growth continues until separation occurs (lower right). The mother cell with the bifurcated prostheca has retained its identity throughout and is now ready to repeat the division cycle.

from other prokaryote appendages (*viz.* flagella, pili (fimbrae), and acellular stalks), being defined as semirigid cellular appendages bound by the cell wall.[36]

The function of prosthecae is not well understood; however, we do know some functions they serve for certain bacteria. For example, in *Hyphomicrobium* and *Rhodomicrobium* they fulfill a reproductive role by mediating bud formation at their distal tips. One could also conjecture that they aid in the dispersal of daughter cells in these genera, or even that they are a mechanism by which these microbes reduce clonal competition by separating, through growth, the mother cells from the daughter cells.

Clearly, prosthecae do not serve a reproductive role for the caulobacters. Poindexter[1] has hypothesized that the caulobacter stalks serve to retard sedimentation of the cells from the aquatic habitats in which they reside and are therefore analogous to the appendages of certain planktonic algae, such as *Scenedesmus* and *Staurastrum*. Pate and Ordal[2] suggested that these structures may have been adapted to increase the surface area of cells, thereby permitting an increase in the rate at which limiting nutrients could be taken up in the nutrient-poor aquatic habitat. Neither function precludes the other, and it could well be that both functions are important for these bacteria.

For the purpose of this article, the budding and prosthecate bacteria are grouped according to morphology and mode of reproduction into the following sections: I. Budding, non-prosthecate forms. II. Prosthecate, non-budding forms. III. Prosthecate, budding forms. IV. Unnamed budding and prosthecate forms. Table 1 lists the bacteria to be included. For additional information on these bacteria, the reader is referred to general review articles[3,4,5] that discuss some or most of these organisms.

GROUP I. BUDDING, NON-PROSTHECATE BACTERIA

Except for their ability to reproduce by budding, the bacteria in this group share few common features. The morphological range spans from spherical to rod-shaped cells, unicellular to multicellular, and appendaged to non-appendaged. Figure 3 shows schematic drawings illustrating in part the range in morphological diversity of the bacteria included in this group.

TABLE 1

REPRESENTATIVES OF THE BUDDING AND PROSTHECATE BACTERIA

Scientific or Descriptive Name	Studied in Pure Culture? (References)	Nutritional Category
Group I. Budding, Non-Prosthecate Bacteria		
Pasteuria-Blastobacter spp.	Yes (3,6,9)	Chemoheterotroph
Planctomyces-Blastocaulis spp.	No (7,11)	Chemoheterotroph (?)
Metallogenium spp.	Yes (21,24,67)	Chemoheterotroph
Nitrobacter winogradskyi	Yes (25)	Chemoautotroph
Rhodopseudomonas acidophila	Yes (26)	Photoheterotroph
Group II. Prosthecate, Non-Budding Bacteria		
Caulobacter spp.	Yes (1)	Chemoheterotroph
Asticcacaulis excentricus	Yes (1)	Chemoheterotroph
Prosthecomicrobium spp.	Yes (36)	Chemoheterotroph
Prosthecochloris aestuarii	Yes (39)	Photoautotroph
Group III. Prosthecate, Budding Bacteria		
Hyphomicrobium spp.	Yes (44,47)	Chemoheterotroph
Hyphomonas polymorpha	Yes (51)	Chemoheterotroph
Rhodomicrobium vannielii	Yes (52)	Photoheterotroph
Rhodopseudomonas palustris (and *R. viridis*)	Yes (57)	Photoheterotroph
Ancalomicrobium adetum	Yes (36)	Chemoheterotroph
Pedomicrobium spp.	No (61)	Chemoheterotroph (?)
Group IV. Unnamed Prosthecate and Budding Bacteria		
Budding microbe from Baltic Sea	Yes (63)	Chemoheterotroph
Non-motile fusiform caulobacter	Yes (64)	Chemoheterotroph
Caulobacter with two prosthecae	Yes (2)	Chemoheterotroph

The nutritional range is equally broad, for photo- and chemoheterotrophs as well as chemoautotrophs are represented. Table 2 lists the major properties to differentiate among the individuals within this group.

Pasteuria-Blastobacter

HISTORY AND HABITAT

In his early studies on phagocytosis, Metchnikoff conducted observations on water fleas of the genus *Daphnia*. These zooplankton harbored parasitic microorganisms in their body cavities. His photomicrographs show that the parasites, which he named *Pasteuria ramosa,* were pear-shaped and aggregated with their narrow poles inward to form rosettes.[6] The bacterium was described as non-motile and reproducing by either longitudinal fission or by the production of endogenously formed polar spores. It was not isolated. Subsequent investigations by Henrici and Johnson[7] and by Zobell and Upham[8] revealed that bacteria of this morphology were common periphytes in fresh and marine waters. These investigators reported, however, that *Pasteuria* divided by budding, a feature not noted by Metchnikoff.

Zavarzin, in his review of the budding bacteria,[3] concluded that the bacteria observed by Henrici and Johnson were entirely different from *Pasteuria ramosa* as described by Metchnikoff. Therefore he proposed

TABLE 2

SPECIES OF BUDDING, NON-PROSTHECATE BACTERIA

Organism	Cell Shape	Motility	Appendages	Morphology of Organism	Other Distinctive Factors
Pasteuria ramosa (or *Blastobacter henricii*)	Oval or pear	Single subpolar flagellum	Absent	Unicellular and rosettes	
Planctomyces bikefii (or *Blastocaulis sphaerica*)	Pear to spherical (1.3-1.5 μm) on appendages	Non-motile (?)	2.7-4.4 μm long, up to 0.3 μm wide	Rosettes	Stalks frequently encrusted with ferric hydroxide
Planctomyces gracilis	Spherical (0.3-0.6 μm) on appendages	?	Up to 11 μm long, 0.3-0.4 μm wide	Rosettes	
Planctomyces crassus	Spherical (1.3-1.7 μm) on appendages	?	Up to 3.3 μm long, 0.7-2.0 μm wide	Rosettes	
Planctomyces guttaeformis	Bell-shaped (2.8-4.7 μm long, 1.0-1.6 μm wide)	?	Wide, possibly prosthecae	Rosettes	A long (*ca.* 20 μm), thin (*ca* 0.1 μm) appendage extends from the non-attaching pole
Planctomyces condensatus	Spherical	?	Very thin and short	Rosettes with central granule	Central granule always encrusted with ferric hydroxide
Metallogenium personatum	Filament; small cocci	?	Absent	Rosettes with radiating filaments; buds formed at tips of filaments	Free-living; entire organisms encrusted with manganese oxide
Metallogenium symbioticum	Tapering filament; small cocci	Yes	Absent	Rosettes with radiating filaments; buds formed laterally on filaments	Mycoplasma; entire organism encrusted with manganese oxide
Nitrobacter winogradskyi	Oval to rod	Single subpolar flagellum	Absent	Unicellular	Nitrite-oxidizing autotroph
Rhodopseudomonas acidophila	Rod to bent rod	Single polar flagellum	Absent	Unicellular and rosettes	Photoheterotroph

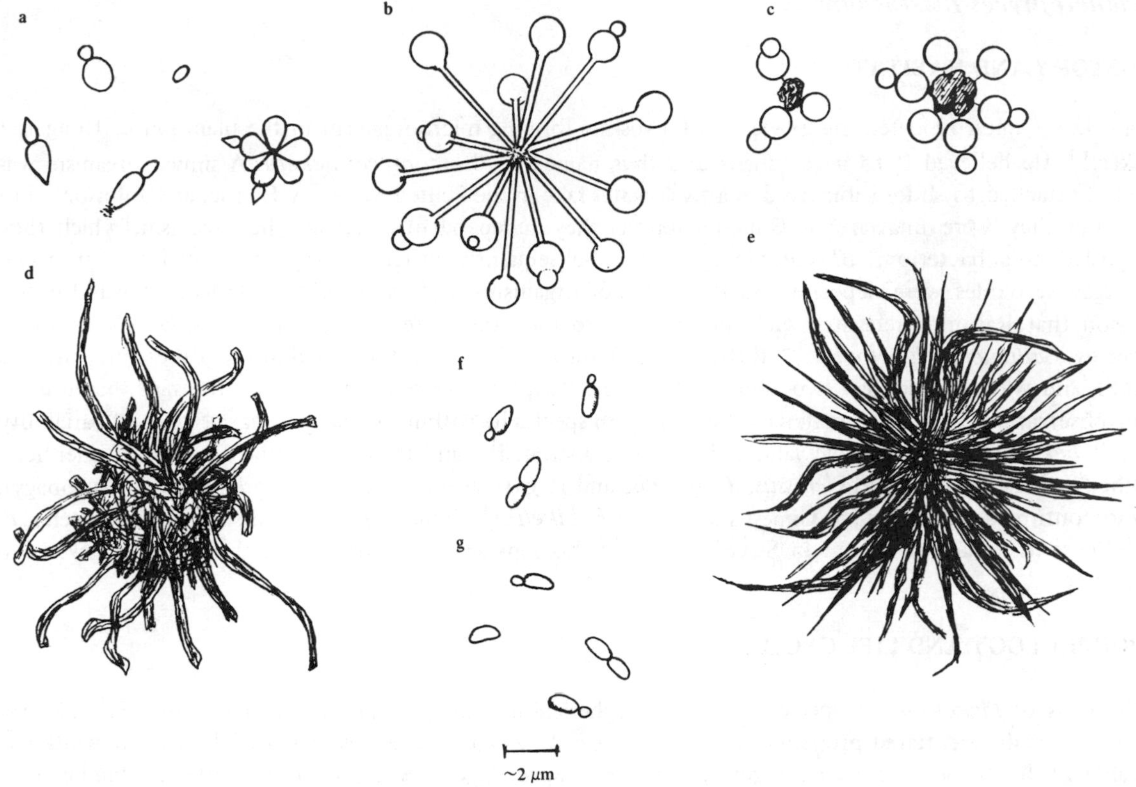

FIGURE 3. Illustrations of Budding, Non-Prosthecate Bacteria. Several cells show the variation in morphology within the *Pasteuria-Blastobacter* group (a). A rosette of *Planctomyces bikefii* is shown in b; the stalks of this microbe may become encrusted with iron hydroxide. *Planctomyces condensatus* is shown with its characteristic iron-encrusted granule at the center of the microcolony (c). The genus *Metallogenium* shows some variation in morphology; the type reported by Guseva[19] consists of a number of manganese-encrusted tubules with a fairly constant diameter (d), whereas the strains studied by Perfil'ev and Zavarzin[3] contain tapering filaments (e). *Nitrobacter winogradskyi* and *Rhodopseudomonas acidophila* are shown in (f) and (g), respectively.

the creation of a new genus and species, *Blastobacter henricii,* to include non-motile, budding bacteria like those reported by Henrici and Johnson and also observed by himself. The first report of the isolation of bacteria of the *Pasteuria-Blastobacter* group was in 1970.[9]

LIFE CYCLE AND MORPHOLOGY

Both of the isolated strains have pear-shaped cells, 0.5 to 2.0 μm in diameter, with holdfast structures located at the narrow pole. Buds are formed at or near the opposite pole of the cell, and, as in *Caulobacter,* the cells aggregate to form rosettes and attach to particulate substrates (Figure 3a). Both strains have motile stages, a feature not reported previously from observations of impure cultures.

NUTRITION AND ISOLATION

The two strains were isolated from pond and lake water samples to which peptone had been added to a final concentration of 0.01%. After these enrichment cultures had incubated two to four weeks at room temperature, portions were streaked onto 1.5% agar plates containing 0.025% each of yeast extract, peptone, and glucose, as well as a mineral-salts solution [Hutner's modified (HMS)[10]] and a vitamin solution.[36] Colonies developed after three to four weeks of incubation at room temperature and were pigmented white and yellow. They were purified on that medium by restreaking. Both strains are chemoheterotrophic.

Planctomyces-Blastocaulis

HISTORY AND HABITAT

In 1924, Gimesi reported the existence of a rosette-forming microorganism in the plankton of Hungarian lakes.[11] He believed it to be a fungus and thus named it *Planctomyces bekefii*. A similar organism was found attached to slides submerged in a fresh-water lake in the United States by Henrici and Johnson,[7] but because they were unaware of Gimesi's report, they independently named the organism, which they regarded as a bacterium, *Blastocaulis sphaerica.* Subsequently other investigators noted that iron and manganese oxides were deposited on the stalks of organisms of this type.[12,13] Indeed, it was for this reason that Rasumov relegated such an organism to the iron bacteria in the genus *Gallionella* as a new species, *Gallionella planctonica.*[14] Ruttner[13] and Wawrik[15] studied the distribution of these organisms in lakes, and Wawrik described two additional species, *Planctomyces subulatus* and *P. stranskae*, based upon her observations of natural materials. Because both species lie within the morphological range of variability for *P. bekefii,* their taxonomic validity has been questioned[17] and they will not be discussed further here. Other species, *Planctomyces crassus, P. gracilis,* and *P. guttaeformis,* have been proposed by Hortobagyi, who confirmed and extended Gimesi's studies of *P. bikefii.*[69] Skuja has described still another species, *P. condensatus,* which he found in Swedish lakes.[16] No representatives of any of these species have been isolated.

MORPHOLOGY AND LIFE CYCLE

The cells of *Planctomyces* species are oval to spherical in shape and are borne at the tip of a stalk. The species are differentiated primarily on the basis of cell size and shape as well as stalk length and width (see Table 2). In all species the cells aggregate with their appendages inward to form rosettes containing up to sixteen cells (Figure 3b). Buds are formed at or near the non-stalked pole of the cell. It is presently not known whether the stalks are prosthecal in nature. Electron-microscopic examinations in our laboratory suggest that this is the case in only one species, *P. guttaeformis;* others appear to be comprised of a fibrous material.

Planctomyces condensatus differs substantially from the aforementioned species (Figure 3c). Its spherical cells (0.7 to 1.7 μm in diameter) are borne upon very thin filaments. Microcolonies contain up to and perhaps in excess of sixteen cells, and rosettes analogous to those of the other species are formed. In contrast to the others, there is a large irregularly shaped granule (1.5 to 4.0 μm in diameter) located at the center of each colony. The granule is always encrusted with ferric hydroxide. Small spherical buds are formed at or near the non-stalked pole of the cell. A form similar to this was reported by Sokolova, who thought it might be a species of *Pasteuria.*[18]

NUTRITION AND PHYSIOLOGY

On the basis of their distribution in lakes, the *Planctomyces-Blastocaulis* group appear to be chemoheterotrophic, mesophilic, and obligately aerobic.

Metallogenium

HISTORY AND HABITAT

Guseva[19] reported large numbers of an iron- and manganese-encrusted microorganism from lake sediments in Russia (Figure 3d). Her findings confirmed and extended the earlier observations of Perfil'ev, who subsequently proposed the name *Metallogenium personatum* for these microorganisms. It was not isolated. Zavarzin[21,22] obtained a monoxenic culture, which he ascribed to the genus as a new species, *M. symbioticum* (Figure 3e). He was unable to grow it away from a fungus with which it was always associated. Subsequently Dubinina isolated the organism by plating it on a medium containing horse serum.[67,68] The requirement for serum coupled with the electron-microscopic evidence led Dubinina to conclude that *Metallogenium* is a cell-wall-less bacterium analogous to the mycoplasmas. If so, it would not belong to the prosthecate and budding bacteria but rather to the Mycoplasmatales discussed elsewhere in the Handbook. Frantsev[23] and Perfil'ev and Gabe[24] regard *Metallogenium* as a very important genus in manganese-cycling in fresh-water lakes.

MORPHOLOGY AND LIFE CYCLE

Because *Metallogenium* is always entirely encrusted with metal oxides, the organism itself has been difficult to observe and study. When freshly collected specimens are viewed with the light microscope, the microcolonies appear as heavily salified brown and yellow filamentous rosettes, some more than 20 μm in diameter. If the manganese encrustations are dissolved by oxalic acid, the organism itself is exposed. It consists of a number of twisted filaments that radiate from the center of the microcolony, tapering outward from a broad base (*ca.* 1.0 μm wide) to a fine tip (*ca.* 0.1 μm).

Zavarzin has followed the growth of *M. symbioticum* by direct observation in the microscope.[21] Round motile buds are formed directly on the lateral surfaces of the filaments. This is in contrast to *M. personatum*, which produces buds apically on its appendages.[24] The buds separate and, through growth and differentiation, develop into the filamentous colonies.

ISOLATION AND NUTRITION

Zavarzin purified *M. symbioticum* using a medium containing 0.01% manganous acetate and distilled water and solidified with 1.5% washed agar (Lieske's acetate agar). Mud samples were inoculated into the molten medium and plates were poured. After incubation for one week, brown colonies of *M. symbioticum* appeared intermixed with a fungus. The organisms were also grown in liquid media containing either 2% purified starch or 2% gum arabic with freshly prepared 0.5% manganous carbonate. The fungus was isolated in pure culture on Czapek's medium, but *Metallogenium* could not be grown by itself. Dubinina finally isolated *Metallogenium* from its symbiont by streaking it on a medium containing 0.1% starch, 1 to 20% horse serum, and saturated with manganese carbonate.[67] Zavarzin regards the organism as an aerobic chemoheterotrophic bacterium, since manganous ion is not required for energy production.

Nitrobacter winogradskyi and Rhodopseudomonas acidophila

Nitrobacter winogradsky and *Rhodopseudomonas acidophila* are usually associated with other groups of bacteria because of their physiological properties. Nevertheless, both have been shown to reproduce by budding, and for this reason deserve some mention in this section.

Although *Nitrobacter winogradskyi* has been known since the latter part of the 19th century, it was not regarded as a budding bacterium until the recent studies of Zavarzin and Legunkova.[25] These scientists reported that this bacterium undergoes a simple life cycle, in which budding is the sole means of reproduction. During logarithmic growth, non-motile pear-shaped mother cells (0.45 x 1.0 μm) produce buds from one pole of the cell. The buds develop flagella in a subpolar position prior to separation from the mother cell. Eventually they become non-motile and capable of forming their own buds, thus repeating the cycle. The physiology of this bacterium, which is a Gram-negative chemoautotrophic nitrite oxidizer, is treated in more detail in the section on autotrophic bacteria. in the Handbook (p.   ).

Rhodopseudomonas acidophila was recently isolated and described by Pfennig as a new species of the budding non-sulfur purple bacteria.[26] Several strains were obtained from a variety of fresh-water sources, using Athiorhodaceae enrichments at low pH (5.2) with succinate as the carbon source. Budding reproduction was verified by following growth in slide cultures. Unlike the other species of *Rhodopseudomonas* that reproduce by budding (Group III, Budding, Prosthecate Bacteria), buds are formed directly on the surface of the mother cell. The bacterium is motile by polar flagella, Gram-negative, and photoheterotrophic. For more on the physiology of this genus, consult the section on phototrophic bacteria.

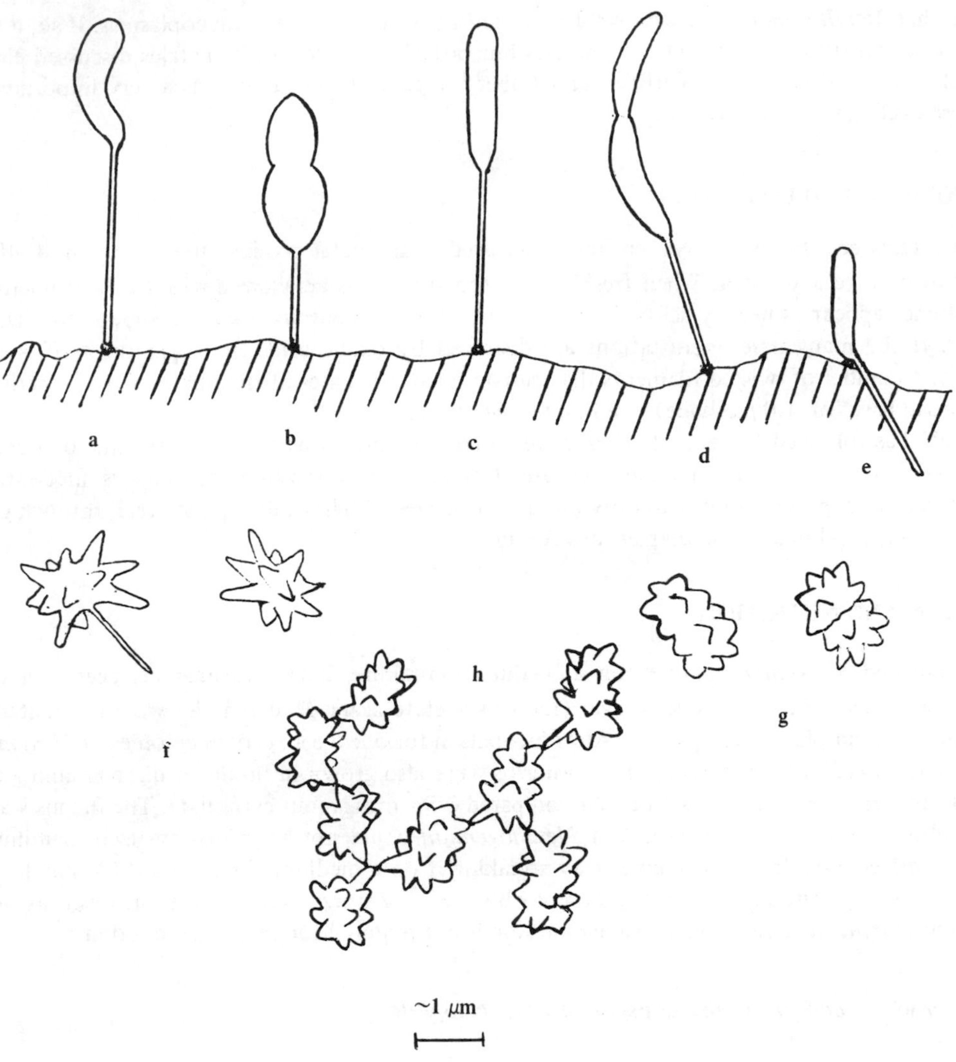

FIGURE 4. Illustrations of Prosthecate, Non-Budding Bacteria. The caulobacters are represented by the vibrioid type (a), the limnoid type (b), the bacteroid type (c), the subvibrioid type (d), all in the genus *Caulobacter,* and the subpolarly stalked type (e), *Asticcacaulis excentricus.* All are attached to a common substratum. *Prosthecomicrobium pneumaticum* (f) has longer appendages than *P. enhydrum* (g). Both are unicellular and heterotrophic, features that distinguish them from their photoautotrophic counterpart, *Prosthecochloris aestuarii* (h).

GROUP II. PROSTHECATE, NON-BUDDING BACTERIA

The bacteria within this group are illustrated in Figure 4. The caulobacters, including the genera *Caulobacter* and *Asticcacaulis*, are distinguished by their single prostheca, which is borne in either a polar (*Caulobacter*) or subpolar (*Asticcacaulis*) position. The genera *Prosthecomicrobium* and *Prosthecochloris* are distinguished morphologically from the caulobacters by the large number of prosthecae that are found on each cell. *Prosthecomicrobium*, a heterotroph, is morphologically indistinguishable from its photoautotrophic counterpart, *Prosthecochloris*.

Caulobacter and *Asticcacaulis*

HISTORY AND HABITAT

In 1905 Mabel Jones[27] reported the isolation of several strains of *Caulobacter* from tap water and sewage in Chicago. She did not name these bacteria, but her photomicrographs leave no doubt about their identity. She, and subsequently Omeliansky,[28] who isolated another strain of *Caulobacter*, misinterpreted the prostheca as a flagellum. Henrici and Johnson[7] correctly recognized that the appendages were stalks, and for this reason proposed the genus name *Caulobacter;* but they mistakenly concluded that the stalks were formed extracellularly by secretion, in a fashion analogous to those of *Gallionella*. The true nature of the stalk was not determined until Houwink[29] observed cells of *Caulobacter* with the electron microscope. He discovered that the cell wall of the bacterium extended *around* the appendage, thereby making it part of the cell.

The caulobacters are regarded primarily as aquatic bacteria since the observations of Henrici and Johnson,[7] who found large numbers in the periphyton of lakes. Water samples from Russian reservoirs[30] and a fresh-water stream in the United States[31] frequently contained from 100 to 10,000 caulobacters per ml. Jannasch and Jones[32] reported large numbers in the periphyton of marine habitats. These bacteria are not restricted entirely to aquatic habitats, though, as Poindexter has isolated them from soil and even from the intestinal tract of millipedes.[1]

LIFE CYCLE AND MORPHOLOGY

Populations of the caulobacters contain two cell types: swarmer cells and stalked cells. Each cell type represents a stage in the life cycle, which is diagrammed for *Caulobacter* in Figure 5a. The stalked cell lengthens as it grows and eventually develops a transverse septum. Differentiation commences at the non-stalked pole of the cell, resulting in the formation of a holdfast and a flagellum at that site. When division is complete, the flagellated swarmer cell separates from the stalked mother cell. The stalked cell can now repeat the cycle while the swarmer cell continues its development. In time the swarmer cell begins to synthesize its own stalk from the same pole at which the flagellum and holdfast are situated. As a result, the holdfast is located at the distal tip of the developing stalk along with the flagellum. When maturity is reached, the fully stalked cell has lost its motility and is prepared to produce a daughter cell of its own. The life cycle of *Asticcacaulis* (Figure 5b) is identical, except that its flagellum and stalk develop in a subpolar position; the holdfast, however, remains at the pole of the cell.

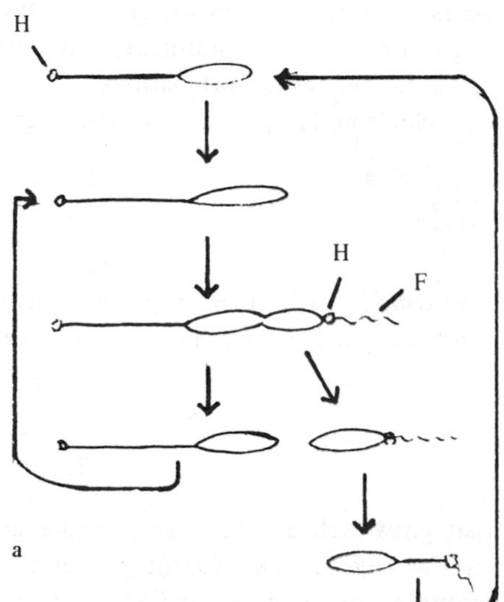

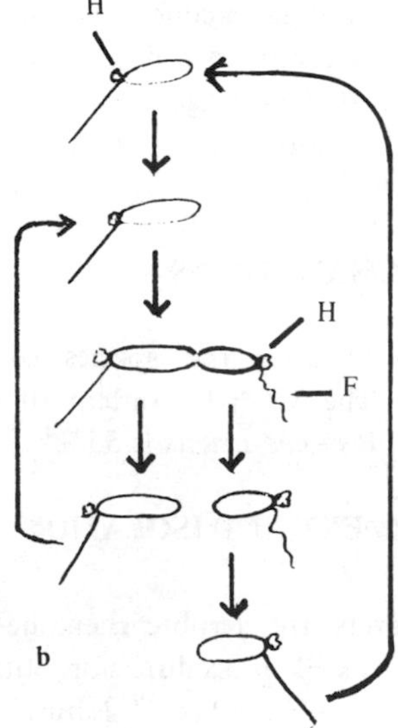

FIGURE 5. Life Cycle of *Caulobacter* and *Asticcacaulis*. In *Caulobacter* (a), the mature stalked cell elongates as growth proceeds until a new cell develops at the pole opposite the stalk. The new cell produces its flagellum (F) and holdfast (H) at the same position at the pole of the cell. Following separation, the newly formed cell continues its growth and development, which culminates in the formation of a stalk at the same locus as the flagellum. Thus the holdfast and flagellum are borne at the tip of the stalk. In time motility is lost, the flagellum disintegrates, and the mature stalked cell is prepared for reproduction. The process is identical in *Asticcacaulis* (b), except both flagellum and stalk develop in a subpolar position.

In pure cultures the caulobacters aggregate by their holdfast structures to form rosettes. In *Caulobacter*, the rosettes appear very much like the seed head of a dandelion, with the stalks meeting at their tips in the center. In *Asticcacaulis*, however, the cells themselves are at the center of the clump and the stalks, lacking holdfasts, extend away from the cluster. Rosettes are rarely seen in natural microbial communities, although the cells frequently attach to particulate materials and other microorganisms, indicating that the specificity of attachment is low.

Cell multiplication in the caulobacters differs substantially from our classical concept of binary transverse fission in bacteria. The cell lacks transverse symmetry throughout the entire division cycle. When division occurs, two morphologically different cells are formed, and the mother cell maintains her identity by virtue of her stalk, the cell envelope of which is completely retained by the mother cell. Although this process more closely resembles budding, it is probable that a substantial portion of the cell wall material from the mother cell is shared in the synthesis of the daughter cell. Thus, reproduction in the caulobacters seems to be a process that is intermediate between binary fission and budding.

The rod-shaped, vibrioid, and liminoid cells of the caulobacters are less than 1.0 μm in diameter (usually *ca.* 0.5 μm) and vary in length from 1.0 to more than 3.0 μm. The stalk diameter is less than 0.1 μm and therefore phase microscopy is recommended for observation, although preparations stained with simple basic stains (prepared with the ammonium oxalate-crystal violet used in Hucker's modification of the Gram stain) are satisfactory for ordinary bright-field microscopy. Though stalks are normally 3.0 μm in length, they may occasionally be much longer, especially under conditions of phosphate limitation; some exceeding 50 μm in length have been observed by the author in stationary-phase cultures. Conversely, some strains produce extremely short stalks, which may actually be imperceptible without the electron microscope. Thin sections reveal that the cell envelope consists of the three basic layers typical of other Gram-negative bacteria; an outer, slightly rippled double track, a middle dense layer, and an inner double-track cell membrane.[2,33]

NUTRITION AND PHYSIOLOGY

The caulobacters are chemoheterotrophs that can use a large number of sugars and organic acids and to a lesser extent sugar alcohols, polysaccharides, and amino acids as carbon sources for growth. Of ten strains tested for growth on a defined medium, all required a single vitamin, either biotin, B_{12}, or riboflavin.[1] Metabolic studies suggest that sugars are catabolized by the Entner-Doudoroff pathway. All strains are aerobic, but some can grow anaerobically with nitrate as an electron acceptor. Many strains grow well at and below 5°C.[1]

TAXONOMIC GROUPS

Table 3 includes the species of *Caulobacter* and *Asticcacaulis* with their important distinguishing features. The G + C content of the DNA of *Caulobacter* ranges from 62 to 67%, whereas that of *Asticcacaulis excentricus* is 55%.[1]

ENRICHMENT AND ISOLATION

Caulobacters are aerobic chemoheterotrophic bacteria that grow well at room temperature and neutral pH. The usual procedure for obtaining pure cultures is the peptone enrichment procedure originally described by Houwink.[29] Natural fresh- and salt-water samples are collected and transferred to beakers (100 ml in 150 ml beakers) containing enough peptone to give a final concentration of 0.01%. The beakers are covered with aluminum foil to reduce excessive contamination (if desired, the process can be conducted aseptically) and incubated at room temperature. The enrichment culture should be examined at daily intervals by removing portions of the surface film with a loop and observing a wet mount with a 100X phase-contrast objective. The proportion of stalked caulobacter cells usually peaks after four to seven days of incubation; attempts to isolate them are most fruitful at that time. Stove recommends streaking on a medium containing 0.05% peptone made up in tap water and solidified with 1.5% agar.[34] Wet mounts of

TABLE 3

SPECIES OF *CAULOBACTER* AND *ASTICCACAULIS*

Species	Cell Shape	Colony Pigmentation	Other Important Distinguishing Factors
Polar Stalk Position: *Caulobacter*			
C. henricii	Vibrioid	Bright yellow	Riboflavin not required
C. henricii auranticus	Vibrioid	Red-golden	Riboflavin not required
C. crescentus	Vibrioid	Colorless	No organic growth factors required
C. intermedius	Vibrioid	Colorless	Biotin required
C. vibrioides	Vibrioid or lemon-shaped	Colorless	Riboflavin required
C. vibrioides limonus	Lemon-shaped	Pale yellow	Riboflavin required
C. subvibrioides	Subvibrioid	Dark orange	
C. subvibrioides albus	Subvibrioid	Colorless	
C. fusiformis	Fusiform	Dark yellow	
C. leidyi	Fusiform	Colorless	
C. bacteroides	Rod-shaped	Colorless to orange	Pentoses used 2% NaCl inhibitory
C. variabilis	Rod-shaped	Colorless	Pentoses not used Stalk position variable 2% NaCl inhibitory
C. halobacteroides	Rod-shaped	Colorless	2% NaCl not inhibitory Amino acids utilized
C. maris	Rod-shaped	Colorless	2% NaCl not inhibitory Amino acids not utilized
Subpolar Stalk Position: *Asticcacaulis*			
A. excentricus	Rod-shaped	Colorless	Biotin only organic growth factor required

Data compiled from Poindexter, J. S., *Bacteriol. Rev., 28,* 231 (1964) and reproduced by permission of the American Society for Microbiology.

TABLE 4

SPECIES OF *PROSTHECOMICROBIUM* AND *PROSTHECOCHLORIS*

Species	Motility	Gas Vacuoles	Oxygen Requirement	Maximum Appendage Length
Prosthecomicrobium				
enhydrum	Yes	No	Obligate aerobe	0.5 μm
pneumaticum	No	Yes	Obligate aerobe	2.0 μm[a]
Prosthecochloris				
aestuarii	No	No[b]	Obligate anaerobe	2.0 μm

[a] Occasionally longer appendages are observed.
[b] Gas-vacuolated strains have been observed in lakes by the author.

individual colonies are examined under oil immersion after the plate has incubated for four to six days at 30°C. Colonies with stalked cells are restreaked for purification on a medium containing 0.1% peptone, 0.05% yeast extract and 0.01% $MgSO_4 \cdot 7H_2O$, prepared with tap water and solidified with 1.0% agar. This medium can also be used for the maintenance of pure cultures, which can be stored in the refrigerator for at least five weeks. For marine cultures, the above media should be supplemented with 3% NaCl or natural marine water should replace the tap water.

Prosthecomicrobium and Prosthecochloris

HISTORY AND HABITAT

Direct electron-microscopic observations of natural materials provided the first evidence that multiple-appendaged bacteria resembling *Prosthecomicrobium* existed. Both soil extracts[35] and materials sedimented from aquatic habitats[36,37] contained forms with several appendages per cell. Heterotrophic isolates of the genus *Prosthecomicrobium* were obtained from a fresh-water pond in the United States.[36] A photoautotrophic isolate was obtained from the mud of a shallow, brackish pond in Russia[38] and subsequently named *Prosthecochloris aestuarii.*[39] As yet no isolates have been obtained from soils or other habitats.

One distribution study showed that there were usually only 0.1 to 1.0 viable *Prosthecomicrobium* cells per ml of water from a polluted river.[31] They were found in considerably smaller numbers than the caulobacters throughout the investigation, suggesting that *Prosthecomicrobium* may be generally less abundant than the caulobacters in fresh-waters. The species included within these genera are listed in Table 4.

MORPHOLOGY AND LIFE CYCLES

These bacteria contain several prosthecae (normally 10 to 20), which extend in all directions from the cell surfaces (Figures 4f, 4g, 4h). The length of the appendages varies somewhat, but is generally less than 2.0 μm, although occasional appendages may exceed that length. Most of the prosthecae are conical in shape, tapering from a wide base (*ca.* 0.2 μm) on the surface of the cell to a blunt tip. Some strains also produce prosthecae of uniform width.

Thin sections indicate that the prosthecae contain typical cytoplasmic constituents, not the membranous material sometimes associated with the stalks of the caulobacters.[33] Those of *Prosthecochloris aestaurii* contain their photosynthetic pigments in chromatophores analogous to *Chlorobium* vesicles.[39] Unlike the caulobacters, holdfast structures are not associated with the prosthecae or with the body of the cell. The cells are coccobacillary in shape, measuring about 0.5 to 1.0 μm in diameter and 1.0 to 2.0 μm in length. The three species discussed below are Gram-negative.

Only one species, *Prosthecomicrobium enhydrum,* is motile. It has a single flagellum, which is located in a polar to subpolar position. The motility exhibited is peculiar but distinctive: cells wind in a circular path and somersault. Motile cells are associated with actively growing populations and are only rarely found in stationary-stage cultures.

Prosthecochloris aestaurii forms multicellular groups due to a tendency of the cells to remain attached after the completion of division.

One species, *Prosthecomicrobium pneumaticum,* produces gas vacuoles that are completely analogous structurally to the gas vacuoles produced in other prokaryotes, namely certain of the blue-green algae, the photosynthetic bacteria, and a few other heterotrophic bacteria.[40] Although the composition of the gas has not been analyzed in this bacterium, Walsby's studies on the gas vacuoles in blue-green algae[65] suggest that the vesicles contain all gases that are present in the cell's microenvironment. The function of the gas vacuoles in these bacteria is not known, but presumably they might enable these aerobes to migrate to the oxygenated surface of stratified lakes.

NUTRITION AND PHYSIOLOGY

Prosthecomicrobium strains are all obligately aerobic and utilize ammonium as a sole source of nitrogen for growth. A variety of carbon sources are used, including a number of sugars, some sugar alcohols, and a few organic acids. All strains require vitamins. *P. enhydrum* requires thiamine, and its growth is stimulated by biotin. *P. pneumaticum* requires biotin, thiamine and B_{12}, and is stimulated by folic acid. The metabolic pathways have not been determined. All strains are mesophilic, growing between 9 and 37°C.

Prosthecochloris aestuarii is an anaerobic, photoautotrophic green sulfur bacterium. Optimum growth occurs from pH 6.7 to 7.0 with 2 to 5% NaCl. Vitamin B_{12} is required for growth. Sulfide and elemental sulfur are used as electron donors in photosynthesis, and, as in *Chlorobium,* sulfur granules are formed when the cells are grown on sulfide. Acetate can be photoassimilated in the presence of sulfide and carbon dioxide. The primary photosynthetic pigment is bacteriochlorophyll c.

ISOLATION

Strains of *Prosthecomicrobium* can be isolated from peptone enrichments as described previously for caulobacters. These bacteria, however, develop more slowly in the enrichments; attempts at isolation are therefore not recommended until the cultures have incubated for three to five weeks at room temperature, providing appreciable numbers have been observed. A variety of reasons might account for their slow development. They normally occur in lower concentrations than the caulobacters and have longer generation times. Their late appearance might also be due in part to succession in the enrichment culture, perhaps initiated by the release of vitamins by the death and lysis of some of the primary organisms. The initial isolation procedure involved selective adsorption of other bacteria from the enrichments by passing the culture through a sterile column containing glass beads. This procedure is described in detail for *Ancalomicrobium* (see Group III), and because of its complexity is not now recommended for *Prosthecomicrobium.* In the new procedure, plates containing 0.01% peptone, distilled water, 20 ml per liter of Hutner's modified mineral salts (HMS[10]), and 10 ml per liter of a vitamin solution[10] solidified with 1.5% agar are streaked or spread with material from the three- to five-week enrichment. Then, after one to two weeks of incubation at room temperature, these bacteria can be located by examining each colony that develops individually in a wet mount with the oil-immersion phase objective. Once found, colonies are restreaked on the same medium until purified. Alternatively, colonies may be restreaked on a richer complex medium containing 0.02% peptone, 0.02% glucose, HMS, and the vitamin solution. After a strain has been isolated, it should be grown on complex media with higher concentrations of glucose (0.1%) and supplemented with peptone and yeast extract (0.025% each) to permit higher growth yields. All strains so far isolated (this also applies to those strains of caulobacters tested) grow on a defined medium that contains 0.025% ammonium sulfate, 0.1% glucose, HMS, and the vitamin solution. Cultures normally reach stationary phase after four to six days of incubation at 30°C and can be stored at refrigerator temperature for three to four months.

Prosthecochloris aestuarii may be isolated with either Larsen's medium[41] or Pfennig's medium.[42] The medium of Larsen consists of the following: 0.1% each of NH_4Cl, KH_2PO_4, and $Na_2S \cdot 9H_2O$; 0.05% $MgCl_2$; 0.2% $NaHCO_3$; and 1.0% NaCl (only for marine isolates). The initial pH of the medium is adjusted to 7.3. Sterile glass-stoppered bottles are filled completely with sterile medium and inoculated with mud samples that contain *Prosthecochloris.* Since B_{12} is required, small amounts of the vitamin should be added to the medium. The enrichments are incubated at 28 to 30°C with continuous illumination from 25- to 50-watt incandescent lamps. Pure cultures can be obtained by adding 2.0% agar to the above-mentioned medium and preparing shake tubes.

GROUP III. BUDDING, PROSTHECATE BACTERIA

Some of the most bizarre unicellular bacteria are found within this group. All undergo morphogenetic changes in their life cycles. The simplest cell forms are undifferentiated oval-, rod-, or pear-shaped buds. As growth proceeds, the bud differentiates to form one or more prosthecae before the organism completes its morphological development. Reproduction occurs solely by bud formation, and this normally happens only after the prosthecae have developed. The buds and prosthecae are synthesized at specific loci on the cell surface; the exact location is a characteristic of the strain or species. Usually the newly formed buds are motile by either polar to subpolar monotrichous or peritrichous flagella. Figure 6 illustrates some of the common cell shapes and colonial groupings characteristic of these genera. Tables 5 and 6 list some important properties of the species that have been studied in pure culture.

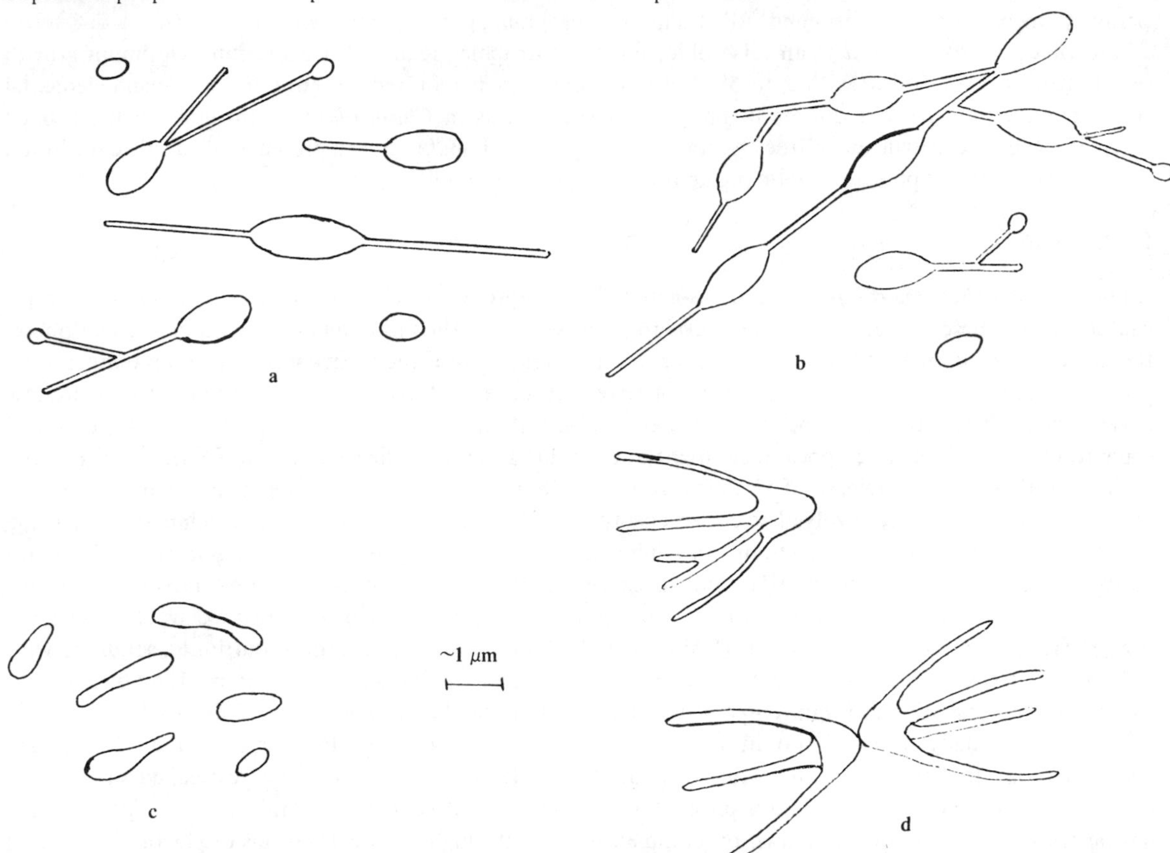

FIGURE 6. Illustrations of Prosthecate Budding Bacteria. Various forms typical of the genus *Hyphomicrobium* are shown in (a). The phototrophic genus *Rhodomicrobium* (b) is similar in appearance to *Hyphomicrobium,* but the cells have a greater tendency to remain together after division, so that multicellular forms are more common. The other photosynthetic budding and prosthecate species are represented by *Rhodopseudomonas palustris* and *R. viridis* (c). Their prosthecae are shorter, wider, and do not have bifurcations. Unlike the other bacteria in this group, *Ancalomicrobium adetum* (d) does not produce prosthecal buds.

TABLE 5

SPECIES OF PHOTOSYNTHETIC BUDDING AND PROSTHECATE BACTERIA

Species	Size of Prosthecae	Cell Diameter	Type of Flagellation	Aerobic Growth in Dark
Rhodomicrobium vannielii	Long (some > 3 μm) and thin (*ca.* 0.2 μm)	1.0–1.2 μm	Peritrichous	No
Rhodopseudomonas palustris (and *R. viridis*)	Short (< 2 μm) and wide (*ca.* 0.5 μm)	0.6–0.8 μm	Single, polar	Yes

TABLE 6

ISOLATES OF NON-PHOTOSYNTHETIC BUDDING AND PROSTHECATE BACTERIA

Species	Morphology	Pellicle Formation	Rosette Formation	Growth on Peptone	Nitrate Anaerobically
Hyphomicrobium					
vulgare	Unicellular rods with prosthecal buds; motile	+	–	–	–
neptunium	Unicellular rods with prosthecal buds; motile	–	+	+	±
Hyphomonas					
polymorpha	Unicellular rods with prosthecal buds; motile	?	?	+	?
Ancalomicrobium					
adetum	Non-prosthecal buds; non-motile	–	–	+	–

Data adapted from Hirsch, P., and Rheinheimer, G., *Arch. Mikrobiol.*, 62, 289 (1968) and reproduced by permission of the publishers.

Hyphomicrobium, Hyphomonas, Rhodomicrobium, and *Rhodopseudomonas*

HISTORY AND HABITAT

When *Hyphomicrobium* was first observed by Rullman[43] and subsequently studied and named by Stutzer and Hartleb,[44] it was interpreted as a nitrifying bacterium because it proliferated in enrichment cultures for these autotrophs. Later Large *et al.* discovered, however, that these bacteria were not living autotrophically[45] and that their close association with the nitrifiers was due instead to their ability to grow oligocarbophilically, using the small amounts of organic material produced by the autotrophs. Kingma-Boltjes[46] was the first to conclude that these bacteria actually reproduced by budding. A number of recent articles concerning the biology of this genus have been published by Hirsch and his collaborators.[47-50]

Hyphomicrobium strains have been isolated from a variety of sources, notably soil and fresh- as well as marine-water habitats. Many strains are associated with the oxidation of iron and manganese compounds, and some have been found in the jellied extracellular integument of certain colonial algae. At least one strain has been obtained from the nasal discharge of a person with sinusitis.[51] This isolate was named *Hyphomonas polymorpha* by Pongratz, who had inferred from the literature that *Hyphomicrobium* was a genus of autotrophs. Morphologically, it is indistinguishable from the genus *Hyphomicrobium.*

The discovery of *Rhodomicrobium vannielli* in 1949[52] was an exciting surprise to microbiologists, for here was a bacterium that could clearly belong to either of two major groups of the Schizomycetes. On the basis of its physiology, it is unquestionably a non-sulfur purple bacterium. An equally strong argument can be made for its placement with the budding bacteria on the basis of its morphology and life cycle. *Rhodomicrobium* strains have been isolated from mud of marine- and fresh-water sources.

Two species of *Rhodopseudomonas* produce abbreviated prosthecae and divide by budding, and for these reasons they are included in this section. The non-sulfur purple bacteria are widely distributed in fresh and marine waters and in muds, and a recent report on their occurrence indicates that they are occasionally found even in drier woodland and grassland soils.[53]

LIFE CYCLE AND MORPHOLOGY

Figure 7 shows a simplified life cycle for *Hyphomicrobium.* Like the caulobacters, its cultures have swarmer cells as well as prosthecate cells. The newly released buds are propelled by a single polar or subpolar flagellum. When a certain stage is reached, the cells usually produce a prostheca from one pole of the cell. The appendage elongates, and when maturity is attained, a bud develops at its distal tip. The bud enlarges, forms its own flagellum, and separates from the mother cell. As the daughter cell repeats the above cycle, the mother (prosthecate) cell proceeds to develop, following one of several available options. It may repeat the former pattern and again produce a bud at its prosthecal tip. Alternatively, another prostheca can be formed either from the same pole or the opposite pole of the cell. Another variation would be the bifurcation of the already existing prostheca, followed by bud formation at its tip. Then, too, buds are occasionally formed directly upon the mother cell without an intermediating prostheca. Thus, the options for development are much more diverse in *Hyphomicrobium* species than in the caulobacters.

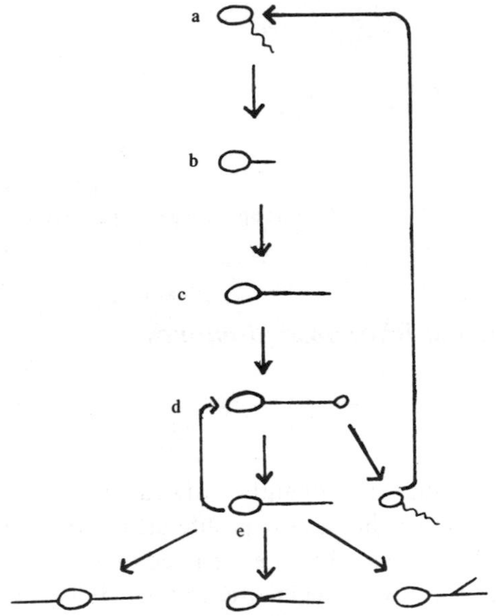

FIGURE 7. Life Cycle of *Hyphomicrobium.* Newly formed buds are uniflagellated with a subpolar flagellum (a). Motility is lost and a prostheca develops (b-c). A bud develops at the tip of the prostheca (d). It becomes motile and separates from the mother cell (e), and may repeat the life cycle. The mature mother cell has several options available. It may produce another bud at the same location as the previous one, or it may produce additional prosthecal appendages, or it may form branches on the already existing prostheca. The next step would probably be the formation of a new bud at the tip of one of the prosthecae.

Rosette formation occurs in one species, *Hyphomicrobium neptunium,* and is used as a characteristic to differentiate it from the other major species, *H. vulgare.*[66] The rosettes that are formed resemble those of *Asticcacaulis* because the holdfast structure is associated with the cells, not the prosthecae. The cells of *Hyphomicrobium vulgare* (Figure 6a) are oval to bean-shaped, measuring 0.5 to 0.75 μm in diameter and 0.6 to 5.0 μm in length. Other species may have much larger cells, 2.5 to 4.0 by 3.0 to 6.0 μm in size.[47] The prosthecal appendages, called hyphae, vary greatly in length but are uniformly 0.2 to 0.3 μm in diameter (about double the diameter of *Caulobacter* stalks). The hyphae do not have septations. Thin sections reveal that the prosthecal core contains the typical constituents characteristic of the cytoplasm; many strains also have lamellar membranes similar in appearance to those of *Rhodomicrobium.*[49] Mother cells invariably have a single refractile granule of poly-β hydroxybutyrate located in their cytoplasm. The *Hyphomicrobium* species are Gram-negative.

The life cycle of *Rhodomicrobium vannielii* closely resembles that of *Hyphomicrobium.* Growth occurs by the production of polar filaments (prosthecae) that develop buds at their tips. Usually the bud remains attached to the mother cell's appendage, resulting in the proliferation of a microcolony of cells. In this regard it differs markedly from the genus *Hyphomicrobium.* Prosthecal branching occurs, so that the microcolony extends to form a three-dimensional network. Motile cells are produced infrequently in younger cultures, but unlike those of *Hyphomicrobium* they have peritrichous flagella.[55] Mature cells are ovoid in shape and about 1.2 by 2.8 μm in size (Figure 6b). The appendages vary considerably in length but maintain a constant diameter of about 0.3 μm. The prosthecae are frequently septate. The photosynthetic membranes as revealed in thin sections have a lamellar configuration,[56] a feature that distinguishes all of the budding photoheterotrophs from their non-budding counterparts with vesicular chromatophores. The organism is Gram-negative.

The two species of *Rhodopseudomonas* that fit in this group, *R. palustris* and *R. virdis,* produce very short, thick and unbranching prosthecae (Figure 6c). In other respects their life cycles are similar to that of *Rhodomicrobium,* except that the motile stage has monotrichous flagella.[57] Some strains aggregate to form rosettes, with the mother cells attaching at the center.

NUTRITION AND PHYSIOLOGY

Hyphomicrobium is an obligate aerobe that grows chemoheterotrophically, utilizing ammonium or nitrate as sole inorganic nitrogen sources. Organic compounds with single carbon atoms, *viz.* formate, methanol, methylamine or urea are preferred as sole carbon sources;[48] some other organic compounds also support growth, but growth rates and yields are much reduced. No vitamins are required by those strains that have been isolated. *Hyphomicrobium* strains are mesophilic, growing at temperatures from about 5 to 40°C. The pH range for growth is 6.0 to 9.5. For some unknown reason, light from the visible spectrum has an inhibitory effect on growth.[48]

Rhodomicrobium vannielii grows well on a defined medium containing ammonium salts as the sole nitrogen source, sodium lactate as the sole carbon source, and the vitamin riboflavin.[58] Other organic hydrogen donors, such as ethanol, propanol, acetate, and butyrate, can also be used, but sugars and sugar alcohols cannot. The organism is mesophilic and obligately anaerobic; unlike other non-sulfur purple bacteria, it has not been grown aerobically in the dark.

The nutrition and physiology of the budding *Rhodopseudomonas* species is similar to that of the other Athiorhodaceae as discussed in the section on phototrophic bacteria.

ENRICHMENT AND ISOLATION

Hyphomicrobium develops in peptone enrichment cultures as described for the caulobacters. Its increase in number, however, is not appreciable until after about three weeks of incubation at room temperature. Pure cultures may be isolated on the 0.01% peptone medium described for *Prosthecomicrobium.* Hirsch[59] recommends enrichment in his mineral medium 337, composed, per liter, of the following: 1.36 g KH_2PO_4; 0.50 g $(NH_4)_2SO_4$; 9.95 mg $CaCl_2 \cdot 2H_2O$; 2.50 mg $MnSO_4 \cdot 4H_2O$; 2.13 g $NaHPO_4$; 0.20 g $MgSO_4 \cdot 7H_2O$; 5.00 mg $FeSO_4 \cdot 7H_2O$; 2.50 mg $NaMoO_4 \cdot 2H_2O$; 1.9% Difco Noble Agar; distilled water to obtain a volume of 1 liter. The pH is adjusted to 7.2 before autoclaving. A sterile 400-ml beaker covered with aluminum foil is filled with the medium and inoculated with 1 to 5 g of soil or 5 to 10 ml of water; it is then incubated in the dark at 30°C for one to four weeks. *Hyphomicrobium* can be isolated from the surface pellicle by streaking plates of medium 337 supplemented with 0.675% methylamine hydrochloride, then incubating in the dark.

Van Niel's procedure for the enrichment of non-sulfur purple bacteria[60] was adapted by Duchow and Douglas[52] for the isolation of *Rhodomicrobium.* The medium contains 0.5% $NaHCO_3$, 0.2% NaCl, 0.1% $(NH_4)_2SO_4$, 0.05% K_2HPO_4, 0.01% $MgSO_4 \cdot 7H_2O$; 0.01% $Na_2S \cdot 9H_2O$, 0.2% ethanol, and distilled water. The bicarbonate and sulfide salts as well as the ethanol are filter-sterilized in concentrated solutions, then added to the autoclaved solution of the other ingredients to make the required volume. The pH is adjusted to 7.0 with sterile H_3PO_4. The medium is distributed in glass-stoppered bottles filled to exclude air, inoculated with mud, and incubated at 25 to 30°C under continuous illumination from 25- to 40-watt incandescent bulbs. When the bacterium develops to significant numbers in the enrichment culture, it can be isolated in the enrichment medium supplemented with 0.2% yeast extract and solidified with 1.5% agar by using the shake culture technique.

Ancalomicrobium

HISTORY AND HABITAT

Only one strain of the genus has been isolated and described.[36] It is apparently distributed widely in natural fresh-waters, but preliminary studies suggest that it may be found in relatively low concentrations at least in temperate zones.[31]

MORPHOLOGY AND LIFE CYCLE

This bacterium superficially resembles *Prosthecomicrobium* in that it has numerous prosthecal appendages radiating from the cell (Figure 6d). It differs, however, in that the appendages are fewer and longer and sometimes form bifurcations. Moreover, it divides by budding (Figure 2). Newly formed buds normally differentiate two to four appendages on the cell surface prior to separation from the mother cell. Unlike those of *Hyphomicrobium,* its prosthecae do not have any reproductive function – buds are formed directly from the surface of the mother cell. Gas vacuoles are produced and are particularly abundant when the cells have entered maximum stationary growth. The cell is pear-shaped, about 1 μm in diameter, and the appendages are about 3.0 μm long, a feature that does not vary significantly with phosphate concentration or carbon source. The diameter of the appendages is about 0.2 μm. Thin sections show that the appendages contain the cytoplasmic constituents typical of the cell proper. The bacterium is non-motile and Gram-negative.

NUTRITION AND PHYSIOLOGY

Ancalomicrobium adetum is a facultative anaerobic chemoheterotroph that can utilize ammonium as sole nitrogen source and a number of mono- and disaccharides as well as sugar alcohols as sole carbon sources. Pantothenic acid is absolutely required for growth, and higher yields are obtained when biotin, folic acid, thiamine, and nicotinic acid are supplied. The bacterium grows anaerobically by the fermentation of sugars. It is mesophilic, growing at temperatures above 6°C and as high as 39°C.

ISOLATION

The strain that has been isolated developed in 0.01% peptone enrichment cultures (see Isolation of *Caulobacter,* Group II) from pond water along with *Prosthecomicrobium.* After the enrichment culture had incubated for four weeks at room temperature, 1 ml was removed and passed through a glass chromatographic column containing glass beads. The column (1 cm I.D.) contained 2″ of autoclaved Ballotini beads (*ca.* 0.2 mm in diameter) and was washed with 100 ml of hot, freshly autoclaved mineral-salts solution. When the column had cooled, the material from the enrichment culture was layered on top of the beads and 1 ml was drained from the column. Alternate drops were then either spread on 0.01% peptone plates solidified with 1.5% agar and prepared with pond water or added to flasks containing the same sterile liquid medium. In this manner some 30 drops were removed from the column and used as inocula. When these cultures had incubated at room temperature for three weeks, they were examined for growth. The first plate in the series that contained colonies had colonies of both *Ancalomicrobium* and *Prosthecomicrobium.* The colonies were purified by restreaking on the 0.01% peptone medium. Both genera grow on the defined medium previously described (see Isolation of *Prosthecomicrobium,* Group II).

Pedomicrobium

This genus was first described by Aristovskaya,[61] who was studying the microorganisms responsible for the oxidation of iron and manganese in organic extracts of Russian soils. The enumeration medium consisted of a fulvic acid fraction prepared by extracting 50 g of soil with 1 liter of 0.1*N* HCl. This extract was filtered, then the filtrate was neutralized to pH 5.0 with NaOH. The gel that formed was washed and combined with water and 1.5% agar. Pour plates were prepared with soil dilutions, and after three to four weeks of incubation small iron- and manganese oxide-encrusted colonies developed (10^4 to 10^5 per gram). When the iron and manganese oxides were dissolved with dilute HCl or oxalic acid respectively, a microorganism resembling *Rhodomicrobium* was exposed. Its growth habit consists of numerous cells interconnected by filaments to form microcolonies. Individual cells may have from one to four branching prosthecae extending from them. Two species were proposed: *Pedomicrobium ferrugineum* for iron oxidizers, and *P. manganicum* for manganese oxidizers. Unfortunately neither species was isolated.

Subsequently, Tyler and Marshall[62] isolated bacteria from manganese encrustations in hydroelectric pipe lines in Tasmania. They identified their isolates as strains of the genus *Hyphomicrobium.* Furthermore, they observed extreme pleomorphy when their isolates were grown under varying cultural conditions. Under some conditions they appeared identical to the classical morphology of *Hyphomicrobium,* whereas under other conditions they closely resembled the morphology described by Aristovskaya for *Pedomicrobium.* For this reason they concluded that *Pedomicrobium* was an invalid genus and should be regarded as a growth form of *Hyphomicrobium.*

GROUP IV. UNNAMED BUDDING OR PROSTHECATE BACTERIA

This article would be incomplete without a brief consideration of some other bacteria that have been isolated and are clearly budding and prosthecate but, for one reason or another, have not as yet been named.

A Budding Bacterium from the Baltic Sea

Ahrens and Moll described a most peculiar bacterium, which they isolated from the Baltic Sea.[63] The bacterium divides by multiple fission to produce a multicellular structure comprised of numerous small ($<0.5\ \mu m$) coccoid subunits. Oval to vibrioid buds are formed on the surface of some of the cells. These buds become uniflagellate swarm cells with lateral flagella. After separation from the mother microcolony they enlarge, become non-motile rods, and divide by multiple fission to produce the coccoid aggregate. The bacterium is Gram-positive and chemoheterotrophic.

A Non-Motile Fusiform Caulobacter

In their classic study on the bacterial periphyton of lakes, Henrici and Johnson[7] noted that there were several types of caulobacters attached to their submerged slides. They referred to one of the varieties as the fusiform type because its cells characteristically had pointed ends. By astute observations of microcolonies of the fusiform type they surmised that the organism was non-motile and, furthermore, that it divided by typical binary transverse fission to produce two symmetrical stalked cells at the time of separation. De Bont *et al.*[64] were the first to report the isolation of such an organism. The cells are about 0.5 μm in diameter, 4 to 10 μm in length, and have pointed ends typical of the fusiform type of Henrici and Johnson. The prostheca is slightly wider than the stalks of the other caulobacters and terminates in a bulbous tip about twice the normal stalk diameter. As predicted by Henrici and Johnson, the life cycle is monomorphic rather than dimorphic. Motility has not been observed, and at the time of division both cells have stalks.

A Caulobacter with Two Prosthecae

Pate and Ordal have isolated one of the strangest caulobacters.[2] It has a motile stage like *Caulobacter* and *Asticcacaulis*, but in the prosthecal stage the cell has two appendages. It more closely resembles *Asticcacaulis,* however, because the prosthecae extend laterally from the cell. Again, as in that genus, the holdfast is located at the pole of the cell, not on the prosthecae.

ACKNOWLEDGMENTS

The author is grateful for the helpful suggestions of E. J. Ordal and H. C. Douglas.

REFERENCES

1. Poindexter, J. S., *Bacteriol. Rev., 28,* 231 (1964).
2. Pate, J. L., and Ordal, E. J., *J. Cell Biol., 27,* 130 (1965).
3. Zavarzin, G. A., *Microbiology USSR* (English translation), *29,* 774 (1961).
4. Starr, M. P., and Skerman, V. B. D., *Annu. Rev. Microbiol., 19,* 407 (1965).
5. Schmidt, J. M., *Annu. Rev. Microbiol., 25,* 93 (1971).
6. Metchnikoff, M. E., *Ann. Inst. Pasteur (Paris), 2,* 165 (1888).
7. Henrici, A. T., and Johnson, D. E., *J. Bacteriol., 30,* 61 (1935).
8. Zobell, C., and Upham, H., *Bull. Scripps Inst. Oceanogr. Univ. Calif., 5,* 243 (1944).
9. Staley, J. T., *Abstracts of the 10th International Congress for Microbiology, Mexico, D. F.,* p. 6. Muñoz, S.A., Mexico City (1970).
10. Van Ert, M., and Staley, J. T., *J. Bacteriol., 108,* 236 (1971).
11. Gimesi, N., *Hydrobiologiai Tanulmányok,* Kiadja A Magyar Ciszterci Rend, Budapest, Hungary (1924).
12. Hortobagyi, T., *Borbasia Nova, 20,* 1 (1944).
13. Ruttner, F., *Arch. Hydrobiol.,* Suppl. *21,* 1 (1952).
14. Razumov, A. S., *Mikrobiologiya, 18,* 5 (1949).
15. Wawrik, F., *Sydowia Ann. Mycol. Ser. II, 6,* 443 (1952).
16. Skuya, H., *Nova Acta Reg. Soc. Sci. Ser. IV, 16,* 3 (1957).
17. Fott, B., and Komárek, J., *Preslia (Praha), 32,* 113 (1960).
18. Sokolova, G. A., *Microbiology USSR* (English translation), *28,* 230 (1959).
19. Guseva, K. A., *Tr. Biol. Sta. "Borok", 2,* 24 (1956).
20. Perfil'ev, B. V., *The Theory and Technique of Continuous-Flow Bacterial Microcultures* (in Russian). Iv. Leningr. gos. Univ. (1959).
21. Zavarzin, G. A., *Microbiology USSR* (English Translation), *30,* 343 (1961).
22. Zavarzin, G. A., *Z. Allg. Mikrobiol., 4,* 390 (1964).
23. Frantsev, A. V., *Tr. Vses. Gidrobiol. Obshchest, 9,* 13 (1959).
24. Perfil'ev, B. V., and Gabe, D. R., *Capillary Methods of Investigating Organisms* (English translation). University of Toronto Press, Toronto, Ontario, Canada (1969).
25. Zavarzin, G. A., and Legunkova, R. M., *J. Gen. Microbiol., 21,* 186 (1959).
26. Pfennig, N., *J. Bacteriol., 99,* 597 (1969).
27. Jones M., *Zentralbl. Bakteriol. Parasitenk. Abt. II, 14,* 459 (1905).
28. Omeliansky, V. L. *Zh. Mikrobiol. Epidemiol. Immunobiol., 1,* 24 (1914).
29. Houwink, A. L., *Nature, 168,* 654 (1951).
30. Belyaev, S. S. *Microbiology USSR* (English translation), *36,* 157 (1967).
31. Staley, J. T., *Appl. Microbiol., 22,* 496 (1971).
32. Jannasch, H. W., and Jones, G. E., *Limnol. Oceanogr., 5,* 432 (1960).
33. Poindexter, J. L. S., and Cohen-Bazire, G., *J. Cell Biol., 23,* 587 (1964).
34. Stove, J. L., *Zentralbl. Bakteriol. Parasitenk. Infektionskr. Hyg. Abt. Orig.,* Suppl. *1,* 95 (1965).
35. Nikitin, D. I., Vasileva, L. V., and Lokmacheva, R. A., *New and Rare Forms of Soil Microorganisms* (in Russian). Science Publishing House, Moscow, U.S.S.R. (1966).
36. Staley, J. T., *J. Bacteriol., 95,* 1921 (1968).
37. Nikitin, D. I., and Kuznetsov, S. I., *Microbiology USSR* (English translation), *36,* 789 (1967).
38. Gorlenko, V. M., *Dokl. Akad. Nauk SSSR* (in Russian), *179,* 1229 (1968).
39. Gorlenko, V. M., *Z. Allg. Mikrobiol., 10,* 147 (1970).
40. Cohen-Bazire, G. Kunisawa, R. and Pfennig, N., *J. Bacteriol., 100,* 1049 (1969).
41. Larsen, H., *J. Bacteriol., 64,* 187 (1952).
42. Pfennig, N., *Zentralbl. Bakteriol. Parasitenk. Infektionskr. Hyg. Abt. Orig.,* Supp. *1,* 179, 503 (1965).
43. Rullmann, W., *Zentralbl. Bakteriol. Parasitenk., 3,* 228 (1897).
44. Stutzer, A., and Hartleb, R., *Mitt. Landwirt. Univ. Breslau, 1,* 75 (1899).
45. Large, P. J., Peel, D., and Quayle, J. R., *Biochem. J., 81,* 470 (1961).
46. Kingma Boltjes, T. Y., *Arch. Mikrobiol., 7,* 188 (1936).
47. Hirsch, P., and Conti, S. F., *Arch. Mikrobiol., 48,* 339 (1964).
48. Hirsch, P., and Conti, S. F., *Arch. Mikrobiol., 48,* 358 (1964).
49. Conti, S. F., and Hirsch, P., *J. Bacteriol., 89,* 503 (1965).
50. Hirsch, P., *Arch. Mikrobiol., 60,* 201 (1968).
51. Pongratz, E., *Schweiz. Z. Allg. Pathol. Bakteriol., 20,* 593 (1957).

52. Duchow, E., and Douglas, H. C., *J. Bacteriol.*, *58,* 409 (1949).
53. Pratt, D. C., and Gorham, E., *Ecology*, *51,* 346 (1970).
54. Leifson, E., *Antonie van Leeuwenhoek J. Microbiol. Serol.*, *30,* 249 (1964).
55. Douglas, H. C., and Wolfe, R. S., *J. Bacteriol.*, *78,* 597 (1959).
56. Boatman, E. S., and Douglas, H. C., *J. Biophys. Biochem. Cytol.*, *11,* 469 (1961).
57. Whittenburg, R., and McLee, A. G., *Arch. Mikrobiol.*, *59,* 324 (1967).
58. Trentini, W. C., *J. Bacteriol.*, *94,* 1260 (1967).
59. Hirsch, P., and Conti, S. F., *Zentralbl. Bakteriol. Parasitenk. Infektionskr. Hyg. Abt. Orig.*, Suppl. *1,* 100 (1965).
60. van Niel, C. B., *Bacteriol. Rev.*, *8,* 1 (1944).
61. Aristovskaya, T. V., *Dokl. Akad. Nauk SSSR* (English Translation), *136,* 111 (1961).
62. Tyler, P. A., and Marshall, K. C., *J. Bacteriol.*, *93,* 1132 (1967).
63. Ahrens, R., and Moo, G., *Arch. Mikrobiol.*, *70,* 243 (1970).
64. de Bont, J. A. M., Staley, J. T., and Pankratz, H. S., *Antonie van Leeuwenhoek J. Microbiol. Serol.*, *36,* 397 (1970).
65. Walsby, A. E. *Proc. Roy. Soc. Ser. B. Biol. Sci.*, *173,* 235 (1969).
66. Hirsch, P., and Rheinheimer, G., *Arch. Mikrobiol.*, *62,* 289 (1968).
67. Dubinina, G. A., *Dokl. Akad. Nauk. SSSR* (English translation), *184,* 87 (1969).
68. Dubinina, G. A., *Z. Allg. Mikrobiol.*, *10,* 309 (1970).
69. Hortobagyi, T., Bot. Közlem., *52/3,* 113 (1965).

BACTERIA WITH ACELLULAR APPENDAGES

DR. JAMES T. STALEY

Gallionella and *Nevskia* are two genera of aquatic bacteria with appendages that can be seen under a light microscope. The appendages of these microorganisms differ from those of the prosthecate bacteria in that they are not bounded by the cell wall. Instead, they are excreted from the cell and are therefore properly termed "acellular" or "extracellular" appendages.

Two other genera, *Blastocaulis* and *Planctomyces,* could also be included here because they, too, have extracellular appendages. However, unlike *Gallionella* and *Nevskia,* which divide by binary transverse fission, they divide by budding and for this reason have been discussed in the section on budding and prosthecate bacteria.

Table 1 lists the *Gallionella* and *Nevskia* species that have been cultivated. For additional information on these bacteria, the reader is referred to the review articles by Schmidt,[1] Starr and Skerman,[2] Zavarzin,[3] and Pringsheim.[4]

TABLE 1

SPECIES OF *GALLIONELLA* AND *NEVSKIA*[a]

Species	Cell Size and Shape	Distinguishing Features
Gallionella ferruginea	Bean-shaped, 0.5-0.7 x 1.2-1.5 μm	Iron-encrusted stalks with 40 or more stalk filaments
Gallionella filamenta	Bean-shaped, 0.4-0.65 x 0.7-1.25 μm	Iron-encrusted stalks with 3 to 8 stalk filaments
Nevskia ramosa	Rod-shaped, 0.7 x 2.4-2.7 μm	Stalk not encrusted; motile cells (1 to 3 polar flagella)

[a]Other species of *Gallionella* have been named (cf. *Bergey's Manual,* 1957). Inasmuch as these have not been cultivated, their validity is questionable. For example, Kucera and Wolfe noted that all the *Gallionella* strains they enriched for and studied fit within the range of variability of *G. ferruginea;*[14] thus, Balashova has reported the only other species that has been cultivated. Balashova also considers *Toxothrix trichogenes* as a member of the genus as *G. trichogenes.*[12]

Gallionella

HISTORY AND HABITAT

Gallionella ferruginea was first observed by Ehrenberg, who described it in 1836.[5] For almost a century it was regarded as a twisted, band-shaped microorganism heavily encrusted with iron. The true nature of the microorganism was finally discovered by Cholodny, who allowed the bacterium to attach and grow on coverslips that he had immersed in iron-bearing waters.[6] After the coverslips had incubated *in situ* for one to two days, he carefully removed them and examined them with a microscope. He noted that the typical band-shaped filaments frequently had small bean-shaped cells attached at their tips. Thus, he interpreted the twisted filament as a stalk that was excreted by the apical cells. Because the cells are easily dislodged from the brittle iron-encrusted stalks by normal sampling procedures, previous investigators had not seen them. Subsequent studies have been hindered by the difficulties encountered in purifying and culturing the bacterium.

Organisms of this genus are widely distributed in iron-containing fresh waters, particularly in cold springs containing ferrous ion; some strains have also been reported from hot springs,[7] and others from marine waters.[8]

MORPHOLOGY AND LIFE CYCLE

Because cultivation has posed such difficult problems, our conception of the genus has been based upon the observation of natural materials, enrichment cultures, and axenic cultures, which are sometimes of dubious purity. For this reason, and also because the cell's morphology is frequently obscured by the heavy deposition of iron, there are conflicting reports on the morphology, mode of reproduction, and life cycle of the bacterium.

Cholodny, whose evidence was derived exclusively from his astute light-microscopic observations of natural materials, described *G. ferruginea* as a bean-shaped bacterium, 0.5 x 1.2 μm, that excretes a stalk from the concave surface of the cell. As the cell grows, it twists, giving the helical shape to the stalk (Figure 1). The stalk contains ferric hydroxide, as indicated by the positive Prussian blue reaction with ferrocyanide. Division occurs by binary transverse fission to produce two daughter cells, each of which develops its own individual stalk after cell separation. In this fashion a microcolony develops numerous long, dichotomously branched filaments.

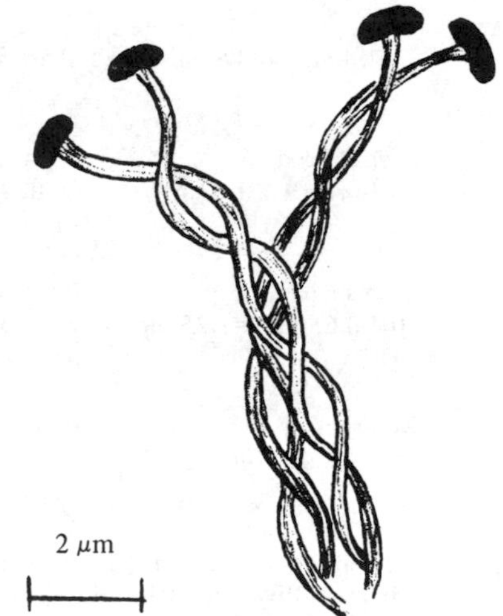

FIGURE 1. *Gallionella ferruginea*. The bean-shaped cells are borne at the tips of the twisted, iron-encrusted stalks.

Contrary to the aforestated view of the genus, the observations of van Beneden,[9] Van Iterson,[10] Zavarzin,[3] and Balashova[11,12] have led these authors to believe that the stalk is either the living part of the cell or that it contains viable elements. To support their arguments, they submit photomicrographs illustrating "sporangia," normal vegetative cells, zoogléal growth stages, and budding cells, all attached to or associated with the stalk.

Electron micrographs of whole cells reveal that the stalk is comprised of numerous fibrils (*ca*. 50 nm wide). These are apparently derived from pores on the concave side of the cell. Thin sections of monocultures of *Gallionella filamenta* by Balashova and Cherni[13] indicate that the stalk fibrils of this species are tubular extensions from the cell. They consist of an electron-dense core surrounded by an electron-transparent area and an external electron-dense layer. The cells themselves have the typical cell-envelope structure of Gram-negative bacteria, i.e., an outer membrane, an intermediate dense layer, and the inner cytoplasmic membrane.

ENRICHMENT AND ISOLATION

The first reproducible enrichment and cultivation techniques were developed by Kucera and Wolfe.[14] Their medium consists of 0.1% ammonium chloride, 0.05% anhydrous dipotassium phosphate, and 0.02% anhydrous magnesium sulfate, each autoclaved individually at tenfold the concentrations given and dispensed aseptically into sterile, distilled water in test tubes plugged with cotton. The ferrous sulfide is prepared by reacting equimolar amounts of ferrous ammonium sulfate and sodium sulfide in boiling distilled water. The precipitate is allowed to settle, then is washed several times by decanting and replacing the supernatant in the stoppered flask with boiling distilled water. Prior to preparation of the medium, 10-ml portions of the ferrous sulfide are sterilized in cotton-plugged 16 x 150 mm test tubes; this is then carefully added to the tubes containing the other ingredients until about 5 to 10% of the total volume is ferrous sulfide. For best results carbon dioxide is bubbled through the medium, then the tubes are stoppered with cork stoppers. A sample of material containing *Gallionella* cells is inoculated into the tubes. Since the medium selects for *Gallionella*, the initial proportion of the cells does not have to be especially high. After several successive transfers, sterile tap water should be added to make up 20% of the volume. Transfers should be made every two to three weeks.

Nunley and Krieg[15] have reported a novel method for obtaining pure cultures. Natural samples containing *G. ferruginea* were concentrated by centrifugation. One to five ml of the sediment was added to a dilution bottle containing 110 ml of the Kucera and Wolfe medium supplemented with 0.5 ml of 40% formaldehyde. This was incubated for one to two days at 25°C, then 1-ml portions were transferred to fresh medium without formaldehyde. In most cases the resulting culture was pure. Pure cultures of *Gallionella ferruginea* have also been reported by Hanert.[16]

NUTRITION AND PHYSIOLOGY

Gallionella ferruginea has a temperature optimum for growth between 20 and 25°C and will not grow above 30°C. Growth is very slow at and below 12°C. The pH optimum is between 6.3 and 6.6. The bacterium is microaerophilic. Ammonium, but not nitrate, can serve as the nitrogen source for growth.[14] Little is known about the metabolic pathways or, in particular, how the iron is used by the organism.

Nevskia

HISTORY AND HABITAT

Nevskia ramosa, the only member in the genus, was first described by Famintzin, who found it in the neuston of a pond in the botanical gardens of Leningrad.[17] It was subsequently reported by Henrici and Johnson from a similar habitat.[18] Babenzien, in his investigations of the microbiology of the neuston, has detected the bacterium at concentrations ranging from 9.1 x 10^5 to 4.9 x 10^6 per ml by direct counting of the neuston of a variety of fresh-water habitats; he has also obtained a pure culture.[19,20]

MORPHOLOGY AND LIFE CYCLE

Nevskia ramosa superficially resembles *Gallionella ferruginea* in that the cells are located at the tips of an extracellular stalk. Indeed, for this reason Krassilnikov included it in the genus *Gallionella.*[21] It differs from all species of *Gallionella,* however, in that the cells are rod-shaped and the stalk is not twisted and does not contain ferric hydroxide. Furthermore, swarm cells with one to three polar flagella are produced. Cells are Gram-negative and measure 0.7 x 2.4-2.7 μm (Figure 2).

Babenzien pictures a life cycle in which the swarm cells leave the neuston and move to the subsurface layers for the duration of their motile stage. The swarm cells do not have any capsular or stalk material.

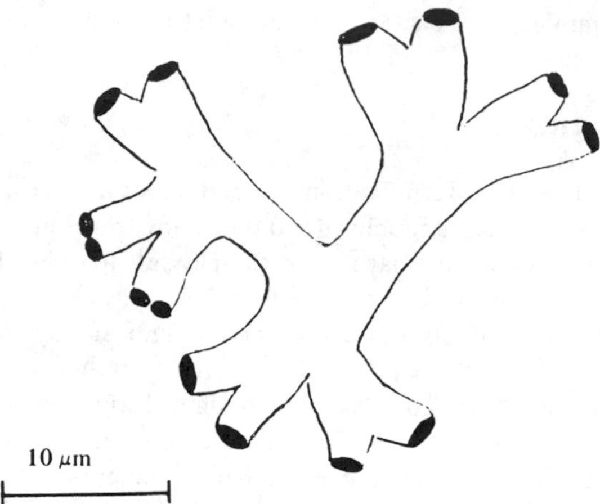

FIGURE 2. A Microcolony of *Nevskia ramosa*. The cells are borne at the tips of the extracellular stalk material; following cell separation, the stalks develop bifurcations, thus accounting for the dichotmously branched structure shown.

When they become non-motile, they produce a capsule at one side of the cell. As the cell grows, the capsule increases in size to become the stalk. After division occurs, each cell synthesizes its own stalk material, resulting in branches in the stalk. The cells are held together in microcolonies by the stalks. Swarm cells are produced at various stages.

If the cells multiply rapidly, as they do under some conditions in pure culture, typical stalks are not formed. Instead, capsular material is deposited over the entire cell surface. Normal stalk formation recurs as the culture enters the stationary phase or at lower temperatures when the growth rate is sufficiently reduced.

ENRICHMENT, ISOLATION, NUTRITION, AND PHYSIOLOGY

Enrichment cultures are prepared by adding sodium lactate at a final concentration of 0.1% to filter-sterilized water from the habitat that contains the bacterium. This is incubated at 20 to 28°C and examined periodically with a microscope. When the numbers become substantial, the bacterium can be purified by extinction dilution techniques in the same medium or by using a micromanipulator.[20]

The bacterium can be maintained on filter-sterilized habitat water, with either sodium lactate or sodium acetate as the added carbon source. Other carbon sources tested, namely glucose, lactose, and DL-β-hydroxybutyric acid, were not used by the bacterium. The temperature range for growth was 15°C to 37°C. Growth occurred between pH 5.5 and 8.0.[20]

REFERENCES

1. Schmidt, J. M., *Ann. Rev. Microbiol., 25,* 93 (1971).
2. Starr, M. P., and Skerman, V. B. D., *Ann. Rev. Microbiol., 19,* 407 (1965).
3. Zavarzin, G. A., *Microbiology, 29,* 774 (1961).
4. Pringsheim, E. G., *Biol. Rev., 24,* 200 (1949).
5. Ehrenberg, C. G., *Ann. Phys., 38,* 213, 455 (1836).
6. Cholodny, N., Die Eisenbakterien, *Pflanzenforschung, Heft 4.* Gustav Fischer, Jena, Germany (1926).
7. Vouk, V., *Archiv Mikrobiol., 36,* 95 (1960).
8. Butkevich, V. S., *Trans. Oceanogr. Inst. Moscow* (in Russian), *3,* 63 (1928).
9. van Beneden, G., *Hydrobiologia, 3,* 1 (1951).
10. Van Iterson, W., *Gallionella ferruginea* Ehrenberg in a Different Light, *Academisch Proefschrift, N. V.* Noord Hollandsche Uitgevers Maatschappij, The Netherlands (1958).

11. Balashova, V. V., *Microbiology, 36,* 879 (1967).
12. Balashova, V. V., *Microbiology, 37,* 590 (1968).
13. Balashova, V. V., and Cherni, N. E., *Microbiology, 39,* 298 (1970).
14. Kucera, S., and Wolfe, R. S., *J. Bacteriol., 74,* 344 (1957).
15. Nunley, J., and Krieg, N. R., *Can. J. Microbiol., 14,* 385 (1968).
16. Hanert, H., *Archiv Mikrobiol., 60,* 348 (1968).
17. Famintzin, A., *Bull. Acad. Imp. St. Petersburg Ser. IV, 34,* 481 (1892).
18. Henrici, A. T., and Johnson, D. E., *J. Bacteriol., 30,* 61 (1935).
19. Babenzien, H. D., *Zentralbl. Bakteriol. Parasitenk. Infektionskr. Hyg. Abteilung I, Suppl. 1,* 111 (1965).
20. Babenzien, H. D., *Z. Allg. Mikrobiol., 7,* 89 (1967).
21. Krassilnikov, N. A., *Diagnostik der Bakterien und Actinomyceten.* Gustav Fischer, Jena, Germany (1959).

TRICHOME-FORMING BACTERIA

DR. ANTONIO H. ROMANO

The trichome-forming bacteria are a group of organisms that form unbranched filaments of cells. The word "trichome" is derived from the Greek "trichoma", meaning hair growth; it is a botanical term used to denote a thread of cells, resulting from division in one plane, that are held together by a common wall layer or sheath.

There are three principal groups of trichome-forming bacteria: (1) the sheath bacteria, which are comprised of a chain of cells enclosed in a sheath; the individual cells are motile by means of polar flagella; (2) large, rigid, cylindrical filaments, actively motile by means of peritrichous flagella; (3) the gliding bacteria, where the entire filament, or a cell abscised from the filament is capable of a gliding motility over solid substrates.

All of these trichome-formers have often been called alga-like and have been considered to be related to blue-green algae. The relationship between bacteria and blue-green algae has been carefully reviewed by Pringsheim.[1] Pringsheim has pointed out that those possessing flagella are not related to blue-green algae, since this structure is entirely absent in blue-green algae; the filamentous gliders are related, however, and Pringsheim actually considered these organisms to be apochlorotic (chlorophyll-less) blue-green algae. Some of the principal characteristics of the filamentous gliders and the blue-green algae to which they are related are listed in Table 1.

The trichome-forming bacteria have also been included in the group that some authors have called "higher bacteria" because they have a greater structural complexity than most bacteria. These morphological features have been recognized in the classification of these forms and have been considered to be sufficiently unique to establish separate orders to include them. Thus, according to the latest edition of *Bergey's Manual of Determinative Bacteriology,*[2] the sheath bacteria are classified in the Chlamydobacteriales, the large rod-shaped filaments with peritrichous flagella in the Caryophanales, and most of the filamentous gliders in the Beggiatoales. A number of types now known to possess gliding motility are classified in other orders, a situation that is not satisfactory. Soriano and Lewin[3] have proposed that all the trichome-forming gliders be classed in a new order, the Flexibacteriales.

No attempt at exhaustive coverage of the trichome-forming bacteria is made here. Many of the forms that have been reported have been described only in samples taken from natural habitats and have never been cultured; others, though cultured, have been studied very little. This report is limited to the better-known forms.

SHEATH BACTERIA

Sphaerotilus

The sheath bacteria are the best-known of the trichome-formers because their principal representative, *Sphaerotilus natans* is a serious interference organism in water supplies. This group of organisms is characterized by the presence of a prominent contiguous, closely fitting sheath, which encloses the chain of rod-shaped cells that comprise the trichome. The sheath is of primary ecological significance to the organism because it allows a means of attachment to solid surfaces. Thus, these organisms gain a selective advantage in streams with appreciable current flow, since they can obtain sufficient nutrient from a large volume of water to achieve massive growth even if the nutrient concentration is very low. They grow in such quantity in streams receiving organic enrichment that massive infestations can result. These infestations are prevalent in streams polluted with paper-mill, cannery, sugar-refining, and brewery wastes and municipal sewage. The organisms attach to submerged rocks, twigs, or debris, and the growth is visible in the water as long grayish or brownish streamers that can be several centimeters long. When these flocs of growth become detached and float downstream, they can foul and clog fishermen's nets and water-intake

TABLE 1

DISTINGUISHING CHARACTERS OF THE PRINCIPAL GENERA OF FILAMENTOUS NON-PHOTOSYNTHETIC ORGANISMS

Structural Properties of Filament				Energy-Yielding Metabolism			
Shape	Cross Section	Motility	Mode of Reproduction	Chemo-autotrophic*	Chemo-heterotrophic	Genus	Structural Counterpart in Filamentous Blue-Green Algae
Straight	Cylindrical	+	Short chains of cells (hormogonia)	+	+(?)	*Beggiatoa*	*Oscillatoria*
Straight	Cylindrical	+	Short chains of cells (hormogonia)	-	+	*Vitreoscilla*	*Oscillatoria*
Straight	Flattened, ribbon-shaped	+	Short chains of cells (hormogonia)	-	+	*Simonsiella*	*Crinalium*
Helical	Cylindrical	+	Short chains of cells (hormogonia)	-	+	*Saprospira*	*Spirulina*
Straight	Cylindrical	-	Single gliding cells	+	?	*Thiothrix*	*Calothrix*
Straight	Cylindrical	-	Single gliding cells	-	+	*Leucothrix*	*Calothrix*

*Oxidation of H_2S

Data taken from Stanier, R. Y., Doudoroff, M., and Adelberg, E. A., *The Microbial World,* 3rd ed., p. 546 (1970). Reproduced by permission of Prentice-Hall, Inc., Englewood Cliffs, New Jersey.

pipes; they can also blanket the bottom of the riverbed and adversely affect fish-hatching and the development of other fauna. Such masses of growth tax the oxygen resources of water courses and detract from the esthetic qualities of the water. Estimates of the amount of *Sphaerotilus* growth carried by the current past a cross section of a polluted river are in the hundreds of tons wet weight per day.

Sphaerotilus natans has also been implicated as a nuisance organism in activated sludge sewage treatment plants. Under certain conditions it grows profusely and causes bulking, a situation in which the sludge does not settle rapidly enough for efficient operation of the unit.

Because of the increasing interest in this organism, a number of good reviews have been written.[4-7]

TAXONOMY

The presence of a prominent sheath enclosing the trichome, which may or may not become encrusted with oxides of iron or manganese, has been considered to be characteristic enough of this group to warrant classification in a separate order, the Chlamydobacteriales. Beyond this, however, the classification has been in a disordered state. The latest edition of *Bergey's Manual of Determinative Bacteriology*[2] lists eight genera with twenty-four species, but few modern workers consider the great majority of these valid. Most descriptions are of organisms sampled from natural habitats that have not been cultured. Pringsheim[8] illuminated the situation considerably when he showed that a number of these organisms, when cultured under a variety of environmental conditions, showed modifications suggesting that one form could be converted to another. Thus, he concluded that the many forms described were variations of one and that there was no justification for retaining more than one genus, *Sphaerotilus.* While there has been argument for retaining the genus *Leptothrix* for forms that show a more pronounced ability to oxidize ferrous salts and an absolute ability to oxidize manganous salts, Rouf and Stokes[9] have provided justification for classifying these forms in a species within the genus, i.e., as *Sphaerotilus discophorus.* Characteristics of *Sphaerotilus* species are shown in Table 2.

TABLE 2

CHARACTERISTICS OF *SPHAEROTILUS* SPECIES

Characteristic	*S. natans*	*S. discophorus*
Diameter of cells	1.2-2.4 μ	0.6-0.6 μ
Vitamin requirement	Cyanocobalamin (vitamin B_{12}),	Cyanocobalamin (vitamin B_{12}), biotin, thiamin*
Deposition of iron oxide in sheath	+	+
Deposition of manganese oxide in sheath	–	+

*Some strains require adenine or guanine.

MORPHOLOGY AND CULTURAL CHARACTERISTICS

The cells are rod-shaped, 1.2 to 2.4 μ wide in the case of *S. natans* and 0.8 to 1.0 μ wide in the case of *S. discophorus,* and 3.8 μ long. They are Gram-negative, non-sporeforming, with a pronounced tendency to deposit poly-*beta*-hydroxybutyric acid in granules,[10] motile by means of a single fascicle of polar flagella,[11,12] and enclosed in a tightly fitting tubular sheath. The sheath is not easily visualized in young cultures because of the tight fit; it is readily seen in older filaments, where portions have been vacated by the cells (Figure 1). Cells can swim from the open end of the sheath, or from broken portions, and colonize a new substrate.

The sheath is composed of a protein-polysaccharide-lipid complex[13] and grows by linear extension of

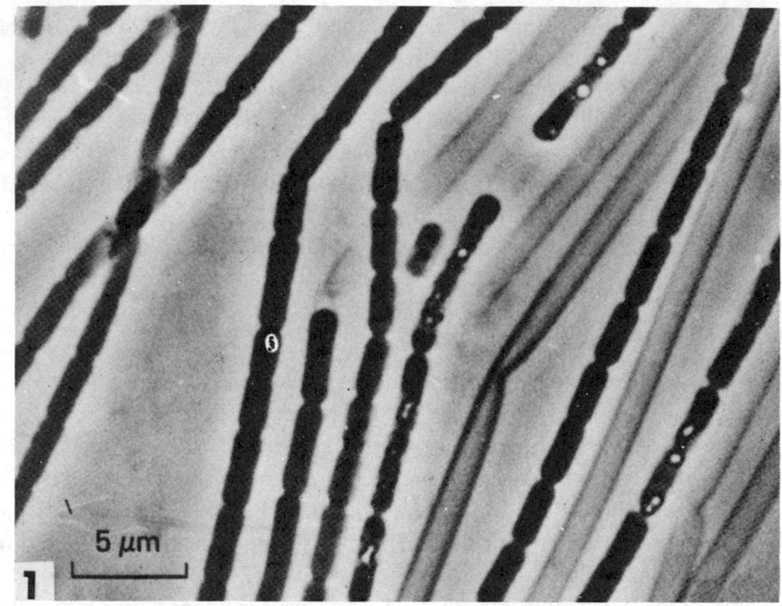

FIGURE 1. Trichomes and Empty Sheaths of *Sphaerotilus natans*. Taken from: van Veen, W. L., and Mulder, E. G., in *Microbiology*, p. 187, J. S. Poindexter (1971). Reproduced by permission of the Macmillan Co., New York.

existing sheath.[14] Surrounding the sheath is a slime layer or capsule that is chemically distinct from the sheath, made up of a polysaccharide containing fucose, glucose, galactose, and glucuronic acid.[15]

Sphaerotilus grows well on a variety of organic culture media, provided that the concentration of nutrients is not too high. Characteristic growth is obtained on glucose-peptone media, such as that of Stokes,[11] containing 0.2% glucose, 0.2% peptone, 0.2% $MgSO_4 \cdot 7H_2O$, 0.005% $CaCl_2$, and 0.001% $FeCl_3 \cdot 6H_2O$ in tap water. Agar colonies are flat and dull, with very irregular edges. Viewed under low power, long curled filaments can be seen along the periphery of the colony. When cultured on the more usual rich laboratory media, there is a dissociation to smooth, glistening colonies of single cells with little or no sheath.[11]

In making a diagnosis of *Sphaerotilus*, it is extremely important that the sheath characteristic of the genus is demonstrated with certainty. Certain species of *Bacillus* that form flat arborescent colonies, are Gram-negative in old cultures, and have a tendency to form chains of cells can be mistaken for *Sphaerotilus;* occasional remnants of cell walls from lysed cells within the chain can resemble a sheath. Such errors in identification have resulted in the occasional reporting of endospores in *Sphaerotilus.*

NUTRITION AND PHYSIOLOGY

There is considerable discrepancy in the literature with regard to specific compounds that can be used by *Sphaerotilus* species as sources of carbon or nitrogen or both (see Reference 6). Many of these discrepancies are doubtlessly due to strain differences and to differences in test procedures and criteria of growth. Nevertheless, some generalizations can be made. A number of carbohydrates (glucose, fructose, mannose, galactose, sucrose, maltose, and mannitol) can be utilized as sources of carbon, as can a number of organic acids (succinate, lactate, fumarate, pyruvate, and acetate) and alcohols (glycerol and ethanol). Many amino acids can serve as nitrogen sources, either singly or in combination. A number of amino acids are toxic, however, at concentrations above 0.36% (alanine, cystine, glycine, tyrosine, and valine); some are toxic at concentrations below 0.02% (DL-serine and DL-valine).[16] This amino acid toxicity probably explains why *Sphaerotilus* does not grow well on rich culture media containing protein hydrolysates. The mechanism of this toxicity is not clear, but it may involve the chelation of metals by amino acids; in addition to the usual metal requirements, *Sphaerotilus* specifically requires calcium for sheath formation.[17] Inorganic ammonium or nitrate salts can be utilized as sources of nitrogen, provided that suitable vitamin supplements are present; *Sphaerotilus natans* requires cyanocobalamin (vitamin B_{12}),[18,19] *Sphaerotilus discophorus* requires, in addition, biotin and thiamin,[9] and some strains of the latter also require adenine or guanine[20] (see Table 2).

Sphaerotilus species will grow over a temperature range from 10 to 40°C, but optimum growth takes place from 25 to 30°C, depending on the strain. All strains require oxygen, although growth will take place at very low oxygen tensions.[11] Optimum pH for growth is in the neutral to slightly alkaline range (pH 7 to 8), though growth will take place over the range from pH 6 to 9.[10,11]

Since the work of Winogradsky in 1888,[21] the pronounced tendency of *Sphaerotilus natans* to deposit oxides of iron in the organic matrix of the sheath, and that of *Sphaerotilus discophorus* to deposit oxides of iron or manganese, has presented the question as to whether these organisms are autotrophic, i.e., whether the energy involved in the inorganic oxidation of ferrous or manganous salts can be utilized for biosynthesis. Actually, the development of the concept of chemoautotrophy by Winogradsky is to a significant extent dependent upon his work on iron oxidation by *Sphaerotilus discophorus* (*Leptothrix ochracea*). However, his work was done with impure cultures that contained organic matter, and no one has as yet succeeded in culturing these organisms in pure culture in the absence of organic materials. Experiments on iron oxidation are complicated by the fact that ferrous salts are rapidly autooxidizable at the pH range in which these organisms grow; this in itself could constitute an argument against the physiological significance of this reaction. However, Johnson and Stokes[22] presented evidence that the oxidation of manganous ion, which is not autooxidizable at the physiological pH range, is catalyzed by an induced enzyme produced by *Sphaerotilus discophorus* only when grown in the presence of Mn^{++} ions. Whether the energy liberated by the oxidation can be utilized by the cells is yet to be established.

BACTERIA FORMING RIGID TRICHOMES WITH PERITRICHOUS FLAGELLA

Caryophanon

The entire trichome of *Caryophanon,* the best-known member of the order Caryophanales, is a rigid rod-shaped structure, measuring up to 3.2 μ in diameter and 30μ in length, rounded at its ends, and vigorously motile by means of numerous peritrichous flagella. It occurs in fresh cow dung.

Caryophanon was first isolated and described by Peshkoff,[a] and fully characterized cytologically by Pringsheim and Robinow.[23] There are two species, *C. latum* and *C. tenue,* distinguished only by the diameter of the trichome. The latter species does not exceed 1.5 μ in diameter; it has been suggested, however, that *C. latum*, under certain conditions of culture, shows sufficient variation in size so that the differentiation of species on this basis alone is not warranted.[24]

The salient morphological feature of this genus is the occurrence of regularly spaced, parallel, transverse lines, which divide the long rods into rows of flattened, discoid compartments (see Figure 2). By specific cell-wall stains and alternative staining of protoplasts with basic dyes it has been demonstrated that the transverse lines that appear dark in light microscopy are septa, consisting of cell wall material, while the clear compartments between them represent the cytoplasm of the individual cellular units.[23] The cytoplasmic areas contain nuclear bodies that can be demonstrated with the standard procedures for staining chromatinic bodies, and because of the relatively large size of *Caryophanon,* this organism has been used frequently to study the organization of bacterial nuclear structures. Occasionally trichomes contain particularly refractile compartments, compressed into a biconcave shape by adjacent compartments; these are dead cells that have become compressed by the turgor of adjoining normal cells.

Trichomes do not grow to indefinite length. Rather, the length of the rods is kept fairly constant due to regular division near the midpoint of the filament by a gradual constriction in the plane of the preformed cross walls. Individual rods may remain attached end to end, however, under certain growth conditions, and form chains more than a hundred microns in length.

Caryophanon does not develop in the intestine, but develops rapidly on cow dung; it reaches its peak of development one to two days after it has been dropped on the field. The organism can be isolated from

[a] Peshkoff, M. A., *J. Gen. Biol.* (Russian), *1,* 598 (1940).

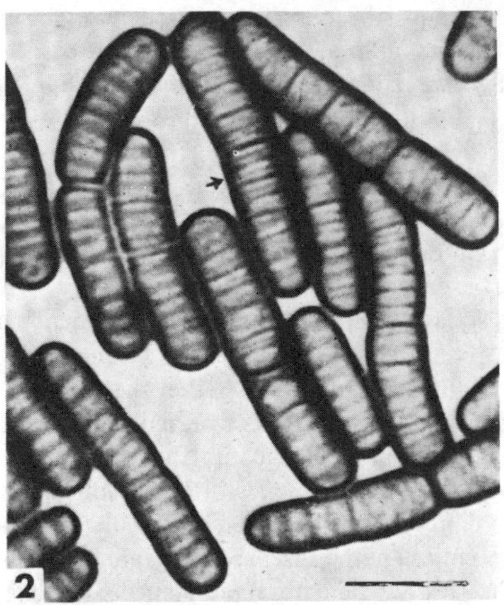

FIGURE 2. Different Stages of Transverse Fission *Caryophanon latum.* Arrow points to cell showing prominent transverse striation. Scale bar represents 5 μ. Taken from: Pringsheim, E. G. and Robinow, C. F., *J. Gen. Microbiol., 1,* 267 (1947). Reproduced by permission of Cambridge University Press.

dung and cultivated on a solid medium containing 0.5% beef extract, 0.5% peptone, and 0.1% sodium acetate, adjusted to a pH of 7.4 to 7.6.[23] It grows over a temperature range of 20 to 37°C and is strictly aerobic. On agar media, it forms convex, circular, undulate colonies with an undulating edge, rarely exceeding 1 to 2 mm in diameter, with smooth or finely granular surface. It exhibits both smooth and rough colonies.

Pringsheim and Robinow[23] failed to obtain good growth of *Caryophanon latum* on liquid media; addition of small amounts of agar (0.1 to 0.2%) stimulated growth considerably, although growth was abnormal compared to growth in solid agar. On the other hand, Provost and Doetsch[25] reported good growth on a defined fluid medium containing acid-hydrolyzed casein, thiamine, biotin, sodium acetate, sodium butyrate, and dipotassium phosphate, when sufficient aeration was provided. These workers detected poly-*beta*-hydroxybutyric acid in large amounts and, in contrast to the findings of the previous workers, reported that *C. latum* is Gram-positive.

THE GLIDING TRICHOME-FORMING BACTERIA

Beggiatoa

Beggiatoa occurs widely in lake, pond, and river muds, in sulfur springs, sewage-polluted streams, and in marine habitats that are characteristically rich in hydrogen sulfide. The organism is of historical importance, since it was the first to be described as autotrophic by Winogradsky[26] in 1887.

MORPHOLOGY AND TAXONOMY

The organism grows in the form of unattached long, colorless, cylindrical trichomes that range in length from 80 μ to well over 1,500 μ.[27] The lengths of individual cells within the trichome range from 7 to 16 μ. The diameters of individual trichomes are fairly uniform throughout their length; most range from 1 to 2.5 μ, although forms as wide as 25 μ have been reported. Cross walls are not easily seen because of the pronounced tendency to deposit sulfur in prominent refractile granules within the cells when grown in the presence of H_2S; poly-*beta*-hydroxybutyric acid inclusions can also be present.[28] Cross walls can be visualized by light microscopy in older filaments that have discharged inclusions,[27,29] or by electron microscopy of thin sections.[30,31] (Figure 3). The cell wall is thin and composed of two layers. Only the inner layer participates in the formation of cross walls;[31] the outer layer appears continuous and, presumably, holds the trichome together. The thinness of the cell wall is further indicated by a high degree of osmotic fragility.

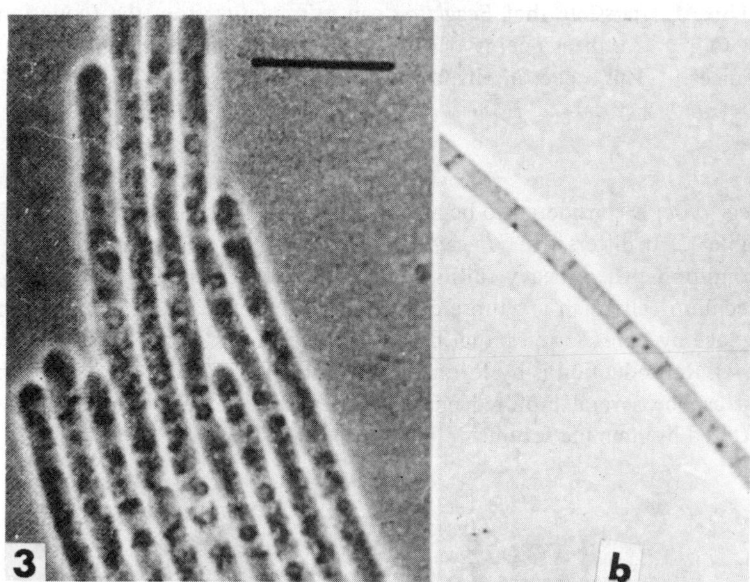

FIGURE 3. *Beggiotoa alba*. a. Edge of microcolony showing trichomes. b. Dying trichome, devoid of sulfur granules, shows cross walls. Scale bar represents 10μ. Taken from: Faust, L. and Wolfe, R. S., *J. Bacteriol., 81*, 103 (1961). Reproduced by permission of the American Society for Microbiology, Washington, D. C.

Flagella are absent. The entire trichome moves by gliding over a solid substrate at a velocity of about 4 μ per second,[29] leaving empty tracks on agar. The movement is both translational and rotational. Microscopic examination of colonies on agar reveals patterns due to movement of the trichomes that are circular or tongue-like.[28,29]

Beggiatoa is considered to be a non-pigmented structural counterpart of the blue-green alga *Oscillatoria* (see Table 1).

The genus *Beggiatoa* has been arbitrarily subdivided into species strictly on the basis of size.[2] This is probably not justified, since considerable variation in growth rate, nutrition, and salt tolerance exists in individuals that lie within the same size range. The type species is *Beggiatoa alba.*

NUTRITION AND PHYSIOLOGY

The question of autotrophy in *Beggiatoa* is as yet not clarified. Since Winogradsky's early work, which was carried out in the presence of small amounts of organic matter, there have been only two reports of successful cultivation in a strictly inorganic medium: the first in 1912 by Kiel,[32] and the second more than 50 years later by Kowallik and Pringsheim,[33] in spite of a number of attempts by other workers in the meantime. The latter report does not include quantitative data regarding growth rates, cell yields, etc., and hence is less than completely convincing.

Beggiatoa grows readily as a heterotroph on a number of dilute organic media containing 0.05 to 0.2%

yeast extract, peptone, or beef extract. It is nonfastidious, since inorganic nitrogen sources such as ammonium salts can be utilized, and most strains show no vitamin requirement. Some amino acids (aspartic acid and glutamic acid) can act as both carbon and nitrogen sources. Acetate is greatly stimulatory to growth[29] and can be used as a sole carbon source.[28] Growth is optimal at pH ranges near neutrality and over a temperature range from 25 to 30°C. Oxygen is required, though many strains are microaerophilic.[27,28]

It is important to note that H_2S has a stimulatory effect on growth even in the presence of organic matter.[27] Moreover, a number of strains require H_2S for growth in the presence of acetate, $(NH_4)_2SO_4$ and mineral salts. Thus it is possible that *Beggiatoa* can gain energy from the oxidation of H_2S, but lacks the capacity to fix CO_2 and utilizes energy derived from the oxidation for the assimilation of acetate or other organic substances.[27] This aspect merits further study.

Vitreoscilla

Vitreoscilla, like *Beggiotoa,* is considered to be an apochlorotic structural counterpart of the blue-green alga *Oscillatoria* (see Table 1). It differs from *Beggiotoa,* however, in that sulfur granules are never found in the cells, even under conditions where heavy sulfur deposits are formed in *Beggiotoa.* Also, the trichomes are more clearly divided into cells than are those of *Beggiotoa.* Individual cells are often barrel-shaped, since cell division takes place by constriction rather than by septation;[34,35] thus the trichomes assume a beaded appearance (Figure 4). Reproduction is by formation of hormogonia (short chains of cells). Trichomes are 1 to 2.5 μ wide, and can be several hundred microns in length. The cells are surrounded by a slime layer, which was demonstrated by immune serum.[35]

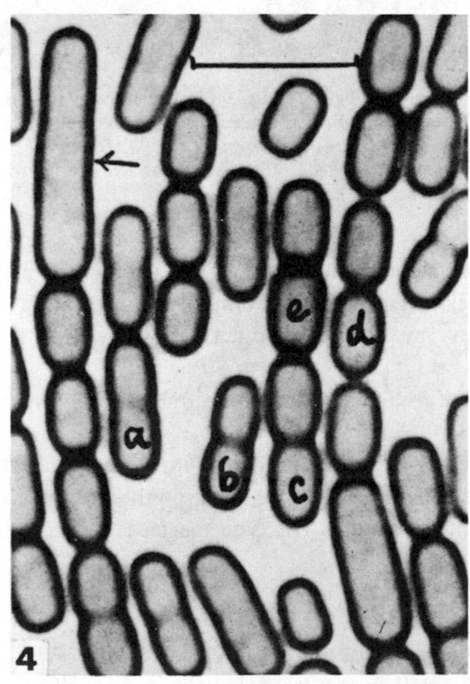

FIGURE 4. *Vitreoscilla* sp., Stained to Show Cell Surface. Intact trichome is shown at left; progressive stages of cell division are shown in a to e. Scale bar represents 5 μ. Taken from: Costerton, J. W. F., Murray, R. G. E., and Robinow, C. F., *Can. J. Microbiol.*, 7, 340 (1961). Reproduced by permission of the National Research Council of Canada, Ottawa, Ontario.

Electron microscopy of thin sections showed a cell wall similar to other Gram-negative bacteria, comprising three layers. The outer layer was found to be consistently folded. Costerton, Murray. and Robinow[35] postulated that the outer layer may be contractile and capable of flexing in a series of orderly waves as a possible explanation of the gliding motility that is exhibited by this organism.[35]

Vitreoscilla is widely distributed in soil, dung, water with decaying plant material, and the algal scum of still waters. Very little is known concerning the nutrition and physiology of *Vitreoscilla.* It has been grown in pure culture on organic media containing 0.2% each of sodium acetate, yeast extract, and tryptone.[35]

Pringsheim[34] has classified organisms that he has isolated within this genus in six species according to characteristics shown in Table 3. Two other species were named provisionally, since they were not cultured.

TABLE 3

CHARACTERISTICS OF SPECIES OF *VITREOSCILLA*

Species	Width, μ	Morphological Features	Culture Appearance
V. beggiatoides	1.2	Straight trichomes; cells longer than wide	Curls and tongues
V. catenula	1.2	Moniliform; readily fragments	Curls and spirals
V. filiformis	1.2	Long, tough trichomes; cells longer than wide	Circular twirls and long loops
V. moniliformis	2.3	Straight trichomes with short or elongate cylindrical cells	Diffuse growth or locks and tongues
V. paludosa	1.8-2	Straight trichomes with long cells	Colonies with concentric structure, no curls
V. stercoraria	1.2-1.5	Trichomes often irregularly bent, moniliform or composed of sausage-shaped cells	Colonies spiral, connected by fine threads

Data taken from Pringsheim, E. G., *J. Gen. Microbiol., 5,* 138 (1951). Reproduced by permission of Cambridge University Press.

Saprospira

Saprospira forms a spiral-shaped trichome; thus it can be considered the apochlorotic relative of the blue-green alga *Spirulina. Saprospira* has been classified among the Spirochaetales because of its spiral form.[2] More recent studies of this organism by Lewin,[36-38] however, have unequivocally demonstrated that it should be excluded from this group. *Saprospira* lacks flagella, an axial filament, and an enveloping membrane. Rather, it has a rigid cell wall and exhibits gliding motility; thus it is more closely related to *Vitreoscilla* and other apochlorotic gliding organisms. Both marine and fresh-water forms of *Saprospira* have been described; the organism occurs in marine muds and decaying vegetation, fresh water, soil, and hot springs.

The filaments measure 0.8 to 1.2 μ in diameter and can be several hundred microns long. Individual cellular units range from 1 to 5 μ, depending upon the species. The spirals are relatively flat, with the width of the helix generally in the range from 1.5 to 2 μ (see Figure 5). The coils in a given filament tend to be regularly spaced, but different filaments within a culture can show a range of wavelengths that is characteristic of different species (*S. grandis,* 4 to 9 μ; *S. albida,* 3 to 9 μ; *S. flammula,* 3 to 4 μ; *S. thermalis,* 7 to 17 μ). Cross walls are not readily observed in young cultures, but they become apparent in dying cultures, when cells tend to separate. Cross walls have also been demonstrated by electron microscopy of longitudinal sections. Cells contain carotenoid pigments, which are principally xanthophylls.[38]

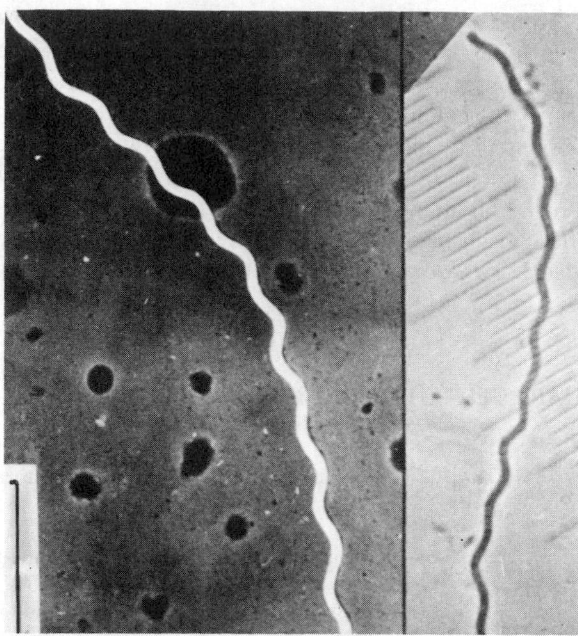

FIGURE 5. Electron Photomicrograph and Phase-Contrast View of *Saprospira grandis*. Scale bar represents 10 μ. Taken from: Lewin, R. A., *Can. J. Microbiol.*, *8*, 556 (1962). Reproduced by permission of the National Research Council of Canada, Ottawa, Ontario.

Saprospira species are strictly aerobic and grow best at pH ranges near neutrality. The marine species (*S. grandis*) grows well in 0.5 to 2X concentrations of sea water supplemented with organic nutrients (tryptone and yeast extract, 0.5% each). Fresh-water species, as represented by *S. thermalis,* do not grow at salinities above one fourth that of sea water. A nutritional study of *S. thermalis*[38] showed that it could utilize glucose, glycerol, or mannose as sources of carbon and energy. Glutamate was a suitable source of nitrogen, as were asparagine, methionine, or threonine; nitrate was not utilized. Ammonium chloride or urea was used by one strain only and did not promote good growth. The amino acids leucine, isoleucine, and valine appear to be absolutely required; in the presence of these three amino acids and of glutamate, growth is further stimulated by tyrosine or phenylalanine. The vitamins thiamine and cobalamin (vitamin B_{12}) are also absolutely required.

Simonsiella and Alysiella

Simonsiella and *Alysiella* are commonly found in the oral cavity of domestic animals. On superficial examination, the trichomes formed by *Simonsiella* resemble those of *Caryophanon* (see Figure 6), and this genus has been classified in the *Caryophanales.* However, there are important differences. First, the trichome of *Simonsiella* is flattened and ribbon-like rather than cylindrical. Second, *Simonsiella* has no flagella; it is motile over an agar surface by gliding, leaving a well-defined track;[39] in this regard, it resembles the blue-green alga *Crinalium* (Table 1).

The related genus *Alysiella* differs from *Simonsiella* in that the former does not form long trichomes, but rather occurs as groups of two or four cells (see Figure 7). Also, terminal cells are not rounded as they are in *Simonsiella.*

The following descriptions are according to Steed.[39]

Simonsiella crassa. Multicellular, unbranched, non-sporing, ribbon-like filaments, 3 to 4 μ wide and 1 to 1.5 μ thick, consisting of closely apposed cells 0.6 μ long, with the free faces of the terminal cells rounded. The filaments appear to divide by constriction into hormogonia-like units about 4.5 μ long, which may remain attached for some time. Filaments may attain a length of 50 μ or more. Phase-dense areas may occur centrally in the intercalary cells and as apical thickenings in the terminal cells. The filaments exhibit gliding motility when the broad face is presented to the solid medium but are immotile and show a pronounced tendency to curl when on their edges. The organism is Gram-negative and basophilic.

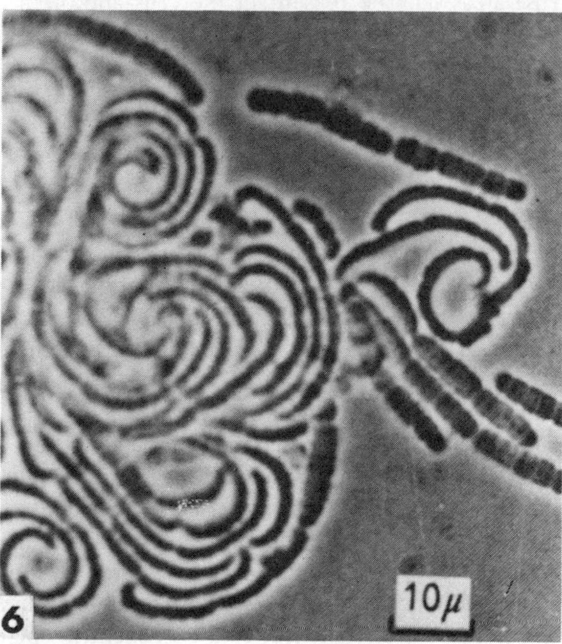

FIGURE 6. Microcolony of *Simonsiella crassa*. Flat filaments coiled at the edges, and filaments moving away from the microcolony. Taken from: Steed, P. D. M., *J. Gen. Microbiol.*, 29, 615 (1962). Reproduced by permission of Cambridge University Press.

FIGURE 7. Microcolony of *Alysiella filiformis*, Showing Phase-Dense Striations (a) and a Lateral View of the Cells (b). Taken from: Steed, P. D. M., *J. Gen. Microbiol.*, *29*, 615 (1962). Reproduced by permission of Cambridge University Press.

Alysiella filiformis. Multicellular, unbranched, non-sporing, ribbon-like filaments whose cells appear to occur in pairs, with a relatively weak linkage between each pair of group of four cells. The filaments are of uniform width throughout; terminal cells are similar to every other cell of the filament, i.e., flat or slightly biconcave. Each cell is 2 to 3 μ wide, about 0.6 μ long, and 0.5 μ thick. Length of the filaments is extremely variable. Phase-dense areas may be present in the cells and often occur along one side of the filament. Gliding motility is exhibited on solid media. The organism is Gram-negative and basophilic.

Both these organisms are aerobic, but they can ferment a number of sugars. They grow optimally at 37° C; their nutritional requirements are complex in keeping with their parasitic habit.

Leucothrix and *Thiothrix*

Leucothrix and *Thiothrix* are more complex than the other trichome-forming gliding bacteria in that there is apical-basal differentiation within the filament. The trichome is not motile, as is the case in *Beggiotoa* and *Vitreoscilla*; only single gliding cells (gonidia) formed at the apical end of the filament are. Moreover, these motile gonidia can aggregate prior to formation of new filaments, so that there is a relatively complex life cycle.

Leucothrix is widely distributed in marine environments and can be readily cultured and isolated by enrichment procedures. *Thiothrix* has been observed in marine and fresh-water environments where hydrogen sulfide is present, but has never been isolated in pure culture. The genus *Leucothrix* has one species, *Leucothrix mucor,* whereas the genus *Thiothrix* has been divided into seven species, based entirely on the diameter of the filaments; this seems hardly justified on the basis of the fragmentary knowledge of this organism. *Leucothrix* has been extensively studied and reviewed by Harold and Stanier[40] and by Pringsheim.[41] The first-mentioned authors have also reviewed literature on *Thiothrix.*[40]

Leucothrix grows in the form of long, unbranched, colorless, slightly tapered filaments that are 3 μ in diameter at the base and 1.5 to 2.0 μ at the apex. Individual cells are short and cylindrical, 1 to 5 μ in length, with a smooth common outer wall. Filaments are variable in length, but may reach 5 mm. There is no prominent outer sheath, although a gelatinous sheath, probably representing surrounding capsular material, has been reported.[41] At the apical end of the filaments, single gonidia are abscised by constriction of the outer wall at the transverse septa, giving rise to a beaded appearance. In liquid media, a number of filaments are characteristically arranged in the form of rosettes, arising from a common holdfast on a solid substratum that is the result of aggregation of number of motile gonidia (Figure 8).

The typical life cycle of *Leucothrix mucor* is illustrated in Figure 9. Gonidia abscised from the tip of a mature filament settle onto a solid surface and glide in jerky fashion over the solid substratum if immersed in liquid. Each gonidium can give rise to a new filament. If there are numerous gonidia in a small area, they glide toward each other, presumably as a manifestation of a chemotactic response, and form a star-like aggregate, with one pole of each of the participating ovoid gonidia in close apposition to one another. A holdfast is then synthesized, and as each gonidium gives rise to a new filament, growing radially, a colony resembling a rosette is formed.

Gonidium formation in a mature filament begins terminally and proceeds toward the base, a situation that on superficial examination might indicate preferential apical cell division. Such is not the case, however; Brock[42] has shown by tritiated-thymidine autoradiography that there is no zone of preferential cell division.

An unusual feature of *Leucothrix mucor* is the ability of the filaments to form knots.[43] This phenomenon takes place in rich media, where growth is rapid. There is faster growth on one side of the filament, resulting in a loop, then growth of the tip through the loop to form a knot. Cells in the knot eventually fuse, forming a bulb that separates from the basal portion of the filament.

Gonidia are not motile on solid media. On agar, a gonidium gives rise to a single filament, which, as it grows, folds many times, forming a flat "thumb-print" colony.

Leucothrix is strictly aerobic and grows optimally at 25°C. It is most abundant in temperate oceanic waters, where tidal or wave action gives good aeration. It does not require vitamins or other complex growth factors. It grows readily on laboratory media containing sugars, other simple organic substances and a salt

FIGURE 8. Young, Developing Rosettes of *Leucothrix mucor*. Scale bar represents 10 μ. Taken from: Harold, R. and Stanier, R. Y., *Bacteriol. Rev.*, *19*, 558 (1955). Reproduced by permission of the American Society for Microbiology, Washington, D. C.

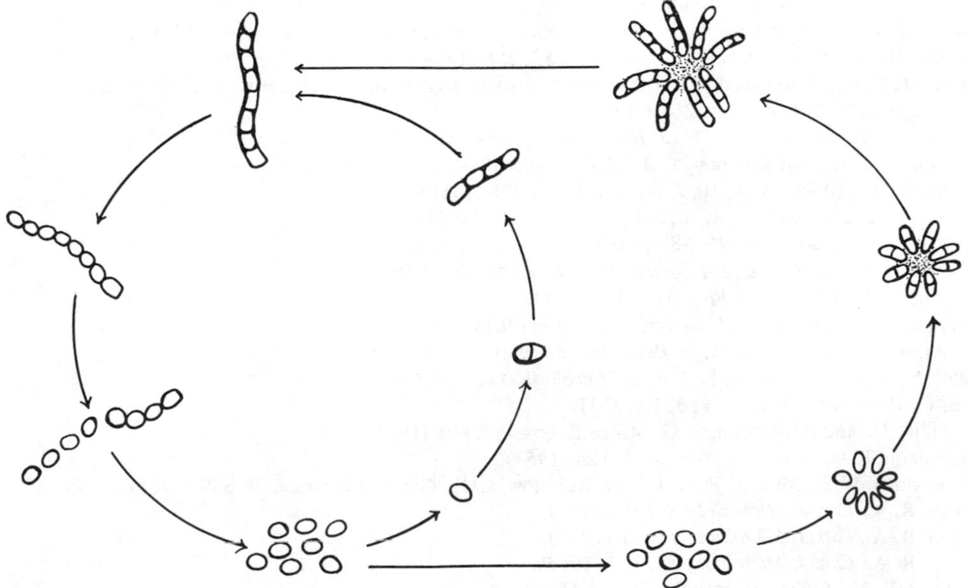

FIGURE 9. Life Cycle of *Leucothrix mucor*. After Brock, T. D., in *Biology of Microorganisms*, p. 610 (1970). Reproduced by permission of Prentice-Hall, Inc., Englewood Cliffs, New Jersey.

concentration of 16 g per liter. Brock[44] has used a synthetic medium containing monosodium glutamate as a sole source of carbon, nitrogen and energy.

Thiothrix is the presumably chemolithotrophic counterpart of *Leucothrix*, occurring in sulfide-containing waters; it presumably oxidizes H_2S and deposits sulfur in extensive intracellular granules. As stated above, knowledge of this organism is fragmentary, since it has not been cultured. Nevertheless, there appears to be little question of its existence, since aggregates of sulfur-containing filaments forming the characteristic rosettes, and even knots, have been observed on occasion in natural substrates, such as sulfur springs.[45]

REFERENCES

1. Pringsheim, E. G., *Bacteriol. Rev., 13*, 47 (1949).
2. Breed, R. S., Murray, E. G. D., and Smith, N. R., *Bergey's Manual of Determinative Bacteriology,* 7th ed. Williams and Wilkins, Baltimore, Maryland (1957).
3. Soriano, S., and Lewin, R. A., *Antonie van Leeuwenhoek J. Microbiol. Serol., 31*, 66 (1965).
4. Harrison, M., and Heukelekian, H., *Sewage Ind. Wastes, 30,* 1278 (1958).
5. Dondero, N. C., *Advan. Appl. Microbiol., 3*, 77 (1961).
6. Phaup, J. D., *Water Res., 2*, 597 (1968).
7. Curtis, E. J. C., *Water Res., 3*, 289 (1969).
8. Pringsheim, E. G., *Phil. Trans. Roy. Soc. London Ser. B Biol. Sci., 233*, 453 (1949).
9. Rouf, M. A., and Stokes, J. L., *Arch. Mikrobiol., 49*, 132 (1964).
10. Rouf, M. A., and Stokes, J. L., *J. Bacteriol., 83*, 343 (1962).
11. Stokes, J. L., *J. Bacteriol., 67*, 278 (1954).
12. Doetsch, R. N., *Arch. Mikrobiol., 54*, 46 (1966).
13. Romano, A. H., and Peloquin, J. P., *J. Bacteriol., 86*, 252 (1963).
14. Romano, A. H., and Geason, D. J., *J. Bacteriol., 88*, 1145 (1964).
15. Gaudy, E., and Wolfe, R. S., *Appl. Microbiol., 10*, 200 (1962).
16. Johnson, A. H., and Stokes, J. L., *Antonie van Leeuwenhoek J. Microbiol. Serol., 31*, 165 (1965).
17. Dias, F. F., Orkend, H., and Dondero, N. C., *Appl. Microbiol., 16*, 1364 (1968).
18. Mulder, E. G., and Van Veen, W. L., *Antonie van Leeuwenhoek J. Microbiol. Serol., 29*, 121 (1963).
19. Orkend, H., and Dondero, N. C., *J. Bacteriol., 87*, 286 (1964).
20. Stokes, J. L., and Johnson, A. H., *Antonie van Leeuwenhoek J. Microbiol. Serol., 31*, 175 (1965).
21. Winogradsky, S., *Bot. Ztg., 46*, 261 (1888).
22. Johnson, A. H., and Stokes, J. L., *J. Bacteriol., 91*, 1543 (1966).
23. Pringsheim, E. G., and Robinow, C. F., *J. Gen. Microbiol., 1*, 267 (1947).
24. Weeks, O. B., and Kelley, L. M., *J. Bacteriol., 75*, 326 (1958).
25. Provost, P. J., and Doetsch, R. N., *J. Gen. Microbiol., 28*, 547 (1962).
26. Winogradsky, S., *Bot. Ztg., 45,* 489 (1887).
27. Scotten, H. L., and Stokes, J. L., *Arch. Mikrobiol., 42*, 353 (1962).
28. Pringsheim, E. G., *Amer. J. Bot., 51*, 898 (1964).
29. Faust, L., and Wolfe, R. S., *J. Bacteriol., 81*, 99 (1961).
30. Morita, R. Y., and Stave, P. W., *J. Bacteriol., 85*, 940 (1963).
31. Maier, S., and Murray, R. G. E., *Can. J. Microbiol., 11*, 646 (1965).
32. Kiel, F., *Beitr. Biol. Pflanz., 11*, 335 (1912).
33. Kowallik, U., and Pringsheim, E. G., *Amer. J. Bot., 53*, 801 (1966).
34. Pringsheim, E. G., *J. Gen. Microbiol., 5*, 124 (1951).
35. Costerton, J. W. F., Murray, R. G. E., and Robinow, C. F., *Can. J. Microbiol., 7*, 329 (1961).
36. Lewin, R. A., *Can. J. Microbiol., 8*, 555 (1962).
37. Lewin, R. A., *Can. J. Microbiol., 11*, 77 (1965).
38. Lewin, R. A., *Can. J. Microbiol., 11*, 135 (1965).
39. Steed, P. D. M., *J. Gen. Microbiol., 29*, 615 (1962).
40. Harold, R., and Stanier, R. Y., *Bacteriol. Rev., 19*, 49 (1955).
41. Pringsheim, E. G., *Bacteriol. Rev., 21*, 69 (1957).
42. Brock, T. D., *J. Bacteriol., 93*, 985 (1967).
43. Brock, T. D., *Science, 144*, 870 (1964).
44. Brock, T. D., and Mandel, M., *J. Bacteriol., 91*, 1659 (1966).
45. Brock, T. D., *Biology of Microorganisms,* Prentice-Hall, Inc., Englewood Cliffs, New Jersey (1970).
46. Stainer, R. Y., Doudoroff, M., and Adelberg, E. A., *The Microbial World,* 3rd ed. Prentice-Hall, Inc., Englewood Cliffs, New Jersey (1970).

THE GENUS *BACILLUS*

DR. RUTH E. GORDON

The rod-shaped bacteria that aerobically form refractile endospores are assigned to the genus *Bacillus.* One spore appears in the spore-bearing cell, although some large rods observed in the intestines of tadpoles but never cultivated *in vitro* were described as forming two spores in the sporangium and assigned to the genus *Bacillus.*[1] The endospores of the bacilli, like those of the clostridia and a few other taxa,[2] are more resistant than the vegetative cells to heat, drying, disinfectants, and other destructive agents, and thus may remain viable for centuries. The basis of the spore's resistance and longevity, its formation, morphology, composition, and stages of germination continue to be subjects of many investigations.[3,4]

With a few exceptions, strains of the genus *Bacillus* form catalase, which, in addition to the aerobic production of spores, distinguishes bacilli from clostridia. The production of catalase also differentiates bacilli from strains of *Sporolactobacillus.*[5] The exceptions, i.e., those strains that produce no catalase or only trace amounts, are strains of *B. larvae, B. lentimorbus, B. popilliae,* and some strains of *B. stearothermophilus.*

The genus *Bacillus* encompasses a great diversity of strains. Some species are strictly aerobic, others are facultatively anaerobic. Strains of *B. polymyxa* fix atmospheric nitrogen.[6-8] Strains of some species grow well in a solution of glucose, ammonium phosphate and a few mineral salts, others need additional growth factors or amino acids, and still others have increasingly complex nutritional requirements.[9,10] Strains of *B. fastidiosus* grow only when uric acid or allantoin is available.[11] Although a pH of 7.0 is suitable for growth of most bacilli, a pH of 9.0 to 10.0 was described[12] as a growth prerequisite for *B. alcalophilus,* and *B. acidocaldarius* was described[13] as growing at pH values of 2.0 to 6.0, with optimum growth at pH 3.0 to 4.0. The bacilli also exhibit great variation in temperatures of growth; some thermophiles grow from a minimum temperature of 45°C to a maximum temperature of 75°C or higher, while some psychrophiles grow at temperatures from −5°C to 25°C. Undoubtedly because of their special requirements for growth, many strains that exist in nature are unknown in the laboratory. On the other hand, because a microorganism has the capacity to adapt to unusual conditions of growth, counterparts of a strain isolated from an unusual source could possibly be found in our culture collections.

The strains of bacilli with ellipsoidal or cylindrical spores that do not appreciably distend the sporangia are generally Gram-positive. Strains with ellipsoidal or round spores that swell the sporangia, however, may be Gram-positive, -variable, or -negative. Strains may be motile or non-motile. Although a few motile strains have been reported as having polar flagella,[14,15] the flagella of the bacilli are generally peritrichous. For a very small number of representative strains the surface configuration of the spores was described as distinctive for some species.[16]

According to the examination of a limited number of strains, the base composition of deoxyribonucleic acid (moles % guanine + cytosine) of the bacilli ranges from 32 to 66.[17,18] Although sporulation and other growth characteristics complicated the techniques of paper chromatography, a study of seventy-five mesophilic strains indicated that strains might be differentiated specifically by the amino acids of their whole cells.[19] For taxonomic purposes, characteristic constituents of the cell lipids may also be useful.[20-22] But the promise of these and other studies of the chemical composition of the cells can only be realized by a comparative examination of many strains. In some species, spore antigens and bacteriophages have provided information that correlates with other properties, but much more work is required before full applications can be determined.[23,24]

Because the bacteria of the genus *Bacillus* are widely distributed in soil, water, and air, and because their spores are so resistant, their control is of great economic concern in the food processing industry and in the preparation of all sterile products. Although *B. anthracis,* which played a memorable role in the history of microbiology, is generally regarded as the only species of the genus that is virulent for man and animals, evidence of infections due to strains of species hitherto accepted as "nonpathogenic" is accumulating.[25-31] At least four species of the genus (*B. thuringiensis, B. larvae, B. popilliae,* and *B. lentimorbus*) are pathogenic for insects. *Bacillus thuringiensis* insecticides, effective against many pests of agricultural crops, forests and stored food, are produced commercially in several countries.[32]

DESCRIPTIONS OF SPECIES

The concept of a microbial species is admittedly man-made, imperfect, and difficult to define.[33] Nevertheless, the specific naming of groups of strains with certain similarities does provide a means of communication among microbiologists. Each name points to a description or definition. Because the rules of nomenclature affix the name of a species to the strain first named and described, old strains as well as freshly isolated ones must be examined and their more stable properties (properties that persist after years of cultivation *in vitro*) incorporated into the species' descriptions. Microorganisms possess many characteristics that are of varying degrees of stability. Therefore, to establish the reliability of a characteristic for describing a species, many strains must be examined.

The most intensively studied species of the genus *Bacillus* are those that are important in medicine or industry and those that are abundantly represented in nature and easily cultivated *in vitro*. As a result, knowledge of the species has accumulated unevenly, and much more work on comparative description of the species remains to be done. With some small modifications in nomenclature, the species presented here are those recognized in the eighth edition of *Bergey's Manual of Determinative Bacteriology*. Histories of the strains and descriptions of media and methods for the determination of properties listed in Tables 1 through 6 and Keys 1 and 2 are available elsewhere.[24,34] The species are arranged in three groups according to shape of the spore and swelling of the sporangium by the spore;[24] the first group is subdivided by diameter of the rod and appearance of its protoplasm (Key 2).

The spores of strains of the species assigned to Group 1 are generally ellipsoidal or cylindrical and do not appreciably distend the sporangia. The cells of *B. megaterium* and *B. cereus* (Table 1) are usually wider than the cells of *B. licheniformis, B. subtilis, B. pumilus, B. firmus*, and *B. coagulans* (Table 2). In addition, the cells of *B. megaterium* and *B. cereus* (Figures 1 and 2), when grown on glucose agar (18 to 24 hours) and lightly stained, are filled with unstained globules, whereas the cells of the other species are not (Figure 3).

Of the properties listed in Table 1, anaerobic growth, Voges-Proskauer and egg-yolk reactions, resistance to lysozyme, and acid production from mannitol are the most useful in separating *B. megaterium* and *B. cereus.* Some characteristics used by other investigators[10,23,35] to delineate these two species are growth in an inorganic ammonium basal solution with glucose, the methyl red test, and production of urease and of acid from raffinose and inulin. An egg-yolk and polymyxin medium was recently proposed for the presumptive identification in foods of *B. cereus,* a recognized cause of food poisoning.[25]

The assignment of *B. anthracis* and *B. thuringiensis* to varietal status (Table 1) is controversial. We

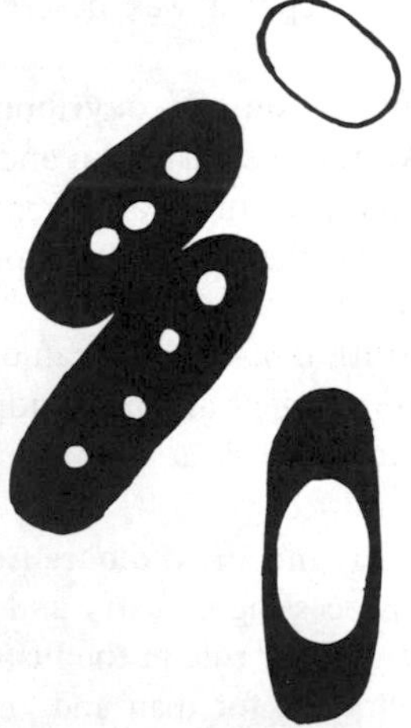

FIGURE 1. *Bacillus megaterium.* A free spore, cells grown on glucose agar, and a sporangium.

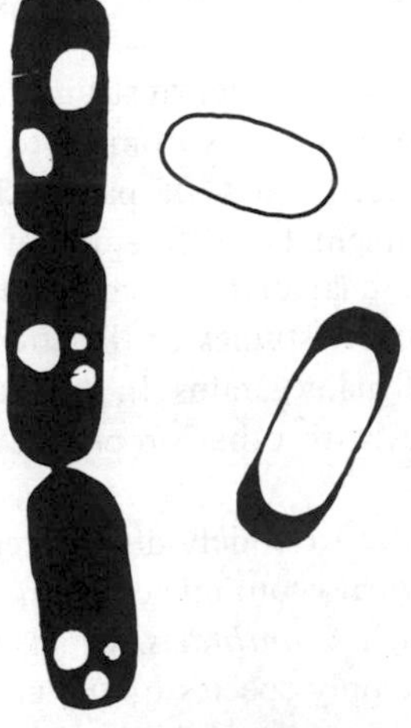

FIGURE 2. *Bacillus cereus.* Cells grown on glucose agar, a free spore, and a sporangium.

FIGURE 3. *Bacillus subtilis.* A sporangium, free spores, and cells grown on glucose agar.

undertook this assignment for the following reasons: (1) our data[24,34] do not provide any stable correlating properties separating *B. anthracis, B. mycoides,* and *B. thuringiensis* from *B. cereus;* (2) data of other investigators support our contention that avirulent strains of *B. anthracis* and *B. thuringiensis* are indistinguishable from strains of *B. cereus;*[36-40] and (3) we believe that a strain should always bear the same species name, despite its loss of unstable properties. In the laboratories of human and veterinary medicine and of insect pathology, however, recognition of virulent strains is admittedly far more important than problems of nomenclature. Criteria for the identification of virulent strains are plentiful.[23,36,41-43] Although the rhizoid strains of *B. mycoides* are currently regarded as unworthy of varietal status,[35,44] in Table 1 the rhizoid strains are presented as a varietal group to demonstrate the similarity between their characteristics and those of the anthrax bacilli.

The strains of the *B. subtilis* group, or the *B. subtilis* spectrum (*B. licheniformis, B. subtilis,* and *B. pumilus*), are morphologically similar and share many physiological properties. In addition to the properties of these three species listed in Table 2, spore antigens, the anaerobic production of gas from nitrate, arginine dihydrolase, reduction of nitrite, and the utilization of acetate and tartrate were reported as useful for the separation of *B. licheniformis* and *B. subtilis.*[45-48] The third member of the *B. subtilis* group, *B. pumilus,* is admittedly very similar to *B. subtilis* and has been given varietal status.[23] However, in addition to the characteristics that separate *B. pumilus* from *B. subtilis* given in Table 2, an examination of the nutritional requirements of twenty-one strains of *B. pumilus* disclosed that all strains required biotin for growth in a solution of glucose and ammonium phosphate, while twenty-six of twenty-seven strains of *B. subtilis* and twelve strains of *B. licheniformis* grew without biotin.[10] The presence of urease has also been used to differentiate between *B. subtilis* and *B. pumilus.*[35]

Despite all attempts to establish the identity of *B. subtilis,* the type species of the genus,[47-50] a Michigan strain of *B. cereus* bearing the label *B. subtilis* is still extant – a warning to all microbiologists against acceptance of a tube's label as identification of the culture.

Although morphologically similar to strains of the *B. subtilis* group, strains of *B. firmus* are distinguishable from this group by their negative V–P reaction, inability to utilize citrate, and their sensitivity to acid (Table 2). Cultures of *B. firmus* form acid from glucose, but are so sensitive to acid that growth ceases at pH 6.0 or above. Strains of *B. lentus* have these same characteristics, but differ from *B. firmus*[51] by their failure to decompose casein and gelatin and by their ability to produce urease. In a study of their nutritional requirements,[9] twenty strains of *B. firmus* and six strains of *B. lentus* showed a gradation in their requirements; the most exacting strains of *B. firmus* and the least exacting strains of *B. lentus* required the same growth factors. Since both *B. firmus* and *B. lentus* are poorly represented in culture collections, examination of many more strains of each is needed.

In contrast to *B. firmus,* strains of *B. coağulans* (Table 2) are aciduric and produce a low pH (4.0 to 5.0) in media containing utilizable carbohydrates. Spoilage of acid foods, such as canned tomatoes, is usually caused by *B. coagulans.* Strains of *B. coagulans* are morphologically variable. The rods of some strains resemble those of the *B. subtilis* group, while those of other strains are longer and slender, and the sporangia may or may not distend the spores (Figure 4). In the presentation of species in Key 2, therefore, *B. coagulans* is intermediate between Groups 1 and 2. Strains of *B. coagulans* also vary in their nutritional requirements, depending on strain, temperature of incubation, and basal medium.[52-54]

Bacillus fastidiosus is one of the two recognized species[44] whose special requirements for growth exclude them from many comparative observations with other recognized species. Because of the limitation of *B. fastidiosus* to media containing uric acid or allantoin, data on this species are not included in Tables 1 to 6. The rods of *B. fastidiosus* are described[44] as large (1.5 to 2.5 by 3 to 6 μ), motile, and Gram-positive in early stages of growth; the spores are oval to cylindrical, terminal or subterminal, and do not distend the sporangia appreciably. Cultures grown on uric acid agar are aerobic, catalase-positive, and mesophilic.

Among the species whose sporangia are swollen by ellipsoidal spores (Figure 5) and whose strains hydrolyze starch (Table 3), *B. polymyxa* is most easily distinguished. Its spores generally have heavily ribbed surfaces; the ribs are longitudinal and parallel; and cross-sections of the spores are star-shaped.[16,55,56] Spore antigens and bacteriophage have been successfully used for identification of its strains.[57,58] If all the species of the genus *Bacillus* were as distinctive as *B. polymyxa,* the taxonomy of the bacilli would be further advanced.

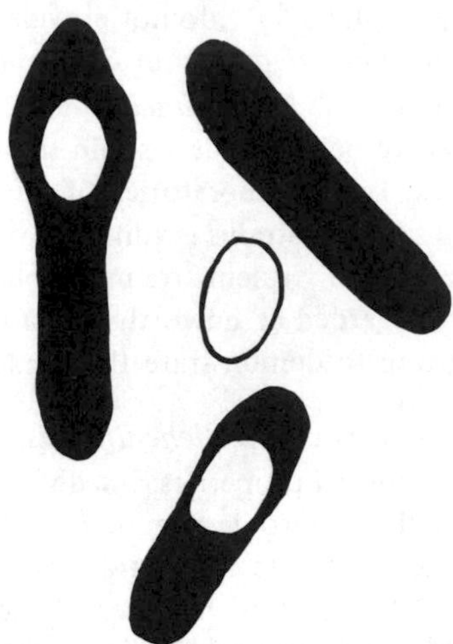

FIGURE 4. *Bacillus coagulans.* Sporangia, a cell, and a free spore.

FIGURE 5. *Bacillus circulans.* A free spore, sporangium, and cells.

Although *B. macerans* resembles *B. polymyxa* in some of its properties, including the morphology of its spores[16] and its active formation of gas from carbohydrates, it is easily separated from *B. polymyxa* (Table 3). At present, however, only gas production and, less satisfactorily, the formation of crystalline dextrins divide *B. macerans* from *B. circulans.* The latter species, aptly labeled a complex rather than a species,[59] encompasses a group of morphologically, nutritionally, physiologically, and chemically heterogeneous strains. Thus far these heterogeneous strains have defied both orderly arrangement within the complex and the satisfactory delineation of *B. circulans* from other species, particularly from *B. macerans, B. stearothermophilus,* and *B. alvei.*

The spores formed by a culture of *B. stearothermophilus* may vary in size. However, in cultures grown on soil extract agar[34] at 45°C to 55°C for three days, spores with diameters wider than the rods predominated. *B. stearothermophilus* is, therefore, assigned to Group 2, Key 2.

In an arrangement (Table 4) of some species according to temperatures of growth of representative strains,[24] *B. stearothermophilus,* the cause of "flat-sours" in canned foods, and *B. coagulans* grow at the higher temperatures. The division between *B. stearothermophilus* and *B. coagulans* is apparently not as clear-cut as this characteristic and the data in Tables 2 and 3 indicate, because other investigators[23,60-64] have described thermophilic strains that share some of the properties of *B. stearothermophilus* with some properties of *B. coagulans.* For example, some of these intermediate strains are as aciduric as strains of *B. coagulans* but grow at 65°C. The existence of these intermediate strains makes it very evident that much more taxonomic work must be done on the thermophilic strains.

The nomenclatural type strain of *B. alvei,* named and described in 1885, was isolated from foul brood of the honeybee, and for a time *B. alvei* was mistakenly thought to be the cause of the disease.[65] Smears of cultures grown on agar usually show the spores arranged side by side in rows (Figure 6). Typical strains of *B. alvei,* as well as some strains of *B. circulans,* are actively motile and may form motile colonies on agar. Although strains of *B. alvei* have been isolated from soil as well as in association with diseases of honeybees, their numbers in culture collections are usually small. Not only are the comparative data on *B. alvei* (Table 3) based on too few strains, but the occurrence of strains seemingly intermediate between *B. alvei* and *B. circulans* makes the reliability of the species description of *B. alvei* uncertain.

Bacillus laterosporus (Table 5) is also sparsely represented in culture collections. A distinctive morphological characteristic of this species is a canoe-shaped[66] or C-shaped[16] parasporal body attached to the spore, with resulting lateral position of the spore in the sporangium. The parasporal body is easily stained and can be seen on the spore after lysis of the sporangium (Figure 7).

In the early 1940's, gramicidin, one of the first antibiotics, was isolated from strains of *B. brevis.* The strains of this species usually form ellipsoidal, central spores that distend the sporangia into spindle shapes;

they resemble strains of *B. laterosporus* in their inability to hydrolyze starch (Table 5). When only inorganic nitrogen is available, cultures of *B. brevis* (86% of ninety-two strains) form acid from glucose (under the conditions of the test); when organic nitrogen is present, production of ammonia masks any action on glucose.

The strains of *B. larvae, B. popilliae,* and *B. lentimorbus* are catalase-negative, grow slowly, and require special media[34] for growth and observation (Table 6). Cultures of *B. larvae*, the cause of American foul brood of the honeybee, grow and sporulate satisfactorily *in vitro.* Cultures of *B. popilliae* and *B. lentimorbus,* causal agents of milky diseases in the larvae of the Japanese beetle and European chafer, can be maintained indefinitely by serial transfer *in vitro*.[67] Infective spores (after ingestion), abundantly produced in the larvae, have not, however, been produced *in vitro*. Insecticides of *B. popilliae* and *B. lentimorbus,* therefore, are not economically feasible.

The spores of *B. sphaericus* are generally round (Figure 8) and terminal, resulting in drumstick-like sporangia (Group 3, Key 2). Examination of twenty-one strains[10] disclosed that nineteen strains grew in a casein basal medium plus thiamin (six strains) or plus thiamin and biotin (thirteen strains). Growth is not enhanced by urea or ammonia. Among the round-spored species, *B. sphaericus* is generally the best represented species in culture collections, and the strains of *B. sphaericus* are quite readily recognized (Table 5). However, since other round-spored strains, including psychrophilic strains,[68] have been reported[44] as differing from *B. sphaericus* in one or more of the distinguishing properties of the species (Table 5), further investigation of more strains may reveal another taxonomically difficult complex.

Bacillus pasteurii is the second recognized species[44] that requires special conditions for growth. The cultures are described as requiring alkaline media containing ammonia (approximately 1% NH_4Cl) or urea (1%). The author of the species[69] characterized freshly isolated strains as decomposing 3 g of urea per liter per hour during their period of maximum growth. The rods of *B. pasteurii* are reported[44] as measuring 0.5 to 1.2 by 1.3 to 4 μ, motile, and Gram-positive. Spores are generally spherical, terminal or subterminal, and swell the sporangia. With the exception of the inability to hydrolyze starch, the physiological properties of the strains are described as variable.

In addition to the species recognized here, many other species names appear in the literature and in our culture collections. Representative strains of a large number of these names have either been lost, found to belong to previously described and named species (see Indexes 1 and 2 in Reference 34 for species names in the synonymy of earlier established species), are believed to be unlike the original strains of their respective species (misnamed), or are temporarily listed as species *incertae sedis* (Table 7). This designation does not imply that the strains bearing these names do not represent distinct species. Some of them undoubtedly do, but the number of available representative strains is insufficient to establish their stable, reliable characteristics. In other instances, time has not permitted a comparison of the representative strains

FIGURE 6. *Bacillus alvei.* Spores and cells.

FIGURE 7. *Bacillus laterosporus.* Spores and a sporangium.

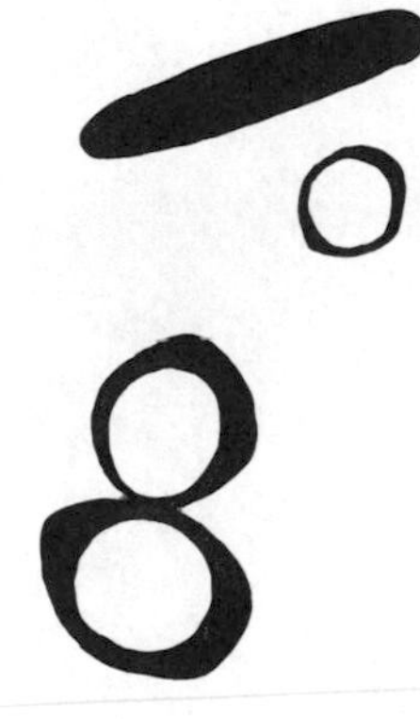

FIGURE 8. *Bacillus sphaericus.* A cell and spores.

with significant numbers of other *Bacillus* species. Information on the comparative examination of strains of some of the species *incertae sedis* (Table 7) is available.[34] The reliability of each characteristic for species description, however, is not established. Despite continued attempts since 1936 to obtain newly described strains and strains bearing species names not represented in our collection, there are unquestionably some we do not have. We should gratefully welcome any representative strains of other *Bacillus* species.

IDENTIFICATION OF STRAINS

For the identification of strains of the genus *Bacillus,* the necessity to compare the unknown strains with a goodly number of named strains of the genus cannot be overstressed. No matter how carefully media and methods are described, the results of a test can be affected by some inadvertently omitted step in technique, changes in the composition of media, regional differences in the water supply, and other factors. By subjecting known strains to a specific test, the investigator may quickly learn whether the test as conducted in his laboratory gives results comparable to those found in the literature. If the results are comparable, the test may be applied to unknown strains; if they are not comparable, the test may be discarded, modified, or replaced by another, provided the usefulness of the modified or new test is established by application to a large number of reference strains. If the investigator cannot spend the time necessary for making a comparison of his unknown strains with available reference strains, he should seek the assistance of someone having a background of experience with the taxon.

For the microbiologist who wishes to acquire a background of knowledge about the taxonomy of the genus *Bacillus*, a mechanical key (Key 1) is presented, to aid in tentatively identifying strains. Although on a much smaller scale, this key resembles *A Guide to the Identification of the Genera of Bacteria*[70] and may be similarly used. In common with other taxonomic keys, Key 1 applies only to the more typical strains of each species.

A more conventional key (Key 2) for the tentative identification of unknown strains is also presented. Key 2 contains more morphological and physiological criteria than Key 1, and the species are arranged in approximately the same order as in the comparative Tables 1 through 6. After tentative identification of an unknown strain as, for example, *B. licheniformis,* observations of the significant properties listed for *B. licheniformis* in Table 2 should then be made. Known reference strains should be used as both positive and negative controls of each observation. The results of these observations would then confirm or disprove the tentative identification of the unknown strain.

Although there undoubtedly remain representative strains of new species to be isolated, described, and named, there are a great many strains belonging to previously established species that were later characterized and named as "new species", a fact that workers should constantly keep in mind.

KEY 1.

TENTATIVE IDENTIFICATION OF TYPICAL STRAINS OF *BACILLUS* SPECIES

Numbers on the right indicate the number (on the left) of the next test to be applied until the right-hand number is replaced by a species name.

1. Catalase: positive → 2
   negative → 16

2. Voges-Proskauer: positive → 3
   negative → 9

3. Growth in anaerobic agar: positive → 4
   negative → 8

4. Growth at 50° C: positive → 5
   negative → 6

5. Growth in 7% NaCl: positive. . . . *B. licheniformis*
   negative. . . *B. coagulans*

6. Acid and gas from glucose (inorganic N_2): positive. . . . *B. polymyxa*
   negative → 7

7. Growth at pH 5.7: positive. . . .*B. cereus*
   negative . . .*B. alvei*

8. Hydrolysis of starch: positive. . . .*B. subtilis*
   negative . . .*B. pumilus*

9. Growth at 65° C: positive. . . .*B. stearothermophilus*
   negative → 10

10. Hydrolysis of starch: positive → 11
    negative →14

11. Acid and gas from glucose (inorganic N_2): positive. . . .*B. macerans*
    negative → 12

12. Utilization of citrate: positive. . . .*B. megaterium*
    negative → 13

13. pH in V-P broth < 6.0: positive. . . .*B. circulans*
    negative . . .*B. firmus*

14. Growth in anaerobic agar: positive. . . .*B. laterosporus*
    negative → 15

15. Acid from glucose (inorganic N_2): positive. . . .*B. brevis*
    negative . . .*B. sphaericus*

16. Growth at 65° C: positive. . . .*B. stearothermophilus*
    negative → 17

17. Decomposition of casein: positive. . . .*B. larvae*
    negative → 18

18. Parasporal body in sporangium: positive. . . .*B. popilliae*
    negative . . .*B. lentimorbus*

KEY 2.

TENTATIVE IDENTIFICATION OF TYPICAL STRAINS OF *BACILLUS* SPECIES

Group 1.

Sporangia not definitely swollen; spores ellipsoidal or cylindrical, central to terminal; Gram-positive.

A. Unstained globules demonstrable in protoplasm of lightly stained cells grown on glucose agar
   1. Strictly aerobic; acetoin not produced: *B. megaterium*
   2. Facultatively anaerobic; acetoin produced: *B. cereus*
      a. Pathogenic to insects: *B. cereus* var. *thuringiensis*
      b. Rhizoid growth: *B. cereus* var. *mycoides*
      c. Causative agent of anthrax: *B. cereus* var. *anthracis*

B. Unstained globules not demonstrable in protoplasm of lightly stained cells grown on glucose agar
   1. Growth in 7% NaCl; acid not produced in litmus milk
      a. Growth at pH 5.7; acetoin produced
         (1) Starch hydrolyzed; nitrates reduced to nitrites
            (a) Facultatively anaerobic; propionate utilized: *B. licheniformis*
            (b) Aerobic[1]; propionate not utilized: *B. subtilis*
         (2) Starch not hydrolyzed; nitrates not reduced to nitrites: *B. pumilus*
      b. No growth at pH 5.7; acetoin not produced: *B. firmus*
   2. No growth in 7% NaCl; acid produced in litmus milk: *B. coagulans*[2]

Group 2.

Sporangia swollen by ellipsoidal spores; spores central to terminal; Gram-positive, -negative, or -variable.

A. Gas formed from carbohydrates
   1. Acetoin produced; dihydroxyacetone formed from glycerol: *B. polymyxa*
   2. Acetoin not produced; dihydroxyacetone not formed: *B. macerans*

B. Gas not formed from carbohydrates
   1. Starch hydrolyzed
      a. Indole not formed
         (1) No growth at 65° C: *B. circulans*
         (2) Growth at 65° C: *B. stearothermophilus*
      b. Indole formed: *B. alvei*
   2. Starch not hydrolyzed
      a. Catalase-positive; survives serial transfer in nutrient broth
         (1) Facultatively anaerobic; pH of cultures in glucose broth less than 8.0: *B. laterosporus*
         (2) Aerobic; pH of cultures in glucose broth 8.0 or higher: *B. brevis*
      b. Catalase-negative; fails to survive serial transfer in nutrient broth
         (1) Nitrates reduced to nitrites; casein decomposed: *B. larvae*
         (2) Nitrates not reduced to nitrites; casein not decomposed
            (a) Sporangium contains a parasporal body; growth in 2% NaCl: *B. popilliae*
            (b) Sporangium does not contain a parasporal body; no growth in 2% NaCl: *B. lentimorbus*

Group 3.

Sporangia swollen; spores generally spherical, terminal to subterminal; Gram-positive, -negative, or -variable.

A. Starch not hydrolyzed; urea or alkaline pH not required for growth: *B. sphaericus*

[1] Do not grow anaerobically in BBL anaerobic agar without glucose or Eh indicator.

[2] Swelling the sporangia by the spores is a variable property of strains of this species and makes the species intermediate between Group 1 and Group 2.

TABLE 1

COMPARISON OF *BACILLUS* SPECIES
[Group 1A, Key 2]

Property	*B. megaterium*	*B. cereus*	*B. cereus* var. *thuringiensis*	*B. cereus* var. *mycoides*	*B. cereus* var. *anthracis*
Rods					
width, μ	1.2–1.5	1.0–1.2	1.0–1.2	1.0–1.2	1.0–1.2
length, μ	2–5	3–5	3–5	3–5	3–5
Gram reaction	+	+	+	+	+
Unstained globules in the protoplasm	+	+	+	+	+
Spores					
ellipsoidal	+	+	+	+	+
round	v	–	–	–	–
central or paracentral	+	+	+	+	+
swelling the sporangium	–	–	–	–	–
Crystalline parasporal bodies	–	–	a	–	–
Motility	a	a	a	–	–
Catalase	+	+	+	+	+
Anaerobic growth	–	+	+	+	+
V-P reaction	–	+	+	+	+
pH in V-P broth	4.5–6.8	4.3–5.6	4.3–5.6	4.5–5.6	5.0–5.6
Temperature of growth, °C					
maximum	35–45	35–45	40–45	35–40	40
minimum[a]	3–20	10–20	10–15	10–15	15–20
Egg-yolk reaction	–	+	+	+	+
Growth in					
0.001% lysozyme	–	+	+	+	+
7% NaCl	+	+	+	a	+
media at pH 5.7	+	+	+	+	+
Acid from					
glucose	+	+	+	+	+
arabinose	a	–	–	–	–
xylose	a	–	–	–	–
mannitol	+	–	–	–	–
Hydrolysis of starch	+	+	+	+	+
Use of citrate	+	+	+	a	b
Reduction of NO_3 to NO_2	b	+	+	+	+

TABLE 1 (Continued)

COMPARISON OF *BACILLUS* SPECIES

Property	*B. megaterium*	*B. cereus*	*B. cereus* var. *thuringiensis*	*B. cereus* var. *mycoides*	*B. cereus* var. *anthracis*
Deamination of phenylalanine, 1 week	a	–	–	–	–
Decomposition of					
casein	+	+	+	+	+
tyrosine	a	+	+	a	–

[a]The lowest temperature tested was 3°C.

Symbol Code

+ = 85 to 100% of the strains positive
a = 50 to 84% of the strains positive
b = 15 to 49% of the strains positive
– = 0 to 14% of the strains positive
v = character inconstant

TABLE 2

COMPARISON OF *BACILLUS* SPECIES

[Group 1B, Key 2]

Property	*B. licheniformis*	*B. subtilis*	*B. pumilus*	*B. firmus*	*B. coagulans*
Rods					
width, μ	0.6-0.8	0.7-0.8	0.6-0.7	0.6-0.9	0.6-1
length, μ	1.5-3	2-3	2-3	1.2-4	2.5-5
Gram reaction	+	+	+	+	+
Unstained globules in protoplasm	–	–	–	–	–
Spores					
ellipsoidal or cylindrical	+	+	+	+	+
central or paracentral	+	+	+	v	v
subterminal or terminal	–	–	–	v	v
swelling the sporangium	–	–	–	–	v
Motility	+	+	+	a	+
Catalase	+	+	+	+	+
Anaerobic growth	+	–	–	–	+
V-P reaction	+	+	+	–	a

TABLE 2 (Continued)

COMPARISON OF *BACILLUS* SPECIES

Property	*B. licheniformis*	*B. subtilis*	*B. pumilus*	*B. firmus*	*B. coagulans*
pH in broth					
V-P	5.0-6.5	5.4-8.0	4.8-5.5	6.0-6.8	4.2-4.8
anaerobic glucose	5.0-5.6	6.2-8.2			
Temperature of growth, °C					
maximum	50-55	45-55	45-50	40-45	55-60
minimum	15	5-20	5-15	5-20	15-25
Egg-yolk reaction	–	–	–	–	–
Growth in					
0.001% lysozyme	–	b	a	–	–
media at pH 5.7	+	+	+	–	+
7% NaCl	+	+	+	+	–
0.02% azide[a]	–	–			+
Acid from					
glucose	+	+	+	+	+
arabinose	+	+	+	b	a
xylose	+	+	+	b	a
mannitol	+	+	+	+	b
Hydrolysis of					
starch	+	+	–	+	+
hippurate, 4 weeks	–	–	+		
Use of					
citrate	+	+	+	–	b
propionate	+	–	–	–	–
Reduction of NO_3 to NO_2	+	+	–	+	b
Decomposition of					
casein	+	+	+	+	b
tyrosine	–	–	–	b	–
Liquefaction of nutrient gelatin, 20°C, 2 weeks	<1.0 cm	1.0 cm or more			

[a]Only strains growing at 55°C or higher were tested.

Symbol Code

+ = 85 to 100% of the strains positive
a = 50 to 84% of the strains positive
b = 15 to 49% of the strains positive
– = 0 to 14% of the strains positive
v = character inconstant

TABLE 3

COMPARISON OF *BACILLUS* SPECIES
[Groups 2A and 2B(1), Key 2]

Property	*B. polymyxa*	*B. macerans*	*B. circulans*	*B. stearothermophilus*	*B. alvei*
Rods					
width, μ	0.6-0.8	0.5-0.7	0.5-0.7	0.6-1.0	0.5-0.8
length, μ	2-5	2.5-5	2-5	2-3.5	2-5
Gram reaction	v	v	v	v	v
Spores					
ellipsoidal	+	+	+	+	+
central or paracentral	v	–	v	–	v
subterminal or terminal	v	+	v	+	v
swelling the sporangium	+	+	+	+	+
Motility	+	+	a	+	+
Catalase	+	+	+	a	+
Anaerobic growth	+	+	a	–	+
V-P reaction	+	–	–	–	+
pH in V-P broth	4.5-6.8	4.5-5.0	4.5-6.6	4.8-5.8	4.6-5.2
Temperature of growth, °C					
maximum	35-45	40-50	35-50	65-75	35-45
minimum	5-10	5-20	5-20	30-45	15-20
Growth in					
0.001% lysozyme	a	–	b	–	+
media at pH 5.7	+	+	b	–	–
0.02% azide				–	
5% NaCl	–	–	a	b	b
10% NaCl	–	–	–	–	–
Acid from					
glucose	+	+	+	+	+
arabinose	+	+	+	b	–
xylose	+	+	+	a	–
mannitol	+	+	+	b	–
Gas from fermented carbohydrates	+	+	–	–	–
Hydrolysis of starch	+	+	+	+	+
Use of citrate	–	b	b	–	–
Reduction of NO_3 to NO_2	+	+	b	a	–

TABLE 3 (Continued)

COMPARISON OF *BACILLUS* SPECIES

Property	*B. polymyxa*	*B. macerans*	*B. circulans*	*B. stearothermophilus*	*B. alvei*
Formation of					
crystalline dextrins	–	+	–		
dihydroxyacetone	+	–	–	–	+
indole	–	–	–	–	+
Decomposition of					
casein	+	–	b	a	+
tyrosine	–	–	–	–	b

Symbol Code:

+ = 85 to 100% of the strains positive
a = 50 to 84% of the strains positive
b = 15 to 49% of the strains positive
– = 0 to 14% of the strains positive
v = character inconstant

TABLE 4

GROWTH TEMPERATURES OF SEVERAL SPECIES OF *BACILLUS*

Species	Number of Strains	Number of Strains Growing at 28°C	33°C	37°C	45°C	50°C	55°C	60°C	65°C	70°C
B. stearothermophilus	87	0	10	73	81	87	87	87	87	45
B. coagulans	73	53	73	73	73	72	66	23	0	
B. subtilis	154	154	154	154	150	105	17	0		
B. brevis	57	57	57	57	38	16	7	0		
B. circulans	55	55	55	51	18	6	1	0		
B. pumilus	65	65	65	65	64	43	0			
B. macerans	13	13	13	13	13	9	0			
B. cereus	50	50	50	50	23	0				
B. sphaericus	42	42	42	42	15	0				

TABLE 5

COMPARISON OF *BACILLUS* SPECIES
[Groups 2B(2a) and 3, Key 2]

Property	*B. laterosporus*	*B. brevis*	*B. sphaericus*
Rods			
width, μ	0.5-0.8	0.6-0.9	0.6-1
length, μ	2-5	1.5-4	1.5-5
Gram reaction	v	v	v
Spores			
ellipsoidal	+	+	–
round	–	–	+
central or paracentral	+	v	–
subterminal or terminal	–	v	+
with "C"-shaped rims	+	–	–
swelling the sporangium	+	+	+
Motility	+	+	+
Catalase	+	+	+
Anaerobic growth	+	–	–
V-P reaction	–	–	–
pH in V-P broth	5.0-6.0	8.0-8.6	7.4-8.6
Temperature of growth, °C			
maximum	35-50	40-60	30-45
minimum	15-20	10-35	5-15
Growth in			
0.001% lysozyme	+	b	a
media at pH 5.7	–	b	b
5% NaCl	a	–	+
10% NaCl	–	–	–
0.02% azide		–	
Acid from			
glucose	+	+	–
arabinose	–	–	–
xylose	–	–	–
mannitol	+	a	–
Hydrolysis of starch	–	–	–
Use of citrate	–	b	b
Reduction of NO_3 to NO_2	+	a	–

TABLE 5 (Continued)

COMPARISON OF *BACILLUS* SPECIES

Property	*B. laterosporus*	*B. brevis*	*B. sphaericus*
Formation of			
dihydroxyacetone	–	–	–
indole	a	–	–
Deamination of phenylalanine, 3 weeks	–	–	+
Decomposition of			
casein	+	+	a
tyrosine	+	+	–

Symbol Code

+ = 85 to 100% of the strains positive
a = 50 to 84% of the strains positive
b = 15 to 49% of the strains positive
– = 0 to 14% of the strains positive
v = character inconstant

TABLE 6

COMPARISON OF *BACILLUS* SPECIES [Group 2B(2b), Key 2]

Property	*B. larvae*	*B. popilliae*	*B. lentimorbus*
Rods			
width, μ	0.5-0.6	0.5-0.8	0.5-0.7
length, μ	1.5-6	1.3-5.2	1.8-7
Gram reaction	+	–	–
Spores			
ellipsoidal	+	+	+
round	–	–	–
central or paracentral	v	v	v
subterminal or terminal	v	v	v
swelling the sporangium	+	+	+
Parasporal body in sporangium		+	–
Motility	a	a	–
Catalase	–	–	–
Anaerobic growth	+	+	+

TABLE 6 (Continued)

COMPARISON OF *BACILLUS* SPECIES

Property	*B. larvae*	*B. popilliae*	*B. lentimorbus*
V-P reaction	–	–	–
pH in V-P broth	5.5-6.2	5.7-6.2	5.9-6.9
Temperature of growth, °C			
maximum	40	35	35
minimum	25	20	20
Growth in			
0.001% lysozyme	+	+	+
media at pH 5.7	–	–	–
nutrient broth	–	–	–
2% NaCl	+	+	–
Acid from			
glucose	+	+	+
trehalose	+	+	+
arabinose	–	–	–
xylose	–	–	–
mannitol	b	–	–
Hydrolysis of starch	–	–	–
Use of citrate	–	–	–
Reduction of NO_3 to NO_2	a	–	–
Formation of			
dihydroxyacetone	–	–	–
indole	–	–	–
Deamination of phenylalanine, 3 weeks	–	–	–
Decomposition of			
casein	+	–	–
gelatin	+	–	–
tyrosine	–	–	–

Symbol Code

+ = 85 to 100% of the strains positive
a = 50 to 84% of the strains positive
b = 15 to 49% of the strains positive
– = 0 to 14% of the strains positive
v = character inconstant

TABLE 7

SPECIES *INCERTAE SEDIS* OF THE GENUS *BACILLUS*

B. acidocaldarius Darland and Brock
B. alcalophilus Vedder
B. aminovorans den Dooren de Jong
B. aneurinolyticus Aoyama
B. anthracoides Flügge
B. apiarius Katznelson
B. badius Batchelor
B. cirroflagellosus ZoBell and Upham
B. cubensis Stührk
B. epiphytus ZoBell and Upham
B. filicolonicus ZoBell and Upham
B. freudenreichii (Miquel) Chester
B. globisporus Larkin and Stokes
B. insolitus Larkin and Stokes
B. laevolacticus Nakayama and Yanoshi
B. lentus Gibson
B. lubinskii Kruse
B. macquariensis Marshall and Ohye
B. macroides Bennett and Canale-Parola
B. maroccanus Delaporte and Sasson
B. medusa Delaporte
B. pacificus Delaporte
B. pantothenticus Proom and Knight
B. psychrophilus Larkin and Stokes
B. psychrosaccharolyticus Larkin and Stokes
B. pulvifaciens Katznelson
B. racemilacticus Nakayama and Yanoshi
B. rarerepertus Schieblich
B. thiaminolyticus Kuno
B. virgula Trevisan
Krusella cascainensis Castellani

REFERENCES

1. Delaporte, B., *Ann. Inst. Pasteur (Paris), 107,* 845 (1964).
2. Cross, T., *J. Appl. Bacteriol., 33,* 95 (1970).
3. Halvorson, H. O., Hanson, R., and Campbell, L. L. (Eds.), *Fifth International Spore Conference, Fontana, Wisconsin, 1971.* American Society for Microbiology, Washington, D. C. (1972).
4. Gould, G. W., and Hurst, A. (Eds.), *The Bacterial Spore.* Academic Press, London, England, and New York (1969).
5. Kitahara, K., and Suzuki, J., *J. Gen. Appl. Microbiol., 9,* 59 (1963).
6. Kalininskaya, T. A., *Mikrobiol.* (USSR), *37,* 923 (1968).
7. Grau, F. H., and Wilson, P. W., *J. Bacteriol., 83,* 490 (1962).
8. Bredemann, G., *Zentralbl. Bakteriol. Parasitenk. Infektionskr. Hyg. Abt. 2, 22,* 44 (1909).
9. Proom, H., and Knight, B. C. J. G., *J. Gen. Microbiol., 13,* 474 (1955).
10. Knight, B. C. J. G., and Proom, H., *J. Gen. Microbiol., 4,* 508 (1950).
11. Dooren de Jong, L. E. den, *Zentralbl. Bakteriol. Parasitenk. Infektionskr. Hyg. Abt. 2, 79,* 344 (1929).
12. Vedder, A., *Antonie van Leeuwenhoek Ned. Tijdschr. Hyg. Microbiol. Serol., 1,* 141 (1934).
13. Darland, G., and Brock, T. D., *J. Gen. Microbiol., 67,* 9 (1971).
14. Mann, E. W., *Southwestern Natur., 13,* 349 (1968).
15. Delaporte, B., and Sasson, A., *C. R. Acad. Sci. Paris Sér. D, 264,* 2344 (1967).
16. Bradley, D. E., and Franklin, J. G., *J. Bacteriol., 76,* 618 (1958).
17. Jones, D., and Sneath, P. H. A., *Bacteriol. Rev., 34,* 40 (1970).
18. Hill, L. R., *J. Gen. Microbiol., 44,* 419 (1966).
19. Jayne-Williams, D. J., and Cheeseman, G. C., *J. Appl. Bacteriol., 23,* 250 (1960).
20. Bulla, L. A., Jr., Bennett, G. A., and Shotwell, O. L., *J. Bacteriol., 104,* 1246 (1970).
21. Shen, P. Y., Coles, E., Foote, J. L., and Stenesh, J., *J. Bacteriol., 103,* 479 (1970).
22. Kaneda, T., *J. Bacteriol., 93,* 894 (1967).
23. Wolf, J., and Barker, A. N., in *Identification Methods for Microbiologists,* p. 93, B. M. Gibbs and D. A. Shapton, Eds. Academic Press, London, England (1968).
24. Smith, N. R., Gordon, R. E., and Clark, F. E., *Agriculture Monograph No. 16.* U.S. Department of Agriculture, Washington, D. C. (1952).
25. Kim, H. U., and Goepfert, J. M., *Appl. Microbiol., 22,* 581 (1971).
26. Coonrood, J. D., Leadley, P. J., and Eickhoff, T. C., *J. Infec. Dis., 123,* 102 (1971).
27. Allen, B. T., and Wilkinson, H. A., III, *Johns Hopkins Med. J., 125,* 8 (1969).
28. Curtis, J. R., Wing, A. J., and Coleman, J. C., *Lancet, 1,* 136 (1967).
29. Lázár, J., and Juresák, L., *Zentralbl. Bakteriol. Parasitenk. Infektionskr. Hyg. Abt. 1 Orig., 199,* 59 (1966).
30. Stopler, T., Camuescu, V., and Voiculescu, M., *Microbiol. Parazitol. Epidemiol., 9,* 457 (1964). English translation, *Rum. (Bucaresti) Med. Rev., 19,* 7 (1965).
31. Farrar, W. E., Jr., *Amer. J. Med., 34,* 134 (1963).
32. Norris, J. R., in *The Bacterial Spore,* p. 485, G. W. Gould and A. Hurst, Eds. Academic Press, London, England, and New York (1969).

33. Cowan, S. T., *J. Gen. Microbiol.*, *67*, 1 (1971).
34. Gordon, R. E., Haynes, W. C., and Pang, C. H-N., *Agriculture Handbook No. 427.* U. S. Department of Agriculture, Washington, D. C. (1972).
35. Lemille, F., de Barjac, H., and Bonnefoi, A., *Ann. Inst. Pasteur (Paris)*, *116*, 808 (1969).
36. Krieg, A., *J. Invertebr. Pathol.*, *15*, 313 (1969).
37. Dowdle, W. R., and Hansen, P. A., *J. Infect. Dis.*, 108, 125 (1961).
38. Lamanna, C., and Eisler, D., *J. Bacteriol.*, *79*, 435 (1960).
39. Bennett, E. O., Peterson, G. E., and Williams, R. P., *Antibiot. Chemother.*, *9*, 115 (1959).
40. Burdon, K. L., *J. Bacteriol.*, *71*, 25 (1956).
41. Weaver, R. E., Brachman, P. S., and Feeley, J. C., in *Diagnostic Procedures for Bacterial, Mycotic, and Parasitic Infections,* 5th ed., p. 354, H. L. Bodily, E. L. Updyke and J. O. Mason, Eds. American Public Health Association, New York (1970).
42. de Barjac, H., and Bonnefoi, A., *C. R. Acad. Sci. Paris, Sér. D, 264*, 1811 (1967).
43. Heimpel, A. M., *Annu. Rev. Entomol.*, *12*, 287 (1967).
44. Gibson, T., and Gordon, R. E., in *Bergey's Manual of Determinative Bacteriology,* 8th ed., Williams and Wilkins, Baltimore, Maryland (in press).
45. Hánákova-Bauerová, E., Kocur, M., and Martinec, T., *J. Appl. Bacteriol.*, *28*, 384 (1965).
46. Kundrat, W., *Zentralbl. Veterinaermed.* Reihe B, *10*, 418 (1963).
47. Gibson, T., *J. Dairy Res.*, *13*, 248 (1944).
48. Lamanna, C., *J. Bacteriol.*, *44*, 611 (1942).
49. *Report of Proceedings, Second International Congress for Microbiology, London, 1936,* p. 28, R. St. John-Brooks, Ed. (1937).
50. Conn, H. J., *J. Infect. Dis.*, *46,* 341 (1930).
51. Gibson, T., *Zentralbl. Bakteriol. Parasitenk. Infektionskr. Hyg. Abt. 2, 92*, 364 (1935).
52. Marshall, R., and Beers, R. J., *J. Bacteriol.*, *94*, 517 (1967).
53. Campbell, L. L., Jr., and Sniff, E. E., *J. Bacteriol.*, *78*, 267 (1959).
54. Humphreys, T. W., and Costilow, R. N., *Can. J. Microbiol.*, *3*, 533 (1957).
55. Murphy, J. A., and Campbell, L. L., *J. Bacteriol.*, *98*, 737 (1969).
56. Holbert, P. E., *J. Biophys. Biochem. Cytol.*, *7*, 373 (1960).
57. Davies, S. N., *J. Gen. Microbiol.*, *5*, 807 (1951).
58. Francis, A. E., and Rippon, J. E., *J. Gen. Microbiol.*, *3*, 425 (1949).
59. Gibson, T., and Topping, L. E., *Proc. Soc. Agr. Bacteriol.*, p. 43 (1938).
60. Baillie, A., and Walker, P. D., *J. Appl. Bacteriol.*, *31*, 114 (1968).
61. Galesloot, T. E., and Labots, H., *Nederlands Melk-en Zuiveltijdschr.*, *13*, 155 (1959).
62. Grinstead, E., and Clegg, L. F. L., *J. Dairy Res.*, *22*, 178 (1955).
63. Marsh, C. L., and Larsen, D. H., *J. Bacteriol.*, *65*, 193 (1953).
64. Gyllenberg, H., *Suomen Maataloustieteellisen Seuran Julkaisuja* (*J. Sci. Agr. Soc. Finland*), No. 73, 1 (1951).
65. Bailey, L., *Nature, 180*, 1214 (1957).
66. Hannay, C. L., *J. Biophys. Biochem. Cytol.*, *3*, 1001 (1957).
67. Haynes, W. C., and Rhodes, L. J., *Bacteriol. Proc.*, p. 10 (1963).
68. Larkin, J. M., and Stokes, J. L., *J. Bacteriol.*, *94*, 889 (1967).
69. Miquel, P., *Ann. Microgr.*, *1*, 506, 552; *2*, 13 (1889).
70. Skerman, V. B. D., *A Guide to the Identification of the Genera of Bacteria,* 2nd ed. Williams and Wilkins, Baltimore, Maryland (1967).

THE CLOSTRIDIA

DR. LOUIS DS. SMITH

The genus *Clostridium* is composed of spore-forming anaerobic rods. Spore-forming anaerobic cocci – the genus *Sporosarcina* – have been described, but we know very little about them. The genus *Desulfotomaculum* also contains anaerobic sporing rods, but these may be differentiated from the clostridia by a combination of three characteristics:[3] (1) Gram-negativity; (2) DNA base composition, GC = 41 to 46%; and (3) cytochrome of the protoheme class. Some members of *Clostridium,* especially the cellulose-fermenters, are Gram-negative; others have DNA with GC = 43 to 46%; in still others, cytochromes have been found. However, the most important characteristic shared by organisms in *Desulfotomaculum* is their ability to reduce sulfate to sulfide, a characteristic uncommon among strains of *Clostridium.* Of the three species of *Desulfotomaculum, D. nigrificans* is thermophilic. The two mesophilic species can be differentiated by the ability of *D. ruminis* to grow in formate and sulfate and the inability of *D. orientis* to do so.

The clostridia vary in tolerance to oxygen, in nutritional requirements, and in optimum and limiting temperature for growth, and in many ways it is difficult to generalize concerning them. *Clostridium carnis, C. histolyticum,* and *C. tertium,* for example, are aerotolerant and will form colonies on freshly poured blood agar medium incubated aerobically. *Clostridium haemolyticum,* on the other hand, is a strict anaerobe[12] and will not grow even on reduced media if the oxygen tension is greater than 0.5%. Some species, such as *C. butyricum,* can grow with ammonia as the nitrogen source and biotin as the only vitamin,[5] whereas others, such as *C. perfringens,* require more than twenty amino acids and vitamins. Some of the psychrophiles, such as *C. putrefaciens,* will not grow at temperatures above 30°C; some of the thermophiles, such as *C. thermosaccharolyticum,* will hardly grow below 50°C. For more information concerning the clostridia than can be given here, see References 19, 23, and 25, as well as *J. Appl. Bacteriol., 28,* No. 1 (1965).

Although the clostridia are usually considered as retaining Gram's stain, not all of them do so equally. Many of the species with terminal spores lose the stain readily, and Gram-positive cells are seldom seen in overnight cultures. Several species with subterminal spores, such as *C. haemolyticum,* also lose stain completely and appear as large Gram-negative rods. Cells of *C. chauvoei* do so only partially and appear mottled or spotted.

It is sometimes necessary to determine whether a strain is an aerotolerant *Clostridium* or a facultative *Bacillus.* Clostridia produce spores anaerobically; bacilli do so aerobically. In addition, clostridia rarely produce catalase, and then only in trace amounts. If spores can be demonstrated in anaerobic culture, the strain should be considered as belonging to *Clostridium,* especially if it does not produce catalase.

The demonstration of spores in clostridia is usually not difficult, because most strains form spores when incubated in a good medium under strictly anaerobic conditions 3 to 8°C below their optimum temperature for growth. The best single medium for demonstrating spore formation is agar made from cooked-meat medium, slanted in tubes, and prepared anaerobically. If a fluid medium is desired, cooked-meat medium may be used.[20] For spore formation by most species, the sporulation medium should not contain fermentable sugar. A few saccharolytic species, however, will not sporulate unless a fermentable carbohydrate is present. A few other species, such as *C. perfringens* and *C. paraperfringens* (*barati*) will not sporulate unless special media are used.

If a strain does not readily form spores, it may often be stimulated to do so by heat-shocking. Heat a freshly inoculated tube of fluid cooked-meat medium at 80°C for 10 minutes, then incubate at a temperature below the optimum for growth. Some clostridial spores, such as those of some strains of *C. botulinum* type E, will not withstand 10 minutes at 80°C; for such strains 10 minutes at 70°C or treatment with 50% ethanol for one hour at room temperature will serve.[10] Although such treatment will usually, but not always, stimulate the production of spores by a strain, it should not be assumed that if a little heat is good, more is better. Many strains will withstand heating at 100°C for an hour or more, but spore production from ensuing cultures will be less than if the heat treatment had been milder, just enough to kill off the vegetative cells. (For a review of some of the factors involved in the production of clostridial spores, see References 18 and 20. For the effect of heat on the properties of cultures, see Reference 17.)

In the identification of the clostridia, the shape – spherical or oval – and the position – terminal or subterminal – of the spore is used. Determination of the shape and position of the spores should be made only with cultures showing fully mature spores, taking no stain and highly refractile. Spores that will, in their mature state, be truly and unmistakably terminal are, in some species, first formed in a subterminal position. As the spore matures, the vegetative material surrounding it disappears, and it attains its position at the end of the rod. Subterminal spores, however, do not do this; they remain in their subterminal to central position until the vegetative portion of the cell is lysed. A few species, such as *C. carnis,* consist of cells that have subterminal spores and cells that have terminal spores, giving the appearance of a mixed culture. For microscopic demonstration of spores, Gram-staining is sufficient. Spore stains are of no real advantage when searching for spores. A phase microscope is better than Gram-staining when mature spores are present, but is less good when the spores are immature and not refractile. Little difficulty will be experienced in determining whether spores are oval or spherical if the spores are mature. Almost all species forming spherical spores do so in the terminal position, giving rise to an unmistakable drumstick configuration.

If strains of clostridia are to be resurrected from old spore suspensions, they should be inoculated into fresh infusion broth containing 0.2% starch, heat-shocked for 10 minutes at 80°C (except for *C. botulinum* type E strains, which should be heated at 70°C), and incubated in an atmosphere containing at least 10% carbon dioxide at a temperature somewhat below the optimum for growth. Thioglycollate should not be used in such media, for it is inhibitory to the outgrowth of some strains;[9,16] its inhibiting effect is much less if the medium contains glucose. The factors governing recovery of severely heated or otherwise damaged spores are complex, for spores of different strains of the same species may behave differently, and spores of the same strain produced in different media may differ in response to the same recovery procedure and the same recovery medium.[21]

When clostridia are to be isolated, it should be remembered that they cover the entire range in their need for anaerobiosis and that some of them are quite fastidious. Consequently, relatively rigorous means of anaerobiosis should be used, or such fastidious strains will be missed. This is important even when handling some of the clostridia usually considered as not highly sensitive, for the most toxigenic variants of some species are fastidious as far as oxygen is concerned, whereas variants of lower toxigenicity are not. Consequently, if inadequate anaerobic methods are used with such strains, only cultures of low or no toxigenicity will be obtained. This seems especially to be the case with *C. botulinum* types C and D.

Two methods of obtaining anaerobiosis can be recommended: the roll-tube method of Hungate as modified by Moore,[10] and the plastic anaerobic glove box of Freter.[1] Freshly poured blood agar plates, incubated in anaerobe jars, may also be successfully used, but if clostridia sensitive to oxygen are to be isolated, the agar should be streaked as soon as it has hardened. Oxygen-sensitive clostridia that will not form colonies on freshly poured and solidified agar may usually be grown by "sloppy streaking", i.e., streaking the fluid blood agar just before it solidifies. Blood agar serves as an excellent solid medium for the clostridia, particularly if 0.05% cysteine and 0.03% dithiothreitol, dithioerythritol, or sodium formaldehyde sulfoxylate is added. For an investigation into the cultivation of exacting species, see Reference 4.

Several media have been devised for the isolation of clostridia from a variety of sources (for an excellent description of methods and media, see Reference 22). For the isolation of clostridia from food, semiselective media containing sulfite and iron are most commonly used. The clostridia reduce sulfite to sulfide, thus causing the blackening of the colonies.[15] Various other substances, inhibitory for other organisms but with little effect on the clostridia, have also been used in efforts to make media that will grow clostridia but no other organisms. Unfortunately, as Gibbs and Fraeme[6] have pointed out, no really satisfactory selective medium for the clostridia is known. Even for an organism as hardy as *C. perfringens,* it has not been possible to devise a medium that suppresses the outgrowth of contaminant bacteria but permits the quantitative recovery of *C perfringens.*[7,8] If a substance is inhibitory for all the facultative bacteria, or even for most of them, it is also inhibitory for at least some of the clostridia. Consequently, media containing crystal violet, sorbic acid, sulfadiazine, neomycin, cycloserine, etc., cannot be trusted for the quantitative isolation of all species of *Clostridium.* When such media are used, non-selective media should be used in parallel. If the limitations of selective media are understood, however, they may be very useful for special purposes. It is well to have a strain of *C. haemolyticum* on hand and occasionally to try obtaining isolated colonies of this organism. If isolated colonies of *C. haemolyticum* can be obtained, the medium and the isolation procedures are adequate for the other clostridia. If isolated colonies of this organism cannot be obtained, some factor is unsatisfactory.

Whatever procedure of incubation and whatever medium is used, after incubation the colonies should be scrutinized with a dissection microscope to make certain that they are well isolated and that the plate is not covered by a thin film of one of the swarming clostridia. Often this film may be just barely perceptible if one is looking carefully for it. When this happens, isolation should be repeated, using a shorter incubation time or increasing the concentration of the agar to 4%. The edge of a swarming area should not be picked with the hope of getting a pure culture of the swarming strain, for bacteria that are swarming often carry other bacteria along with them. An isolated colony should be picked to a tube of cooked-meat–glucose medium, incubated overnight, and used to inoculate the differential media listed in Table 1 for identification.

TABLE 1

DIFFERENTIAL MEDIA FOR THE CLOSTRIDIA

Medium	Purpose
Glucose broth	Fermentation, chromatography, toxin
Lactose broth	Fermentation
Maltose broth	Fermentation
Mannitol broth	Fermentation
Mannose broth	Fermentation
Salicin broth	Fermentation
Sucrose broth	Fermentation
Trehalose broth	Fermentation
Gelatin	Liquefaction
Milk	Digestion
Cooked-meat medium	Motility, indol, digestion
Cooked-meat slant	Sporulation (incubate at 30°C)
Molten deep agar	Aerobic or anaerobic growth
Blood agar plate	Colony description, hemolysis
Blood agar plate, aerobic	Aerotolerance
Egg yolk agar plate	Lecithinase, lipase

The cultural characteristics of the most commonly occurring species are listed in Tables 2 and 3. Among the species that will be encountered most often will be *C. perfringens, C. bifermentans, C. sordellii, C. sporogenes, C. innocuum,* and *C. subterminale.* If pathologic specimens from animals are being examined, *C. novyi, C. chauvoei,* and *C. septicum* will also be encountered fairly frequently. It will be convenient to be familiar with the morphology, colonial aspects and outstanding cultural characteristics of these organisms.

In addition to the determination of the usual cultural characteristics, it is most helpful to determine, with the aid of a gas chromatograph, the principal fermentation products that each strain produces. This may be done from the culture in peptone-yeast-extract-glucose medium[10] after incubation for three days or longer. The major products of fermentation in this medium are listed in Tables 4 and 5. Determination of the fermentation products is a valuable aid when quick identification of a strain is desired, and is always helpful in confirming conclusions arrived at by consideration of other data.

The fluorescent-antibody technique has also been found to be of considerable value in the rapid identification of some of the pathogenic species. *Clostridium chauvoei, C. septicum, C. novyi, C. tetani,* and *C. botulinum* have been studied by this method.[2] *Clostridium chauvoei* strains apparently form a single antigenic group when tested by this method, while *C. septicum* strains form two groups. Strains of *C. novyi* of whatever type share antigens. However, antiserum prepared against *C. haemolyticum* (*C. novyi* type D)

TABLE 2

CULTURAL CHARACTERISTICS OF CLOSTRIDIA WITH SUBTERMINAL SPORES

Species	Gelatin Digestion	Milk Digestion	Lecithinase Production	Lipase Production	Indol Production	Glucose Fermentation	Lactose Fermentation	Maltose Fermentation	Mannitol Fermentation	Mannose Fermentation	Salicin Fermentation	Sucrose Fermentation	Toxin Production
C. sordelli[a]	+	+	+	–	+	+	–	+	–	v	–	–	+
C. bifermentans[a]	+	+	+	–	+	+	–	+	–	v	–	–	–
C. sporogenes[b]	+	+	–	+	–	+	–	v	–	–	–	–	–
C. botulinum ABF (proteolytic)	+	+	–	+	–	+	–	v	–	–	–	–	+
C. hystolyticum[c]	+	+	–	–	–	–	–	–	–	–	–	–	+
C. subterminale[c]	+	+	–	–	–	–	–	–	–	–	–	–	–
C. limosum	+	+	+	–	–	–	–	–	–	–	–	–	±
C. perfringens	+	–	+	–	–	+	+	+	–	+	–	+	+
C. novyi A[b]	+	–	+	+	–	+	–	+	–	–	–	–	+
C. novyi B	–	+	+	–	–	+	–	+	–	+	–	–	+
C. haemolyticum	+	–	+	–	+	+	–	–	–	v	–	–	+
C. botulinum CD[b]	+	–	–	+	–	+	–	+	–	+	–	–	+
C. botulinum BEF[b] (non-proteolytic)	+	–	–	+	–	+	–	+	–	+	–	+	+
C. difficile	+	–	–	–	–	+	–	–	+	+	v	–	+
C. septicum	+	–	–	–	–	+	+	+	–	+	+	–	+
C. chauvoei	+	–	–	–	–	+	+	+	–	+	–	+	+
C. paraperfringens	–	–	+	–	–	+	+	+	–	+	+	+	–
C. fallax	–	–	–	–	–	+	+	+	+	+	+	+	–
C. butyricum	–	–	–	–	–	+	+	+	–	+	+	+	–
C. carnis[d]	–	–	–	–	–	+	v	+	–	+	+	+	–
C. beijerinckii	–	–	–	–	–	+	+	+	–	+	–	+	–

[a] *C. sordelli* produces urease; *C. bifermentans* does not.
[b] Toxin tests are required for identification.
[c] *C. histolyticum* is aerotolerant; *C. subterminale* is not.
[d] *C. carnis* is aerotolerant; spores are subterminal to terminal.

and absorbed with *C. novyi* strains of the other types has been reported as satisfactory for identifying *C. haemolyticum* in tissue smears.[13] The indirect-fluorescence method has also been used to demonstrate spores of *C. haemolyticum* in biopsy specimens.[24]

TOXIN PRODUCTION AND TESTING

For the identification of some species, and for types within some species, it is necessary to determine the production of toxin and to identify the toxin that is produced. For example, it is not possible to distinguish, on cultural characteristics, between strains of *C. sporogenes* and proteolytic strains of *C. botulinum* of types A, B, and F. *Clostridium botulinum* forms a potent paralytic neurotoxin, whereas *C. sporogenes* does not, and the demonstration of such a toxin is sufficient to remove a strain from *C. sporogenes.* Determination of a type within the species *C. botulinum, C. novyi* or *C. perfringens* can be carried out only by identifying the toxin by neutralization with type-specific antitoxin.

All of the pathogenic clostridia produce toxin. All clostridial toxins are produced in the interior of the cell, but some, the "extracellular" toxins, readily diffuse out and are found in maximum amount in the culture fluid shortly after the end of the log phase of growth. Others, the "protoplasmic" toxins, do not

TABLE 3

CULTURAL CHARACTERISTICS OF CLOSTRIDIA WITH TERMINAL SPORES

Species	Gelatin Digestion	Milk Digestion	Lecithinase Production	Lipase Production	Indol Production	Glucose Fermentation	Lactose Fermentation	Maltose Fermentation	Mannitol Fermentation	Mannose Fermentation	Salicin Fermentation	Sucrose Fermentation	Trehalose Fermentation	Toxin Production
C. cadaveris	+	+	–	–	+	+	–	–	–	–	–	–	–	–
C. lentoputrescens	+	+	–	–	+	–	–	–	–	–	–	–	–	–
C. putrificum	+	+	–	–	–	+	–	–	–	–	–	–	–	–
C. tetani	+	–	–	–	v	–	–	–	–	–	–	–	–	+
C. malenominatum	–	–	–	–	+	–	–	–	–	–	–	–	–	–
C. tertium[a]	–	–	–	–	–	+	+	+	+	+	+	+	+	–
C. carnis[a]	–	–	–	–	–	+	v	+	–	+	+	+	–	+
C. paraputrificum[a]	–	–	–	–	–	+	+	+	–	+	+	+	–	–
C. ramosum[a]	–	–	–	–	–	+	+	+	v	+	+	+	+	–
C. innocuum	–	–	–	–	–	+	–	–	+	+	+	+	+	–
C. cochlearium	–	–	–	–	–	–	–	–	–	–	–	–	–	–

[a] *C. tertium* and *C. carnis* are aerotolerant; *C. paraputrificum* and *C. ramosum* are not.

TABLE 4

FERMENTATION PRODUCTS OF CLOSTRIDIA WITH SUBTERMINAL SPORES

Major Products	Species
Acetic acid	*C. histolyticum, C. limosum*
Acetic and butyric acids; smaller amounts of propionic acid	*C. perfringens, C. septicum, C. chauvoei, C. botulinum* types B, E, F (non-proteolytic), *C. butyricum, C. fallax, C. paraperfringens, C. carnis, C. beijerinckii*
Propionic and butyric acids, with or without acetic acid	*C. botulinum* types C, D, *C. novyi, C. haemolyticum*
Acetic, isobutyric, and isovaleric acids	*C. subterminale*
Acetic, isobutyric, sometimes butyric, isovaleric and isocaproic acids	*C. sordellii, C. bifermentans*
Acetic and butyric acids, with smaller amounts of propionic, isobutyric, isovaleric, and valeric acids; with or without valeric acid and ethanol, butyanol, isoamyl alcohol	*C. sporogenes, C. botulinum* types A, B, F (proteolytic), *C. difficile*

TABLE 5

FERMENTATION PRODUCTS OF CLOSTRIDIA WITH TERMINAL SPORES

Major Products	Species
Acetic and formic acids	*C. ramosum*
Butyric acid	*C. cochlearium, C. lentoputrescens*
Acetic and butyric acids	*C. innocuum*
Acetic and butyric acids; smaller amount of propionic acid	*C. malenominatum, C. terium, C. carnis, C. paraputrificum*
Acetic and butyric acids, butanol; with or without ethanol	*C. tetani, C. cadaveris*
Acetic, isobutyric, butyric, and isovaleric acids	*C. putrificum*
Acetic, propionic, isobutyric, and isovaleric acids, and ethyl, propyl, butyl, and isoamyl alcohols	*C. glycolicum*

readily diffuse out and reach their maximum in the culture fluid only after lysis has occurred. The toxins of all types of *C. botulinum,* the neurotoxin of *C. tetani,* and the alpha toxin of *C. novyi* are all protoplasmic toxins. Almost all the other clostridial toxins are extracellular. The enteropathogenic toxin of *C. perfringens* differs from the rest; it is formed intracellularly and is released only at the time of sporulation.

Some toxins appear to be synthesized by the clostridia in fully active form. Others are synthesized as less active prototoxins and attain full toxicity only after exposure to certain proteolytic enzymes. In some cases proteolytic enzyme is also synthesized by the organism producing the prototoxin, and activation is spontaneous; in other cases the bacteria do not synthesize proteolytic enzyme, or not enough of it to fully activate the prototoxin, and treatment with trypsin or some similar enzyme is required for full toxic activity.

The neurotoxins of all serological types of *C. botulinum* seem to be synthesized as prototoxins, but only that of type E and those of some strains of other non-proteolytic types of this species require trypsin activation. The situation with regard to *C. tetani* neurotoxin is not clear. The *epsilon* and *iota* toxins of *C. perfringens* definitely require activation for full toxin activity. Trypsin activation cannot be carried out routinely with all cultures, for all clostridial toxins can be inactivated by proteolytic enzymes if the treatment is sufficiently rigorous.

If a strain under investigation produces easily demonstrable toxin, its identification is relatively easy. More work is required if toxin is not easily demonstrable, for it may be necessary to incubate for different lengths of time and to examine culture fluid that has been treated with trypsin as well as culture fluid that has not been so treated. Both young (six to twelve hours incubation) and old (three to five days incubation) cultures should be examined. A portion of each should be used for toxicity tests (0.5 ml injected intraperitoneally into each of two mice) without treatment with trypsin, and another portion after trypsin treatment. Trypsin treatment is done by adjusting the pH of the culture fluid to 6.0-6.2, adding a solution of commercial trypsin (Difco 1:250, for example) to a final concentration of 0.1%, then incubating at 37°C for one hour. Too long or too rigorous treatment with trypsin will first activate and then inactivate prototoxin; consequently, it is best to have the hydrogen ion concentration of the fluid below neutrality, where the action of trypsin is slower than under alkaline conditions.

When species- or type-specific antitoxin is available, 0.2 ml of antitoxin should be added to 1.3 ml of culture fluid, mixed, and allowed to stand for 30 minutes at room temperature before injecting 0.5 ml intraperitoneally or 0.2 ml intravenously into each of two mice. Culture fluid that has been mixed with normal serum in a similar fashion should be injected into a second pair of mice. The mice should be observed for two days. If the mice receiving the mixture of culture fluid and antitoxin die after the control mice do, the experiment should be repeated, diluting the culture fluid 1:10 with broth. With most toxigenic strains the control mice will die overnight; only occasionally will death occur on the second day. It is not necessary to sterilize clostridial culture fluid by filtration when intraperitoneal or intravenous injection is used, for thorough centrifuging suffices for the removal of the great majority of the organisms, and infection is not encountered.

The pathogenic clostridia and their *lethal* toxins are listed in Table 6.

TABLE 6

LETHAL CLOSTRIDIAL TOXINS

Species	Designation	Mode of Action
C. botulinum	Neurotoxin	Prevents release of acetylcholine at motor end plate
C. carnis	None	Hemolytic: lethal activity unknown
C. chauvoei	Alpha	Necrotizing:[a] hemolytic; leucocidic
C. difficile	None	Lethal activity unknown
C. haemolyticum	Identical with the beta toxin of *C. novyi*	Hemolytic: necrotizing: phospholipase
C. histolyticum	Alpha	Lethal activity unknown
	Beta	Collagenase
C. novyi	Alpha	Necrotizing
	Beta	Necrotizing: hemolytic: phospholipase
C. perfringens	Alpha	Necrotizing: hemolytic: phospholipase
	Beta	Necrotizing
	Epsilon	Necrotizing
	Iota	Necrotizing: increases capillary permeability
	Enterotoxin	Diarrheal: erythemal
C. septicum	Alpha	Necrotizing: hemolytic: leucocidic
C. sordellii	None	Lethal activity unknown
C. tetani	Tetanospasmin	Neurotoxic: action in central nervous system not known
	Tetanolysin	Hemolytic: edema inducing

[a] Necrotizing properties were determined by intracutaneous inoculation of guinea pigs. The true lethal action of toxins designated as necrotizing is not known.

REFERENCES

1. Aranki, A., Syed, S. A., Kenney, E. B., and Freter, R., Isolation of Anaerobic Bacteria from Human Gingiva and Mouse Cecum by Means of a Simplified Glove-Box Procedure. *Appl. Microbiol., 17,* 568 (1969).
2. Batty, I., and Walker, P. D., Colonial Morphology and Fluorescent-Labelled Antibody Staining in the Identification of Species of the Genus *Clostridium. J. Appl. Bacteriol., 28,* 112 (1965).
3. Campbell, L. L., and Postgate J. R., Classification of the Spore-Forming Sulfate-Reducing Bacteria. *Bacteriol. Rev., 29,* 359 (1965).
4. Collee, J. G., Rutter, J. M., and Watt, B., The Significantly Viable Particle: A Study of the Subculture of an Exacting Sporing Anaerobe. *J. Med. Microbiol., 4,* 271 (1971).
5. Cummins, C. S., and Johnson, J. L., Taxonomy of the Clostridia: Wall Composition and DNA Homologies in *Clostridium butyricum* and Other Butyric Acid-Producing Clostridia. *J. Gen. Microbiol., 67,* 33 (1971).
6. Gibbs, B. M., and Fraeme, B., Methods for the Recovery of Clostridia from Foods. *J. Appl. Bacteriol., 28,* 95 (1965).
7. Harmon, S. M., Kautter, D. A., and Peeler, J. T., Comparison of Media for the Enumeration of *Clostridium perfringens. Appl. Microbiol., 21,* 922 (1971).
8. Harmon, S. N., Kautter, D. A., and Peeler, J. T., Improved Medium for Enumeration of *Clostridium perfringens. Appl. Microbiol., 22,* 688 (1971).
9. Hibbert, H. R., and Spencer, R., An Investigation of the Inhibitory Properties of Sodium Thioglycollate in Media for the Recovery of Clostridial Spores. *J. Hyg., 68,* 131 (1970).
10. Holdeman, L. V., and Moore, W. E. C., *Anaerobic Laboratory Manual.* Virginia Polytechnic Institute Educational Foundation, Blacksburg, Virginia (1972).
11. Johnson, R., Harmon, S., and Kautter, D., Method to Facilitate the Isolation of *Clostridium botulinum* Type E. *J. Bacteriol., 88,* 1521 (1964).
12. Loesche, W. J., Oxygen Sensitivity of Various Anaerobic Bacteria. *Appl. Microbiol., 19,* 723 (1969).
13. McCain, C. S., Isolation and Identification of *Clostridium haemolyticum* in Cattle in Florida. *Amer. J. Vet. Res., 28,* 878 (1967).
14. Moore, W. E. C., Cato, E. P., and Holdeman, L. V., Fermentation Characteristics of *Clostridium* Species. *Int. J. Syst. Bacteriol., 16,* 383 (1966).
15. Mossel, D. A. A., Enumeration of Sulfite-Reducing Clostridia Occurring in Foods. *J. Sci. Food Agr., 10,* 662 (1959).
16. Mossel, D. A. A., and Beerens, H., Studies on the Inhibitory Properties of Sodium Thioglycollate on the Germination of Wet Spores of Clostridia. *J. Hyg., 66,* 269 (1968).
17. Nishida, S., Yamagishi, T., Tamai, K., Sanada, I., and Takahashi, K., The Effects of Heat Selection on Toxigenicity, Cultural Properties, and Antigenic Structures of Clostridia. *J. Infect. Dis., 120,* 507 (1969).
18. Perkins, W. E., Production of Clostridial Spores. *J. Appl. Bacteriol., 28,* 1 (1965).
19. Prévot, A. R., Turpin, A., and Kaiser, P., *Les Bactèries Anaérobies.* Dunod, Paris, France (1967).
20. Roberts, T. A., Sporulation of Mesophilic Clostridia. *J. Appl. Bacteriol., 30,* 430 (1967).
21. Roberts, T. A., Recovering Spores Damaged by Heat, Ionizing Radiation, or Ethylene Oxide. *J. Appl. Bacteriol., 33,* 74 (1970).
22. Shapton, D. A., and Board, R. G., *Isolation of Anaerobes.* Academic Press, New York and London (1971).
23. Smith, L. DS., and Holdeman, L. V., *The Pathogenic Anaerobic Bacteria.* Charles C Thomas, Springfield, Illinois (1968).
24. Van Kampen, K. R., and Kennedy, O. C., Experimental Bacillary Hemoglobinuria: Intrahepatic Detection of Spores of *Clostridium haemolyticum* by Immunofluorescence in the Rabbit. *Amer. J. Vet. Res., 29,* 2173 (1968).
25. Willis, A. T., *Clostridia of Wound Infection.* Butterworths, London, England (1969).

VIBRIOS AND SPIRILLA

DR. RITA R. COLWELL

Short curved or straight cells, single or united into spirals, that grow well and rapidly on the surfaces of standard culture media can be readily isolated from salt- and fresh-water samples. These heterotrophic organisms vary in their nutritional requirements; some occur as parasites and pathogens for animals and for man. These short curved, asporogenous, Gram-negative rods, members of the genera *Vibrio* and *Spirillum*, are most commonly encountered in the marine or fresh-water habitat. Distinguishing species of the genus *Vibrio* from those of *Spirillum* can often be accomplished by examining Gram and flagella stains of carefully prepared specimens; *Spirillum* species are frequently seen as rigid, helical cells with a single or several turns, motile by means of bipolar polytrichous flagella, whereas *Vibrio* species are short rods with a curved axis, motile by means of a single polar flagellum. However, *Vibrio* species may be short, straight rods (1.5–3.0 μm x 0.5 μm), or may be S-shaped or spiral-shaped when individual cells are joined. Possession of two or more flagella in a polar tuft has also been demonstrated in *Vibrio* species. Thus, to identify and classify vibrios and spirilla, physiological and biochemical taxonomic tests should be made. *Vibrio* species are facultatively anaerobic, with both a respiratory (oxygen-utilizing) and a fermentative metabolism. *Spirillum* species are aerobic or microaerophilic, with a strictly respiratory metabolism (oxygen is the terminal electron acceptor). Marine vibrios and spirilla require 1 to 3% NaCl for growth.

In the seventh edition of *Bergey's Manual,*[1] published in 1957, the genera *Vibrio* and *Spirillum* are placed in the Family Spirillaceae, Suborder Pseudomonadineae, Order Pseudomonadales, Class Schizomycetes. The Spirillaceae are defined therein as simple cells, curved or spirally twisted rods that, after transverse division, frequently remain attached to each other to form chains of spirally twisted cells. The cells are described as rigid, usually motile by means of a single flagellum (rarely two) or a tuft of polar flagella, Gram-negative, and most frequently isolated from water, although some species are pathogenic for higher animals and man.

The borderline between the genus *Vibrio* and the genus *Pseudomonas* has been considered in the past to be somewhat vague. However, evidence that sharpens the demarcation between the two genera has been gathered. *Pseudomonas* species are oxidative and possess overall DNA base compositions in the range from 57 to 67% G+C, whereas *Vibrio* species are fermentative, producing acid in carbohydrate-containing media without formation of gas, and possess a DNA moles % G+C in the range from 39 to 49%. Separation of *Spirillum* and *Pseudomonas* species, unfortunately, remains less clear-cut, and subsequent work may well show that those *Spirillum* species with a G+C ranging from 57 to 67% are more appropriately grouped with *Pseudomonas.*

Genus *Vibrio* Pacini, 1854

The type species for the genus *Vibrio* is *Vibrio cholerae* Pacini, 1854. A total of thirty-four species of *Vibrio* are listed in *Bergey's Manual;*[1] however, the two strictly anaerobic species have been removed from the genus, since no cultures of obligately anaerobic *Vibrio* species are extant, and the microaerophilic species, including *Vibrio fetus* and *Vibrio bubulus,* have been transferred to the genus *Campylobacter* by Sebald and Véron,[2] who observed the DNA base composition of *Campylobacter* species to be in the range from 30 to 34% G+C.

Of the twenty-nine species of *Vibrio* listed in *Bergey's Manual,*[1] those not attacking carbohydrates have been transferred to the genus *Commamonas.* Davis and Park[3] described *Commamonas* as Gram-negative rod-like bacteria that give no reaction on carbohydrate and an alkaline reaction in the Hugh and Leifson test.[4] Other test results described were as follows: negative in indole, cholera red, methyl red, phenylalanine deaminase, Moeller's lysine, arginine, ornithine, and Voges-Proskauer tests; positive in urease, catalase, and oxidase tests; no liquefaction of gelatin, no utilization of citrate, no evidence of growth in KCN medium, no pigment or fluorescence under ultraviolet light, and possession of lophotrichous flagella. Sebald and Véron,[2] confirming the observations of Davis and Park assigned, on the basis of DNA G+C composition (64%), *Vibrio percolans, Vibrio cyclosites, Vibrio neocistes,* and *Vibrio alcaligenes* to the genus *Commamonas.* Colwell and Liston[5] observed that *Vibrio cuneatus* produced a green-fluorescent pigment, and subsequent DNA studies[6] confirmed the conclusion that this species should be assigned to the genus *Pseudomonas.*

The C27 organisms of Ferguson and Henderson,[7] previously assigned to *Aeromonas* by Ewing, Hugh and Johnson,[8] to *Plesiomonas* by Habs and Schubert[9] and by Eddy and Carpenter,[10] and to *Fergusonia* by Sebald and Véron,[2] should be included in the genus *Vibrio,* according to Hendrie, Shewan and Véron.[11] Thus, the number of characterized and defined species resident in the genus *Vibrio* can be reduced to six.

The description of the genus *Vibrio*, as amended by the Subcommittee on Taxonomy of Vibrios International Committee on Nomenclature of Bacteria, is concise and provides a good working definition.

> "Gram-negative, asporogenous rods which have a single, rigid curve or which are straight. Motile by means of a single, polar flagellum. Produce indophenol oxidase and catalase. Ferment glucose without gas production. Acidity is produced from glucose by the Embden-Meyerhof glycolytic pathway. The guanine plus cytosine in the DNA of *Vibrio* species is within the range of 40 to 50 moles per cent."

The type species of the genus, *Vibrio cholerae* Pacini 1854, can be succinctly described as follows: producing L-lysine and L-ornithine decarboxylases; L-arginine dihydrolase and hydrogen sulfide (Kligler iron agar) are not produced. The guanine-plus-cytosine (G+C) in the DNA of *Vibrio cholerae* is approximately 48 ± 1%. *Vibrio cholerae* includes strains that may or may not elicit the cholera-red (nitroso-indole) reaction, may or may not be hemolytic, may or may not be agglutinated by Gardner and Venkatraman O group I antiserum, and may or may not be lysed by Mukerjee *Vibrio cholerae* bacteriophages I, II, III, IV, and V.[12]

Vibrio cholerae strains possess a common H antigen and can be serologically grouped into thirty-nine serotypes according to their O antigens, as described in Reference 13. Strains agglutinated by Gardner and Venkatraman O group I antiserum are in serotype I and are the principal cause of cholera in man. An M antigen can obscure the agglutinability of mucoid strains of *Vibrio cholerae.* Furthermore, *Vibrio cholerae* strains in the R (rough) form cannot be serotyped. General reviews on cholera are provided in References 14 to 16. Isolation and diagnosis procedures are outlined in these reviews and in References 17 and 18.

In the past, many cholerae-like vibrios have been given separate species status because they were isolated from patients suffering diarrhea, not cholera, or from water and foods. *Vibrio proteus, Vibrio metschnikovii, Vibrio berolinensis, Vibrio albensis* and *Vibrio paracholerae* can be considered to be biotypes of *Vibrio cholerae.* The so-called non-agglutinable (NAG) or non-cholera (NCV) vibrios, including *Vibrio eltor,* have been shown by Citarella and Colwell,[19] using the techniques of DNA/DNA reassociation measurements, to be related to *Vibrio cholerae* at the species level.

Since its initial isolation by Fujino and Fukumi[20] in 1953, *Vibrio parahaemolyticus* has received considerable attention. *Vibrio parahaemolyticus* (syn. *Pasteurella parahaemolytica, Pseudomonas enteritis, Oceanomonas parahaemolytica*) is the causative agent of food poisoning arising from ingestion of contaminated seafood and can be isolated from the marine environment. A review of its identification and classification is provided in References 21 and 22.

The species *Vibrio alginolyticus* is presently in dispute; some investigators advocate its synonymy with *Vibrio parahaemolyticus,* and others consider it to be a separate species.

Other species of *Vibrio* that should be considered are the following: *Vibrio anguillarum* (syn. *Vibrio piscium, Achromobacter ichthyodermis, Pseudomonas ichthyodermis, Vibrio piscium* var. *japonicus, Vibrio ichthyodermis*), isolated from diseased conditions in marine- and fresh-water fish; *Vibrio marinus,* found in sea water and associated with marine animals;[23,24] *Vibrio costicolus,* a species tolerating salt concentrations from 2 to 23%, with an optimal concentration of 6 to 12%; and *Vibrio shigelloides* (syn. *Pseudomonas shigelloides, Pseudomonas michigani, Aeromonas shigelloides, Plesiomonas shigelloides,* and *Fergusonia shigelloides*), found in feces of man and in the mesenteric lymph nodes of healthy dogs.

Distinctions can be made among the species of the genus *Vibrio,* and useful differentiating characteristics are given in Table 1. Some species of *Vibrio* may demonstrate sheathed flagella, i.e., a flagellum with a central core and an outer sheath. "Round bodies" or sphaeroplasts are commonly present during various stages of growth,[25,26] and fimbriae (pili) have been observed in strains of *Vibrio cholerae.*[27]

TABLE 1

FEATURES USEFUL IN DIFFERENTIATING AND CHARACTERIZING SPECIES OF THE GENUS *VIBRIO*

Characteristic	*Vibrio cholerae*	*Vibrio parahaemolyticus*	*Vibrio anguillarum*	*Vibrio marinus*	*Vibrio costicolus*	*Vibrio shigelloides*
Rod shape	+	+	+	+	+	+
Motility	+	+	+	+	+	+
Single polar flagellum	+	+	+	v	+	–
Lophotrichous flagella	–	–	–	v	–	+
Gram reaction	–	–	–	–	–	–
Diffusible pigment	–	–	–	–	–	–
Luminescence	–	–	–	–	–	–
Pathogenicity for man or animals	+	+	+	–	–	+
DNA base composition (% G+C)	46–49	44–46	44–45	40–44	50	51–52
Indole reaction	+	+	+	–	–	+
Methyl red reaction	+	+	+	+	–	+
Voges-Proskauer reaction	+	v	+	–	+	–
Citrate utilization	+	+	–	–	–	–
Citrulline utilization	–	–	+	–	nt	nt
Sensitivity						
0/129	+	+	+	+	v	v
novobiocin	+	+	+	+	+	–
penicillin, 10 units	+	–	–	–	nt	nt
polymyxin, 300 units	v	–	v	+	nt	nt
streptomycin, 10 μg	+	–	–	+	nt	nt
Growth						
in 0% NaCl	+	–	+	+	–	+
in 1% NaCl	+	+	+	+	–	+
in 7% NaCl	v	+	v	+	+	–
in 10% NaCl	–	+	–	–	+	–
at 5°C	–	–	+	+	+	–
at 20°C	+	+	+	+	+	+
at 37°C	+	+	v	–	v	+
at 42°C	+	+	–	–	–	–

TABLE 1 (Continued)

FEATURES USEFUL IN DIFFERENTIATING AND CHARACTERIZING SPECIES OF THE GENUS *VIBRIO*

Characteristic	*Vibrio cholerae*	*Vibrio parahaemolyticus*	*Vibrio anguillarum*	*Vibrio marinus*	*Vibrio costicolus*	*Vibrio shigelloides*
Acid production						
from arabinose	–	+	–	–	–	–
from inositol	–	–	–	v	–	v
from mannitol	+	+	+	+	v	nt
from mannose	+	+	+	+	+	–
from salicin	–	–	–	v	–	v
from sucrose	+	v	+	v	+	–
Gelatin liquefaction	+	+	+	+	v	–
Hydrolysis						
casein	+	+	+	+	–	–
starch	+	+	+	v	–	–
Tween 80®	+	+	+	+	+	–
H_2S production (on lead acetate agar)	–	–	–	–	–	v
Lecithinase (egg yolk)	+	+	nt	nt	v	–
Arginine dihydrolase	–	–	+	–	+	+
Lysine decarboxylase	+	+	–	+	–	+
Ornithine decarboxylase	+	+	–	–	–	+
Hemolysis	+	+	v	–	–	v

Symbol Code

+ = positive or present
– = negative or absent
v = reaction varied among the strains tested
nt = not tested

Data compiled from several sources, including Reference 11, 21, 23, 24, and 28 and unpublished data obtained by the author.

Genus *Spirillum* Ehrenberg, 1832

The genus *Spirillum* has been less sharply defined than the genus *Vibrio. Spirillum* Ehrenberg, 1832, as described in *Bergey's Manual,*[1] includes crescent-shaped to spiral cells that are frequently united into spiral chains of cells, which are not embedded in zoogloeal masses. The spiral cells are usually motile by means of polar flagellation, i.e., possessing a tuft of polar flagella at one or both ends of the cells. The cells form either long screws or portions of a turn. Volutin granules are usually present. Spirilla are either aerobic growing well on ordinary culture media, or microaerophilic.

Nine species are listed in *Bergey's Manual,*[1] but at least thirty species have been described in the literature. The type species is *Spirillum undula* (Müller), 1832. Species described in *Bergey's Manual*[1] of which no representative isolates are available include the following: *Spirillum undula; Spirillum tenue; Spirillum viginianum; Spirillum minus,* and *Spirillum kutscheri.* Of the remaining species, *Spirillum Serpens, Spirillum itersonii,* and *Spirillum volutans* are recognized as *bona fide* species of the genus *Spirillum.* However, it must be noted that, in a numerical taxonomy study carried out by Colwell,[21] *Spirillum itersonii,* ATCC strain 11331, was found to be a green-fluorescent pigment-producing organism clustering with *Pseudomonas* species, including *Pseudomonas fluorescens.* The overall G+C of this organism and that of *Spirillum serpens,* subsp. *serpens,* ATCC strain 11330, both 64%, suggest that green-fluorescent pigment-producing strains of *Spirillum* with a G+C content of more than 50% may, in fact, be strains of *Pseudomonas* with spiral-shaped (as opposed to straight-rod) morphology.

In general terms, however, the genus can be described as comprising rigid, helical, Gram-negative cells, motile with bipolar polytrichous flagella. Spirilla possess a strictly respiratory metabolism, with oxygen as the terminal electron acceptor; they are oxidase- and catalase-positive, and usually phosphatase-positive. H_2S is usually produced from cysteine, but indol, sulfatase, and amylase are not produced. Other reactions usually negative are hydrolysis of casein, hippurate, and gelatin, and production of urease and acid from sugars. Though species cannot utilize sugars, various organic acids, alcohols, or amino acids can suffice as sole carbon sources. Most *Spirillum* species do not demonstrate amino acid, vitamin, purine, or pyrimidine requirements, and ammonium ion can usually be utilized as a sole nitrogen source. Growth below 10°C or above 45°C is uncommon. Salt requirements for marine species range from 1 to 3% NaCl.

Table 2 lists differentiating characteristics of four *Spirillum* species. *Spirillum* species other than these have been described elsewhere (see References 28 to 30). It must be emphasized that the taxonomy of the genus *Spirillum* is presently under study in several laboratories and will doubtless be altered in the near future.

DIFFERENTIATION FROM OTHER BACTERIA

Characteristics useful for distinguishing vibrios and spirilla from several genera with which they are frequently confused are given in Table 3.

TABLE 2

DIFFERENTIATING CHARACTERISTICS OF SELECTED *SPIRILLUM* SPECIES

Characteristic	*Spirillum volutans*	*Spirillum anulus*	*Spirillum itersonii*	*Spirillum serpens*
Size	1.4–1.7 nm* x 14–60 nm	1.4–1.5 nm x 7–15 nm	0.4–0.6 nm x 2.0–7.0 nm	0.0–1.0 nm x 5–35 nm
Motility (lophotrichous)	+	+	+	+
Fluorescent pigment	–	–	+	v
Oxidase	+	+	+	+
Catalase	+	+	+	+
Phosphatase	+	+	+	+
Urease	–	–	–	–
Phenylalanine deaminase	–	–	–	–
DNAse	–	+	+	+
RNAse	–	+	+	+
Gelatin hydrolysis	–	–	–	–
Esculin hydrolysis	–	–	+	–
Nitrate reduction	–	–	+	–
Acid from carbohydrates	–	–	v	–
Growth				
in 1% bile	–	–	+	+
in 3% NaCl	–	–	–	–
at 10°C	–	–	–	–
at 30–32°C	+	+	+	+
at 42°C	–	–	–	–
Oxygen requirements	Obligately microaerophilic	Strict aerobe	Strict aerobe	Strict aerobe
DNA base composition	38%	58%	64%	64%

* Diameter; the other measurements represent width by length.

Symbol Code

+ = positive or present
– = negative or absent
v = variable among the strains tested

TABLE 3

DIFFERENTIATION OF RELATED GENERA FREQUENTLY ISOLATED FROM THE SAME SOURCE IN NATURE

Characteristic	*Vibrio*	*Aeromonas*	*Photobacterium*	*Pseudomonas*	*Spirillum*
Morphology	Straight or curved rod	Straight rod	Straight rod	Straight rod	Helical
Diffusible pigment	None	None[a]	None	None or green-fluorescent	None or green-fluorescent
Motility	+	+	+	+	+
Flagella	Polar	Polar	Polar	Polar	Lophotrichous
Carbohydrate metabolism	Fermentative	Fermentative	Fermentative	Respiratory	Respiratory
Gas production from carbohydrates	–	v	+	–	–
Luminescence	v	–	+	–	–
Oxidase	+	+	v	+	+
0/129 sensitivity	+	–	+	–	–
"Round bodies" or "cysts" produced	+	–	–	–	+

[a] Species of *Aeromonas* may produce a brown pigment.

Symbol Code

+ = positive or present
– = negative or absent
v = variable

REFERENCES

1. Breed, R. S., Murray, E. G. D., and Smith, N. R., *Bergey's Manual of Determinative Bacteriology*, 7th ed. Williams and Wilkins, Baltimore, Maryland (1957).
2. Sebald, M., and Véron, M., *Ann. Inst. Pasteur Lille, 105,* 897 (1963).
3. Davis, G. H. G., and Park, R. W. A., *J. Gen. Microbiol., 27,* 101 (1962).
4. Hugh, R., and Leifson, E., *J. Bacteriol., 66,* 24 (1953).
5. Colwell, R. R., and Liston, J., *J. Bacteriol., 82,* 1 (1961).
6. Colwell, R. R., and Mandel, M., *J. Bacteriol., 87,* 1412 (1964).
7. Ferguson, W.W., and Henderson, N. D., *J. Bacteriol., 54,* 178 (1947).
8. Ewing, W. H., Hugh, R., and Johnson, J. G., *Studies on the Aeromonas Group.* Communicable Disease Center, Atlanta, Georgia (1961).
9. Habs, H., and Schubert, R. H. W., *Zentralbl. Bakteriol. Parasitenk. Infektionskr. Hyg. Abt. I. Orig., 186,* 316 (1962).
10. Eddy, B. R., and Carpenter, K. P., *J. Appl. Bacteriol., 27,* 96 (1964).
11. Hendrie, M. S., Shewan, J. M., and Véron, M., *Int. J. Syst. Bacteriol., 21,* 25 (1971).
12. Hugh, R., and Feeley, J. C., Report (1966–1970) of the Subcommittee on Taxonomy of Vibrios to the International Committee on Nomenclature of Bacteria. *Int. J. Syst. Bacteriol., 22,* 123 (1972).

13. Sakazaki, R., Kazunichi, T., Gomez, C. Z., and Sen, R., *Jap. J. Med. Sci. Biol., 23,* 13 (1970).
14. Pollitzer, R., *Cholera,* W.H.O. Monograph Series No. 43, p. 1019. World Health Organization, Geneva, Switzerland (1959).
15. Felsenfeld, O., *Bacteriol. Rev., 28,* 72 (1964).
16. Felsenfeld, O., *Bull. World Health Organ., 34,* 161 (1966).
17. Burrows, W., and Pollitzer, R., *Bull. World Health Organ., 18,* 275 (1958).
18. Carpenter, K. P., Hart, J. M., Hatfield, J., and Wicks, G., Identification Methods for Microbiologists (B. M. Gibbs and D. A. Shapton, Eds.), *Soc. Appl. Bacteriol. Tech. Ser., 2,* 8 (1968).
19. Citarella, R. V., and Colwell, R. R., *J. Bacteriol., 104,* 434 (1970).
20. Fujino, T., and Fukumi, H. (Eds.), *Vibrio parahaemolyticus,* 2nd ed. (in Japanese). Naya Shoten, Tokyo, Japan (1967).
21. Colwell, R. R., *J. Bacteriol., 104,* 410 (1970).
22. Vanderzant, C., Nickelson, R., and Parker, J. C., *J. Milk Food Technol., 33,* 161 (1970).
23. Colwell, R. R., and Morita, R. Y., *J. Bacteriol., 88,* 831 (1964).
24. Bianchi, M. A. G., *Arch. Mikrobiol., 77,* 127 (1971).
25. Felter, R. A., Kennedy, S. F., Colwell, R. R., and Chapman, G. B., *J. Bacteriol., 102,* 552 (1969).
26. Kennedy, S. F., Colwell, R. R., and Chapman, G. B., *Can. J. Microbiol., 16,* 1027 (1970).
27. Tweedy, J. M., Park, R. W. A., and Hodgkiss, W., *J. Gen. Microbiol., 51,* 235 (1968).
28. Williams, M. A., *Int. Bull. Bacteriol. Nomencl. Taxon.,* 9, 137 (1959).
29. Watanabe, N., *Bot. Mag. (Tokyo), 72,* 77 (1959).
30. Terasaki, Y., *Bull. Suzugamine Wom. Coll. Natur. Sci., Suppl. 8–9,* p. 1 (1962).

THE MYCOPLASMATALES

DR. SHMUEL RAZIN

Mycoplasmas are minute prokaryotic organisms, highly pleomorphic, varying in shape from spherical structures with diameters of about 125 to 250 nm to slender branched filaments of uniform diameter, and ranging in length from a few μm to 150 μm. Their mode of reproduction is still controversial. According to Freundt,[1] the cells, during growth, go through a cycle of morphological changes, commencing with a small spherical particle or elementary body, from which one or more short, thin filaments emerge. The filaments lengthen and branch frequently, so that in the first 12 to 18 hours of growth a mycelial structure develops (hence the name *Mycoplasma*, fungus form). Regularly spaced spheres of uniform size and shape then form within the filaments. Subsequently constrictions develop between the spheres, dividing the filament into a chain of coccoid elements. By disintegration of the chain the spherical elements are liberated and are considered to be new elementary bodies. On the other hand, several authors, in particular Furness,[2] claim that mycoplasmas reproduce by simple binary fission. However, there is no essential difference between the two schools of thought. Recent studies have shown that the replication of the mycoplasma genome is semiconservative, like that of the typical prokaryotic genome of the eubacteria.[3] The various mycoplasmas may differ in the stage at which division of the cytoplasm takes place. It may either be synchronized with the replication of the genome, resulting in characteristic binary fission, or it may be delayed, leading to the formation of multinucleate filaments. The degree of synchronization of genome replication and cytoplasmic division may be a characteristic of the *Mycoplasma* species, though it also varies within the same species with the age of the culture and the growth conditions. Morphology and mode of reproduction can, therefore, not be taken as definite criteria for distinguishing between different mycoplasmas.

Electron microscopy of thin sections of mycoplasmas reveals an extremely simple ultrastructure, supporting the idea that these may be the simplest and most primitive organisms capable of autonomous growth. The mycoplasma cell is built of only three organelles: the cell membrane, the ribosomes, and the prokaryotic nucleus. In contrast to bacteria, the mycoplasmas have no cell wall. This is a characteristic of outstanding importance, to which the mycoplasmas owe many of their peculiarities, for example, their morphological instability, osmotic sensitivity, tendency to penetrate and grow in the depth of solid media, resistance to antibiotics interfering with cell wall synthesis, susceptibility to lysis by detergents, alcohols, specific antibody and complement, and others. The mycoplasmas have, therefore, been set apart from bacteria and assigned to a new class, *Mollicutes* (*mollis* = soft; *cutis* = skin).

The vast majority of mycoplasmas require cholesterol, or some related sterols, for growth.[4] Cholesterol has been shown to be an important constituent of their plasma membrane.[5] The mycoplasmas are Gram-negative, have no flagella, and are usually non-motile, except for a few species for which gliding motility has been described. No resting stages are known. A rough comparison of some of their characteristics with those of other groups of microorganisms is given in Table 1. Figure 1 outlines the steps to be taken for the identification of mycoplasmas as recommended by the ICSB Subcommittee on the Taxonomy of Mycoplasmatales.[6]

Although most mycoplasma species are facultative anaerobes, they grow better aerobically. Several mycoplasmas of primate origin prefer anaerobic conditions, however, and require an atmosphere composed of 95% N_2 and 5% CO_2 for growth (see Table 2). The "fried-egg" colony, consisting of an opaque, granular central zone that grows down into the medium and a flat, translucent peripheral zone on the medium surface, is an important characteristic of mycoplasmas. The colonies are usually very small, their diameters ranging from 10 to 600 μm, but occasionally they come up to 4 mm in diameter. A "fried-egg" colony is also typical of the L-phase variants of bacteria. These variants, however, which are usually induced in the laboratory by a single exposure to antibiotics affecting cell wall synthesis, revert to the bacterial phase and lose the "fried-egg" colonial form when the antibiotic is omitted from the growth medium. It is, therefore, recommended to subculture each new isolate suspected of being a mycoplasma at least five consecutive times on media not containing penicillin or other antibacterial agents, to check for non-reversion to a bacterial form.

TABLE 1

CHARACTERISTICS OF MYCOPLASMAS COMPARED WITH THOSE OF OTHER MICROORGANISMS

Characteristic	Mycoplasmas	Bacteria	Rickettsiae	Chlamydiae[a]	Viruses	Fungi(Yeasts)	Protozoa
Growth in cell-free medium	+	+	–[b]	–	–	+	some + some –
Absence of cell wall	+	–	–	–	+	–	some + some –
Smallest forms 200 nm or less	+	–	–	–	+	–	–
Contain DNA and RNA	+	+	+	+	–	+	+
Metabolic systems present	+	+	+	+	–	+	+
Sterol requirement	+[c]	–	–	–	–	some + some –	some + some –
Growth inhibited by antibody alone	+	–	+	+	+	–	–
Growth inhibited by antibiotics	+	+	+	+	–	+	–

[a]Previously termed *Bedsonia* or psittacosis-lymphogranuloma; TRIC group of agents.
[b]Except for *R. quintana.*[8]
[c]Most of the species.[4]

Data taken, with a few modifications, from: Taylor-Robinson, D., Addey, J. P., Hare, M. J., and Dunlop, E. M. C., *Brit. J. Vener. Dis., 45,* 265 (1969). Printed by permission from the Editor of the publication and the authors of the article.

Another important step in mycoplasma identification is cloning of the culture to establish its purity. For this purpose an agar medium is inoculated with a filtrate of a broth culture of the organism. Filtration through a membrane filter with the smallest pore diameter possible breaks up any clumps of organisms and results in colonies originating from single cells. An isolated colony is then transferred to the broth and the cloning procedure is repeated at least twice more. The cloned cultures are examined by light and electron microscopy to verify the characteristic mycoplasma morphology and the absence of a cell wall.

The mycoplasmas have recently been divided into two families according to their dependence on cholesterol for growth: *Mycoplasmataceae* are dependent on cholesterol, and *Acholeplasmataceae* are not.[25] Growth response to cholesterol should, therefore, be determined in order to place the new isolate in one of these families.[4] Each family comprises only one genus, *Mycoplasma* and *Acholeplasma*, respectively. Classification of any isolate within the family depends on biochemical and serological tests. Some of these tests are designated by the Subcommittee on the Taxonomy of Mycoplasmatales as obligatory; others are optional (see Figure 1).

The obligatory biochemical tests include tests for the breakdown of glucose and arginine. Most mycoplasmas utilize either glucose or arginine as major sources of energy. The carbohydrate-forming strains mostly catabolize glucose by homolactic or heterolactic glycolytic pathways, the major end products being lactic acid and, to a smaller extent, pyruvic acid, acetic acid, and acetylmethylcarbinol. Breakdown of carbohydrates is, therefore, indicated by the production of acid and the color change of a pH indicator incorporated into the carbohydrate-containing medium. The breakdown of carbohydrates other than glucose, such as mannose, mannitol, lactose, xylose, sorbitol, glycerol, cellobiose, saccharose, salicin, fructose and galactose, is of some diagnostic value, and these determinations were accordingly included among the optional tests. In most species the respiratory pathways seem to be flavin-terminated, so that heme compounds (cytochromes, catalase, etc.) are absent.[5]

With very few exceptions, the non-fermentative mycoplasmas contain the arginine dihydrolase pathway:

(a) arginine $\xrightarrow{\text{arginine deiminase}}$ citrulline + NH_3

(b) citrulline + Pi $\xrightarrow{\text{ornithine transcarbamylase}}$ ornithine + carbamyl phosphate

(c) carbamyl phosphate + ADP $\xrightarrow[\text{Mg}^{2+}]{\text{carbamyl phosphokinase}}$ ATP + NH_3 + CO_2

This pathway supplies the organism with ATP.[26] The liberated ammonia raises the pH of the medium, furnishing the basis for the detection of arginine hydrolysis in the laboratory test.

Hydrolysis of urea is a property of a group of mycoplasmas known as the T-mycoplasmas. The T stands for "tiny", referring to the small size of their colonies. Recent studies by M. C. Shepard and C. D. Lunceford (personal communication) indicate that the T-mycoplasmas actually require urea for growth. This nutritional requirement and other unique properties of the T-mycoplasmas (see Tables 2 and 3) have led to the recent suggestion, now under consideration by a committee of experts, that these strains should be classified as a new genus within the family Mycoplasmataceae.

FIGURE 1.
RECOMMENDED STEPS IN THE IDENTIFICATION OF MYCOPLASMAS
According to the Recommendations of the Subcommittee on the Taxonomy of Mycoplasmatales*

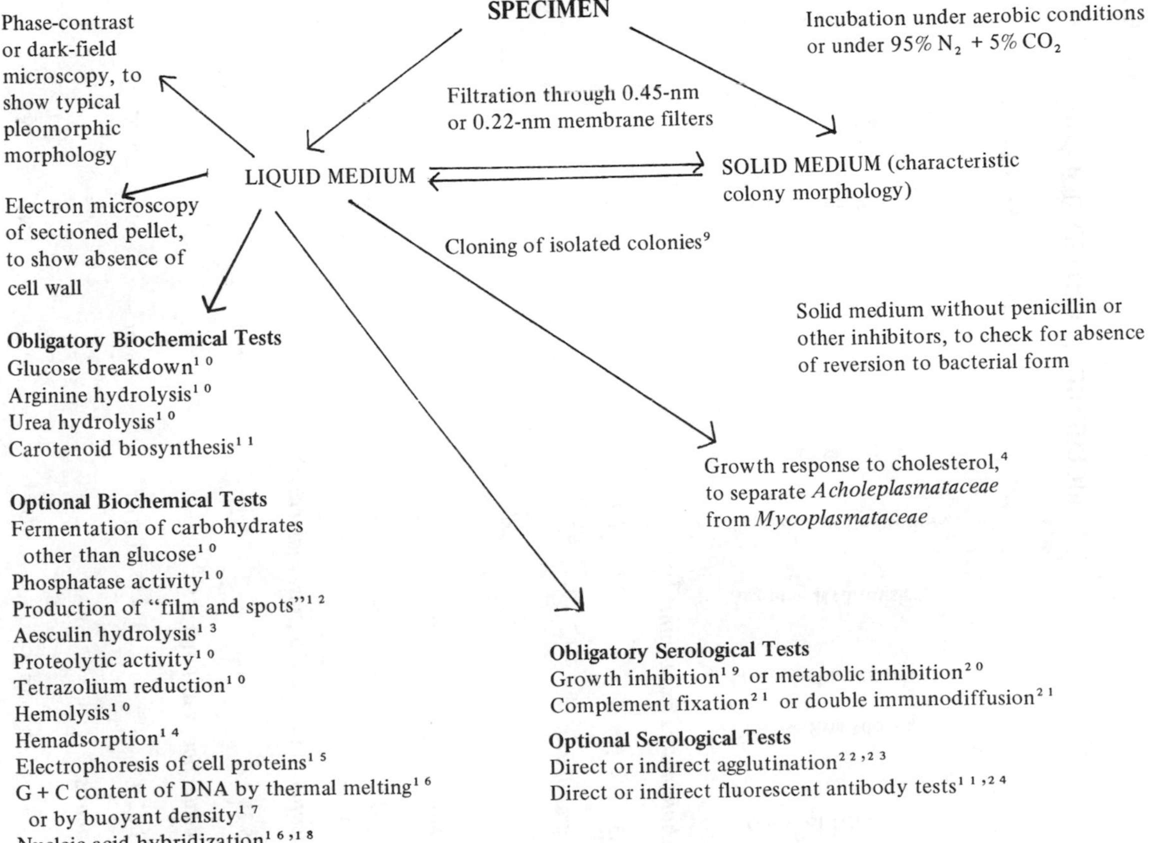

*Methods are given in the references cited.

TABLE 2

BIOCHEMICAL AND PHYSIOLOGICAL CHARACTERISTICS OF MYCOPLASMAS

(Ae = aerobic; An = anaerobic; NT = not tested or not known)

%G + C of DNA	Glucose Breakdown	Mannose Breakdown	Arginine Hydrolysis	Tetrazolium Reduction, Ae/An	Phosphatase	Film and Spots	Gelatin Hydrolysis	Coagulated-Serum Digestion	Hemadsorption	Preferred Atmosphere	Rate of Growth	Colony Morphology	Special Growth Factors	Inhibition by Erythromycin, μg/ml
Acholeplasma axanthum[27,28]														
31.3	+	–	–	+/+	–	–	NT	–	–	Ae	Moderate	Regular[a]	None	NT
A. granularum[10,16,17,28,29,97]														
30.5-32.4	+	–	–	±/+	–	–	–	–	–	Ae	Moderate	Regular, yellow	None	0.03
A. laidlawii[10,28,29,30,97]														
31.7-35.7	+	–	–	±/+	± or –	–	+ or –	–	–	Ae	Rapid	Regular, yellow	None	0.03-0.13
Mycoplasma agalactiae*, var. *agalactiae[28,29]														
33.5-34.2	–	–	–	+/+	+	+ or –	–	–	+	Ae	Moderate	Regular	None	200-512
M. agalactiae*, var. *bovis[28,29]														
32.7-32.9	–	–	–	+/+	+	+ or –	NT	NT	–	Ae	Moderate	Regular	None	NT
M. anatis[29]														
NT	+	+	–	–/+	+	+	NT	NT	–	Ae	Slow	Regular	None	NT

M. arginini[29,31]														
28.6	–	–	+	–/+	–	–	NT	NT	–	Ae	Slow	Regular	None	NT
M. arthritidis[10,17,30,97]														
30.0-33.7	–	–	+	–/–	+	–	+	–	–	Ae	Moderate	Regular	None	>50
M. bovigenitalium[10,17,29,30]														
28.0-32.0	–	–	–	–/+	+	+	–	–	+	Ae	Moderate	Regular	DNA on primary isolation	
M. bovirhinis[10,17,29,30]														
24.5-25.7	+	–	–	+/+	–	–	–	+	NT	Ae	Moderate	Regular	None	NT
M. canis[10,16,17,29,97]														
28.5-29.1	+	–	–	–/+	–	–	+ or –	–	+	Ae	Moderate	Regular	None	>50
M. conjunctivae[98]														
NT	+	+	–	+/+	–	–	NT	–	–	Ae	Moderate	Regular	None	NT
M. dispar[32]														
28.5	+	NT	–	NT	NT	NT	NT	NT	NT	NT	Slow	Lacy, no definite nipple	Very exacting	NT
M. edwardii[10,33]														
29.2	+	–	–	–/+	–	+ (slow)	–	–	NT	Ae	Moderate to rapid	Regular and large	None	NT
M. feliminutum[29,34]														
NT	–	NT	–	–/+	–	–	NT	NT	NT	Ae	Slow	Small	None	NT
M. felis[10,29,30,35]														
25.0-25.4	+	–	–	–/+	+	+	–	–	NT	Ae	Moderate	Regular	None	Resistant[b]

TABLE 2 (Continued)

BIOCHEMICAL AND PHYSIOLOGICAL CHARACTERISTICS OF MYCOPLASMAS

(Ae = aerobic; An = anaerobic; NT = not tested or not known)

%G + C of DNA	Glucose Breakdown	Mannose Breakdown	Arginine Hydrolysis	Tetrazolium Reduction, Ae/An	Phosphatase	Film and Spots	Gelatin Hydrolysis	Coagulated-Serum Digestion	Hemadsorption	Preferred Atmosphere	Rate of Growth	Colony Morphology	Special Growth Factors	Inhibition by Erythromycin, μg/ml
M. fermentans[10,28,29,30]														
27.5-29.1	+	–	+	–/+	– or +	+	–	–	–	An	Moderate	Regular	None	10-512
M. gallinarum[10,17,29,30]														
26.3-28.0	–	–	+	+/+	–	+	–	–	–	Ae	Moderate	Regular	None	40-1,000
M. gallisepticum[10,17,29,30]														
31.6-35.7	+	+	–	+/+	–	–	–	–	+	Ae	Moderate to rapid	Central small nipple	None	0.02-2.0
M. gateae[10,30,35]														
28.4-28.6	–	–	+	–/±	–	–	–	–	NT	Ae	Moderate	Regular	None	Resistant[b]
M. hominis[10,17,29,30]														
27.3-29.3	–	–	+	–/–	–	–	–	–	–	Ae	Moderate	Regular	None	500-1,000
M. hyorhinis[10,16,29]														
27-28	+	–	–	+/±	+	–	–	–	–	Ae	Moderate	Regular	None	>512

M. hyosynoviae(M. suidaniae)[36,37]														
NT	–	–	+	–/–	–	+	NT	–	–	Ae	Slow to moderate	Regular	Growth is enhanced by mucin	NT
M. iners[10,29]														
28.9-29.6	–	–	+	–/–	–	+ (on egg yolk medium)	–	–	–	Ae	Moderate	Regular	None	NT
M. maculosum[16,17,29,38]														
26.5-29.6	–	–	+	–/+	+	+	–	–	–	Ae	Moderate	Regular	None	NT
M. meleagridis[10,17,29]														
28.0-28.5	–	–	+	–/+	+	–	NT	NT	– or +	Ae	Moderate	Regular	None	NT
M. mycoides*, var. *capri[29,97]														
23.6-25.8	+	+	–	+/+	–	–	+	+	–	Ae	Rapid	Large	None	0.05-4.0
M. mycoides*, var. *mycoides[16,28,29,97]														
26.1-26.8	+	+	–	+/+	–	–	+	±	–	Ae	Moderate to rapid	Regular	None	0.03-4.0
M. neurolyticum[10,28,29,30,97]														
22.8-26.5	+	+	–	–/+	–	–	–	–	–	Ae	Relatively rapid	Regular	None	50
***M. orale*, type 1**[10,29,30]														
24.0-28.2	–	–	+	–/–	–	–	–	–	+ (chicken cells only)	An	Moderate	Regular	None	25-512
***M. orale*, type 2**[10,29,30]														
24-28	–	–	+	–/+	+	–	–	–	–	An	Slow to moderate	Small, minute central nipple	None	25-512

TABLE 2 (Continued)

BIOCHEMICAL AND PHYSIOLOGICAL CHARACTERISTICS OF MYCOPLASMAS
(Ae = aerobic; An = anaerobic; NT = not tested or not known)

%G + C of DNA	Glucose Breakdown	Mannose Breakdown	Arginine Hydrolysis	Tetrazolium Reduction, Ae/An	Phosphatase	Film and Spots	Gelatin Hydrolysis	Coagulated-Serum Digestion	Hemadsorption	Preferred Atmosphere	Rate of Growth	Colony Morphology	Special Growth Factors	Inhibition by Erythromycin, μg/ml
M. orale, type 3[29,39]														
NT	–	–	+	–/–	–	NT	NT	NT	+ (chick-en cells only)	An	Moderate	Small	Growth is stimulated by L-cyste-ine	NT
M. pneumoniae[10,28,29,30]														
38.6-40.8	+	+	–	+/+	–	–	–	–	+	Ae	Slow	No central nipple on isolation	None	0.001-1.0
M. primatum[99]														
28.6	–	–	+	–/–	+	–	–	–	–	95% N_2 + 5% CO_2	Moderate	Regular	None	NT
M. pulmonis[10,29,30]														
27.5-28.3	+	+	–	–/+	–	+	–	–	+	Ae	Moderate	Nipple is less defined	None	>40
M. salivarium[10,29,30]														
27.0-31.5	–	–	+	–/±	–	+	–	–	–	An	Moderate	Regular	None	>512

M. spumans [10,17,29,30]														
28.4-29.1	–	–	+	–/–	+	–	–	–	+	Ae	Moderate	Coarse on primary culture	None	NT
M. suipneumoniae (M. hyopneumoniae) [40,41]														
NT	+	NT	NT	NT	NT	NT	NT	NT	NT	Ae	Very slow	Minute, no central nipple	Satellitism with *Staphylococcus aureus* gastric mucin	NT
M. synoviae [29]														
34.2	+	NT	NT	NT	NT	+	NT	NT	–	Ae	Slow	Regular	NADH is required for growth; growth is enhanced by cysteine	NT
T-strains [29,42]														
27.7-28.5	–	–	–	–/–	+	NT	NT	NT	–	10% CO_2 + 90% N_2	Rapid; optimal at pH 5.5 to 6.5	Tiny	Growth is stimulated by urea	0.8-3.0

[a] Regular colony morphology means the typical "fried-egg" colony.
[b] Determined by growth inhibition zones around discs impregnated with 2 μg of erythromycin.[7]

TABLE 3

DIFFERENCES BETWEEN T-STRAIN MYCOPLASMAS AND *MYCOPLASMA HOMINIS*

Property	T-Strain Mycoplasmas	Mycoplama hominis	References
Colony morphology	Usually 15 μm to 30 μm in diameter, without a peripheral zone	Typical "fried-egg" colonies, more than 200 μm in diameter	43
Optimal pH for growth	6.0 ± 0.5	7.0 ± 1.0	44
Growth-curve characteristics	Steep logarithmic rise up to 10^7 CFU/ml in 24 hours, and rapid decline	Slower logarithmic rise reaching over 10^9 CFU/ml in 24 to 48 hours, and slow decline	45
Urea hydrolysis	+	–	43
Arginine breakdown	–	+	43
Sensitivity to erythromycin	Sensitive to 0.8-3.0 μg/ml	Resistant to 500 μg/ml	46
Sensitivity to lincomycin	Resistant to 200 μg/ml	Sensitive to 5 μg/ml	47
Sensitivity to thallium acetate	Sensitive to 1:500-1:2000	Resistant to 1:500	43

The ability to synthesize carotenoid pigments is confined to *Acholeplasma laidlawii* and *Acholeplasma granularum* and provides a good means for distinguishing them from other *Acholeplasma* and *Mycoplasma* species (see Table 4).

The optional biochemical tests listed in Figure 1 are very useful in the classification of the mycoplasmas, as may be seen in Table 2, which summarizes the biochemical and physiological characteristics of the *Acholeplasma* and *Mycoplasma* species established so far. Some of the data, however, may not reflect possible strain variations, since they were obtained only with the type or representative strains.

The tetrazolium reduction test is based on the ability of many mycoplasmas to reduce 2,3,5-triphenyltetrazolium chloride. More strains are capable of reducing tetrazolium under aerobic than under anaerobic conditions. Phosphatase activity is determined according to the ability or inability of the mycoplasma to hydrolyze phenolphthalein diphosphate incorporated in the growth medium.[10] Film and spots are indicative of the intensity of the lipolytic activity of the organisms, a property that has some diagnostic value. During growth of certain mycoplasmas on media containing horse serum or egg yolk emulsion, a characteristic wrinkled pearly film appears on the medium surface together with tiny black spots beneath and around the colonies. The film contains cholesterol and phospholipids, whereas the spots consist of calcium and magnesium salts of fatty acids liberated by the mycoplasma lipases.[12]

Hydrolysis of the glucoside aesculin is recommended for differentiation between *Acholeplasma granularum* and the other two *Acholeplasma* species (see Table 4). Hydrolysis is expressed by the development of a black color in a growth medium containing aesculin and ferric citrate. The test is positive only with *A. laidlawii* and *A. axanthum*; all other mycoplasmas tested so far were negative.[13] Proteolytic activity, usually determined by the ability to liquefy gelatin or coagulated serum, may serve as another criterion for distinguishing mycoplasmas (see Table 2).

TABLE 4

DIFFERENTIATION AMONG *ACHOLEPLASMA* SPECIES

Acholeplasma Species	Carotenoid Biosynthesis[48]	Aesculin Hydrolysis[13]
A. axanthum	–	+
A. granularum	+	–
A. laidlawii	+	+

The optional tests also include hemolysis of sheep and guinea pig red blood cells by mycoplasmas (see Figure 1), which can be tested by covering colonies with a thin layer of blood agar[10] or by inoculating concentrated suspensions of organisms onto blood agar.[49] Weak to strong *alpha* and *beta* hemolysis was shown almost throughout to result from the production of peroxide by the organisms.[49] Since hemolytic activity is shared by most mycoplasmas, and since its degree and type seem to depend on minor differences in the technique used, demonstration of the property *per se* is of secondary importance and was accordingly omitted from Table 2. Of greater diagnostic value is the ability of certain mycoplasma species to adsorb erythrocytes or other animal cells (see Table 2). This property can be tested microscopically by determining the adsorption of erythrocytes or other animal cells to mycoplasma colonies.[14]

More sophisticated methods for mycoplasma classification are those based on molecular genetics. Determination of the G + C content of the mycoplasma DNA by thermal melting, buoyant density or chemical analysis may be of great diagnostic value. The data presented in Table 2 show that the G + C content ranges from 23 to 40%, reflecting the marked genetic heterogeneity of mycoplasmas.[28] The most typical aspect is the low G + C content of mycoplasma DNA, which in many strains is as low as 23 to 24%, much less than in bacterial DNAs. Since a similar base composition does not always imply genetic identity, the nucleotide sequence in the DNA strands must also be determined. Various nucleic acid hybridization techniques have helped to establish the identity of unknown mycoplasma strains.[28] A simpler approach to genetic classification, though less direct, is the comparison of the electrophoretic patterns of cell proteins in polyacrylamide gels.[15] Since the synthesis of cell proteins is genetically directed, the electrophoretic patterns are likely to reflect the genetic identity or non-identity of microorganisms.[5]

Serological tests form an essential part of any identification and classification of a new mycoplasma isolate. The new isolate should be compared serologically with other named species – ideally with all of them. The minimum requirement is that it should differ antigenically from all species having the same habitat and sharing the same general biological properties. The obligatory serological tests according to the recommendations of the Subcommittee (see Figure 1) include either the growth or the metabolism inhibition test together with any one of the less specific complement fixation and double-immunodiffusion tests. The growth inhibition test, based on inhibition of growth on agar around discs saturated with the specific antiserum,[19] is the most specific serological test known so far, but it requires highly potent sera, which are not always available. The metabolism inhibition test is based on the specific antibodies that inhibit certain metabolic activities of the mycoplasmas, such as glucose fermentation or arginine and urea hydrolysis.[20] Complement fixation test with whole-cell antigens or immunodiffusion tests with extracts of mycoplasma cells[21] have the advantage of showing antigenic relationships between strains, and one of them should, therefore, be included in the battery, to supplement the growth or metabolism inhibition test. Further optional serological tests are the simple agglutination test and the direct identification of mycoplasma colonies on agar by specific fluorescent antibodies.[24] The usefulness of the last test can hardly be overestimated. It is very rapid, specific, and the only one capable of distinguishing between a mixture of colonies of different serotypes on the same plate – a most important feature, since clinical material very frequently contains more than one mycoplasma species or serotype.

Table 5 shows the habitats and describes the pathogenicity of mycoplasmas. It is evident from the table that mycoplasmas are most prevalent in nature. Plants and insects can apparently be added to the long list

TABLE 5

HABITAT AND PATHOGENICITY OF MYCOPLASMAS

Type Strain	Natural Host	Site of Recovery and Material for Isolation	Disease Manifestation	Experimental Pathology
Acholeplasma axanthum[27]				
S-743 (ATCC 25176)	Not known	Tissue cultures of murine leukemia cell lines	Not known	Not studied
A. granularum[29,41]				
BTS39 (ATCC 19168)	Swine (?)	Nasal cavity of swine Synovial fluid	Possibly swine arthritis (may be due to *M. hyosynoviae*)	The organism does not multiply, cause lesions, or produce death in embryonated hen's eggs. Not known to infect any of the common laboratory or domestic animals other than swine
A. laidlawii[29,50,51]				
PG8 (ATCC 23226)	Cow, swine, man, monkey, chicken, etc.	Tissue cultures, soil, compost, sewage, genitalia of cattle	Usually regarded as saprophytes, but may be associated with infections of the oviducts and infertility in cattle, and with burns in man	Not studied
Mycoplasma agalactiae*, var. *agalactiae[29]				
PG2	Goat, sheep	Mammary glands, lymph nodes, joints Milk, synovial fluid	Contagious agalactia, characterized by arthritis, mastitis and keratitis	Goats and sheep are susceptible to experimental infection by subcutaneous inoculation. The inflammatory lesions are localized in the udders of females and in 10 to 20% of cases in the joints

***M. agalactiae*, var. *bovis* (*M. bovimastitidis*)**[29,52,53,54]				
Donetta	Cattle	Cow udders, sometimes joints Milk or exudate from udders, semen, synovial fluid	Bovine mastitis, sometimes complicated by septic arthritis	Mastitis is experimentally produced by inoculation of the organism into the udder. Endometritis and salpingo-oophoritis, sometimes associated with impaired fertility, have been produced by inoculation of cultures or infectious semen into the uterus of a heifer. Histopathologic examination shows a characteristic eosinophilic cell response. Some strains were cytopathic to bovine embryo tissue cultures. Involvement of a toxin in the pathogenesis is claimed.
M. anatis[55]				
1340	Duck	Sinuses and air sacs (single isolation)	Not known	Not studied
***M. arginini* (*M. leonis*)**[31,56]				
G230 (ATCC 23838)	Cattle, sheep, goat, chamois, mouse, lion	Brain tissue, joint fluid, cell cultures	Not known	Not pathogenic to mice.
***M. arthritidis* (*M. hominis*, type 2)**[29,57]				
PG6 (Preston; ATCC 19611)	Rat	Joint fluid, infected tissue	Polyarthritis of rats, submandibular abscesses, ocular lesions, infection of the middle ear, purulent rhinitis, lung lesions	Widespread infection, characterized by suppurative polyarthritis, conjunctivitis and urethritis, can be produced by intravenous inoculation of virulent strains. Subcutaneous inoculation with agar produces localized abscesses and septicemia in rats. Localized arthritis is produced by inoculation into the footpads. The organisms are non-pathogenic to monkeys, rabbits, and guinea pigs; they may or may not be pathogenic to mice.

TABLE 5 (Continued)

HABITAT AND PATHOGENICITY OF MYCOPLASMAS

Type Strain	Natural Host	Site of Recovery and Material for Isolation	Disease Manifestation	Experimental Pathology
M. bovigenitalium[29,58]				
PG11 (B2; ATCC 19852)	Cattle	Common inhabitant of the lower genital tract in both males and females	Bovine mastitis; possible cause of vaginal disorders and seminal vesiculitis	Mastitis and seminal vesiculitis, characterized by eosinophilic cell response, can be experimentally produced. Intravenous inoculation into calves produced general infection with low fever, mild diarrhea, and arthritic lesions characterized by eosinophilic granulomas of the joint capsules. Marked cytopathogenic effects in monkey, calf, and pig cell cultures, characterized by enlargement of the cells, appearance of intracytoplasmic inclusions, and partial destruction of the cell layer.
M. bovirhinis[29,58]				
PG43 (5M331; ATCC 19884)	Cattle	Common inhabitant of the upper respiratory tract or joints Lung material, milk, sinovial fluid	Possible cause of respiratory disease and mastitis in cattle	Mastitis could-be experimentally produced with some of the strains. Cytopathogenicity for bovine and canine embryo kidney tissue cultures has been reported.
M. canis[59]				
PG14 (C55; ATCC 19525)	Dog	Common inhabitant of the upper respiratory tract and genital tract	Non-pathogenic, as far as known	Not studied
M. conjunctivae[98]				
HRC 581 (ATCC 25834)	Goat, sheep	Conjunctival tissue	Associated with 'pink eye' in goats and sheep	Not studied

***M. dispar*[32]**				
462/2 (NCTC; Colindale 10125)	Cattle	Normal habitat as yet not defined, but has been isolated from pneumonic lung	Possible cause of pneumonia	Not studied
***M. edwardii*[59]**				
PG24 (C21; ATCC 23462)	Dog	Upper respiratory tract and genital tract Pneumonic lung	Significance in respiratory disease is not clear	Not studied
***M. feliminutum*[29,34]**				
Ben	Cat	Oral cavity	Not known (so far only one strain has been isolated)	Not studied
***M. felis*[29,35]**				
CO (ATCC 23391)	Cat	Common inhabitant of the oral and nasal cavities, conjuntivae, and lower genital tract	May be associated with conjunctivitis, but is usually regarded as non-pathogenic	Not studied
***M. fermentans*[7,60,61]**				
PG18 (G; ATCC 19989)	Man	Genitourinary tract, oropharynx (relatively rare) Urethral or cervical scrapings	Not known; possible implication in rheumatoid arthritis requires more study	Produces chromosomal aberrations in tissue cells in culture. Inhibits migration of leukocytes in patients with rheumatoid arthritis.
***M. gallinarum*[55]**				
PG16 (Fowl; ATCC 19708)	Chicken	Upper respiratory tract	Not pathogenic	Not studied

TABLE 5 (Continued)

HABITAT AND PATHOGENICITY OF MYCOPLASMAS

Type Strain	Natural Host	Site of Recovery and Material for Isolation	Disease Manifestation	Experimental Pathology
M. gallisepticum[55,62]				
PG31 (X95; ATCC 19610)	Chicken, turkey	Respiratory system, air sacs, ovaries, eggs	Chronic respiratory disease in chickens, infectious sinusitis in turkeys, encephalitis associated with polyarthritis of the cerebral arteries in turkeys	The chronic respiratory disease in chickens can be experimentally produced. The neurological disease and characteristic pathology can be produced in turkeys by intravenous inoculation of high doses of washed organisms, but not of cell-free filtrates; the cells apparently contain a toxic component.
M. gateae[29,35]				
CS (ATCC 23392)	Cat	Common inhabitant of the upper respiratory tract, conjunctivae, and genital mucosa	Not known	Not studied
M. hominis[7,60,63,64,65]				
PG21 (H50; ATCC 23114)	Man, monkey (?)	Genitourinary tract (common), oropharynx (rare) Urethral or cervical exudates or scrapings, prostatic secretions, rectum, urinary sediments, products of abortion (rare), pleural fluid or abscess, throat swab, sputum	Usually considered commensal, but may be potentially pathogenic and cause salpingitis, tubo-ovarian abscesses, empyema, exudative pharyngitis (not definite), septicemia (rare) or abortion (disputable)	Produces chromosomal aberrations in tissue cells in culture.

M. hyorhinis [29,41,66]				
BTS7 (ATCC 17981)	Swine	Common inhabitant of the nasal cavity Nasal secretions	Many be incriminated in swine penumonia; septicemia associated with arthritis and polyserositis has been reported in young pigs	Young pigs are susceptible to experimental infection, characterized by a febrile reaction that is followed by arthritis and polyserositis. The organisms produce irregular mortality and occasionally pericardial and peritoneal lesions in chick embryos; most strains are cytopathogenic to primary calf, swine, monkey kidney and fetal human lung diploid cell lines.
***M. hyosynoviae* (*M. suidaniae*)** [29,36,37]				
S16 (ATCC 25591) M60	Swine	Joints, respiratory tract Synovial fluid, nasal secretions, tonsillar material, lungs of pigs with catarrhal pneumonia	Association with arthritis or pneumonia has not been established so far	Experimental inoculation of aerosols failed to produce any respiratory disease. On intra-articular inoculation, the synovial fluid was sometimes increased, showing a proliferation of lymphoreticular cells.
M. iners [55]				
PG30 (M; ATCC 19705)	Chicken, turkey	Respiratory tract	Not known	Causes joint lesions when inoculated into chick embryos.
M. maculosum [59,67]				
PG15 (C27; ATCC 19327)	Dog	Genitourinary tract, upper respiratory tract	Not known	Not studied
M. meleagridis [55,68]				
N (17529; ATCC 25294)	Turkey	Respiratory tract and urogenital tract of turkeys; also present in semen, ovaries, and eggs	Air sacculitis in turkeys	Air sacculitis can be experimentally produced in turkeys; the suppurative type, complicated by purulent pneumonia, pericarditis and peritonitis, was produced by combined infection with *Escherichia coli*.

TABLE 5 (Continued)

HABITAT AND PATHOGENICITY OF MYCOPLASMAS

Type Strain	Natural Host	Site of Recovery and Material for Isolation	Disease Manifestation	Experimental Pathology
M. mycoides*, var. *capri[29,58,69,70]				
PG3	Goat	Lung, spleen, liver Exudate, blood, synovial fluid	Caprine pleuropneumonia, fatal edema, cellulitis, septicemia, polyarthritis; disease manifestations vary among strains	Pneumonia can be experimentally produced in goats by intratracheal inoculation or exposure to nebulized cultures. Subcutaneous inoculation produces extensive and often fatal cellulitis and spreading edematous lesions in goats and sheep. Rabbits and mice are susceptible only to organisms injected with mucin. The organisms stop ciliary activity in tracheal-organ cultures of chick embryos, apparently by secretion of H_2O_2.
M. mycoides*, var. *mycoides[58,71,72]				
PG1	Cattle	Respiratory tract Pleural exudate, infected lung	Contagious bovine pleuropneumonia	Cattle can be experimentally infected by nasal instillation of lung material. Goats and sheep are susceptible to subcutaneous incubation, which causes inflammatory lesions. Laboratory animals are susceptible to organisms injected with mucin. A toxin that causes tissue necrosis is produced in diffusion chambers implanted in rabbit peritoneum
M. neurolyticum[29,57,62]				
Sabin type A (PG39; ATCC 19988)	Mouse	Upper respiratory tract	"Rolling" disease, epidemic conjunctivitis	Produces a neurotoxin that, on intravenous injection into young mice, produces characteristic symptoms of "rolling" disease, neuropathological lesions, and pulmonary hemorrhages. Less characteristic neurological symptoms are elicited in young rats.

				Hamsters, guinea pigs, and chickens are not susceptible to the toxin. Production of the neurotoxin in tissue cultures, as well as cytopathic effects on repeated passage of the organisms, has been observed in a variety of cell lines. The neurotoxin is a true exotoxin of protein nature, with a molecular weight in excess of 200,000.
***M. orale*, type 1 (*M. pharyngis*)**[60]				
CH19299 (ATCC 23714)	Man	Oropharynx (very common) Throat swab, sputum	Apparently not pathogenic	Not studied
***M. orale*, type 2**[60,73,74]				
CH20247 (ATCC 23636	Man, monkey	Oropharynx (common in monkeys, rare in man) Throat swab	Apparently not pathogenic	Not studied
***M. orale*, type 3**[9]				
DC333 (ATCC 25293)	Man	Oropharynx (rare)	Apparently not pathogenic	Not studied
M. pneumoniae[60,75,76,77]				
FH (ATCC 15531)	Man	Respiratory tract, oral region, middle ear (rare) Sputum, infected lung	Atypical penumonia, febrile upper respiratory infection, bullous myringitis, various cutaneous and neurological conditions (not definite)	Pneumonia can be experimentally produced in man, in hamsters, and in cotton rats. The organism produces distinct cytopathology, including loss of ciliary activity, in the epithelia of hamster tracheae in organ cultures.
M. primatum[99]				
Navel (ATCC 15497)	Monkey, man	Respiratory tract and genitourinary tract of monkey, infected umbilicus of man	Not known	Not studied

TABLE 5 (Continued)

HABITAT AND PATHOGENICITY OF MYCOPLASMAS

Type Strain	Natural Host	Site of Recovery and Material for Isolation	Disease Manifestation	Experimental Pathology
***M. pulmonis* (*M. histotropicum*, *M. mergenhagen*)[29,57,78,79]**				
PG34 (Ash; ATCC 19612)	Mouse, rat, rabbit (?)	Very common inhabitant of the respiratory tract	Infectious catarrh, sometimes complicated by otitis media and bronchopneumonia; also associated with arthritis in rats and mice	Rhinitis, otitis media, and characteristic pneumonic lesions have been produced in gnotobiotic mice. On intraperitoneal inoculation into female mice, the organisms show a predilection for the ovaries and oviducts and may produce inflammation of the reproductive system. Intravenous inoculation may produce arthritis in mice.
***M. salivarium*[60,73,74]**				
PG20 (ATCC 23064)	Man, monkey	Very common in the oral cavity, particularly in the gingival sulci	Apparently not pathogenic	Not studied
		Throat swab, sputum, gingival debris		
***M. spumans*[59,67]**				
PG13 (C48; ATCC 19526)	Dog	Genitourinary tract, upper respiratory tract	Not known	Not studied
		Lung tissue		
***M. suipneumoniae* (*M. hyopneumoniae*)[29,41,80]**				
J (ATCC 25934) VMR-11 (ATCC 25095)	Swine	Respiratory tract	Enzootic pneumonia of pigs	Pneumonia was experimentally produced by intranasal inoculation of both conventional and gnotobiotic pigs.
		Pneumonic lung		

M. synvoiae[55,81]				
WVU1853	Chicken, turkey	Joints, respiratory tract	Infectious synovitis in chickens and turkeys; pericarditis and myocarditis usually appear at the chronic phase of the disease	The natural disease manifestation can be experimentally produced by inoculation of the organisms through the footpad of the chickens.
T-strains[7,82,83,84,85]				
T-960	Man, monkey, cattle, dog	Genitourinary tract (common in man), oropharynx (uncommon in man) Urethral or vaginal scrapings, seminal fluid, throat swab	Suspected as an agent of non-gonococcal urethritis in man, but recent evidence does not support this	Not studied

of animals known to harbor mycoplasmas. Mycoplasma-like bodies have been detected in electron micrographs of plants suffering from yellows diseases and in insects known to transmit them.[86] The electron micrographs leave little doubt that these were mycoplasmas, which was corroborated by the successful treatment of the diseased plants and vectors with tetracyclines and chloramphenicol and the ineffectiveness of penicillin,[86] but establishment of their etiology must await their isolation, cultivation, identification, and plant inoculation. Several reports claiming the cultivation of mycoplasmas from plants on bacteriological media have recently appeared, but they cannot be accepted as definite until confirmed by other laboratories.

It has been inferred from the highly exacting nature of mycoplasmas that they would exhibit strict host specificity, but several recent findings, reflected in the data presented in Table 5, seem to contradict this assumption. Of special interest is the isolation of *A. laidlawii* strains from a variety of hosts, which casts doubt on the saprophytic nature ascribed to them because, as distinct from the so-called parasitic mycoplasmas, they were originally isolated from sewage and soil. It may well be that *A. laidlawii* strains are commensals or parasites in animals and reach sewage or soil through their excreta. Nevertheless, it may well be that some as yet unknown mycoplasmas or mycoplasma-like organisms can lead a truly saprophytic life. Thus, not long ago, organisms resembling mycoplasmas in their dimensions, their low G + C content of DNA and the absence of a cell wall, were isolated from a hot coal refuse pile and from hot springs. These organisms, which grow optimally at 59°C and at pH values between 1 and 2, were named *Thermoplasma acidophilum.*[87] Though they appear to fit into the class *Mollicutes*, it is debatable whether they should be assigned to the order of Mycoplasmatales. Accordingly, they were, in the meantime, omitted from the tables presented in this chapter.

The disease manifestations and experimental pathology caused by the different *Mycoplasma* species are described in detail in Table 5. Mycoplasma infections have been associated with diseases of the respiratory tract, arthritis, mastitis, and involvement of mucous membranes. The pathologic response of animal tissues of mycoplasma infections is basically the same as that to bacterial infections. Little is known about the mechanisms of mycoplasma pathogenicity. As can be seen from Table 5, the symptoms of only one mycoplasma infection, the "rolling" disease caused by *M. neurolyticum* in mice, can be directly attributed to a potent neurotoxin excreted by the organism. For a detailed discussion of the possible mechanisms for mycoplasma pathogenicity, the reader is referred to recent review and books.[5,88,89]

Table 6 presents data on the sensitivity of several pathogenic mycoplasmas to antibiotics. The prokaryotic nature of mycoplasmas is reflected in their marked sensitivity to antibiotics affecting protein

TABLE 6

SENSITIVITY OF MYCOPLASMAS TO ANTIBIOTICS

Antibiotic	Minimum Inhibitory Concentration, μg/ml		
	M. pneumoniae	*M. gallisepticum*	*M. mycoides* var. *mycoides*
Chloramphenicol	1.5-10.0[90,91]	0.1-8.0[93]	0.5-8.0[93]
Chlorotetracycline	1.5-6.2[90,92]	0.1-2.0[93]	0.5[93]
Erythromycin	0.12-0.3[92]	0.02-0.5[93]	0.1-0.4[29]
Kanamycin	6.25[90] 5.0[91] 100[92]	0.5-40.0[93]	2-200[93]
Nystatin	>50[92]	200[93]	200[93]
Polymyxin B	>100[92]	1,000[93]	>100[94]
Spiramycin	0.6[92]	0.02-2.0[93]	0.1[93,94]
Streptomycin	25-100[90;92]	0.5-40.0[93]	10-1,000[93,94]
Tylosin	0.3[92]	0.004-0.02[93]	0.004[93]

and nucleic acid synthesis in bacteria, such as the tetracyclines and chloramphenicol. Resistance of mycoplasmas to streptomycin develops rapidly and is exhibited by many strains. The mycoplasmas are particularly sensitive to tylosin, a drug widely used in the treatment of mycoplasmosis in poultry. Since the sensitivity of mycoplasmas to erythromycin varies with the species, it can be helpful in the differentiation and selection of mycoplasmas (see Tables 2 and 3). As stated, the wall-less mycoplasmas are completely resistant to those antibiotics and drugs that specifically inhibit bacterial cell wall synthesis.

Detailed information on the biology and pathogenicity of mycoplasmas may be found in two recent books.[88,89] For excellent reviews of diagnostic procedures, see Purcell and Chanock,[60] Lemcke and Leach,[95] and Crawford.[96]

ACKNOWLEDGMENTS

The author is indebted to Dr. D. G. ff. Edward, Dr. E. A. Freundt, and Dr. J. G. Tully for helpful comments and suggestions.

REFERENCES

1. Freundt, E. A., in *The Mycoplasmatales and the L-Phase of Bacteria*, p. 281, L. Hayflick, Ed. Appleton-Century-Crofts, New York (1969).
2. Furness, G., *J. Infect. Dis., 122*, 146 (1970).
3. Morowitz, H. J., in *The Mycoplasmatales and the L-Phase of Bacteria*, p. 405, L. Hayflick, Ed. Appleton-Century-Crofts, New York (1969).
4. Razin, S., and Tully, J. G., *J. Bacteriol., 102*, 306 (1970).
5. Razin, S., *Annu. Rev. Microbiol., 23*, 317 (1969).
6. ICSB Subcommittee on the Taxonomy of Mycoplasmatales, *Proposals for Minimum Standards for Descriptions of New Species of Mycoplasmatales Class Mollicutes Int.J. Syst. Bacteriol.*, (in press).
7. Taylor-Robinson, D., Addey, J. P., Hare, M. J., and Dunlop, E. M. C., *Brit. J. Vener. Dis., 45*, 265 (1969).
8. Mason, R. A., *J. Bacteriol., 103*, 184 (1970).
9. Fox, H., Purcell, R. H., and Chanock, R. M., *J. Bacteriol., 98*, 36 (1969).
10. Aluotto, B. B., Wittler, R. G., Williams, C. O., and Faber, J. E., *Int. J. Syst. Bacteriol., 20*, 35 (1970).
11. Tully, J. G., and Razin, S., *J. Bacteriol., 95*, 1504 (1968).
12. Fabricant, J., and Freundt, E. A., *Ann. N.Y. Acad. Sci., 143,* 50 (1967).
13. Williams, C. O., and Wittler, R. G., *Int. J. Syst. Bacteriol.*, (in press).
14. Sobeslavsky, O., Prescott, B., and Chanock, R. M., *J. Bacteriol., 96*, 695 (1968).
15. Razin, S., *J. Bacteriol., 96*, 687 (1968).
16. McGee, Z. A., Rogul, M., and Wittler, R. G., *Ann. N.Y. Acad. Sci., 143*, 21 (1967).
17. Kelton, W. H., and Mandel, M., *J. Gen. Microbiol., 56*, 131 (1969).
18. Somerson, N. L., Reich, P. R., Chanock, R. M., and Weissman, S. M., *Ann. N.Y. Acad. Sci., 143*, 9 (1967).
19. Clyde, W. A., Jr., *J. Immunol., 92*, 958 (1964).
20. Purcell, R. H., Wong, D., Chanock, R. M., Taylor-Robinson, D., Canchola, J., and Valdesuso, J., *Ann. N.Y. Acad. Sci., 143*, 664 (1967).
21. Kenny, G. E., *J. Bacteriol., 98*, 1044 (1969).
22. Argaman, M., and Razin, S., *J. Gen. Microbiol., 55*, 45 (1969).
23. Morton, H. E., *J. Bacteriol., 92*, 1196 (1966).
24. Del Giudice, R. A., Robillard N. F., and Carski, T. R.,*J. Bacteriol., 93,* 1205 (1967).
25. Edward, D. G. ff., and Freundt, E. A., *J. Gen. Microbiol., 62*, 1 (1970).
26. Schimke, R. T., Berlin, C. M., Sweeney, E. W., and Carroll, W. R., *J. Biol. Chem., 241*, 2228 (1966).
27. Tully, J. G., and Razin, S., *J. Bacteriol., 103*, 751 (1970).
28. Neimark, H. C., *J. Gen. Microbiol., 63*, 249 (1971).
29. Freundt, E. A., in *Bergey's Manual*, 8th ed. (In preparation).
30. Williams, C. O., Wittler, R. G., and Burris, C., *J. Bacteriol., 99*, 341 (1969).
31. Barile, M. F., Del Giudice, R. A., Carski, T. R., Gibbs, C. J., and Morris, J. A., *Proc. Soc. Exp. Biol. Med., 129*, 489 (1968).
32. Gourlay, R. N., and Leach, R. H., *J. Med. Microbiol., 3*, 111 (1970).
33. Tully, J. G., Barile, M. F., Del Giudice, R. A., Carski, T. R., Armstrong, D., and Razin, S., *J. Bacteriol., 101*, 346 (1970).
34. Heyward, J. T., Sabry, M. Z., and Dowdle, W. R., *Amer. J. Vet. Sci., 30*, 615 (1969).
35. Cole, B. C., Golightly, L., and Ward, J. R., *J. Bacteriol., 94*, 1451 (1967).
36. Ross, R. F., and Karmon, J. A., *J. Bacteriol., 103*, 707 (1970).

37. Friis, N. F., *Acta Vet. Scand.*, *11*, 487 (1970).
38. Yamamoto, R., Bigland, C. H., and Ortmayer, H. B., *J. Bacteriol.*, *90*, 47 (1965).
39. Fox, H., Purcell, R. H., and Chanock, R. M., *J. Bacteriol.*, *98*, 36 (1969).
40. Goodwin, R. F. W., Pomeroy, A. P., and Whittlestone, P., *J. Hyg. (Camb.)*, *65*, 85 (1967).
41. Switzer, W. P., in *The Mycoplasmatales and the L-Phase of Bacteria*, p. 607, L. Hayflick, Ed. Appleton-Century-Crofts, New York (1969).
42. Bak, A. L., and Black, F. T., *Nature*, *219*, 1044 (1968).
43. Shepard, M. C., in *The Mycoplasmatales and the L-Phase of Bacteria*, p. 49, L. Hayflick, Ed. Appleton-Century-Crofts, New York (1969).
44. Shepard, M. C., and Lunceford, C. D., *J. Bacteriol.*, *89*, 265 (1965).
45. Manchee, R. J., and Taylor-Robinson, D., *J. Bacteriol.*, *100*, 78 (1969).
46. Shepard, M. C., Lunceford, C. D., and Baker, R. S., *Brit. J. Vener. Dis.*, *42*, 21 (1966).
47. Csonka, G. W., and Spitzer, R. J., *Brit. J. Vener. Dis.*, *45*, 52 (1969).
48. Tully, J. G., and Razin, S., *J. Bacteriol.*, *98*, 970 (1969).
49. Cole, B. C., Ward, J. R., and Martin, C. H., *J. Bacteriol.*, *95*, 2022 (1968).
50. Hoare, M., *Vet. Rec.*, *85*, 351 (1969).
51. Markham, J. G., and Markham, N. P., *J. Bacteriol.*, *98*, 827 (1969).
52. Hirth, R. S., Plastridge, W. N., and Tourtellotte, M. E., *Amer. J. Vet. Res.*, *28*, 97 (1967).
53. Mosher, A. H., Plastridge, W. N., Tourtellotte, M. E., and Helmboldt, C. F., *Amer. J. Vet. Res.*, *29*, 517 (1968).
54. Karbe, E., and Mosher, A. H., *Zentralbl. Veterinaermed.*, *15*, 817 (1968).
55. Fabricant, J., in *The Mycoplasmatales and the L-Phase of Bacteria*, p. 621, L. Hayflick, Ed. Appleton-Century-Crofts, New York (1969).
56. Leach, R. H., *Vet. Rec.*, *87*, 319 (1970).
57. Tully, J. G., in *The Mycoplasmatales and the L-Phase of Bacteria*, p. 571, L. Hayflick, Ed. Appleton-Century-Crofts, New York (1969).
58. Cottew, G. S., and Leach, R. H., in *The Mycoplasmatales and the L-Phase of Bacteria*, p. 527, L. Hayflick, Ed. Appleton-Century-Crofts, New York (1969).
59. Barile, M. F., Del Giudice, R. A., Carski, T. R., Yamashiroya, H. M., and Verna, J. A., *Proc. Soc. Exp. Biol. Med.*, *134*, 146 (1970).
60. Purcell, R. H., and Chanock, R. M., in *Diagnostic Procedures for Viral and Rickettsial Infections*, 4th ed., p. 786, E. H. Lennette and N. J. Schmidt, Eds. American Public Health Association, New York (1969).
61. Williams, M. H., Brostoff, J., and Roitt, I. M., *Lancet*, *2*, 277 (1970).
62. Thomas, L., *Yale J. Biol. Med.*, *40*, 444 (1968).
63. Allison, A. C., and Paton, G. R., *Lancet*, *2*, 1229 (1966).
64. Harwick, H. J., Purcell, R. H., Iuppa, J. B., and Fekety, F. R., Jr., *J. Infect. Dis.*, *121*, 260 (1970).
65. Tully, J. G., and Smith, L. G., *J. Amer. Med. Assoc.*, *204*, 827 (1968).
66. Schulmann, A., Estola, T., and Garry-Andersson, A. S., *Zentralbl. Veterinaermed.*, *17*, 549 (1970).
67. Armstrong, D., Tully, J. G., Yu, B., Mortonon, V., Friedman, M. H., and Steger, L., *Infect. Immun.*, *1*, 1 (1970).
68. Saif, Y. M., Moorhead, P. D., and Bohl, E. H., *Amer. J. Vet. Res.*, *31*, 1637 (1970).
69. Smith, G. R., *J. Comp. Pathol.*, *77*, 21 (1967).
70. Cherry, J. D., and Taylor-Robinson, D., *Nature*, *228*, 1099 (1970).
71. Shifrine, M., and Moulton, J. E., *J. Comp. Pathol.*, *78*, 383 (1968).
72. Lloyd, L. C., *J. Pathol. Bacteriol.*, *92*, 225 (1966).
73. Del Giudice, R. A., Carski, T. R., Barile, M. F., Yamashiroya, H. M., and Verna, J. E., *Nature*, *222*, 1088 (1969).
74. Madden, D. L., Hildebrandt, R. J., Monif, G. R. G., London, W. T., Sever, J. L., and McCullough, N. B., *Lab. Anim. Care*, *20*, 467 (1970).
75. Smith, C. B., Chanock, R. M., Friedewald, W. T., and Alford, R. H., *Ann. N.Y. Acad. Sci.*, *143*, 471 (1967).
76. Steele, J. C., Gladstone, R. M., Thanasophon, S., and Fleming, P. C., *Lancet*, *2*, 710 (1969).
77. Collier, A. M., Clyde, W. A., Jr., and Denny, F. W., *Proc. Soc. Exp. Biol. Med.*, *132*, 1153 (1969).
78. Barden, J. A., and Tully, J. G., *J. Bacteriol.*, *100*, 5 (1969).
79. Deeb, B. J., and Kenny, G. E., *J. Bacteriol.*, *93*, 1416 (1967).
80. Hodges, R. T., Betts, A. O., and Jennings, A. R., *Vet. Rec.*, *84*, 268 (1969).
81. Kerr, K. M., and Olson, N. O., *Avian Dis.*, *14*, 291 (1970).
82. Fowler, W., and Leeming, R. J., *Brit. J. Vener. Dis.*, *45*, 287 (1969).
83. Mardh, P.-A., and Weström, L., *Acta Pathol. Microbiol. Scand., Ser. B.*, *78*, 367 (1970).
84. Hare, M. J., Dunlop, E. M. C., and Taylor-Robinson, D., *Brit. J. Vener. Dis.*, *45*, 282 (1969).
85. Shepard, M. C., *J. Amer. Med. Assoc.*, *211*, 1335 (1970).
86. Whitcomb, R. F., and Davis, R. E., *Annu. Rev. Entomol.*, *15*, 405 (1970).
87. Darland, G., Brock, T. D., Samsonoff, W., and Conti, S. F., *Science*, *170*, 1416 (1970).
88. Hayflick, L. (Ed.), *The Mycoplasmatales and the L-Phase of Bacteria*. Appleton-Century-Crofts, New York (1969).
89. Sharp, J. T. (Ed.), *The Role of Mycoplasmas and L-Forms of Bacteria in Disease*. Charles C Thomas, Springfield, Illinois (1970).

90. Stewart, S. M., Burnet, M. E., and Young, J. E., *J. Med. Microbiol., 2*, 287 (1969).
91. Slotkin, R. I., Clyde, W. A., Jr., and Denny, F. W., *Amer. J. Epidemiol., 86*, 225 (1967).
92. Arai, S., Yoshida, K., Izawa, A., Kumagai, K., and Ishida, N., *J. Antibiot., Ser. A, 19*, 118 (1966).
93. Newnham, A. G., and Chu, H. P., *J. Hyg. (Camb.), 63*, 1 (1965).
94. Omura, S., Lin, Y. C., Yajima, T., Nakamura, S., Tanaka, N., and Umezawa, H., *J. Antibiot., Ser. A, 20*, 241 (1967).
95. Lemcke, R. M., and Leach, R. H., in *Identification Methods for Microbiologists*, Part B, p. 132, B. M. Gibb and D. A. Shapton, Eds. Academic Press, London, England (1968).
96. Crawford, Y. E., in *Manual of Clinical Microbiology*, p. 251, J. E. Blair, E. H. Lennette, and J. P. Truant, Eds. American Society for Microbiology, Washington, D.C. (1970).
97. Ogata, M., Atobe, H., Kushida, H., and Yamamoto, K., *J. Antibiot. (Tokyo), 24*, 443 (1971).
98. Barile, M. F., Del Giudice, R. A., and Tully, J. G. (submitted for publication in *Infect. Immun.*).
99. Del Giudice, R. A., Carski, T. R., Barile, M. F., Lemcke, R. M., and Tully, J. G., *J. Bacteriol, 108,* 439 (1971).

THE CHLAMYDIAE

DR. LESLIE A. PAGE

Chlamydiae are pathogenic bacteria that multiply only within the cytoplasm of vertebrate host cells by a developmental cycle that is unique among microorganisms. They are Gram-negative, non-motile, coccoidal organisms, 0.2 to 1.5 μm in diameter, classified in the genus *Chlamydia*, family Chlamydiaceae, order Chlamydiales. The intracellular developmental cycle begins after phagocytosis of the small (0.2 to 0.5 μm), infectious form of the organism, called elementary body (EB). Within a cytoplasmic vesicle, the EB reorganizes into a larger (0.8 to 1.5 μm), non-infectious form, called initial body, that multiplies by fission. Numerous daughter cells are formed, and these again reorganize, becoming small, electron-dense EB's which, when released from damaged host cells, survive extracellularly to repeat the cycle in other host cells.

Elementary bodies contain compact nuclear material (RNA/DNA ratio is 1:1) and ribosomes and are bounded by a rigid trilaminar cell wall that is chemically similar to that of Gram-negative bacteria.[1] Initial bodies contain a loose network of nuclear fibrils with an RNA/DNA ratio of 2:1 and have a thin, fragile cell wall.[2] Because of this nucleic fibrilar network, they are sometimes called reticulate bodies. Although initial bodies are the intracellular vegetative form of the organism, they have limited ability to survive extracellularly.

METABOLIC CHARACTERISTICS AND CLASSIFICATION

Chlamydiae depend on host cells for certain growth factors. Though chlamydiae are capable of independent enzymatic activities, such as catabolism of glucose, pyruvate and glutamate, when provided with essential organic and inorganic cofactors, they have no apparent ability to produce high-energy compounds (e.g., ATP) for energy storage and utilization.[3] Thus, they have been characterized as "energy parasites".[4]

Some strains are capable of synthesizing folates and glycogen, thereby providing biochemical criteria for distinguishing two species: *C. trachomatis* and *C. psittaci.*[5,6] Strains of *C. trachomatis* synthesize folates, and hence their growth is inhibited by sulfa compounds; they also produce a glycogen matrix surrounding the organisms in their intracytoplasmic microcolony. On the other hand, strains of *C. psittaci* fail to synthesize either folates or glycogen, and their growth is not inhibited by sulfa compounds. This division of the chlamydial strains into two species is supported by evidence that hybridization of DNA strands between the two species is 10% or less (see Table 1).[7]

ANTIBIOTIC SENSITIVITIES AND DISINFECTION

Multiplication of chlamydiae in host cells is inhibited by tetracyclines, chloramphenicol, erythromycin, and 5-fluorouracil. Some strains are sensitive to penicillin and D-cycloserine. Most strains appear to be insensitive to streptomycin, vancomycin, kanamycin, neomycin, and bacitracin. Infectivity is rapidly destroyed by dilute solutions of quaternary ammonium detergents and lipid solvents, but is relatively unaffected by dilute acids, alkalies, alcohols, phenol, creosyl, or oxidants such as hypochlorite or permanganate.[8] The organisms are moderately resistant to low-energy ultraviolet treatment, to drying, and to lyophilization. They are inactivated in 5 to 30 minutes at 56°C when suspended in beef heart infusion broth as a 10% infected tissue homogenate.[9]

DISEASES CAUSED BY CHLAMYDIAE

Chlamydiae produce cytopathology and are the etiologic agents of a variety of diseases of man and other animals. Strains of *C. trachomatis* cause well-known diseases of the ocular and urogenital tracts in humans[10] (see Table 2), and strains of *C. psittaci* cause numerous diseases of man and animals, manifested primarily as pneumonitis, arthritis, placentitis (leading to abortion), or enteritis[11] (see Table 3).

TABLE 1.

BIOCHEMICAL PROPERTIES DIFFERENTIATING TWO SPECIES OF *CHLAMYDIA*

Chlamydia trachomatis	*Chlamydia psittaci*
1. Forms compact intracytoplasmic microcolonies that produce glycogen and phospholipids, which stain differentially with iodine–potassium iodide solution.	1. Intracytoplasmic microcolonies are less compact, with developing microorganisms distributed throughout the cytoplasm of the host cell. Glycogen or phospholipids detectable by iodine staining of infected cultures are apparently not formed.
2. Synthesizes essential folates, and growth in chicken embryos is therefore inhibited by sodium sulfadiazine at the level of 1 mg per embryo.	2. Growth in chicken embryos is not sensitive to sodium sulfadiazine at the level of 1 mg per embryo.
3. Degree of DNA hybridization with strains of *C. psittaci* is 10% or less.	3. Degree of DNA hybridization with strains of *C. trachomatis* is 10% or less.

Note:

Strains of chlamydiae now considered to be members of the single species *C. psittaci* were formerly known as any of eight species of *Miyagawanella (Bergey's Manual of Determinative Bacteriology,* 7th ed., 1957). They have also been variously labeled as *Bedsonia, Rickettsiformis, Rakeia,* and *Ehrlichia.*[4]

Chlamydiae are widely distributed in nature.[12] They have occasionally been isolated from mites and ticks, but arthropod transmission of chlamydial disease has not been proven.[13]

ANTIGENIC STRUCTURE

Every chlamydial strain possesses common group antigens, which are lipopolysaccharides. These antigens are resistant to heat, phenol, and proteolytic and nucleolytic enzymes, but are inactivated by periodate and lecithinase. They are solubilized in aqueous solutions by treatment of chlamydiae suspensions with deoxycholate or sodium lauryl sulfate.[14,15] Some of these antigens are common to all strains, but certain groups of specific antigens present in a number of strains indicate whether the strain's natural host was man, other mammals, or birds.[16] Clusters of strain-specific antigens are found in chlamydial cell walls.

TINCTORIAL CHARACTERISTICS

Chlamydiae, like rickettsiae, stain red with the Gimenez or Macchiavello procedure, or purple with Giemsa stain. The Gimenez procedure is preferred for staining chlamydiae propagated in yolk sacs of infected chicken embryos.[17] Giemsa stain or Macchiavello's method may be preferred for staining chlamydiae in infected animal tissues or exudates. Intracellular chlamydiae may be clearly observed in fresh wet-mounts of infected cell suspension or exudates by using a microscope equipped with phase-contrast optics. Purified suspensions of chlamydiae, free of host cells, are Gram-negative, but various forms of intracellular chlamydiae stain irregularly by Gram's method.

DIAGNOSIS OF CHLAMYDIOSIS

Whatever the host species, conclusive proof of chlamydial infection or disease usually rests upon isolation

TABLE 2

DISEASES AND INFECTIONS CAUSED BY VARIOUS STRAINS OF *CHLAMYDIA TRACHOMATIS*

Disease	Natural Host	Principal Effects and Transmission Routes
Trachoma	Man	Progressive conjunctivitis, primarily of upper eyelid, with hyperemia, exudation, follicular hypertrophy; neovascularization of cornea may lead to opacity and pannus formation. Transmitted by contamination of conjunctiva with infectious exudate.
Inclusion conjunctivitis	Man	Conjunctivitis, primarily of lower eyelid, which tends to heal spontaneously; organisms may spread to mucous membranes of genitalia. Transmitted by contamination of eyelid with infectious material.
Non-gonococcal urethritis, proctits	Man	Inflammation and exudation in urethra of males and in cervix and anus of females; infection may spread by contact contamination to conjunctiva. Transmitted venereally.
Lymphogranuloma venereum	Man	Effects on lymphatic tissue of iliac and inguinal region result in lymphoadenopathy with suppuration, occasional elephantiasis of penis or scrotum, or rectal strictures in females. Transmitted venerally.
Murine pneumonitis	Mouse	Mild Pneumonitis, rarely fatal; endemic in some mouse colonies. Probably transmitted via aerosol of respiratory ejecta.

and identification of the etiologic agent. Clinical history, evidence of typical lesions, and positive serology (or skin test, where appropriate) greatly assist confirmation of frank chlamydiosis, but they are not conclusive by themselves. Overt clinical trachoma may be a diagnostic exception. Evidence of a fourfold or greater rise in serologic titer between acute and convalescent phases of infection is usually diagnostic of recent infection, but in cases of primary brucellosis or Q-fever this evidence may occasionally be misleading. The serology of viral diseases in animals may be obscured by chlamydial antibody titer rises due to incidental subclinical intestinal infection with chlamydiae of low virulence. On the other hand, concurrent salmonellosis or trichomoniasis in birds greatly enhances otherwise benign chlamydial infection, producing high mortality. In domestic herbivores, chlamydiae of modest virulence combined with pasteurellae, mycoplasmata, or parainfluenza viruses may cause a "shipping fever" syndrome.

ISOLATION AND IDENTIFICATION OF CHLAMYDIAE

Choice of tissue for examination depends on the site of principal damage caused by chlamydial infection (see Table 4). Contaminated specimens may be treated with antibiotics known not to affect chlamydial growth.

All strains of chlamydiae may be isolated in developing chicken embryos inoculated by the yolk-sac route on the 6th to 7th day of development. Strains of *C. trachomatis* are sensitive to temperatures above 37°C, and embryos inoculated with homogenates of specimens from humans should, therefore, be incubated at 35°C. Embryos inoculated with specimens from infected birds or mammals may be incubated

TABLE 3

DISEASES AND INFECTIONS CAUSED BY VARIOUS STRAINS OF *CHLAMYDIA PSITTACI*

Disease	Natural Hosts	Principal Effects, Epidemiology, and Transmission Route
Psittacosis, ornithosis	Wild and domestic birds	Lethargy, hyperthermia, anorexia, abnormal excretions, lowered egg production. Fibrinous airsacculitis, pericarditis, peritonitis, perihepatitis, splenomeglay, hepatopathy. Mortality 0 to 40%, depending on the virulence of the organism. Endemic in psittacine and in columbine species. Transmitted by air-borne route to domestic birds and man.
Psittacosis	Man	Malaise, headache, hyperthermia, anorexia, cough. Pneumonitis, splenitis, occasional meningitis. Mortality is less than 1% in antibiotic-treated cases, and 20% in untreated cases. World-wide distribution. Transmitted by air-borne route from birds to man and from man to man.
Pneumonitis	Cats, sheep, cattle, goats, pigs, horses, rabbits	Conjunctival and/or nasal mucopurulent discharge, lethargy, anorexia, labored breathing, hyperthermia. Conjunctivitis, pneumonitis. Rarely fatal, unless complicated by concurrent infection with viruses or other bacteria, especially *Mycoplasma* and *Pasteurella.*
Polyarthritis	Lambs, calves, pigs	Lameness, swollen carpal, tarsal, and stifle joints, hyperthermia, anorexia, lethargy. Fibrinous synovitis, tendonitis, occasional hepatopathy. Mortality is variable. Widespread among lambs and calves in the western U.S. and among pigs in Europe.
Placentitis (leading to abortion)	Cattle, sheep, pigs, goats, rabbits, mice	Transient hyperthermia, chlamydemia, inflammation and necrosis of placentome, abortion of fetus late in gestation. Fetal hepatopathy, edema, ascites, vascular congestion, tracheal petechia. Periodically epidemic in California and Oregon cattle. Endemic in sheep throughout the world.
Encephalomyelitis	Calves	Lethargy, incoordination, weakness, hyperthermia, anorexia, diarrhea, paralysis. Fibrinous perihepatitis, pericarditis, ascites. Endemic in cattle in the mid-western and western U.S.
Conjunctivitis	Sheep, cattle, pigs, cats, guinea pigs	Vacular congestion and edema of conjunctiva, mucopurulent discharge, hyperthermia in cats. Follicular conjunctivitis, keratitis, pannus formation. May be related to pneumonitis strains. Probably transmitted by air-borne route and contact contamination.
Fatal enteritis	Snowshoe hare, muskrat	Bizarre behavior, diarrhea. Enteritis, splenomegaly, focal necrosis in liver. High mortality in hares caused extensive deaths in snowshoe hares in Canada between 1959 and 1961. Transmission route is unknown; muskrats may be reservoirs.

TABLE 3 (Continued)

DISEASES AND INFECTIONS CAUSED BY VARIOUS STRAINS OF *CHLAMYDIA PSITTACI*

Disease	Natural Hosts	Principal Effects, Epidemiology, and Transmission Route
Enteritis	Cattle, sheep	Diarrhea, weakness, and death in newborns. Enteritis. Epidemiology in relation to widespread subclinical infection of adult animals is not known.
Subclinical intestinal infection	Cattle, sheep	No clinical signs in adults; may cause transient hyperthermia and diarrhea in newborns. Widespread in cattle and sheep throughout the U.S. Probably transmitted by ingestion of feces-contaminated feed.

Note:
Clinical signs and gross lesions vary widely in cases of mild or chronic chlamydial infections. Variations in effects may also be caused by differences in natural resistance of individuals, virulence of the organisms, dosages, presence of other pathogens, and other stress factors.

Data adapted from: Page, L. A., The Chlamydioses: Diseases Caused by Organisms of the Genus *Chlamydia,* in *Standard Methods for Veterinary Microbiology.* American Veterinary Medical Association, Chicago, Illinois (in press). Printed by permission of the publishers.

at 37 to 39°C. The latter temperature enhances the growth rate of most *C. psittaci* strains.[18] Depending on the virulence and growth rate, chlamydiae multiply to numbers sufficient to kill the embryo in 4 to 12 days. If the inoculum contains few organisms, several blind passages of infectious material in eggs may be necessary to establish an embryo death pattern.

Alternative isolation hosts are listed in Table 4. These species include primates, mice, guinea pigs, and mouse lung cell cultures ("L" cell line, McCoy cell line). Lesions observed in these hosts are described in detail elsewhere.[19,20]

Identification of chlamydiae in yolk sacs of chicken embryos or in tissues of other experimental hosts depends on demonstration of the presence of chlamydial group antigen. Infected yolk sacs are triturated in phosphate-buffered saline (pH 7.2) to make a 20 to 25% suspension and boiled for 30 minutes; when the suspension has cooled, phenol is added to a final concentration of 0.5%. The suspension should then be reacted in serial dilution in Veronal® buffer against a constant dilution of chlamydial antiserum in a CF test. The dilution of antiserum should contain four to eight CF units of antibody per chlamydial group antigen. If the yolk-sac suspension has a CF titer of 1:32 or more against antiserum and is negative against normal serum, the yolk sac can then be assumed to contain organisms of the genus *Chlamydia*. It has been calculated[6,21] that an approximately-10,000-embryo LD_{50} of chlamydiae will fix one unit of complement against four units of antiserum in a CF test.[a] Confirmatory examination of infected yolk sacs may be made after staining yolk-sac smears by the Gimenez method. Chlamydiae stain red, whereas other bacteria stain blue against a greenish background.

Chlamdiae in cell cultures and animal tissues may also be identified microscopically by using specific fluorescent-antibody (FA) methods, but reliable FA preparations are not available for general use. Most FA preparations are products of individual laboratory effort and are used experimentally. FA preparations that are specific for certain strains of organisms, e.g., those that cause trachoma, have been used successfully to distinguish serotypes.

[a] Yolk sacs containing large numbers of bacteria of the genera *Herellea* or *Bacterioides* may fix complement with chlamydial antisera also.

TABLE 4

PREFERRED SPECIMENS, MICROSCOPY, AND HOSTS FOR ISOLATION OF CHLAMYDIAE
(CE = chicken embryo; MLCC = mouse lung cell culture)

Disease	Preferred Specimen*	Microscopy, Stain or Method†	Principal (Alternative) Isolation Hosts; Incubation Temperature, °C
Trachoma	Conjunctival scraping	Giemsa stain, iodine stain, fluorescent antibody	CE (MLCC); 35. Primates, conjunctival scarification
Conjunctivitis; urethritis or cervicitis	Conjunctival scraping; urethral or cervical swab	Fluorescent antibody, Giemsa stain, iodine stain	CE (MLCC); 35. Primates, conjunctival scarification
Lymphogranuloma venereum	Bubos aspirate	Inconclusive	CE (MLCC); 35. Mice, intracerebral route
Human pneumonitis	Sputum, blood; lungs, spleen at autopsy	Inconclusive	CE (MLCC); 37-39. Mice, intraperitoneal route
Avian pneumonitis	Airsac or pericadial exudate, spleen, liver	Giemsa stain, Macchiavello stain, phase-contrast microscopy of fresh exudate	CE (MLCC); 37-39. Mice, intraperitoneal route
Mammalian pneumonitis	Tracheal exudate, lung	Giemsa stain, Macchiavello stain	CE; 37-39. Guinea pig, intraperitoneal route
Mammalian polyarthritis	Synovial fluid of carpal, tarsal, or stifle joints	Phase-contrast microscopy of fresh exudate, fluorescent antibody	CE; 37-39. Guinea pig, intraperitoneal route
Mammalian placeentitis	Hyperemic areas of placenta, placentomes, fetal liver	Giemsa stain, Macchiavello stain	CE; 37-39. Guinea pig, intraperitoneal route
Mammalian enteritis	Surface mucus of formed feces, hyperemic areas of intestinal wall	Inconclusive	CE; 37-39 after centrifugation and antibiotic treatment to remove contaminating bacteria

* For isolation of chlamydiae, all specimens should be treated with a combination of antibiotics, e.g., streptomycin sulfate, vancomycin and kanamycin sulfate, 1 mg of each per ml of phosphate-buffered saline.

† Positive specimens contain mononuclear cells with intracytoplasmic microcolonies of chlamydiae.

Chlamydial species are identified on the basis of characteristics described in Table 1. The test for glycogen production is performed on infected cell monolayer cultures (and uninoculated control cultures) by staining them with a 5% iodine-potassium iodide solution after chlamydial growth has occurred.[5] Iodine-positive microcolonies of chlamydiae are dark-tan against a light-tan background. The test for sulfa sensitivity is performed by inoculating decimal dilutions of a suspension of chlamydiae by the yolk-sac route into 6-day-incubated embryos, using at least 12 embryos per dilution. Half of the embryos in each series of dilutions should be given a second yolk-sac inoculation, using a solution containing sufficient sodium sulfadiazine so that each embryo receives 1 mg. The eggs are then incubated for 14 days, the deaths recorded, and the LD_{50} calculated for each series. If the chlamydiae are sensitive to sulfadiazine, there should be at least a 2-log difference between the LD_{50} calculation for the series with sulfadiazine and that for the series without sulfadiazine.[6]

SEROLOGY

The method most commonly used at present for detecting antibodies in sera of infected individuals is the CF tests.[16] Group- and strain-specific antigens have been prepared from yolk-sac- or cell-culture-propagated organisms. However, more rapid and simpler serologic methods should be available in the near future for general use. Antigens have been developed for use in capillary-tube agglutination,[22] double diffusion in gel,[23] and immunofluorescence[24] tests for antibodies.

Elevations of chlamydial antibodies generally signify current infection. Incidences of serologic titers of 1:64 among a herd or flock of domestic animals in excess of 50% also indicate current infection. The incidence of seropositive tests among apparently healthy populations of humans, cattle, or sheep may run as high as 25%, reflecting the ubiquity of chlamydiae in nature and the possibility of widespread natural infections that escape clinical notice because of their mildness.

REFERENCES

1. Manire, G. P., and Tamura, A., Preparation and Chemical Composition of the Cell Walls of Mature Infectious Dense Forms of Meningopneumonitis Organisms. *J. Bacteriol., 94,* 1178 (1967).
2. Tamura, A., and Manire, G. P., Preparation and Chemical Composition of the Cell Membranes of Developmental Reticulate Forms of Meningopneumonitis Organisms. *J. Bacteriol., 94,* 1184 (1967).
3. Weiss, E., and Wilson, N. N., Role of Exogenous Adenosine Triphosphate in Catabolic and Synthetic Activities of *Chlamydia psittaci. J. Bacteriol., 97,* 719 (1969).
4. Moulder, J. W., The Relation of the Psittacosis Group (Chlamydiae) to Bacteria and Viruses. *Annu. Rev. Microbiol., 20,* 107 (1969).
5. Gordon, F. B., and Quan, A. L., Occurrence of Glycogen in Inclusions of the Psittacosis-Lymphogranuloma venereum-Trachoma Agents. *J. Infect. Dis., 115,* 86 (1965).
6. Page, L. A., Proposal for the Recognition of Two Species in the Genus *Chlamydia* Jones, Rake and Stearns, 1945. *Int. J. Syst. Bacteriol., 18,* 51 (1968).
7. Kingsbury, D. T., and Weiss, E., Lack of Deoxyribonucleic Acid Homology Between Species of Genus *Chlamydia. J. Bacteriol., 96,* 1421 (1968).
8. Nabli, B., and Tarizzo, M. L., The Effect of Antiseptics and Other Substances on TRIC Agents. *Amer. J. Ophthalmol., 63,* 1441 (1967).
9. Page, L. A., Thermal Inactivation Studies on a Turkey Ornithosis Virus. *Avian Dis., 3,* 67 (1959).
10. Jawetz, E., Agents of Trachoma and Inclusion Conjunctivitis. *Annu. Rev. Microbiol., 18,* 301 (1964).
11. Storz, J., *Chlamydia and Chlamydia-Induced Diseases.* Charles C Thomas, Springfield, Illinois (1971).
12. Meyer, K. F., The Host Spectrum of Psittacosis-Lymphogranuloma venereum (PL) Agents. *Amer. J. Ophthalmol., 63,* 1224 (1967).
13. Eddie, B., Radovsky, F. J., Stiller, D., and Kumada, N., Psittacosis-Lymphogranuloma venereum (PL) Agents *(Bedsonia, Chlamydia)* in Ticks, Fleas, and Native Mammals in California. *Amer. J. Epidemiol., 90,* 449 (1969).
14. Jenkin, H. M., Preparation and Properties of Cell Walls of the Agent of Meningopneumonitis. *J. Bacteriol., 80,* 639 (1960).
15. Benedict, A. A., and McFarland, C., Direct Complement Fixation Test for Diagnosis of Ornithosis in Turkeys. *Proc. Soc. Exp. Biol. Med., 92,* 768 (1956).
16. Fraser, C. E. O., Analytical Serology of the Chlamydiaceae, in *Analytical Serology of Microorganisms,* Vol. 1, p. 257. John Wiley and Sons, New York (1969).
17. Gimenez, D. F., Staining Rickettsiae in Yolk-Sac Cultures. *Stain Technol., 39,* 135 (1964).
18. Page, L. A., Influence of Temperature on Multiplication of Chlamydiae in Chicken Embryos. *Excerpta Med. Int. Congr. Ser., Proc. Int. Trachoma Conf., Boston, 1970,* p. 40 (1971).
19. Wang, S. P., A Microimmunofluorescence Method: Study of Antibody Response to TRIC Organisms in Mice. *Excerpta Med. Int. Congr. Ser., Proc. Int. Trachoma Conf., Boston, 1970,* p. 273 (1971).
20. Page, L. A., Interspecies Transfer of Psittacosis-LGV-Trachoma Agents: Pathogenicity of Two Avian and Two Mammalian Strains for Eight Species of Birds and Mammals. *Amer. J. Vet. Res., 27,* 397 (1966).
21. Schachter, J., Recommended Criteria for the Identification of Trachoma and Inclusion Conjunctivitis Agents. *J. Infect. Dis., 122,* 105 (1970).
22. Mason, D. M., A Capillary-Tube Agglutination Test for Detecting Antibodies Against Ornithosis in Turkey Serum. *J. Immunol., 83,* 661 (1959).
23. Collins, A. R., and Barron, A. L., Demonstration of Group- and Species-Specific Antigens of Chlamydial Agents by Gel Diffusion. *J. Infect. Dis., 121,* 1 (1970).
24. Schachter, J., Studies on *Bedsonia* Antigens and Studies on Some Human *Bedsonia* Infections. Ph.D. thesis, University of California, Berkeley (1965).

THE RICKETTSIALES

DR. PAUL FISET, DR. WILLIAM F. MYERS AND DR. CHARLES L. WISSEMAN, JR.

The order Rickettsiales comprises small rod-shaped or coccoid organisms parasitic of man or animals. They are considered Gram-negative, although the Gram stain is not the stain of choice. Some are obligate intracellular parasites, others may grow in a cell-free environment. Rickettsiales have both RNA and DNA. They exist in natural vertebrate hosts and are usually transmitted from host to host by more or less specific arthropod vectors. Most are capable of causing inapparent infections in their vertebrate and arthropod hosts.

The seventh edition of *Bergey's Manual* lists four families in the order Rickettsiales: Rickettsiaceae, Chlamydiaceae, Bartonellaceae and Anaplasmataceae. The next edition will probably classify the family Chlamydiaceae in a new, monofamilial order, the Chlamydiales, which are discussed by Dr. L. A. Page on page 131 of this volume. The major characteristics of the Rickettsiales are described in Table 1.

For reasons of convenience the present section will be divided into two parts: the family Rickettsiaceae, and the families Bartonellaceae and Anaplasmataceae.

Rickettsiaceae

The family Rickettsiaceae comprises the following important genera: *Rickettsia, Coxiella, Rochalima,* and *Ehrlichia.* All except *Ehrlichia* are potential human pathogens. All except *Rochalima* are obligate intracellular parasites. All are transmitted to their natural hosts by arthropods, although *Coxiella* is usually transmitted by aerosol. All replicate by binary fission, and all have a cell wall. Although most of the Rickettsiaceae are considered obligate intracellular parasites, they are capable of a large variety of independent metabolic activities. Except for *Ehrlichia,* all genera present certain hazards to laboratory workers and should be handled carefully. All Rickettsiaceae are susceptible to varying but clinically attainable concentrations of chloramphenicol and tetracycline, and to a lesser degree to erythromycin.

CHEMICAL COMPOSITION

Most of the chemical-composition data available were derived from studies of a limited number of members of this group, primarily *Rickettsia prowazeki, R. typhi (R. mooseri),* and *Coxiella burneti.* However, there is no reason to believe that other members of this group are different. From the data available, all Rickettsiaceae would seem to contain DNA, RNA, proteins, polysaccharides, lipids, phospholipids, and coenzymes. In addition, their cell walls are similar to those of Gram-negative bacteria and contain amino acids, muramic acid, diaminopimelic acid, hexoses, and hexosamines.

MORPHOLOGY

Rickettsiaceae vary in morphology from small rods to coccoid forms ranging in size from approximately 0.5 to 2.0 μm in length and 0.2 to 0.6 μm in width. Under certain, as yet poorly defined, conditions some may assume filamentous form.

Although very few electron microscopic studies have been carried out, the data available indicate that the ultrastructures of rickettsiae resemble that of bacteria. The rickettsial cell wall and cytoplasmic membrane appear to be trilamellar structures. In addition, an amorphous capsular substance has been demonstrated in *Rickettsia prowazeki;* there is also evidence for the presence of cytoplasmic membranous organelles. Ultrathin sections of other rickettsiae suggest the presence of ribosome-like structures and nuclear material similar to that of bacteria. In *Coxiella burneti,* typical bacterial ribosomes have been demonstrated.

TABLE 1

GENERAL CHARACTERISTICS OF RICKETTSIALES

Families and Genera	Characteristics
Rickettsiaceae	Most are human pathogens. All have a true cell wall. They replicate by binary fission, mostly within endothelial and reticuloendothelial cells. They are not associated with erythrocytes.
1. *Rickettsia*	Most show a close association with arthropod vectors, which are required for transmission to natural vertebrate hosts, including man. All are potential human pathogens. All are labile and are readily inactivated outside the natural host. All are considered obligate intracellular parasites. Several species are known.
2. *Coxiella*	Similar to the genus *Rickettsia,* except: (a) they are highly resistant to physical and chemical agents in an extracellular environment; (b) arthropod vectors are not usually involved in natural transmission to vertebrate hosts; (c) they undergo an antigenic phase variation. Only one species is known.
3. *Rochalima*	Similar to the genus *Rickettsia,* except that these organisms are not obligate intracellular parasites. Man is the only known vertebrate host. Only one species is known.
4. *Ehrlichia*	They have many of the characteristics of the genus *Rickettsia,* except that they are non-pathogenic for man. Three species are known, the most important being *E. canis,* the agent of tropical canine pancytopenia.
Bartonellaceae	Only one genus is pathogenic for man. Some lack a true cell wall. They replicate by binary fission on or inside erythrocytes.
1. *Bartonella*	Human pathogens growing in the endothelial cells of man, the only known vertebrate host. They grow in cell-free medium. The organisms have cell walls. Only one species is known.
2. *Haemobartonella*	These organisms are non-pathogenic for man. Multiple species are defined on the basis of vertebrate-host association. They are closely associated with erythrocytes, but are rarely found free in plasma. No demonstrable growth occurs in host tissue other than erythrocytes. The organisms rarely produce disease in natural hosts without splenectomy. They have not been grown in cell-free medium. *Haemobartonella* have no cell wall. There is no differentiated nuclear structure, although both RNA and DNA are reported to be present.
3. *Eperythrozoon*	Similar to *Haemobartonella.* Differentiation between these two genera is difficult and somewhat arbitrary. There are minor differences in morphology, ring forms being predominant with *Eperythrozoon.* Although associated with red cells, numerous organisms are found free in plasma.
4. *Grahamella*	Morphologically similar to *Bartonella.* They grow inside erythrocytes. Multiple species are defined on the basis of host association. Although parasitemia can be demonstrated, no clinical disease is manifested in natural vertebrate hosts, even after splenectomy. The organisms grow in cell-free medium. They possess true cell walls. *Grahamella* are not infectious to man.
Anaplasmataceae	Non-pathogenic for man. The organisms lack a true cell wall. They replicate inside erythrocytes within a vacuole (marginal body).
1. *Anaplasma*	The only genus in this family. The growth cycle is different from that of all other members of the Rickettsiales; it resembles that of the Chlamydiales. Three species are known to infect cattle and sheep. Anaplasmosis in cattle varies from asymptomatic infection to severe anemia. The organisms have not been grown in cell-free medium.

BIOLOGICAL PROPERTIES

Tables 2 to 6 present detailed descriptions of the biological properties of all Rickettsiaceae other than *Ehrlichia. Ehrlichia* was excluded from the tables because of the paucity of information available on the biological characteristics of this group of organisms. Suffice it to say that the genus *Ehrlichia* comprises three species: *E. bovis, E. ovina,* and *E. canis. Ehrlichia canis* is the only species that has been studied to any significant extent in recent years. It causes tropical canine pancytopenia. It is prevalent in Southeast Asia, Central America, the Caribbean, Florida, and Texas. The usual laboratory animals are refractory, but experimental infection is readily achieved in dogs (beagles and German shepherds), though clinical manifestations vary with the breed. The organism has not been grown in chick embryos or in tissue cultures other than canine monocytes. The dog tick, *Rhipicephalus sanguineus,* seems to be the natural vector, in which the organism undergoes transovarial and transtadial transmission. Although the natural vertebrate hosts are not really known, foxes, coyotes, and jackals are suspected.

The Rickettsiaceae are grouped on the basis of antigenic relationships.

Bartonellaceae and Anaplasmataceae

The family Bartonellaceae includes the genera *Bartonella, Haemobartonella, Eperythrozoon,* and *Grahamella;* the family *Anaplasmataceae* contains only the genus *Anaplasma.* The genus *Bartonella* is the only one of the group that infects man. The other Bartonellaceae produce infections in lower vertebrates, frequently with little or no clinical symptoms. *Anaplasma marginale* produces a severe infectious anemia of cattle, anaplasmosis, a disease of economic importance in the USA and elsewhere. A common characteristic of all these agents is their location on or within erythrocytes.

Human bartonellosis is manifested in two distinctly different clinical forms: Oroya fever, an acute hemolytic anemia, and verruga peruana, a relatively benign, chronic type of disease. The latter disease syndrome seems to reflect a partial immunity to the agent and is the usual form of bartonellosis seen in cases of reinfection. The asymptomatic carrier state is also probably an expression of a partial immunity. Latent infections seem to be a common characteristic of both human and animal bartonellosis. Persisting non-apparent infections and partial immunity seem to be a common theme for all the Bartonellaceae and Anaplasmataceae.

Only *Bartonella* and *Grahamella* have been cultured in cell-free media and possess true cell walls. Arthropod vectors are involved in the transmission of most of these agents. Virtually nothing is known about their metabolism or chemical and antigenic composition. The classification of this group rests on somewhat tenuous grounds, and the status of these organisms will undoubtedly change as more is learned about their biological properties. Tables 7 and 8 describe the known biological properties of the Bartonellaceae and Anaplasmataceae.

TABLE 2

HUMAN DISEASES AND GEOGRAPHIC DISTRIBUTION OF RICKETTSIACEAE

Species	Human Diseases	Geographic Distribution
Typhus Group: Genus *Rickettsia*		
R. prowazeki	Epidemic typhus: the only epidemic rickettsial disease; outbreaks occur under conditions of overcrowding and louse infestation. Brill-Zinsser disease: recrudescence of infections in individuals who had typhus years earlier.	Worldwide; endemic on all continents except Australia, with occasional epidemic outbreaks in Africa, Central and South America, Eastern Europe, and Asia.
R. typhi (mooseri)	Murine typhus: less severe than epidemic typhus.	Worldwide.
R. canada	Serologic evidence for a Rocky Mountain spotted feverlike disease; no isolation from man at this time.	Unknown (only one isolation reported in Ontario, Canada, from rabbit ticks).
Spotted Fever Group: Genus *Rickettsia*		
R. rickettsi	Rocky Mountain spotted fever (RMSF); severe.	Western hemisphere; the organisms are transmitted by different species of ixodid ticks in different areas.
R. conori	Boutonneuse fever, Kenya tick typhus, South African tick typhus, Indian tick typhus (may be caused by other species): less severe than RMSF.	Mediterranean area, East and South Africa, Pakistan, and India; the organisms are transmitted by different species of ixodid ticks in these different areas.
R. sibirica	Siberian or North Asian tick typhus: less severe than RMSF.	Siberia, Armenia, and Central Asian Republics of USSR, possibly also in Czechoslovakia.
R. australis	Queensland tick typhus: less severe than RMSF	Australia.
R. akari	Rickettsial pox: mild	Urban infections in USA and USSR (house mouse); sylvatic infections in Korea (field mouse).

TABLE 2 (continued)

HUMAN DISEASES AND GEOGRAPHIC DISTRIBUTION OF RICKETTSIACEAE

Species	Human Diseases	Geographic Distribution
Scrub Typhus: Genus *Rickettsia*		
R. tsutsugamushi (orientalis)	Scrub typhus: mild to severe, depending on the strain.	Soviet Far East, Japan, Chinese mainland, Pacific islands, Southeast Asia to Northern Australia, and westward to West Pakistan.
Q-Fever: Genus *Coxiella*		
C. burneti	Unrecognized infections: mostly mild. Q fever: moderately severe. Hepatitis: moderately severe. Subacute endocarditis:severe.	Worldwide.
Trench Fever: Genus *Rochalima*		
R. quintana	Trench fever, probably many unrecognized infections.	Probably worldwide; same distribution as epidemic typhus.

TABLE 3

NATURAL VERTEBRATE HOSTS AND VECTORS OF RICKETTSIACEAE

Species	Natural Vertebrate Hosts	Vectors Transmitting within the Natural Cycle	Vectors Transmitting to Man	Infection in Vectors
Typhus Group: Genus *Rickettsia*				
R. prowazeki	Man.	*Pediculus humanus humanus* (human body louse).	*Pediculus humanus humanus.*	Growth in gut epithelium; lice die 8 to 10 days after infection; organisms excreted in feces.
R. typhi (mooseri)	*Rattus rattus* and *R. norvegicus.*	*Xenopsylla cheopis* (rat flea); *Polyplax spinulosa* (rat louse)?	*Xenopsylla cheopis; Pulex irritans* (human flea)? *Pediculus humanus humanus*?	Permanent infection; fleas do not die of infection; growth in gastric epithelium; organisms excreted in feces.
R. canada	Unknown.	*Haemaphysalis leporispalustris* (rabbit tick).	Unknown.	Experimental infection of *Haemaphysalis leporispalustris* indicates transovarial and transtadial transmission.
Spotted Fever Group: Genus *Rickettsia*				
R. rickettsi	Various rodents; rabbits; dog?	Northwest USA: *Dermacentor andersoni.* Eastern USA and Canada: *Dermacentor variabilis; Haemaphysalis leporispalustris.* South and Southwest USA: *Amblyoma americanum.* Mexico and South America: *Rhipicephalus sanguineus; Amblyoma cajenense.*	Same as in the zoonotic cycle, except for *Haemaphysalis leporispalustris,* which does not bite man.	Persistent infection in ticks without overt pathology; transovarial and transtadial transmission; growth in all tissues of ticks; organisms excreted in saliva and feces.

TABLE 3 (Continued)

NATURAL VERTEBRATE HOSTS AND VECTORS OF RICKETTSIACEAE

Species	Natural Vertebrate Hosts	Vectors Transmitting within the Natural Cycle	Vectors Transmitting to Man	Infection in Vectors
Spotted Fever Group:Genus *Rickettsia* (continued)				
R. conori	Primarily dogs, but also several species of rodents.	Mediterranean area: *Rhipicephalus sanguineus.* Kenya: *Haemaphysalis leachi; Rhipicephalus simus.* South Africa: *Amblyoma hebreum; Haemaphysalis aegypticum.* India: *Ixodes ricinus.*	Same as in the zoonotic cycle.	Same as for *R. rickettsi.*
R. sibirica	Domestic animals, rodents.	Various species of ticks: *Dermacentor nuttali; D. sylvarum; D. marginatus; D. pictus; Haemaphysalis concina; H. punctata.*	Same as in the zoonotic cycle.	Same as for *R. rickettsi.*
R. australis	Several species of marsupials.	*Ixodes holocyclus* (tick of marsupials).	Same as in the zoonotic cycle.	Same as for *R. rickettsi.*
R. akari	*Mus musculus* (house mouse); *Microtus fortis pellicus* (sylvatic in Korea).	The mite *Allodermanyssus sanguineus.*	Same as in the zoonotic cycle.	Transmission by biting; transovarial transmission.
Scrub Typhus:Genus *Rickettsia*				
R. tsutsugamushi (orientalis)	Various rodents and other small mammals.	The trombiculid mites: *Leptotrombidium akamushi; L. deliense; L. pallidum; L. scutellare,* others.	Same as in the zoonotic cycle.	Distribution of the parasites in the vector is not well known; transmission by biting; transovarial and transtadial transmission.

TABLE 3 (Continued)

NATURAL VERTEBRATE HOSTS AND VECTORS OF RICKETTSIACEAE

Species	Natural Vertebrate Hosts	Vectors Transmitting within the Natural Cycle	Vectors Transmitting to Man	Infection in Vectors
Q-Fever:Genus *Coxiella*				
C. burneti	Various domestic animals: sheep, goats, cattle; various other animals: birds, rodents, marsupials.	Direct transmission by aerosol; several species of ticks may be involved in some animal cycles.	Aerosol.	In ticks transovarial and transtadial transmission; organisms are excreted in saliva and feces.
Trench Fever:Genus *Rochalima*				
R. quintana	Man.	*Pediculus humanus humanus.*	*Pediculus humanus humanus.*	Lice do not die of infection but seem to remain infected for life; growth is extracellular in the gut; organisms excreted in feces.

TABLE 4

EXPERIMENTAL INFECTIONS IN LABORATORY ANIMALS AND SEROLOGIC REACTIONS OF RICKETTSIACEAE

	Experimental Infections		Serologic Reactions	
Species	Guinea Pig	Mouse	With Rickettsial Antigens	With Proteus[a]
Typhus Group:Genus *Rickettsia*				
R. prowazeki	High susceptibility to infection; mild disease with fever; scrotal swelling (rare).	Low susceptibility to infection; acute toxic death with large inoculum.	A "soluble" antigen that is group-specific in complement fixation reaction is released by ether treatment; after ether treatment rickettsial bodies exhibit species-specific activity in complement fixation and in agglutination tests.	OX-19, OX-2
R. typhi (mooseri)	Organisms more virulent than *R. prowazeki,* usually causing scrotal swelling.	High-dose infection causes death in 3 to 8 days; acute toxic death with large inoculum.		OX-19, OX-2
R. canada	Subclinical infection shown by serologic conversion.	Subclinical infection shown by serologic conversion; acute toxic death with large inoculum.		Not determined.
Spotted Fever Group:Genus *Rickettsia*				
R. rickettsi	High susceptibility to infection; most strains are virulent, causing scrotal swelling and necrosis as well as death.	Low susceptibility to infection; acute toxic death with large inoculum.	Group-specific "soluble" antigen; species-specific rickettsia-associated antigen.	OX-19, OX-2
R. conori	Organisms less virulent than *R. rickettsi,* usually causing fever and scrotal swelling, but not scrotal necrosis and death.	Low susceptibility to infection; acute toxic death with large inoculum.		OX-19, OX-2
R. sibirica	Infection similar to that of *R. rickettsi.*	Low susceptibility to infection; acute toxic death with large inoculum.		OX-19, OX-2.

TABLE 4 (Continued)

EXPERIMENTAL INFECTIONS IN LABORATORY ANIMALS AND SEROLOGIC REACTIONS OF RICKETTSIACEAE

	Experimental Infections		Serologic Reactions	
Species	**Guinea Pig**	**Mouse**	**With Rickettsial Antigens**	**With Proteus**[a]
Spotted Fever Group:Genus *Rickettsia* (continued)				
R. australis	Infection similar to that of *R. conori.*	Infection produced by intraperitoneal route; acute toxicity has not been demonstrated.	Group-specific "soluble" antigen; species-specific rickettsia-associated antigen.	OX-19, OX-2.
R. akari	Infection similar to that of *R. conori.*	High susceptibility to infection; acute toxicity has not been demonstrated.		Negative.
Scrub Typhus:Genus *Rickettsia*				
R. tsutsugamushi (orientalis)	Except for a few strains, the organisms usually cause no apparent infection.	High susceptiblity to infection; strains vary from almost avirulent to lethal; death occurs 7 to 14 days after inoculation with lethal strains; acute toxic death has been demonstrated with only one strain.	At least three serotypes are recognized, but there probably are many more; the rickettsial surface behaves like a mosaic with predominant "type-specific" antigens; a group-specific, ether-released, water-soluble antigen has been described.	OX-K.
Q-Fever:Genus *Coxiella*				
C. burneti	Animal of choice for primary isolation; considerable variation in virulence with different strains; fever and enlargement of spleen.	High susceptibility to infection; acute toxic death has not been produced.	Only one serotype; two antigenic phases; phase I is found in nature; surface phase I polysaccharide is lost on adaptation to chick embryos → phase II.	Negative.

TABLE 4 (Continued)

EXPERIMENTAL INFECTIONS IN LABORATORY ANIMALS AND SEROLOGIC REACTIONS OF RICKETTSIACEAE

	Experimental Infections		Serologic Reactions	
Species	**Guinea Pig**	**Mouse**	**With Rickettsial Antigens**	**With Proteus**[a]
Trench Fever: Genus *Rochalima*				
R. quintana	Infection not produced; experimental infections have been produced only in man and in monkeys.	Infection not produced; acute toxic death has not been produced.	Only one recognized serotype; no ether-released "soluble" antigen.	Negative.

[a] Weil-Felix reaction (Proteus agglutination).

TABLE 5

CULTIVATION AND STABILITY OF RICKETTSIACEAE

Species	Embryonated Hen's Egg (Yolk Sac)	Cell Culture	Optimal Temperature	Stability
Typhus Group:Genus *Rickettsia*				
R. prowazeki	Generation time about 6 hours; peak titer just prior to death; rickettsial growth stops after embryo death; optimal yield about 10^9–10^{10} infectious units per egg.	Plaques (1 mm) in 7 days with primary chick embryo cell culture; other cultures employed: BS-C-1, guinea pig kidney, chick entodermal cells, *Hyalomma* tick cells, L-cells, and colubrid snake cells.	35°C	Labile; readily inactivated at 56°C; unstable at room temperature, but may survive for months in dried vector feces in a cool, dry climate; stable at −76°C and under lyophilization; labile between −5°C and −20°C; stabilized in SPG (sucrose phosphate-glutamate) medium.
R. typhi (mooseri)	Same as *R. prowazeki.*	Plaques (1 mm) in 7 days with primary chick embryo cell culture; other cultures employed: *Hyalomma* tick cells.	35°C	Same as *R. prowazeki.*
R. canada	Grows well in yolk sac after adaptation.	Plaques (0.75 mm) in 7 days with primary chick embryo cell culture.	35°C	Same as *R. prowazeki.*
Spotted Fever Group:Genus *Rickettsia*				
R. rickettsi	Embryo may die early due to toxicity; rickettsial multiplication continues about 2 days after death.	Plaques (2 mm) in 6 days with primary chick embryo cell culture; other cultures employed: 14 pf (rat fibroblast) and guinea pig scrotal tissue explant.	32°C	Labile; similar to the organisms of the typhus group, except not stabilized by SPG; some stability is conferred by glutathione and by microaerophilic conditions.
R. conori	Same as *R. rickettsi.*	Plaques (2 mm) in 5 days with primary chick embryo cell culture; other cultures employed: guinea pig kidney, pig embryo kidney, and *Hyalomma* tick cells.	32°C	Same as *R. rickettsi.*

TABLE 5 (Continued)

CULTIVATION AND STABILITY OF RICKETTSIACEAE

Species	Embryonated Hen's Egg (Yolk Sac)	Cell Culture	Optimal Temperature	Stability
Spotted Fever Group:Genus *Rickettsia* (continued)				
R. sibirica	Same as *R. rickettsi.*	Plaques (2 mm) in 5 days with primary chick embryo cell culture.	32°C	Same as *R. rickettsi.*
R. australis	Same as *R. rickettsi.*	Plaques (2 mm) in 5 days with primary chick embryo cell culture.	32°C	Same as *R. rickettsi.*
R. akari	Same as *R. rickettsi.*	Plaques (3 mm) in 5 days with primary chick embryo cell culture; other cultures employed: *Hyalomma* tick cells.	32°C	Same as *R. rickettsi.*
Scrub Typhus:Genus *Rickettsia*				
R. tsutsugamushi (orientalis)	Peak titer just prior to death; titer declines rapidly after death; poor growth and low yield.	Plaques (1 mm) in 10 to 17 days with primary chick embryo cell culture; other cultures employed: 14 pf, guinea pig kidney, pig embryo kidney, MB-III, and L-929.	35°C	Very labile; inactivated at 37°C and at room temperature in several hours; stabilized by SPG, albumin, or protein; stable at −70°C; inactivated by lyophilization.
Q-Fever:Genus *Coxiella*				
C. burneti	Slow growth rate, may require three passages; generation time about 12 hours; yield about $10^{10}-10^{11}$ infectious particles per egg.	Plaques (0.75 mm) in 16 days with primary chick embryo cell culture; other cultures employed: L-cells, guinea pig kidney, pig embryo kidney, chick fibroblast, *Antheraea* (moth) and *Aedes* (mosquito) cell lines, Detroit-6, H-Ep-2, and human amnion.	35°C	Very stable; survives for months in dried materials at room temperature; may survive heating at 60°C for 30 minutes.

TABLE 5 (Continued)

CULTIVATION AND STABILITY OF RICKETTSIACEAE

Species	Embryonated Hen's Egg (Yolk Sac)	Cell Culture	Optimal Temperature	Stability
Trench Fever: Genus *Rochalima*				
R. quintana	Poor growth; will grow in cell-free medium (blood agar with 5% carbon dioxide atmosphere).	Has been grown with some difficulty; will grow in cell-free medium.	32–34°C	Stability similar to that of *R. prowazeki;* organisms remain viable for several months in dried louse feces in a cool, dry climate.

TABLE 6

ENERGY PRODUCTION AND BIOSYNTHESIS OF RICKETTSIACEAE

Species	Energy Production	Biosynthesis
Typhus Group: Genus *Rickettsia*		
R. prowazeki	Oxygen uptake with glutamate, glutamine, pyruvate, and succinate; demonstration of dicarboxylic portion of citric acid cycle; adenosine diphosphate → adenosine triphosphate; glutamic dehydrogenase.	Glutamate-oxaloacetate transaminase; glycine-^{14}C, methionine-^{14}C, or valine-^{14}C + amino acid mixture → protein-^{14}C; acetate-^{14}C → lipid-^{14}C.
R. typhi (mooseri)	Oxygen uptake with glutamate, glutamine, pyruvate, and succinate; demonstration of dicarboxylic portion of citric acid cycle; glutamine → glutamate; asparagine → aspartic acid; glutamic dehydrogenase.	Glutamate-oxaloacetate transaminase; methionine-^{14}C + amino acid mixture → protein-^{14}C; *Chlorella* protein hydrolysate-^{14}C → protein-^{14}C.
R. canada	Unknown.	Unknown.
Spotted Fever Group: Genus *Rickettsia*		
R. rickettsi	Carbon dioxide-^{14}C from all carbon atoms of glutamate; carbon dioxide-^{14}C production reduced by addition of unlabeled citric acid cycle intermediates.	Unknown.
R. conori	Unknown.	Unknown.
R. sibirica	Unknown.	Unknown.
R. australis	Unknown.	Unknown.
R. akari	Unknown.	Unknown.
Scrub Typhus: Genus *Rickettsia*		
R. tsutsugamushi (orientalis)	Unknown.	Unknown.

TABLE 6 (continued)

ENERGY PRODUCTION AND BIOSYNTHESIS OF RICKETTSIACEAE

Species	Energy Production	Biosynthesis
Q-Fever: Genus *Coxiella*		
C. burneti	Oxygen uptake with pyruvate, glutamate, succinate, α-ketoglutarate, oxaloacetate, fumarate, malate, and serine; glutamate, malate, and isocitrate dehydrogenase; citrate synthase; hexokinase; glucose-6-phosphate dehydrogenase, adenosine diphosphatase, adenosine triphosphatase.	Glycine + formaldehyde → serine; ornithine + carbamyl phosphate → citrulline; aspartate + carbamyl phosphate → ureidosuccinate; glutamate-oxaloacetate transaminase; DNA-dependent RNA polymerase; leucine-^{14}C or phenylalanine-^{14}C + amino acid mixture → protein-^{14}C; algal hydrolysate-^{14}C → protein-^{14}C.
Trench Fever: Genus *Rochalima*		
R. quintana	Oxygen uptake with succinate, glutamine, glutamate, α-ketoglutarate, pyruvate; demonstration of dicarboxylic portion of citric acid cycle.	Glutamate-oxaloacetate transaminase.

TABLE 7

NATURAL INFECTION AND HOST RANGE, MODE OF TRANSMISSION, AND GEOGRAPHIC DISTRIBUTION OF BARTONELLACEAE AND ANAPLASMATACEAE

Species	Natural Infection and Host Range	Mode of Transmission	Geographic Distribution
Bartonellaceae			
Bartonella			
bacilliformis[a]	Man is the only known host. Clinical manifestations: acute hemolytic anemia (Oroya fever), and subacute benign wart-like skin eruptions (verruga peruana). In both diseases the parasite is found in reticuloendothelial cells. In Oroya fever, the parasite is found on the surface of red blood cells; at the peak of the disease, 90% of the red blood cells are parasitized. If untreated, Oroya fever has a mortality of about 40%. In endemic areas, asymptomatic carriers represent 5 to 10% of the population.	Known vectors are *Phlebotomus verrucarum* and *P. columbianus* (sand flies). The geographic distribution of the vectors agrees with that of the diseases. Sand flies feed at night. The status of the parasite in the vectors is unknown.	The mountainous regions of Peru, Ecuador, and Columbia at altitudes from 2,500 to 8,000 feet, but not elsewhere.
Haemobartonella			
muris[b,c]	Asymptomatic infection in apparently normal rats. Acute anemia develops after splenectomy, with the appearance of large numbers of the parasites on erythrocytes; this may result in death. Surviving animals are resistant to reinfection.	Transmission by *Polyplax spinulosa* (rat louse). Infection is induced by biting; feces are apparently non-infective.	Not determined, but probably the same as that of rats.
felis	Infectious anemia of cats occurs without splenectomy, loss of weight, weakness, and dyspnea. In untreated cases the mortality is high.	Transmission probably by cat bite. No vectors have been incriminated.	Probably worldwide.
Eperythrozoon			
coccoides[b]	Asymptomatic infection in apparently normal mice splenectomy in infected mice leads to moderate anemia. Several species of wild rats and mice are susceptible, but the common laboratory mouse is most susceptible.	*Polyplax serrata* (mouse louse) is the natural vector.	Worldwide.

TABLE 7 (continued)

NATURAL INFECTION AND HOST RANGE, MODE OF TRANSMISSION, AND GEOGRAPHIC DISTRIBUTION OF BARTONELLACEAE AND ANAPLASMATACEAE

Species	Natural Infection and Host Range	Mode of Transmission	Geographic Distribution
Bartonellaceae (continued)			
Grahamella			
talpae[b,c]	Infection without clinical symptoms in a variety of animals, including shrews, voles, mice, desert rats, gerbils, hamsters, and squirrels. Splenectomy has virtually no effect.	Transmission by fleas; contamination of the bite by flea feces causes the infection.	Worldwide.
Anaplasmataceae			
Anaplasma			
marginale[d]	Anaplasmosis in cattle varies from nônapparent infection to severe anemia; it is more severe in older animals. In deer the infection is widespread and asymptomatic. Dear seem to be a natural source of infection for cattle and are probably important in maintaining the parasite in nature.	Many species of ticks have been experimentally infected, but *Dermacentor andersoni* and *D. occidentalis* seem to be the most important vectors in the USA. Both transtadial and transovarial transmission occur. Horse flies seem to be important insect vectors, but transmission is strictly mechanical. Mosquitoes may be involved.	Probably worldwide.

[a] The only species in the genus *Bartonella*.
[b] Prototype species; many species have been described, based on host relationships.
[c] Definite speciation is not well established.
[d] Prototype species; other species, of lesser importance, are *A. centrale* (cattle) and *A. ovis* (sheep).

TABLE 8

EXPERIMENTAL INFECTIONS, CULTIVATION, AND MORPHOLOGY OF BARTONELLACEAE AND ANAPLASMATACEAE

Species	Experimental Infections	Cultivation	Morphology
Bartonellaceae			
Bartonella			
bacilliformis	The experimental host range is limited to primates. Clinical manifestations are localized verrugas following intradermal or subcataneous inoculation. Experimental Oroya fever has been produced in man only.	Growth becomes visible in semi-solid agar enriched with serum and hemoglobin in 10 days at 28°C. The organisms are obligate aerobes. No hemolysin has been demonstrated. In tissue cultures, growth is intra- and extracellular. In chick embryos, growth occurs in chorioallantoic fluid and in the yolk sac, with no erythrocyte association.	The parasites are found on the surface of red cells, with as many as ten organisms per cell at the peak of Oroya fever. They are extremely pleomorphic and may exist as spheres, rods, and ring-like bodies, 0.3–3.0 x 0.2–0.5 μm. In culture, there is a predominance of rods in the early growth phase. Cell walls and flagella have been demonstrated, but the latter are found in culture only.
Haemobartonella			
muris	Experimental infection can only be demonstrated in splenectomized rats.	Several reports of growth in cell-free media and in chick embryos have been confirmed.	Coccoid bodies, 0.3–0.5 μm are found in blood; under light microscopy, chains of cocci may resemble rods. The parasites are found attached to or within the red blood cells. No cell wall has been demonstrated; there is a single limiting membrane. The organisms appear to divide by binary fission.
felis	The organisms are transmitted only to cats. Recovery from the disease leads to a carrier state.	Reports of growth in cell-free media are conflicting.	Ring forms, 0.5–1.0 μm. Four to twelve parasites, sometimes in chains, are found on the surface of red cells. No cell wall has been observed.
Eperythrozoon			
coccoides	Transmission to intact mice leads to subclinical infection and a carrier state. Infection of splenectomized mice leads to moderate anemia and parasitemia. Concomitant infection with mouse hepatitis virus leads to fatal hepatitis.	No growth has been reported in cell-free media. Adaptation is required for growth on chick embryos.	The organisms have a single limiting membrane; no cell wall has been demonstrated. They are ring forms with clear centers, 0.5 μm in diameter; some are rods, 0.6–1.2 μm in length.

TABLE 8 (continued)

EXPERIMENTAL INFECTIONS, CULTIVATION, AND MORPHOLOGY OF BARTONELLACEAE AND ANAPLASMATACEAE

Species	Experimental Infections	Cultivation	Morphology
Bartonellaceae (continued)			
Grahamella			
talpae	Persistent parasitemia may be produced in a large variety of animals without overt clinical manifestation.	The organisms will grow in semi-solid medium containing blood or hemoglobin. They are obligate aerobes. Growth occurs in about 10 days at temperatures between 20°C and 37°C. In rat embryo cell cultures, growth occurs both intra- and extracellularly.	In blood smears, the parasites appear as bacilliform bodies or, more rarely, as coccoids within the erythrocytes; the rods measure 0.5–1.0 x 0.2 μm. The organisms are non-motile and appear to divide by binary fission. They possess true cell walls.
Anaplasmataceae			
Anaplasma			
marginale	The usual laboratory animals are refractory to infection. A large variety of ungulates are susceptible to experimental infection with infected blood, washed erythrocytes, or organ suspensions administered by inoculation.	No growth has been demonstrated in a wide variety of cell-free media, in cell cultures, or in chick embryos.	The organisms grow within the erythrocyte. Dense spherical bodies, known as marginal bodies (0.3–1.0 μm in diameter) and possessing a limiting membrane, contain subunits (initial bodies). The latter are 0.3–0.4 μm and are enclosed in a double membrane and, possibly, an outer envelope-like membrane. Dense aggregates of fine granular material are seen within the initial bodies. The initial bodies are the infectious units. They divide by binary fission.

REFERENCES

1. Huxsoll, D. L., Hildebrandt, P. K., Nims, R. M., and Walker, J. S., Tropical Canine Pancytopenia, in *Current Veterinary Therapy,* Vol. 4, pp. 677–679, R. W. Kirk, Ed. W. B. Saunders Co., Philadelphia, Pennsylvania (1971).
2. Kobayashi, Y., Nagai, K., and Tachibana, N., Purification of Complement-Fixing Antigens of *Rickettsia orientalis* by Ether Extraction, *Amer. J. Trop. Med. Hyg., 18,* 942 (1969).
3. McDade, J. E., Stakebake, J. R., and Gerone, P. J., Plaque Assay System for Several Species of Rickettsiae, *J. Bacteriol., 99,* 910 (1969).
4. Ormsbee, R. A., Q-Fever Rickettsia, in *Viral and Rickettsial Infections of Man,* 4th ed., pp. 1144–1160, F. L. Horsfall and I. Tamm, Eds. J. B. Lippincott Co., Philadelphia, Pennsylvania (1965).
5. Ormsbee, R. A., Rickettsiae (As Organisms), *Annu. Rev. Microbiol., 23,* 275 (1969).
6. Paretsky, D., Biochemistry of Rickettsiae and Their Infected Hosts, with Special Reference to *Coxiella burneti, Zentralbl. Bakteriol. Parasitenk. Infektionskr. Hyg. Abt. Orig., 206,* 283 (1968).
7. Peters, D., and Wigand, R., Bartonellaceae, *Bacteriol. Rev., 19,* 150 (1955).
8. Philip, C. B., The Rickettsiales, in *Bergey's Manual of Determinative Bacteriology,* 7th ed. Williams and Wilkins Co., Baltimore, Md. (1957).
9. Smadel, J. E., and Elisberg, B. L., Scrub Typhus Rickettsiae, in *Viral and Rickettsial Infections of Man,* 4th ed., pp. 1130–1143, F. L. Horsfall and I. Tamm, Eds. J. B. Lippincott Co., Philadelphia, Pennsylvania (1965).
10. Snyder, J. C., Typhus Fever Rickettsiae, in *Viral and Rickettsial Infections of Man,* 4th ed., pp. 1059–1094, F. L. Horsfall and I. Tamm, Eds. J. B. Lippincott Co., Philadelphia, Pennsylvania (1965).
11. Tanaka, H., Hall, W. T., Sheffield, J. B., and Moore, D. H., Fine Structure of *Haemobartonella muris* as Compared with *Eperythrozoon coccoides* and *Mycoplasma pulmonis, J. Bacteriol., 90,* 1735 (1965).
12. Weinman, D., and Ristic, M., *Infectious Blood Diseases of Man and Animals.* Academic Press, New York (1968).
13. Weiss, E., Comparative Metabolism of Rickettsiae and Other Host-Dependent Bacteria, *Zentralbl. Infektionskr. Bakteriol. Parasitenk. Hyg. Abt. Orig., 206,* 292 (1968).
14. Weiss, E., and Moulder, J. W., Taxonomy of the Rickettsiae, in *Bergey's Manual of Determinative Bacteriology,* 8th ed. Williams and Wilkins Co., Baltimore, Md. (1973).
15. Wike, D. A., Tallent, G., Peacock, M. G., and Ormsbee, R. A., Studies of the Rickettsial Plaque Assay Technique, *Infect. Immun., 5,* 715 (1972).
16. Wisseman, C. L., Jr., Some Biological Properties of Rickettsiae Pathogenic for Man, *Zentralbl. Bakteriol. Parasitenk. Infektionskr. Hyg. Abt. Orig., 206,* 299 (1968).
17. Woodward, T. E., and Jackson, E. B., Spotted Fever Rickettsiae, in *Viral and Rickettsial Infections of Man,* 4th ed., pp. 1095–1129. J. B. Lippincott Co., Philadelphia, Pennsylvania (1965).
18. Zdrodovskii, P. F., and Golinevich, H. M., *The Rickettsial Diseases,* p. 629. Pergamon Press, New York (1960).

THE SPIROCHAETALES

DR. ROBERT M. SMIBERT

Members of the order Spirochaetales are slender, flexuous, unicellular, helically coiled organisms 5 to 500 μm long and 0.1 to 3 μm wide; they have one or more complete turns in the helix. The organisms are Gram-negative, but are best observed by dark-field microscopy or phase-contrast microscopy. They are motile, with a rapid whirling about the long axis of the cell, flection, and movement in a corkscrew or serpentine fashion. Spirochetes consist of an outer cell envelope and an inner protoplasmic cylinder. Between the cell envelope and the protoplasmic cylinder are axial fibrils that are inserted into the cylinder wall at each end of the cell and extend along the cell towards the opposite end of the cell. Bizarre forms of spirochetes may be seen. Bullae, for example, are swellings of the cell envelope. In old cultures, spirochetes may coil up, forming spheres or coccoid forms that may break up into granules. The coccoid forms and granules are also found in cultures treated with chemicals such as penicillin.[1] Spirochetes are both aerobic and anaerobic. They are found free-living or saprophitic in nature or are parasitic. Some species are pathogenic for man and animals.

The genera of Spirochaetales may be divided into two groups on the basis of their relationship to oxygen. Anaerobic genera include *Spirochaeta, Borrelia, Treponema,* and *Cristispira;* the obligate aerobic genus is *Leptospira.*

Spirochaeta are found free-living in nature. Some are anaerobes, others are facultative anaerobes. They are not very strict anaerobes.

Borrelia cause relapsing fever in man and are transmitted by lice and ticks. They are not very strict anaerobes.

Treponema are found in the intestines, oral cavity, and genital tract of man and animals. Some are pathogens. They are anaerobes and range from strict anaerobes to those that will tolerate some oxygen.

Cristispira are found in the intestinal tract of molluscs. They are the largest of the spirochetes and have a characteristic lateral ridge ("crista") made by a large bundle – fifty to several hundred – of axial fibrils. They have not been cultured.

Leptospira are thin, tightly coiled spirochetes that usually have one or both ends hooked. Some are pathogenic to man and animals, others are saprophytes. Leptospires are the only spirochetes that are obligate aerobes.

MORPHOLOGY AND ULTRASTRUCTURE

The different genera of spirochetes have a similarity that makes it desirable to consider their morphology from a comparative viewpoint.[2-21] All spirochetes are helically coiled. There is a note in the literature, however, stating that *Treponema pallidum* may be a flat wave twisted into one to five different planes.[22] Spirochetes have an outer cell envelope – usually made up of three layers – that is from 80 to 140 Å thick, a space between the envelope and the protoplasmic cylinder, a protoplasmic cylinder made up of three layers about 80 to 100 Å thick, and an axial fibril (filament) system inserted by terminal bulbs into each end of the protoplasmic cylinder. The axial fibril system runs along the cell from one end to the other in the space in between the inner surface of the cell envelope and the outer surface of the protoplasmic cylinder. Spirochetes do differ in the structure and number of fibrils in the system. The axial fibril in the leptospires has also been called an axistyle, and the diameter of the structure appears to be wider than that of fibrils of other spirochetes. A cross section of leptospira axial filaments viewed under electron microscopy appears to have a fiber bundle containing twelve to fifteen individual fibers;[2] the axial fibril is encased in a sheath. *Spirochaeta, Treponema, Borrelia,* and *Cristispira* have one or more loose axial fibrils without an outer covering that binds them all together.

Spirochaeta

Spirochaeta are 5 to 500 μm long and 0.2 to 0.75 μm wide. They are motile, free-living in nature, and are usually found in H_2S-containing fresh- or sea-water mud as well as in sewage and polluted waters. Those species cultivated are either strict or facultative anaerobes and do not require animal sera in the media. The guanidine-plus-cytosine (GC) content of their DNA ranges from 56 to 67%. Two axial fibrils are usually found in the known species, one inserted at each end of the protoplasmic cylinder.[23] So far, five species have been recognized in the genus.

Spirochaeta plicatilis is the type species of the genus. It is 0.5 to 0.75 μm wide and 100 to 200 μm long and has not been cultivated.

Spirochaeta stenostrepa is 0.2 to 0.3 μm wide and 15 to 45 μm long. Colonies, 2 mm in diameter, are subsurface, white, spherical and fluffy.[23,24] The organism is strictly anaerobic, catalase negative, and grows at temperatures from 15 to 40°C; optimal temperature is between 35 and 37°C; optimal pH is between 7.0 and 7.5. Carbohydrates are fermented by the Embden-Meyerhof pathway and act as energy sources; amino acids are not fermented. An unidentified growth factor is present in yeast extract. Biotin, riboflavin and Vitamin B_{12} are either required or stimulatory; carbon dioxide is not required for growth. Generation time is six hours. The HL antigen, which is genus-specific for leptospires, is not present. The GC content of the DNA is 60.2%.

Spirochaeta zuelzerae is 0.2 to 0.35 μm wide and 8 to 16 μm long.[23,25] Colonies are subsurface, white, fluffy, and spherical; they have a tendency to spread in the agar medium. The organism is strictly anaerobic, catalase-negative, and grows at 20°C but not at 45°C. Optimal temperature is between 37 and 40°C; optimal pH is between 7 and 8. Carbon dioxide is required for growth. A fermentable carbohydrate is needed as an energy source. Cells contain a protein antigen that reacts with syphilitic serum. The GC content of the DNA is 56.1%.

Spirochaeta aurantia is 0.3 μm wide and 5 to 35 μm long.[23,26] Colonies measuring 1 mm in diameter are subsurface, white, and fluffy when grown anaerobically. When grown aerobically, colonies ranging from 2 to 4 mm in diameter are partially subsurface, colored yellow to orange, and round with slightly irregular edges. The pigment is a carotenoid; two fractions have been tentatively identified as *trans*-lycopene and as an isomer of lycopene. The organism is a facultative anaerobe, catalase-positive, reduces nitrate to nitrite, and grows at 15°C but not at 37°C. Optimal temperature is 30°C; optimal pH is between 7.0 and 7.3. The organism uses carbohydrates, but not amino acids, as energy sources. There is no growth without a fermentable carbohydrate in the medium. Thiamin and biotin are required; adenine, guanine, and uracil are not required. Generation time is 3.8 hours. Leptospiral HL antigen is not present. The GC content of the DNA is 66.8%.

Spirochaeta litoralis is 0.4 to 0.5 μm wide and 5.5 to 7 μm long.[27] Colonies, 1 to 5 mm in diameter, are subsurface, cream-colored, spherical, and fluffy. The organism is obligately anaerobic and obligately marine. Optimal concentration of NaCl is between 0.2 and 0.3*M*. Optimal temperature is 30°C; optimal pH is between 7 and 7.5. Fermentable carbohydrates are required as an energy source. A rubredoxin was found, but no ferredoxin. The GC content of the DNA is 50.6%.

Carbohydrates fermented by species of *Spirochaeta* are shown in Table 1. End products of glucose fermentation are presented in Table 2. Enzymes found in cell extracts of *S. stenostrepa* are in Table 3.

CHEMICAL COMPOSITION AND NUTRITION

Chloroform–methanol-extractable lipids of *Spirochaeta zuelzerae* consist of 9% neutral lipids, 37% glycolipids, and 54% phospholipids.[28] Glycolipids consist of 27% glucosyldiglyceride and 7% lysoglucosyldiglyceride. The phospholipids consist of 31% phosphatidylglycerol, 16% cardiolipin, and 5% lysocardiolipin. An unidentified glycolipid, 3%, and an unidentified phospholipid, 2%, were found. The neutral lipids contained 7% fatty aldehydes and 1% fatty acids; the remaining 1% was an unidentified fraction. Of the fatty acids of *S. zuelzerae,* 30% were straight-chain saturated fatty acids, 47% were branched-chain fatty acids, and 23% were monounsaturated acids. The saturated and unsaturated straight-chain acids were $C_{12}{:}0$, $C_{14}{:}0$, $C_{14}{:}1\Delta^5$, $C_{14}{:}1\Delta^7$, $C_{15}{:}0$, $C_{16}{:}0$, $C_{16}{:}1\Delta^7$, $C_{16}{:}1\Delta^9$, $C_{18}{:}1\Delta^9$,

TABLE 1

CARBOHYDRATES FERMENTED BY SPECIES OF *SPIROCHAETA*[23-27]

Carbohydrate	*S. stenostrepa*	*S. aurantia*	*S. zuelzerae*	*S. litoralis*
Glucose	+	+	+	+
Galactose	+	+	+	+
Mannose	+	+	+	+
Fructose	+	+	–	+
Sucrose	+	+	–	+
Lactose	+	+	–	+
Maltose	+	+	+	+
Cellobiose	+	+	+	+
Ribose	+	–		–
Arabinose	+	+	+	+
Xylose	+	+	+	+
Trehalose		+	+	+
Starch			+	
Dextrin		+		
Rhamnose		+	–	+
Raffinose		–	–	+
Inulin		+	–	+
Mannitol		+	–	–
Sorbitol		–	–	–
Sorbose		–	–	–
Glycerol		+		–
Dulcitol		–		–
Fucose				+
Lactate		–		+
Pyruvate		–		+
Acetate		–		

Code

+ = sugar fermented
– = sugar not fermented
blank space = no information available

Note

Ethanol, allantoin, uric acid, succinate, orotic acid, fumarate, and α-ketoglutarate are not attacked by *S. aurantia* and *S. litoralis.*

and $C_{18}:1\Delta^{11}$. Branched-chain acids were iso-C_{13}:0, iso-C_{14}:0, iso-C_{15}:0, and iso-C_{16}:0. Fatty aldehydes present were similar to the fatty acid spectrum. Slightly different results on fatty acids present in *S. zuelzerae* have been reported in Reference 29.

S. zuelzerae grows in a fatty-acid-free medium. It can synthesize all its fatty acids, fatty aldehydes and lipids *de novo* from glucose or acetate.[28] Monoenoid fatty acids are synthesized by the anaerobic pathway, in which β-hydroxy fatty acylthioesters of medium chain length will undergo β,γ-dehydration to yield Δ^3 monoenoic acylthioesters that are then chain-elongated.[28,30]

TABLE 2

END PRODUCTS OF GLUCOSE FERMENTATION BY SPECIES OF *SPIROCHAETA*[23–27]

Product	*S. stenostrepa*	*S. aurantia*	*S. litoralis*	*S. zuelzerae*	Strain Z4
Acetate	20.4	69.2	37.5	+	94.8
Lactate	8.2	1.0	6.5	+	56.8
Formate	–	5.2	2.8	–	10.7
Pyruvate	–	–	0.3	–	–
Succinate	–	–	–	+	26.3
Acetoin	–	tr	–	–	–
Diacetyl	–	tr	–	–	–
Ethanol	146.2	151.0	109.5	–	10.5
CO_2	187.5	165.3	127.5	+	72.7
H_2	27.2	107.7	74.10	+	186.9

Code

The values given represent μmoles of product per 100 μg of glucose
tr = product present in trace amounts
+ = product present
– = product not found

Note

Strain Z4 is similar to *S. stenostrepa*, but not identical.

TABLE 3

ENZYMES FOUND IN CELL EXTRACTS OF *SPIROCHAETA STENOSTREPA*[24]

Enzyme		Enzyme	
Hexokinase	+	Pyruvate kinase	+
Glucosephosphate kinase	+	Phosphogluconate dehydrogenase	+
Phosphofructokinase	+	Glucose-6-phosphate dehydrogenase	–
Fructosediphosphate aldoase	+	Gluconokinase	–
Glyceraldehydephosphate dehydrogenase	+	Glycerol kinase	–
Triosephosphate isomerase	+	Glycerolphosphate dehydrogenase	–
Phosphoglyceromutase	+	Glyceroldehydrogenase	–
Phosphopyruvate hydratase	+		

Code

+ = enzyme present
– = enzyme not present

Borrelia

Borrelia are spirochetes that are 5 to 20 μm long and 0.2 to 0.5 μm wide. They are motile and have twelve to fifteen axial fibrils at each end of the cell. There are three to ten spirals to the cells, the average being five to seven spirals. The organisms are loosely coiled; the amplitude of their spiral is about 1 μm. *Borrelia* cause relapsing fever in man and a similar condition in animals; they are transmitted by lice or ticks.

The taxonomy and classification of *Borrelia* is poor and confusing. Until they are cultivated *in vitro* and systematically studied, the classification will remain in its present state. Classification of *Borrelia* is now accomplished mainly by naming the species of *Borrelia* after the species of host vector. Species of *Borrelia* and their vectors are in Table 4. Type species of the genus is *B. anserina.*

TABLE 4

CLASSIFICATION OF *BORRELIA*

Organism[a]	Vector	Reservoir	Geographical Distribution
Louse-Borne *Borrelia*			
B. recurrentis	*Pediculus humanus*		
(obermeieri)			Cosmopolitan
[berbera]			North Africa
,*[carteri]*			India
[novyi]			U.S.A.
[kocki]			North America
[aegyptica]			
Tick-Borne *Borrelia*			
B. hispanica [subsp. *maroccana*] [subsp. *mansouria*]	*Ornithodoros erraticus erraticus*		North Africa, Spain, Portugal
B. crocidurae group	*Ornithodoros erraticus sonrae*	Rodents	Central Africa, Turkey, Middle East
B. microti	*Ornithodoros erraticus sonrae*	Rodents	Middle East, Iran
B. merionesi	*Ornithodoros erraticus sonrae*	Rodents	West Africa
B. dipodilli	*Ornithodoros erraticus sonrae*	Gerbils	East Africa, Kenya
B. duttonii	*Ornithodoros moubata*		Tropical East and South Africa, Madagascar, Senegal, Arabia
	saviguyi		
B. granigeri	*Ornithodoros granigeri*		Kenya

TABLE 4 (Continued)

CLASSIFICATION OF *BORRELIA*

Organism[a]	Vector	Reservoir	Geographical Distribution
Tick-Borne *Borrelia* (continued)			
B. persica	*Ornithodoros*		
	tholozani	Rodents	Eastern Mediterranean
[uzbekistana]	*papillipis*	Rodents	Arabian peninsula, Iran
[sogdiana]	*crossi*	Rodents	Central Asia
[babylonensis]	*asperns*	Rodents	
B. caucasica	*Ornithodoros verrucosus*	Rodents	Caucasus
B. latyschewii	*Ornithodoros tartakouskyi*	Rodents, dogs	Caucasus
B. venezuelensis (neotropicalis)	*Ornithodoros rudis (venezuelensis)*	Rodents	Central and South America
B. turkmenica	*Ornithodoros cholodkouskyi*		
B. mazzotti	*Ornithodoros talaje*	Rodents	Central and South America, Texas
B. parkerii	*Ornithodoros parkeri*	Rodents	Western U.S.A., Canada
B. turicatae	*Ornithodoros turicata*	Rodents	Central and South America, Western U.S.A., Texas, Canada
B. hermsii	*Ornithodoros hermsi*	Rodents	Western U.S.A.
B. brasiliensis	*Ornithodoros brasiliensis*	Rodents	South America
B. dugesi	*Ornithodoros dugesi*	Rodents	Central America, Mexico
Unnamed species	*Ornithodoros talaje*		Central and South America, Western U.S.A., Canada
Tick-Borne Animal-Pathogenic *Borrelia*			
B. theileri	*Rhipicephalus evertsi*		
B. anserina	*Argas*		
	miniatus		
	persicus		
	reflexus		

[a] Names in parentheses are synonyms; names in brackets are possible subspecies or synonyms.

Note:

Blank spaces: unknown, or information not available.

CULTIVATION

Until recently, *Borrelia* species have not been cultivated *in vitro*. Most culture media used by early investigators maintained the organisms, but did not allow unlimited subculturing. In 1971, Kelly[31] reported the cultivation of *B. hermsi* in artificial medium. *Borrelia parkerii* and *B. turicatae* were also cultivated by Kelly. These results were confirmed in the author's laboratory.[32] Table 5 shows the composition of the orginal medium as well as modifications used in the author's laboratory. In preparing the media, certain precautions are essential to obtain media that are not toxic to the organisms. Glassware must be cleaned by methods satisfactory for tissue culture. The media must be filter-sterilized by pressure filtration, using nitrogen, and filtered only once, using a membrane filter. Other methods of filtration make the media incapable of supporting growth. All ingredients should be made with double-distilled water or with distilled water passed through an ion-exchange resin.

TABLE 5

CULTURE MEDIA FOR *BORRELIA*[31,32]

Constituents	Kelly	Smibert-1	Smibert-2
Basal Medium[a]			
$Na_2HPO_4 \cdot 7H_2O$	26.52	–	–
$NaH_2PO_4 \cdot H_2O$	1.03	1.03	1.03
Na_2HPO_4	–	12.96	12.96
NaCl	1.20	1.20	1.20
KCl	0.85	0.85	0.85
$MgCl_2 \cdot 6H_2O$	0.68	0.68	0.68
Glucose	12.75	12.75	12.75
Bacto-proteose peptone No.2	5.95	5.95	–
Peptone M (Pfizer)	–	–	20
Bacto-tryptone	2.55	2.55	–
Sodium pyruvate	1.06	1.06	1.06
Sodium citrate dihydrate	0.47	0.47	0.47
N-Acetyl-D-glucosamine	0.53	0.53	0.60
Asparagine	–	2.5	2.5
Glutamine	–	2.5	2.5
Distilled water	1 liter	1 liter	1 liter
Culture Medium[b]			
Basal medium, stored at –20°C	80	80	80
Sodium bicarbonate solution	4	4	4
Bovine serum albumin (fraction V) solution	34	34	34
Distilled water	2	2	2
Rabbit serum, non-hemolyzed	10	10	10
Gelatin solution, autoclaved	40	40	40
Phenol red	1.7 mg	1.7 mg	1.7 mg

[a] Amounts given represent grams per liter unless otherwise specified.

[b] Amounts given represent milliliters unless otherwise specified.

TABLE 6

LABORATORY ANIMALS CAPABLE OF BEING INFECTED BY *BORRELIA*[33-37]

Organism	Mice	Rats	Guinea Pigs
B. recurrentis	Young only; disease mild; blood positive for 3 to 5 days	Young only; blood positive for 1 week	Some young; adults not infected
B. hispanica	Young only; disease mild; blood positive for 2 to 5 days	Young only; disease mild; blood positive for 1 to 2 weeks	Disease severe
B. crocidurae group	Young only; disease of long duration	Young only; disease of long duration	Usually not infected
B. duttonii	Old only; disease of long duration	Old only; disease of long duration	Young only; disease of long duration
B. persica	Long incubation required; disease mild	Long incubation required; disease mild	Disease severe
B. latyshevyi	Sick; only few cells in blood	Not infected	Not infected
B. venezuelensis	Disease mild; blood positive for 1 to 2 weeks	Disease mild; blood positive for 1 to 2 weeks	Disease mild
B. mazzottii	Disease mild; blood positive for 1 to 2 weeks	Disease mild; blood positive for 1 to 2 weeks	Not infected
B. turicatae	Disease mild; few cells in blood	Disease mild; few cells in blood	Young only
B. parkerii	Disease mild; few cells in blood	Disease mild; few cells in blood	Young only
B. hermsi	Disease mild	Disease mild	Disease mild
B. brasiliensis	Disease mild; blood positive for 1 to 2 weeks	Disease mild; blood positive for 1 to 2 weeks	Disease mild; blood positive for 1 to 2 weeks
B. anserina	Not infected	Not infected	Not infected
Agent 277F	Not infected	Not infected	Not infected

Instructions for preparing the culture medium described in Table 5 are given below. Double-distilled water is used for all solutions. Glassware must be cleaned to meet the requirements for tissue cultures. Serum does not have to be inactivated. The final pH of the medium should be between 7.6 and 7.8.

1. Prepare the sodium bicarbonate solution by dissolving 4.5 g of $NaHCO_2$ in 95.5 ml of distilled water. Make the solution fresh each time medium is made.

2. Prepare the bovine albumin solution by adding 10 g of bovine serum albumin (fraction V) to 90 ml of distilled water. Adjust the pH to 7.8.

3. Prepare the gelatin solution by dissolving 7 g of Bacto-gelatin in 93 ml of distilled water. Autoclave the solution.

4. With the exception of the gelatin solution, mix all the ingredients required for the culture medium. Sterilize the mixture by pressure filtration through a 0.22 μ membrane filter. Do not use vacuum filtration.

5. Add the gelatin solution. Fill 8.5-ml aliquots of the medium into 13 x 100 mm screw-cap tubes, closing the caps tightly.

6. When inoculating medium, leave 0.5 ml of air space, closing the cap tightly.

A spirochete called Agent 277F, isolated from *Haemaohysulis temporispalustris,* was cultivated in medium 10, which consisted of pooled chorioallantoic and amniotic fluids from twelve-day-old chicken embryos, fortified with 0.2% casamino acids, and crystallized bovine plasma albumin.[33] The medium was adjusted to pH 6.5 and filter-sterilized. Red blood cells (5% final concentration) from thirteen-day-old chicken embryos were added to the medium. The red cells, however, were found not to be essential for growth.

Laboratory animals that can be infected by different species of *Borrelia* are listed in Table 6. Mice are susceptible to *Borrelia* infecting man; young mice are usually more susceptible than old mice. Monkeys are susceptible to most *Borrelia.* Young rabbits may be infected by some species, but are more useful for antibody production. Embryonating chicken eggs have been shown to cultivate several species of *Borrelia,* such as *B. recurrentis, B. duttonii, B. anserina,* and Agent 277F.

The wild-animal reservoir of *B. hermsi* has been studied, and it was shown that *B. hermsi* can infect laboratory mice, rats, guinea pigs, and monkeys as well as rodents found in nature.[38] Table 7 offers additional information regarding this and other species of *Borrelia.*

Borrelia anserina infects ducks, turkeys, chickens, and geese. Turkey poults, one to three weeks old, inoculated intramuscularly or subcutaneously had spirochetes in their blood in 24 to 48 hours.[39] Young turkeys, ten to sixteen weeks old, rarely had spirochetes in their blood before 48 hours. Adult turkeys, seven months old, had spirochetes in their blood in 48 hours; the organisms could be found there for four days. Spirochaetemia will usually last only about three to four days in birds of any age. Six-month-old chickens and two-year-old hens infected with *B. anserina* have a spirochaetemia that usually clears up in five to six days with the disappearance of the organism from blood and tissues.[40,41] In general, spirochetes appear in the blood of chickens 24 to 48 hours after inoculation, depending on their age and on the amount of inoculum; their numbers increase rapidly, so that many cells are present in the blood by 72 to 120 hours. The spirochetes decline rapidly in number after 120 hours and disappear.

Vaccines against human *Borrelia* have been studied. They protected mice only against the homologous strain. *Borrelia anserina* vaccines protected chickens against challenge and were 98.5% effective.[41]

TABLE 7

WILD ANIMALS INFECTED BY *BORRELIA* EITHER NATURALLY OR IN THE LABORATORY[35-38]

Animal	*B. recurrentis*	*B. duttonii*	*B. hispanica*	*B. crocidurae* Group	*B. persica*	*B. turicatae*	*B. hermsi*
Gerbil	+			+			
Dog		+	+	+		+	
Horse		+		+			
Goat		+		+			
Sheep		+		+	+		
Hedgehog		–	+	+	+		
Rat			+				
Jackal			+				
Fox			+			+	
Bat			+	+	±		
Weasel			+				
Porcupine			+				
Wild mouse			+	+	+		
Pig			–			+	
Donkey			–				
Cat			–			+	
Shrew				+			
Hamster				+			
Bandicoot				+			
Vole							+
Flying squirrel							–
Cotton rat						+	
Squirrel	+						
Wild rat					+		
Pine squirrel							+
Chipmunk							+
Ground squirrel							–
Wood rat							–
Deer mouse							–

Code

+ = animal infected either naturally or in the laboratory; – = animal not infected; blank space = no information

METABOLISM

Biochemical activities of *Borrelia* are not well understood. Only a few reports are available. The studies have been hampered by the inability to cultivate *Borrelia in vitro.* Cell-free extracts of *B. recurrentis* utilized glucose via the Embden-Meyerhof pathway.[42,43] *Borrelia novyi* was found to ferment glucose with formation of lactic acid (65%) and CO_2 (10%). *Borrelia recurrentis* fermented glucose with production of lactic acid. Hexokinase, phosphoglucoisomerase, phosphofructokinase, aldolase, phosphoglyceraldehyde dehydrogenase, triose-phosphoglycerate kinase, phosphoglyceromutase, enolase, pyruvate kinase, and a DPN-dependent lactic dehydrogenase were found in cell-free extracts of *B. recurrentis.*[43-46] Current studies in the author's laboratory show that *B. hermsi, B. parkeri,* and *B. turicatae* ferment glucose with production of lactic acid, the latter being the only acid end product.

IMMUNITY

Man and animals develop an immunity against *Borrelia* during the first attack of relapsing fever. Relapse strains or variants develop *in vivo* causing relapse attacks.. The relapse variants are antigenically different from the attack strain. Virulence of a relapse strain is usually lower than that of the original attack strain. How and why *Borrelia* can change its antigenic characteristics remains a mystery, but it seems that the relapse occurring in the infection is an immunological phenomenon resulting from the genetic capability of the organism to undergo antigenic variations. The number of antigenic variants differs from strain to strain and from species to species. *Borrelia recurrentis* has been reported to have nine phases, designated as A to I.[37] Phase A was the attack variant, and B the first relapse phase; phases C to H were found in the second relapses. In man, phases A and B are the ones most frequently found.

Studies in rats showed *B. hermsi* to have four major serotypes, named O, A, B, and C.[34,47] In rats infected with the O serotype, the first relapse contained serotype A, the second serotype B, and the third serotype C; thus the relapse serotypes appeared in sequential order of A, B, and C. When a relapse serotype (A, B, or C) was used to initiate infection of rats, there was a tendency in the relapse to revert to the serotype that preceded them in infections started with the O serotype.

ANTIGENS

Antigens of some species of *Borrelia* have been studied. Three antigenic factors were isolated from *B. turicatae* and *B. parkerii* and were labeled A, B, and C.[48] The B antigen was shared by both species of *Borrelia* and may be considered genus-specific; antigens A and C were strain- and relapse-specific. For more detailed information in reviews on *Borrelia* see References 35, 36, 37, and 49.

Treponema

Treponemes are 5 to 20 μm long and 0.09 and 0.5 μm wide. They have one or more axial fibrils inserted at each end of the cell. Members of this genus are chemoorganotrophs with a fermentative metabolism; some ferment glucose, others ferment amino acids. They are anaerobes and are catalase- and oxidase-negative. Those that can be cultured require serum or one or more volatile fatty acids added to the culture medium. Treponemes are found in the oral cavity, in the intestinal tract, and in the genital regions of man and animals. Some species are pathogenic. The G+C content of the DNA of cultivatable species ranges from 32 to 50 moles %. The type species of the genus is *Treponema pallidum.*

Classification of treponemes is difficult. At our current state of knowledge they may be divided into two groups. The first group consists of pathogens propagated in laboratory animals but not cultured *in vitro;* the second group comprises organisms that have been cultivated *in vitro.*

The pathogens in the first group are *T. pallidum,* the cause of syphilis; *T. pertenue,* the cause of yaws; *T. carateum,* the cause of pinta; *T. paraluis-cuniculi* (*T. cuniculi* is an illegitimate name), the cause of rabbit syphilis, which is reviewed in Reference 50; and two organisms without names. One of the unnamed species is the cause of non-venereal endemic syphilis in man, the other (treponeme FB) causes a natural disease in primates. A classification system of these non-cultivatable pathogens is based on the lesions produced in laboratory animals (Table 8). It has been suggested that the treponeme of endemic syphilis is a subspecies or variant of *T. pallidum* and that treponeme FB is a subspecies or variant of *T. pertenue.*[56]

Like any bacteria, the cultivatable treponemes can be grown in pre-reduced media and their phenotypic characteristics studied. Table 9 shows the classification and some characteristics of these treponemes.[57-60] Gas chromatography is very useful in determining fatty acids and alcohols that are end products of the metabolism of treponemes (Table 10). *Treponema phagedenis* comprises the Reiter treponeme, English Reiter and all the Kazan strains. *Treponema refrigens* comprises the Noguchi and avirulent cultivated Nichols strains as well as strains previously labeled *T. calligyrum* and *T. minutum. Treponema denticola* (the correct designation for *T. microdentium,* which is an illegitimate name) includes strains formerly named *T. microdentium, T. commondonii* and *T. ambiguum.*

TABLE 8

SOME BIOLOGIC CHARACTERISTICS OF PATHOGENIC TREPONEMES[51–55]

Organism	Rabbit	Hamster	Mouse	Guinea Pig	Chimpanzee
T. pallidum	+	–	–	+/–	+
T. pertenue	+	+	–	–	+
T. carateum	–	–	–	–	+
T. paraluis-cuniculi	+	–	–	+	+
Endemic syphilis	+	+	–	+	+
Treponema FB	+	+			+
					+

Code

+ = produces skin lesions; – = no skin lesions; +/– = only rarely produces skin lesions.

TABLE 9

SOME CHARACTERISTICS OF TREPONEMES[57–60]

Species	Glucose	Lactose	Fructose	Sucrose	Mannitol	Galactose	Cellobiose	Maltose	Mannose	Trehalose	Indol	H_2S	1% Glycine, gr.	Lactate Used	Esculin Hydrolysis	Cell Diameter, μ
T. phagedenis biotype Reiter	+	+	+	–	+	v	–	–	+	v	+	w	+	–	–	0.25–0.35
T. phagedenis biotype Kazan	+	+	+	–	+	+	–	–	+	v	+	w	+	–	+	0.25–0.35
T. refringens biotype *refringens*	–	–	–	–	–	–	–	–	–	–	+	+	–	–	+	0.25–0.35
T. refringens biotype *calligyrum*	–	–	–	–	–	–	–	–	–	–	+	+	+	–	+	0.25–0.35
T. denticola biotype *denticola*	–	–	–	–	–	–	–	–	–	–	+	+	–	–	+	0.15–0.25
T. denticola biotype *comondonii*	–	–	–	–	–	–	–	–	–	–	–	+	v	–	+	0.15–0.25
T. oralis	–	–	–	–	–	–	–	–	–	–	+	+		+		0.15–0.25
T. scoliodontum	–	–	–	–	–	–	–	–	–	–	–	–	–	–	–	0.10–0.15
T. macrodentium	+	–	+	+	–	v	v	+	–	–	–	+		–	–	0.15–0.25
T. vincentii	–	–	–	–	–	–	–	–	–	–	+	+	–	–	–	0.25–0.35
T. hyodysenteriae[a]																
																0.35–0.45

[a] Associated with swine dysentery.

Code

+ = positive reaction or weak acid formation without gas; – = negative reaction or no acid formation; v = variable results (some strains +, some –); w = weak reaction; blank space = no data available

TABLE 10

END PRODUCTS OF FERMENTATION OF TREPONEMES[57-59]

Fatty Acid	*T. phagedenis*	*T. refringens*	*T. denticola*	*T. oralis*	*T. scoliodontum*	*T. macrodentium*	*T. vincentii*
Acetic acid	+	+	+	+	+	+	+
Propionic acid	±	t	±	+	+	–	–
Isobutyric acid	–	–	–	–	+	–	–
n-Butyric acid	+	t	t	–	–	–	+
Ethanol	±	–	–	–	–	–	±
n-Butanol	±	–	–	–	–	–	±
n-Propanol	±	–	–	–	–	–	±
Lactic acid	±	±	+	–	t	+	t
Succinic acid	t	t	±		t		t

Code

+ = major end product; ± = minor end product that is usually, but not always, found; t = trace amounts sometimes found; blank = no data available

COLONIES

The capability of growing microorganisms as isolated colonies is necessary for obtaining pure cultures that can be used for taxonomic, biochemical, and antigenic studies. Socransky, MacDonald and Sawyer[61] were able to colonize some oral treponemes on both streaked and poured plates. Others have also reported successful colonization of treponemes.[57,58,62-64]

Treponemes require either short-chain volatile fatty acids or long-chain fatty acids incorporated into culture media. Short-chain acids can be provided by rumen fluid from cattle and other ruminants or by artificial mixtures of fatty acids. Long-chain fatty acids are supplied to treponemes from animal sera. Rabbit, horse, sheep, and cattle serum have been used in culture media at a concentration of 10 to 12%. Rabbit serum is more satisfactory than other animal sera. Bovine serum albumin can replace whole serum.[65] However, oral treponemes called *T. dentium (T. denticola)* were found to be able to grow in a medium supplemented with α-globulin or α^2-globulin.[66] Albumin gave poor growth, as did β-globulin, γ-globulin, transferrin, and ceruloplasmin.

Most treponemes require long-chain fatty acids in albumin. The albumin acts as a detoxifying carrier for the long-chain fatty acids, which by themselves are toxic at concentrations necessary for growth. The Reiter and Kazan strains of *T. phagedenis* were found to use a variety of long-chain fatty acids (Table 11). Some results were probably influenced by contamination of the albumin with lipid. In another study, using lipid-poor albumin thioglycollate medium, the Reiter and Kazan strains of *T. phagedenis* were shown to require a pair of fatty acids for growth (Table 12). One fatty acid was saturated and the other unsaturated. The saturated fatty acid needed a chain length of at least fourteen carbon atoms, whereas the unsaturated fatty acid needed a chain length of fifteen or more carbon atoms and one, two, or three double bonds. The pair of fatty acids could be replaced only with *trans*-18-carbon monounsaturated fatty acid. Long-chain fatty acids were incorporated unchanged into the lipid of the Kazan-5 strain, so that the fatty acids in the medium ended up unchanged in the lipid of serum-requiring treponemes. Kazan-5 and Reiter strains could not synthesize fatty acids and could not modify the chain length, nor reduce or desaturate long-chain fatty acids.[28,68] The combination of a saturated and a *cis*-unsaturated fatty acid probably provides the right "fluidity" for the cell membrane of the flexuous treponemes. Treponemes requiring short-chain volatile fatty acids include human oral isolates as well as strains from the intestinal tract and rumen of animals (Table 13).

TABLE 11

EFFECT OF FATTY ACIDS ON GROWTH OF SOME SERUM-REQUIRING TREPONEMES[65,67,68]

Fatty Acid[a]	Growth	Fatty Acid[a]	Growth
Caprylic acid	±	Heptylic acid	–
Lauric acid	±	Pelargonic acid	–
Tridecylic acid	±	Capric acid	–
Palmitic acid	±	Undecylic acid	–
10-Methylhexadecanoic acid	±	Myristic acid	–
Stearic acid	±	Pentadecylic acid	–
cis-6-Octadecanoic acid	+	Margaric acid	–
trans-6-Octadecanoic acid	+	2-Methylhexadecanoic acid	–
cis-8-Octadecanoic acid	+	Nonadecylic acid	–
trans-8-Octadecanoic acid	+	Arachidic acid	–
Elaidic acid	+	Acrylic acid	–
cis-10-Octadecanoic acid	+	Erotonic acid	–
Vaccenic acid	+	Sorbic acid	–
Linoleic acid	+	Undecylic acid	–
Linolenic acid	+	Undecylenyl alcohol	–
8-Octadecynoic acid	+	Myristyl alcohol	–
9-Octadecynoic acid	+	Cetyl alcohol	–
10-Octadecynoic acid	±	Oleyl alcohol	–
Erucic acid	±	Tween 80	+
Ricinoleic acid	+	Sodium oleate	+
Oleic acid methyl ester	+	TEM-4T[b]	+
Oleic acid ethyl ester	+	TEM-4C[c]	–
Arachidonic acid methyl ester	±	TEM-4S[d]	–
Formic acid	–	C_{16}:0[e]	–
Acetic acid	–	*cis*-9-C_{18}:1[e]	–
Propionic acid	–	*trans*-9-C_{16}:1[e]	–
Butyric acid	–	*trans*-9-C_{18}:1[e]	+
Valeric acid	–	*trans*-11-C_{18}:1[e]	+
Caproic acid	–	*trans*-9,12-C_{18}:2[e]	–

[a] The strain tested on all compounds was the Reiter strain of *T. phagedenis*. The medium contained bovine serum albumin.

[b] TEM-4T = diacetyl tartaric acid ester of tallow monoglycerides.

[c] TEM-4C = diacetyl tartaric acid ester of cotton seed oil monoglycerides.

[d] TEM-4S = diacetyl tartaric acid ester of soybean oil monoglycerides.

[e] Also tested against the Kazan-5 strain.

Code

+ = growth; ± = slight growth; – = no growth

TABLE 12

EFFECT OF COMBINATIONS OF FATTY ACIDS ON TREPONEMES[68]

Fatty Acids	Growth		Fatty acids	Growth	
	Reiter[a]	Kazan-5[a]		Reiter[a]	Kazan-5[a]
$C_{16}{:}0$	–	–	*cis*-9-$C_{14}{:}1$ + $C_{16}{:}0$	–	–
cis-9-$C_{18}{:}1$	–	–	*cis*-9-$C_{16}{:}1$ + $C_{16}{:}0$	+	+
$C_4{:}0$ + *cis*-9-$C_{18}{:}1$	–	–	*cis*-9-$C_{18}{:}1$ + $C_{16}{:}0$	+	+
$C_5{:}0$ + *cis*-9-$C_{18}{:}1$	–	–	*cis*-11-$C_{18}{:}1$ + $C_{16}{:}0$	+	+
$C_6{:}0$ + *cis*-9-$C_{18}{:}1$	–	–	*cis*-9,12,-$C_{18}{:}2$ + $C_{16}{:}0$	+	+
$C_7{:}0$ + *cis*-9-$C_{18}{:}1$	–	–	*cis*-9,12,15-$C_{18}{:}3$ + $C_{16}{:}0$	+	+
$C_8{:}0$ + *cis*-9-$C_{18}{:}1$	–	–	*cis*-5,8,11,14-$C_{20}{:}4$ + $C_{16}{:}0$	–	–
$C_9{:}0$ + *cis*-9-$C_{18}{:}1$	–	–	*trans*-9-,$C_{18}{:}1$		+
$C_{10}{:}0$ + *cis*-9-$C_{18}{:}1$	–	–	*trans*-9-$C_{18}{:}1$ + $C_8{:}0$		+
$C_{11}{:}0$ + *cis*-9-$C_{18}{:}1$	–	–	*trans*-9-$C_{18}{:}1$ + $C_{10}{:}0$		+
$C_{12}{:}0$ + *cis*-9-$C_{18}{:}1$	–	–	*trans*-9-$C_{18}{:}1$ + $C_{12}{:}0$		+
$C_{13}{:}0$ + *cis*-9-$C_{18}{:}1$	–	–	*trans*-9-$C_{18}{:}1$ + $C_{14}{:}0$		+
$C_{14}{:}0$ + *cis*-9-$C_{18}{:}1$	+	+	*trans*-9-$C_{18}{:}1$ + $C_{15}{:}0$		–
$C_{15}{:}0$ + *cis*-9-$C_{18}{:}1$		+	*trans*-9-$C_{18}{:}1$ + $C_{16}{:}0$		–
$C_{16}{:}0$ + *cis*-9-$C_{18}{:}1$	+	+	*trans*-9-$C_{18}{:}1$ + $C_{17}{:}0$		–
$C_{17}{:}0$ + *cis*-9-$C_{18}{:}1$		+	*trans*-9-$C_{18}{:}1$ + $C_{18}{:}0$		–
$C_{18}{:}0$ + *cis*-9-$C_{18}{:}1$		+	*trans*-9-$C_{18}{:}1$ + *cis*-9-$C_{18}{:}1$		+

[a] Medium contained 2% lipid-poor bovine albumin.

Code

+ = growth; – = no growth; blank space = no data available

TABLE 13

EFFECT OF FATTY ACIDS ON GROWTH OF RUMEN FLUID-REQUIRING TREPONEMES[29,69–72]

Fatty Acid(s)	Growth	Fatty Acid(s)	Growth
Acetic acid	–	Palmitic acid	–
Propionic acid	–	Formic acid	–
Isobutyric acid	+[a]	Isobutyric + *n*-valeric acid	+[b]
n-Butyric acid	–	Isobutyric + isovaleric acid	–
Isovaleric acid	–	Isobutyric + acetic acid	–
n-Valeric acid	–	Isobutyric + propionic acid	–
DL-2-Methylbutyric acid	–	Isobutyric + caproic acid	–
Caproic acid	–	Acetic + propionic acid	–
Caprylic acid	–	*n*-Butyric + isobutyric acid	–
2-Methyl valerate	–	*n*-Butyric + *n*-valeric acid	–
Oleic acid	–	*n*-Butyric + acetic acid	–
Linoleic acid	–	*n*-Butyric + propionic acid	–
Lauric acid	–	*n*-Butyric + caproic acid	–

[a] Strains of *T. macrodentium*.[69,70]

[b] Isolates from animal intestinal tract or rumen.

Code

+ = growth; – = no growth

COMPOSITION AND METABOLISM

Chemical composition of some treponemes has been investigated. Most information was obtained on the Reiter and Kazan strains of *T. phagedenis* and least information on *T. pallidum.* Cellular fatty acids of treponemes that have been examined indicate that the main cellular fatty acids of treponemes are C-16:0, C-18:1, C-18:2, and C-18:0.[29,73] Other acids found ranged from C-10:0 to C-20:0. Another study showed Reiter treponeme to contain as major cell lipids C-16:0, C-18:0, C-18:1 Δ^9, and C-18:2 Δ^9,Δ^{12} fatty acids.[32] The fatty aldehydes in Reiter treponeme corresponded to the fatty acids present in cells. Fatty aldehydes were in rather high concentration, 25 μM/100 mg of lipid.

Lipids of treponemes have been studied[28,73-76] and are listed in Table 14. All treponemes studied so far have a monogalactosyldiglyceride, whereas *Spirochaeta zuelzerae* has monoglucosyldiglyceride. In Kazan-5, lipids comprise 18 to 20% of the dry weight of the cells, which is a larger amount of lipid than is found in most bacterial cells. Glycolipid and phospholipid make up 90 to 95% of the total lipid, and free fatty acids and aldehydes make up the remaining 5 to 10%. The monogalactosyldiglyceride found in Kazan-5 treponeme was identified as I-(O-*a*-D-galactopyranosyl)-2,3-diglyceride by Livermore and Johnson;[75] another investigator[76] stated that the compound isolated from the Reiter treponeme had the *beta* linkage.

The Reiter treponeme contains 1.8% DNA and 10.2% RNA.[81] Guanine + cytosine content of the DNA was 38 to 40% in the Reiter treponeme, 37 to 40% in *T. refringens*, 36 to 37% in *T. denticola*, 39% in *T. macrodentium*, and 37% in *T. oralis*.[23,58,59,81]

The metabolism of amino acids has been investigated in a few strains (Table 15).

TABLE 14

LIPIDS FOUND IN *TREPONEMA* AND *SPIROCHAETA*[28,73-76]

Lipid	Reiter[28]	Reiter[76]	Reiter[73]	Kazan-5[73]	Noguchi[73]	Nicholas[a][73]	*S. zuelzerae*[28]
Phosphatidylcholine	+	+		+			–
Lysophosphatidylcholine	±	–		–			–
Phosphatidylglycerol	±	+		–			+
Cardiolipin	±	+		±			+
Lysocardiolipin	±	–		–			±
Unidentified phospholipid	±	±		–			±
Monogalactosyldiglyceride	+	+	+	+	+	+	–
Lysogalactosyldiglyceride	±	–		–			–
Monoglucosyldiglyceride	–	–		–			+
Lysoglucosyldiglyceride	–	–		–			+
Unidentified glycolipid	±	±		±			±
Phosphatidylethanolamine	–	–		+			–

[a] Avirulent cultivated Nichols strain.

Code

+ = present; ± = small amount found; – = none found; blank space = not reported

Note

Neutral lipids are also present, as well as free fatty acids and aldehydes.

TABLE 15

AMINO ACIDS USED BY TREPONEMES[77-80]

Amino Acid	Used as Energy Source: Reiter	Used as Energy Source: *T. denticola*	Dissimilated[a] Reiter
Lysine	–	–	–
Histidine	+	–	+
Arginine	+	–	+
Aspartic acid	–	–	–
Threonine	+	–	+
Serine	+	+	
Glutamate	+	–	+
Proline	–	–	–
Glycine	–	+	–
Alanine	–	+	–
Valine	–	–	–
Methionine	–	–	–
Isoleucine	–	–	–
Leucine	–	–	–
Tyrosine	–	–	–
Phenylalanine	–	–	–
Cysteine	–	+	+

[a] Deaminated or decarboxylated by cell-free extracts.

Code

+ = fermented or dissimilated by cell-free extracts; – = not used; blank space = no data available

ENZYMES

Some enzyme systems of treponemes have been investigated.[78-80,82-93] The Reiter treponeme has been reported to have cytochromes.[82,83] *Treponema denticola* ferments glucose by the Embden-Meyerhof pathway;[80] amino acids were also fermented and serve as the major energy source for the organism. Only a small amount of the end products were produced from glucose. The Reiter, Kazan-2, Kazan-4, and Kazan-5 strains had acetokinase, phosphotransacetylase, and β-galactosidase activity, but no aceto-CoA-kinase activity. The pathogenic Nichols strain of *T. pallidum* did not have acetokinase, aceto-CoA-kinase, phosphotransacetylase, or β-galactosidase activity.[84,89,90] Proteolytic activity of treponemes has been reported in Table 16.[91-93]

ANTIGENS

The antigenic relationships of treponemes have received some study;[94-103] they are shown in Table 17. The Reiter and Kazan treponemes are very similar, but not identical. The Kazan-2 strain has been shown to have two phenol-water-extractable polysaccharide antigens not possessed by the Reiter strain, as well as three other antigens found in sonically disrupted cells.[96] *Treponema phagedenis* and the Reiter treponeme are also very closely related, having at least six antigens in common.[98,99] The Reiter strain has at least one antigen not found in *T. phagedenis*. *Treponema refringens* has four or five antigens in common with *T. calligyrum*, but both organisms also have some individual antigens not found in the other. The Nichols and Noguchi strains are identical. *Treponema minutum* shares a weak antigen with *T. refringens, T. calligyrum* and the Reiter strain. *Treponema pallidum* has only one antigen in common with all of the cultivated treponemes;[98-100] this common antigen is probably the same one shared by all treponemes and *Spirochaeta zuelzerae*. Larger amounts of this antigen are found on the Reiter strain and *T. phagedenis,* while smaller amounts are found on *T. refringens, T. calligyrum,* and *T. minutum.*[98,99]

TABLE 16

PEPTIDASE ACTIVITY OF THE REITER TREPONEME[86,88,91–93]

	Activity		Activity
Hyaluronidase	+	N-Acetyl-DL-glutamate	–
Proteolysis	+	N-Acetyl-DL-leucine	–
Glycylglycine	+	N-Acetyl-DL-methionine	–
Glycyldiglycine	+	L-Leucyl-β-naphthylamide	+
Glycyltriglycine	+	L-Tyrosyl-β-naphthylamide	+
Glycyl-L-leucine	+	L-Arginyl-β-naphthylamide	+
Glycyl-L-phenylalanine	+	L-Alanyl-β-naphthylamide	+
Glycyl-L-tyrosine	+	L-Pyrrolidonyl-β-naphthylamide	+
L-Leucylglycine	+	α-L-Glutamyl-β-naphthylamide	+
L-Prolylglycine	+	γ-L-Glutamyl-α-naphthylamide	+
L-Glutamine	+	Glycyl-β-naphthylamide	+
L-Asparagine	+	L-Histidyl-β-naphthylamide	+
Glutathione	+	L-Cystinyldi-β-naphthylamide	+
γ-DL-Glutamylglycylglycine	+	L-Propyl-β-naphthylamide	+

Code

+ = activity found; – = activity not found

Note

T. denticola strains and *T. vincentii* have been reported as not having hyaluronidase or chondroitinase activity; another report states that oral treponemes had hyaluronidase activity.

TABLE 17

SEROLOGICAL RELATIONSHIPS OF TREPONEMES[94–101]

Serogroup / Organism	*T. phagedenis* Reiter Kazan Kazan 2–8	*T. refringens* Nichols Noguchi *T. calligyrum*	Kroó	*T. denticola* MRB FM N-9	*T. vincentii* N-9	*T. minutum*
T. phagedenis	+	–		–	–	∓
Reiter	+	–	–	–	–	∓
English Reiter	+	–		–	–	∓
Kazan	+	–	–	–	–	
Kazan 2	+	–		–	–	
Kazan 4	+	–		–	–	
Kazan 5	+	–		–	–	
Kazan 8	+	–		–	–	
T. refringens	–	+		–	–	∓
Nichols	–	+	–	–	–	
Noguchi	–	+	–	–	–	
T. calligyrum	–	+		–	–	∓
T. minutum	–	–		–	–	+
Kroó	–	–	+	–	–	
T. denticola MRB	–	–		+	–	
T. denticola FM	–	–		+	–	
T. denticola N-39	–	–		+	–	
T. vincentii N-9	–	–		–	+	

Code

+ = cross reaction; ∓ = weak cross reaction; – = no cross reaction; blank space = no data available

Note

Most strains will agglutinate each others' sera, due to possession of a common antigen on cells. Most of the data given above were derived from sera absorbed with Reiter treponeme to remove the common antigen.

REACTION TO ANTIBIOTICS

Antibiotic susceptibility of treponemes has received some attention.[94,104–109] Sensitivity of pathogenic treponemes has been established by observing immobilization of cells in the presence of the antibiotic and comparing the percentage of immobilized cells with a non-antibiotic control. The relationship of immobilization of cells to inhibition of growth may be fallacious; a better index of susceptibility might be the use of a cultivated treponeme. The minimal inhibitory concentrations of some cultivated treponemes are listed in Tables 18 and 19. All strains were resistant to cycloserine (500–1000 μg/ml), polymyxin-B (500–1000 units/ml), nitrofurazone (100–1000 μg/ml), sulfathiazole (1000 μg/ml), sulfaquinoxaline (1000

TABLE 18

INHIBITORY CONCENTRATIONS[a] OF ANTIBIOTICS FOR SOME STRAINS OF TREPONEMES[104,105]

	T. phagedenis				*T. denticola*		*T. vincentii*	Rumen Fluid	
Antibiotic	**Reiter**	**Kazan**	*T. refringens*	**Nichols**	**FM**	**T-32**	**N-9**	**Oral**	**Intestinal**
P	1	0.1	0.1	1	0.1	0.1	0.1	100	10
A	1	1	1	0.1	0.1	0.1	1	100	10
N	1	1	10	1	1	0.1	1	100	10
Ox	1	1	10	0.1	0.1	0.1	1	100	10
Cl	1	1	1	0.1	0.1	0.1	0.1	100	10
KP	1	0.1	0.1	1	0.1	0.1	0.1	500	10
Ce	10	10	0.01	0.1	0.1	0.1	0.1	1	1
No	100	10	100	500	100	10	1	100	1
Van	1	1	1	0.1	1	10	1	10	100
Bac	1	0.1	0.1	0.1	1	0.1	0.1	1	10
E	0.1	0.1	0.1	0.01	0.01	0.01	0.1	0.01	0.1
Ty	100	10	10	1	100	10	1	1	10
Lin	100	10	1	1	10	10	10	10	10
Tet	1	1	1	1	1	1	1	1	100
Chl	1	10	10	10	10	1	1	10	100
Oxt	10	10	1	1	1	1	1	1	100
DM	1	1	1	1	1	0.1	0.1	1	100
Doxy	1	0.1	0.1	1	1	0.1	0.1	1	100
Me	1	1	0.1	1	1	0.1	0.1	1	100
Chlo	500	500	100	100	100	100	100	100	100
S	100	500	10	10	10	100	10	500	1000
DHS	500	100	1	1	1	100	100	100	1000
K	1000	1000	100	1	100	100	100	100	100
Gen	500	–	–	1	10	–	–	–	–
Neo	500	500	10	1	10	10	100	100	100
Vio	1000+	1000+	100	10	10	100	1000	100	500
Tyr	500	100	100	10	100	100	500	10	500

[a] Concentrations are given in μg of antibiotic per ml of medium.

Code

P = penicillin G; A = ampicillin; N = nafcillin; Cl = cloxacillin; KP = K-phenoxymethyl penicillin; Ce = cephalothin; No = novobiocin; Van = vancomycin; Bac = bacitracin; E = erythromycin; Ty = tylosing; Lin = lincomycin; Tet = tetracycline; Chl = chlorotetracycline; Oxt = oxytetracycline; DM = demethylchlorotetracycline; Doxy = doxycycline; ME = methacycline; Chlo = chloramphenicol; S = streptomycin; DHS = dihydrostreptomycin; K = kanamycin; Gen = gentamycin; Neo = neomycin; Vio = viomycin; Tyr = tyrothricin; – = not tested.

Note

Clindamycin is not inhibitory at 0.5 μg/ml; higher concentrations were not tested.

TABLE 19

INHIBITORY CONCENTRATIONS[a] OF CHEMICALS FOR SOME STRAINS OF TREPONEMES[104,105]

Antibiotic	*T. phagedenis* Reiter	*T. phagedenis* Kazan	*T. refrigens*	*T. refringens* Nichols	*T. denticola* FM	*T. denticola* T-32	*T. vincentii*	Rumen Fluid[b] Oral	Rumen Fluid[b] Intestinal
Fur	100	100	100	100	100	100	10	100	500
USNIC	1000	1000	100	100	100	100	500	100	10
5-FL	500	1000	1000	1000	100	100	100	100	1000
Tell	100	100	100	100	100	10	10	100	100
Thall	100	500	100	500	500	100	100	100	100
BG	500	500	100	100	100	100	100	100	100
CV	100	500	100	100	500	500	100	100	100

[a] Concentrations are given in μg of chemical per ml of medium.

[b] Oral strains require fluid for growth, intestinal strains require rumen fluid for growth.

Code

Fur = furazolidone; USNIC = usnic acid; 5-FL = 5-fluorouracil; Tell = K-tellurite; Thall = thallium acetate; BG = brilliant green; CV = crystal violet

μg/ml), sulfadiazine (1000 μg/ml), succinylsulfathiazole (1000 μg/ml), nalidixic acid (500–1000 μg/ml), methenamine mandelate (500–1000 μg/ml), 5-aminouracil (1000 μg/ml), 5-iodouracil (1000 μg/ml), and lysostaphin (1000 μg/ml). The bactericidal concentrations of the penicillins are 10 to 1000 times higher than the inhibitory concentrations.[105] Synergism of antibiotics against treponemes shows that there is lowering of the bactericidal concentrations for combinations of erythromycin and penicillin, erythromycin and tetracycline, erythromycin and vancomycin, cephalothin and tetracycline, and bacitracin in combination with erythromycin, cephalothin, penicillin, tetracycline, or vancomycin.[106]

Natural or induced resistance of treponemes to penicillin has not been found in syphilis or in the laboratory.[105] L-forms of treponemes have been reported by some workers,[20,110] while others have not been able to induce stable cultivatable L-forms.[1]

For current review and texts on treponemes, see References 50, 56, 94, 102, 103, 109, 111, 112.

Leptospira

Leptospires are aerobic spirochetes whose cells are flexuous, motile, tightly coiled, and have a single axial fibril (axistyle). Some are pathogenic for man and animals and others are saprophytes found in water.

Classification is based on serologic analysis of the antigens of leptospires and divides the pathogenic leptospires into serogroups, serotypes, and subserotypes (Table 20). Recent ideas on the taxonomy of leptospires have resulted in reducing all species names previously used to the taxonomic level of serotypes and retaining only two species names: *L. interrogans* for the pathogens, and *L. biflexa* for the water forms or saprophytes. Twenty-eight water leptospires were grouped by their antigens into sixteen serogroups.[113]

Some very interesting work[114] that has a great influence on the taxonomy of leptospires has been done with G + C ratios and DNA homology (Table 21). Serologic relationships of the strains studied did not correlate with homology groups, but sensitivity to 8-azaguanine and 2,6-diaminopurine, growth at 13°C, and lipase production did correlate with the four homology groups (Table 22). Further work is needed with DNA–DNA homology, using more strains to represent all serogroups and serotypes.

TABLE 20

SEROTYPES OF PATHOGENIC LEPTOSPIRES OF THE SPECIES *LEPTOSPIRA INTERROGANS*

Serogroup	Serotype	Subserotype
icterohaemorrigiae	*icterohaemorrigiae*	*icterohaemorrigiae*
	icterohaemorrigiae	incomplete
	naom	
	mankarso	
	sarmin	
	birkini	*birkini*
	birkini	*smithii*
	nadamdari	
javanica	*javanica*	
	poi	
	coxus	
canicola	*canicola*	
	schueffneri	
	benjamin	
	jousis	
	sumneri	
	malaya	
ballum	*ballum*	*ballumensis*
	ballum	*castellonis*
pyrogenes	*pyrogenes*	
	zanone	
	abramnii	
	biggis	
	hamptoni	
cynopteri	*cynopteri*	
	butembo	
sentot	*sentot*	
autumnalis	*autumnalis*	
	autumnalis	*autumnalis*
	bangkingang	*rachmati*
	mooris	

Serogroup	Serotype	Subserotype
djasiman	*djasiman*	
australis	*australis*	
	muenchen	
	esposito	
pomona	*pomona*	
grippotyphosa	*grippotyphosa*	
hebdomadis	*hebdomadis*	
	medanensis	
	wolffii	
	hardjo	
	mini	*mini*
	mini	*szwajizak*
	kremastos	
	kabura	
	jules	
	haemolytica	*haemolytica*
	haemolytica	*ricardi*
	worsfoldi	
	sejroe	
	saxhoebing	
	borincana	
bataviae	*bataviae*	
	paidjan	
semaranga	*semaranga*	
andamanan	*andamanan*	
hyos	*hyos*	*hyos*
	hyos	*bakeri*
celledoni	*celledoni*	*celledoni*
	celledoni	*whitcombi*

CULTIVATION

Leptospires are usually cultured in dilute media supplemented with animal sera or albumin and Tween 80.

TABLE 21

PERCENT G + C IN THE DNA OF PATHOGENIC AND *BIFLEXA* LEPTOSPIRES

Complex	Serotype	Strain	Percent G + C as Determined by: Tm	Buoyant Density
Pathogens (*L. interrogans*)	*autumnalis*	Akiyami A	35.5 ± 1.3[a]	35.4 ± 0.95
	australis	Ballico	36.7 ± 0.72	35.5 ± 0.95
	ballum	Mus 127		39.0 ± 1.57
	bataviae	Van Tienen	36.6 ± 0.58	35.3 ± 1.35
	canicola	Hond Utrecht	36.7 ± 0.9	
	hyos	Mitis Johnson	40.2 ± 0.73	
	copenhageni	M-20	36.6 ± 0.70	35.4 ± 0.69
	icterohaemorrhagiae	RGA	35.5 ± 1.3	35.4 ± 0.95
	javanica	Veldrat Bataviae 46	40.4 ± 0.75	39.9 ± 1.32
	javanica	TR-73[b]		37.7 ± 1.0
	celledoni	Celledoni		38.3 ± 0.30
	pomona	Pomona	36.0 ± 1.38	36.0 ± 0.79
	pomona	Cornelli CB	36.0 ± 1.09	35.4 ± 1.85
	pyrogenes	Salinem		34.2 ± 2.04
Unclassified		Turtle strain A-183[c]		36.0 ± 1.8
		Turtle strain A-284[c]		35.0 ± 0.46
L. biflexa	*andamana*	Correo	39.1 ± 1.03	39.4 ± 0.32
	patoc	Patoc 1	39.0 ± 0.7	38.3 ± 1.0
	sao-paulo	Sao Paulo		37.8 ± 0.87
	undetermined	CDC		38.0 ± 0.59

[a] Twice the standard deviation of the mean.

[b] Isolated from rodents in Thailand.

[c] Isolated from cloacae of turtles in Illinois by L. E. Hanson.

Data taken from Haapala, D. K., Rogul, M., Evans, L. B., and Alexander, A. D., *J. Bacteriol.*, *98*, 421 (1969). Reproduced by permission of the copyright owners, American Society for Microbiology.

TABLE 22

SELECTED PHENOTYPIC CHARACTERISTICS OF GENETIC GROUPS OF *LEPTOSPIRA*

Genetic Group	Strains Used in DNA Duplex Studies	% GC	Pathogenicity	13°C	Growth[a] DAP[b]	8-Azaguanine[b]	Lipase Production[a]
I	*australis* (Ballico), *bataviae* (Van Tienen), and *pomona* (Pomona)	36 ± 1	+	–	–	–	+
II	*javanica* (Veldbat Bataviae 46), *celledoni* (Celledoni), and *hyos* (Mitis Johnson)	39 ± 1	+	–	+	–	–
III	*patoc* (Patoc I), *sao-paulo* (Sao Paulo), and *andamana* (Correo)	39 ± 1	–	+	+	+	+
IV	*biflexa* type (CDC)	39 ± 1	–	+	+	+	+

[a] Data of Johnson and Harris,[20,21] Kmety and Bakoss,[25] Fuzi and Czoka,[17] and Parnas *et al.*,[39] published in the reference cited below.

[b] DAP (2,6-diaminopurine) concentration is 5 to 10 μg/ml; 8-azaguanine concentration is 200 μg/ml.

Data taken from Haapala, D. K., Rogul, M., Evans, L. B., and Alexander, A. D., *J. Bacteriol.*, *98*, 421 (1969). Reproduced by permission of the copyright owners, American Society for Microbiology.

NUTRITION AND METABOLISM

Horse, cattle, fetal-calf, sheep, pig, and guinea pig sera have been used, but the most successful is pooled rabbit serum. Some rabbit sera have been found to be inhibitory for leptospires.[115,116] The inhibitory fraction, a so-called natural antibody, was found to be a β-macroglobulin (IgM), which is heat-labile (2 hours at 65°C) and is reduced by 2-mercaptoethanol. This serum fraction seems to act in conjunction with complement and lysozyme. Formalin-treated cells have a Z-antigen, which absorbs the toxic serum fraction. The Z-antigen is formalin-stable and may be associated with virulence.[116]

The active growth-promoting fraction in serum is albumin. Globulin fractions are inactive; however, the globulin fraction of serum combined with serum ultrafiltrate and either soluble starch or the weakly basic resin Amberlite IR-45 is active in promoting growth. Amberlite IR-45 alone did not support growth. The weakly acidic resin Amberlite ICR-50, the strongly acidic resin Dowex 50, the strongly basic resin Dowex-1, and charcoal did not support growth when combined with globulin and serum ultrafiltrate.[117] The PPLO serum fraction supports growth, as does albumin with oleic acid or with Tween 80 and Tween 60.

While the active fraction of serum is albumin, it has been shown that leptospires actually require long-chain fatty acids contained in the serum and that serum protein (albumin) acts to inhibit the toxicity of the fatty acids to the organism.[118-121] Fatty acids can be used as the sole source of carbon and energy. There is no growth when lipid-free albumin is used to supplement media.

Table 23 shows the lipids and fatty acids that can be used by leptospires or that suuport growth of these organisms in media containing albumin.[120,122-126] Saprophitic leptospires (serotype *semaranga*) grew in albumin medium containing saturated or unsaturated fatty acids with chain lengths of C_{12} to C_{18}.[126] Pathogenic serotypes *canicola* and *ballum* grew in media with only unsaturated fatty acids that had fourteen carbon atoms or with saturated fatty acids that had twelve to eighteen carbon atoms.

Cellular lipids of some parasitic and saprophytic leptospires have been investigated; they make up 18 to 26% of the dry weight of the organisms.[127] The total lipid content was composed of 60 to 70% phospholipids. The remaining lipid content consisted of free long-chain fatty acids. The phospholipids found were phosphotidylethanolamine, phosphatidylglycerol, and diphosphatidylglycerol. Traces of lysophatidylethanolamine were often found. The major fatty acids were hexadecanoic, hexadecenoic, and octadecanoic acids. Leptospires cannot chain-elongate fatty acids, but they are capable of β-oxidation of fatty acids.[124,126,128]

Various enzymes of leptospires have been investigated; they are listed in Table 24. Baseman and Cox[131] have investigated the terminal electron transport system in leptospires. Spectral evidence showed cytochromes of the a, c, and c_1 types in serotypes *pomona* and *shueffneri* and in the water isolate B-16. Cytochrome b was not found. Absorption peaks showed that cytochrome oxidase of the O type was present in all strains; cytochrome oxidase of the a_1 or a_3 type and a pigment were found only in the two pathogenic strains.

Serotype *pomona* oxidizes glucose; its growth is increased by addition of glucose to the medium.[140] Baseman and Cox[141] investigated the metabolic pathway of glucose for serotypes *pomona* and *shüeffneri* and for the water isolate B-16. Figure 1 shows their proposed pathway of *Leptospira* for the breakdown of glucose.

The leptospires do not require preformed purines or pyrimidines in culture media. Some encouraging work has been done with purine analogues. Johnson and Harris[142] found that leptospires could be divided into three groups on the basis of purine analogue sensitivity and lipase activity. Group 1, the parasitic leptospires, cannot grow in media containing 10 μg of 2,6-diaminopurine or 200 μg of 8-azaguanine per ml; the organisms have lipase (triolein) activity. Group 2, also parasitic leptospires, grows in media containing 2,6-diaminopurine but not 8-azaguanine; these organisms do not have lipase or triolein activity. Group 3 consists of saprophytic leptospires that have lipase activity and grow in media containing both purine analogues. *Leptospira biflexa* was insensitive to 8-azaguanine at concentrations varying from 25 to 600 μg per ml. Twenty pathogenic serotypes were sensitive to 8-azaguanine.[143] Another purine analogue, 6-mercaptopurine, also inhibited pathogenic strains. The pyrimidine analogues 5-fluorouracil and 5-bromonuracil did not inhibit either pathogenic or saprophytic leptospires; 5-fluorouracil has been used successfully in isolation media to inhibit contaminating bacteria.

TABLE 23

LIPIDS AND FATTY ACIDS UTILIZED OR HYDROLYZED BY SAPROPHYTIC AND PATHOGENIC LEPTOSPIRES[120,122-126]

Lipid or Fatty Acid(s)	Concentration	Used[a]
2-Octadecanoic acid	200 μg/ml	–
3-Octadecanoic acid	200 μg/ml	+
4-Octadecanoic acid	200 μg/ml	+
5-Octadecanoic acid	200 μg/ml	–
6-Octadecanoic acid	200 μg/ml	+
7-Octadecanoic acid	200 μg/ml	–
9-Octadecanoic (oleic) acid	200 μg/ml	+
10-Octadecanoic acid	200 μg/ml	–
11-Octadecanoic acid	200 μg/ml	+
12-Octadecanoic acid	200 μg/ml	–
13-Octadecanoic acid	200 μg/ml	–
14-Octadecanoic acid	200 μg/ml	–
15-Octadecanoic acid	200 μg/ml	+
16-Octadecanoic acid	200 μg/ml	+
17-Octadecanoic acid	200 μg/ml	–
2- + 9-Octadecanoic acid	200 μg/ml	+
3- + 9-Octadecanoic acid	200 μg/ml	+
Arachidic acid	1–4 mg/100 ml	–
Arachidonic acid	1–4 mg/100 ml	–
Butyric acid	1–4 mg/100 ml	–
Capric acid	1–4 mg/100 ml	–
Caproic acid	1–4 mg/100 ml	–
Caprylic acid	1–4 mg/100 ml	–
Lauric acid	1–4 mg/100 ml	+/–[b]
Linoleic acid	1–4 mg/100 ml	–
Linolenic acid	1–4 mg/100 ml	–
Propionic acid	1–4 mg/100 ml	–
Ricinoleic acid	1–4 mg/100 ml	–
Undecanoic acid	1–4 mg/100 ml	–
Egg lecithin	2 mg/100 ml	+
Egg-yolk fat	2 mg/100 ml	+
Monoolein	2 mg/100 ml	+/–
Olive oil	2 mg/100 ml	+
Peanut oil	2 mg/100 ml	+
TEM-4T[c]	5 mg/100 ml	+
Tween 60	14 mg/100 ml	+/–
Tween 80	28 mg/100 ml	+
Aleuric acid		–
Capric acid		–
Caproic acid		–
Caprylic acid		–
Heptadecanoic acid		+
Methylacetic acid		–
Methylcholesterol		–
Methyloleic acid		–
Methylpalmitic acid		–
Myristic acid		+/–

TABLE 23 (Continued)

LIPIDS AND FATTY ACIDS UTILIZED OR HYDROLYZED BY SAPROPHYTIC AND PATHOGENIC LEPTOSPIRES[120,122-126]

Lipid or Fatty Acid(s)	Concentration	Used[a]
Myristoleic acid		+/–
Palmitic acid		+/–
Palmitoleic acid		+
Pentadecanoic acid		+/–
Stearic acid		+/–
Tridecanoic acid		+/–
Triglycerides of butyric acid		–
Triolein		+/–
Tween 20		+/–
Tween 40		+/–
Acetic + capric acid		–
Acetic + caproic acid		–
Acetic + caprylic acid		–
Acetic + lauric acid		–

[a] Where two results are shown, data were obtained from different reports.

[b] +/– = saprophyte/pathogen; for differences in fatty acid utilization by saprophytic and pathogenic leptospires see Reference 126.

[c] TEM-4T = diacetyl tartaric acid ester of tallow monoglycerides.

TABLE 24

ENZYMES OF LEPTOSPIRES[128,130-141]

Enzyme	Present
Catalase	+
Hemolysin	+
Oxidase	+
Lipase	+
Transaminase	+
Fumarase	+
Enolase	+
Aconitase	+
NADH oxidase	+
Phosphoglyceromutase	+
Phosphofructokinase	+
α-Ketoglutarate dehydrogenase	+
Malate dehydrogenase	+
Transaldolase	+
Fructose-1,6-diphosphate aldolase	+
Condensing enzyme	+
Acyl CoA dehydrogenase	+
Triosephosphate isomerase	+
Isocitrate dehydrogenase	+
Glyceraldehyde 3-phosphate dehydrogenase	+
Acyl dehydrogenase	+

Enzyme	Present
Enoyl hydrase	+
β-Hydroxyacyl dehydrogenase	+
Succinate dehydrogenase	+
Acyl thiolase	+
Acyl CoA synthetase	+
Glucokinase	–
Phosphoglucoisomerase	–
Glucose-6-phosphate dehydrogenase	–
Pyruvate kinase	–
Lactate dehydrogenase	–
Acetokinase	–
Phosphotransacetylase	–
Alcohol dehydrogenase	–
Phosphoriboisomerase	+
D-Ribulose 5-phosphate 3-epimerase	+
Phosphatase	+
Malic dehydrogenase	+
α-Glycerophosphate dehydrogenase	+
6-Phosphogluconic dehydrogenase	+
Naphthylamidase	+
Esterase	+
Phospholipase C	+
Aminopeptidase	+
Transketolase	+

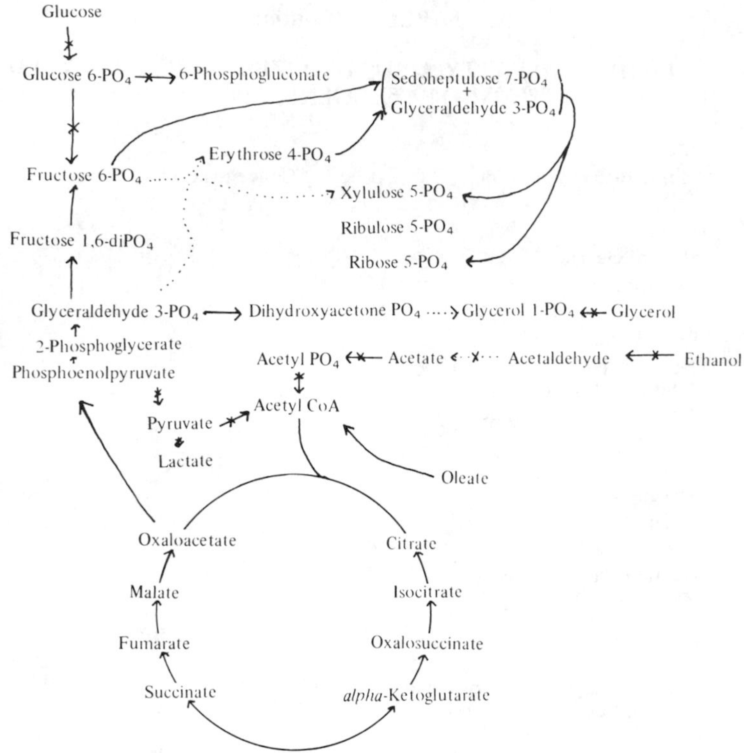

FIGURE 1. Proposed Metabolic Pathway of *Leptospira.* (→ = enzyme present in leptospiral extract; –x→ = enzyme absent from leptospiral extract; - - → = enzyme not assayed, but presence of enzyme suspected; - -x- → = enzyme not assayed, but presence of enzyme doubted).

From: Baseman, J. B., and Cox, C. D., *J. Bacteriol., 97,* 992 (1969). Reproduced by permission of the copyright owners, American Society for Microbiology, Washington, D.C.

ANTIGENS

Antigens of leptospires have been studied. Extraction of leptospires with 70% ethanol yields two water-soluble antigens; the P-antigen is type specific, whereas the S-antigen is genus specific.[144] Extraction of serotype *pomona* with 50% ethanol and 0.15*M* NaCl also yielded two fractions,[145,146] an alcohol fraction and a saline S fraction.

Cox discovered the HL-antigen (hemolytic-lysis antigen), which is genus-specific for leptospires. HL-antigen-sensitized sheep erythrocytes are hemolyzed in serum containing leptospiral antibodies in the presence of complement.[147] The HL-test is genus-specific and, therefore, a good screening test for leptospirosis in man and animals.

An erythrocyte-sensitizing substance (ESS) similar to the HL-antigen has been isolated from leptospires. The sensitized-erythrocyte lysis (SEL) test detects ESS antibodies in people with leptospirosis. It has broad reactivity against most serogroups of leptospires.[148]

A leptospire type-specific antigen (TM) has been isolated by extraction with 90% phenol, enzyme treatment and fractionation by ethanol precipitation.[149] It had only one antigenic component.

COLONIES

Leptospires can be grown as colonies on solid agar medium.[150] They grow into the agar medium, giving several types of colonies.[150–156]

For additional information on Leptospires, see References 157 to 159.

Cristispira

Cristispira have very coarse spirals. They are 28 to 120 μm long and 0.5 to 3 μm wide. Cells have fifty to several hundred axial fibrils, which form a lateral ridge of "crista" that may be seen by dark-field or phase microscopy. *Cristispira* are found in molluscs. They have not been cultured. Type species of the genus is *C. pectinis* Gross, 1910.[160] Another organism found in molluscs, *C. hartmannii*, smaller than *C. pectinis*, has fewer axial fibrils and may be considered a *Treponeme*. Very little is known about *Cristispira*. For more information on *Cristispira*, see Reference 15.

REFERENCES

1. Abramson, I. J., Ph.D. Thesis, *Effect of Antimicrobial Agents on Treponemes.* Virginia Polytechnic Institute and State University, Blacksburg, Virginia (1971).
2. Ritchie, A. E., and Ellinghausen, H. C., *J. Bacteriol., 89,* 223 (1965).
3. Nauman, R. K., Holt, S. C., and Cox, C. D., *J. Bacteriol., 98,* 264 (1969).
4. Anderson, D. L., and Johnson, R. C., *J. Bacteriol., 95,* 2293 (1968).
5. Pillot, J., and Ryter, A., *Ann. Inst. Pasteur (Paris), 108,* 791 (1965).
6. Listgarten, M. A., and Socransky, S. S., *Arch. Oral Biol., 10,* 127 (1965).
7. Jackson, S., and Black, S. H., *Arch. Mikrobiol., 76,* 308 (1971).
8. White, F. H., and Simpson, C. F., *J. Infect. Dis., 115,* 123 (1965).
9. Simpson, C. F., and White, F. H., *J. Infect. Dis., 109,* 243 (1961).
10. Bharier, M. A., and Rittenberg, S. C., *J. Bacteriol., 105,* 413 (1971).
11. Holt, S. C., and Canale-Parola, E., *J. Bacteriol., 96,* 822 (1968).
12. Bladen, H. A., and Hampp, E. G., *J. Bacteriol., 87,* 1180 (1964).
13. Listgarten, M. A., and Socransky, S. S., *J. Bacteriol., 88,* 1087 (1964).
14. Jepsen, O. B., Hovind Hougen, K., and Birch-Anderson, A., *Acta Pathol. Microbiol. Scand., 74,* 241 (1968).
15. Ryter, A., and Pillot, J., *Ann. Inst. Pasteur (Paris), 109,* 552 (1965).
16. Bharier, M. A., and Rittenberg, S. C., *J. Bacteriol., 105,* 422 (1971).
17. Bharier, M. A., and Rittenberg, S. C., *J. Bacteriol., 105,* 430 (1971).
18. Jackson, S., and Black, S. H., *Arch. Mikrobiol., 76,* 308 (1971).
19. Hovind Hougen, K., and Birch-Anderson, A., *Acta Pathol. Microbiol. Scand., 79,* 37 (1971).
20. Ovcinnikov, N. M., and Delktorskij, V. V., *Brit. J. Vener. Dis., 47,* 315 (1971).
21. Yanagihara, Y., and Mifuchi, I., *J. Bacteriol., 95,* 2403 (1968).
22. Cox, C. D., *J. Bacteriol., 109,* 943 (1972).
23. Canale-Parola, E., Udris, Z., and Mandel, M., *Arch. Mikrobiol., 63,* 385 (1968).
24. Hespell, R. B., and Canale-Parola, E., *J. Bacteriol., 103,* 216 (1970).
25. Veldhamp, H., *Antonie van Leeuwenhoek J. Microbiol. Serol., 26,* 103 (1960).
26. Breznak, J. A., and Canale-Parola, E., *J. Bacteriol., 97,* 386 (1968).
27. Hespell, R. B., and Canale-Parola, E., *Arch. Mikrobiol., 74,* 1 (1970).
28. Meyer, H., and Meyer, F., *Biochim. Biophys. Acta, 231,* 93 (1971).
29. Cohen, P. G., Moss, C. W., and Farshtchi, D., *Brit. J. Vener. Dis., 46,* 10 (1970).
30. Meyer, H., and Meyer, F., *Biochim. Biophys. Acta, 176,* 202 (1969).
31. Kelly, R., *Science, 173,* 443 (1971).
32. Smibert, R. M., *Personal Communication* (1972).
33. Pickens, E. G., Gerhoff, R. K., and Burgdorfer, W., *J. Bacteriol., 95,* 291 (1968).
34. Coffey, E. M., and Eveland, W. C., *J. Infect. Dis., 117,* 29 (1967).
35. Southern, P. M., and Sanford, J. P., *Medicine, 48,* 129 (1969).
36. Felsenfeld, O., *Bacteriol. Rev., 29,* 46 (1965).
37. Felsenfeld, O., *Borrelia Strains, Vectors, Human and Animal Borreliosis.* Warren H. Green, Inc., St. Louis, Missouri (1971).

38. Burgdorfer, W., and Mavros, A. J., *Infect. Immun.*, *2*, 256 (1970).
39. McNeil, E., Hinshaw, W. R., and Kissling, R. E., *J. Bacteriol.*, *57*, 191 (1949).
40. Dhanhov I., Soumrov, I., Lozera, T., and Penev, P., *Zentralbl. Veterinaermed. Reihe B*, *17*, 544 (1970).
41. Packchanian, H., and Smith, J. B., *Tex. Rep. Biol. Med.*, *28*, 287 (1970).
42. Jepson, W. F., *Nature*, *160*, 874 (1947).
43. Fulton, J. D., and Smith, P. J. C., *J. Biochem.*, *76*, 491 (1960).
44. Smith, P. J. C., *J. Biochem.*, *76*, 500 (1960).
45. Smith, P. J. C., *J. Biochem.*, *76*, 508 (1960).
46. Smith, P. J. C., *J. Biochem.*, *76*, 514 (1960).
47. Coffey, E. M., and Eveland, W. C., *J. Infect. Dis.*, *117*, 23 (1967).
48. Felsenfeld, O., Decker, W. J., Wohlhieter, J. A., and Rafyi, Z., *J. Immunol.*, *94*, 805 (1965).
49. Davis, G. E., *Annu. Rev. Microbiol.*, *2*, 305 (1943).
50. Smith, J. L., and Pesitsky, B. R., *Brit. J. Vener. Dis.*, *43*, 117 (1967).
51. Paris-Hamelin, A., Vaisman, A., and Dunoyer, F., *Bull. WHO*, *38*, 308 (1968).
52. Vaisman, A., Paris-Hamelin, A., Dunoyer, F., and Dunoyer, M., *Bull. WHO*, *36*, 339 (1967).
53. Fribourg-Blanc, A., and Mallaret, H. H., *WHO/VDT/Res.*, *68*, 135 (1968).
54. Sepetjian, M., Tissot Guerraz, F., Salassola, D., Thirolet, T., and Monier, J. C., *Bull. World Health Organ.*, *40*, 141 (1969).
55. Kuhn, U. S. G., III, Medina, R., Cohen, P. G., and Vegas, M., *Brit. J. Vener. Dis.*, *46*, 311 (1970).
56. Treponematoses Research, *WHO Tech. Rep. Ser.*, No. 455 (1970).
57. Holdeman, L. V., and Moore, W. E. G. (Eds.), *Anaerobe Laboratory Manual*. Virginia Polytechnic Institute and State University, Blacksburg, Virginia (1972).
58. Smibert, R. M., *WHO/VDT/Res.*, *71*, 242 (1971).
59. Socransky, S. S., Listgarten, M. A., Hubersak, C., Cotmore, J., and Clark, H., *J. Bacteriol.*, *98*, 878 (1969).
60. Harris, D. L., Glock, R. D., Christensen, C. R., and Kinyon, J. M., *Vet. Med. Small Anim. Clin.*, *67*, 61 (1972).
61. Socransky, S. S., MacDonald, J. B., and Sawyer, S., *Arch. Oral Biol.*, *1*, 171 (1959).
62. Hardy, P. H., Lee, Y. C., and Nell, E. E., *J. Bacteriol.*, *86*, 616 (1963).
63. Christiansen, A. H., *Acta Pathol. Microbiol. Scand.*, *60*, 234 (1964).
64. Hanson, A. W., and Cannefax, G. R., *Brit. J. Vener. Dis.*, *41*, 163 (1965).
65. Oyama, V. I., Steinman, H. G., and Eagle, H., *J. Bacteriol.*, *65*, 609 (1953).
66. Socransky, S. S., and Hubersak, C., *J. Bacteriol.*, *94*, 1795 (1967).
67. Power, D. A., and Pelczar, M. J., *J. Bacteriol.*, *77*, 789 (1959).
68. Johnson, R. C., and Eggerbraten, L. M., *Infect. Immun.*, *3*, 723 (1971).
69. Socransky, S. S., Loesche, W. J., Hubersak, C., and MacDonald, J. B., *J. Bacteriol.*, *88*, 200 (1964).
70. Hardy, P. H., and Munro, C. O., *J. Bacteriol.*, *91*, 27 (1966).
71. Wegner, G. H., and Foster, E. M., *J. Bacteriol.*, *85*, 53 (1963).
72. Sachan, D. S., and Davis, C. L., *J. Bacteriol.*, *98*, 300 (1969).
73. Váczi, L., Király, K., and Réthy, H., *Acta Microbiol. Acad. Sci. Hung.*, *13*, 79 (1966).
74. Johnson, R. C., Livermore, B. P., Jenkins, H. M., and Eggerbraten, L. M., *Infect. Immun.*, *2*, 606 (1970).
75. Livermore, B. P., and Johnson, R. C., *Biochim. Biophys. Acta*, *210*, 315 (1970).
76. Coulon-Morelec, M. J., Dupouey, P., and Marechal, J., *C. R. Acad. Sci. (Paris)*, *269*, 854 (1969).
77. Allen, G. L., Johnson, R. C., and Peterson, D., *Infect. Immun.*, *3*, 727 (1971).
78. Barban, S., *J. Bacteriol.*, *68*, 493 (1954).
79. Barban, S., *J. Bacteriol.*, *69*, 274 (1955).
80. Hespell, R. B., and Canale-Parola, E., *Arch. Mikrobiol.*, *78*, 234 (1971).
81. Rathlev, T., and Pfau, C. J., *Arch. Biochem. Biophys.*, *106*, 343 (1964).
82. Kawata, T., *J. Gen. Appl. Microbiol.*, *13*, 405 (1967).
83. Kawata, T., *Jap. J. Bacteriol.*, *22*, 590 (1967).
84. Ajello, F., *G. Microbiol.*, *15*, 17 (1967).
85. Bucca, M. A., *J. Vener. Dis. Inform.*, *32*, 16 (1951).
86. Hussey, M. S., and Nowminski, W. W., *Tex. Rep. Biol. Med.*, *7*, 73 (1949).
87. Tauber, H., Cannefax, G. R., Hanson, A. W., and Russell, H., *Exp. Med. Surg.*, *20*, 324 (1962).
88. Berger, U., *Zentralbl. Bakteriol. Parasitenk. Infektionskr. Hyg. Abt. I Orig.*, *165*, 563 (1956).
89. Ajello, F., *G. Microbiol.*, *17*, 107 (1969).
90. Ajello, F., *WHO/VDT/Res.*, *71*, 240 (1971).
91. Hampp, E. G., Mergenhagen, S. E., and Omata, R. R., *J. Dent. Res.*, *38*, 979 (1959).
92. Omata, R. R., and Hampp, E. G., *J. Dent. Res.*, *40*, 171 (1961).
93. Szewczuk, A., and Metzger, M., *Arch. Immunol. Ther. Exp.*, *18*, 643 (1970).
94. Turner, T. B., and Hollander, D. H., Biology of the Treponematoses, *WHO Monogr. Ser.*, No. 35 (1957).
95. Meyer, P. E., and Hunter, E. F., *J. Bacteriol.*, *93*, 784 (1967).
96. Eagle, H., and Germuth, F. G., Jr., *J. Immunol.*, *60*, 223 (1948).
97. Christiansen, A. H., *Acta Pathol. Microbiol. Scand.*, *60*, 123 (1964).
98. Dupouey, P., *Ann. Inst. Pasteur (Paris)*, *105*, 725 (1963).

99. Dupouey, P., *Ann. Inst. Pasteur (Paris), 105,* 949 (1963).
100. Kiraly, K., Jobbagy, A., and Kovats, L., *J. Invest. Dermatol., 48,* 98 (1967).
101. Pillot, J., Dupouey, P., and Faure, M., *Ann. Inst. Pasteur (Paris), 98,* 734 (1960).
102. Wallace, A. L., and Harris, A., Reiter Treponeme: A Review of the Literature, *Bull. WHO, 36,* Suppl. 2 (1967).
103. Willcox, R. R., and Guthe, R., *Treponema pallidum*: A Bibliographical Review of the Morphology, Culture and Survival of *T. pallidum* and Associated Organisms, *Bull. WHO, 35* (1960).
104. Abramson, I. J., and Smibert, R. M., *Brit. J. Vener. Dis., 47,* 407 (1971).
105. Abramson, I. J., and Smibert, R. M., *Brit. J. Vener. Dis., 47,* 413 (1971).
106. Abramson, I. J., and Smibert, R. M., *Brit. J. Vener. Dis., 48,* 113 (1972).
107. Fitzgerald, R. J., and Hampp, E. G., *J. Dent. Res., 31,* 20 (1952).
108. Berger, U., *Arch. Hyg. Bakteriol., 140,* 605 (1956).
109. Rosebury, T., *Microorganisms Indigenous to Man.* Blakiston Division, McGraw-Hill Book Co., New York (1962).
110. Ovcinnikov, N. M., Delektorskij, V. V., and Ustimenko, L. M., *Vestn. Dermatol. Venerol., 44,* 53 (1970).
111. Miller, J. N., Falcone, V. H., Golden, B., Israel, C. W., Kuhn, U. S. G., and Smibert, R. M., *Spirochetes in Body Fluids and Tissues: Manual of Investigative Methods.* Charles C Thomas, Springfield, Illinois (1971).
112. Smith, J. L., *Spirochetes in Late Seronegative Syphilis, Penicillin Notwithstanding.* Charles C Thomas, Springfield, Illinois (1969).
113. Henneberry, R. C., and Cox, C. D., *J. Bacteriol., 96,* 1419 (1968).
114. Haapala, D. K., Rogul, M., Evans, L. B., and Alexander, A. D., *J. Bacteriol., 98,* 421 (1969).
115. Rhu, E., *Bull. Inst. Zool. Acad. Sin. (Taipei), 3,* 1 (1964).
116. Faine, S., and Carter, J. N., *J. Bacteriol., 95,* 280 (1968).
117. Johnson, R. C., and Wilson, J. B., *J. Bacteriol., 80,* 406 (1960).
118. Ellinghausen, H. C., and McCullough, W. G., *Amer. J. Vet. Res., 26,* 45 (1965).
119. Ellinghausen, H. C., and McCullough, W. G., *Amer. J. Vet. Res., 26,* 39 (1965).
120. Helprin, J. J., and Hiatt, C. W., *J. Infect. Dis., 100,* 136 (1957).
121. Stalheim, O. H. V., *J. Bacteriol., 92,* 946 (1966).
122. Jenkin, H. M., Anderson, L. E., Halman, R. T., Ismaie, I. A., and Tunstone, F. P., *J. Bacteriol., 98,* 1026 (1969).
123. Johnson, R. C., and Gary, N. D., *J. Bacteriol., 85,* 976 (1963).
124. Stalheim, O. H. V., and Wilson, J. B., *J. Bacteriol., 88,* 55 (1964).
125. Bertok, L., and Kemens, F., *Acta Microbiol. Acad. Sci. Hung., 7,* 251 (1960).
126. Johnson, R. C., Harris, V. G., and Walby, J. K., *J. Gen. Microbiol., 55,* 399 (1969).
127. Johnson, R. C., Livermore, B. P., Walby, J. K., and Jenkin, H. M., *Infect. Immun., 2,* 286 (1970).
128. Henneberry, R. C., and Cox, C. D., *Can. J. Microbiol., 16,* 41 (1970).
129. Henneberry, R. C., Baseman, J. B., and Cox, C. D., *Antonie van Leeuwenhoek J. Microbiol. Serol., 36,* 489 (1970).
130. Rao, P. J., Larson, A. D., and Cox, C. D., *J. Bacteriol., 88,* 1045 (1964).
131. Baseman, J. B., and Cox, C. D., *J. Bacteriol., 97,* 1001 (1969).
132. Patel, V., Goldberg, H. S., and Blenden, D. C., *J. Bacteriol., 88,* 877 (1964).
133. Green, S. S., and Goldberg, H. S., *J. Bacteriol., 93,* 1739 (1967).
134. Chorvath, B., and Fried, M., *J. Bacteriol., 102,* 879 (1970).
135. Berg, R. N., Green, S. S., Goldberg, H. S., and Blenden, D. C., *Appl. Microbiol., 17,* 467 (1969).
136. Green, S. S., Goldberg, H. S., and Blenden, D. C., *Appl. Microbiol., 15,* 1104 (1967).
137. Burton, G., Blenden, D. C., and Goldberg, H. S., *Appl. Microbiol., 19,* 586 (1970).
138. Stalheim, O. H. V., *Amer. J. Vet. Res., 32,* 843 (1971).
139. Markovetz, A. J., and Larson, A. D., *Proc. Soc. Exp. Biol. Med., 101,* 638 (1959).
140. Ellinghausen, H. C., *Amer. J. Vet. Res., 29,* 191 (1969).
141. Baseman, J. B., and Cox, C. D., *J. Bacteriol., 97,* 992 (1969).
142. Johnson, R. C., and Harris, V. G., *Appl. Microbiol., 16,* 1584 (1968).
143. Johnson, R. C., and Rogers, P., *J. Bacteriol., 88,* 1618 (1964).
144. Rothstein, N., and Hiatt, C. W., *J. Immunol., 77,* 257 (1956).
145. Schricker, R. L., and Hanson, L. E., *Amer. J. Vet. Res., 24,* 854 (1963).
146. Schricker, R. L., and Hanson, L. E., *Amer. J. Vet. Res., 24,* 861 (1963).
147. Cox, C. D., *Proc. Soc. Exp. Biol. Med., 90,* 610 (1955).
148. Sharp, C. F., *J. Pathol. Bacteriol., 77,* 349 (1958).
149. Shinagawa, M., and Yanagawa, R., *Infect. Immun., 5,* 12 (1972).
150. Cox, C. D., and Larson, A. D., *J. Bacteriol., 73,* 587 (1957).
151. Smibert, R. M., *Can. J. Microbiol., 15,* 127 (1968).
152. Stalheim, O. H. V., and Wilson, J. B., *J. Bacteriol., 86,* 482 (1963).
153. Fujikura, T., *Jap. J. Vet. Res., 28,* 63 (1966).
154. Armstrong, J. C., and Goldberg, H. S., *Amer. J. Vet. Res., 21,* 311 (1960).
155. Fujikura, T., *Jap. J. Vet. Res., 28,* 297 (1966).
156. Fujikura, T., *Jap. J. Microbiol., 10,* 79 (1966).
157. Turner, L. H., *Trans. R. Soc. Trop. Med. Hyg., 61,* 842 (1967).

158. Turner, L. H., *Trans. R. Soc. Trop. Med. Hyg.*, *62,* 880 (1968).
159. Turner, L. H., *Trans. R. Soc. Trop. Med. Hyg.*, *64,* 623 (1970).
160. Kuhn, D. A., *Int. J. Syst. Bacteriol.*, *20,* 301 (1970).

Additional New References

161. Joseph, R., and Canale-Parola, E., *Arch. Mikrobiol.*, *81,* 146 (1972).
162. Breznak, S. A., and Canale-Parola, E., *Arch. Mikrobiol.*, *83,* 261 (1972).
163. Breznak, S. A., and Canale-Parola, E., *Arch. Mikrobiol.*, *83,* 278 (1972).
164. Smibert, R. M., and Claterbaugh, R. L., Jr., *Can. J. Microbiol.*, *18,* 1073 (1972).

THE MYXOBACTERALES [(FRUITING) MYXOBACTERALES][a]

DR. MARTIN DWORKIN

The Myxobacterales are Gram-negative rods that are characterized by two properties – one unusual and the other unique among the bacteria. The unusual feature concerns the motility of the vegetative rods, which occurs in the absence of flagella or any other visible organelle of locomotion. Gliding movement takes place when the cells are in contact with an interface, and seems similar to that of some of the *Cyanophyta* as well as other groups of gliding bacteria (e.g., *Beggiatoa*). While it seems likely that the laying down of slime tracks is in some way associated with the motility, there are no insights presently available into the process.[6] The most unique visible characteristic of the myxobacteria is their ability to form fruiting bodies under the appropriate physical and nutritional conditions. The vegetative cells migrate to form aggregates of cells, which are then transformed into more or less elaborate structures called fruiting bodies. The physiology of the cells within the mature fruiting body is, in most cases, not well understood. In the genus *Myxococcus*, the cells in the fruiting body differ from vegetative cells in having a much lower metabolic rate[42] and a greater resistance to environmental stresses.[40] The life cycles of two typical fruiting myxobacteria are illustrated in Figures 1 and 2.

FRUITING BODIES

It has been frequently observed that the nature of the nutritional milieu plays a determining role as to whether the swarm continues to develop vegetatively or enters the fruiting stage; i.e., high concentrations of nutrients seem to favor vegetative development, while lower concentrations appear to induce the fruiting stage. In a defined medium that supported exponential growth of *M. xanthus,* it was possible to induce the conversion from vegetative growth to fruiting-body formation by the specific elimination of tryptophan and phenylalanine from the complete amino acid salts medium.[8] On the other hand, under conditions where the components of the medium were limiting (and which may more accurately reflect the natural environment), the elimination of any one of a variety of required amino acids was shown to induce fruiting body formation.[12]

The nature and regulation of the process that leads to the aggregation of the individual vegetative cells remains unclear. While a number of investigators have provided evidence that is consistent with the production of a chemotactic substance,[3,11,13,19,45] no such material has yet been identified.

The complexity of the mature fruiting structure varies from the simple fruiting body of *Myxococcus,* which is a mound of individual resting cells held together in a slimy matrix, to the stalked, cyst-bearing fruiting bodies of *Chondromyces* and *Stigmatella.* The fruiting body is usually brightly pigmented and represents the resistant, resting stage of the myxobacterial life cycle. Under appropriate conditions of moisture and nutrition, the individual resting cells will germinate and give rise to the vegetative rods. A variety of fruiting bodies are illustrated in Figures 3 to 11.

THE VEGETATIVE CELL

The existence of two morphologically distinct vegetative cell types among the myxobacteria was pointed out by Helena and Seweryn Krzemieniewski[15] in 1928. These are exemplified by *Chondromyces* and *Sorangium* whose cells are relatively thick and blunt-ended (0.8-1.2 x 2.5-6 μ), and by *Cystobacter* and *Stigmatella*, whose cells are slender, spindle-shaped rods (0.65-0.75 x 3.5-8.5 μ).

[a] The Myxobacterales previously included both fruiting and non-fruiting genera. It is now quite evident that the non-fruiting genera (*Cytophaga* and *Sporocytophaga*) belong to a different order. Thus the qualification "fruiting" is no longer necessary.

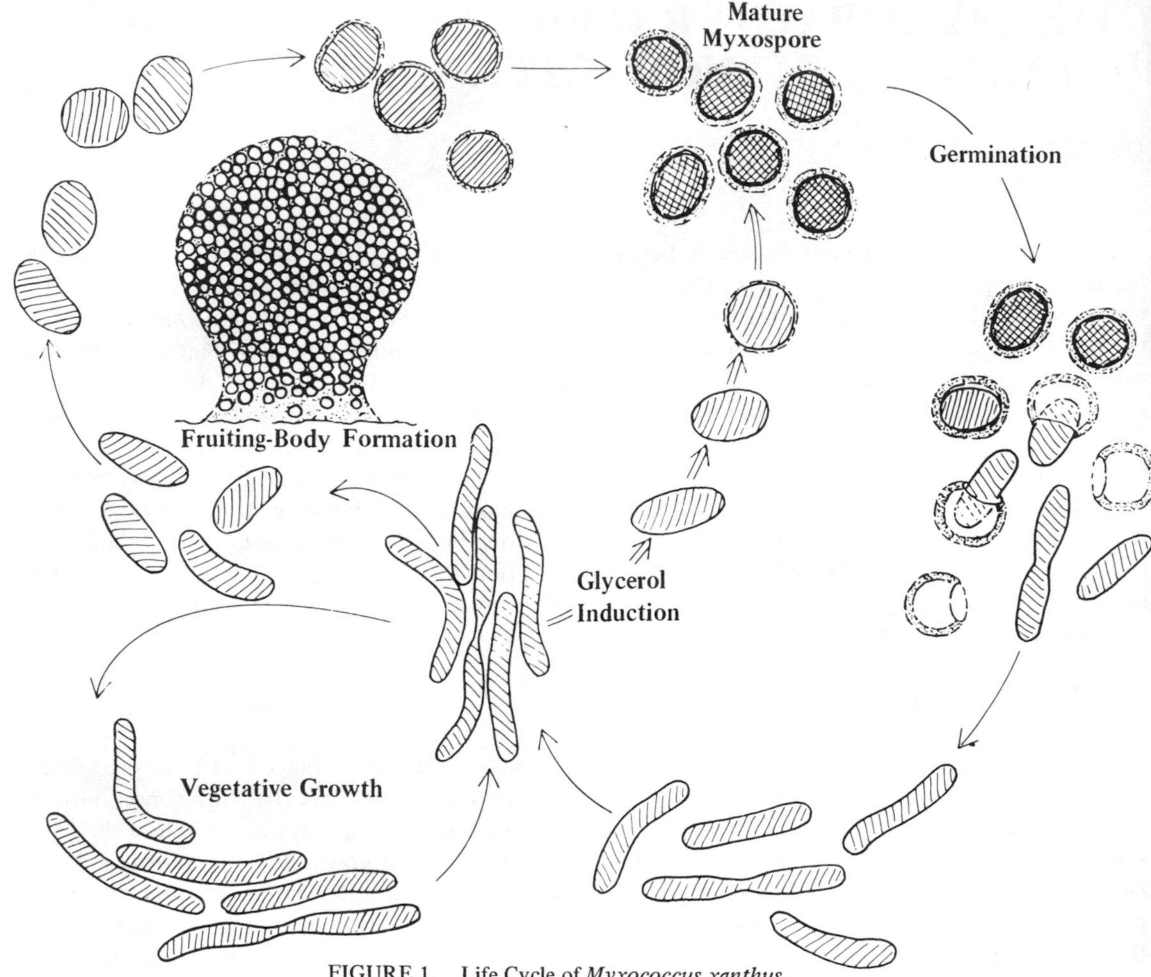

FIGURE 1. Life Cycle of *Myxococcus xanthus*.

The fine structure of two fruiting myxobacteria has been examined in some detail. These are *Myxococcus xanthus*[41] and *Stigmatella aurantiaca.*[35] The generalization that emerges from these and other investigations is that there are no obvious fundamental differences between the fine structure of the vegetative cells of myxobacteria and of the Gram-negative eubacteria. Furthermore, chemical analyses[43] have detected the presence of a typical peptidoglycan component in the cell wall of *M. xanthus.* It is, however, present as a smaller percentage of the total-cell dry weight than is the case for *Escherichia coli.* The earlier suggestions that the cell wall of the myxobacteria differs in some fundamental way from that of eubacteria seems not to be reflected in any basic difference in the peptidoglycan.

THE RESTING CELL

The term "microcyst" has been used to refer to all types of resting cells formed by the myxobacteria. The term is somewhat misleading and could profitably be replaced by the term "myxospore." "Myxospores" are here defined as resting cells that appear during the myxobacterial life cycle, usually within the fruiting body, and that are substantially more resistant than the corresponding vegetative cells.

The myxobacteria form two types of resting cells, with some intermediate forms. These include at one extreme the round, capsulated, optically refractile, resistant cells formed by the Myxococcaceae; at the other extreme are most of the remainder of the myxobacteria – the somewhat shortened rods, which may or may not be optically refractile and which usually lack a heavy capsule. In the latter category are some such as the Archangiaceae, where the cell itself is the unit-resting structure, and others such as the

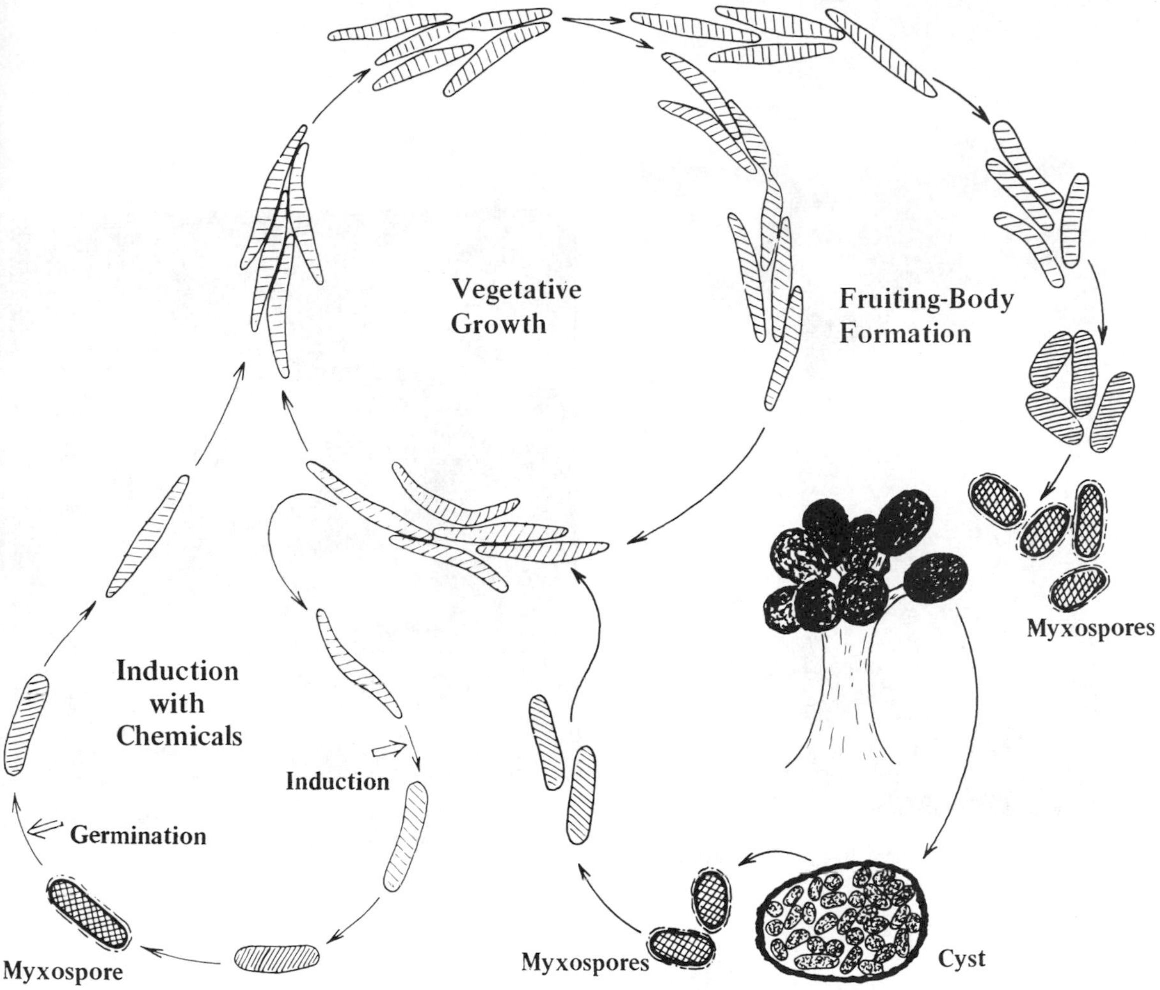

Vegetative
Growth

Fruiting-Body
Formation

Myxospores

Induction
with
Chemicals

Induction

Germination

Myxospore

Myxospores

Cyst

FIGURE 2. Life Cycle of *Stigmatella aurantiaca*.

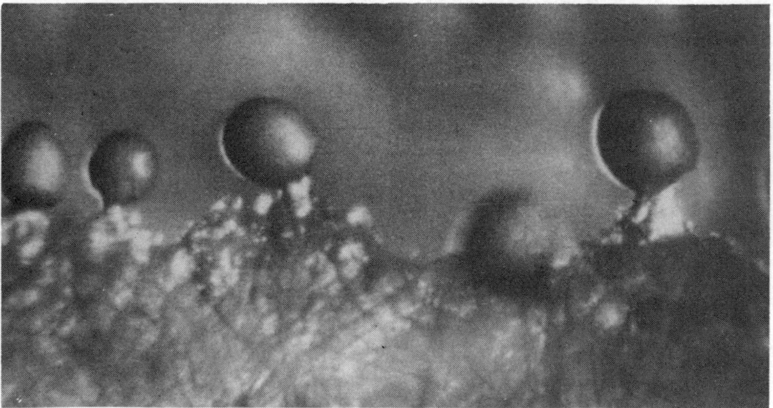

FIGURE 3. Fruiting Bodies of *Myxococcus fulvus*. The fruiting bodies have no stalks but are perched on bits of cellulose. Magnification about 300X.

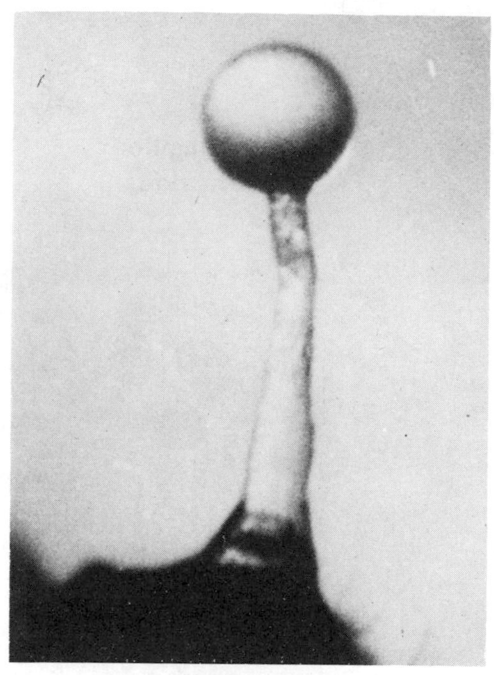

FIGURE 4. Fruiting Body of *Myxococcus stipitatus*. Magnification 400X.

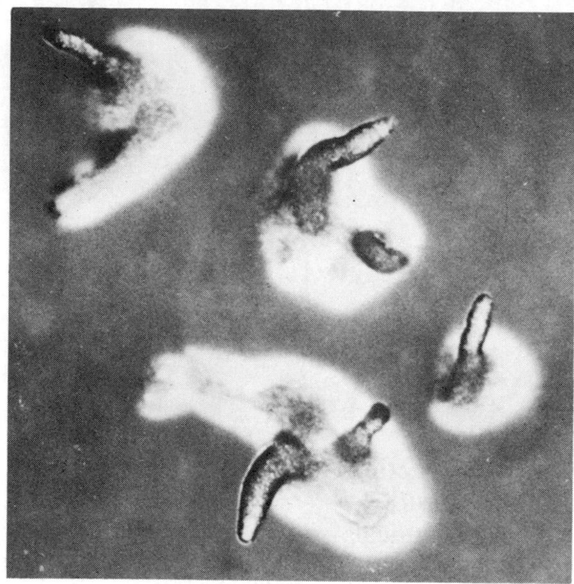

FIGURE 5. Fruiting Bodies of *Chondrococcus macrosporus*. Magnification 145X.

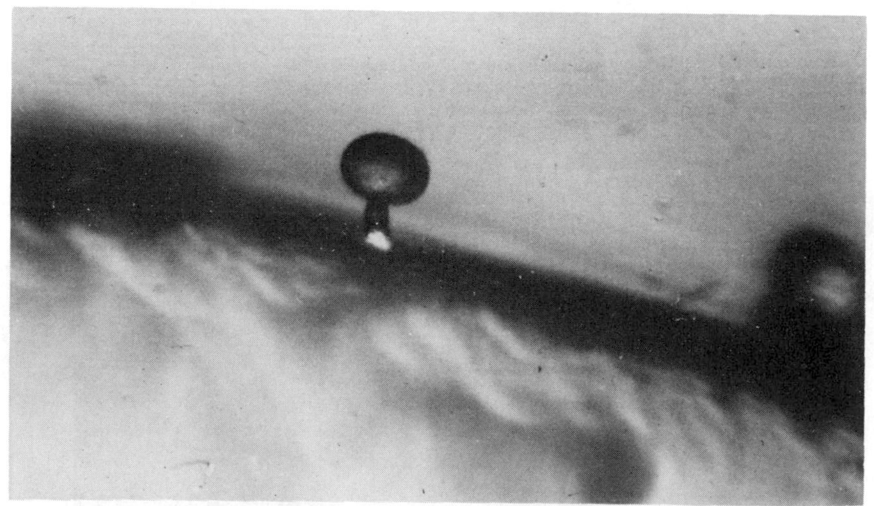

FIGURE 6. Fruiting Body of *Podangium* spp. Magnification about 250X.

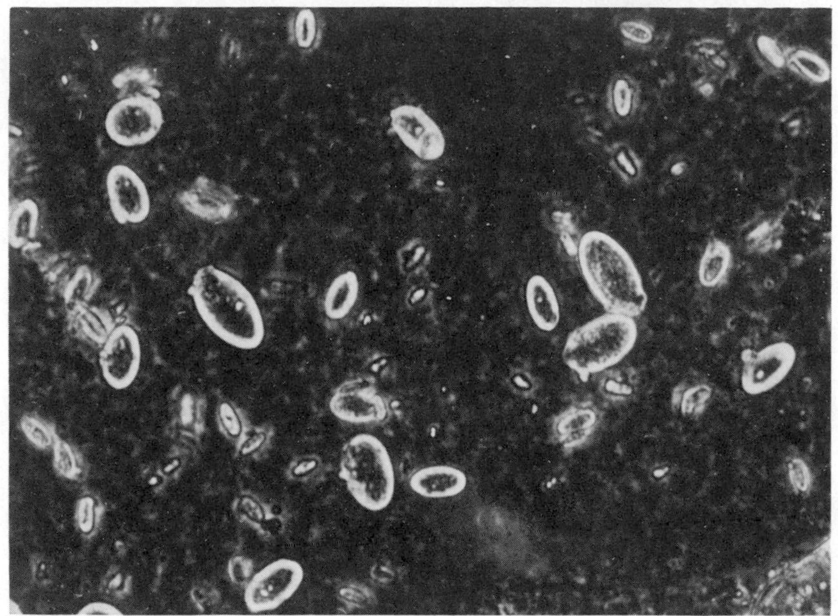

FIGURE 7. Cysts of *Nannocystis exedens*. Note the extreme variability in cyst size. Magnification 550X.

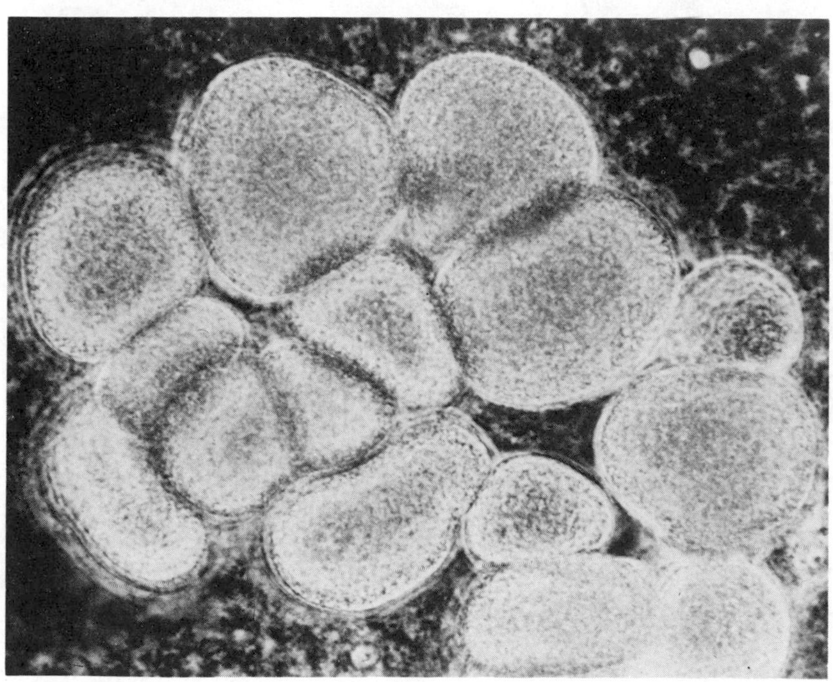

FIGURE 8. Fruiting Body of *Sorangium* spp. Magnification 525X.

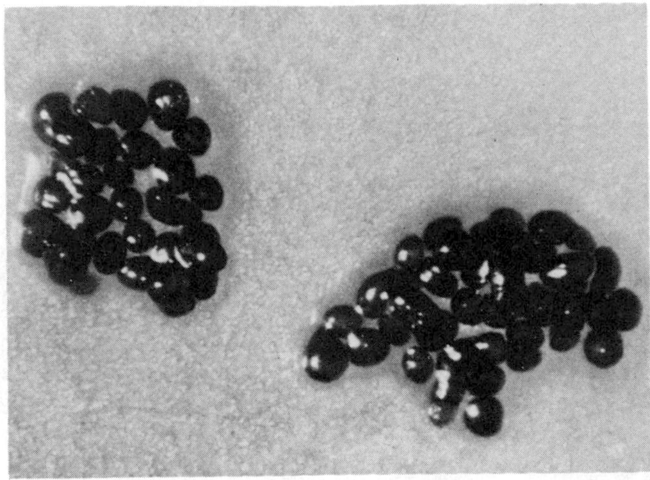

FIGURE 9. Fruiting Bodies of *Polyangium fuscum*. Magnification 90X.

FIGURE 10. Fruiting Body of *Chondromyces apiculatus*. Magnification about 300X.

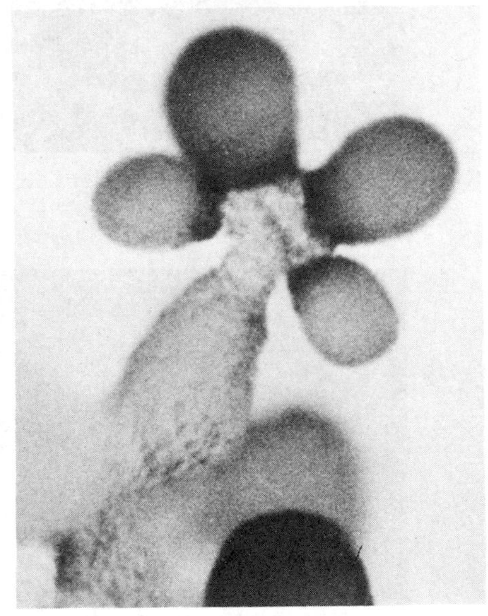

FIGURE 11. Fruiting Body of *Stigmatella aurantiaca*. Magnification about 550X.

Cystobacteraceae, where the unit-resting structure is a cyst containing the shortened vegetative cells. The resistance properties of the resting cells of *Myxococcus* have been examined in some detail.[40] These cells, while lacking the dramatic resistance of bacterial endospores, are substantially more resistant to heat, desiccation, UV irradiation, and physical disruption than the corresponding vegetative cells. In the case of the rod-shaped resting cells that are enclosed in a cyst, the degree of protection conferred by the cyst is not clear. However, resting cells of *Stigmatella aurantiaca,* freed from the cyst, are completely resistant to desiccation and somewhat more resistant than vegetative cells to sonic vibration.[34]

The fine structure of the myxospores of *Myxococcus xanthus*[1,41] and of *Stigmatella aurantiaca*[35] has been examined. These reports confirm the earlier notion, based on light-microscope examinations, that the resting cells are formed by the conversion of single, entire vegetative cells with no visible breakdown and resynthesis of major structural components of the cell.

Dworkin and Gibson[9] reported a technique for inducing the rapid conversion of vegetative cells of *M.*

xanthus to myxospores in liquid media without the intervening formation of fruiting bodies. This technique, which consists of adding 0.5*M* glycerol to exponentially growing cells, has been shown to be effective with members of the genera *Stigmatella,*[35] *Cystobacter, Archangium* and *Chondrococcus*, as well as with other species of *Myxococcus.*[33]

NUTRITION AND CULTIVATION

While there are some exceptions, there are three generalizations that one may make about the nutrition of the myxobacteria. They invariably can hydrolyze one or more varieties of insoluble macromolecules; they are frequently unable to metabolize sugars; and they usually require an organic source of nitrogen. As a rule, they can be grown on a medium containing enzymatic hydrolysate of protein (e.g., Casitone or Tryptone) and a few simple salts. In the case of *Myxococcus xanthus,* whose nutrition has been examined in the greatest detail, growth under these conditions is exponential, with a minimum generation time of about 3.5 hours. *Myxococcus xanthus* can also be grown on a defined medium containing amino acids plus salts.[7,12,44] While some myxobacteria will not grow in liquid media in a dispersed state, many can be grown as true suspensions with the proper manipulation of salt and nutrient concentration, or occasionally by patient selection of dispersed-growing strains.

The nature of the growth on solid media is often a function of the nutrient concentration. At relatively high concentrations of Casitone (>0.1%) the cells grow vegetatively, forming colonies on the agar surface. At lower concentrations or on water agar the cells will usually aggregate and form fruiting bodies.

While there have been reports of cellulose decomposition by a number of genera of the myxobacteria, it has been demonstrated unequivocally only in some members of the genus *Sorangium.*[16] These myxobacteria can often be grown on a simple medium containing glucose, salts and inorganic nitrogen.

ISOLATION

There are three principal techniques for isolating myxobacteria from nature. These make use of the following generalizations: myxobacteria occur primarily in soil and on rotting plant material, or on bark; most of the myxobacteria will cause lysis of, grow, and fruit on living bacteria; dung pellets are a convenient and natural source of these food bacteria. The first method was worked out by the Krzemieniewskis.[14] Petri dishes are lined with filter paper and filled with soil; and the soil is then packed with sterile rabbit dung pellets. The plates are moistened until the soil is almost saturated, then incubated at 28 to 30°C. Visible fruiting bodies should appear on the dung pellets within one week. The second technique is a modification of that used by Singh.[37] Patches of living *Sarcina lutea* are placed on the surface of water agar plates and inoculated with a small amount of soil; the plates are then incubated at 28 to 30°C. The myxobacteria will lyse the *Sarcina,* swarm out, and usually form fruiting bodies. Where fruiting-body formation does not occur under these conditions, the myxobacteria can be recognized by the characteristic swarm pattern. The swarm will usually appear within two or three days, and the fruiting bodies within a week.

Where wood or bark is the natural habitat (certain myxobacteria, such as *Stigmatella aurantiaca,* can be isolated from other sources only with difficulty), the sample is placed in petri dishes on filter paper moistened with distilled water and incubated at 28 to 30°C for one to two weeks. Fruiting bodies will appear on the wood or bark.

For the isolation of cellulose-decomposing myxobacteria, filter paper moistened with a salts-glucose medium is inoculated with soil, then incubated at 28 to 30°C for two to four weeks. Fruiting bodies will eventually appear. It is often useful to include 10 μg of Actidione per ml of medium, to supress the growth of fungi.

Obtaining a pure culture is sometimes complicated by contaminants remaining entrapped in the copious slime formed by the myxobacteria. In such cases it is often useful to homogenize the suspension with fine glass beads, to facilitate the separation of myxobacteria and contaminants.[21] The suspension may then be streaked on a plate containing a lawn of *Sarcina* or on a nutrient plate, and the leading edge of the myxobacterial swarm picked. After a number of such transfers, pure cultures are usually obtained. In general, however, once a medium that will support growth of the myxobacterium is available, the purification process should utilize two properties of the organism: one, its ability to swarm and thus frequently leave the contaminant behind; and two, if fruiting bodies are obtained, the resistance of the myxospore to treatments such as sonic vibration or heating to 50 to 60°C. For a review and compilation of enrichment and isolation procedures, the reader is referred to References 17, 24, 29.

ECOLOGY

The fruiting myxobacteria are frequent and ubiquitous inhabitants of soil and decaying wood and bark. Their function in the cycle of matter is undoubtedly associated with their ability to hydrolyze such insoluble macromolecules as starch, peptidoglycan, protein, and cellulose. Certain of the myxobacteria may be associated with particular ecological niches; e.g. *Sorangium* seems to be a frequent cellulose-decomposing inhabitant of desert soils,[28] and *Stigmatella aurantiaca* and *Chondromyces apiculatus* are usually isolated from decaying soft wood. Most of the other myxobacteria are, however, widely distributed.

The lack of a reliable, accurate method for counting myxobacteria in soil has prevented any detailed ecological examination of the role of these organisms in nature. Nor has there been any systematic investigation of the relation between normal environmental fluctuations and the developmental cycle of the myxobacteria.

INTERMEDIARY METABOLISM

The ability of *Myxococcus xanthus* to grow in dispersed liquid culture offered the opportunity to examine the intermediary metabolism of the organism. *Myxococcus xanthus* has been shown to have a tricarboxylic acid cycle and most of the enzymes required for glycolysis and gluconeogenesis. These data, coupled with the absence of hexokinase and the inability of the organism to incorporate glucose, suggest that the flow of carbon in this organism (and perhaps in most non-cellulose-decomposing myxobacteria) involves the entry of amino acids into the tricarboxylic acid cycle and their subsequent oxidation. Biosynthetically, the amino acids are converted *via* gluconeogenesis to carbohydrate. In addition to certain qualitative metabolic changes, the Q_{O_2} of myxospores of *M. xanthus* is considerably lower than that of vegetative cells.[42]

BACTERIOPHAGE

The only member of the fruiting myxobacteria for which a bacteriophage has been isolated is *Myxococcus xanthus.* Morphologically, the virus is a T-even type with double-stranded DNA and not unusual in any obvious sense. The myxospores are completely resistant to infection. Neither a closely related species, *Myxococcus fulvus,* nor members of the genera *Cytophaga* and *Sporocytophaga* were sensitive to the bacteriophage.[5]

DEVELOPMENTAL BIOLOGY

The myxobacteria manifest two developmental processes: one, a colonial morphogenesis, involving the formation of a more or less differentiated multicellular structure from individual vegetative rods; and two, a cellular morphogenesis, where the vegetative rods are converted to myxospores within the fruiting body. Under appropriate conditions of moisture and nutrition, the myxospores will germinate, giving rise once again to vegetative cells. This process varies from the considerable change manifested by the Myxococcaceae, where the myxospores have drastically altered morphology, metabolic properties and resistance, to the slightly shortened rods within the cysts of *Chondromyces*. The discovery that it was possible to induce the formation of myxospores of *Myxococcus xanthus* directly, without the intervening formation of fruiting bodies,[9] has made it possible to subject the process to rigorous biochemical scrutiny (e.g., see References 2, 36, 42, 44, and 46).

MYXOBACTERIAL COMMUNALITY

The ability of the myxobacteria to aggregate and form the elaborate multicellular fruiting bodies is often viewed as a sudden shift during the life cycle from a unicellular to a multicellular mode. Careful and detailed observation of these organisms, however, leads to a contrary view; that is, a communal association among the cells, far from being expressed only during fruiting-body formation, is a pervasive feature of the organisms, expressed at all stages of the life cycle. Examination of a time-lapse photomicrograph of myxobacteria reveals that the cells often move as coordinated swarms, pulsating rythmically or moving concertedly.[31,32] While cells can ordinarily move either as a coordinated swarm or singly, a mutant has been isolated that will move only as part of a swarm.[4] Furthermore, myxospores of *Myxococcus xanthus* will germinate in water only if the cell density exceeds about 10^9 cells per ml. The signal is apparently an extracellular material excreted by myxospores suspended in water, which, at a critical concentration, can trigger germination.[30] This would guarantee that resting cells suddenly immersed in water would not germinate unless there were sufficient cells to constitute a swarm. This organized communality is certainly the simplest and most primitive known.

These features of myxobacteria, along with the obvious technical facility with which the procaryotes can be experimentally manipulated, make them a most suitable system for investigating basic developmental processes.

TAXONOMY

In 1947, Soriano suggested that the non-fruiting myxobacteria (*Cytophaga* and *Sporocytophaga*) be removed from the Myxobacterales and united with other Gram-negative gliding bacteria in the order Flexibacterales.[38] This suggestion received experimental support when Mandel and Leadbetter[20] and McCurdy and Wolf[23] showed that, while DNA of the non-fruiting myxobacteria had a mole % G + C range of 34 to 43, all of the fruiting myxobacteria showed a strikingly different and narrow range of mole % G + C values (67 to 71). Further support for this view was provided by an Adansonian analysis of the two groups. Comparison between the two groups showed a matching coefficient of 56% S, whereas within the groups of fruiting myxobacteria S values ranged from 82% to 94%.[23] The approach of Soriano and Lewin,[39] transferring the nonfruiting myxobacteria to the Flexibacterales, has been followed in the forthcoming (eighth) edition of *Bergey's Manual.*

The taxonomy of the fruiting myxobacteria is in somewhat of a state of flux. One taxonomic revision has been presented by Dr. H. McCurdy in Reference 25 and in personal communications; at least one other is being prepared by Dr. H. Reichenbach (personal communication). These two revisions shall be presented here, along with some general and interpretive comments:

Taxonomic Scheme Proposed by Dr. H. McCurdy

I. Vegetative cells tapered; microcysts (slime-encapsulated myxospores) produced
   - A. Microcysts spherical or oval: Myxococcaceae[b]
     - Type species: *Myxococcus fulvus*
   - B. Microcysts rod shaped
     - 1. Microcysts not in sporangia: Archangiaceae[c]
       - Type species: *Archangium gephyra*
     - 2. Microcysts in sporangia: Cystobacteraceae
       - Type species: *Cystobacter fuscus*
       - (a) Sporangia sessile: *Cystobacter*

[b] Includes only *Myxococcus*, with which *Chondrococcus* has been recombined; *Angiococcus; species incertae sedis;* number of species reduced to five.

[c] Probably should be discarded; will likely be reduced to *Archangium gephyra*, with the remainder relegated to *species incertae sedis.*

(b) Sporangia borne on stalked fruiting bodies
i. Sporangia solitary: *Melittangium*
ii. Sporangia usually in clusters: *Stigmatella*

II. Vegetative cells of uniform diameter, with blunt, rounded ends; myxospores resemble vegetative cells: Polyangiaceae[d]

Type species: *Polyangium vitellinum*

A. Sporangia sessile, solitary and in groups: *Polyangium*
B. Sporangia in stalks: *Chondromyces*

A more detailed discussion of the data on which this scheme is based may be found in References 24 to 27.

Taxonomic Scheme Proposed by Dr. H. Reichenbach

I. Vegetative cells cylindrical, with blunt, rounded ends: SORANGINEAE

A. Only one family: Sorangiaceae

Fruiting bodies always consist of cysts with definite cyst walls; myxospores do not differ morphologically from vegetative cells.

1′ Fruiting bodies are single cysts, oval or spherical in shape, varying considerably in size and shape within a single culture; myxospores are spherical or cube-shaped, as are the vegetative cells in old cultures: *Nannocystis*

1″ Fruiting bodies consist of densely packed, more or less polygonal cysts; some of these organisms decompose cellulose: *Sorangium*

1‴ Fruiting bodies consist of groups of large oval cysts enclosed in a tough vitreous envelope: *Polyangium*

1⁗ Fruiting bodies consist of clusters of cysts borne on a branched or unbranched slime stalk: *Chondromyces*

1′′′′′ Fruiting bodies as with *Chondromyces,* but the cysts of each cluster are fused at their bases, so that the whole cluster is actually one cyst: *Synangium*

II. Vegetative cells slender, with ends more or less tapered, often spindle- or boat-shaped; myxospores always differ morphologically from vegetative cells: CYSTOBACTERINEAE

A. Myxospores spherical or oval, optically refractile (fruiting bodies without cyst walls): Myxococcaceae

1′ Fruiting bodies more or less spherical, soft-slimy: *Myxococcus*

1″ Fruiting bodies horn-like, often branched, tough, cartilaginous: *Chondrococcus*

B. Myxospores short rods, bean-shaped or oval; fruiting bodies without a wall, irregularly shaped pads or meandering ridges, sometimes with horn-like processes; internally, coiled tube-like structures may often be discerned; of tough cartilaginous consistency: Archangiaceae, with only one genus, *Archangium*

C. Myxospores short rods, oval or spherical, fruiting bodies consist always of cysts with definite cyst walls: Cystobacteraceae

1′ Fruiting bodies groups of (large) cysts, often covered by a common sheet of translucent slime: *Cystobacter*

1″ Fruiting bodies groups of (small) disk-shaped cysts; myxospores spherical: *Angiococcus*

1‴ Fruiting bodies (small) cysts borne individually on delicate slime stalks: *Melittangium*

1⁗ (Relatively large) single cysts on top of coarse slime stalks: *Podangium (erectum)*

1′′′′′ Clusters of cysts on top of slime stalks: *Stigmatella*

[d] Included are all of the old Sorangiaceae and all *Polyangium* except *Cystobacter fuscus, C. ferrugineus* (= *indivisum*), and *C. minus,* which are removed. Also included are *Chondromyces.*

COMMENTS

Both of these authors agree that the vegetative cell shape is the first dichotomous character in the group, with Reichenbach assigning subordinal rank at this level of division (SORANGINEAE for the thick, fat rods and CYSTOBACTERINEAE for the thin, tapered rods). McCurdy has incorporated the genus *Chondrococcus* into *Myxococcus*, whereas Reichenbach retains generic status for both. Both authors have agreed on the establishment of a new family, Cystobacteraceae, to include *Stigmatella*, *Cystobacter* (containing what were previously some of the species of *Polyangium*), and *Podangium*, which McCurdy has chosen to redefine as *Mellitangium*. Both authors have retained the Archangiaceae, although McCurdy feels that its legitimacy as a valid family is highly questionable. Finally, those members of the SORANGINEAE that form fruiting bodies consisting of densely packed, polygonal cysts have been retained in the genus *Sorangium* by Reichenbach and transferred to *Polyangium* by McCurdy. Only time will determine which of the two schemes has the greater utility and more truly reflects the evolutionary relationships among the myxobacteria.

For a more detailed description of the species of the myxobacteria, the reader is referred to the eighth edition of *Bergey's Manual.* The reader is cautioned, however, not to view the taxonomic arrangement presented in *Bergey's Manual* as representing the consensus of most workers in the field. It is an interim proposal, representing only one point of view. A more detailed review of the properties of the myxobacteria can be found in References 10 and 18.

ACKNOWLEDGMENT

The author is grateful to Dr. Hans Reichenbach for permission to use his drawing of the life cycle of *Stigmatella aurantiaca* (Figure 2) and his photographs of fruiting bodies (Figures 3 to 11).

REFERENCES

1. Bacon, K., and Eiserling, F. A., A Unique Structure in Microcysts of *Myxococcus xanthus. J. Ultrastruct. Res., 21,* 378 (1968).
2. Bacon, K., and Rosenberg, E., Ribonucleic Acid Synthesis During Morphogenesis in *Myxococcus xanthus. J. Bacteriol., 94,* 1883 (1967).
3. Bonner, J. T., in *Morphogenesis,* p. 168. Atheneum, New York (1963).
4. Burchard, R. P., Gliding Motility Mutants of *Myxococcus xanthus. J. Bacteriol., 104,* 940 (1970).
5. Burchard, R. P., and Dworkin, M., A Bacteriophage for *Myxococcus xanthus*: Isolation, Characterization and Relation of Infectivity to Host Morphogenesis. *J. Bacteriol., 91,* 1305 (1966).
6. Doetsch, R. N., and Hageage, G. J., Motility in Procaryotic Organisms: Problems, Points of View, and Perspectives. *Biol. Rev., 43,* 317 (1968).
7. Dworkin, M., Nutritional Requirements for Vegetative Growth of *Myxococcus xanthus. J. Bacteriol., 84,* 250 (1962).
8. Dworkin, M., Nutritional Regulation of Morphogenesis in *Myxococcus xanthus. J. Bacteriol., 86,* 67 (1963).
9. Dworkin, M., and Gibson, S. M., A System for Studying Microbial Morphogenesis: Rapid Formation of Microcysts in *Myxococcus xanthus. Science, 146,* 243 (1964).
10. Dworkin, M., Biology of the Myxobacteria. *Annu. Rev. Microbiol., 20,* 75 (1966).
11. Fluegel, W., Myxobacterial Chemotaxis. *Bacteriol. Proc.,* p. 48 (1963).
12. Hemphill, H. E., and Zahler, S. A., Nutritional Induction and Suppression of Fruiting in *Myxococcus xanthus* FBa. J. *Bacteriol., 95,* 1018 (1968).
13. Jennings, J., Association of a Steroid and a Pigment with a Diffusible Fruiting Factor in *Myxococcus virescens. Nature, 190,* 190 (1961).
14. Krzemieniewska, H., and Krzemieniewski, S., Über die Verbreitung der Myxobakterien im Boden. *Acta Soc. Bot. Pol., 5,*102 (1927).
15. Krzemieniewska, H., and Krzemieniewski, S., Zur Morphologie der Myxobakterienzelle. *Acta Soc. Bot. Pol., 5,* 46 (1928).
16. Krzemieniewska, H., and Krzemieniewski, S., Über die Zersetzung der Zellulose durch Myxobakterien. *Bull. Acad. Pol. Sci. Lett. Sci. Math. Natur., 1,* 33 (1937).
17. Kühlwein, H., and Reichenbach, H., Anreicherung und Isolierung von Myxobakterien. *Zentralbl. Bakteriol. Parasitenk. Infektionskr. Hyg. Abt. Orig. Suppl. 1,* 57 (1965).
18. Larpent, J. P., *De la cellule à l'organisme. Acrasiales, myxomycetes, myxobactériales.* Masson et Cie, Paris, France (1970).

19. Lev, M., Demonstration of a Diffusible Fruiting Factor in Myxobacteria. *Nature, 173,* 501 (1954).
20. Mandel, M., and Leadbetter, E. R., Deoxyribonucleic Acid Base Composition of Mycobacteria. *J. Bacteriol., 90,* 1795 (1965).
21. McCurdy, H. D., Jr., A Method for the Isolation of Myxobacteria in Pure Culture. *Can. J. Microbiol., 9,* 282 (1963).
22. McCurdy, H. D., Jr., and Wolf, S., Deoxyribonucleic Acid Base Compositions of Fruiting Myxobacterales. *Can. J. Microbiol., 13,* 1707 (1967).
23. McCurdy, H. D., Jr., and Wolf, S., Studies on the Taxonomy of Fruiting Myxobacterales. *Bacteriol. Proc.,* p. 39 (1967).
24. McCurdy, H. D., Jr., Studies on the Taxonomy of the Myxobacterales. I. Record of Canadian Isolates and Survey of Methods. *Can. J. Microbiol., 15,* 1453 (1969).
25. McCurdy, H. D., Jr., Studies on the Taxonomy of the Myxobacterales. II. *Polyangium* and the Demise of the Sorangiaceae. *Int. J. Syst. Bacteriol., 20,* 283 (1970).
26. McCurdy, H. D., Jr., Studies on the Taxonomy of the Myxobacterales. III. *Chondromyces* and *Stigmatella. Int. J. Syst. Bacteriol., 21,* 40 (1971).
27. McCurdy, H. D., Jr., Studies on the Taxonomy of the Myxobacterales. IV. *Melittangium. Int. J. Syst. Bacteriol., 21,* 50 (1971).
28. Peterson, J. E., The Myxobacterium *Sorangium cellulosum* as a Major Component of North American Desert Soils. *Bacteriol. Proc.,* p. 39 (1967).
29. Peterson, J. E., Isolation and Maintenance of Myxobacteria, in *Methods of Microbiology,* Vol. 3b, p. 185, J. R. Norris and D. W. Ribbons, Eds. Academic Press, London and New York (1969).
30. Ramsey, W. S., and Dworkin, M., Microcyst Germination in *Myxococcus xanthus. J. Bacteriol., 95,* 2249 (1968).
31. Reichenbach, H., Schwarmentwicklung und Morphogenese bei Myxobakterien – *Archangium, Myxococcus, Chondrococcus, Chondromyces. Film C 893,* Institut für den Wissenschaftlichen Film, Göttingen, West Germany (1965).
32. Reichenbach, H., Rhythmische Vorgänge bei der Schwarmentfaltung von Myxobakterien. *Ber. Deut. Bot. Ges., 78,* 102 (1965).
33. Reichenbach, H., and Dworkin, M., Unpublished Data (1968).
34. Reichenbach, H., and Dworkin, M., Studies on *Stigmatella aurantiaca* (Myxobacterales). *J. Gen. Microbiol., 58,* 3 (1969).
35. Reichenbach, H., Voelz, H., and Dworkin, M., Structural Changes in *Stigmatella aurantiaca* During Myxospore Induction. *J. Bacteriol., 97,* 905 (1969).
36. Rosenberg, E., Katarski, M., and Gottlieb, P., Deoxyribonucleic Acid Synthesis During Exponential Growth and Microcyst Formation in *Myxococcus xanthus. J. Bacteriol., 93,* 1402 (1967).
37. Singh, B. N., Myxobacteria in Soils and Composts; Their Distribution, Number and Lytic Action on Bacteria. *J. Gen. Microbiol., 1,* 1 (1947).
38. Soriano, S., The Flexibacterales and Their Systematic Position. *Antonie van Leeuwenhoek J. Microbiol. Serol., 12,* 215 (1947).
39. Soriano, S., and Lewin, R. A., Gliding Microbes: Some Taxonomic Considerations *Antonie van Leeuwenhoek J. Microbiol. Serol., 31,* 66 (1965).
40. Sudo, S. Z., and Dworkin, M., Resistance of Vegetative Cells and Microcysts of *Myxococcus xanthus. J. Bacteriol., 98,* 883 (1969).
41. Voelz, H., and Dworkin, M., Fine Structure of a Fruiting Myxobacterium during Morphogenesis. *J. Bacteriol., 84,* 943 (1962).
42. Watson, B. F., and Dworkin, M., Comparative Intermediary Metabolism of Vegetative Cells and Microcysts of *Myxococcus xanthus. J. Bacteriol., 96,* 1465 (1968).
43. White, D., Dworkin, M., and Tipper, D. J., Peptidoglycan of *Myxococcus xanthus*: Structure and Relation to Morphogenesis. *J. Bacteriol., 95,* 2186 (1968).
44. Witkin, S., and Rosenberg, E., Induction of Morphogenesis by Methionine Starvation in *Myxococcus xanthus*: Polyamine Control. *J. Bacteriol., 103,* 641 (1970).
45. Zahler, S. A., and McVittie, A., Chemotaxis in *Myxococcus. Nature, 194,* 1299 (1962).
46. Zusman, D., and Rosenberg, E., Deoxyribonucleic Acid Synthesis During Microcyst Germination in *Myxococcus xanthus. J. Bacteriol., 96,* 981 (1968).

THE ACTINOMYCETALES

Introduction

DR. HUBERT A. LECHEVALIER AND DR. LEO PINE

Actinomycetes are filamentous, branching bacteria that are widely distributed in nature. They are mainly found in soil, where they play a major role in the decomposition of organic matter. Some species cause diseases of animals or plants; a few species are parasitic and have been isolated only from their animal hosts. Some actinomycetes are of medical and industrial importance as producers of antibiotics, and others are used industrially as agents of chemical transformations. For general reviews on actinomycetes see References 1 to 6.

Although branched filaments may also be observed in some budding bacteria, mycoplasmata, and highly pleomorphic organisms found in the genera *Actinobacillus* and *Streptobacillus,*[7] these are not classified in the Actinomycetales. Basically, members of the Actinomycetales may be considered to be those bacteria that form a well-defined coherent mycelium. But this structure may be so rudimentary and transient in the genera *Actinomyces* and *Mycobacterium* that it may go unnoticed or be essentially nonexistent. Indeed, in the genus *Bifidobacterium* no mycelium has been described. Thus, no unambiguous morphological feature serves to unite all the members of this widely diversified group.

Actinomycetales may form a substrate mycelium only, or both aerial and substrate mycelia, or an aerial mycelium only (*Sporichthya*). While some species are holocarpic, eucarpic members may show highly complex mycelial structures with conidia and sporangia.[8] Motility, when observed, is due to flagella. Conidia and sporangiospores may be variously arranged and/or variously structured with surface hairs, spines, or ridges.

Although generic separation may be made on morphological differences in the case of highly evolved types, classification of the most primitive forms rests primarily on oxygen requirements, fermentative abilities, formation of catalase and respiratory cytochrome systems, and cell-wall and lipid composition. Most actinomycetes are Gram-positive, but part of their thallus (depending on the stages of growth) may be Gram-negative; members of the genus *Mycoplana* are Gram-negative.

Phylogenetically there is very little reason to classify organisms on the basis of ecology. However, a realistic appraisal of research pertinent to the actinomycetes shows that two major groups may be recognized: 1) parasitic forms found in the mucosal cavities of man and animals, and 2) soil forms. It is recognized that among the so-called soil forms parasitic actinomycetes may be found; similarly, some soil forms may be closely related to the animal parasites.

As a group, the parasitic actinomycetes are presently contained within the family Actinomycetaceae and are unified by the following characteristics: they are Gram-positive; anaerobic to aerobic, but generally microaerophilic, they ferment glucose under anaerobic or aerobic conditions to form acids, which can account for 50 to 100% of the substrate carbon; they form a loosely bound or transient mycelium, but do not sporulate or form motile cells; at best, they have a limited oxidative system, by which oxygen is utilized for growth.

Conversely, the soil actinomycetes may be viewed as Gram-positive, sometimes acid-fast organisms having an aerobic metabolism that mediates growth and does not result in large accumulation of acid from carbohydrate substrates. Well-recognizable coherent mycelium is usually formed, often with specialized structures and spores.

As a practical device for presentation of the various genera, the following review on actinomycetes is divided into two sections: soil or oxidative actinomycetes, and parasitic or fermentative actinomycetes. At one extreme of the soil actinomycetes are oxidative mycelial organisms, whereas at the opposite pole of the parasitic actinomycetes are fermentative diphtheroids. Between these two extremes there are no voids, but a population of intermediate forms.

Soil or Oxidative Actinomycetes

DR. HUBERT A. LECHEVALIER

Analyses of cell-wall preparations of soil or oxidative actinomycetes, some of which are animal pathogens (see Tables 7, 8, and 9), reveal that they fall into four main groups as indicated in Tables 1 to 5.

Strains of *Oerskovia, Agromyces,* and *Mycoplana* do not produce aerial mycelia. Oerskoviae and mycoplanae form branched filaments, which break up into motile segments. Strains of *Agromyces* are microaerophilic to aerobic catalase-negative actinomycetes with no special morphological features, but they are abundantly distributed in soil.

It is possible to determine by one-way chromatography of whole-cell hydrolysates whether an actinomycete contains major amounts of L- or of *meso*-DAP.[1] On the basis of paper chromatography of sugars present in whole-cell hydrolysates of actinomycetes containing major amounts of *meso*-DAP, it is possible to separate them into four groups.[2,3] As can be seen in Table 6, whole-cell sugar pattern A indicates the presence of Type IV wall, and whole-cell sugar pattern D indicates the presence of Type II wall. Organisms with Type III wall can be separated into two groups, depending on the presence or absence of madurose (see Tables 4 and 6).

A distinction between the genera *Mycobacterium* and *Nocardia* can be made on the basis of the type of mycolic acid produced.[4] Mycobacteria contain true mycolic acids, giving C_{22} to C_{26} fatty acids upon pyrolysis. Nocardiae contain related nocardomycolic acids, which are smaller molecules and give smaller fatty acids (C_{12} to C_{18}) upon pyrolysis.

A tentative phylogenetic sketch of some of the better-known aerobic actinomycetes is given in Figure 1. Chemical criteria are incorporated in the scheme. Table 7 lists species of actinomycetes of special importance, following the same generic order as in Tables 2 to 5. Tables 8 and 9 list properties of zoopathogenic members of the genera *Nocardia, Actinomadura, Streptomyces,* and *Mycobacterium.*

TABLE 1

MAJOR CONSTITUENTS OF CELL WALLS OF SOIL ACTINOMYCETES[a]

Cell Wall Type	Major Constituents	Reference
Streptomyces or Type I	L-DAP[b], glycine	5,6
Micromonospora or Type II	*meso*-DAP[c], glycine; hydroxy-DAP may also be present	5,6
Actinomadura or Type III	*meso*-DAP	5,6
Nocardia or Type IV	*meso*-DAP, arabinose, galactose	5,6
Oerskovia	Lysine, aspartic acid, galactose	7
Agromyces	DAB[d], glycine	8
Mycoplana[e]	*meso*-DAP; also many amino acids	9

[a] All cell-wall preparations contain major amounts of alanine, glutamic acid, glucosamine, and muramic acid.
[b] DAP = 2,6-diaminopimelic acid.
[c] No differentiation is made between *meso*-DAP and D-DAP.
[d] DAB = 2,4-diaminobutyric acid.
[e] These microorganisms are Gram-negative.

FIGURE 1.
PHYLOGENETIC SKETCH OF SOME AEROBIC ACTINOMYCETES

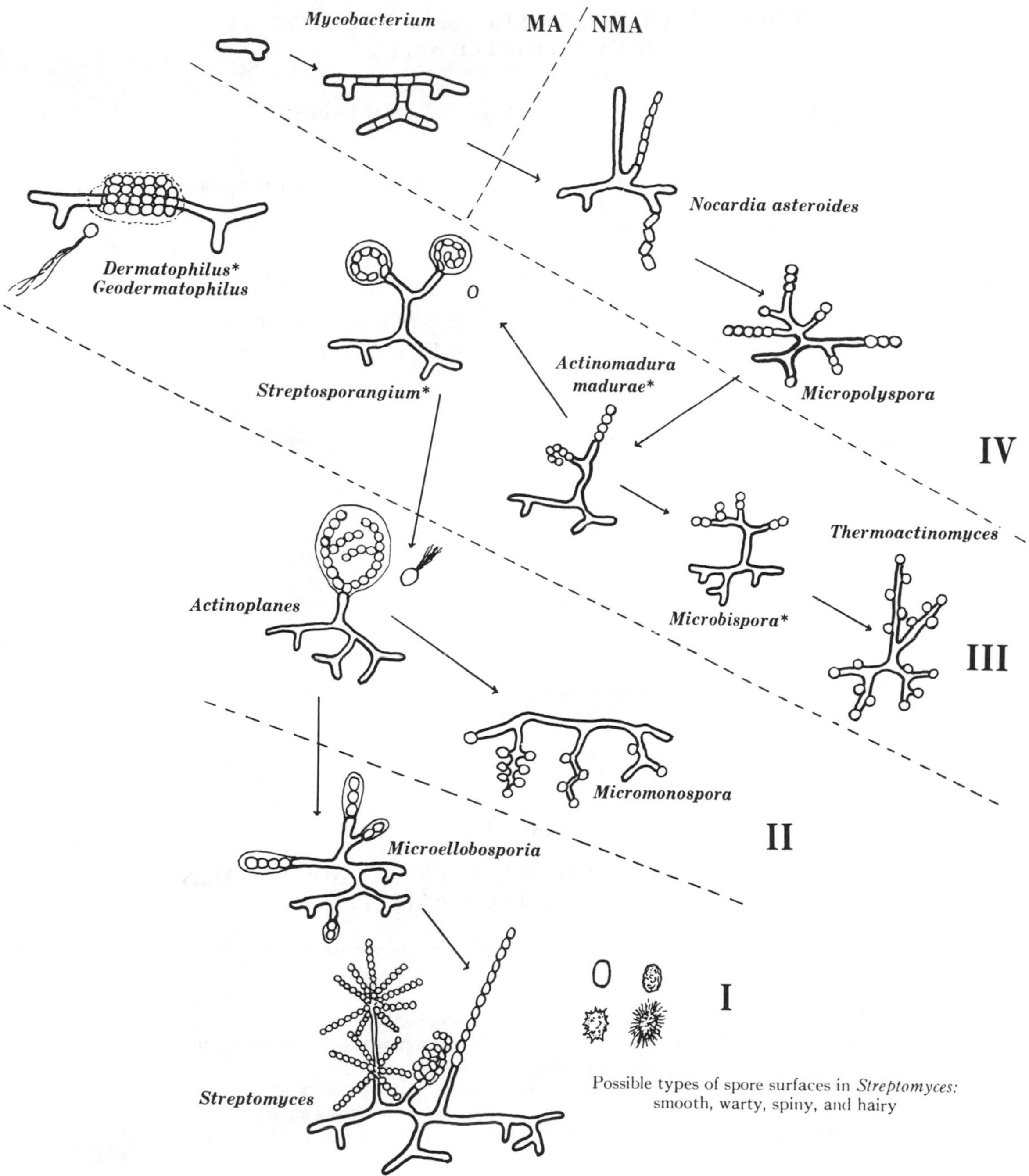

Legend

Roman numerals refer to cell-wall types (see Table 1).
* = madurose-containing.
MA = contains mycolic acid; NMA = contains nocardomycolic acid.

Reproduced from: *Laval Méd., 38,* 740 (1967), by permission of the copyright owners, *Laval Médical,* Ecole de Médicine, Université Laval, Quebec, Canada.

TABLE 2

MORPHOLOGIC CHARACTERISTICS OF ACTINOMYCETALES WITH A TYPE I CELL WALL[a]

Generic Name	Morphologic Characteristics
Streptomyces[b]	Aerial mycelium with chains (usually long) of nonmotile conidia
Streptoverticillium[c]	Same as *Streptomyces,* but the aerial mycelium bears verticils consisting of at least three side branches, which may be chains of conidia or hold sporulating terminal umbels
Chainia	Same as *Streptomyces,* but sclerotia are also formed
Actinopycnidium	Same as *Streptomyces,* but pycnidia-like structures are also formed
Actinosporangium	Same as *Streptomyces,* but spores accumulate in drops
Elytrosporangium	Same as *Streptomyces,* but merosporangia are also formed
Microellobosporia	No chains of conidia; merosporangia with nonmotile spores are formed
Sporichthya	No substrate mycelium is formed; aerial chains of motile, flagellated conidia are held to the surface of the substratum by holdfasts[12]
Intrasporangium	No aerial mycelium; substrate mycelium forms terminal and subterminal vesicles[13]

[a] All aerobic.
[b] This genus includes the most common soil forms and most of the important producers of antibiotics. See Reference 10 for methods of characterization.
[c] See Reference 11.

TABLE 3

MORPHOLOGIC CHARACTERISTICS OF ACTINOMYCETALES WITH A TYPE II CELL WALL[a]

Generic Name	Morphologic Characteristics
Micromonospora	Aerial mycelium absent, conidia single
Actinoplanes	Globose to lageniform sporangia; globose spores with one polar tuft of flagella
Amorphosporangium	Same as *Actinoplanes,* but the sporangia are often very irregular; sporangiospores are usually nonmotile
Ampullariella	Lageniform to globose sporangia; rod-shaped spores with one polar tuft of flagella
Dactylosporangium	Claviform sporangia, each with one chain of spores with one polar tuft of flagella

[a] All aerobic, except for some micromonosporae; cell-wall composition of anaerobic micromonosporae is not known.

TABLE 4

MORPHOLOGIC CHARACTERISTICS OF ACTINOMYCETALES WITH A TYPE III CELL WALL[a]

Generic Name	Morphologic Characteristics
Actinomadura	Short chains of conidia on the aerial mycelium (*madurae*-type[b]); very long chains of conidia on the aerial mycelium (*dassonvillei*-type)[2]
Microbispora[b]	Longitudinal pairs of conidia on the aerial mycelium
Thermoactinomyces	Single spores are formed on the aerial and substrate mycelia
Actinobifida	Same as *Thermoactinomyces*, but the sporophores are dichotomously branched
Streptosporangium[b]	Globose sporangia containing nonmotile spores
Spirillospora[b]	Globose sporangia with rod-shaped spores, each with a subpolar tuft of flagella
Planomonospora[b]	Cylindrical sporangia, each containing one motile spore with one polar tuft of flagella
Dermatophilus[b]	Hyphae dividing in all planes, forming packets of cocci motile by means of a tuft of flagella; pathogenic to animals
Geodermatophilus	Same as *Dermatophilus*, but a soil form[14]

[a] All aerobic.
[b] Madurose-containing organisms (see Table 6).

TABLE 5

MORPHOLOGIC CHARACTERISTICS OF ACTINOMYCETALES WITH A TYPE IV CELL WALL[a]

Generic Name	Morphologic Characteristics
Mycobacterium	Filamentation is usually limited, and aerial mycelium is usually not formed; filaments fall easily apart into rods and cocci
Nocardia	Filamentation is abundant, and aerial mycelium is often formed; chains of conidia may be formed
Micropolyspora	Short chains of globose conidia are formed on both the aerial and the substrate mycelia
Pseudonocardia	Long, cylindrical conidia in chains on the aerial mycelium
Thermomonospora	Mainly single spores are formed on the aerial mycelium

[a] All aerobic.

TABLE 6

CELL WALL TYPES AND WHOLE-CELL SUGAR PATTERNS OF AEROBIC ACTINOMYCETES CONTAINING *meso*-DIAMINOPIMELIC ACID[a]

Cell Wall[5,6]		Whole-Cell Sugar Pattern[2,3]	
Type	Distinguishing Major Constituents[b]	Type	Diagnostic Sugars
II	Glycine	D	Xylose, arabinose
III	None	B	Madurose[c]
		C	None
IV	Arabinose, galactose	A	Arabinose, galactose

[a]No differentiation is made between *meso*-DAP and D-DAP.
[b]All cell-wall preparations contain major amounts of alanine, glutamic acid, glucosamine, and muramic acid.
[c]Madurose = 3-O-methyl-D-galactose.[15]

TABLE 7

IMPORTANT SPECIES OF AEROBIC ACTINOMYCETES

Organism	Importance	Reference
Streptomyces		
antibioticus	Production of actinomycin	16
aureofaciens	Production of chlortetracycline and tetracycline	16
erythreus	Production of erythromycin	16
fradiae	Production of neomycin	16
griseus	Production of streptomycin, cyclo-heximide and candicidin	16
griseus (scabies)	Cause of potato scab	17
nodosus	Production of amphotericin B	16
noursei	Production of nystatin	16
rimosus	Production of oxytetracycline	16
somaliensis	Found in mycetomas	18
venezuelae	Production of chloramphenicol	16
Micromonospora		
echinospora	Production of gentamicin	19
purpurea	Production of gentamicin	19
Actinomadura		
madurae	Found in mycetomas	18
pelletieri	Found in mycetomas	18
Thermoactinomyces		
vulgaris	Cause of pneumonitis (farmer's lung)	20
Dermatophilus		
congolensis	Cause of streptothricosis of animals	18
Mycobacterium		
avium	Cause of tuberculosis	21
bovis	Cause of tuberculosis	21
farcinicum	Cause of bovine farcy	22
fortuitum	Causes abscesses in man; grows more rapidly and is more filamentous than most pathogenic mycobacteria	21
kansasii	Cause of tuberculosis	21
leprae	Associated with leprosy; has never been cultured on laboratory media	21
marinum	Cause of skin lesions in man	21
paratuberculosis	Cause of Johne's disease of cattle	23
tuberculosis	Cause of tuberculosis	21
Nocardia		
asteroides	Cause of deep-seated nocardiosis; also found in mycetomas	18
brasiliensis	Cause of deep-seated nocardiosis; also found in mycetomas	18
caviae	Cause of deep-seated nocardiosis; also found in mycetomas	18
farcinica	See *Mycobacterium farcinicum*	
Micropolyspora		
faeni	Cause of pneumonitis (farmer's lung)	24

TABLE 8

DISTINCTIVE PROPERTIES OF MYCOBACTERIAL SPECIES

Code

Column 1: *M. abscessus*
Column 2: *M. avium*
Column 3: *M. bovis*
Column 4: *M. flavescens*
Column 5: *M. fortuitum*
Column 6: *M. gastri*
Column 7: *M. intracellulare*
Column 8: *M. kansasii*
Column 9: *M. marianum* (*scrofulaceum*)
Column 10: *M. marinum*
Column 11: *M. terrae*
Column 12: *M. tuberculosis*
Column 13: *M. ulcerans*
Column 14: *M. xenopei*
Column 15: "Aquae" strains
Column 16: "V" (*M. triviale*)
Column 17: other Group IV[a]

Property	1	2	3	4	5	6	7	8	9	10	11	12	13	14	15	16	17
Specimen Source																	
See footnote[b]	+								+	+			+				
Growth Rate[c]																	
At 45°C	–	+			–		–							S			V
At 37°C	R	S	S	M	R	S	S	S	S	∓	S	S	–	S	S	S	R
At 31°C										M			S				
At 24°C	R	±	–	M	R	S	±	S	S	M	S	–	–	–	S	S	R
Colony																	
Always rough on 7H10	–	–	+	–	–	–	–	–	–	–	–	+	+	–	–	+	–
Usually smooth, thin		+					+		+								
Branched filamentous extension	–			–	+									+			V
Pigment																	
Photochromogen	–	–	–		–	–	–	+		+	–	–	–			–	–
Scotochromogen	–	–	–	+	–	–	–	–	+	–	–	–	–	+	+	–	V
Tests																	
Niacin	V	–	–			–	–				–	+	+			–	
Susceptible to isoniazid (1 μg/ml)	–	–	+	+	–	+	–	+	–	+	–	+	–	+	–	–	–
Susceptible to T_2H (10 μg/ml)		–	+			–	–				–	–	–			–	
Agglutination tests available	+	+			+		+	+	+	+							
Tween* hydrolysis (5 days)	–	–		+	∓	+	–	+	–	+	+		–	–	+	+	+
Tellurite reduction (3 days)	–	+		–	+	–	+		–		–			–	–	–	V
Nitrate reduction	–	–	–	+	+	–	–	+	–	–	+	+	–	–	–	+	+
Arylsulfatase (3 days)	+			+	+	–					–					+	–
MacConkey agar	±			–	+	–					–					–	–
Iron uptake				–	+	–			–		–			–	–	–	+

TABLE 8 (Continued)

DISTINCTIVE PROPERTIES OF MYCOBACTERIAL SPECIES

Property	1	2	3	4	5	6	7	8	9	10	11	12	13	14	15	16	17
Tests (continued)																	
68°C catalase	+	+	–	+	+	–	+				+	–	+	+		+	+
>45 mm catalase	+	–		+	+	–	–	+	+		+			–		+	+
5% NaCl tolerance	+			±	+	–					–					+	+

*Tradename, Atlas Chemical Industries, Inc.
[a]Group IV of Runyon = growing fully in less than one week.
[b]A plus sign indicates that the nature of the source of the specimen (as from a superficial body area) is information contributing to the identification.
[c]S = slow; M = moderate; R = rapid; V = variable.
[d]Thiophene-2-carboxylic acid hydrazide.

Data taken from: Blair, J. E., Lennette, E. H., and Truant, J. P., Eds., *Manual of Clinical Microbiology,* p. 122, American Society for Microbiology, Bethesda, Maryland, 1970. By permission of the copyright owners.

TABLE 9

PHYSIOLOGICAL CHARACTERISTICS OF PATHOGENIC NOCARDIA, ACTINOMADURA, AND STREPTOMYCES SPECIES

Species	Decomposition[a] of							
	Casein	Tyrosine	Xanthine	Starch	Gelatin	BCP[b] Milk	Urea	Acid from Arabinose and Xylose
Nocardia								
asteroides	–	–	–	–[c]	–[d]	–[e]	+	–
brasiliensis	+	+	–	–[c]	+	+	+	–
caviae	–	–	+	–[c]	–	–[e]	+	–
farcinica	–	–	–	–	–	–	–	
Actinomadura								
madurae	+	+	–	+	+	+	–	+
pelletieri	+	+	–	–	+	+	–	–
Streptomyces								
paraguayensis	+	+	+	–	+	+	+	
somaliensis	+	+	–	±	+	+	–	–

[a]Within two weeks at 27°C.
[b]BCP = bromcresol purple.
[c]About 50% of strains give a positive test result when a different method is used.
[d]Some strains reportedly liquefy certain gelatin media.
[e]Usually turns alkaline.

Data taken from: Blair, J. E., Lennette, E. H., and Truant, J. P., eds., *Manual of Clinical Microbiology,* p. 138, American Society for Microbiology, Bethesda, Maryland, 1970. By permission of the copyright owners.

Parasitic or Fermentative Actinomycetes

DR. LEO PINE

Twelve species of organisms distributed among five genera have been proposed for inclusion in the family Actinomycetaceae[1] (see Tables 1 and 2). Of the species listed, the taxonomical status of *Actinomyces eriksonii* remains to be firmly established; on the basis of glucose fermentation, it may well belong in the genus *Bifidobacterium.* The genus *Agromyces*[2] is microaerophilic and fermentative and lacks both catalase and a cytochrome system. It belongs to the Actinomycetaceae, but it is discussed as one of the soil actinomycetes (see page 204). All species listed in Table 2, except *Actinomyces humiferus,* have been isolated from human or animal sources, and several cause invasive disease.

In general, there are two microcolony types. Colonies showing soft, smooth convex surfaces with entire edges are associated with *Actinomyces bovis, A. eriksonii, A. odontolyticus,* or *Bifidobacterium bifidum.* All other species form rough-edged mycelial microcolonies that are composed of loosely attached branching cells or of true mycelium giving a spider-like appearance to the colony. The macrocolony, which may take 7 to 14 days to develop, may exhibit a wider range of structures. Regardless of the microcolony type, the macrocolony developing from it may vary from a smooth-surfaced convex colony to one resembling a cauliflower or a classical molar tooth. For excellent photographs of the various morphological forms, the reader is directed to References 3 to 8. Diagnostically, the source of the material and cellular or colonial morphology do not identify genera; nevertheless, they are strongly suggestive and often highly characteristic.

Differentiation between species can be made on the basis of the physiological characteristics listed in Table 3. When acid is formed, it is not accompanied by the formation of gas. The amount of acid produced depends on the rate of growth and on the period of incubation. In view of the slow growth rates of some strains, fermentation tubes should be incubated no less than 14 days. *Arachnia propionica,* for example, does not ferment glycerol readily, but some strains may do so in incubation periods longer than 7 days. Similarly, the formation of red colonies by some strains of *Actinomyces odontolyticus* requires up to 21 days on blood agar.

Serologically, each of the *Actinomyces* species may be identified by the specific-fluorescent-antibody technique; *A. bovis, A. israelii, A. viscosus,* and *A. odontolyticus,* each show two relatively specific species serotypes.[1,9-12]

Relationships between the effects of carbon dioxide and oxygen on growth are complex (see Table 4). In general, carbon dioxide is required for growth, and substantial amounts may be fixed into the fermentation products. Some species show a twofold increase in cell yields when grown aerobically in shake culture. Regardless of their ability to grow aerobically, cultures of *Actinomyces israelii, A. naeslundii,* and *Arachnia propionica* are maintained anaerobically. Of the species that have no catalase, only one, *Actinomyces humiferus,* reportedly requires oxygen for growth. *Actinomyces viscosus, Bacterionema matruchotii,* and *Rothia dentocariosa,* which are catalase-positive, are also maintained aerobically. Anaerobic *Bacterionema* species have been described.[13,14]

The anaerobic and aerobic fermentations of glucose by the various species of the family Actinomycetaceae are given in Table 5, with fermentations reported for closely related genera. The fermentations are of three types: a propionic acid fermentation (*Arachnia* and *Bacterionema*),[5,15-17] a homolactic acid fermentation (*Actinomyces* and *Rothia*),[18-20] and a heterolactic acid fermentation (*Bifidobacterium*).[21-23]

Just as the species are divided into groups forming lactic acid or propionic acid (Table 1), they may also be divided into two groups based on cell-wall composition (see Table 6). Thus the two genera fermenting glucose to propionic acid are the only members having diaminopimelic acid in the cell walls; all other species have lysine or lysine + ornitihine as the dibasic component of the cell-wall murein. Within this latter group a further division can be made; species can be divided into those having glucose and rhamnose or those primarily having galactose in the polysaccharide of the cell wall. If one considers oxygen relationship, cell-wall composition, and morphology, the actinomycetes exhibit a phylogenetic progression from a strictly homofermentative, catalase-negative, anaerobic bifurcated type of organism (*Actinomyces bovis*) that has lysine in its cell wall to a strict aerobe that has catalase and a cytochrome system,[24,25] diaminopimelic acid in its cell wall, and a specialized mycelial structure that produces spores (Streptomycetes). An examination of the G + C ratios of the DNA of various actinomycetes (see Volume II) also shows a general progression from about 57% for some Actinomycetaceae to about 75% for the Streptomycetes.

TABLE 1

CHARACTERISTICS DELINEATING GENERA OF ACTINOMYCETACEAE[a]

Oxygen Relationship	Catalase	Cell Wall	Fermentation	Major Products Formed From Glucose
Actinomyces				
Anaerobe, facultative anaerobe	+, –	Lysine + ornithine	Homolactic	Formate + acetate + lactate + succinate
Arachnia				
Facultative anaerobe	–	L-DAP, glycine	Homolactic	Acetate + propionate
Bacterionema				
Aerobe, facultative anaerobe	+, –	*meso*-DAP	Not reported	Lactate + propionate
Bifidobacterium				
Anaerobe	–	Ornithine	Heterolactic	Acetate + lactate
Rothia				
Aerobe	+	Lysine	Homolactic	Lactate

[a] Members of the family Actinomycetaceae are Gram-positive, non-acid-fast, non-motile, non-spore-forming, diphtheroid or branched rod-like bacteria that ferment carbohydrates, producing volatile and non-volatile acids.

TABLE 2

GENERA AND SPECIES OF FERMENTATIVE ACTINOMYCETES (FAMILY ACTINOMYCETACEAE)

Organism	Synonym	Habitat	Disease	References
Actinomyces				
bovis		Cattle	Lumpy jaw, actinomycosis	26–31
eriksonii		Man	Pulmonary and subcutaneous abscess	32
humiferus		Soil	None	33
israelii		Oral cavity, dental calculus, tonsillar crypts of man	Lacrimal canaliculitis, actinomycosis	6,8,26,31, 33
naeslundii		Oral cavity, dental calculus, tonsillar crypts	Actinomycosis	6,35–37
odontolyticus		Human dental caries	None	38,39
suis		Swine	Swine mammary actinomycosis	40,41
viscosus	Hamster organism *Odontomyces viscosus*	Oral cavity of hamster, human dental calculus	Subginigival plaque, peridontal disease	20,42–45
Arachnia				
propionica	*Actinomyces propionicus*	Human tissues	Actinomycosis, lacrimal canaliculitis	5,17,36,46, 47
Bacterionema				
matruchotii	*Leptotrichia*	Oral cavity of man and primates	None	13,14,16,48, 49
Bifidobacterium				
bifidum	*Lactobacillus bifidus, Actinomyces parabifidus, Lactobacillus parabifidus, Actinobacterium bifidum*	Stool of milk-fed infants, adult stool, rat feces, bovine rumen, turkey liver	None	50–56
Rothia				
dentocariosa	*Nocardia salivae*	Oral cavity of man	None	4,57–59

TABLE 3

FERMENTATIVE AND METABOLIC CHARACTERISTICS OF ACTINOMYCETACEAE

(A = acid produced; An = anaerobically; Ae = aerobically; NR = not reported)

Organism	Catalase	Nitrate Reduction	Casein and Gelatin Hydrolysis	Starch Hydrolysis	Glucose	Starch	Mannitol	Ribose	Xylose	Glycerol	Color on Blood Agar	Oxygen Requirement
Actinomyces												
bovis	–	–	–	+	A	A	–	–	A(–)[a]	A(–)[a]	White	$+CO_2$ An
eriksonii	–	–	–	+	A	A	A	A	A	–	White	An
humiferus	–	–	+[b]	+	A[c]	A	A	–	A	±	White	Ae
israelii	–	+(–)[a]	–	–	A	–	A	A	A	–	White	$+CO_2$ An
naeslundii	–	+	–	–	A	–	–	–	–	–	White	$+CO_2$ An
odontolyticus	–	+	–	–	A	–	–	–	A(–)[a]	A	Dark Red	An
suis	–	NR	–	+	A	A	A(–)[a]	NR	A	A	Brown to red on blood	An
viscosus	+	+	–	+	A	±	–	–	–		White	$+CO_2$ An
Arachnia												
propionica	–	+			A	–	A	–	–	–(A)[a]	White	An
Bacteriònema												
matruchotii	+	+	–	+(–)[a]	A	NR	–	NR	–	–	White	Ae
Bifidobacterium												
bifidum	–	–	–	–	A	–	–	NR	–	–	White	An
Rothia												
dentocariosa	+	+	–	–	A	–	–	–	–	A	White	Ae

[a] A few strains of the species give the reaction shown in parentheses.

[b] Strong + for casein, ± for gelatin.[33]

[c] Fermentation of glucose is limited; fructose is readily fermented.[33]

Data compiled from: Slack, J. M., *Report to the International Committee on Bacterial Nomenclature, Subgroup on Taxonomy of Microaerophilic Actinomycetes.* Department of Microbiology, West Virginia University, Morgantown, West Virginia (1969).

TABLE 4

RELATIONSHIPS BETWEEN ANAEROBIOSIS, CARBON DIOXIDE, OXYGEN, AND GROWTH IN SPECIES OF THE FAMILY ACTINOMYCETACEAE

(na = not applicable; an = anaerobic; ae = aerobic)

Code

Column 1: *Actinomyces bovis*
Column 2: *Actinomyces eriksonii*
Column 3: *Actinomyces humiferus*
Column 4: *Actinomyces israelii*
Column 5: *Actinomyces naeslundii*
Column 6: *Actinomyces viscosus*
Column 7: *Arachnia propionica*
Column 8: *Bacterionema matruchotii*
Column 9: *Bifidobacterium bifidum*
Column 10: *Rothia dentocariosa*

Conditions of Growth	1	2	3	4	5	6	7	8	9	10
Net fixation of CO_2	+	+	?	+	+	+	–	–	±	–
CO_2 required for anaerobic growth	+	?	?	+	+	+	–	–	+	na
Anaerobic growth increased by CO_2	+	–	–	+	+	+	–	?	–	na
Aerobic growth	+	–	+	+	+	+	+	+	–	+
Aerobic growth requires CO_2	+	na	–	+	–	+	–	–	na	–
Aerobic growth (in CO_2) is two or more times greater than anaerobic growth	–	na	?	+	+	+	–	+	na	+
Aerobic growth changes fermentation, signifying complete or partial oxidation of substrate	na	na	+	+	+	+	+	+	na	+
Aerobic formation of CO_2	na	na	?	+	+	+	+	+	na	+
Catalase	–	–	–	–	–	+	–	+	–	+
Maintenance of stock cultures	an	an	ae	an	an	ae	an	ae	an	ae

Data compiled from References 5, 7, 16, 20, 33, 60, 61, and 62.

TABLE 5

ANAEROBIC AND AEROBIC FERMENTATIONS OF ACTINOMYCETACEAE AND RELATED ORGANISMS

Organisms	Fermentations
Anaerobic Cultures	
Actinomyces bovis, israelii, naeslundii, viscosus	1 glucose → 2 lactate
	3.5 glucose + 3 CO_2 → 3 formate + 3 acetate + 3 succinate + 1 lactate
	1 glucose + 2 malate → 2 formate + 2 acetate + 2 succinate
Arachnia propionica	18 glucose → 10 CO_2 + 2 formate + 10 acetate + 23 propionate + 1 succinate + 1 lactate
Bifidobacterium bifidum	10 glucose + CO_2 → 1 formate + 13 acetate + 10 lactate + 1 succinate
	10 glucose → 15 acetate + 10 lactate
Propionibacterium acnes	10 glucose → 5 CO_2 + 4 acetate + 12 propionate + 2 succinate + 1 lactate
Propionibacterium pentosaceum	10 glucose → 6 CO_2 + 2 acetate + 14 propionate + 2 succinate
Anaerobic Microaerophilic Cultures	
Actinomyces humiferus	1 glucose → 2 lactate
Bacterionema matruchotii	10 glucose → 6 propionate + 13 lactate
Aerobic Shake Cultures	
Actinomyces israelii, naeslundii; Arachnia propionica; Propionibacterium acnes, arabinosum	1 glucose + O_2 → 2 acetate + 2 CO_2
Actinomyces viscosus	10 glucose → 10 lactate + 2 succinate + 1 formate + 1 acetate
Bacterionema matruchotii	10 glucose → 3 acetate + 4 propionate + 5 lactate
Rothia dentocariosa	10 glucose → 15 lactate + traces of formate, acetate and succinate

Data compiled from References 5, 18, 33, 36, 60, 61, 62, 63, and 64.

TABLE 6

CELL-WALL COMPOSITION OF ACTINOMYCETACEAE AND RELATED SPECIES[a]

Organism	Catalase	Aspartic Acid	Lysine	Ornithine	Glycine	DAP	Galactose	Mannose	Glucose	Rhamnose	Fucose	Deoxytalose	Arabinase
Actinomyces													
bovis													
ATCC 13683	–	+	+					+	+	+	+	+	
P2R	–		+					+	+	+	+	+	
13R	–		+					+	+	+	+	+	
eriksonii	–	+	+				+				+		
humiferus	–	+	+	+					±	+	±		
israelii													
ATCC 12102, Type 1	–		+	+			+						
ATCC 12102, Type 2	–	–	+	+			+			+			
israelii (296)													
ATCC 10049	–	±	+	?			+	+		+			
naeslundii (279)													
ATCC 12104	–	±	+	+				+	+	+	+	+	
odontolyticus	–		+	+			+		+	±	±	±	
viscosus	+		+	+			+	+	+				
Arachnia													
propionica	–				+	L	+		+				+
Bacterionema													
matruchotii	+					*meso*	+	+	+				
Bifidobacterium													
bifidum	–			+			+		+				
corynebacterium													
Propionibacterium													
species	+				+	L	+		+				
acnes	+				+	L	+		+				
diphtheriae	+					*meso*	+						+
Rothia													
dentocariosa	+		+				+						

[a] All strains have alanine and glutamic acid.

Data compiled from References 2, 5, 26, 35, 57, 65–70.

REFERENCES

Introduction

1. Lechevalier, H. A., and Lechevalier, M. P., *Annu. Rev. Microbiol., 21,* 71 (1967).
2. Prauser, H., *The Actinomycetales.* Gustav Fischer, Jena, Germany (1970).
3. Waksman, S. A., *The Actinomycetes.* Ronald Press, New York (1967).
4. Williams, S. T., Davies, F. L., and Cross, T., in *Identification Methods for Microbiologists,* B, p. 111, B. M. Gibbs and D. A. Shapton, Eds. Academic Press, London, England (1968).
5. Pine, L., and George, L. K., *Int. J. Syst. Bacteriol., 15,* 143 (1965).
6. Pine, L., *Int. J. Syst. Bacteriol., 20,* 445 (1970).
7. Lechevalier, H. A., and Pramer, D., *The Microbes.* J. B. Lippincott, Philadelphia (1971).
8. Lechevalier, H. A., Lechevalier, M. P., and Holbert, P. E., *J. Bacteriol., 92,* 1228 (1966).

Soil or Oxidative Actinomycetes

1. Becker, B., Lechevalier, M. P., Gordon, R. E., and Lechevalier, H. A., *Appl. Microbiol., 12,* 421 (1964).
2. Prauser, H., (Ed.), *The Actinomycetales.* Gustav Fischer, Jena, Germany (1970).
3. Lechevalier, M. P., *J. Lab. Clin. Med., 71,* 934 (1968).
4. Lechevalier, M. P., Horan, A. C., and Lechevalier, H. A., *J. Bacteriol., 105,* 313 (1971).
5. Becker, B., Lechevalier, M. P., and Lechevalier, H. A., *Appl. Microbiol., 13,* 236 (1965).
6. Yamaguchi, T., *J. Bacteriol., 89,* 444 (1965).
7. Sukapure, R. S., Lechevalier, M. P., Reber, H., Higgins, M. L., Lechevalier, H. A., and Prauser, H., *Appl. Microbiol., 19,* 527 (1970).
8. Gledhill, W. E., and Casida, L. E., *Appl. Microbiol., 18,* 340 (1969).
9. Higgins, M. L., Lechevalier, M. P., and Lechevalier, H. A., *J. Bacteriol., 93,* 1446 (1967).
10. Shirling, E. B., and Gottlieb, D., *Int. J. Syst. Bacteriol., 16,* 313 (1966).
11. Locci, R., Baldacci, E., and Petrolini-Baldan, B., *G. Microbiol., 17,* 1 (1969).
12. Lechevalier, M. P., Lechevalier, H. A., and Holbert, P. E., *Ann. Inst. Pasteur, 114,* 277 (1968).
13. Lechevalier, H., and Lechevalier, M. P., *J. Bacteriol., 100,* 522 (1969).
14. Luedemann, G. M., *J. Bacteriol., 96,* 1848 (1968).
15. Lechevalier, M. P., and Gerber, N. N., *Carbohyd. Res., 13,* 451 (1970).
16. Waksman, S. A., and Lechevalier, H. A., *Antibiotics of Actinomycetes.* Williams and Wilkins, Baltimore (1962).
17. Waksman, S. A., *The Actinomycetes.* Ronald Press, New York (1967).
18. Gordon, M. A., in *Manual of Clinical Microbiology,* p. 137, J. E. Blair, E. H. Lennette, and J. P. Truant, Eds. American Society for Microbiology, Bethesda, Maryland (1970).
19. Lechevalier, H. A., and Lechevalier, M. P., *Annu. Rev. Microbiol., 21,* 71 (1967).
20. Wenzel, F. J., Emanuel, D. A., and Lawton, B. R., *Amer. Rev. Resp. Dis., 95,* 652 (1967).
21. Runyon, E. H., Kubica, G. P., Morse, W. C., Smith, C. R., and Wayne, L. G., in *Manual of Clinical Microbiology,* p. 112, J. E. Blair, E. H. Lennette, and J. P. Truant, Eds. American Society for Microbiology, Bethesda, Maryland (1970).
22. Asselineau, J., Lanéele, M. A., and Chamoiseau, G., *Rev. Elevage Med. Vet. Pays Trop., 22,* 205 (1969).
23. Siegmund, O. H., and Eaton, L. G., (Eds.), *The Merck Veterinary Manual,* 3rd ed. Merck and Company, Inc., Jersey (1967).
24. Lacey, J., *J. Gen. Microbiol., 41,* 406 (1965).

Parasitic or Fermentative Actinomycetes

1. Slack, J. M., *Report to the International Committee on Bacterial Nomenclature, Subgroup on Taxonomy of Microaerophilic Actinomycetes.* Department of Microbiology, West Virginia University, Morgantown, West Virginia (1969).
2. Gledhill, W. E., and Casida, L. E., *Appl. Microbiol., 18,* 340 (1969).
3. Brock, D. W., and Georg, L. K., *J. Bacteriol., 97,* 589 (1969).
4. Brown, J. M., Georg, L. K., and Waters, L. C., *Appl. Microbiol., 17,* 150 (1969).
5. Buchanan, B. B., and Pine, L., *J. Gen. Microbiol., 28,* 305 (1962).
6. Howell, A., Jr., Murphy, W. C., Paul, F., and Stephan, R. M., *J. Bacteriol., 78,* 82 (1959).
7. Pine, L., Howell, A., Jr., and Watson, S. J., *J. Gen. Microbiol., 23,* 403 (1960).
8. Slack, J. M., Langried, S., and Gerencser, M. A., *J. Bacteriol., 97,* 873 (1969).
9. Blank, C. H., and Georg, L. K., *J. Lab. Clin. Med., 71,* 283 (1968).
10. Brock, D. W., and Georg, L. K., *J. Bacteriol., 97,* 581 (1969).
11. Lambert, F. W., Brown, J. M., and Georg, L. K., *J. Bacteriol., 94,* 1287 (1967).
12. Slack, J. M., and Gerencser, M. A., *J. Bacteriol., 103,* 266 (1970).
13. Gilmour, M. N., *Bacteriol. Rev., 25,* 142 (1961).
14. Gilmour, M. N., and Beck, P. H., *Bacteriol. Rev., 25,* 152 (1961).
15. Allen, S. H. G., Kellermeyer, R. W., Stjernholm, R. L., and Wood, H. G., *J. Bacteriol., 87,* 171 (1964).
16. Howell, A., Jr., and Pine, L., *Bacteriol. Rev., 25,* 162 (1961).
17. Pine, L., and Hardin, H., *J. Bacteriol., 78,* 165 (1959).
18. Buchanan, B. B., and Pine, L., *J. Gen. Microbiol., 46,* 225 (1967).
19. Buyze, G., van den Hamer, J. A., and DeHaan, P. G., *Antonie van Leeuwenhoek J. Microbiol. Serol., 23,* 345 (1951).
20. Howell, A., Jr., and Jordan, H. V., *Sabouraudia, 3,* 93 (1963).
21. DeVries, W., and Stouthamer, A. H., *J. Bacteriol., 93,* 574 (1967).
22. DeVries, W., Gerbrandy, S. J., and Stouthamer, A. H., *Biochem. Biophys., 136,* 415 (1967).
23. Scardovi, V., and Trovatelli, L. D., *Ann. Microbiol. Enzimol., 15,* 19 (1965).
24. Domnas, A., and Grant, N. G., *J. Bacteriol., 101,* 652 (1970).
25. Hein, A. H., Silver, W. S., and Birk, Y., *Nature, 180,* 608 (1957).

26. Cummins, C. S., *J. Gen. Microbiol., 28,* 35 (1962).
27. Erikson, D., *Med. Res. Counc. (Gt. Brit.) Spec. Rep. Ser.,* No. 240, 1 (1940).
28. Harz, C. O., *Deut. Z. Tiermed., 5,* 125 (1879).
29. Silberschmidt, W., *Zentralbl. Bakteriol. Parasitenk. Infektionskr. Hyg. Abt. I. Orig., 27,* 486 (1900).
30. Silberschmidt, W., *Z. Hyg. Infektionskr., 37,* 345 (1901).
31. Thompson, L., *Proc. Staff Meet. Mayo Clinic, 25,* 81 (1950).
32. Georg, L. K., Roberstad, G. W., Brinkman, J. A., and Hicklin, M. D., *J. Infect. Dis., 115,* 88 (1965).
33. Gledhill, W. E., and Casida, L. E., *Appl. Microbiol., 18,* 114 (1969).
34. Wolff, M., and Israel, J., *Virchows Arch. Abt. A Pathol. Anat., 126,* 11 (1891).
35. Pine, L., and Boone, C. J., *J. Bacteriol., 94,* 875 (1967).
36. Pine, L., Hardin, H., Turner, L., and Roberts, S. S., *Amer. J. Ophthalmol., 49,* 1278 (1960).
37. Thompson, L., and Lovestadt, S. A., *Proc. Staff Meet. Mayo Clinic, 26,* 169 (1951).
38. Batty, I., *J. Pathol. Bacteriol., 75,* 455 (1958).
39. Georg, L. K., and Coleman, R. M., Comparative Pathogenicity of Various Actinomyces Species, in *The Actinomycetales,* p. 35, H. Prauser, Ed. Gustav Fischer, Jena, Germany (1970).
40. Gräser, R., *Zentralbl. Bakteriol. Parasitenk. Infektionskr. Hyg. Abt. I. Orig., 184,* 478 (1962).
41. Gräser, R., *Zentralbl. Bakteriol. Parasitenk. Infektionskr. Hyg. Abt. I. Orig., 188,* 251 (1963).
42. Georg, L. K., Pine, L., and Gerencser, M. A., *Int. J. Syst. Bacteriol., 19,* 291 (1969).
43. Gerencser, M. A., and Slack, J. M., *Appl. Microbiol., 18,* 80 (1969).
44. Howell, A., Jr., *Sabouraudia, 3,* 81 (1963).
45. Jordan, H. V., and Keyes, P. H., *Arch. Oral Biol., 9,* 401 (1967).
46. Gerencser, M. A., and Slack, J. M., *J. Bacteriol., 94,* 109 (1967).
47. Pine, L., and Georg, L. K., *Int. J. Syst. Bacteriol., 19,* 267 (1969).
48. Cock, D. J., and Bown, W. H., *J. Periodontal Res., 2,* 36 (1967).
49. Gilmour, M. N., Howell, A., Jr., and Bibby, B. J., *Bacteriol. Rev., 25,* 131 (1961).
50. Harrison, A. P., and Hansen, J. A., *J. Bacteriol., 60,* 543 (1954).
51. Orla-Jensen, J., Orla-Jensen, A. D., and Winther, O., *Zentralbl. Bakteriol. Parasitenk. Infektionskr. Hyg. Abt. II. Naturwiss., 93,* 321 (1936).
52. Pine, L., and Georg, L. K., *Int. J. Syst. Bacteriol., 15,* 143 (1965).
53. Prévot, A. R., *Traité de systématique bactérienne,* Vol. 2. Dunnod, Paris, France (1961).
54. Puntoni, V., *Ann. Ig. Sper., 47,* 157 (1937).
55. Weiss, J. E., and Rettger, L. F., *J. Bacteriol., 35,* 17 (1938).
56. Weiss, J. E., and Rettger, L. F., *J. Infect. Dis., 62,* 115 (1938).
57. Davis, G. H. G., and Freer, J. H., *J. Gen. Microbiol., 23,* 163 (1960).
58. Onisi, M., *Shikagaku Zasshi, 6,* 273 (1949).
59. Roth, G. D., and Thurn, A. N., *J. Dent. Res., 41,* 1279 (1962).
60. Buchanan, B. B., and Pine, L., *Sabouraudia, 3,* 26 (1963).
61. Inchikawa, Y., *Chem. Abstr., 51,* 11458 (1957).
62. Pine, L., and Howell, A., Jr., *J. Gen. Microbiol., 15,* 428 (1956).
63. Moore, W. E. C., and Cato, E. P., *J. Bacteriol., 85,* 870 (1963).
64. Wood, H. G., and Werkman, C. H., *Biochem. J., 30,* 618 (1936).
65. Becker, B., Lechevalier, M. P., and Lechevalier, H. A., *Appl. Microbiol., 13,* 236 (1965).
66. Cummins, C. S., in *The Actinomycetales,* p. 29, H. Prauser, Ed. Gustav Fischer, Jena, Germany (1970).
67. Cummins, C. S., and Harris, H., *J. Gen. Microbiol., 14,* 583 (1956).
68. Cummins, C. S., and Harris, H., *J. Gen. Microbiol., 18,* 173 (1958).
69. Cummins, C. S., Glendenning, O. M., and Harris, H., *Nature, 180,* 337 (1957).
70. Deweese, M. S., Gerencser, M. A., and Slack, J. M., *Appl. Microbiol., 16,* 1713 (1968).

HETEROTROPHIC RODS AND COCCI

Introduction

DR. HUBERT A. LECHEVALIER*

In the preceding chapters on bacteria, information has been given about bacteria that are either autotrophic or have striking morphological properties. On the following pages we are presenting a modicum of information about rods and cocci that utilize preformed organic matter and that have not been previously discussed. In the absence of a new edition of *Bergey's Manual,* we are simply separating these organisms on the basis of (1) morphology (rods *vs* cocci), (2) staining properties (Gram-positive *vs* Gram-negative), and (3) oxygen requirement (aerobic *vs* anaerobic). Such groupings are not very satisfactory, of course, since some bacteria are coccoid rods, some are Gram-variable, and some are microaerophilic. In any case, for better or for worse, we have grouped the remaining bacteria as follows:

The sections on Enterobacteriaceae, the genus *Pseudomonas* and Cocci have been written by Dr. Carl F. Clancy. These sections have their own lists of references. In the rest of the text, information is based on the consultation of the following general references, with recent references occasionally being given in the text.

GENERAL REFERENCES

1. Blair, J. E., Lennette, E. H., and Truant, J. P. (Eds.), *Manual of Clinical Microbiology.* American Society for Microbiology, Bethesda, Maryland (1970).
2. Breed, R. S., Murray, E. G. D., and Smith, N. R. (Eds.), *Bergey's Manual of Determinative Bacteriology.* Williams and Wilkins, Baltimore, Maryland (1957).
3. Davis, B. D., Dulbecco, R., Eisen, H. N., Ginsberg, H. S., and Wood, W. B., *Microbiology.* Hoeber Medical Division, Harper and Row, New York (1968).
4. Prévot, A. R., Turpin, A., and Kaiser, P., *Les Bactéries Anaérobies.* Dunod, Paris, France (1967).
5. Skerman, W. B. D., *A Guide to the Identification of the Genera of Bacteria,* 2nd ed. Williams and Wilkins, Baltimore, Maryland (1967).

* The author gratefully acknowledges the kind assistance of Dr. D. Kronish, Warner-Lambert Research Institute.

Aerobic Gram-Negative Rods

DR. HUBERT A. LECHEVALIER

SMALL MOTILE RODS OR VIBRIOS CAUSING PHAGE-LIKE LYSIS OF BACTERIA

Bdellovibrio

Vibrios or rods, 0.35 X 1–2 μm. They are motile with a single thick (28 nm) sheathed flagellum; speed of travel is estimated at 100 cell lengths per second. The organisms are capable of attachment to bacterial cells (usually Gram-negative) that they can penetrate. Multiplication takes place within the bacterial host cell (host-independent strains are known). Release of the extracellular form is effected by lysis of the host cell. Type species: *B. bacterovorus,* found in soil, water, and sewage (Starr, M. P., and Seidler, R. J., *Annu. Rev. Microbiol., 25,* 649–678, 1971).

CURVED RODS FORMING HALF-CIRCLES, CIRCLES AND/OR COILS

Microcyclus

Slightly curved rods about 1 μm long, forming horseshoe-like structures, closed ring-like cells, or even curved corkscrew-shaped filaments. They are non-motile and are found in fresh-water bodies and soil. The type species, *M. aquaticus,* is colorless; the growth of another species, *M. flavus,* is yellow (Raj, H. D., *Int. J. Syst. Bacteriol., 20,* 61–81, 1970).

RODS CAPABLE OF FIXING ATMOSPHERIC NITROGEN IN LABORATORY MEDIA

There are numerous reports of microorganisms that are capable of fixing atmospheric nitrogen non-symbiotically. Blue-green algae, fungi, and some bacteria, such as species of *Azotobacter, Azotomonas* and *Klebsiella* (Mahl, M. C., Wilson, P. W., Fife, M. A., and Ewing, W. H., *J. Bacteriol., 89,* 1482–1487, 1965), have been implicated. The most studied and probably the most important bacterial non-symbiotic nitrogen fixers belong to the genus *Azotobacter.*

Azotobacter

Spherical to rod-shaped, these organisms may exceed 2 μm in diameter. They are non-motile, or motile with a single or several peritrichous flagella. In older cultures, the cell wall thickens and a melanin pigment is deposited as "cysts" are formed (Cagle, G. D., and Vela, G. R., *J. Bacteriol., 107,* 315–319, 1971). *Azotobacter* are found in soil and water. The type species is *A. chroococcum.*

A. agilis. The cells are almost spherical. A slightly yellowish pigment is produced. The genus *Azotococcus* has been proposed to harbor this species.

A. indicus. Bicellular organisms or unicellular organisms with polar refringent bodies. The genus *Beijerinckia* has been proposed for this organism and similar species (Hilger, F., *Ann. Inst. Pasteur (Paris), 109,* 406–423, 1965).

Azotomonas

Rods that may be up to 5 μm long. They are motile by means of one to three polar flagella. Sugars are utilized with the production of acid or acid and gas. *Azotomonas* are found in soil. The type species is *A. insolita.*

RODS PRODUCING NODULES AND MALFORMATIONS OF ROOTS AND PLANTS

Bacteria of the genus *Rhizobium* have the capacity to induce the formation of nitrogen-fixing nodules on the root of leguminous plants. Those of the genus *Agrobacterium* produce crown-gall tumors on many higher plants. In addition, bacteria from the genera *Chromobacterium, Phyllobacterium, Klebsiella* and *Xanthomonas* may form nodules on the leaves of plants of the Myrsinaceae and Rubiaceae families (Horner, H. T., and Lersten, N. R., *Int. J. Syst. Bacteriol., 22,* 117–122, 1972).

Rhizobium

Rods that may be up to 3 μm long. They are non-motile, or motile with flagella that may be either single or peritrichous. The organisms are aerobic and capable of forming nitrogen-fixing nodules on the roots of legumes; they do not fix nitrogen on laboratory media. They are pleomorphic (T- and Y-shaped) in the host plant. Different species and different subspecies have different host ranges. The type species is *R. leguminosarum* (Dixon, R. O. D., *Annu. Rev. Microbiol., 23,* 137–158, 1969).

Agrobacterium

Rods that may be up to 3 μm long. They are usually motile with flagella that may be either single or peritrichous. Growth is good on nutrient agar. Most species are pathogenic to plants and produce hyperplasia. The type species is *A. tumefaciens.*

RODS WITH FASTIDIOUS GROWTH REQUIREMENTS

Pasteurella

These coccobacilli sometimes occur in pairs. They are facultatively anaerobic. Carbohydrates are fermented without production of gas. Enriched media, such as blood or serum agar, may be needed for growth, especially for primary isolation. *Pasteurellae* are non-hemolytic. They are parasitic on warm-blooded animals. Four species of *Pasteurella* are of special importance.

P. multocida. The cause of fowl cholera and of hemorrhagic septicemia of rabbits, rats, horses, sheep, dogs, cats, chinchilla, and swine. It is also often found to be associated with pneumonia in various animals.

P. (Yersinia) pestis. The cause of plague in man. It infects a number of rodents and is transmitted from rat to rat and from rat to man by fleas. The bacilli introduced by the flea bite enter the dermal lymphatics and are transported to the regional lymph nodes, where they form enlarged tender buboes; in the terminal stages bacteremia develops. A pneumonic form of the disease, with mortality approaching 100% in untreated patients, often develops under crowded conditions; early treatment with streptomycin, chloramphenicol or tetracycline is effective.

P. (Yersinia) pseudotuberculosis. The only one of these four species of *Pasteurella* to be flagellated; this property is observed only when growth takes place between 18 and 37°C. The organism is the cause of pseudotuberculosis of rabbit, chinchilla, and other animals. The tubercular lesions bear a superficial relationship to those formed during the course of tuberculosis.

P. (Francisella) tularensis. This organism causes tularemia, a disease of wild rodents, especially rabbits. It is transmitted to man by contact with infected tissues and is thus most common in hunters. The bites of ticks and deer flies can also transmit the infection to man, where it takes many forms, such as ulceroglandular, oculoglandular, pneumonic, or typhoidal. The latter two forms are the most serious, with mortality rates of about 30% in untreated patients. Streptomycin is the drug of choice.

Bordetella

These coccobacilli usually occur in pairs. They grow on media containing blood, but not on ordinary meat infusion; they also grow on media containing charcoal, suggesting that toxic factors may be adsorbed. They are hemolytic. *Bordetella* cause respiratory-tract infections of warm-blooded animals. Three species are known: the non-motile *B. pertussis* and *B. parapertussis* found in man, and the motile *B. bronchiseptica,* which is associated with animals.

B. pertussis. The cause of whooping cough, an infection of the mucous membranes of the respiratory tract mainly occurring in children. The bacterium is very delicate, and its cultivation requires special media, such as a peptone–glycerol–potato-extract–blood medium. Immunization is highly successful, but chemotherapy is not very effective.

B. parapertussis. This organism differs from *B. pertussis* in producing larger colonies on solid media, in splitting urea, and in utilizing citrate as a sole source of carbon. It causes a mild form of whooping cough.

B. bronchiseptica. This species is also urease- and citrate-positive. It may be found in healthy rodents, such as rabbits; in these animals it may cause bronchopneumonia. One common problem in the care of laboratory animals is the passage of this pathogen from a healthy carrier to a sensitive cagemate.

Brucella

These non-motile coccobacilli grow slowly on tryptose-containing laboratory media; growth is often enhanced by increasing the carbon dioxide in the atmosphere. The organisms cause brucellosis, a disease of cattle, swine, sheep, and goats, and occasionally affecting horses. *Brucellae* have an affinity for the mammary glands and the reproductive organs of their animal hosts and often cause contagious abortion in animals (Bang's disease). Brucellosis is acquired by man through contact with diseased tissues or by ingestion of contaminated milk. The human disease, called Malta fever or undulant fever, is debilitating, but usually self-limiting. Treatment with streptomycin and the tetracyclines is of assistance, but eradication of the bacteria is difficult, due to their intracellular location. Immunological control has not been successful. Three species of *Brucella* are usually recognized: *B. melitensis,* mainly found in goats; *B. abortus,* principally affecting cows; and *B. suis,* whose chief host is the hog. Differences between the species are based on quantitative physiological differences rather than qualitative ones.

Haemophilus

These coccobacilli are facultatively anaerobic. They are strict parasites of vertebrates, but not always pathogenic. The organisms require blood for growth on laboratory media; individual species have been found to require more specifically, singly or in combination, hemin (X-factor), phosphopyridine nucleotide (V-factor), diphosphothiamine, or adenosine. Some species are hemolytic. Hemolysis may be of the α-type (green) or of the β-type (colorless). At least a dozen species are recognized.

H. influenzae. The most important human pathogen of this genus. Its virulence is directly related to a capsule that contains a polysaccharide, which is useful for serotyping. Young children are frequently carriers of this bacterium, but they usually develop natural immunity by the age of 10 years. *Haemophilus influenzae* usually strikes children in the upper respiratory tract, where it may cause serious obstructive laryngitis. Bacteremia may occur, leading to a meningitis that is usually caused by serotype "b" strains. Chloramphenicol in combination with sulfadiazine constitutes the treatment of choice.

H. ducreyi. The most difficult species to cultivate. It causes soft chancre or chancroid, which usually can be treated effectively with sulfonamides

H. aegyptius. This species produces purulent conjunctivitis.

H. parainfluenzae. Most often non-pathogenic, but may cause bacterial endocarditis.

Other species of *Haemophilus* have been isolated from cattle, dogs, swine, ferret, fowl, and even trout.

Actinobacillus

These coccobacilli may be pleomorphic on some media, even forming filaments. They are non-motile, facultatively anaerobic, and grow best on serum or blood agars in the presence of 10% carbon dioxide. The organisms are avirulent for the usual laboratory animals, but they cause diseases of domestic animals. Two species are of importance.

*A. **lignieresii.*** The cause of actinobacillosis, a disease similar to actinomycosis but mainly affecting soft tissues rather than bones; it occurs most frequently in cattle, occasionally in sheep, and rarely in man. In cattle, the proliferation of granule-containing abscesses leads to a "wooden tongue". Treatment is partially surgical and partially systemic with iodides or antibiotics.

*A. **(Malleomyces) mallei.*** The cause of equine glanders, a disease that may occur in one of three forms: (1) nasal, (2) pulmonary, and (3) cutaneous; this third form is also called farcy and should not be confused with bovine farcy, which is caused by *Mycobacterium (Nocardia) farcinicum. Felidae* and man may be affected, and the condition is usually fatal. Glanders is characterized by the formation of nodules or tubercles that tend to break down, forming ulcers. The prognosis is unfavorable, but sulfadiazine has been effective. *A. mallei* has been the object of an extensive study by D. H. Evans (*Can. J. Microbiol., 12,* 609–652, 1966). See also page 239, *Pseudomonas.*

Calymmatobacterium (Donovania) granulomatis

The causative agent of granuloma inguinale of man; it is not pathogenic for laboratory animals. This microaerophilic organism is considered by many to be closely related to *Klebsiella pneumoniae* and to be transmitted by sexual contacts. The initial lesion commonly appears around the genitalia and forms an ulcer, which spreads over a large area. Treatment with the tetracyclines, chloramphenicol or streptomycin is effective. *Calymmatobacterium granulomatis* is a non-motile coccobacillus, best isolated by inoculation on media containing unheated embryonic egg yolk. Upon cultivation it becomes less fastidious.

Mima, Herellea, Acinetobacter and *Moraxella*

A number of coccoid bacteria, often occurring as diplococci and with a tendency to pleomorphism that in some cases leads to filament formation, have been isolated repeatedly from clinical specimens and have received a number of names. Often they were given one of the four generic names listed above. They can be separated into two main groups as follows:

oxidase-positive–*Moraxella*
oxidase-negative–*Acinetobacter*

Moraxella

Moraxella bovis. The cause of infectious keratoconjunctivitis (pink-eye) of cattle, which is characterized by photophobia, lacrimation, and eventually ulceration of the cornea. Chloramphenicol, the tetracyclines, and nitrofurazone are beneficial, especially if treatment is started early.

Moraxella lacunata. Isolated from cases of human conjunctivitis.

Acinetobacter

The Acinetobacters can be separated into two main species: (1) the carbohydrate oxidizers, *Herellea (Acinetobacter) vaginicola,* and (2) the nonoxidizers, *Mima (Acinetobacter) polymorpha.* They are widely distributed in nature and have been recovered from every part and fluid of the human body, as well as from soaps, milk, foods, water, and also from animal sources.

The taxonomy of all these organisms is still in a state of flux (Baumann, P., Doudoroff, M., and Stanier, R. Y., *J. Bacteriol., 91,* 58–73, 1520–1541, 1968; Samuels, S. B., Pittman, B., Tatum, H. W., and Cherry, W. B., *Int. J. Syst. Bacteriol., 22,* 19–38, 1972).

Streptobacillus

These highly pleomorphic rods are less than 1 μm wide, often forming long, curved and looped filaments that may be 150 μm long. In older cultures, spheroid granule-containing bodies that may be 15 μm wide may occur. The organisms are facultatively anaerobic. Blood serum or ascitic fluid is required for growth.

S. moniliformis. The cause of a rat-bite fever in man. Following the bite by a rodent, there is normal healing of the wound, which is followed after approximately ten days by a prostrating fever with chills, vomiting, headache, and a cutaneous eruption. Painful arthritic swellings may occur. In untreated cases, mortality may reach 10%. Penicillin, streptomycin, or chlortetracycline have been used successfully for treatment.

RODS TO COCCI GROWING WELL ON ORDINARY LABORATORY MEDIA WITH 12 TO 30% SALT ADDED

Halobacterium

These pleomorphic rods may be up to 8 μm long and are usually associated with coccoid forms. They are polarly flagellated when motile. The organisms are colorless, yellow, orange, or bright red. They are found in tidal pools, especially in the tropics, and in salt ponds, brines, salted fish, salted hides, etc. The type species is *H. salinarium.*

RODS, NON-MOTILE OR FLAGELLATED, GROWING WELL ON ORDINARY LABORATORY MEDIA

Alcaligenes

These rods may be up to 3 μm long. They are non-motile, or motile with polar or peritrichous flagella. Some species are facultatively anaerobic. No acid or gas is produced from glucose or lactose; litmus milk is turned strongly alkaline. *Alcaligenes* are found in the intestinal tract of animals and in milk or water. See also Tables 1 and 2 of *Pseudomonas.* (pp. 240–241).

A. faecalis. Considered an opportunistic pathogen in the intestine and feces of man. It may cause urinary-tract infections and more serious diseases in debilitated patients.

A. viscolactis. Causes ropiness of milk.

Achromobacter

These rods may be up to 8 μm long, aerobic to facultatively anaerobic, and non-motile or peritrichously flagellated. No gas is produced from carbohydrates; acid may be formed by some species; litmus milk is not strongly alkalinized. The organisms are found in soil, water, and other natural habitats. *Achromobacter*-like bacteria are occasionally isolated from clinical specimens of body fluids (blood, pus). The type species is *A. liquefaciens* (Tulecke, W., Orenski, S. W., Taggart, R., and Colavito, L., *J. Bacteriol., 89,* 905–906, 1965).

Agarbacterium

These rods may be up to 4 μm long, non-motile, or motile with peritrichous flagella. They are aerobic or facultatively anaerobic. No gas is produced from carbohydrates, but acid may be formed. In the presence of an appropriate concentration of salt, *Agarbacterium* will grow well on meat-infusion agar, with liquefaction of the agar; the color of the growth varies from buff to orange. The organisms are found in sea water and on algae. The type species is *A. aurantiacum.*

Beneckea

These rods may be up to 2 μm long. They have a single polar sheathed flagellum when grown in liquid medium; many strains are peritrichously flagellated when grown on solid media. The organisms are facultatively anaerobic. Glucose is fermented with the production of acid, but not gas; chitin is hydrolyzed. *Beneckeae* are found in sea water or brackish water and in soil. The type species is *B. labra* (Baumann, P., Baumann, L., and Mandel, M., *J. Bacteriol., 107,* 268–302, 1971).

Alginomonas and *Alginobacter*

These rods, known to decompose alginic acid, have been placed in the genus *Alginomonas* if they are polarly flagellated (type species *A. nonfermentans*), or in the genus *Alginobacter* if they are peritrichously flagellated (type species *A. acidofaciens*). Members of these genera can be isolated from soil or sea water. They are presumably associated with the presence of algae.

Chromobacterium

These violet-chromogenic rods may be up to 6 μm long and polarly or peritrichously flagellated. They produce the indole pigment violacein. These bacteria grow on ordinary peptone media. Found in soil and water, they may be opportunistically pathogenic.

The two subgroups of *Chromobacterium* have the following characteristics (Sneath, P. H. A., *Iowa State J. Sci., 34,* 243, 1960):

I. Growth at 37°C but not at 4°C. Production of hydrogen cyanide. Turbidity from egg-yolk. Production of acid from trehalose. No hydrolysis of esculin. Production of acid from arabinose, xylose, or mannitol. Moderately proteolytic, hemolytic, and chitinolytic.

Chromobacterium violaceum (mesophils)

II. Growth at 4°C but not at 37°C. No production of hydrogen cyanide. No turbidity from egg-yolk. No production of acid from trehalose. Hydrolysis of esculin. Production of a small amount of acid from arabinose, xylose, and mannitol. Poorly proteolytic, haemolytic, and chitinolytic.

Chromobacterium lividum (psychrophils)

Flavobacterium

These rods may be more than 4 μm long, aerobic to facultatively anaerobic, and non-motile or peritrichously flagellated. They produce yellow or orange pigments that are not water-soluble; often they require salt for typical pigmentation. The organisms are usually found in soil and water. The type species, *F. aquatile,* is found in water containing calcium carbonate.

F. meningosepticum. Can be isolated from water, sinks, nursing bottles, etc. It is one of the flavobacteria that can be deadly to children. Infants may succumb rapidly to general sepsis; if they survive the original acute phase, meningitis may develop. Adults are rarely affected. The morphology of these pathogens is distinctive: they are elongate, thin rods with bulbous ends.

Thermus

These rods measure up to 10 μm in length; filaments up to 200 μm are also formed, especially in freshly isolated cultures; large spheres (up to 20 μm in diameter), composed of rods enclosed within the outer layer of the cell wall, are produced in old cultures. The organisms are non-motile. Growth is colorless to bright orange; the temperature minimum for growth is 40°C, the optimum 70 to 72°C. *Thermus* species are found in hot springs and in hot tap water. The type species is *T. aquaticus* (Brock, T. D., and Freeze, H., *J. Bacteriol., 98,* 289–297, 1969; Brock, T. D., and Edwards, M. R., *J. Bacteriol., 104,* 509–517, 1970; Ramaley, R. F., and Hixson, J., *J. Bacteriol., 103,* 527, 528, 1970).

Escherichia, Shigella, Salmonella, Arizona, Citrobacter, Enterobacter, Providencia, Edwardsiella, Klebsiella, Aerobacter, Serratia, Pectobacterium, Proteus

See Enterobacteriaceae, page 230.

Erwinia

The genus *Erwinia* is used to harbor plant-pathogenic bacteria that have the morphological and fermentative properties of Enterobacteriaceae. They may be implicated in some animal infections. The type species, *E. amylovora,* attacks a large number of species of *Rosaceae,* including pear and apple trees. Most species of *Erwinia* are named after a host plant; for example, *E. carotovora, E. chrysanthemi* (Graham, D. C., *Annu. Rev. Phytopathol., 2,* 13–42, 1964).

Acetobacter

These rods may be 2 μm long. They are non-motile or monotrichously or peritrichously flagellated. *Acetobacter* oxidize ethanol to acetic acid, and acetate or lactate to carbon dioxide and water. They are found on fruits, vegetables, and in fermented beverages. The type species is *A. aceti.* One species, *A. xylinum,* produces cellulose fibers (Gibson, E. J., and Colvin, J. R., *Can. J. Microbiol., 14,* 93–95, 1968).

Acetomonas (Gluconobacter)

These rods may be up to 3 μm long, non-motile, or motile with one to eight polar flagella. They are obligately aerobic. *Acetomonas* oxidize ethanol to acetic acid, but not acetate and lactate to carbon dioxide; many species produce 2- and 5-ketogluconic acid. The organisms are found on fruits and in fermented beverages. The type species are *A. suboxydans* or *G. oxydans,* depending on the generic name selected (de Ley, J., and Frateur, J., *Int. J. Syst. Bacteriol., 20,* 83–95, 1970). Acetomonads, as well as the acetobacters (see above), are of industrial importance (Prescott, S. C., and Dunn, C. G., *Industrial Microbiology,* McGraw-Hill, New York, 1959).

Aeromonas

These rods are up to 3 μm long, generally monotrichous, but occasionally with a tuft of polar flagella or non-motile. They are facultatively aerobic. Carbohydrates are utilized with the production of acid and gas. *Aeromonas* are mostly found in water and may be pathogenic to aquatic animals. The type species is *A. punctata,* which is found in water and may cause an infection of carps (Schubert, R. H. W., *Int. J. Syst. Bacteriol., 18,* 1–7, 1968).

A. salmonicida. A non-motile species that causes furunculosis of *Salmonideae.*

A. hydrophila. A monotrichous species. Some cells are slightly curved. It is found in water, sewage, foods, etc. This organism is pathogenic for reptiles and amphibians; it has been isolated from a variety of human clinical specimens.

Another species, *A. shigelloides,* is now placed in the genus *Vibrio.*

Photobacterium

Photobacteria are coccobacilli or rods that tend to show rudimentary branching on media containing glucose and asparagine. They are facultatively aerobic. The organisms are luminescent, especially in the presence of 3 to 5% salt. Isolated from sea water or from marine animals, photobacteria may be pathogenic for some animals; they have been found to be associated with phosphorescent organs of marine animals. The type species is *P. phosphoreum.* Recent studies have indicated that some luminescent bacteria might be better placed in the genus *Vibrio;* for example, *V. fischeri* (Hendrie, M. S., Hodgkiss, W., and Shewan, J. M., *Int. J. Syst. Bacteriol., 21,* 217–221, 1971). The genus *Lucibacterium* was proposed for peritrichously flagellated luminous bacteria (Hendrie, M. S., Hodgkiss, W., and Shewan, J. M., *J. Gen. Microbiol., 64,* 151–169, 1970). Some strains of *Lucibacterium* share with strains of *Beneckea* the property of having a thick (sheathed) polar flagellum and thin (unsheathed) lateral flagella when they are peritrichous.

Zooglea

These rods may be up to 4 μm long. When motile, they are monotrichous. *Zooglea* are oxidase-positive and strictly aerobic. They deposit extracellularly a gelatinous matrix that may be cellulosic. The organisms are found in water containing decomposing organic matter, such as trickling filters. The type species is *Z. ramigera* (Crabtree, K., and McCoy, E., *Int. J. Syst. Bacteriol., 17,* 1–10, 1967).

In nature, strains of *Zooglea* form mucilaginous masses that have either dactylate or filiform projections. From such masses one can isolate not only strains of *Zooglea,* but also other bacteria that do not form mucilage (Unz, R. F., and Dondero, N. C., *Can. J. Microbiol., 13,* 1671–1694, 1967).

Xanthomonas

These rods may be up to 4 μm long. They are polarly flagellated, usually monotrichous. The organisms are usually aerobic, but some strains are facultatively anaerobic. Growth is yellow; acid – but usually no gas – is produced when organisms are grown on sugars. *Xanthomonas* are usually isolated from plant necroses. Most so-called species bear the name of a host. The type species is *X. hyacinthi.*

Pseudomonas

See page 239 for a comprehensive discussion of this genus by Dr. Carl F. Clancy.

GLIDING RODS

A number of gliding bacteria have already been discussed in the chapters "Trichome-Forming Bacteria" (page 57) and "The Fruiting Myxobacteria" (Page 191). In this section gliding rods that are pigmented with carotenoids will be briefly reviewed. These may be placed in the family Cytophagaceae, also called Flexibacteraceae (Lewin, R. A., *J. Gen. Microbiol., 58,* 189–206, 1969).

Cytophaga

These are flexible rods that may be up to 50 μm long. They are found in rotting vegetable matter, both on land and in aqueous environments. They are very capable decomposers of polysaccharides such as cellulose, chitin, agar, alginates, etc. The genus *Sporocytophaga* harbors similar organisms, which are capable of forming microcysts. The latter are spherical or oval and occur among the rods in the colonies. Differentiation between the genera *Cytophaga* and *Flavobacterium* (page 227) may not always be easy to make (Mitchell, T. G., Hendrie, M. S., and Shewan, J. M., *J. Appl. Bacteriol., 32,* 40–50, 1969; Weeks, O. B., *J. Appl. Bacteriol., 32,* 13–18, 1969).

Microscilla

These flexible cells are usually 20 to 100 μm long. They are unable to attack cellulose, but may liquefy alginate and gelatin. Microscillae are usually isolated from marine samples. In addition to enzymatic differences, microscillae are differentiated from members of the genus *Cytophaga* by their longer cells and their more rapid motion.

Flexibacter

Flexibacters are similar in morphology to the microscillae, but their cells are shorter (up to 50 μm long). They can liquefy gelatin, but not cellulose, agar or alginate. These organisms are usually isolated from fresh-water habitats, including hot springs.

Enterobacteriaceae

DR. CARL F. CLANCY

Some Enterobacteriaceae exist as saprophytes in nature, others are parasitic for plants, causing blights and soft rots; many exist as commensals in the intestinal tracts of animals, and still others may cause serious infectious disease.

The group includes motile organisms with peritrichous flagella as well as non-motile varieties. Most species ferment glucose with production of acid, reduce nitrate to nitrite, and lack a cytochrome C oxidase. Their biochemical and serological characteristics have been studied intensively in efforts to arrive at a logical classification. Five tribes and about twelve genera are currently recognized as distinguishable entities; however, so many intermediate strains exist that it is difficult to express them in a single formal classification.

Many of the genera to be discussed are susceptible to lysis by specific bacteriophages, and many fundamental principles of phage–host cell relationships have been elucidated by studies with *Escherichia coli* and its various phages. Another characteristic of these enteric bacteria is the ability to produce bacteriocins, metabolites that are lethal for other strains of the same or closely related species. Bacteriocins are named after the producing strain; for example, colicins are produced by *Escherichia coli.* They may play a role in the maintenance of normal flora in the intestinal tract, although *in-vivo* activity has not been thoroughly investigated. Endotoxins, composed of lipopolysaccharides, occur in the cell wall of nearly all the enteric bacteria. The endotoxins of *E. coli, Salmonella, Shigella* and *Serratia* have been studied intensively and are discussed in Volume II of the Handbook.

Whenever these Gram-negative rods are implicated as the etiologic agent of infection, the question arises as to the best antimicrobial agent, if any, to be used for therapy. With the exception of typhoid fever and shigellosis, no reference will be made in the following pages to specific chemotherapy for each individual infectious agent. Most are sensitive, in varying degrees, to ampicillin, cephalothin, chloramphenicol, tetracycline, streptomycin, sulfonamide, etc., but in any specific instance the isolated strain must be tested to determine its particular antimicrobial-sensitivity pattern.

The more important biochemical reactions of these organisms, which enable them to be separated into twelve genera, are shown in Table 1. Antigenic analysis and serotyping are also used as taxonomic tools and will be discussed under the several genera for which they are most valuable.

Escherichia

The only important species of this genus, *Escherichia coli,* is a common inhabitant of the lower bowel of man and animals. It is often present in man in concentrations of 10^7 or more viable organisms per gram of fecal material. It is commonly looked for in water supplies and food as an indicator of fecal pollution. It may be motile or non-motile, always ferments glucose, and about 90% of strains ferment lactose in one to two days. When studied by the IMViC (indol, methyl red, Voges-Proskauer, and citrate) reactions, it produces indol from tryptophane and is methyl-red-positive. Acetyl methyl carbinol is not formed, and citrate is not utilized as a sole source of carbon.

TEST or SUBSTRATE	ESCHERICHIEAE		EDWARD-SIELLEAE	SALMONELLEAE			KLEBSIELLEAE									PROTEEAE					
								Enterobacter							*Pectobacterium*	*Proteus*				*Providencia*	
										hafniae		*liquefaciens*									
	Escherichia	*Shigella*	*Edwardsiella*	*Salmonella*	*Arizona*	*Citrobacter*	*Klebsiella*	*cloacae*	*aerogenes*	37C	22C	37C	22C	*Serratia*	25C	*vulgaris*	*mirabilis*	*morganii*	*rettgeri*	*alcalifaciens*	*stuartii*
INDOL	+	-or +	+	-	-	-	- or +	-	-	-	-	-	-	-	- or +	+	-	+	+	+	+
METHYL RED	+	+	+	+	+	+	-	-	-	+ or -	-	+ or -	- or +	- or +	+ or -	+	+	+	+	+	+
VOGES - PROSKAUER	-	-	-	-	-	-	+	+	+	+ or -	+	- or +	+ or -	+	- or +	-	- or +	-	-	-	-
SIMMONS'S CITRATE	-	-	-	d	+	+	+	+	+	(+) or -	d	+	+	+	d	d	+ or (+)	-	+	+	+
HYDROGEN SULFIDE (TSI)	-	-	+	+	+	+ or -	-	-	-	-	-	-	-	-	-	+	+	-	-	-	-
UREASE	-	-	-	-	-	d^w	+	+ or -	-	-	-	d	-	d^w	d^w	+	+	+	+	-	-
KCN	-	-	-	-	-	+	+	+	+	+	+	+	+	+	+ or -	+	+	+	+	+	+
MOTILITY	+ or -	-	+	+	+	+	-	+	+	+	+	d	+	+	+ or -	+	+	+	+	+	+
GELATIN (22C)	-	-	-	-	(+)	-	-	(+)or -	- or (+)		-		+	+	+ or (+)	+ or (+)	+	-	-	-	-
LYSINE DECARBOXYLASE	d	-	+	+	+	-	+	-	+	+	+	+ or -	+	+	-	-	-	-	-	-	-
ARGININE DIHYDROLASE	d	- or (+)	-	(+) or +	+ or (+)	d	-	+	-	-	-	-	-	-	- or +	-	-	-	-	-	-
ORNITHINE DECARBOXYLASE	d	d[1]	+	+	+	d	-	+	+	+	+	+	+	+	-	-	+	+	-	-	-
PHENYLALANINE DEAMINASE	-	-	-	-	-	-	-	-	-	-	-	-	-	-	-	+	+	+	+	+	+
MALONATE	-	-	-	-	+	d	+	+ or -	+ or -	+ or -	+ or -	-	-	-	- or +	-	-	-	-	-	-
GAS FROM GLUCOSE	+	-[1]	+	+	+	+	+	+	+	+	+	+	+	+ or -[3]	- or +	+ or -	+	d	- or +	+ or -	-
LACTOSE	+·	-[1]	-	-	d	d	+	+	+	- or (+)	- or (+)	d	(+)	- or (+)	d	-	-	-	-	-	-
SUCROSE	d	-[1]	-	-	-	d	+	+	+	d	d	+	+	+	+	+	d	-	d	d	d
MANNITOL	+	+ or -	-	+	+	+	+	+	+	+	+	+	+	+	+	-	-	-	+ or -	-	d
DULCITOL	d	d	-	d[2]	-	d	- or +	- or +	-	-	-	-	-	-	-	-	-	-	-	-	-
SALICIN	d	-	-	-	-	d	+	+ or (+)	+	d	d	+	+	+	+	d	d	-	d	-	+
ADONITOL	-	-	-	-	-	-	+ or -	- or +	+	-	-	d	d	d	-	-	-	-	d	+	-
INOSITOL	-	-	-	d	-	-	+	d	+	-	-	+	+	d	-	-	-	-	+	-	+
SORBITOL	+	d	-	+	+	+	+	+	+	-	-	+	+	+	-	-	-	-	d	-	d
ARABINOSE	+	d	-	+[2]	+	+	+	+	+	+	+	+	+	-	+	-	-	-	-	-	-
RAFFINOSE	d	d	-	-	-	d	+	+	+	-	-	+	+	-	+ or (+)	-	-	-	-	-	-
RHAMNOSE	d	d	-	+	+	+	+	+	+	+	+	-	-	-	d	-	-	-	+ or -	-	-

(1) Certain biotypes of *Shigella flexneri* produce gas; *S. sonnei* cultures ferment lactose and sucrose slowly and decarboxylate ornithine. (2) *Salmonella typhi, S. cholerae-suis, S. enteritidis* bioser. Paratyphi A and Pullorum, and a few others ordinarily do not ferment dulcitol promptly; *S. cholerae-suis* does not ferment arabinose. (3) Gas volumes produced by cultures of *Serratia, Proteus,* and *Providencia* are small.

The results of serological studies on *E. coli* have shown it to have three different classes of antigens.[1] The "O antigens" are somatic antigens, stable to heat at 100°C or 121°C. "K antigens" are somatic antigens that occur as sheaths, envelopes or capsules and may act to inhibit "O" agglutination; they are designated as L, A, or B, depending upon their heat lability at 100° or 120°C. The "H antigens" are flagellar and are heat-labile at 100°C. By application of serotyping, as done with *Salmonella,* it has been possible to characterize many strains of *E. coli* in relation to these three antigens.

Evidence has now accumulated to show that many outbreaks of diarrheal disease in which no *Salmonella* or *Shigella* can be isolated are associated with certain serotypes of *E. coli.* A particular serotype of *E. coli* (O 11:B 4) was first noted in 1945 by Bray[2] to be associated with infantile diarrhea; many subsequent reports confirm that one of several serotypes of *E. coli* may be isolated from the stool of infants with epidemic diarrhea of the newborn. Commercial typing sera that contain appropriate antibody against the O (somatic) antigen and B (envelope) antigen are available for the identification of ten different serotypes known to be associated with infantile diarrhea. Some of these same serotypes are also found in diarrhea of adults.

Occasionally biotypes of *E. coli* are encountered that are anaerogenic and non-motile. They were once thought to be *Shigella,* but more recently have been referred to as the *Alkalescens-Dispar* group. They do have O antigens in common with *Escherichia* and are now included in this genus.

Escherichia coli is the etiologic agent of a variety of other types of human and animal infections. It is one of the most common causes of cystitis and other infections of the urinary tract. It is also one of the organisms most commonly recovered from peritonitis following rupture of the appendix or other bowel perforation, and from pneumonia following aspiration of intestinal contents. It may cause pyogenic wound infections, especially in lesions that are fecally contaminated.

Escherichia coli has also served as a "model" organism for many types of studies in bacterial physiology, cytology by electron microscopy, genetics, chromosome mapping, etc. It grows well on simple media, has a generation time of about twenty minutes, can be grown in large quantities, and generally lends itself to laboratory manipulation.

Shigella

Members of this genus, the etiologic agents of bacillary dysentery, are widespread throughout the world, especially in tropical areas. They are non-motile Gram-negative rods, anaerogenic, and ferment dextrose but not lactose. Their reactions in the IMViC tests are the same as those of *Escherichia coli,* except that not all strains produce indol. Some species also ferment mannitol, a characteristic that is useful in the identification of the species (Table 2). Group A (*S. dysenteriae*) organisms do not ferment mannitol, whereas Group B (*S. flexneri*), Group C (*S. boydii*), and Group D (*S. sonnei*) are nearly all fermenters of mannitol. Group A has ten serotypes based on O antigens that are serologically related to each other, and Groups B and C are composed of twelve to fifteen serotypes. Group D (*S. sonnei*) has only one serotype.

It should be noted (3) that the O antigens of ten *Shigella* serotypes are identical with those of *Escherichia coli* or intermediate coliform bacilli. Furthermore, seventeen other *Shigella* serotypes also bear reciprocal O antigen relationships with *E. coli* and coliform intermediates. This is of some practical importance in the identification of *Shigella* species, especially if one is dealing with an isolate that is actually an anaerogenic, slow lactose-fermenting *E. coli.* It is important to analyze carefully the results of both biochemical and serological tests.

The natural habitat of *Shigella* is the lower bowel of man. Isolation of the organisms from monkeys has been reported, but they are rarely isolated from other animal species. Transmission of bacillary dysentery usually occurs directly by contact with a human case or carrier or by ingestion of food or water that is fecally contaminated. Dysentery is characterized by diarrhea, fever, cramps, and sometimes vomiting; the watery stools characteristically contain blood, mucus, and many pus cells. The incubation period is usually less than seven days. In contrast to typhoid fever, the organisms in *Shigella* infections are localized in the lower bowel. They cannot be cultured from blood, but are readily isolated from feces during illness and for several weeks afterward. The carrier state may then persist for months, and occasionally for a year or more. Clinical severity of the disease may be affected by the age, state of nutrition, and general well-being of the

TABLE 2

GENUS *SHIGELLA*

Serologic Group	Mannitol	Species
A (1)	Negative	*S. dysenteriae*
A (2–10)	Negative	*S. schmitzii, S. arabinotarda,* etc.
B	Positive	*S. flexneri* and miscellaneous serotypes
C	Positive	*S. boydii* and miscellaneous serotypes
D	Positive	*S. sonnei*

patient. Infants and debilitated elderly persons may be severely affected. In temperate zones the disease may be self-limiting, whereas in the tropics, under conditions of crowding, severe epidemics occur.

An attack of dysentery does not confer protection against subsequent attacks, although persons living in areas where the disease is endemic do not continue to have recurrent attacks. Antibody formation after an attack of dysentery cannot be demonstrated by conventional serologic techniques, but if measured by the hemagglutination technique,[4] titers of 1/40 or higher may be demonstrated.

Shigella dysenteriae type 1 (the Shiga bacillus) produces a disease that is considerably more severe than that caused by other organisms in the group. It has been shown to produce a powerful exotoxin, which damages the blood vessels of the central nervous system of rabbits, producing symptoms of flaccid paralysis. In rats and hamsters, no neurological symptoms appear, but the blood vessels of various other organs are affected. The true role of the toxin in human disease is not known. Disease due to the Shiga bacillus has been quite rare for many years, except for endemic foci in Asia, but in 1969 an epidemic, first noted in Guatemala,[5] spread throughout Central America.

Specific treatment of bacillary dysentery with chloramphenicol or tetracycline results in a rapid decline in symptoms. Earlier attempts at chemotherapy with sulfonamides, however, resulted in bacteriostasis of the organisms, with subsequent emergence of drug-resistant strains. Studies on the antibiotic sensitivity of *Shigella* strains in Japan have resulted in the discovery of the RTF (Resistance Transfer Factor, see Volume IV). It was found that, by means of transfer of an episome through conjugation, a *Shigella* strain could acquire resistance to an antibiotic from a resistant strain of *Escherichia coli.*

Salmonella

Salmonella organisms are usually motile and produce both acid and gas from glucose. Lactose, sucrose, and salicin are not fermented, but hydrogen sulfide is usually produced. The organisms do not produce acetyl methyl carbinol, rarely liquefy gelatin, and do not hydrolyze urea. They are found in man and lower animals, especially in the intestinal tract. Members of this group cause typhoid fever, gastroenteritis, and sepsis in man, as well as similar infections in lower animals.

The typhoid bacillus, now designated as *Salmonella typhi,* was the first member of this group to be described and isolated.[6,7] As bacteriologists began to study the etiology of enteric fevers, it became apparent that many of the organisms isolated had the common characteristics mentioned above, but attempts to identify and classify these organisms by the usual biochemical tests failed. By use of antigenic analysis, White[8] and Kauffmann[9] established some taxonomic order; they showed that each organism was a serologic entity with characteristic somatic and flagellar antigens, of which some were common to other members of the group.

The O (*ohne,* German for "without") or somatic antigens are present in both motile and non-motile bacilli and are resistant to boiling at 100°C and to treatment with alcohol. The H (*Hauch,* German for "haze") or flagellar antigens are only found in motile cultures and are destroyed at 100°C or by treatment with alcohol or dilute acids. At least two species, *S. typhi* and *S. paratyphi* C, have, in addition to O and H antigens, an antigen designated as Vi. This substance is an envelope antigen and was originally called Vi to denote association with a virulent strain. The virulence concept has not been borne out; however, presence of this antigen in a freshly isolated culture of *S. typhi* may interfere with its agglutination by antiserum for the specific O-antigenic factors.

The Kauffmann-White scheme is a systematic tabulation of the exact antigenic structure of the thousand-odd species of *Salmonella* now described. Somatic (O) antigens are designated with arabic numerals, as may be seen in Table 3; for example, *S. typhimurium* cells contain O-antigenic factors 1, 4, 5, and 12.

The O antigen is a constitutive part of the bacterial cell wall; it is a complex of lipopolysaccharide and protein and is closely associated with endotoxin. The polysaccharide complex, which determines the serologic specificity of a serotype, may be composed of five to eight different monosaccharides.[10,11] All *Salmonella* O antigens contain five common sugars: D-galactose, D-glucosamine, D-glucose, heptose, and ketodeoxyoctonic acid. It is believed that specific *Salmonella* polysaccharides possess a common core of these five sugars, to which specific side chains composed of the sugars characteristic of the serotype in question are attached. For example, the group D organism, *S. typhi,* possesses the O factors 9 and 12; in addition to the five basal sugars, the polysaccharide of *S. typhi* contains mannose, rhamnose, and tyvelose, which presumably accounts for the specificity of factors 9 and 12.

Flagellar H antigens are divided into specific (Phase 1) factors, designated by lower-case letters, and non-specific (Phase 2) factors, designated by arabic numerals. These two entities differ in that Phase 1 antigens are shared by only a few serotypes, whereas the non-specific antigens are common to many different serotypes. Thus it may be noted (Table 3) that the flagellar antigens of *S. typhimurium* are as follows: Phase 1 – i; Phase 2 – 1, 2. The specificity of the flagellar antigens is dependent on the proteins present in the flagella.

The Kauffmann-White schema, as developed, has become an aid in the serologic identification of organisms isolated from disease in man and animals. Before the multitude of serotypes was comprehended,

TABLE 3

ANTIGENIC STRUCTURES OF REPRESENTATIVE *SALMONELLA* AS DEPICTED IN THE KAUFFMANN-WHITE SCHEMA

Group	Species	O Antigens	H Antigens	
			Phase 1	Phase 2
A	*S. paratyphi* A	1, 2, 12	a	
B	*S. paratyphi* B (*schottmuelleri*)	1, 4, 5, 12	b	1, 2
	S. typhimurium	1, 4, 5, 12	i	1, 2
	S. derby	1, 4, 5, 12	f,g	
C_1	*S. cholerae-suis*	6, 7	c	1, 5
C_2	*S. newport*	6, 8	e,h	1, 2
D	*S. typhi*	9, 12	d	
	S. enteritidis	1, 9, 12	g,m	
	*S. pullorum**	9, 12		

* *S. pullorum*, one of the few non-motile species, has no flagella, and hence no H antigens.

each was given a species name, usually the name of the area of isolation (e.g., *S. rutgers*). It now seems apparent that spontaneous mutations occur in nature and that genetic recombinations may occur by transduction and conjugation. This may explain the multiplicity of serotypes and the overlapping patterns of antigenicity.

Salmonella typhi,[a] the etiologic agent of typhoid fever, is found only in man or in food or water fecally contaminated by a human carrier. The disease is an acute febrile illness of about three weeks' duration. The cellular response to *S. typhi* infection, a neutropenia, is unlike that of most bacterial diseases. *Salmonella typhi* may be isolated from the blood, urine, and feces of the patients at various stages in the disease. Such isolates are often subjected to bacteriophage typing for epidemiological purposes to correlate with isolations from carriers or other cases of typhoid.

Chloramphenicol is the drug of choice for patients with typhoid fever. It has been found that treatment with other drugs (e.g., tetracycline), though showing activity *in vitro* against *S. typhi*, appeared to exert little or no effect on the course of the disease. More recently, ampicillin has been shown to be useful in the therapy of typhoid fever. After an attack of typhoid fever, the patient may become a carrier of the organism for weeks and months; occasionally the carrier state continues for years, and these people serve as a reservoir for the maintenance of the disease. Typhoid fever now occurs usually in small, localized epidemics.

A vaccine for immunization against *Salmonella* infection has been available since the early days of prophylactic immunization. The usual preparation, a mixture, contains killed cells of *S. typhi*, *S. paratyphi* A, and *S. paratyphi* B. Although widely used for many years, no experimental proof of its efficacy can be demonstrated; however, it is recommended for persons living in areas where typhoid fever occurs.

Several other organisms (*S. paratyphi* A, *S. paratyphi* B) produce a disease clinically similar to typhoid fever, but less severe. These organisms, like the typhoid bacillus, are harbored almost exclusively by man and are spread by carriers to susceptible individuals.

Gastroenteritis due to *Salmonella* is probably the most commonly occurring form of infection caused by these organisms. The course of the disease is three to seven days and is usually self-limited. Patients are carriers for weeks after the infection, but eventually most clear spontaneously. The strains of *Salmonella* causing gastroenteritis are usually animal pathogens or organisms maintained in animal species. For example, the *Salmonella* most commonly isolated from cases of human gastroenteritis is *S. typhimurium*, the causative agent of mouse typhoid. Some of the more common sources of *Salmonella* dissemination are listed below:

1. Contaminated water and food, especially poultry products, such as eggs, egg powder, improperly cooked meat (e.g., turkey roll), etc.

2. Animal reservoirs; direct contamination of food, such as fecal contamination by mice carrying *S. typhimurium*, or direct contact with animlas (e.g., handling of infected pet turtles). .

3. Animal feed prepared from contaminated meat scraps, tankage, etc.

4. Human cases, subclinical human infections, and carriers.

Certain Group C *Salmonella*, especially *S. cholerae-suis*, are known to cause disease unrelated to the gastrointestinal tract. Abscesses may develop anywhere in the body, sometimes resulting in bacteremia. The source of infection in these patients is not clear. *Salmonella cholerae-suis* is found in swine, associated with hog cholera, but not the causal agent; it is rarely isolated from the feces of humans or animals.

Miscellaneous Enterobacteriaceae

A transitional group of Gram-negative rods should be mentioned, which has certain biochemical and serologic characteristics in common with *Salmonella*, *Shigella*, and also with other members of the Enterobacteriaceae. They include *Arizona*, *Citrobacter*, *Enterobacter hafniae*, and *Providencia* (Table 1), as

[a] *S. typhi*, *S. pullorum*, and *S. gallinarum* ferment glucose, but are anaerogenic.

well as a more recently described genus, *Edwardsiella.*[12] This somewhat heterogeneous group of organisms is referred to as the "Paracolon Group", a term of convenience that has no taxonomic standing.

These organisms are of importance mainly because they complicate the process of isolating *Salmonella* and *Shigella* from stool, since they either do not ferment lactose or ferment it only slowly. Thus, on stool culture media, which depend on non-fermentation of lactose to indicate suspect colonies, they mimic the appearance of the true pathogens.

The *Arizona* strains are quite closely related to *Salmonella* and have numerous O and H antigens in common with the latter. They are found in lower animals, especially reptiles, and occasionally produce *Salmonella*-like gastroenteritis in man.

Representative *Citrobacter* strains (formerly called *Escherichia freundii*) ferment lactose and produce large amounts of hydrogen sulfide. However, many strains attack this key carbohydrate slowly or not at all. This group of bacteria, which also has some antigens in common with *Salmonella,* has also been termed "Bethesda-Ballerup."

Enterobacter hafniae is included in the tribe Klebsielleae. It does not produce hydrogen sulfide and is mentioned here because it usually does not ferment lactose. The *Providencia* strains are closely related to *Proteus* in their biochemical reactions. *Serratia* strains are also lactose-negative or ferment the sugar slowly, but they will be discussed under a separate heading.

In summary, this heterogeneous group of organisms is of nuisance value in the search for *Salmonella* and *Shigella* from cases of enteric fever. Although the true role of these bacteria in the etiology of gastroenteritis is uncertain, they may cause urinary-tract or other infections if they somehow gain access to a part of the body normally sterile; e.g., *Edwardsiella tarda*[13] has been shown to have caused a case of bacterial meningitis.

Klebsielleae

Members of the tribe Klebsielleae are widespread in their biologic activities; they are found as human pathogens, as commensals in man and animals, in soil and water, and as plant pathogens. Most genera ferment glucose, lactose, and sucrose and utilize citrate, but do not produce indol, hydrogen sulfide, or phenylalanine deaminase. Those in the genus *Klebsiella* are non-motile, whereas most of the *Enterobacter* strains are motile.

Classification of these organisms into an orderly scheme has been difficult. Historically, organisms arising from the respiratory tract were designated as *Klebsiella,* whereas those from urine, stool, soil, water, etc., were called *Aerobacter aerogenes.* As time went on, it became apparent that application of critical biochemical tests showed some strains of *A. aerogenes* to be indistinguishable from *Klebsiella.* Julianelle,[14] using a capsular antigen–antibody reaction, showed that at least three serotypes of *Klebsiella* (A, B, and C) could be recognized on the basis of the specificity of the polysaccharide in the capsule. Subsequently Kauffmann[15] prepared a schematic representation (Table 4) of the antigenic structure of *Klebsiella* based on their somatic O and capsular K antigens.

It may be seen in Table 4 that seventy-two serotypes have been described. Identification of organisms on this basis is not practical for the routine diagnostic laboratory, but Eichoff, Steinhauer and Finland[16] studied 257 strains of *Klebsiella* from human sources and found that serotypes 1, 3, 4, and 5 were most frequently encountered from the respiratory tract, whereas type 2 and untypable strains were more often found in urine. In the same study, the motile species *Enterobacter cloacae* and *E. aerogenes* were more commonly, but not exclusively, isolated from urine. These latter two species are distinct from *Klebsiella* in being motile, as are *Serratia* and some strains of *Enterobacter liquefaciens.*

Klebsiella is the causative agent of Friedlander's pneumonia, a fulminating type of disease that has a high case fatality rate if untreated. Since the advent of chemotherapy, however, the disease has been less serious. It occurs primarily in chronic alcoholics or physically debilitated individuals. Those patients recovering from the disease may have chronic lung abscesses and cavities that persist over a period of slow convalescence.

TABLE 4

DIAGNOSTIC ANTIGENIC SCHEMA OF THE GENUS *KLEBSIELLA*

O Group	Capsule Type
1	1, 2, 3, 7, 8, 10, 12, 16, 19, 20, 21, 22, 23, 24, 26, 29, 30, 32, 34, 37, 39, 41, 44, 45, 46, 47, 62
2	2, 3, 4, 5, 6, 8, 27, 28, 35, 43, 59
3	11, 25, 31, 33, 48, 49, 50, 51, 53, 54, 55, 58
4	15, 42
5	57, 61
Ungrouped	9, 13, 14, 17, 18, 36, 38, 40, 52, 56, 60, 63–72

Data taken from: Kaufmann, F., *The Bacteriology of Enterobacteriaceae.* Williams and Wilkins, Baltimore, Maryland (1966). Reproduced by permission of the copyright owners.

Serratia

Serratia marcescens has long been used as a laboratory stock culture to hand out to students in introductory microbiology. It is a small Gram-negative rod or coccobacillus that does not ferment lactose or does so only slowly. It is found in water, soil, and occasionally in humans, but not nearly as constantly as *Enterobacter aerogenes.* The colonies have orange-red pigmentation, which varies in color intensity and shade depending on conditions of cultivation. The red pigment, prodigiosin, has been characterized chemically and is discussed in Volume III.

Serratia marcescens has been used as a tracer organism to follow the natural spread of bacteria from person to person, instruments to person, etc., because the colonies are easily recognized on subculture and the organism was once considered a harmless saprophyte. More recently it has been recognized that *Serratia* can be the cause of infections in various parts of the human body. Some strains are non-pigmented, and it is probable that in the past many of these were reported as *Aerobacter*. It is now believed that this organism may spread in epidemic form, causing nosocomial infections in hospitals, similar to that of penicillin-resistant staphylococci in the newborn and to that of *Salmonella derby* gastroenteritis.[17]

Enterobacter agglomerans

A group of organisms that were described at one time in the genus *Erwinia,* and later as *Pectobacterium,* are now classified as *Enterobacter agglomerans.* They are found in soil and invade plant tissues, causing dry necroses, galls, wilts, and soft rots. Certain of the species produce pectinolytic enzymes, which is manifested in their ability to produce plant diseases. The biochemical reactions of these organisms are listed in Table 1, under the heading *Pectobacterium.*

Occasionally these bacilli are isolated from human sources.[18,19] Their role as human pathogens is doubtful. In one instance, however, in which they were accidentally introduced into the blood stream of patients via contaminated bottles of intravenous glucose solution, they apparently caused serious illness and death.

The Proteus Group

Organisms of this genus are motile and ferment dextrose but not lactose. They produce a strong urease and phenylalanine deaminase. Four species (Table 1) are recognized; two of these, *P. vulgaris* and *P. mirabilis*, produce large amounts of hydrogen sulfide, liquefy gelatin, and are so motile that they swarm over the surface of a moist agar plate. *Proteus morganii* and *P. rettgerii* do not liquefy gelatin, nor do they produce hydrogen sulfide; they are motile, but do not tend to swarm. The habitat of *Proteus* is the intestinal tract of man and lower animals; in nature, the organisms occur in locales that are fecally contaminated.

The O, H, and K antigens of the *Proteus* group have been studied, but they are of little importance, since identification on the basis of biochemical reactions is usually satisfactory. Of more practical interest is the fact that certain strains of *Proteus* (OX19, OX2 and OXK) have an O-antigenic component in common with certain rickettsiae. This cross reaction is used in the Weil-Felix reaction, in which the serum of patients convalescing from epidemic typhus agglutinates *Proteus* OX19 antigen. Similarly, sera of patients who have had scrub typhus agglutinate the OXK antigen. The phenomenon is interesting, but not as specific for diagnostic purposes as the complement fixation test with purified rickettsial antigens.

Proteus species are commonly isolated from urinary-tract infections; they are sometimes difficult to eliminate because of their resistance to the more commonly used antimicrobial agents. Like other Enterobacteriaceae, they may cause pyogenic infections in other parts of the body when accidentally introduced.

REFERENCES

1. Edwards, P. R., and Ewing, W. H., *The Identification of Enterobacteriaceae*, 3rd ed. Burgess Publishing Co., Minneapolis, Minnesota (1972).
2. Bray, J., Isolation of Antigenically Homogeneous Strains of *Bact. coli neapolitanum* from Summer Diarrhea of Infants, *J. Pathol. Bacteriol.*, *57*, 239 (1945).
3. Ewing, W. H., Serological Relationships between *Shigella* and Coliform Cultures, *J. Bacteriol.*, *66*, 333 (1953).
4. Neter, E., Westphal, O., Lüderitz, O., and Gorzynski, E. A., The Bacterial Hemagglutination Test for the Demonstration of Antibodies to Enterobacteriaceae, *Ann. N.Y. Acad. Sci.*, *66*, 141 (1956).
5. Mata, L. J., Gangarosa, E. J., Carceres, A., Perera, D. R., and Mejicanos, M. L., Epidemic Shiga Bacillus Dysentery in Central America. I. Etiologic Investigations in Guatemala, 1969, *J. Infect. Dis.*, *122*, 170 (1970).
6. Eberth, C. J., Die Organismen in den Organen bei Typhus abdominalis, *Virchow's Arch.*, *81*, 58 (1880).
7. Gaffky, G., *Mitt. Kais. Ges. Amt.*, *2*, 372 (1884).
8. White, P. B., *Spec. Rep.* Ser. *Med. Res. Counc. London,* No. 103 (1926).
9. Kauffmann, F., Der Antigen-Aufbau der Typhus–Paratyphus Gruppe. *Z. Hyg. Infektionskr.*, *111*, 740 (1930).
10. Kauffmann, F., Lüderitz, O., Stierlin, H., and Westphal, O., Zur Chemie der O-Antigene von Enterobacteriaceae. I. Analyse der Zuckerbausteine von *Salmonella* O-Antigenen, *Zentralbl. Bakteriol. Parasitenk. Infektionskr. Hyg. Abt. I Orig.*, *178*, 442 (1960).
11. Lüderitz, O., Staub, A. M., and Westphal, O., Immunochemistry of O and R Antigens of *Salmonella* and Related Enterobacteriaceae, *Bacteriol. Rev.*, *30*, 192 (1966).
12. Ewing, W. H., McWhorter, A. C., Escobar, M. R., and Lubin, A. H., *Edwardsiella,* a New Genus of Enterobacteriaceae, Based on a New Species, *E. tarda*, *Int. Bull. Bacteriol. Nomencl. Taxon.*, *149*, 33 (1965).
13. Sonnenwirth, A. C., and Bozena, A. K., Meningitis Due to *Edwardsiella tarda.* First Report of Meningitis Caused by *E. tarda*, *Amer. J. Clin. Pathol.*, *49*, 92 (1968).
14. Julianelle, L. A., A Biological Classification of *Encapsulatus pneumoniae* (Friedlander's Bacillus), *J. Exp. Med.*, *44*, 113 (1926).
15. Kauffman, F., *The Bacteriology of Enterobacteriaceae.* Williams and Wilkins Co., Baltimore, Maryland (1966).
16. Eickhoff, T. C., Steinhauer, B. W., and Finland, M., The *Klebsiella-Enterobacter-Serratia* Division. Biochemical and Serologic Characteristics and Susceptibility to Antibiotics, *Ann. Intern. Med.*, *65*, 1163 (1966).
17. Sweeney, F. J., and Randall, E. L., Clinical and Epidemiological Studies of *Salmonella derby* Infections in a General Hospital, in *Proceedings of the National Conference on Salmonellosis, 1964*, pp. 130–139. U.S. Department of Health, Education and Welfare, Washington, D.C. (1964).
18. Ewing, W. H., and Fife, M. A., *Enterobacter agglomerans.* U.S. Department of Health, Education and Welfare, Washington, D.C. (1971).
19. von Gravenitz, A., and Strouse, A., Isolation of *Erwinia* spp. from Human Sources. *Antonie van Leeuwenhoek J. Microbiol. Serol.*, *32*, 429 (1966).

Pseudomonas

DR. CARL. F. CLANCY

Bergey's Manual of Determinative Bacteriology describes 149 species of the genus *Pseudomonas*. However, as with species designations in other groups of organisms, many are based on minor points of difference, which may vary under different conditions of growth and nutrition.

Most species are motile with polar flagella; straight rods or occasionally coccoid in shape. They grow well on the usual culture media, and many strains produce characteristic pigment. All species except *Pseudomonas maltophilia* are recognized as having a cytochrome C oxidase present when tested with tetramethyl-*p*-phenylenediamine, a characteristic that distinguishes them from the Enterobacteriaceae.

The habitat of these organisms is usually considered to be soil and water; but once they are established in areas somewhat removed from their natural habitat, they are difficult to eradicate. For example, they have been known to contaminate the cooling bath of an atomic reactor, and they have been found in "antiseptic" solutions of benzalkonium chloride, used for rinsing contact lenses.

By use of Adansonian analysis, Rhodes[1,2] simplified the taxonomy and reduced the number of species. Lysenko[3] simplified the genus into what are designated as eighteen species centers. The use of Adansonian classification also enabled Colwell and Liston[4] to define four main groups:

1. A marine group that requires sea water for growth or enhancement of growth; many members of this group are psychrophilic.
2. A group whose members are mainly mesophilic; some are curved rods and may produce a yellow or yellow-green diffusible fluorescent pigment.
3. Fluorescent pigment-producing strains; one subgroup is psychrophilic, another is mesophilic (*Pseudomonas aeruginosa* strains), and still another includes *Vibrio* forms.
4. Non-pigmented and mesophilic organisms (including *Vibrio tyrogenus, Pseudomonas ovalis* and *P. denitrificans*).

A number of different pigments are produced by various pseudomonads; many of these are phenazines.[5] Pyocyanin, elaborated by *Pseudomonas aeruginosa,* has been studied intensively. It has antibiotic properties, but is much too toxic to be used for antimicrobial therapy. Other pigments produced are chlororaphin and oxychlororaphin, phenazine *a*-carboxylic acid, and indigoidine. Brown pigments have been referred to as "melanins," but their nature is unknown. The substances responsible for fluorescence have been described as a pterine that is blue and a yellow flavin. Usually the pigments are water-soluble and diffuse throughout the medium.

A review[6] of the entire taxonomy of the pseudomonads that is based on biochemical, physiological and nutritional characteristics proposes thirteen different species designations; eight of these had been described by previous workers. The organisms are arranged in four main groups as seen in Tables 1 and 2, somewhat different from those designated by Colwell and Liston. The fluorescent group includes *P. aeruginosa, P. fluorescens,* and *P. putida.* Two other groups are designated as *Acidovorans* and as *Alcaligenes.* A fourth group, *Pseudomallei,* includes *P. pseudomallei* and *P. mallei.* These two organisms are almost identical, except that *P. mallei* is non-motile.

The fluorescent group is of special interest in medical microbiology, because *P. aeruginosa* and *P. fluorescens* can produce a wide variety of infections, although the former is considered to be the more pathogenic of the two. Clinical laboratories often report these organisms as *P. aeruginosa* and do not attempt to apply the differential tests of incubation at 4° and 41°C, flagella staining, denitrification, and utilization of trehalose, inositol and geraniol as sole sources of carbon.

Pseudomonas aeruginosa produces a bacteriocin, designated as pyocin. This substance, produced by one strain, is inhibitory for certain other strains of the species, but not all. The phenomenon has been suggested[7] as a means of typing strains for epidemiological purposes, since its application is somewhat simpler than phage-typing or serotyping. An improved technique[8] has been suggested, in which unknown

TABLE 1

GENERAL CHARACTERS OF DIAGNOSTIC VALUE FOR DIFFERENTIATING SPECIES OF AEROBIC PSEUDOMONADS

	Fluorescent Group				Pseudomallei Group*		Acidovorans Group		Alcaligenes Group				
Character	*P. aeruginosa*	*P. fluorescens*	*P. putida*	*P. multivorans*	*P. pseudomallei*	*P. mallei*	*P. acidovorans*	*P. testosteroni*	*P. alcaligenes*	*P. pseudoalcaligenes*	*P. stutzeri*	*P. maltophilia*	*P. lemoigneit*†
Number of flagella	1	>1	>1	>1	>1	0	>1	>1	1	1	1	>1	1
Poly-β-OH-butyrate as cellular reserve	–	–	–	+	+	+	+	+	–	V	–	–	+
Pigments													
fluorescent	+	+	+	–	–	–	–	–	–	–	–	–	–
phenazine	+	V	–	V	–	–	–	–	–	–	–	–	–
Methionine requirement	–	–	–	–	–	–	–	–	–	–	–	+	–
Denitrification	+	V	–	–	+	V	–	–	–	–	+	–	–
Growth													
at 4°C	–	+	V	–	–	–	–	–	–	–	–	–	–
at 41°C	+	–	–	V	+	+	–	–	+	+	V	–	–
Extracellular hydrolases													
gelatin	+	+	–	+	+	+	–	–	+	V	+	+	–
poly-β-OH-butyrate	–	–	–	–	+	+	–	V	–	–	–	–	+
starch	–	–	–	–	+	V	–	–	–	–	+	–	–
Oxidase reaction	+	+	+	+	+	+	+	+	+	+	+	–	+
Arginine dihydrolase	+	+	+	–	+	+	–	–	+	V	–	–	–
Cleavage mechanism for diphenols	ortho						meta		None		ortho	None	

* Data of Redfearn *et al.* (1966). † Data of Delafield *et al.* (1965). V = variable.

Taken from: Stanier, R. Y., Palleroni, N. J., and Doudoroff, M., *J. Gen. Microbiol.*, *43*, 159–271 (1966). Reproduced by permission of Cambridge University Press, New York.

strains may be "epidemiologically fingerprinted" by testing for pyocin-sensitivity and pyocin production against known standard strains.

Although *Pseudomonas* are not particularly invasive, once they are established as infective agents, they are very difficult to eradicate. *Pseudomonas aeruginosa,* the "blue pus" organism, produces a dramatic effect when it infects a burn: the pigment of the organism, mixed with blood and tissue detritus, makes the involved area literally blue. *Pseudomonas* also have a predilection for the urinary tract. They have been known to cause meningitis, and sometimes invade the eye after removal of cataracts or after trauma. The end result of this type of infection is usually loss of sight in the involved eye.

TABLE 2

NUTRITIONAL CHARACTERS OF DIAGNOSTIC VALUE FOR DIFFERENTIATING SPECIES OF AEROBIC PSEUDOMONADS

	Fluorescent Group			Pseudomallei Group*			Acidovorans Group		Alcaligenes Group				
Utilization as Sole Carbon and Energy Source of	*P. aeruginosa*	*P. fluorescens*	*P. putida*	*P. multivorans*	*P. pseudomallei*	*P. mallei*	*P. acidovorans*	*P. testosteroni*	*P. alcaligenes*	*P. pseudoalcaligenes*	*P. stutzeri*	*P. maltophilia*	*P. lemoigneit*†
D-Fucose	–	–	–	+	+	+	–	–	–	–	–	–	–
D-Glucose	+	+	+	+	+	+	–	–	–	–	+	+	–
Trehalose	–	+	–	+	+	+	–	–	–	–	–	+	–
Cellobiose	–	–	–	+	+	+	–	–	–	–	–	+	–
Maltose	–	–	–	–(+)	+	V	–	–	–	–	+	+	–
Starch	–	–	–	–	+	V	–	–	–	–	+	–	–
Inositol	–	+	–	+	+	+	V	–	–	–	–	–	–
Mannitol	+	+	–(+)	+	+	+	+	–	–	–	V	–	–
Geraniol	+	–	–	–	–	–	–	–(+)	–	–	–	–	–
2-Ketogluconate	+	+	+	+	+	V	–	–	–	–	–	–	–
Maleate	–	–(+)	–	–	–	–	+	–	–	–	–	–	–
Glycollate	–	–(+)	–(+)	V	–	–	+	+	–	–	+	–	–
DL-Lactate	+	+	+	+	+	+	+	+	+	+	+	+	–
Pelargonate	+	+	+	+	+	V	–	–	+	+	+	–	–
Adipate	+	V	–	+	+	V	+	+	–	–	–(+)	–	–
m-Hydroxybenzoate		–(+)	–(+)	+	–	–	+	+	–	–	–	–	–
Testosterone	–	–(+)	–(+)	+	–	–	–	+	–	–	–	–	–
Acetamide	+	–	V	v	–	–	+	–	–	–	–	–	–
Arginine	+	+	+	+	+	+	–	–	+	+	–	–	–
Valine	+	+	+	v	+	V	–	–	–	–	+	–	–
Norleucine	–	–	–	–	–	–	+	+	–	–	–	–	–
D-Tryptophan	–	–	–	–	–	–	+	–	–	–	–	–	–
δ-Aminovalerate	+	V	+	+	+	V	+	–	+	+	–	–	–
Betaine	+	+	+	+	+	+	–	–	–	+	–(+)	–	–
Putrescine	+	+	+	+	+	V	–	–	–	+	+	–	–

* Data of Redfearn *et al.* (1966). † Data of Delafield *et al.* (1965). + = positive; –(+) = usually negative, occasionally positive; – = always negative.

Taken from: Stanier, R. Y., Palleroni, N. J., and Doudoroff, M., *J. Gen. Microbiol., 43,* 159–271 (1966). Reproduced by permission of Cambridge University Press, New York.

Infection with other pseudomonads occurs rarely. Occasional reports[11,12] are noted of isolations of *P. acidovorans, P. alcaligenes, P. maltophilia, P. multivorans, P. putida, P. stutzeri,* and others, from human infections, especially from the urinary tract. However, they may be commonly isolated if environmental studies are made.

Chemotherapy for *Pseudomonas* infections is rather difficult, because the organisms are often resistant to the commonly used antibiotics. Some strains are moderately sensitive to streptomycin, and many strains are moderately sensitive to colistin, gentamycin, and polymyxin B. These latter three drugs may be used topically, but all have varying degrees of toxicity when applied parenterally.

A substituted penicillin, carbenicillin, has recently been shown[9] to be moderately active against *Pseudomonas* strains. It has the advantage over the above-mentioned drugs that its toxicity is almost negligible. Levels equivalent to the minimal lethal concentration of carbenicillin are difficult to attain in blood, but they can be readily attained in urine.

It should be noted that *P. aeruginosa* is notoriously resistant to quaternary ammonium compounds. Solutions of these, such as benzalkonium chloride, are often used for chemical disinfection in the laboratory and hospital; if allowed to stand for days, they may become contaminated with pseudomonads and serve as a reservoir to contaminate the environment further; they should, therefore, be discarded immediately after use. Recent studies of these organisms suggest that their ability to grow in the benzalkonium chloride may be due to the presence of ammonium acetate buffer in the commercial product.[10]

Pseudomonads have been isolated from a wide variety of plants, usually from blighted leaves, spots on fruit, necrotic spots in mushrooms, etc. They are said to be pathogenic, but their presence may just be that of an opportunistic organism growing in a favorable environment.

THE *Pseudomallei* Group

Pseudomonas pseudomallei is the causative agent of melioidosis. The organism has been isolated from infected domestic animals, rodents, and man, although the disease is relatively rare, even in southeast Asia, where the incidence is highest. The source of infection, when it occurs, is probably soil or water. *Pseudomonas pseudomallei* is probably an accidental pathogen, like *P. aeruginosa,* which may set up an infectious process after introduction into a portal of entry in man or animals. The organism has long been recognized to be a typical *Pseudomonas* on the basis of cultural and biochemical properties. Results of recent studies[13] confirm the basis for the original classification and, in addition, suggest that *Actinobacillus mallei* is closely related and should be included in this genus.

The latter organism causes glanders in horses, donkeys, and mules. Other animal species are susceptible, but rarely acquire the disease naturally. Humans occasionally acquire the infection from an infected animal. As is the case with *Brucella* and *Pasteurella tularensis,* laboratory infections may arise in workers studying this species. Glanders has almost been eradicated from the U.S., Canada and Europe by destruction of infected animals and application of effective quarantine measures. It should be pointed out that glanders is transmitted from animal to animal rather than by invasion of the host by an opportunistic environmental bacterium, such as *P. pseudomallei.*

Generically, the glanders bacillus has variously been designated as *Bacillus, Pfeifferella,* and *Malleomyces*, in addition to the currently accepted genus, *Actinobacillus*. The proposal to include the glanders bacillus in the genus *Pseudomonas* has some merit, even though the organism has no flagella. If one compares *P. pseudomallei* and *P. mallei* in Tables 1 and 2, it is obvious that the two organisms are very similar in their characteristics and reactions. The application of DNA base composition as a taxonomic tool is well recognized and has been determined for many genera. The values between 57 and 70 mols % guanine + cytosine (GC) have been suggested[14] as a range that would include all *Pseudomonas* species. Comparison of the buoyant densities in cesium chloride and the DNA base composition showed that *P. pseudomallei* had a buoyant density of 1.7281 ± 0.0007 and a guanine + cytosine value of 69.5 ± 0.7 mols %, while for *P. mallei* the buoyant density was 1.7276 ± 0.0010 and the GC value was 69.0 ± 1.0 mols %. These values are obviously close enough to justify inclusion of the glanders bacillus in the pseudomonads on the basis of DNA base composition.

REFERENCES

1. Rhodes, M. E., *J. Gen. Microbiol., 21,* 221 (1959).
2. Rhodes, M. E., *J. Gen. Microbiol., 25,* 331 (1961).
3. Lysenko, O., *J. Gen. Microbiol., 25,* 379 (1961).
4. Colwell, R. R., and Liston, J., *J. Bacteriol., 82,* 1 (1961).
5. DeLey, J., *Annu. Rev. Microbiol., 18,* 17 (1964).
6. Stanier, R. Y., Palleroni, N. J., and Doudoroff, M., *J. Gen. Microbiol., 43,* 159 (1966).
7. Holloway, B. W., *J. Pathol. Bacteriol., 80,* 448 (1960).
8. Farmer, J. J., III, and Herman, L. G., *Appl. Microbiol., 18,* 760 (1969).
9. Knudsen, E. T., Robinson, E. N., and Sutherland, R., *Brit. Med. J., 3,* 75 (1967).
10. Adair, F. W., Geftic, S. G., and Gelzer, J., *Appl. Microbiol., 18,* 299 (1969).
11. Pickett, M. J., and Pedersen, M. M., *Amer. J. Clin. Pathol., 54,* 164 (1970).
12. Pedersen, M. M., Marso, E., and Pickett, M. J., *Amer. J. Clin. Pathol., 54,* 178 (1970).
13. Redfearn, M. S., Palleroni, N. J., and Stanier, R. Y., *J. Gen. Microbiol., 43,* 293 (1966).
14. Mandel, M., *J. Gen. Microbiol., 43,* 273 (1966).

Anaerobic Gram-Negative Rods

DR. HUBERT A. LECHEVALIER

Non-motile, non-sporing strains of anaerobic bacteria used to be separated into genera on the basis of morphology as follows:

Bacteroides	– rods with rounded ends
Fusobacterium	– rods with pointed ends
Sphaerophorus	– pleomorphic rods
Dialister	– very small rods
Leptotrichia	– very large rods

The presence of many strains with intermediate forms makes purely morphological differentiation difficult. Another approach has been to combine morphology with consideration of the *major* products of fermentation of glucose:

Bacteroides	– production of acetic and succinic acids, often with propionic and formic acids; organisms usually not pleomorphic
Fusobacterium	– production of butyric acid; rods with pointed ends or pleomorphic
Leptotrichia	– production of lactic acid; large rods
Dialister	– very small rods, 0.15 to 0.6 μm in diameter

In the above system of classification, the generic name *Sphaerophorus* is not used.

On the basis of morphology and fermentative abilities, one can separate the motile, non-sporing anaerobes as follows:

A. Peritrichously flagellated

a. producing acetic and succinic acids . . . *Bacteroides*
b. producing butyric acid . . . . *Fusobacterium*

B. Flagella in tufts

a. fermenting glucose to alcohol . . . . . . *Zymomonas*
b. not as above

aa. polar tufts . . . . *Spirillum* (see page 97)
bb. lateral tufts . . . . *Selenomonas*

C. One single flagellum

a. non-fermentative . . . . *Vibrio* (see page 101)
b. fermentative

aa. producing butyric acid . . . . . . *Butyrivibrio*
bb. producing succinic acid . . . . . . *Succinivibrio*

Bacteroides

Members of this genus are abundantly distributed in the intestinal flora of man. They are peritrichously flagellated when motile. The type species, *B. fragilis,* is one of the most common. Other species are also part of the normal flora of the human gastrointestinal, respiratory and genitourinary tracts. In general, *Bacteroides* are harmless bacteria that may become opportunistic pathogens; they are not highly invasive, and usually the infections they cause respond to the administration of the tetracyclines.

Fusobacterium

When motile, these cells are peritrichously flagellated. The type species, *F. fusiforme,* is a normal inhabitant of the mouth and upper respiratory tract. The tapered rods may reach 16 μm in length and are found associated with *Borrelia vincenti* in Vincent's angina (trench mouth), an acute ulcerating disease of the oropharynx.

Fusobacterium (Sphaerophorus) necrophorus. Also normally present in the upper respiratory and gastrointestinal tracts of man. It is more pleomorphic than *F. fusiforme* and has been implicated in numerous human and animal infections.

Leptotrichia

These rods may be as long as 10 μm. Some cells may have two somewhat pointed ends, others may be pointed at one end and blunt at the other. Formation of chains of cells may exceed 100 μm in length. Young cells are Gram-positive; those one-day-old and older are Gram-positive, but ultramicroscopic examination reveals a Gram-negative cell wall structure. Freshly isolated strains are strictly anaerobic; tolerance to oxygen has been observed after continued cultivation on laboratory media. *Leptotrichia* are found in the oral cavity of man or in the urogenital region; they are not known to be pathogenic. The type species is *L. buccalis* (Hofstad, T., *Int. J. Syst. Bacteriol., 20,* 175–177, 1970).

Dialister

These very small rods (up to 0.6 μm) are capable of passing through some of the bacteriological filters; their isolation is helped by the presence of ascitic fluid or fresh tissue in the medium. They are isolated from nasopharynx washings. The type species is *D. pneumosintes.*

Zymomonas

These rods may be 5 μm long. They are motile, with tufts of flagella, when young. The organisms are anaerobic, developing a certain tolerance to air in the presence of some sugar. Glucose is fermented with the production of ethanol and some lactic acid. Zymomonads are found in plant juices and fermented beverages. The type species is *Z. mobilis.*

Selenomonas

These curved cells may be up to 12 μm long. They are motile by a tuft of flagella, which is attached to the concave side of the cell. The organisms are strictly anaerobic. They ferment sucrose to lactate, acetate, propionate, and formate. Selenomonads are difficult to isolate in culture or media that are not enriched with blood or rumen juice; they may require one or more of the following acids: isobutyric, *n*-valeric, isovaleric, and/or DL-2-methylbutyric. These bacteria are found in the cecum of guinea pigs, in the human buccal cavity, and in the rumen of herbivorous mammals. The type species is *S. palpitans* (Prins, R. A., *J. Bacteriol., 105,* 820–825, 1971).

Butyrivibrio

These curved rods may be up to 5 μm long, with monotrichous flagellation. Glucose is fermented with the production of butyric, formic, and lactic acids. The organisms are found in the rumen of herbivorous mammals and in the intestinal tracts of other animals, including man. The type species is *B. fibrisolvens* (Bryant, M. P., and Small, N., *J. Bacteriol., 72,* 16–21, 1956).

Succinivibrio

These curved rods may be up to 5 μm long, with monotrichous polar flagellation. Glucose is fermented with the production of succinic and acetic acids. The organisms are found in the rumen of herbivorous mammals. The type species is *S. dextrinosolvens* (Bryant, M. P., and Small, N., *J. Bacteriol., 72,* 22–25, 1956; Hungate, R. E., Bryant, M. P., and Mah, R. A., *Annu. Rev. Microbiol., 18,* 131–166, 1964).

The Cocci

DR. CARL F. CLANCY

Most of the cocci discussed in this section are members of the family Micrococcaceae or the tribe Streptococceae of the family Lactobacillaceae. These two groups include nearly all spherical bacteria that have been described, with the exception of a few species of photosynthetic and chemoautotrophic bacteria previously discussed, and members of the genera *Siderocapsa, Lampropedia,* and relatives.

MICROCOCCACEAE

The family Micrococcaceae is composed of spherically shaped Gram-positive organisms, classified into six different genera. The characteristic for assignment to a genus is the arrangement of cells as seen in the microscope. The cells may appear singly, in pairs and clusters, in tetrads, or in packets of eight, and the particular arrangement is thought to be a reflection of whether cell division is in one, two, or three planes. Most species are Gram-positive, but occasionally some species are described as Gram-negative or variable. Micrococcaceae are widespread in nature and, with the exception of the staphylococci, usually not parasitic.

Micrococcus, Gaffkya, Sarcina, Methanococcus and *Peptococcus*

Three of these genera – *Micrococcus, Gaffkya,* and *Sarcina* – are found frequently in our environment. They are rarely isolated from disease processes. The genus *Micrococcus* includes about sixteen species; microscopically they appear arranged singly, in pairs, and in clusters. Several species may be isolated from dairy products and dairy utensils. They appear occasionally as air-borne contaminants on culture plates, often as yellow-, orange-, or red-pigmented colonies.

Cocci that regularly appear in tetrads are included in the genus *Gaffkya.* The type species *G. tetragena* is occasionally isolated from human sources, but rarely, if ever, produces infection.

Cocci that appear in packets of eight are *Sarcina,* found in air, dust, water, soil. *Sarcina lutea, S. flava,* and *S. aurantiaca* are notable for yellow and orange colonies. One unusual species, *S. ureae,* is not only motile but forms spores; because of this anomaly, in spite of its coccal shape, *S. ureae* might well be classified with the aerobic spore formers.

The genus *Methanococcus* has only two species: *M. mazei,* and *M. vannielii.* Both have been isolated from soil. They are anaerobic and have the ability to convert formate, acetate, and butyrate into methane and carbon dioxide and undoubtedly are responsible, at least in part, for the methane present in marsh gas.

Another genus, *Peptococcus,* is composed of anaerobic micrococci found in or on the human body. One species has been found in soil, but most have been recovered from blood, urine, intestinal tract, genitalia, tonsils, etc. Except when isolated from blood or other foci in pure culture, the pathogenicity of *Peptococcus* species is not clearly defined. The organisms may be involved in mixed pyogenic infections.

Staphylococcus

The staphylococci are commonly found in and on the human and animal body. *Staphylococcus epidermidis* (formerly *albus*) is usually found on the skin and external nares; it is usually not considered pathogenic, although it may cause such diseases as subacute bacterial endocarditis, meningitis, etc., after accidental introduction into the body. The colonies on agar are white, shiny, butyrous, and about 3 mm in diameter after an incubation period of 48 hours. The organism grows at room temperature, but more rapidly at 37°C.

Staphylococcus aureus is the cause of a wide variety of human and animal infections. It is the common cause of furunculosis and pyogenic wound infections, and may produce an infection in almost any part of the body into which it is introduced. It is one of the common causes of mastitis in cattle. Colonies on agar are initially a dirty-white color that may develop into a golden color after two to three days of incubation; it should be noted that the pigmentation is not the lemon-yellow color seen with some of the other micrococci. There is considerable variation from strain to strain in the amount of golden pigmentation that develops, even after several days of incubation.

Differentiation between *S. epidermidis* and *S. aureus* is not too difficult, since the former has a white colony, does not ferment mannitol, nor produce the enzyme coagulase. *Staphylococcus aureus,* on the other hand, ferments mannitol and produces coagulase. Occasional strains may deviate from the usual pattern.

Toxins and Enzymes

Staphylococcus aureus, when grown on blood-agar plates, usually produces a zone of clear hemolysis after two days of incubation. The size of the zone is enhanced by cooling the plates to 5°C for several hours. Four antigenically distinct hemolysins, designated as *alpha-, beta-, gamma-,* and *delta-,* have been shown to be involved in this red-cell lysis.

The *alpha*-hemolysin, a protein of molecular weight of about 44,000, is active against rabbit and sheep erythrocytes. It produces necrosis on intracutaneous injection into rabbits, and small amounts are lethal for rabbits and mice. This hemolysin is produced mainly by strains isolated from human sources. It should be noted that the *alpha*-hemolysin produces a clear type of hemolysis that should not be confused with the reaction produced on blood agar by *alpha*-streptococci. The *beta*-hemolysin is active against sheep erythrocytes and is produced mainly by strains from animal sources. *In-vitro* activity is best demonstrated by preliminary incubation of blood-broth tubes at 37°C, then holding overnight at 4°C, the so-called "hot-cold lysis". This hemolysin is much less toxic on injection into mice or rabbits. The *gamma-* and *delta*-hemolysins are active against erythrocytes of a wider species of animals, including man. They are less toxic for experimental animals than the *alpha*-hemolysin. It is believed that these four toxins play a part in the pathogenesis of staphylococcal disease, but their true role is unknown.

Staphylococcal enterotoxin[1] is produced by relatively few strains of *S. aureus,* usually those in phage-lytic group II. There are three types of enterotoxin, each antigenically distinct from the others and from the hemolysins.

Leucocidin, a substance produced by many strains of *S. aureus,* can be shown to destroy white blood cells of a variety of animal species. It is distinct from the other metabolites of staphylococci, proteinaceous in nature, and antigenic.

A number of enzymes are produced by staphylococci, such as hyaluronidase (spreading factor), staphylokinase, proteinases, lipases, coagulase, and penicillinase.

Coagulase can be demonstrated in cultures of nearly all strains of *S. aureus* that are freshly isolated from human infections. This enzyme has a thrombokinase-like activity that initiates the conversion of fibrinogen into fibrin. Coagulase does not require calcium ions to initiate the conversion of fibrinogen into fibrin; however, it does require a factor called CRF (coagulase reacting factor), which is present in high concentrations in horse, human, and rabbit plasma. The coagulase test, as performed in the laboratory, is usually carried out using human or rabbit plasma in a tube to which a broth culture of the organism is added; after a suitable incubation period, a typical fibrin clot forms if the enzyme is present. A slide test may be carried out by adding plasma to a uniform suspension of the organism and rocking the mixture for a few minutes; the bacterial cells become coated with fibrin as it is formed, then clump together, appearing much the same as in a rapid slide agglutination test.

The production of the enzyme penicillinase, by staphylococci has considerable clinical importance. The enzyme acts by hydrolyzing the unstable β-lactam ring of penicillin G, thereby yielding inactive penicilloic acid. Penicillinase-producing strains are resistant to penicillin G when tested *in vitro,* and infections caused by them do not respond to treatment with penicillin G. Penicillinase is much less active against the semisynthetic penicillins, due to the steric effect of the bulky side chains substituted in the acyl group.

Serologic Relationships

Identification of staphylococci by serologic tests has been described.[2] There are two main groups: the pathogenic Group A, and the non-pathogenic Group B. The specificity depends on the polysaccharides in the cell wall, which are precipitated by antiserum prepared against the whole cells. Subsequent work has shown that immune serum prepared from pathogenic staphylococci agglutinates many strains of *S. aureus,* but not *S. epidermidis.* Many strains are inagglutinable, however, and serologic studies seem, therefore, of little value.

Bacteriophage Typing

One of the first descriptions of bacteriophage lysis was that of Twort,[3] who noted the moth-eaten appearance of colonies of bacteria (presumably *S. epidermidis*) cultured from calf vaccine. Six years later, d'Herelle[4] reported on the isolation of the lytic principle from *S. aureus* isolated from a human boil. A system of phage typing[5] has been developed that enables the characterization of cultures of *S. aureus* according to their lysis by twenty to thirty different phage types. Phage typing is useful for epidemiological studies, especially in tracking down healthy carriers of *S. aureus* in institutional outbreaks, such as occur in maternity and newborn units. It is now recognized that certain phage types, such as 80/81 and 52A/79, are epidemic types and are repeatedly isolated from staphylococcal infections in many parts of the world.

Nature of Staphylococcal Disease

The reservoir that maintains staphylococcal disease in the population is man, either as hosts with actual infections or as healthy carriers. *Staphylococcus epidermidis* can be isolated routinely from the skin and nares of most people, but occasionally, for reasons not known, some individuals harbor *S. aureus* in the nares.

The most common type of infection is that of suppurative wound infection following injury. This type of infection may occur in any part of the body; the source of the organism is not always clear. Staphylococcal infections tend to become walled-off abscesses, which eventually drain or are incised by the surgeon's scalpel. A particularly annoying type of infection is recurrent furunculosis, which usually occurs in youths and young adults. Furuncles develop without apparent break in the skin, proceed to drain or are incised, and finally heal; subsequently, another furuncle develops at a different site. This condition may continue for weeks and months. At one time vaccination with a killed autogenous vaccine (prepared from the patient's own organism) was tried, but proved to have little therapeutic effect. Application of chemotherapy also seems to have no effect in controlling the disease.

Chronic osteomyelitis may be a sequela to pyogenic infection with *S. aureus.* It often persists for years and is especially refractory to chemotherapy.

In recent years, staphylococcal gastroenteritis has been found to occur occasionally in patients who receive intensive antibiotic therapy. It is believed that, when the normal flora of the lower bowel is altered, an antibiotic-resistant staphylococcus may be established in the gut, resulting in gastroenteritis. The disease is often severe, especially in patients who already have another disease.

Staphylococcal food poisoning is an intoxication due to ingestion of food that contains preformed enterotoxin. Cream-filled pastries, puddings, and other foods that are only lightly heated in cooking often are the vehicle, although almost any food contaminated by a carrier and improperly refrigerated may cause this disease. The incubation period is short, two to six hours, and the course of the disease is limited to about twenty-four hours. The patient develops nausea, then violent diarrhea and vomiting, which is followed by prostration. Recovery occurs in one to two days, except in debilitated persons.

NEISSERIACEAE

The family Neisseriaceae is composed of Gram-negative cocci of two genera. In one genus, *Neisseria,* the organisms appear as Gram-negative diplococci and are facultative organisms in relation to their free-oxygen requirements; in the other genus, *Veillonella,* the cocci appear in pairs, clusters, and short chains and are strict anaerobes. Organisms from both groups live in or on man and other animals.

Neisseria

The *Neisseria* are Gram-negative "biscuit"-shaped diplococci; one side of each cell of the pair is flattened against the side of the other cell, and the pair always appears as one unit. In cultures three to four days old or older they may "balloon out" in size and have a tendency to become Gram-positive; in smears of body fluids and young cultures they usually have the typical diplococcus arrangement and are uniformly Gram-negative. All species are found in the human body, with the exception of *N. caviae,* which has been found in the pharynx of guinea pigs. All *Neisseria* produce an oxidase that acts on dimethyl- or tetramethyl-*p*-phenylene diamine; when flooded with the reagent, colonies turn pink to purple to black because of the oxidation reaction.

Neisseria catarrhalis is commonly found in sputum and throat cultures of normal individuals. The colony is dirty white in color, grows to a diameter of 2 to 3 mm in two days, and often is rough. *Neisseria sicca* has the same habitat, but colonies grow in a rather lacy fashion likened, by some, to cartwheels. Several other species (e.g., *N. flava*) have dirty-yellow pigmentation and may also be readily isolated from sputa or throat swabs. These different species are similar to one another in having rough colonies that are difficult to emulsify. They are a well-recognized entity in the oral flora and are not associated with any disease processes. They grow at 22°C.

Neisseria meningitidis (formerly *intracellularis*), the meningococcus, is the cause of a bacterial meningitis and meningococcemia. This Gram-negative diplococcus is maintained in the population in the throats of healthy carriers and produces its disease only when conditions are favorable for the development of the disease in a susceptible host. Colonies of this species grow to be two to three mm in diameter and are usually gray and mucoid on primary isolation. Good growth occurs on blood agar or chocolate agar at 37°C and is enhanced by incubation in an atmosphere of 5% carbon dioxide. When grown in a special cystine-casein semisolid agar with appropriate CHO substrate, glucose and maltose are fermented, but not sucrose. The amount of acid produced is not large.

Meningococci are separated into four serologic groups (A, B, C, and D) on the basis of antigenically specific polysaccharides present in the capsules. A rapid slide agglutination test may be carried out with a freshly isolated culture by use of appropriate antisera that have been absorbed, to ensure group specificity. Encapsulated cells demonstrate the quellung reaction when mixed with homologous antiserum. Serologic studies show that Group A organisms are usually isolated from epidemic cases, whereas Group B and Group C strains are usually recovered from sporadic cases. Group D organisms are rarely encountered.

A potent endotoxin can be extracted from meningococcal cells; it is believed to play a part in the pathogenesis of the infection. When injected into experimental animals, the endotoxin produces vascular damage; the lesions resemble those seen in meningococcemia. Analysis[6] of endotoxin from an organism of the serologic Group C showed it to contain 20% lipid, composed of cephalin, fatty acids, and plasmalogen; the polysaccharide constituents were identified as galactose, glucose, glucosamine, and sialic acid. The lipids, when separated from the endotoxin, fail to show lethal activity and are non-pyrogenic.

Meningococcal Disease

Man is the only natural host for meningococci, and the disease is maintained and spread by carriers harboring the organism in their nasopharynx. The relationship between the carrier state and clinical disease is not understood, since studies have shown a carrier rate of 25% in a population where only sporadic cases of the disease may occur. Meningococcal disease attacks children and young adults, and epidemics may break out in institutions or military induction camps. Development of the disease seems to be related in some way to physical stress; in military units the carrier rate may rise well over 50% during epidemic periods.

Invasion of the blood by the organism may produce chronic or acute meningococcemia; this may result in the development of petechiae in the skin, with subsequent progress to purpurae several centimeters in diameter, due to hemorrhage. Positive cultures may be recovered from the hemorrhaged areas, and it is even possible to see the Gram-negative diplococci in smears of blood from the purpurae. Cultures of blood drawn at this time are usually positive.

Invasion of the central nervous system from the blood results in development of the meningeal form of the disease. Smears of spinal fluid sediment reveal the presence of Gram-negative diplococci, and cultures made at this time should be positive. It should be emphasized that the meningococcus undergoes autolysis rather quickly, and bacteriological studies should, therefore, be carried out immediately after collection of the specimen.

Chemotherapy and Prophylaxis

Initially, all strains of meningococci were believed to be sensitive to sulfonamides. Clinical cures were reported, and mass prophylaxis with sulfonamides was shown to curb epidemics in closed communities, such as military installations. However, the situation is different today; more than 70% of the cases are caused by sulfonamide-resistant meningococci.[7]

Penicillin is effective in the treatment of clinical cases, but on a large scale it does not lend itself to prophylaxis of the carrier state. A recent study[8] has shown that application of oral Rifampin, a semisynthetic antibiotic, was effective in clearing the carrier state. Another interesting approach to the carrier problem has been the immunization[9] of potential carriers with a high-molecular-weight polysaccharide prepared from *N. meningitidis,* Group C. Follow-up studies[10] showed the results of the vaccination to have a degree of efficiency in the prevention of subsequent meningococcal disease.

Neisseria gonorrhoeae

Neisseria gonorrhoeae, the gonococcus, is the causative agent of gonorrhea, a venereal disease, and of several other conditions that are usually sequela to the venereal infection. Like the meningococcus, the organism only infects man, and transmission of the disease is practically always by person-to-person contact.

Microscopically, the organism is indistinguishable from many other *Neisseria*; it is, however, more fastidious in its growth requirements. The gonococcus will not grow on blood agar, but it will grow on chocolate agar to which a yeast extract supplement has been added; the latter provides glutamine, cocarboxylase, and glutathione, which have been shown to be essential for growth. Certain amino acids in peptone and agar are somewhat toxic for the gonococcus, but the serum in the chocolate agar adsorbs or neutralizes these toxic factors. For primary isolation, appropriate concentrations of crystal violet in the chocolate agar will reduce the growth of Gram-positive organisms that might be present in urethral or cervical swabs. It is also necessary to incubate plates for primary isolations from clinical material in an atmosphere of about 5% added carbon dioxide. Colonies, about 1 to 2 mm in diameter, grow in two days, are colorless to gray, shiny and smooth. The oxidase test has been used to identify *Neisseria* colonies on plates from urethral or cervical swabs, where contaminating organisms make recognition of gonococcal colonies difficult. The gonococcus ferments glucose, but not maltose and sucrose; the sugar fermentations are of value in confirming the identity of an isolate.

Laboratory Diagnosis

The gonococcus is even more delicate than the meningococcus and will not withstand adverse conditions, such as drying; hence, it is important to be sure that smears and cultures are made promptly after collection of specimens. Cultures should be made on chocolate agar, of course, as discussed above. Smears from urethral discharge are prepared by rolling the swab onto a glass slide, to avoid destruction of leucocytes. The finding of Gram-negative intracellular diplococci in such smears is almost specific identification for the gonococcus, provided that the history and clinical findings are coincident.

In the past, because of the presence of a multiplicity of other organisms, the demonstration of gonococci in cervical swabs was so difficult that laboratory studies were often not carried out and the diagnosis was made on a clinical basis. Since the advent of fluorescent-antibody techniques for identification of microorganisms, appropriately labeled antigonococcal serum has also been used for identification of *N. gonorrhoeae*. Slides may be prepared directly from the swab or a culture for fluorescent-antibody studies.

Gonococcal Disease

Gonorrhea in males usually appears two to eight days after exposure and is manifested by urethritis with purulent discharge. In females the onset of disease may also be urethritis, but the infection often progresses to involvement of the ovaries and other organs and is then called "pelvic inflammatory disease" (P.I.D.).

Among the complications of gonorrhea are gonococcal arthritis and ophthalmia neonatorum. Prophylaxis for the latter, a disease of newborns, consists of washing the eyes of the infant with a solution of penicillin immediately after birth.

When sulfonamides first became available they were used successfully in the treatment of gonorrhea. During World War II, certain strains of the gonococcus appeared, which were resistant to the drug. Later, penicillin became the drug of choice, and many strains were sensitive to 0.06 units per ml or less.[11] Subsequent studies showed that each year the sensitivity to penicillin is decreasing; as of 1969[12] more than 65% of routine isolations were resistant to doses larger than 0.05 units of penicillin per ml, and some even to doses larger than 3.0 units per ml. Penicillin is still the drug used most often; when given in high doses, with addition of probenecid to slow down excretion, it provides an effective therapeutic measure.

In spite of the fact that effective antimicrobial agents have been available since about 1940, the incidence of gonorrhea is increasing at an alarming rate. From 1954 to 1962 the total of reported cases in the United States was around 250,000 per year; this has been rising constantly, so that by 1971 a total of almost 700,000 cases was reported.

Veillonella

This group of anaerobic Gram-negative cocci named after Adrien Veillon, who first described the type species, is included in the Neisseriaceae. The cocci appear in pairs, short chains, and clusters. They are very small; most species are less than 1.0 μ in diameter. Six species of *Veillonella* are recognized; they have been isolated from oral cavity, lungs, lower bowel, and urogenital tract of man and animals, and sometimes from abscesses or other pyogenic infections. Prévot[13] considers them to have a degree of natural pathogenicity, capable of causing abscesses, etc., in various parts of the body where they are naturally found. It should be pointed out that, because of the toxicity of free oxygen, many strict anaerobes, such as *Veillonella*, are missed with the usual techniques for routine cultures. In fact, it is almost a research procedure to isolate some of the strict anaerobes found in the gastrointestinal tract.

STREPTOCOCCEAE

In the tribe Streptococceae are the Gram-positive cocci that occur singly, in pairs, and often in chains. Three genera–*Diplococcus, Streptococcus*, and *Pediococcus*–are homofermentative, producing mainly lactic acid from carbohydrate fermentation. Another genus, *Leuconostoc*, is heterofermentative, producing carbon dioxide, ethanol, acetic acid, and lactic acid from carbohydrates. A fifth genus, *Peptostreptococcus*, includes strict anaerobes. These cocci are found in nature, in humans and lower animals. Several species produce serious infectious diseases.

Diplococcus

Diplococcus pneumoniae, the common cause of lobar pneumonia in man, appears as lanceolate-shaped diplococci in which the pair comprises the unit. The organisms are, of course, Gram-positive, but they may be readily decolorized by the use of too much ethanol or acetone in the Gram-staining procedure. Their usual habitat is the upper respiratory tract of healthy carriers. Pneumococci are delicate, die off readily away from the body, and often cannot be subcultured if they are held longer than two or three days in the incubator. They will not grow on ordinary culture media unless blood is added; in blood broth they grow with a light uniform turbidity. On primary isolation on a blood agar plate, they appear after one day as tiny colonies, which enlarge after two days to about 1 mm in diameter. The typical appearance is flat and glassy, with a central depression. Growth is much better in a partial carbon dioxide atmosphere. Some strains show a marked tendency to autolyze; if left in the incubator for several days, the entire colony may disappear.

Pneumococci possess a polysaccharide capsule, which accounts for the glassy appearance of the colonies. Certain serotypes (types 3 and 8) have unusually large capsules, a property that is reflected in large colonies. Serological studies on capsular polysaccharide have made it possible to establish a system of classification based on the serologic specificity of the capsular material. As initially developed, there were only a few serotypes; but as time went on, new ones were discovered, until more than 100 serotypes have been described. Prior to the antimicrobial era, serotherapy of pneumococcal infection was successfully used; however, the therapeutic antiserum had to be of the same serotype as the organism that infected the patient.

Pneumococcal Disease

The pneumococcus is the most common cause of lobar pneumonia and, as a complication, may produce empyema or pericarditis. It also may cause mastoiditis and otitis media, and is a common agent of bacterial meningitis.

The laboratory diagnosis of pneumococcal infection depends on demonstration of the organism by smear and confirmation by culture. A smear of body fluid (e.g., spinal fluid) and a Gram stain may be sufficient to show the presence of Gram-positive lanceolate-shaped diplococci. The most specific test that can be done to identify the pneumococcus is the Neufeld quellung reaction. If an encapsulated pneumococcus is mixed with its homologous serum (i.e., antibodies against the specific capsular polysaccharide), a precipitation reaction occurs and the capsule can then be seen surrounding the organism. Visualization under the microscope is enhanced by adding a drop of methylene blue and a cover slip. Commercially prepared sera are available for the serotyping procedure.

Identification of the culture depends on the appearance of typical flat, glassy colonies on blood-agar plates in one to two days. The blood in the agar is turned green, an effect similar to that produced by *alpha*-hemolytic streptococci. Confirmation tests that may be done are fermentation of inulin and bile solubility tests. The pneumococcus ferments inulin, and the cells are readily lysed by 10% sodium taurocholate. However, a simple test, and one that is easier to interpret, is the cuprein hydrochloride (Optochin®) sensitivity test. This compound is active against pneumococci, but not against *alpha*-hemolytic streptococci. Disks impregnated with cuprein hydrochloride may be placed on a plate that has been freshly streaked with the unknown organism; after incubation, there will be a zone of inhibition around the disk if the organism is a pneumococcus, whereas *alpha*-hemolytic streptococci will not be inhibited. The Neufeld quellung reaction also may be used for identification of young cultures of pneumococci on primary isolation.

Pneumococci are sensitive to many antimicrobial agents. Penicillin is usually the drug of choice, since the pneumococcus is extremely sensitive to this antibiotic.

Streptococcus

The streptococci are Gram-positive cocci that appear under the microscope in pairs and chains. Chains can best be visualized in smears from fluid media; strains that grow "rough" usually have longer chains than smooth-growing strains. A basic character for classification is the reaction produced on the surface of blood-agar plates:[14] *alpha*-hemolytic streptococci turn the medium (actually the hemoglobin) green in the vicinity of the colony; *beta*-hemolytic streptococci lyse the red blood cells, creating a clear zone around the colony; *gamma*-hemolytic streptococci have no effect, and the medium remains unchanged. Four basic groups of streptococci are now recognized: (1) hemolytic streptococci, (2) viridans streptococci, (3) enterococci, and (4) lactic streptococci.

HEMOLYTIC STREPTOCOCCI

These organisms are, for the most part, human and animal pathogens. Most appear as *beta*-hemolytic streptococci when grown on blood agar. *Streptococcus pyogenes* was recognized very early in microbiology as a causative agent of human infection. It is maintained in the throats of carriers, usually children. While streptococcal disease is not as rampant as in the pre-antibiotic era, it continues to smolder and breaks out as clinical disease from time to time.

Streptococcus pyogenes grows only poorly in ordinary media, but addition of blood gives much better growth. Colonies on blood agar are of three different types – mucoid, matt, and glossy – and about 1 to 2 mm in diameter. The hemolysin produced diffuses into the medium and lyses the red cells, creating a clear area surrounding each colony. In broth there is usually a floccular or granular sediment, due to long chains of cocci settling to the bottom of the tube. The classification of hemolytic streptococci was considerably simplified when Lancefield[15] showed that an antigen, now called the C carbohydrate, could be extracted from cells of *S. pyogenes.* The antigen was specific and antigenically identical in all strains studied. Thus, the human strains from scarlet fever, streptococcal sore throat, erysipelas, puerperal sepsis, wound infections, etc., became designated as serologic Group A. It is now recognized that the C carbohydrate is in the cell wall. Elaboration of studies of the C carbohydrate of streptococci from many sources has shown at least thirteen different serologic groups.

Another antigenically active component of streptococci is the M protein, present in the cell wall. Within Group A, more than forty antigenically distinct serotypes are recognized (on the basis of the M antigen) and are used for epidemiologic purposes in studying the spread of streptococcal disease.

S. pyogenes Metabolites

Erythrogenic toxin is produced by many Group A strains and is responsible for the rash of cases of scarlet fever. Injection of toxin into the skin (Dick test) produces an erythematous reaction in susceptible children, whereas in convalescents it is negative, presumably due to neutralization of the toxin by antibody. It is of interest that erythrogenic strains are lysogenic,[16] similar to toxigenic *Corynebacterium diphtheriae.*

Streptolysin O lyses red blood cells, but does so only under anaerobic conditions. It is antigenic and stimulates antibody production during infection. This reaction has clinical application in that a significant increase in Streptolysin O antibody level indicates a recent streptococcal infection.

Streptolysin S is stable in air and is cell-bound, although it can be extracted from the bacterial cells. It is apparently non-antigenic.

Streptokinase (originally called fibrinolysin) is an enzyme that initiates the lysis of fibrin. It acts to convert plasminogen to plasmin, a protease found in blood plasma. It has had some use in human medicine as an agent to lyse fibrin deposits, but is of limited value because of its immunogenicity.

Hyaluronidase, originally called "spreading factor", is an enzyme that acts on hyaluronic acid; the latter is found in the streptococcal capsule and is very similar to the hyaluronate in the ground substance of connective tissue. Culture filtrates injected intradermally into animals enhance the spreading of such inert substances as India ink.

Streptococcal Disease

The usual streptococcal infections have been listed above; however, these organisms can produce infection and disease in any part of the body to which they gain entrance. Two diseases that are sequelae of streptococcal infection are acute glomerulonephritis and rheumatic fever. The pathogenesis of these conditions is essentially unknown, except that they are known to follow an acute streptococcal infection at a time when the patient has developed antibodies against the disease.

Acute glomerulonephritis follows a streptococcal infection by about a week. The glomeruli of the kidney are damaged, but the patients usually undergo spontaneous recovery. It is interesting that the great majority of cases of this disease follow infection by serotype 12, Group A streptococci. In acute rheumatic fever, the latent period after streptococcal infection is about three weeks. Patients, usually children, show carditis, migratory polyarthritis, and fever for many weeks; there is great danger of permanent damage to the heart. In this disease there may be recurrent attacks following subsequent streptococcal infections; prophylactic penicillin is often recommended to ward off further streptococcal infections. Both of these conditions may be an expression of autoimmune phenomenon.

Laboratory Diagnosis

Streptococcus pyogenes grows out readily from throat or wound swabs when planted on blood agar plates. The colonies and hemolytic activity are most typical after two days of incubation. Positive identification may be accomplished in 5 to 6 hours by inoculation of a swab into broth, followed by preparation of a smear with appropriate fluorescent-antibody and microscopic examination; fluorescent-antibody studies may even be done on the initial swab, if large numbers of organisms are present. Another simple test, though less specific, is to subculture the isolate on a blood-agar plate, apply a filter paper disk containing the antibiotic bacitracin, and look for a zone of inhibition of growth. Most strains of *S. pyogenes* are sensitive to bacitracin, whereas most other *beta*-hemolytic streptococci are bacitracin-resistant. Serologic tests for grouping streptococci are too cumbersome for routine diagnostic purposes.

Streptococcus pyogenes is very sensitive to penicillin, and apparently no antibiotic-resistant mutants have appeared, even though the drug has been vigorously applied for nearly thirty years. Most strains are also sensitive to the tetracyclines, chloramphenicol, etc., as well as to sulfonamides.

Other Hemolytic Streptococci

Streptococcus agalactiae of the Lancefield Group B streptococci is indigenous to cattle and a common cause of mastitis, as are *S. dysgalactiae,* a Group C organism, and *S. überis,* a viridans streptococcus. Some Group C strains cause human infections, others infect horses (*S. equi*) and a wide variety of domestic animals.

VIRIDANS STREPTOCOCCI

The viridans streptococci are usually *alpha*-hemolytic, and occasionally non-hemolytic. Some species (*S. salivarius, S. mitis*) are a part of the normal human oral flora; others have been found in milk and in the bovine alimentary tract. The human species are usually commensals, except when they are accidentally introduced by trauma (e.g., vigorous tooth brushing, tooth extraction, etc.) into the blood stream of persons with valvular heart disease. In such situations these benign organisms can attach themselves to the heart valves, grow, and produce vegetation, causing development of subacute bacterial endocarditis (SBE) in the patient. At one time a uniformly progressive, fatal disease, endocarditis can now usually be treated effectively with penicillin or other antimicrobials. Viridans streptococci do not have a C carbohydrate antigen in the cell wall and therefore cannot be classified on a serologic basis.

ENTEROCOCCI

The enterococci are quite different from the other streptococci in many ways. They are so consistently a part of the bowel flora of humans and animals that they have sometimes been used as indicators of fecal pollution in the bacteriological examination of water. Some strains grow so smoothly that it is difficult to demonstrate the presence of chains. Colonies on agar plates are 2 to 3 mm in diameter in two days, colorless to gray, and usually butyrous. Hemolytic activity is not critical for identification, but two of the four species are *beta*-hemolytic and the other two are non-hemolytic. Enterococci differ from other human streptococci in their ability to grow at 10°C in broth that contains 6.5% sodium chloride. They have a specific C carbohydrate antigen and are designated as Lancefield Group D.

Enterococci have the same capabilities to cause subacute bacterial endocarditis as some of the viridans group organisms, often as a sequela to prostatectomy or bowel surgery. They are penicillin-resistant, but sensitive to some degree to a wide spectrum of other antibiotics. Enterococci are also a cause of cystitis and other urinary-tract infections.

LACTIC STREPTOCOCCI

Streptococcus lactis and *S. cremoris* are found in milk and milk products, although *Bergey's Manual* lists their habitat as "probably of plant origin". These organisms play an important role in the souring of milk and are often used as starter cultures for the preparation of some cheeses. Some strains of *S. lactis* produce an antibiotic, nisin, that acts against other Gram-positive bacteria.

Miscellaneous Streptococceae

Pediococcus is a genus of microaerophilic Gram-positive cocci that appear singly, occasionally in chains, and often as tetrads. They are found in beer, wine, sauerkraut, and other fermenting materials. They are homofermentative, producing racemic lactic acid from carbohydrates. Due to the formation of lactic acid, they are sometimes responsible for beer spoilage.

Organisms in the *Leuconostoc* genus are normally spherical, but under certain conditions they appear as rods. They are microaerophilic, growing better under anaerobic conditions, and homofermentative, producing carbon dioxide, ethanol, acetic and lactic acid, from carbohydrates. In sauerkraut and ensilage fermentation, *Leuconostoc* species produce some of the lactic acid that functions as a preservative of the foodstuff. Since these organisms can convert sucrose to polysaccharide dextrans, they have been utilized in the commercial manufacture of those products. Two species are notable for having unusual vitamin requirements. *Leuconostoc mesenteroides* requires nicotinic acid, and *L. citrovorum* requires folic acid (citrovorum factor) for growth. These organisms are often used for assay of these two vitamins.

The peptostreptococci are anaerobic streptococci found in the oral cavity, lower bowel, and vagina of humans and other animals. Some species become microaerophilic on culture in the laboratory. They have been isolated in pure culture and in mixed culture from a variety of pyogenic wounds and abscesses. There are also reports of isolations from the blood of patients with subacute bacterial endocarditis and from urinary-tract infections. Peptostreptococci are also found in the rumen of herbivores and undoubtedly play a role in that bacteriological hodgepodge.

MISCELLANEOUS COCCI

Siderocapsa

Members of this genus are heterotrophic cocci that are distributed at random in a primary capsule. A secondary capsule unites a number of primary capsules, thus forming a mucilaginous colony; iron or manganese compounds are stored in the secondary capsule. Individual cocci may be up to 2 μm in diameter. These organisms are mainly found on the surface of growing plants in iron-rich waters. The genus *Siderococcus* harbors similar cocci, which are not encapsulated; the metal oxides are merely deposited on the cells of the cocci.

Lampropedia

The coccoid cells of the *Lampropedia* adhere in rectangular arrays, one cell thick, forming sheets of cells. These organisms have been found in stagnant water, on rotting vegetation, and in the intestine of numerous herbivorous animals.[17]

REFERENCES

1. Dack, G. M., *Food Poisoning,* 3rd ed. University of Chicago Press, Chicago, Illinois (1956).
2. Julianelle, L. A., and Wieghard, C. W., The Immunological Specificity of Staphylococci. I. The Occurrence of Serological Types, *J. Exp. Med., 62,* 11 (1935).
3. Twort, F. W., An Investigation on the Nature of Ultramicroscopic Viruses, *Lancet, 2,* 1241 (1915).
4. d'Herelle, F., *Le Bacteriophage: Son Rôle dans l'Immunitè. Monographie de l'Institut Pasteur.* Masson et Cie, Paris, France (1921).
5. Wentworth, B. B., Bacteriophage Typing of Staphylococci, *Bacteriol. Rev., 27,* 253 (1963).
6. Mergenhagen, S. E., Martin, G. R., and Schiffman, E., Studies on an Endotoxin of a Group C, *Neisseria meningitidis, J. Immunol., 90,* 312 (1963).
7. *Meningococcal Infections, Morbidity and Mortality Weekly Reports,* Vol. 18, p. 135, Communicable Disease Center, Atlanta, Georgia (1969).
8. Deal, W. B., and Sanders, E., Efficacy of Rifampin in Treatment of Meningococcal Carriers, *N. Engl. J. Med., 281,* 641 (1969).
9. Gotschlich, E. C., Goldschneider, I., and Artenstein, M. S., Human Immunity to the Meningococcus. V. The Effect of Immunization with Meningococcal Group C Polysaccharide on the Carrier State, *J. Exp. Med., 129,* 1385 (1969).
10. Artenstein, M. S., Gold, R., Zimmerly, J. G., Luyle, F. A., Schneider, H., and Harkins, C., Prevention of Meningococcal Disease by Group C Polysaccharide Vaccine, *N. Engl. J. Med., 282,* 417 (1970).
11. Love, B. D., and Finland, M., Šusceptibility of *Neisseria gonorrhoeae* to Eleven Antibiotics and Sulfadiazine, *Arch. Intern. Med., 95,* 66 (1955).
12. Martin, J. E., Jr., Lester, A., Price, E. V., and Schmale, J. D., Comparative Study of Gonococcal Susceptibility to Penicillin in the United States, 1965-1969, *J. Infect. Dis., 122,* 459 (1970).
13. Prévot, A. R., *Manual for the Classification and Determination of the Anaerobic Bacteria,* 1st American ed. Lea and Febiger, Philadelphia, Pennsylvania (1966).
14. Brown, J. H., *The Use of Blood Agar for the Study of Streptococci,* Monograph No. 9, Rockefeller Institute for Medical Research, New York (1919).
15. Lancefield, R. C., A Serologic Differentiation of Human and Other Groups of Hemolytic Streptococci, *J. Exp. Med., 57,* 571 (1933).
16. Zabriskie, J. B., The Role of Temperate Bacteriophage in the Production of Erythrogenic Toxin by Group A Streptococci, *J. Exp. Med., 119,* 761 (1964).
17. Starr, M. P., and Skerman, V. B. D., Bacterial Diversity: The Natural History of Selected Morphologically Unusual Bacteria, *Annu. Rev. Microbiol., 19,* 407 (1965).

Anaerobic Non-Sporulating Gram-Positive Rods[a, b]

DR. HUBERT A. LECHEVALIER

Lactobacillus

Lactobacilli are catalase-negative, , aerobic to anaerobic (often microaerophilic), non-motile rods that grow poorly on meat-infusion agar if carbohydrates and yeast extract are not added. Rods may appear singly or in chains. Colonies are usually non-pigmented. Glucose is fermented either to lactic acid (homofermentative

[a] One should keep in mind that some anaerobic spore formers (e.g., *Clostridium perfringens*) do not sporulate on the usual culture media that support their growth.

[b] The anaerobic *Actinomyces* and *Corynebacterium* are discussed elsewhere in this volume.

strains, subgenus *Lactobacillus*) or to a mixture of lactic acid, acetic acid, alcohol, and carbon dioxide (heterofermentative strains, subgenus *Saccharobacillus*). In addition, a distinction can be made between strains on the basis of the optical rotation of the lactic acid produced (Cato, E. P., and Moore, W. E. C., *Can. J. Microbiol., 11,* 319–324, 1965). Lactate is not metabolized. The type species is *L. caucasicus,* which occurs symbiotically with yeasts in kefir. Lactobacilli are found in fermenting plant juices, in milk products, including cheddar cheese, and in the intestine of milk-fed animals, including man (Seyfried, P. L., *Can. J. Microbiol., 14,* 313–318, 1968).

Bifidobacterium

The bifid bacteria are anaerobic, non-motile, pleomorphic rods. The cells are often irregularly shaped, club-shaped, or show "Y" branching (Kojima, M., Suda, S., Hotta, S., and Hamada, K., *J. Bacteriol., 95,* 710–711, 1968). Glucose is fermented through the fructose-6-phosphate shunt, with acetic and lactic acids as chief end products. The organisms are catalase-negative and do not reduce nitrate. Bifidobacteria show some relationships to the actinomycetes, the lactobacilli, and the corynebacteria. They are isolated from the intestine and other natural cavities of warm-blooded animals (see "Parasitic or Fermentative Actinomycetes", page 212). The type species is *B. bifidum.*

Cillobacterium

Strictly anaerobic, Gram-positive rods that do not form spores and are motile by peritriochous flagella belong to the genus *Cillobacterium.* They are found in the intestinal tract (including the rumen) of warm-blooded animals, in clinical specimens, and may also be isolated from soil. The type species is *C. moniliforme.*

Catenabacterium, Eubacterium, and *Ramibacterium*

These are strictly anaerobic, Gram-positive, non-motile rods, which grow abundantly on peptone media containing no carbohydrates. Generic designation is mainly on a basis of morphology and cellular arrangement. The distinction among the various genera is not always clear. Ramibacteria are diphtheroid in morphology, i.e., delicate, club-shaped rods that often do not stain uniformly; the cells often effect a "Chinese character arrangement". Eubacteria are boldly staining thick rods that occur singly, in pairs, or in short chains. Catenabacteria are delicate thin rods that occur in long filaments. Members of these three genera are found in the intestinal tract and in lesions of warm-blooded animals.

Propionibacterium, Butyribacterium, and *Zymobacterium*

Anaerobic Gram-positive, non-sporing, non-motile rods that do not grow well on meat-infusion agar without the addition of carbohydrates may belong not only to the genera *Lactobacillus* and *Bifidobacterium,* discussed above, but may also belong to the genera *Propionibacterium, Butyribacterium* and *Zymobacterium.*

Propionibacterium. Propionibacteria are coccoid to pleomorphic rods that are catalase-positive and anaerobic to aerotolerant. Anaerobically they produce propionic acid, acetic acid, and carbon dioxide from several carbohydrates and from lactic acid (see Table 5 of "Anaerobic Actinomycetes," page 217). Propionic acid bacteria are found in dairy products, especially cheeses of the Gruyère and Edam types. The type species is *P. freudenreichii* (Malik, A. C., Reinbold, G. W., and Vedamuthu, E. R., *Can. J. Microbiol., 14,* 1185–1191, 1968).

Butyribacterium. The generic name *Butyribacterium* is used to refer to anaerobic Gram-positive rods that produce butyric acid, acetic acid, and carbon dioxide from glucose and other carbohydrates as well as from lactic acid.

Zymobacterium. Rods placed in the genus *Zymobacterium* are anaerobic to microaerophilic and ferment glucose in a yeast-like manner with the production of ethyl alcohol.

Too few isolates of these last two types of anaerobic bacteria have been studied to establish their role in nature.

Aerobic Gram-Positive Rods

DR. HUBERT A. LECHEVALIER

This section includes bacteria grouped in the 7th edition of *Bergey's Manual* in the families Brevibacteriaceae and Corynebacteriaceae. Both families harbor Gram-positive rods, and the distinction between these families is not clear. In general, the Brevibacteriaceae are non-pathogenic, non-pleomorphic coccobacilli to rods found in soil and other natural substrates, whereas the Corynebacteriaceae are often pleomorphic and often associated with diseases.

Brevibacteriaceae

The family Brevibacteriaceae is composed of only two genera, *Brevibacterium* and *Kurthia*.

Brevibacterium. This genus was introduced primarily to provide a taxonomic shelter for Gram-positive, non-sporing, non-acid-fast rods that have no clear-cut affinity to other better-defined genera. The rods are usually short, non-motile, and may have reddish to brown pigment. Growth in the presence of glucose usually results in the formation of acid. The organisms are aerobic to facultatively anaerobic. The type species, *B. linens*, is widely distributed in dairy products, foods, and soil.

Kurthia. These rods may be up to 12 μm long, forming filaments when grown in liquid media, and are motile with peritrichous flagella. They are facultatively anaerobic. Carbohydrates are not utilized. *Kurthia* are found in decomposing organic matter. The type species is *K. zopfii*.

Saprophytic Corynebacteriaceae

The saprophytic Corynebacteriaceae fall mainly into the genera *Microbacterium, Cellulomonas,* and *Arthrobacter*.

Microbacterium. These rods may be up to 3 μm long, are irregularly shaped, and arranged in an angular fashion. Growth is good on meat-infusion agar. Production of lactic acid from carbohydrates is weak. The organisms are found in dairy products. The types species is *M. lacticum*.

Cellulomonas. (Small pleomorphic rods, which may be up to 2.5 μm long.) They may be motile with one or more peritrichous flagella. The organisms are Gram-variable. Acid, but no gas, is produced from glucose and lactose. Active decomposers of cellulose, *Cellulomonas* are found in soil and decomposing vegetable matter. The type species is *C. biazotea*.

Arthrobacter. In complex organic media, the cells undergo a change in morphology; they turn from spheres to rods when inoculated into fresh medium and revert to spheres at the end of exponential growth. They are mostly non-motile. In chemically defined media, it has been possible to grow these organisms either as rods or as spheres (Krulwich, T. A., *et al., J. Bacteriol., 94,* 734–750, 1967). The rods may be up to 7 μm long. These Gram-variable, heterotrophic organisms are typically found in soil, on plants, in activated sludge, and in milk products. The type species is *A. globiformis*.

Pathogenic Corynebacteriaceae

The pathogenic Corynebacteriaceae are members of the following three genera: *Erysipelothrix, Listeria,* and *Corynebacterium*.

Erysipelothrix. *Erysipelothrix rhusiopathiae (insidiosa),* the single member of the genus, is the cause of swine erysipelas, non-suppurative arthritis of lambs and calves, acute septicemias of turkeys, ducks and geese, septicemia in mice, and erysipeloid in man. The latter is usually a self-limiting wound infection and is especially common in veterinarians, hunters, fishermen, and all those who handle meat and fish. In addition, this organism may cause a number of ill-defined infections. *E. rhusiopathiae* is widely distributed in nature. It may contaminate insecticidal solutions, resulting in wound infections in treated animals. These slender beaded rods grow to 2.5 μm in length and may form filaments up to 15 μm long. They are Gram-variable, non-motile, facultatively anaerobic or microaerophilic, and require serum for optimal growth. Carbohydrate reactions are variable; hydrogen sulfide is produced, indole is not formed, and hemolysis is seen on 10% horse blood agar. This species is resistant to kanamycin, neomycin, and vancomycin; they are sensitive to penicillin, and less sensitive to the tetracyclines and streptomycin.

Listeria. *Listeria monocytogenes* is composed of short (up to 2 μm) rods that usually occur in pairs and are motile with a characteristic tumbling motion. They are peritrichously flagellated. Facultatively anaerobic, they produce acid, but no gas, from carbohydrates. *Listeria monocytogenes* causes numerous infections, including encephalitis and meningoencephalitis in adult ruminants, septicemia with focal hepatic necrosis in monogastric animals, septicemia with myocardial degeneration in fowls, abortion in sheep, goats and cows, and meningoencephalitis in man. In the human newborn it causes granulomatosis infantiseptica, which results from intrauterine infection and is characterized by numerous necrotic foci located throughout the body. A striking monocytic blood response occurs in human listeriosis, which once led to the belief that *L. monocytogenes* was the cause of infectious mononucleosis. The organism grows well on infusion media without added carbohydrate or blood at 25°C or 37°C, but poorly at 42°C. *Beta*-hemolysis is seen on sheep blood agar and, slowly, on human or rabbit blood agar. On McBride's agar illuminated with oblique light, colonies have a characteristic blue-green color. Indole, hydrogen sulfide, and urease are not produced, nitrates are not reduced, and citrate is not utilized. The organism is catalase-positive, oxidase-negative, and Voges-Proskauer-variable, depending on the procedure. *Listeria monocytogenes* is sensitive to the tetracyclines, penicillin, and erythromycin.

Corynebacterium. Quite diverse organisms are currently placed in the genus *Corynebacterium*. They are rods that are considered somewhat pleomorphic and usually cause either animal or plant diseases (Veldkamp, H., *Annu. Rev. Microbiol., 24,* 209–240, 1970).

Corynebacterium diphtheriae is the type species of the genus. It is the cause of diphtheria, a human disease characterized by the formation of pseudomembranes in the throat. A toxin produced by the bacterium can be shown to interfere with protein synthesis. Invasiveness (virulence) and toxinogenicity are separate properties, the latter conferred by a gene that is carried by a temperate phage. Early antitoxin therapy, coupled with the use of penicillin, tetracyclines or erythromycin, constitutes effective treatment. Prevention is by immunization with diphtheria toxoid.

Barksdale (*Bacteriol. Rev., 34*, 378–422, 1970) proposed the following description of *C. diphtheriae*:

> "Facultatively aerobic, gram-positive to gram-variable, nonsporulating, nonmotile, rodlike, tapered bacteria. Actively growing cells appear as doublets tapered from their septal ends. Club-shaped phenotypes occur in old cultures and on inadequate media. Intracellular polyphosphate granules, formed on serum slants rich in phosphate, can be revealed by staining with the metachromatic dyes, Toluidine Blue and methylene blue. Most strains ferment glucose, maltose, and dextrin; fewer ferment starch, still fewer sucrose. GC content is about 55%. Cell walls are distinguished by having meso-a, ϵ-DAP in conjunction with arabinogalactan, corynemycolic, and corynemycolenic acids and trehalose (dimycolate). O (polysaccharide) antigen cross-reacts with O antigens of *Mycobacterium* and *Nocardia*. Specific K antigens (protein) are basis for serotyping. Corynebacteriophages may be used for further typing members of the genus and for distinguishing corynebacteria from mycobacteria and nocardias. Certain lysogenic strains harboring prophages carrying the *tox* gene produce the immunologically distinct protein, diphtherial toxin, molecular weight 64,000, 4.2*S*. Subunits of toxin, 2.5*S*, obtained by treatment with dithiothreitol, bring about the ribosylation of the mammalian translocase, transferase II. Most strains produce a neuraminidase (sialidase) which cleaves neuraminlactose to lactose and N-acetyl-neuraminic acid. Neotype: *Corynebacterium diphtheriae*, strain(s) $C7_s(-)^{tox-}$ and $C7_s(\beta)^{tox+}$."

Although the appearance of stained smears of *C. diphtheriae* is considered highly characteristic, identification should not be made by morphologic criteria. Appearance on blood–tellurite agar, hemolysis, glucose and sucrose fermentation, and *in-vivo* or *in-vitro* virulence tests should be included.

It seems reasonable to limit membership in the genus *Corynebacterium* to organisms similar to *C. diphtheriae*, that is, organisms that have cell walls of Type IV (see Table 1 of "Soil or Oxidative Actinomycetes") and contain corynomycolic acids similar to, but of smaller molecular weight than nocardomycolic and true mycolic acids (see Volume II, Lipids). Examples of such organisms are described below.

Corynebacterium pseudotuberculosis (ovis) causes lymphadenitis of sheep, a chronic disease in which the pathogen is believed to enter through skin abrasions that occur at the time of shearing. It also causes ulcerative lymphangitis in horses, which resembles a cutaneous form of glanders. The organism can also infect cattle, goats, deer, and rabbits. It has been isolated from the lymph nodes and appendix of man. *Corynebacterium renale* causes an inflammatory disease of the urinary tract of cattle.

Corynebacterium bovis is found in freshly drawn cow's milk (Cummins, C. S., *J. Bacteriol., 105,* 1227–1228, 1971). *Corynebacterium pseudodiphthericum (hofmannii)* is found in the throat of normal human beings, and *C. xerosis* is a skin and conjunctiva inhabitant; neither is known to cause any disease. *Corynebacterium fascians* is pathogenic to plants. *Corynebacterium equi* causes purulent pneumonia in young horses and is also found in tuberculous-like lesions of swine; it is now considered to belong to the *Rhodochrous* group of "nocardiae" (Gordon, R. E., *J. Gen. Microbiol., 43,* 329–343, 1966).

Corynebacterium pyogenes has a cell wall quite different from that of *C. diphtheriae,* since it contains no DAP. Its metabolic activities are similar to those of lactobacilli. It is involved in numerous infections of animals, either alone or in combination with other pathogens.

The plant-pathogenic corynebacteria *C. tritici*, *C. betae*, *C. flaccumfaciens*, and *C. poinsettiae* are also very different from *C. diphtheriae*; they also have cell walls without DAP. Their generic assignment is still to be made. *Corynebacterium michiganense*, also a plant pathogen, contains the L-isomer of DAP, as does the ubiquitous, motile *C. aquaticum*, which is found widely distributed in nature.

The anaerobic to aerotolerant *C. acnes* is now placed in the genus *Propionibacterium* (see Tables 5 and 6 of "Parasitic or Fermentative Actinomycetes").

FUNGI

THE PHYCOMYCETES

DR. EVERETT S. BENEKE*

The term "Phycomycete" has been used to include a group of diverse organisms considered to be the lower fungi. Myxomycetes, Acrasiales, and Labyrinthulales are not included in the Phycomycetes. Some of the Phycomycetes occur in an aquatic habitat, others are found in soil, on organic matter, or parasitic on plants and animals; a few are obligate parasites.

The vegetative phase of these fungi is either unicellular (without budding) or filamentous. The filamentous hyphae may be aseptate with numerous nuclei or, occasionally, septate. The cell wall is typically chitinized, except for cellulose in the Oömycetes. Most of the organisms produce their spores in sporangia. Many produce motile spores or motile gametes. In some of the higher forms the sporangia function as conidia or are replaced by conidia in asexual reproduction. Sexual reproduction is by fusion of gametes or by the formation of gametangia. For more detailed information on the Phycomycetes, References 1 to 5 should be consulted.

The lower fungi have been classified into five or six classes, depending on the system followed. The use of the class Phycomycetes in the modern classification systems is no longer recognized. The classes of the lower fungi are listed in Table 1 according to the classification systems of Ainsworth (1971), Bessey (1950), Gäumann (1964), and Alexopoulos (1962).[1,2,7,8] Alexopoulos considers four of the six classes as aquatic fungi: the Chytridomycetes, producing posteriorly uniflagellate zoöspores; the Hypochytridiomycetes, forming anteriorly uniflagellate zoöspores; the Oömycetes, which produce biflagellate zoöspores; and the obligate endoparasites, the Plasmodiophoromycetes, which have unequal biflagellate zoöspores. The classification used by Ainsworth is similar, except the class Plasmodiophoromycetes is classified with the Myxomycota (the slime molds). The class Zygomycetes is a well-limited group of terrestrial fungi that produce non-motile spores known as sporangiospores, or conidia and a zygospore by gametangial copulation. The class Trichomycetes is a specialized group of fungi attached by a basal cell to the digestive tract or the external cuticle of arthropods. Sporangiospores and resting spores are reproductive characteristics of this group.

In making a comparison of the six classes of lower fungi (including the Plasmodiophoromycetes), one observes an assemblage of diverse types of organisms. The presence or absence of flagellated cells and uniflagellate or biflagellate cells has been the basis used for classification.[2,9] Table 2 gives the key characteristics of the classes of the lower fungi.

Each of the six classes (under the older system one class, Phycomycetes) contains a number of orders. The Chytridiomycetes are typically found in aquatic habitats or in soil. A few parasitize and kill algae. Representatives of the two genera *Synchytrium* and *Physoderma* parasitize plants. *Allomyces* and *Blastocladiella* have been used extensively as research tools.[10-12] Chytridiomycetes include the following three orders.[9]

Chytridiales: The thallus lacks mycelium and converts entirely into a reproductive structure (holocarpic), or it has rhizomycelium (eucarpic) and converts into one (monocentric) or more (polycentric) reproductive structures. Zoöspores have a single conspicuous oil globule.[15]

Blastocladiales: True mycelium, usually with a basal cell and tapered rhizoids; thick-walled, often punctate resistant sporangia; sexual reproduction by planogametes; alternation of generations in some species. Zoöspores have no conspicuous oil globule.

Monoblepharidales: Delicate much-branched mycelium; sexual reproduction by fusion of a free-

* The author wishes to thank Kerry O'Donnell for the drawings of Figures 1 to 15.

swimming male gamete with a female aplanogamete, forming an oöspore. Zoöspores have no conspicuous oil globule. No resistant sporangia.

Representative genera of the three orders are listed in Table 3 together with the chief morphologic characteristics for each genus. Some are illustrated in Figures 1 to 5.

TABLE 1
CLASSES OF LOWER FUNGI[a]

Ainsworth (1971)	Bessey (1950)	Gäumann (1964)	Alexopoulos (1962)
KINGDOM FUNGI			**KINGDOM PLANTAE**
			Mycota
Myxomycota			Myxomycotina
1. Acrasiomycetes			
2. Hydromyxomycetes			
3. Myxomycetes	**Mycetozoa**		Myxomycetes
			Eumycotina
4. Plasmodiophoromycetes			Plasmodiophoromycetes
		Archimycetes	
Eumycota			
	Class Phycomyceteae	Phycomycetes	
Mastigomycotina			
5. Chytridiomycetes			Chytridiomycetes
6. Hyphochytridiomycetes			Hyphochytridiomycetes
7. Oömycetes		Order Oömycetes	Oömycetes
Zygomycotina			
8. Zygomycetes		Order Zygomycetes	Zygomycetes
9. Trichomycetes			Trichomycetes

[a] Division names end in *-mycota*, subdivision names in *-mycotina;* the names of classes end in *-mycetes* in classification systems in 1962 and thereafter.

TABLE 2

CHARACTERISTICS OF THE LOWER FUNGI

Class	Characteristics[2,4,9]
Chytridiomycetes (Water molds)	Posteriorly uniflagellate (whiplash) zoöspores; coenocytic thallus/ mycelium. The zygote usually develops into a resting spore or sporangium.
Hyphochytridiomycetes (Water molds)	Anteriorly uniflagellate (tinsel) zoöspores; coenocytic thallus.
Oömycetes (Water molds, blight, downy mildews)	Biflagellate zoöspores; flagella nearly equal, one whiplash, the other tinsel type, borne in sporangia of various types; sporangia may function as a spore; coenocytic thallus/mycelium; heterogametangia form oöspores.
Plasmodiophoromycetes (Club root organisms)	Biflagellate zoöspores; flagella unequal in length, both whiplash type; plasmodium; resting spores.
Zygomycetes (Pin molds, bread mold)	Sporangiospores; coenocytic hyphae; zygospores.
Trichomycetes (Parasites or commensals of arthropods)	Sporangiospores usually form; coenocytic hyphae; resting spores.

The Hyphochytridiomycetes are fresh-water or marine chytrid-like fungi. The anteriorly uniflagellate zoospores have tinsel-type flagella. These organisms are found as parasites on fungi and algae, or as saprobes on plant and insect debris. The one order in this class consists of a small number of genera, including such representative members as *Anisolpidium, Rhizidiomyces,* and *Hypochytrium.* For further information on this small class see References 2 and 9.

Members of the class Oömycetes are usually aquatic, although some Saprolegniales and Peronosporales grow in soil. The Lagenidiales are usually parasitic on algae, fungi, and microscopic aquatic animals. A few species in the Saprolegniales attack economically important plants, and a few others cause a serious fish disease. The Peronosporales include some of the most destructive parasites known. This class contains four orders.[9]

Saprolegniales: Holocarpic or eucarpic; hyphae, when present, without constrictions; zoöspores formed in the sporangia, diplanetic, monoplanetic, or (rarely) aplanetic; oöspores in oögonium.[18,21]

Leptomitales: Eucarpic; hyphae constricted, oögonium usually with a single oöspore; similar to the Saprolegniales in many characteristics.

Lagenidiales: Holocarpic, reniform zoöspores formed in a sporangium or in a vesicle, monoplanetic; thick-walled resting spore.

Peronosporales: Eucarpic; reniform zoöspores formed in a sporangium or in a vesicle, monoplanetic; sporangia may function as a spore; oöspheres in oögonium usually have periplasm.

TABLE 3

MORPHOLOGIC CHARACTERISTICS OF REPRESENTATIVE GENERA IN THE CHYTRIDIOMYCETES

Genus	Morphologic Characteristics[2,9]
Order Chytridiales	
Olpidium	Thallus holocarpic, inoperculate, endobiotic; single sporangium; does not fill host cell; scattered, one discharge tube.
Rozella	Thallus holocarpic, inoperculate, endobiotic; single sporangium, fused walls; fill host cell; hypertrophy of host.
Achlyogeton	Thallus holocarpic, inoperculate, endobiotic; series of sporangia forms in host.
Synchytrium	Thallus holocarpic, inoperculate, endobiotic; forms a sorus with sporangia or resting spores.
Rhizophydium	Thallus eucarpic, monocentric; sporangia and resting spores epibiotic; tapering rhizoids endobiotic.
Entophlyctis	Thallus eucarpic, monocentric; sporangia and resting spores endobiotic; rhizoids arise from sporangia.
Rhizidium	Thallus eucarpic, monocentric; sporangia form from enlarged encysted zoöspores; branched rhizoids on single axis on sporangium anastomose to form resting spore.
Cladochytrium	Thallus eucarpic, polycentric; sporangia and rhizoids usually endobiotic, at times apophysate.
Physoderma	Thallus eucarpic, monocentric; sporangia epibiotic; resting spores on separate thallus, endobiotic, polycentric.
Chytridium	Thallus eucarpic; operculate sporangia; develops from all or part of encysted zoöspore; resting spore in substrate.
Nowakowskiella	Thallus eucarpic, polycentric, endobiotic; strongly tapering, branched rhizoids, septate if operculate sporangium delimited.
Order Blastocladiales	
Blastocladiella[13]	Thallus a short basal swelling or an elongate, unbranched basal cell with rhizoids; apex with a zoösporangium or thick-walled resistant sporangium.
Blastocladia[14]	Thallus simple lobed or branched basal cell; secondary axis may be present; no pseudosepta; setae in some species; zoosporangia and thick-walled resistant sporangia.
Allomyces	Thallus with unbranched basal cell, dichotomously branched, pseudoseptate; no setae; gametothalli may develop; sporothalli with zoösporangia and thick-walled resistant sporangia.

TABLE 3 (Continued)

MORPHOLOGIC CHARACTERISTICS OF REPRESENTATIVE GENERA IN THE CHYTRIDIOMYCETES

Genus	Morphologic Characteristics[2,9]
Order Blastocladiales (continued)	
Coelomomyces	Thallus without wall or rhizoids; parasitic in aquatic larvae of insects; zoösporangia; resistant sporangia thick-walled; ornate.[16]
Catenaria	Walled thallus, catenulate with rhizoids; cross walls delimit sporangia; resistant sporangia or sterile isthmuses.
Order Monoblepharidales	
Gonapodya	Mycelial contents reticulate, with constrictions and pseudosepta; zygote, motile for a period, forms smooth-walled oöspores.
Monoblepharis	Mycelial contents reticulate, no constrictions or pseudosepta; zygote, endogenous or at mouth of oögonium, encysts; oöspores usually bullate.

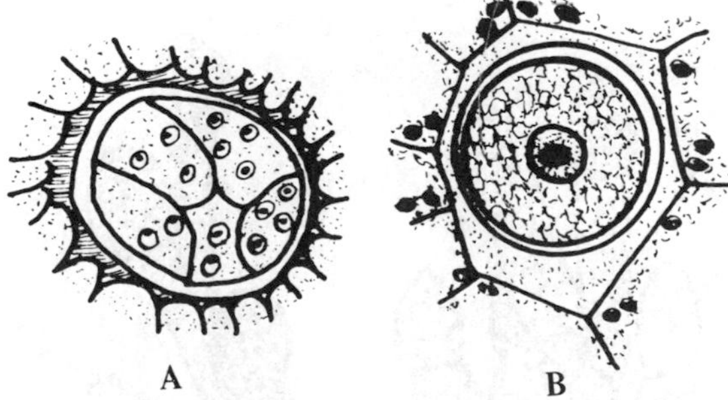

FIGURE 1. *Synchytrium.* A. Sorus with sporangia. B. Resting spores. Modified drawing of Figure 11 in Bessey, E. A., *Morphology and Taxonomy of Fungi,* p. 51 (1950). Reproduced by permission of the copyright owner, McGraw-Hill Book Co., New York.

FIGURE 2. *Rhizophydium.* A. Sporangium. B. Resting spore. Modified drawings of Figures 15B and 15C in Bessey, E. A., *Morphology and Taxonomy of Fungi,* p. 57 (1950). Reproduced by permission of the copyright owner, McGraw-Hill Book Co., New York.

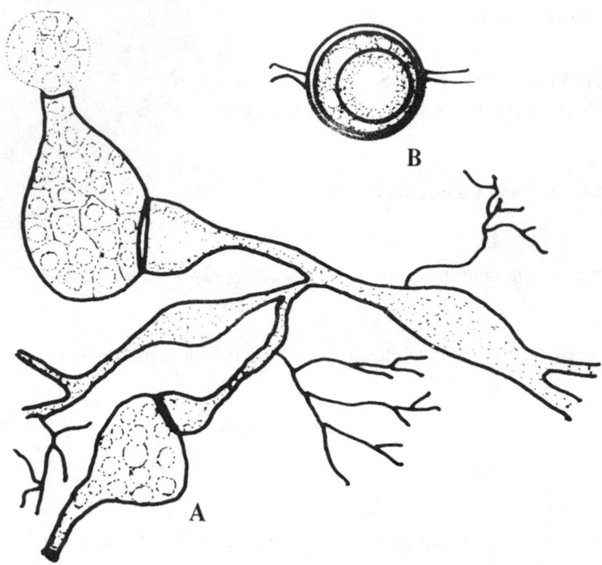

FIGURE 3. *Nowakowskiella.* A. Sporangia. B. Resting spore. Modified drawings of Figures 17D and 17E in Bessey, E. A., *Morphology and Taxonomy of Fungi,* p. 63 (1950). Reproduced by permission of the copyright owners, McGraw-Hill Book Co., New York.

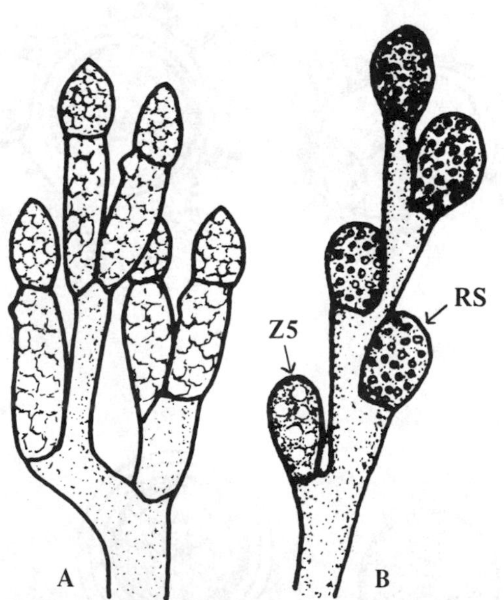

FIGURE 4. *Alloymyces.* A. Gametothallus with male and female sporangia. B. Sporothallus, zoösporangium (ZS), resistant sporangia (RS). Modified drawings of Figures 28D and 28G in Bessey, E. A., *Morphology and Taxonomy of Fungi,* p. 89 (1950). Reproduced by permission of the copyright owners, McGraw-Hill Book Co., New York.

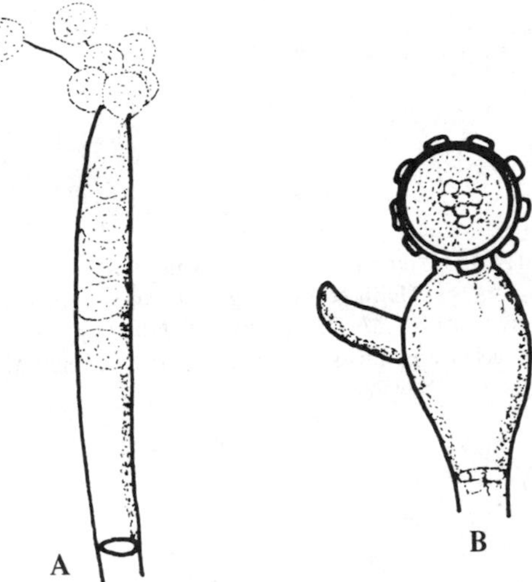

FIGURE 5. *Monoblepharis.* A. Sporangium. B. Oögonium, oöspore.

Representative genera of the four orders are listed in Table 4 together with the chief morphologic characteristics for each genus. Some are illustrated in Figures 6 to 11.

The Plasmodiophoromycetes occur as obligate endoparasites in vascular plants, algae, and fungi. The somatic phase, the plasmodium, which gives rise to the sporangium with anteriorly biflagellate zoöspores or resting spores, produces an abnormal enlargement of the host cells. The class Plasmodiophoromycetes has a single order, the Plasmodiophorales.

Plasmodiophorales: Thallus multinuclear; naked protoplast or plasmodium in host cells; zoöspores with unequal flagella, anterior, whiplash; resting spores.

Selected genera in this order are listed in Table 5 with the chief morphologic characteristics indicated for each genus. The genus *Plasmodiophora* is illustrated in Figure 12.

TABLE 4

MORPHOLOGIC CHARACTERISTICS OF REPRESENTATIVE GENERA IN THE OÖMYCETES

Genus	Morphologic Characteristics[2,9]
Order Lagenidiales	
Olpidiopsis	Thallus holocarpic, endobiotic, one-celled; does not fill host; resting spore from conjugation of thalli; zoösporangia; zoöspores differentiated within.
Lagenidium	Thallus holocarpic, endobiotic, multi-celled, unbranched or branched in one or more host cells; irregular segments; zoöspores form in vesicle; resting spores.
Order Saprolegniales	
Ectrogella	Thallus holocarpic, endobiotic; zoöspores encyst outside sporangium; parasitic in diatoms; diplanetic zoöspores, resting spore.
Thraustochytrium	Thallus eucarpic, monocentric; epibiotic sporangium; endobiotic part is a rhizoidal system; sporangia burst, release biflagellate zoöspores; resting spore.
Saprolegnia	Thallus eucarpic; zoöspores diplanetic; sporangia renewed by internal proliferation.[17]
Leptolegnia	Thallus eucarpic; zoöspores diplanetic; sporangia have one row of spores; one oöspore.
Achlya	Thallus eucarpic; all primary zoöspores encyst at orifice, forming a hollow sphere after sporangial discharge.[19]
Aphanomyces	Thallus eucarpic; zoöspores in a single row in sporangium, encysting at orifice; one oospore.[20]
Thraustotheca	Thallus eucarpic; zoöspores encyst within sporangium; secondary zoöspores emerge after sporangium deliquesces.
Dictyuchus	Thallus eucarpic; zoöspores encyst within sporangium; secondary zoöspores emerge, leaving a network of cyst walls; one oöspore.
Geolegnia	Thallus eucarpic; encysted spores in single row; aplanetic; one oöspore.

TABLE 4 (Continued)

MORPHOLOGIC CHARACTERISTICS OF REPRESENTATIVE GENERA IN THE OÖMYCETES

Genus	Morphologic Characteristics[2,9]
Order Leptomitales	
Leptomitus	Hyphae constricted; no basal cells; dichotomous branches; diplanetic zoöspores in undifferentiated sporangium; no oöspore.
Apodachlya	Hyphae constricted, branched, diplanetic; pedicellate sporangia; oöspores.
Araiospora	Hyphae constricted; basal cell stout; sporangia smooth or spiny-walled; oöspore has reticulate or cellular outer wall.
Order Peronosporales	
Pythium	Differentiated or undifferentiated sporangia formed directly on hyphae; germinate directly or form vesicle with zoöspores; one oöspore.[22]
Phytophthora	Differentiated sporangia; zoöspores usually differentiate in sporangium; sporangiophore formed; one oöspore.[23]
Zoophagus	Undifferentiated sporangia; hyphae form short lateral branches adapted to capturing rotifers; one oöspore.
Plasmopora	Sporangiophores branched at right angles; sporangia form zoöspores, encyst, germinate.[7]
Bremia	Sporangiophores dichotomously branched, with acute angles; sporangia germinate by germ tube or form zoöspores; disks at tip of branches.[7]
Peronospora	Sporangiophores dichotomously branched, with acute angles; sporangia germinate by germ tube.[7]
Albugo	Sporangia catenulate on subepidermal sporangiophores; intercellular hyphae.[7]

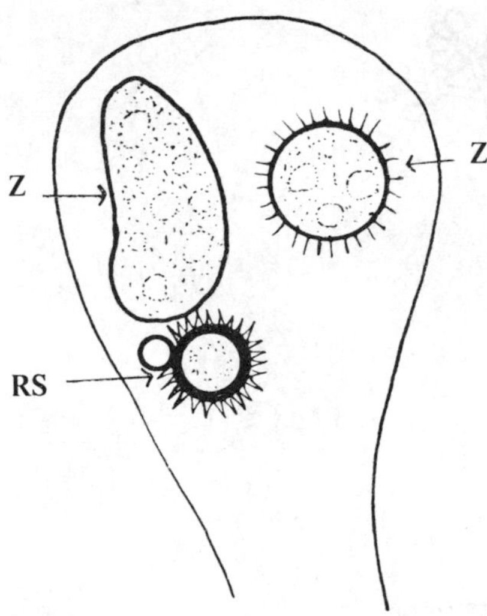

FIGURE 6. *Olpidiopsis.* Smooth- and spiny-walled zoösporangia (Z) and resting spores (RS).

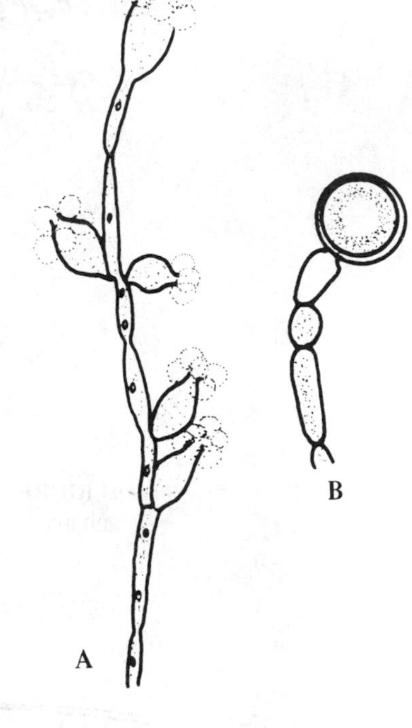

FIGURE 8. *Apodachlya.* A. Sporangia. B. Oögonium, oöspore. Modified drawings of Figures 69C and 69E in Sparrow, F. K., Jr., *Aquatic Phycomycetes,* p. 875 (1960). Reproduced by permission of the copyright owners, University of Michigan Press, Ann Arbor, Michigan.

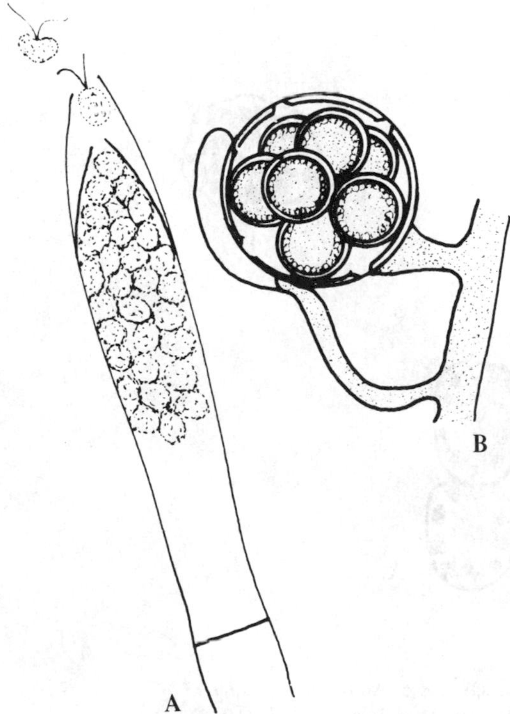

FIGURE 7. *Saprolegnia.* A. Sporangium. B. Oögonium, oöspores.

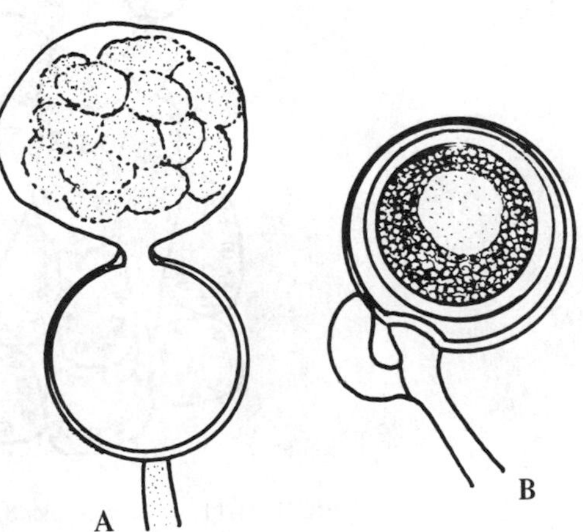

FIGURE 9. *Pythium.* A. Sporangium. B. Oögonium, oöspore.

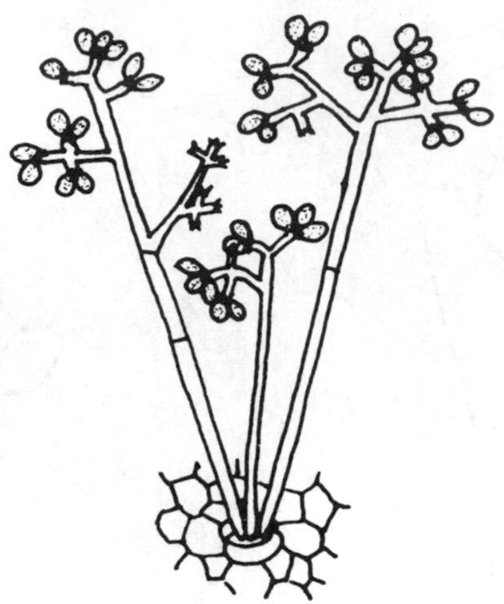

FIGURE 10. *Plasmopara.* Sporangiophore.

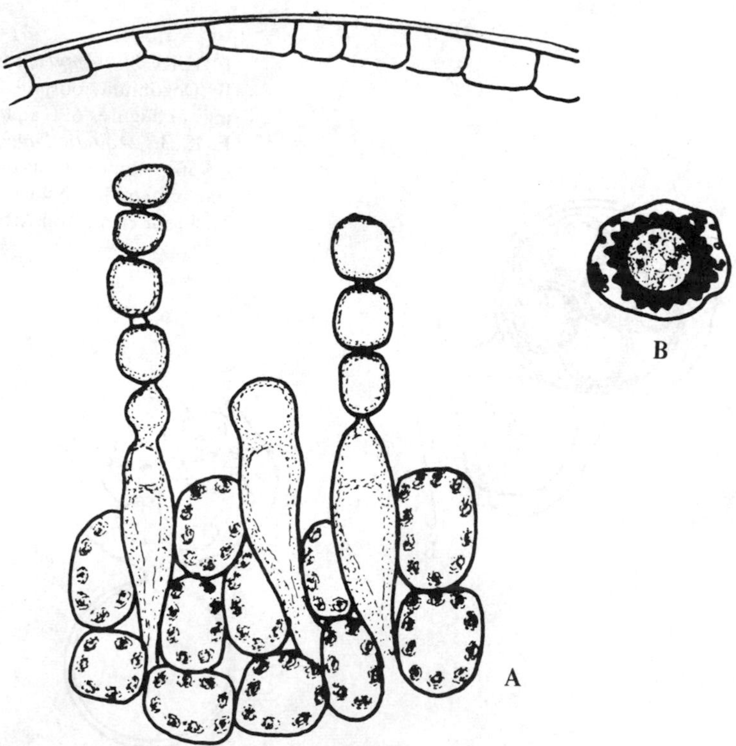

FIGURE 11. *Albugo.* A. Sporangiophore, sporangia. Modified drawing of Figure 61B in Alexopoulos, C. J., *Introductory Mycology,* p. 168 (1962). Reproduced by permission of the copyright owners, John Wiley and Sons, New York. B. Oögonium, oöspore. Modified drawing of Figure 52-7 in Gäuman, E. A., and Dodge, C. W., *Comparative Morphology of Fungi,* p. 85 (1928). Reproduced by permission of the copyright owners, McGraw-Hill Book Co., New York.

TABLE 5

MORPHOLOGIC CHARACTERISTICS OF REPRESENTATIVE GENERA IN THE PLASMODIOPHOROMYCETES

Genus	Morphologic Characteristics[2,24]
Order Plasmodiophorales	
Plasmodiophora	Plasmodium endobiotic; zoösporangia few or numerous; resting spores not united.
Tetramyxa	Plasmodium endobiotic; no zoösporangia; resting spores usually in tetrads or dyads.
Sorodiscus	Plasmodium endobiotic; no zoösporangia; cystosorus or aggregate of resting spores predominantly disk-shaped, two-layered.
Spongospora	Plasmodium endobiotic; zoösporangia oval to irregular; cystosorus oval to spherical, spongy, canals in center.

The Zygomycetes produce a zygospore, the chief characteristic of the group. The zygospore results from the fusion of two gametangia. Characteristic sporangia or conidia are produced. Motile cells are lacking. The group contains saprobes, plant and animal parasites, and obligate parasites on other zygomycetes. This class includes three orders: Mucorales, Entomophthorales, and Zoöpagales, which are characterized below.

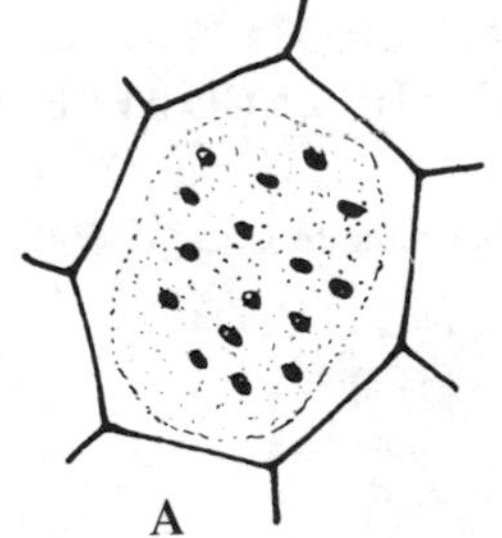

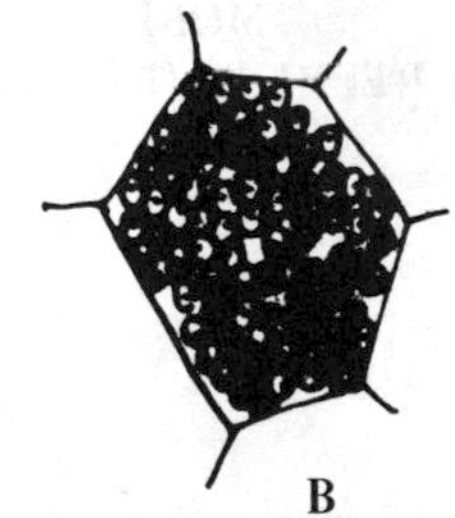

FIGURE 12. *Plasmodiophora*. A. Plasmodium. B. Resting spores. Modified drawings of Figures 64M and 64P in Alexopoulos, C. J., *Introductory Mycology*, p. 179 (1962). Reproduced by permission of the copyright owners, John Wiley and Sons, New York.

Mucorales: Usually saprobic, weakly parasitic in plants; a few are endoparasitic in vertebrates and man. Sporangia have one or more spores or conidia. Septa in older hyphae or at base of reproductive structures.

Entomophthorales: Usually parasitic on lower animals. Modified sporangia function as conidia, forcibly discharged.

Zoöpagales: Usually parasitic on lower animals. True conidia, not forcibly discharged. Fusion of hyphal tips form zygospores.

Representative genera well known or of importance in the three orders are listed in Table 6 together with the chief morphologic characteristics for each genus. Examples are shown in Figures 13, 14, and 15.

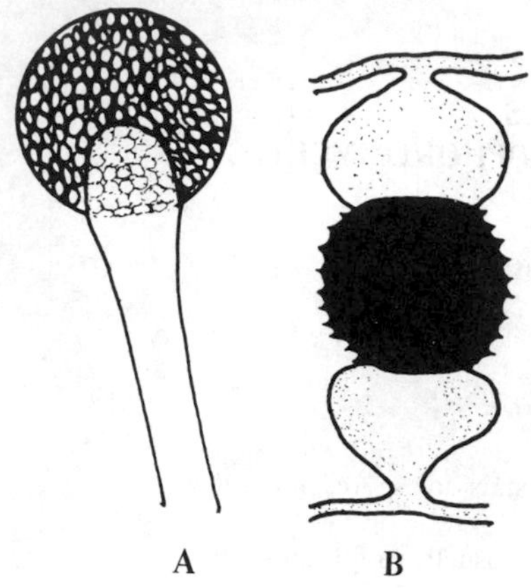

FIGURE 13. *Rhizopus.* A. Sporangium. B. Zygospore.

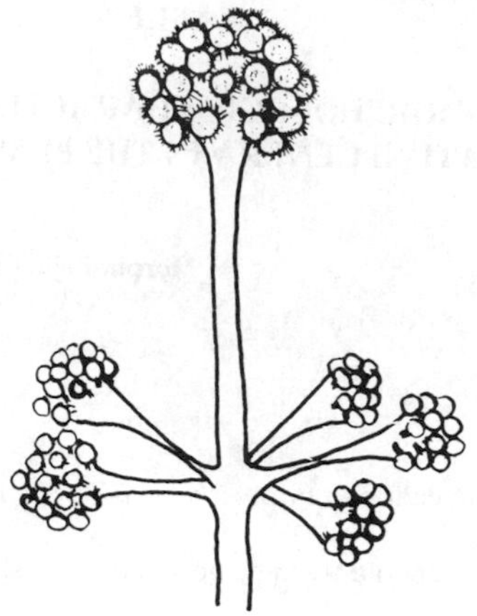

FIGURE 14. *Cunninghamella.* Conidiophore, conidia. Modified drawing of Figure 54 in Zycha, H., and Siepman, R., *Mucorales,* p. 147 (1969). Reproduced by permission of the copyright owners, Verlag von J. Cramer, Lehre, West Germany.

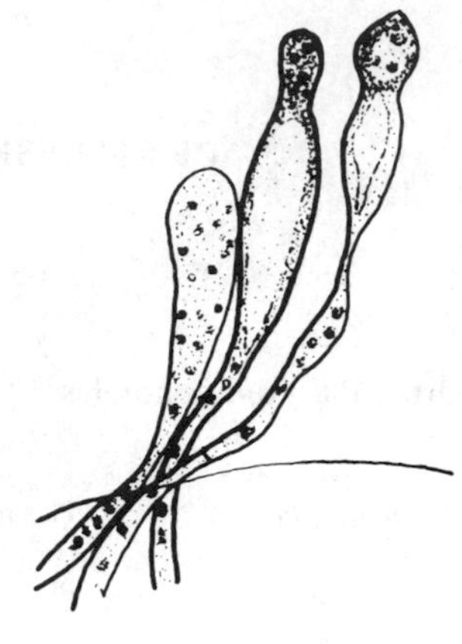

FIGURE 15. *Entomophthora.* Conidiophore.

TABLE 6

MORPHOLOGIC CHARACTERISTICS OF REPRESENTATIVE GENERA IN THE ZYGOMYCETES

Genus	Morphologic Characteristics[2,25-27]
Order Mucorales	
	Family Mucoraceae
	Columella present; sporangeal membrane thin, fugacious; zygospores rough, not tong-like.
Absidia	Sporangia pyriform; sporangiophores borne on stolons, not typically opposite rhizoids.
Rhizopus	Sporangiophores on stolons opposite rhizoids.
Circinella	Sporangiophores branched, all branches bearing sporangia circinately.
Phycomyces	Sporangiophores large, over 80 mm, with a metallic luster, unbranched; zygospore suspensor with branched finger-like projections.
Zygorhynchus	Zygospores on short side branches of sporangiophores; homothallic.
	Family Pilobolaceae
	Columella present; sporangial wall cutinized; sporangium violently discharged from sporangiophore; zygospores smooth-walled, tong-like suspensors.
Pilobolus	Discharge of sporangia violent, with a subsporangial swelling.
	Family Thamnidiaceae

TABLE 6 (Continued)

MORPHOLOGIC CHARACTERISTICS OF REPRESENTATIVE GENERA IN THE ZYGOMYCETES

Genus	Morphologic Characteristics[2,25-27]
Order Mucorales (continued)	
	Terminal sporangium with columella, multispored or with spines or neither on tip of sporangiophores; branches have sporangiola with one to a few spores.
Helicostylum	Terminal sporangia; pyriform to globose sporangiola borne circinately; heterothallic.
Thamnidium	Terminal sporangia; globose sporangiola borne dichotomously.
	Family Syncephalastraceae
	Sporangiophores branched, with terminal, globose vesicles; merosporangia over surface; wall evanescent; spores uniseriate; zygospores rough-walled.
Syncephalastrum	Sporangiophore with globose vesicle; merosporangia.
	Family Piptocephalidaceae
	Sporangiophores may branch; merosporangia may be deciduous; wall may be evanescent; obligate parasites on fungi, mostly Mucorales.
Piptocephalis	Dichotomously branched sporangiophores; deciduous vesicles; obligate parasites.
Syncephalis	Sporangiophores not branched; vesicles not deciduous; obligate parasites.
	Family Kickxellaceae
	Deciduous unispored sporangiola borne on phialids, which in turn are on sporocladia that are on sporangiophores.
Kickxella	Sporocladia elongate, septate, verticillate.
Spirodactylon	Sporocladia elongate, septate; spores ellipsoidal; fertile region of sporangiophore coiled.
	Family Mortierellaceae
	No columella; sporangia, sporangiola, and conidia borne singly; tips of sporangiophores not swollen; zygospores in thick hyphal matrix.
Haplosporangium	Sporangia with one or two sporangiospores.
Mortierella	Sporangia with several sporangiospores.
	Family Cunninghamellaceae
	Only conidia develop; zygospores formed as in *Mucor.*
Cunninghamella	Conidiophores usually ramose; vesicle globose at each branch apex; zygospores formed as in *Mucor.*
	Family Choanephoraceae
	Sporangia with persistent walls or conidia dark-colored, borne on swollen tips; zygospores smooth-walled.

TABLE 6 (Continued)

MORPHOLOGIC CHARACTERISTICS OF REPRESENTATIVE GENERA IN THE ZYGOMYCETES

Genus	Morphologic Characteristics[2,25-27]
Order Mucorales (continued)	
Blakeslea	Conidia absent, but sporangiola present.
Choanephora	Conidia present, but sporangiola absent.
Order Entomophthorales	
Entomophthora	Conidiophore extruded through the body wall; conidia forcibly discharged; parasitic on insects (flies); mycelium may be septate.[4,7]
Conidiobolus	Conidiophore discharges conidia violently; conidia function as sporangia, forming spores; septate hyphae; saprophytic or weakly parasitic on fruiting bodies of higher fungi (rarely insects).[4,7]
Basidiobolus	Sporangiophores produce pear-shaped sporangia, which are violently discharged; and later, on germination, form spores.[4,7]
Order Zoöpagales	
Cochlonema	Conidiophores produce chains of spindle-shaped conidia; thick, spiral haustorial thallus in cell of an amoeba.[7,28,29]
Cystopage	No conidia; reproduction by chlamydospores; effuse mycelium adheres to rhizopods or nematodes with penetrating haustoria.[7,28,29]

The Trichomycetes are parasites or commensals inside of the bodies of arthropods. Their hyphae are long, unbranched, or branched, and usually attached to the cuticle lining inside the gut of the host. Asexual reproduction is by sporangia, with spores linearly arranged, or by arthrospores. Sexual reproduction results in a thick-walled resting spore, analagous to a zygospore in some genera. Four orders are recognized.[30]

Amoebidales: Spores produced internally; amoeboid cells may be produced; entire unbranched thallus functions as one sporangium.

Eccrinales: Spores produced internally; no amoeboid cells produced; thallus unbranched; spores produced singly in a terminal series of sporangia.

Asellariales: Spores produced externally; unbranched thallus reproduces by arthrospores (or fragmentation).

Harpellales: Spores produced exogenously; thallus simple or branched.

Representative genera in the four orders of the Trichomycetes, the location in the host, and the types of hosts are listed in Table 7. The genus *Enterobryus* is illustrated in Figure 16.

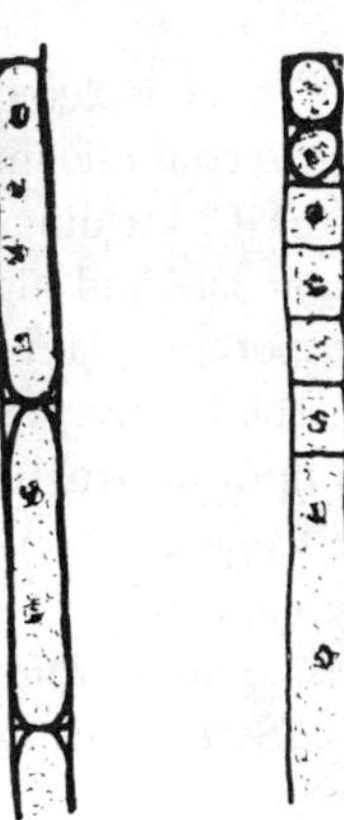

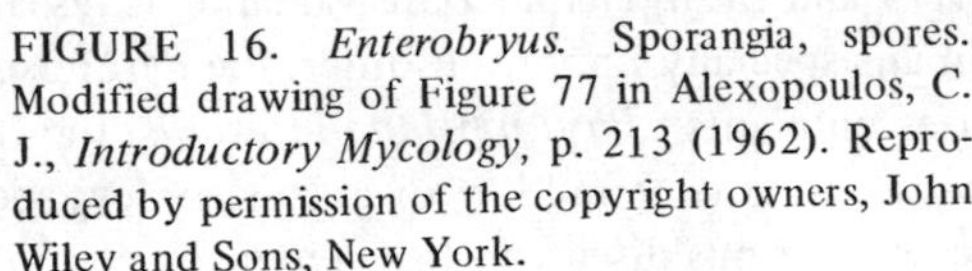

FIGURE 16. *Enterobryus.* Sporangia, spores. Modified drawing of Figure 77 in Alexopoulos, C. J., *Introductory Mycology,* p. 213 (1962). Reproduced by permission of the copyright owners, John Wiley and Sons, New York.

TABLE 7

SELECTED GENERA OF TRICHOMYCETES[30]

Genus	Location in Host	Types of Hosts
Order Eccrinales		
Enterobryus	Hindgut (and foregut)	Millipeds, beetles, crabs
Parataeniella	Hindgut	Isopods (pill bugs, sow bugs)
Arundinula	Foregut and hindgut	Crustaceans (hermit crabs, crayfish)
Order Amoebidiales		
Amoebidium	External	Minute aquatic crustaceans and various immature aquatic insects
Order Asellariales		
Asellaria	Hindgut	Intertidal and terrestrial isopods
Order Harpellales		
Harpella	Midgut (peritrophic membrane)	Black-fly larvae
Smittium (Rubetella)	Hindgut	Diptera larvae (black flies, mosquitoes, midges)

IMPORTANT SPECIES

The six classes of fungi that are considered under the older classification as one class, Phycomycetes, range from organisms of little direct economic importance to some of the most important parasites on plants, the downy mildews. Most of the Chytridiomycetes are of little direct economic importance. Some live as saprobes, others parasitize animals and algae. Members of the genera *Synchytrium* and *Physoderma* are parasitic on economically important plants. *Allomyces* and *Blastocladiella* are valuable as research tools in the study of morphogensis. The Oömycetes contain an especially important order, the Peronosporales; this group includes some of the most destructive parasites on plants: *Phytophthora infestans,* the cause of late blight of potatoes, and *Plasmopara viticola,* the cause of downy mildew of grapes. A few species in the Saprolegniales are especially important as the cause of a serious disease in fish. There are several important plant diseases caused by Plasmodiophoromycetes, including club root in cabbage and related plants, caused by *Plasmodiophora brassicae.* Several of the Zygomycetes are parasitic on fungi, green plants, and animals or grow as saprobes.

Some of the more important species of the class Phycomycetes are listed in Table 8. They are usually grouped under the six classes under some current classification systems outlined in Table 2.

BIOCHEMICAL APPROACH TO CLASSIFICATION

In recent years various biochemical properties have been found to be useful in evaluating relationships among the major classes and subclasses of fungi. In some cases biochemical tests may differentiate a taxon of lower rank.

The fungal cell wall polysaccharides comprise hexoses, amino sugars, hexuronic acids, methylpentoses, and pentoses. By selection of dual combinations of these polysaccharides, which appear to be the main components of the cell wall, Bartnicki-Garcia[38] proposed a classification of the fungi based on a minimum of eight cell wall categories. In this classification system a close correlation can be seen between the cell wall chemistry and the conventional morphological classification of the fungi, as indicated in Table 9.

The four classes of fungi formerly grouped together in the class Phycomycetes, namely Hyphochytridiomycetes, Chytridiomycetes, Oömycetes, and Zygomycetes, have different cell wall composition. This is one basis to justify the dissolution of the class Phycomycetes. The term "Phycomycete" is being retained to denote a coenocytic fungus.[38]

The fungi synthesize lysine by either the 2,6-diaminopimelic acid route (DAP) or by the 2-aminoadipic acid (AAA) route.[39] The Oömycetes and the Hyphochytridiomycetes are the only fungi with the DAP pathway, which is the same as that in vascular plants, green algae, blue-green algae, and bacteria. The rest of the fungi, including the Chytridiomycetes, Zygomycetes, and all higher fungi, have the AAA pathway almost exclusively. Table 10 illustrates the lysine pathways in the classes of the lower fungi.

Other biochemical studies have been made to further assess the taxonomic and phylogenic relationships of the lower fungi. LeJohn[40] noted that the uridylates, uridine nucleotide sugars, and uridine nucleotide amino sugars function as allosteric activators of DPN-linked glutamic dehydrogenases obtained from cellulosic fungi that do not produce chitin in the cell wall. These glutamic dehydrogenases also retained the ability to interact with five other activators found in all members of the Oömycetes. Alexopoulos and Storch[41] made an extensive study of the DNA in the fungi, including the Oömycetes and Zygomycetes. They noted that the Zygomycetes as a group were characterized by having a lower G+C (guanine and cystosine) DNA content than the Oömycetes.

In an examination of Golgi dictyosomes in the lower fungi, Moore[34] reported that all the biflagellates and all the uniflagellates except the Blastocladiales showed evidence of this organelle, but not the non-flagellates.

Various schemes of fungal phylogeny have been proposed for evolution of the lower and the higher fungi. For further information on the phylogeny and classification of the lower fungi see References 7, 8, 12, 34, 38, and 39.

TABLE 8

IMPORTANT SPECIES OF THE SIX CLASSES OF PHYCOMYCETES (FORMERLY ONE CLASS)

Generic Name	Importance	References
Class Chytridiomycetes		
Synchytrium endobioticum	Black wart of potato	2, 32
Physoderma zeae-maydis	Maize brown spot	32
Rhizophidium couchii	Parasitic on *Spirogyra*	2, 9, 31
Coelomomyces dodgei	Parasitic on larvae of mosquito *(Anopheles crucians)*	9
Allomyces macrogynus	In soil; for morphogenesis studies	33
Blastocladiella emersonii	On silica-gel agar, fresh-water pond; for morphogenesis studies	12
Class Hyphochytridiomycetes		
Rhizidiomyces apophysatus	Parasitic on oögonia of *Achlya* and *Saprolegnia*	9
Class Plasmodiophoromycetes		
Plasmodiophora brassicae	Club root of cabbage, crucifers	2
Spongospora subterranea	Powdery scab of potatoes	2
Class Oömycetes		
Olpidiopsis sp.	Parasitic in host cells of aquatic fungi and algae	9
Lagenidium rabenhorstii	Parasitic in host cells of algae; may kill *Spirogyra*	9
Aphanomyces euteiches	Root rot of peas	2
Saprolegnia parasitica	Pathogenic to fish	2

TABLE 8 (Continued)

IMPORTANT SPECIES OF THE SIX CLASSES OF PHYCOMYCETES

Generic Name	Importance	References
Class Oömycetes (continued)		
Leptomitus lacteus	In large masses in water with high organic content	9
Achlya bisexualis	Hormones, sexual mechanism	35
Zoophagus insidians	Capturer of rotifers	9
Pythium debaryanum	Common cause of "damping off" of seedlings and soft rot	36
Phytophthora infestans	Late blight of potato and tomato	2, 32
Albugo candida	White rusts of crucifers	2, 32
Plasmopara viticola	Downy mildew of grape	2, 32
Peronospora destructor	Downy mildew of onion	2, 32
Bremia lactucae	Downy mildew of lettuce	2, 32
Rhizopus stolonifera	Decay of organic matter, soft rot of sweet potato, common bread mold	2, 32
Mucor sp.	Cosmopolitan in decay of organic matter	2, 3
Phycomyces sp.	For study of phototropism	1, 2
Choanephora cucurbitarum	Blossom blight, fruit rot of cucurbits	1, 2
Rhizopus arrhizus	May cause phycomycosis in animals and humans	37
Basidiobolus ranarum	Causes subcutaneous phycomycosis in man	4, 37
Entomophthora muscae	Parasitic on house flies	2, 7

TABLE 9

CELL WALL COMPOSITION AND TAXONOMY OF THE PHYCOMYCETES[a]

Cell Wall Category	Class	Representative Genera
II. Cellulose–glucan[b]	Oömycetes	*Phytophthora, Pythium, Saprolegnia*
III. Cellulose–chitin	Hyphochytridiomycetes	*Rhizidiomyces*
IV. Chitosan–chitin[c]	Zygomycetes	*Mucor, Phycomyces, Zygorhinchus*
V. Chitin–glucan[b]	Chytridiomycetes	*Allomyces, Blastocladella*
VIII. Polygalactosamine–galactan	Trichomycetes	*Amoebidium*

[a] Modified from Bartnicki-Garcia, S., in *Phytochemical Phylogeny,* p. 81, J. B. Harborne, Ed. (1970). Reproduced by permission of the copyright owners, Academic Press, Inc., New York.

[b] Incompletely characterized.

[c] Glucan has been demonstrated in the spores of *Mucor rouxii.*

TABLE 10

DISTRIBUTION OF LYSINE PATHWAYS AMONG LOWER FUNGI[a]

Lysine Pathway	Class	Representative Genera
DAP (diaminopimelic acid)	Oömycetes	*Achlya, Sapromyces, Sirolpidium, Pythium, Thraustotheca*
	Hyphochytridiomycetes	*Hyphochytrium, Rhizidiomyces*
AAA (aminoadipic acid)	Chytridiomycetes	*Rhizophlyctis, Phlyctochytrium, Allomyces, Monoblepharella*
	Zygomycetes	*Rhizopus, Cunninghamella, Syncephalastrum*

[a] Modified from Vogel, H. J., in *Evolving Genes and Proteins,* p. 25, V. Pryson and H. J. Vogel, Eds. (1963). Reproduced by permission of the copyright owners, Academic Press, New York.

REFERENCES

1. Ainsworth, G. C., and Bisby, G. R., *A Dictionary of the Fungi,* 6th ed. Commonwealth Mycological Institute, Kew, Surrey, England (1971).
2. Alexopoulos, C. J., *Introductory Mycology,* 2nd ed. John Wiley and Sons, New York (1962).
3. Burnett, J. H., *Fundamentals of Mycology.* Edward Arnold, London, England (1968).
4. Webster, J., *Introduction to Fungi.* Cambridge University Press, London, England (1970).
5. Hawker, L. E., *Fungi: An Introduction.* Hutchinson University Library, London, England (1968).

6. Hawker, L. E., and Linton, A. H., *Microorganisms: Function, Form and Environment.* American Elsevier Publishing Co., Inc., New York (1971).
7. Bessey, E. A., *Morphology and Taxonomy of Fungi.* The Blakiston Co., Philadelphia, Pennsylvania (1950).
8. Gäumann, E. A., *Die Pilze: Grundzüge ihrer Entwicklungsgeschichte und Morphologie,* 5th ed. Birkhauser, Basel, and Stuttgart, Germany (1964).
9. Sparrow, F. K., Jr., *Aquatic Phycomycetes.* University of Michigan Press, Ann Arbor, Michigan (1960).
10. Emerson, R., in *Aspects of Synthesis and Order in Growth,* p. 171, D. Rudnick, Ed. Princeton University Press, Princeton, New Jersey (1955).
11. Turian, G., *Nature, 196,* 493 (1962).
12. Cantino, E. C., in *The Fungi: An Advanced Treatise,* Vol. 2, p. 283, G. C. Ainsworth and A. S. Sussman, Eds. Academic Press, New York (1966).
13. Cantino, E. C., and Lovett, J. S., in *Advances in Morphogenesis,* Vol. 3, p. 33, M. Abercrombie and J. Brachet, Eds. Academic Press, New York (1964).
14. Emerson, R., and Wilson, C. M., *Mycologia, 46,* 393 (1954).
15. Miller, C. E., in *Mycological Studies Honoring John N. Couch,* p. 100, W. J. Koch, Ed. University of North Carolina Press, Chapel Hill, North Carolina (1968).
16. Couch, J. N., *J. Elisha Mitchell Sci. Soc., 78,* 835 (1962).
17. Scott, W. W., and O'Bier, A. H., *Progr. Fish-Cult., 24,* 3 (1962).
18. Coker, W. C., and Matthews, V. D., *N. Amer. Flora, 2,* 15 (1937).
19. Johnson, T. W., *The Genus Achlya: Morphology and Taxonomy,* The University of Michigan Press, Ann Arbor, Michigan (1956).
20. Scott, W. W., *Va. Agr. Exp. Sta. Tech. Bull., 151,* 1 (1956).
21. Dick, M. W., *J. Gen. Microbiol., 42,* 257 (1966).
22. Middleton, J. T., *Mem. Torrey Bot. Club, 20,* 1 (1943).
23. Waterhouse, G. M., in *Miscellaneous Publications,* Vol. 12, p. 1. Commonwealth Mycological Institute, Kew, Surrey, England (1956).
24. Karling, J. S., *The Plasmodiophorales.* Published by author, New York (1942).
25. Hesseltine, C. W., *Mycologia, 47,* 344 (1955).
26. Benjamin, R. K., *Aliso, 4,* 321 (1959).
27. Gilman, J. C., *A Manual of Soil Fungi,* 2nd ed. Iowa State University Press, Ames, Iowa (1957).
28. Drechsler, C., *Mycologia, 27,* 6 (1935).
29. Drechsler, C., *Mycologia, 33,* 248 (1941).
30. Manier, J. F., and Lichtwardt, R. W., *Ann. Sci. Nat. Bot. Ser., 12, 9,* 519 (1968).
31. Couch, J. N., *J. Elisha Mitchell Sci. Soc., 47,* 245 (1932).
32. Walker, J. C., *Plant Pathology.* McGraw-Hill Book Co., New York (1950).
33. Emerson, R., and Wilson, C. M., *Mycologia, 46,* 393 (1954).
34. Moore, R. T., *Recent Advances in Microbiology,* A. Perez-Miravete and D. Pelaez, Eds. Tenth International Congress for Microbiology, Mexico City, Mexico (1971).
35. Barksdale, A. W., *Science, 166,* 831 (1969).
36. Middleton, J. T., *Mem. Torrey Bot. Club, 20,* 1 (1943).
37. Beneke, E. S., and Rogers, A. L., *Medical Mycology Manual,* 3rd ed. Burgess Publishing Co., Minneapolis, Minnesota (1970).
38. Bartnicki-Garcia, S., in *Phytochemical Phylogeny,* p. 81, J. B. Harborne, Ed. Academic Press, New York and London, England (1970).
39. Vogel, H. J., in *Evolving Genes and Proteins,* p. 25, V. Pryson and H. J. Vogel, Eds. Academic Press, New York and London, England (1963).
40. LeJohn, H. B., *Biochem. Biophys. Res. Commun., 42,* 538 (1971).
41. Alexopoulos, C. J., and Storch, R., *Bacteriol. Rev., 34,* 126 (1970).

THE ASCOMYCETES

DR. WILLARD A. TABER AND RUTH ANN TABER

INTRODUCTION

The filamentous Ascomycetes[a] are fungi that produce sexual spores by free cell formation following karyogamy and meiosis in a specialized sac-like structure called an ascus. The spore is an ascospore. The ascospore contains one or more haploid nuclei and upon germination produces multinucleated septate assimilative filaments called hyphae or mycelia (singular, hypha and mycelium). The septum or cross wall, which is perforated, provides structural strength, but does not prevent movement of cytoplasm, nutrients, nuclei, or mitochondria. The cross walls divide the hypha up into cell-like compartments. The term "cell" may not be appropriate for these compartments, because the cytoplasm can be continuous from one to another.

An Ascomycete may consist of three developmental states: (a) the assimilative hyphae, which take up nutrients and exchange gases through the external envelope made up of a cytoplasmic membrane and cell wall and growing by extension of the hyphal tip; (b) the ascigerous or perfect state, in which nuclear pairing, fusion, and meiosis occur; and (c) the imperfect state, which consists of specialized hyphae upon which asexual disseminating spores are formed. Both the perfect and imperfect states may develop from the same hyphal mass. Many Ascomycetes possess all three states. Others consist only of assimilative hyphae and the imperfect state. Individuals of the latter category are referred to as Fungi Imperfecti or Deuteromycetes (Deuteromycotina). Such fungi are given a name based on the form of the asexual state. If such a fungus is later found to produce the perfect state, it is also named according to the structure and form of the perfect state, and, according to Botanical Nomenclatural Rules, the perfect-state name takes precedence. An example is *Penicillium vermiculatum* Dangeard (author of the species), which is in fact the imperfect state of *Talaromyces vermiculatus* (Dang.) Benjamin.[327,328]

NUMBER OF ASCOMYCETES

The number of species of Ascomycetes is not known with certainty, because some probably have not yet been discovered, and, no doubt, some described species are identical with others already described (synonyms). There are probably 15,000 species of Ascomycetes distributed in 1,950 genera.[1,266] Fifteen thousand species of Fungi Imperfecti have been described, a third of which are known to be Ascomycetes.[1,266]

LIFE CYCLE

The life cycle (Figure 1) of a representative Ascomycete begins with the haploid (1n) ascospore. The ascospore is a sexual spore, because its nucleus is the product of the fusion of two nuclei, followed by meiosis (Figure 2). When the two nuclei that fuse originate from a mycelium that has developed from one uninucleate spore (or a multinucleate spore in which all nuclei are genetically identical), the species is homothallic. When the nuclei must originate from two mycelia each of which developed from genetically different spores (that is, different mating types), the species is heterothallic.

Most heterothallic species produce both male and female structures (or female structures and morphologically indistinct male elements) on one mycelium; however, the male nucleus cannot fuse with the female nucleus (self-sterile). Self-sterility forces outbreeding and is effected by genetic control at one locus. This regulation is called the "incompatibility factor" and is usually represented by the alleles *A* and *a*; nuclear fusion takes place only between nuclei representing these two factors, never between *A* and *A* or *a* and *a*. Exceptions to this hermaphroditic, self-sterile condition can be found in the most unusual

[a] Most yeasts are Ascomycetes and are discussed on pp. 351–389. The class Ascomycetes is now often referred to as the subdivision Ascomycotina.[308] Others[189] place most fungi in the Eumycotina and place Ascomycetes in the class Ascomycetes.

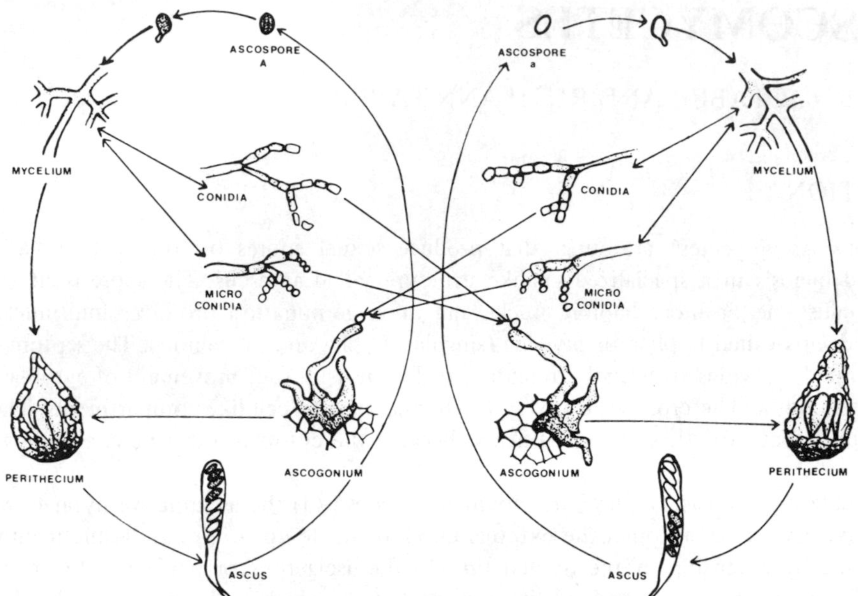

FIGURE 1. Life Cycle of *Neurospora crassa*.

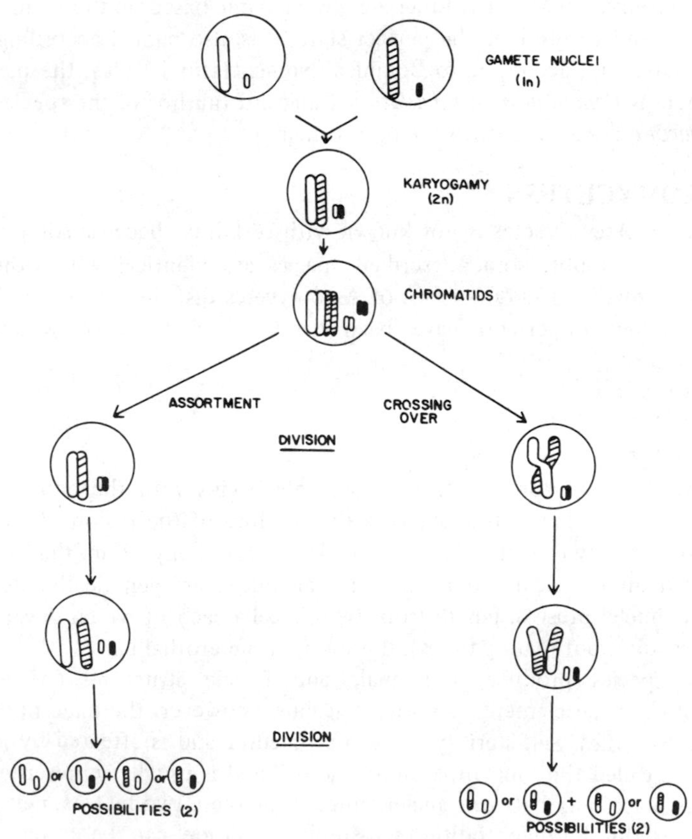

FIGURE 2. Random Assortment and Crossing Over.

Laboulbeniales (non-mycelial) and Ascosphaera, which are sexually dimorphic (either male or female organ on a given thallus). In some heterothallic hermaphroditic species the male organ is not morphologically distinct, and its function may be assumed by assimilative hyphae or conventional asexual spores. The life cycle of an heterothallic hermaphroditic Ascomycete, *Neurospora crassa* Shear & Dodge, is shown in Figure 1. Ascospores of mating types *A* and *a* germinate and develop into mycelium bearing both female sex organs (ascogonium and the receptor of the male nucleus, the trichogyne) and the male organ (microconidium). The male cell of each strain can fertilize the ascogonium of the other strain (in both strains an asexual conidium can also function as a male cell). Both nuclei continue to divide in the growing ascogonium, with the result that for a short time a multinucleate state exists. As the perithecial wall develops about the ascogonium, ascogenous hyphae develop from the tops or sides of the ascogonium. Eventually the tip of the ascogenous hypha bends into a crook ("crozier"); the second-from-last compartment (antipenultimate cell) receives one of each of the two kinds of nuclei and develops into an ascus initial (Figure 3). The two genetically unlike nuclei fuse (2n or zygote nucleus), and the diploid nucleus produces four nuclei by meiosis. A third division by conventional mitosis produces a total of eight nuclei. Astral rays, possibly emanating from a centriole, are believed to arc around the nucleus and surrounding cytoplasm and cut out the new ascospore. The ascospore wall is laid down by deposition of material on this membrane.

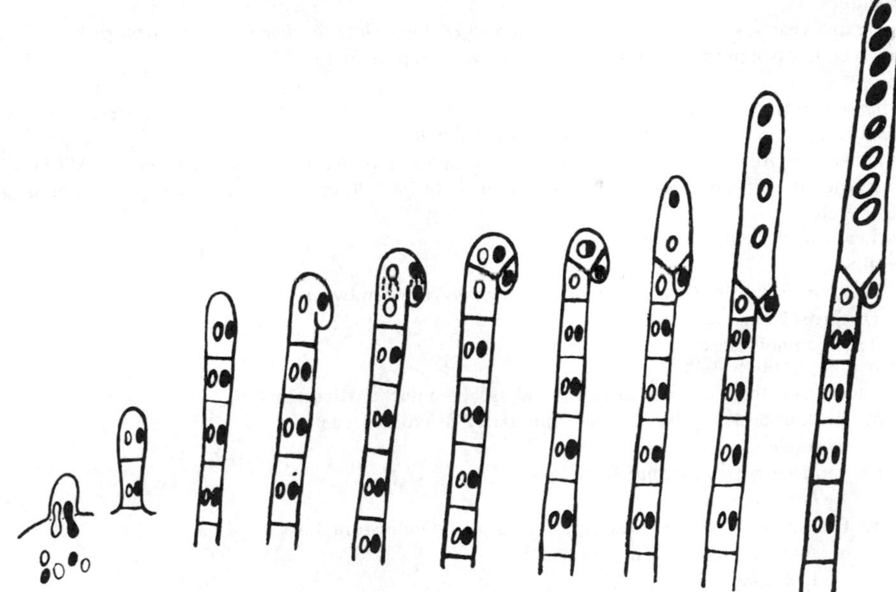

FIGURE 3. Formation of an Ascus from an Ascogenous Hypha.

TAXONOMY

Taxonomically, the term Ascomycete refers to a class of fungi united by the presence of ascospores produced within an ascus. The ascus is usually, but not always, borne on ascogenous hyphae which develop from an ascogonium (Figures 1 and 3). The class comprises three subclasses: Hemiascomycetes, Euascomycetes, and Loculoascomycetes.[189,191] These subclasses are organized into classifications that are constructed on morphological, if not phylogenetic, similarities or relationships. For a different interpretation of the simpler Hemiascomycetes, see References 194 and 341. The class Ascomycetes is divided into the following major groups: order (ending in -ales), family (ending in -aceae), genus, and species.

For the purpose of defining the various kinds of Ascomycetes, a key has been compiled (Table 1) from

TABLE 1.

KEY TO ASCOMYCETE GROUPS

I. Ascus not in an ascocarp and not arising from an ascogonium: HEMIASCOMYCETES
 1. Copulation of cells produces zygotes, which become asci: Endomycetales
  2. Ascus with more than eight spores: Ascoideaceae
   Example:
    Dipodascus, on plant exudate
  2. Ascus usually with eight spores or less
   3. Mycelium present
    4. Cell fusion of uninucleate gametangia: Endomycetaceae
     Examples:
      Eremascus
      Endomyces
    4. Cell fusion of gametes produced in gametangium: Spermophthoraceae
     Examples:
      Eremothecium
      Ashbya
      Nematospora, on plants
   3. Mycelium absent: Saccharomycetaceae
    Examples:
     Sexual yeasts, such as *Saccharomyces cerevisiae* and *Schizosaccharomyces octosporus*
 1. Ascus produced in open in ascogenous-like cell on leaves: Taphrinales
  Example:
   Taphrina deformans (peach leaf curl)

II. Ascus produced from ascogonium inside ascocarp or hyphal ball
 1. Ascus wall usually unitunicate (one layer); rare exception when ascocarp is an apothecium: EUASCOMYCETES
  2. Ascus borne in ascocarp and usually not from a definite hymenial (fertile) layer; ascus globose or clavate (Plectomycetes)
   3. Ascocarp with a stalk: Onygenales
    Families:
     Onygenaceae (e.g., *Onygena equina,* on shed horns or hooves)
     Dendrosphaeriaceae
     Trichocomoidaceae
   3. Ascocarp without a stalk
    4. Ascocarp with an ostiole (aperture) and usually a neck: Microascales
     5. Neck or beak papillate and not fimbriate: Microascaceae
      Example:
       Microascus manganii, on skin
       Petriella spp.
     5. Beak long and more or less fimbriate at tip: Ophiostomataceae
      6. Ascocarp usually dark; on wood
       Examples:
        Ceratocystis ulmi (Dutch elm disease); asexual phase is *Graphium*
        C. fagacearum (oak wilt)
        C. fimbriata (black rot of sweet potato)
        C. pilifera (blue stain of lumber)
        C. minor (blue stain of lumber)
      6. Ascocarp pale; on dung (coprophilous); considered by some to be a member of Melanosporaceae: *Sphaeronaemella fimicola*
    4. Ascocarp without an ostiole (a cleistothecium): Eurotiales
     5. Globose asci in spheres contained in sporocyst: Ascosphaeraceae
      Example:
       Ascosphaera apis (bee hive fungus)
     5. Ascus in ascocarp with membrane wall (peridium) or scattered in cottony ball of mycelium
      6. Ascocarp a cottony ball or weft of mycelium; some have distinct peripheral appendages: Gymnoascaceae
       Examples:
        Gymnascus rosea in soil
        Byssochlamys, in canned food
        Arthroderma spp. (athlete's foot) } perfect stages of some *Trichophyton*
        Nannizza spp. (athlete's foot) } and *Microsporum* spp.

TABLE 1 (Continued)

KEY TO ASCOMYCETE GROUPS

6. Ascocarp has a membranous wall or a "wall" of closely woven hyphae: Eurotiaceae
    7. Ascospore usually girdled with ridge
        8. Asexual stage is *Aspergillus*
            9. Ascocarp wall a single layer of cells, usually yellow
               Example:
               Eurotium
            9. Ascocarp wall more than one cell layer thick, with peripheral hyphae and Hülle cells
               Example:
               Emericella
            9. Ascocarp wall one or more cell layers thick, but not as above: *Sartorya*
        8. Asexual stage is *Penicillium*
            9. Ascocarp wall made up of hyphae
               Example:
               Talaromyces
            9. Ascocarp wall membranous (pseudoparenchymatous)
               Example:
               Carpenteles
    7. Ascospore not girdled
       Examples:
       Thielavia
       Anixiopsis

2. Ascus borne in ascocarp, with definite hymenial layer or tufts; ascus usually not dissolving; paraphyses usually present
    3. Ascus operculate (lid, not just pore): Pezizales
        4. Receptacle cup-shaped, usually not stalked
            5. Ascus tip stains blue with iodine
                6. Ascus cylindrical; ascospores in a single series (column): Pezizaceae
                    7. Ascus stains blue at upper end
                        8. Spores not globose
                            9. Apothecia at first subterranean, then superficial
                               Example:
                               Sarcosphaera
                            9. Apothecia on top of ground, saucer-shaped
                               Example:
                               Peziza
                        8. Spores globose
                           Example:
                           Plicaria
                    7. Ascus does not stain
                       Example:
                       Otidea
                6. Ascus more oval than above: Pseudoascobolaceae
                   Example:
                   Ascodesmis
            5. Ascus or ascus tip does not stain blue with iodine
                6. Apothecium very small, housing one ascus: Thelebolaceae
                   Example:
                   Thelebolus
                6. Apothecium larger, but down to 1 mm in some cases, with asci in hymenial layer or fascicle
                    7. Large asci extending above hymenial layer: Ascobolaceae
                       Examples:
                       Ascobolus
                       Pyronema
                    7. Asci not extending above hymenial layer
                        8. Ascospores symmetrical, smooth; apothecium large, yellow to tan: tribe Otidaea of Pezizaceae
                           Example:
                           Otidea (see above)

TABLE 1 (Continued)

KEY TO ASCOMYCETE GROUPS

8. Apothecium large, fibrous; on wood: Sarcoscyphaceae
   Examples:
   Urnula
   Plectania
   Sarcoscypha
8. Apothecium small, soft, often with bristles: Humariaceae
   Examples:
   Aleuria
   Lamprospora
   Humaria

4. Receptacle with heavy trunk-like stalk (compare with Sarcoscyphaceae above)
   5. Ascospores elliptical, without oil droplets: Morchellaceae
      Examples:
      Morchella esculenta (morel)
      Verpa conica
   5. Ascospores elliptical to slender, with oil drops: Helvellaceae
      Examples:
      Helvella
      Gyromitra

3. Ascus inoperculate (pore or splitting, but no lid)
   4. Ascus with pore, in cleistothecium (compare with Eurotiales, globose ascus and no pore); above ground
      5. White mycelium; on leaf surface: Erysiphales
         One family:
         Erysiphaceae (powdery mildews)
         6. One ascus per cleistothecium
            7. Appendages mycelium-like: *Sphaerotheca*
            7. Appendages dichotomously branched: *Podosphaera*
         6. Several asci per cleistothecium
            7. Appendages mycelium-like: *Erysiphe graminis*, on grass
            7. Appendages rigid
               8. Bulbous base on appendage: *Phyllactinia*
               8. No bulbous base
                  9. Tips of appendage dichotomously branched: *Microsphaera*
                  9. Tips coiled: *Uncinula*
      5. White mycelium; inside leaf; conidia solitary: *Leveillula*
      5. Dark mycelium: Meliolales
         Two families:
         Meliolaceae
         Example:
         Meliola
         Englerulaceae
   4. Ascus without pore; ascocarp closed; subterranean: Tuberales
      Four families:
      Pseudotuberaceae
      Geneaceae
      Eutuberaceae
      Terfeziaceae (some are truffles)
      Examples:
      Asci arranged in hymenium; fruit body convoluted: *Gyrocratera*
      Asci not in hymenium; tissue marbled: *Tuber*
   4. Ascus opening by apical pore or deliquescing; perithecium or apothecium
      5. Asci and ascospores very long, the latter thread-like
         6. Asci in isolated apothecium: Ostrapales
            One family:
            Ostrapaceae
            Examples:
            Apothecia superficial and stalked: *Vibrissea*
            Apothecia immersed in plant tissue, a disk opening by a slit: *Ostrapa*

TABLE 1 (Continued)

KEY TO ASCOMYCETE GROUPS

6. Perithecia imbedded in fungal stroma, which may arise from a sclerotium: Clavicipitales
   One family:
   Clavicipitaceae
   7. Stroma on stalk arising from sclerotium; parasitic on grasses
      Examples:
      Claviceps purpurea
      C. paspali
   7. Stroma on stalk arising from an insect or subterranean Ascomycete host
      8. Ascospores breaking into cells: *Cordyceps militaris*, on moth pupae
      8. Ascospores not breaking into cells: *Ophiocordyceps* spp.
   7. Stroma sessile
      8. Stroma forms cylinder around grass stem: *Epichloe*
      8. Stroma soft, small, light-colored; on leaves: *Oomyces*

5. Asci not long and cylindrical; spores sometimes long, but not thread-like
   6. Ascus borne in apothecium, unitunicate; asci not stained blue by iodine
      7. Apothecia-stroma with dark covering (clypeus), opening by a slit; immersed in plant tissue: Phacidiales (Families: Cryptomycetaceae, Phacidiaceae, Hypodermataceae)
         8. Flattened, stroma-like body; on leaf tissue; several hymenia in stroma
            Example:
            Rhytisma acerinum (tar spot of maple)
         8. Hymenia developed singly
            9. Apothecia immersed in bark: *Colpoma*
            9. Apothecia in leaves and herbaceous stems
               10. Ascospores filiform: *Lophodermium*
               10. Ascospores elliptical, clavate, or cylindrical
                   11. Ascospores rod-shaped; clypeus black: *Hypoderma*
                   11. Ascospores clavate; spores *ca.* half the length of ascus: *Hypodermella*
            9. Apothecia circular, opening by teeth; in leaf tissue: *Coccomyces*
      7. Apothecia superficial, no dark covering layer: Helotiales
         8. Asci dissolving, leaving dry spore mass in saucer-shaped ascocarp: Caliciaceae
            Example:
            Roesleria
         8. Asci persistent, with a pore
            9. Asci in palisades on bark (like *Taphrina*, but with ascal pore): Ascocorticiaceae
               Example:
               Ascocorticium
            9. Asci in hymenium, bearing paraphyses
               10. Ascocarp upright, club- or finger-shaped or with a mushroom-like head: Geoglossaceae
                   11. Fruit body black or deep-purple to black, woody
                       12. Spines on hymenial area: *Trichoglossum*
                       12. No spines on hymenial area
                           13. Ascospores brown
                               Example:
                               Geoglossum ("devils tongue", "earth tongue")
                           13. Ascospores colorless
                               Example:
                               Corynites
                   11. Fruit body yellow, tan, or green; ascospores hyaline
                       12. Fertile head gelatinous
                           Example:
                           Leotia
                       12. Fertile head not gelatinous
                           13. Ascocarp greenish: *Microglossum*
                           13. Ascocarp yellow or brown
                               14. Ascospores multiseptate
                                   15. Fertile head flattened (spathulate): *Spathularia*
                                   15. Fertile head hemispherical: *Cudonia*

TABLE 1 (Continued)

KEY TO ASCOMYCETE GROUPS

14. Ascospores non- to one- septate: *Mitrula*

10. Ascocarp saucer-shaped or submerged in plant tissue, usually not on ground
  11. Apothecia small, near-sessile, waxy; Orbiliaceae
    Examples:
      Orbilia
      Hyalina
  11. Hymenium not waxy
    12. Apothecium tissue cartilaginous to fleshy; excipulum of globose cells: Dermateaceae
      Example:
        Small dark cups breaking through plant tissue: *Dermea*, *Diplocarpon rosae*, on roses (usually as imperfect *Marssonina*)
    12. Excipulum of elongated cells
      13. Apothecia with pronounced hairs: Hyaloscyphaceae
        Example:
          Mollisina
      13. Apothecia smooth or downy; arising from sclerotium
        14. Apothecium on stalk arising from sclerotium: Sclerotiniaceae
          Examples:
            Sclerotinia (*Monilinia*), on stone fruits and on soil (*Monilia* and *Botrytis* imperfect form)
        14. Apothecium not arising from sclerotium; excipulum of parallel hyphae: Helotiaceae
          Examples:
            Cudoniella
            Hymenoscyphus

6. Asci in a perithecium
  7. Paraphyses absent
    8. Perithecial cavity a locule, but asci unitunicate: Coronophorales
      One family:
        Coronophoraceae
      Example:
        Coronophora
    8. Perithecia (and stroma, if present) bright-colored or white, fleshy and soft, sometimes membranous; apical paraphyses present: Hypocreales
      One family:
        Hypocreaceae
      9. Perithecia immersed in compact stroma; superficial or erumpent
        10. Ascospores one-celled
          Examples:
            Balzania
            Selinia
        10. Ascospores more-than-one-celled
          11. Ascospores two-celled
            Examples:
              Hypocrea
              Podostroma
          11. Ascospores more-than-two-celled
            Example:
              Thyronectria
      9. Perithecia not in compact stroma; immersed in substrate or superficial, with or without subiculum or basal stroma
        10. Perithecia immersed in plant substrate
          11. Ascospores one-celled
            Example:
              Hyponectria
          11. Ascospores more-than-one-celled
            Example:
              Nectriella

TABLE 1 (Continued)

KEY TO ASCOMYCETE GROUPS

10. Perithecia superficial, with or without subiculum or basal stroma
  11. Perithecia superficial
    Example:
      Neocosmospora
  11. Perithecia in or on subiculum or basal stroma
    12. Perithecia on subiculum
      Example:
        Protocrea
    12. Perithecia on basal stroma
      13. Ascospores one-celled
        Example:
          Allantonectria
      13. Ascospores more-than-one-celled
        Example:
          Nectria

7. Paraphyses present, but may disappear; perithecia yellowish-brown to black (color may be due to spores): Sphaeriales
  Exception:
    Endothia parasitica (chestnut blight), with light spores and reddish stroma
  8. Asci dissolving
    9. Ascocarps light; ascospores emerging in a dark mass: Melanosporaceae
      10. Ascocarps smooth, beaked
        Example:
          Melanospora
      10. Ascocarps smooth, not beaked
        Example:
          Sphaeroderma
    9. Ascocarps dark and hairy: Chaetomiaceae
      10. Appendages not jointed: *Chaetomium*
      10. Appendages jointed
        Example:
          Ascotricha
  8. Asci not dissolving in ascocarp
    9. Ascospores hyaline, uniseptate, with appendages; aquatic
      Example:
        Loramyces (see also Sphaeriaceae)
    9. Ascospores hyaline, non-septate, large; on dung: Seliniaceae
      Example:
        Selinia (see also Hypocreales)
    9. Ascospores not large and not with long appendages
      10. Perithecium immersed in tissue or stroma; ascospores hyaline
        11. Ascus pore stains blue with iodine: Amphisphaeriaceae
          Example:
            Amphisphaeria
        11. Ascus pore not stained blue with iodine: Polystigmataceae
          Example:
            Glomerella cingulata, on pears, apples, grapes
      10. Ascospores dark, non-septate
        11. Perithecia in dark stroma: Xylariaccae (spores flatter on one side)
          12. Perithecium immersed in stalked, upright-stick, or club-like stroma
            13. Stroma black; perithecia not on tip only: *Xylaria*
            13. Stroma gray to flesh-colored, conical; like head of nail; ascocarps inserted in head of nail; on dung: *Poronia* (*Podosordaria*)
          12. Stroma sessile, cushion- or knob-like
            13. Stroma tissue concentrically zoned; on wood; large (bigger than 1 cm): *Daldinia*
            13. Stroma tissue not zoned: *Hypoxylon*
          12. Stroma prostrate, dark or black; crusty layer
            13. Perithecia tubular, densely crowded: *Camarops*

TABLE 1 (Continued)

KEY TO ASCOMYCETE GROUPS

13. Perithecia flask-like to subglobose
 14. Perithecia large, with ascospores longer than 25 μ: *Ustilina*
 14. Ascospores usually less than 20 μ long
  15. Perithecia just under stroma surface; ostiole may protrude: *Hypoxylon* (see above for cushion forms)
  15. Perithecia at bottom of stroma; long neck connects to surface
   16. Perithecia in one horizontal layer; stroma concave: *Nummularia*
   16. Perithecia distributed at different levels: *Bolinia*
11. Perithecia free, but with a subiculum, clypeus, or rudimentary stroma
 12. Not on dung; ascospores have no appendages
  13. On surface of wood; clustered on dark subiculum: *Rosellinia*
  13. On burnt ground or immersed in plant tissue; not clustered, or if so, under clypeus
   14. Ascospores with four germination pores: *Amphisphaerella*
   14. Ascospores without such pores
    15. Ascospore with colorless basal appendage: *Entosordaria*
    15. Ascospore without such an appendage
     16. Perithecia without a neck: *Anthostomella*
     16. Perithecia with a neck: *Anthostoma*
 12. Beneath surface of dung: *Hypocopra*
11. Perithecia free; spores may have appendages: Lasiosphaeriaceae
 12. Coprophilous, but isolatable from soil
  13. Ascospores with colorless appendages
   14. Ascospores oblong; perithecium with hairs: *Lasiosordaria*
   14. Ascospores ovoid; perithecium dark, with clustered neck hairs: *Podospora*
   Examples:
    P. curvula
    P. anserina
  13. Ascospores without appendages
   14. Ascospores furrowed: *Coniochaeta*
   14. Ascospores not furrowed: *Sordaria*
   Example:
    S. fimicola
 12. Not coprophilous
  13. Parasitic on certain Basidiomycetes; perithecia hairy; spore dark: *Helminthosphaeria*
  13. Not parasitic; on wood, plants, or lichens
   14. Ascospores brown, elliptical
    15. On lichens; perithecia without ostiole; spores held in matrix: *Synaptospora*
    15. Ascospores not clustered (not held in matrix)
     16. Ascospores smooth; perithecia hairy; on wood and dung: *Coniochaeta*
     16. Ascospores with pits on wall; perithecia dark to black: *Gelasinospora*
     16. Ascospores with striations; perithecia smooth, pear-shaped, black: *Neurospora*
   14. Ascospores large, colorless
    15. Ascospores with dark, swollen cell; perithecia black, elliptical: *Bombardia*
    15. Ascospores without dark, swollen cell: *Lasiosphaeria*
10. Ascospores colorless to pale yellow, not cylindrical
 11. Perithecia in stroma; necks congregated; emerging as erumpent disk with ostiole holes; ascospores sausage-shaped: Diatrypaceae
 Examples:
  Diatrype
  Valsa

TABLE 1 (Continued)

KEY TO ASCOMYCETE GROUPS

11. Perithecia not in stroma, or if so, ascospores not sausage-shaped but septate
    12. Perithecia on surface, dark; scattered light-colored spores: Sphaeriaceae
        Examples:
            Loramyces
            Niesslia
    12. Perithecia immersed in wood; no stroma; long neck to surface: Ceratostomataceae
        Example:
            Ceratostomella
    12. Perithecia small; in or on host tissue; stroma may be present
        13. Ascospores in two columns in long-stalked ascus, or spores colored: *Diaporthaceae*
            Examples:
                Valsaria, in stroma
                Gnomonia ulmea, on elm leaves
                Diaporthe
        13. Ascospores in one column; hyaline; uniseptate when mature: *Endothella* (see Polystigmataceae)

1. Ascus wall bitunicate; asci in cavity (locule) of stroma (ascostroma); not a true perithecium; uniloculate stroma may resemble perithecium; pseudoparaphyses may be present, but not paraphyses: LOCULOASCOMYCETES
    2. Small stroma; on plants, man, or insects; spherical asci scattered inside; Myriangiales
        Three families:
            Elsinoeaceae (*Elsinoe fawcetti*, on citrus; stroma just under surface, ascospores multiseptate, one ascus per cavity of stroma)
            Myriangiaceae (*Myriangium*, on insects)
            Piedraiaceae (*Piedraia hortai*, on human hair; "black piedra")
    2. Stroma oblong with slit opening (hysterothecium), or irregular stroma
        3. Irregular stroma: lichens
        3. Hysterothecium elongated; forked, star-shaped on wood: Hysteriales
            4. Hysterothecium higher than wide
                Examples:
                    Actidium
                    Lophium
            4. Hysterothecium flatter, and wider than high
                5. Ascospores pale to hyaline
                    Examples:
                        Glonium
                        Gloniopsis
                5. Ascospores brown or brownish
                    Examples:
                        Hysterium
                        Hysterographium
    2. Stroma flattened (thyriothecium), shield-like, often with central pore or slit; small "fly specks"; on leaves and stems; more than one ascus per cavity: Microthyriales (Hemisphaeriales)
        3. Upper surface of shield made of hyphae radiating outward: Microthyriaceae
            Example:
                Microthyrium
        3. Upper surface without radiating pattern: Hemisphaeriaceae
            Example:
                Microthyriella
    2. Asci in pseudothecia (false perithecia)
        3. Pseudoparaphyses present: Pleosporales
            4. Pseudothecia with slit-like ostiole: Lophiostomataceae
                Examples:
                    Lophiosphaeria
                    Lophiostoma
            4. Pseudothecia with circular ostiole
                5. Ascospores one-septate

TABLE 1 (Continued)

KEY TO ASCOMYCETE GROUPS

6. Ascospores hyaline: Pleosporaceae (in part)
   Example:
   Herpotrichia
6. Ascospores hyaline to gray-greenish: Venturiaceae
   Example:
   Venturia inaequalis, on apple leaves
6. Ascospores brown: *Trichodelitschia* and Verrucariaceae (lichens)

5. Ascospores two- or more-septate, some with transverse walls
   6. Ascospores hyaline to brown: Pleosporaceae (in part)
      Examples:
      Pleospora
      Pyrenophora
      Sporormia
      Leptosphaeria
   6. Ascospores becoming greenish: Herpotrichiellaceae
      Example:
      Berlesiella
5. Ascospores non-septate, hyaline: Botryosphaeriaceae
   Example:
   Botryosphaeria (imperfect, *Dothiorella*)

4. Pseudothecia with no ostiole: Perisporiaceae
   Example:
   Perisporium

3. Pseudoparaphyses absent: Dothideales
   4. Several to many locules per stroma; erumpent to subepidermal: Dothideaceae
      5. Ascospores hyaline
         Examples:
         Scirrhia, on cherry
         Dibotryon morbosum, on cherry
      5. Ascospores becoming yellow to brown
         Example:
         Dothidea
   4. Stroma small; immersed in tissue, with few asci per locule; cavities orderly arranged in stroma: Pseudosphaeriaceae
      5. Ascospores muriform
         Example:
         Pseudoplea (*Leptosphaerulina*)
      5. Ascospores one-septate: *Wettsteinina*
   4. Stroma very small, resembling perithecium, but without paraphyses: Mycosphaerellaceae
      5. Ascospore cells equal
         6. Ascospores one-septate
            7. Asci eight-spored
               Example:
               Mycosphaerella (leaf spot of strawberry and pear)
            7. Asci many-spored: *Rehmielopsis*
         6. Ascospores with several cross-septations: *Sphaerulina*
         6. Ascospores also with longitudinal septations: *Saccothecium*
      5. Ascospore cells unequal: *Guignardia bidwellii*, on grape leaves
   4. Pseudothecia on leaves, "sooty moulds", growing on secretions of insects: Capnodiaceae
      Example:
      Capnodium

III. Minute, non-mycelial ascomycetes living on exoskeletons of certain insects; approximately 1500 species: Laboulbeniales

Data compiled from References 187, 189-191, 195, 267, 288, and 296.

several existing authoritative sources (References 186, 187, 189-195, 267, 288, 289, and 296), all of which should be consulted for more detailed taxonomic discussions. This key is not intended to lead to the identification of all genera, since only a few of the 1,950 genera could be included. All of the major orders and families are represented, however.

Hemiascomycetes

The Hemiascomycetes do not produce asci in an ascocarp, but form asci directly from zygotes or ascogenous cells. In the laboratory, on agar media, they usually grow as moist, yeast-like colonies consisting of budding cells and (or) pseudomycelium.[290] The remaining Ascomycetes, those producing an ascocarp, are divided into two subclasses, the Euascomycetes and Loculoascomycetes.

Euascomycetes

The Euascomycetes have unitunicate asci, i.e., ascal walls that do not separate into two walls upon maturation,[291,292] and an ascocarp wall, which develops below (Discomycetes) or around (Pyrenomycetes) the ascogonium. By contrast, the Loculoascomycetes develop ascogonia in a stroma (see below). Unitunicate asci may discharge their ascospores through a special apical opening with a lid called an operculum (Pezizales), through a simple pore, a slit, or merely through rupture or disintegration of the ascus wall (Figure 4). The unitunicate ascus and the presence of paraphyses characterize most of the Euascomycetes. Paraphyses are sterile hyphae, which grow upward from the subhymenial layer that produces the asci. The Euascomycetes are divided into those bearing globose to clavate asci scattered within an ascocarp (Plectomycetes) and those usually bearing cylindrical asci in a definite fertile layer, the hymenium, or in tufts. The Plectomycete ascocarp may be either a closed ascocarp, called a cleistothecium (Eurotiales), a cottony ball of hyphae with asci scattered throughout (Gymnoascaceae), or an ascocarp with an opening, or ostiole, and usually a neck (e.g., Microascaceae and Ophiostomataceae). Members of the Microascales superficially resemble flask-shaped Euascomycetes that bear asci and paraphyses in a fertile layer; they are included in the Plectomycetes, however, because they bear globose asci scattered in the ascocarp. The Myriangiales[293] bear spherical asci, but are placed in the Loculoascomycetes because they form in a locule. Ascomycetes with a distinct fertile layer of asci may bear them in a cleistothecium (such as the Erysiphales), perithecium (flask-shaped or globose ascocarp with an ostiole), or apothecium (cup-shaped with exposed hymenium). Apothecial ascomycetes are referred to as Discomycetes and, according to the treatment of Seaver,[192,193] may be shaped like a stalked or stalkless saucer (as Pezizaceae and Helotiales), stalked saddle (Helvellaceae), or stalked, sponge-like conical head (Morchellaceae). The Pezizales are currently being revised,[294] and the current literature in journals such as *Mycologia* and *Taxon* should be followed for proposed changes. A forthcoming book[345] is to have taxonomic keys to various fungi. Figure 5 illustrates the fruiting bodies of ascomycetes.

The bright-colored Euascomycetes, the Hypocreales, have recently been treated comprehensively by Rogerson.[288]

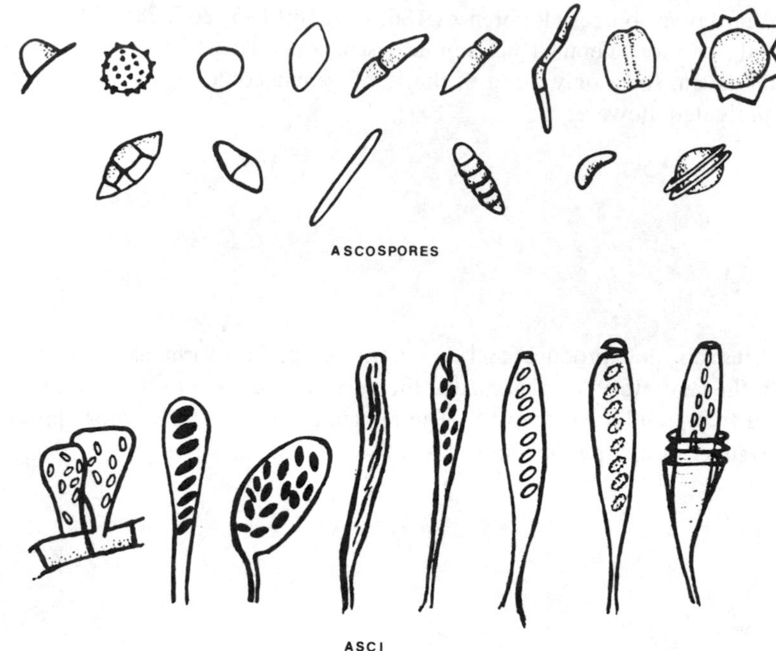

FIGURE 4. Representative Types of Ascospores and Asci.

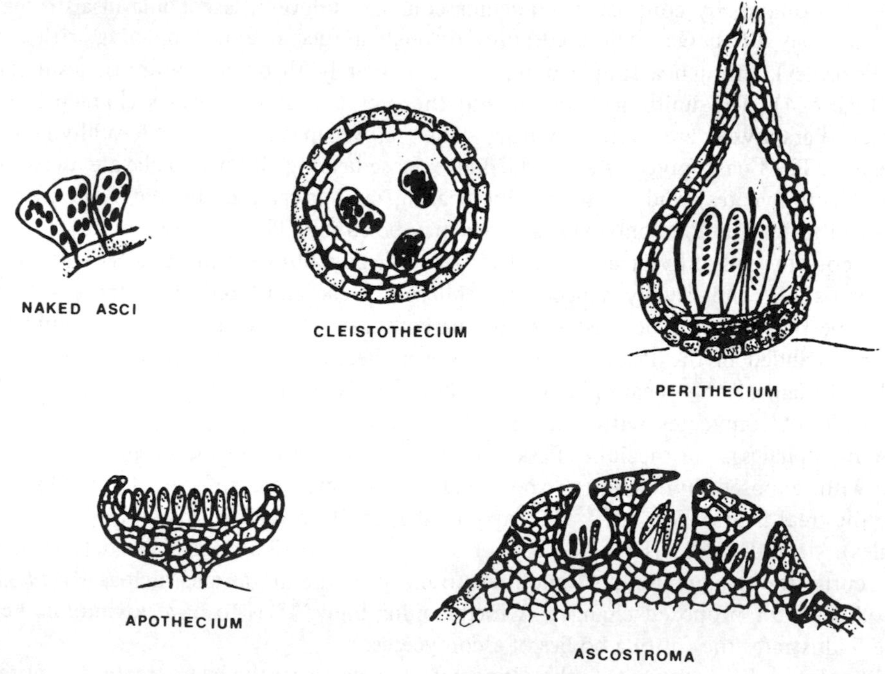

FIGURE 5. Ascomycete Fruiting Bodies.

Loculoascomycetes

The Loculoascomycetes, as defined by Luttrell,[296] possess bitunicate asci, which develop in a cavity, or locule, in stromatic tissue. Luttrell[187,295,296,306] described four different patterns of development of locules in stromatic tissue; the Elsinöe type, the Pseudosphaeria type, the Dothidea type, and the Pleospora type. The asci of the Elsinöe type arise in monascal locules and remain separated by relatively unaltered stromal tissue. The asci of the Pseudosphaeria type arise individually in the stromatic tissue (monascal locules), with the result that some stromatic tissue is often left between locules; such strands of tissue are termed interthecial strands. The asci of the Dothidea type arise in a group, or fascicle, and expand into the disintegrating stromatic tissue, forming the locule with asci inside. The asci of the Pleospora type grow up between previously formed vertical hyphae originating in the roof of the locule (pseudoparaphyses).

The bitunicate nature of the ascus may be difficult to determine in many cases, and Dennis[191] has suggested several other characteristics usually associated with bitunicate asci that help in their identification: (a) short-stalked ascus; (b) ascus stalk sharply differentiated from the ascus body; (c) ascus thick-walled at the tip, with a depression on the inner wall; (d) non-staining of ascus wall with iodine; (e) multicellular and usually dark-spored; and (f) ascocarps usually small and closed. See Reference 185 for a discussion of the morphology of the ascocarp.

Cultures of some of the Loculoascomycetes and immersed Euascomycetes cannot be identified according to a key such as that given in Table 1 because they do not develop identifiable structures on laboratory media. In many cases such fungi may be identified only if they can be observed on their natural substrate.

ASSIMILATIVE HYPHAE

Fine and Gross Structure

The assimilative mycelium consists of multinucleated cytoplasm, partially divided into sections by perforated septa (Figure 6) and contained within a profusely branched system of tubes bounded by the wall and cytoplasmic membrane. Mycelium may exist as masses of hyphae, or the hyphae may be compacted into tissues. Tissue of loosely woven hyphae is referred to as prosenchyma, and tissue formed by cells that round up and fuse is referred to as pseudoparenchyma (Figure 7).

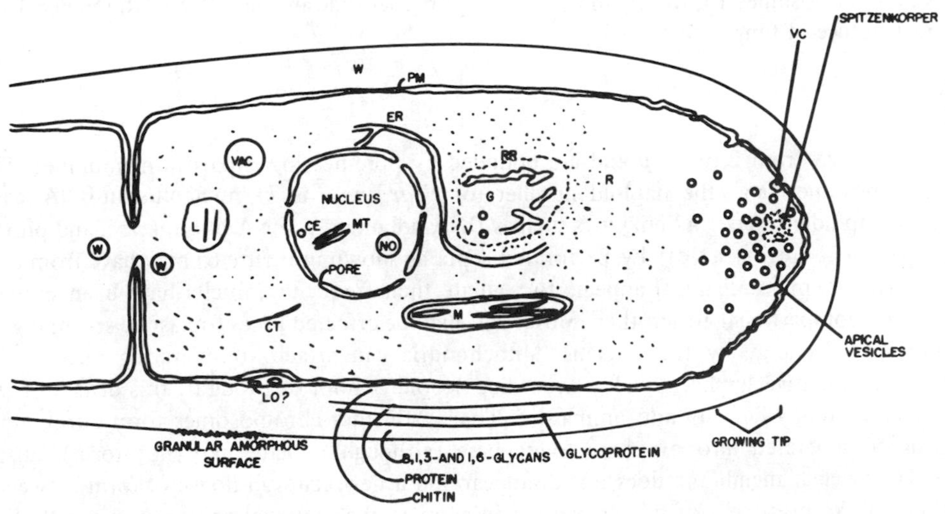

CE = centriole-like body
CT = cytoplasmic microtubule
ER = endoplasmic reticulum
G = smooth-surfaced Golgi cisternae
L = lipid body
LO = lomasomes
M = mitochondrion
MT = microtubule
NO = nucleolus

PM = plasma or cytoplasmic membrane
R = ribosomes
RR = ribosome-rare zone of exclusion
V = vesicle
VAC = vacuole
VC = vesicle continuous with plasma membrane
W = cell wall
(W) = Woronin body

FIGURE 6. Fine Structure of Hyphal Tip.

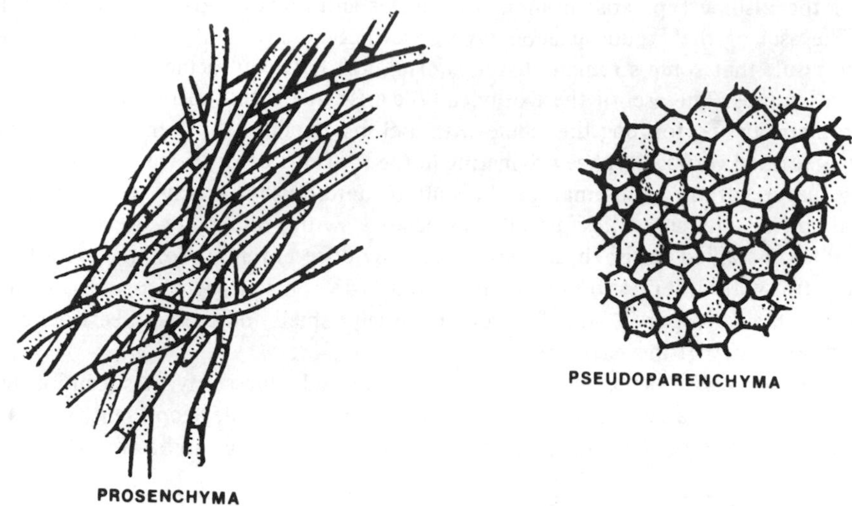

FIGURE 7. Tissue-like Formations of Hyphae.

Ascomycetes are eukaryons. Some of the cellular organelles and inclusions are the following: (a) nucleus bounded by a pore-bearing membrane and containing chromosomes (which, however, apparently do not contain histones[2]), nucleolus, centriole-like plate,[3] and microtubules,[4,5] (b) mitochondria; (c) endoplasmic reticulum, which may bear ribosomes (rough ER); (d) Golgi body-like cisternae in a ribosome-free area,[5] (e) Spitzenkörper region at the hyphal tip,[5,6] (f) secretory vesicles, whose membranous wall, by intercalation, becomes new cytoplasmic membrane and whose contents become cell wall;[5,7] (g) septum; (h) Woronin bodies, which occur near septal pores and may regulate translocation through the pore;[8] (i) lipid globules; (j) vacuoles; (k) non-structural polymers, such as glucans and polyphosphates; (l) cytoplasmic membrane; (m) cell wall; (n) ribosomes; (o) lomasomes;[9,10,11] (p) perhaps mesosomes.[12] See Reference 4 for a review of the ultrastructure of fungi.

Nucleus

Nuclei range in size from 2 to 25 μ and are bounded by pore-bearing tripartite membranes. The nucleus contains chromosomes (e.g., the haploid number for *Neurospora* is 7), nucleolus, m-RNA, centriole-like body, possibly spindle fibers, and enzymes such as DNA polymerase, RNA polymerase, and phosphatases.[3] A cell compartment (region set off by perforated septa as shown in Figure 6) may have from one to many nuclei, and while some mechanism appears to regulate their frequency, nuclei have been shown to freely move from one compartment to another. Movement may be effected by cytoplasmic streaming or possibly by expenditure of energy by the nucleus. Mitochondria can attach to migrating nuclei.[13] Nuclei of vegetative or assimilative hyphae divide. According to one school of thought, this division (i.e., mitosis) resembles closely that of plants and animals.[13] That is, distinct chromosomes form during division, and division can be separated into prophase, metaphase (although a plate may not form), anaphase, and telophase. The nuclear membrane does not disappear and a new cell wall does not form between the sister nuclei. Thus, unlike most cells of higher plants and animals, many nuclei can exist in one "cell". Another school[14,15] holds that somatic-cell division is not a typical mitosis. The chromosomes are said to line up with the long axis of the mycelium and to divide longitudinally while being connected by a thread. Spindle fibers maintain mutual contact of chromosomes destined to occupy one nucleus, and cytoplasmic streaming, rather than contraction of fibers, separates chromosomes going into nuclei. Meiosis is in essence similar to that occurring in plants and animals.[16,17] Centrioles and spindle fibers play conventional roles.

DNA and GC Content

Fungi contain double-stranded DNA both in the mitochondrion and in the nucleus. The mitochondrion of

Neurospora contains 0.2-1.8 x 10^{-16} grams, and the haploid nucleus contains 900 x 10^{-16} grams DNA.[3] Guanine + cytosine base ratios vary widely in prokaryons (i.e., bacteria and blue-green algae), ranging from 25 to 75%. Bacteria that are taxonomically similar possess similar ratios. The range in fungi is much smaller,[18] from 34 to 63%. The GC content of the Phycomycetes is less than 50%, while that of the filamentous Ascomycetes, according to a survey of 90 isolates representing 69 species, was greater than 50%, with the mode being near 52%.[18] As might be expected, the GC ratios of Fungi Imperfecti, almost all of which probably either are Ascomycetes or were derived from them, were similar, ranging from 48.5 to 60%.

Ribosomes

Ascomycetes contain ribosomes consisting of approximately 53% RNA and 47% protein.[19] The ribosomes of Ascomycetes are 80S, as are those of other eukaryons. Alignment on endoplasmic reticulum does not appear to occur commonly. Polyribosomes are demonstrable (up to hexamers) in growing cells, but not in conidia;[20] thus, informational RNA is not stored in resting cells. Ribosomal RNA of *Neurospora* consists of 19S and 28S RNA, whereas that of *Aspergillus niger* (a Deuteromycete with no known perfect stage) consists of 10S, 15S, and 19S.[19] Approximately 73% of all *A. niger* RNA is ribosomal and is 53% G + C.[19] The GC ratio of soluble RNA is 58%.

Mitochondria

Mitochondria in Ascomycetes no doubt develop from preexisting mitochondria and, as other mitochondria, may have originally derived from some early prokaryon parasite.[21] The mitochondria of *Neurospora* possess ribosomes and messenger RNA and carry out chloromycetin-sensitive protein synthesis on mitochondrial ribosomes (cytoplasmic protein synthesis is cycloheximide-sensitive). It has been estimated[3] that approximately 18% of mitochondrial DNA codes for ribosomes, whereas nuclear DNA does not possess any genes for mitochondrial ribosomes. Certain mitochondrial proteins, however, are coded for from chromosomal DNA. Mitochondrial DNA is at least in part circular, thus resembling that of prokaryons.

Cell Wall

Through an ingenious use of depolymerizing enzymes, involving their application in various sequences and combinations to cell wall preparations from *Neurospora crassa*, Hunsley and Burnett[22] found that the inner layer of wall consists of chitin (*beta*-linked polymer of N-acetyl glucosamine). Immediately external to this is a protein layer, then a glycoprotein layer, and finally, on the outside, a layer of *beta*-1,3- and 1,6-glucans. In contrast, a Phycomycete contained cellulose rather than chitin. Phycomycetes also contain uronic acids. A general statement may be made, then, that cell walls of Ascomycetes contain chitin, protein, 1,3- and 1,6-*beta*-linked polymers of hexoses, and perhaps lipid. The walls of some Ascomycetes definitely contain cellulose, although the content is quite small. Using X-ray diffraction, cellulose and chitin were identified in walls of *Ceratocystis ulmi* (Buis.) Moreau and in *C. olivacea* (Math.) Hunt;[23] chitin, cellulose, mannan, and a *beta*-glucan were identified in *C. ulmi*, using infra-red spectroscopy.[24] Gorin and Spencer[25,26] found that walls of *C. ulmi* contained a 1,6-linked *alpha*-D-mannopyranose main chain, substituted in the 3 position mainly by a single-unit *alpha*-L-rhamnopyranosyl side chain and probably by a small proportion of O-L-rhamnopyranosyl-^{1,4}-L-rhamnosyl side chain. The related *C. brunnea* Dav. contained a D-gluco-D-mannan. These compounds may be an integral part of the cell wall, or they may pass through the cell wall, or they may be synthesized near the surface. The outer wall surface of *Neurospora* was found to be granular, whereas interior layers were reticulated;[22] the outer surface was frayed or at least not distinct.[27] Electron-microscope examination of thin sections of hyphae treated with snail-gut enzymes before fixation with glutaraldehyde and $KMnO_4$ showed microfibrils in the wall (these fibrils were not detected when using osmium tetroxide). The fibrils of the chitin layer are arranged randomly in the plane of the wall. Manocha and Colvin[27] reported that the wall consisted of two phases: chitin microfibrils, and a mixture of amorphous glucan and protein, which fills the interstices between the chitin microfibrils. The protein does not contain hydroxyproline or cysteine, but is rich in aspartic and glutamic amides. The protein is viewed as having skeletal functions and contributing at least in part to fibrils imbedded in the wall. Pores 40 to 70 Å in diameter course throughout the wall, and protein forms a part of the boundaries of these pores. *Neurospora* wall may also contain some galactosamine, but this may be due to infection with the double-stranded RNA fungal virus. Lemke[28] has found such viruses in *Penicillium* and noted the correlation between infection and the presence of galactosamine. *Neurospora* walls may also contain some glucuronic acid.[29] Conidial walls have similar composition.[30]

The sugar content of Ascomycete walls may vary from species to species. In general, the cell walls appear to contain glucose, mannose, and galactose (and galactosamine?) but lack xylose and fucose. Basidiomycetes (mushrooms) on the other hand, contain mannose, xylose and fucose, but not galactose or galactosamine.[29] Interestingly, the walls of many ascomycetous yeasts have only traces of chitin and consist mainly of mannans, glucans, and glycoproteins.[31] The pink yeasts, *Sporobolomyces* and *Rhodotorula*, which are believed to be imperfect forms of Basidiomycetes, contain chitin, mannans, galactose, fucose, and *gamma*-aminobutyric acid.[29] Mutations that affect mycelial shape also affect cell wall composition.[29] Certain marine and terrestrial species of *Leptosphaeria* have similar walls.[32] These walls are approximately 50% glucose polymers, with lesser amounts of mannose and galactose. Glucosamine residues constitute approximately 10% of cell wall weight. Protein is present, and 3 to 10% of the wall weight remains after ashing. Analyses of cell walls of certain Fungi Imperfecti, which probably are, or were derived from, Ascomycetes, reveal additional differences in composition. For instance, the cell wall of *Pithomyces chartarum* (Berk. & Curt.) M. B. Ellis, the fungus causing facial eczema of sheep in New Zealand[33] and found in soils,[34,35] consists of 20% protein, 40% bound hexoses, 10% bound glucosamine, 10% lipid, and 5% ash.[33] The following constituents were identified: glucose, galactose, mannose, glucosamine, aspartic acid, threonine, serine, glutamic acid, proline, glycine, alanine, valine, methionine, isoleucine, leucine, tyrosine, phenylalanine, lysine, histidine, arginine, and tryptophan. The isolated cell wall of *Aspergillus niger* consists chiefly of neutral carbohydrate (73 to 83%) and hexosamine (9 to 13%), with smaller amounts of lipid (2 to 7%) and protein (0.15 to 2.5%).[36] Phosphorus constituted less than 0.1% of wall weight. The acetyl content was 3 to 3.4%, which corresponded to 1 mole per mole hexosamine. Sixteen amino acids and six sugars were identified. The sugars, all in D-configuration except possibly a trace of L-galactose, were glucose, galactose, mannose, arabinose, glucosamine, and galactosamine. At least part of the glycan existed as *alpha*-1,3-linkages. Purified cell-wall preparations of the imperfect fungus *Penicillium chrysogenum* consisted of at least two layers and contained mannose, galactose, glucosamine, glucose, xylose, and rhamnose in molar ratios of 1:3:4.5:9:0.5:0.5.[37] Approximately 2% of the wall was protein.

The septum of most Ascomycetes is a simple wall with one central pore (0.05 to 0.5 μ in diameter). Nuclei and cytoplasm move freely through the pore. *Fusarium solani* (Mart.) Sacc. (*Hypomyces solani* Rke. & Berth, but see Reference 332 for synonymy with *Nectria haematococca* Berk. & Br.) possesses septa with many pores, as do fungi in lichen associations. Woronin bodies (electron-dense spherical organelles) may be found near the pores of many Ascomycetes. Their function is not clearly understood, but they are thought to act as pore plugs (Figure 8). Bracker[4] has concluded that Woronin bodies are unique to Ascomycetes. Carrol[38] described two other types of pore structures in the fruiting structure of *Ascodesmis*, one involving striate formations in the pores of paraphyses and the other a dome-shaped structure in the basal pore of the asci. Furtado[39] described an elaborate pore in the ascogenous elements of *Sordaria*. There has been an attempt to correlate such ultrastructural features with taxonomic treatments of the fungi.[4]

Membranes

Apparently the cytoplasmic, nuclear, and endoplasmic reticulum membranes are tripartite. They show a dark-light-dark banding with the electron microscope. Little is known concerning the structure of the cytoplasmic membrane. Presumably it is composed of protein, phospholipid, and ergosterol or other steroids. It may exist as a bimolecular leaflet consisting of two layers of phospholipid sandwiched between two outer layers of protein,[40] as in the protein-lipid monolayer complex as proposed for mitochondria,[41,42] or as a bimolecular leaflet coated with protein and phospholipid.[42] In *Saccharomyces*

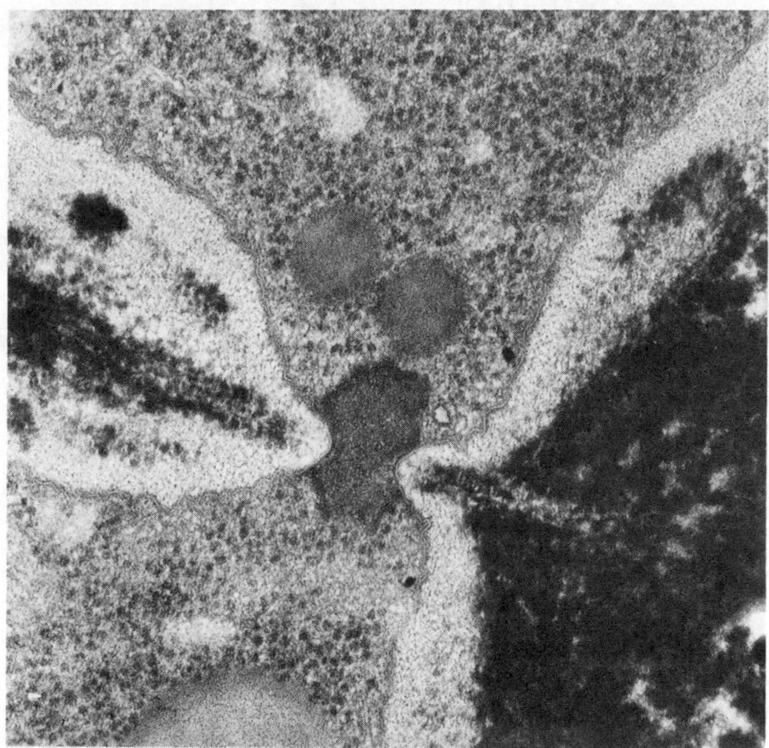

FIGURE 8. Woronin Bodies Near and Passing Through (or Plugging) a Septal Pore.

cerevisiae, which is an ascomycetous yeast, the cytoplasmic membrane is approximately 49% protein and 45% lipid;[43] 84% of the lipids are lecithin and cephalins. The major fatty acids of the *S. cerevisiae* cell envelope are: C16:1, C18:1, and C16, with C18:1 predominating in the membrane[44] and C16:1 predominating in the cell wall. The main phospholipids are phosphatidylcholine, phosphatidylethanolamine, phosphatidyl inositol, and phosphatidyl serine. The envelope, which consists of the wall and cytoplasmic membrane, also contains lipase and phospholipase, with most of both compounds in the membrane.[45] The cytoplasmic membrane of hyphae is thought to be derived from secretory vesicles cut off from the Golgi-like body or endoplasmic reticulum.[5,7] These vesicles (Figure 3) are thought to migrate toward the growing apex, where the vesicle wall becomes part of the growing cytoplasmic membrane by intercalation and the vesicle contents become part of the growing cell wall.[5,7] A visual difference was reported in the thickness of the ER or Golgi-like membrane and the cytoplasmic membrane, with intermediate differences occurring as the membrane units became secretory vesicles and migrated away from their origin toward the growing hyphal tip.[4]

CELL CONSTITUENTS AND PRODUCTS

Gross Composition

The chemical composition of the hyphae varies with species, distance from the hyphal tip,[46] age, aeration, redox potential, and composition of the medium. Mycelium of the Ascomycete *Claviceps purpurea* consists (in percent of dry weight) of approximately 31% protein, 10% polyols, 8% trehalose and polysaccharides, 11% lipid, 18% cell wall, 4% RNA and DNA, 3% polyphosphate and orthophosphate, 9% amino acids, and 5% ash (not necessarily additive).[47-49] For convenience, cell constituents and metabolism can be divided into the categories listed in Table 2.

TABLE 2

CELL COMPOSITION AND PRODUCTS

Types of Cell Constituents and Metabolism	Examples
Molecules of structural cell and essential enzymes	Cell wall, membranes, protein, DNA, RNA, dehydrogenases
Intermediates and co-factors	Pyruvate, B-vitamins, amino acids, NAD, sugar phosphates
Overflow products (conventional metabolites produced in excess by "sick" strains)	Citrate, amino acids, riboflavin
Shunt metabolism products (not integral part of cell and not intermediates)	
Primary shunt products (present in most or all fungi, stable synthesis)	Polyols, neutral lipids, polyphosphates, trehalose, probably non-structural polysaccharides
Secondary shunt products (produced only by certain strains, rare, unstable synthesis)*	Antibiotics, alkaloids, toxins (mostly cyclic)
Degradation products and unique H-acceptor	Amines, ethylene and ethanol, lactic acid
Epicellular and extracellular enzymes	Cellulase, amylase, lipase

*Synthesis may be induced by fungal viruses.[28] They may be regulatory products synthesized in excess,[50,51] may keep cell metabolism active,[52] or may result from metabolic lesion.[53]

Small-Molecular-Weight Constituents

Lipids. Fungi contain sterols, phospholipids, free fatty acids, esterified sterols, diglycerides, and neutral triglycerides. The latter exist in globules more or less free in the cytoplasm and may constitute 10 to 50% of dry weight. Composition of the lipid fraction of various Ascomycetes, and of others for comparison, is given in Table 3. Apparently the branched fatty acids found in some Gram-positive bacteria are not present in filamentous fungi. Unusual fatty acids are synthesized, however, as evidenced by the presence of ricinoleic acid in *Claviceps purpurea*, the fungus synthesizing ergot alkaloids used in reducing post-partum hemorrhaging, treating migraine headaches, and synthesizing LSD (lysergic acid diethylamide). Both the sclerotium occurring on infected rye plants and mycelium cultured *in vitro* contain 20 to 30% lipids,[47,54] and ricinoleic acid (12-hydroxy-9-octadecenoid acid) constitutes 20 to 30% of the lipids.[54] Not all hydroxy groups are free, however, as the acid is esterified to glycerol and acetylated to normal fatty acids through the hydroxy group.[54] In addition to the fatty acids listed in Table 3 and the phospholipids listed previously, Ascomycetes or other fungi contain in the membrane the following sterols: ergosterol, ergosterol peroxide, and cerevisterol.[57] For further data see References 58, 59, 60, and 61.

Carbohydrates and Polyols. Probably all filamentous fungi synthesize the non-reducing disaccharide trehalose (*alpha*-D-glucosido-*alpha*-D-glucoside), which may be a food reserve or endogenous nutrient or may function to reduce the active groups of hexoses in the cell. Phosphorylated trehalose has been isolated from *Saccharomyces cerevisiae*[62] and from bacteria.[63] Spores of the imperfect fungus *Myrothecium*

TABLE 3

COMPARISON OF FATTY ACIDS OF ASCOMYCETES WITH THOSE OF OTHER FUNGI

Species	Fatty Acids, molar percent												
	C12:0	C12:1	C14:0	C14:1	C16:0	C16:1	C16:2	C18:0	C18:1	C18:2	*alpha* C18:3	*gamma* C18:3	C20:1
Ascomycetes[55]													
Botryosphaeria ribis	–	–	1.3	–	31.7	–	–	20.6	15.3	–	19.8	–	11.3
*Botrytis cinerea**	3.4	–	1.6	1.8	19.0	1.3	–	3.8	11.0	16.4	41.7	–	–
*Cephalosporium subverticillatum**	0.9	–	1.2	0.5	21.6	2.7	–	6.8	28.3	35.0	3.1	–	–
Chaetomium globosum	–	–	0.4	–	19.2	1.7	–	8.3	17.0	46.4	7.0	–	–
Claviceps purpurea[54]†													
sclerotia, USA			0.7		28.0	3.7		6.4	19.6	17.4			
sclerotia, UK			0.1		19.9	6.5		4.3	22.5	14.3			
mycelia, USA			0.2		19.5	6.6		3.3	38.0	32.4			
mycelia, UK			1.0		23.4	6.0		3.6	30.1	14.5			
*Cylindrocarpon radicola**	3.5	–	1.9	1.2	22.4	1.7	–	7.9	24.6	25.5	9.2	–	2.1
Glomerella cingulata[56]			1.1		43.7	2.2		5.8	26.4	19.8	1.2		
Mycosphaerella musicola	1.9	–	1.3	0.8	16.0	2.3	–	5.6	16.8	48.7	5.4	–	1.2

TABLE 3 (Continued)

COMPARISON OF FATTY ACIDS OF ASCOMYCETES WITH THOSE OF OTHER FUNGI

Species	Fatty Acids, molar percent												
	C12:0	C12:1	C14:0	C14:1	C16:0	C16:1	C16:2	C18:0	C18:1	C18:2	*alpha* C18:3	*gamma* C18:3	C20:1
Ascomycetes (continued)													
Nectria ochroleuca	–	–	2.3	–	18.8	5.0	–	12.4	9.8	43.1	8.6	–	–
Neurospora crassa	2.8	–	2.3	1.5	18.2	2.5	–	7.9	11.1	42.3	7.7	–	3.7
Pyronema domesticum	0.4	–	0.4	–	13.6	–	–	21.2	29.3	35.1	–	–	–
Taphrina deformans	2.1	–	2.9	1.9	21.2	6.7	–	4.6	51.5	7.0	2.1	–	–
Basidiomycetes[55]													
Corticium solani	5.5		2.2	1.3	13.4	2.2	–	7.3	22.3	28.4	14.9		2.5
Rhizoctonia lamellifera	3.6		2.6	1.6	21.4	2.6	–	9.6	12.3	29.9	13.5		–
Stilbum lacalloxanthum	0.6		0.5	0.5	21.8	4.3	1.9	3.7	21.5	41.4	3.8		–
Fungi Imperfecti[55]													
Aspergillus flavus	0.5		0.2	0.2	12.6	–		17.6	37.0	31.9	–		–
niger	1.8		0.8	0.7	14.2	2.0		8.2	27.8	35.6	8.9		–

Penicillium													
chrysogenum	3.9		3.1	–	12.8	–	–	11.9	18.8	43.1	6.4	–	–
notatum	0.6		0.8	0.7	19.6	3.0	–	5.7	13.8	53.5	2.3		–
Phycomycetes[55]													
Mucor													
javanicus	–	–	3.2	0.4	20.3	4.7		8.3	26.8	12.6		13.7	
Pythium													
debaryanum‡	–	–	10.8	2.4	21.3	17.9		4.5	21.3	15.8		18.3	
Rhizopus													
arrhizus	1.0	–	0.7	1.3	19.9	3.5		7.4	40.6	15.8		9.8	
stolonifer	4.9	–	3.6	1.9	20.8	3.9		5.2	34.3	7.8		15.6	
Saprolegnia													
litoralis	–	–	14.5	4.3	16.1	3.9		14.2	30.2	14.1		2.8	

* Imperfect forms of Ascomycetes.

† Molar percent of ricinoleic acid containing glycerides was as follows: sclerotia, USA, 24.1; mycelium, USA, none; sclerotia, UK, 32.3; mycelium, UK, 21.5.

‡ Contained 10.4% of C20 to C22 unsaturated fatty acids.

verrucaria (Albert & Schwein) Ditmar are 20% trehalose (dry weight); trehalase is present, but it may not be involved in utilization,[64] although it is claimed to be for *Aspergillus oryzae* (Ahlb) Cohn.[65] Trehalose content of *Neurospora*,[66] *Claviceps*,[47,48] and *Sclerotinia*[67] has been reported. Mycelium of *Neurospora crassa* utilizes lipids as an endogenous carbon source, whereas ascospores utilize trehalose; activation of trehalase is involved in the germination of ascospores.[66] Probably all fungi synthesize one or more non-cyclic polyols, and *meso*-erythritol, D-threitol, D-arabinitol, L-arabinitol, xylitol, sorbitol (D-glucitol), D-mannitol, dulcitol (D-galacticol), and glycerol have been isolated.[68,69] *Claviceps purpurea* synthesizes mannitol and a pentitol, and the imperfect fungus *Penicillium chrysogenum* Thom produces mannitol, erythritol, and glycerol.[70] Polyols may be present in combined forms, as in riboflavin, esters of fatty acids and phosphate, and as constituents of polysaccharides.[68,69] Mannitol is derived from fructose, and its formation involves NAD or NADP and free sugar or phosphorylated sugar, depending upon the species. Presumably, hexitols enter glycolysis, and pentitols and tetritols enter the pentose phosphate cycle.[68,69] Kinases and phosphatases may be involved. Polyols may be endogenous nutrients, hydrogen sinks in the re-oxidation of $NADH_2$ or $NADPH_2$, or chemically unreactive derivatives of sugars. Glycogen is produced by some fungi,[71] and *Claviceps purpurea* synthesizes a glucan of *beta*-D-glucopyranosyl units containing (1,3)-linkages with some (1,6)-branches.[72]

Polyphosphates. Probably all fungi synthesize linear inorganic polyphosphates, which are formed by anhydrous linkage between orthophosphate, and contain up to 165 phosphate residues.[73] Their formula is $(M_{n+2}P_nO_{3n+1})$, where M is a monovalent ion.[74] They appear to function as phosphate pools, exchanging with exogenous orthophosphate and endogenous ADP[75] and RNA.[76] They also have been implicated in sugar transport[77] and possibly could complex with metal ions. Polyphosphates have been isolated from *Neurospora*,[76] *Saccharomyces cerevisiae* Hansen,[73] *Claviceps purpurea* (Fr.) Tul.,[47] and the imperfect fungus *Aspergillus niger* van Tiegh.[75,78]

Secondary Metabolites. Many ascomycetes synthesize secondary metabolites, which are small-molecular-weight compounds that are not integral parts of the cell and are not intermediates. Examples of the hundreds known are given in Table 4. Maximal yields are generally obtained either during restricted growth or after growth (as increase in total biomass) is over.[79,310,311] High aeration is required. The compilations of Benedict and Brady[80] reveal that 16% of 776 known fungal secondary metabolites are synthesized by Ascomycetes. If *Aspergillus spp.* and *Penicillium spp.* are considered Ascomycetes (as they must be), 54% of known fungal secondary metabolites are synthesized by Ascomycetes. Synthesis is at least in part under chromosomal control.[81] Synthetic capacity is very unstable; this is due to mutation and, perhaps, to some cytoplasmic state that is unstable (see discussion of heterokaryosis below). Secondary metabolism may be influenced or brought about by the double-stranded RNA fungal virus recently characterized.[28] For further details of secondary metabolism consult References 53, 79, 81, 82, 83, 84, 85, 86, 87, 88, 89, 90, 91, 92, 301, and 313. Product formation can be manipulated by inducing the heterokaryotic state[87,137] (see pages 316 to 317, where sexuality is discussed).

Toxins. Certain chemicals produced by some of the ascomycetes have been shown to be toxic to plants and/or animals (Table 5); however, plant toxins produced by fungi are not, in general, toxic to animals.[333]

Besides those listed in the table, many Deuteromycetes with ascomycetous affinities produce toxic compounds. Examples are *Fusarium oxysporum* f. sp. *lycopersici* (lycomarasmin),[93] *Penicillium* spp. (tremortin A and B),[94] *Penicillium rubrum* (rubratoxins),[95] *Aspergillus flavus* (aflatoxins, see Table 4), *Aspergillus ochraceus* (ochratoxin A, B, and C),[96,97] *Pithomyces chartarum* (sporidesmin, see Table 4),[98] *Stachybotrys atra* (stachybotryotoxin),[99] *Alternaria* spp. (alternaric acid), *Aspergillus clavatus* (ascladiol),[100] *Penicillium roqueforti* (toxin 1, 2, and 3),[101] *Fusarium tricinctum* [8-(3-methylbutyloxy)-diacetoxyscirpenol],[102] *Piricularia oryzae* (picolinic acid, $C_6H_5NO_2$, and piricularin),[103] and *Periconia circinata* toxin, a polypeptide.[339,340]

GROWTH

Growth of filamentous Ascomycetes can be expressed as increase in length of hyphae when growth occurs on solid media, and as increase in total biomass (dry weight, 100°C) when growth occurs in liquid media. Increase in weight due to neutral triglycerides, polyols and oligosaccharides probably should not be

TABLE 4.
SECONDARY METABOLITES

Related to	Example	Fungus
Tryptophan	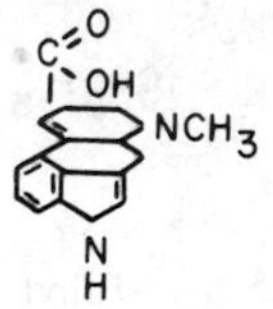 Lysergic acid	*Claviceps purpurea*
Peptides	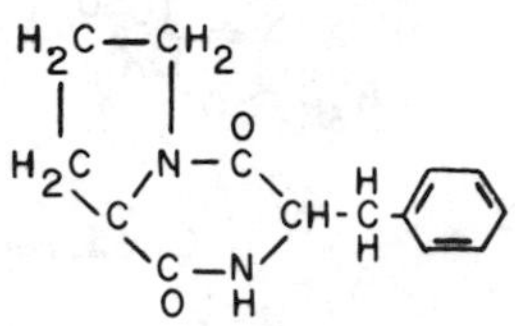	*Rosellinia necatrix*
	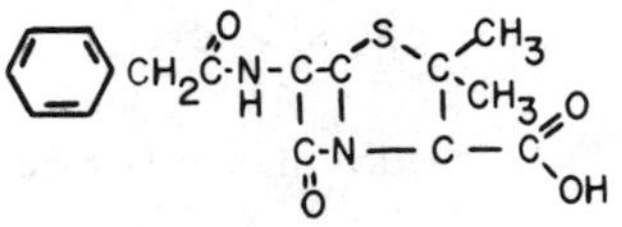 Penicillin G	*Penicillium chrysogenum*
	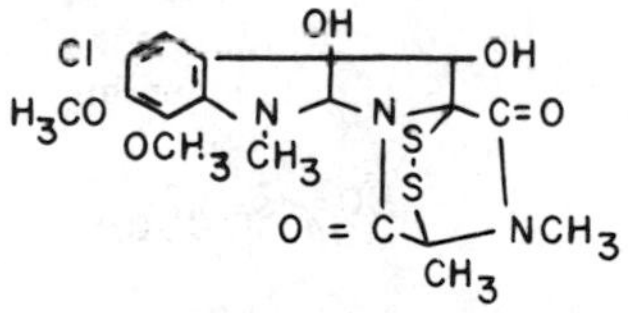 Sporidesmin	*Pithomyces chartarum*
Heterocyclic terpenes	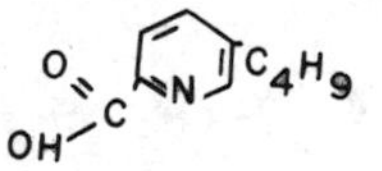 Fusaric acid	*Nectria cinnabarina*
	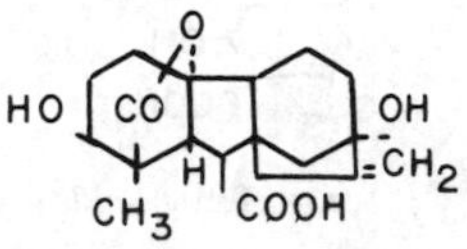 Gibberellin A_1	*Gibberella fujikuroi*
Monosaccharides	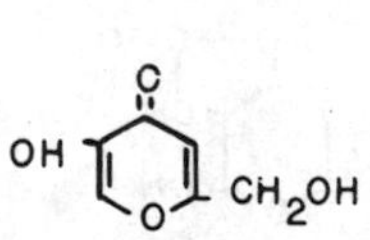 Kojic acid	*Aspergillus oryzae*

TABLE 4 (Continued). SECONDARY METABOLITES

Related to	Example	Fungus
Fatty acids	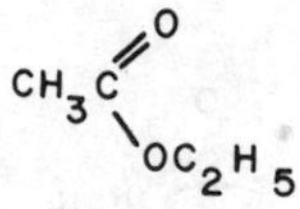 $CH_3C(=O)OC_2H_5$ Ethyl acetate	*Ceratocystis moniliformis*
Tetronic acid	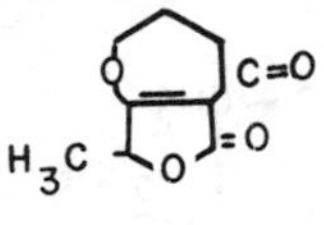 Carolic acid	*Penicillium charlesii*
Lactones	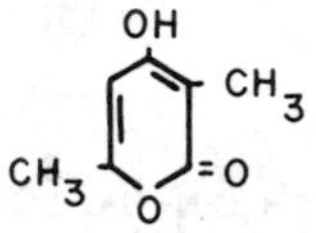 Methyl triacetic lactone	*Penicillium stipitatum*
Polyenes	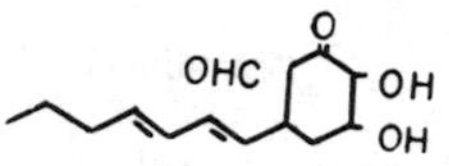 Frequentin	*Penicillium brefeldianum*
Tropolones	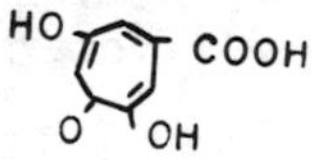 Stipitatic acid	*Penicillium stipitatum*
Benzene	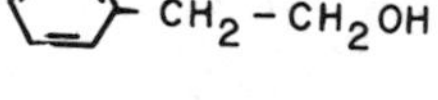 C_6H_5–CH_2-CH_2OH Phenethyl alcohol	*Gibberella fujikuroi*
Resorcinol	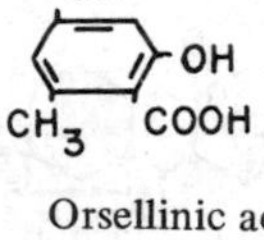 Orsellinic acid	*Chaetomium cochlioides*
Phenol	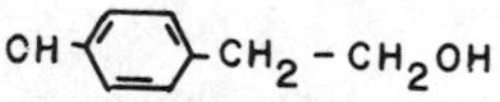 OH–C_6H_4–CH_2-CH_2OH Tyrosol	*Gibberella fujikuroi*
Hydroquinone	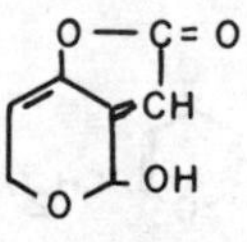 Patulin	*Byssochlamys nivea*

TABLE 4 (Continued). SECONDARY METABOLITES

Related to	Example	Fungus
Benzofuran	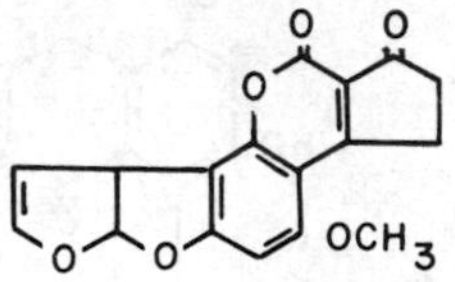 Aflatoxin B_1	*Aspergillus flavus*
Quinones	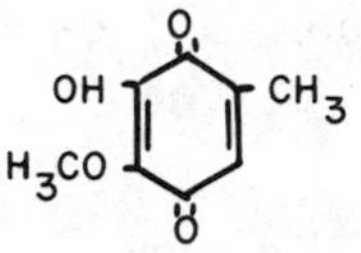 Fumigatin	*Aspergillus fumigatus*
Bibenzoquinones	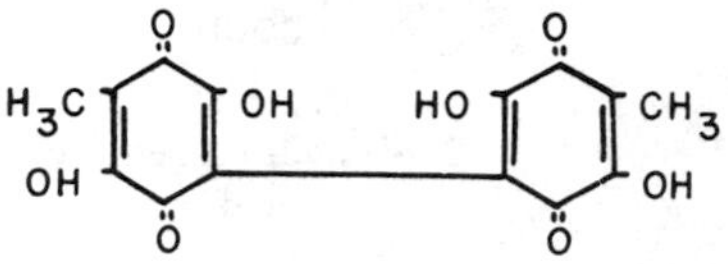 Oosporein	*Chaetomium aureum*
Naphthoquinones	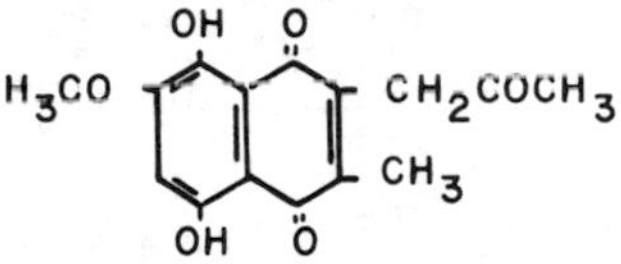 Javanicin	*Nectria haematococca*
Binaphthoquinone	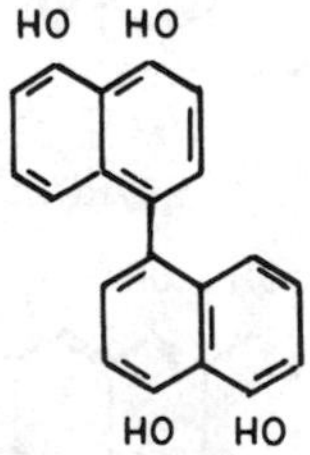	*Daldinia concentrica*
Monoanthraquinones	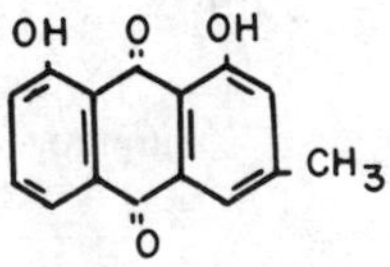 Chrysophanol	*Chaetomium elatum*

TABLE 4 (Continued). SECONDARY METABOLITES

Related to	Example	Fungus
Bianthraquinones	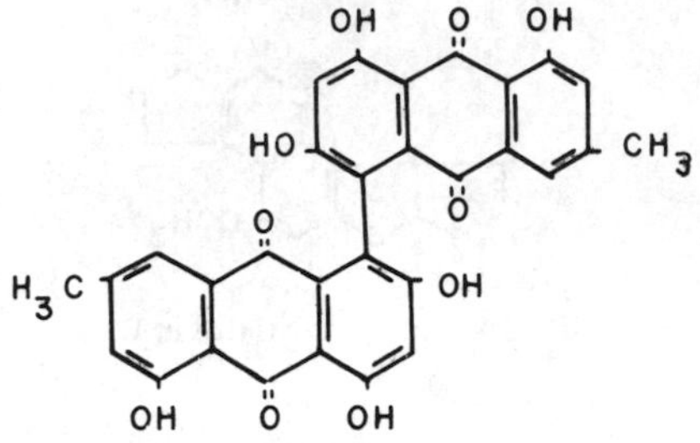 Skyrin	*Endothia parasitica*
Pyrocatechol	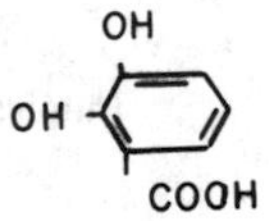 2,3-Dihydroxybenzoic acid	*Claviceps paspali*
Hydroxyhydroquinone	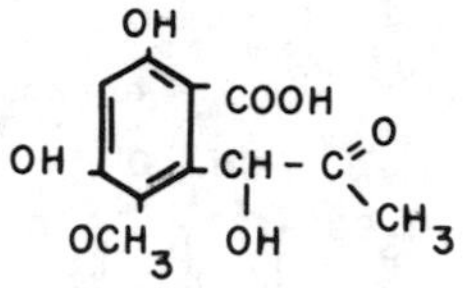 Ustic acid	*Paecilomyces victoriae*
Pyrogallol	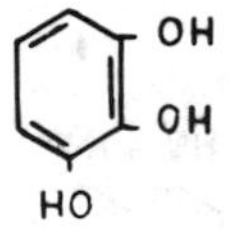 Pyrogallol	*Penicillium urticae*
Methylenequinones	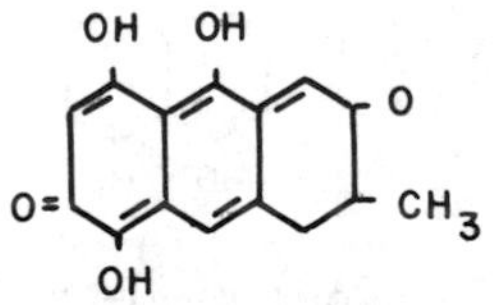 Purpurogenone	*Penicillium purpurogenum*
Benzophenone	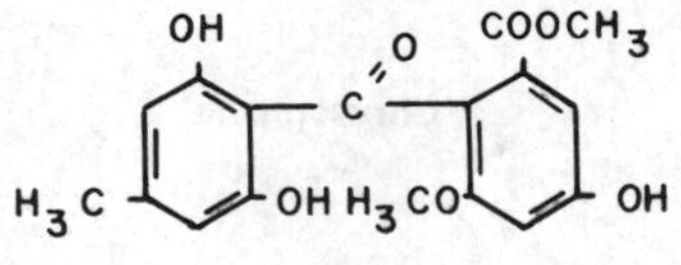 Sulochrin	*Penicillium frequentans*

TABLE 4 (Continued). SECONDARY METABOLITES

Related to	Example	Fungus
Spirans	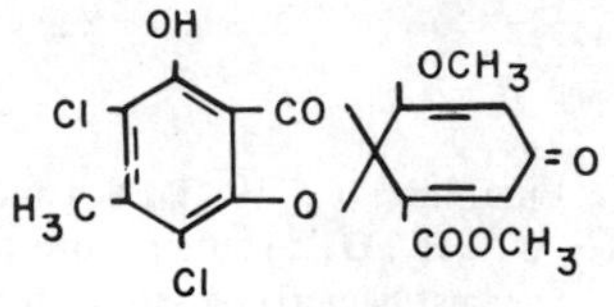 Geodin	*Aspergillus terreus*
Diphenyl ether	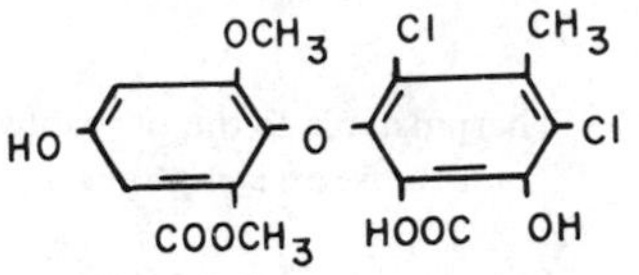 Geodin hydrate	*Aspergillus terreus*
γ-Pyrone	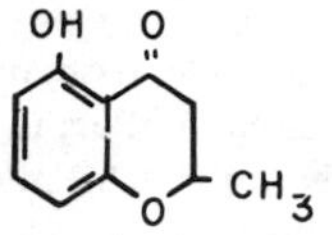	*Daldinia concentrica*
Azaphilones	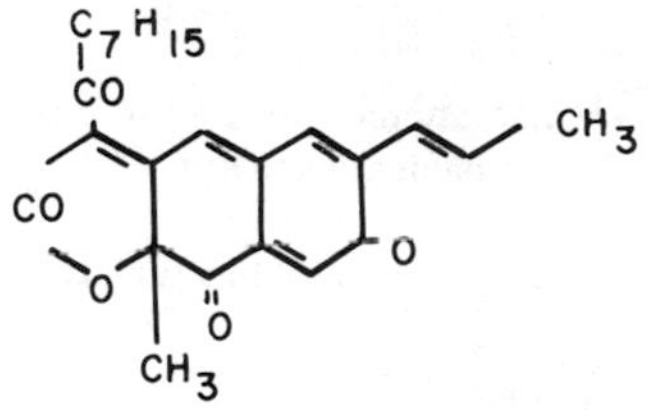 Monascorubrin	*Monascus purpureus*

TABLE 5
EXAMPLES OF TOXINS PRODUCED BY ASCOMYCETES

Ascomycete	Toxic Compounds	References
Calonectria		
graminicola [*Fusarium nivale* (Fr.) Ces.]	Fusarenon ($C_{18}H_{25}O_8N_3$) and nivalenol ($C_{15}H_{20}O_7$); neither shows antimicrobial activity against bacteria, yeasts, or fungi; fusarenon blocks cell division of *Tetrahymena*	118
Ceratocystis		
fagacearum (Bretz) Hunt	Thermostable toxin, probably a polysaccharide; causes disease symptoms in oak treated with toxin	117
fimbriata Ellis & Halst.	Ipomeamarone; sesquiterpenoid; toxic to animal liver	302
	4-Ipomeanol; fatal lung edema of cattle	302
ulmi (Buis.) C. Moreau	Thermostable toxin, probably a polysaccharide; causes elm leaflets to curl and dry at the margins, similar to Dutch elm disease symptoms	116
Chaetomium		
cochlioides Pall	Chaetomin (tentative formula $C_{16}H_{18}N_2O_5S_2$); antibacterial activity	98,124
globosum Kunze *ex* Fr.	Corn invaded by this fungus was toxic to rats	123
Claviceps		
paspali Stevens & Hall	Ergot alkaloids; poisonous to cattle	87
purpurea Fr. (Tul.)	Ergot alkaloids (lysergic acid derivatives); contracts smooth muscles, producing gangrene	87
Cochliobolus		
miyabeanus (Ito & Kurib.) Drechs. (*Helminthosporium oryzae* Breda de Haan)	Ophiobolin[108] = cochliobolin;[109] vivotoxin; treatment of rice leaves with 10 μg/ml solution causes necrotic spots resembling those of disease	108,109
	C_{25}-terpenoid; abnormal growth at low concentrations	103,110
sativus (Ito & Kurib.) Drechs. (*Helminthosporium sativum* Pam., King & Bakke = *Bipolaris sorokineana*)	Sesquiterpenoid dialdehyde (helminthosporal); relatively non-specific	107
victoriae Nelson (*Helminthosporium victoriae* Meehan and Murphy)	Victorin, which on treatment with alkali yields two toxins: victoxinine (tricyclic secondary amine $C_{17}H_{29}NO$) and a pentapeptide (aspartic acid, leucine, valine, glycine, and glutamic acid); the toxin inhibits susceptible roots at 0.0002 μg/ml concentration and is highly selective	103

TABLE 5 (Continued)
EXAMPLES OF TOXINS PRODUCED BY ASCOMYCETES

Ascomycete	Toxic Compounds	References
Endothia		
parasitica (Murr.) Anders	Diaporthin ($C_{13}H_{14}C_5$) and skyrin ($C_{30}H_{18}O_{10}$), an anthraquinone (see Table 4); they change water permeability of cells of certain plants; diaporthin induces symptoms of chestnut blight, but neither toxin has been isolated from diseased tissue; diaporthin is active at $10^{-11}M$, skyrin at $10^{-6}M$	104,105,106
Erysiphe		
graminis D. C.	Crude spore extracts produce "green islands" on detached unwounded leaf surfaces	125
Gibberella		
baccata (*Fusarium lateritium*, var. *fructigenum*)	Baccatin A, a mixture of enniatin A and B ($C_{24}N_{42}O_6N_2$ and $C_{22}H_{38}O_6N_2$)	104
fujikuroi Wr. (*Fusarium moniliforme* Sheld.)	Fusaric acid (5-*n*-butylpyridine-2-carboxylic acid, see Table 4); respiratory inhibition; role in pathogenicity (wilt) is uncertain; detected in many species	112
zeae (Schw.) Petch (*Fusarium graminearum* Schwabe)	Zearalenone, one of the enantiomorphs of 6-(10-hydroxy-6-oxo-*trans*-1-undecenyl)-β-resorcylic acid lactone; estrogenic in rats; vomiting, diarrhea in pigs; emetic	113
Hypomyces		
solani Rke & Berth. [*Nectria haematococca* Berk. & Br., *Fusarium solani* (Mart.) Sacc.]	Four toxic pigments (naphthazarin derivatives), including novarubin, were isolated from cultures and from diseased pea; they affect water permeability of cell membranes; T-2 toxin; solaniol (3α,8α-dihydroxy-4 β, 15-diacetoxy-12,13-epoxy-Δ9-trichothecene; diacetooxyscirpenol; fusarenone-X	111,331
Leptosphaeria		
avenaria G. F. Weber	Brown pigment; causes leaf spot of oats when sprayed on plants	119
Leptosphaerulina		
briosiana (Poll.) Graham & Luttrell	Polypeptide-like, with acid and amino groups; soluble in water, insoluble in organic solvents; produces lesions on alfalfa, barley, strawberry, and bean	120
Nectria		
cinnabarina (Tode *ex* Fr.) Fr.	Two toxins produced in Richards' solution: one is acidic and produces necrosis, the other is alkaline and produces wilting	115
radicicola Gerlach & Nilsson	Radicicol; similar in structure to zearalenone	114

TABLE 5 (Continued)
EXAMPLES OF TOXINS PRODUCED BY ASCOMYCETES

Ascomycete	Toxic Compounds	References
Rosellinia		
necatrix Prill	*p*-Hydroxyphenylacetic acid; concentrations of 1:2,000 to 1:20,000 produce rot of soybean	121
	Dioxopiperazines; they affect elongation of rice seedlings at concentrations of 1:2,500 to 1:100,000	121
Sclerotinia		
sclerotiorum (Lib.) de Bary	When growing on celery, produces two furocoumarins that have a strong synergistic action with ultraviolet light (340-380 nm) and cause skin lesions following exposure to sunlight; the toxins inhibit mitosis and influence copper ions in liver	122
Trichoderma		
viride Pers. ex Fr. (asexual stage of several Hypocreales)	Trichotoxin A; a cyclic polypeptide ($C_{58}H_{95}O_{17}N_{15}$); intraperitoneal injection in mice (LD_{50} = 4.36 mg/kg) causes lethargy and death in 1 hour to 3 days; some antimicrobial activity	334

considered growth, since these are probably not essential and their amount varies with age and composition of the medium. Growth takes place at the hyphal tip (probably in the first 0.1 mm[16,46] by increase in cell content and deposition of the membrane and cell-wall material at the tip (see discussion of fine structure). Regions more remote from the tip are more heavily vacuolated. Reference 16 gives details on hyphal-tip extension. Growth is normally exponential and may be expressed as growth constants, i.e., $k = \frac{\log_{10} X_1 - \log_{10} X_0}{0.434 \times t_1 - t_0}$ and (three-dimensional growth) $k = \frac{\sqrt[3]{X_1} - \sqrt[3]{X_0}}{t_1 - t_0}$, where X_1 is weight at time t_1 and X_0 is inoculum weight or earlier weight at time t_0.[126,127,128,129] Growth can be linear if some environmental or nutritional condition restricts synthesis or uptake. Growth can also be expressed as the time required to double the weight, or as increase in hyphal length.[16] Growth in liquid culture can be divided into the conventional lag phase, phase of increase, exponential (log) phase, phase of decrease, and stationary phase. Bu'Lock[92] divides it into trophophase (constant cell composition) and idiophase (post-trophophase metabolism). References 126 and 127 offer relevant data concerning chemical composition of the phases.

Steady state growth can be maintained by culturing in a chemostat or turbidistat, employing drip addition of rate-limiting concentration of nutrients or nutrient addition to constant turbidity by use of a photocell.[90,130] Growth on solid media assumes a disk shape on the surface and is called a colony. Some hyphae will penetrate the substrate, and some will extend into the air, where sporophores may develop on those hyphae receiving nutrients from submerged hyphae. Increases in colony diameter may be linear; the circular form occurs because of branching and subbranching of hyphae behind every tip. Colony growth finally ceases due to exhaustion of limiting nutrient, accumulation of toxic wastes, formation of a repressor, or translocation of essential growth hormone[131] away from the growing tips. Assimilative or vegetative hyphae of certain Ascomycetes (e.g., *Claviceps* and *Sclerotinia*) can develop into a firm, dry body of definite shape known as a sclerotium.[132] The sclerotium can function as a resting or overwintering body capable of germination. It is not necessarily long-lived, however, since *Claviceps* sclerotia usually do not survive beyond a year. The sclerotium can be differentiated into tissue-like areas; *Claviceps purpurea* sclerotia have a purple outer rind and a white internal medulla.

Germination of sclerotia of *Claviceps* and *Sclerotinia* leads to development of stromatic tissue, which is the site of nuclear fusion and ascospore formation. Stromata are tissues that resemble sclerotia but house either perithecia or asci in locules. They may be flat and resemble a layer of tar on wood, or they may be globose and even resemble perithecia.

Individual terminal or intercalary hyphal "cells" or compartments can balloon out and develop thick walls. Such cells are termed chlamydospores. They usually do not readily separate from the mycelium. Terminal or lateral chlamydospores of regular shape are usually considered to be spores and are called aleuriospores.[133] Chlamydospores formed within the substrate no doubt result from development under unfavorable nutritional or environmental conditions. See Reference 329 for discussion of the chlamydospore.

HETEROKARYOTIC STATE

Hyphae within a mass of mycelium, hyphae from different individuals of the same species, and perhaps occasionally those of unrelated species, can fuse (anastamose) by the growing together of hyphal tips followed by migration of nuclei. If the introduced nuclei are different and survive and multiply, the fungus is said to be an heterokaryon. *Neurospora* forms stable heterokaryons only when fusing hyphae are the *same* mating type (i.e., $A + A$, or $a + a$).[134] A suppressor mutation has been produced,[136] however, that suppresses inhibition of heterokaryon formation by unlike mating types. (Recall that sexuality requires interaction between *unlike* mating types.) Genes governing fusion *per se* exist; however, these genes have no influence on mating compatibility.[135,136] Heterokaryotic mycelium may possess genetic traits different from those of either "parent" and may synthesize amounts or types of metabolites different from either parent.[137] Internal heterokaryons may develop from mutations in a given nucleus and perpetuation of it and the parent nucleus. This phenomenon has been invoked to explain decrease in penicillin yields by strains of the imperfect fungus *Penicillium chrysogenum* Thom; constant reisolation from spores or hyphal tips is required to maintain high yields. Heredity differences may reside in cytoplasmic organellae, such as mitochondria, and in that case the fungus is said to be an heteroplasmon.

METABOLISM

Those Ascomycetes examined possess the glycolytic pathway, pentose phosphate cycle, Krebs cycle, and aerobic electron transport system, including coenzyme Q_{10},[138] which resembles that of mammals.[139] Some also possess a glucose-repressible, acetate-inducible glyoxylate bypass,[140] although others, for example, *Claviceps purpurea,* will not grow on acetate. Lysine no doubt is synthesized from *alpha*-aminoadipic acid rather than from diaminopimelic acid (the precursor in bacteria and certain primitive Phycomycetes,[141] and the aromatic amino acids are derived from the shikimic pathway. See References 16, 139, 142, and 143-145 for comprehensive discussions of fungal metabolism. Metabolic uniqueness in Ascomycetes is expressed by biogenesis of ethylene,[146] of large quantities of polyols, polyphosphate, and trehalose, of secondary metabolites, such as pigments, antibiotics, toxins, and alkaloids (see earlier discussions of metabolites and toxins), and of cell walls (see discussion of the life cycle).

METABOLIC REGULATION

Much of the present understanding of metabolic regulation has resulted from studies of the prokaryons *Escherichia coli* and *Salmonella typhimurium.* While it would not be safe to extrapolate from these to the eukaryotic Ascomycetes, it has been established that repression,[147-150] negative feedback or endproduct inhibition,[151-154,158] suppressor genes,[135] and induction[149,155-157] exist among Ascomycetes.

NUTRITION AND TRANSPORT

Utilization of nutrients in metabolism must be preceded by uptake or transport of the nutrients. The yeast *S. cerevisiae* takes up sugars by facilitated diffusion (a protein carrier transports sugar across the membrane into the cell, but the sugar is not concentrated), which does not involve phosphorylation during transport.[157] Amino acids are taken up by active transport (protein carrier and/or permease, plus expenditure of energy, which allows concentration of substrate).[159-162] The filamentous Ascomycetes *Neurospora crassa* and *Aspergillus nidulans* take up sugars by active transport, which, unlike that of some bacteria,[163] does not require phosphorylation.[164-167] *Neurospora* has two transport systems: one constitutive, and one suppressible by high sugar concentration.[165,166] *Neurospora* takes up amino acids by four different active transport mechanisms:[159] system I transports L-amino acids; system II transports basic, neutral, and acidic DL-amino acids; system III transports basic L-amino acids; and system IV transports L-methionine. *Claviceps purpurea,* when grown on glucose, takes up succinic acid when undissociated, and then only by diffusion.[168] For discussions on transport see References 169-176.

Some Ascomycetes will grow on a completely synthetic medium, containing, for example, the following (in grams per liter): glucose, 15; NH_4NO_3, 0.8; KH_2PO_4, 1.6; K_2HPO_4, 2.4; $MgSO_4 \cdot 7H_2O$, 0.3; $FeSO_4 \cdot 7H_2O$, 0.005; NaCl, 0.005; $ZnSO_4 \cdot 7H_2O$, 0.004; $MnSO_4 \cdot H_2O$, 0.0028; $CaCl_2 \cdot 2H_2O$, 0.006; $CuCl_2 \cdot 2H_2O$, 0.0002; $CoCl_2$, 0.00002; and $(NH_4)_6Mo_7O_{24} \cdot 4H_2O$, 0.002. Some Ascomycetes require biotin (*ca.* 5μg per liter), or thiamine (*ca.* 10 μg), or both. The optimal amount depends upon the amounts of carbon source, etc. Members of the Ophiostomataceae require pyridoxal and are stimulated by biotin and thiamine.[177,304] See References 16, 145, and 178 for discussions of nutrition.

Ascomycetes normally are aerobic (see, however, Reference 179) and at least half of the carbon source is respired to CO_2.[49,87] Many Ascomycetes can hydrolyze starch and lipids; fewer can hydrolyze cellulose (e.g., *Chaetomium* and certain imperfect fungi, such as *Trichoderma* and *Myrothecium,* are particularly high producers of cellulase). Ascomycetes probably are not the principal hydrolyzers of lignin and chitin. Some filamentous Ascomycetes can ferment sugars (i.e., use an intermediate as acetaldehyde to oxidize $NADH_2$). See Reference 16 for a discussion.

REPRODUCTIVE PHASE

Sexuality

Sex involves plasmogamy (cell fusion, or at least cell contact and nuclear transfer), karyogamy (nuclear fusion), which temporarily establishes the 2n state, and meiosis, which through two divisions reduces the diploid nucleus to four haploid nuclei. These nuclei – or, much more frequently, their mitotic product – become the nuclei of the sexual spore, the ascospore. Thus the haploid state of the mycelium-to-be is established in the ascus.

Haploidy refers to the state of having only one chromosome of a type per nucleus or of having only one gene per trait. By contrast, higher plants and animals are diploid (except for gametes) and have two genes per trait per nucleus (e.g., two genes for eye color, each on separate but homologous chromosomes). The section dealing with the life cycle and Figures 1 and 2 describe sexual structures and nuclear processes. The survival value of sexuality is that gamete nuclear fusion allows exchange of genes and whole chromosomes, with the result that new progeny may have combinations of genes, or traits, different from those of either parent, and the new combination may be advantageous in a changing environment. Following fusion, homologous chromosomes (bearing the same type of genes) line up side by side, and each, after doubling in content, splits into two chromatids (tetrad when paired, see Figure 2). Each pair of chromatids is held together by a centromere of the chromosome. In the simplest process, each chromatid pair separates from the tetrad (Figure 2, left side) and becomes the nuclear material for one nucleus. A second, mitotic, division separates the pair in each of these two nuclei, with the final formation of four haploid nuclei. If some chromosomes from one parent and some from the other move into the same nucleus, the progeny will have some traits of each parent. This process is called random assortment of whole chromosomes, because there is an equal chance that a chromosome of a homologous pair will be included in any one nucleus. A second opportunity for recombination occurs at the tetrad state, where sticking between non-sister chromatids can cause an exchange of segments of chromosomes as they part from one another (Figure 2, right side). This is termed crossing over. These two processes can produce nuclei for progeny that have traits different from those of either parent. In most instances mitosis produces eight nuclei from the above four, and these eight become the nuclei of the eight ascospores (Figure 1).

Since most Ascomycetes are haploid, every gene is, or can be on induction, expressed, and consequently new traits will be expressed. Of course, if both gamete nuclei originated from the same nuclear clone (homothallism), the same traits would exist whether or not the above process had occurred. Outbreeding is no doubt usually advantageous in nature, and approximately fifty species of Ascomycetes are known to have a genetic process that prevents self-breeding.[180] This process is regulated in almost all heterothallic Ascomycetes by the "incompatibility factor" and involves two alleles for mating types, *A* or *a*, at one locus. Only nuclei containing *unlike* factors (i.e., *A* and *a*) can fuse in the ascogonium. An exception to this rule is *Podospora anserina* (Rabenh.) Winter, which usually bears four binucleate ascospores, each spore containing both mating types; this is an example of secondary homothallism. Approximately 3% of the spores are small and uninucleate,[180] and perithecia form only between crosses of certain of these ascospores. Here apparently four unlinked loci are involved,[300] each with two alleles. Sexuality occurs between opposite mating types (+ and −) only when the other loci are the same (i.e., $+c_1v_1ab$ crossed with $-c_1v_1ab$). Esser[180] terms this "heterogenic incompatibility" (unlike alleles prevent sexuality).

Most Ascomycetes are homothallic; that is, mycelium derived from one uninucleate ascospore, or from a multinucleate ascospore with identical nuclei, can produce ascocarps and ascospores. This situation allows potentially undesirable inbreeding, which, however, can be overcome by formation of heterokaryotic mycelium by mutation and perpetuation of both parent and mutant nuclei and by anastomosis between hyphae of the same species but with different alleles. Such heterokaryotic hyphae can have traits of both nuclei and thus behave much as if recombination had produced new combinations of traits within one nucleus. True genetic recombination can be effected by parasexuality as well as by conventional sexuality. Parasexuality[182,183] is the rare fusion of two haploid nuclei in ordinary assimilative hyphae. While the result can be a stable diploid (often with large 2n asexual spores), mitotic crossing over followed by haploidization also occurs, thus producing haploid nuclei with new gene combinations. Haploidization can occur by non-disjunction and, in some cases, by formation of hyperhaploids and stepwise loss of chromosomes until the haploid state is attained.[183]

Inheritance

Ascomycetes are usually haploid, and therefore all nuclear genes are normally expressed; that is, "the geneotype is, or can be, the phenotype". They possess conventional chromosomes (although histoneless[2]); *Neurospora* has seven and *Aspergillus nidulans* has eight. Mendelian inheritance exists, but normally is not complicated by a dominance-recessive state. Dominance-like expression of a trait may exist, however, as a result of the heterokaryotic state.[81] Ascomycetes also have mitochondria with circular DNA and its genes, and possibly also other forms of cytoplasmic inheritance.

The recently found fungus viruses (see Reference 28) may be a source of inheritance if they can interact with host DNA and transfer sections of DNA to other hosts, as is the case with certain bacteriophages.

The female thallus contributes more "heredity" to the new progeny than the male; the latter contributes only nuclear DNA to the ascogonium, whereas the female contributes both nuclear and mitochondrial DNA, and perhaps other inheritable material as well. Thus the "poky" respiratory deficiency of *Neurospora* is passed on only when the mutation resides in the thallus bearing the ascogonium in a given cross.

Genetics can be used in commercial-product formation to increase yields or to synthesize new products. If the product of a heterothallic Ascomycete is under nuclear gene control, heredity can be manipulated by crossing various strains, with or without prior induction of mutation in one or both. Mitochondrial changes can be sought if the product is at least in part under mitochondrial gene control, i.e., making certain that the change is in the "female" thallus. If the culture is homothallic or does not readily form sexual spores, changes in expression of heredity can be effected (a) by parasexuality, producing either new recombinants or stable diploids, (b) by forming stable heterokaryons,[87,137] (c) by isolating desired homokaryons from hyphal tips or uninucleate asexual spores of heterokaryons, (d) by producing mutations in homokaryons, and, perhaps, (e) by inducing changes, using fungal viruses. Sermonti[81] discusses the genetics of industrial microbiology.

The Ascocarp

Ascogenous hyphae are contained within the ascocarp in all Ascomycetes except the Hemiascomycetes. In the Euascomycetes the body is in the form of a cleistothecium, perithecium, or apothecium (Figure 5) and is believed by some to be developed in response to sexual stimulation during formation of the ascogenous hyphae. In the Loculoascomycetes the stroma forms before the ascogonium (Figure 5) develops and, therefore, probably is not developed in response to a sexual stimulus (see References 16 and 185 for discussions of possible stimulation).

The ascocarp, with or without stroma, is composed of pseudoparenchymatous or prosenchymatous tissue (Figure 7). Ascospores are either ejected forcibly from the ascocarp if the asci remain intact, or issue from the ascocarp as a fluid mass if the asci dissolve while inside the ascocarp.

ECOLOGY

Terrestrial Ascomycetes

Terrestrial Ascomycetes, for the most part, live on dung (coprophilous), wood (lignicolous), grass (graminicolous), insects (insecticolous), rocks (saxicolous; for example, in lichen associations), leaves and stems (foliicolous and caulicolous), on the ground (terricolous), on bark (corticolous), in the soil, in and on roots (mycorrhizal, saprophytic, and parasitic), and in and on animals other than insects. Since there are 1,950 genera and 15,000 species of Ascomycetes, only a cursory consideration of them and their activities can be presented.

SOIL FUNGI

Soil is an heterogeneous material, consisting mainly of inorganic colloids and associated ions, salts, sand, pebbles, gas, and humus and other organic matter. Soil is the temporary site of dead animal bodies, plant debris, and living and dead plant roots. Microorganisms observed to be growing on these need not necessarily be soil microbes. For example, roots of living plants can harbor specific fungi, as can the soil particles in direct contact with the root surface;[229] many such fungi will not be found growing remote from roots. Spores of Ascomycetes growing on substrates above ground or on litter on the ground may find their way into the soil, but may not be capable of growing there. Consequently, it is somewhat difficult to ascertain whether a given soil isolate actually grows there unless it has developed from a hyphal fragment actually collected from the soil. Microscopic examination of soil, retrieval of hyphal fragments and plating the fragments on laboratory media is the only certain way of establishing that a given Ascomycete grows in soil. Probably the fungi that are most active in soil, or at least most abundant by weight, are slow-growing Ascomycetes and Basidiomycetes (mushrooms), which are not usually isolated by the plating procedures because the plates are overrun by fast-growing fungi that may or may not be capable of growing in soil.

True soil Ascomycetes, or course, would not necessarily be continuously growing in soil. Unless a soil is rich in all nutrients and is continuously moist, aerated, and of the correct temperature, it will not support continuous growth. Thus fungi live a discontinuous existence, growing when a given limiting factor, such as water or carbon source, is replenished, and either dying or rapidly forming resistant spores or other bodies when some factor becomes limiting. While no wholly satisfactory definition of a soil fungus exists, a meaningful one would probably include the condition that it be capable of growing *in situ* in unamended soil. Soil amendment can be useful, however, in isolating apparent soil fungi capable of utilizing a specified nutrient. For example, one can add sterile cellulose or chicken feathers (keratin) to soil, retrieve them a few days or weeks later, then isolate "soil fungi" capable of using these. Athlete's foot fungi can be isolated from soil in this manner;[230] other serious human pathogens can be isolated by injecting saline suspensions of soil (containing antibacterial antibiotics) into mice and isolating the fungi from spleen, liver, or other organs.[231] No doubt most Ascomycetes grow in the upper few inches of soil. Viable Ascomycete spores can be isolated from much deeper soil, but these may have been merely washed down through cracks and fissures of the soil.

While it is difficult to generalize, a soil fungus would probably have to possess one or more of the following attributes: (a) ability to grow rapidly on substrates of low concentration, (b) ability to utilize substrates not readily used by other microbes, (c) tolerance of CO_2, (d) ability to grow in low O_2 tension, (e) resistance to antimicrobial products and hydrolytic enzymes elaborated by other microbes, (f) ability to rapidly synthesize spores and resistant bodies, and, perhaps, (g) ability to form antimicrobial metabolites or hydrolytic enzymes. Bloomfield and Alexander found that fungal walls containing certain pigments were resistant to degradation in the soil.[232] Soil fungistasis, which is the ability of soil to prevent fungal spore germination, has long been known to exist and appears to be due to rapid removal of limiting nutrients by rapid-growing bacteria.[252] Some Ascomycetes isolated from the soil environment are listed in Table 6.

TABLE 6

SOME ASCOMYCETES ISOLATED FROM SOIL

Ascomycete	Reference
Amauroascus	
niger Schroet.	233
Arachniotus	
candidus (Eidam) Schroet.	242
citrinum Massee & Salm.	233
terrestris Raillo	233
Ascobolus	
stercorarius (Bullard) Schroet.	233
Auxarthron	
thaxeri (Kuehn) Orr & Chu.	236
Bulgaria	
iniquans Fr.	233
Byssochlamys	
nivea Westl.	242
Caloscypha	
fulgens (Pers.) Boud.	238
Carpenteles	
javanicum (van Beyma) Shear	243
Chaetoceratostoma	
longirostre Farrow	249
Chaetomidium	
minutum Cain	241
Chaetomium	
bostrychodes Zopf	235
caprinum Bain.	235
cochliodes Palliser	235
contusum van Warm.	244
crispatum Fuckel	235
cristatum Ames	235
flavigenum van Warm.	244
flavum Omvik	235
Chaetomium (continued)	
funicolum Cooke	235
globosum Kunze	235
gracile Udagawa	235
homopilatum Omvik	235
humicolum van Warm.	244
indicum Corda	235
lentum van Warm.	244
longirostre (Farrow) Farrow	235
magnum Bain.	235
murorum Corda	235
nigricolor Ames	235
olivaceum Cooke & Ellis	235
riburogenum van Warm.	244
spirale Zopf	235
subterraneum Swift & Povah (*C. globosum* ?)	233, 234
trilaterale Chivers	235
Coniochaeta	
pulveracea (Ehr.) Munk	242
tetraspora Cain	241
Cordyceps	
militaris (Fr.) Link. (conidial stage)	247
Cyathipodia	
corium (Wegerb.) Boud.	238
Dialonectria	
brassicae (Ellis & Sacc.) Cooke	233
galligena (Bresaud.) Petch	233
Eidamella	
deflexa (Berk.) Benj.	242
Emericellopsis	
humicola (Cain) Gilman	242
minima Stolk	233
terricola van Beyma	233

TABLE 6 (Continued)

SOME ASCOMYCETES ISOLATED FROM SOIL

Ascomycete	Reference
Fimetaria	
fimicola (Roberge) Griff. & Seaver	233
macrospora (Auersw.) Griff. & Seaver	233
sylvatica (Dasz.) Griff. & Seaver	233
Gelasinospora	
cerealis Dowding	233
Geopyxis	
carbonaria (Fr.) Sacc.	238
Gymnoascus	
reesil Baran.	233
setosus Eidam	233
subumbrinus A. L. Smith & Rams	233
Gyromitra	
esculenta (Pers.) Fr.	238
Humarina	
convexula (Pers.) Seaver	233
Leptosphaeria	
artemisiae (Fuckel) Auersw.	242
Levispora	
terricola Rout.	251
Magnusia	
nitida Sacc.	233
Melanomma	
sylvanum Sacc. & Speg.	233
Melanospora	
chionea (Fr.) Wint.	245
Microascus	
cirrosus Curzi	240
intermedium Emm. & Dodge	245
trigonosporus Emm. & Dodge	242
Microsporum	
gypseum (Bodin) Gui. & Grig. (imperfect; see Table 12)	239
Muellerella	
nigra Rout.	239
Monosporium	
apiospermum Sacc. (imperfect; see Table 12)	239
Myxotrichum	
chartarum Kunze	233
conjugatum Kuehn	233
Nectria	
haematococca Berk. & Br.	248
Neocosmospora	
vasinfecta E. F. Smith	246
Neottiella	
hetieri Boud.	238
Neurospora	
africana Huang & Backus	237
crassa Shear & Dodge	237
dodgei Nelson & Novak	237
erythrea (Moll.) Shear & Dodge	237
galapagosensis Mahoney & Backus	237
intermedia Tai	237
lineolata Fred. & Ueck	237
phoenix (Kunze) Dennis	237
sitophila Shear & Dodge	237
terricola Goch. & Backus	237
tetrasperma Shear & Dodge	237
Perisporium	
funiculatum Preuss	233
vulgare Corda	233
Peziza	
bufonia Pers.	238
ostracoderma Korf	242
sylvestris (Boud.) Sacc. & Trott.	238
violacea Pers.	238
Pleospora	
calvescens (Fr.) Tul.	242
Pleurage	
setosa Wint.	233
verruculosis Jensen	233
Plicaria	
leiocarpa (Currey) Boud.	238
Pseudoarachniotus	
citrinus (Massee & Salm.) Kuehn & Goos	242
marginisporus Kuehn & Orr	236
reticulatus Kuehn & Goos	242
roseus Kuehn	242
Pseudogymnoascus	
roseus Raillo	234
vinaceus Raillo	233
Pseudonectria	
diparietospora Miller	250

TABLE 6 (Continued).

SOME ASCOMYCETES ISOLATED FROM SOIL

Ascomycete	Reference	Ascomycete	Reference
Pyronema		*Thielavia*	
omphalodes (Bull.) Fuckel	233	*terricola* (Gil. & Abb.) Emm.	233, 250
		variospora Cain	241
Sphaerostilbe			
repens Berk. & Br.	248	*Trichophyton*	
		mentagrophytes (Robin) Blanch (imperfect; see Table 12)	239
Sporormia			
fasciculata Jensen	233		
intermedia Auersw.	233	*Trichosphaeria*	
minima Auersw.	233	*pilosa* (Pers.) Fuckel	233
Strickeria		*Westerdykella*	
mutabilis (Quel.) Wint.	242	*ornata* Stolk	233
Talaromyces			
wortmanni (Kloeck.) Benj.	243		

Continuation of life depends on recycling of carbon and phosphorus, fixing, if not recycling, of nitrogen, and "recycling" space. Microbes effect most of these changes. Ascomycetes certainly play a significant role in all but nitrogen fixation, and they may even be active in this role. For a discussion of soil fungi see Reference 252.

LITTER FUNGI

Many Ascomycetes grow on litter of forests and fields; their number is too numerous to list. Many are Discomycetes. Most of these are large enough that the ascocarps can be seen with the eye or hand lens. For listing and descriptions see References 192, 193, and 282.

COPROPHILOUS FUNGI

Herbivore dung constitutes a unique ecosystem with a more or less unique flora. Countless species of fungi are ingested along with plant material, but only a few survive and grow in dung. Since some of these fungi are seldom, if ever, found free in soil, they are not merely air contaminants or soil fungi contacting the dung after deposition. Enzymes and the same temperature as that of the herbivore may be required in some cases to force spore germination.[253] A fungal succession occurs on dung; fast-growing "sugar-utilizing" Phycomycetes appear first, are followed by cellulose-utilizing Ascomycetes, and finally by lignin-utilizing Basidiomycetes. Webster[253] points out, however, that the succession is not as distinct as was once believed. While we have observed such a succession on horse dung, we have also observed both the Phycomycete *Pilobolus* and a minute Basidiomycete (mushroom) on horse dung within three days of incubation in a moist chamber. Coprophiles, in general, appear to have a high optimal growth temperature. Temperature may play a critical role, since there are few coprophiles on the dung of cold-blooded reptiles and amphibians.[253] Some genera and species of coprophilous Ascomycetes are given in Table 7. Some fungi that can grow on other substances may either be deposited with dung or contact it after deposition and grow on it. Note the presence of *Chaetomium* on different substrates (Tables 6, 7, and 13).

MYCORRHIZAL FUNGI

Mycorrhizal fungi that exist in or on roots do not cause more harm than good. The fungus no doubt

TABLE 7

SOME GENERA AND SPECIES OF COPROPHILOUS ASCOMYCETES

Ascomycete	Reference
Anixiopsis	
stercoraria (Hansen) Hansen	261
Ascobolus	
furfuraceus Pers. *ex* Fr.	262
stercorarius (Bull.) Schroet.	261
Ascodesmis	
macrospora Obr.	257
microscopica (Crou.) LeGal	257
nigricans van Tiegh.	257
procina Seaver	257
sphaerospora Obr.	257
Chaetomium	
africanum Ames	235
alba-arenulum Ames	235
ampullare Chivers	235
anguipilium Ames	235
angustum Chivers	235
aterrimum Ell. & Ev.	235
atrobrunneum Ames	235
aureum Chivers	235
bostrychodes Zopf	235
brasiliense Batista & Pont.	235
cancroideum Tschudy	235
caprinum Bain.	235
cochliode Pall.	235
congoensis Ames	235
contortum Peck	235
convolutum Chivers	235
crispatum Fuckel	235
cuniculorum Fuckel	235
dolichotrichum Ames	235
elatum Kunze *ex* Fr.	235
fusiforme Chivers	235
indicum Corda	235
molicellum Ames	235
murorum Corda	235
ochraceum Tschudy	235
quadrangulatum Chivers	235
semen-citrulli Sergejeva	235
simile Massee & Salm.	235
spinosum Chivers	235
spirale Zopf	235
subspirale Chivers	235
torulosum Bainier	235
trigonosporum (March.) Chivers	235
trilaterale Chivers	235
Cheilymenia sp.	256
Coniochaeta	
discospora (Auersw.) Cain	261
leucoplaca (Berk. & Rav.) Cain	261
Coprobia sp.	253
Coprotinia sp.	253
Delitschia	
auerswaldii Fuckel	261
marchalii Berl. & Vogel	261
Gelasinospora	
tetrasperma	261
Gymnoascus	
reesii Baran.	261
Lasiobolus sp.	253
pilosus (Fr.) Sacc.	262
Lophotrichus	
ampullus Benj.	235
brevirostratus Ames	235
martinii Benj.	235
Martinia sp.	253
Melanospora	
fimbriata (Rost.) Petch	261
Microascus	
doguetii Moreau	240
longirostris Zukal	240
schumacheria (Hans.) Curzi	240
Nectria sp.	255
Phaeotrichum	
circinatum Cain	259
hystricinum Cain & Barr	259
Podosordaria sp.	253
leporina (Ellis & Everh.) Dennis	254
Preussia	
dispersa (Clum.) Cain	258
fleischhakii (Auersw.) Cain	258
funiculata (Preuss) Fuckel	258
indica (Chattop & Das Gupta) Cain	258
isomera Cain	258
multispora (Saito & Mino) Cain	258

TABLE 7 (Continued)

SOME GENERA AND SPECIES OF COPROPHILOUS ASCOMYCETES

Ascomycete	Reference
Preussia (continued)	
nigra (Rout.) Cain	259
punctata (Auersw.) Cain	258
purpurea Cain	258
terricola Cain	258
typharum (Sacc.) Cain	258
vulgaria (Corda) Cain	258
Saccobolus sp.	253
Sordaria	
arctica Cain	261
fimicola (Rob.) Ces. & de Not.	261
humana (Fuckel) Wint.	262
leucoplaca (Berk. & Rav.) Ellis & Everh.	262
macrospora Auersw.	261
maxima Niessl.	261
tetraspora Wint.	261
Sphaeronaemella	
fimicola March.	260
Sporormia	
australis Speg.	
corynespora Niessl.	261
fimetaria de Not.	261
intermedia Auersw.	261
leporina Niessl.	261
octomera Auersw.	261
polymera Cain	261
venusta Cain	261
vexans Auersw.	261
Thelebolus sp.	253
Triangularia	
angulospora Cain & Farrow	241
bambusae (van Beyma) Boedijn	241
obliqua Cain	241
Trichodelitschia	
bisporula (Crouan) Munk	261

benefits by receiving organic nutrients from the plant, and the plant benefits by the increased surface area that the hyphae present to soil water and dissolved nutrients. While Ascomycetes are generally thought not to be mycorrhizal,[263] recent accounts suggest there may be exceptions.[264,265]

HYPOGEAN FUNGI

Hypogean fungi not only grow beneath the surface of soil, but also fruit there. Some are Phycomycetes, some are Basidiomycetes, but most are Ascomycetes. The famed truffles of epicurean delight are Tuberales of the Ascomycetes. Specially trained pigs and dogs are used to locate them. A one-ounce can containing three to four French truffles costs approximately $3.50 in the USA. They are also collected and sold in Mexico.[265] Most hypogean Ascomycetes are Tuberales. There are 140 species in 34 genera, with some 50 species in the genus *Tuber.*[266] Some examples are given here: *Tuber aestivum* Vitt., *Tuber magnatum* Pico, *Terfezia* spp. (12 species), *Elaphomyces granulatus* Fr., *Elaphomyces reticulatus* Vitt., *Genea asperula* Trappe & Guzman, *Geopora cooperi* Harkn., *Tuber murinum* Hesse, *Tuber brumale* Vitt., *T. mesentericum* Vitt., *T. panniferum* Tul., and *T. lapideum* Matt. See References 189, 267-272 for accounts of these interesting and unique fungi. These fungi grow under oaks and other trees and may be mycorrhizal.

Ascomycetes Living In Or On Plants

Many plants are attacked by fungi. It is interesting that plants appear to be the preferred hosts of fungi, whereas animals are the preferred hosts of bacteria. A recent U.S. government survey[199] estimates that the total annual loss in value caused by plant diseases is $3,251,114,000. The Ascomycetes are responsible for a significant portion of this loss. According to Ainsworth, 23% of 1,288 fungal plant pathogens are Ascomycetes.[298]

Ascomycetes may live on the surface of the plants, e.g., *Capnodia,* or actually enter the plant, either through direct penetration of the epidermis or gain entrance through natural openings such as lenticels and stomata or wounds. Several Ascomycetes have been responsible for almost exterminating certain plants. Examples are *Endothia parasitica,* the cause of chestnut blight, and *Ceratocystis ulmi,* the cause of Dutch elm disease. Tables 8, 9, 10 and 11 list Ascomycetes that cause various important plant diseases. In addition, several Ascomycetes cause losses of foods in storage. Examples of such Ascomycetes are *Nectria ipomoeae* Halst., *Ceratocystis fimbriata,* and *Diaporthe batatatis* Harter & Field on sweet potato, and *Pleospora* on apple. Certain species of *Sclerotinia* cause watery soft rots of vegetables.

TABLE 8

ASCOMYCETES CAUSING TREE DISEASES

Fungus	Disease	Tree
Appendicullella		
pinicola (Dearn.) Piroz. & Shoem.	Black mildew	Conifers
Ascocalyx		
asiaticus Groves	Leaf spot	Fir
tennuisporus Groves	Leaf spot	Fir
Atropellis		
arizonica Lohm. & Cash	Canker	Ponderosa pine
pinicola Zeller & Goodding	Canker	Pine
piniphila (Weir) Lohm. & Cash	Canker	Ponderosa pine, lodge pole pine
tingens Lohm. & Cash	Canker	Hard pine
Bifusella		
linearis (Peck) Höhnel	Needle blight	Pine
Botryosphaeria		
laricis (Wehm.) Arx & E. Müll.	Disease	Tamarack
ribis (Tode *ex* Fr.) Gross. & Dug.	Canker	Willow
tsugae Funk	Dieback	Western hemlock
Caliciopsis		
pinea Peck	Canker	Eastern pine
Cenangium		
ferruginosum Fr. *ex* Fr.	Dieback	Pine
Ceratocystis		
coerulescens (Münch) Bakshi	Stain	Spruce
fagacearum (Bretz) Hunt	Oak wilt	Black oak, red oak
fimbriata Ellis & Halst.	Disease	Cacao
ulmi (Buisman) C. Moreau	Dutch elm disease	American elm
Ciborinia		
whetzelii (Seaver) Seaver	Ink spot	Aspen
Cucurbidothis		
pithyophila (Fr.) Petr.	Branch disease	Conifers
Cucurbitaria		
naucosa (Fr.) Fuckel	Canker	European elm
Davisomycella		
ampla (J. J. Davis) Darker	Needle cast	Jack pine

TABLE 8 (Continued)

ASCOMYCETES CAUSING TREE DISEASES

Fungus	Disease	Tree
Delphinella		
balsameae (Waterm.) E. Müll.	Tip blight	Balsam fir
Dermea		
pseudotsugae Funk	Dieback	Douglas fir
Diaporthe		
alleghaniensis R. H. Arn.	Canker	Yellow birch
lokoyae Funk	Dieback	Douglas fir
Dibotryon		
morbosum (Schw.) Theiss & Syd.	Black knot	Cherry
Didymascella		
thujina (Durand) Maire	Cedar leaf blight	Cedar
Didymosphaeria		
oregonensis Goodding	Canker	Alder
Dothidella		
ulei P. Henn.	Leaf blight	Rubber plant
Elsinoe		
fawcettii Bitanc. & Jenk.	Citrus scab	Citrus
Elytroderma		
deformans (Weir) Darker	Needle cast	Ponderosa pine
Endothia		
parasitica (Murr.) P. J. & H. W. Anders	Chestnut blight	American and European chestnut
Epipolaeum		
abietis (Dearn.) Shoem.	Sooty mold	Fir
Eutypella		
parasitica Davidson & Lorenz	Canker	Maple, box elder
Glomerella		
cingulata (Stonem.) Spauld. & Schrenk	Anthracnose	Chestnut, apple
Gnomonia		
veneta Kleb.	Leaf, twig blight	Sycamore, oak
ulmea (Schw.) Thüm	Leaf spot	Elm
Gnomoniella		
tubiformis (Fr.) Sacc.	Leaf spot	Alder
Guignardia		
aesculi (Peck) Stewart	Leaf spot	Horse chestnut
Herpotrichia		
juniperi (Duby) Petrak.	Brown felt blight	Conifers
Hypodermella		
laricis Tub.	Needle disease	Western larch

TABLE 8 (Continued)

ASCOMYCETES CAUSING TREE DISEASES

Fungus	Disease	Tree
Hypoxylon		
atropunctatum (Schw. *ex* Fr.) Cooke	Dieback	Oak
mammatum (Wahl.) J. H. Miller	Canker	Maple, aspen
mediterraneum (DeNot) J. H. Miller	Canker	Oak, eucalyptus
Isthmiella		
quadrispora Ziller	Needle blight	Alpine fir
faullii (Darker) Darker	Needle blight	Balsam fir
Lachnellula		
pini (Brunch.) Dennis	Pine canker	Pine
pseudotsugae (Hahn) Dennis	Canker	Douglas fir
willkommii (Hartig) Dennis	Larch canker	Larch
Leptosphaeria		
coniothyrium (Fuckel) Sacc.	Canker	British willow
Leveillula		
taurica (Lev.) Arn.	Powdery mildew	Mesquite
Linospora		
gleditschiae J. H. Miller & Wolf	Leaf spot	Honey locust
Lirula		
nervisequia (D.C. *ex* Fr.) Darker	Needle blight	Fir
Lophodermella		
concolor (Dearn.) Darker	Needle blight	Lodge pole pine
Lophodermium		
pinastri (Schrad. *ex* Fr.) Chev.	Needle cast	Conifers
Maurodothina		
dothideoides (Ellis & Everh.) Piroz. & Shoem.	Black mildew	Conifers
farrae Piroz. & Shoem.	Black mildew	Conifers
Melanconis		
juglandis (Ellis & Everh.) Graves	Dieback	Butternut
Microsphaera		
alni (Wallr.) Salm.	Powdery mildew	Oak, many others
Mollisia		
grisleae (Syd.) J. H. Miller & Burton	Dieback	Pine
Monilinia		
fructicola (Wint.) Honey	Brown rot	Stone fruits
laxa (Aderh. & Ruhl.) Honey	Brown rot	Stone fruits
Morenoella		
quercina (ellis & Martin) Theiss	Leaf spot	Oak
Mycosphaerella		
dendroides (Cooke) Demaree & Cole	Leaf spot	Pecan
fraxinicola (Schw.) House	Leaf spot	Ash
musicola Leach	Leaf spot	Banana
ulmi Kleb.	Leaf spot	Elm

TABLE 8 (Continued).

ASCOMYCETES CAUSING TREE DISEASES

Fungus	Disease	Tree
Nectria		
cinnabarina (Tode *ex* Fr.) Fr.	Dieback	Hardwoods
coccinea (Pers. *ex* Fr.) Fr., var. *faginata* Lohm., A. M. Wats & Ayers	Beech disease	Beech
galligena Bres.	Canker	Aspen
Neofabraea		
populi G. E. Thomps.	Canker	Aspen, poplar
Neopeckia		
coulteri (Peck) Sacc.	Brown felt blight	Pine
Ophiobolus		
heveae Petch	Leaf spot	Rubber plant
Phacidium		
infestans Karst.	Snow blight	Conifers
Phaeocryptopus		
gaeumannii (Rohde) Petr.	Needle cast	Douglas fir
Phyllactinia		
guttata (Fr.) Lev.	Powdery mildew	Hardwoods
Physalospora		
glandicola (Schw.) N. E. Stevens	Canker	Oak
miyabeana Fukushi	Black canker	Hardwoods
Ploioderma		
lethale (Dearn.) Darker	Needle cast	Eastern pine
Rhabdocline		
pseudotsugae Syd.	Needle cast	Douglas fir
weirii Parker & Reid	Needle cast	Douglas fir
Rhizinia		
undulata Fr.	Rhizinia root rot	Conifers
Rhytidiella		
moriformis Salasky	Rough bark	Balsam fir, poplar
Rhytisma		
acerinum (Pers. *ex* St. Amans) Rr.	Tar spot	Red maple, silver maple
Scirrhia		
pini Fund & Parker	Needle blight	Pine
Scleroderris		
abieticola Zeller & Goodding	Canker	Fir
abietina (Lagerb.) Gremmen	Canker	Pine
Sclerotinia		
fructigena Aderh. & Ruhl.	Brown rot	Stone fruits
kerneri Wettstein	Needle disease	Fir
Sphaerotheca		
lanestris Harkn.	Powdery mildew	Oak

TABLE 8 (Continued).

ASCOMYCETES CAUSING TREE DISEASES

Fungus	Disease	Tree
Sphaerulina		
taxi Cooke & Massee	Blight	Pacific yew
Systremma		
ulmi (Duval *ex* Fr.) Theiss & Syd.	Leaf spot	Elm
Taphrina		
aesculi (Patt.) Gies.	Blister	Buckeye
bacteriosperma Johans	Blister, curl	Birch
boycei Mix	Blister	Birch
confusa (Atk.) Gies.	Leaf blister	Choke cherry
coryli Nishida	Blister, leaf curl	Hazel
deformans (Berk.) Tul.	Leaf blister	Hardwoods
farlowii Sadeb.	Leaf curl	Black cherry
flava Farl.	Blister	Birch
Japonica Kusano	Leaf curl	Red alder
populina Fr.	Yellow leaf blister	Poplar
ulmi (Fuckel) Johans	Blister	American elm
virginica	Blister	Hop hornbeam
Thyronectria		
austro-americana (Speg.) Seeler	Canker	Honey locust
Tympanis		
confusa Nyl.	Tympanis canker	Pine
Uncinula		
circinata Cooke & Peck	Powdery mildew	Maple
clintonii Peck	Powdery mildew	Basswood
flexuosa Peck	Powdery mildew	Buckeye, elm
macrospora Peck	Powdery mildew	American elm
salicis (D.C. *ex* Merat.) Wint.	Powdery mildew	Poplar, willow
Urnula		
craterium (Schw.) Fr.	Canker	Oak
Valsa		
sordida Nits.	Cytospora canker	Poplar, willow
Venturia		
inaequalis (Cooke) Wint.	Leaf spot	Apple
Virgella		
robusta (Tub.) Darker	Needle blight	Fir
Xylaria		
digitata (L. *ex* Fr.) Grev.	Xylaria root rot	Apple, hardwoods
polymorpha Pers. *ex* Grev.	Decay	Hardwoods

Data compiled in cooperation with Dr. Al Funk, Forest Pathologist, and Daphyne P. Lowe, Department of Fisheries and Forestry, Canadian Forestry Service, Victoria, British Columbia.[303]

TABLE 9

IMPORTANT ASCOMYCETE DISEASES OF FRUITS AND VEGETABLES

Fungus	Disease	Fruit and/or Vegetable
Botryosphaeria		
melathroa Berk. & Curt.	Dieback	Hawthorne
ribis G. & D.	Dieback	Currant
Calonectria		
crotalariae (Loos) Bell & Sobers	Necrosis	Peanut
Ceratocystis		
fimbriata Ellis & Halst.	Black rot	Sweet potato
Diaporthe		
batatatis Harter & Field	Dry rot	Sweet potato
phaseolarum (Cooke & Ell.) Sacc.	Pod blight	Lima bean, soybean
vexans Gratz	Fruit rot	Eggplant
Erysiphe		
cichoracearum D.C.	Powdery mildew	Cucurbits
polygoni D.C.	Powdery mildew	Legumes
Glomerella		
cingulata (Ston.) Spauld. & Schrenk	Anthracnose	Pepper, bean, gourds, soybean, grape, tomato, eggplant
Gnomonia		
fragariae Kleb.	Leaf spot	Strawberry
Guignardia		
bidwellii (Ellis) V. & R.	Black rot	Grape
vaccinii Shear	Rot, leaf spot	Cranberry
Hypomyces		
solani Rke. & Berth.	Root rot	Bean
Leptosphaeria		
coniothyrium (Fuckel) Sacc.	Cane blight	Raspberry
Leptosphaerulina		
crassiasca (Sechet) Jackson & Bell	Leaf spot	Peanut
Leveillula sp.	Root rot	Tomato, eggplant, carrot, many others
Mycosphaerella		
arachidicola Jenkins	Leaf spot	Peanut
brassicicola (Duby) Lind.	Ring spot	Crucifers
berkeleyii Jenkins	Leaf spot	Peanut
fragariae (Tul.) Lind.	Leaf spot	Strawberry
louisianae Plak.	Leaf spot	Berries
pinodes (Berk. & Blox.) Vestergr.	Foot rot	Pea
rubi Roark	Leaf spot	*Rubus*
Nectria		
haematococca Berk. & Br.	Wilt	Pea, squash
Physalospora		
obtusa (Schw.) Cooke	Canker	Pistachio nut

TABLE 9. (Continued)

IMPORTANT ASCOMYCETE DISEASES OF FRUITS AND VEGETABLES

Fungus	Disease	Fruit and/or Vegetable
Pleospora		
lycopersici Ellis & Emm. March.	Leaf spot, rot	Tomato, etc.
Pseudopeziza		
ribis Kleb.	Leaf and stem spot	Currant, gooseberry
Rosellinia		
necatrix (Prill.) Berl.	Root rot	Grapevine
Sclerotinia		
sclerotiorum (Lib.) DBy.	Soft rot	Many vegetables
Sphaerotheca		
macularis (Wall. *ex* Fr.) W. B. Cooke	Powdery mildew	Strawberry
mors-uvae (Schw.) Berk.	Powdery mildew	Gooseberry, currant
Uncinula		
necator (Schw.) Burr.	Powdery mildew	Grape

TABLE 10

IMPORTANT ASCOMYCETE DISEASES OF ORNAMENTALS

Fungus	Disease	Ornamental
Botryosphaeria		
ribis Gross. & Dug.	Canker	Allspice, mulberry
Diaporthe		
linearis (Nees *ex* Fr.) Nits.	Stem blight	Forsythia, delphinium, butterfly bush, aster, cosmo, gardenia, etc.
stewartii Harr.	Stem blight	As *D. linearis*
umbrina Jenk.	Brown canker	Rose
Didymellina		
macrospora Kleb.	Leaf spot	Iris
Diplocarpon		
rosae (Fr.) Wolf	Black spot	Rose
Erysiphe		
cichoracearum D.C.	Mildew	Begonia, bellflower, sunflower, gaillardia, crysanthemum, aster, etc.
polygoni D.C.	Mildew	California poppy
Fabraea		
maculata Atk.	Blight	Cotoneaster, hawthorne, apple

TABLE 10 (Continued).

IMPORTANT ASCOYMCETE DISEASES OF ORNAMENTALS

Fungus	Disease	Ornamental
Glomerella		
cincta (B. & C.) Spauld. & Schrenk	Leaf spot	Begonia, orchid, dracaena, rose, snapdragon, *Dieffenbachia*
cingulata (Stonem.) Spauld. & Schrenk	Anthracnose	Privet, sweet pea, rubber plant, iris, croton, clematis, agave, camellia, etc.
spp.	Leaf spot, canker	Pansy, *Nandina,* lily, ivy, dogwood, boxwood
Hyponectria		
buxi (D.C. *ex* Fr.) Sacc.	Leaf blight	Boxwood
Keithia		
thujina Dur.	Leaf spot	Arbor vitae
Leptosphaeria		
arunci Zeller	Canker	Goatsbeard
coniothyrium (Fuckel) Sacc.	Blight	Rose
Meliola		
ambigua Pat. & Gaill.	Black mildew	Camellia, trumpet flower, lantana
Microsphaera		
alni (Wallr.) Salm.	Powdery mildew	Privet, bittersweet, lilac, *Ceanothus,* holly, trailing arbutus, etc.
Mycosphaerella		
andrewsii (Sacc.) J. J. Davis	Leaf spot	Prickly pear, kalanchoe, mulberry, rubber plant, gentian, rose, mountain laurel, etc.
arachnoidea Wolf	Leaf spot	As *M. andrewsii*
bolleana Higgins	Leaf spot	As *M. andrewsii*
ligulicola Bak, Dim. & Dav.	Petal blight	Chrysanthemum
opuntiae (Ellis & Everh.) Dearn.	Leaf spot	As *M. andrewsii*
rosae Fuckel	Leaf spot	As *M. andrewsii*
Nectria		
ditissima Tul.	Canker	*Broussonetia,* allspice, acacia, magnolia
Physalospora		
dracaenae Sheld.	Tip blight	Dracaena
Physalospora (Botryosphaeria)		
obtusa (Schw.) Cooke	Leaf spot	Crabapple, allspice, cotoneaster
Pseudonectria		
pachysandricola B. O. Dodge	Canker or leaf blight	Pachysandra
rousseliana (Mont.) Seaver	Canker or leaf blight	Boxwood

TABLE 10 (Continued).

IMPORTANT ASCOMYCETE DISEASES OF ORNAMENTALS

Fungus	Disease	Ornamental
Sclerotinia sp.	Stem blight or black slime	Peony
gladioli Drayton	Dry rot	Gladioli
narcissicola Gregory	Smoulder	Narcissus
polyblastis Gregory	Fire	Narcissus
sclerotiorum (Libert.) de Bary	Stem rot	Cornflower, daisy, snapdragon, camellia, columbine, sunflower, calendula, forget-me-not, hyacinth, orchid
Sphaerotheca		
humuli (D.C.) Burr.	Powdery mildew	Rose
pannosa (Wallr.) Lev.	Powdery mildew	Rose
Taphrina		
filicina Rostr. *ex* Johans	Leaf blister	Fern

TABLE 11.

IMPORTANT ASCOMYCETE DISEASES OF CEREALS, GRASSES, AND FORAGE CROPS

Fungus	Disease	Cereals, Grasses, and Forage Crops
Calonectria		
nivalis Schaffnit	Snow mould	Cereals
Claviceps		
paspali Stevens & Hall	Ergot	Dallis grass
purpurea (Fr.) Tul.	Ergot	Cereals, grasses
Cochliobolus		
carbonum Nelson	Leaf spot	Grasses
heterostrophus Drechs.	Leaf spot	Maize, etc.
sativus (Ito & Kurib.) Drechs. *ex* Dast.	Spot blotch	Barley, etc.
stenospilus Mats. & Yamam.	Brown stripe	Sugar, cane
victoriae Nelson	Foot rot	Cereals
Epichloe		
typhina (Pers. *ex* Fr.) Tul.	Choke	Grasses
Erysiphe		
graminis D.C. *ex* Merat	Powdery mildew	Many cereals and grasses
Gibberella		
fujikuroi (Saw.) Wollenw.	Bakanae disease, pink ear rot	Rice, corn
zeae (Schw.) Petch	Seedling blight	Cereals
Gnomonia		
andropogonis Ellis & Everh.	Leaf spot	Andropogon

TABLE 11 (Continued).

IMPORTANT ASCOMYCETE DISEASES OF CEREALS, GRASSES, AND FORAGE CROPS

Fungus	Disease	Cereals, Grasses, and Forage Crops
Leptosphaeria		
avenariae G. F. Weber	Speckled blotch	Oat
herpotrichoides de Not.	Blight	Wheat
maculans (Desm.) Ces. & de Not.	Black leg	Rape, mustard
sacchari van Breda	Ring spot	Sugar cane
salvinii Catt.	Stem rot	Rice
tritici (Gar.) Pass.	Blight	Wheat
Leptosphaerulina		
briosiana (Poll.) Graham & Luttrell	Leaf spot	Alfalfa, clover
Mycosphaerella		
carinthiaca Jaap	Midvein spot	Clover
Ophiobolus		
graminis (Sacc.) Sacc.	Take-all	Wheat
Phyllachora		
graminis Pers.	Black spot	Grasses
trifollii Pers.	Black spot	Clover
Pleospora		
herbarum (Fr.) Rabenh.	Target spot	Alfalfa
hyalospora Ellis & Everh.	Leaf spot	Alfalfa
Pseudopeziza		
medicaginis (Lib.) Sacc.	Leaf spot	Alfalfa
trifollii (Biv.-Bern) Fuckel	Leaf spot	Clover
Pyrenophora		
graminea Ito & Kurib.	Leaf stripe	Barley
teres Drechs.	Net blotch	Barley
Sclerotinia		
borealis	Snow mould	Grasses
homeocarpa F. T. Bennett	Dollar spot	Grasses
trifoliorum Erikss.	Rot	Clover

See References 196, 197, and 198 for further information.

Many seeds harbor Ascomycetes that have grown in the coat during maturation and are far enough below the surface to be out of reach of such common surface-sterilization agents as 10% chlorox or mercury compounds. Many seed-borne Ascomycetes are *Chaetomium* species which are also present in dung, soil, and marine environment. Their presence in various seeds is documented in References 283-285. The seed-coat flora is probably quite specific, but many of the fungi probably do not damage the seed unless the seeds are stored under moist conditions.[286] Both the ascigerous aspergilli (especially the *Aspergillus glaucus* series) and penicillia are notoriously active in the deterioration of stored seed, since many of these fungi can grow at reduced moisture levels.[314] In addition, many ascomycetous plant pathogens, such as *Cochliobolus* and *Pyrenophora*, may be transmitted through seed. Contaminating fungi growing on moist seed can result in the production of toxins active against livestock and no doubt also against man (see the section on toxins).

Ascomycetes Living On Animals

ASCOMYCETES AND MAN

Few fungi attack man. *Histoplasma capsulatum* Darlina, *Coccidioides immitis* Rix. & Gilchrist, and *Cryptococcus neoformans* (Sanf.) Vuill. are the most deadly human-pathogenic fungi, and some or all may be imperfect forms of Ascomycetes.[278] The known Ascomycetes that attack man are listed in Table 12 along with their imperfect forms. According to Ainsworth,[298] 13.6% of 205 fungi pathogenic to vertebrates are Ascomycetes. These fungi live in soil and on plant material, and the incidence of some of them increases if the soil regularly receives bird and bat droppings,[230,239,279] which are sources of nutrients.

ASCOMYCETES AND INSECTS

One genus of the Ascomycetes, *Cordyceps,* is well known for its parasitism of insects, such as moth pupae, and of arachnids, such as spiders. Main[217] has described many species, and Kobayasi[218] has recorded 137 species attacking insects, 37 of which are found in North America. We have encountered hundreds of the orange stromata of *Cordyceps militaris* protruding an inch or more above the buried pupae of moths in hemlock-rhododendron-maple-beech forests of West Virginia following long rain spells. Such spontaneous outbreaks may constitute significant biological control of insects. Madelin[219] lists the following genera of ascomycete parasites: *Calonectria, Cordyceps, Ascosphaera, Hypocrella, Myriangium, Nectria, Ophiocordyceps, Podonectria, Sphaerostilbe,* and *Torrubiella.* All members of the most unusual Laboulbeniales live on insect skeletons and perhaps should be considered parasitic.

Two Deuteromycetes related to Ascomycetes, *Aspergillus flavus* Link and *A. parasiticus* Speare, attack insects. These two species grow in plant fruits and synthesize aflatoxin, the most potent inducer of liver cancer known. The former attacks silk worm larvae,[220] grasshoppers,[221] honey bees,[222] corn borers,[223] desert locusts,[224] bed bugs,[225] and cockroaches.[226] Interestingly, *Aspergillus flavus* can invade the human lung, where it grows as hyphae and produces conidiophores.

Ambrosia fungi are Ascomycetes (e.g., *Ascoidea, Dipodascus,* and *Endomycopsis*) that grow in tunnels made by ambrosia beetles. These fungi may be a source of food for both larvae and adults.[297]

Certain attine ants cultivate fungi by growing them on leaf parts that have been carried to the nests. The ants fertilize these leaf pieces by adding their feces to them. While the identity of these fungi is unknown, they may be Basidiomycetes. Martin[216] has worked out the biochemistry of this relationship, which concerns in part cellulase production by the fungi.

For further discussion of fungi on insects see References 227, 228, and 338.

ASCOMYCETES AND OTHER ANIMALS

Certain dermatophytes, i.e., Gymnoascaceae, live on the skin and hairs of both wild and domestic animals. In some cases it has been shown that the fungus can be transmitted to man. Children are particularly likely to pick up certain dermatophytes from dogs and cats. *Nannizzia persicolor* has been isolated from 53% of 127 bank voles and 25% of 113 field voles, indicating that these animals can serve as a reservoir from which man can become infected.[298]

Aquatic Ascomycetes

FRESH-WATER ASCOMYCETES

While many strictly aquatic Phycomycetes exist, only a few fresh-water Ascomycetes are known.[214] Several families of Ascomycetes are represented, and consequently it is assumed that aquatic species probably evolved from various terrestrial species. Few features suggest adaptation to the aquatic environment, although the ascospores are usually enveloped in a gelatinous sheath and spores abound in

TABLE 12

HUMAN-PATHOGENIC ASCOMYCETES

Ascomycete	Imperfect Form	Disease
Ajellomyces	*Blastomyces*	
dermatitidis McD. & Louis	*dermatitidis*	Blastomycosis
Allescheria	*Monosporium*	
boydii Shear	*apiospermum*	Mycetoma
Arthroderma	*Trichophyton*	
benhamiae Ajello & Cheng	*mentagrophytes*	Dermatomycosis, skin, hair, nails
gertleri Bohme	*vanbreuseghemii*	Dermatomycosis
simii Stock, Mack. & Austw.	*simii*	Dermatomycosis
Emmonsiella	*Histoplasma*	
capsulata Kwon-Chung[330]	*capsulatum*	Histoplasmosis
Leptosphaeria		
senegalensis Segret, Bay., Darker & Camain.		Mycetoma
*Microascus**	*Scopulariopsis*	
cinereus (Emile-Weil & Gaud.) Curzi	*cinerea*	Skin
manganii (Loub.) Curzi	*albo-flavescens*	Skin
trigonosporus Emm. & Dodge	sp.	Skin
Nannizzia	*Microsporum*	
cajetani Ajello	*cookei*	Dermatomycosis
fulva Stock	*fulvum*	Dermatomycosis
grubyia Georg, Ajello, Fried & Brink	*vanbreuseghemii*	Dermatomycosis
gypsea Stock	*gypseum*†	Dermatomycosis
incuravata Stock	*gypseum*	Dermatomycosis
obtusa Daws. & Gentles	*nanum*	Dermatomycosis
persicolor Stock	*persicolor*	Dermatomycosis
Neotestudina		
rosatii Segret & Dest.		Mycetoma
Pichia	*Candida*	
guillermondii Wick. (yeast)	*guillermondii*	Dermatomycosis
Piedraia		
hortai (Brum) Fon. & Leao		Black piedra
Sartorya	*Aspergillus*	
fumigata Vuill.	*fumigatus*	Aspergillosis

* Microascus may not be pathogenic.

† Imperfect form of two ascomycetes.

Data compiled from References 230, 240, and 287, and from personal communication with Dr. L. Georg.

lipid (which would decrease density).[214] *Loramyces juncicola* Weston, *Ceriospora caudae-suis* Ing, *Ophiobolus typhae* Feltg., *Pleospora scirpicola* (D.C. *ex* Fr.) Karst., *Vibrissea truncorum* (Alb. & Schw.) Fr., and *Apostemidium guernisaci* (Cr.) Boud. are examples of fresh-water Ascomycetes. See Reference 215 for a discussion of the fresh-water environment.

MARINE ASCOMYCETES

Most marine fungi are Ascomycetes (see References 200-203 and 346 for taxonomic treatments). Most of those described to date are listed in Table 13.

Most marine Ascomycetes occur on submerged wood or on algae, where they may be mainly epiphytes. Some have been found at depths of 3,000 meters,[200] and while they may grow there, they could be spores that have settled but not yet died. Few physiological or morphological traits unite the marine Ascomycetes,

TABLE 13

SOME MARINE ASCOMYCETES

Ascomycete*	Substrate	Reference
Amphisphaeria		
biturbinata (Dur. & Mont.) Sacc.	*Posidonia*	200
maritima Linder	Wood	200
posidoniae (Dur. & Mont.) Ces. & de Not.	*Posidonia*	200
Antennospora		
quadricornuta (Cribb & Cribb) Johnson	Wood	200
Arenariomyces		
trifurcatus Höhnk	Wood	200
Banhegyia		
uralensis (Naovmoff) Kohlm.	Driftwood	209
Ceriosporopsis		
cambrensis Wilson	Wood	200
halima Linder	Driftwood	209
hamata Höhnk	Submerged test blocks	202
Chadefaudia		
calyptrata Kohlm.	Wood	200
marina G. Feld.	*Rhodymenia*	
Chaetomium		
erectum Skolko & Groves	Submerged test blocks	202
globosum Kunze	Submerged test blocks	202
Chaetospheria		
chaetosa Kohlm.	Driftwood	209
Corollospora		
comata (Kohlm.) Kohlm.		212
maritima Wedermann	Driftwood	209
pulchella Kohlm., Schmidt & Nair	Mangrove roots	210
trifurcata (Hohnk) Kohlm.	Sea foam	209
Didymella		
magnei G. Feld.	*Rhodymenia*	200
Didymosamarospora		
euryhalina Johnson & Gold	*Juncus*	200
Didymosphaeria		
enalia Kohlm.	Roots, breakwater wood	210
fucicola Suth.	*Fucus*	200
maritima (Crouan) Sacc.	*Atriplex*	200
pelvetiana Suth.	*Pelvetia*	200

TABLE 13 (Continued).

SOME MARINE ASCOMYCETES

Ascomycete*	Substrate	Reference
Endoxyla		
cirrhosa (Pers.) Muller & Arx†		213
Gnomonia		
longirostris Cribb & Cribb	*Avicennia*	200
marina Cribb & Cribb	*Avicennia*	200
salina Jones	Driftwood	202
Guignardia		
alaskana Reed	*Prasiola*	200
prasiolae (Wint.) Reed	*Prasiola*	200
ulvae Reed	*Ulva*	200
Haligena		
elaterophora Kohlm.	Driftwood	209
Haloguinardia sp.	Driftwood	212
decidua Cribb & Cribb	Sargassum	200
irritans (Setch. & Est.) Cribb & Cribb	*Cystoseira*	200
longispora Cribb & Cribb	Sargassum	200
tumefaciens (Cribb & Herb.) Cribb & Cribb	Sargassum	200
Halosphaeria		
appendiculata Linder	Driftwood	209
circumvestita Kohlm.	Submerged test blocks	203
mediosetigera (Cribb & Cribb) Johnson		204
torquata Kohlm.	Wood	200
*tubulifera*Kohlm.	Driftwood	212
Herpotrichiella		
ciliomaris Kohlm.	Driftwood	200
Hydronectria		
tethys J. & E. Kohlm.	Submerged mangrove	210
Hypoderma		
laminariae Suth.	*Laminaria*	200
Keissleriella		
elaphospora J. & E. Kohlm.	Bark of prop root *Rhizophora*	210
Kymadiscus		
haliotrephus J. & E. Kohlm.	Wood or prop root	210
Lentescospora		
submarina Linder	Wood	200
Leptosphaeria		
albopunctata (West.) Sacc.	Driftwood	202
australiensis (Cribb & Cribb) G. C. Hughes	Prop root	210
avicenniae J. & E. Kohlm.	Submerged bark	210
discors Sacc. & Ellis		
fuscella (Berk. & Br.) Ces. & de Not †		213
halima Johnson	Wood, *Spartina*	200
macrosporidium Jones	Marsh grass, *Spartina*	202
marina Ellis & Everh.	*Spartina*	200

TABLE 13 (Continued)

SOME MARINE ASCOMYCETES

Ascomycete*	Substrate	Reference
Leptosphaeria (continued)		
maritima (Cooke & Plowr.) Sacc.	*Juncus*	200
orae-maris Linder	Driftwood	209
pelagica Jones	Marsh grass	202
typharum (Rabenh.) Karst.	Marsh grass	202
Lindra		
inflata Wilson	Wood	200
marinera Meyers	Foam	210
thalassiae Oraurt	Turtle grass	211
Lignincola		
laevis Höhnk	Mangrove root	210
Lulworthia		
attenuata Johnson	Wood	200
conica Johnson	Wood	200
floridana Meyers	Submerged test blocks	202
fucicola Suth.	Submerged test blocks	203
grandispora Meyers	Prop root	210
halima (Diehl & Mounce) Cribb & Cribb	In *Zostera*	200
kniepii Kohlm.	Algae	209
medusa (Ellis & Everh.) Cribb & Cribb	Driftwood	203
opaca (Linder) Cribb & Cribb	Submerged test blocks	203
purpurea (Wilson) Johnson	Timber, ropes, driftwood	202
rotunda Johnson	Wood	200
rufa (Wilson) Johnson	Submerged test blocks	202
submersa Johnson	Wood	200
Manglicola		
guatemalensis J. & E. Kohlm.	Bark of mangrove	210
Marinospora		
calyptrata (Caval) Caval	Driftwood	212
Massariella		
maritima Johnson	Wood	200
Melanopsamma		
cystophorae (Cribb & Herb.) Meyers	*Cystophora*	200
tregoubovii Olliv.	*Aglaxonia* and others	200
Microthelia		
maritima (Linder) Kohlm.	Driftwood	209
Mycophycophila		
corallinarum Kohlm.	Algae	209
Mycosphaerella		
ascophylli Cott.	*Ascophyllum*	200
pelvetiae Suth.	*Pelvetia*	200
pneumatophorae Kohlm.	Bark of *Avicennia*	210
Nautosphaeria		
cristaminuta Jones	Driftwood	209

TABLE 13 (Continued)

SOME MARINE ASCOMYCETES

Ascomycete*	Substrate	Reference
Nectria		
penicilloides Ranzoni†		213
Ophiobolus		
australiensis Johnson & Sparrow	*Avicennia*	200
kniepii Ade & Bauch	*Lithophyllum*	200
laminariae Suth.	In *Laminaria*	200
littoralis (Crouan) Sacc.	*Agrostis*	200
maritimus (Sacc.) Sacc.	*Zostera* leaves	200
salina Meyers	*Andropogon*	200
Orcadia		
ascophylli Suth.	*Ascophyllum*	200
pelvetiana Suth.	*Pelvetia*	200
Paraliomyces		
lentiferus Kohlm.	Wood	200
Peritrichospora		
comata Kohlm.	Driftwood	200
cristata Kohlm.	Driftwood	200
integra Linder	Submerged test blocks	202
lacera Linder	Wood	200
Pharcidia		
balani (Wint.) Bauch (lichen)	Sea shells	209
pelvetiae Suth.	*Pelvetia*	200
Phyllachlorella		
oceanica Ferd. & Winge	Sargassum	200
Placostroma		
laminariae (Rostrup) Meyers	*Laminaria*	200
pelvetiae (Suth.) Meyers	*Pelvetia*	200
Plectolitus		
acanthosporum Kohlm.	In wood	200
Pleospora		
herbarum, var. *herbarum* Wehm.	Wood	202
laminariana Suth.	*Laminaria*	200
pelagica Johnson	*Spartina*	200
pelvetiae Suth.	*Pelvetia*	200
Remispora		
cucullata Kohlm.	Intertidal wood	210
hamata (Höhnk) Kohlm.	Driftwood	212
maritima Linder	Driftwood	212
ornata Johnson & Cav.	Driftwood	212
pilleata Kohlm.	Driftwood	212
quadriremis (Höhnk) Kohlm.	Driftwood	200
salina (Meyers) Kohlm.	Intertidal wood	210
stellata Kohlm.	Wood	200
Rosellinia		
laminariana Suth.	*Laminaria*	200

TABLE 13 (Continued).

SOME MARINE ASCOMYCETES

Ascomycete*	Substrate	Reference
Samarosporella		
pelagica Linder	Wood	200
Sphaerulina		
codicola E. Y. Dawson	*Codium*	200
orae-maris Linder	Wood	200
pedicellata Johnson	Wood	200
Stimatea		
pelvetiae Suth.	*Pelvetia*	200
Taeniolella		
audis (Sacc.) S. J. Hughes†		213
Thallassoascus		
tregoubovii Olliv. (see *Melanopsamma*)	Algae	209
Torpedospora		
ambispinosa Kohlm.	Driftwood	200
radiata Meyers	Driftwood, bamboo	209
Trailia		
ascophylli Suth.	*Ascophyllum*	200
Zignoella		
calospora Pat.	*Castanea*	200
cubensis Har. & Pat.	*Stypocaulon*	200
enormis Pat. & Har.	Driftwood	212

* See Reference 201 for possible revisions.

† Terrestrial fungi living in brackish water.

although Johnson[200] has pointed out that most produce deliquescing asci that might favor spore dispersal by submerged ascocarps. Interestingly, like marine bacteria, they appear to have less tolerance for high salt concentrations than their terrestrial counterparts. Probably none of the marine Ascomycetes require sea water; they probably have evolved, or have become adapted, from diverse terrestrial species and families. Some marine Ascomycetes grow best at alkaline pH (sea water pH is *ca.* 8.3) and some at pH 4.4; some grow best at temperatures not much above 20°C and others grow most rapidly at higher temperatures; some require thiamine, and perhaps biotin, whereas others do not.[200,204] Sea water contains vitamins.[205-208] Probably most marine Ascomycetes synthesize cellulase and/or pectinases. Spores appear to be common in sea water, since many marine Ascomycetes have been isolated by suspending sterile blocks of wood in the sea for a few weeks at a time;[202,203] fungal succession may occur on test blocks and natural substrates, but if so, it is probably not very complicated.[200] Ascomycetes do not appear to be common in ocean beach sand.[200]

Lichens consist of colonies of Ascomycetes (for the most part) and algae living in such constant form that they are given names. The blue-green, orange, or red crusty growths on rocks and trees are lichens, and their marine counterparts exist (see Table 14). Almost all live on rocks (saxicolous); some remain submerged, while others are moistened by spray at high tide.

TABLE 14

MARINE LICHENS

Ascomycete	Reference
Calcoplaca	
marina Wedd.	200
thallinicola (Wedd.) DR.	200
Lecanora	
actophila Wedd.	200
atra (Huds.) Ach.	200
Lichina	
confinis (Mull.) Ag.	200
pygmaea (Lightf.) Ag.	200
Parmelia	
saxatilis (L.) Ach.	200
Ramalina	
siliquosa (Huds.) A. L. Smith	200
Rhizocarpon	
constrictum Malme	200
Verrucaria	
ditmarsica Erichs	200
maura Wg.	200
microspora Nyl.	200
mucosa Wg.	200
Xanthorina	
parietina (L.) Th. Fr.	200

Ascomycetes in the Atmosphere

Dry spores of certain Ascomycetes growing on soil and plant materials are carried into the atmosphere, where some remain viable. Most of the viable spores are dark-pigmented. The spores can reinfect other plant or animal materials in other locations and also be a source of allergic reactions in humans. See Reference 299 for discussions.

DISEASES OF ASCOMYCETES

VIRUSES

Although it was believed until recently that fungi were not hosts of viruses, it has now been established that some Fungi Imperfecti (some of which may be Ascomycetes) are attacked by RNA viruses.[273,274,344] The following are examples: *Penicillium stoloniferum* Thom, *P. funiculosum* Thom, *P. chrysogenum* Thom, *P. cyaneofulvum* Biourge, *P. brevicompactum* Dierck, *P. citrinum* Thom, *P. variabile* Sopp., *Aspergillus foetidus* (Naka) Thom & Raper, *Aspergillus niger* van Tiegh., *Stemphylium* sp. and *Cephalosporium* spp.[28] These viruses may be of great importance, because they induce synthesis in mammalian cells of interferon, which prevents attack by some viruses. If not dangerous to man, these fungal viruses could be used to induce interferon synthesis. The virus of *P. chrysogenum* is double-stranded RNA.[28]

Fungi

The truffle-like hypogeous *Elaphomyces* is attacked by the Ascomycetes *Cordyceps agariciformia* (Bolt) Seaver and *C. capitata* (Holm. *ex* Fr.) Lk. *Trichoglossum* sp. is attacked by *Micropyxis* sp., and *Tuber* spp. are attacked by *Battarrina* sp. *Cordyceps* also attacks the ergot Ascomycetes, *Claviceps* spp.[218] Several Phycomycetes attack other phycomycetes, but only *Syncephalis wynneae* Thaxter attacks an ascomycete.[219] See References 217-219, 262, 275-277, and 315 for revelant literature. Certain *Hypomyces* parasitize mushrooms.[191]

FUNGI IN PULP MILLS

Fungi of various classes create problems in the manufacture of pulp and paper. Many have been identified.[280,281] Wang[280] records the Ascomycetes *Byssochlamys fulva* Oll. & Smith, *Chaetomium indicum* Cda., *Eurotium repens* de Bary, *Leptosphaeria tenera* (Ellis) Sacc., *Sartorya fumigata* Vuill., and *Talaromyces vermiculatus* (Dang.) Benj.

ASCOMYCETES IN WASTE DISPOSAL

Ascomycetes probably do not play a significant role in domestic-sewage-disposal plant operations, although they may be present in the aerated process. Such fungi have been used in developing processes for specific waste-product removal.[335]

ASCOMYCETES AND MYCOPHAGY

Certain fleshy Ascomycetes are collected from nature and eaten, although most fungi collected for consumption are Basidiomycetes (mushrooms). Perhaps the most sought-after of all fleshy fungi are the morels, which are Ascomycetes of the genus *Morchella.* Other examples are *Verpa* spp., certain species of *Helvella* and *Gyromitra* (caution – some of these are poisonous[189]) and *Tuber* spp. (the truffles). See References 336 and 337 for discussion of the use of fungi for food.

COLLECTION, ISOLATION, AND PRESERVATION OF ASCOMYCETES

Fruit bodies of Ascomycetes occurring on plant material and dung can be stored in boxes or other containers containing a few crystals of *p*-dichlorobenzene (commercial moth balls) after the material has been gently dried to prevent mold growth. See References 318, 342, and 343 for methods of isolating human-pathogenic fungi and References 316, 317, 342, and 343 for methods of isolating soil- and plant-inhabiting fungi. Living cultures can be preserved by storage under sterile heavy-duty mineral oil, in sterile soil, lyophilized, or submerged in liquid nitrogen.[319-324] These methods will preserve many Ascomycetes for several years.

The morphological characteristics, and often pigmentation, of Ascomycetes growing on agar media in petri plates can be preserved indefinitely by careful drying and storage.[325,342,343] Microscopic characteristics of Ascomycetes can be preserved on microscope slides for at least eighteen years, in the authors' experience, by mounting mycelium in lactophenol blue (phenol, 20 ml; lactic acid, 20 ml; glycerol, 40 ml; water, 20 ml; and cotton blue, 50 mg) and ringing the cover slip several times, on separate days, with ordinary nail polish. The slides must be stored in horizontal position. Also see References 326, 347 and 348.

ACKNOWLEDGMENT

The authors thank Dr. E. S. Luttrell, Department of Plant Pathology, University of Georgia, Athens, for his critical review and helpful suggestions.

REFERENCES

1. Ainsworth, G. C., in *The Fungi,* Vol. 3, p. 505, G. C. Ainsworth and A. S. Sussman, Eds. Academic Press, New York (1968).
2. Leighton, T. J., Dill, B. C., Stock, J. J., and Phillips, C., *Proc. Nat. Acad. Sci. U.S.A., 68,* 677 (1971).
3. Dupraw, E. J., *DNA and Chromosomes.* Holt, Rinehart and Winston, New York (1970).
4. Bracker, C. E., *Annu. Rev. Phytopathol., 5,* 343 (1967).
5. Grove, S. N., and Bracker, C. E., *J. Bacteriol., 104,* 989 (1970).
6. Girbardt, M., *Protoplasma, 6,* 413 (1969).
7. Marchant, R., Peat, A., and Banbury, G. H., *New Phytol., 66,* 623 (1967).
8. Reichle, R. E., and Alexander, J. V., *J. Cell Biol., 24,* 489 (1965).
9. Girbardt, M., *Arch. Mikrobiol., 39,* 351 (1961).
10. Moore, R. T., and McAlear, J. H., *Mycologia, 53,* 194 (1961).
11. Heath, I. B., and Greenwood, A. D., *J. Gen. Microbiol., 62,* 129 (1970).
12. Leiva, S., and Carbonell, L. M., *J. Gen. Microbiol., 62,* 43 (1970).
13. Robinow, C. F., and Bakerspigel, A., in *The Fungi,* Vol. 1, p. 119, G. C. Ainsworth and A. S. Sussman, Eds. Academic Press, New York (1965).
14. Weijer, J., and Weisberg, S. H., *Can. J. Genet. Cytol., 8,* 361 (1966).
15. Weisberg, S. H., and Weber, J., *Can. J. Genet. Cytol., 10,* 699 (1968).
16. Burnett, J. H., *Fundamentals of Mycology.* St. Martin's Press, New York (1968).
17. Olive, L. J., in *The Fungi,* Vol. 1, p. 143, G. C. Ainsworth and A. S. Sussman, Eds. Academic Press, New York (1965).
18. Storck, R., and Alexopoulos, C. J., *Bacteriol. Rev., 34,* 126 (1970).
19. Moyer, B. C., and Storck, R., *Arch. Biochem. Biophys., 104,* 193 (1964).
20. Henney, H. R., and Storck, R., *Proc. Nat. Acad. Sci. U.S.A., 51,* 1050 (1964).
21. Margulis, L., *Origin of Eukaryotic Cells.* Yale University Press, New Haven, Connecticut (1971).
22. Hunsley, D., and Burnett, J. H., *J. Gen. Microbiol., 62,* 203 (1970).
23. Rosinski, M. A., and Campana, R. J., *Mycologia, 56,* 738 (1964).
24. Michell, A. J., and Scurfield, G., *Trans. Brit. Mycol. Soc., 55,* 488 (1970).
25. Gorin, P. A. J., and Spencer, J. F. T., *Carbohyd. Res., 13,* 339 (1970).
26. Spencer, J. F. T., and Gorin, P. A. J., *Mycologia, 63,* 387 (1971).
27. Manocha, J. J., and Colvin, J. R., *J. Bacteriol., 94,* 202 (1970).
28. Lemke, P. A., and Ness, T. M., *J. Virol., 6,* 813 (1970).
29. Bartnicki-Garcia, S., *Annu. Rev. Microbiol., 22,* 87 (1968).
30. Mahadelin, P. R., and Mahaskar, V. R., *Indian J. Exp. Biol., 8,* 207 (1970).
31. Nickerson, W. J., *Bacteriol. Rev., 27,* 305 (1963).
32. Szaniszlo, P. J., and Mitchell, R., *J. Bacteriol., 106,* 640 (1971).
33. Russell, D. W., Sturgeon, R. J., and Ward, W., *J. Gen. Microbiol., 36,* 289 (1964).
34. Taber, R. A., Pettit, R. E., Taber, W. A., and Dollahite, J. W., *Mycologia, 60,* 727 (1968).
35. Gregory, P. H., and Lacey, M. E., *Trans. Brit. Mycol. Soc., 47,* 25 (1964).
36. Johnston, L. R., *Biochem. J., 96,* 651 (1965).
37. Hamilton, P. B., and Knight, S. G., *Arch. Biochem. Biophys., 99,* 282 (1962).
38. Carroll, G. C., *Ph.D. Thesis.* University of Texas, Austin, Texas (1966).
39. Furtado, J., *Mycologia, 63,* 104 (1971).
40. Stein, W. D., *The Movement of Molecules Across Cell Membranes.* Academic Press, New York (1967).
41. Vanderkool, G., and Green, D. E., *Bioscience, 21,* 409 (1971).
42. Racker, E., Seminar, Texas A & M University, College Station, Texas (1971).
43. Kotyk, A., and Janacek, K., *Cell Membranes and Transport.* Academic Press, New York (1970).
44. Suomolainen, H. S., and Nurminen, T., *Chem. Phys. Lipids, 4,* 247 (1970).
45. Nurminen, T., and Suomolainen, H. S., *Biochem. J., 118,* 759 (1970).
46. Zalokar, M., *Amer. J. Bot., 46,* 602 (1959).
47. Taber, W. A., *Appl. Microbiol., 12,* 321 (1964).
48. Taber, W. A., and Siepmann, R., *Appl. Microbiol., 14,* 120 (1966).
49. Taber, W. A., *Mycologia, 60,* 345 (1968).
50. Katz, E., in *Antibiotics,* Vol. 2, p. 276, D. Gottlieb and P. D. Shaw, Eds. Springer-Verlag, New York (1967).
51. Lingappa, B. T., and Lingappa, Y., *Nature, 214,* 516 (1967).
52. Bu'Lock, J. D., and Powell, A. J., *Experientia, 21,* 55 (1965).
53. Birkinshaw, J. H., in *The Fungi,* Vol. 1, p. 179, G. C. Ainsworth and A. S. Sussman, Eds. Academic Press, New York (1965).
54. Morris, C. J., and Hall, S. W., *Lipids, 1,* 188 (1966).
55. Shaw, R., *Biochim. Biophys. Acta, 98,* 230 (1965).
56. Jack, R. C. M., *J. Bacteriol., 91,* 2101 (1966).

57. Starrott, A. N., and Madhosing, C., *Can. J. Microbiol., 13,* 1351 (1967).
58. Wirth, J. C., and Anand, S. R., *Can. J. Microbiol., 10,* 23 (1964).
59. McKeen, W. E., *Can. J. Microbiol., 16,* 1041 (1970).
60. Brody, S., and Nyl, J. F., *J. Bacteriol., 104,* 780 (1970).
61. Leegwater, D. C., and Craig, B. M., *Can. J. Biochem. Physiol., 40,* 858 (1962).
62. Savioja, T., and Miettinen, J. K., *Acta Chem. Scand., 20,* 2451 (1966).
63. Narumi, K., *J. Biol. Chem., 242,* 2233 (1967).
64. Mandels, G. R., Vitols, R., and Parrish, F. W., *J. Bacteriol., 90,* 1589 (1965).
65. Horikoshi, K., and Ikeda, Y., *J. Bacteriol., 91,* 1883 (1966).
66. Hill, E. P., and Sussman, A. S., *J. Bacteriol., 88,* 1556 (1964).
67. Letourneau, D., *Mycologia, 58,* 934 (1966).
68. Lewis, D. H., and Harley, J. L., *New Phytol., 64,* 238 (1965).
69. Lewis, D. H., and Harley, J. L., *New Phytol., 64,* 256 (1965).
70. Taber, W. A., and Tertzakian, G., *Appl. Microbiol., 13,* 590 (1965).
71. Cochrane, V. C., *Physiology of Fungi,* p. 42. John Wiley and Sons, New York (1958).
72. Perlin, A., and Taber, W. A., *Can. J. Chem., 41,* 2278 (1963).
73. Liss, E., and Langen, P., *Biochem. Z., 333,* 193 (1960).
74. Schmidt, G., in *Phosphorus Metabolism,* Vol. 1, p. 443, W. D. McElroy and B. Glass, Eds. Johns Hopkins Press, Baltimore, Maryland (1951).
75. Nishi, A., *J. Biochem. (Tokyo), 48,* 758 (1960).
76. Harold, F. M., *Biochim. Biophys. Acta, 45,* 172 (1960).
77. van Steveninck, J., and Dawson, E. C., *Biochim. Biophys. Acta, 150,* 47 (1968).
78. Mann, T., *Biochem. J., 38,* 345 (1944).
79. Brar, S. S., Giam, C. S., and Taber, W. A., *Mycologia, 60,* 806 (1968).
80. Benedict, R. G., and Brady, L. R., in *Fermentation Advances,* p. 63, D. Perlman, Ed. Academic Press, New York (1969).
81. Sermonti, G., *Genetics of Antibiotic-Producing Microorganisms.* Interscience Publications, John Wiley and Sons, New York (1969).
82. Bu'Lock, J. D., *The Biosynthesis of Natural Products.* McGraw-Hill, New York (1965).
83. Bu'Lock, J. D., *Essays in Biosynthesis and Microbial Development.* John Wiley and Sons, New York (1967).
84. Demain, A. L., *Lloydia, 31,* 395 (1968).
85. Vining, L. C., *Can. J. Microbiol., 16,* 473 (1970).
86. Vining, L. C., and Taber, W. A., in *Biochemistry of Industrial Microorganisms,* p. 341, C. Rainbow and A. H. Rose, Eds. Academic Press, London, England (1963).
87. Taber, W. A., *Lloydia, 30,* 39 (1967).
88. Wallen, L. L., Stodola, F. H., and Jackson, R. W., *Type Reactions in Fermentation Chemistry (U.S.D.A. ARS T1-13).* U.S. Government Printing Office, Washington, D.C. (1959).
89. Shibata, S., Natori, S., and Udagawa, S., *List of Fungal Products.* Charles C Thomas, Springfield, Illinois (1964).
90. Perlman, D. (Ed.), *Fermentation Advances.* Academic Press, New York (1969).
91. Miller, M. W., *The Pfizer Handbook of Microbial Metabolites.* Blakiston Division, McGraw-Hill, New York (1961).
92. Bu'Lock, J. D., and Barr, J. G., *Lloydia, 31,* 342 (1968).
93. Gaumann, F., *Advan. Enzymol., 11,* 401 (1951).
94. Hou, C. T., Ciegler, A., and Hesseltine, C. W., *Appl. Microbiol., 21,* 1101 (1971).
95. Townsend, R. J., Moss, M. O., and Peck, H. M., *J. Pharm. Pharmacol., 18,* 471 (1966).
96. Van Der Merve, K. P., Steyn, P. S., and Fourie, L., *J. Chem. Soc.,* p. 7083 (1965).
97. Purchase, I. F. H., and Nel, W., in *Biochemistry of Some Food-Borne Microbial Toxins,* p. 153, R. I. Mateles and G. N. Wogan, Eds. M.I.T. Press, Cambridge, Massachusetts (1967).
98. Taylor, A., in *Biochemistry of Some Food-Borne Microbial Toxins,* p. 67, R. I. Mateles and G. N. Wogan, Eds. M.I.T. Press, Cambridge, Massachusetts (1967).
99. Forgas, J., in *Mycotoxins in Food Stuffs,* p. 87, G. Wogan, Ed. M.I.T. Press, Cambridge, Massachusetts (1964).
100. Tanabe, H., and Suzuki, T., in *Toxic Microorganisms, Proceedings of the First U.S.-Japan Conference,* p. 127, M. Herzberg, Ed. U.S. Government Printing Office, Washington, D.C. (1970).
101. Kanota, K., in *Toxic Microorganisms, Proceedings of the First U.S.-Japan Conference,* p. 129, M. Herzberg, Ed. U.S. Government Printing Office, Washington, D.C. (1970).
102. Bamburg, J. R., Riggs, N. V., and Strong, F. M., *Tetrahedron, 24,* 3329 (1968).
103. Mirocha, C. J., and Uritani, I. (Eds.), *The Dynamic Role of Molecular Constituents in Plant Parasite Interaction.* Bruce Publishing Co., St. Paul, Minnesota (1967).
104. Gaumann, E., Naef-Roth, S., and Kern, H., *Phytopathol. Z., 37,* 145 (1960).
105. Gaumann, E., *Endeavour, 13,* 198 (1954).
106. Gaumann, E., and Obrist, W., *Phytopathol. Z., 37,* 145 (1960).
107. de Mayo, P., Spencer, E. Y., and White, R. W., *J. Amer. Chem. Soc., 84,* 494 (1962).
108. Nakamura, M., and Oku, H., *Ann. Takamine Lab., 12,* 266 (1960).

109. Orsenigo, M., *Phytopathol. Z., 29,* 189 (1957).
110. Nazoe, S., Morisaku, M., Tsuda, K., Iitaka, Y., Takahashi, N., Tamura, S., Ishibashi, K., and Shirasaka, M., *J. Amer. Chem. Soc., 87,* 4768 (1965).
111. Owens, L. D., *Science, 165,* 18 (1969).
112. Kuo, M. S., and Scheffer, R. P., *Phytopathology, 54,* 1041 (1964).
113. Urray, W. H., Wehrmeister, H. L., Hodge, E. B., and Hidy, P. H., *Tetrahedron Lett.,* No. 27, p. 3109 (1966).
114. Mirrington, B. N., Ritchie, E., Shoppee, C. W., Taylor, C. W., and Sternhell, S., *Tetrahedron Lett.,* No. 7, p. 365 (1964).
115. Kobel, F., *Phytopathol. Z., 18,* 157 (1951).
116. Feldman, A. W., Caroselli, N. E., and Howard, F. L., *Phytopathology, 40,* 341 (1950).
117. Boyer, M. G., *Diss. Abstr., 18,* 1564 (1958).
118. Morovka, N., and Tatsuno, T., in *Toxic Microorganisms, Proceedings of the First U.S.-Japan Conference,* p. 114, M. Herzberg, Ed. U.S. Government Printing Office, Washington, D.C. (1970).
119. Poole, D. D., and Murphy, H. C., *Phytopathology, 42,* 16 (1952).
120. Sundheim, H., and Wilcoxson, R. D., *Phytopathology, 55,* 546 (1965).
121. Chen, Y. S., *Bull. Agr. Chem. Soc. Jap., 24,* 372 (1960).
122. Perone, V. B., Scheel, L. D., and Meitus, R. A., *J. Invest. Dermatol., 42,* 267 (1964).
123. Christiansen, C. M., Nelson, G. H., Mirocha, C. J., Bates, F., and Dorworth, C. E., *Appl. Microbiol., 14,* 774 (1966).
124. Waksman, S. A., and Bugie, E., *J. Bacteriol., 48,* 527 (1944).
125. Bushnell, W. R., and Allen, P. J., *Plant Physiol., 37,* 50 (1962).
126. Borrow, A., Jeffreys, E. G., Kessell, R. H. J., Lloyd, E. C., Lloyd, P. B., and Nixon, I. J., *Can. J. Microbiol., 7,* 227 (1961).
127. Borrow, A., Brown, C., Jeffreys, E. G., Kessel, R. H. J., Lloyd, E. C., Lloyd, P. B., Rothwell, A., Rothwell, B., and Swait, J. C., *Can. J. Microbiol., 10,* 407 (1964).
128. Emerson, S., *J. Bacteriol., 60,* 22 (1950).
129. Taber, W. A., and Siepmann, R., *Appl. Microbiol., 13,* 827 (1965).
130. Malek, I., Beran, K., and Hospodka, J. (Eds.), *Biogenesis of Antibiotics.* Academic Press, New York (1962).
131. Gottlieb, D., *Mycologia, 63,* 619 (1971).
132. Butler, G. M., in *The Fungi,* Vol. 2, p. 83, G. C. Ainsworth and A. S. Sussman, Eds. Academic Press, New York (1966).
133. Barron, G. L., *The Genera of Hyphomycetes from Soil.* Williams and Wilkins, Baltimore, Maryland (1968).
134. Beadle, G. W., and Coonradt, V. L., *Genetics, 29,* 291 (1944).
135. Garnjobst, L., *Amer. J. Bot., 42,* 444 (1955).
136. Newmeyer, D., *Can. J. Genet. Cytol., 12,* 914 (1970).
137. Amici, A. M., Cotti, T. S., Spalla, C., and Tognoli, L., *Appl. Microbiol., 15,* 611 (1967).
138. Anderson, J. A., Sun, F. K., McDonald, J. K., and Cheldelin, V. H., *Arch. Biochem. Biophys., 107,* 37 (1964).
139. Lindenmeyer, A., in *The Fungi,* Vol. 1, p. 301, G. C. Ainsworth and A. S. Sussman, Eds. Academic Press, New York (1965).
140. Sjorgren, R. F., and Romano, A. H., *J. Bacteriol., 93,* 1638 (1967).
141. Hoare, D. S., and Work, E., *Biochem. J., 61,* 562 (1955).
142. Blumenthal, H. J., in *The Fungi,* Vol. 1, p. 229, G. C. Ainsworth and A. S. Sussman, Eds. Academic Press, New York (1965).
143. Niederpruem, D. J., in *The Fungi,* Vol. 1, p. 269, G. C. Ainsworth and A. S. Sussman, Eds. Academic Press, New York (1965).
144. Zalokar, M., in *The Fungi,* Vol. 1, p. 377, G. C. Ainsworth and A. S. Sussman, Eds. Academic Press, New York (1965).
145. Cochrane, V. W., *Physiology of Fungi,* p. 99. John Wiley and Sons, New York (1958).
146. Ilag, L., and Curtis, R. W., *Science, 163,* 1357 (1968).
147. Cherest, H., Surdin-Kerjan, Y., and de Robichow-Szulmajster, H., *J. Bacteriol., 106,* 758 (1971).
148. Magee, P. T., and Hereford, L. M., *J. Bacteriol., 98,* 857 (1969).
149. Hynes, M. J., *J. Bacteriol., 103,* 482 (1970).
150. Williams, L. G., and Davis, R. H., *J. Bacteriol., 103,* 335 (1970).
151. Atkinson, D. E., *Annu. Rev. Biochem., 35,* 85 (1966).
152. Tucci, A. F., *J. Bacteriol., 99,* 624 (1969).
153. Gross, S. R., *Proc. Nat. Acad. Sci. U.S.A., 54,* 1538 (1965).
154. Bussey, H., and Umbarger, H. E., *J. Bacteriol., 103,* 277 (1970).
155. Lacroute, F., *J. Bacteriol., 95,* 824 (1968).
156. Turner, J. R., Sorsoli, W. R., and Matchett, W. H., *J. Bacteriol., 103,* 364 (1970).
157. Kuo, S., and Cirillo, V., *J. Bacteriol., 103,* 679 (1970).
158. Eberhart, B. M., and Beck, R. J., *J. Bacteriol., 101,* 408 (1970).
159. Pall, M. L., *Biochim. Biophys. Acta, 203,* 139 (1970).
160. Thwaites, W. M., and Pendyla, L., *Biochim. Biophys. Acta, 192,* 455 (1969).

161. Grenson, M., and Hennaut, C., *J. Bacteriol., 105,* 477 (1971).
162. DeBusk, C. G., and DeBusk, A. G., *Biochim. Biophys. Acta, 104,* 139 (1965).
163. Kundig, W., Ghosh, S., and Roseman, S., *Proc. Nat. Acad. Sci. U.S.A., 52,* 1067 (1964).
164. Brown, C. E., and Romano, A. H., *J. Bacteriol., 100,* 1198 (1969).
165. Schneider, R. P., and Wiley, W. R., *J. Bacteriol., 106,* 487 (1971).
166. Schneider, R. P., and Wiley, W. R., *J. Bacteriol., 106,* 479 (1971).
167. Scarborough, G. A., *J. Biol. Chem., 245,* 1694 (1970).
168. Taber, W. A., *Mycologia, 63,* 290 (1971).
169. Kotyk, A., and Janacek, K. (Eds.), *Cell Membrane and Transport.* Academic Press, New York (1970).
170. Kalckar, H. M., *Science, 174,* 557 (1971).
171. Tosteson, D. L. (Ed.), *The Molecular Basis of Membrane Function.* Prentice-Hall, Englewood Cliffs, New Jersey (1969).
172. Kaback, H. R., *Annu. Rev. Biochem., 39,* 561 (1970).
173. Pardee, A. B., *Science, 162,* 632 (1968).
174. Berlin, R. D., *Science, 168,* 1539 (1970).
175. Cirillo, V. P., *Annu. Rev. Microbiol., 15,* 197 (1961).
176. Koch, A. L., *Biochim. Biophys. Acta, 79,* 177 (1964).
177. Wikberg, E., *Physiol. Plant., 12,* 100 (1959).
178. Lilly, V. G., and Barnett, H. L., *Physiology of the Fungi.* McGraw-Hill, New York (1951).
179. Tabak, H. H., and Cooke, W. B., *Mycologia, 60,* 115 (1968).
180. Esser, K., in *The Fungi,* Vol. 2, p. 661, G. C. Ainsworth and A. S. Sussman, Eds. Academic Press, New York (1966).
181. Emerson, J., in *The Fungi,* Vol. 2, p. 513, G. C. Ainsworth and A. S. Sussman, Eds. Academic Press, New York (1966).
182. Pontecorvo, G., and Sermonti, G., *J. Gen. Microbiol., 11,* 94 (1954).
183. Roper, J. A., in *The Fungi,* Vol. 2, p. 589, G. C. Ainsworth and A. S. Sussman, Eds. Academic Press, New York (1966).
184. Jinks, J. L., in *The Fungi,* Vol. 2, p. 619, G. C. Ainsworth and A. S. Sussman, Eds. Academic Press, New York (1966).
185. Booth, C., in *The Fungi,* Vol. 2, p. 133, G. C. Ainsworth and A. S. Sussman, Eds. Academic Press, New York (1966).
186. Miller, J. H., *Mycologia, 41,* 99 (1949).
187. Luttrell, E. S., Taxonomy of the Pyrenomycetes. *Univ. Mo. Stud., 24,* 1 (1951).
188. Chadefaud, M., and Emberger, L., *Les Végétaux Non Vasculaires,* Vol. 1, Masson, Paris, France (1960).
189. Alexopoulos, C. J., *Introductory Mycology.* John Wiley and Sons, New York (1966).
190. Martin, G. W., Key to the Families of Fungi, in *Dictionary of Fungi,* 5th ed., p. 497, G. C. Ainsworth, Ed. Commonwealth Mycological Institute, Kew, Surrey, England (1961).
191. Dennis, R. W. G., *British Ascomycetes.* J. Cramer, Lehre, Germany (1968).
192. Seaver, F. J., *The North American Cup Fungi (Inoperculates).* Published by author, New York (1951).
193. Seaver, F. J., *The North American Cup Fungi (Operculates).* Published by author, New York (1928).
194. von Arx, J. A., *Pilzkunde.* J. Cramer, Lehre, Germany (1967).
195. von Arx, J. A., *The Genera of Fungi Sporulating in Pure Culture.* J. Cramer, Lehre, Germany (1970).
196. Chupp, C., and Sherf, A. F., *Vegetable Diseases and Their Control.* Ronald Press, New York (1960).
197. Pirone, P. P., Dodge, B. O., and Rickett, H. W., *Diseases and Pests of Ornamental Plants.* Ronald Press, New York (1960).
198. U.S. Department of Agriculture, *Agriculture Handbook No. 165, Index of Plant Disease in the United States.* U.S. Government Printing Office, Washington, D.C. (1960).
199. Agriculture Research Service, USDA, *Agriculture Handbook No. 291, Losses in Agriculture.* U.S. Government Printing Office, Washington, D.C. (1965).
200. Johnson, T. W., and Sparrow, F. K., *Fungi in Oceans and Estuaries.* J. Cramer, Lehre, Germany (1961).
201. Kohlmeyer, J., and Kohlmeyer, E., *Icones Fungorum Maris,* Vols. 1-7. J. Cramer, Lehre, Germany (1964-1969).
202. Jones, E. B. G., *Trans. Brit. Mycol. Soc., 45,* 93 (1962).
203. Jones, E. B. G., *Trans. Brit. Mycol. Soc., 46,* 135 (1963).
204. Sguros, P. L., and Simms, J., *Can. J. Microbiol., 9,* 585 (1963).
205. Vishniac, H. S., and Riley, G., Vitamin B_{12} and Thiamin in Long Island Sound: Patterns of Distribution and Ecological Significance, *Preprints International Oceanographic Congress,* p. 942. American Association for the Advancement of Science, Washington, D.C. (1959).
206. Anita, N. J., *Can. J. Microbiol., 9,* 403 (1963).
207. Belser, W. L., *Proc. Nat. Acad. Sci. U.S.A., 45,* 1533 (1959).
208. Oppenheimer, C. (Ed.), *Symposium on Marine Microbiology.* Charles C Thomas, Springfield, Illinois (1961).
209. Kohlmeyer, J., *Trans. Brit. Mycol. Soc., 50,* 137 (1967).
210. Kohlmeyer, J., and Kohlmeyer, E., *Mycologia, 63,* 831 (1971).
211. Meyers, S. A., and Simms, J., *Can. J. Bot., 43,* 379 (1965).

212. Cavaliere, A. R., *Mycologia, 60,* 475 (1968).
213. Hughes, G. C., *Ph.D. Dissertation: Ecological Aspects of Some Lignicolous Fungi in Estuarine Waters.* Florida State University, Tallahassee, Florida (1960).
214. Ingold, C. T., *Dispersal in Fungi.* Clarendon Press, Oxford, England (1953).
215. Sparrow, F. K., in *The Fungi,* Vol. 3, p. 41, G. C. Ainsworth and A. S. Sussman, Eds. Academic Press, New York (1968).
216. Martin, M. M., *Science, 169,* 16 (1970).
217. Main, E. B., *Mycologia, 50,* 169 (1958).
218. Kobayasi, Y., *Sci. Rep. Tokyo Bunrika Daigaku, 5,* 53 (1941).
219. Madelin, M. F., in *The Fungi,* Vol. 3, p. 227, G. C. Ainsworth and A. S. Sussman, Eds. Academic Press, New York (1968).
220. Aoki, J., *J. Sericult. Sci. Jap., 30,* 43 (1961).
221. Lepesme, P., *Bull. Soc. Hist. Natur. Agr., 29,* 372 (1938).
222. Burnside, C. E., *USDA Technical Bulletin 149.* U.S. Government Printing Office, Washington, D.C. (1930).
223. Toumanoff, A., *Int. Corn Borer Invest. Sci. Rep.,* p. 74 (1928).
224. Abbas, H. M., *Agr. Pakistan, 10,* 195 (1959).
225. Cockbain, J. A., and Hastie, A. C., *J. Insect Pathol., 3,* 95 (1961).
226. Beal, R. H., and Kais, A. G., *J. Insect Pathol., 4,* 488 (1962).
227. Steinhaus, E. A. (Ed.), *Insect Pathology.* Academic Press, New York (1963).
228. Steinhaus, E. A., *Insect Microbiology.* Comstock Publishing Associates, Ithaca, New York (1946).
229. Garrett, S. D., *Soil Fungi and Soil Fertility.* Pergamon Press, Oxford, England (1963).
230. Ajello, L., *Sabouraudia, 6,* 147 (1968).
231. Emmons, C. W., *Pub. Health Rep., 64,* 892 (1949).
232. Bloomfield, B. J., and Alexander, M., *J. Bacteriol., 93,* 1276 (1967).
233. Gilman, J. C., *A Manual of Soil Fungi,* 2nd ed. Iowa State College Press, Ames, Iowa (1957).
234. Taber, W. A., *Proc. Iowa Acad. Sci., 58,* 209 (1951).
235. Ames, L. M., *A Monograph of the Chaetomiaceae. U.S. Army Research and Development Series No. 2.* Army Research Office, Natick, Massachusetts (1961).
236. Orr, G. F., and Kuehn, H. H., *Mycologia, 63,* 191 (1971).
237. Frederick, L., Uecker, F. A., and Benjamin, C. R., *Mycologia, 61,* 1077 (1969).
238. Maas, J. L., and Stuntz, D. E., *Mycologia, 61,* 1106 (1969).
239. Kishomoto, R. A., and Baker, G. E., *Mycologia, 61,* 537 (1969).
240. Barron, G. L., Cain, R. F., and Gilman, J. C., *Can. J. Bot., 39,* 1610 (1961).
241. Cain, R. F., and Farrow, W. M., *Can. J. Bot., 34,* 689 (1956).
242. Ranzoni, F. V., *Mycologia, 60,* 356 (1968).
243. Gochenaur, S. E., and Backus, M. P., *Mycologia, 59,* 893 (1967).
244. van Warmelo, K. T., *Mycologia, 58,* 846 (1966).
245. Joffe, A. Z., and Borut, S. Y., *Mycologia, 58,* 692 (1966).
246. Hodge, C. S., *Mycologia, 54,* 221 (1962).
247. Rall, G., *Mycologia, 57,* 872 (1965).
248. Goos, R. D., *Mycologia, 55,* 142 (1963).
249. Goos, R. D., *Mycologia, 52,* 877 (1960).
250. Miller, J. H., Giddens, J. E., and Foster, A. A., *Mycologia, 49,* 779 (1957).
251. Routien, J. B., *Mycologia, 49,* 188 (1957).
252. Parkinson, D., and Waid, J. S. (Eds.), *The Ecology of Soil Fungi.* Liverpool University Press, Liverpool, England (1960).
253. Webster, J., *Trans. Brit. Mycol. Soc., 54,* 161 (1970).
254. Koehn, R. D., *Mycologia, 63,* 441 (1971).
255. Walkey, D. G. A., and Harvey, F., *Trans. Brit. Mycol. Soc., 48,* 35 (1965).
256. Kar, A. K., and Pal, K. P., *Mycologia, 60,* 1086 (1968).
257. Obrist, W., *Can. J. Bot., 39,* 943 (1961).
258. Cain, R., *Can. J. Bot., 39,* 1633 (1961).
259. Cain, R. F., *Can. J. Bot., 34,* 675 (1956).
260. Cain, R. F., and Weresub, L. K., *Can. J. Bot., 35,* 119 (1957).
261. Cain, R. F., *Can. J. Bot., 35,* 255 (1957).
262. Hanlin, R. T., *Ga. Exp. Sta. N.S., 175,* 1 (1963).
263. Harley, J. L., in *The Fungi,* Vol. 3, p. 139, G. C. Ainsworth and A. S. Sussman, Eds. Academic Press, New York (1968).
264. Trappe, J. M., *Proceedings of the First National American Conference on Mycorrhizae, USDA, 1971.* U.S. Government Printing Office, Washington, D.C. (in press).
265. Trappe, J. M., and Guzman, G., *Mycologia, 63,* 317 (1971).

266. Ainsworth, G. C., *Dictionary of the Fungi,* 5th ed. Commonwealth Mycological Institute, Kew, Surrey, England (1961). Also see 6th ed. (1971).
267. Bessey, E. A., *Morphology and Taxonomy of the Fungi.* Blakiston Co., Philadelphia, Pennsylvania (1950).
268. Ramsbottom, J., *Mushrooms and Toadstools.* Collins, London, England (1953).
269. Hawker, L. E., *Trans. Brit. Mycol. Soc., 38,* 73 (1955).
270. Burdsall, H. H., *Mycologia, 60,* 496 (1968).
271. Lange, M., *Dan. Bot. Ark., 16,* 1 (1956).
272. Szemere, L., *Die unterirdischen Pilze des Karpatenbeckens.* Akad. Kiaao, Budapest, Hungary (1965).
273. Ellis, L. F., and Kleinschmidt, W. J., *Nature (London), 215,* 649 (1967).
274. Banks, G. T., Buck, K. W., Chain, E. B., Darbyshire, J. E., and Himmelweit, F., *Nature (London), 222,* 89 (1969).
275. Hansford, C. G., *Mycological Paper #15,* Commonwealth Mycological Institute, Kew, Surrey (1969).
276. Nicot, J., *Bull. Soc. Mycol. Fr., 78,* 221 (1962).
277. Seeler, E. V., Jr., *Farlowia, 1,* 119 (1943).
278. Conant, N. F., Smith, D. T., Baker, R. D., Callaway, J. L., and Martin, D. S., *Manual of Clinical Mycology.* Saunders, Philadelphia, Pennsylvania (1954).
279. Shimoto, R. A., and Baker, G. E., *Mycologia, 61,* 537 (1969).
280. Wang, C. J. K., *Fungi of Pulp and Paper in New York.* State University of New York, College of Forestry at Syracuse University, Syracuse, New York (1965).
281. Brewer, D., *TAPPI J. Tech. Assoc. Pulp Pap. Ind., 43,* 609 (1960).
282. Brandsberg, J. W., *Mycologia, 61,* 373 (1969).
283. Skolko, A. J., and Groves, J. W., *Can. J. Res., 26,* 269 (1948).
284. Skolko, A. J., *Can. J. Bot., 31,* 779 (1953).
285. Taber, W. A., and Heacock, R. A., *Can. J. Microbiol., 8,* 137 (1962).
286. Anonymous, *Seeds. The Yearbook of Agriculture.* U.S. Government Printing Office, Washington, D. C. (1961).
287. Ajello, L., and Cheng, S., *Mycologia, 59,* 689 (1967).
288. Rogerson, C. T., *Mycologia, 62,* 865 (1970).
289. Martin, G. W., *Outline of the Fungi.* Wm. C. Brown Co., Dubuque, Iowa (1950).
290. Nickerson, W. J., *Bacteriol. Rev., 27,* 305 (1963).
291. Greenhalgh, G. N., and Evans, L. V., *Trans. Brit. Mycol. Soc., 50,* 183 (1967).
292. Reeves, F. B., *Mycologia, 63,* 204 (1971).
293. Miller, J. H., *Mycologia, 41,* 91 (1949).
294. Korf, R. P., *Taxon, 19,* 782 (1970).
295. Luttrell, E. S., *Trans. Brit. Mycol. Soc., 48,* 135 (1965).
296. Luttrell, E. S., *Mycologia, 47,* 511 (1955).
297. Batra, R., *Mycologia, 59,* 976 (1967).
298. Ainsworth, G. C., in *The Fungi,* Vol. 3, p. 211, G. C. Ainsworth and A. S. Sussman, Eds. Academic Press, New York (1968).
299. Gregory, P. H., *The Microbiology of the Atmosphere.* Interscience Publications, John Wiley and Sons, New York (1961).
300. Esser, K., Heterogenic Incompatibility in *Incompatibility of Fungi,* p. 6, K. Esser and J. R. Raper, Eds. Springer-Verlag, Berlin, Germany (1965).
301. Broadbent, D., *Bot. Rev., 32,* 219 (1966).
302. Boyd, M. R., Wilson, B. J., and Harris, T. M., *Chem. Eng. News, 49,* 25 (1971).
303. Lowe, D. P., *Check List and Host Index of Bacteria, Fungi and Mistletoes of British Columbia.* Canadian Forestry Service, Department of Fisheries and Forestry, Victoria, B.C., Canada (1969).
304. Harris, J. L., and Taber, W. A., *Mycologia, 62,* 152 (1970).
305. Harris, J. L., *Mycologia, 62,* 1130 (1970).
306. Luttrell, E. S., *Phytopathology, 55,* 828 (1965).
307. Corlett, M., *Can. J. Bot., 49,* 39 (1971).
308. Talbot, P. H. B., *Priniciples of Fungal Taxonomy.* Macmillan, London, England (1971).
309. Miller, J. H., *A Monograph of the World Species of Hypoxylon.* University of Georgia Press, Athens, Georgia (1961).
310. Diendoerfer, F. H., in *Advances in Applied Microbiology,* Vol. 2, p. 321, W. W. Umbreit, Ed. Academic Press, New York (1960).
311. Gaden, E. L., Jr., *Chem. Ind. (London),* p. 154 (1959).
312. Aiba, S., Humphrey, A. E., and Millis, N. F., *Biochemical Engineering.* Academic Press, New York (1965).
313. Blakebrough, N. (Ed.), *Biochemical and Biological Engineering Science,* Vol. 1. Academic Press, London, England (1967).
314. Christensen, C. M., and Kaufmann, H. H., *Grain Storage. The Role of Fungi in Quality Loss.* University of Minnesota Press, Minneapolis, Minnesota (1969).
315. Barnett, H. L., *Mycologia, 56,* 1 (1964).
316. Johnson, L. F., Curl, E. A., Bond, J. H., and Fribour, H. A., *Methods for Studying Soil Microflora-Plant Disease Relationships.* Burgess Publishing Co., Minneapolis, Minnesota (1959).

317. Johnson, L. F., and Curl, E. A., *Methods for Research on the Ecology of Soil-Borne Plant Pathogens.* Burgess Publishing Co., Minneapolis, Minnesota (1971).
318. Ajello, L., Georg, L. K., Kaplan, W., and Kaufman, L., *Laboratory Manual for Medical Mycology.* U.S. Department of Health, Education and Welfare, Communicable Disease Center, Atlanta, Georgia (1963).
319. Norris, J. R., and Ribbons, D. W. (Eds.), *Methods in Microbiology,* Vol. 1. Academic Press, New York (1969).
320. Martin, S. M. (Ed.), *Culture Collections: Perspectives and Problems.* University of Toronto Press, Toronto, Ont., Canada (1963).
321. Ellis, J. J., and Roberson, J. A., *Mycologia, 60,* 399 (1968).
322. Wellman, A. M., and Walden, D. K., *Can. J. Microbiol., 10,* 585 (1964).
323. Hwang, S., *Appl. Microbiol., 14,* 784 (1966).
324. Little, G. N., and Gordon, M. A., *Mycologia, 59,* 733 (1967).
325. Pollack, F. G., *Mycologia, 59,* 541 (1967).
326. Dade, H. A., and Waller, S., *Mycological Paper #27.* Commonwealth Mycological Institute, Kew, Surrey, England (1949).
327. Hennebert, G. L., Pleomorphism in Fungi Imperfecti, in *Taxonomy of Fungi Imperfecti,* p. 202, B. Kendrick, Ed. University of Toronto Press, Toronto, Ont., Canada (1971).
328. Muller, E., Imperfect-Perfect Connections in Ascomycetes, in *Taxonomy of Fungi Imperfecti,* p. 184, B. Kendrick, Ed. University of Toronto Press, Toronto, Ont., Canada (1971).
329. Kendrick, B. (Ed.), in *Taxonomy of Fungi Imperfecti,* pp. 225-252. University of Toronto Press, Toronto, Ont., Canada (1971).
330. Kwon-Chung, K. J., *Science, 177,* 368 (1972).
331. Ishii, K., Sakai, K., Ueno, Y., Tsunoda, H., and Enomoto, M., *Appl. Microbiol., 22,* 718 (1971).
332. Booth, C., *The Genus Fusarium,* p. 46. Commonwealth Mycological Institute, Kew, Surrey, England (1971).
333. Main, C. E., and Hamilton, P. B., *Appl. Microbiol., 23,* 193 (1972).
334. Hou, C. T., Ciegler, A., and Hesseltine, C. W., *Appl. Microbiol., 23,* 183 (1972).
335. Anonymous, *Use of Fungi Imperfecti in Waste Control.* U.S. Department of the Interior, Federal Water Quality Administration, Superintendent of Documents, Washington, D.C. (1970).
336. Christensen, C. M., *Common Edible Mushrooms.* Charles T. Branford Co., Newton Center, Massachusetts (1964).
337. Gray, W. D., *The Use of Fungi as Food and in Food Processing.* Chemical Rubber Co., Cleveland, Ohio (1970).
338. Burges, H. D., and Hussey, N. W. (Eds.), in *Microbial Control of Insects and Mites,* p. 125. Academic Press, New York (1971).
339. Ciegler, A., Kadis, S., and Ajl, S. J. (Eds.), *Microbial Toxins,* Vol. 6, Fungal Toxins. Academic Press, New York (1971).
340. Kadis, S., Ciegler, A., and Ajl, S. J. (Eds.), *Microbial Toxins,* Vol. 7, Algal and Fungal Toxins. Academic Press, New York (1971).
341. von Arx, J. A., and Muller, E., *Keys to the Genera of Amerospored and Didymospored Pyrenomycetes.* Commonwealth Mycological Institute, Kew, Surrey, England (1969).
342. Staff, *Herb. I.M.I. Handbook.* Commonwealth Mycological Institute, Kew, Surrey, England (1960).
343. Staff, *Plant Pathologists Handbook.* Commonwealth Mycological Institute, Kew, Surrey, England (1968).
344. Hollings, M., and Stone, O. M., *Annu. Rev. Phytopathol., 9,* 93 (1971).
345. Ainsworth, G. C., and Sussman, A. S. (Eds.), *The Fungi,* Vol. 4. Academic Press, New York (in press).
346. Kohlmeyer, J., and Kohlmeyer, E., *Synoptic Plates of Higher Marine Fungi,* 3rd ed. J. Cramer, Lehre, Germany (1971).
347. Kohlmeyer, J., and Kohlmeyer, E., *Mycologia, 64,* 666 (1972).
348. Cunningham, J. L., *Mycologia, 64,* 906 (1972).

YEASTS
General Survey

DR. JAMES D. MACMILLAN AND DR. HERMAN J. PHAFF

INTRODUCTION

Although the term yeast is used extensively in scientific literature, it does not represent a taxonomic designation which can be rigorously defined. Historically, the word itself originated from ancient words describing the visible changes occurring in fermenting liquids. During the 19th century when the biological basis for the alcoholic fermentation became firmly established, the organisms responsible (such as *Saccharomyces cerevisiae*) were described as single, hyaline, round to oval budding cells capable of forming ascospores. After Pasteur related the fermentative process to the ability of yeasts to grow in the absence of air, it was assumed that all yeasts could grow anaerobically. As the years passed, other organisms were discovered which were similar but not identical in morphological and physiological properties and the definition for yeast was expanded to include them. For example, although most yeast cells are hyaline some such as *Rhodotorula* produce red or yellow carotenoid pigments. Budding, too, is a characteristic not common to all yeasts; species of the genus *Schizosaccharomyces* multiply exclusively by fission. Species of several yeast genera such as *Endomycopsis* and *Rhodosporidium* are capable of forming true mycelium. Even fermentation is not a universal characteristic since many species are incapable of anaerobic growth. Many yeasts are apparently unable to form ascospores or other kinds of sexually produced spores and are, therefore, assigned to the Fungi Imperfecti. Some budding yeasts produce spores which are forcibly discharged from the tips of sterigmata. Although these so-called ballistospores resemble basidiospores, they do not result from a sexual process. Recently, other budding yeasts (*Rhodosporidium* and *Leucosporidium*) have been shown to produce sexual spores known as sporidia; these yeasts have life cycles similar to the Ustilaginales of the Basidiomycetes.

With all these variations, it is little wonder that it has been difficult to state a precise definition of yeasts on the basis of general morphological and physiological considerations. According to Lodder,[6] "Yeasts may be defined as microorganisms in which the unicellular form is conspicuous and which belong to the fungi." This simple definition is perhaps the only one possible in view of the heterogeneous nature of this group of organisms. In spite of this, the world's leading yeast taxonomists apparently have had little difficulty in deciding what are yeasts. The most recent taxonomic study of 4,300 strains led to a 1,385-page tome[6] describing 39 genera and 349 species. The major part of the information presented here was obtained from this study.

There are four groups of yeasts:

I. Ascomycetous yeasts: yeasts capable of forming ascospores in asci, considered to be primitive ascomycetes. Twenty-two genera.

II. Basidiomycetous yeasts: yeasts having life cycles similar to those of the order Ustilaginales of the Basidiomycetes. Four genera.

III. Ballistosporogenous yeasts: yeasts in the family Sporobolomycetaceae that forcibly discharge spores by the drop excretion mechanism. Morphologically, ballistospores resemble basidiospores but they are generally considered to possess an asexual rather than a sexual means of reproduction. Three genera.

IV. Asporogenous yeasts: yeasts incapable of producing ascospores, ballistospores, or sporidia. (Some species produce asexual spores called endospores.) Since sexual life cycles do not occur or have not been observed so far, these yeasts are members of the Fungi Imperfecti. Twelve genera.

In addition to the above well-recognized categories, there are a number of microorganisms with yeast-like properties which will also be included in this section.

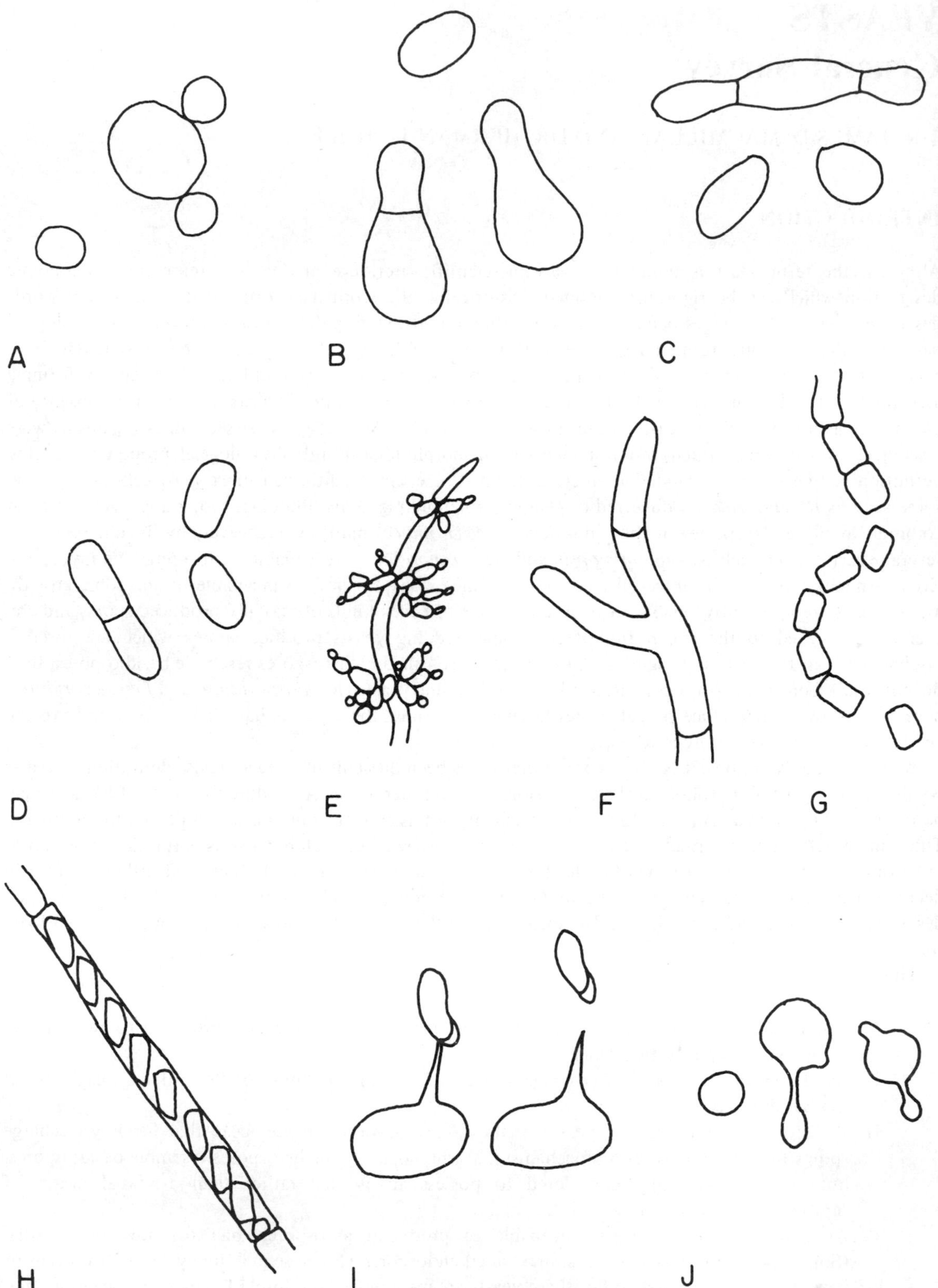

FIGURE 1. Kinds of Asexual Reproduction Known to Occur in Yeasts. A. Multilateral budding. B. Monopolar budding. C. Bipolar budding, bud fission. D. Fission. E. Pseudomycelium. F. Mycelium. G. Arthrospores. H. Endospores. I. Ballistospores. J. Conidia.

The classification of yeasts is based on morphological, cultural, sexual, and physiological characteristics. Some of these properties are shown in Tables 1 to 6.

YEAST MORPHOLOGY

To a considerable extent, the morphology exhibited by a particular yeast is directly associated with the mechanism it employs for asexual reproduction. Various types of asexual or vegetative reproduction are illustrated in Figure 1. Although the majority of yeasts reproduce by budding, fission occurs in some and in others, there is a combination of the two processes. The following terms are used to describe various types of budding: multilateral – buds occurring at different sites on the surfaces of cells; bipolar – buds formed exclusively at the two opposite poles of a cell; monopolar – buds restricted to one pole only; bud-fission – a so-called combination of budding and fission in which a broadly based bud emerges from a yeast cell, usually polarly. Later a cross wall forms by centripetal growth across the base of the bud.

Members of several genera including *Endomycopsis* and *Nematospora* produce true mycelium. The hyphae of some of these filamentous types disarticulate into cells called arthrospores. In some budding yeasts the buds do not detach themselves from one another. The chains of cells that result are reminiscent of filamentous fungal growth and are, therefore, termed pseudomycelia.

In addition to budding, members of the Sporobolomycetaceae characteristically reproduce by production of ballistospores. When agar plates containing colonies of these yeasts are inverted for a short time over fresh plates, spores are forcibly discharged onto the lower plates. Upon germination, these form new colonies that are "mirror images" of the original ones. Members of one genus, *Sterigmatomyces,* reproduce asexually by the formation of single conidia at the ends of sterigmata. These conidia, however, are not forcibly discharged.

Some yeast species, such as members of the genus *Metschnikowia,* form chlamydospores. They are thick-walled, nondeciduous, intercalary or terminal asexual spores formed by rounding off of a cell or cells. According to van der Walt,[10] the chlamydospores of *Candida albicans* may have a special role as a structure in which meiosis takes place (gonotocont) and are, therefore, not truly chlamydospores at all.

Some members of the genera *Trichosporon* and *Oosporidium* produce endospores in older cultures, These spores are vegetative cells which are produced within the confines of other cells or hyphae.[2]

The morphology of vegetative yeast cells is sometimes very useful taxonomically, particularly when the cell shape is characteristic of an entire genus. For example, *Trigonopsis* has triangularly shaped cells; *Pityrosporum* is flask shaped (i.e., because of monopolar budding); *Nadsonia, Hanseniaspora,* and *Saccharomycodes* are apiculate or lemon shaped because of the build-up of scar material at the poles due to bipolar budding. *Brettanomyces* and *Dekkera* are frequently ogival (i.e., pointed at one end). Other words describing cell shape, such as spheroidal, ellipsoidal, ovoidal, cylindrical, elongate, etc., are usually self-explanatory. Examples of typical shapes are shown in Figure 2. Reliance on cell shape or size for yeast identification can, however, be misleading since considerable variation may occur within a single culture. As yeast cells age they become scarred and misshapen from budding and, therefore, appear considerably different from younger cells.

FIGURE 2. Various Shapes of Yeast Cells. From left to right: spheroidal; ovoidal; cylindrical; ogival; triangular; flask-shaped; apiculate.

Taken from: Phaff, H. J., Miller, M. W., and Mrak, E. M., *The Life of Yeasts* (1966). Reproduced by permission of Harvard University Press, Cambridge, Massachusetts.

TABLE 1

MORPHOLOGICAL PROPERTIES OF ASCOMYCETOUS YEASTS

Genus	Number of Species	Cell Shape[a]	Budding[b]	Pseudomycelium[c]	Mycelium[c]	Pellicle[c]	Ascospore Shape[a]	Ascospore Number	Special Morphological Characteristic
Citeromyces	1	A,B	ML	–	–	–	A	1	Asci have thick walls; ascospore walls are warty
Coccidiascus	1	A,H	ML		–		I	8	Spores are fusiform arranged in a helix
Debaryomyces	8	A,B,L	ML	(±)	–	(M)	A,C,D	1–4	Bud and mother cell or two independent cells conjugate
Dekkera	2	A,B,J,K,L	ML	+	–	(+)	E	1–4	Cells frequently ogival
Endomycopsis	10	Various	ML	M	+	(F)	A,E,F,G	1–4	Asci at hyphal tips or intercalary
Hanseniaspora	3	M,C	BP	F	–	–	E,N,O	1–4	Apiculate cells
Hansenula	25	A,B,H,K	ML	±	F	±	E,P,A,F	1–4	Colonies slimy, butyrous, or chalky dull
Kluyveromyces	18	A,B,K,L	ML	M	–	(F)	Q,R,S,A,T	1–many	One species produces numerous ascospores
Lipomyces	3	B,A	ML	–		–	B,U	1–16	"Active" buds may conjugate or directly form asci; spores brown
Lodderomyces	1	A,B,K	ML	+	–	–	S,T	1–2	Large, elongate ascospores
Metschnikowia	5	A,B,V,K	ML	(M)	–	F	W	1–2	Elongate asci; chlamydospores in some species
Nadsonia	2	M,C,L	BP,BF	–	–	(±)	A,D	1–2	Bud and mother cell conjugate
Nematospora	1	Various (heteromorphic)	ML	M	+	–	X	8	Large asci containing two bundles of four spindle-shaped spores

TABLE 1 (Continued)

MORPHOLOGICAL PROPERTIES OF ASCOMYCETOUS YEASTS

Genus	Number of Species	Cell Shape[a]	Budding[b]	Pseudomycelium[c]	Mycelium[c]	Pellicle[c]	Ascospore Shape[a]	Ascospore Number	Special Morphological Characteristic
Pachysolen	1	B,A	ML	±	–	–	Y	4	Ascus develops at tip of a long tube
Pichia	35	A,B,H,K	ML	M	F	±	A,E,F,D	1–4	Colonies slimy, pasty, or chalky dull
Saccharomyces	41	A,B,K,L	ML	(±)	–	–	A,T	1–4	Colonies pasty, semi-glossy
Saccharomycodes	1	M,L	BP,BF	(±)	–	–	A	4	Large cells; spores conjugate in ascus; they have a very narrow ledge
Saccharomycopsis	1	C,K	ML	+	–	–	C,K	1–4	Double-walled spore
Schizosaccharomyces	4	A,K	–	–	±	–	A,C	4–8	Cells reproduce by fission
Schwanniomyces	4	C,A,L,K	ML	–	–	–	O,D	1–2	Meiosis bud
Wickerhamia	1	C,L,M	BP,BF	–	–	–	Z	1–16	Spores shaped like sporting cap (usually one or two per ascus)
Wingea	1	A,B	ML	–	–	(±)	U	1–4	Lens-shaped spores

[a] + = all species possess property; – = no species possess property; ± = approximately half of species or strains possess property; M = most species possess property; F = few species possess property; above symbols in ( ) = property primitive or weakly exhibited.

[b] BP = bipolar budding; ML = multilateral budding; BF = bud fission.

[c] A = spheroidal or globose; B = ellipsoidal; C = ovoidal; D = warty; E = hat-shaped; F = saturn-shaped; G = sickle-shaped; H = oblong; I = fusiform; J = ogival; K = cylindroidal; L = elongate; M = apiculate or lemon-shaped; N = helmet-shaped; O = walnut-shaped; P = hemispheroidal; Q = crescentiform; R = reniform; S = oblong with obtuse end; T = prolate-ellipsoidal; U = oblate-ellipsoidal or lenticular; V = pyriform; W = needle-shaped without appendage; X = spindle-shaped with appendage; Y = hemispheroidal with narrow ledge; Z = cap-shaped.

TABLE 2

PHYSIOLOGICAL PROPERTIES OF ASCOMYCETOUS YEASTS

Genus	Number of Species	Fermentation[a]	Nitrate Utilization[a]	Growth without Vitamins[a]	Acid Production	Cycloheximide Resistance	Growth at 37°C	Growth on 50% Glucose	Growth in 10% NaCl	Special Physiological Property
Citeromyces	1	+	+	±	(+)	–	–	+	+	
Coccidiascus	1									Observed in drosophila – not cultivated
Debaryomyces	8	(M)	–	±			F	+		High salt tolerance
Dekkera	2	+	±	–	+	+	+	–		Produces acetic acid
Endomycopsis	10	(M)	F	F			(±)	(F)		Several species grow well on starch
Hanseniaspora	3	+	–	–	+		±	±		All species have an absolute requirement for inositol and pantothenate
Hansenula	25	M	+	F	F		±		F	Some species produce phosphomannans or sphingolipids
Kluyveromyces	18	+	–	–		M	±	F		Red (non-carotenoid) pigments may be formed in sporulating cultures
Lipomyces	3	–	–	±	–		±	–		Cells produce lipids in high concentration
Lodderomyces	1	(+)	–	±	–	+	+	(+)		Usually utilize higher paraffins
Metschnikowia	5	M	–	–			F	M		Some species produce pulcherrimin
Nadsonia	2	+	–	±	±		–	–		Maximum growth less than 26°C
Nematospora	1	+	–	–			+	–		Plant parasitic – produces riboflavin
Pachysolen	1	+	+	–			+		+	Produces a slimy extracellular phosphomannan
Pichia	35	M	–	F		F	±	F		One species common in olive brines; some species form phosphomannan
Saccharomyces	41	+	–	±		F	±	F		All species ferment strongly
Saccharomycodes	1	+	–	–	–		–	–		
Saccharomycopsis	1	(+)		–	–		+	–		Inhabits digestive tract of rabbits; grows only between 30 and 40°C; requires CO_2 and organic nitrogen
Schizosaccharomyces	4	+	–	–	±		+	M		No chitin is present in the cell wall
Schwanniomyces	4	+	–	–	(±)		±	–		All species utilize soluble starch
Wickerhamia	1	+	–	–	+	+	–	+		Produces extracellular riboflavin
Wingea	1	+	–	(±)		–	(±)	+		

[a] + = all species possess property; – = no species possess property; ± = approximately half of species of strains possess property; M = many species possess property; F = few species possess property; above smbols in ( ) = property slowly or weakly exhibited

TABLE 3

MORPHOLOGICAL PROPERTIES OF ASPOROGENOUS YEASTS

Genus	Number of Species	Cell Shape	Budding[a]	Pseudomycelium[b]	Mycelium[b]	Arthrospores[b]	Endospores[b]	Conidia[b]	Chlamydospores[b]	Pigments[b]	Pellicle[b]	Special Morphological Characteristic
Brettanomyces	7	Ogival, ellipsoidal, spheroidal, elongate	ML	+	–	–	–	–	–	–	(±)	Ogival cells
Candida	81	Spheroidal, ovoidal, cylindrical, elongate	ML	+	F	–	–	–	F	–	±	All produce pseudomycelium; may produce chlamydospores
Cryptococcus	17	Spheroidal, ovoidal, elongate	ML	(F)	–	–	–	–	F	±	F	Cells of most species have capsules
Kloeckera	4	Apiculate, ovoidal, elongate	BP	±	–	–	–	–	–	–	–	Apiculate cells
Oosporidium	1	Various	ML	+		–	+	–	–	+	–	Produces asexual endospores; chains of cells
Pityrosporum	3	Flask-shaped, spheroidal, ellipsoidal	MP,BF	(F)	(F)	–	–	–	–	–	–	Flask-shaped cells
Rhodotorula	9	Spheroidal, ovoidal, elongate	ML	(F)	(F)	–	–	–	F	+	–	Growth pink to orange, due to carotenoid pigments

TABLE 3 (Continued)

MORPHOLOGICAL PROPERTIES OF ASPOROGENOUS YEASTS

Genus	Number of Species	Cell Shape	Budding[a]	Pseudomycelium[b]	Mycelium[b]	Arthrospores[b]	Endospores[b]	Conidia[b]	Chlamydospores[b]	Pigments[b]	Pellicle[b]	Special Morphological Characteristic
Schizoblastosporion	1	Ellipsoidal, cylindrical, flask-shaped	BP,BF	±	–	–	–	–	–	–	–	Flask-shaped cells
Sterigmatomyces	3	Spheroidal, ovoidal	–	–	–	–	–	+	–	–	+	Multiplies by cells on stalks (conidia)
Torulopsis	36	Spheroidal, ovoidal, elongate	ML	(F)	–	–	–	–	–	–	F	Few produce rudimentary pseudomycelium
Trichosporon	8	Various	ML	+	+	+	F	–	–	–	±	Some species produce endospores (asexual)
Trigonopsis	1	Triangular, ellipsoidal		–	–	–	–	–	–	–	+	Triangular-shaped cells bud at apices

[a] + = all species possess property; – = no species possess property; ± = approximately half of species or strains possess property; M = most species possess property; F = few species possess property; above symbols in ( ) = property primitive or weakly exhibited.

[b] BP = bipolar budding; ML = multilateral budding; BF = bud fission.

TABLE 4

PHYSIOLOGICAL PROPERTIES OF ASPOROGENOUS YEASTS

Genus	Number of Species	Fermentation[a]	Nitrate[a]	Starch Synthesis[a]	Growth without Vitamins[a]	Acid Production[a]	Growth at 37°C[a]	Growth on 50% Glucose[a]	Growth in 10% NaCl[a]	Assimilate Inositol[a]	Gelatin Liquefaction[a]	Special Physiological Property
Brettanomyces	7	+	±		–	+	+	–		–		Produce acetic acid, resistant to cycloheximide; growth is slow
Candida	81	M	±		(M)		M		±	F		
Cryptococcus	17	–	±	M	F	–	F	–		+	F	Inositol assimilated
Kloeckera	4	+	–	–	–	±	–	±		–	(±)	Absolute requirement for inositol and pantothenate
Oosporidium	1	–	+	(+)	–		–	–				
Pityrosporum	3	–					+					Requires lipids for growth
Rhodotorula	9	–	±	–	F	–	±	–		–	F	Carotenoid pigments
Schizoblastosporion	1	–	–		–		–	–		–		
Sterigmatomyces	2	–	±	–	–		–			–		
Torulopsis	36	M	±	–	(±)	F	M	±	±	–		
Trichosporon	8	(F)	F	F	±	–	M	–		±		
Trigonopsis	1	–	–	–	–		+	–		–		

[a] + = all species possess property; – = no species possess property; ± = approximately half of species of strains possess property; M = many species possess property; F = few species possess property; above smbols in ( ) = property slowly or weakly exhibited

TABLE 5

MORPHOLOGICAL PROPERTIES OF BASIDIOMYCETOUS AND BALLISTOSPOROGENOUS YEASTS

Genus	Number of Species	Cell Shape	Budding[a]	Pseudomycelium[b]	Mycelium[b]	Pellicle Formation[b]	Teliospore Formation[b]	Ballistospore Formation[b]	Pigment Formation[b]	Special Morphological Characteristic
Basidiomycetous Yeasts										
Leucosporidium	6	Ovoidal, elongate	B	M	+	–	+	–	–	Perfect form of certain *Candida*-like organisms
Rhodosporidium	4	Spheroidal, ovoidal, elongate	B	±	+	–	+	–	+	Perfect form of *Rhodotorula*
Aessosporon	2	Ovoidal, elongate, cylindrical	B	+	±	+	+	+	+	Asymmetrical ballistospores; perfect form of *Sporobolomyces*
Filobasidium	2	Ovoidal, elongate	B	+	+	–	–	–	–	Forms *Tilletia*-like basidia with sessile basidiospores
Ballistosporogenous Yeasts										
Bullera	3	Spheroidal, ovoidal	B	–	–	–	–	+	±	Symmetrical ballistospores
Sporidiobolus	2	Ovoidal, elongate, cylindrical	B	(±)	+	+	?	+	+	Asymmetrical ballistospores; brown thick-walled chlamydospores
Sporobolomyces	9	Ovoidal, elongate	B	±	±	±	–	+	M	Asymmetrical ballistospores

[a] B = budding

[b] + = all species possess property; – = no species possess property; ± = approximately half of species or strains possess property; M = most species possess property; above symbols in ( ) = peroperty primitive or weakly exhibited

TABLE 6

PHYSIOLOGICAL PROPERTIES OF BASIDIOMYCETOUS AND BALLISTOSPOROGENOUS YEASTS

Genus	Number of Species	Fermentation	Nitrate Utilization[a]	Growth without Vitamins[a]	Growth on 50% Glucose[a]	Growth at 17°C[a]	Growth at 19°C[a]	Growth at 30°C[a]	Growth at 37°C[a]	Acid Production[a]	Starch Production[a]	Urea Hydrolysis[a]	Gelatin Liquefaction[a]
Basidiomycetous Yeasts													
Leucosporidium	6	(±)	M	F	–	+	(±)	F	–	F	M	+	M
Rhodosporidium	4	–	+	±	–	+	+	+	(±)	–	–	+	±
Aessosporon	2	–	+	+	–	+	+	+	–	–	–	+	–
Filobasidium	2	±	±	–	–	+	+	+	(+)	(+)	+		–
Ballistosporogenous Yeasts													
Bullera	3	–	±	±	–	+	+	–	–	–	±	+	–
Sporidiobolus	2	–	+	+	–	+	+	±	±	–	–	+	(F)
Sporobolomyces	9	–	±	M	–	+	+	±	F	–	–	+	F

[a] + = all species possess property; – = no species possess property; ± = approximately half of species of strains possess property; M = many species possess property; F = few species possess property; above smbols in ( ) = property slowly or weakly exhibited

CULTURAL CHARACTERISTICS

The cultural characteristics on solid or liquid media sometimes are sufficiently unique to be of taxonomic value. Distinctive growth on solid media such as malt agar may be a manifestation of hyphal or pseudohyphal growth or due to formation of carotenoids or pulcherrimin, etc. Growth in stationary liquid media results in the formation of a sediment, ring, islets, or pellicle – properties that are readily identifiable and of some value in species characterization.

SEXUAL CHARACTERISTICS

Ascomycetous Yeasts

Yeasts which produce ascospores are either homothallic or heterothallic. Homothallic refers to yeasts (or fungi) in which sexual reproduction can take place with identical nuclei undergoing fusion. Heterothallic refers to the opposite case, in which the fusing nuclei are not identical because they originate from opposite mating types. Life cycles are further characterized on the basis of the ploidy of the vegetative reproductive stage, which is either haploid or diploid or a mixture of these two phases. Forms with higher ploidy also have been found to exist. In the life cycle of a typical haploid yeast, budding cells are predominantly haploid and ascus formation occurs immediately following conjugation of the cells. The process of spore formation occurs so rapidly after conjugation that the shape of the ascus that results is reminiscent of the shape assumed by the two cells as they are engaged in the conjugation process. For example, in some haploid strains of *Saccharomyces* and *Schizosaccharomyces,* plasmogamy is initiated at the tips of tube-like outgrowths. Shortly after the union, karyogamy occurs, followed by meiosis, and spores are formed in asci that are characteristically elongated or dumbbell-shaped. Upon germination, the haploid spores produce haploid budding vegetative cells and the cycle is complete. The life cycle of such a haploid yeast is shown in Figure 3a.

In diploid yeasts, conjugation followed by karyogamy takes place shortly after spore germination. Spore formation, however, is delayed. The diploid cells or zygotes may bud for many generations, producing additional diploid vegetative cells until conditions are finally suitable for meiosis. Ascospores are formed within the cells and the asci are usually about the same size and shape as the original diploid budding vegetative cells from which they were derived (Figure 4).

The life cycles described above are typical ones for haploid and diploid ascomycetous yeasts. Other life cycles are simply variations on these two themes. Three variations are known to occur among yeasts that spend their vegetative life as haploid cells (Figures 3a to c):

(a) Conjugation tubes and dumbbell-shaped asci are formed as previously described. Examples: *Schizosaccharomyces pombe* and *Saccharomyces elegans.*

(b) A haploid cell produces a special bud called a "meiosis" bud which does not separate from the original cell. The nucleus of the mother cell divides mitotically and both nuclei migrate into the daughter bud, where karyogamy and meiosis take place. The four nuclei migrate back into the original cell and spore formation occurs. Usually, two of the four nuclei do not produce ascospores. (The life cycle depicted was for members of the genus *Schwanniomyces.* Similar life cycles may occur in species of *Hansenula, Pichia,* and some *Saccharomyces.*)

(c) Two gametes fuse. (The case depicted is *Nadsonia,* where a mother cell and daughter cell conjugate.) The dikaryotic cell thus formed does not become the ascus. Instead, karyogamy and ascospore formation occur in another specialized bud that grows out of the opposite end of the dikaryotic structure. (Other examples can be found for this type of life cycle in *Endomycopsis* and the yeast-like fungus *Eremascus.*)

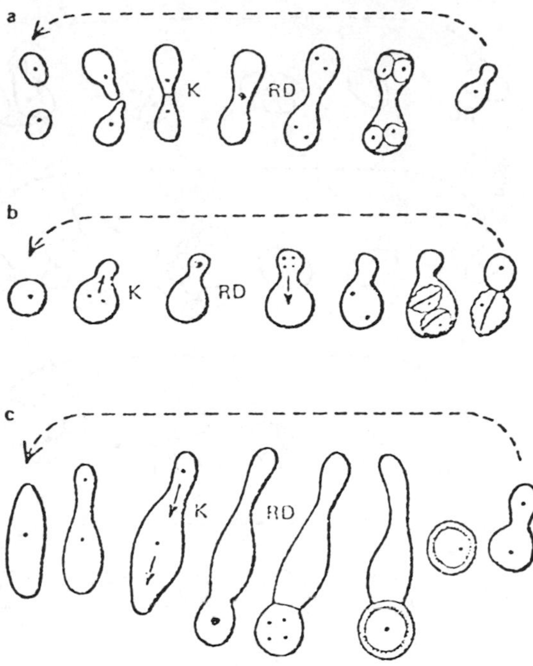

FIGURE 3. Life Cycles of Haploid Yeasts. The heavier dots represent diploid nuclei; the lighter dots are haploid nuclei. K = karyogamy. RD = reduction division.

Taken from: Phaff, H. J., Miller, M. W., and Mrak, E. M., *The Life of Yeasts* (1966). Reproduced by permission of Harvard University Press, Cambridge, Massachusetts.

There are four kinds of life cycles for yeasts that grow vegetatively as diploid cells (Figures 4a to d):

(a) Two ascospores conjugate directly in the ascus and the first bud from this zygote is a diploid. Example: *Saccharomycodes ludwigii.*
(b) Ascospores may germinate and bud as haploid cells for a short time prior to conjugation. Example: strains of *Saccharomyces cerevisiae.*
(c) Some spores germinate, bud for a while, and then one of the cells fuses with an ungerminated spore. Example: strains of *Saccharomyces cerevisiae.*
(d) The nucleus in a swelling ascospore divides into two haploid nuclei that fuse prior to germination of the spore into a diploid cell (*Saccharomyces chevalieri* and some species of *Hanseniaspora*).

It is noteworthy that various yeasts cannot be categorized as strictly haploid or strictly diploid. In some yeast cultures, both haploid and diploid vegetative cells may exist together. For a more complete discussion of life cycles in yeasts, see Phaff, Miller, and Mrak[7] and Fowler.[4]

Ascomycetous yeasts are members of a primitive group known as the Hemiascomycetidae. Also included in this subclass are mycelial saprobes, such as *Dipodacus aggregatus,* which grows in tree exudates, and other yeast-like organisms, such as *Taphrina deformans,* that cause leaf curl disease in certain plants. Species of *Taphrina* reproduce by budding in culture but form mycelium in the host plant. The ascomycetous yeasts have been differentiated from other members of the Hemiascomycetidae partly on the basis of number of spores per ascus. Most yeasts produce one to four ascospores. By comparison, *Dipodacus* produces an indefinite number of ascospores (up to 100). This differentiation has not stood up taxonomically since some species in the genera *Lipomyces* and *Kluyveromyces* can be multispored and produce more than 16 spores per ascus. The formation of more than four spores per ascus has been

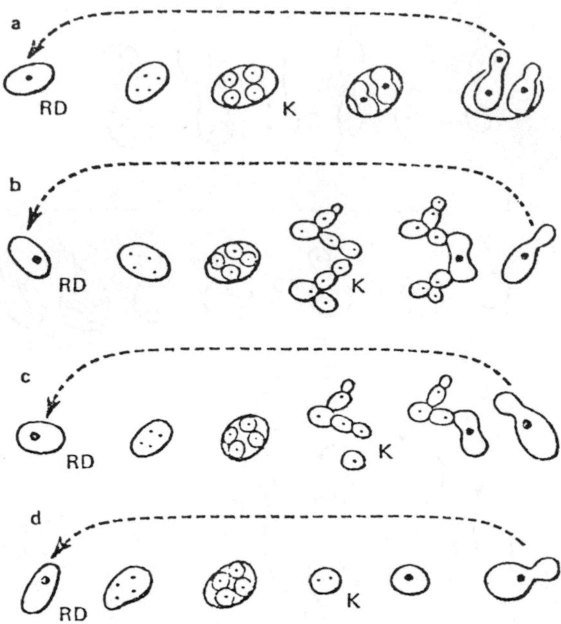

FIGURE 4. Life Cycles of Diploid Yeasts. The heavier dots represent diploid nuclei; the lighter dots are haploid nuclei. K = karyogamy. RD = reduction division.

Taken from: Phaff, H. J., Miller, M. W., and Mrak, E. M., *The Life of Yeasts* (1966). Reproduced by permission of Harvard University Press, Cambridge, Massachusetts.

attributed to postmeiotic mitoses within the ascus. Ascomycetous yeasts usually do not produce the well-developed true mycelia which are characteristic of many other organisms in the Hemiascomycetidae.

The taxonomy used in Lodder[6] assigns ascomycetous yeasts to a single order, Endomycetales, three of the four families of which contain yeast genera (Figure 5). Admittedly, this is not the final word since considerable information is lacking for a natural phylogenetic classification of yeasts.

Considerable variation in the shapes of ascospores is encountered among different yeast species. Spore morphology is a fairly consistent property which is useful in species identification. In some cases all of the species in an entire genus have essentially the same spore shape. Typical shapes of spores are shown in Figure 6.

Basidiomycetous Yeasts

It has long been inferred that certain yeasts such as *Rhodotorula* and *Sporobolomyces* are related to the Basidiomycetes. It was not until 1967, however, when Banno[1] reported a sexual life cycle in certain members of the genus *Rhodotorula,* that this suspicion was confirmed. The sexual stage of *Rhodotorula* was named *Rhodosporidium.* Three additional genera, *Leucosporidium, Filobasidium,* and *Aessosporon,* have been described with basidiomycete-like life cycles. Yeasts in the first two of these newly described genera were formerly members of the imperfect genus *Candida* and the last one of *Sporobolomyces,* respectively, before their sexual stages were elucidated. Members of this group of yeasts can be either homothallic or heterothallic. Characteristically, they form thick-walled diploid teliospores. In the heterothallic strains, haploid budding cells of compatible mating types conjugate and give rise to a dikaryotic mycelial phase which exhibits clamp formation. Teliospores are produced either terminally or

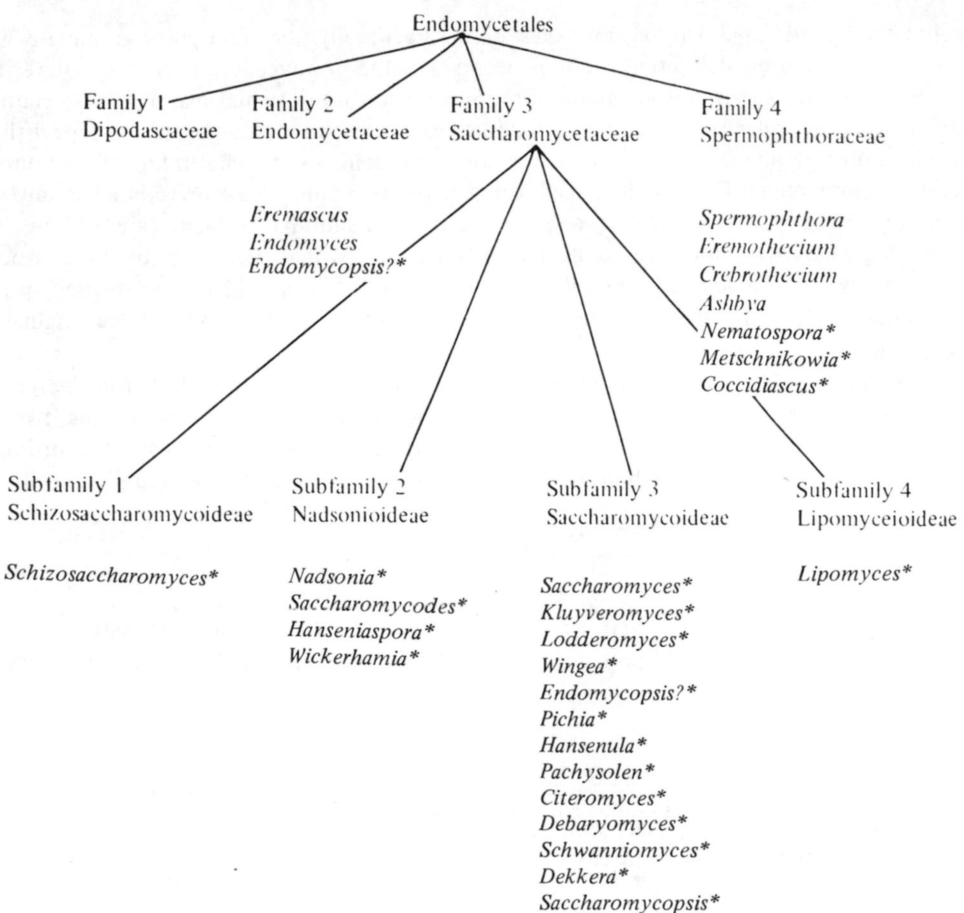

FIGURE 5. Classification of the Ascomycetous Yeasts According to Lodder.[6] Asterisked genera are yeasts.

Taken from: Lodder, J. (Ed.), *The Yeasts – A Taxonomic Study* (1970). Reproduced by permission of American Elsevier Publishing Co., Inc., New York.

FIGURE 6. Shapes of Various Ascospores Produced by Yeasts. From top, left to right: spheroidal; ovoidal; reniform; crescent- or sickle-shaped; hat-shaped; helmet-shaped; spheroidal with warty surface; walnut-shaped; saturn-shaped; spheroidal with spiny surface; needle-shaped without appendage; spindle-shaped with appendage.

Taken from: Phaff, H. J., Miller, M. W., and Mrak, E. M., *The Life of Yeasts* (1966). Reproduced by permission of Harvard University Press, Cambridge, Massachusetts.

within the hyphal strands, and karyogamy takes place. Eventually the teliospores germinate with the formation of a promycelium. Reduction division occurs and the promycelium becomes septate, forming four cells on which sporidia (basidiospores) are born. Segregation into original mating types occurs during formation of sporidia. Sporidia can reproduce as budding yeast cells and can conjugate and repeat the cycle. Sometimes teliospores germinate without reduction division, giving rise to what is known as a uninucleate self-sporulating budding phase. These budding cells can give rise to a uninucleate mycelium (usually without clamp connections), which produces teliospores, promycelia, and sporidia. In many cases the life cycles of the self-sporulating phase have not been completely worked out, so that the ploidy of these structures is unknown. In the case of *Rhodosporidium sphaerocarpum* (Figure 7), sporidia from the self-sporulating phase are presumably haploid since they mate with the appropriate haploid cells of the original mating types of this species.

Aessosporon salmonicolor formerly was classified as a ballistosporogenous yeast *Sporobolomyces.* It was found to produce teliospores which germinated into a nonseptate promycelium bearing two to four sporidia.[11] This homothallic yeast does not produce a dikaryotic mycelium phase. According to the postulated life cycle, ballistospores and buds are produced by both haploid- and diploid-phase cells.

Asporogenous Yeasts.

Asporogenous is probably an inappropriate term for describing this group of yeasts since certain strains produce arthrospores and other endospores or conidia. None, however, produce sexually derived spores

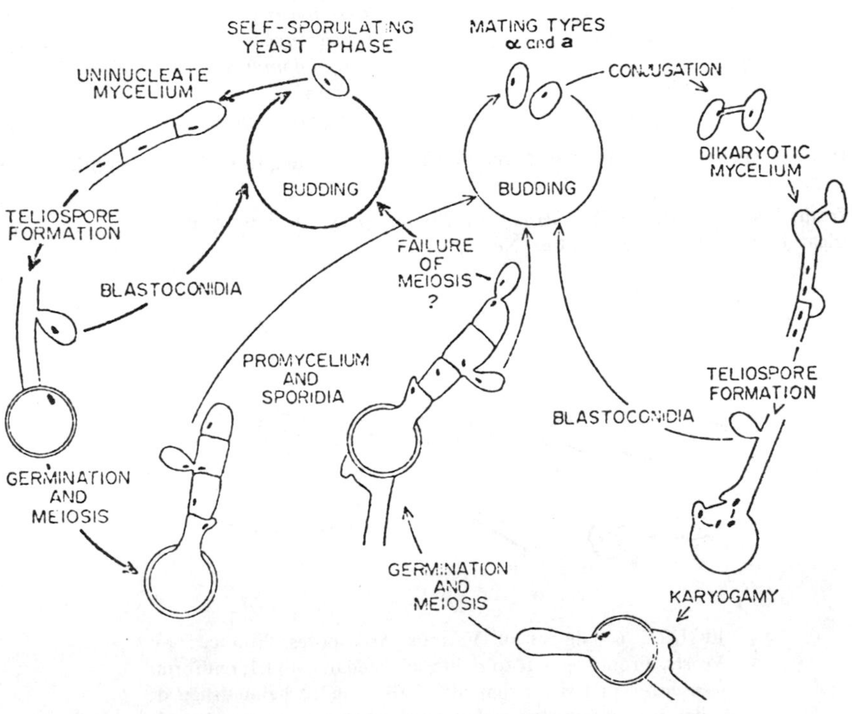

FIGURE 7. Life Cycle of *Rhodosporidium sphaerocarpum.*

Taken from: Fell, J. W., Phaff, H. J., and Newell, S. Y., Genus 2, *Rhodosporidium* Banno, in *The Yeasts – A Taxonomic Study,* pp. 801–814, J. Lodder, Ed. (1970). Reproduced by permission of American Elsevier Publishing Co., Inc., New York.

such as ascospores or sporidia and none produce forcibly discharged spores (ballistospores). All genera reproduce either by budding or by fission except one, *Sterigmatomyces,* which is, perhaps, not a yeast because it reproduces exclusively by formation of an unusual type of conidia.

Some genera are essentially the same as ascomycetous genera except for the absence of formation of ascospores. For example, *Dekkera* is the perfect form of the asporogenous genus *Brettanomyces,* and *Hanseniaspora,* that of *Kloeckera.* It is natural that the asporogenous group of yeasts has become a catchall for those yeasts in which sexuality does not exist or simply has never been observed. As more information is obtained about specific strains, however, sexual processes sometimes are discovered and the participating strains are removed from this group. (For example, sexually active strains of *Rhodotorula* are now called *Rhodosporidium.*)

Other forms of sexuality that do not include production of ascospores or sporidia have been suggested as occurring in certain members of this group. For example, van der Walt[12] reported that a kind of sexual cycle occurs in both *Candida albicans* and *Cryptococcus albidus.* Conjugation took place in certain homothallic strains. Zygotes can form chlamydospores or they can give rise to a diploid generation which produces large, dormant, multispored cells. Ascospores are not produced, but the haploid generation probably is reestablished from the conidia found on these multinucleate cells. According to van der Walt these multispored cells might be similar to basidia. Lodder,[6] however, suggests that this could be true for *Cryptococcus albidus* but not for *Candida albicans,* which is an ascomycete-like yeast on the basis of DNA base composition, cell wall structure, and serological analysis.

Wickerham[13] has proposed that a sort of parasexual cycle is exhibited by certain asporogenous yeasts. He coined the term "protosexual" to describe species in which diploid vegetative cells give rise to haploid cells without the production of fruiting bodies or sexual spores. Unfortunately, the strains in which this type of sexuality was observed later were found to form ascospores; therefore, it is not known whether there are exclusively "protosexual" yeasts, as originally proposed by Wickerham.

PHYSIOLOGICAL CHARACTERISTICS

Physiological characteristics generally are of limited value in taxonomic delimitation of genera; however, they are essential in the identification of species. The most important properties are those related to fermentation and assimilation of carbon sources, nitrogen utilization, vitamin and temperature requirements, and ability to grow in high sugar or salt concentrations. Susceptibility to fungal antibiotics also has been employed. Recently, criteria for delimiting relationships among various yeasts have been based on immunological properties, analysis of base composition of deoxyribonucleic acid, and proton magnetic resonance spectra of cell wall components.

Many yeasts are capable of fermenting various carbon sources. There are probably no strictly anaerobic yeasts and the same carbon sources which a yeast can ferment can also be assimilated oxidatively under appropriate conditions. The reverse is not true, in that ability to assimilate a certain carbon source does not mean that it necessarily can be fermented. All fermentative yeasts are able to ferment glucose, producing ethanol and CO_2. Generally, fermentation tests are conducted in tubes containing a basal medium (such as 0.5% yeast extract) and 2% sugar and inverted vials (Durham tubes) are used for gas collection.

Carbon assimilation tests can be performed either in liquid or on solid medium, using a chemically defined basal medium such as yeast nitrogen base (Difco) which contains ammonium sulfate as a nitrogen source and all vitamins, amino acids, and trace elements known to be required by yeasts. In the auxanographic method an agar basal medium is seeded with a suspension of the yeast and plated. Small amounts of various carbon sources are then placed in specific locations on the dry agar surface. Plates are incubated for a few days and growth occurs on those carbon sources which can be assimilated. Replicate plating methods have been designed for use when large numbers of yeast strains are to be tested.

Similar procedures have been employed for testing the ability to utilize various nitrogen sources. In this case, however, the basal medium lacks a nitrogen source and contains glucose as a carbon source (e.g., Yeast Carbon Base, Difco). Almost all yeasts are able to utilize peptone, asparagine, ammonium sulfate, and urea. The ability to utilize potassium nitrate, sodium nitrite, aliphatic amines, or amino acids is useful in the characterization of various yeast isolates.

A more complete description of these methods and other physiological tests such as splitting of arbutin, growth in vitamin-free media, growth in media of high osmolarity, growth at elevated temperatures, acid production, starch production hydrolysis of urea, fat splitting, pigment formation, ester production, cycloheximide resistance, and gelatin liquefaction have been published by van der Walt.[10]

YEAST-LIKE ORGANISMS

The literature abounds with descriptions of yeast-like organisms which yeast taxonomists have either refused to accept or which have at one time been accepted and later rejected. These include borderline genera such as *Endomyces* (and its imperfect form *Geotrichum*), which would be similar to the yeast *Schizosaccharomyces* if it did not produce such extensive mycelium and gametangia, and related to *Endomycopsis* if it were also to reproduce by budding. Likewise, *Ashbya* might be a yeast if budding were not so "rare and atypical." The so-called "black yeasts" (*Pullularia, Aureobasidium*) produce mycelial phases and budding yeast phases but, because of the production of a black pigment, are not accepted as yeasts. Other organisms are definitely not yeasts, yet they resemble them in colony morphology. For example, *Prototheca* is actually a colorless alga, probably related to *Chlorella.* The spheroidal to ellipsoidal cells of these organisms do not bud, but instead reproduce vegetatively by partitioning of the protoplasm into two or several irregularly shaped spore-like bodies (aplanospores). Other genera that can be confused with yeasts are *Taphrina, Eremothecium, Dipodascus,* and *Eremascus.* For brief descriptions of these genera and reasons for their exclusion from the yeasts, consult Lodder.[6]

YEASTS AND MAN

Throughout the last several thousand years man has capitalized on the products formed by yeast cells. In addition to leavened bread, beer, wine, sake, and other forms of potable alcohol, yeasts have provided man with other products such as glycerol, enzymes, coenzymes, and vitamins. Today there is considerable interest in employing yeast cells for upgrading waste materials into utilizable forms of single cell protein. Yeasts have been convenient tools for biochemists, physiologists, geneticists, cell biologists, and other scientists. Indeed, the initiation of the fields of biochemistry and nutrition was based on the discovery of enzymes and vitamins in yeast cells.

The main genera of commercial significance are *Saccharomyces* and *Candida.* Frequently, species were named on the basis of the fermentation they were associated with, e.g., *Saccharomyces sake* and *Saccharomyces vini.* The morphological and physiological properties of many of these species are very similar, and separate names have not been justified from a taxonomic point of view. Special characteristics of certain yeast strains are extremely desirable for various industrial processes; therefore, the strain differentiation may be more important than species classification.

A partial listing of yeast species either used or suitable for the manufacture of various products is shown in Table 7. Other commercial products isolated from *Saccharomyces cerevisiae* include alcohol dehydrogenase, hexokinase, L-lactate dehydrogenase, glucose-6-phosphate dehydrogenase, glyceraldehyde-3-phosphate dehydrogenase, inorganic pyrophosphatase, coenzyme A, oxidized and reduced diphosphopyridine nucleotides, and the mono-, di-, and triphosphates of adenosine, cytidine, guanosine, and uridine.[8]

Under certain circumstances, yeasts may cause infections of man and animals.[5] Treatment of patients with bactericidal antibiotics as well as the nutritional status of the infected individual is often of prime importance for the opportunistic invasion by yeasts. Table 8 is a listing of important pathogenic yeasts. Diseases range from superficial infections of cutaneous and mucosal sites to serious systemic diseases involving the viscera and circulatory fluids.

Yeasts have also been a detriment to man in causing food spoilage. Yeasts ordinarily do not compete well in mixed populations and, therefore, cause spoilage under conditions which are adverse for the growth of other organisms. Bacteria usually outgrow yeasts at pH values in the neutral or slightly acidic range, but below pH 5, yeasts are readily able to compete. A listing of important food spoilage yeasts is shown in Table 9. An extensive discussion of food spoilage yeasts was published by Walker and Ayres.[9]

TABLE 7
INDUSTRIALLY IMPORTANT YEAST SPECIES*

Product	Yeast Employed
Baker's yeast	*Saccharomyces cerevisiae*
Food yeasts	*Candida utilis* *Candida tropicalis* *Kluyveromyces fragilis* *Saccharomyces carlsbergensis* *Saccharomyces cerevisiae*
Protein (from hydrocarbons)	*Candida lipolytica* *Candida tropicalis* *Trichosporon japonicum*
Liquors and industrial alcohol	*Saccharomyces cerevisiae* (and others)
Lager beer	*Saccharomyces carlsbergensis*
Ale	*Saccharomyces cerevisiae*
Sakē	*Saccharomyces cerevisiae*
Wine	*Saccharomyces cerevisiae* *Saccharomyces fermentati* *Saccharomyces bayanus* Others
Yeast autolysates and extracts	*Saccharomyces cerevisiae* *Candida utilis*
Lipid	*Rhodotorula gracilis* (syn. *R. glutinis*) *Candida utilis* *Metchnikowia pulcherrima* *Metchnikowia reukaufii*
Invertase	*Saccharomyces cerevisiae* *Candida utilis*
Amylase	*Endomycopsis fibuligera* *Endomycopsis capsularis*
Lactase	*Kluyveromyces fragilis* *Kluyveromyces lactis* *Candida pseudotropicalis*
Uricase	*Candida utilis*
Polygalacturonase	*Kluyveromyces fragilis*
Ergosterol	*Saccharomyces cerevisiae* *Saccharomyces carlsbergensis* *Saccharomyces bayanus*
Ribonucleic acid	*Candida utilis* *Candida tropicalis*
Riboflavins	*Eremothecium ashbyi*
Lysine	*Candida utilis*
Cystine	*Rhodotorula gracilis*
Methionine	*Rhodotorula gracilis*

*These data were compiled from appropriate sections of Reference 8.

TABLE 8

YEASTS PATHOGENIC ON HUMANS*

Candida albicans	Candidiasis: superficial infections of various sites on the skin and in the digestive, genital, urinary, and respiratory tracts (for example, oral thrush, chronic paronychia, vaginitis, etc.), which can lead to deep-seated infections of individual viscera and the bloodstream.
Candida tropicalis	Associated with vulvovaginitis, and other forms of candidiasis; also free-living.
Candida stellatoidea	Associated with vulvovaginitis, also thrush; occasionally other infections.
Candida pseudotropicalis	Frequently respiratory infections; rarely cutaneous lesions and septicaemia; also free-living. Imperfect form of *Kluyveromyces fragilis.*
Candida parapsilosis	Cutaneous, mucosal, and deep-seated lesions, especially on heart valves.
Candida guilliermondii	Implicated in infections of nails, heart, and blood; also free-living.
Candida krusei	Implicated in infections of vagina, mouth, heart, and blood; also occurs as a saprobe in the intestinal tract of certain warm-blooded animals or free-living.
Cryptococcus neoformans	Cryptococcosis: infections may occur in the skin, lungs, or other parts of the body; frequently the central nervous system is involved, causing a form of meningitis.
Torulopsis glabrata	Possible pathogen isolated from urinary tract, skin, epidermal scales, nails, and genital secretions.
Torulopsis famata	Possible pathogen isolated from various skin lesions and genital secretions.
Pityrosporum orbiculare	Related to pityriasis versicolor (fine scales, lesions on skin).
Trichosporon cutaneum	Causes white piedra (soft white nodules on hair); also free-living in a large range of ecological niches.
Trichosporon capitatum	Possibly associated with respiratory complaints, such as bronchitis and asthma.

* Information in this table was obtained from Gentles and Touche.[5]

TABLE 9

TYPICAL FOOD SPOILAGE YEASTS*

Species	Source of Isolation
Brettanomyces spp.	Beer, wine, cucumber brines
Candida catenulata	Frankfurters
Candida guilliermondii	Dates
Candida krusei	Fresh figs, dates, tomatoes, food brines, pickle brines
Candida lipolytica	Frankfurters, margarine
Candida pseudotropicalis	Cream, cheese, butter, sour milk

TABLE 9 (Continued)

TYPICAL FOOD SPOILAGE YEASTS*

Species	Source of Isolation
Candida vini	Wine, beer
Candida zeylanoides	Frankfurters, beef
Citeromyces matritensis	Condensed milk, fruit in syrup
Debaryomyces hansenii	Sausage, frankfurters, food brines, meat brines, sakē, lunch meats, tomato puree
Endomycopsis fibuligera	Spoiled starchy foods
Hanseniaspora uvarum	Tomatoes, fresh figs, grapes, etc.
Hanseniaspora valbyensis	Dates, fresh figs, fresh fruits
Hansenula subpelliculosa	Cucumber brines, foods of high sugar content
Kloeckera apiculata	Tomatoes, strawberries, fruit juices, fresh figs
Kluyveromyces fragilis	Fresh figs, yogurt and other dairy products
Kluyveromyces lactis	Cream, cheese, milk
Pichia kluyveri	Tomatoes
Pichia membranefaciens	Food brines, wine, olive brines
Rhodotorula spp.	Oysters, crabs, dairy products, olive brines
Saccharomyces bailii	Wine, apple juice, salad dressing, mayonnaise
Saccharomyces bisporus	Honey, maple sugar, cucumber brines, salted beans
Saccharomyces carlsbergensis	Figs
Saccharomyces chevalieri	Fruit juice, wine
Saccharomyces rouxii	Honey, maple syrup, citrus juice, sugar cane juice, dates, prunes, dried figs, jelly, strawberry juice, candied fruits, cucumber brines
Saccharomyces uvarum	Maple sugar, sugar cane juice, honey, sugar beet juice, fruit juice, lemonade
Schizosaccharomyces octosporus	Prunes, figs, raisins
Schizosaccharomyces pombe	Sugar cane juice
Torulopsis lactis-condensi	Condensed milk, food brines
Torulopsis sphaerica	Cream, butter

* Most information in this table was obtained from Walker and Ayres.[9] Many names of the yeasts listed there were changed here to conform to the taxonomy listed in Lodder,[6] from which additional examples of food spoilage yeasts were also obtained. Incorrect nomenclature may be responsible for unlikely habitats in a few instances.

REFERENCES

1. Banno, I., Studies on sexuality of *Rhodotorula, J. Gen. Appl. Microbiol., 13,* 167 (1967).
2. Carmo-Sousa, L. do, Proceedings of the Second International Symposium on Yeasts. Vydavatel'sto Slovenskej Akademie Vied, Bratislava, Czechoslovakia, p. 87 (1966).
3. Fell, J. W., Phaff, H. J., and Newell, S. Y., Genus 2, *Rhodosporidium* Banno, in *The Yeasts – A Taxonomic Study,* pp. 801–814, J. Lodder, Ed. North Holland, Amsterdam, The Netherlands (1970).
4. Fowler, R. R., Life Cycles in Yeasts, in *The Yeasts,* Vol. I, pp. 461–471. *The Biology of Yeasts,* A. H. Rose and J. S. Harrison, Eds. Academic Press, London, England, and New York (1969).
5. Gentles, J. C., and Touche, C. J. L., Yeasts as Human and Animal Pathogens, in *The Yeasts,* Vol. I, pp. 107–182, *The Biology of Yeasts,* A. H. Rose and J. S. Harrison, Eds. Academic Press, London (1969).
6. Lodder, J., Ed., *The Yeasts – A Taxonomic Study,* North Holland, Amsterdam, The Netherlands (1970).
7. Phaff, H. J., Miller, M. W., and Mrak, E. M., *The Life of Yeasts,* Harvard University Press, Cambridge, Massachusetts (1966).
8. Rose, A. H., and Harrison, J. S., Eds., *The Yeasts,* Vol. III, *Yeast Technology,* Academic Press, London (1970).
9. Walker, H. W., and Ayres, J. C., Yeasts as Spoilage Organisms, in *The Yeasts,* Vol. III, pp. 463–527, A. H. Rose and J. S. Harrison, Eds. Academic Press, London, England, and New York (1970).
10. van der Walt, J. P., Criteria and Methods Used in Classification, in *The Yeasts – A Taxonomic Study,* Chapter II, pp. 34–113, J. Lodder, Ed. North Holland, Amsterdam, The Netherlands (1970).
11. van der Walt, J. P., The Perfect and Imperfect States of *Sporobolomyces salmonicolor, Antonie van Leeuwenhoek, J. Microbiol. Serol., 36,* 49 (1970).
12. van der Walt, J. P., Sexually Active Strains of *Candida albicans* and *Cryptococcus albidus, Antonie van Leeuwenhoek, J. Microbiol. Serol., 33,* 246 (1967).
13. Wickerham, L. J., A Preliminary Report on a Perfect Family of Exclusively Protosexual Yeasts, *Mycologia, 56,* 253 (1964).

Description of Various Genera and Species of Special Interest

DR. HERMAN J. PHAFF AND DR. JAMES D. MACMILLAN

Below is a listing of the various yeast genera in alphabetical order, adapted from Lodder,[22] including a brief discussion of those species thought to be of special interest. Where sufficient data are available, information on the range of base composition of nuclear deoxyribonucleic acid (expressed as mole % guanine + cytosine) is reported also. Data currently available indicate that for yeasts (in contrast to the higher fungi), organisms with a GC% greater than approximately 49% belong to the basidiomycetes, whereas those with less than about 49% are hemiascomycetes. Yeasts belonging to the imperfect fungi (Deuteromycetes) may represent either ascomycetous or basidiomycetous fungi and their range of % GC extends all the way from about 30 to 70%. In the borderline area (% GC ± 49%), when a sexual stage has not been demonstrated, decisions are not always possible with respect to a basidiomycetous or ascomycetous origin. Some support for one or the other may be gathered from the observation that basidiomycetous yeasts (as far as is known) usually produce urease when grown on urea and they have a laminated cell wall consisting of a number of layers, especially in older cells. Ascomycetous yeasts are normally urease negative and their cell wall consists of only two layers.[21]

ASCOSPOROGENOUS YEASTS

Citeromyces Santa Maria

Diagnosis. Cells reproduce vegetatively by budding. No true or pseudomycelium is produced. *Citeromyces* is heterothallic but may be isolated from nature in the bisexual, diploid form. Such cells form one or two spheroidal, warty spores per ascus and do not show evidence of conjugation. Heterothallic haploid forms also occur in nature. For sporulation to occur, these must be mixed with the opposite mating type. The thick-walled asci do not rupture at maturity. Nitrate is assimilated. The genus belongs to the family *Saccharomycetaceae.*

Speciation. Only one species is known – *C. matritensis* (Santa Maria) Santa Maria. The type was isolated from fruit preserved in syrup. Other strains have been isolated from sweetened condensed milk and (in our own laboratory) from a concentrated tacky gum from a tree in Hawaii. This organism appears to be quite rare. The % GC was found to be 45.5%.[27] Its relationship to other yeasts is uncertain. For further information, see Wickerham.[75]

Coccidiascus Chatton

Diagnosis. Budding yeast; cells are spherical, oblong, or slightly curved. After two cells conjugate, banana-shaped asci are formed which contain eight fusiform spores per ascus. Spores are screwed in a helix.

Speciation. The only species of this genus, *C. legeri,* was observed in the intestinal tract of *Drosophila funebris,* but could not be cultivated. It is placed in the family *Spermophthoraceae.* For further details, see Chatton.[8]

Debaryomyces Lodder et Kreger-van Rij nom. conserv.

Diagnosis. Cells are haploid and conjugation before sporulation occurs most commonly between a cell and a bud-like appendage. Most species form asci with one or two spores which have warty walls; a few species form four warty spores per ascus. Vegetative cells are often spherical or globose, but in some species they are elongate. Budding is multilateral. Fermentation is weak, slow, or absent and nitrate is not assimilated. Some strains, however, can assimilate nitrite.

Speciation. The history of species concept in this genus is complicated and several of the presently or previously accepted species have been variously classified in the genera *Saccharomyces* or *Pichia.*[17] Eight species are currently accepted, four of which have spheroidal to short oval cells and the other four possess long oval cells.

Results on DNA base composition (C. W. Price, unpublished results and Reference 34) indicate that the species with small spheroidal cells have a GC range of 36.6 to 38.4% and the four species with long oval cells show a range from 33.0 to 35.5% GC. One species of the small-celled group, *D. tamarii,* has a GC content of 46.2% and appears, therefore, only distantly related to the other members of the genus. Species of *Debaryomyces* are probably related to some species of *Saccharomyces* (the so-called *Torulaspora* group), the round-spored species of *Pichia,* and possibly to those of *Schwanniomyces.*[31,34]

By far the most common species is *D. hansenii,* a highly salt-tolerant yeast. It is commonly isolated from seawater, salted meat products (e.g., wieners, salami), and other salt-preserved products. It can grow slowly in media containing 18 to 20% NaCl, but grows even better in salt-free media. For further details, see Kreger-van Rij.[17]

Dekkera van der Walt

Diagnosis. These are multilaterally budding diploid yeasts, which are slow growing in malt extract, produce a characteristic aroma and, under aerobic conditions, high concentrations of acetic acid. Spores are hat-shaped, easily liberated from the ascus. Cells are of various shapes but often elongate and pointed at one end (ogival). Cultures are short-lived due to acid production, and are cycloheximide-resistant. Vitamins are required for growth.

Speciation. The two species now recognized were formerly placed in the asporogenous form genus *Brettanomyces.* Sporulation was discovered by placing the cells on yeast extract – malt extract agar enriched with vitamins. *Dekkera intermedia* is found as a spoilage organism of table wines and *D. bruxellensis* (in the past) was associated with the secondary fermentation of English stout or Belgian lambic beers. *Dekkera bruxellensis* has a 36.1% GC,[40] but that of *D. intermedia* is not known. The genus may be related to *Pichia* or *Hansenula.* For further information, see Reference 67.

Endomycopsis Dekker

Diagnosis. Vegetative growth consists of abundant true (septate) mycelium (containing septal pores) with blastospores (buds); pseudomycelium and individual budding cells also occur. Occasionally there is disarticulation of the mycelium into arthrospores. Asci are formed at the tips of hyphae, or at the point of juncture of two cells of a pseudo- or true mycelium, or intercalarily, and occasionally after gametangial fusion by gametangia originating from distant cells. The ascospores, one to four per ascus, may be spheroidal-, hat-, Saturn-, or sickle-shaped (depending on the species) and are usually liberated from the ascus soon after maturation. Fermentation is slow, weak, or absent; respiration constitutes the main metabolic activity. Assimilation of nitrate is positive or negative.

Speciation. The type species of the genus is *E. capsularis,* isolated in 1903 from soil in the Swiss Alps. Kreger-van Rij[18] accepted 10 species which were separated on the basis of number and shape of the ascospores, sugar fermentation, and assimilation of 5 carbon sources and of nitrate. Recently, van der Walt has discovered several additional species in the galleries of xylophagous insects. The sources from which strains of the various species have been isolated are quite varied.[18] Some appear to be associated with ambrosia beetles which attack broad-leafed trees (*E. monospora, E. platypodus*), others, with bark beetles that attack coniferous trees (*E. bispora*), or with slime fluxes of broad-leafed trees (*E. javanensis*). Some species exhibit a strong starch-splitting ability and may be associated with spoiled, starchy food or plant material (*E. burtonii, E. capsularis,* and *E. fibuligera*). Wickerham et al.[74] demonstrated the production of extracellular amylases in *E. fibuligera* and suggested the use of this yeast for the direct fermentation of starchy substrates.

Information on the base composition of the nuclear DNA is not complete at this time. The range now extends from 31.5% GC for *E. javanensis* to 43.4% GC for *E. capsularis.*[27,31] For further details, see Kreger-van Rij.[18]

Hanseniaspora Zikes

Diagnosis. The cells are diploid, lemon-shaped (apiculate), and reproduce by bipolar budding. Spores are either helmet-shaped (two or four per ascus and released upon maturity) or spheroidal with a subequatorial ledge (one or two per ascus and not released at maturity). Haploid spores diploidize during germination by fusion of two haploid nuclei. All species ferment glucose and none assimilates nitrate. All strains require inositol and pantothenate and growth is stimulated by biotin, niacin, thiamine, and pyridoxine.

Speciation. Species of *Hanseniaspora* represent the sporogenous stages of the form genus *Kloeckera.* They occur very commonly during early stages of natural wine fermentation and on all kinds of fermenting spoiled fruit, including tomatoes. Their alcohol tolerance is low, probably not over 4 to 6%. Phaff[43] accepted three species. However, Nakase and Komagata[32] have shown by base composition studies that *H. guilliermondii* is not a synonym of *H. valbyensis* and thus four species currently are known. They are separated by spore morphology, carbon assimilation reactions, and growth or no growth at 37°C. The range of GC % is from about 27 to 38 for the 4 species. Species of *Hanseniaspora* may be used for the microbiological assay of inositol or of pantothenic acid. See Reference 29 for further details.

Hansenula H. et P. Sydow

Diagnosis. Cells are haploid or diploid, varying from spheroidal to cylindrical and reproducing by multilateral budding. Pseudomycelium and true hyphae occur in some species. Ascospores are hat- or Saturn-shaped, hemispheroidal or spheroidal, and are usually released at maturity. Both heterothallic and homothallic species are known. Depending on the species, sugars may or may not be fermented, pellicle formation on liquid media is variable, vitamin requirement is variable, ester formation in aerobic cultures is variable, and growth on solid media varies from slimy to dull and dry. All species assimilate nitrate.

Speciation. Wickerham[76] accepted 25 species, but several additional species have been described recently. Interspecific hybridization is rare or absent. He has postulated a number of phylogenetic lines suggesting a development from primitive species, containing a capsular phosphomannan, which are associated with bark beetles infesting coniferous trees, to free-living species which lack capsules. Common species of the first group include *H. capsulata, H. holstii,* and *H. wingei,* while free-living species are represented by *H. anomala* (the type species of the genus) and *H. subpelliculosa.* Some species have only been isolated from soil (*H. saturnus, H. californica*), and *H. polymorpha* is common in wild species of *Drosophila* in California and in insect frass occurring in broad-leafed trees. *Hansenula* species are related to those of *Pichia* and *Pachysolen.* The base composition of the *Hansenula* species has been studied by Nakase and Komagata;[35] it ranged from 32% GC for *H. ciferrii* to 50% GC for *H. henricii.* These authors proposed four (what appear to be natural) groups in *Hansenula,* based on GC contents, antigenic structure, PMR spectra of cell wall mannans, and criteria employed in the current system of yeast taxonomy. For further details, see Wickerham.[76]

Kluyveromyces van der Walt emend. v.d. Walt

Diagnosis. Cells are spheroidal to elongate, and reproduce by multilateral budding; pseudomycelium are formed by some species. Spores are one to numerous per ascus, bean-shaped, crescentiform, or spheroidal with a smooth surface; they are rapidly released from the ascus when mature. All species require vitamins and ferment glucose; some ferment other sugars also. Nitrate is not assimilated. Most species are not inhibited by cycloheximide.

Speciation. Originally, the genus contained only two multispored species, *K. polysporus* (the type species) and *K. africanus.* Later, van der Walt emended the diagnosis to include species with up to four spores per ascus, which were formerly placed in the genus *Saccharomyces,* and he now accepts 18 species.[68] They are separated on the basis of number and shape of ascospores, fermentation and assimilation reactions, and sensitivity to cycloheximide. Some species show interfertility and identical base compositions (GC%), suggesting that the number of valid species may be less than 18. The base composition of the various species ranges from 35.3 to 47.4% GC and several natural groups appear to exist.[23] A number of species have been isolated frequently from the crops of wild species of *Drosophila* (*K. drosophilarum, K. dobzhanskii, K. phaseolosporus,* and *K. wickerhamii*), while others able to ferment lactose may be isolated from dairy products (*K. fragilis, K. lactis, K. bulgaricus,* and *K. cicerisporus*). *K. fragilis* is a versatile species. It is the source of the enzyme lactase,[57] it can ferment the polyfructoside inulin,[62] and it excretes a constitutive polygalacturonase.[42] Its temperature range for growth ranges from about 5 to 46°C. See Reference 68 for further details.

Lipomyces Lodder et Kreger-van Rij

Diagnosis. The cells are capsulated globose to ellipsoidal that reproduce by multilateral budding. Stationary phase cells often contain large lipid globules. Occasionally, the buds have a broad base. Before ascospore formation, cells develop one or more protuberances or appendages, which fill with ascospores directly or after conjugation between appendages. Ascospores are dark brown and number from 4 to less, to 8 or 16 per ascus, depending on strain or species. Spores are ellipsoidal and their surface is smooth or it may bear denticles or longitudinal ridges; asci rupture at maturity. Fermentative ability is lacking; growth is generally slimy. Nitrate is not assimilated.

Speciation. Slooff[58] recognized three species, which were separated on the basis of ascospore morphology, carbon compound assimilation, and synthesis of a starch-like compound at low pH. An additional criterion was the presence or absence of galactose in the polysaccharide of the capsule. In nitrogen-deficient media, the cells become filled with large lipid globules (50 to 60% of the dry weight may be lipid). Strains of all three species have been isolated exclusively from soil in various geographic areas. *Lipomyces starkeyi,* the type species of the genus, is the only species whose base composition (47.6% GC) has been studied.[40]

Lodderomyces van der Walt

Diagnosis. Cells are diploid, spheroidal to cylindrical, reproducing by multilateral budding or pseudomycelium formation. Ascospores are oblong with obtuse ends, one or two per ascus, not liberated at maturity. Nitrate is not utilized.

Speciation. The only species in this genus, *L. elongisporus,* was originally described by Recca and Mrak as *Saccharomyces elongisporus.* It had been isolated from orange juice concentrate. Van der Walt established a separate genus for it on the basis of the unusual morphology of the ascospores and on the ability of the species to assimilate paraffinic hydrocarbons. On the basis of a superficial similarity in morphological and physiological properties, he suggested that *L. elongisporus* is the perfect form of the asporogenous species *Candida parapsilosis.* Although Meyer and Phaff[26] showed that the two species have the same base composition (39.7 to 40% GC), they later found (Meyer, S. A., and Phaff, H. J., Proceedings of the 1st Specialized Symposium on Yeasts, Smolenice Castle, Czechoslovakia, June 1–4, 1971, in press) that the two organisms lacked significant homology between their DNA's and they rejected the proposed relationship between the two species. For further details, see van der Walt.[69]

Metschnikowia Kamienski

Diagnosis. Cells are generally diploid, spheroidal, ellipsoidal, pear-shaped, or cylindrical; they reproduce by multilateral budding and may form a rudimentary pseudomycelium. Asci are much larger than the vegetative cells and are club-shaped, sphaeropedunculate, or ellipsoidopedunculate, containing one or two needle-shaped ascospores, pointed at one or both ends. Fermentative capacity is present or absent; nitrate is not utilized; there is no growth in the absence of added vitamins.

Speciation. This genus has had an interesting history. The first representative of the genus was observed in 1884 by Metschnikoff as a parasite in sporulating condition in the body cavity of a fresh-water crustacean *Daphnia magna.* It was not until 1961 that the first pure culture of *Metschnikowia* was isolated by van Uden and Castelo-Branco from marine materials of the Pacific Ocean. It is not clear why these aquatic species were not cultured earlier since they do not exhibit special nutritional requirements. In 1968, Pitt and Miller made the discovery that two well-known species belonging to the imperfect genus *Candida* are able to go through a sexual cycle and to produce characteristic needle-shaped ascospores if they are grown on very dilute vegetable juice at 12 to 21°C. *Candida pulcherrima,* known since 1901, occurs commonly on decomposing fruits visited by bees and other insects, while *C. reukaufii,* discovered in 1918, is common in the nectar of certain flowers. Miller and van Uden[30] accepted five species. *M. bicuspidata,* the type of the genus (with three varieties), *M. krissii,* and *M. zobellii* have an aquatic habitat (some may be pathogenic to crustaceans); *M. reukaufii* and *M. pulcherrima* are terrestrial in habitat. The base composition of the various species and varieties[27] ranges from 42.2% GC for *M. reukaufii* to 48.3% GC for *M. pulcherrima,* a relatively narrow range as compared to many other yeast genera. There appears to be no relation between base composition and habitat. See Reference 30 for further details.

Nadsonia Sydow

Diagnosis. Cells are haploid, lemon-shaped, oval, or elongate; they reproduce by bipolar budding. Buds are not separated by constriction but by formation of a distinct septum across the neck of the bud. Spores are formed after conjugation between a bud and the mother cell. The nuclei then move into a bud opposite the first one and after karyogamy and meiosis the second bud becomes the ascus, containing one or, rarely, two spores. They are dark brown, spheroidal with convoluted surface, and they contain a prominent lipid globule. Both species ferment glucose; they do not assimilate nitrate and the maximum temperature for growth is about 26 to 27°C.

Speciation. Two species were recognized by Phaff.[44] *N. fulvescens,* the type species of the genus, was isolated in 1911 from the exudate of an oak in Russia. No additional isolates of this species have been reported. The second species, *N. elongata,* is separated from *N. fulvescens* on the basis of sugar fermentations and assimilations. The first isolate came from the exudate of a birch in Russia in 1913.

Subsequent isolates have been obtained from exudates of broad-leafed trees in Europe, the Eastern U.S., and Japan. Despite extensive ecological surveys, it has not been found along the Pacific Coast of North America. *Nadsonia* species seem to have a specific habitat in exudates of broad-leafed trees, but only in certain geographic regions. No data are available on the base composition of the DNA of the two species. A high chitin content of the cell wall and the apparent absence of normal cell wall mannan suggest that *Nadsonia* is not closely related to other yeast genera with lemon-shaped cells (*Hanseniaspora, Saccharomycodes*). For further information, see References 29 and 44.

Nematospora Peglion

Diagnosis. The cells are strongly polymorphic, reproducing by multilateral budding. Pseudo- and true septate mycelia may be formed. The asci are cylindrical and much larger than the vegetative cells. They contain usually eight spindle-shaped spores provided with a whip-like appendage at one end. The spores, usually in two bundles of four, are rapidly liberated from the ascus at maturity. Glucose is fermented; nitrate is not assimilated.

Speciation. Do Carmo-Sousa[5] accepted only a single species, *N. coryli,* which had been isolated from the tissue of diseased hazelnuts in 1897. Similar mild plant diseases caused by *Nematospora* yeasts have been reported from tomatoes, beans, cotton bolls, oranges, and coffee berries. Apparently, the spores are introduced into the plant tissue by hemipterous insects – in particular, bugs. Some isolates have been named after the host plant from which they were isolated, but they are now considered to belong to the single species *N. coryli.* Its nuclear DNA contains 40.2% GC.[40] All strains of the species require inositol for growth. For further details, see Reference 5.

Pachysolen Boidin et Adzet

Diagnosis. Cells are haploid or diploid, spheroidal to ellipsoidal, reproducing by multilateral budding. Colonies are mucoid, due to the presence of a capsular phosphomannan, or butyrous. Asci are formed as follows: a cell produces a long tube that bears at its distal end a globose compartment containing four helmet-shaped ascospores. At maturity, the tube becomes thick-walled. The compartment remains thin-walled and the spores are released. In haploid cultures the asci have a small conjugant cell attached near the cell that initiated the tube; in diploid strains such a cell is absent. Glucose is fermented, esters are synthesized, pellicles are not formed on liquid media, and nitrate is assimilated.

Speciation. Wickerham[77] recognized a single species, *P. tannophilus,* which Boidin and Adzet had isolated in 1957 from concentrated tanning liquor of various broad-leafed trees in France. *Pachysolen* is related to *Hansenula* physiologically, but differs by the shape of the ascus and by the mechanism of spore formation. *P. tannophilus* has a base composition of 42.4% GC.[40]

Pichia Hansen

Diagnosis. Cells are haploid or diploid, homothallic or heterothallic, spheroidal to cylindrical, reproducing by multilateral budding. Pseudomycelium formation is present or absent; occasionally there is a very limited true mycelium. Pellicles on liquid media are present or absent. Spores are hat-shaped, spheroidal, or Saturn-shaped. Spore wall is usually smooth but if warts are present they are exclusively formed by the outer spore wall. There are one to four spores per ascus, and the latter usually ruptures or lyses at maturity. Glucose fermentation is present or absent. Nitrate is not assimilated.

Speciation. *Pichia* is a large genus, related through some of its species to *Hansenula* and through others to *Endomycopsis, Pachysolen,* and *Debaryomyces.* Kreger-van Rij recognized 35 species[19] and a number of additional species have been described recently. Some of the oldest known species are free-living or associated with food fermentations or food spoilage. Examples are *P. membranaefaciens* (the type of the genus), *P. etchelsii, P. fermentans, P. kluyveri,* and *P. vini.* Many other and several of the more recently described species were isolated from exudates of broad-leafed trees, e.g., *P. pastoris, P. angophorae, P. fluxuum, P. trehalophila, P. salictaria,* and *P. quercuum.* Others occur in insect frass of coniferous trees (*P.*

scolyti, P. pinus, and *P. toletana*) or of broad-leafed trees (*P. acaciae, P. media,* and *P. stipitis*), and one species came from insect frass of a cycad (*P. wickerhamii*). Several were isolated from tanning liquor (*P. chambardii, P. rhodanensis,* and *P. strassburgensis*). It is clear that many species are associated with trees and the insects that attack them. The range of GC values for the species that have been studied (27 out of 35) is large and varies from about 29% GC for *P. kluyveri* to 50% GC for *P. rhodanensis.*[33] The genus appears to be very complex and heterogeneous since Nakase and Komagata[33] were unable to establish well-defined subdivisions of the genus in which correlations could be established between base composition, antigenic cell wall structure, PMR spectra of the cell wall mannan, or by the system of Boidin and co-workers. For further information, see References 19 and 33.

Saccharomyces Meyen emend Reess

Diagnosis. Cells are diploid or haploid, hetero- or homothallic, spheroidal to elongate, reproducing by multilateral budding; pseudomycelium may be formed but no true mycelium. Ascospores (one to four per ascus) are spheroidal to ellipsoidal, with a smooth surface; asci do not release the spores at maturity. Some species show interspecific fertility between their spores. Glucose is fermented; nitrate, lactose, and higher paraffinic hydrocarbons are not assimilated.

Speciation. *Saccharomyces,* first recognized in 1838 by Meyen, is the oldest known genus of the yeasts. It once contained a number of species which have subsequently been removed and placed into other more appropriate genera (e.g., *S. pastori* to *Pichia pastoris; S. fragilis* to *Kluyveromyces fragilis*). The genus still contains certain haploid species which at one time were placed in separate genera, e.g., *S. rosei*, which was *Torulaspora rosei* and *S. rouxii,* which was *Zygosaccharomyces rouxii*). Van der Walt,[70] who accepted 41 species, recognized the diversity of the species by placing them into three well-defined groups and a fourth group consisting of phylogenetically heterogeneous species. The first group consists of diploid species, which are interfertile, containing, among others, *S. cerevisiae* (used for baking, distillery fermentations, and top (or ale) fermentation, *S. carlsbergensis* (used for lager beer fermentation), and *S. bayanus* (some strains of which form the film or "flor" during Spanish sherry fermentation). The second group, consisting of haploid species which show conjugation between independent cells before sporulation, includes the highly sugar-tolerant species *S. rouxii* and *S. bisporus,* which may cause incipient spoilage of honey, syrups, dried fruits and other foods containing 50 to 60% sugar. The third group shows mother-daughter cell conjugation (meiosis buds) and contains species formerly placed in the genus *Torulaspora* (*S. rosei, S. fermentati* and others). These species may be related to *Debaryomyces. Saccharomyces* species are separated by the type of life cycle, assimilation and fermentation reactions, sugar tolerance, ethylamine assimilation, and cycloheximide sensitivity. Some attention is paid to cell size as well. Base composition studies[38] have shown a range of 32.4 to 43.7% GC. The 31 species examined can be placed into 3 groups, with GC values of 32.4 to 33.7, 38.3 to 40.7, and 42.7 to 43.7% GC. Van der Walt's[70] first and third groups proved to be natural taxonomic groups based on GC content. Only a small intraspecific variation in GC content was found.[38] For further details, see References 38 and 70.

Saccharomycodes Hansen

Diagnosis. Cells are large, lemon-shaped or elongate, and diploid, which reproduce by bipolar budding. Buds are separated by a septum which is easily seen in the light microscope (bud fission). Spores (two to four per ascus) are spheroidal, with an indistinct subequatorial ledge. Spores usually conjugate in pairs inside the ascus. Glucose is fermented, pellicles are absent on liquid media, and nitrate is not assimilated.

Speciation. Phaff[45] recognized only a single species, *S. ludwigii,* which Hansen had isolated in 1889 from the exudate of an oak in Denmark. We recently studied another strain isolated from oak exudate in Arizona. Sporadic isolations from grapes or wine fermentations have been reported. Strains of *Saccharomycodes* are relatively rare. The base composition of *S. ludwigii* has been reported as 38.3% GC.[40] The genus may be related to *Hanseniaspora.* For further information, see Reference 45.

Saccharomycopsis Schionning

Diagnosis. Cells are large, diploid, long-oval to cylindrical, in pairs (solid media) or forming pseudomycelia (liquid media). Ascospores are oval to cylindrical, one to four per ascus (not tightly packed); spores are not liberated at maturity. Upon germination an exosporium is evident. Growth occurs only between 30 and 40°C, in the presence of high concentrations of gaseous CO_2, and provided media rich in amino acids are used. Sporulation, however, occurs only at low temperature (18 to 22°C). Glucose is slowly fermented. The cells are short-lived.

Speciation. Only a single species is known, *S. guttulata.*[46] It occurs in the intestinal tract and the stomach of domestic rabbits, where it was first observed in 1845. Because of its unusual growth requirements, the first pure cultures were not obtained until 1956. At 37°C, the cells are short-lived; after a few days they become strongly granular and die. Storage is best when sporulated cultures are placed at 5°C; under such conditions, growth will resume at 37°C after 4 to 6 months of storage. Gaseous requirements for optimal growth have been defined.[3] A mycelial variant, isolated from a wild jack rabbit, has been described recently.[4] This variant is dimorphic in its behavior and may transform from a filamentous into a budding form.

Schizosaccharomyces Lindner

Diagnosis. Cells are haploid, cylindrical to short ovoidal; reproduction is by fission; rudimentary true hyphae may be formed, breaking up into arthrospores. Asci are formed after conjugation of two haploid cells. Spores are globose to ovoidal, four to eight per ascus (depending on the species); asci usually rupture at maturity. Spore wall is smooth or convoluted, varying with the species. The cell wall is of unusual composition. The usual mannan is replaced by a galactomannan, chitin is absent, and a significant fraction of the wall is made up of an α-1,3-linked glucan. Glucose is fermented by all species and nitrate is not assimilated. Inositol is an essential requirement for growth in synthetic media.

Speciation. Four species and one variety were accepted by Slooff.[59] The species are separated on the basis of numbers of spores per ascus, occurrence of bean-shaped spores, and fermentation reactions with various sugars. The base composition of *S. pombe* and *S. octosporus* DNA is the same – 42% GC.[27] The other two species have not been studied yet for DNA base composition. The oldest species, *S. pombe* (the type species of the genus), had been isolated in 1893 from a native African fermented beverage. It has been isolated subsequently at various times from fermented beverages and fruits in subtropical and tropical countries. It has been extensively studied genetically. *S. octosporus* appears to prefer substrates of high sugar content, such as sun-dried raisins, prunes, raw cane sugar, and the like. *S. malidevorans* is notable for its ability to decarboxylate malic acid during fermentation and transform it to lactic acid. The fourth species is *S. japonicus* and its variety *versatilis*, strains of which are occasionally found in spoiled wines. For further details, see Reference 59.

Schwanniomyces Klöcker

Diagnosis. Cells are haploid, usually ovoidal, reproducing by multilateral budding. Before sporulation the nucleus divides; after karyogamy the diploid nucleus moves into a bud-like appendage, the meiosis bud, where reduction division occurs. One or two of the haploid nuclei, after returning to the mother cell, become the nucleus of one or, rarely, two ascospores. The spores have an equatorial ledge, a warty or convoluted surface, and contain a conspicuous lipid globule. They resemble somewhat the appearance of walnuts. Isogamic conjugation occurs rarely. Spores are not released from the ascus upon maturity. Glucose is fermented; nitrate is not assimilated.

Speciation. Phaff[47] accepted four species, which were separated on the basis of fermentation and assimilation reactions. All species of the genus have been isolated exclusively from soil samples collected in various parts of the world. The type species of the genus is *S. occidentalis,* isolated by Klöcker in 1909 from a soil sample from the island of St. Thomas in the Virgin Islands. It is noteworthy that all species of the genus grow well on soluble starch and inulin as sole carbon sources. The DNA base composition (cf.

Reference 27) is known only for *S. occidentalis* (34.5% GC). The other species of the genus are *S. alluvius, S. persoonii,* and *S. castellii.* For further information, see Reference 47.

Wickerhamia Soneda

Diagnosis. Cells are diploid, reproducing by bipolar budding on a broad base, apiculate to elongate. Spores are cap-shaped; the crown deflects to one side of a sinuous brim, giving the appearance of a sporting cap; usually 1 or 2 spores per ascus, which are liberated at maturity (reports of up to 16 spores per ascus have not been confirmed). Glucose is fermented; nitrate is not assimilated.

Speciation. The genus contains a single species, *W. fluorescens,* isolated in 1959 from the dung of a wild squirrel in Japan. It was first thought to be a member of the asporogenous genus *Kloeckera,* but spores were recognized subsequently. It may excrete a yellowish pigment in liquid media which fluoresces in ultraviolet light. The genus was separated from other apiculate yeast genera on the basis of the shape of the ascospores. The DNA of *W. fluorescens* was shown to have 37.6% GC.[40] *W. fluorescens* has shown itself an unusually fruitful organism for the study of the mitotic process[24] and for ultrastructural studies of the membrane systems of this yeast.[2] For additional details, see Reference 48.

Wingea van der Walt

Diagnosis. Cells are haploid, spheroidal to ellipsoidal; reproduction is by multilateral budding. Before sporulation, cells develop protuberances or appendages. Presumably, these are involved in conjugation and meiosis. Occasionally, two independent cells conjugate. Spores (one to four per ascus) are lens-shaped and light brown in color. Glucose is slowly fermented. Nitrate is not utilized.

Speciation. The single species of the genus, *W. robertsii,* was originally considered a species of *Pichia.* It had been isolated in 1959 from the pollen pabulum provided for the larval stage of the South African carpenter bee, *Xylocopa caffra.* Because of the unique lens-shaped spores, van der Walt established a separate genus for this species. Four strains have been isolated, all from the above habitat, which appears to be specific for this yeast. For further information, see Reference 71.

BASIDIOMYCETOUS YEASTS BELONGING TO THE ORDER USTILAGINALES

Leucosporidium Fell, Statzel, Hunter et Phaff

Diagnosis. Cells are haploid, ovoidal to elongate, reproduce by budding, usually show a well-developed pseudomycelial growth, and occasionally a true mycelium. Growth on solid media is nonpigmented. Two haploid cells of opposite mating type may form a zygote which remains binuclear and develops into a septate binucleate mycelium with clamp connections at the cross walls. The mycelium develops intercalary or terminal, thick-walled, teliospores with a granular content. Karyogamy takes place in the teliospores, which germinate to produce thin-walled, one- to four-celled promycelia. Meiosis occurs in the promycelium and haploid sporidia are formed laterally or terminally. They are not forcibly discharged. A self-sporulating stage, not involving conjugation between haploid cells, is known for some of the species and strains. All species assimilate nitrate; glucose may or may not be fermented. Asexual ballistospores are not produced.

Speciation. When the genus was described in 1969 (cf. Reference 13), 7 species were accepted. However, *L. capsuligenum* was transferred recently[56] to the genus *Filobasidium* (see below), leaving six species in the genus. Prior to the discovery of the sexual life cycle, the species were assigned to the genus *Candida.* Except for some marine isolates of *L. scottii* (the type species of the genus), strains of all species show a decided thermophobic tendency, having maximum growth temperatures of 18 to 22°C. Most of the strains have been isolated from antarctic or arctic samples which were continually exposed to low temperatures, but specific ecological niches are not known. The base composition of three of the species has been reported, based on the haploid *Candida* phase: *L. scottii* 59.0, *L. gelida* 56.1, and *L. frigida* 54.1% GC.[39] A recent study of the mitotic behavior of *L. scottii*[25] has shown that, upon budding, the nucleus moves into the bud, accompanied by partial disintegration of the nuclear membrane. This is followed by a return of one of the daughter nuclei to the mother cell and reformation of the nuclear membrane. This behavior is very different from mitosis in ascomycetous species. For fuller details, see Reference 13.

Rhodosporidium Banno

Diagnosis. The uninucleate cells are haploid, spheroidal to elongate, reproducing by budding. Appropriate mating types may conjugate, developing a dikaryotic mycelium with clamp-connections at the septa. Thick-walled teliospores develop on the mycelium where karyogamy takes place. During germination, meiosis occurs and a thin-walled, transversely septate promycelium is formed on which the bud-like lateral, or terminal sporidia are formed. The sporidia are haploid, function as basidiospores, but are not forcibly discharged. They reproduce by budding, reestablishing the yeast stage.

A self-sporulating uninucleate yeast phase is also known, which can go through the above cycle without conjugation between two cells. Asexual ballistospores are not produced. Glucose is not fermented. Growth on solid media is orange to pink, due to the synthesis of carotenoid pigments.

Speciation. The species of *Rhodosporidium* were known in their asexual stage as strains of *Rhodotorula*. In 1967, Banno discovered the sexual stage by mixing a number of strains of *Rhodotorula glutinis*. Those that mated and showed cross-fertility were named *Rhodosporidium toruloides*, the type species of the genus. A second species, *Rh. sphaerocarpum*, was discovered by Newell and Fell[12] by mixing a number of other strains of *Rhodotorula glutinis*, which they isolated from antarctic seawater. The new species showed no cross-fertility with *Rh. toruloides* and formed smooth, spherical teliospores rather than the irregular, angular teliospores of *Rh. toruloides*. Another recently described species, *Rhodosporidium diobovatum*,[41] again corresponds in its imperfect stage to *Rh. glutinis*. It also differs in teliospore morphology and is not interfertile with the other species. A fourth species, *Rhodosporidium malvinellum*, corresponds to another species of *Rhodotorula*, *R. graminis*.[10] The base compositions of the four species[40] are as follows: *Rh. diobovatum* 67.1 to 67.3, *Rh. sphaerocarpum* 64.9 to 65.4, *Rh. toruloides* 60.2 to 61.2, and *Rh. malvinellum* 50.5% GC. For further information, see References 10, 12, 40, and 41.

Filobasidium Olive

Diagnosis. Cells are uninucleate, haploid, ovoidal to elongate; reproduction is by budding, sometimes forming pseudomycelium. Upon mixing of compatible mating types a binucleate, septate mycelium with clamp connections is formed. The mycelium ultimately forms long, slender basidia which are nonseptate. At the somewhat swollen tip, a cluster (about five to nine) of thin-walled sessile basidiospores is formed; upon germination they produce blastospores. Fermentation of glucose is present or absent; nitrate assimilation is positive or negative. A starch-like compound is synthesized.

Speciation. Two species are recognized.[56] They are thought to belong to the family *Tilletiaceae* of the *Ustilaginales*. *F. floriforme* is the type species of the genus and was described by Olive in 1968. The second species is *F. capsuligenum*. This species was placed in *Leucosporidium* by Fell and Phaff (cf. Reference 13), who interpreted the basidia of the sexual stage as teliospores. It was transferred to *Filobasidium* by Rodrigues de Miranda[56] on the basis of its *Tilletia*-like basidia. The two species were differentiated by fermentation and assimilation reactions as well as by morphology of the basidia and basidiospores. Base compositions of their DNA are not known.

Aessosporon van der Walt

Diagnosis. Vegetative cells reproduce by budding and by the formation of asymmetrical ballistospores on sterigmata. Pseudo- and true mycelium may be formed. Basidiocarps are lacking; the teliospores are smooth, spheroidal to ellipsoidal, thick-walled, and filled with lipid globules. They germinate, forming nonseptate promycelia on which one to four budding sporidia are produced. Glucose is not fermented, starch-like compounds are not formed, and carotenoid pigments may be synthesized.

Speciation and Discussion. The genus *Aessosporon* was established in 1970 by van der Walt.[73] Although the complete life cycle has not been worked out in detail, it appears that meiosis occurs during germination of the teliospores. The teliospores germinate to produce basidia and sporidia characteristic of the family *Tilletiaceae*. Both self-sporulating forms (as in the case of *A. salmonicolor*) and forms with haploid mating types (as in *A. odorus*) appear to exist. In the latter case, conjugation is followed by formation of a dikaryotic, septate mycelium with clamp connections.[1] *Sporobolomyces* is the imperfect form of *Aessosporon*. A sexual stage has been demonstrated thus far for only two species and only for some strains of *Sporobolomyces salmonicolor* and of *Sporobolomyces odorus*. For further details, see References 1 and 73.

YEAST-LIKE GENERA BELONGING TO THE FAMILY *SPOROBOLOMYCETACEAE* (DEUTEROMYCETES)

Bullera Derx

Diagnosis. Cells are spheroidal to ovoidal; they reproduce by budding and by the formation of ballistospores, which are borne on sterigmata of various lengths. The asexual ballistospores are symmetrical, spheroidal, ovoidal, or apiculate, and are forcefully discharged. Pseudomycelium or true mycelium is lacking. Growth on solid media is nonpigmented to slightly yellowish. Glucose is not fermented.

Speciation. Phaff[49] recognized three species, of which one, *B. grandispora*, is no longer available in culture. *B. alba*, the type species of the genus, was repeatedly isolated in Canada around 1929 from wheat and oat straw infected with rust. Only a few strains have been isolated since that time. *B. tsugae* was isolated from bark beetle frass of a hemlock tree in Oregon. The DNA base composition of *B. alba* is 54.4% GC.[40] Strains of *Bullera* tend to lose their ballistosporulating ability. Such asporogenous strains are very similar to species of *Cryptococcus*. For further information, see Reference 49.

Sporidiobolus Nyland

Diagnosis. Reproduction in young cultures on complex media occurs by budding and by formation of asymmetrical ballistospores on sterigmata. Ballistospores may bud or reproduce in turn by ballistosporulation. In older cultures or on media that stimulate filamentous growth (e.g., corn meal agar), a sparsely septate mycelium with clamp connections may form. Brown, thick-walled chlamydospores are produced terminally or intercalarily on the mycelium. Arthrospores are absent. Glucose is not fermented. A pink carotenoid pigment is synthesized.

Speciation. There is insufficient evidence at this time for a sexual life cycle and for this reason *Sporidiobolus* is retained for the present in the *Deuteromycetes* until more information is available to support the postulated haplophase and diplophase of this genus, as well as the exact stage where karyogamy and meiosis occur. The genus presently contains two species (*S. johnsonii*, the type, and *S. ruinenii*), which are separated on the basis of α-glucosidase activity. Strains of *Sporidiobolus* occur on leaves of tropical plants as well as plants growing in temperate zones. For further information, see Reference 50.

Sporobolomyces Kluyver et van Niel

Diagnosis. Vegetative reproduction is by budding and in some species also by the formation of pseudomycelium and true mycelium with cross walls. Cells are ovoidal to elongate. Some of the vegetative cells develop asymmetrical ballistospores on sterigmata. The ballistospores are sickle- to kidney-shaped and they are forcefully discharged. Most of the species produce pink growth, due to the synthesis of carotenoid pigments. Glucose is not fermented. Starch-like compounds are not synthesized.

Speciation. The nine species accepted by Phaff[51] are separated by nitrate and sugar assimilation reactions, formation of true mycelium, and by carotenoid synthesis. The type species of the genus is *S. roseus*, a species commonly found on the surface of the leaves of grasses and many other plants. The DNA base composition has been determined for 6 of the 9 species and ranges from 51.5 to 65% GC.[64] The values for particulate species are not meaningful, due to the large intraspecific variation in GC content, indicating that the present basis for species separation is inadequate. *Sporobolomyces singularis* is an interesting yeast since it lends itself unusually well for *in-vivo* transgalactosylation and transglucosylation reactions.[14–16] For further details, see Reference 51.

GENERA OF YEASTS BELONGING TO THE FAMILY *CRYPTOCOCCACEAE* (DEUTEROMYCETES)

This is a heterogeneous group of genera and species, some of which are of ascomycetous and some of basidiomycetous origin. None of the various species, however, has a known sexual stage. In some cases this may be due to the fact that sexually compatible strains have not yet been discovered.

Brettanomyces Kufferath et van Laer

Diagnosis. For the characteristics of the genus *Brettanomyces,* see the ascomycetous genus *Dekkera,* of which *Brettanomyces* represents the imperfect form genus.

Speciation. As only certain strains of *B. bruxellensis* and *B. intermedius* were found to produce ascospores, these two species were also retained in *Brettanomyces,* together with five other species.[72] *Brettanomyces bruxellensis* is the type species of the genus. All species produce large amounts of acetic acid under aerobic conditions and are, therefore, short-lived; they are resistant to cycloheximide, require more than the usual concentration of thiamine for optimal growth, and their natural occurrence seems to be limited to products of the fermentation industries (beer and wine) or to materials on or around their premises. More recently, two additional species have been reported as spoilage yeasts of soft drinks.[78] The species of the genus are separated on the basis of fermentation and assimilation reactions, including nitrate utilization. For further information, see Reference 72.

Candida Berkhout

Diagnosis. Cells vary from ovoidal to elongate, sometimes irregularly shaped. Reproduction by budding; in some species budding is clearly multilateral, but in others it appears to be mono- or bipolar. In the latter case, budding is not on a broad base. All species have in common that a pseudomycelium with blastospores is formed under certain conditions of growth, although this property is not always distinct. Chlamydospores are formed by some species. True mycelium may be formed but arthrospores are absent. Physiological properties are variable (nitrate assimilation, glucose fermentation, starch synthesis, capsule formation, cycloheximide sensitivity, etc.). Pigments are not synthesized.

Speciation. Van Uden and Buckley accepted 81 species,[65] but nearly 40 more species have been described since 1970. The indiscriminate introduction of new species, sometimes on very flimsy grounds, has made species identification in the genus *Candida* a very difficult and frustrating task. Many of these species will undoubtedly turn out to be metabolic variants of one and the same species. For example, DNA-DNA homology studies have shown[28] that *C. obtusa* and *C. lusitaniae* (differentiated by galactose metabolism) constitute the same species. This is also true for *C. salmonicola* and *C. sake* (differentiated by cellobiose utilization). In contrast, DNA base composition studies of strains of a single "species" have shown that in some instances several distinct species may have been combined. The base composition of many species of *Candida* has been determined.[39] The range is from about 30 to 63% GC, including species of ascomycetous origin (less than about 50% GC) and basidiomycetous species (GC $\geqslant$ 50%). In spite of the above critique on speciation, several species are well defined. A few will be discussed.

Candida albicans may be pathogenic to susceptible individuals. It causes thrush, vaginitis, and may become systemic. *Candida parapsilosis,* although it may be free-living, has been repeatedly isolated as the cause of heart valve infection, especially in drug addicts. *Candida guilliermondii* and *C. tropicalis* have been isolated in a number of instances from clinical material but they are also found as free-living organisms. They probably are not important pathogens. *Candida utilis* is important for the production of single cell protein (e.g., from sulfite waste liquor of the paper industry). *Candida tenuis* is commonly found in association with bark beetles which attack coniferous trees. *Candida diddensii* has its habitat in ocean water or marine products. For further information, see Reference 65.

Cryptococcus Kützing emend Phaff et Spencer

Diagnosis. Cells are ovoidal to occasionally elongate, or polymorphic, reproducing by budding. Cells of most strains are surrounded by a capsule consisting of an acidic heteropolysaccharide. The thickness of the capsule varies with growing conditions. Most species synthesize a starch-like compound when the pH of the medium turns low. All species assimilate inositol as a sole source of carbon. Growth is often slimy and in liquid media a mucous pellicle may be formed. Glucose is not fermented. Nitrate assimilation varies with the species. Carotenoid pigments are synthesized by some of the species.

Speciation. Lack of a sexual cycle and fermentation has led to a somewhat arbitrary selection of a number of carbon sources for assimilation tests. In addition, nitrate assimilation, growth at 37°C, starch synthesis, and pigment production have been used for species differentiation. These criteria have proven unsatisfactory since data on DNA base composition have shown a large intraspecific variation in several species.[36,64] Unless sexual cycles are discovered or better criteria are found based on physiological properties, the classification of species in the genus *Cryptococcus* is entirely arbitrary, with the possible exception of *C. neoformans,* the type species of the genus. This yeast may cause serious infections of the central nervous system. Cryptococcal meningitis is a frequently fatal, but relatively rare disease. *Cryptococcus laurentii* is worldwide in distribution as a saprophyte of both terrestrial and marine origin. Another common species is *C. albidus.* Several *Cryptococcus* species have features in common with species of the basidiomycetous genus *Tremella,* of which they may represent the yeast phase. Information on base composition of *Tremella* species supports such a relationship.[40] The red yeast *C. infirmo-miniatus* was placed only provisionally in *Cryptococcus.*[52] Its high GC content (68%) confirms its unusual position in *Cryptococcus.* For further details, see Reference 52.

Kloeckera Janke

Diagnosis. The species of *Kloeckera* have all the properties of species of *Hanseniaspora* of the ascomycetous yeasts, except ascosporulation. The reader is referred to the diagnosis of *Hanseniaspora* for further details.

Speciation. Four species, including two varieties, were accepted by Phaff.[53] They were differentiated by fermentation and assimilation reactions of several sugars. The various strains show no other salient characteristics. Nakase and Komagata[32] showed that *K. corticis* and *K. africana* had very similar GC contents (37.3 to 38%), corresponding to *H. osmophila. K. javanica* and its variety *lafarii* had GC contents of 33.7 to 34.6% not corresponding to a perfect stage. *Kloeckera apiculata* may represent the imperfect stage of either *H. uvarum* or *H. guilliermondii* (GC% 31.0 to 31.7). Other strains of *K. apiculata* with a lower GC% (27 to 27.5%) correspond to *H. valbyensis. Kloeckera brevis,* a strain used for detailed studies of its cell wall mannan,[55] represents a synonym of one of the forms of *K. apiculata.* For further details, see References 29 and 53.

Oosporidium Stautz

Diagnosis. Cells are of various shapes, from spheroidal to elongate, reproducing by multilateral budding on a broad base; they usually remain attached in the form of chains. True septa may be present but there are no arthrospores. Asexual endospores are formed by protoplasmic cleavage. Glucose is not fermented. Pink or orange-yellow pigments, apparently of a noncarotenoid nature, are synthesized.

Speciation. Only one species is known, *O. margaritiferum,* which occurs in slime fluxes of broad-leafed trees. The organism is related to the genus *Trichosporon* (in particular to *T. pullulans*), but in the absence of arthrospores it is placed in a separate form genus. *Oosporidium margaritiferum* grows slowly in complex media and poorly in synthetic media. For this reason it may be difficult to isolate this species from slime fluxes (in competition with other organisms), although it is easily recognized (if present) by microscopy due to the characteristic chain formation of the cells in its natural habitat. For further details, see Reference 6.

Pityrosporum Sabouraud

Diagnosis. Cells are globose to elongate, reproducing mainly by monopolar budding on a broad base. Repeated budding produces added wall layers which are visible at the location of budding, forming a "collarette," from which young buds emerge. Ridges occur on the inside of the cell wall, running obliquely to the long axis of the cell. In culture, little or no pseudomycelium or mycelium is formed. Growth is optimal at about 36°C and is stimulated for most strains by natural oils or fatty materials. Glucose is not fermented.

Speciation and relation to *Malassezia*. Evidence has gradually been forthcoming that a dimorphic fungus is responsible for the various skin disorders (such as tinea versicolor and possibly dandruff) from which cultures of *Pityrosporum* have been isolated. Apparently on the scalp both the yeast phase and the mycelial phase (*Malassezia*) can occur, but in culture the organism grows only as the yeast phase (*Pityrosporum*). Lipids are required or stimulate growth in culture. Three species were recognized by Slooff,[60] strains of all of which had been isolated from skin disorders of warm-blooded animals or man. *Pityrosporum ovale* (referred to as the "bottle bacillus" in older literature) is commonly isolated from cases of human dandruff. *Pityrosporum canis* is considered a synonym of *P. pachydermatis* and is common in the ear wax of dogs. For further information, see Reference 60.

Rhodotorula Harrison

Diagnosis. Cells are ovoidal to elongate, reproducing by budding. Occasionally, strains of some species may form chlamydospore-like cells, suggesting a relation to *Rhodosporidium.* Rudimentary pseudo- or true hyphae of various length occur in some strains. Red or yellow carotenoid pigments are synthesized in young malt agar cultures. Glucose is not fermented; inositol is not assimilated; starch-like substances are not synthesized. Capsule formation varies from conspicuous to lacking, depending on the species and carbon source in the medium.

Speciation. Members of the genus *Rhodotorula* are widespread in nature and they are particularly common in marine waters, the atmosphere, and on pasture grasses. As in *Cryptococcus,* the species of *Rhodotorula* are arbitrarily separated on the basis of nitrate and sugar assimilation reactions and vitamin requirements. Data on DNA base composition[36,64] show large intraspecific variations for some of the species, indicating their heterogeneity. Further evidence for this lies in the discovery that three separate species of the basidiomycetous genus *Rhodosporidium* have *Rhodotorula glutinis* (the type species of the genus) as their imperfect stage. Eight additional species (including four varieties) were accepted by Phaff and Ahearn.[54] Base composition studies[36] support the synonymy of several *Rhodotorula* species proposed by Phaff and Ahearn,[54] whose publication should be consulted for further details.

Schizoblastosporion Ciferri

Diagnosis. Cells reproduce by bipolar budding on a broad base. Glucose is not fermented. Pseudomycelium is not formed.

Speciation. There is a single species, *S. starkeyi-henricii,* which Ciferri described in 1930 based on a single strain isolated from soil in the U.S. by Starkey and Henrici. Subsequent strains have been isolated in various parts of the world from soils, which appear to be its normal habitat. The genus shows some resemblance to *Kloeckera* and to *Pityrosporum.* However, species of the former genus ferment glucose and those of the latter reproduce by unipolar budding. The DNA base composition of *S. starkeyi-henricii* has not yet been determined. For further details, see Reference 20.

Sterigmatomyces Fell

Diagnosis. Cells are spheroidal to ovoidal; they develop one or more sterigma- or conidiophore-like structures of variable length at the end of which a conidium or new cell develops. When the daughter cell is mature it may produce sterigmata in turn, often resulting in short branched groups of cells connected by narrow tubes. The cells appear to separate by a septum formed in the middle of the sterigma. Cells are nonpigmented, do not synthesize starch-like compounds, and do not ferment glucose.

Speciation. Two species have been described by Fell[11] which were separated by the property that *S. halophilus* assimilates nitrate and *S. indicus* does not. Most of the strains of these two species were isolated from seawater. A third species,[63] *S. elviae,* was isolated from human skin eczema in Finland. Its DNA base composition (51.5% GC) and positive urease reaction[40] suggest that this organism may represent the yeast phase of a basidiomycetous fungus. *Sterigmatomyces elviae* differs from *S. indicus* in several carbon assimilation reactions. See Reference 11 for further details.

Torulopsis Berlese

Diagnosis. Cells are globose to ovoidal, rarely elongate, reproducing by multilateral budding. Pseudomycelium is not formed or is rudimentary. Cells are nonpigmented. Starch-like compounds are not formed. Capsule formation, nitrate assimilation, and glucose fermentation are variable, depending on the species.

Speciation. Van Uden and Vidal-Leiria[66] assigned 36 species to this form genus but numerous additional species have been described in the last few years. Since few conspicuous morphological details characterize the genus, speciation is based mainly on assimilation and fermentation reactions and maximum growth temperature. The DNA base composition of 33 species and 2 varieties has been reported[37] to extend from 32.4 (*T. pintolopesii*) to 60% GC (*T. magnoliae*). The genus, therefore, like several other form genera of the *Cryptococcaceae,* is very heterogeneous and no prominent groups of species can be delimited on the basis of GC content. Additional criteria are needed in order to develop a sounder system for the classification of these species. Several species have been shown to have a specific habitat. *Torulopsis glabrata* is a medically important species which is often associated with infections of the urinary tract. Certain free-living species have frequently been confused with *T. glabrata. T. stellata* is a common species on fermenting and spoiled fruit. *T. pintolopesii* (growth range 25 to 40°C) occurs in the intestinal tract of mice, rats, and other small rodents. It multiplies in the highly acid environment constituted by the mucosa of the stomach. Some species have been related to certain ascomycetous yeasts.

Trichosporon Behrend

Diagnosis. True mycelium and arthrospores are always present, and are usually abundant. The pseudomycelium is more or less well developed. Budding cells are of various shapes. Asexual endospores are formed in strains of some species. Glucose fermentation is weak, latent, or absent. Nitrate assimilation varies with the species.

Speciation. Some of the former species of the genus in which ascospores have been discovered were transferred to the genus *Endomycopsis.* Do Carmo-Sousa[7] accepted seven species, all of which (except *T. penicillatum*) have high GC contents in the nuclear DNA.[9] The range extends from 55 to 64% GC, suggesting that these organisms are of basidiomyceous origin. *Trichosporon penicillatum,* a common species in slime fluxes of broad-leafed trees, has a 45% GC and may represent an asporogenous form of the ascomycetous genus *Endomycopsis* or of a related taxon. *Trichosporon pullulans* is also commonly found in exudates of broad-leafed trees. It is nitrate-positive and has a rather low maximum growth temperature (23 to 27°C, depending on the strain). *T. cutaneum,* the type species of the genus, is widely distributed in nature. It assimilates a large number of different carbon compounds including aromatics. It is a prominent and active species in aerobic sewage digestion systems. Some strains occur commonly in the digestive tract of warm-blooded animals and some have been isolated from clinical specimens or were found associated with skin lesions. Intraspecific variation in DNA base composition, maximum growth temperature (29 to 41°C), carbon assimilation pattern, and other properties indicate that *T. cutaneum* represents several species. For further information, see References 7, 27, and 40.

Trigonopsis Schachner

Diagnosis. Cells are triangular or ellipsoidal, the proportion of each depending on the medium composition. Tetrahedral and rhombohedral cells occur also. Budding occurs at the corners of the triangular and similar cells and is multilateral on ellipsoidal cells. Pseudomycelium formation is lacking. Glucose is not fermented, and nitrate is not assimilated.

Speciation. Only a single species, *T. variabilis,* is known, which was isolated in 1929 from beer in Munich, Germany. Another strain was isolated later from grape must in Brazil. Nakase and Komagata[40] reported that this yeast is urease-negative and that its DNA base composition was 46.1% GC. It appears to be a rare species. For further details, see Slooff.[61]

Descriptions of the Various Genera and Species of Special Interest

1. Bandoni, R. J., Lobo, K. J., and Brezden, S. A., Conjugation and Chlamydospores in *Sporobolomyces odorus, Can. J. Bot., 49,* 683 (1971).
2. Bauer, H., A Freeze-Etch Study of Membranes in the Yeast *Wickerhamia fluorescens, Can. J. Microbiol., 16,* 219 (1970).
3. Buecher, E. J., and Phaff, H. J., Growth of *Saccharomycopsis* Schiönning under Continuous Gassing, *J. Bacteriol., 104,* 133 (1970).
4. Buecher, E. J., and Phaff, H. J., Dimorphism in a New Isolate of *Saccharomycopsis* Schiönning, *Can. J. Microbiol., 18,* 901 (1972).
5. Carmo-Sousa, L. do, Genus *Nematospora,* in *The Yeasts – A Taxonomic Study,* pp. 44–447, J. Lodder, Ed. North Holland, Amsterdam, The Netherlands (1970).
6. Carmo-Sousa, L. do, Genus *Oosporidium,* in *The Yeasts – A Taxonomic Study,* pp. 1161–1166, J. Lodder, Ed. North Holland, Amsterdam, The Netherlands (1970).
7. Carmo-Sousa, L. do, Genus *Trichosporon,* in *The Yeasts – A Taxonomic Study,* pp. 1309–1352, J. Lodder, Ed. North Holland, Amsterdam, The Netherlands (1970).
8. Chatton, E., *Coccidiascus legeri* n.g., n.sp. Levure Ascosporée Parasite des Cellules Intestinales de Drosophila funebris Fabr. *C. R. Soc. Biol. Paris, 75,* 117 (1913).
9. Dupont, P. F., and Hedrick, L. R., Deoxyribonucleic Acid Base Composition and Numerical Taxonomy in the Genus *Trichosporon, J. Gen. Microbiol., 66,* 349 (1971).
10. Fell, J. W., Yeasts with Heterobasidiomycetous Life Cycles, in *Recent Trends in Yeast Research,* Vol. 1,pp. 49–67, D. G. Ahearn, Ed. Spectrum University of Georgia, Athens, Georgia (1970).
11. Fell. J. W., Genus *Sterigmatomyces,* in *The Yeasts – A Taxonomic Study,* pp. 1229–1234, J. Lodder, Ed. North Holland, Amsterdam, The Netherlands (1970).
12. Fell, J. W., Phaff, H. J., and Newell, S. Y., Genus *Rhodosporidium,* in *The Yeasts – A Taxonomic Study,* pp. 803–814, J. Lodder, Ed. North Holland, Amsterdam, The Netherlands (1970).
13. Fell, J. W., and Phaff, H. J., Genus *Leucosporidium,* in *The Yeasts – A Taxonomic Study,* pp. 776–802, J. Lodder, Ed. North Holland, Amsterdam, The Netherlands (1970).
14. Gorin, P. A. J., Spencer, J. F. T., and Phaff, H. J., The Structures of Galactosyl-lactose and Galactobiosyl-lactose produced from Lactose by *Sporobolomyces singularis, Can. J. Chem., 42,* 1341 (1964).
15. Gorin, P. A. J., Spencer, J. F. T., and Phaff, H. J., The Synthesis of β-Galacto- and β-Glucopyranosyl Disaccharides by *Sporobolomyces singularis, Can. J. Chem., 42,* 2307 (1964).
16. Gorin, P. A. J., Horitsu, K., and Spencer, J. F. T., Formation of O-β-D-Glucopyranosyl and O-β-D-Galacto-pyranosyl-myo-inositols by Glycosyl Transfer, *Can. J. Chem., 43,* 2259 (1965).
17. Kreger-van Rij, N. J. W., Genus *Debaryomyces,* in *The Yeasts – A Taxonomic Study,* pp. 129–156, J. Lodder, Ed. North Holland, Amsterdam, The Netherlands (1970).
18. Kreger-van Rij, N. J. W., Genus *Endomycopsis,* in *The Yeasts – A Taxonomic Study,* pp. 166–208, J. Lodder, Ed. North Holland, Amsterdam, The Netherlands (1970).
19. Kreger-van Rij, N. J. W., Genus *Pichia,* in *The Yeasts – A Taxonomic Study,* pp. 455–554, J. Lodder, Ed. North Holland, Amsterdam, The Netherlands (1970).
20. Kreger-van Rij, N. J. W., Genus *Schizoblastosporion,* in *The Yeasts – A Taxonomic Study,* pp. 1224–1228, J. Lodder, Ed. North Holland, Amsterdam, The Netherlands (1970).
21. Kreger-van Rij, N. J. W., and Veenhuis, M., A Comparative Study of the Cell Wall Structure of Basidiomycetous and Related Yeasts, *J. Gen. Microbiol., 68,* 87 (1971).
22. Lodder, J., Ed., *The Yeasts – A Taxonomic Study,* North Holland, Amsterdam, The Netherlands (1970).

23. Martini, A., Phaff, H. J., and Douglass, S. A., Deoxyribonucleic Acid Base Composition of Species in the Yeast Genus *Kluyveromyces* van der Walt emend. v. d. Walt, *J. Bacteriol., 111,* 481 (1972).
24. Matile, P., Moor, H., and Robinow, C. F., Yeast Cytology, in *The Yeasts,* Vol. I, pp. 220–297, A. H. Rose, and J. S. Harrison, Eds. Academic Press, London (1969).
25. McCully, E. K., and Robinow, C. F., Mitosis in Heterobasidiomycetous Yeasts. I. *Leucosporidium scottii (Candida scottii), J. Cell Sci., 10,* 857 (1972).
26. Meyer, S. A., and Phaff, H. J., Deoxyribonucleic Acid Base Composition in Yeasts, *J. Bacteriol., 97,* 52 (1969).
27. Meyer, S. A., and Phaff, H. J., Taxonomic Significance of the DNA Base Composition in Yeasts, in *Recent Trends in Yeast Research,* Vol. I, pp. 1–29, D. G. Ahearn, Ed. Spectrum University of Georgia, Athens, Georgia (1970).
28. Meyer, S. A., Ph.D. Dissertation, *DNA Base Composition and Homology in Candida Species and Related Yeasts,* pp. 1–149. University of California, Davis, California (1970).
29. Miller, M. W., and Phaff, H. J., A Comparative Study of the Apiculate Yeasts, *Mycopath. Mycol. Appl., 10,* 113 (1958).
30. Miller, M. W., and van Uden, N., Genus *Metschnikowia,* in *The Yeasts – A Taxonomic Study,* pp. 408–429, J. Lodder, Ed. North Holland, Amsterdam, The Netherlands (1970).
31. Nakase, T., and Komagata, K., Taxonomic Significance of Base Composition of Yeast DNA, *J. Gen. Appl. Microbiol., 14,* 345 (1968).
32. Nakase, T., and Komagata, K., Significance of DNA Base Composition in the Classification of the Yeast Genera *Hanseniaspora* and *Kloeckera, J. Gen. Appl. Microbiol., 16,* 241 (1970).
33. Nakase, T., and Komagata, K., Significance of DNA Base Composition in the Classification of the Yeast Genus *Pichia, J. Gen. Appl. Microbiol., 16,* 511 (1970).
34. Nakase, T., and Komagata, K., Significance of DNA Base Composition in the Classification of the Yeast Genus *Debaryomyces, J. Gen. Appl. Microbiol., 17,* 43 (1971).
35. Nakase, T., and Komagata, K., Further Investigation on the DNA Base Composition of the Genus *Hansenula, J. Gen. Appl. Microbiol., 17,* 77 (1971).
36. Nakase, T., and Komagata, K., Significance of DNA Base Composition in the Classification of the Yeast Genera *Cryptococcus* and *Rhodotorula, J. Gen. Appl. Microbiol., 17,* 121 (1971).
37. Nakase, T., and Komagata, K., Significance of DNA Base Composition in the Classification of the Yeast Genus *Torulopsis, J. Gen. Appl. Microbiol., 17,* 161 (1971).
38. Nakase, T., and Komagata, K., Significance of DNA Base Composition in the Classification of the Yeast Genus *Saccharomyces, J. Gen. Appl. Microbiol., 17,* 227 (1971).
39. Nakase, T., and Komagata, K., Significance of DNA Base Composition in the Classification of the Yeast Genus *Candida, J. Gen. Appl. Microbiol., 17,* 259 (1971).
40. Nakase, T., and Komagata, K., DNA Base Composition of Some Species of Yeast and Yeast-like Fungi, *J. Gen. Appl. Microbiol., 17,* 363 (1971).
41. Newell, S. Y., and Hunter, I. L., *Rhodosporidium diobovatum* sp.n., *J. Bacteriol., 104,* 503 (1970).
42. Phaff, H. J., α-1,4-Polygalacturonide Glycanohydrolase from *Saccharomyces fragilis,* in *Methods in Enzymology,* Vol. 8, Complex Carbohydrates, *pp. 636–641,* E. F. Neufeld and V. Ginsburg, Eds. Academic Press, New York (1966).
43. Phaff, H. J., Genus *Hanseniaspora, The Yeasts -- A Taxonomic Study,* pp. 209–225, J. Lodder, Ed. North Holland, Amsterdam, The Netherlands (1970).
44. Phaff, H. J., Genus *Nadsonia,* in *The Yeasts – A Taxonomic Study,* pp. 430–439, J. Lodder, Ed. North Holland, Amsterdam, The Netherlands (1970).
45. Phaff, H. J., Genus *Saccharomycodes,* in *The Yeasts – A Taxonomic Study,* pp. 719–724, J. Lodder, Ed. North Holland, Amsterdam, The Netherlands (1970).
46. Phaff, H. J., Genus *Saccharomycopsis,* in *The Yeasts – A Taxonomic Study,* pp. 725–732, J. Lodder, Ed. North Holland, Amsterdam, The Netherlands (1970).
47. Phaff, H. J., Genus *Schwanniomyces,* in *The Yeasts – A Taxonomic Study,* pp. 756–766, J. Lodder, Ed. North Holland, Amsterdam, The Netherlands (1970).
48. Phaff, H. J., Genus *Wickerhamia,* in *The Yeasts – A Taxonomic Study,* pp. 767–771, J. Lodder, Ed. North Holland, Amsterdam, The Netherlands (1970).
49. Phaff, H. J., Genus *Bullera,* in *The Yeasts – A Taxonomic Study,* pp. 815–821, J. Lodder, Ed. North Holland, Amsterdam, The Netherlands (1970).
50. Phaff, H. J., Genus *Sporidiobolus,* in *The Yeasts – A Taxonomic Study,* pp. 822–830, J. Lodder, Ed. North Holland, Amsterdam, The Netherlands (1970).
51. Phaff, H. J., Genus *Sporobolomyces,* in *The Yeasts – A Taxonomic Study,* pp. 831–862, J. Lodder, Ed. North Holland, Amsterdam, The Netherlands (1970).
52. Phaff, H. J., and Fell, J. W., Genus *Cryptococcus,* in *The Yeasts – A Taxonomic Study,* pp. 1088–1145, J. Lodder, Ed. North Holland, Amsterdam, The Netherlands (1970).
53. Phaff, H. J., Genus *Kloeckera,* in *The Yeasts – A Taxonomic Study,* pp. 1146–1160, J. Lodder, Ed. North Holland, Amsterdam, The Netherlands (1970).

54. Phaff, H. J., and Ahearn, D. G., Genus *Rhodotorula,* in *The Yeasts – A Taxonomic Study,* pp. 1187–1223, J. Lodder, Ed. North Holland, Amsterdam, The Netherlands (1970).
55. Raschke, W. C., and Ballou, C. E., Immunochemistry of the Phosphomannan of the Yeast *Kloeckera brevis, Biochemistry, 10,* 4131 (1971).
56. Rodrigues de Miranda, L., *Filobasidium capsuligenum* nov. comb., *Antonie van Leeuwenhoek, J. Microbiol. Serol., 38,* 91 (1972).
57. Szabo, G., and Davies, R., Studies on the β-Galactosidase Activity of *Saccharomyces fragilis, J. Gen. Microbiol., 37,* 99 (1964).
58. Slooff, W. Ch., Genus *Lipomyces,* in *The Yeasts – A Taxonomic Study,* pp. 379–402, J. Lodder, Ed. North Holland, Amsterdam, The Netherlands (1970).
59. Slooff, W. Ch., Genus *Schizosaccharomyces,* in *The Yeasts – A Taxonomic Study,* pp. 733–755, J. Lodder, Ed. North Holland, Amsterdam, The Netherlands (1970).
60. Slooff, W. Ch., Genus *Pityrosporum,* in *The Yeasts – A Taxonomic Study,* pp. 1167–1186, J. Lodder, Ed. North Holland, Amsterdam, The Netherlands (1970).
61. Slooff, W. Ch., Genus *Trigonopsis,* in *The Yeasts – A Taxonomic Study,* pp. 1353–1357, J. Lodder, Ed. North Holland, Amsterdam, The Netherlands (1970).
62. Snyder, H. E., and Phaff, H. J., The Pattern of Action of Inulinase from *Saccharomyces fragilis* on Inulin, *J. Biol. Chem., 237,* 2438 (1962).
63. Sonck, C. E., and Yarrow, D., Two New Yeast Species Isolated in Finland, *Antonie van Leeuwenhoek J. Microbiol. Serol., 35,* 172 (1969).
64. Storck, R., Alexopoulos, C. J., and Phaff, H. J., Nucleotide Composition of Deoxyribonucleic Acid of Some Species of *Cryptococcus, Rhodotorula* and *Sporobolomyces, J. Bacteriol., 98,* 1069 (1969).
65. Uden, N. van, and Buckley, H., Genus *Candida,* in *The Yeasts – A Taxonomic Study,* pp. 893–1087, J. Lodder, Ed. North Holland, Amsterdam, The Netherlands (1970).
66. Uden, N. van, and Vidal-Leiria, M., Genus *Torulopsis,* in *The Yeasts – A Taxonomic Study,* pp. 1235–1306, J. Lodder, Ed. North Holland, Amsterdam, The Netherlands (1970).
67. Walt, J. P. van der, Genus *Dekkera,* in *The Yeasts – A Taxonomic Study,* pp. 157–165, J. Lodder, Ed. North Holland, Amsterdam, The Netherlands (1970).
68. Walt, J. P. van der, Genus *Kluyveromyces,* in *The Yeasts – A Taxonomic Study,* pp. 316–378, J. Lodder, Ed. North Holland, Amsterdam, The Netherlands (1970).
69. Walt, J. P. van der, Genus *Lodderomyces,* in *The Yeasts – A Taxonomic Study,* pp. 403–407, J. Lodder, Ed. North Holland, Amsterdam, The Netherlands (1970).
70. Walt, J. P. van der, Genus *Saccharomyces,* in *The Yeasts – A Taxonomic Study,* pp. 555–718, J. Lodder, Ed. North Holland, Amsterdam, The Netherlands (1970).
71. Walt, J. P. van der, Genus *Wingea,* in *The Yeasts – A Taxonomic Study,* pp. 772–775, J. Lodder, Ed. North Holland, Amsterdam, The Netherlands (1970).
72. Walt, J. P. van der, Genus *Brettanomyces,* in *The Yeasts – A Taxonomic Study,* pp. 863–892, J. Lodder, Ed. North Holland, Amsterdam, The Netherlands (1970).
73. Walt, J. P. van der, The Perfect and Imperfect States of *Sporobolomyces salmonicolor, Antonie van Leeuwenhoek J. Microbiol. Serol., 36,* 49 (1970).
74. Wickerham, L. J., Lockwood, L. B., Pettijohn, O. G., and Ward, G. E., Starch Hydrolysis and Fermentation by the Yeast *Endomycopsis fibuliger, J. Bacteriol., 48,* 413 (1944).
75. Wickerham, L. J., Genus *Citeromyces* Santa Maria, in *The Yeasts – A Taxonomic Study,* pp. 121–127, J. Lodder, Ed. North Holland, Amsterdam, The Netherlands (1970).
76. Wickerham, L. J., Genus *Hansenula,* in *The Yeasts – A Taxonomic Study,* pp. 216–315, J. Lodder, Ed. North Holland, Amsterdam, The Netherlands (1970).
77. Wickerham, L. J., Genus *Pachysolen,* in *The Yeasts – A Taxonomic Study,* pp. 448–454, J. Lodder, Ed. North Holland, Amsterdam, The Netherlands (1970).
78. Yarrow, D., and Ahearn, D. G., *Brettanomyces abstinens* sp.n., *Antonie van Leeuwenhoek J. Microbiol. Serol., 37,* 296 (1971).

THE BASIDIOMYCETES

MARY P. LECHEVALIER

Basidiomycetes, the most evolved class of fungi, are characterized by the formation of basidiospores on the outside of spore-bearing structures known as basidia (Figure 1). Such organisms as rusts, smuts, and jelly fungi, as well as mushrooms, puffballs, shelf and coral fungi, stinkhorns, earthstars, and birds-nest fungi, are basidiomycetes. Morphologically fascinating, some are the delight of gastronomists, others the despair of farmers, nurserymen, foresters and industrialists. As plant parasites (rusts, smuts, and higher forms), as destroyers of lumber and wooden structures, they cause an untold amount of destruction. At the same time, such organisms serve a useful purpose in the degradation of vegetable matter, besides providing food for mushroom fanciers.

Basidiospores are usually uninucleate and haploid and are usually borne on sterigmata (Figure 1, A, b). The mycelium is uninucleate to begin with, then becomes binucleate through plasmogamy, with karyogamy and meiosis occurring only in the basidium. Clamp connections (Figure 1, A, d), characteristic of the mycelium of many basidiomycetes, play a role in the production of the binucleate condition. The vast majority of basidiomycetes are heterothallic. The sexual compatibility of such organisms depends not on the differentiation of distinct sexual forms but on "incompatibility factors." These factors regulate both the recognition of a potential mate by a given strain as well as subsequent morphogenesis.[1] The fruiting bodies (basidiocarps), except for those of most rusts and smuts, are quite complex; they may be open or closed, varying in shape and size from exceedingly tiny to approximately 100 cm.[2] The cell walls of the basidiomycetes that have been examined are of the chitin–glucan type; L-fucose is also found in the wall.[3] Among the economically important products of the basidiomycetes are such enzymes as cellulases and ligninases, such toxins as the amanitins, phalloidin, ibotenic acid, and atropine, and hallucinogens such as psilocybine. A review of the methods used in the study of basidiomycetes can be found in Reference 4.

Basidiomycetes may be classified as follows.[5]

A. Basidia septate or deeply divided, or consisting of a probasidium (teliospore) or cyst that germinates to give rise to a promycelium (epibasidium, Figure 1, B, C, D), basidiospores often germinating by the formation of conidia . . . HETEROBASIDIOMYCETIDAE

AA. Basidia not septate or deeply divided, but of various shapes; basidiospores usually germinating to form a mycelium (Figure 1, A) . . . HOMOBASIDIOMYCETIDAE

HETEROBASIDIOMYCETIDAE

A. Basidiocarp usually well developed; saprophytic or parasitic on plants or insects . . . Tremellales (jelly fungi)

The nine families of this order are distinguished on the basis of the morphology of the basidiocarps, epibasidia, probasidia, hymenium, and basidiospores. None has great economic importance.[6] Into this order fall the septobasidia, which form a lichenoid growth in association with scale insects.[7]

AA. Basidiocarp lacking, except in the Graphiolaceae (Ustilaginales); teliospores present *en masse* or scattered on or in host tissue; plant parasites.

B. Promycelium becoming transversely septate, forming four cells; each cell produces a basidiospore borne on a sterigma (Figure 1, C) . . . Uredinales (rusts)

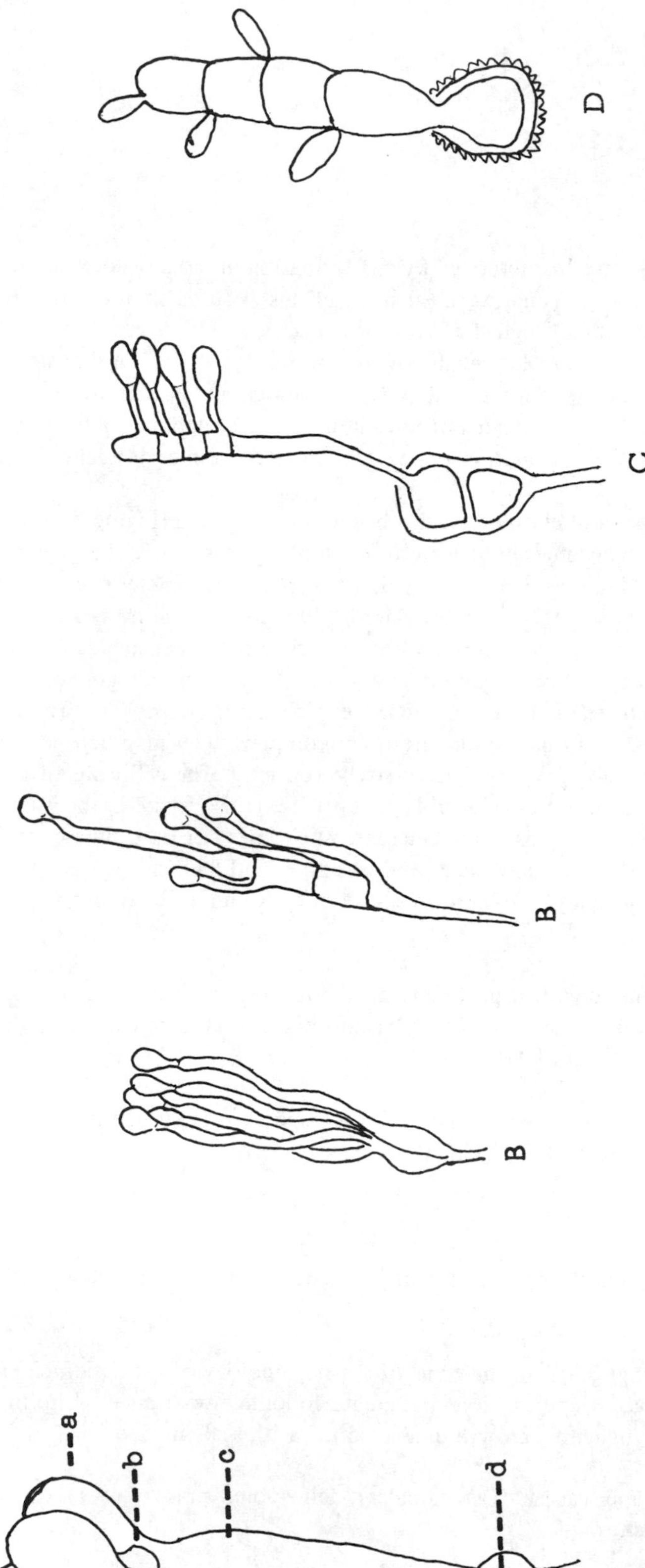

FIGURE 1. Types of Basidia. A. Homobasidiomycetaceae, Agaricales; (a) basidiospore; (b) sterigma; (c) basidium; (d) clamp connection. B. Heterobasidiomycetaceae, Tremellales. C. Heterobasidiomycetaceae, Uredinales. D. Heterobasidiomycetaceae, Ustilaginales.

A group of economically important fungi, the rusts cause tremendous damage not only to plants cultivated as food, such as cereals, vegetables, and fruits, but also to flowers, ornamentals, ferns, and trees, such as white pine. Rusts display an exceedingly complex life cycle, which often involves more than one host. Some of the economically important species and their hosts (in parentheses) are the following: *Puccinia graminis* (various subspecies attack different small grains, such as wheat, oat, barley, and rye); *P. sorghi* (corn); *Cronartium ribicola* (white pine); *Gymnoconia peckiana* (blackberry, black raspberry); and *Uromyces dianthi* (carnation).[8]

BB. Promycelium septate or not; basidiospores sessile (Figure 1, D) . . . Ustilaginales (smuts)

Like the rusts, to which they are closely related, the smuts cause great crop damage. Some important members of this group and their hosts (in parentheses) are the following: *Ustilago avenae* (oats); *U. maydis* (corn); *Tilletia caries* (wheat); and *Urocystis cepulae* (onion).[9]

HOMOBASIDIOMYCETACEAE

Basidiomycetes classified here have a club-shaped basidium that is not divided (Figure 1, A).

A. Basidiocarp not formed; hymenial layer formed at the surface of parasitized plant tissues . . . Exobasidiales

Members of this group, although they attack certain ornamental plants, do not cause damage that is as important as that caused by the rusts and smuts.[10]

AA. Basidiocarp formed, ranging from a fine network of hyphae to an intricate fruit body.

B. Hymenium present and exposed before spores mature . . . Hymenomycetes[a]

Hymenomycetes[a]

C. Hymenium unilateral or not restricted to any particular surface; hymenial surface smooth, ridged, warty, spiny, or porous, rarely bearing gills; basidiocarp of porous or lamellate forms not soft or putrescent . . . Polyporales (Aphyllophorales)

D. Hymenium smooth, rough, or corrugate.

E. Basidiocarp typically cobweb-like, membranous, leathery, or hard, varying from forms simply resting on the substratum and facing outwards (resupinate) to forms having a cap and stalk . . . Thelephoraceae

The Thelephoraceae include many rather primitive forms that resemble crusts or shelves on the surface of dead wood. Other forms resemble densely branched plants; still others approach a mushroom-like shape. *Sparassis radicata,* which resembles a buff-colored lettuce plant, is highly prized by mushroom eaters (mycophagists). This fungus is parasitic on conifers. Other parasites in this group include *Pellicularia filamentosa,* better known as *Rhizoctonia solani,* which causes black scurf of potatoes, and members of the

[a] Not officially recognized as a taxonomic entity.

genus *Stereum,* which attack many economically important trees, such as oak and apple.[11-15]

EE. Basidiocarps mostly pileate and fleshy, sometimes gelatinous.

F. Basidiocarps club-shaped, branched or not; hymenium amphigenous . . . Clavariaceae (fairy clubs)[12,16,17]

FF. Basidiocarps funnel-shaped or mushroom-like; hymenium inferior, composed of ridges or gill-like folds . . . Cantharellaceae[18]
Members of these two families are often brightly colored. Many are edible; the famous "chanterelle," *Cantharellus cibarius,* is considered among the most delectable mushrooms.

DD. Hymenium spiny or porous, rarely lamellate.

E. Hymenial covering consisting of warts, spines, or teeth, which point downward . . . Hydnaceae (tooth fungi)

Among the edible members of this family is *Hericium corralloides,* a striking white-yellow fungus with many fine, long teeth.[12,19-21]

EE. Hymenium lining interior of pits or tubes.

F. Pits shallow, fertile on ridges . . . Meruliaceae

Merulius lacrymans, the dry rot fungus, is classed here.[22,23]

FF. Tubes (pores) deep, or if shallow, sterile on ridges; basidiocarp not soft and putrescent . . . Polyporaceae (polypores or bracket fungi)

Some mycologists split off certain groups normally included here to form new families, such as the Ganodermataceae (containing *Ganoderma applanatum,* a parasite that attacks beech and other trees) and *Fistulinaceae* (containing *Fistulina hepatica,* the "beefsteak fungus"). Because of their tough consistency, polypores do not harbor many forms considered edible. A few that are soft when young, such as *Polyporus sulfureus,* are regarded as good to eat. Many species are wood destroyers. The basidiocarps of certain types are annual; those of other types are perennial and may live to be quite old (more than 50 years).[12,24-27]

CC. Hymenium borne on gills (lamellae), or, if lining the interior of tubes, soft and putrescent basidiocarp . . . Agaricales (see below)

BB. Hymenium present or absent; basidiocarp closed at least until spores are released from basidia . . . Gasteromycetes

GASTEROMYCETES[28,29]

C. Hymenium present in early stages.

D. Gleba powdery; spores usually pale and small . . . Lycoperdales (common puffballs and earthstars)

All puffballs are edible as long as the interior of the sporocarp is pure white. Since certain poisonous forms of agarics (e.g., some *Amanita*) have an immature stage resembling a puffball, all puffballs that will be cooked should be cut to be sure no organized structure can be discerned in the interior.

DD. Gleba fleshy, waxy, or slimy when mature.

E. Gleba slimy and fetid, exposed . . . Phallales (stinkhorns)

EE. Gleba usually fleshy or waxy; if slimy or fetid, not exposed . . . Hymenogastrales

CC. Hymenium lacking or indistinct.

D. Gleba waxy; peridioles formed . . . Nidulariales (birds-nest fungi)

DD. Gleba powdery; spores usually large and dark . . . Sclerodermatales (earthballs and earthstars)

Agaricales

One can easily recognize two main groups of Agaricales: (1) those with pores, the boletes, and (2) those with lamellae, the agarics.

Boletes

The boletes, or Boletaceae, have been the object of three recent books,[30-32] in which three different systems of classification and nomenclature are used; thus no attempt will be made here to characterize genera.

The boletes have, as indicated above, fleshy, readily decaying basidiocarps. Many species form mycorrhizal associations with trees. This explains why some species are found only where certain trees are growing. *Boletus edulis* is one of the most delectable mushrooms. Some tropical boletes may be dangerous to eat. Some of the species from temperate regions are not very tasty, and still other species can produce intestinal disorders in some individuals and not in others.

Agarics

Among the criteria used in the classification of the agarics, the following may be cited: gross and microscopic morphology, size, structure, pigments, habitat and associations, chemistry of certain reactions, taste, and odor. Reviews of the status of chemotaxonomy in the classification of homobasidiomycetes have recently been published,[33,34] as well as a review covering the chemical tests used in the classification of agarics.[35] An effort to apply anatomy and morphology of cultures grown *in vitro* to taxonomy has been made.[36] Some morphologic characteristics of the agarics are illustrated in Figures 2, 3, and 4.

The classification below, based on the system of Kühner and Romagnesi,[37] is not followed by all mycologists. It is, however, a useful guide for the identification of agarics. Although species descriptions given in their text (not included here) cover mushrooms found in western and central Europe, many of these are similar or identical to forms found in North America. Useful guides to some American species include Kauffman's classic text[38] and Graham's largely ignored handbook,[39] both now available as reprints.

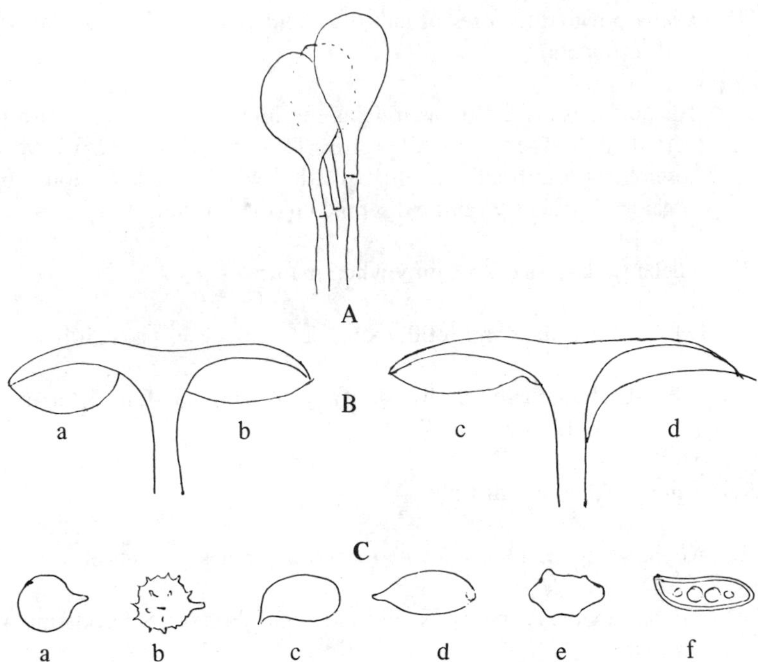

FIGURE 2. Morphologic Characteristics of the Agarics. A. Basal cells of the Amanitaceae. B. Gills of the Agaricales; (a) free; (b) adnate; (c) sinuate; (d) decurrent. C. Spore types; (a) round; (b) spiny; (c) oval-ellipsoid; (d) with germ pore; (e) angular; (f) bolete-like.

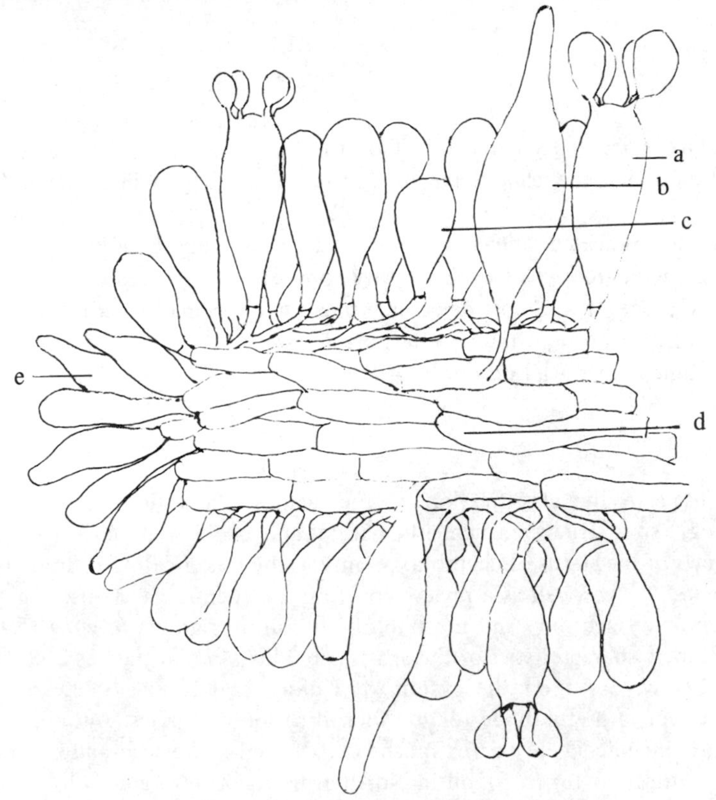

FIGURE 3. Section through a Gill. (a) basidium; (b) cystidium; (c) basidiole; (d) trama; (e) sterile cell.

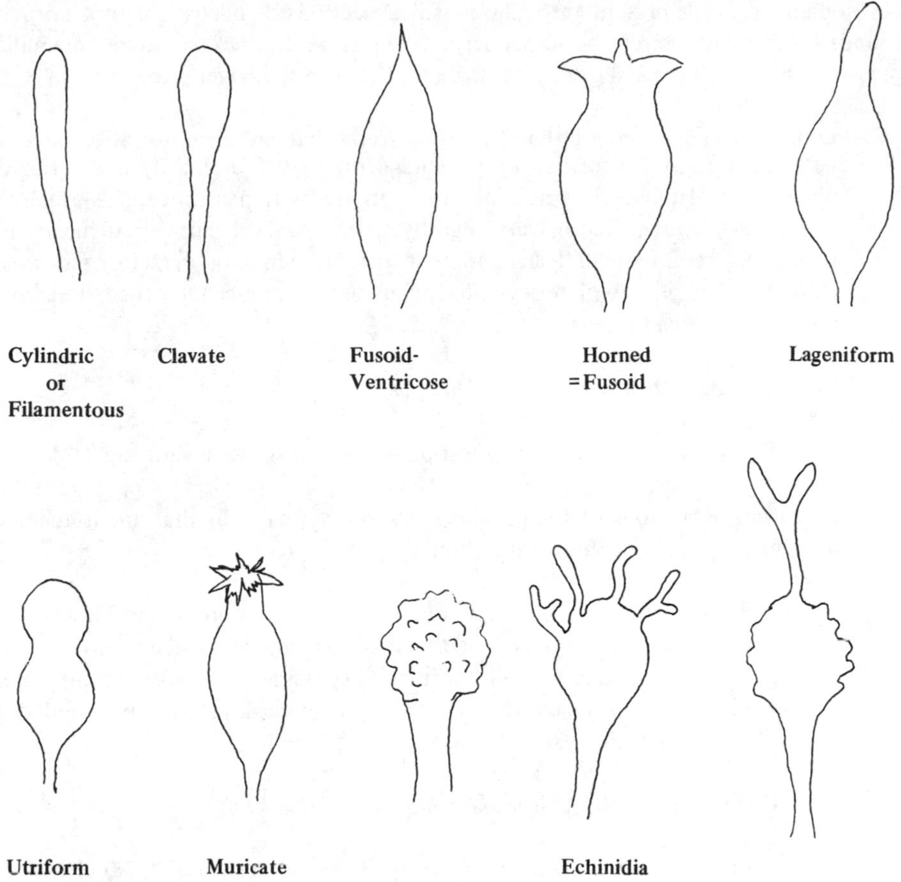

FIGURE 4. Some Types of Cystidia and Other Sterile Cells.

A. Stipe not horny, flesh not leathery, readily decomposing; gills not mucilaginous; spore prints, when dark, never olivaceous, and when yellow, always clear, never brownish; spore surface or ornaments amyloid. Spores never violet when viewed microscopically . . . FIRST GROUP OF FAMILIES

FIRST GROUP OF FAMILIES

B. Radially sectioned pileus shows groups of isodiametric cells resembling plant parenchyma; basidiocarp usually terrestrial and squat; stipe central, without volva, annulus or cortina, confluent with the pileus; pileus may be depressed or wrinkled, but never hygrophanous or striate by transparence; clear or milk-like sap may be present; sectioned gills may show rounded cells; flesh tastes mild or acrid-peppery; spores white, cream, or yellow (rarely pink) in mass, colorless or pale when viewed microscopically, round or elliptical, with amyloid ornaments; cystidia almost always present . . . Russulaceae

Genera: *Lactarius, Russula*

Among the species of *Lactarius, L. deliciosus* is highly prized by gourmets. The number of species of russulae is very large, and determinations are difficult to make. *R. emetica* is well known for its very acrid taste.

BB. No isodiametric cells present throughout radical sections of pileus; gills not normally showing rounded cells; taste rarely acrid-peppery; when stipe is thick or squat, no milk-like sap is present; gills always show shorter intermediates between the larger ones.

C. Longitudinal sections of the stipe show cells that are abruptly attenuated to a narrow base (Figure 2,A); terrestrial forms with a central stipe, ordinarily with a ring or a cortina; pileus never depressed, hygrophanous, or striate by transparence; gills usually some shade of white, crowded, free or very slightly adnate, never decurrent, at most with a narrow ridge (Figure 2,B); gill trama bilateral; spores white in mass or glaucous, round or elliptical; normally no pleurocystidia; pileus surface non-cellular (occasional scales may be cellular) . . . Amanitaceae

Genera: *Amanita, Amanitopsis, Limacella*

The genus *Amanita* contains the most poisonous mushrooms known.[40,41]

CC. Longitudinal sections of the basidiocarp show hyphal cells that are attached end-to-end, forming long chains of basal mycelium.

D. Gill trama inversely bilateral; stipe central, dry, more or less fleshy, and separable from pileus; no annulus or cortina; pileus campanulate or convex, not depressed; gills free, crowded, usually whitish, becoming rose; spore print rose or a rosy terra-cotta; spores pale when viewed microscopically, smooth, rounded or elliptical; pleuro- and cheilocystidia almost always present . . . Volvariaceae

Genera: *Volvaria, Volvariella, Pluteus*

Volvaria volvacea, known as the "padi straw mushroom", is cultivated as food in China and in certain other areas in the East.[42]

DD. Gill trama different; when the spore print is rose or reddish, the gills are not free, or the stipe is either confluent with the pileus or has an annulus.

E. When the spores are rose-colored, they are rounded and smooth or present very minor irregularities.

F. Mushrooms with an annulus or a cortina; gills not deliquescing; stipe separable from cap, or gills free; terrestrial forms putrefying, with a central thick or fleshy non-viscid stipe, ordinarily without volva; pileus with dry cuticle, never umbilicate, hydrophanous or striate by transparence; crowded gills; flesh white, sometimes changing color when exposed to air or on aging; spores essentially smooth; cheilocystidia frequent; pleurocystidia ordinarily lacking; gill trama not bilateral, but regular or intermingled . . . Lepiotaceae

Genera: *Lepiota, Cystoderma, Psalliota (Agaricus)*

To the genus *Psalliota* belongs the mushroom most widely cultivated in the western world, *Ps. bispora.* The production of this mushroom in the United States alone was estimated in 1965 at 170 million pounds annually.[43,44]

FF. Not as above.

G. Spore print dark; spores black, dull brown, or dark brown-red when viewed microscopically; spore color changes in concentrated sulfuric acid; stipe central, whitish in young specimens; pileus often brownish or grayish, frequently striate by transparence or striate-wrinkled; flesh often thin; gills never decurrent, occasionally mottled in appearance; spores smooth with a germ pore (Figure 2,C,d); pleurocystidia present; cheilocystidia frequently present . . . Coprinaceae

Genera: *Coprinus, Drosophila*

Coprinus species characteristically deliquesce to give an inky-black fluid. Certain species of this genus have been cultivated *in vitro* and serve as tools in genetic studies.

GG. Spore color does not change in concentrated sulfuric acid. See SECOND GROUP OF FAMILIES, below.

EE. Spore print rose-colored or terra cotta; individual spores polygonal when viewed microscopically from any angle; basidiocarps usually terrestrial; stipe not viscid, central in young specimens, usually central in mature specimens; no volva or ring; cortina rare; pileus confluent with stipe; gills decurrent or adnexed, usually rose-colored when mature . . . Rhodophyllaceae

Genera: *Rhodophyllus* (subgenera *Claudopus, Eccilia, Leptonia, Nolanea, Entoloma*)[45]

AA. Not as above . . . SECOND GROUP OF FAMILIES

SECOND GROUP OF FAMILIES

A. Spore print frankly colored, but not rose-colored or reddish; echinidia (brush cells, Figure 4) lacking.

B. Gills similar to the pores of the Boletaceae: of soft-mucilaginous consistency and easily separable from the cap; or spores similar in shape and color to those of the boletes: fusoid in shape (Figure 2,C,f), and yellow, brownish, bistre, or olive-colored when viewed microscopically; stipe usually well-developed, fleshy and thick (at least 5–8 mm in diameter), ordinarily solid, and more or less central (when it is eccentric, the gills are usually decurrent); cap fleshy, non-hygrophanous, and opaque; no volva . . . Boletineae (gilled boletes)

Genera: *Phylloporus, Gomphidius, Paxillus*[30]

BB. Basidiocarps not having characteristics of boletes.

C. Stipe easily distinguished from pileus; gills rarely decurrent, or, if decurrent, showing cheilocystidia; young specimens have a veil that unites the cap and stipe . . . Naucoriaceae

Genera: *Inocybe,*[46] *Cortinarius, Rozites, Hebeloma, Gymnopilus,*[47] *Phaeocollybia, Galera*[48] *Naucoria, Dryophila* (subgenera *Pholiota*[49] and *Flammula*), *Geophila* (subgenera *Stropharia, Psilocybe,* and *Hypholoma*), *Panaeolus, Agrocybe, Bolbitius, Conocybe*

The genus *Psilocybe* contains most of the mushrooms known to be hallucinogenic. *Psilocybe mexicana* and others (the "teonanacatl" of the Mexican Indians) are thought to be sacred by certain tribes.[50] Members of the genus *Cortinarius* are characterized by a distinct cortina or veil, frequently of cobwebby appearance, visible in young specimens. Certain *Inocybe* species have strikingly polygonal spores.

CC. Stipe lacking, rudimentary or scarcely distinguishable from the pileus in mature basidiocarps, or gills decurrent, with fertile edges; spores never blackish or violet when viewed microscopically and lacking germ pores. See F or FF, below.

AA. Spores not as above, usually colorless when viewed microscopically; spore prints may be more or less rose-colored, reddish or white.

D. Gills waxy, normally thick, and distant; basidia slender and elongated (4 to 8 times longer than wide), often showing marked fatty inclusions; terrestrial forms not leathery, but readily putrifying; stipe central, never horny or black and velvety at the same time; spores white in mass, smooth, elliptical or round, not amyloid; cap cuticle typically "non-cellular" (see glossary); no echinidia on the pileus surface or gill edge; normally no cheilocystidia; clamp connections present as a rule . . . Hygrophoraceae

Genus: *Hygrophorus*[51]

DD. Gills membranous, or spores or basidia not as above; stipe not viscid, except in certain forms having amyloid spores.

E. Stipe of mature forms never fibrous-fleshy, but cartilaginous or horny, central, and relatively slender; spore prints usually white, not rose or reddish; cap rarely umbilicate, gills rarely decurrent; cap cuticle neither scaly nor velvety, except in *Crinipellis;* no striking cortina; normally no ring . . . Marasmiaceae

Genera: *Xeromphalina, Crinipellis, Collybia, Marasmius, Deliculata, Mycena*

Two monographs on *Mycena* exist: one (52) is oriented toward European species, the other (53) to North American forms. Species of *Marasmius* are characterized by the fact that, comparatively speaking, they resist decomposition in nature, probably because of the cyanides they form. *Marasmius oreades,* often found on lawns, is delicious.

EE. Stipe never threadlike, never long and thin and pure white at the same time, never horny, always continuous with the pileus; gills of unequal length or branched; pileus cuticle usually not hymeniform.

F. Species readily decomposing; flesh formed only of thin-walled hyphae; pileus not leathery; stipe usually central; gills never reaching the stipe base; lignicolous forms lacking elongated cylindrical spores; most species terrestrial . . . Tricholomaceae

Genera: *Rhodotus, Macrocystidia, Rhodopaxillus, Clitopilus, Ripartites, Nyctalis, Lyophyllum, Laccaria Hygrophoropsis, Clitocybe (Armillaria), Omphalia, Biannularia, Leucopaxillus, Melanoleuca, Tricholoma, Leucocortinarius, Hebelomina*

One of the most destructive agarics is *Clitocybe (Armillaria) mellea,* the "honey fungus". It attacks conifers, shrubs, broad-leafed trees, logs, and occasionally even such plants as potatoes and strawberries. It is edible. Some of the genera of the Tricholomaceae are monospecific, representing forms that cannot be classified easily elsewhere.

FF. Species not as above; when the spore print is rose-colored or reddish, the species is lignicolous or epiphytic, the spores lack striate markings, and the hyphae show clamp connections; clamp connections usually present in all other forms; sporocarp tough, sessile or pleuropodial . . . Pleurotaceae

Genera: *Leptoglossum, Crepidotus,*[54] *Schizophyllum, Phyllotopsis, Chaetocalathus, Pleurotellus, Pleurotus, Geopetalum, Panus, Lentinus, Lentinellus, Panellus.*

Schizophyllum commune, cosmopolitan in distribution, has the unique character of having split-edged gills. It is widely used in genetic studies because of its ease of maintenance *in vitro.*[55]

SELECTED FIELD GUIDES

The larger basidiomycetes and ascomycetes are easily seen by the layman. A number of field books are available, designed to help identify fungi with a minimal amount of knowledge of mycology. The following is a list of such books that are considered by the author to have special merit.

Graham, V. O., *Mushrooms of the Great Lakes Region.* Dover Publications, New York (1970).

This is a republication of a book first published in 1944. It has an excellent key to genera and is probably the most useful book of its type.

Lange, M., and Hora, F. B., *Mushrooms and Toadstools.* E. P. Dutton, New York (1963).

This is a translation and adaptation of a Danish text. Color illustrations are numerous and excellent. Together with Graham's book, it should constitute the "basic" library of the amateur.

Pomerleau, R., *Mushrooms of Eastern Canada and the United States.* Les Editions Chantecler Ltd, Montreal, P. Q., Canada (1951).

An excellent book for the mycophagist. In addition to black-and-white photographs, it contains five plates of beautiful water colors by Henry Jackson.

Christensen, C. M., *Common Fleshy Fungi.* Burgess Publishing Co., Minneapolis, Minnesota (1946).

A good, general, basic field book.

Christensen, C. M., *Common Edible Mushrooms.* University of Minnesota Press, Minneapolis, Minnesota (1943).

A good, inexpensive booklet for the casual mycophagist.

Maublanc, A., *Les Champignons Comestibles et Vénéneux,* 6th edition (two volumes). P. Lechevalier, Paris, France (1971).

Volume 1 contains general statements and keys. Volume 2 contains excellent color plates and descriptions of species.

Smith, A. H., *The Mushroom Hunter's Field Guide.* University of Michigan Press, Ann Arbor, Michigan (1967).

A revised and enlarged collection of excellent photographs of mushrooms, some in color. Like many books of this type, it suffers from the fact that it covers too few species.

Hesler, L. R., Mushrooms of the *Great Smokies.* University of Tennessee Press, Knoxville, Tennessee (1960).

Contains fine black-and-white photographs of selected mushrooms.

von Frieder, L., *Mushrooms of the World.* Bobbs-Merrill Co., Inc., New York (1969).

Presumably translated from German, this book is printed in Italy and contains some of the most beautiful color plates of mushrooms the author has seen.

Pilát, A., and Ušák, O., *Atlas Hub.* Státní Pedagogické Nakladatelstór, Prague, Czechoslovakia (1961).

A collection of beautiful water colors of mushrooms. Text in Czech.

Miller, O. K., *Mushrooms of North America.* Dutton, New York (1972).

This is a new and yet untested field guide just published in the U.S.A. Printed in Japan, it has beautiful reproductions of color photographs; some of these are mislabeled in the first printing.

REFERENCES

1. Koltin, V., Stamberg, J., and Lemke, P. A., *Bacteriol. Rev., 36,* 156 (1972).
2. Smith, A. H., in *The Fungi,* Vol. 2, pp. 151-177, G. C. Ainsworth and A. S. Sussman, Eds. Academic Press, New York (1966).
3. Bartnicki-Garcia, S., *Annu. Rev. Microbiol., 22,* 87 (1968).
4. Booth, C. (Ed.), Vol. 4 of *Methods in Microbiology,* J. R. Norris and D. W. Ribbons, Eds. Academic Press, New York (1971).
5. Martin, G. W., in *Dictionary of the Fungi,* 5th ed., pp. 497-517, G. C. Ainsworth, Ed. Mycological Institute, Kew, Surrey, England (1963).
6. Martin, G. W., *Revision of the North American Tremellales.* J. Cramer, Lehre, Germany (1969).
7. Couch, J. N., *The Genus Septobasidium.* University of North Carolina Press, Chapel Hill, North Carolina (1938).

8. Arthur, J. C., and Cummins, G. B., *Manual of the Rusts in the United States and Canada.* Hafner, New York (1962).
9. Fischer, G. W., and Holton, C. S., *Biology and Control of the Smut Fungi.* Ronald Press, New York (1957).
10. Saville, D. B. O., *Can. J. Bot., 37,* 641 (1959).
11. Burt, E. A., *The Thelephoraceae of North America.* Hafner, New York (1966).
12. Bourdot, H., and Galzin, A., *Hyménomycètes de France.* J. Cramer, Lehre, Germany (1969).
13. Corner, E. J. H., *A Monograph of Thelephora.* J. Cramer, Lehre, Germany (1968).
14. Slysh, A. R., The Genus *Peniphora* in New York State and Adjacent Regions, *Technical Publication No. 83.* New York State College of Forestry at Syracuse University, Syracuse, New York (1960).
15. Overholts, L. O., The Genus *Stereum* in Pennsylvania, *Bull. Torrey Bot. Club, 66,* 515 (1939).
16. Corner, E. J. H., *A Monograph of Clavaria and Allied Genera.* Oxford University Press, Oxford, England (1950); supplement, J. Cramer, Lehre, Germany (1970).
17. Petersen, R. H., *The Genus Clavulinopsis in North America.* Hafner, New York (1968).
18. Smith, A. H., The Cantharellaceae of Michigan, *Mich. Bot., 7,* 143 (1968).
19. Coker, W. C., and Beers, A. H., *The Stipitate Hydnums of the Eastern United States.* J. Cramer, Lehre, Germany (1970).
20. Harrison, K. A., The Stipitate Hydnums of Nova Scotia, *Can. Dep. Agr. Publ. 1099* (1961).
21. Hall. D., and Stuntz, D. E., *Mycologia, 63,* 1099 (1971); *64*, 15 (1972); *64*, 560 (1972).
22. Burt, E. A., *Merulius* in North America, *Ann. Mo. Bot. Gard., 4,* 305 (1917); *6*, 143 (1919).
23. Ginns, J. H., The Genus *Merulius, Mycologia, 60,* 1211 (1968); *61*, 357 (1969); *62*, 238 (1970); *63*, 219 (1971); *63*, 800 (1971).
24. Overholts, L. O., *The Polyporaceae of the United States, Alaska and Canada.* University of Michigan Press, Ann Arbor, Michigan (1967).
25. Lowe, J. L., The Polyporaceae of New York State (Except *Poria*), *Technical Publication No. 60.* New York State College of Forestry at Syracuse University, Syracuse, New York (1942).
26. Lowe, J. L., The Genus *Poria, Technical Publication No. 90.* New York State College of Forestry at Syracuse University, Syracuse, New York (1966).
27. Lowe, J. L., Polyporaceae of North America; The Genus *Fomes, Technical Publication No. 80.* New York State College of Forestry at Syracuse University, Syracuse, New York (1957).
28. Coker, W. C., and Couch, J. N., *The Gasteromycetes of Eastern United States and Canada.* J. Cramer, Lehre, Germany (1969).
29. Smith, A. H., *Puffballs and Their Allies in Michigan.* University of Michigan Press, Ann Arbor, Michigan (1951).
30. Leclair, A., and Essette, H., *Les Bolets.* P. Lechevalier, Paris, France (1969).
31. Snell, W. H., and Dick, E. A., *The Boleti of Northeastern North America.* J. Cramer, Lehre, Germany (1970).
32. Smith, A. H., and Thiers, H. D., *The Boletes of Michigan.* University of Michigan Press, Ann Arbor, Michigan (1970).
33. Tyler, V. E., Jr., in *Evolution in the Higher Basidiomycetes,* pp. 29–62, R. H. Petersen, Ed. University of Tennessee Press, Knoxville, Tennessee (1971).
34. Arpin, N., and Fiasson, J. L., in *Evolution in the Higher Basidiomycetes,* pp. 63–98, R. H. Petersen, Ed. University of Tennessee Press, Knoxville, Tennessee (1971).
35. Watling, R., in *Methods in Microbiology,* pp. 567–597, C. Booth, Ed. Academic Press, New York (1971).
36. Miller, O. K., in *Evolution in the Higher Basidiomycetes,* pp. 197–215, R. H. Petersen, Ed. University of Tennessee Press, Knoxville, Tennessee (1971).
37. Kühner, R., and Romagnesi, H., *Flore Analytique des Champignons Supérieurs.* Masson et Cie, Paris, France (1953).
38. Kauffman, C. H., *The Agaricaceae of Michigan.* Dover Publications, New York (1971).
39. Graham, V. O., *Mushrooms of the Great Lakes Region.* Dover Publications, New York (1970).

40. Heim, R., *Les Champignons Toxiques et Hallucinogènes.* Editions Boubée et Cie., Paris, France (1963).
41. Pomerleau, R., *Natur. Can., 93,* 861 (1966).
42. Gray, W. D., *Crit. Rev. Food Technol., 1,* 225 (1970).
43. Worgan, J. T., in *Progress in Industrial Microbiology,* Vol. 8, p. 76. CRC Press, Cleveland, Ohio (1968).
44. Essette, H., *Les Psalliotes.* P. Lechevalier, Paris, France (1964).
45. Hesler, L. R., *Entoloma in Southeastern North America.* J. Cramer, Lehre, Germany (1967).
46. Heim, R., *Le Genre Inocybe.* P. Lechevalier, Paris, France (1931).
47. Hesler, L. R., *North American Species of Gymnopilus.* Hafner, New York (1969).
48. Smith, A. H., and Singer, R., *A Monograph of the Genus Galerina.* Hafner, New York (1964).
49. Smith, A. H., and Hesler, L. R., *The North American Species of Pholiota.* Hafner, New York (1968).
50. Heim, R., and Wasson, R. G., *Les Champignons Hallucinogènes du Mexique.* Museum National d'Histoire Naturelle, Paris, France (1958).
51. Hesler, L. R., and Smith, A. H., *North American Species of Hygrophorus.* University of Tennessee Press, Knoxville, Tennessee (1963).
52. Kühner, R., *Le Genre Mycena.* P. Lechevalier, Paris, France (1938).
53. Smith, A. H., *North American Species of Mycena.* University of Michigan Press, Ann Arbor, Michigan (1947).
54. Hesler, L. R., and Smith, A. H., *North American Species of Crepidotus.* Hafner, New York (1965).
55. Niederpruem, D. J., and Jersild, R. A., *Crit. Rev. Microbiol., 1,* 545 (1972).

DEUTEROMYCETES (FUNGI IMPERFECTI)

DR. BARRY B. HUNTER AND DR. H. L. BARNETT

The Deuteromycetes are a heterogenous group of fungi in which sexual stages (perfect stages) are not known or rarely found and reproduction is limited to the production of conidia or sclerotia. These fungi are commonly called imperfect fungi, and technically Deuteromycetes or Fungi Imperfecti. The absence of the sexual phase makes it impossible to assign these fungi to a natural taxon; thus, description and taxonomic relationships must be based upon non-sexual features.

The conidial states of the imperfect fungi are like the conidial states of known Ascomycetes, and it is likely that most of these septate fungi really belong to this class, although a few are known to be Basidiomycetes. Alexopoulos[1] considers the Fungi Imperfecti as conidial states of Ascomycetes or, more rarely, as Basidiomycetes whose sexual states have not been discovered or no longer exist. Likewise, the discovery by Pontecorvo and Roper[2] of the parasexual cycle, which is operational in many of these fungi, suggests that certain Fungi Imperfecti may never have possessed a sexual state.

When sexual states are found in Deuteromycetes, these species may properly be placed in the Ascomycetes or Basidiomycetes. However, many mycologists prefer to retain the imperfect name, as well as the correct taxonomic designation, as a synonym (see Table 1). Likewise, most mycologists now concur that the Deuteromycetes constitute a valid and important group of fungi, even though true systematic relationships are often obscure.

TABLE 1

SEXUAL AND CONIDIAL STATES OF SELECTED HUMAN- AND PLANT-PATHOGENIC FUNGI

Sexual State	Conidial State	Common Name
Arthroderma benhamiae	*Trichophyton mentagrophytes*	Athlete's foot (Tinea pedis)
Ceratocystis ulmi	*Graphium ulmi*	Dutch elm disease
Cochliobolus hetero strophus	*Helminthosporium maydis*	Southern corn blight
Gibberella roseum	*Fusarium roseum*	Seedling blight of small grains
Gymnoscus demonbreunii	*Histoplasma capsulatum*	Histoplasmosis
Monilinia fructicola	*Monilia cinerea* var. *americana*	Brown rot of stone fruits
Nannizzia fulva	*Microsporum fulvum*	Ringworm (Tinea capitis)
Sartorya fumigata	*Aspergillus fumigatus*	Aspergillosis
Thanatephorus cucumeris	*Rhizoctonia solani*	Brown patch on turf grasses, potato scurf, etc.
Venturia inaequalis	*Spilocaea pomi*	Apple scab

Excluding the asporogenous yeasts, such as *Candida, Rhodotorula, Cryptococcus* and *Trichosporon,* the thallus of Deuteromycetes is filamentous, consisting of branching septate hyphae, most often multinucleate and rarely possessing clamp connections. All imperfect fungi are eukaryotic (i.e., the nuclear material is surrounded by a membranous nuclear envelope) and have cytoplasm replete with organelles as in higher plant cells, plasmalemmas, and cell walls. Compared to most other eukaryotic plant and animal cells, fungal cells are less complex and smaller. The septa of the imperfects, like Ascomycetes, are usually perforated, and their nuclei and cytoplasm can pass freely from one cell to another.[3] *Botrytis cinerea* has been shown to have a distinct central pore separating two simple plates that are connected to the longitudinal cell walls; this constitutes a typical Ascomycete-Deuteromycete septum.[4] Plasmodesmata have been found in complete septa of *Geotrichum candidum,*[5] and multiperforate septa have been observed in *Fusarium solani.*[6] The fungal cell composition and ultrastructure will be discussed later.

CONIDIAL STATES

There are two distinct ways in which conidia are produced on conidiophores: (1) separate and loosely arranged on the cottony hyphae; (2) enclosed within asexual fruiting structures. The former are often referred to as hyphomycetes and include the greatest number of imperfect fungi. The conidiophores of the hyphomycetes may be simple, variously branched, fused together (synnema), or aggregated on a stroma (sporodochia). These fungi are placed in the order Moniliales and are separated into families according to color of mycelium, conidia and conidiophores, and the presence of synnemata or sporodochia.

Imperfect fungi that produce conidia enclosed within variously shaped pycnidia belong to the order Sphaeropsidales. The pycnidial wall is pseudoparenchymatous and the conidiophores are produced from cells inside the pycnidium. Externally, the pycnidia closely resemble the variously shaped perithecia, and it is often necessary to crush the structure and examine the contents for asci or conidia before correct identification can be made.

In plants, the development of conidia subepidermally or subcortically, often surrounded by a gummy mass, constitutes an acervulus. At maturity the structure becomes erumpent, with conidia being dispersed in droplets of water. Some mycologists interpret the acervulus as a group of closely compact conidiophores growing on a stroma or stroma-like aggregation of hyphae, not associated with a host. Accordingly, the acervulus does not form readily in culture, often appearing more like a sporodochium. Some fungi produce fruiting structures intermediate between the pycnidium and the acervulus, making identification and classification complex. Illustrations of the different fruiting structures and descriptions of the form orders are given below.

ULTRASTRUCTURE OF THE CELL

In recent years there have been many investigations of the fine structure of the fungal cell; however, information on the ultrastructure of cells of Fungi Imperfecti is scarce. Nevertheless, enough research studies have been undertaken to make some generalizations on the fine structure of propagules of selected imperfect fungi.

In most instances it has been ascertained that the ultrastructure of Deuteromycete cells is similar to that of Ascomycete cells. Most imperfect fungi, like Ascomycetes, have a central pore in their septum (simple pore), whereas *Rhizoctonia solani* was shown to have a dolipore septum, which is common in Homobasidiomycetidae and the Tremelalles of Heterobasidiomycetidae.[3] Figures 1 and 2 illustrate the typical Ascomycete-Deuteromycete septum and Basidiomycete septum.

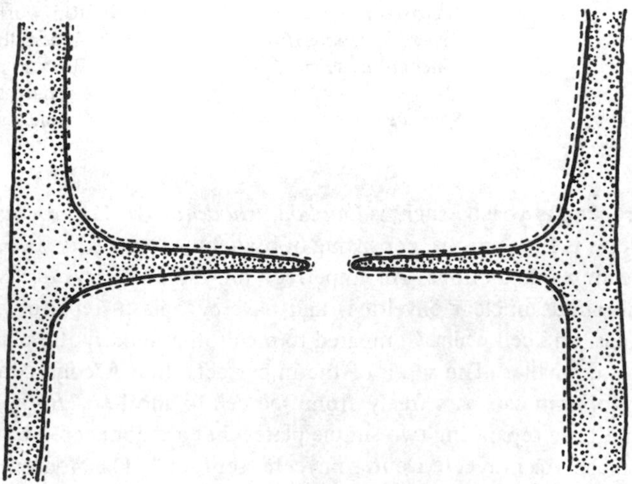

FIGURE 1. Ascomycete-Deuteromycete Septum

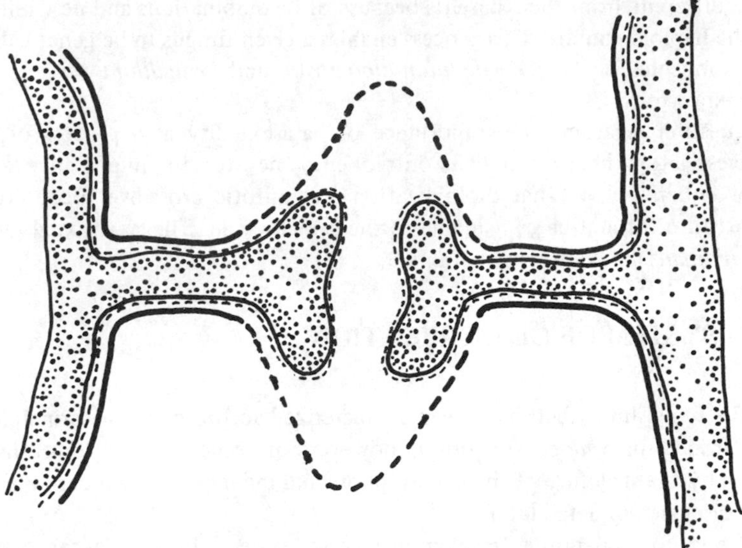

FIGURE 2. Basidiomycete Septum

The organelle composition and morphology of the imperfect fungal cell is similar to that of all eumycophytic fungi. Hyphae and conidia of *Verticillium albo-atrum* and *V. nigrescens* have the typical organelle composition of an imperfect fungal cell.[7] These two fungi have uninucleate cells containing mitochondria, endoplasmic reticulum, lomasomes (unique to fungi), ribosomes, and Woronin bodies, all enclosed within the plasmalemma. Lysosomes, vacuoles, Golgi bodies, and microtubules were not evident in cells of the two fungi, but they have been observed in cells of other imperfect fungi.

Table 2 provides information regarding the organelle composition of cells of various fungal structures of selected imperfect fungi. The organelle composition varied slightly when conidium, sclerotium, haustorium, and hyphal cell were compared.

THE PARASEXUAL CYCLE

Numerous imperfect fungi are now known to possess functional parasexual cycles. This process enables these fungi to have segregation and recombination of diploid nuclei at mitosis, which provides features of sexuality as a part of the parasexual cycle.

The most important aspects of the parasexual cycle are: (1) heterokaryosis; (2) fusion of unlike nuclei in the hypae; and (3) segregation and recombination at mitosis.[11]

Parasexuality was first observed in 1952 by Pontecorvo and Roper at the University of Glasgow while working with *Aspergillus nidulans.* They found that *A. nidulans* could form heterokaryotic mycelium through anastomosis of hyphae with different nuclear composition or, to a lesser extent, by mutation of one or more nuclei in a homokaryotic mycelium. At times fusion of like or unlike nuclei occurred, the first resulting in a homozygous diploid nucleus and the latter in a heterozygous diploid nucleus. Approximately one heterozygous diploid nucleus formed for every 1,000 haploid nuclei present.[12] Mitotic cross-over resulting in different combinations and new linkages occurred infrequently during multiplication of the diploid nucleus.

One method by which diploid nuclei have been detected was analysis of DNA phosphorus content of *A. nidulans* conidia. The diploid nuclei had approximately twice the phosphorus content that the haploid nuclei had, and the conidia were larger and of a different color. During growth of a diploid strain, chromosomes were successively lost by non-disjunction until the haploid condition was restored (haploidization). This condition could be observed by the presence of sectors in the mycelium that produced haploid conidia. These conidia were isolated and grew into haploid mycelium. Some of the haploid strains were different from their parents because of recombinations and new linkage groups, which were sorted out in the haploid conidia. This process enables a given fungus to be genetically different, and it can occur in Deuteromycetes, such as *Verticillium albo-atrum* and *Penicillium chrysogenum,* in which no sexual structures are known.

There is some question regarding the importance of parasexuality as a process of increasing genetic variation. Some investigators believe that its infrequency negates its importance as an evolutionary mechanism, whereas others believe that diploidization and mitotic cross-over occur frequently in some fungi and could substitute for and be as valuable as the sexual cycle. The parasexual cycle and the sexual cycle of *Aspergillus nidulans* are compared in Table 3.

THE SACCARDO SYSTEM OF CLASSIFICATION

The major taxa of the Fungi Imperfecti have been characterized according to the principles of classification established by Saccardo[13] in *Sylloge fungorum;* however, in recent years interest has been shown in classifying these fungi by the structure of the conidiophore and manner in which conidia are produced. This latter system will be discussed in detail later.

The primary criteria for separating large groups (form orders) by the Saccardoan system are (1) conidiophores and conidia occurring free and distributed over the mycelium, Hyphomyceteae (Moniliales); (2) conidiophores and conidia contained within a pycnidium, Sphaeropsideae (Sphaeropsidales); (3) conidiophores and conidia produced on layers of the substrate, Melanconieae (Melanconiales). A fourth group of fungi is Mycelia Sterilia, which produce no conidia or reproductive cells. Each of the sporulating form orders can be divided into one or more form families, based upon such characteristics as color, shape, and consistency of the pycnidium in the Sphaeropsidales, or color of conidia and presence of synnemata or sporodochia in the Moniliales.

The tremendous number of form genera of Deuteromycetes, estimated at more than 1,400, necessitates a practical means of separation. The system that is accepted and most widely used is that of Saccardo,[14] in which the color and morphology of the conidia are the basis for separation of form genera into "sections". The sections are subdivided as follows:

Amerosporae	Conidia are one-celled, spherical or ovoid to short cylindrical.
Hyalosporae	Hyaline conidia.
Phaeosporae	Colored conidia, ranging from light brown to black.
Didymosporae	As Amerosporae, except that the conidia are two-celled.
Hyalodidymae	Hyaline conidia.
Phaeodidymae	Colored conidia.
Phragmosporae	Conidia are transversely septate, consisting of three or more cells.
Hyalophragmiae	Hyaline conidia.
Phaeophragmiae	Colored conidia.
Dictyosporae	Conidia are multicellular, consisting of both transverse and longitudinal septa.
Hyalodictyae	Hyaline conidia.
Phaeodictyae	Colored conidia.
Scolecosporae	Conidia are slender, thread- to worm-like, one- to several-celled, hyaline or colored.
Helicosporae	Conidia are cylindrically spiraled, one- to several-celled, hyaline or colored.
Staurosporae	Conidia are radially lobed, star- or cross-shaped or branched, one- to several-celled, hyaline or colored.

TABLE 2

ORGANELLES IN CELLS OF VARIOUS IMPERFECT FUNGAL STRUCTURES AS DETERMINED BY ELECTRON MICROSCOPY

Organelles	*Verticillium albo-atrum* Conidium[7]	*Macrophomina phaseoli* Sclerotium[8]	Imperfect Form of *Erysiphe graminis* Haustoria[9]	*Rhizoctonia solani* Hyphal Cell[10]
Lipid droplets	–	+	+	–
Nucleus	+	+	+	+
	(one)	(one to three per cell)	(one)	(one)
Mitochondria	+	+	+	+
Lysosomes	—	—	—	—
Woronin bodies	+	+	—	—
Endoplasmic reticulum	+	+	+	+
Lomasomes	+	+	—	+
Ribosomes	+	+	—	—
Microtubules	—	—	—	—
Plasmalemma	Unit membrane	Unit membrane	Unit membrane	Unit membrane
Vacuoles	—	—	+	+
Golgi bodies	—	—	—	—

TABLE 3

A SUMMARIZED COMPARISON OF THE SEXUAL AND PARASEXUAL CYCLES IN *ASPERGILLUS NIDULANS*

Sexual Cycle	Parasexual Cycle
1. Heterokaryosis	1. Heterokaryosis
2. Nuclear fusion in specialized structures to yield "selfed" and hybrid zygotes.	2. Rare nuclear fusion in vegetative cells. Heterozygotes selected by color and/or nutrition. Homozygotes, if formed, are not detected
3. Zygote persists through one nuclear generation only.	3. Zygote may persist through many mitotic divisions.
4. Recombination at meiosis: cross-over at 4-strand stage in all chromosome pairs, random assortment of members of each chromosome pair, and reduction to haploid state.	4. Recombination by rare "accidents" of mitosis: (a) mitotic cross-over at 4-strand stage, usually only one exchange in a single chromosome arm; (b) haploidization, usually probably via aneuplidy; independent of cross-over random assortment of chromosome pairs.
5. Products of meiosis readily recognized and isolated.	5. Recombinants occur among vegetative cells. Recognized by use of suitable markers.

Data taken from: Roper, J. A., The Parasexual Cycle, in *The Fungi*, Vol. 2, pp. 589-617, G. C. Ainsworth and A. S. Sussman, Eds., Academic Press, New York, 1966. Reproduced by permission of the publishers.

Figure 3 illustrates the conidial types.

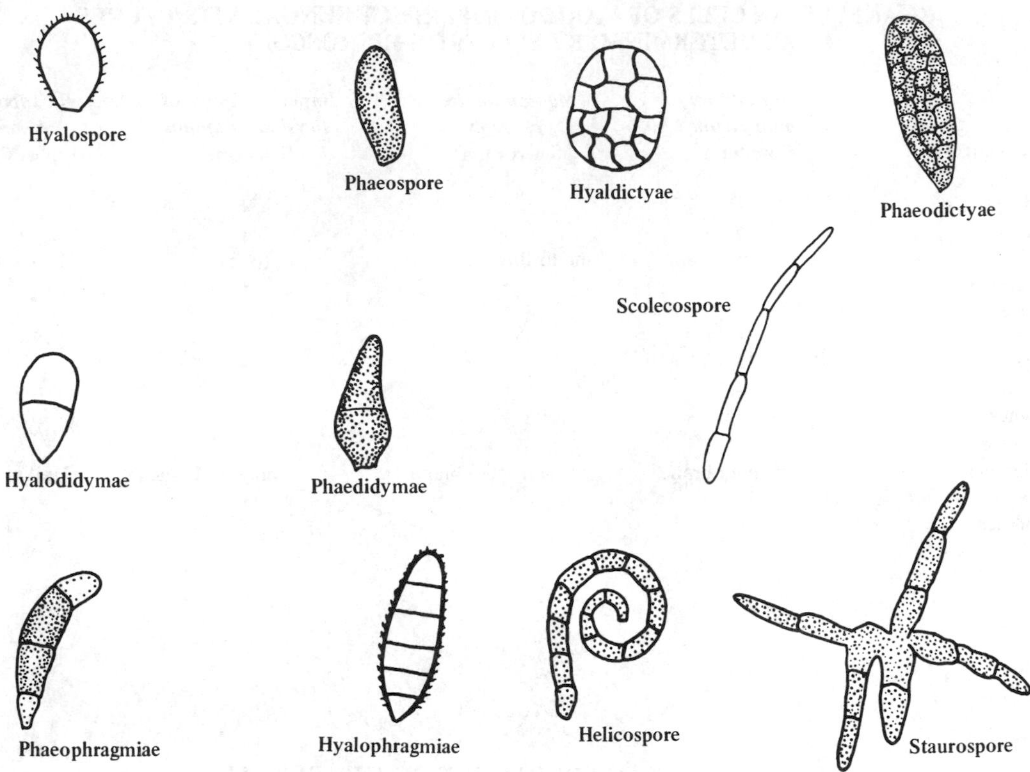

FIGURE 3. Conidial Types of Fungi Imperfecti.

Key to the Form Orders of Fungi Imperfecti

1A	Conidia produced	2
1B	Conidia absent	Mycelia Sterilia
2A	Conidia produced in pycnidia or acervuli	3
2B	Conidia produced on conidiophores distributed throughout the mycelium	Moniliales
3A	Conidia produced in pycnidia	Sphaeropsidales
3B	Conidia produced in acervuli	Melanconiales

Form Order Mycelia Sterilia

This diversified array of fungi is classified into approximately twenty genera, of which the best known are *Rhizoctonia* and *Sclerotium;* both genera have species that are plant-pathogenic.

No form families exist in this order because of the heterogeneity of the form genera. Identification is predicated upon mycelial and sclerotial structures. Several species are thought to have affinities with Basidiomycetes because of the presence of clamp connections; *Rhizoctonia solani* has been shown to have a Basidiomycetes' sexual state, *Thanatephorus cucumeris.* Mycelial and sclerotial characteristics of *R. solani* are shown in Figure 4.

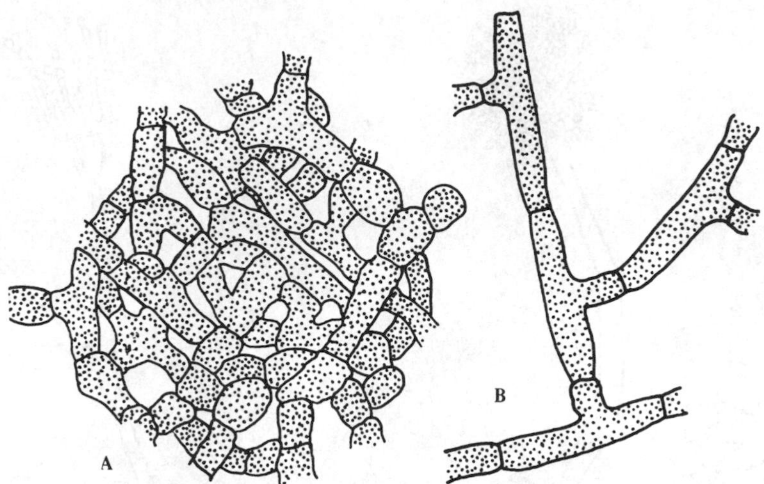

FIGURE 4. *Rhizoctonia solani.* A. Section of loose sclerotium. B. Cells of mycelium.

Form Order Moniliales

This form order of the Fungi Imperfecti contains the greatest number of species; Alexopoulos[1] and Bessey[15] concluded that there are approximately 10,000 form species in the Moniliales. Important in industry, these fungi cause many serious plant diseases and most of the fungal diseases of man; some of them will be discussed later in that context.

There are four form families in the Moniliales, and they will be discussed as they are presented in the third edition of *Illustrated Genera of Imperfect Fungi.*[16]*

Moniliaceae	Both conidia and conidiophores, if present, are hyaline or brightly colored; conidiophores are single or in loose clusters.
Dematiaceae	Either conidia or conidiophores, or both, have distinct dark pigment; conidiophores are single or in loose clusters.
Stilbaceae	Conidiophores are united into synnemata.
Tuberculariaceae	Conidiophores are united into sporodochia.

MONILIACEAE

This form family of the Moniliales contains the greatest number of species. Form genera are delimited by one or more of the following characteristics: septation; conidiophore appearance and branching; conidial morphology; true and pseudo mycelium; the manner in which conidia are produced; presence of chlamydospores and their morphology; conidia produced in chains or in a head; presence or absence of mucilage; conidial number and arrangement at apex of conidiophore; conidia produced on conidiophore or mycelium; and exogenous or endogenous production of conidia.

Many of the more common fungi belong to the Moniliaceae. Species of *Aspergillus* (Figure 5) and *Penicillium* (Figure 6) seem to be ubiquitous. They are common laboratory contaminants and are often isolated from soil and air. *A. fumigatus* is an opportunistic pathogen of man and other animals, causing aspergillosis, a disease of the lungs. *P. chrysogenum* and closely related species are the fungi that produce the antibiotic penicillin, which has saved countless human beings from death due to bacterial diseases. Other species of *Penicillium* are responsible for molds of food and clothing products. Species of

* Reproduced by permission of Burgess Publishing Co., Minneapolis, Minnesota (Copyright 1971).

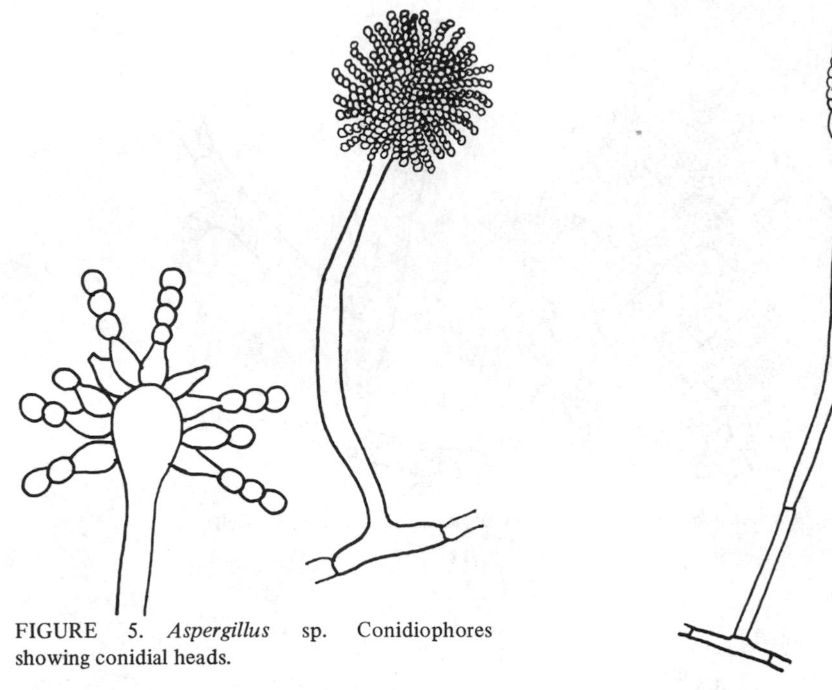

FIGURE 5. *Aspergillus* sp. Conidiophores showing conidial heads.

FIGURE 6. *Penicillium* sp. Conidiophores bearing catenulate conidia.

Gliocladium closely resemble *Penicillium*, but differ at maturity by having the spore mass encompassed by mucilage. *G. roseum* is one of the many fungi that exhibits two conidial types, the second one being a *Verticillium* type (Figure 7). The form genus *Verticillium* is named for the arrangement of whorled branches of the conidiophore. *V. albo-atrum*, a destructive plant pathogen, is responsible for many of the wilt diseases of higher plants. *Monilia* is another common imperfect fungus that produces branched conidia chains (Figure 8). *Monilinia fructicola*, the perfect state of *Monilia cinerea*, var. *americana*, causes brown rot of peach and other stone fruits. *Cylindrocladium* is a plant-pathogenic fungus that possesses one- to several-septate cylindrical conidia as well as an ancillary vesicle inflated at the apex (Figure 9). This vesicle is considered by Sobers and Seymour[17] and by Morrison and French[18] as the most significant taxonomic structure for delimiting species of *Cylindrocladium*. *Candida albicans* is a filamentous yeast that produces conidia by budding from single cells or from the mycelium (Figure 10). This fungus is mostly saprophytic, but at times it can become pathogenic to man, affecting mucous membranes, skin, nails, and lungs.

DEMATIACEAE

The members of the Dematiaceae differ from Moniliaceae in having light-brown to darkly pigmented conidia or conidiophores. In practice, however, it is often very difficult to distinguish a hyaline species from a lightly pigmented species.

Many of the common fungi are found in Dematiaceae. *Stachybotrys*, a soil-borne saprophyte (Figure 11), has pigmented conidiophores as well as dark amerospores that often slime down to form glistening beads. *Hormodendrum* is a common air-borne fungus now considered synonymous with *Cladosporium* (Figure 12). Both have much-branched conidiophores as well as branched conidial chains and one- or two-celled conidia. *C. fulvum* causes a serious plant disease in tomatoes grown in greenhouses. *Aureobasidium*, a filamentous black yeast (Figure 13), is somewhat similar to *Candida* of the Moniliaceae, but the mycelium becomes black and shiny with age and bears lateral blastophores, subhyaline or dark. *A.*

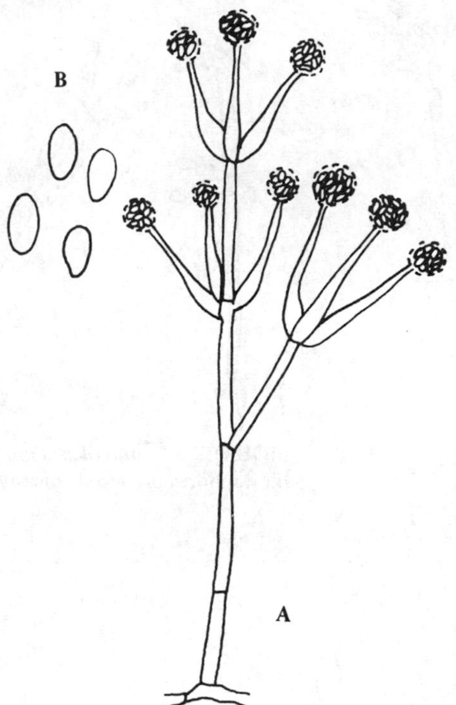

FIGURE 7. *Verticillium* sp. A. Conidiophores bearing clustered conidia in mucilage. B. Conidia.

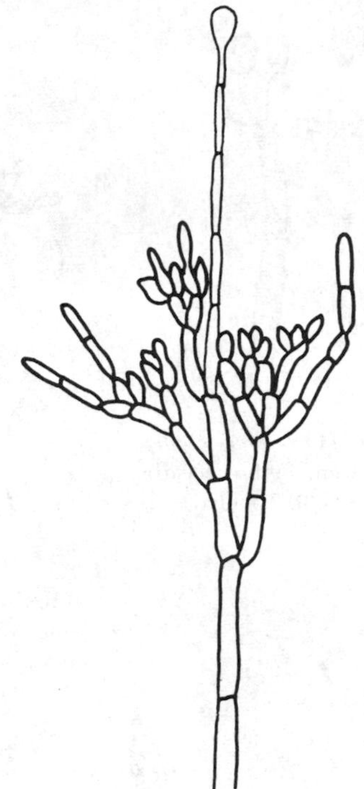

FIGURE 9. *Cylindrocladium scoparium*. Conidiophore bearing a vesicle and conidia.

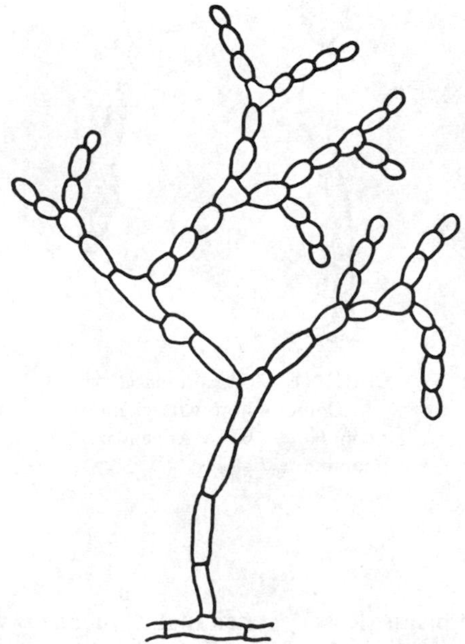

FIGURE 8. *Monilia sitophila*. Conidiophore with branched conidia.

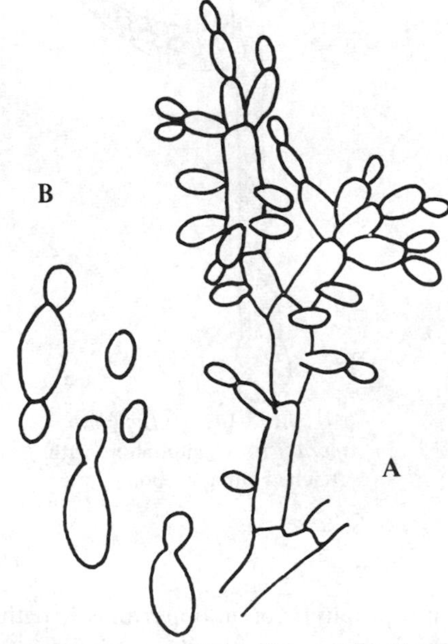

FIGURE 10. *Candida albicans*. A. Hyphae with attached blastospores. B. Budding conidia.

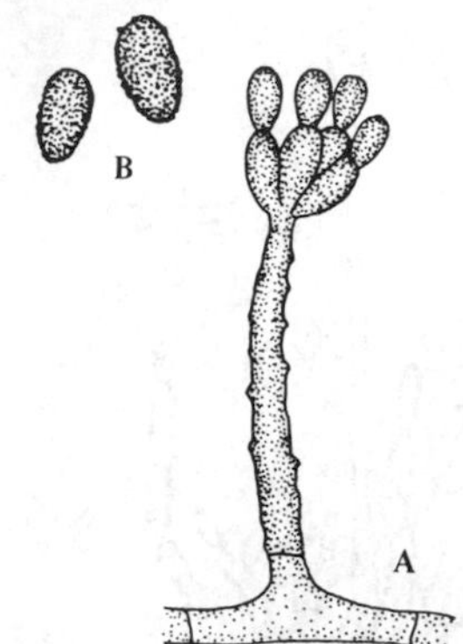

FIGURE 11. *Stachybotrys atra*. A. Conidiophore bearing phaeospores. B. Conidia.

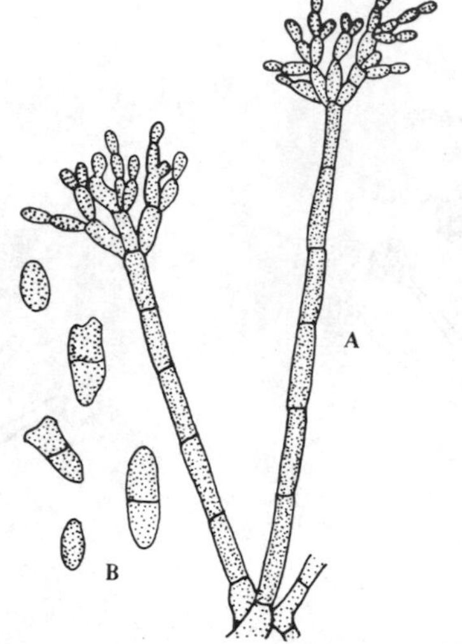

FIGURE 12. *Cladosporium herbarum*. A. Simple conidiophores bearing branched chain of conidia. B. Pleiomorphic conidia.

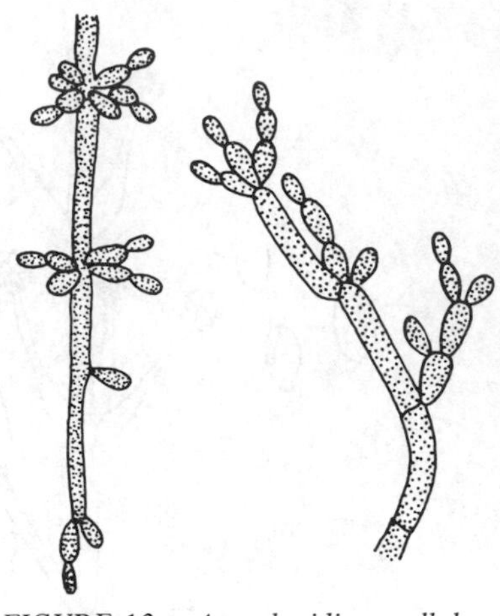

FIGURE 13. *Aureobasidium pullulans.* Blastophores produced from hyphae.

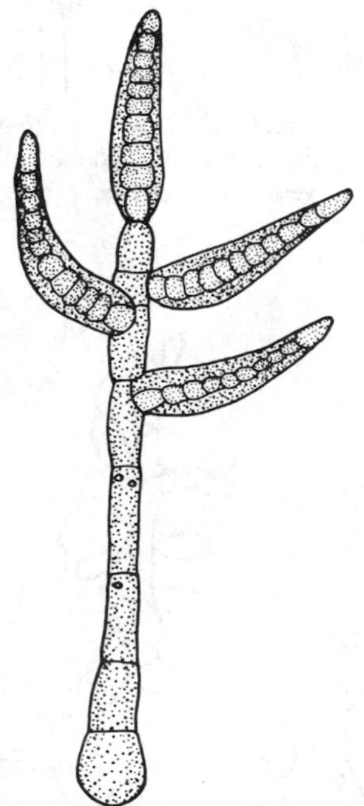

FIGURE 14. *Helminthosporum* sp. Conidiophore with attached phragmospores.

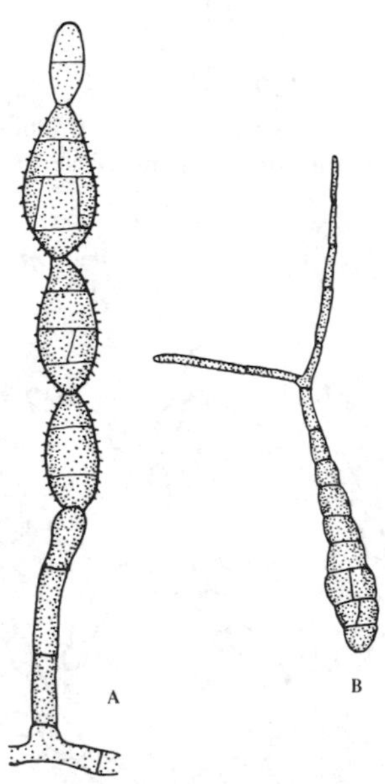

FIGURE 15. *Altenaria* sp. A. Conidiophore with chain of conidia. B. Appendaged conidium.

pullulans is saprophytic or an opportunistic pathogen of plants. Ross[19] stated that *A. pullulans* was a major agent in the deterioration of painted surfaces. In the form genus *Helminthosporium,* the dark conidia are cylindrical, multiseptate, and usually have rounded ends (Figure 14). Many species of *Helminthosporium* are parasitic on the leaves of grasses. The conidia of *Bipolaris* and *Dreschlera* are very similar to those of *Helminthosporium,* differing only in mode of formation. The most commonly encountered fungus of this family is *Alternaria,* which produces large muriform conidia, often borne acropetally in chains (Figure 15).

This fungus is readily isolated from air, soil, and decaying vegetation. *Alternaria solani* causes a destructive disease, known as early blight, in potatoes and tomatoes.

STILBACEAE

The fungi of this family have conidiophores united in columns, known as synnemata or coremia, which produce conidia on the upper portions. Best known of this family is probably *Graphium ulmi* (Figure16). The synnema is tall and has a rounded mass of light-colored conidia embedded in mucilage. This fungus is the imperfect state of *Ceratocystis ulmi,* the cause of the destructive Dutch elm disease.

TUBERCULARIACEAE

The presence of a sporodochium distinguishes fungi of this form family from other Moniliales. However, the production of sporodochia often varies with the cultural conditions employed, thus making identification of many of these fungi difficult. Some species of *Fusarium* produce sporodochia (Figure 17). The tremendous variability of the fusaria makes a species identification difficult. Tousson and Nelson[20] have compiled a pictorial key to simplify this problem. *F. moniliforme* and other species produce both macronidia and micronidia and may also produce chlamydospores. The fusaria cause many different types of plant diseases, usually by plugging the vascular system of the host or through toxic secretions. Another form genus, *Epicoccum* (Figure 18), is a common saprophyte found both on decaying wood and in soil. This fungus has dark sporodochia, from which compact or loose conidiophores give rise to dark, globose dictyospores.

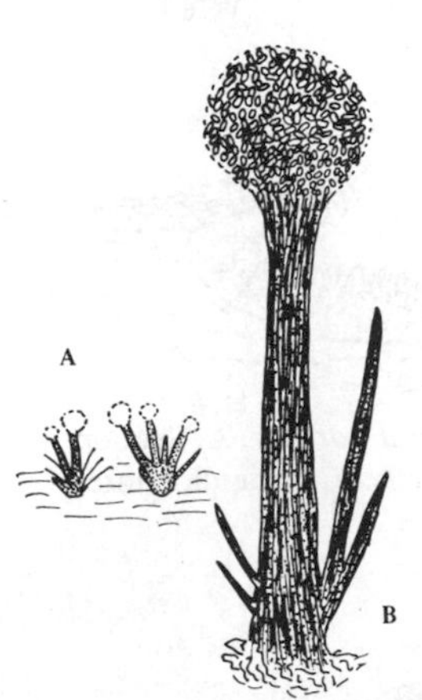

FIGURE 16. *Graphium* sp. A. Habit of synnemata. B. Synnema and mucilaginous conidial head.

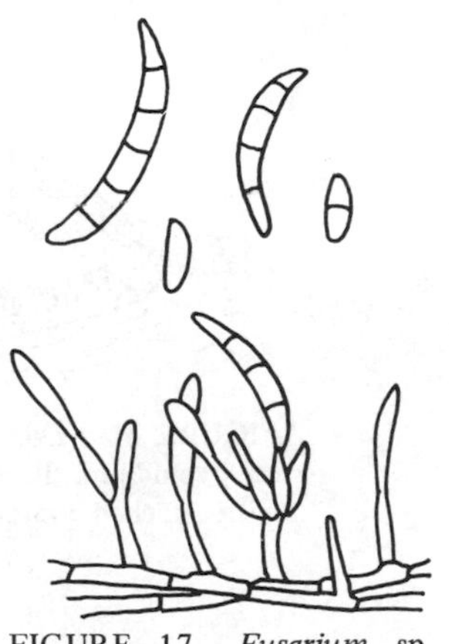

FIGURE 17. *Fusarium* sp. A. Variable conidiophores. B. Macronidia and micronidia.

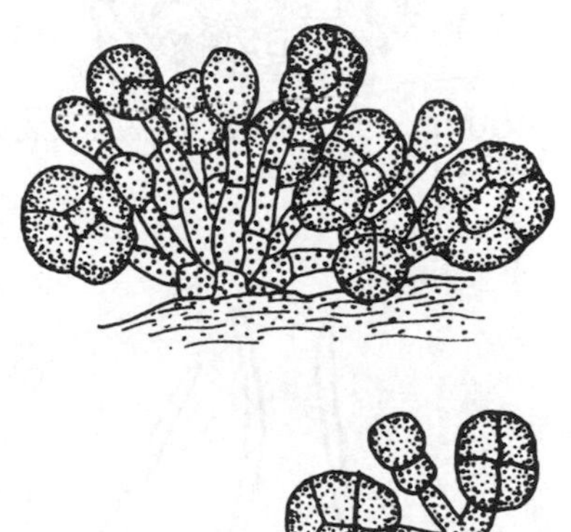

FIGURE 18. *Epicoccum* sp. Conidiophores and conidia.

Form Order Sphaeropsidales

There are four form families in the form order Sphaeropsidales, and these can be differentiated as follows: Sphaeropsidaceae – dark pycnidia, leathery to carbonous, which may or may not be produced on a stroma, usually having a circular opening; Zythiaceae – physical characteristics are the same as those of the Sphaeropsidaceae, but the pycnidia are bright-colored and waxy; Leptostromataceae – pycnidia are more fully developed in the upper half rather than in the basal portion; Excipulaceae – pycnidia are cup- or saucer-shaped.

FORM FAMILY SPHAEROPSIDACEAE

Most of the members of this form family are saprophytic, but many are parasites of plants, and a few are parasitic on insects and on other fungi. The most common form genera are *Phoma* (Figure 19), *Phyllosticta, Sphaeropsis, Coniothyrium,* and *Septoria.* Most species of these five genera are parasites on stems and leaves of plants. *Phoma* and *Phyllosticta* are morphologically so similar that distinctions among them are purely arbitrary. Both have dark, erumpent pycnidia enclosing short conidiophores that produce hyaline non-septate conidia. Two other form genera that closely resemble *Phoma* are *Sphaeropsis* and *Coniothyrium;* however, their conidia are dark, and those of *Coniothyrium* are smaller than those of *Sphaeropsis. S. malorum* is pathogenic to leaves and fruits of apples. Its perfect state is *Physalospora obtusa. Septoria* (Figure 20) is a form genus containing approximately 1,000 species, most of them plant-pathogenic, whose names are based on the host on which they were found. This scheme has led to great confusion and has brought about species names that are scientifically invalid. Nevertheless, until research is undertaken to clarify these inconsistencies, the numerous species of *Septoria* will persist. The pycnidia of *Septoria* are dark, globose, ostiolate, and erumpent; they enclose short conidiophores bearing long, thin scolecospores. Most species cause leaf spots in higher plants, and *S. apii* is responsible for a serious disease in celery, late blight. *Dothischiza* (Figure 21) and *Protostroma* (Figure 22) illustrate two other variations in pycnidial morphology.

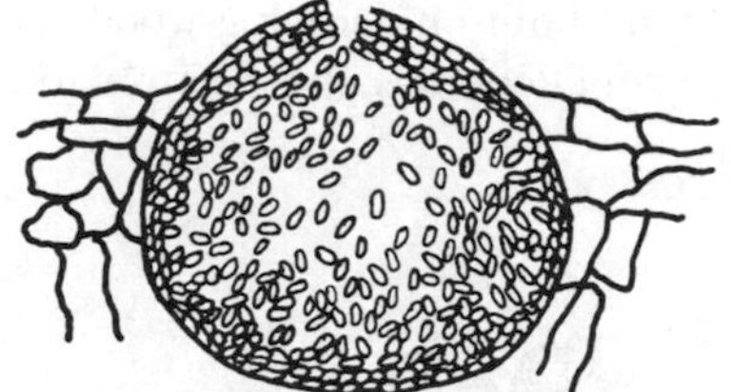

FIGURE 19. *Phoma lingam.* Cross section of pycnidium.

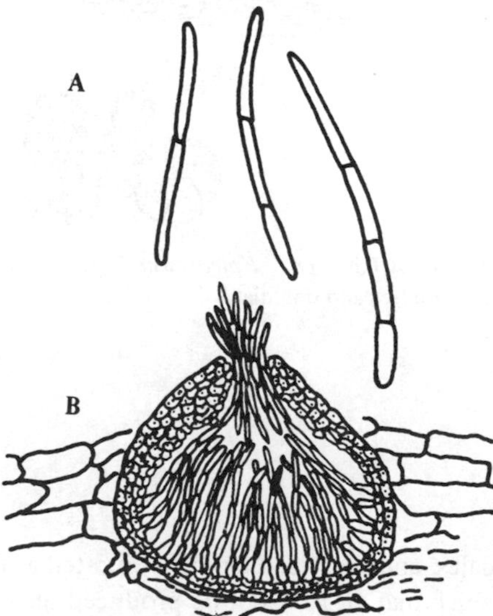

FIGURE 20. *Septoria* sp. A. Conidia emerging from pycnidium. B. Scolecospores.

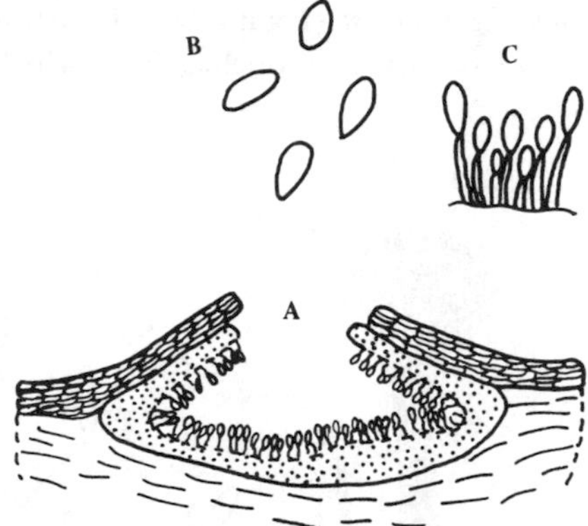

FIGURE 21. *Dothischiza populae.* A. Section of pycnidium. B. Conidia. C. Conidiophores with attached conidia.

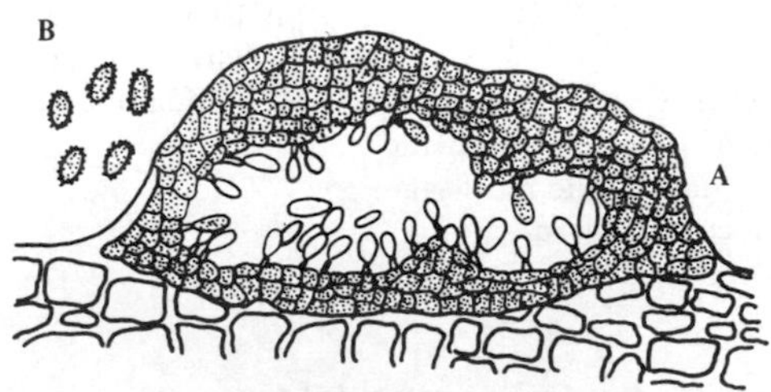

FIGURE 22. *Protostroma hyphaeneae.* A. Section through pycnidium. B. Conidia.

Form Order Melanconiales

The acervulus is the characteristic structure of all fungi in this form order, and all species are placed in one form family, Melanconiaceae. The form genera *Gloeosporium* and *Colletotrichum* are very similar in appearance, except that the latter has prominent dark setae associated with the conidiophores. These conidia are hyaline, one-celled, and ovoid to oblong. Under certain cultural conditions, however, the setae of *Colletotrichum* fail to form, and the validity of distinguishing *Colletotrichum* from *Gloeosporium* by this characteristic is therefore questionable. Both of these fungi have known perfect states of *Glomerella,* an Ascomycete; both are parasitic on various species of higher plants. *C. lindemuthianum* is illustrated in Figure 23. Another form genus is *Pestalotia,* which produces multiseptate conidia with pointed ends and apical appendages (Figure 24). Species of *Pestalotia* can be either parasitic or saprophytic. The form genus *Cylindrosporium* is similar to *Gloeosporium,* but its conidia are long and straight to curved. It causes leaf spots on many higher plants. *C. hiemalis* is destructively pathogenic to cherry, causing a leaf-spot disease. Many species of *Cylindrosporium* have *Mycosphaerella,* an Ascomycete, as the perfect state.

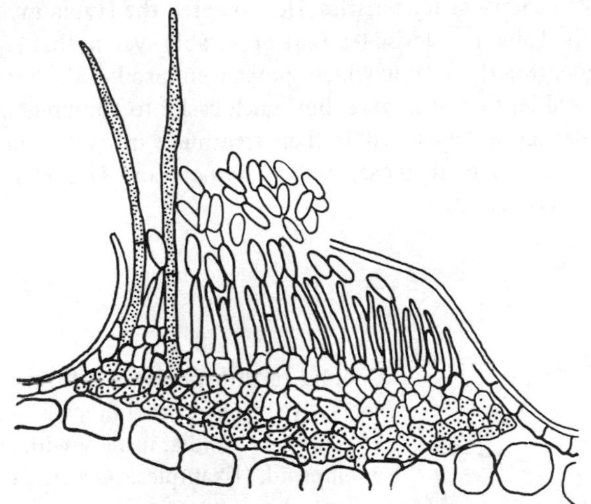

FIGURE 23. *Colletotrichum lindemuthianum.* Section through the acervulus.

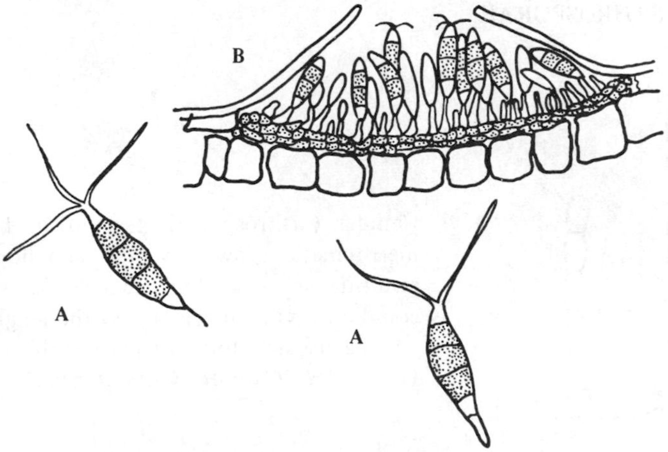

FIGURE 24. *Pestalotia macrotricha.* A. Conidia with apical appendages. B. Section through the acervulus.

THE HUGHES-TUBAKI-BARRON SYSTEM

The absence of natural taxonomic characteristics in the Saccardian system has brought about problems in the identification and classification of the imperfect fungi. Separation of genera on the basis of dubious characteristics such as pigmentation and conidial septation has led to increased dissatisfaction with this artificial scheme. The disparities of the Saccardian system have revealed the need for a modern, systematic approach that is predicated on fundamental taxonomic characteristics of conidiophore structure and conidium development. This, however, is not to imply that the Saccardian system is totally without value; its simplicity and comprehensiveness have aided the plant pathologist and mycologist in describing, identifying, and classifying microfungi for many years.

Vuillemin[21,22] and Mason[23] suggested that spore development rather than spore characteristics should serve as the basis for identifying Fungi Imperfect. Later, Hughes,[24] Tubaki,[25] Barron,[26] and Barnett and Hunter[16] employed conidiophore morphology and conidial development as the primary characteristics for classifying Hyphomycetes. It was Hughes[24] who gave the greatest impetus to the new system based on conidium development and who relegated conidial septation, pigmentation, synnemata, sporodochia, and presence of slime to supplementary characteristics. He separated the Hyphomycetes into eight sections, and Tubaki later added a ninth. Tubaki[27] devised a more applicable system that segregated the imperfect fungi into six major groups named for the way in which spores were produced. Barron[26] has presented a system for soil Hyphomycetes based on that of Hughes, but much easier to comprehend and to work with. Barnett and Hunter[16] followed Barron and included in their treatment many genera of imperfect fungi in other habitats. The following nine series are those used in Barnett and Hunter's book. An imperfect fungus representative of each series is illustrated.

Series ARTHROSPORAE

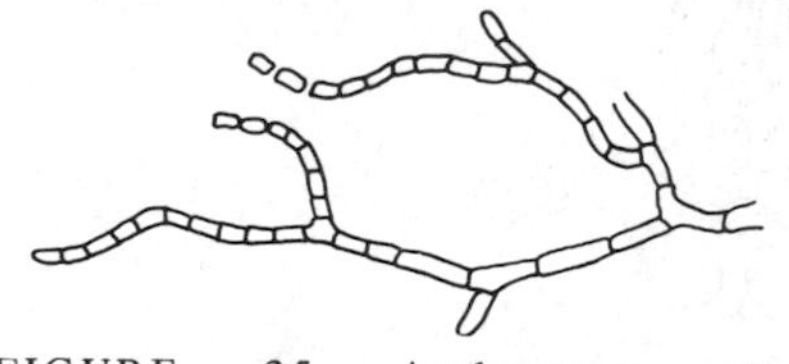

FIGURE 25. Arthrospore of *Geotrichum albidum.*

Conidia (arthrospores) are formed following segmentation of vegetative hyphae or branches of conidiophores (see Figure 25); mature conidia, usually with truncate ends, are cylindrical or ellipsoid. (Examples: *Geotrichum, Oidodendron.*)

Series MERISTEM ARTHROSPORAE

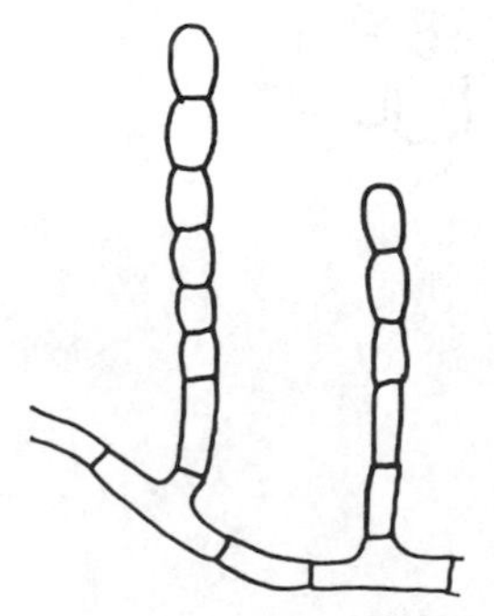

FIGURE 26. Meristem arthrospore of *Oidium monilioides.*

Conidia (arthrospores) develop in basipetal succession by meristematic growth of the conidiophore (see Figure 26), resulting in a gradual change from conidiophore cell to conidium, with no change in the length of the conidiophore; conidia usually, but not necessarily, hang together in chains. (Examples: *Oidium, Basipetospora.*)

Series ALEURISPORAE

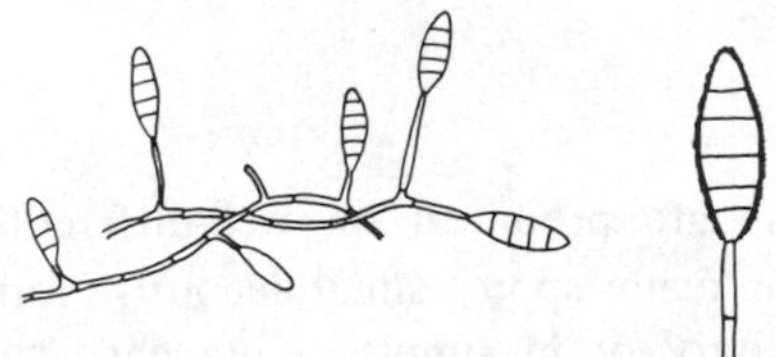

FIGURE 27. Aleurispore of *Microsporum gypseum*.

Conidia (aleuriospores) are usually single and apical on conidiophores or sporogenous cells (see Figure 27), often thick-walled and pigmented (but may be hyaline), not easily deciduous, or deciduous by means of a special cell at the apex of the conidiophore; accessory spore stages are often present. (Examples: *Sepedonium, Microsporum.*)

Series ANNELLOSPORAE

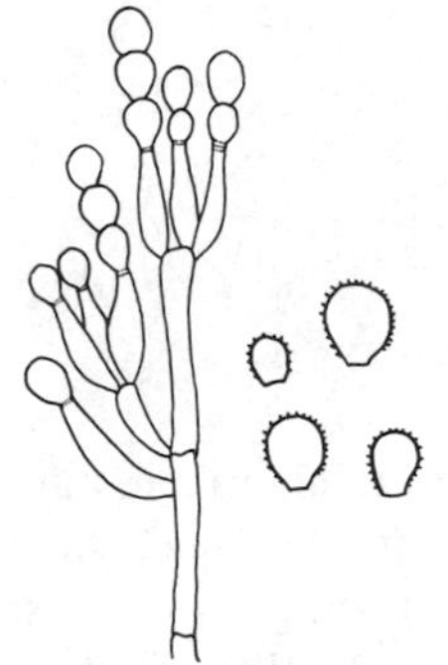

FIGURE 28. Annellospore of *Scopulariopsis* sp.

The first conidium (annellospore, similar to the aleurispore) is terminal, with successive conidia produced on the new apex of the sporogenous cell (see Figure 28), which increases slightly in length by proliferation through the previous spore scar; successive scars appear as annelations (rings) near the apex of the sporogenous cell and are often difficult to see. (Examples: *Scopulariopsis, Spilocaea.*)

Series BLASTOSPORAE

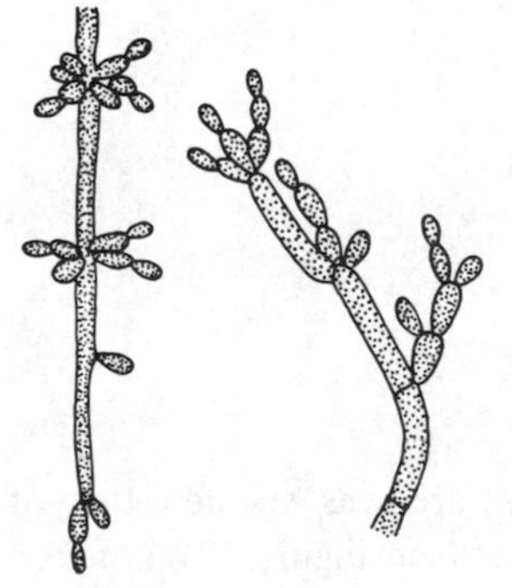

FIGURE 29. Blastospore of *Aureobasidium pullulans*.

Conidia (blastospores) develop as buds from simple or branched conidiophores, or directly from vegetative cells or previous spores (see Figure 29), often forming simple or branched acropetalous chains that are easily broken into separate spores. (Examples: *Aureobasidium, Monilia.*)

Series BOTRYOBLASTOSPORAE

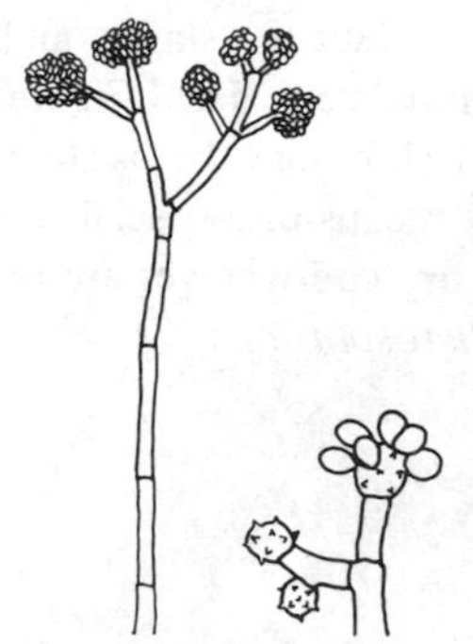

FIGURE 30. Botryblastospore of *Botrytis cinerea*.

Conidia (blastospores) are produced on well-differentiated swollen cells that bear many spores simultaneously, forming clusters or heads, solitary or in simple or branched chains; mature conidia are easily deciduous, revealing minute denticles on the spore-bearing cells. (Examples: *Oedocephalum, Botrytis.*)

Series POROSPORAE

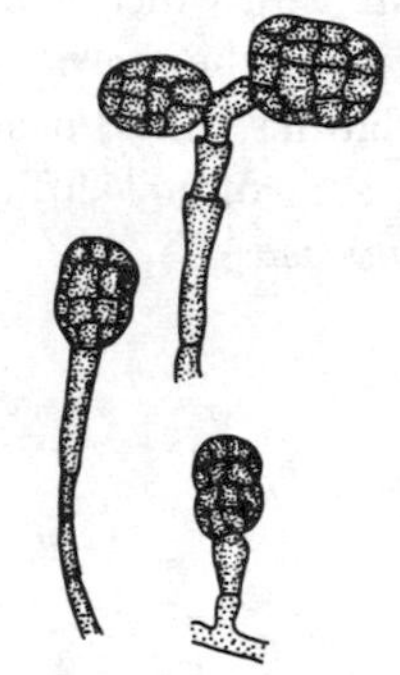

FIGURE 31. Porospore of *Stemphylium sarcinaeforme*.

Conidia (porospores), often thick-walled, are developed through pores in the wall at the apex or side of the conidiophore (see Figure 31), single or, in some genera, formed on successive new growing tips of the conidiophore, which are produced by proliferation at the apex or by sympodial extension from beneath the previous conidium. (Examples: *Helminthosporium, Stemphylium.*)

Series SYMPODULOSPORAE

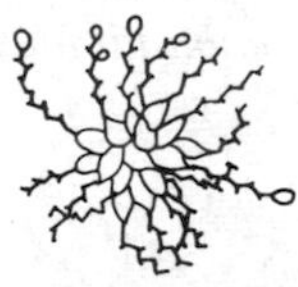

FIGURE 32. Sympodulospore of *Beauveria bassiana*.

Conidia (sympodulospores) arise as swollen tips of conidiophores or sporogenous cells (see Figure 32), not from pores in the wall, single or formed successively on new growing tips of the sporogenous cell, which increases in length by sympodial extension from beneath the previous conidium; sometimes there is little evidence of the increase in length, but the cell may increase slightly in size, bearing conidia of different ages. (Examples: *Fusicladium, Beauveria.*)

TABLE 5

CLASSIFICATION, NUTRITION, HABITAT, AND GENERIC CHARACTERISTICS OF SOME COMMON FUNGI IMPERFECTI

Form Genus	Form Family	Nutrition and Habitat	Conidial Description and Generic Characteristics
Papulospora		Decaying wood, soil.	Conidia lacking; compact clusters of bulbils and light mycelium.
Beauveria	Moniliaceae	Parasitic on insects.	Hyalosporae.[a] Sympodulospore.[b] Conidiophores often inflated at base and tapering to a slender zigzagged apex.
Oedocephalum	Moniliaceae	Dead plant parts.	Hyalosporae.[a] Botryoblastospore.[b] Simple conidiophores with swollen apex bearing conidial head; light mycelium.
Trichoderma	Moniliaceae	Soil; wood; parasitic on other fungi.	Hyalosporae.[a] Phialospore.[b] Greatly branched conidiophores bearing phialides; conidia borne in small clusters at end of phialides, green in mass; light mycelium.
Nigrospora	Dematiaceae	Living plant parts; soil or decaying material.	Phaeosporae.[a] Aleuriospore.[b] Short, dark conidiophores, simple, bearing black conidium at end of conidiophore.
Spilocaea	Dematiaceae	Leaves; plant parts.	Phaeosporae.[a] Annellospore.[b] Simple conidiophores; conidia pyriform and attached to conidiophore at broad end.
Stemphylium	Dematiaceae	Living plant parts; decaying plant material; soil.	Phaeodictyae.[a] Porospore.[b] Simple, dark conidiophores producing conidia on successive new growing tips.
Trichurus	Stilbaceae	Decaying plant material.	Phaeosporae.[a] Annellospore.[b] Dark synnemata with thin stalk; setae present among conidiophores in expended apical portion of synnemata; conidia catenulate.
Myrothecium	Tuberculariaceae	Living plant parts; decaying plant material; soil.	Phaeosporae.[a] Phialospore.[b] Light- to dark-colored sporodochia infrequently bearing setae; conidiophores much-branched and lightly colored.
Ascochyta	Sphaeropsidaceae	Primarily on living leaves.	Hyalodidymae.[a] Pycnidia ostiolate, single, dark, and embedded in plant tissue.
Diplodia	Sphaeropsidaceae	Living plant parts; decaying plant parts.	Phaeodidymae.[a] Pycnidia ostiolate, single, black, and embedded in plant tissue; conidiophores thin and simple.
Melanconium	Melanconiaceae	Living plant parts; decaying plant parts.	Phaeosporae.[a] Subepidermal or subcortical black acervuli; conidiophores short and simple.
Pestalotia	Melanconiaceae	Mostly on living plant parts.	Phaeophragmiae.[a] Dark subepidermal acervuli; simple conidiophores; conidia have hyaline pointed end cells with hyaline appendages.

[a]Conidial description according to Saccardo.
[b]Conidial description according to Barnett and Hunter.

Series PHIALOSPORAE

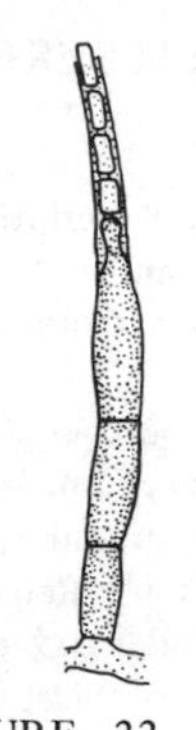

FIGURE 33. Phialospore of *Chalara quercina.*

Conidia (phialospores) are formed successively from the open apex of the conidiophore or sporogenous cell (phialide), which ordinarily does not increase in length (see Figure 33); conidia often collect in a drop of mucilage or slime at the apex of the phialide or remain attached in basipetal chains; conidiophores are simple or branched; in a few genera the simple conidiophore proliferates through the apex and forms new phialides. (Examples: *Chalara, Phialophora.*)

Table 4 shows the relationships of various systems based primarily on conidium development. The new classification schemes presented should not be construed as the panacea for all the difficulties inherent in the Saccardo system, but as a system that will alleviate many of the problems. Often the experts disagree on the type of spore formed, annelations on the conidiophores (annellophores) are frequently difficult to distinguish, and the differences between blastospores and porospores are not readily determined by inexperienced workers.

Characteristics of several other imperfect fungi are described in Table 5.

TABLE 4

A COMPARISON OF PROPOSED SCHEMES FOR SOIL HYPHOMYCETES

Subramanian[28]	Hughes[24]	Barron[26]	Tubaki[27]
	Section IA	Blastosporae }	Blastosporae
Torulaceae	Section IB	Botryoblastosporae }	
	Section II	Sympodulosporae	Radulasporae
Bactridiaceae	Section III	Aleurisporae } Annellosporae }	Aleuriosporae
Tuberculariaceae	Section IV	Phialosporae	Phialosporae
Coniosporiaceae	Section V	Meristem arthrosporae	
Helminthosporiaceae	Section VI	Porosporae	Porosporae
Geotrichaceae	Section VII	Arthrosporae	Arthrosporae
	Section VIII	Meristem blastosporae	

PATHOGENIC FUNGI OF MAN

Human diseases incited by fungi are called mycoses. Of the tens of thousands of known fungi, there are fewer than sixty species that are pathogenic to man; most of these pathogens belong to the Deuteromycetes and Ascomycetes. Human fungal diseases can be divided into two categories: cutaneous mycoses, caused by dermatophytes (mostly imperfect fungi) growing in or on the skin, hair and nails, and systemic mycoses, involving the internal organs.

Cutaneous Mycoses

Dermatophytes (skin plants) are distributed throughout the world, with certain species more numerous than others in some areas. The pathogenic species are found primarily in three genera: *Epidermophyton,*

Microsporum, and *Trichophyton.* Clinically it is impossible to distinguish the three fungi when skin material is examined. More information regarding the nature of the fungi can be obtained when they attack hair. Certain fungi penetrate into the hair, but extensive mycelial growth and spore formation is external (ectothrix). Endothrix (internal) hair infections are characterized by scarce hyphae elements on the surface of the hair and chains of arthrospores formed inside the hair. The only other infection of the hair is neoendothrix, in which spores can be produced on the inside or outside of the hair.

Tinea is the medical term for superficial fungal infections of the skin, hair, and nails. The infection is called tinea pedis when the fungus occurs on the feet, tinea capitis when it occurs on the head, and tinea corporis when it occurs on the body. Any of the pathogens that can produce the first two diseases can also incite tinea corporis. A few of the dermatophytes are found in the soil, and rubbing soil onto the skin can produce infection in man. However, most infections are transferred from animals to man or from man to man; similarly, man can infect animals such as dogs and cats. One *Epidermophyton* species and one *Microsporum* species will be discussed as representative of most superficial human mycoses.

Epidermophyton floccosum

Epidermophyton floccosum most frequently causes tinea cruris, an infection of the groin or perianal region. This fungus is also prominent in mycoses of the nails (tinea unguium). The usual mode of transfer is by direct contact with infected persons or their clothing. In tinea cruris the skin becomes thick and tough, with darker pigmentation than normal. This disease usually doesn't respond to topical therapy, but oral griseofulvin cures the disease in about four to eight weeks.

Epidermophyton floccosum, unlike species of *Microsporum* and *Trichophyton,* does not produce microconidia. The macroconidia are clavate, two- to six-celled, and occur in groups. Colonies are white, later becoming yellowish green, with or without aerial mycelium.

Microsporum canis

Ringworm of the scalp (tinea capitis) results from *Microsporum canis* and other dermatophytes. Man acquires this disease most often by contact with infected animals, although man-to-man transmission is possible. The fungus invades the hair (ectothrix infection), causing the hair to fall off or break off at the scalp line. Many patches may coalesce to produce a large, ragged bald spot on the head. Hairs that fluoresce designate infected areas of the scalp. Oral griseofulvin is usually successful in curing ringworm of the scalp.

Microsporum canis is characterized by large, multicelled, thickwalled, spindle-shaped macronidia. Micronidia are few; they are single-celled and emanate directly from the hyphae. Colonies grow rapidly on agar media, bearing white aerial mycelium; submerged mycelium is bright yellow to orange and later becomes dark.

Deep Mycoses

The consequences of systemic fungal infections are much greater than those of the superficial injuries of the dermatophytes, and often fatal. Usually the infection is initiated in the lungs, but may spread to visceral organs, skin, or other parts of the body. Most systemic fungal infections result when the body's resistance is low due to sickness, age, or other debilitating conditions.

The incidence of systemic mycoses seems to be rising, perhaps as a result of improved diagnostic procedures, or for reasons not yet known. Treatment of systemic infections is predicated upon the positive identity of the fungus, usually through isolation techniques. Chemotherapeutic agents useful in treating deep-seated fungal diseases are few, although nystatin cycloheximide and amphotericin B are effective in certain cases. *Cryptococcus neoformans* and *Histoplasma capsulatum* are imperfect fungi that can cause systemic diseases and are described as representative examples.

Cryptococcus neoformans

Cryptococcus neoformans, the pathogen causing Cryptococcosis, is widely distributed in nature and, for the most part, probably saprophytic. It is readily air-borne, and infected individuals most likely contract the organism in this manner. The lung is the primary site of infection, with consequent metastasis to other parts of the body. The fungus produces no toxic metabolites that are injurious to humans. However, damage is caused by the growth of the yeast-like cells, which crowd the normal cells, usually in the central nervous system, to the extent that they do not function. If this occurs in the brain, the person infected may become paralyzed, go blind, and eventually die. If the disease is diagnosed promptly and amphotericin B is administered, approximately 80% of the patients recover.

Cryptococcus neoformans is a false yeast, which reproduces only by budding and usually has no hyphae. Cells are round to ovoid, aggregating into cream-colored colonies that later become light brown.

Histoplasma capsulatum

Histoplasmosis is a systemic fungal disease caused by *Histoplasma capsulatum.* This fungus is soil-borne, especially in areas enriched with bird feces. Air-borne spores are usually responsible for entry into the body. This disease is endemic throughout the Mississippi valley, and occasionally it has reached epidemic proportions, though it is not transmissible from man to man or from animals to man. Fortunately, most infected individuals recover.

In man, *Histoplasma capsulatum* is found in its yeast-like form. Most infections occur in the lungs, spleen, lymph glands, kidney, and liver. This fungus is unique because it develops inside white blood cells and is released upon the death of these cells. Histoplasmosis is usually localized and subsides with little permanent damage to man. However, in about 1% of infected individuals the fungus becomes disseminated, resulting in a high mortality rate. The antibiotic amphotericin B has been effective in curing the disseminated phase.

In host tissue or in culture at 37°C, *Histoplasma capsulatum* is yeast-like, reproducing only by budding; in culture at 24°C, a mycelial phase occurs. Conidia are one-celled, round to pear-shaped, and are produced directly from the hyphae or from short conidiophores. Colony growth is slow; the mycelium is white and later becomes light brown.

Common Fungal Diseases

Additional imperfect fungi and the symptoms of the diseases they cause are described in Table 6.

PATHOGENIC FUNGI OF PLANTS

Numerous plant diseases are caused by imperfect fungi. The conidial state, with or without a sexual state, is usually responsible for the dissemination of the fungus. Thus, several crops of conidia can be produced in one growing season, with concomitant spread of the fungus to distant areas. A recent example of this occurred in the summer of 1970, when the southern corn blight, a disease caused by *Helminthosporium maydis,* was spread from corn-growing areas of the South to the corn fields in the North, Midwest and East.

Plant-pathogenic fungi can invade their hosts by mechanical and/or chemical means. Infection (establishment of a food relationship between parasite and host) is followed by invasion of the tissue, which usually adversely affects the host by causing disease. The disease can result from one or more factors initiated by the pathogen, such as depletion of molecular nutrients of the host, secretion of one or more enzymes that soften or dissolve host tissue, production of toxins that kill host cells, or release of growth-regulating chemicals that can induce hypertrophy of host tissue. Polysaccharides and toxins are often important factors in various wilt diseases of higher plants. Victoria blight of oats, caused by *Helminthosporium victoriae,* tomato wilt, caused by *Fusarium oxysporum* f. *lycopersici,* and blue mold (soft rot) of apples and pears, caused by *Penicillium expansum,* are common plant diseases and will be described as representative of the imperfect fungi.

Victoria Blight of Oats

The Victoria blight of oats is a disease relegated only to oat varieties of Victoria parentage. The variety

TABLE 6

SOME COMMON CHARACTERISTICS OF FUNGAL PATHOGENS OF MAN

Disease	Symptoms	Fungal Characteristics
Aspergillus fumigatus		
Aspergillosis systemic	Pulmonary infections, otomycoses.	Rapid growth on agar; non-septate conidia are abundant and varying in color.
Candida albicans		
Candidiasis systemic	Infection may occur on any region of the body, but is usually relegated to nails, vulvovagina, or between folds of skin. Systemic infections also occur.	Colonies are cream-colored, occur rapidly, but grow slowly. They may appear as yeast-like cells or have a true mycelium producing hyaline blastospores.
Cladosporium wernickii		
Tinea nigra cutaneous	Surface infection of the skin, most often the hands.	Colonies are typically slow-growing, black, and often yeast-like; conidia are slightly pigmented and one- or non-septate.
Geotrichum		
Geotrichosis	Pulmonary and oral infections and possibly diarrhea.	Rapid colonial growth, white to cream-colored; mycelium forms into arthrospores.
Microsporum audounii		
Tinea corporis, tinea pedis cutaneous	Ramification through the skin of body and feet. The organism is also present on hair.	Slow colonial growth; conidia are usually of brownish color and, infrequently, phragmosporous.
Phialophora pedrosi		
Chromo-mycosis cutaneous	Infection is mostly confined to skin and subcutaneous tissue of the extremities and the head. The organism also causes granulomas.	Slow-growing colonies that become olive green in old culture; conidia are hyaline to lightly pigmented and non-septate.
Sporotrichum schenckii		
Sporotrichosis systemic	The site of infection usually becomes nodular and progresses to the lymphatic channels.	Moderate colonial growth, at first white, later becoming brown to black; single-celled hyaline conidia are borne in clusters.
Trichophyton rubrum		
Tinea capitis, tinea pedis, tinea corporis cutaneous	The organism infects nails, skin, and hair. Occasionally it is recovered from deep lesions.	Colonies are slow-growing; initially the mycelium is velvety to powdery; microconidia, single-celled and hyaline, are plentiful; macroconidia, hyaline and phragmosporous, are rare.

Victoria was crossed with other varieties to insure resistance to certain rust and smut diseases, and this group became the dominant line by 1946. However, in 1944, after extensive distribution, *Helminthosporium victoriae,* a heretofore unknown pathogen, incited an epiphytotic disease of oats, although the disease was not described until 1946. The pathogen is usually a saprophyte or a weak pathogen of many grasses, but when it invades susceptible oat plants, it produces a powerful toxin that can act at some distance from the site of infection. Infection is usually on the stem near the soil line or on the leaf, but the resultant blight strikes the leaves, and soon the entire plant is destroyed. The initial symptoms of the disease are yellow to orange-red stripes on the leaves. Within four to five days a rapidly spreading necrosis encompasses the whole plant, and later dark-olive conidia appear on certain necrotic areas. The best control of this disease is planting of oat varieties other than Victoria or those of Victoria parentage.

Tomato Wilt

Fusarium oxysporum f. *lycopersici* can cause serious wilting of greenhouse and field tomatoes; the latter condition is of greater importance in the southern United States. Once this fungus becomes established in the soil, it is virtually impossible to eradicate. Local spread is primarily by transplants, drainage water, air-borne soil, and infested tools; wide distribution is the result of transplants. The wilting of the plant is related to three toxins, all of which are produced by *F. oxysporum* f. *lycopersici;* however, mycelium occlusion in the vessels, collapsed vessels, degradation products accumulating in vessels, and tyloses, all are an important part of the wilt syndrome. In the field, the disease is primarily restricted to the warmer climates and the warm sandy soils of the temperate regions. Susceptible varieties become stunted and wilt when attacked by the pathogen; however, if soil and air temperatures during most of the season are not high, even susceptible varieties can produce good yields. Because soil sterilization is impractical in the field, control must be predicated on the use of resistant tomato varieties as well as on planting only healthy seed or transplants. Greenhouse tomatoes should only be planted in sterilized soil.

Blue Mold of Apples and Pears

Penicillium expansum is responsible for blue mold of apples and pears and is omnipresent wherever apples and pears are grown or shipped. The disease occurs mostly in transit or during storage, but it is not uncommon for injured fruit to rot while on the tree, due to the blue-mold fungus. *P. expansum* is only a weak parasite and depends upon injury before invasion can occur. Uninjured fruit adjacent to decaying apples will remain uninfected. In its early stages, blue mold is characterized by a light color and a soft, watery texture of the diseased area. A moldy odor is noticeable and is imparted to adjacent unrotted fruits, making them equally unmarketable. Once the rot begins, it spreads rapidly and soon involves the entire fruit. The rotted area is slightly sunken and usually yellow to light brown; it is usually free of wrinkles. Mats of gray-blue fruiting structures frequently cover the surface of the fruit. The best control of the blue-mold disease is prevention of injury to the fruit and discard of all rotted fruit. It is also important to keep fruit at low temperatures from the time of harvest until it reaches the consumer.

Other Plant Diseases

There are many other plant diseases caused by imperfect fungi. Some of them are briefly characterized in Table 7.

FACTORS AFFECTING GROWTH AND SPORULATION

The Fungi Imperfecti are adapted to live under a wide variety of environmental and nutritional conditions. Conidia of some species may survive for years in a dry or cold environment and may germinate immediately when favorable conditions return. The conditions that favor or inhibit growth and sporulation of a given fungus are usually correlated with its habitat. For example, *Bispora,* which grows on decaying wood, is limited in growth only by temperature and moisture, whereas *Septoria lycopersici* (leaf spot of tomato)

TABLE 7

SOME COMMON CHARACTERISTICS OF PLANT-PATHOGENIC FUNGI

Disease	Signs and/or Symptoms
Alternaria dauci	
Blight of carrot	Irregular necrotic areas on leaves, usually confined to older leaves.
Botrytis allii	
Onion neck rot	Tissue becomes soft and sunken. Mycelium may develop on tissue.
Cercospora oryzae	
Leaf spot of rice	Reddish to dark-brown linear spots on leaves.
Cladosporium carpophilum	
Peach scab	Numerous dark, circular lesions on fruit; they occur in greater numbers on the upper face of the fruit and may coalesce.
Dothichiza populea	
Canker of poplar	Sunken area (cankers) can occur on trunk, limbs, and twigs; diseased bark is usually darker than healthy tissue.
Helminthosporium sativum	
Spot blotch of barley	Linear brown spots with definite margins, often coalescing to give a striped appearance to the leaf.
Myrothecium roridum	
Fruit rot of tomatoes	Numerous brown irregular spots that darken with age and often become sunken.
Pyricularia oryzae	
Blast of rice	Occurs on leaves, culms, and flowers; spots are long on young leaves, and small and circular on older leaves. The spots often coalesce at the base of the leaf blade.
Sclerotium rolfsii	
Root rot of various plants	Rot of roots and underground parts, resulting in severe wilting and death of the plants. Small brown sclerotia are present.
Septoria musiva	
Septoria canker of poplar	Cankers form around wounds, lenticels, petioles, and stipules; the fungus spreads from the lateral branches to the main stem, and ultimately girdles the main stem and kills the tree.

grows and sporulates only on the living leaves. In fact, the dissemination of conidia of plant pathogens is limited to the growing season of the host plant, and the production of conidia is therefore favored by these conditions. Such adaptation has led to the survival of the imperfect fungi that exist today. Some of the responses of imperfect fungi to factors of environment and nutrition will be discussed briefly, and occasional examples will be given.

Temperature

Temperature is, of course, a universal factor, affecting all organisms; together with moisture, it must be within a favorable range if the organism is to grow and reproduce. It is customary to list three cardinal temperatures, which describe the response of the organism: minimum, ,optimum, and maximum. A selection of cardinal temperatures is given in Table 8. These are not strictly points, but rather ranges that are influenced by other factors.

TABLE 8

CARDINAL TEMPERATURES (°C) OF SELECTED FUNGI IMPERFECTI

Fungus	Minimum	Optimum	Maximum
Alternaria solani	2	7	45
Aspergillus fumigatus	<20	35	50
Botrytis cinerea	0	20	30
Diplodia zeae	10	30	35
Epicoccum nigrum	< 5	25	35
Fusarium moniliforme	< 5	25	40
Helminthosporium sativum	< 5	25-30	35
Humicola grisea, var. *thermoides*	24	38-46	56
Microsporum gypseum		25-30	35
Rhizoctonia solani	2	25-30	35
Trichophyton mentagrophytes	8	30	40
Trichothecium roseum	<10	30	35
Verticillium albo-atrum	5	25	35

Moisture

Most fungi can grow in liquid solutions of nutrients, provided that enough oxygen is present, but many imperfect fungi do not require liquid water for their development. Some of the plant pathogens (*Botrytis cinerea, Penicillium expansum*) cause rot of plant parts and obtain their moisture from the decomposing plant cells. Species of *Aspergillus, Penicillium, Alternaria, Cladosporium* and *Aureobasidium* are among the common causes of molds and mildew on moist cloth, paper, leather, wood, and even painted surfaces. Stored grains may be attacked especially by species of *Aspergillus* and *Penicillium* if the moisture content is greater than 14 to 18%. *Aspergillus glaucus* and its close relatives are notorious for their ability to grow under conditions of severe physiological drought.

In contrast to the above group are the many imperfect species that are checked by relatively dry conditions. Spores of most of these require liquid water or a saturated atmosphere for good germination.

Survival of spores is another matter. Dry spores remain viable much longer than do moist spores.

Light

Visible light or ultraviolet radiation may affect some of the imperfect fungi in different ways. One of the easiest to demonstrate is the directional growth of the conidiophores of *Aspergillus giganteus, A. clavatus* and *Penicillium claviforme* when cultures receive unilateral illumination. The conidiophores grow toward the source of white light, irrespective of the position of the culture.

SPORULATION

Sporulation of a great number of imperfect fungi is either induced (i.e., light is necessary) or favored (increased) by exposure of mycelium to radiation. Different species respond to different wavelengths of radiation: ultraviolet, near ultraviolet, blue (the most common), a wide band of blue-green-yellow, and far red. The red band is seldom effective if any radiation is required. There is some evidence that a few imperfects respond to far-red radiation and perhaps also to a shorter band of wavelengths. Table 9 gives a few examples.

TABLE 9

EFFECTS OF RADIATION ON SPORULATION

Species	Effective Wavelength Region	Response
Alternaria solani	UV	Sporulation induced (conidia)
Ascochyta	UV	Sporulation induced (pycnidia)
Botrytis cinerea	UV	Sporulation induced (conidia)
Cylindrocladium spp.	UV, near UV, and blue	Sporulation favored (conidia)
Cylindrocladium citri	Blue to far red	Sporulation induced (conidia)
Dendrophoma obscurans	Blue	Sporulation induced (pycnidia)
Epicoccum nigrum	UV	Sporulation induced (conidia)
Helminthosporium vagans	Near UV	Sporulation induced (conidia)
Septoria oudemansii	UV	Sporulation induced (pycnidia)
Trichoderma viride	Blue	Sporulation favored (conidia)

White visible light may be as effective as any given color if the intensity is nearly equal. The intensity of white light required to induce sporulation by *Epicoccum nigrum* varied inversely with the duration. One exposure of mycelial cultures on agar to sunlight (7,000 ft-c) for 15 minutes induced the production of about as many spores as a single exposure of 24 hours at 50 ft-c or of 6 hours at 100 ft-c. Spores were produced only in that zone of young hyphae at the time of the exposure. While the inhibitory effects of ultraviolet radiation are well known, there are few substantial examples of inhibition of imperfect fungi by visible light.

pH

Most imperfect fungi are favored by a slightly acid substrate, the optimal range being between pH 5.0 and pH 6.0. However, most achieve fair to good growth over a very wide range, from about pH 3.0 to pH 8.0. Certain species are able to tolerate even greater ranges: *Aspergillus niger,* 2.8 to 8.8; *Aspergillus oryzae,* 1.6 to 9.3; *Penicillium italicum,* 1.9 to 9.3; *Fusarium oxysporum,* 1.8 to 11.1; *Botrytis cinerea,* <2.8 to 7.4; *Rhizoctonia solani,* 2.5 to 8.5.

It must be emphasized that a fungus alters the pH of its substrate as it grows, and the extent of the change depends on the composition of the substrate as well as on the nature of the organism.

Carbon Sources

The requirement of fungi for carbon is greater than that for any other nutrient. The cosmopolitan nature of many of the imperfect fungi suggests that they have the ability (necessary enzymes) to utilize the carbon from many different sources; among these, cellulose is the most abundant utilizable source. Seldom does a fungus find a pure single carbon source in nature, and preferential utilization of one source may occur.

In order to determine the ability of specific fungi to utilize single sources of carbon, experiments must be conducted in the laboratory under controlled conditions, using a medium that is complete for all nutrients except carbon. Table 10 lists the relative growth of selected imperfect fungi on several sugars. Growth on glucose, fructose, and mannose is approximately the same for all fungi.

TABLE 10

RELATIVE GROWTH OF SELECTED IMPERFECT FUNGI ON SOME COMMON SUGARS

(+++ = good; ++ = fair; + = poor; 0 = not utilized)

Species	Glucose, Fructose, or Mannose	Galactose	Sorbose	Xylose	Maltose	Sucrose	Lactose
Alternaria solani	+++	+++	+	+++	+++	++	++
Aspergillus clavatus	+++	++	+	+++	+++	+++	++
Aspergillus niger	+++	++	++	+++	+++	+++	+
Botrytis cinerea	+++	++	++	++	+++	+++	++
Colletotrichum lindemuthianum	+++	+++	0	+	++	++	++
Cordana pauciseptate	+++	+++	0	+++	+	+	+
Dendrophoma obscurans	+++	+++	+	++	++	++	++
Fusarium lycopersici	+++	+++	++	+++	+++	+++	++
Helminthosporium sativum	+++	++	+	++	+++	+++	++
Melanconium fuligenum	+++	++	+	++	+++	++	0
Penicillium expansum	+++	++	+++	+++	+++	+++	+
Rhizoctonia solani	+++	+++	0	+++	+++	+++	++
Sphaeropsis malorum	+++	+++	+	++	+++	+++	+
Stysanus stemonitis	+++	+++	0	++	++	++	++
Thielaviopsis basicola	+++	+++	0	+++	+	+++	0

Vitamins

The majority of imperfect fungi are probably self-sufficient for their vitamins; i.e., they are able to synthesize the vitamins they require from other compounds. Some species, however, are unable to synthesize certain vitamins and must obtain them from the substrate. In nature, this means products of other organisms. Such deficiencies in synthetic ability can be determined only by cultivation in suitable synthetic media with and without added vitamins. The deficiency is more common for thiamine than for other vitamins. A deficiency may be single or multiple, complete or partial. Some selected species of imperfect fungi and their known deficiencies, if any, are listed in Table 11.

Deficiency for a new growth factor (in the nature of a B vitamin) has been discovered in four mycoparasites (parasites on other fungi). The growth factor is present in many Ascomycetes, imperfect fungi and Basidiomycetes and has been extracted and partially purified. The four mycoparasites deficient for this growth factor – *Calcarisporium parasiticum, Gonatobotrys simplex, Gonatobotryum fuscum,* and *Gonatorrhodiella highlei* – make no growth on any common medium without the addition of this substance.

Nitrogen Sources

In nature, the organic material in the substrate would usually furnish the nitrogen for growth of fungi. Most of the imperfect fungi that have been studied are able to utilize nitrate nitrogen as well as ammonium nitrogen and amino acids as sources of nitrogen. Growth rate on inorganic sources is often less than on a mixture of amino acids or on a complex organic nitrogen source. If one merely desires to cultivate fungi on a laboratory medium, yeast extract or casein hydrolysate can be used as an excellent nitrogen source for almost all of the imperfect fungi. If one wishes to study the relative rate of utilization of various sources, however, single amino acids and single organic sources must be used. Asparagine, aspartic acid, and glutamic acid are among the best amino acids. Not enough study of the nitrogen requirements of the imperfect species has been made to justify a detailed report here.

TABLE 11

VITAMIN DEFICIENCIES OF SELECTED IMPERFECT FUNGI
(0 = no known deficiency; synthesizes its own vitamins)

Species	Deficiencies for Vitamins	Species	Deficiencies for Vitamins
Amblyosporium botrytis	0	*Gliocladium roseum*	0
Alternaria solani	0	*Graphium* spp.	Pyrodoxine
Arthrobotrys musiformis	Thiamine, biotin	*Helminothosporium sativum*	0
Aspergillus spp.	0	*Histoplasma capsulatum*	0
Botrytis cinerea	0	*Penicillium* spp.	0
Cladosporium herbarum	0	*Pyricularia oryzae*	Thiamine, biotin
Colletotrichum lindemuthianum	0	*Sclerotium rolfsii*	Thiamine
Cylindrocladium floridanum	Thiamine	*Sepedonium chrysopermum*	0
Cylindrocladium scoparium	0	*Stachybotrys atra*	Biotin
Dendrophoma obscurans	Thiamine	*Stysanus stemonitis*	0
Diplodia macrospora	Biotin	*Thielaviopsis basicola*	Thiamine
Fusarium spp.	0	*Trichophyton* spp.	0 or thiamine

Inorganic Salts and Microelements

Natural organic compounds often furnish all of the inorganic salts necessary for growth. However, if one desires to culture imperfect fungi on synthetic or semisynthetic media, it is necessary to add certain compounds. Monobasic potassium phosphate (KH_2PO_4) and magnesium sulfate ($MgSO_4 \cdot 7H_2O$) will supply potassium, phosphorus, magnesium and sulfur. The microelements Fe, Zn, Mn, Cu, and Ca are frequently added to synthetic media. A mixture of salts containing these elements is usually used. So few studies of the requirements of imperfect fungi for microelements have been made that little general information can be given here.

Suggested Media for Growth of Imperfect Fungi

A simple, easy-to-prepare medium for general cultivation of a wide variety of fungi in the laboratory is a glucose-yeast extract medium of the following composition:

glucose: 5 g
yeast extract: 1 g
distilled water: 1 liter
agar: 18–20 g

No adjustment of the pH is necessary. The medium is autoclaved for 15 minutes at 15 psi (120°C), then poured into sterile petri dishes.

For more precise studies, the following synthetic medium is recommended:

glucose: 5–20 g, as desired
KNO_3: 1–2 g (or 2 g of glutamic acid)
KH_2PO_4: 1.0 g
$MgSO_4 \cdot 7H_2O$: 0.5 g
Fe, Zn, Mn, Ca: 1 ml of solution
Thiamine: 100 μg
Biotin: 5 μg
Pyridoxine: 100 μg
distilled water: 1 liter

The pH must be adjusted to 6.0-6.5. If desired, agar may be added.

Imperfect Fungi That Parasitize On Other Fungi

In the long evolutionary development of fungi living in close association, some species have emerged as parasites on other species. The imperfect fungi that are parasites may obtain their nutrients in two ways: (1) by first killing the cells of the host fungus (destructive or necrotrophic parasites), or (2) by absorbing nutrients from living host cells without killing them (biotrophic parasites).

The species of the first group are composed of a number of soil- or wood-inhabiting fungi, which vary greatly in their destructive ability. A few common species in that group are *Trichoderma viride, Gliocladium roseum, Cephalosporium* spp., and *Fusarium* spp.

Only four species are known in the second group. They absorb nutrients through special cells that make close contact with the host hyphae without penetration, and they cause little or no harm. These four species are unlike all other fungi in that they do not grow on common laboratory media unless a water extract of certain fungi containing a special growth factor is added.

Nematode-Trapping Imperfect Fungi

A few imperfect fungi that inhabit rich soil or decaying wood have the ability to produce sticky knobs, loops, or constricting rings adapted for capturing and holding small nematodes. A captured nematode may struggle violently, but seldom breaks free after it is securely caught. Hyphae from the fungus finally penetrate the nematode's body and consume it for food. On common agar media, the production of traps is stimulated by the presence of nematodes.

The nematode-trapping fungi mostly belong to the genera *Arthrobotrys, Dactylella,* and *Monacrosporium.* Although there has been some success in limited experiments in reducing the number of plant-parasitic nematodes by this method, few of the extensive experiments have been promising as a means of controlling the plant-parasitic nematodes on crop lands.

Imperfect Fungi In Industry

Industry takes advantage of the synthetic abilities of the imperfect fungi in many ways by cultivating selected species and isolates under carefully controlled conditions. Species of *Aspergillus* and *Penicillium,* for example, are used in the synthesis of several organic acids: gallic acid, citric acid, gluconic acid, itaconic acid, and kojic acid.[29]

Several antibiotics, of which the most notable is penicillin, are synthesized by imperfect fungi, produced in industry by selected isolates of the *Penicillium notatum-chrysogenum* group.[30] Other imperfect fungi are involved in the synthesis of less important antibiotics.

In the area of production and processing of food, species of *Penicillium* are used in the ripening process of certain cheeses, particularly *P. roqueforti* for Roquefort cheese and *P. camemberti* for Camembert cheese.

The people of the Orient have developed many fermented foods by using a number of fungi, including species of *Aspergillus* and yeasts. Soybeans and rice are the common raw materials, and *A. oryzae* is one of the most common fungi used.[31]

The use of certain yeasts, such as *Torulopsis utilis,* as a food yeast has been proposed. Various low-cost natural carbohydrate substances and ammonium salts can be used as substrates. The resulting yeast cells are high in proteins and vitamins. It is possible that this yeast might be used as a protein supplement in the human diet as well as in animal feeds.[29]

Species of *Aspergillus, Penicillium* and *Alternaria,* as well as species of some other genera, are found on many manufactured products and sometimes cause heavy loss. Cloth, paper, and leather goods stored in a moist environment commonly show mildew and are often covered by the green, blue, or yellow spores of *Aspergillus* and *Penicillium*. Growth of these fungi in moist climates is not uncommon on electrical wiring, photographic film, and even on microscope or camera lenses, where etching of glass may occur because of the acids produced by the fungi.

REFERENCES

1. Alexopoulos, C. J., *Introductory Mycology.* John Wiley and Sons, New York (1962).
2. Pontecorvo, G., and Roper, J. A., Genetic Analysis Without Sexual Reproduction by Means of Polyploidy in *Aspergillus nidulans* (abstr.), *J. Gen. Microbiol., 6,* 7 (1952).
3. Bracker, C. E., The Ultrastructure of Fungi. *Annu. Rev. Phytopathol., 5,* 343 (1967).
4. Hawker, L. E., and Hendy, R. J., An Electron-Microscope Study of Germination of Conidia of *Botrytis cinerea. J. Gen. Microbiol., 33,* 43 (1963).
5. Kirk, B. T., and Sinclair, J. B., Plasmodesmata Between Hyphal Cells of *Geotrichum candidum. Science, 153,* 1646 (1966).
6. Reichle, R. E., and Alexander, J. V., Multiperforate Septations, Woronin Bodies and Septal Plugs in *Fusarium. J. Cell Biol., 24,* 489 (1965).
7. Buckley, P. M., Wyllie, T. D., and De Vay, J. E., Fine Structure of Conidia and Conidium Formation in *Verticillium albo-atrum* and *V. nigrescens. Mycologia, 61,* 240 (1969).
8. Wyllie, T. D., and Brown, M. F., Ultrastructural Formation of Sclerotia of *Macrophomina phaseoli. Phytopathology, 60,* 524 (1970).
9. Bracker, C. E., Ultrastructure of the Haustorial Apparatus of *Erysiphe graininis* and Its Relationship to the Epidermal Cells of Barley. *Phytopathology, 58,* 12 (1968).
10. Bracker, C. E., and Butler, E. E., The Ultrastructure and Development of Septa in Hyphae of *Rhizoctonia solani. Mycologia, 55,* 35 (1963).
11. Roper, J. A., The Parasexual Cycle, in *The Fungi,* Vol. 2, p. 589, G. C. Ainsworth and A. S. Sussman, Eds. Academic Press, New York (1966).
12. Pontecorvo, G., *Trends in Genetic Analysis.* Columbia University Press, New York (1958).
13. Saccardo, P. A., *Sylloge fungorum,* Vol. 4. Published by the author in Pavia, Italy (1886).
14. Saccardo, P. A., *Sylloge fungorum,* Vol. 14. Published by the author in Pavia, Italy (1899).
15. Bessey, E. A., *Morphology and Taxonomy of Fungi.* Hafner Publishing Co., New York (1964).
16. Barnett, H. L., and Hunter, B. B., *Illustrated Genera of Imperfect Fungi, 3rd ed.* Burgess Publishing Co., Minneapolis, Minnesota (1972).
17. Sobers, E. K., and Seymour, C. P., *Cylindrocladium floridanum* sp. n. Associated with Decline of Peach Trees in Florida. *Phytopathology, 57,* 389 (1967).
18. Morrison, R. H., and French, D. W., Taxonomy of *Cylindrocladium floridanum* and *C. scoparium. Mycologia, 61,* 957 (1969).
19. Ross, R. T., Microbiology of Paint Films. *Advan. Appl. Microbiol., 5,* 217 (1963).
20. Tousson, T. A., and Nelson, P. E., *A Pictorial Guide to the Identification of Fusarium Species.* Pennsylvania State University Press, University Park, Pennsylvania (1968).
21. Vuillemin, P., Les conidiophores. *Bull. Soc. Sci. Nancy, 11,* 129 (1910).
22. Vuillemin, P., Les aleurispores. *Bull. Soc. Sci. Nancy, 12,* 151 (1911).
23. Mason, E. W., Annotated Account of Fungi Received at the Imperial Mycological Institute, List II *Commonw. Mycol. Inst. Mycol. Pap.,* No. 2 (1928).
24. Hughes, S. J., Conidiophores, Conidia, and Classification, *Can. J. Bot., 31,* 577 (1953).
25. Tubaki, K., Studies on the Japanese Hyphomycetes. V. Leaf and Stem Group, with a Discussion of the Classification of Hyphomycetes and Their Perfect Stages. *J. Hattori Bot. Lab., 31,* 142 (1958).
26. Barron, G. L., *The Genera of Hyphomycetes from Soil.* Williams and Wilkins, Baltimore, Maryland (1968).
27. Tubaki, K., Taxonomic Study of Hyphomycetes. *Annu. Rep. Inst. Ferment. Osaka, 1,* 25 (1963).
28. Subramanian, C. V., A Classification of the Hyphomycetes. *Curr. Sci. (India), 31,* 409 (1962).
29. Gray, W. D., *The Relation of Fungi to Human Affairs.* Holt, Rinehart and Winston, New York (1959).
30. Raper, K. B., and Thom, C., *Manual of the Penicillia.* Williams and Wilkins, Baltimore, Maryland (1949).
31. Hesseltine, C. S., A Millenium of Fungi, Food, and Fermentation. *Mycologia, 57,* 149 (1965).

LICHENS

DR. ROGER A. ANDERSON

Lichens are specific structural and functional entities formed by the association of certain algae and fungi. The association is so close-knit that, in nature, lichens look and behave like distinct autonomous organisms. Yet they clearly have a dual nature and, in fact, represent one of the most successful experiments in mutualism, a type of symbiosis.[1-3]

Lichens constitute a biological group of "organisms", not a taxonomic group. The fungi involved in lichen formation are polyphyletic and represent three classes – Ascomycetes, Basidiomycetes, and Deuteromycetes. Lichen algae are also polyphyletic in the sense that they are derived from four divisions – Cyanophyta, Chlorophyta, Phaeophyta, and Xanthophyta. There are as many types of lichens as there are species of lichenized fungi. Well over 20,000 species have been described, but the number of "good" species is estimated to be about 15,000 in about 400 genera.[2] Basidiomycetes and Deuteromycetes (Fungi Imperfecti) account for only several small genera; the vast majority of lichen fungi therefore are Ascomycetes. The lichenized Ascomycetes come from only six of the more than twenty orders in the class. These are the orders Sphaeriales, Lecanorales, and Caliciales of the subclass Ascomycetidae, and the orders Myriangiales, Pleosporales, and Hysteriales of the subclass Loculoascomycetidae. The orders Lecanorales and Caliciales are comprised primarily of lichenzied members, whereas the others are a mixture of lichenized and non-lichenized species.

At least twenty-eight genera of algae are known to serve as primary symbionts in lichen associations (see Table 1).[1,4] With the exception of one genus of brown algae (Phaeophyta) and one genus of yellow-green algae (Xanthophyta), which are associated with a few species of the pyrenomycete genus *Verrucaria,* all known lichen algae belong to either the Cyanophyta or Chlorophyta. *Trebouxia* (green) is the most common phycobiont, being represented in about 50% of the known lichens.[5] *Trentepohlia* (green) and *Nostoc* (blue-green) are the next most common genera of lichen algae.[4] Few, if any, of the genera of algae found in lichens are restricted to a lichenized state. Their behavior in culture suggests that nearly all lichen algae can, and probably do, exist in a free-living state in nature, although some, for example *Trebouxia,* may be rare outside of lichens.[1] Lichen fungi, however, seem to be almost entirely restricted to a lichenized existence. Possible exceptions are indicated by conidial stages produced by lichen fungi in culture, which are identified as, or similar to, free-living imperfect fungi.[2]

MORPHOLOGY

The gross morphology of lichens is determined by the lichen fungus, except in a few simple lichens where the overall outline of the thallus is that of the filamentous algal component. In these few associations the alga makes up the bulk of the thallus; the fungus is present only as a thin hyphal sheath over the algal cells. In more complex lichen associations the fungus makes up the bulk of the thallus, perhaps as much as 90 to 95% of it, and is responsible for the characteristic growth form and color of the thallus. Thus the algae constitute a small portion of the thallus in most lichens and are confined to a thin layer beneath the cortex or outer surface of the thallus. Below the algal layer is the medulla, often a thick layer composed of loosely arranged hyphae. For a general introduction to the internal and external morphology of lichens see Reference 2. Reference 9 is a more comprehensive treatment.

REPRODUCTION

Methods of lichen reproduction have been recently reviewed.[2] Two basic methods are evident: vegetative reproduction, and natural resynthesis. Vegetative reproduction involves the formation of vegetative diaspores of various shapes and sizes, which contain both components of the lichen. The most common reproductive unit in this category is probably the unspecialized thallus fragment, which, when separated from a lichen thallus, can "germinate" into a new one. Many lichens also produce specialized diaspores,

TABLE 1

ALGAL GENERA SERVING AS PRIMARY SYMBIONTS AND REPRESENTATIVE ASSOCIATED LICHENIZED FUNGI

Algal Division	Algal Genus	Representative Associated Fungi
Cyanophyta	*Anacystis*[6]	*Peltula*
	Calothrix	*Lichina, Porocyphus*
	Chroococcus	*Phylliscum*
	Dichothrix	*Placynthium*
	Gloeocapsa	*Peccania, Psorotichia*
	Hyella	*Arthopyrenia*
	Nostoc	*Collema, Leptogium, Pannaria, Peltigera*
	Scytonema	*Coccocarpia, Pannaria*
	Stigonema	*Ephebe, Spilonema*
Chlorophyta	*Cephaleuros*	*Strigula*
	Chlorella	*Calicium, Lecidea*
	Chlorosarcina	*Lecidea*
	Coccobotrys	*Lecidea, Verrucaria*
	Coccomyxa	*Peltigera, Solorina*
	Dilabifilum[a]	*Verrucaria*
	Gloeocystis	*Gyalecta*
	Hyalococcus	*Dermatocarpon*
	Leptosira	*Thrombium*
	Myrmecia	*Bacidia, Catillaria, Lecidea, Psoroma*
	Phycopeltis	*Arthonia, Opegrapha*
	Pleurococcus	*Endocarpon, Lecidea*
	Pseudochlorella	*Lecidea*
	Stichococcus	*Calicium, Chaenotheca*
	Trebouxia	*Alectoria, Caloplaca, Cladonia, Lecanora, Parmelia, Umbilicaria*
	Trentepohlia	*Arthonia, Graphis, Opegrapha, Pyrenula*
	Trochiscia	*Polyblastia*
Phaeophyta	*Petroderma*[8]	*Verrucaria*
Xanthophyta	*Heterococcus*	*Verrucaria*

[a] *Pseudopleurococcus* pro parte; see Reference 7.

Data taken from: Ahmadjian, V., *The Lichen Symbiosis,* Blaisdell Publishing Co., Waltham, Massachusetts (1967). Reproduced by permission of the copyright owners.

known as soredia and isidia. Soredia are small (25 to 100 μ), non-corticate, powdery particles that consist of a few algal cells surrounded by fungal hyphae.[2] Isidia are small, corticate, clavate or minutely coralloid units produced on the thallus surface. Both soredia and isidia can develop into new thalli under appropriate conditions.

Natural resynthesis of lichens involves synthesis of a new thallus by means of recombining appropriate algal cells and fungal hyphae. Ascospores resulting from sexual reproduction by the lichen fungus become dispersed and germinate to form a small prothallus. If the prothalline hyphae encounter a suitable alga, an association of fungal hyphae and algal cells will develop and ultimately form the typical lichen. Evidence for such natural resyntheses is largely circumstantial, yet there is little doubt that they do occur.[1,2] Published reports that confirm the direct observation of the process in nature are lacking;[3] yet, experimental evidence indicates that new lichen associations can arise from separated algal and fungal components.[1,10,11] Supportive evidence comes from the observation that many lichens, particularly crustose ones, lack any obvious means of producing vegetative diaspores but do produce ascocarps and ascospores.[2,12] The extent of reproduction by natural resynthesis among lichens is unknown, and no reliable estimates are available. However, Degelius[12] has determined that 11% of the fifty-six species of lichens that occur on *Fraxinus excelsior* twigs in southern Scandinavia reproduce exclusively by this means. He also found that 77% of the fifty-six species were capable of reproducing by means of fragmentation, 32% formed soredia, and 11% produced isidia. Table 2 provides further information on the frequency of isidia and soredia in selected lichens. Estimates on the frequency of thallus fragments in the taxa or groups listed have not been made.

Numerical abundance of lichens and geographic distribution tends to be correlated with the mode of reproduction. Although exceptions are known, lichens that produce vegetative diaspores are often more widespread and abundant than lichens that reproduce by resynthesis.[2]

PHYSIOLOGICAL ATTRIBUTES

From a functional point of view, lichens behave like autonomous organisms, even though they are composed of two distinctly different kinds of organisms. They are, in a sense, a functional whole and therefore have a physiology. The physiological characteristics of lichens are reviewed comprehensively in References 1, 2, and 3. The present review covers only some of the major aspects of lichen physiology and some pertinent recent information.

Water is a critical limiting factor in the physiology of lichens. Lichens, like other poikilohydric organisms, are unable to control their water content; they have no special means of absorbing or conserving water. Their water content thus fluctuates with that of the immediate surrounding environment. When no liquid water is available and relative humidity is low, the water content of the thallus is low, perhaps 2 to 15% of the thallus dry weight.[3] When liquid water is available, such as from dew, rain, or snowmelt, lichens quickly absorb it and their water content increases to as much as 100 to 300% or more of the dry weight.[3] Lichens can also absorb water vapor; when placed in a humid atmosphere, air-dry lichens slowly absorb water until their water content reaches an equilibrium value. When the relative humidity is 100%, the water content usually ranges between 50 and 75% of the maximum value, and at 95% relative humidity, values are in the range of 30 to 50% of the saturated water content.[1,3,17,18]

The water content of the thallus affects the rates of respiration and photosynthesis in a way that varies from lichen to lichen. With increasing water content, respiration increases until it reaches a maximum value.

TABLE 2

PERCENTAGE OF ISIDIATE AND SOREDIATE SPECIES IN SELECTED LICHENS

Genus, Family, or Group	Number of Species	% Isidiate	% Sorediate	Reference
Anaptychia, worldwide	79	6	16	2
Cetrelia, worldwide	14	14	21	13
Cladonia, North America	116	0	33	14
Collema, Europe	35	46	0	2
Foliicolous lichens, worldwide	236	2	1	2
Heppiaceae, North America	13	0	23	6
Lecanora (in part), worldwide	118	0	1	2
Lecidea, United States,	88	0	3	2
Europe	235	0	9	2
Leptogium, North America	43	26	0	15
Lichinaceae, worldwide	32	0	0	2
Maronea, worldwide	13	0	0	2
Ochrolechia, worldwide	59	10	30	16
Parmelia subgenus *Amphigymnia*, worldwide	106	23	32	2
Physcia, North America	47	9	49	2
Platismatia, worldwide	10	50	20[a]	13

[a] These species are also isidiate.

For the majority of lichens tested, this value is reached at a water content of 80 to 95% of saturation and maintained up to complete saturation.[3,19] As the lichen thallus dries out, respiration decreases rapidly; however, measurable amounts of respiration have been noted in air-dry thalli.[1] Optimal rates of photosynthesis are usually reached when water contents range between 65 and 90% of saturation.[1,3,19] At water contents above and below this value, the photosynthetic rate decreases. For many lichens photosynthesis stops when water contents decrease below 25 to 30% of saturation.[18,19] Since respiration and photosynthesis show different patterns of response to water content, it is to be expected that the rate of net CO_2 assimilation in lichens will vary considerably with changes in thallus water content.

In a recent study, respiration rates were measured for eight lichens in aqueous solution, using an oxygen electrode.[20] Values at 35°C ranged from 2.80 to 63.0 μl 0_2 absorbed per gram water-saturated thallus per minute, somewhat higher than rates reported by earlier investigators. Respiration rates increased to a maximum at 40°C for about half the species tested; maxima for the others were between 30 and 40°C. Respiration decreases at lower temperatures. In a few lichens respiration persists at measurable levels at temperatures of -24° to -26°C,[21,22] although in some antarctic lichens it is not evident below -10°C.[23]

Rates of photosynthesis are much lower for lichens than for higher plants, probably due primarily to the fact that lichens have fewer photosynthetic units and less chlorophyll per unit surface area than vascular plant leaves.[1,3] Optimal temperatures for photosynthesis vary with different lichens, season, habitat, and light intensity. Maximal rates of photosynthesis in temperate-zone lichens are reached in the temperature range between 10° and 20°C at a light intensity of 1600 foot-candles.[1] Cold-zone lichens often exhibit a capacity for photosynthesis at very low temperatures, with apparent CO_2 absorption noted at temperature minima between -22° and -24° C;[24] in these lichens, though respiration becomes inhibited at lower temperatures, photosynthesis is less affected and thus a positive metabolic balance can be maintained more easily at low temperatures.[23]

Lichens are commonly thought to be highly resistant to extremes in environmental conditions, particularly with respect to temperature and the water factor. Lichens have withstood freezing down to -183°C (liquid oxygen)[21] and -198°C (liquid nitrogen);[23] return to normal levels of respiration and photosynthesis is more rapid if freezing and thawing are accomplished slowly.[23] The extreme-cold resistance of lichens is thought to be due to their ability to undergo rapid dehydration. This also accounts for the heat resistance of lichens. The limits of heat resistance of air-dry thalli range from 70° to 101°C, whereas the limits in moist thalli range from 35° to 46°C.[25]

Drought resistance in lichens has been little studied. Present evidence indicates that the capacity to resist drought is correlated with habitat preference; lichens from aquatic or cool, moist habitats seem less resistant than those from warm, dry habitats.[25,26] For many lichens the limits of drought resistance greatly exceed what they experience in nature.[3,25] Thus, though an aquatic lichen may exhibit permanent damage from a 24-hour drought, many terrestrial lichens can survive desiccation periods of many weeks' duration, the most resistant even more than a year.[25,26]

Lichens are also distinguished by some aspects of their nutrition and metabolism. Nutrient absorption can probably occur over the whole thallus surface; it also occurs within the thallus in the exchange between components. In view of the typically nutrient-poor habitats where lichens grow, it is not surprising to note that they also possess highly efficient mechanisms for the accumulation of a wide variety of organic and inorganic nutrients.[3]

A major feature of the nutrition of lichens is the production of carbohydrates by the alga and their transfer to the fungus. The types of carbohydrates produced by the lichen algae and utilized by the fungus have been recently discussed[27-29] and are listed in Table 3. In all cases a single carbohydrate is released by a given alga. In the five genera of green algae, the mobile carbohydrate is always a polyol – either ribitol, erythritol, or sorbitol. In lichens with blue-green algae the mobile carbohydrate is apparently glucose, although Hill[29] has recently suggested that *Nostoc,* the blue-green alga associated with *Peltigera polydactyla,* releases a glucan that is quickly hydrolyzed by an extracellular enzyme produced by the fungus; the free glucose thus formed is absorbed by the fungus.

The rate at which carbohydrate is transferred from the alga to the fungus varies with the kind of alga. Experimental studies to determine the rate of movement have involved use of ^{14}C, which becomes incorporated into photosynthate. The rate of movement from alga to fungus corresponds to the rate of ^{14}C

TABLE 3

MOBILE CARBOHYDRATES MOVING FROM ALGA TO FUNGUS IN TWENTY-SEVEN LICHENS

Fungal Symbiont	Algal Symbiont		Mobile Carbohydrate
	Phylum	Genus	
Lecanora conizaeoides	Chlorophyta	*Trebouxia*	Ribitol
Parmelia saxatilis			
Pseudevernia furfuracea			
Umbilicaria pustulata			
Xanthoria aureola			
Dermatocarpon hepaticum		*Myrmecia*	Ribitol
Lobaria amplissima			
Lobaria laetevirens			
Lobaria pulmonaria			
Peltigera aphthosa		*Coccomyxa*	Ribitol
Solorina saccata			
Gyalecta cupularis		*Trentepohlia*	Erythritol
Lecanactis stenhammarii			
Roccella fuciformis			
Roccella phycopsis			
Dermatocarpon fluviatile		*Hyalococcus*	Sorbitol
Dermatocarpon miniatum			
Collema auriculatum	Cyanophyta	*Nostoc*	Glucose
Leptogium sp.			
Lobaria scrobiculata			
Peltigera canina			
Peltigera horizontalis			
Peltigera polydactyla			
Sticta fuliginosa			
Sticta sp.			
Cephalodia of *Lobaria amplissima*			
Cephalodia of *Peltigera aphthosa*			
Cephalodia of *Solorina saccata*			
Lichina pygmaea		*Calothrix*	?Glucose/?Glucosan
Coccocarpia sp.		*Scytonema*	Glucose

Data taken from: Richardson, D. H. S., Hill, D. H., and Smith, D. C., *New Phytol., 67,* 469 (1968). Reproduced by permission of the copyright owners, Blackwell Publications Ltd., Oxford, England.

movement, which presumably is related to the rate of carbohydrate movement. The rate may range from very slow to very fast. Under identical experimental conditions, the rate of ^{14}C movement from *Nostoc* to *Peltigera polydactyla* is twelve times faster than from *Trebouxia* to *Xanthoria aureola.*[30] In general, ^{14}C movement in lichens with *Nostoc* is "fast" (20 to 40% of the total fixed ^{14}C is released by the alga in three hours); lichens with *Trebouxia* are described as "slow" (2 to 4% of the fixed ^{14}C is released in three hours, 20 to 45% in twenty-four hours); in *Trentepohlia* lichens movement is "very slow" (1 to 2% in three hours; 5 to 10% in twenty-four hours).[27] Regardless of the form of the carbohydrate transferred or its rate of movement, it accumulates in the lichen fungus as polyols; mannitol is always formed, and arabitol is also commonly formed.[28]

Carbohydrates are obviously not the only nutrients moving between the components of lichens. The inorganic elements required by lichen algae are in most cases first absorbed by the fungus and subsequently transported to the algal cells. Lichen fungi frequently require either biotin or thiamine, or both. These vitamins, and others, are synthesized by the algal component and partly released to the fungus. In addition, nitrogen fixation has been demonstrated in some genera of blue-green lichen algae, and some transport of nitrogenous compounds undoubtedly occurs.[1-3]

One of the most outstanding physiological characteristics of most lichens is their slow rate of growth. For many lichens the annual radial increase in thallus size is less than one millimeter. The least amount of yearly growth for a lichen has been reported for *Rhizocarpon geographicum* (growing in the Colorado Rockies), with an average diameter increase of 14 mm per 100 years for the first century and a rate of 3.3 mm per 100 years thereafter.[31] The largest increment of radial growth in crustose and foliose lichens has been recorded for *Peltigera praetextata*, which had a rate of 45 mm per year.[17] Most values are between 0.5 and 8 mm radial increase per year.[2] Among the reasons for the slow growth rates of lichens are the following:[1] (1) low net CO_2 assimilation rates; (2) slow rates of protein synthesis and breakdown; (3) nutrient-poor substrates; and (4) existence in habitats that allow only brief, and sometimes infrequent, periods of optimal metabolic activity.

Coinciding with the slow rate of growth is the great longevity of many lichens. Some of the oldest lichens apparently reach the age of 1,000 to 4,500 years,[32] possibly even 6,000 years; the last-mentioned value is based on extrapolation of a growth curve intended for use up to 3,000 years only.[31]

REFERENCES

1. Ahmadjian, V., *The Lichen Symbiosis.* Blaisdell Publishing Co., Waltham, Massachusetts (1967).
2. Hale, M. E., *The Biology of Lichens.* Edward Arnold, London, England (1967).
3. Smith, D. C., *Biol. Rev., 37,* 537 (1962).
4. Ahmadjian, V., *Phycologia, 6,* 127 (1967).
5. Ahmadjian, V., in *Evolutionary Biology,* Vol. 4, p. 166, Th. Dobzhansky, M. K. Hecht, and W. C. Steere, Eds. Appleton-Century-Crofts, New York (1970).
6. Wetmore, C., *Ann. Mo. Bot. Gard., 57,* 158 (1970).
7. Tschermak-Woess, E., *Oesterr. Bot. Z., 118,* 443 (1970).
8. Wynne, M. J., *Univ. Calif. Publ. Bot., 50,* 1 (1969).
9. Ozenda, P., in *Handbuch der Pflanzenanatomie,* Vol. 6, p. 1, W. Zimmermann and P. Ozenda, Eds. Gebrüder Borntraeger, Berlin, Germany (1963).
10. Ahmadjian, V., *Science, 151,* 199 (1966).
11. Ahmadjian, V., and Heikkila, H., *Lichenologist (Oxf.), 4,* 259 (1970).
12. Degelius, G., *Acta Horti Gotob., 27,* 11 (1964).
13. Culberson, W. L., and Culberson, C. F., *Contrib. U.S. Natl. Herb., 34,* 449 (1968).
14. Thomson, J. W., *The Lichen Genus Cladonia in North America.* University of Toronto Press, Toronto, Ontario, Canada (1967).
15. Sierk, H., *Bryologist, 67,* 245 (1964).
16. Verseghy, K., Beih. zur *Nova Hedwigia, Suppl. 1,* 1 (1962).
17. Butin, H., *Biol. Zentralbl., 73,* 459 (1954).
18. Reid, A., *Flora Allg. Bot. Ztg. (Jena), 149,* 345 (1960).
19. Reid, A., *Biol. Zentralbl., 79,* 129 (1960).
20. Baddeley, M., Ferry, B., and Finegan, E., *Lichenologist (Oxf.), 5,* 18 (1971).
21. Scholander, P., Flagg, W., Hock, R., and Irving, L., *J. Cell. Comp. Physiol., 42, Suppl. 1,* 1 (1953).
22. Lange, O., *Ber. Deut. Bot. Ges., 75,* 351 (1963).
23. Ahmadjian, V., in *Antarctic Ecology,* Vol. 2, p. 801, M. W. Holdgate, Ed. Academic Press, New York (1970).
24. Lange, O., *Planta (Berlin), 64,* 1 (1965).
25. Lange, O., *Flora (Jena), 140,* 39 (1953).
26. Reid, A., *Biol. Zentralbl., 79,* 657 (1960).
27. Richardson, D. H. S., Hill, D. J., and Smith, D. C., *New Phytol., 67,* 469 (1968).
28. Smith, C., Muscatine, L., and Lewis, D., *Biol. Rev., 44,* 17 (1969).
29. Hill, D. J., *New Phytol., 71,* 31 (1972).
30. Richardson, D. H. S., and Smith, D. C., *New Phytol., 67,* 61 (1968).
31. Benedict, J., *J. Glaciol., 6,* 817 (1967).
32. Beschel, R., in *Geology of the Arctic,* p. 1044, G. O. Raasch, Ed. University of Toronto Press, Toronto, Ontario, Canada (1961).

ALGAE

A GUIDE TO THE LITERATURE ON ALGAE

DR. HUBERT A. LECHEVALIER*

The word "alga" was used by the Romans much as we do at present. In addition, figuratively, it meant an object of little value.

Algae can be defined, by exclusion, as oxygen-evolving photosynthetic organisms that are not bryophytes or vascular plants (Table 1). Their thallus, like that of the fungi, is not differentiated into roots, stems, and leaves. They constitute a heterogeneous group and, like the fungi, are more easily recognized than defined.

One can recognize two large groups of algae: the *pocaryotic* blue-green algae or *Cyanophyta,* which are closely related to the bacteria, and the *eucaryotic true algae,* which show relationships to higher forms of life. Some forms, like the euglenoids, can be classified with the protozoa. Algae can be separated into phyla as follows:

TABLE 1

Some Characteristics of Photosynthetic Organisms

	Green Plants and Algae	**Green Bacteria (Chlorobacteriaceae)**	**Purple Bacteria (Thiorhodaceae and Athiorhodaceae)**
Source of reducing power	H_2O	H_2S, other reduced inorganic compounds	H_2S, other reduced inorganic compounds, organic compounds
Photosynthetic oxygen evolution	Yes	No	No
Principal source of carbon	CO_2	CO_2	CO_2 or organic compounds
Relation to oxygen	Aerobic	Strictly anaerobic	Strictly anaerobic or facultatively anaerobic
Major form of chlorophyll	Chlorophyll	*Chlorobium*-chlorophyll	Bacteriochlorophyll

Taken from Lechevalier, H. A., and Pramer, D., *The Microbes,* p. 339. J. B. Lippincott, Philadelphia (1971).

SYNOPSIS OF THE ALGAL PHYLA

1 Cells without chloroplasts or chromatophores; pigments blue-green, olive-green, or purplish, distributed throughout the entire protoplast (although cells may be somewhat less colored in the central region); wall usually thin (often showing as a membrane only) and generally with a mucilaginous sheath (wide or narrow, watery or firm, and definite); food reserve in the form of glycogen or a starch-like substance; iodine test for starch negative; no motile cells present. . . . .Blue-Green Algae. . . .CYANOPHYTA

1 Cells with chloroplasts or with chromatophores, the pigments not distributed throughout the protoplast; cell wall clearly evident (with rare exceptions *Pyramimonas*; stored food not in the form of glycogen; iodine test for starch positive or negative. . .2

*The kind assistance of Dr. Peter Edwards, Rutgers University, is gratefully acknowledged.

2 Cells with grass-green chloroplasts (but see some species of *Euglena,* or the filamentous alga, *Trentepohlia,* which, although possessing chlorophyll, have the green color masked by an abundance of the red pigment, haematochrome). . .3

2 Cells with chloroplasts or chromatophores some other color, gray-green, brown, violet-green, or yellow-green, sometimes purplish. . .5

3 Free-swimming, unicellular; with numerous ovoid, star-shaped, or plate-like chloroplasts which are grass-green; food stored as clearly evident grains of insoluble paramylum (sticks, or plates); iodine test for starch negative; one or two (rarely three) coarse flagella attached at the apex in a gullet; eye-spot or red pigment spot usually evident. . . .Euglenoids. . . .EUGLENOPHYTA

3 Organisms not as above. . .4

4 Unicellular, without an eye-spot; chloroplasts numerous discs usually radially directed at the periphery of the cell; motile by means of two flagella inserted in an apical reservoir; trichocyst organelles usually present just within the cell wall; food reserve oil. . . .CHLOROMONADOPHYTA

4 Unicellular, colonial, or filamentous; swimming or not swimming (although often free-floating); when swimming using 2 to 4 fine flagella attached at the apex of the cell but not in a colorless reservoir; chloroplasts one to several, usually with a conspicuous pyrenoid (starch-storing granule); iodine test for starch positive. . . .Green Algae. . . .CHLOROPHYTA

5 Chromatophores light olive-brown to dark brown; nearly all marine, essentially filamentous, but occurring mostly as thalli of macroscopic size (brown sea weeds); stored food in the form of laminarin and alcohol; starch test with iodine negative. . . .Brown Algae. . . .PHAEOPHYTA

5 Plants marine or fresh-water, but not occurring as brown thalli of macroscopic size. . .6

6 Chromatophores yellow-green to yellow- or golden-brown; food in the form of leucosin or oil; starch test with iodine negative; plants unicellular, colonial or filamentous; sometimes swimming with apically attached flagella; many forms (especially the diatoms) with the cell wall impregnated with silicon. . . .Yellow-green Algae. . . .CHRYSOPHYTA

6 Chromatophores not yellow-green or pale green, but dark golden brown, gray-green, violet-green; food in the form of oil or starch-like carbohydrates; iodine test for starch mostly negative. . .7

7 Unicellular, with dark, golden-brown chromatophores; swimming by means of two laterally attached flagella; a conspicuous eye-spot usually present; many forms with the cell wall composed of polygonal plates. . . .Dinoflagellates. . . .PYRRHOPHYTA

7 Organisms unicellular or filamentous, not swimming by means of laterally attached flagella; chromatophores brown, green, bluish, violet-green, or gray-green. . .8

8 Chromatophores violet or gray-green, sometimes bluish-green in fresh water, red in marine forms; occurring as filamentous thalli of both macroscopic and microscopic size; food stored in the form of starch-like carbohydrates; starch test with iodine negative. . . .Red Algae. . . .RHODOPHYTA

8 Chromatophores one or two golden-brown (rarely blue) bodies; organisms unicellular; swimming by means of subapically attached flagella; food reserve in the form of starch-like carbohydrates; iodine test positive in some. . . .CRYPTOPHYTA

(This class of the algae has several characteristics in common with Dinoflagellates and in some systems of classification is included under the Pyrrhophyta.)

Taken from Prescott, G. W., *How to Know The Fresh-Water Algae,* 2nd ed., pp. 19–20. Wm. C. Brown, Dubuque, Iowa (1954).

Some general characteristics of the various phyla are given in Table 2, and a list of general references on algae follows.

GENERAL REFERENCES

1. Boney, A. D., *Biology of Marine Algae,* Hutchinson Educational, London (1966).
2. Chapman, V. J., *The Algae,* Macmillan, London (1962).
3. Dawson, E. Y., *Marine Botany,* Holt, Rinehart and Winston, New York (1966).
4. Dawson, E. Y., *How to Know the Seaweeds,* Wm. C. Brown, Dubuque, Iowa (1956).
5. Fritsch, F. E., *The Structure and Reproduction of the Algae,* Vol. I and Vol. II, Cambridge University Press (1935, 1945).
6. Godward, M. B. E., *The Chromosomes of the Algae,* Edward Arnold, London (1966).
7. Hutner, S. H., and Provasoli, L., Nutrition of Algae, *Annu. Rev. Plant Physiol., 15, 37*(1964).
8. Japanese Society of Plant Physiologists, Ed., *Studies on Microalgae and Photosynthetic Bacteria,* University of Tokyo Press (1963).
9. Kumar, H. D., and Singh, H. N., *A Textbook on Algae,* East-West Press, New Delhi (1971).
10. Lewin, R. A., Ed., *Physiology and Biochemistry of Algae,* Academic Press, New York (1962).
11. Palmer, C. M., *Algae in Water Supplies,* U.S. Public Health Service Publication No. 657 (1959).
12. Prescott, G. W., *How to Know the Fresh-Water Algae,* Wm. C. Brown, Dubuque, Iowa (1954).
13. Prescott, G. W., *The Algae: A Review,* Houghton Mifflin, Boston, Massachusetts (1968).
14. Rosowski, J. R., and Parker, B. C., Eds., *Selected Papers in Phycology,* Department of Botany, University of Nebraska (1971).[a]
15. Round, F. E., *The Biology of the Algae,* Edward Arnold, London (1965).
16. Smith, G. M., *The Fresh-Water Algae of the United States,* 2nd ed., McGraw-Hill, New York (1950).
17. Smith, G. M., Ed., *Manual of Phycology,* Chronica Botanica, Waltham, Massachusetts (1951).

[a] Available from: The University Bookstore, 14th and R Street, Lincoln, Nebraska.

TABLE 2

Characteristics of Algae

	Pigments*				
Phylum	**Chlorophyll**	**Other Major Pigments**	**Cell Wall**	**Food Reserve**	**Chloroplast and Nuclear Membranes**
Cyanophyta		Phycobilins	Mucopolymer	Amylopectin	–
Chlorophyta	*b*	–	Cellulose	Starch	+
Euglenophyta	*b*	–	–	Paramylum and oil	+
Chrysophyta	*c*	Xanthophylls	Silicon impregnated	Chrysolaminarin	+
Phaeophyta	*c*	Xanthophylls	Cellulose and algin	Laminarin and mannitol	+
Pyrrophyta	*c*	Xanthophylls	Cellulose	Starch and oil	+
Rhodophyta	*d*	Phycobilins	Cellulose	Starch	+

*All contain chlorophyll *a*- and β-carotene.

Taken from Lechavalier, H. A., and Pramer, D., *The Microbes,* 300. J. B. Lippincott Co., Philadelphia (1971).

1. Cyanophyta (Blue-green Algae)

Plants are unicellular or more often colonial. The colony may be an orderly or a random agglomeration of cells; it may be a simple or a branched (branching may be "false") filament, with procaryotic cytology. Meiosis and sexual reproduction are unknown. Gas vacuoles may be present; photosynthesis is accompanied by evolution of oxygen. The photosynthetic apparatus (thylakoid) is composed of two membranes joined at their ends and enclosing a space. Flagella are not formed but a gliding motion may occur. Some species at least are able to fix atmospheric nitrogen and may form vegetative spores capable of germinating (akinetes) and other differentiated cells called "heterocysts," which may play a role in nitrogen fixation.

BIBLIOGRAPHY

From Lang, N. J., and Waaland, J. R., *Selected Papers in Phycology,* pp. 755–759, J. R. Rosowski and B. C. Parker, Eds. Department of Botany, University of Nebraska, Omaha (1971).

A. Reviews and General Treatments

1. Allsopp, A., Phylogenetic Relationships of the Procaryota and the Origin of the Eucaryotic Cell, *New Phytol., 68,* 591 (1969).
2. Barghoorn, E. S., and Tyler, S. A., Microorganisms from the Gunflint Chert, *Science, 147,* 563 (1965).
3. Echlin, P., and Morris, I., The Relationships between Blue-green Algae and Bacteria, *Biol. Rev., 40,* 143 (1965).
4. Gusev, M. V., The Blue-green Algae, *Microbiology, 30,* 897 (1961); translation from *Mikrobiologiya, 30,* 1108–28.
5. Holm-Hansen, O., Ecology, Physiology, and Biochemistry of Blue-green Algae, *Ann. Rev. Microbiol., 22,* 47 (1968).
6. Jackson, D. F., Ed., *Algae, Man, and the Environment,* Proceedings International Symposium, sponsored by Syracuse University and New York State Science and Technological Foundation. Syracuse University Press, Syracuse, New York (1968).
7. Knight, B. C. J. G., and Charles, H. P., Organization and Control in Prokaryotic and Eukaryotic Cells, *Symp. Soc. Gen. Microbiol.,* No. 20. Press, Cambridge, England (1970).
8. Lang, N. J., The Fine Structure of Blue-green Algae, *Ann. Rev. Microbiol., 22,* 15 (1968).
9. Margulis, L., *Origin of Eukaryotic Cells,* Yale University Press, New Haven (1970).
10. Pavoni, M., Blaualgenliteratur aus den Jahren 1960-1966, *Schweiz. Z. Hydrol., 29,* 226 (1967).
11. Pringsheim, E. G., The Relationship between Bacteria and Myxophyceae, *Bacteriol. Rev., 13,* 47 (1949).
12. Raven, P. H., A Multiple Origin for Plastids and Mitochondria, *Science, 169,* 641 (1970).
13. Schopf, J. W., Precambrian Microorganisms and Evolutionary Events Prior to the Origin of Vascular Plants, *Biol. Rev., 45,* 319 (1970).
14. Spearing, J. K., Studies on the Cyanophycean Cell. I. Vital Staining – A Study in the Production of Artifacts, *La Cellule, 61,* 243 (1961).

B. Cultivation

15. Allen, M. B., The Cultivation of the Myxophyceae, *Arch. Mikrobiol., 17,* 34 (1952).
16. Allen, M. M., Simple Conditions for the Growth of Unicellular Blue-green Algae on Plates, *J. Phycol., 4,* 1 (1968).
17. Allen, M. M., and Stanier, R. Y., Selective Isolation of Blue-green Algae from Water and Soil, *J. Gen. Microbiol., 51,* 203 (1968).

C. Cyanophages

18. Luftig, R., and Haselkorn, R., Studies on the Structure of Blue-green Algae Virus LPP-1, *Virology, 34,* 664 (1968).
19. Safferman, R. S., Morris, M., Sherman, L. A., and Haselkorn, R., Serological and Electron Microscopic Characterization of a New Group of Blue-green Algal Viruses (LPP-2), *Virology, 39,* 775 (1969).
20. Safferman, R. S., Schneider, I. R., Steere, R. L., Morris, M., and Diener, T. O., Phycovirus SM-1: A Virus Infecting Unicellular Blue-green Algae, *Virology, 37,* 386 (1969).

D. Development

21. Lazaroff, N., Photoinduction and Photoreversal of the Nostocacean Developmental Cycle, *J. Phycol., 2,* 7 (1966).
22. Wolk, C. P., Physiological Basis of the Pattern of Vegetative Growth of a Blue-green Alga, *Proc. Natl. Acad. Sci. USA, 57,* 1246 (1967).

E. Ecology and Symbiosis

23. Castenholz, R. W., Thermophilic Blue-green Algae and the Thermal Environment, *Bacteriol. Rev., 33,* 476 (1969).
24. Gorham, P. R., Toxic Waterblooms of Blue-green Algae, in *Biological Problems in Water Pollution,* 3rd Seminar, 1962, 37, U.S. Department of Health, Education, and Welfare, Public Health Service Pub. No. 999-WP-25 (1965).
25. Jackim, E., and Gentile, J., Toxins of a Blue-green Alga: Similarity to Saxitoxin, *Science, 162,* 915 (1968).
26. Hall, W. T., and Claus, G., Ultrastructure Studies on the Cyanelles of *Glaucocystis nostochinearum* Itzigsohn, *J. Phycol., 3,* 57 (1967).
27. Richardson, F. L., and Brown, T. E., *Glaucosphaera vacuolata,* Its Ultrastructure and Physiology, *J. Phycol., 6,* 165 (1970).
28. Various authors, in *Environmental Requirements of Blue-green Algae,* Proceedings of a Symposium sponsored by the University of Washington and the Federal Water Pollution Control Administration, U.S. Dept. of Interior, Corvallis, Oregon (1967).

F. Gas Vacuoles

29. Cohen-Bazire, G., Kunisawa, R., and Pfennig, N., Comparative Study of the Structure of Gas Vacuoles, *J. Bacteriol., 100,* 1049 (1969).
30. Jones, D. D., and Jost, M., Isolation and Chemical Characterization of Gas-vacuole Membranes from *Microcystis aeruginosa* Kuetz. emend. Elenkin, *Arch. Mikrobiol., 70,* 43 (1970).
31. Waaland, J. R., and Branton, D., Gas Vacuole Development in a Blue-green Alga, *Science, 163,* 1339 (1969).
32. Waaland, J. R., Waaland, S. D., and Branton, D., Gas Vacuoles: Light Shielding in Blue-green Algae, *J. Cell. Biol., 48,* 212 (1971).
33. Walsby, A. E., The Permeability of Blue-green Algal Gas Vacuoles to Gas, *Proc. R. Soc. Lond. B., 173,* 235 (1969).

G. Genetics and Mutagens

34. Asato, Y., and Folsome, C. E., Temporal Genetic Mapping of the Blue-green Alga *Anacystis nidulans, Genetics, 65,* 407 (1970).

35. Bazin, M., Sexuality in a Blue-green Alga: Genetic Recombination in *Anacystis nidulans, Nature, 218,* 282 (1968).
36. Kumar, H. D., Action of Mutagenic Chemicals on *Anacystis nidulans, Z. Allg. Mikrobiol., 9,* 137 (1969).
37. Pikálek, P., An Attempt to Find Genetic Recombination in *Anacystis nidulans, Nature, 215,* 666 (1967).
38. Shestakov, S. V., and Khyen, N. T., Evidence for Genetic Transformation in Blue-green Alga *Anacystis nidulans, Mol. Gen. Genet., 107,* 372 (1970).
39. Singh, R. N., and Sinha, R., Genetic Recombination in a Blue-green Alga *Cylindrospermum majus* Kütz, *Nature, 207,* 782 (1965).

H. Heterocysts and Nitrogen Fixation

40. Fay, P., Cell Differentiation and Pigment Composition in *Anabaena cylindrica, Arch. Mikrobiol., 67,* 62 (1969).
41. Fay, P., Stewart, W. D. P., Walsby, A. E., and Fogg, G. E., Is the Heterocyst the Site of Nitrogen Fixation in Blue-green Algae? *Nature, 220,* 810 (1968).
42. Singh, R. N., *Role of Blue-green Algae in Nitrogen Economy of Indian Agriculture,* Indian Council of Agriculture Research, New Delhi (1961).
43. Singh, R. N., and Tiwari, D. N., Frequent Heterocyst Germination in the Blue-green Alga *Gloeotrichia ghosei* Singh, *J. Phycol., 6,* 172 (1970).
44. Stewart, W. D. P., Algal Fixation of Atmospheric Nitrogen, *Plant Soil, 32,* 555 (1970).
45. Stewart, W. D. P., and Lex, M., Nitrogenase Activity in the Blue-green Alga *Plectonema boryanum* Strain 594, *Arch. Mikrobiol., 73,* 250 (1970).
46. Stewart, W. D. P., Haystead, A., and Pearson, H. W., Nitrogenase Activity in Heterocysts of Filamentous Blue-green Algae, *Nature, 224,* 226 (1969).
47. Wolk, C. P., and Simon, R. D., Pigments and Lipids of Heterocysts, *Planta (Berlin), 86,* 92 (1969).
48. Wyatt, J. T., and Silvey, J. K. G., Nitrogen Fixation by *Gloeocapsa, Science, 165,* 908 (1969).

I. Physiology and Biochemistry

49. Ahmad, M. R., and Winter, A., Studies on the Hormonal Relationships of Algae in Pure Culture. I. The Effect of Indole-3-Acetic Acid on the Growth of Blue-green and Green Algae, *Planta (Berlin), 78,* 277 (1968).
50. Ahmad, M. R., and Winter, A., Studies on the Hormonal Relationships of Algae in Pure Culture. II. The Effect of Potential Precursors of Indole-3-Acetic Acid on the Growth of Several Freshwater Blue-green Algae, *Planta (Berlin), 81,* 16 (1968).
51. Ahmad, M. R., and Winter, A., Studies on the Hormonal Relationships of Algae in Pure Culture. III. Tryptamine as an Intermediate in the Conversion of Tryptophan to Indole-3-Acetic Acid by the Blue-green Alga *Chlorogloea fritschii, Planta (Berlin), 88,* 61 (1969).
52. Berns, D. S., Immunochemistry of Biliproteins, *Plant Physiol., 42,* 1569 (1967).
53. Berns, D. S., and Edwards, M. R., Electron Micrographic Investigations of c-Phycocyanin, *Arch. Biochem. Biophys., 110,* 511 (1965).
54. Cohen-Bazire, G., and Sistrom, W. R., The Procaryotic Photosynthetic Apparatus, in *The Chlorophylls,* pp. 313–341, L. P. Vernon and G. R. Seely, Eds. Academic Press, New York (1966).
55. Edelman, M., Swinton, D., Schiff, J. A., Epstein, H. T., and Zeldin, B., Deoxyribonucleic Acid of the Blue-green Algae (Cyanophyta), *Bacteriol. Rev., 31,* 315 (1967).
56. Fogg, G. E., The Comparative Physiology and Biochemistry of the Blue-green Algae, *Bacteriol. Rev., 20,* 148 (1956).
57. Fogg, G. E., The Physiology of an Algal Nuisance, *Proc. R. Soc. Lond. B., 173,* 175 (1969).
58. Halfen, L. N., and Castenholz, R. W., Gliding in a Blue-green Alga: A Possible Mechanism, *Nature, 225,* 1163 (1970).

59. Holt, S. A., Nature, Properties and Distribution of Chlorophylls, in *Chemistry and Biochemistry of Plant Pigments,* pp. 3–28, T. W. Goodwin, Ed. Academic Press, New York (1965).
60. Kenyon, C. N., and Stanier, R. Y., Possible Evolutionary Significance of Polyunsaturated Fatty Acids in Blue-green Algae, *Nature, 227,* 1164 (1970).
61. O'hEocha, C., Phycobilins, in *Chemistry and Biochemistry of Plant Pigments,* pp. 175–195, T. W. Goodwin, Ed. Academic Press, New York (1965).
62. Stewart, W. D. P., and Pearson, H. W., Effects of Aerobic and Anaerobic Conditions on Growth and Metabolism of Blue-green Algae, *Proc. R. Soc. Lond. B., 175,* 293 (1970).
63. Thomas, J., Absence of the Pigments of Photosystem II of Photosynthesis in Heterocysts of a Blue-green Alga, *Nature, 228,* 181 (1970).
64. Walsby, A. E., Mucilage Secretion and the Movements of Blue-green Algae, *Protoplasma, 65,* 223 (1968).

J. Taxonomy and Morphology

65. Desikachary, T. V., *Cyanophyta,* Indian Council of Agricultural Research, New Delhi (1959).
66. Drouet, F., *Revision of the Classification of the Oscillatoriaceae,* Monograph 15, Academy of Natural Science, Philadelphia (1968).
67. Drouet, F., and Daily, W. A., Revision of the Coccoid Myxophyceae, *Butler Univ. Bot. Stud., 12,* 1 (1956).
68. Geitler, L., Cyanophyceae, in *Kryptogamenflora von Deutschland, Österreich und der Schweiz.,* Vol. 14, pp. 1–1056, L. Rabenhorst, Ed. (1930–1932).
69. Geitler, L., Schizophyceen, *Handbuch der Pflanzenanatomie,* Vol. 6 (1960).
70. Kantz, T., and Bold, H. C., Phycological Studies. IX. Morphological and Taxonomic Investigations of *Nostoc* and *Anabaena* in Culture. The University of Texas Publication No. 6924 (1969).

K. Ultrastructure

71. Allen, M. M., Ultrastructure of the Cell Wall and Cell Division of Unicellular Blue-green Algae, *J. Bacteriol., 96,* 842 (1968).
72. Bisalputra, T., Brown, D. L., and Weier, T. E., Possible Respiratory Sites in a Blue-green Alga *Nostoc sphaericum* as Demonstrated by Potassium Tellurite and Tetranitro-blue Tetrazolium Reduction, *J. Ultrastruct. Res., 27,* 182 (1969).
73. Edwards, M. R., Berns, D. S., Ghiorse, W. C., and Holt, S. C., Ultrastructure of the Thermophilic Blue-green Alga, *Synechococcus lividus* Copeland, *J. Phycol., 4,* 283 (1968).
74. Gantt, E., and Conti, S. F., Ultrastructure of Blue-green Algae, *J. Bacteriol., 97,* 1486 (1969).
75. Jensen, T. E., Fine Structure of Developing Polyphosphate Bodies in a Blue-green Alga, *Plectonema boryanum, Arch. Mikrobiol., 67,* 328 (1969).
76. Jost, M., Die Ultrastruktur von *Oscillatoria rubescens* D.C., *Arch. Mikrobiol., 50,* 211 (1965).
77. Lamont, H. C., Shear-oriented Microfibrils in the Mucilaginous Investments of Two Motile Oscillatoriacean Blue-green Algae, *J. Bacteriol., 97,* 350 (1969).
78. Lang, N. J., and Fay, P., The Heterocysts of Blue-green Algae. II. Details of Ultrastructure, *Proc. R. Soc. Lond. B., 176,* in press (1971).
79. Leak, L. V., Studies on the Preservation and Organization of DNA-Containing Regions of a Blue-green Alga, a Cytochemical and Ultrastructural Study, *J. Ultrastruct. Res., 20,* 190 (1967).
80. Lefort, M., Sur le chromatoplasma d'une Cyanophycee endosymbiotique: *Glaucocystis nostochinearum* Itzigs, *C.R. Acad. Sci. Paris, 261,* 233 (1965).
81. Miller, M. M., and Lang, N. J., The Fine Structure of Akinete Formation and Germination in *Cylindrospermum, Arch. Mikrobiol., 60,* 303 (1968).
82. Pankratz, H. S., and Bowen, C. C., Cytology of Blue-green Algae. I. The Cells of *Symploca muscorum, Am. J. Bot., 50,* 387 (1963).

83. Ris, H., Ultrastructure and Molecular Organization of Genetic Systems, *Can. J. Genet. Cytol., 3,* 95 (1961).
84. Ris, H., and Singh, R. N., Electron Microscope Studies on Blue-green Algae, *J. Biophys. Biochem. Cytol., 9,* 63 (1961).
85. Wildon, D. C., and Mercer, F. V., The Ultrastructure of the Heterocyst and Akinete of the Blue-green Algae, *Arch. Mikrobiol., 47,* 19 (1963).

2. Chlorophyta (Green Algae)

Plants are unicellular, colonial, or filamentous, and are swimming, floating, or attached and stationary. Cells contain plastids in which chlorophyll (grass-green) is predominant, and in which there is usually a shiny, starch-storing body, the pyrenoid. Pigments are chlorophyll, xanthophyll and carotene. The starch test with iodine is positive (in almost every instance); the nucleus is definite (although often small and inconspicuous). The cell wall is usually relatively thick and definite, composed of cellulose and pectose; swimming cells or motile reproductive elements are furnished with two to four flagella of equal length attached at the anterior end; sexual reproduction is by iso-, aniso-, and by heterogametes.

A. BIBLIOGRAPHY

C. Parker, Eds. Department of Botany, University of Nebraska, Omaha (1971), for the Chlorophyceae (grass-green algae).

General References to the Chlorophyta

1. Ahmadjian, V., *The Lichen Symbiosis,* Blaisdell Publishing Co., Waltham, Massachusetts (1967).
2. Bourrelly, P., *Les Algues d'Eau Douce,* Tome I: Les Algues Vertes, Éditions N. Boubeé et Cie, Paris (1966).
3. Chapman, V. J., The Chlorophyta, *Oceanogr. Mar. Biol. Annu. Rev., 2,* 193 (1964).
4. Christensen, T., Alger. 2. Udgave, in *Botanik, Systematisk Botanik,* Bd. II, Nr. 2. T. Böcher, M. Lange, and T. Sørensen, Munksgaard, Copenhagen (1966).
5. Dodge, J. D., A Review of the Fine Structure of Algal Eyespots, *Br. Phycol. J., 4,* 199 (1969).
6. Frederick, J. F., and Klein, R. M., Eds., Phylogenesis and Morphogenesis in the Algae, *Ann. N.Y. Acad. Sci., 175,* 413 (1970).
7. Fritsch, F. E., *The Structure and Reproduction of the Algae,* Vol. I. Cambridge University Press, Cambridge, England (1935).
8. Fritsch, F. E., Studies in the Comparative Morphology of the Algae. IV. Algae and Archegoniate Plants, *Ann. Bot. N. S., 9,* 1 (1945).
9. Godward, M. B. E., *The Chromosomes of the Algae.* Edward Arnold Ltd., London (1966).
10. Griffiths, D. J., The Pyrenoid, *Bot. Rev., 36,* 29 (1970).
11. Klein, R. M., and Cronquist, A., A Consideration of the Evolutionary and Taxonomic Significance of Some Biochemical, Micromorphological, and Physiological Characters of the Thallophytes, *Quart. Rev. Biol., 42,* 105 (1967).
12. Lewin, R., Ed., *Physiology and Biochemistry of Algae.* Academic Press, New York (1962).
13. Manton, I., Some Phyletic Implications of Flagellar Structure in Plants, in *Advances in Botanical Research,* Vol. 2, pp. 1 – 34, R. D. Preston, Ed. Academic Press, New York (1965).
14. McLaughlin, J. J. A., and Zahl, P. A., Endozoic Algae, in *Symbiosis,* Vol. I, pp. 257 – 297, S. M. Henry, Ed. Academic Press, New York (1966).
15. Papenfuss, G. F., Classification of the Algae, in *Century of Progress in the Natural Sciences, 1853 – 1953,* pp. 115 –224, California Academy of Sciences, San Francisco, California (1955).
16. Raven, P. H., A Multiple Origin for Plastids and Mitochondria, *Science, 169,* 641 (1970).
17. Round, F. E., The Taxonomy of the Chlorophyta, *Br. Phycol. Bull., 2,* 224 (1963).

18. Schopf, J. W., Precambrian Microorganisms and Evolutionary Events Prior to the Origin of Vascular Plants, *Biol. Rev., 45,* 319 (1970).
19. Smith, G. M., *Fresh-water Algae of the United States,* 2nd ed. McGraw-Hill Book Co., New York (1950).
20. Smith, G. M., *Cryptogamic Botany,* Vol, 1, *Algae and Fungi,* 2nd ed. McGraw-Hill Book Co., New York (1955).
21. Stebbins, G. L., The Comparative Evolution of Genetic Systems, in *Evolution After Darwin,* Vol. 1, *The Evolution of Life,* pp. 197–226, S. Tax, Ed. University of Chicago Press, Chicago, Illinois (1960).
22. Wiese, L., Algae, in *Fertilization,* Vol. 2, pp. 135–188, C. B. Metz and A. Monroy, Eds. Academic Press, New York (1969).

Prasinophyceae

23. Chihara, M., The Life History of *Prasinocladus ascus* as Found in Japan, with Special Reference to the Systematic Position of the Genus, *Phycologia, 3,* 19 (1963).
24. Craigie, J. S., McLachlan, J., Ackman, R. G., and Tocher, C. S., Photosynthesis in algae. III. Distribution of Soluble Carbohydrates and Dimethyl-β-propriothetin in Marine Unicellular Chlorophyceae and Prasinophyceae, *Can. J. Bot., 45,* 1327 (1967).
25. Manton, I., Rayns, D. G., Ettl, H., and Parke, M., Further Observations on Green Flagellates with Scaly Flagella: the Genus *Heteromastix Korshikov, J. Mar. Biol. Assoc. U.K., 45,* 241 (1965).
26. Parke, M., and den Hartog-Adams, I. Three Species of *Halosphaera., J. Mar. Biol. Assoc. U.K., 45,* 537 (1965).
27. Parke, M., and Manton, I., Observations on the Fine Structure of Two Species of *Platymonas* with Special Reference to Flagellar Scales and the Mode of Origin of the Theca, *J. Mar. Biol. Assoc. U.K., 45,* 743 (1965).
28. Provasoli, L., Yamasu, T., and Manton, I., Experiments on the Resynthesis of Symbiosis in *Convoluta roscoffensis* with Different Flagellate Cultures, *J. Mar. Biol. Assoc. U.K., 48,* 465 (1968).
29. Ricketts, T. R., The Pigments of the Prasinophyceae and Related Organisms, *Phytochemistry, 9,* 1835 (1970).

Volvocales – Tetrasporales

30. Brown, R. M., Jr., Johnson, C., Sr., and Bold, H. C., Electron and Phase-contrast Microscopy of Sexual Reproduction in *Chlamydomonas moewusii, J. Phycol., 4,* 100 (1968).
31. Cavalier-Smith, T., Electron Microscopic Evidence for Chloroplast Fusion in Zygotes of *Chlamydomonas reinhardi, Nature, 228,* 333 (1970).
32. Darden, W. H., Sexual Differentiation in *Volvox aureus, J. Protozool., 13,* 239 (1966).
33. Goldstein, M. E., Speciation and Mating Behavior in *Eudorina, J. Protozool., 11,* 317 (1964).
34. Johnson, U. G., and Porter, K. R., Find Structure of Cell Division in Chlamydomonas *reinhardi:* Basal Bodies and Microtubules, J. Cell. Biol., 38, 403 (1968).
35. Lang, N. J., Electron Microscopy of the Volvocaceae and Astrephomenaceae, *Am. J. Bot., 50,* 280 (1963a).
36. Lang, N. J., Electron-microscopic Demonstration of Plastids in *Polytoma, J. Protozool., 10,* 333 (1963b).
37. Lembi, C. A., and Walne, P. L., Interconnections between Cytoplasmic Microtubules and Basal Bodies of Tetrasporalean Pseudocilia, *J. Phycol., 5,* 202 (1969).
38. McCracken, M. D., and Starr, R. C., Induction and Development of Reproductive Cells in the K-32 Strains of *Volvox rousseletii, Arch. Protistenk., 112,* 262 (1970).
39. Pocock, M. A., *Haematococcus* in Southern Africa, *Trans. R. Soc. S. Afr., 36,* 5 (1959).
40. Ringo, D. L., Flagellar Motion and Fine Structure of the Flagellar Apparatus in *Chlamydomonas, J. Cell. Biol., 33,* 543 (1967).

41. Ris, H., and Plaut, W., Ultrastructure of DNA-containing Areas in the Chloroplast of *Chlamydomonas, J. Cell. Biol., 13,* 383 (1962).
42. Sager, R., and Ramanis, Z., A Genetic Map of Non-Mendelian Genes in *Chlamydomonas, Proc. Natl. Acad. Sci. U.S.A., 65,* 593 (1970).
43. Stein, J. R., Growth and Mating of *Gonium pectorale* (Volvocales) in Defined Media, *J. Phycol., 2,* 23 (1966).

Chlorococcales-Chlorosarcinales

44. Arnott, H. J., and Brown, R. M., Jr., Ultrastructure of the Eyespot and its Possible Significance in Phototaxis of *Tetracystis excentrica, J. Protozool., 14,* 529 (1967).
45. Brown, R. M., Jr., and Arnott, H. J., Structure and Function of the Algal Pyrenoid. I. Ultrastructure and Cytochemistry during Zoosporogenesis of *Tetracystis excentrica, J. Phycol., 6,* 14 (1970).
46. Brown, R. M., Jr., and Lester, R. N., Comparative Immunology of the Algal Genera *Tetracystis* and *Chlorococcum, J. Phycol., 2,* 60 (1965).
47. Fott, B., and Nováková, M., A Monograph of the Genus *Chlorella.* The Fresh Water Species, in *Studies in Phycology,* pp. 10 – 74, B. Fott, Ed. E. Schweizerbart'sche Verlags, Stuttgart, Germany (1969).
48. Groover, R. D., and Bold, H. C., *Phycological Studies.* VIII. The Taxonomy and Comparative Physiology of the Chlorosarcinales and Certain Other Edaphic Algae. University of Texas Publ. No. 6907 (1969).
49. Herndon, W., Studies on Chlorosphaeracean Algae from Soil, *Am. J. Bot., 45,* 298 (1958).
50. Kies, L., Oogamie bei *Eremosphaera viridis* De Bary, *Flora, 157 B,* 1 (1967).
51. Krauss, R. W., Mass Culture of Algae for Food and other Organic Compounds, *Am. J. Bot., 49,* 425 (1962).
52. Pocock, M. A., *Hydrodictyon:* A Comparative Biological Study, *J. S. Afr. Bot., 26,* 167 (1960).
53. Shihira, I., and Krauss, R. W., *Chlorella, Physiology and Taxonomy of Forty-one Isolates.* University of Maryland, College Park, Maryland (1965).
54. Starr, R. C., *A Comparative Study of Chlorococcum meneghini and Other Spherical Zoospore-producing Genera of the Chlorococcales.* Indiana University Press, Bloomington, Indiana (1955).
55. Thomas, D. L., and Brown, R. M., Jr., New Taxonomic Criteria in the Classification of *Chlorococcum* Species. III. Isozyme Analysis, *J. Phycol., 6,* 293 (1970).
56. Trainor, F. R., and Burg, C. A., Motility in *Scenedesmus dimorphus, Scenedesmus obliquus,* and *Coelastrum microporum, J. Phycol., 1,* 14 (1965).

Ulotrichales-Chaetophorales-Ulvales-Schizogoniales

57. Bliding, C., A Critical Survey of European Taxa in Ulvales. I. *Capsosiphon, Percursaria, Blidingia, Enteromorpha, Opera Bot., 8,* 1 (1963).
58. Bliding, C., A Critical Survey of European Taxa in Ulvales. II. *Ulva, Ulvaria, Monostroma, Kornmannia, Bot. Not., 121,* 535 (1968).
59. Blinn, D. W., The Influence of Sodium on the Development of *Ctenocladus circinnatus* Borzi (Chlorophyceae), *Phycologia, 9,* 49 (1970).
60. Bravo, L., Studies on the Life History of *Prasiola meridionalis, Phycologia, 4,* 177 (1965).
61. Cox, E. R., and Bold, H. C., *Phycological Studies.* VII. Taxonomic Investigations of *Stigeoclonium.* University of Texas Publ. No. 6618 (1966).
62. Dube, M. A.. On the Life History of *Monostroma fuscum* (Postels et Ruprecht) Wittrock, *J. Phycol., 3,* 64 (1967).
63. Friedmann, I., Structure, Life-history, and Sex Determination of *Prasiola stipitata* Suhr, *Ann. Bot. N. S., 23,* 571 (1959).
64. Geitler, L., Entwicklungsgeschichtliche Untersuchungen an *Coleochaete*-Arten, *Österr. Bot. Zeit., 109,* 495 (1962).

65. Kornmann, P., Die *Ulothrix*-Arten von Helgoland I. *Helgolaender Wiss. Meeresunters, 11,* 27 (1964).
66. Mattox, K. R., and Bold, H. C., *Phycological Studies.* III. The Taxonomy of Certain Ulotrichacean Algae. University of Texas Publ. No. 6222 (1962).
67. McBride, G. E., Cytokinesis and Ultrastructure in *Fritschiella tuberosa* Iyengar, *Arch. Protistenk., 112,* 365 (1970)
68. Nichols, H. W., and Bold, H. C., *Trichosarcina polymorpha* gen. et sp. nov., *J. Phycol., 1,* 33 (1965).
69. Thompson, R. H., Sexual Reproduction in *Chaetosphaeridium globosum* (Nordst.) Klebahn (Chlorophyceae) and Description of a Species New to Science, *J. Phycol., 5,* 285 (1969).
69a. Ramanathan, K. R., *Ulotrichales.* Indian Council of Agricultural Research, New Delhi, India (1964).

Siphonous Orders of Chlorophyta

70. Brachet, J., and Bonotto, S., Eds., *Biology of Acetabularia.* Academic Press, New York (1970).
71. Burr, F. A., and West, J. A., Light and Electron Microscopic Observations on the Vegetative and Reproductive Structures of *Bryopsis hypnoides, Phycologia, 9,* 17 (1970).
72. Egerod, L. E., An Analysis of the Siphonous Chlorophycophyta, with Special Reference to the Siphonocladales, Siphonales, and Dasycladales of Hawaii. University Calif. Publ. Bot., 25, 325 (1952).
73. Gibor, A., *Acetabularia:* A Useful Giant Cell, *Sci. Am., 215,* 118 (1966).
74. Hoek, C. van den, *Revision of the European Species of Cladophora.* E. J. Brill, Leiden, The Netherlands (1963).
75. Hustede, H., Entwicklungsphysiologische Untersuchungen über den Generationswechsel zwischen *Derbesia neglecta* Berth. und *Bryopsis halymeniae, Bot. Mar., 6,* 134 (1964).
76. Jónsson, S., Recherches sur les Cladophoracées Marines (Structure Reproduction, Cycles Comparés, Conséquences Systématiques), *Ann. Sci. Nat. Bot.* (Ser. 12), *3,* 25 (1962).
77. Kleinig, H., Carotenoids of Siphonous Green Algae: A Chemotaxonomic Study, *J. Phycol., 5,* 281 (1969).
78. Kornmann, P., Der Lebenzyklus von *Acrosiphonia arcta, Helgolaender Wiss. Meeresunters., 11,* 110 (1964).
79. Preston, R. D., Plants without Cellulose, *Sci. Am., 218,* 102 (1968).
80. Puiséux-Dao, S., *Acetabularia and Cell Biology,* Springer-Verlag, New York (1970).
81. Rietma, H., A New Type of Life History in *Bryopsis* (Chlorophyceae, Caulerpales), *Acta Bot. Neerl., 18,* 615 (1970).
82. Sears, J. R., and Wilce, R. T., Reproduction and Systematics of the Marine Alga *Derbesia* (Chlorophyceae) in New England, *J. Phycol., 6,* 381 (1970).
83. Taylor, D. L., Chloroplasts as Symbiotic Organelles in the Digestive Gland of *Elysia viridis* (Gastropoda: Opiisthobranchia), *J. Mar. Biol. Assoc. U.K., 48,* 1 (1968).
84. Valet, G., Contribution à l'étude des Dasycladales. 1. Morphogenèse. 2. Cytologie et reproduction. 3. Révision systématique, *Nova Hedwigia, 16,* 21 (1968); *17,* 551 (1969).
85. Ziegler, J. R., and Kingsbury, J. M., Cultural Studies on the Marine Green Alga *Halicystis parvula-Derbesia tenuissima.* I. Normal and Abnormal Sexual and Asexual Reproduction, *Phycologia, 4,* 105 (1964).

Oedogoniales

86. Cook, P. W., Growth and Reproduction of *Bulbochaete hiloensis* in Unialgal Culture, *Trans. Am. Microsc. Soc., 81,* 384 (1962).
87. Hoffman, L. R., Cytological Studies of *Oedogonium.* I. Oospore Germination in *O. foveolatum, Am. J. Bot., 52,* 173 (1965).
88. Hoffman, L. R., and Manton, I., Observations on the Fine Structure of *Oedogonium.* II. The Spermatozoid of *O. cardiacum, Am. J. Bot., 50,* 455 (1963).
89. Pickett-Heaps, J. D., and Fowke, L. C., Cell Division in *Oedogonium.* I. Mitosis, Cytokinesis and Cell Elongation, *Aust. J. Biol. Sci., 22,* 857 (1969).

90. Rawitscher-Kunkel, E., and Machlis, L., The Hormonal Integration of Sexual Reproduction in *Oedogonium, Am. J. Bot., 49,* 177 (1962).

Zygnematales

91. Biebel, P., The Sexual Cycle of *Netrium digitus, Am. J. Bot., 51,* 697 (1964).
92. Cook, P. W., Host Range Studies of Certain Phycomycetes Parasitic on Desmids, *Am. J. Bot.,50,* 580 (1963).
93. Fowke, L. C., and Pickett-Heaps, J. D., Cell Division in *Spirogyra.* I. Mitosis. II. Cytokinesis, *J. Phycol., 5,* 240; *5,* 273 (1969).
94. Hoshaw, R. W., Biology of the Filamentous Conjugating Algae, in *Algae, Man and the Environment,* pp. 135–184, D. F. Jackson, Ed., Syracuse University Press, Syracuse, New York (1968).
95. Kies, L., Über die Zygotenbildung bei *Micrasterias papillifera* Bréb, *Flora, 157B,* 301 (1968).
96. Lippert, B. E., Sexual Reproduction in *Closterium moniliferum* and *Closterium ehrenbergii, J. Phycol., 3,* 182 (1967).
97. Pickett-Heaps, J. D., and Fowke, L. C., Mitosis, Cytokinesis, and Cell Elongation in the Desmid, *Closterium littorale, J. Phycol., 6,* 189 (1970).
98. Starr, R. C., Zygospore Germination in *Cosmarium botrytis* var. *subtumidum, Am. J. Bot., 42,* 577 (1955).
99. Taylor, A. O., and Bonner, B. A., Isolation of Phytochrome from the Alga *Mesotaenium* and Liverwort *Sphaerocarpos, Plant Physiol., 42,* 762 (1967).
100. Teiling, E., Evolutionary Studies on the Shape of the Cell and of the Chloroplast in Desmids, *Bot. Not., 105,* 264 (1952).

B. BIBLIOGRAPHY

From Wood, R. D., and Harrington, D., *Selected Papers in Phycology,* pp. 812–815, J. R. Rosowski and B. C. Parker, Eds. Department of Botany, University of Nebraska, Omaha (1971), for the Charophyceae (stoneworts).

1. Anderson, R. G., and Lommasson, R. C., Some Effects of Temperature on the Growth of *Chara zeylanica, Butler Univ. Bot. Stud., 13,* 113 (1958).
2. Barton, R., Autoradiographic Studies on Wall Formation in *Chara, Planta, 82,* 302 (1968).
3. Blinks, L. R., The Relations of Bioelectrical Phenomena to Ionic Permeability and to Metabolism in Large Plant Cells, *Cold Spring Harbor Symp. Quant. Biol., 8,* 224 (1940).
4. Bold, H. C., *Morphology of Plants,* 2nd ed. Harper and Brothers, Publishers, New York (1967).
5. Carr, D. J., and Ross, M. M., Studies on the Morphology and Germination of *Chara gymnopitys* A. Braun. II. Factors in Germination, *Port. Acta Biol. Ser. A. – Morfol. Fisiol. Genet. Biol. Geral., A, 8(½),* 41 (1963).
6. Collander, R., The Permeability of *Nitella* Cells to Non-electrolytes, *Physiol. Plant., 7,* 420 (1954).
7. Conkin, J. E., Conkin, B. M., Sawa, T., and Kern, J. M., Middle Devonian *Moellerina greenei* Zone and Suppression of the Genus Weikkoella Summerson, 1958, *Micropaleontology, 16,* 399 (1970).
8. Corillion, R., Les Charophycées de France et d'Europe Occidentale, *Trav. Lab. Bot. Fac. Sci. Angers, 11* and *12,* 1 (1957).
9. Daily, F. K., Some Observations on the Occurrence and Distribution of the Characeae of Indiana, *Indiana Acad. Sci., 68,* 95 (1958).
10. Daily, F. K., Oospore Variation in Culture as Applied to the Taxonomy of *Chara, Torrey Bot. Club Bull., 91,* 281 (1964).
11. Dainty, J., and Hope, A. B., The Electric Double Layer and the Donnan Equilibrium in Relation to Plant Cell Walls, *Aust. J. Biol. Sci., 14,* 541 (1960).
12. Dambska, I., Charophyta – Ramienice, *Flora Slodkowodna Polski (Warsaw), 13,* 1 (1964).

13. Findlay, G. P., and Hope, A. B., Ionic Relations of Cells of *Chara australis.* IX. Analysis of Transient Membrane Currents, *Aust. J. Biol. Sci., 17,* 400 (1964).
14. Forsberg, C., Phosphorus, a Maximum Factor in the Growth of Characeae, *Nature, 201,* 517 (1964).
15. Forsberg, C., Nutritional Studies of *Chara* in Axenic Culture, *Physiol. Plant., 18,* 275 (1965).
16. Fridvalszky, L., Nagy, T., and Lovas, B., Feny-es elektronmikroszkopos vizsgalatok a csillarkamoszatok sejtfalan, *Bot. Kozlemenyek, 51,* 211 (1964).
17. Fritsch, F. E., *The Structure and Reproduction of the Algae,* Vol. I. University Press, Cambridge (1935).
18. Goebel, K., Die Deutung der Characean-Antheridien. Ein Versuch., *Flora, 124,* 491 (1930).
19. Grambast, L., Tendances Evolutives dans le Phylum des Charophytes, *C. R. Acad. Sci., 249,* 557 (1959).
20. Green, P. B., Structural Characteristics of Developing *Nitella* Internodal Cell Walls, *J. Biophys. Biochem. Cytol., 4,* 505 (1963).
21. Green, P. B., Pathways of Cellular Morphogenesis. A Diversity in *Nitella, J. Cell. Biol., 27,* 343 (1965).
22. Green P. B., Growth Physics in *Nitella:* A Method for Continuous *in vivo* Analysis of Extensibility Based on a Micromanometer Technique for Turgor Pressure, *Plant Physiol., 43,* 1169, (1968).
23. Green, P. B., and King, A., A Mechanism for the Origin of Specifically Oriented Textures in Development with Special Reference to *Nitella* Wall Texture, *Aust. J. Biol. Sci., 19,* 421 (1966).
24. Griffin, D. G., and Proctor, V. W., Population Study of *Chara zeylanica* in Texas, Oklahoma and New Mexico, *Am. J. Bot., 51,* 120 (1964).
25. Groves, J., and Bullock-Webster, G. R., *The British Charophyta,* Vol. I and II. Ray Society, London, England (1920–1924).
26. Guerlesquin, M., Sur quelques cas de variabilité naturelle du stock chromosomique haploïde dans les spermatocytes des genres *Tolypella, Nitellopsis,* et *Chara* (Charophycées), *Bull. Soc. Sci. Anjou. n.s., 1965,* 53 (1965).
27. Hampson, M. A., Uptake of Radioactivity by Aquatic Plants and Location in the Cells; Uptake of Cerium-144 by the Freshwater Plant *Nitella opaca, J. Exp. Bot., 18,* 34 (1967).
28. Harrington, D., Anderson, R. G., Bird, G. W., and Mai, W. F., Occurrence of *Monhystrella plectoides* (Nematoda: Monhysteridae) within the White Oogonia of *Chara zeylanica, Can. J. Bot., 45,* 973 (1967).
29. Haughton, P. M., Sellen, D. B., and Preston, R. D., Dynamic Mechanical Properties of the Cell Wall of *Nitella opaca, J. Exp. Bot., 19,* 1 (1968).
30. Hoagland, D. R., and Broyer, T. F., Accumulation of Salt and Permeability in Plant Cells, *J. Gen. Physiol., 25,* 865 (1942).
31. Hofmeister, W., Ueber die Stellung der Moose im System, *Flora, 35,* 1 (1852).
32. Horn af Rantzien, H., Recent Charophyte Fructifications and Their Relations to Fossil Charophyte Gyrogonites, *Ark. Bot., Ser. 2, 4,* 165 (1959).
33. Hotchkiss, A. T., Chromosome Numbers in Characeae from the South Pacific, *Pac. Sci., 19,* 31 (1965).
34. Imahori, K., On the Significance of the Masamune's Line, from the Viewpoint of Characeous Distribution, *J. Geobot., 8,* 63 (1960).
35. Imahori, K., and Iwasa, K., Pure Culture and the Chemical Regulation of the Growth of Charophytes, *Phycologia, 4,* 127 (1965).
36. Karczmarz, K., *Charophyceae Poloniae Exiscati Reg. Lublinensis (Polonia orientalis),* Fasc. 1: No. 1–10. Lublini, Poland (1965).
37. Karling, J. S., A Preliminary Account of the Influence of Light and Temperature on Growth and Reproduction in *Chara fragilis, Torrey Bot. Club Bull., 51,* 469 (1924).
38. Kasaki, H., On the Distribution of *Nitellopsis* (Charophyta), *Acta Phytotax. Geobot., 1962,* 258 (1962).
39. Kishimoto, U., Hyperpolarizing Response in *Nitella* Internodes, *Plant Cell Physiol., 7,* 429 (1966).
40. Krause, W. von, Zur Characeenvegetation der Oberrheinebene. Characeae in the Upper Plain of the Rhine, *Arch. Hydrobiol. Suppl., 35,* 252 (1969).

41. Liebert, V. E., and Jahnke, E., Untersuchungen über die Apikaldominanz bie *Chara*-Arten sowie ihre Beeinfluszbarkeit durch Auxin (Indolessigsaure) und Antiauxin (p-Chlorphenoxyisobuttersaure), *Biol. Zentralbl.*, *84*, 25 (1965).
42. Littlefield, L., and Forsberg, C., Absorption and Translocation of Phosphorus-32 by *Chara globularis* Thuill, *Physiol. Plant.*, *18*, 291 (1965).
43. Lokke, D. H., and van Sant, J. F., Upper Pennsylvanian Charophyta from Kansas, *J. Paleontol.*, *40*, 971 (1966).
44. MacRobbie, E. A. C., The Nature of the Coupling between Light Energy and Active Transport in *Nitella translucens*, *Biochim. Biophys. Acta*, *94*, 64 (1965).
45. Maeda, M., and Imahori, K., Light Effects on the Morphogenesis of Charophytes, *Sci. Rep. Coll. Gen. Ed., Osaka Univ.*, *16*, 37 (1967).
46. McCracken, M. D., Proctor, V. W., and Hotchkiss, A. T., Attempted Hybridization between Monoecious and Dioecious Clones of *Chara*, *Am. J. Bot.*, *53*, 937 (1966).
47. Moestrup, O., The Fine Structure of Mature Spermatozoids of *Chara corallina*, with Special Reference to Microtubules and Scales, *Planta*, *93*, 295 (1970).
48. Ophel, I. L., Some Ecological Effects of Substances Produced by the Characeae, *Proc. Okla. Acad. Sci.*, *29*, 15 (1948).
49. Osterhout, W. J. V., Physiological Studies of Single Plant Cells, *Biol. Rev.*, *6*, 369 (1931).
50. Osterhout, W. J. V., Movements of Water in Cells of *Nitella*, *J. Gen. Physiol.*, *32*, 553 (1949).
51. Peck, R. E., Fossil Charophytes, *Bot. Rev.*, *19*, 209 (1953).
52. Peck, R. E., and Eyer, J. A., Pennsylvanian, Permian, and Triassic Charophyta of North America, *J. Paleontol.*, *37*, 835 (1963).
53. Peck, R. E., and Morales, G. A., The Devonian and Lower Mississippian Charophytes of North America, *Micropaleontology*, *12*, 303 (1966).
54. Pickard, W. F., Correlation between Electrical Behavior and Cytoplasmic Streaming in *Chara braunii*, *Can. J. Bot.*, *47*, 1233 (1969).
55. Pickett-Heaps, J. D., Ultrastructure and Differentiation in *Chara* sp. I. Vegetative Cells, *Aust. J. Biol. Sci.*, *20*, 539 (1967).
56. Pickett-Heaps, J. D., Ultrastructure and Differentiation in *Chara* sp. II. Mitosis, *Aust. J. Biol. Sci.*, *20*, 883 (1967).
57. Pickett-Heaps, J. D., Ultrastructure and Differentiation in *Chara* sp. III. Formation of the Antheridium, *Aust. J. Biol. Sci.*, *21*, 255 (1968).
58. Pickett-Heaps, J. D., Ultrastructure and Differentiation in *Chara (fibrosa)*. IV. Spermatogenesis, *Aust. J. Biol. Sci.*, *21*, 655 (1968).
59. Printz, H., Chlorophyceae, in *Die Natürlichen Pflanzenfamilien*, 2nd ed., Vol. 3, A. Engler and K. A. E. Prantl, Eds. (1927).
60. Probine, M. C., and Preston, R. D., Cell Growth and Structure and Mechanical Properties of the Wall in Internodal Cells of *Nitella opaca*. II. Mechanical Properties of the Walls, *J. Exp. Bot.*, *13*, 111 (1962).
61. Proctor, V. W., Viability of *Chara* Oospores Taken from Migratory Water Birds, *Ecology*, *43*, 528 (1962).
62. Proctor, V. W., Storage and Germination of *Chara* Oospores, *J. Phycol.*, *3*, 208 (1967).
63. Proctor, V. W., Taxonomy of *Chara braunii:* An Experimental Approach, *J. Phycol.*, *6*, 317 (1970).
64. Remusat, R. L., Contribuicão ao estudio das Characeae, para o combate a Esquistossomose: communicacão especial, com novas observacoes e estudos sistematicos e morfologicos das Characeae, *An. Soc. Bot. Brasil*, *11*, 21 (1962).
65. Sarma, Y. S. R. K., and Kahn, M., Chromosome Numbers in some Indian Species of *Nitella*, *Chromosoma (Berlin)*, *15*, 246 (1964).
66. Sarma, Y. S. R. K., and Kahn, M., Chromosome Numbers in some Indian Species of *Chara*, *Phycologia*, *4*, 173 (1965).
67. Sawa, T., Cytotaxonomy of the Characeae: Karyotype Analysis of *Nitella opaca* and *Nitella flexilis*, *Am. J. Bot.*, *52*, 962 (1965).

68. Shen, E. Y. F., Microspectrophotometric Analysis of DNA in *Chara zeylanica, J. Cell Biol., 35,* 377 (1967).
69. Shimizu, A., Formation of Surface Membrane of the Endoplasmic Drop Isolated *in vitro* from the *Nitella* (Charophyta) Internode, *Sci. Rep. Coll. Gen. Ed., Osaka Univ., 14,* 21 (1965).
70. Smith, F. A., Rates of photosynthesis in Characean Cells. II. Photosynthesis $^{14}CO_2$ Fixation and ^{14}C-Bicarbonate Uptake by Characean Cells, *J. Exp. Bot., 19,* 207 (1968).
71. Smith, G. M., *The Fresh-water Algae of the United States,* 2nd ed. McGraw-Hill, New York (1950).
72. Spanswick, R. M., Measurements of Potassium Ion Activity in the Cytoplasm of the Characeae as a Test for the Sorption Theory, *Nature, 218,* 357 (1968).
73. Spear, D. G., Barr, J. K., and Barr, C. E., Localization of Hydrogen Ion and Chloride Ion Fluxes in *Nitella, J. Gen. Physiol., 54,* 397 (1969).
74. Sundaralingham, V. S., The Cytology and Spermatogenesis in *Chara zeylanica* Willd, *J. Indian Bot. Soc., Iyengar Commem. Vol., Silver Jubilee Session Allahabad, 1946,* 289 (1946).
75. Turner, F. R., An Ultrastructural Study of Plant Spermatogenesis. Spermatogenesis in *Nitella, J. Cell. Biol., 37,* 370 (1968).
76. Umrath, K., Versuche an *Nitella* zur Frage einer electroosmotichen Komponete der Saugkraft, *Protoplasma, 61,* 229 (1966).
77. Vaidya, B. S., Studies of some Environmental Factors Affecting the Occurrence of Charophytes in Western India, *Hydrobiologia, 29,* 256 (1967).
78. Vouk, V., and Benzinger, F., Some Preliminary Experiments on Physiology of Charophyta, *Act. Bot. Inst. Univ. Zagreb, 4,* 64 (1929).
79. Wetzel, R. G., and McGregor, D. L., Axenic Culture and Nutritional Studies of Aquatic Macrophytes, *Am. Midland Nature, 80,* 52 (1968).
80. Wood, R. D., Stability and Zonation of Characeae, *Ecology, 31,* 642 (1950).
81. Wood, R. D., Preliminary Report on Characeae of Australia and the South Pacific, *Torrey Bot. Club Bull., 89,* 150 (1962).
82. Wood, R. D., Charophytes of North America. Available from Bookstore, University of Rhode Island, Kingston, Rhode Island (1967).
83. Wood, R. D., and Imahori, K., *A Revision of the Characeae,* Vol. I and II. J. Cramer, Weinheim, Germany (1964–1965).

3. Chrysophyta (Yellow-Green or Yellow-Brown Algae)

Plants are unicellular or colonial, and are rarely filamentous. Pigments are contained in chromatophores in which yellow or brown often predominates; chlorophyll, carotene, and xanthophyll are also present (some chromatophores appearing pale green or yellow-green). Food storage is in the form of oil or leucosin, the latter often giving the cell a metallic luster. The starch test with iodine is negative; the wall is relatively thick and definite, pectic in composition, often impregnated with silicon (especially in the diatoms), and sometimes built in two sections which overlap in the midregion. Motile cells and swimming reproductive cells are furnished with two flagella of unequal length, or with a single flagellum; rhizopodial (pseudopodial or amoeboid) extensions of the cell are not uncommon in some families.

A. BIBLIOGRAPHY

Chrysophyta, excluding *Bacillariophyceae* (diatoms) from Norris, R. E., and Blankley, W. F., *Selected Papers in Phycology,* pp. 780–783, J. R. Rosowski and B. C. Parker, Eds. Department of Botany, University of Nebraska, Omaha (1971).

1. Aaronson, S., and Baker, H., A Comparative Study of Two Species of *Ochromonas, J. Protozool., 6,* 282 (1959).
2. Aaronson, S., and Bensky, B., Effect of Aging of a Cell Population on Lipids and Drug Resistance in *Ochromonas danica, J. Protozool., 14,* 76 (1967).
3. Allen, M. B., Nannoplankton and the Carbon Cycle in Tropical Waters, *Stud. Trop. Oceanogr. Inst. Mar. Sci. Univ. Miami, 5,* 273 (1967).
4. Allen, M. B., Structure, Physiology, and Biochemistry of the Chrysophyceae, *A. Rev. Microbiol., 23,* 29 (1969).
5. Antia, N. J., Kalmakoff, J., Growth Rates and Cell Yields from Axenic Mass Culture of Fourteen Species of Marine Phytoplankters, *Fish. Res. Board Can., Ms. Rep. Ser., 203,* (1965).
6. Belcher, J. H., A Morphological Study of the Phytoflagellate *Chrysococcus rufescens* Klebs in Culture, *Br. Phycol. J., 4,* 105 (1969).
7. Belcher, J. H., and Swale, E. M. F., *Chromulina placentula* sp. nov. (Chrysophyceae), a Freshwater Nannoplankton Flagellate, *Br. Phycol. Bull., 3,* 257 (1967).
8. Bernard, F., and Lecal, J., Plancton unicellular recolte dans l'ocean Indien par le Charcot (1950) et le Norsel (1955–56), *Bull. Inst. Oceanogr. (Monaco), 57,* 1166 (1960).
9. Birkenes, E., and Braarud, T., Phytoplankton in the Oslo Fjord during a "*Coccolithus huxleyi*-summer," *Avh. Norske Videnskaps – Akad. Oslo Mat. – Naturvidensk K. L., 2,* 1 (1952)
10. Black, M., The Fine Structure of the Mineral Parts of the Coccolithophoridae, *Proc. Linnean Soc. Lond., 174,* 41 (1963).
11. Blankley, W. F., Heterotrophic Growth and Calcification in Coccolithophorids, *Abstr. XI Int. Bot. Congr., 16* (1969).
12. Boney, A. D., and Burrows, A., Experimental Studies on the Benthic Phases of Haptophyceae. I. Effects of some Experimental Conditions on the Release of Coccolithophorids, *J. Mar. Biol., Assoc. U.K., 46,* 295 (1966).
13. Bourrelly, P., Recherches sur les Chrysophycées: Morphologie, phylogénie, systématique, *Revue Algol., Mém. Hors Sèr., 1,* 1 (1957).
14. Bourrelly, P., Chrysophycées et Phylogénie, *Deut. Bot. Ges. N. F., 1,* 32 (1962).
15. Bourrelly, P., Loricae and Cysts in the Chrysophyceae, *Ann. N.Y. Acad. Sci., 108,* 421 (1963).
16. Bourrelly, P., *Les algues d'eu douce. Initiation à la systèmatique,* Tome II: Les algues jaunes et brunes. Chrysophycées, Phéophycées, Xanthophycées et Diatomées. Boubée et Cie, Paris (1968).
17. Braarud, T., Deflandre, G., Halldal, P., and Kamptner, E., Terminology, Nomenclature, and Systematics of the Coccolithophoridae, *Micropaleontology, 1,* 157 (1955).
18. Bradley, D. E., The Ultrastructure of the Flagella of Three Chrysomonads with Particular Reference to the Mastigonemes, *Exp. Cell Res., 41,* 162 (1966).
19. Brown, R. M., Jr., Franke, W. W., Kleinig, H., Falk, H., and Sitte, P., Scale Formation in Chrysophycean Algae. I. Cellulosic and Noncellulosic Wall Components Made by the Golgi Apparatus, *J. Cell Biol., 45,* 246 (1970).
20. Casselton, P. J., Chemo-organotrophic Growth of Xanthophyceae Algae, *New Phytol., 65,* 134 (1966).
21. Chapman, D. J., and Haxo, F. T., Chloroplast Pigments of Chloromonadophyceae, *J. Phycol., 2,* 89 (1966).
22. Ettl, H., Ein Beitrag zur Systematik der Heterokonten, *Bot. Notiser, 109,* 411 (1956).
23. Fauré-Frémiet, E., and Rouiller, C., Le flagelle interne d'une Chrysomonadale: *Chromulina psammobia, C. R. Acad. Sci. Paris, 244,* 2655 (1957).
24. Fjerdingstad, E. J., Ultrastructure of the Collar of the Choanoflagellate *Codonosiga botrytis* (Ehrenb.), *Z. Zellforsch., 54,* 499 (1961).
25. Fott, B., Taxonomy of *Mallomonas* Based on Electron Micrographs of Scales, *Preslia, 34,* 69 (1962).
26. Fott, B., Hologamic and Agamic Cyst Formation in Loricate Chrysomonads, *Phykos, 3,* 15 (1964).
27. Fournier, R. C., Observations of Particulate Organic Carbon in the Mediterranean Sea and Their Relevance to the Deep-living Coccolithophorid *Cyclococcolithus fragilis, Limnol. Oceanogr., 13,* 693 (1968).

28. Gaarder, K. R., Comments on the Distribution of Coccolithophorids in the Oceans, in *The Micropaleontology of Oceans,* pp. 97–103, B. M. Funnell and W. R. Riedel, Eds. Cambridge University Press, Cambridge, England (1970).
29. Gayral, P., and Haas, C., Étude comparée des genres *Chrysomeris* Carter et *Giraudyopsis* P. Dang. position systématique des Chrysomeridaceae (Chrysophyceae), *Rev. Gén. Bot., 76,* 659 (1969).
30. Geitler, L., and Schiman-Czeika, H., Über das sogenannte Palmella-stadium von *Phaeothamnion confervicola, Österr. Bot. Z., 118,* 293 (1970).
31. Gibbs, S. P., Nuclear Envelope-chloroplast Relationships in Algae, *J. Cell Biol., 14,* 488 (1962).
32. Gold, K., Pfister, R. M., and Liguori, V. R., Axenic Cultivation and Electron Microscopy of Two Species of Choanoflagellida, *J. Protozool., 17,* 210 (1970).
33. Gooday, G. W., Aspects of the Carbohydrate Metabolism of *Prymnesium parvum, Arch. Mikrobiol., 72,* 9 (1970).
34. Green, J. C., and Jennings, D. H., A Physical and Chemical Investigation of the Scales Produced by the Golgi Apparatus within and Found on the Surface of the Cells of *Chrysochromulina chiton* Parke et Manton, *J. Exp. Bot., 18,* 359 (1967).
35. Greenwood, A. D., Observations on the Structure of the Zoospores of *Vaucheria,* II. *J. Exp. Bot., 10,* 55 (1959).
36. Heynig, H., Beiträge zur Taxonomie und Ökologie der Gattung *Chrysococcus* Klebs (Chrysophyceae), *Arch. Protistenk., 110,* 259 (1967).
37. Hibberd, D. J., and Leedale, G. F., Eustigmatophyceae – A New Algal Class with Unique Organization of the Motile Cell, *Nature, 225,* 758 (1970).
38. Hilliard, D. K., Seasonal Variation in Some *Dinobryon* Species (Chrysophyceae) from a Pond and a Lake in Alaska, *Oikos, 19,* 28 (1968).
39. Hovasse, R., and Joyon, L., Contribution à l'étude de la Chrysomonadine *Hydrurus foetidus, Rev. Algol., 5,* 66 (1960).
40. Ignatiades, L., The Relationship of the Seasonality of Silicoflagellates to Certain Environmental Factors, *Bot. Marina, 13,* 44 (1970).
41. Isenberg, H. D., Douglas, S. D., Lavine, L. S., and Weissfellner, H., Laboratory Studies with Coccolithophorid Calcification, *Stud. Trop. Oceanogr. Inst. Mar. Sci. Univ. Miami, 5,* 155 (1967).
42. Jahn, T. L., Landman, M. D., and Fonseca, F. R., The Mechanism of Locomotion of Flagellates. II. Function of the Mastigonemes of *Ochromonas, J. Protozool., 11,* 291 (1964).
43. Kristiansen, J., Lorica Structure in *Chrysolykos* (Chrysophyceae), *Bot. Tidsskr., 64,* 162 (1969).
44. Leedale, G. F., Leadbeater, B. S. C., and Massalski, A., The Intracellular Origin of Flagellar Hairs in the Chrysophyceae and Xanthophyceae, *J. Cell Sci., 6,* 701 (1970).
45. Loeblich, A. R., Jr., and Tappan, H., Annotated Index and Bibliography of the Calcareous Nannoplankton V, *Phycologia, 9,* 157 (1970).
46. Loeblich, A. R., III, Loeblich, L. A., Tappan, H., and Loeblich, A. R., Jr., Annotated Index of Fossil and Recent Silicoflagellates and Ebridians with Descriptions and Illustrations of Validly Proposed Taxa, *Geol. Soc. Am. Mem.,* p. 106 (1968).
47. Lund, J. W. G., Unsolved Problems in the Classification of the Nonmotile Chrysophyceae with References to Those in Parallel Groups, *Preslia, 34,* 140 (1962).
48. Manton, I., Further Observations on the Fine Structure of the Haptonema in *Prymnesium parvum, Arch. Mikrobiol., 49,* 315 (1964).
49. Manton, I., and Harris, K., Observations on the Microanatomy of the Brown Flagellate *Sphaleromantis tetragona* Skuja with Special Reference to the Flagellar Apparatus and Scales, *J. Linnean Soc. (Bot.), 59,* 397 (1966).
50. Manton, I., and Leedale, G. V., Observations on the Microanatomy of *Coccolithus pelagicus* and *Cricosphaera carterae,* with Special Reference to the Origin and Nature of Coccoliths and Scales, *J. Mar. Biol. Assoc. U.K., 49,* 397 (1969).
51. Massalski, A., and Leedale, G. F., Cytology and Ultrastructure of the Xanthophyceae. I. Comparative Morphology of the Zoospores of *Bumilleria sicula* Borzi and *Tribonema vulgare* Pascher, *Br. Phycol. J., 4,* 159 (1969).

52. Mattox, K. R., and Williams, J. P., Plastid Pigments of the Xanthophyceae, *J. Phycol., 1,* 191 (1965).
53. McLachlan, J., Some Considerations of the Growth of Marine Algae in Artificial Media, *Can. J. Microbiol., 10,* 769 (1964).
54. McLaughlin, J. J. A., Euryhaline Chrysomonads: Nutrition and Toxigenesis in *Prymnesium parvum* with Notes on *Isochrysis galbana* and *Monochrysis lutheri, J. Protozool., 5,* 75 (1958).
55. Mignot, J.-P., Structure et ultrastructure de quelques Chloromonadines, *Protistologica, 3,* 5 (1967).
56. Nival, R., Sur le cycle de *Dictyocha fibula* Ehrenberg dans les eaux de surface de la rade de Villefranche-sur-Mer, *Cahiers Biol. Mar., 6,* 67 (1965).
57. Norris, R. E., Neustonic Marine Craspedomonadales (Choanoflagellates) from Washington and California, *J. Protozool., 12,* 589 (1965).
58. Paasche, E., Biology and Physiology of Coccolithophorids, *Annu. Rev. Microbiol., 22,* 71 (1968).
59. Padilla, G. M., Bragg, R. J., and Kennedy, J. R., Jr., Characteristics and Cellular Localization of the Hemolytic Toxin from the Euryhaline Flagellate *Prymnesium parvum,* in *Drugs from the Sea,* pp. 185–201, H. D. Freudenthal,Ed., Marine Technology Society (1968).
60. Parke, M., Some Remarks Concerning the Class Chrysophyceae, *Br. Phycol. Bull., 2,* 47 (1961).
61. Parsons, T. R., Stephens, K., and Strickland, J. K. H., On the Chemical Composition of Eleven Species of Marine Phytoplankters, *J. Fish. Res. Board Can., 18,* 1 (1961).
62. Pascher, A., Heterokonten, in *L. Rabenhorst's Kryptogamen-Flora von Deutschland, Österreich und der Schweiz,* 2nd ed., Vol. 2, R. Kolkwitz, Ed. Akademische Verlagsgesellschaft, Leipzig, Germany (1937–1939).
63. Peterson, J. B., and Hansen, J. B., On the Scales of Some *Synura* Species, *Biol. Medd. Dan. Vid. Selsk., 23,* 1 (1958).
64. Pintner, I. J., and Provasoli, L., Nutritional Characteristics of Some Chrysomonads, in *Symposium on Marine Microbiology,* pp. 114–121, C. H. Oppenheimer, Ed. Charles C Thomas, Springfield, Illinois (1963).
65. Pintner, I. J., and Provasoli, L., Heterotrophy in Subdued Light of 3 *Chrysochromulina* Species, *Bull. Misaki Mar. Biol. Inst., Kyoto Univ., 12,* 25 (1968).
66. Pringsheim, E. G., On the Nutrition of *Ochromonas, Quart. J. Microscop. Sci., 93,* 71 (1952).
67. Pringsheim, E. G., *Farblose Algen. Ein Beitrag zur Evolutionsforschung,* pp. xii and 471. Gustav Fischer Verlag, Stuttgart, Germany (1963).
68. Rahat, M., and Spira, Z., Specificity of Glycerol for Dark Growth of *Prymnesium parvum, J. Protozool., 14,* 45 (1967).
69. Ricketts, T. R., Chlorophyll c in Some Members of the Chrysophyceae, *Phytochemistry, 4,* 725 (1965).
70. Scagel, R. V., Bandoni, R. J., Rouse, G. E., Schofield, W. B., Stein, J. R., and Taylor, T. M. C., *An Evolutionary Survey of the Plant Kingdom,* pp. xii and 658. Wadsworth, Belmont, California (1966).
71. Schiller, J., Coccolithineae, in *L. Rabenhorst's Kryptogamen-Flora von Deutschland, Österreich und der Schweiz,* 2nd ed., Vol. 10, Abt. 2, pp. 89–273. R. Kolkwitz, Ed. Akademische Verlagsgesellschaft, Leipzig, Germany (1930).
72. Schnepf, E., and Deichgräber, G.; Uber die Feinstruktur von *Synura petersenii* unter besonderer Berücksichtigung der Morphogenese ihrer Kieselschuppen, *Protoplasma, 68,* 85 (1969).
73. Schwarz, E., Beiträge zur Entwicklungsgeschichte der Protophyten. IX. Der Formwechsel von *Ochrosphaera neapolitana, Arch. Protistenk., 77,* 434 (1932).
74. Shilo, M., Formation and Mode of Action of Algal Toxins, *Bacteriol. Rev., 31,* 180 (1967).
75. von Stosch, H. A., Chryosphyta, in *Vegetative Fortpflanzung, Parthenogenese und Apogamie bei Algen,* Sekt. E, pp. 637–692, H. Ettl, D. G. Müller, K. Neumann, H. A. von Stosch, and W. Weber, Eds.; in *Handbuch der Pflanzenphysiologie,* Vol. 18, Sexualität, Fortpflanzung, Generationswechsel, pp. 597–776, W. Ruhland, Ed. Springer-Verlag, Berlin, Germany (1967).
76. Thomas, D. M., and Goodwin, T. W., Nature and Distribution of Carotenoids in the Xanthophyta (Heterokontae), *J. Phycol., 1,* 118 (1965).
77. Van Valkenburg, S. D., and Norris, R. E., The Growth and Morphology of the Silicoflagellate *Dictyocha fibula* Ehrenberg in Culture, *J. Phycol., 6,* 48 (1970).

78. Whittle, S. J., and Casselton, P. J., The Chloroplast Pigments of Some Green and Yellow-green Algae, *Br. Phycol. J., 4,* 55 (1969).

B. BIBLIOGRAPHY

Bacillariophyceae (diatoms) from Hostetter, H. P., and Stoermer, E. F., *Selected Papers in Phycology,* pp. 787–790, J. R. Rosowski and B. C. Parker, Eds. Department of Botany, University of Nebraska, Omaha (1971).

A. Reviews and General Treatments

1. Conger, P. S., Significance of Shell Structure in Diatoms, *Annual Report of the Smithsonian Institution,* pp. 325–44 (1936).
2. Gran, H. H., Diatomeen, in *Nordisches Plankton,* Vol. 19, pp. 1–146, K. Brandt and C. Apstein, Eds. (1905).
3. Hendey, N. I., An Introductory Account of the Smaller Algae of British Coastal Waters, Part 5. Bacillariophyceae (Diatoms), *Fishery Investigations,* London, Ser. 4 (1964).
4. Lauterborn, R., *Untersuchungen über Bau, Kernteilung und Bewegung der Diatomeen.* W. Engelmann, Leipzig, Germany (1896).
5. Lewin, J. C., and Guillard, R. R. L., Diatoms, *Annu. Rev. Microbiol., 17,* 373 (1963).
6. Patrick, R., and Reimer, C. W., *The Diatoms of the United States,* Vol. 1, Monograph 13, Academy of Natural Sciences of Philadelphia (1966).
7. Pfitzer, E., Untersuchungen über Bau und Entwicklung der Bacillariaceen (Diatomeen), in *Bot. Abh. Gebiet Morphol. Physiol.,* Vol. 1, Teil 2, R. Hanstein, Ed. (1871).
8. Taylor, F. F., *Notes on Diatoms.* Guardian Press, Bournemouth, England (1929).
9. Wornardt, W. W., Jr., Diatoms, Past, Present, Future, *Proceedings of the 1st International Conference on Planktonic Microfossils,* Vol. 2, pp. 690–714, Geneva, Switzerland. E. J. Brill, Leiden, The Netherlands (1967).

B. Morphology

10. Holmes, R. W., and Reimann, B. E. F., Variation in Valve Morphology during the Life Cycle of the Marine Diatom *Coscinodiscus concinnus, Phycologia, 5,* 233 (1966).
11. Kolbe, R. W., Zur Phylogenie des Raphe-Organs der Diatomeen: *Eunotia (Amphicampa) eruca* Ehr., *Bot. Not., 109,* 91 (1956).
12. Müller, O., Kammern und Poren in der Zellwand der Bacillarien, *Ber. Deutsch. Bot. Ges., 19,* 195 (1901).

C. Physiology, Biochemistry, and Nutrition

13. Coombs, J., and Volcani, B. E., Studies on the Biochemistry and Fine Structure of Silica Shell Formation in Diatoms. Chemical Changes in the Wall of *Navicula pelliculosa* during Its Formation, *Planta, 82,* 280 (1968).
14. Darley, W. M., and Volcani, B. E., Role of Silicon in Diatom Metabolism. A Silicon Requirement for Deoxyribonucleic Acid Synthesis in the Diatom *Cylindrotheca fusiformis* Reimann and Lewin, *Exp. Cell Res., 58,* 334 (1969).
15. Guillard, R. R. L., B_{12} Specificity of Marine Centric Diatoms, *J. Phycol., 4,* 59 (1968).
16. Höfler, K., Uber die Permeabilität der Diatomee *Caloneis obtusa, Protoplasma, 52,* 5 (1960).
17. Kates, M., and Volcani, B. E., Lipid Components of Diatoms, *Biochim. Biophys. Acta, 116,* 264 (1966).

18. Lewin, J. C., Silicon Metabolism in Diatoms. V. Germanium Dioxide, a Specific Inhibitor of Diatom Growth, *Phycologia, 6,* 1 (1966).
19. Lewin, J. C., and Lewin, R. A., Auxotrophy and Heterotrophy in Marine Littoral Diatoms, *Can. J. Microbiol., 6,* 127 (1960).
20. Thomas, W. H., and Dodson, A. N., Effects of Phosphate Concentration on Cell Division Rates and Yield of a Tropical Oceanic Diatom, *Biol. Bull., 134,* 199 (1968).

D. Movement

21. Drum, R. W., and Hopkins, J. T., Diatom Locomotion: An Explanation, *Protoplasma, 62,* 1 (1966).
22. Gordon, R., and Drum, R. W., A Capillarity Mechanism for Diatom Gliding Locomotion, *Proc. Natl. Acad. Sci. U.S.A., 67,* 338 (1970).
23. Palmer, J. D., and Round, F. E., Persistent, Vertical-migration Rhythms in Benthic Microflora. VI. The Tidal and Diurnal Nature of the Rhythm in the Diatom *Hantzschia virgata, Biol. Bull., 132,* 44 (1967).

E. Ultrastructure

24. Drum, R. W., and Pankratz, H. S., Pyrenoids, Raphes, and Other Fine Structure in Diatoms, *Am. J. Bot., 51,* 405 (1964).
25. Helmcke, J. G., and Krieger, W., *Diatomeenschalen im Elektronen-mikroskopischen Bild,* Teil I, pl. 1–102 (1953); Teil II, pl. 103–200 (1954); Teil III, pl. 201–300 (1961); Teil IV, pl. 301–413 (1963); Teil V, pl 414–513 (1964); Teil VI, pl. 514–613 (1966). J. Cramer, Lehre, Germany (1953–1966).
26. Manton, I., Kowallik, K., and von Stosch, H. A., Observations on the Fine Structure and Development of the Spindle at Mitosis and Meiosis in a Marine Centric Diatom (*Lithodesmium undulatum*). I. Preliminary Survey of Mitosis in Spermatogonia, *J. Microscop., 89,* 295 (1969); II. The Early Meiotic Prophases in Male Gametogenesis, *J. Cell Sci., 5,* 271 (1969); III. The Later Stages of Meiosis I, in *J. Cell Sci., 6,* 131 (1970); IV. The Second Meiotic Division and Conclusion, *J. Cell Sci., 7,* 407 (1970).
27. Stoermer, E. F., Pankratz, H. S., and Bowen, C. C., Fine Structure of the Diatom *Amphipleura pellucida.* II. Cytoplasmic Fine Structure and Frustule Formation, *Am. J. Bot., 52,* 1067 (1965).
28. Wornardt, W. W., Jr., Diatom Research and the Scanning Electron Microscope, *Beih. Nova Hedwigia, 31,* 355 (1970).

F. Reproduction

29. Geitler, L., Comparative Studies on the Behavior of Allogamous Pennate Diatoms in Auxospore Formation, *Am. J. Bot., 56,* 718 (1969).
30. Klebahn, H., Beiträge zur Kenntnis der Auxosporenbildung, *Jahrb. Wiss. Bot., 29,* 595 (1896).
31. Liebisch, W., Experimentelle und kritische Untersuchungen über die Pektinmembran der Diatomeen unter besonderer Berücksichtigung der Auxosporenbildung und der Kratikularsustände, *Z. Naturwiss. Abteil. (Bot.), 22,* 1 (1929).
32. Rao, V. N. R., and Desikachary, T. V., MacDonald-Pfitzer Hypothesis and Cell Size in Diatoms, *Beih. Nova Hedwigia, 31,* 485 (1970).
33. von Stosch, H. A., Oogamy in a Centric Diatom, *Nature, 165,* 531 (1950).
34. von Stosch, H. A., Manipulierung der Zellgrösse von Diatomeen im Experiment, *Phycologia, 5,* 21 (1965).

G. Ecology

35. Cholnoky, B. J., *Die Ökologie der Diatomeen in Binnengewässern.* J. Cramer, Lehre (1968).
36. Hostetter, H. P., and Hoshaw, R. W., Environmental Factors Affecting Resistance to Desiccation in the Diatom *Stauroneis anceps, Am. J. Bot., 57,* 512 (1970).
37. Jφrgensen, E. G., Diatom Periodicity and Silicon Assimilation, *Dansk Bot. Ark., 18,* 6 (1957).
38. Lund, J. W. G., Observations on Soil Algae. I. The Ecology, Size and Taxonomy of British Soil Diatoms, *New Phytol., 44,* 196 (1945).
39. Lund, J. W. G., Studies on *Asterionella formosa* Hass. II. Nutrient Depletion and the Spring Maximum, *J. Ecol., 38,* 1 (1950).
40. Maynard, N. G., Aquatic Foams as an Ecological Habitat, *Z. Allg. Mikrobiol., 8,* 119 (1968a).
41. Maynard, N. G., Significance of Air-borne Algae, *Z. Allg. Mikrobiol., 8,* 225 (1968b).
42. Revill, D. L., Stewart, K. W., and Schlichting, H. E., Jr., Passive Dispersal of Viable Algae and Protozoa by Selected Craneflies and Midges, *Ecology, 48,* 1023 (1967).
43. Steemann Nielsen, E., and Jφrgensen, E. G., The Adaptation of Plankton Algae. I. General Part, *Physiol. Plant., 21,* 401 (1968).
44. Stockner, J. G., and Benson, W. W., The Succession of Diatom Assemblages in the Recent Sediments of Lake Washington, *Limnol. Oceanogr., 12,* 513 (1967).

H. Taxonomy

45. Boyer, C. S., *The Diatoms of Philadelphia and Vicinity.* Lippincott, Philadelphia, Pennsylvania (1916).
46. Cleve, P. T., Synopsis of the Naviculoid Diatoms. I., *K. Svenska Vet.-Akad. Handl., Ny Foljd., 26*(2), 1 (1894); II., *27*(3), 1 (1895).
47. Cleve-Euler, A., Die Diatomeen von Schweden und Finnland, I., *K. Svenska Vet.-Akad. Handl., Fjärde Serien., 2*(1), 1 (1951); II., *4*(1), 1 (1953); III., *4*(5), 1 (1953); IV., *5*(4), 1 (1955); V., *3*(3), 1 (1952).
48. Cupp., E. E., Marine Plankton Diatoms of the West Coast of North America, *Bull. Scripps Inst. Oceanogr., 5,* 1 (1943).
49. Ehrenberg, C. G., *Mikrogeologie. Das Erden und felsen schaffende Wirken des unsichtbar kleinen selbstandigen Lebens auf der Erde.* Leopold Voss, Leipzig, Germany (1854–1856).
50. Hustedt, F., Die Kieselalgen Deutschlands, Osterreichs und der Schweiz, in *Kryptogamen-Flora von Deutschland, Österreich und der Schweiz,* Vol. VII. Teil 1: Liefr. 1, pp. 1–272; Liefr. 2, pp. 273–464; Liefr. 3, pp. 465–608; Liefr. 4, pp. 609–784; Liefr. 5, pp. 785–920; Teil 2: Liefr. 1, pp. 1–176; Liefr. 2, pp. 177–320; Liefr. 3, pp. 321–432; Liefr. 4, pp. 433–576; Liefr. 5, pp. 577–736; Liefr. 6, pp. 737–845; Teil 3: Liefr. 1, pp. 1–160; Liefr. 2, pp. 161–348; Liefr. 3, pp. 349–556; Liefr. 4, pp. 557–816, L. Rabenhorst, Ed. Geest und Portig K.-G., Leipzig, Germany (1927–1966).
51. Karsten, G., Untersuchungen über Diatomeen, III. *Flora (Marburg), 83,* 203 (1897).
52. Kützing, F. T., *Die kieselschaligen Bacillarien oder Diatomeen.* Kohne, Nordhausen (1844).
53. Mills, F. W., *An Index to the Genera and Species of the Diatomaceae and Their Synonyms, 1816–1932,* pp. 1–525; pp. 527–1444; pp. 1444–1726. Wheldon and Wesley, London, England (1933–1935).
54. Peragallo, H., and Peragallo, M., *Diatomées marines de France et des districts maritimes voisins.* M. J. Tempère, Grez-sur-Loing, France (1908).
55. Schmidt, A., *Atlas der Diatomaceenkunde,* Taf. 1–472, R. Reisland, Leipzig, Germany (1874–1959).
56. Van Heurck, H. F., *Synopsis des diatomées de Belgique,* Atlas, pl. 1–30 & bis 22 (1880); Atlas, pl. 31–77 & bis 53 (1881); Atlas, pl. 78–103 & bis 82, bis 83, ter 83, bis 95 (1882); Atlas, pl. 104–132 (1883). Anvers, J. Ducaju & Cie. Table Alphabétique des Noms Génériques et Spécifiques et des Synonyms contenus dans l'Atlas. J. F. Dieltjens, Anvers. 120 pp. (1884). Texte & Suppl. pl. A, B, C. Mtin. Brouwers & Co., Anvers. 120 pp. (1885).
57. VanLandingham, S. L., *Catalog of the Fossil and Recent Genera and Species of Diatoms and Their Synonyms,* J. Cramer, Lehre. pp. 1–493 + i-xi; pp. 494–1086; pp. 1087–1756 (1969); (continuing) (1967–1969).

58. Zabelina, M. M., Kiselev, I. A., Proschkina-Lavrenko, A. I., and Sheshukova, V. S., Diatomovyje Vodorosli, *Opredelitel Presnovodnych Vodoroslei S.S.S.R.*, Vypusk 4. Isdanije Sovetskaja Nauka, Moskva, U.S.S.R. (1951).

I. Fossil

59. Grunow, A., Beiträge zur Kenntnis der fossilen Diatomeen Österreich-Ungarns, in *Beiträge zur Paläontologie Österreich-Ungarns und des Orients,* Vol. II., Heft IV, pp. 136–59, E. von Mojsisovics and M. Neumayr, Eds. (1882).
60. Hanna, G. D., Diatom Deposits, *Bull. Natl. Res. Div. Mines, 154,* 281 (1951).
61. Pantocsek, J., *Beiträge zur Kenntnis der fossilen Bacillarien Ungarns,* Teil I, 30 Taf.; Teil II, 30 Taf.; Teil III, 42 Taf. J. Platzko, Nagy-Tapolcsány (1886–1892).

4. Euglenophyta (Euglenoids)

Cells are solitary, swimming by one (usually) or by two (rarely three) flagella; a gullet is present in the anterior end of the cell in many members, as is also a red pigment (eye) spot; chloroplasts are few to many variously shaped green bodies (a few relatives are colorless). A chlorophyll-like pigment predominates, but carotene is also present. The nucleus is large and centrally located. The food reserve is in the form of an insoluble starch-like substance, paramylum, which is negative to the starch test with iodine and fatty substances. The cell membrane is in the form of a pellicle, rigid or plastic, frequently striated. Sexual reproduction is unknown.

BIBLIOGRAPHY

From Leedale, G. F., and Walne, P. L., *Selected Papers in Phycology,* pp. 800–802, J. R. Rosowski and B. C. Parker, Eds. Department of Botany, University of Nebraska, Omaha (1971).

1. Arnott, H. J., and Walne, P. L., Observations on the Fine Structure of the Pellicle Pores of *Euglena granulata, Protoplasma, 64,* 330 (1967).
2. Batra, P. P., and Tollin, G., Phototaxis in *Euglena.* I. Isolation of the Eyespot Granules and Identification of the Eyespot Pigments, *Biochim. Biophys. Acta, 79,* 371 (1964).
3. Blum, J. J., and Padilla, G. M., Studies on Synchronized Cells. The Time Course of DNA, RNA, and Protein Synthesis in *Astasia longa, Exp. Cell. Res., 28,* 512 (1962).
4. Brawerman, G., and Eisenstadt, J. M., Deoxyribonucleic Acid from the Chloroplasts of *Euglena gracilis, Biochim. Biophys. Acta, 91,* 477 (1964a).
5. Brawerman, G., and Eisenstadt, J. M., Template and Ribosomal Ribonucleic Acids Associated with the Chloroplasts and the Cytoplasm of *Euglena gracilis, J. Mol. Biol., 10,* 403 (1964b).
6. Brody, M., Brody, S. S., and Levine, J. H., Fluorescence Changes During Chlorophyll Formation in *Euglena gracilis* (and other organisms) and an Estimate of Lamellar Area as a Function of Age, *J. Protozool., 12,* 465 (1965).
7. Bruce, V. G., and Pittendrigh, C. S., Temperature Independence in a Unicellular "Clock," *Proc. Natl. Acad. Sci. U.S.A., 42,* 676 (1956).
8. Buetow, D. E., Ed., *The Biology of Euglena,* Vol. I. and II. Academic Press, New York (1968).
9. Buetow, D. E., and Buchanan, P. J., Oxidative Phosphorylation in Mitochondria Isolated from *Euglena gracilis, Biochim. Biophys. Acta, 96,* 9 (1965).
10. Christen, H. R., Zur Taxonomie der farblosen Eugleninen, *Nova Hedwigia, 4,* 437 (1963).
11. Cook, J. R., Photo-inhibition of Cell Division and Growth in Euglenoid Flagellates, *J. Cell Physiol., 71,* 177 (1968).

12. Diehn, B., Action Spectra of the Phototactic Responses in *Euglena, Biochim. Biophys. Acta, 177,* 136 (1969).
13. Edelman, N., Cowan, C. A., Epstein, H. T., and Schiff, J. A., Studies of Chloroplast Development in *Euglena.* VIII. Chloroplast-associated DNA, *Proc. Natl. Acad. Sci. U.S.A., 52,* 1214 (1964).
14. Edmunds, L. N., and Funch, R. R., Circadian Rhythm of Cell Division in *Euglena:* Effects of a Random Illumination Regimen, *Science, 165,* 500 (1970).
15. Gibor, A., Effect of Ultraviolet Irradiation on DNA Metabolism of *Euglena gracilis, J. Protozool., 16,* 190 (1970).
16. Gibbs, S. P., The Fine Structure of *Euglena gracilis* with Special Reference to Chloroplasts and Pyrenoids, *J. Ultrastruct. Res., 4,* 127 (1960).
17. Hall, R. P., A Note on Behavior of the Chromosomes in *Euglena, Trans. Am. Microsc. Soc., 56,* 288 (1937).
18. Hollande, A., Etude cytologique et biologique de quelques flagellés libres. Volvocales, Cryptomonadines, Eugléniens, Protomastigines. *Arch. Zool. Exp. Gen., 83,* 1 (1942).
19. Holwill, M. E. J., The Motion of *Euglena viridis:* The Role of Flagella, *J. Exp. Biol., 44,* 579 (1967).
20. Huber-Pestalozzi, G., *Das Phytoplankton des Süsswassers.* 4. *Euglenophyceen.* E. Schweizerbart'sche Verlagsbuchhandlung, Stuttgart, Germany (1955).
21. Krichenbauer, H., Beitrag zur Kenntnis der Morphologie und Entwicklungeschichte der Gattungen *Euglena* und *Phacus, Arch. Protistenk., 90,* 88 (1937).
22. Lackey, J. B., Studies on the Life Histories of Euglenida. IV. A Comparison of the Structure and Division of *Distigma proteus* Ehrb. and *Astasia dangeardi* Lemm., a Study in Phylogeny, *Biol. Bull. (Woods Hole), 67,* 145 (1934).
23. Leedale, G. F., Nuclear Structure and Mitosis in the Euglenineae, *Arch. Mikrobiol., 32,* 32 (1958).
24. Leedale, G. F., The Evidence for a Meiotic Process in the Euglenineae, *Arch. Mikrobiol., 42,* 237 (1962).
25. Leedale, G. F., Euglenida/Euglenophyta, *Annu. Rev. Microbiol., 21,* 31 (1967a).
26. Leedale, G. F., *Euglenoid Flagellates.* Prentice-Hall, Inc., Englewood Cliffs, New Jersey (1967b).
27. Leedale, G. F., Meeuse, B. J. D., and Pringsheim, E. G., Structure and Physiology of *Euglena spirogyra.* I and II. III–VI., *Arch. Mikrobiol., 50,* 68; 133 (1965).
28. Mast, S. O., Motor Response in Unicellular Animals, in *Protozoa in Biological Research,* pp. 271–351, G. N. Calkins and F. M. Summers, Eds. Columbia University Press, New York (1941).
29. Michajow, W., Euglenoidina (Flagellata) – Parasites of Cyclopidae (Copepoda), *Acta Protozool., 5,* 181 (1968).
30. Mignot, J.-P., Quelques particularités de l'ultrastructure d'*Entosiphon sulcatum* (Duj.) Stein, flagellé Euglénien, *C.R. Acad Sci. Paris, 257,* 2530 (1963).
31. Mignot, J.-P., Ultrastructure des Eugléniens. I. Étude de la cuticle chez differentes éspecies, *Protistologica, 1,* 5 (1965a).
32. Mignot, J.-P., Étude ultrastructurale des Eugléniens. II. A, Dictyosomes et dictyocinése chez *Distigma proteus* Ehrbg. B, Mastigonèmes chez *Anisonema costatum* Christen, *Protistologica, 1*(2), 17 (1965b).
33. Mignot, J.-P., Structure et ultrastructure de quelques Euglenomonadines, *Protistologica, 2,* 51 (1966).
34. Parenti, F., Brawerman, G., Preston, J., and Eisenstadt, J. M., Isolation of Nuclei from *Euglena gracilis, Biochim. Biophys. Acta, 195,* 234 (1970).
35. Pochmann, A., Struktur, Wachstum und Teilung der Körperhülle bei den Eugleninen, *Planta, 42,* 478 (1953).
36. Pochmann, A., Untersuchungen über Plattenbau und Spiralbau, über Wachstum und Zerteilung der Paramylon-kórner, *Ost. Bot. Z., 104,* 321 (1956).
37. Pringsheim, E. G., Contributions to Our Knowledge of Saprotrophic Algae and Flagellata. III. *Astasia, Distigma, Menoidium* and *Rhabdomonas, New Phytol., 41,* 171 (1942).
38. Pringsheim, E. G., Taxonomic Problems in the Euglenineae, *Biol. Rev., 23,* 46 (1948).
39. Pringsheim, E. G., Observations on Some Species of *Trachelomonas* Grown in Culture, *New Phytol., 52,* 93; 238 (1953a).
40. Pringsheim, E. G., Salzwasser-Eugleninen, *Arch. Mikrobiol., 18,* 149 (1953b).

41. Pringsheim, E. G., Notiz über *Colacium* (Euglenaceae), *Ost. Bot. Z., 100,* 270 (1953c).
42. Pringsheim, E. G., Contributions toward a Monograph of the Genus *Euglena, Nova Acta Leopold., 18,* 1 (1956).
43. Pringsheim, E. G., *Farblose Algen. Ein Beitrag zur Evolutionsforschung.* Gustav Fischer, Stuttgart, Germany (1963).
44. Pringsheim, E. G., and Hovasse, R., The Loss of Chromatophores in *Euglena gracilis, New Phytol., 47,* 52 (1948).
45. Pringsheim, E. G., and Pringsheim, O., Experimental Elimination of Chromatophores and Eye-spot in *Euglena gracilis, New Phytol., 51,* 65 (1952).
46. Provasoli, L., Hutner, S. H., and Schatz, A., Streptomycin-induced Chlorophyll-less Races of *Euglena, Proc. Soc. Exp. Biol. Med., 69,* 279 (1948).
47. Sagan, L., Ben-Shaul, Y., Epstein, H. T., and Schiff, J. A., Studies of Chloroplast Development in *Euglena.* XI. Radioautographic Localization of Chloroplast DNA, *Plant Physiol., 40,* 1257 (1965).
48. Schwelitz, F. D., Evans, W. R., Mollenhauer, H. H., and Dilley, R. A., The Fine Structure of the Pellicle of *Euglena gracilis* as Revealed by Freeze Etching, *Protoplasma, 69,* 341 (1970).
49. Stern, A. I., Schiff, J. A., and Epstein, H. T., Studies of Chloroplast Development in *Euglena.* VI. Light Intensity as a Controlling Factor in Development, *Plant Physiol., 39,* 226 (1964).
50. Walne, P. L., and Arnott, H. J., The Comparative Ultrastructure and Possible Function of Eyespots: *Euglena granulata* and *Chlamydomonas eugametos, Planta, 77,* 325 (1967).
51. Wolken, J. J., *Euglena,* 2nd ed. Appleton-Century-Crofts, New York (1967).

5. Cryptophyta (Cryptophyceae of some authors)

Cells are solitary or colonial, mostly swimming by means of two, often laterally placed or subapical flagella. Chromatophores are large and brown, or rarely blue, often with pyrenoids. The food reserve is in the form of starch or oil. The membrane is firm but relatively thin. Sexual reproduction is unknown.

BIBLIOGRAPHY

From Cox, E. R., and Zingmark, R. G., *Selected Papers in Phycology,* p. 810, J. R. Rosowski and B. C. Parker, Eds. Department of Botany, University of Nebraska, Omaha (1971).

1. Anderson, E., A Cytological Study of *Chilomonas paramecium* with Particular Reference to the So-called Trichocysts, *J. Protozool., 9,* 380 (1962).
2. Antia, N., and Chorney, V., Nature of the Nitrogen Compounds Supporting Phototropic Growth of the Marine Cryptomonad *Hemiselmis virescens, J. Protozool., 15,* 198 (1968).
3. Antia, N., Cheng, J., and Taylor, F., The Heterotrophic Growth of a Marine Photosynthetic Cryptomonad (*Chroomonas salina*), *Proceedings of the International Seaweed Symposium, 6,* 17 (1969).
4. Butcher, R. W., An Introductory Account of the Smaller Algae of British Coastal Waters. IV. Cryptophyceae, *Fish. Invest. Minist. Agric. Fish. Food B (Brit.) Ser. IV,* Her Majesty's Stationery Office (1967).
5. Chapman, D. J., Three New Carotenoids Isolated from Algae. *Phytochemistry, 5,* 1331 (1966).
6. Dodge, J. D., The Ultrastructure of *Chroomonas mesostigmatica* Butcher (Cryptophyceae), *Arch. Mikrobiol., 69,* 266 (1969).
7. Fritsch, F. E., *The Structure and Reproduction of the Algae,* Volume I. Cambridge University Press, Cambridge, England (1961).
8. Gantt, E., Edwards, M. R., and Provasoli, L., Chloroplast Structure in the Cryptophyceae. Evidence for Phycobiliproteins within Intrathylakoidal Spaces, *J. Cell. Biol., 48,* 280 (1971).
9. Hibberd, D. J., Greenwood, A. D., and Griffiths, H. B., Observations on the Ultrastructure of the Flagella and Periplast in the Cryptophyceae, *Br. Phycol. J., 6,* in press.

10. Lucas, I. A. N., Observations on the Ultrastructure of Representatives of the Genera *Hemiselis* and *Chroomonas* (Cryptophyceae), *Br. Phycol., J., 5,* 29 (1970).
11. Lucas, I. A. N., Observations on the Fine Structure of the Cryptophyceae. I. The Genus *Cryptomonas, J. Phycol., 6,* 30 (1970).
12. Pringsheim, E. G., Zur Kenntnis der Cryptomonaden des Susswassers, *Nova Hedwigia, 16,* 367 (1968).
13. Schuster, F. L., The Gullet and Trichocysts of *Cyathomonas truncata, J. Exp. Cell. Res., 49,* 277 (1968).

6. Pyrrhophyta (Dinoflagellates)

Cells are solitary or (rarely) filamentous, mostly swimming by means of two flagella, one commonly wound about the cell in a transverse furrow, and one extended posteriorly from the point of flagellar attachment in a longitudinal furrow. Cells are dorsiventrally flattened and differentiated, the longitudinal furrow extending along the ventral surface. The cell wall, if present, is firm and often composed of regularly arranged polygonal plates (as in the so-called armored or thecate Dinoflagellates). Pigments are chlorophyll, carotene, four xanthophylls, brown phycopyrrin, and red peridinin (the latter sometimes predominating) contained within chromatophores. The food reserve is starch or a starch-like substance, and oil. A pigment (eye) spot is often present, and sexual reproduction is unknown.

BIBLIOGRAPHY

From Cox, E. R., and Zingmark, R. G., *Selected Papers in Phycology,* pp. 805–808, J. R. Rosowski and B. C. Parker, Eds. Department of Botany, University of Nebraska, Omaha (1971).

Ultrastructure and Cytology

1. Bouck, B., and Sweeney, B., The Fine Structure and Ontogeny of Trichocysts in Marine Dinoflagellates, *Protoplasma, 61,* 205 (1966).
2. Bouligand, Y., Soyer, M.-O., and Puiseux-Dao, S., La structure fibrillaire et l'orientation des chromosomes chez les Dinoflagellates, *Chromosoma, 24,* 251 (1968).
3. de Cao Vien, M., Sur l'existence de formes à chromosomes géants chez le Péridinien libre *Amphidinium carteri, C.R. Acad. Sci. Paris, 267,* 2309 (1968).
4. Chunosoff, L., and Hirshfield, H. I., Nuclear Structure and Mitosis in the Dinoflagellate *Gonyaulax monilata, J. Protozool., 14,* 157 (1967).
5. Cox, E. R., and Arnott, H. J., The Ultrastructure of the Theca of *Ensiculifera loeblichii* sp. nov, in *Contributions in Phycology,* B. C. Parker and R. M. Brown, Eds. The Allen Press, Lawrence, Kansas, in press.
6. Dodge, J. D., The Dinophyceae, in *The Chromosomes of the Algae,* St. Martin's Press, New York (1966).
7. Dodge, J. D., The Fine Structure of Chloroplasts and Pyrenoids in some Marine Dinoflagellates, *J. Cell Sci., 3,* 41 (1968).
8. Dodge, J. D., A Review of the Fine Structure of Algal Eyespots, *Br. Phycol. J., 4,* 199 (1969).
9. Dodge, J. D., and Crawford, R. M., Fine Structure of the Dinoflagellate *Amphidinium carteri* Hulburt, *Protistologica, 4,* 231 (1968).
10. Dodge, J. D., and Crawford, R. M., Observations on the Fine Structure of the Eyespot and Associated Organelles in the Dinoflagellate *Glenodinium foliaceum, J. Cell Sci., 5,* 479 (1969).

11. Dodge, J. D., and Crawford, R. M., The Fine Structure of *Gymnodinium fuscum* (Dinophyceae), *New Phytol., 68,* 613 (1969).
12. Dodge, J. D., and Crawford, R. M., The Morphology and Fine Structure of *Ceratium hirundinella* (Dinophyceae), *J. Phycol., 6,* 137 (1970).
13. Dodge, J. D., and Crawford, R. M., A Survey of Thecal Fine Structure in Dinophyceae, *Bot. J. Linn. Soc., 63,* 53 (1970).
14. Greuet, C., Organisation ultrastructurale de l'ocelle de deux peridiniens warnowiidae, *Erythropsis pavillardi* Kofoid et Swezy et *Warnowia pulchra* Schiller, *Protistologica, 4,* 209 (1968).
15. Kalley, J. P., and Bisalputra, T., *Peridinium trochoideum:* The Fine Structure of the Theca as Shown by Freeze-etching, *J. Ultrastr. Res., 31,* 95 (1970).
16. Kevin, M. J., Hall, W. T., McLaughlin, J. J. A., and Zahl, P. A., *Symbiodinium microadriatricum* Freudenthal, a Revised Taxonomic Description and Ultrastructure, *J. Phycol., 5,* 341 (1969).
17. Kowallik, K., The Crystal Lattice of the Pyrenoid Matrix of *Prorocentrum micans, J. Cell Sci., 5,* 251 (1969).
18. Kubai, D. F., and Ris, H., Division in the Dinoflagellate *Gyrodinium cohnii* Schiller, *J. Cell. Biol., 40,* 508 (1969).
19. Leadbeater, B., and Dodge, J. D., An Electron Microscope Study of Nuclear and Cell Division in a Dinoflagellate, *Arch. Mikrobiol., 57,* 239 (1967).
20. Soyer, M.-O., Sur l'existence d'un axe chromosomien chez certains dinoflagellés, *C.R. Acad. Sci. Paris, 265,* 1206 (1967).
21. Soyer, M.-O., Étude cytologique ultrastructurale d'un dinoflagellé libre, *Noctiluca miliaris* Suriray: Trichocystes et inclusions paracristallines, *Vie Milieu, 19,* 305 (1968).
22. Soyer, M.-O., L'enveloppe nucléaire chez *Noctiluca miliaris* Suriray (Dinoflagellata). I. Quelques données sur son ultrastructure et son évolution au cours de la sporogenèse, *J. Microsc., 8,* 569 (1969).
23. Soyer, M.-O., Observations ultrastructurales sur la condensation sporogénétique des chromosomes chez *Noctiluca miliaris* S. (Dinoflagellé, Noctilucidae), *C.R. Acad. Sci. Paris, 271,* 1003 (1970).
24. Swift, E., and Remsen, C. C., The Cell Wall of *Pyrocystis* spp. (Dinococcales), *J. Phycol., 6,* 79 (1970).
25. Taylor, D. L., In Situ Studies on the Cytochemistry and Ultrastructure of a Symbiotic Marine Dinoflagellate, *J. Mar. Biol. Assoc. U.K., 48,* 349 (1968).
26. Zingmark, R. G., Ultrastructural Studies on Two Kinds of Mesocaryotic Dinoflagellate Nuclei, *Am. J. Bot., 57,* 586 (1970).

Morphology and Taxonomy

27. Balech, E., Tintinnoinea y dinoflagellata del Pacifico segun material de las expediciones Norpac y Downwind del Instituto Scripps de Oceanografia, *Rev. Mus. Argent. Cienc. Nat. (Zool.), 7,* 1 (1962).
28. Balech, E., Dinoflagelados nuevos o interesantes del Golfo de Mexico y Caribe, *Rev. Mus. Argent. Cienc. Nat. (Hidrobiol.), 2,* 77 (1967).
29. Buchanan, R. J., Studies at Oyster Bay in Jamaica, West Indies. IV. Observations on the Morphology and Asexual Cycle of *Pyrodinium bahamense* Plate, *J. Phycol., 4,* 272 (1968).
30. Bursa, A., The Genus *Prorocentrum* Ehrenberg. Morphodynamics, Protoplasmic Structures and Taxonomy, *Can. J. Bot., 37,* 1 (1959).
31. Evitt, W. R., and Davidson, S. E., Dinoflagellate Studies. I. Dinoflagellate Cysts and Thecae, *Stanford Univ. Publ. Sci., 10,* 3 (1964).
32. Graham, H. W., Studies in the Morphology, Taxonomy, and Ecology of the Peridiniales, *Publ. Carneg. Inst., 542,* 1 (1942).
33. Halim, Y., Dinoflagellates of the South-east Caribbean Sea (East Venezuela), *Int. Rev. Hydrobiol., 52,* 701 (1967).
34. Kofoid, C. A., and Swezy, O., The Free-Living Unarmoured Dinoflagellates, *Mem. Univ. Calif. Berkeley, 5,* 1 (1921).

35. Lebour, M. V., *The Dinoflagellates of the Northern Seas,* Marine Biological Association of the United Kingdom. Plymouth, England (1925).
36. Loeblich, A. R., Jr., and Loeblich, A. R., III, Index to the Genera, Subgenera, and Sections of the Pyrrhophyta, *Studies in Tropical Oceanography,* No. 3. University of Miami Institute of Marine Science (1966).
37. Schiller, J., Dinoflagellate (Peridineae), in Rabenhorst's *Kryptogamen-Flora von Deutschland, Österreich und der Schweiz,* Vol. 10. Akademischer Verlag (1933–1937).
38. Sournia, A., Le genre *Ceratium* (Péridinien planctonique) dans le canal de Mozambique. Contribution á une Révision Mondiale, *Vie Milieu, 18,* 375 (1967).
39. Steidinger, K., and Williams, J., *Memoirs of the Hourglass Cruises.* II. *Dinoflagellates,* pp. 1–251. Marine Research Laboratories, Florida Department of Natural Resources (1970).
40. Subrahmanyan, R., The Dinophyceae of the Indian Seas. Genus *Ceratium* Schrank, *Mar. Biol. Assoc. India, 1,* 129 (1968).

Life Histories

41. de Cao Vien, M., Sur l'existence de phénomènes sexuels chez un Péridinien libre, l'*Amphidinium carteri, C.R. Acad. Sci. Paris, 2640,* 1006 (1967).
42. de Cao Vien, M., Sur la germination du zygote et sur un mode particular de multiplication végétative chez le Péridinien libre *Amphidinium carteri, C.R. Acad. Sci. Paris, 2670,* 701 (1968).
43. Freudenthal, H. D., *Symbiodinium* gen. nov. and *Symbiodinium microadriaticum* sp. nov., a Zooxanthella: Taxonomy, Life-cycle, and Morphology, *J. Protozool., 9,* 45 (1962).
44. von Stosch, H. A., Zum problem der sexuellen fortpflanzung in der Peridineengattung *Ceratium, Helgolaender Wiss. Meeresunters, 10,* 140 (1964).
45. von Stosch, H. A., Sexualität bei *Ceratium cornutum* (Dinophyta), *Naturwissenschaften, 52,* 112 (1965).
46. Wall, D., Guillard, R. R. L., and Dale, B., Marine Dinoflagellates Cultured from Resting Spores, *Phycologia, 6,* 83 (1967).
47. Wall, D., and Dale, B., Modern Dinoflagellate Cysts and Evolution of the Peridiniales, *Micropaleontology, 14,* 265 (1968).
48. Wall, D., and Dale, B., The "hystrichosphaerid" Resting Spore of the Dinoflagellate *Pyrodinium bahamense* Plate 1906, *J. Phycol., 5,* 140 (1969).
49. Wall, D., Guillard, R. R. L., Dale, B., Swift, E., and Watable, N., Calcitic Resting Cysts in *Peridinium trochoideum* (Stein) Lemmermann, an Autotrophic Marine Dinoflagellate, *Phycologia, 9,* 151 (1970).
50. Wilson, W. B., Forms of the Dinoflagellate *Gymnodinium breve* Davis in culture, *Contrib. Mar. Sci., 12,* 120 (1967).
51. Zingmark, R. G., Sexual Reproduction in the Dinoflagellate *Noctiluca miliaris* Suriray, *J. Phycol., 6,* 122 (1970).

Ecology

52. Allen, W. E., Twenty Years Statistical Studies of Marine Plankton Dinoflagellates of Southern California, *Am. Midl. Nat., 26,* 603 (1941).
53. Balech, E., The Change in the Phytoplankton Population off the California Coast, *Calif. Coop. Oceanic Fish. Invest., 7,* 127 (1958).
54. Dragovich, A., Kelly, J. A., and Goodell, H. G., Hydrological and Biological Characteristics of Florida's West Coast Tributaries, *Fish. Bull., 66,* 463 (1968).
55. Eppley, R. W., Holm-Hansen, O., and Strickland, J. D. H., Some Observations on the Vertical Migration of Dinoflagellates, *J. Phycol., 4,* 333 (1968).
56. Hasle, G. R., Photactic Vertical Migrations in Marine Dinoflagellates, *Oikos, 2,* 162 (1950).
57. Hulburt, E. M., Competition for Nutrients by Marine Phytoplankton in Oceanic, Coastal, and Estuarine Regions, *Ecology, 51,* 475 (1970).

58. Le Fevre, J., and Grall, J. R., On the Relationship of *Noctiluca* Swarming off the Western Coast of Brittany with Hydrological Features and Plankton Characteristics of the Environment, *J. Exp. Mar. Biol. Ecol., 4,* 287 (1970).
59. Ryther, J. H., Ecology of Autotrophic Marine Dinoflagellates with Reference to Red Water Conditions, in *The Luminescence of Biological Systems,* pp. 387–414, F. H. Johnson, Ed. American Association for the Advancement of Science, Washington, D.C. (1955).
60. Seliger, H. H., Carpenter, J. H., Loftus, M., and McElroy, W. D., Mechanisms for the Accumulation of High Concentrations of Dinoflagellates in a Bioluminescent Bay, *Limnol. Oceanogr., 15,* 234 (1970)

Physiology and Biochemistry

61. Aldrich, D. V., Ray, S. M., and Wilson, W. B., *Gonyaulax monilata:* Population Growth and Development of Toxicity in Cultures, *J. Protozool., 14,* 636 (1967).
62. Bentley-Mowat, J. A., and Reid, S. M., Effect of Gibberellins, Kinetin and Other Factors on the Growth of Unicellular Marine Algae in Culture, *Bot. Mar., 12,* 185 (1969).
63. Biggley, W. H., Swift, E., Buchanan, R. J., and Seliger, H. H., Stimulable and Spontaneous Bioluminescence in the Marine Dinoflagellates, *Pyrodinium bahamense, Gonyaulax polyedra,* and *Pyrocystis lunula, J. Gen. Physiol., 54,* 96 (1969).
64. Forward, R. B., Change in the Photoresponse Action Spectrum of the Dinoflagellate *Gyrodinium dorsum* Kofoid by Red and Far-red Light, *Planta, 92,* 248 (1970).
65. Forward, R. B., and Davenport, D., The Circadian Rhythm of a Behavioral Photoresponse in the Dinoflagellate *Gyrodinium dorsum, Planta, 92,* 259 (1970).
66. Francis, D., On the Eyespot of the Dinoflagellate *Nematodinium, J. Exp. Biol., 47,* 495 (1967).
67. Hashimoto, Y., Okaichi, T., Dang, L. D., and Noguchi, T., Glenodinine, an Ichthytoxic Substance Produced by a Dinoflagellate, *Peridinium polonicum, Bull. Jap. Soc. Sci. Fish., 34,* 528 (1968).
68. Honjo, T., and Hanaoka, T., Diurnal Fluctuations of Photosynthetic Rate and Pigment Contents in Marine Phytoplankton, *J. Oceanogr. Soc. Jap., 25,* 182 (1969).
69. Jahn, T., Harmon, W., and Landman, M., Mechanisms of Locomotion in Flagellates. I. *Ceratium, J. Protozool., 10,* 358 (1963).
70. Kelly, M., and Katona, S., An Endogenous Diurnal Rhythm of Bioluminescence in the Natural Population of Dinoflagellates, *Biol. Bull., 131,* 115 (1966).
71. Mandelli, E. F., Carotenoid Pigments of the Dinoflagellate *Glenodinium foliaceum* Stein, *J. Phycol., 4,* 347 (1968).
72. Mandelli, E. F., Carotenoid Interconversion in Light-dark cultures of the Dinoflagellate *Amphidinium klebsii, J. Phycol., 5,* 382 (1969).
73. Muscatine, L., and Cernichiari, E., Assimilation of Photosynthetic Products of Zooxanthellae by a Reef Coral, *Biol Bull., 137,* 506 (1969).
74. Nevo, Z., and Sharon, N., The Chemical Structure of the Cell Wall of the Dinoflagellate Alga, *Peridinium westii, Isr. J. Chem., 5,* 139 (1967).
75. Sweeney, B. M., and Hastings, J. W., Characteristics of the Diurnal Rhythm of Luminescence in *Gonyaulax polyhedra, J. Cell. Comp. Physiol., 49,* 115 (1957).
76. Sweeney, B. M., and Hastings, J. W., Rhythmic Cell Division in Populations of *Gonyaulax polyhedra, J. Protozool., 5,* 217 (1958).
77. Swift, E., and Taylor, W. R., Bioluminescence and Chloroplast Movement in the Dinoflagellate *Pyrocystis lunula, J. Phycol., 3,* 77 (1967).

7. Rhodophyta (Red Algae)

Plants have simple or branched filaments (unicellular in one questionable form). Pigments are contained within chromatophores, and are chlorophyll, xanthophyll, carotene, phycocyanin, and phycoerythrin, in the freshwater forms appearing blue-green, gray-green, or violet (not red). The food reserve is in the form of a special starch (floridean), which is negative to the iodine test for starch. Walls are relatively thick and often mucilaginous, sometimes furnished with pores through which protoplasmic extensions occur. Sexual reproduction is by heterogametes, but the male elements drift and do not swim. Thalli are often of macroscopic size.

BIBLIOGRAPHY

From Hommersand, M. H., and Searles, R. B., *Selected Papers in Phycology*, pp. 763–767, J. R. Rosowski and B. C. Parker, Eds. Department of Botany, University of Nebraska, Omaha (1971).

1. Abbott, I. A., Studies in Some Foliose Red Algae of the Pacific Coast. III. Dumontiaceae, Weeksiaceae, Kallymeniaceae, *J. Phycol., 4,* 180 (1968).
2. Adey, W. H., The Crustose Corallines of the Northwestern North Atlantic, Including *Lithothamnium lemoineae* n. sp., *J. Phycol., 6,* 225 (1970).
3. Allsopp, A., Phylogenetic Relationships of the Protocaryota and the Origin of the Eucaryotic Cell, *New Phytol., 68,* 591 (1969).
4. Ardré, F., Remarques sur la structure des *Pterosiphonia* (Rhodomélacées, Céramiales) et leurs rapports systématiques avec les *Polysiphonia, Rev. Algol., 9,* 37 (1967).
5. Bisalputra, T. and Bisalputra, A., The Occurrence of DNA Fibrils in Chloroplasts of *Laurencia spectabilis, J. Ultrastruct. Res., 17,* 14 (1967).
6. Boillot, A., Sur l'alternance de générations hétéromorphes d'une Rhodophycée, *Halarachnion ligulatum* (Woodward) Kützing (Gigartinales, Furcellariacées), *C.R. Acad. Sci. Paris, 261,* 4191 (1965).
7. Boillot, A., Sur le développement des carpospores de *Naccaria wiggii* (Turner) Endlicher et d'*Atractophora hypnoides* Crouan (Naccariacées, Bonnemaisoniales), *C.R. Acad. Sci. Paris, 264D,* 257 (1967).
8. Bourrelly, P., *Les Algues d'eau douce.* III. Les Algues bleues et rouges, les Eugléniens, Peridiniens et Cryptomonadines. N. Boubée et Cie, Paris (1970).
9. Brown, D. L., and Weier, T. E., Chloroplast Development and Ultrastructure in the Freshwater Red Alga *Batrachospermum, J. Phycol., 4,* 199 (1968).
10. Brown, D. L., and Weier, T. E., Ultrastructure of the Freshwater Alga *Bactrachospermum.* I. Thin-section and Freeze-etch Analysis of Juvenile and Photosynthetic Filament Vegetative Cells, *Phycologia, 9,* 217 (1970).
11. Cabioch, J., Sur l'importance de phénomènes cytologiques pour la systématique et la phylogénie des Corallinacées (Rhodophycées, Cryptonemiales), *C.R. Acad. Sci. Paris, 271D,* 296 (1970).
12. Chadefaud, M., La morphologie des végétaux inférieures: données fondamentales et problèmes, *Mém. Soc. Bot. Fr., 115,* 5 (1968).
13. Chapman, V. J., *Seaweeds and Their Uses,* 2nd ed. Methuen, London (1970).
14. Chen, L. C.-M., Edelstein, T., and McLachlan, J., *Bonnemaisonia hamifera* Hariot in Nature and in Culture, *J. Phycol., 5,* 211 (1969).
15. Chiang, Y.-M., Observations on the Development of the Carposporophyte of *Scinaia pseudojaponica* Yamada et Tanaka (Nemaliales, Chaetangiaceae), *J. Phycol., 6,* 289 (1970).
16. Chiang, Y.-M., Morphological Studies of Red Algae of the Family Cryptonemiaceae, *Univ. Calif. Publ. Bot., 58* (1970).
17. Chihara, M., Life Cycle of the Bonnemaisoniaceous Algae in Japan. II., *Sci. Rep. Tokyo Kyoiku Daigaku, 11B,* 27 (1962).
18. Cortel-Breeman, A. M., and van den Hoek, C., Life-history Studies on Rhodophyceae I. *Acrosymphyton purpuriferum* (J. Ag.) Kyl, *Acta Bot. Neerl., 19,* 265 (1970).
19. Dawson, E. Y., Marine Red Algae of Pacific Mexico. Part 4. Gigartinales, *Pac. Natur., 2,* 190 (1961).
20. Dawson, E. Y., *Marine Botany.* Holt, Rinehart & Winston, Inc., New York (1966).

21. Denizot, M., Les Algues Floridées Encrotantes (à l'exclusion des Corallinacées), *Lab. Crypt. Mus. Nat. Hist. Natur. Paris* (1968).
22. Dixon, P. S., The Rhodophyceae: Some Aspects of Their Biology. II., *Oceanogr. Mar. Biol. Annu. Rev.*, Godward, Ed. E. Arnold, Longon (1966).
23. Dixon, P. S., The Rhodophyceae: Some Aspects of Their Biology II., *Oceanogr. Mar. Biol. Annu. Rev., 7,* 111 (1970).
24. Doty, M. S., and Abbott, I. A., Studies in Helminthocladiaceae. III. *Liagoropsis, Pac. Sci., 18,* 441 (1964).
25. Dring, M. J., Phytochrome in Red Alga *Porphyra tenera, Nature, 215,* 1411 (1967),
26. Edelstein, T., The Life History of *Gloiosiphonia capillaris* (Hudson) Carmichael, *Phycologia, 9,* 55 (1970).
27. Edwards, P., Field and Cultural Observations on the Growth and Reproduction of *Polysiphonia denudata* from Texas, *Br. Phycol. J., 5,* 145 (1970).
28. Evans, J. V., Electron Microscopical Observations on a New Red Algal Unicell, *Rhodella maculata* gen. nov., sp. nov., *Br. Phycol. J., 5,* 1 (1970).
29. Fan, K. C., Morphological Studies of the Gelidiales., *Univ. Calif. Publ. Bot., 32,* 315 (1961).
30. Feldmann, J., Les cycles de reproduction des algues et leurs rapports avec la phylogénie, *Rev. Cytol. Biol. Vég., 13,* 1 (1952).
31. Feldmann, G., Sur le cycle haplobiontique du *Bonnemaisonia asparagoides* (Woodw.) Ag., *C. R. Acad. Sci. Paris, 262D,* 1695 (1966).
32. Fredrick, J. F., Evolution of Polyglucoside-synthesizing Isozymes in the Algae. Phylogenesis and Morphogenesis in the Algae, *Ann. N. Y. Acad. Sci., 175,* 524 (1970).
33. Frei, E., and Preston, R. D., Non-cellulosic Structural Polysaccharides in Algal Cell Walls. II. Association of Xylan and Mannan in *Porphyra umbilicalis, Proc. R. Soc. B., 160,* 314 (1964).
34. Fritsch, F. E., *The Structure and Reproduction of the Algae,* Vol. 2. Cambridge University Press, London (1945).
35. Gantt, E., and Conti, S. F., Phycobiliprotein Localization in Algae, in Energy Conversion by the Photosynthetic Apparatus, *Brookhaven Symp. Biol., 19,* 393 (1967).
36. Gantt, E., Edwards, M. R., and Conti, S. F., Ultrastructure of *Porphyridium aerugineum,* a Blue-green Colored Rhodophytan, *J. Phycol., 4,* 65 (1968).
37. Giraud, A., and Magne, F., La place de la méiose dans le cycle de développement de *Porphyra umbilicalis, C. R. Acad. Sci. Paris, 267D,* 586 (1968).
38. Goedheer, J. C., Energy Transfer from Carotenoids to Chlorophyll in Blue-green, Red and Green Algae and Greening Bean Leaves, *Biochim. Biophys. Acta, 172,* 252 (1969).
39. Guérin-Dumartrait, E., Sarda, C., and Lacourly, A., Sur la structure fine du chloroplaste de *Porphyridium* sp. (Lewin), *C. R. Acad. Sci. Paris, 270D,* 1977 (1970).
40. L'Hardy-Halos, M.-Th., Recherches sur les Céramiacées (Rhodophycées-Céramiales) et leur morphogénèse, *Rev. Gén. Bot., 77,* 211 (1970).
41. van den Hoek, C., and Cortel-Breeman, A. M., Life History Studies on Rhodophyceae III. *Scinaia complanata* (Collins) Cotton, *Acta Bot. Neerl., 19,* 457 (1970).
42. Hollenberg, G. J., An Account of the Species of *Polysiphonia* of the Central and Western Tropical Pacific Ocean. I. *Oligosiphonia, Pac. Sci., 22,* 56; II. *Polysiphonia, Pac. Sci., 22,* 198 (1968).
43. Hollenberg, G. J., and Abbott, I. A., New Species of Marine Algae from California, *Can. J. Bot., 46,* 1235 (1968).
44. Hommersand, M. H., The Morphology and Classification of Some Ceramiaceae and Rhodomelaceae, *Univ. Calif. Publ. Bot., 35,* 165 (1963).
45. Hudson, P.R., and Wynne, M. J., Sexual Plants of *Bonnemaisonia geniculata* (Nemaliales), *Phycologia, 8,* 207 (1969).
46. Johansen, H. W., Morphology and Systematics of Coralline Algae with Special Reference to *Calliarthron, Univ. Calif. Publ. Bot., 49,* (1969).
47. Jones, W. E., and Smith, R. M., The Occurrence of Tetraspores in the Life History of *Naccaria wiggii* (Turn.) Endl., *Br. Phycol. J., 5,* 91 (1970).

48. Knaggs, F. W., A Review of Florideophycidean Life Histories and of the Culture Techniques Employed in Their Investigation, *Nova Hedwigia Z. Kryptogamenko, 18,* 293 (1969).
49. Kylin, H., *Die Gattungen der Rhodophyceen.* Lund: CWK Gleerups, (1956).
50. Lee, R. E., and Fultz, S. A., Ultrastructure of the *Conchocelis* Stage of the Marine Red Alga *Porphyra leucosticta, J. Phycol., 6,* 22 (1970).
51. Levring, T., Hoppe, H. A., and Schmid, O. J., *Marine Algae, A Survey of Research and Utilization.* Cram, de Gruyter & Co., Hamburg, Germany (1969).
52. Lichtlé, C., and Giraud, G., Aspects ultrastructuraux particuliers au plaste du *Batrachospermum virgatum* (Sirdt) – Rhodophycée – Nemalionale, *J. Phycol., 6,* 281 (1970).
53. McBride, D. L., and Cole, K., Ultrastructural Characteristics of the Vegetative Cell of *Smithora naiadum* (Rhodophyta), *Phycologia, 8,* 177 (1969).
54. Magne, F., Recherches caryologiques chez les Floridées (Rhodophycées), *Cahiers Biol. Mar., 5,* 461 (1965).
55. Magne, F., Sur le déroulement et le lieu de la méiose chez les Lémanéacées (Rhodophycées, Nemalionales), *C. R. Acad. Sci.* Paris, *265D,* 670 (1967).
56. Magne, F., Le cycle de développement des Rhodophycées, *Botaniste, 50,* 297 (1967).
57. Magne, F., Méiose sans tétrasporocystes chez les Rhodophycées, *Proc. Int. Seaweed Symp., 6,* 251 (1969).
58. Majak, W., Craigie, J. S., and McLachlan, J., Photosynthesis in Algae. I. Accumulation Products in the Rhodophyceae, *Can. J. Bot., 44,* 541 (1966).
59. Martin, M. T., A Review of Life-histories in the Nemalionales and some Allied Genera, *Br. Phycol. J., 4,* 145 (1969).
60. Mikami, H., A Systematic Study of the Phyllophoraceae and Gigartinaceae from Japan and its Vicinity, *Sci. Pap. Inst. Algol. Res. Fac. Sci. Hokkaido Univ., 5,* 181 (1965).
61. Neushul, M., A Freeze-etching Study of the Red Alga *Porphyridium, Am. J. Bot., 57,* 1231 (1970).
62. Nichols, H. W., and Lissant, E. K., Developmental Studies of *Erythrocladia* Rosenvinge in Culture, *J. Phycol., 3,* 6 (1967).
63. Norris, R. E., Morphological Studies on the Kallymeniaceae, *Univ. Calif. Publ. Bot., 28,* 251 (1957).
64. Nozawa, Y., Systematic Account of the Red Algal Genus *Rhodopeltis, Pac. Sci., 24,* 99 (1970).
65. O'hEocha, C., Biliproteins of Algae, *Annu. Rev. Plant Physiol., 16,* 415 (1965).
66. Papenfuss, G. F., A Review of the Present System of Classification of the Florideophycidae, *Phycologia, 5,* 247 (1966).
67. Percival, E., and McDowell, R. H., *Chemistry and Enzymology of Marine Polysaccharides.* Academic Press, London (1967).
68. Peyriére, M., Evolution d'lappareil de Golgi au cours de la tetrasporogenese de *Griffithsia flosculosa* (Rhodophycée), *C. R. Acad. Sci. Paris, 270D,* 2071 (1970).
69. Pringle, J. D., and Austin, A. P., The Mitotic Index in Selected Red Algae In Situ. II. A Supralittoral Species, *Porphyra lanceolata, J. Exp. Mar. Biol. Ecol., 5,* 113 (1970).
70. Provasoli, L., Media and Prospects for the Cultivation of Marine Algae, in *Culture and Collections of Algae, Proceedings of the U.S. – Japan Conference at Hakone, Japan, September 1966,* pp. 63–75, A. Watanabe and A. Hattori, Eds. Japan Society of Plant Physiology (1968).
71. Ramus, J., The Development Sequence of the Marine Red Alga *Pseudogloiophloea* in Culture, *Univ. Calif. Publ. Bot., 52,* 1 (1969).
72. Ramus, J., Pit Connection Formation in the Red Alga *Pseudogloiophloea, J. Phycol., 5,* 56 (1969).
73. Richardson, F. L., and Brown, T. E., *Glaucosphaera vacuolata,* its Ultrastructure and Physiology, *J. Phycol., 6,* 165 (1970).
74. Richardson, N., Studies on the Photobiology of *Bangia fuscopurpurea, J. Phycol., 6,* 215 (1970).
75. Richardson, N., and Dixon, P. S., The *Conchocelis* phase of *Smithora naiadum* (Anders.) Hollenb., *Br. Phycol. J., 4,* 181 (1969).
76. Richardson, N., and Dixon, P. S., Culture Studies on *Thuretellopsis peggiana* Kylin, *J. Phycol., 6,* 154 (1970).

77. Saito, Y., On Morphological Distinctions of Some Species of Pacific North American *Laurencia, Phycologia, 8,* 85 (1969).
78. Scagel, R. F., The genus *Dasyclonium* J. Agardh., *Can. J. Bot., 40,* 1017 (1962).
79. Schnepf, E., and Brown, R. M., Jr., On Relationships between Endosymbiosis and the Origin of Plastids and Mitochondria, in *Origin and Development of Cell Organelles,* pp. 299–322, W. Urspring and C. Reinert, Eds. Springer-Verlag, Berlin (1971).
80. Schotter, G., Recherches sur les Phyllophoracées, *Bull. Inst. Océanogr. (Monaco), 67,* 1 (1968).
81. Searles, R. B., Morphological Studies of Red Algae of the Order Gigartinales, *Univ. Calif. Publ. Bot., 43* (1968).
82. Siegelman, H. W., Chapman, D. J., Cole, W. J., The Bile Pigments of Plants, in *Porphyrins and Related Compounds,* pp. 107–20, G. Goodwin, Ed. Biochemical Society Symposium No. 28. Academic Press, London (1968).
83. Smith, G. M., *Cryptogamic Botany,* Vol. 1, *Algae and Fungi,* 2nd ed. McGraw-Hill, New York (1955).
84. Sommerfeld, M. R., and Nichols, H. W., Developmental and Cytological Studies of *Bangia fuscopurpurea* in Culture, *Am. J. Bot., 57,* 640 (1970).
85. Sparling, S. R., The Structure and Reproduction of some Members of the Rhodymeniaceae, *Univ. Calif. Publ. Bot., 29,* 313 (1957).
86. von Stosch, H. A., The Sporophyte of *Liagora farinosa* Lamour, *Br. Phycol. Bull., 2,* 486 (1965).
87. von Stosch, H. A., Observations on *Corallina, Jania* and other Red Algae in Culture, *Proc. Int. Seaweed Symp., 6,* 389 (1969).
88. Svedelius, N., Are the Haplobiontic Florideae to be Considered Reduced Types? *Sven. Bot. Tidskr., 50,* 1 (1956).
89. Turvey, J. R., and Williams, E. L., The Structures of some Xylans from Red Algae, *Phytochemistry, 9,* 2383 (1970).
90. West, J., The Life Histories of *Rhodochorton purpureum* and *R. tenue* in Culture, *J. Phycol., 5,* 12 (1969).
91. Wollaston, E. M., Morphology and Taxonomy of Southern Australian Genera of Crouaniae Schmitz (Ceramiaceae, Rhodophyta), *Aust. J. Bot., 16,* 217 (1968).
92. Wynne, M. J., *Platysiphonia decumbens* sp. nov., a New Member of the *Sarcomenia* Group (Rhodophyta) from Washington, *J. Phycol., 5,* 190 (1969).
93. Wynne, M. J., Marine Algae of Amchitka Island (Aleutian Islands). I. Delesseriaceae, *Syesis, 3,* 95 (1970).

8. Chloromonadophyta (Chloromonads)

An obscure and little-understood group, it is composed of a few genera and species; cells swim, and have one or two flagella, apically attached. Chromatophores are green, with chlorophyll (supposedly) predominating, but with an abundance of xanthophyll also present. The food reserve is in the form of oils or a fat. Contractile vacuoles and a reservoir are in the anterior end of the cell. Cell contents with trichocysts are radiately arranged just within the cell membrane (in the genus *Gonyostomum*); sexual reproduction is unknown.

BIBLIOGRAPHY

Hollande, A., Classe des Chloromonadinés, in *Traite de Zoologie,* Vol. I, Part 1, pp. 227–237, P. P. Grassé, Ed. Masson, Paris (1952).

9. Phaeophyta (Brown Algae)

This phylum is mostly marine, including the brown sea weeds (kelps). It is essentially filamentous (some microscopic) but mostly robust and leathery. Pigments include chlorophylls, ccarotene, xanthophyll, and fucoxanthin (predominating brown pigment). The food reserve is soluble carbohydrates including alcohol (mannitol); reproduction is asexual by kidney-shaped zoospores with two lateral flagella or sexual by iso-, aniso-, or heterogametes.

BIBLIOGRAPHY

From Druehl, L. D., and Wynne, M. J., *Selected Papers in Phycology*, pp. 793–796, J. R. Rosowski and B. C. Parker, Eds. Department of Botany, University of Nebraska, Omaha (1971).

1. Bisalputra, T., and Bisalputra, A. A., Chloroplast and Mitochondria DNA in *Egregia menziesii, J. Cell Biol., 33,* 511 (1967).
2. Bisalputra, T., and Bisalputra, A. A., The Ultrastructure of Chloroplast of a Brown Alga *Sphacelaria* sp. I. Plastid DNA Configuration–the Chloroplast Genophore, *J. Ultrastr. Res., 29,* 151 (1969).
3. Blood, C. T., The Many Applications of Alginates, *CIBA Rev., 1,* 19 (1969).
4. Bouck, G. B., Fine Structure and Organelle Associations in Brown Algae, *J. Cell Biol., 26,* 523 (1969).
5. Bouck, G. B., Extracellular Microtubules. The Origin, Structure, and Attachment of Flagellar Hairs in *Fucus* and *Ascophyllum* antherozoids, *J. Cell Biol., 40,* 446 (1969).
6. Bryan, G. W., The Absorption of Zinc and Other Metals by the Brown Seaweed *Laminaria digitata, J. Mar. Biol. Assoc. U.K., 49,* 225 (1969).
7. Burrows, E. M., An Experimental Assessment of some of the Characters Used for Specific Delimitation in the Genus *Laminaria, J. Mar. Biol. Assoc. U.K., 44,* 137 (1964).
8. Cardinal, A., Etude sur les Ectocarpacees de la Manche, *Beih. Nova Hedwigia Z. Kryptogamenkd, 15* (1964).
9. Chapman, D. J., and Tocher, R. D., Occurrence and Production of Carbon Monoxide in some Brown Algae, *Can. J. Bot., 44,* 1438 (1966).
10. Cole, K., Chromosome Numbers in the Phaeophyceae, *Can. J. Genet. Cytol., 9,* 519 (1967).
11. Cole, K., Ultrastructural Characteristics in some Species in the Order Scytosiphonales, *Phycologia, 9,* 275 (1970).
12. Drew, E. A., Uptake and Metabolism of Exogenously Supplied Sugars by Brown Algae, *New Phytol., 68,* 35 (1969).
13. Druehl, L. D., Distribution of Two Species of *Laminaria* as Related to some Environmental Factors, *J. Phycol., 3,* 103 (1967).
14. Druehl, L. D., Taxonomy and Distribution of Northeast Pacific Species of *Laminaria, Can. J. Bot., 46,* 539 (1968).
15. Druehl, L. D., The Pattern of Laminariales Distribution in Northeast Pacific, *Phycologia, 9,* 237 (1970).
16. Druehl, L. D., and Hsiao, S. I. C., Axenic Culture of Laminariales in Defined Media, *Phycologia, 8,* 47 (1969).
17. Earle, S. A., Phaeophyta of the Eastern Gulf of Mexico, *Phycologia, 7,* 71 (1969).
18. Edelstein, T., Chen, L. C-M., and McLachlan, J., The Life Cycle of *Ralfsia clavata* and *R. borneti, Can. J. Bot., 48,* 527 (1970).
19. Evans, L. V., Cytological Studies in the Laminariales, *Ann. Bot., N. S., 29,* 541 (1965).
20. Evans, L. V., Distribution of Pyrenoids Among some Brown Algae, *J. Cell Sci., 1,* 449 (1966).
21. Evans, L. V., Chloroplast Morphology and Fine Structure in British Fucoids, *New Phytol., 67,* 173 (1968).
22. Fulcher, R. G., and McCully, M. E., Histological Studies on the Genus *Fucus.* IV. Regeneration and Adventive Embryony, *Can. J. Bot., 47,* 1643 (1969).
23. Hoek, C. van den., and Flinterman, A., The Life-history of *Sphacelaria furcigera* Kütz. (Phaeophyceae), *Blumea, 16,* 193 (1968).
24. Hollenberg, G. J., An Account of the Ralfsiaceae (Phaeophyta) of California, *J. Phycol., 5,* 290 (1969).

25. Jaasund, E., Aspects of the Marine Algal Vegetation of North Norway, *Bot. Gothoburg,* 4 (1965).
26. Jaasund, E., Marine Algae in Tanzania. IV, *Bot. Mar., 13,* 71 (1970).
27. Jaffe, L. F., Localization in the Developing *Fucus* Egg and the General Role of Localizing Currents, in *Advances in Morphogenesis,* pp. 295-326, M. Abercrombie, Ed. Academic Press, New York (1968).
28. Jones, N. S., and Kain, J. M., Subtidal Algal Colonization Following the Removal of *Echinus, Helgoländer Wiss. Meeresunters, 15,* 460 (1967).
29. Kain, J. M., The Biology of *Laminaria hyperborea.* V. Comparison with Early Stages of Competitors, *J. Mar. Biol. Assoc. U.K., 49,* 455 (1969).
30. Knight, M., and Parke, M., A Biological Study of *Fucus vesiculosus* L., and *F. serratus L., J. Mar. Biol. Assoc. U.K., 29,* 439 (1950).
31. Kuckuck, P., Ectocarpaceen-Studien I. *Hecatomema, Chilionema, Compsonema, Helgoläender Wiss. Meeresunters., 4,* 317 (1953).
32. Kuckuck, P., Ectocarpaceen-Studien VII. *Giffordia, Helgoläender Wiss. Meeresunters., 8,* 119 (1961).
33. Levring, T., Hoppe, H. A., and Schmid, O. J., Marine Algae. A Survey of Research and Utilization, *Botanica Marina Handbooks,* Vol. 1. Cramer, de Gruyter & Co., Hamburg, Germany (1969).
34. Liddle, L. B., Reproduction *A Zonaria farlowii.* I. Gametogenesis, Sporogenesis, and Embryology, *J. Phycol., 4,* 298 (1968).
35. Lindauer, V. W., Chapman, V. J., and Aiken, M., The Marine Algae of New Zealand. II. Phaeophyceae, *Nova Hedwigia Z. Kryptogamenko, 3,* 129 (1961).
36. Loiseaux, S., Notes on Several Myrionemataceae from Calfornia using Culture Studies, *J. Phycol., 6,* 248 (1970).
37. Lund, S., The Marine Algae of East Greenland. I. Taxonomical Part, *Medd. Grønland, 156,* 1 (1959).
38. Lüning, K., Growth of Amputated and Dark-exposed Individuals of the Brown Alga *Laminaria hyperborea, Mar. Biol., 2,* 218 (1969).
39. Mathieson, A. C., Morphology and Life History of *Phaeostrophion irregulare* S. et G., *Nova Hedwigia Z. Kryptogamenko, 13,* 293 (1967).
40. McCully, M. E., Histological Studies on the Genus *Fucus.* I. Light Microscopy of the Mature Vegetative Plant, *Protoplasma, 62,* 287 (1966).
41. McCully, M. E., Histological Studies on the Genus *Fucus.* II. Histology of the Reproductive Tissues, *Protoplasma, 66,* 205 (1968).
42. Misra, J. N., *Phaeophyceae in India.* Indian Council of Agricultural Research, New Delhi, India (1966).
43. Moss, B., The Apical Meristem in *Fucus, New Phytol., 66,* 67 (1967).
44. Moss, B., The Transition from Vegetative to Fertile Tissue in *Fucus vesiculosus, Br. Phycol. Bull., 3,* 567 (1968).
45. Moss, B., Apical Meristems and Growth Control in *Himanthalia elongata* (S. F. Gray), *New Phytol., 68,* 387 (1969).
46. Müller, D. B., Generationswechsel, Kernphasenwechsel und Sexualität der Braunalge *Ectocarpus siliculosus im Kulturversuch, Planta* (Berlin),
siliculosus im Kulturversuch, *Planta (Berlin), 75,* 39 (1967).
47. Müller, D. G., Jaenicke, L., Donike, M., and Akintobi, T., Sex Attractant in a Brown Alga: Chemical Structure, *Science, 171,* 815 (1971).
48. Nakamura, Y., Development of Zoospores in *Ralfsia*-like Thallus, with Special Reference to the Life Cycle of the Scytosiphonales, *Bot. Mag. Tokyo, 78,* 109 (1965).
49. Nicholson, N. L., Field Studies on the Giant Kelp *Nereocystis, J. Phycol., 6,* 177 (1970).
50. Nizamuddin, M., Phytogeography of the Fucales and Their Seasonal Growth, *Bot. Mar., 13,* 131 (1970).
51. North, W. J. (principal investigator), *Kelp Habitat Improvement Project Annual Reports* (1962–1969).
52. Norton, T. A., and Burrows, E. M., Studies on Marine Algae of the British Isles. 7. *Saccorhiza polyschides* (Lightf.) Batt., *Br. Phycol. J., 4,* 19 (1969).
53. Norton, T. A., and South, G. R., Influence of Reduced Salinity on the Distribution of Two Laminarian Algae, *Oikos, 20,* 320 (1969).

54. Paine, R. T., and Vadas, R. L., The Effects of Grazing by Sea Urchins *Strongylocentrotus* spp. on Benthic Algal Populations, *Limnol. Oceanogr., 14,* 710(1969).
55. Papenfuss, G. F., Phaeophyta, in *Manual of Phycology,* pp. 119–158, G. M. Smith, Ed. Chronica Botanica, Waltham, Massachusetts (1951).
56. Parker, B. C., Translocation in the Giant Kelp *Macrocystis.* I. Rates, Direction, and Quality of C^{14}-labeled Organic Products and Fluorescein, *J. Phycol., 1,* 41 (1965).
57. Parker, B. C., Translocation in *Macrocystis.* III. Composition of Sieve Tube Exudate and Identification of the Major C^{14}-labeled Products, *J. Phycol., 2,* 38 (1966).
58. Parker, B. C., and Dawson, E. Y., Non-calcareous Marine Algae from California Miocene Deposits, *Nova Hedwigia Z. Kryptogamenko, 10,* 273 (1965).
59. Parker, B. C., and Huber, J., Translocation in *Macrocystis.* II. Fine Structure of the Sieve Tubes, *J. Phycol., 1,* 172 (1965).
60. Quatrano, R. S., Rhizoid Formation in *Fucus* Zygotes: Dependence on Protein and Ribonucleic Acid Synthesis, *Science, 162,* 468 (1968).
61. Roberts, M., Studies on the Marine Algae of the British Isles. 3. The Genus *Cystoseira, Br. Phycol. Bull., 3,* 345 (1967).
62. Russell, G., The Genus *Ectocarpus* in Britain. I. The Attached Forms, *J. Mar. Biol. Assoc. U.K., 46,* 267 (1966).
63. Scagel, R. F., The Phaeophyceae in Perspective, *Oceanogr. Mar. Biol. Annu. Rev., 4,* 123 (1966).
64. South, G. R., and Burrows, E. M., Studies on Marine Algae of the British Isles. 5. *Chorda filum* (L) Stackh., *Br. Phycol. Bull., 3,* 379 (1967).
65. Sundene, O., Interfertility between Forms of *Laminaria digitata, Nytt Mag. Bot., 6,* 121 (1958).
66. Sundene, O., Growth in the Sea of *Laminaria digitata* Sporophytes from Culture, *Nytt Mag. Bot., 9,* 5 (1962).
67. Tatewaki, M., Formation of a Crustaceous Sporophyte with Unilocular Sporangia in *Scytosiphon lomentaria, Phycologia, 6,* 62 (1966).
68. Torrey, J. G., and Galun, E., Apolar Embryos of *Fucus* Resulting from Osmotic and Chemical Treatment, *Am. J. Bot., 57,* 111 (1970).
69. Widdowson, T. B., A Taxonomic Study of the Genus *Hedophyllum* Setchell, *Can. J. Bot., 43,* 1409 (1965).
70. Widdowson, T. B., I. A Taxonomic Revision of the Genus *Alaria* Greville. II. A Statistical Analysis of Variation in the Brown Alga *Alaria, Syesis, 4,* in press, (1971).
71. Wilce, R. T., Studies in the Genus *Laminaria.* III. A Revision of the North Atlantic Species of the Simplices Section of *Laminaria, Bot. Gothoburg, 3,* 247 (1965).
72. Wilce, R. T., Heterotrophy in Arctic Sublittoral Seaweeds: An Hypothesis, *Bot. Mar., 10,* 185 (1967).
73. Wilce, R. T., Webber, E. E., and Sears, J. R., *Petroderma* and *Porterinema* in the New World, *Mar. Biol., 5,* 119 (1970).
74. Womersley, H. B. S., The Morphology and Taxonomy of *Cystophora* and Related Genera (Phaeophyta), *Aust. J. Bot., 12,* 53 (1964).
75. Wynne, M. J., Life History and Systematic Studies of Some Pacific North American Phaeophyceae (Brown Algae), *Univ. Calif. Publ. Bot., 50,* (1969).
76. Yabu, H., Early Development of Several Species of Laminariales in Hokkaido, *Mem. Soc. Fish., Hokkaido Univ., 12,* 1 (1964).
77. Ziegler, H., and Ruck, I., Untersuchungen über die Feinstruktur des Phloems. III. Die "Trompetzellen" von *Laminaria*-Arten, *Planta (Berlin), 73,* 62 (1967).

ACKNOWLEGEMENT

The bibliographies in this chapter were taken from *Selected Papers in Phycology,* © 1971, Department of Botany, University of Nebraska, Lincoln, Nebraska, and are reproduced by permission of the copyright owners. The above publication is available from the University Bookstore, 14th and R Streets, Lincoln, Nebraska.

PROTOZOA

OUTLINE OF PROTOZOA

DR. S. H. HUTNER

The protozoa (*proto* = first; *zoa* = animals) are the animals that explored unicellularity. Animals, as opposed to plants, are primarily specialized to ingest particulate food. According to this definition, tapeworms and other parasites that live on dissolved rather than particulate food are not proper animals, but the word "primarily" implies that these parasites evolved from phagotrophic, i.e., particle-ingesting ancestors.

Algal flagellates such as *Euglena* are discussed in texts of protozoology, although *Euglena* is not known to ingest particles. But *Euglena* is unmistakably kin to voracious predators such as *Peranema.* The same protozoan may combine phagotrophy and photosynthesis – an especially common pattern in chrysomonads. Motility does not separate plants and animals; many photosynthetic flagellates swim well toward light, but elaborate specializations for motility are more common among phagotrophs.

A widely accepted classification of protozoa was advanced by a committee of the Society of Protozoologists.[1] The following outline is, in the main, in accordance with it, though it is outdated in some respects. Permission for extensive quotation was granted by the editor of the *Journal of Protozoology.*

SUBPHYLUM I. SARCOMASTIGOPHORA

Superclass I. Mastigophora

Flagellates are regarded as ancestral, as shown by simplicity of structure and photosynthesis in several groups. Primitivity of one flagellate group, the cryptomonads, is assumed from the fact that its members share photosynthetic bile pigments (phycobilins) with red and blue-green algae. A major phyletic gap, therefore, separates red algae (non-flagellated, eukaryotic) + blue-green algae (non-flagellated, prokaryotic) and the flagellates (eukaryotic). A missing link may be *Glaucocystis,* which has phycobilins, flagella (though functionless), and a chloroplast reminiscent of that in red algae.[2,3] The eyespot is emerging as a valuable taxonomic character (Table 1).

CLASS 1. PHYTAMASTIGOPHOREA (PHYTOFLAGELLATES)

Typically these organisms have chloroplasts; if chloroplasts are lost secondarily, the relationship to pigmented forms is clearly evident. Commonly they have only one or two emergent flagella; ameboid forms are found in some groups. The carotenoids of the various phytoflagellate groups are characteristic; chlorophytes have pigments like those of higher plants; euglenids also have the carotenoids diadinoxanthin and diatoxanthin. For a survey of algal carotenoids, see Reference 4; for details, see Reference 5. Satisfactory agreement between the zoological and botanical classifications of the phytoflagellates has not yet been reached.

ORDER 1. CHRYSOMONADIDA

A vast group, yellow to brown because of abundant yellow carotenoids of the fucoxanthin series. In common with other brown-pigmented groups of algae (diatoms, phaeophycean seaweeds) and dinoflagellates, they contain chlorophyll *c*, which is lacking in the chlorophytes (green algae and higher plants). They usually have one, two, or three flagella. Relations to brown seaweeds and Xanthophycene are briefly reviewed in Reference 6.

The group includes the following widely studied species:

Ochromonas malhamensis. Phagotrophic and weakly photosynthetic. The only microorganism known to share the pattern of B_{12} requirement of higher animals (no response to pseudo B_{12}'s; B_{12}, i.e., cyanocobalamin, is spared by methionine; propionate oxidation is stimulated by B_{12}).

TABLE 1

EYESPOT TYPES IN FLAGELLATES[a]

Type A	
Eyespot part of a chloroplast, not associated with flagella	Chlorophyceae Prasinophyceae Cryptophyceae
Type B	
Eyespot part of a chloroplast, closely associated with a flagellum	Chrysophyceae Phaeophyceae (brown seaweeds)
Type C	
Eyespot independent of chloroplasts but adjacent to flagella	Euglenophyceae Dinophyceae
Other Types	
Type D: eyespot independent of chloroplasts but adjacent to flagellar bases and a lamellar body	Some Dinophyceae
Type E: eye-like structure or ocellus	Some dinophycean genera, e.g., *Erythropsis, Nematodium*

[a] Botanical names of taxa are used here to illustrate this alternative system.

Ochromonas danica. Vigorously phagotrophic and photosynthetic. At least 95% reduction of the photosynthetic apparatus occurs on growth in the dark. Photosynthesis is regained in light; hence, this is a convenient organism for determining which cellular constituents are located in the chloroplast. Its other vitamin requirements, biotin and thiamine, are applied in determining biotin and thiamine in clinical material. For use of *O. malhamensis* and *O. danica* in vitamin assays, see Reference 7. *Ochromonas danica* has novel sulfatides, including sulfonolipids, and also halogenated lipids.[8,9]

Suborder *Euchrysomonadina*

In the Euchrysomonadina, the flagellate stage, e.g., as in *Ochromonas,* is dominant. *O malhamensis,* under poorly understood circumstances, forms a vase-like lorica that encloses most of the body; therefore some phycologists (i.e., algologists) assign them to another genus, *Poteriochromonas; P. stipita* has been shown to be virtually identical in metabolic characteristics to *O. malhamensis.* Such loricas, borne on branching stalks, characterize *Dinobryon,* common in fresh-water reservoirs, where it causes off-odors when abundant.

Prymnesium. A pest in brackish fish ponds, notably in Israel, where it occasionally forms blooms. It produces a gill-paralyzing, fish-killing toxin.[10] The group containing *Prymnesium* has a unique attachment filament, the haptonema. Pure cultures of *Prymnesium* are available.

Other Noteworthy Chrysomonads. These include *Hydrurus foetidus,* which forms immense gelatinous masses (the palmelloid state) in cold streams; *Phaeocystis,* at times so abundant in the North Atlantic that its masses clog fishermen's nets; and the common fresh-water colonial genera *Synura* and *Uroglena.* Colorless chrysomonads, e.g., *Monas,* are common, some definitely ameboid. Siliceous scales and bristles characterize *Mallomonas,* common in brackish waters.

Suborder (2). Rhizochrysidina

The ameboid stage is dominant. Example: the chloroplastidic *Chrysomoeba radians.*[11] (Hibberd, D. J. 1971. *Brit. Phycol. J.* 6:207-23).

ORDER 2. SILICOFLAGELLIDA

Members of this group have an internal siliceous skeleton. They are marine organisms and are of some importance as index fossils. None has been cultivated.

ORDER 3. COCCOLITHOPHORIDA

Cosmopolitan in marine plankton, some species are notoriously abundant in polluted estuaries, e.g., the Oslo fjord. They have elaborately sculptured calcareous scales, coccoliths, formed in Golgi apparatus and extruded to the surface of the cell;[12] hence coccolithophorids, e.g., *Hymenomonas,* serve as models of calcification in dentition.[13] Coccoliths are important geological markers. Pure cultures of coccolithophorids are available. Coccolithophorida probably should be a suborder of chrysomonads.

ORDER 4. HETEROCHLORIDA

Typically these organisms have two very unequal flagella, two to several yellow-green chromatophores, and brown-pigmented chloroplasts. Ameboid forms are frequent. Food reserves typically consist of lipids; some reports of glycogen and leucosin have been made. Cyst walls are siliceous.

Included by phycologists in the Xanthophyceae, the Heterochlorida are a neglected group, mainly found in fresh water. The difficulty in distinguishing between chrysomonads and heterochloridians is illustrated by a discussion of the marine *Olisthodiscus lutens,* which combines xanthophycean and chrysomonad characters.[14]

ORDER 5. CRYPTOMONADIDA

Cryptomonadida have two flagella, usually originating in a depression; their bodies are compressed. Typically they have two chromatophores, often brown, but also red, olive-green, or blue-green. Ameboid forms are absent. Food reserves consist of starch and amyloid substances.

Cryptomonads, because of the simplicity of their photosynthetic apparatus and, as noted, possession of bile pigments, may be the most archaic extant flagellates. The colorless fresh-water *Chilomonas paramecium,* available pure (axenic), is a much-studied, hardy "acetate flagellate", i.e., it grows well on acetate as sole source of carbon and energy and is resistant to acetate toxicity. Olive-green *Cryptomonas* is common in fresh waters. Isolates of the following predominantly marine genera, in pure culture, have been used to study phycobilins: *Chroomonas, Rhodomonas, Hemiselmis.* The phagotrophic *Cyathomonas* is available in bacterized culture.

ORDER 6. DINOFLAGELLIDA

Members of this group have two flagella, typically one transverse and one trailing. The body is usually grooved transversely and longitudinally, forming girdle and sulcus, each containing a flagellum. Chromatophores are usually yellow or dark brown, and occasionally green or blue-green. Many species thecate. Food reserves consist of starch and lipids.

The dinoflagellates, a large group, are fresh-water and marine organisms. They are among the most abundant of all flagellates because they are photosynthetic symbionts of photic-zone corals and many other invertebrates, e.g., some jellyfish or giant clams. Isclates of symbionts, *Symbiodinium,* are available axenic.[15,16]

At times dinoflagellates form vast blooms. Where the bloom-forming species is toxic, marine animals may be killed en masse by a deadly alkaloid, saxitoxin, of incompletely known structure, which may be passed up the food chain, especially in shellfish, and so may be fatal to man. Saxitoxin paralyzes the Na^+ pump in nerve and muscle, as does the pharmacologically virtually identical but chemically different tetrodotoxin from puffer fish. The red tides in Florida are caused by *Gymnodinium breve,* and those in Nova Scotia by *G. tamarensis.*[17,18]

Dinoflagellates are unique among eukaryotes in lacking histones. The chromosomes in some groups are permanently condensed;[19] (Rae, P.M.M. 1970. *J. Cell Biol.* 46:106-13); for other unique nuclear features see Reference 20. Dinoflagellates are rich in a unique $C_{22:6}$ polyenic fatty acid[21] and probably in unique carotenoids. They exemplify transitions between algae and protozoa, since many of the photosynthesizers are also phagotrophic. There are colorless saprophytic forms, e.g., *Gyrodinium cohnii* (pure cultures available); colorless phagotrophs, e.g., *Oxyrrhis marina,* which requires ubiquinone and cholesterol among other growth factors;[22] and intensely luminescent and voracious forms, such as *Noctiluca* (see Reference 23 for structure). Small dinoflagellates are responsible for the phosphorescence of the phosphorescent bays in Puerto Rico and Jamaica. Some dinoflagellates have elaborate eye-like ocelli. Dinoflagellate spores may serve as index fossils.[24] Many dinoflagellates are parasitic; *Oodinium* may cause serious losses in fresh-water aquarium fish because of dermal ulcerations. Some photosynthetic dinoflagellates are available in pure culture; among the hardiest is *Amphidinium carterae,* but less so than the colorless *Gyrodinium (Crypthecodinium) cohnii,*[25] which has been used for preparing labeled $C_{22:6}$ acid.[26]

Suborder (1). Adinina

Two flagella are inserted at the apex, one directed forward and the other laterally; girdle and sulcus are absent. This is considered a primitive character. *Exuviella* and *Prorocentrum* species are in pure culture.

Suborder (2). Dinifera

Girdle and sulcus with their flagella are present; thecate (cell wall arranged in plates) and athecate forms exist. Electron microscopy reveals that many athecate (naked) forms (*Gymnodinium,* etc.) have delicate plates and hence intergrade with thecate forms. The swarm cells of dinoflagellates, as in *Symbiodinium,* are gymnodiniod, i.e., naked and simple in structure and with the characteristic girdle. *Peridinium* and *Ceratium* are common in fresh waters and in the ocean. Reference 27 offers a monograph with many color plates depicting the range of dinoflagellates from simple photosynthetic forms to highly evolved predators.

ORDER 7. EBRIIDA

No chloroplasts; internal siliceous skeleton. These organisms are rare.

ORDER 8. EUGLENIDA

Typically one or two flagella emerge from the anterior reservoir; the green chromatophores are of various shapes, absent in some species. Motility is shown by typically "metabolic" contortions of the body, but there is no ameboid movement. Paramylum (a glucan) is the food reserve.[28,29] Cultural details for *Euglena gracilis* can be found in References 30, 31, and 32.

Suborder (1). Euglenina

Euglena gracilis. This species includes strains that, while fully photosynthetic, are also intensely heterotrophic, attaining dense populations in the dark with near-total dwindling of the photosynthetic apparatus, which, however, is quickly reassembled in light. The photosynthetic apparatus is easily permanently destroyed by antibiotics of the streptomycin family, by mutagens, various drugs, UV irradiation, and heat treatment, giving rise to "bleached" (permanently albinized) strains. Like *Ochromonas danica,* the photosynthetic apparatus and metabolism of *E. gracilis* are receiving much attention. *Euglena gracilis* is widely used to assay B_{12} in blood and natural waters.

Euglena mutabilis (klebsii). Common in extremely acid waters; it crawls rather than swims (rudimentary flagellum).

Common Genera. A common colorless counterpart of *Euglena* is *Astasia.* The two-flagellated photosynthetic brackish or marine *Eutreptia* is common in polluted estuaries; an isolate is in pure culture.

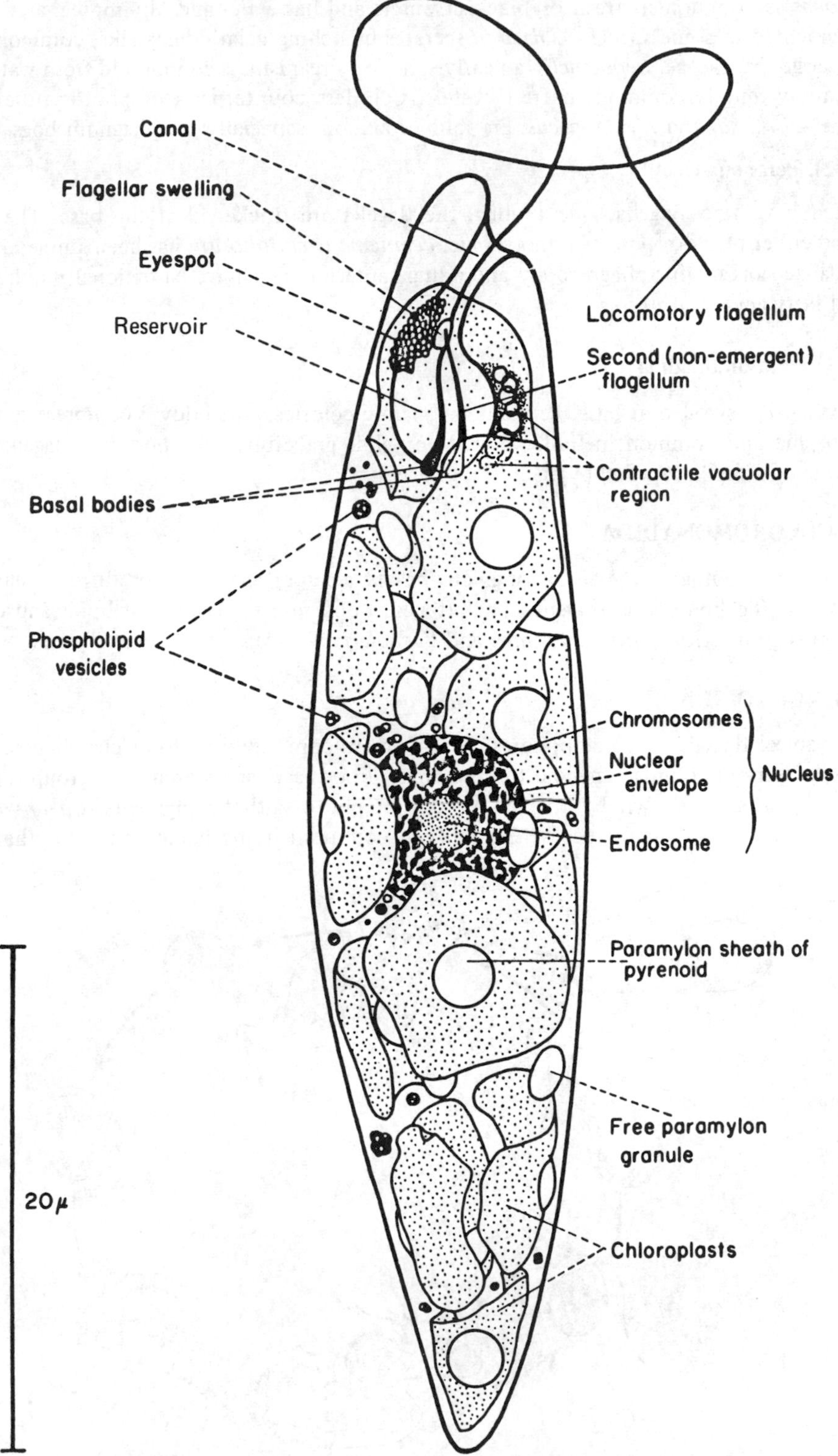

FIGURE 1. *Euglena gracilis*. Taken from: Leedale, G. F., *Euglenoid Flagellates,* p. 22 (1967). Reproduced by permission of Prentice-Hall, Inc., Englewood Cliffs, New Jersey.

Trachelomonas is common in fresh or brackish waters and has a Fe- and Mn-impregnated rigid cell wall, highly ornamented in some species. *Colacium* secretes branching gelatinous stalks, commonly attached to turtles and large Crustaceae. *Lepocinclis,* a nearly spherical organism, is common in fresh waters. *Phacus* has a leaf-like body and is common in fresh waters. Colorless counterparts of photosynthetic forms, e.g., *Heteronema, Distigma,* and *Hyalophacus,* are fairly common, especially in sphagnum bogs.

Suborder (2). Peranematina

Peranematina have two flagella, one trailing; the flagella are thickened at the base. The organisms are colorless and either phagotrophic or osmotrophic. *Peranema trichophorum* has been studied in pure culture. It has specialized organs for phagotrophy and a huge anterior flagellum. Nutritional requirements include unsaturated fatty acids and sterols.

Suborder (3). Petalomonadina

One or two flagella, swollen at base. Petalomonadina are colorless; the body is compressed and rigid. They are phagotrophic and common in fresh water. Many glide gracefully on a posterior flagellum, which acts like the runner of a sled, propelled by an anterior flagellum.

ORDER 9. CHLOROMONADIDA

Two flagella, one trailing; the flagella originate beside a superficial cleft or furrow. Many have green chromatophores; the body is dorso-ventrally flattened. Food reserves consist of lipids and glycogen. Two members of this group, *Gonyostomum semen* and *Vacuolaria virescens,* occur in acid bogs.

ORDER 10. VOLVOCIDA

Two to four apical flagella. Chromatophores are leaf-green, commonly shell- or cup-shaped. Some species are colorless; solitary or colonial ameboid forms are rare. Food reserve is starch. This group is clearly related to the higher plants, as shown by the composition of photosynthetic pigments. *Chlamydomonas* (two flagella, see Figure 2) has many species. Sexuality is common; homothallic and heterothallic strains are

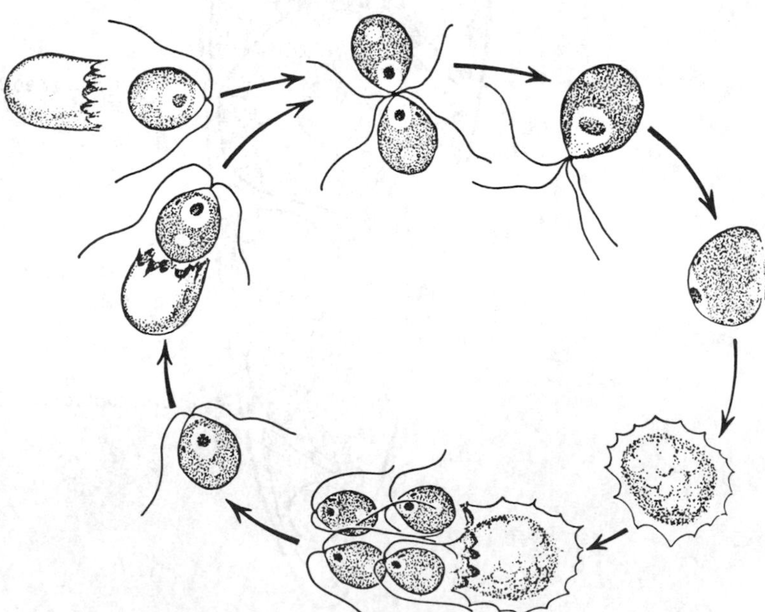

FIGURE 2. Mating and Life Cycle in *Chlamydomonas reinhardi.* Taken from: Curtis, H., *The Marvelous Animals: An Introduction to the Protozoa,* p. 68, Natural History Press (1968). Copyright (c) 1968 by Helena Curtis. Reproduced by permission of Doubleday and Company, New York.

available. The heterothallic strains (i.e., those with separate mating types, + and -) *C. reinhardi (C. reinhardti)* are frequently used to study the genetic control of the chloroplast. The organism grows in the dark on acetate and hence survives many mutations that partly or wholly eliminate photosynthesis. Whether a character is controlled by nuclear (i.e., Mendelian) or extranuclear genes is determinable by crosses. Reviews on Volvocida are given in References 33, 34, and 35.

Chlamydomonas mundana is used to study photoassimilation, i.e., light-promoted assimilation of organic compounds, especially acetate.[36]

Photosynthetic unicellular volvocines have many colorless counterparts. *Polytoma,* essentially a colorless *Chlamydomonas,* is an established research object. *Polytomella,* the four-flagellated colorless counterpart of *Carteria,* is prized for its abundant growth in strongly acid media on simple substrates, e.g., lower fatty acids. Other noteworthy unicellular forms in pure culture include the following: *Haematococcus pluvialis,* commonly forming brick-red incrustations in cement bird baths (allied species occur in potholes just above the tidal zone); *Dunaliella salina,* abundant at end stages of concentration of seawater in producing solar salt, has a high electrolyte requirement; the less halophilic marine but extremely hardy *D. tertiolecta* is widely used as a model marine flagellate.

Colonial forms range from *Gonium* and *Pandorina* to *Eudorina* to *Volvox;* many are in pure culture; all need vitamin B_{12}. *Chlamydobotrys* and *Astrephomene* are used to study photoassimilation because of pronounced heterotrophy. Various *Volvox* species are under scrutiny because they secrete diffusible factors eliciting differentiation of sexual colonies.[37]

ORDER PRASINOPHYCEAE

Mainly marine planktonic, these organisms are not yet assimilated into the protozoological taxonomic scheme. They comprise forms previously included in the Volvocida (Chlorophyceae) or Chrysophyceae (Chrysomonadida), having scales or hairs on the surface of the flagella. Prominent genera are *Pryamimonas, Nephroselmis, Platymonas,* and *Micromonas* (one flagellum, one mitochondrion, and one chloroplast – macromorphologically the simplest known flagellate).

CLASS 2. ZOOMASTIGOPHOREA (ANIMAL-LIKE FLAGELLATES)

Chromatophores are absent. The organisms have one to many flagella; additional organelles may be present in mastigonts; ameboid forms, with or without flagella, are found in this class. Sexuality is known in a few groups. Many are parasitic.

ORDER 1. CHOANOFLAGELLIDA

A single anterior flagellum is surrounded posteriorly by a delicate collar. Some species are loricate. Attached forms may be with or without peduncle. Solitary and colonial exist. The organisms are free-living and fairly common, especially in inshore waters. Their resemblance to sponge choanocytes elicits speculation as to their being forebears of sponges. Few laboratory studies have been made; no cultures are available.

ORDER 2. BICOSOECIDA

Members of this group have two flagella, one free, the other attaching the posterior end of the organism to shell. The organisms are free-living. They are common, but seldom abundant, especially in fresh water. No laboratory studies have been made.

ORDER 3. RHIZOMASTIGIDA

Pseudopodia and one to four flagella (many in one family) are present simultaneously or at different times in trophozoites. Most species are free-living. The rare, free-living *Mastigamoeba* is a spectacular giant ameba with a giant flagellum. *Histomonas meleagridis* causes a serious disease, blackhead, in turkeys and young chickens; highly pleomorphic, its tissue forms are ameboid without flagella; in cultures (grown with bacteria because pure cultures are not achieved) or in the lumen of the coeca it has flagella, generally four. The organism is transmitted via the eggs of the cecal nematode *Heterakis gallinarum.* It probably belongs in the order Trichomonadida, since it has an axostyle, pelta, and parabasal body;[38] it shares with trichomonads the high sensitivity to certain nitro-heterocyclic drugs, e.g., metronidazole.

ORDER 4. KINETOPLASTIDA

Kinetoplastida have one to four flagella. They contain a kinetoplast, a DNA-rich, specialized region of the mitochondrion, which appears to be single, continuous, and more or less branched. Most species are parasitic.

Suborder (1). Bodonina

Typically two unequal flagella, one directed anteriorly, the other posteriorly; there is no undulating membrane. The free-living, kinetoplastidic *Bodo* is ubiquitous in fresh and marine waters (one bacterized species is available from culture collections). *Icthyobodo (Costia) necatrix* is pathogenic to fish.

Suborder (2) Trypanosomatina

One flagellum, free or attached to the body by an undulating membrane; all species are parasitic. Generic distinctions are diagrammed in Figure 3.

The trypanosomatids are a vast, well-defined group, parasitizing insects, annelids, and less commonly, other invertebrates; many parasitize vertebrates. Trypanosomatids are divided into "lower" Trypanosomatidae, i.e., forms lacking a vertebrate host; the main genera are *Leptomonas, Herpetomonas, Crithidia, Blastocrithidia,* and *Phytomonas.* Higher Trypanosomatidae comprise *Leishmania, Trypanosomas,* and *Endotrypanum.* Representatives of all well-known genera of lower Trypanosomatidae, except *Phytomonas* (from lactiferous plants), have been grown in defined media and are amenable to biochemical

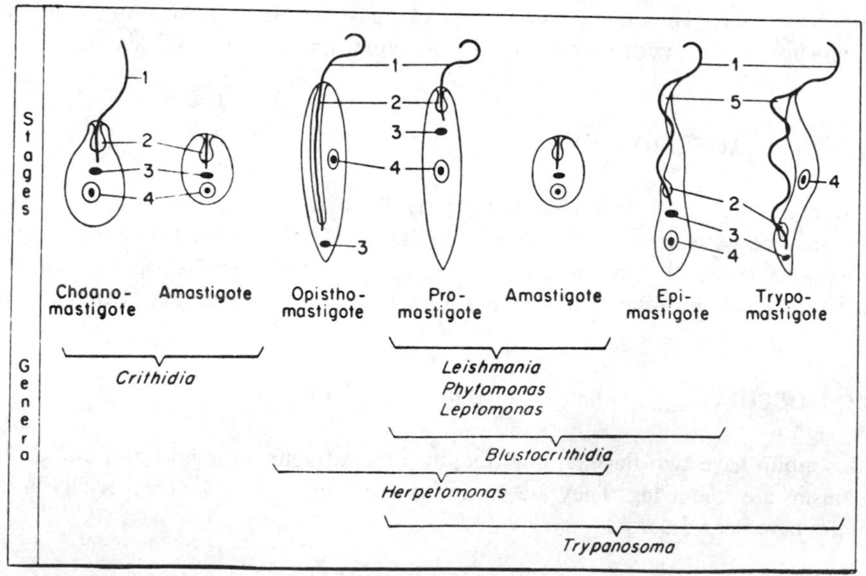

FIGURE 3. Generic Distinctions in Tryptanosomatidae. 1. Flagellum. 2. Flagellar pocket or reservoir. 3. Kinetoplast. 4. Nucleus. 5. Undulating membrane. Taken from: Cosgrove, W. B., in *Developmental Aspects of the Cell Cycle,* p. 2, I. L. Cameron, G. M. Padilla and A. M. Zimmerman, Eds. (1971). Reproduced by permission of the author and the publishers, Academic Press, New York.

experimentation, especially *Crithidia fasciculata.* A defined medium for *Crithidia fasciculata* is given in Table 2.

A heme requirement is probably phyletic, but may be obscured, as in *Crithidia oncopelti,* by rickettsia-like endosymbionts.[39] A requirement for biopterin, also probably phyletic, permits a unique microbiological assay for biopterin (including its coenzymic di- and tetrahydro derivatives).

A listing of useful drug solvents for Crithidia is given in Table 3.

The complex anterior structures of *Crithidia fasciculata* are shown in Figure 4.

Herpetomonas is unique in that the flagellum traverses much of the body in a long, narrow canal extending under some conditions posteriorly to the nucleus. *Blastocrithidia* links lower trypanosomatids to

TABLE 2

DEFINED MEDIUM FOR *CRITHIDIA FASCICULATA*

Ingredient	g/liter	Ingredient	mg/liter
Nitrilotriacetic acid	0.3	Fe	9.0
1,2,3,4-Cyclopentanetetra-	2.5	Mn	7.5
carboxylic acid	0.30	Zn	7.5
NH_4HCO_3	0.40	Mo	3.0
KH_2PO_4	0.60	Cu	0.6
$MgCO_3$	2.0	V	0.3
NaCl	0.02	Co	0.15
$CaCO_3$	0.06	B	0.15
Adenine	0.02	Ni	0.15
Guanine	0.3	Cr	0.15
L-Arginine HCl	0.04	Orotic acid	6.0
L-Cysteine HCl	0.30	Uracil	3.0
L-Histidine (free base)	0.50	Thymine	1.0
L-Isoleucine	0.75	Nicotinamide	0.75
L-Leucine	0.30	Ca pantothenate	0.45
L-Lysine HCl	0.10	Na riboflavine $PO_4 \cdot 2H_2O$	0.75
L-Methionine	0.22	Pyridoxamine·2HCl	0.4
L-Phenylalanine	0.25	Thiamine HCl	0.45
L-Threonine	0.15	Folic Acid	0.45
L-Tryptophan	0.08	Hemin	14.0
L-Tyrosine	0.25		
L-Valine	0.2	**Ingredient**	**μg/liter**
Tween 80	0.5		
Substrate (sucrose,		Biotin	3.5
sorbitol, asparagine, etc.)		Biopterin	0.18

pH adjusted to 4.2–4.6 with Tris

Notes

Nearly all media for trypanosomids are adjusted to pH 7.1–7.8 because of ordinarily intense glycolytic acid formation. This can be obviated by heavy buffering of the medium, here with cyclopentanetetracarboxylate, and use of shallow layers of medium. Pyrimidines are dispensable; the purines represent an absolute requirement. Biopterin is spared by folic acid. The heme requirement is met by hemin dissolved in a viscous, hydrophilic, alkaline solvent, e.g., triethanolamine. Tween 80 (polyoxyethylene sorbitan monoleate) promotes growth, perhaps by hindering polymerization and precipitation of hemin. The medium is designed to be assembled as a dry mix with the exception of Tween and hemin solutions. For details on the assembly of trace elements see Reference 40.

Data taken from: Tamburro, K. M., and Hutner, S. H., *J. Protozool., 18,* 668 (1971). Reproduced by permission of B. M. Honigberg, Editor, *Journal of Protozoology.*

the trypanosomes; its undulating membrane, unlike that of the blood forms of trypanosomes, is inserted before rather than behind the nucleus; hence the organism resembles in morphology the trypanosomes in culture or in the invertebrate vector (the epimastigote form). Most *Leptomonas* isolates are virtually indistinguishable from culture and insect-vector form of *Leishmania. Leptomonas karyophilus* may destroy the macronucleus of *Paramecium,* then break out to infect others.

Much interest attends the kinetoplast-mitochondrion organelle because of the unique concentration of mitochondrial DNA in the kinetoplast, comprising as much as 20% of the total cell DNA. The kinetoplast is approximately 90% destroyed by such DNA-binding drugs as acriflavin and ethidium, leading to

TABLE 3

USEFUL DRUG SOLVENTS AND CRYOPROTECTANTS FOR *CRITHIDIA*[a]

Compound	Maximum Tolerated Concentration for Growth, mg/ml
1,2,4-Butanetriol	60
1,4-Cyclohexanediol	60
Dimethylsulfoxide	40
Ethylene glycol	80
Glycerol	100
1,2,6-Hexanetriol	80
1,2-Propanediol	60

[a] Based on O'Connell, K. M., *et al., J. Protozool., 15,* 719 (1968).

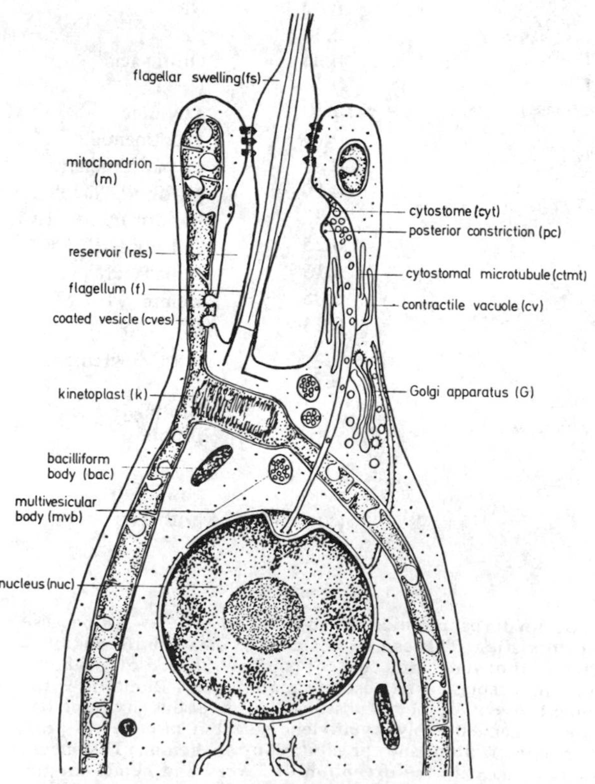

FIGURE 4. Anterior Half of *Crithidia fasciculata.* Taken from: Brooker, B. E., The Fine Structure of *Crithidia, Z. Zellforsch.,* 116, 534 (1971). Reproduced by permission of Springer Verlag, New York.

permanently dyskinetoplastic lines. Some *Trypanosoma* species are naturally dyskinetoplastic, e.g., *T. equinum;* dyskinetoplastic races of *T. brucei*-type trypanosomes may arise during drug treatment and persist in the bloodstream or tissue fluids. No dyskinetoplastic trypanosomatid has been maintained in culture, unlike the somewhat similar *petite* mutations in yeasts.

The other important trypanosomatid pathogens are described below.

Leishmania. Distinguished by intracellular multiplication in the vertebrate host, notably in cells of the macrophage series, with disappearance of the flagella (amastigote morphology). The usual vectors are sandflies (*Phlebotomus*). Proof of leishmanial habit, i.e., of intracellularity in vertebrates, is lacking. Culture forms are leptomonad in structure. The epidemiology of leishmaniases is discussed in Reference 41.

L. tropica. The agent of kala-azar or visceral leishmaniasis, this species occurs in Mediterranean countries, India, parts of Africa, and Brazil. It is nearly always fatal if untreated; it has devastated India. Most strains are sensitive to antimonials and/or diamidines, e.g., stilbamidine.

L. braziliensis. The cause of espundia in South and Central America; deep ulcerations of the nasopharynx may cause death. Some strains are rather drug resistant.

L. mexicana. Found in Central American countries, it causes "chiclero" disease and dermal lesions.

L. enriettii. Causes skin lesions in guinea pigs; it is extensively studied as a model for dermal leishmaniasis. The organism is remarkably temperature sensitive; lesions occur only in the cool regions of the body (nose and toes).

L. tarentolae. Isolated from lizards. Defined media are available.

Trypanosoma. Found in all classes of vertebrates, this genus is divided into two groups. Group I, *Stercoraria:* infection is "contaminative" by the feces of an invertebrate vector ("posterior station"; presumably primitive). Group II, *Salivaria:* infection is produced by injection of metacyclic trypanosomes in the proboscis or salivary gland ("anterior station") of an insect vector, with some exceptions where an invertebrate vector is lacking. Stercorarians retain cytochromes in a vertebrate host; salivarians lose cytochromes in a warm-blooded host; their degenerate mitochondria make them obligately glycolytic despite intense aerobiosis. For comprehensive monographs, including a detailed, critical chapter on chemotherapy, consult References 43 and 44; diagrams of life cycles and color photomicrographs of stained blood and fecal smears of these and other protozoan pathogens are found in Reference 45.

Important Stercorarian Trypanosomes

T. lewisi. Common in rats, this species is cultivable on blood agar (e.g., NNN agar). It is transmitted by fleas, which are hosts to the intracellular phase.[42] The organism is much studied as a model of non-virulent trypanosome infection.

T. cruzi. The agent of Chagas' disease of Central and South America (7 million estimated cases). Intracellular in the cardiac muscle, the organism is easily grown on blood agar. It is transmitted by reduviid hemipterans ("kissing" bugs); no drug effective against this infection is known. Many mammals are reservoir hosts. The species is discussed in Reference 46.

T. rangeli. Common in man in some areas of the American tropics, apparently non-pathogenic. Unlike *T. cruzi,* it has no intracellular stage. The organism is noteworthy because it is unique among stercorarians in transmission by the salivary route. It may kill the reduviid vector by overwhelming growth in hemocoel. A culture is available.

T. conorhini. Morphologically like *T. cruzi,* this species is isolated from reduviids. Apparently it is non-infective for man. A culture is available.

T. theileri. Common in cattle in the U.S.A., Canada, Scotland, and presumably world-wide. Apparently it is non-pathogenic. Some strains are cultivable, though with difficulty.

Amphibian Trypanosomes. *T. rotatorium* and *T. mega* have been cultivated. Leeches are the vector.

Important Salivarian Trypanosomes

T. brucei Group (includes *T. rhodesiense* and *T. gambiense*). The cause of African sleeping sickness and of nagana in domestic animals, species of this group are found in wild-game reservoirs. The vector is the tsetse fly (*Glossina*). These organisms are classical objects of chemotherapy with arsenic, and later with suramin, diamidines, and Berenil (diaceturate). Cultivated strains (insect-vector form is then assumed) generally become non-infective; infective strains defy cultivation.

T. evansi. This species occurs in many animals (equine, sheep, etc.) in North Africa, south and southeast Asia, Indonesia, and Central and South America. It is transmitted "mechanically" by tabanids and other blood-sucking insects.

T. equinum. Like *T. vivax,* but, as noted, lacks kinetoplast.

T. equiperdum. Causes dourine, a venereal disease of equines (no invertebrate vector). Infections in mice are rapidly fatal; the organism is used for standard model infection in screening trypanocides.

T. congolense. Pathogenic to wild game and domestic animals. Infection is treated with ethidium or quinapyramine (Antrycide).

T. simiae. The cause of fulminating disease in pigs; the vector is *Glossina.*[47]

T. vivax. In cattle and domestic animals, except dogs. Vectors are *Glossina* in Africa, and principally tabanids in Central and South America and in some West Indian islands.

Endotrypanum. This genus, confined to sloths, parasitizes red cells, imitating the malaria organism. Pure cultures are available.[48]

Family Cryptobiidae

Two flagella, one free, the other forming an undulating membrane. Some Cryptobia species are blood parasites of fish and may devastate trout and salmon hatcheries; they have defied cultivation. Other members of the group occur in reproductive organs of invertebrates, mainly molluscs.

ORDER 5. RETORTAMONADIDA

Two or four flagella, one twined posteriorly and associated with a ventrally located cytostomal area; the cytostome is bordered by a fibril. Members of this group are parasitic. *Chilomastic mesnili* is occasionally found in the human intestine; the clinical significance is unknown. The organism is not cultivated. *Retortamonas (Embadomonas) intestinalis* occurs in the cecum of man, and probably in primates generally; it is probably a harmless commensal.

ORDER 6. DIPLOMONADIDA

The bodies of these organisms are bilaterally symmetrical and have two karyomastigonts, each with four flagella and a set of accessory organelles. Most species are parasitic. *Giardia lamblia* inhabits the duodenum and jejunum and may cause enteritis and malnutrition, especially in children; it is spread by cysts; chloroquine and quinacrine are used for therapy. This species has been cultivated with a yeast. *Hexamita* species are occasionally pathogenic to oysters and fish.

ORDER 7. OXYMONADIDA

Oxymonadida have one or more karyomastigonts, each with three flagella. They are intestinal parasites of cockroaches and termites and strictly anaerobic. Almost no experimental work has been done with these organisms.

ORDER 8. TRICHOMONADIDA

Trichomonads typically have four to six flagella, one recurrent; an undulating membrane, if present, is associated with the recurrent flagellum. Both axostyle and parabasal bodies are present. True cysts and sexuality are unknown. Nearly all members of this order are parasitic.

Family Monocercomonadidae

Lacking the undulating membrane and costa found in the Trichomonadida, these flagellates may be ancestral to the trichomonads. Some *Monocercomonas* species have been cultivated. *Monocercomonas, Hexamastix, Protrichomonas,* and *Chilomitus* occur in the intestinal tracts of domestic animals.

Family Trichomonadida

A large family, which is discussed in detail (subfamilies, etc.) in Reference 49. Only a few *Trichomonas* species are mentioned here. *Trichomonas* is of biochemical interest because it lacks mitochondria (a rarity among non-protozoal eukaryotes); some species are so anaerobic that they produce copious H_2. The physiology of trichomonads is exhaustively reviewed in Reference 51. Little later work, aside from chemotherapy, is available.

T. foetus. A cause of infectious abortion in cattle, has enzymes that split off and destroy haptenic carbohydrate determinants of some human blood groups (e.g., groups H and Le^a).[50] This species is relatively easy to cultivate; it is not as stringently anaerobic as *T. vaginalis.*

T. vaginalis. The source of common venereal disease of man, causing superficial inflammatory lesions in the vagina, and sometimes of the urethra and prostate. Metronidazole, a nitroimidazole, is remarkably effective – even by mouth – against *T. vaginalis* infections.

T. gallinae. The cause of serious losses of pigeons and turkeys, because of ulcerative lesions in the upper digestive tract; the organism may invade the liver. This species is readily cultivable.

ORDER 9. HYPERMASTIGIDA

Nearly all members of this group occur in termites, and some in the wood roach *Cryptocercus.* Some have attained fantastic complexity, along with remarkable sexual phenomena in which induction of sexuality is governed by hormones of the host; hence, some are potential assay organisms for insect hormones. Conspicuous centrioles and chromosomes are present in some species. In some termites the organisms are conspicuously symbiotic, ingesting and digesting wood splinters. Details of the life cycles of Hypermastigida have been described by L. R. Cleveland in various publications; his classical motion pictures of sexual and mitotic phenomena amply document the often spectacular cytological phenomena (conspicuous centrioles, complex gametic behavior). Experimental work on these flagellates has recently been virtually nil because of technical obstacles, especially extreme sensitivity to O_2 and difficulty in establishing laboratory colonies of *Cryptocercus* to ensure a dependable supply of their protozoa.

Superclass II. Opalinata

Opalinata have numerous cilia-like organelles over the entire body surface; cytostomes are absent. They have two to many nuclei of one type. All are parasitic. *Opalina* and *Zelleriella* are ubiquitous in tadpoles. These hardy-seeming intestinal parasites have not been cultivated. An account of the coordination of their life cycles with those of their hosts is given in Reference 52. Affinities of opalinids to ciliates and other protozoa are obscure.

Superclass III. Sarcodina

Locomotion is usually by pseudopodia, filopodia, or reticulopodia. The group includes many shelled species. All are phagotrophs. Flagella, when present, are restricted to the developmental stages.

CLASS 1. RHIZOPODEA

Locomotion is associated with formation of characteristic lobopodia, filopodia, or reticulopodia.

Subclass (1). Lobosia

Pseudopodia are typically lobose, rarely filiform or anastomosing.

ORDER 1. AMOEBIDA

Amoebida are naked and typically uninucleate. The majority are free-living; many are parasitic. Small soil and water forms include *Acanthamoeba, Hartmannella, Mayorella,* and *Vahlkampfia.* Many of these, mainly *Acanthamoeba* and *Hartmannella,* are easily grown in defined media. *Naegleria gruberi,* abundant in soils and fresh-water muds, has a flagellate stage (as does *Tetramitus*). Some strains of this species cause a fulminating meningitis in bathers in certain waters, for which the quite toxic polyene antibiotic amphotericin B seems to be the only available drug. Some strains of *Acanthamoeba* (*Hartmannella*) are also pathogenic, but non-fatally. References 45 and 53 offer reviews of these organisms. Other reviews on ameboflagellates can be found in References 54 and 55.

The flagellate ↔ ameba transformation is much studied but poorly understood, because easily reproducible defined media for *Naegleria* and *Tetramitus* are not available. Encystation and excystation in *Hartmannella* (*Acanthamoeba*) is studied as an instance of controllable cellular differentiation. Reviews of encystment are given in References 56 and 57.

Other Noteworthy Genera. The large classroom ameba (*Ameba proteus*) is by some assigned to the genus *Chaos.* Its nutritional requirements are poorly known; it is reared on mixed live food, e.g., *Chilomonas* + *Tetrahymena* + bacteria. Some postulated life cycles may include its parasites. Most cultivated lines have one or more types of endobionts (the word "symbionts" seems inappropriate, since endobiont functions are unknown). *Pelomyxa* are multinucleate and may have as many as one thousand nuclei; individuals ranging in size from 0.5 to 1.0 mm are common. These organisms are reviewed in References 58 and 59.

Intestinal Amebas

Entamoeba histolytica. The cause of amebic dysentery, existing world-wide. Infection is caused by resistant cysts. The trophozoite is actively motile. This organism may cause liver abcesses as well as intestinal ulcerations. It may be diagnosed immunologically or by fecal examination for four-nucleate cysts. Infection is treated with emetine, iodohydroxyquinolines, and recently with metronidazole or niridazole. Cultivation generally takes place in artificial symbiosis with a bacterium or a flagellate, e.g., *Crithidia fasciculata,* and also axenically in very complex media.

E. hartmanni. This species is non-pathogenic. Its cysts are smaller.

E. coli. Common in man and monkeys, has eight nuclei.

E. invadens. A parasite of reptile intestine, highly pathogenic in zoo animals. Its cysts have four nuclei almost identical to those of *E. histolytica.* It has been grown axenically.[60]

E. moshkovskii. Recovered from sewage, this species is found in several countries. Trophozoites and cysts are almost identical to those of *E. histolytica's.*

Iodoamoeba bütschlii. Common in pigs, and occasionally in man, this organism is non-pathogenic. Its cysts have a single nucleus.

Dientamoeba fragilis. This non-pathogenic organism occurs in the mouth; it is common in some populations.

ORDER 2. ARCELLINIDA

The body is enclosed by a test or a rigid membrane; pseudopodia extrude through a definite aperture. Members of this group are free-living; they are ubiquitous in fresh waters. Common genera include *Arcella* (brown test) and *Centropyxis* and *Difflugia* (test formed of cemented sand grains or other debris). Reference 61 provides a classic magnificent monograph.

ORDER 3. ACONCHULINIDA

Naked, with filopodia. The representative genus is *Penardia.*

ORDER 4. GROMIDA

Pseudopodia are reticulose; the test has a distinct aperture. Uniflagellate gametes are present in some forms. Representative genera include *Euglypha* (common among mosses; has imbricated plates); *Gromia* (common in inshore waters; may reach a size of 2 mm; some recent descriptions are available; *Paulinella* (*P. chromatophora* has chloroplasts resembling blue-green algae).

ORDER 5. ATHALAMIDA

Naked, with reticulose pseudopodia arising from any position.

ORDER 6. FORAMINIFERIDA

The test has one to many chambers; pseudopodia protrude from an aperture, wall perforations, or both. Reproduction alternates with sexual and asexual generations, of which one may be secondarily repressed; gametes are usually flagellate, rarely ameboid; nuclear dimorphism occurs in the developmental stages of some species.

Foraminifera are extensively studied as index fossils by petroleum geologists and oceanographers. Vast areas of the ocean floor are covered by *Globigerina* ooze; calcareous skeletons of foraminifera compose the white cliffs of Dover.[62] Complex life cycles have been described but not yet subjected to systematic laboratory analysis.

ORDER 7. XENOPHYOPHORIDA

Multinucleate in hollow-tube system, with a pseudopodial network.

ORDER 8. ACRASIDA

Myxamebae typically aggregate before spore formation but do not fuse to form a true plasmodium. The flagellate stage is absent; sexuality is unconfirmed. The organisms are free-living.

This group is under intensive study because of amenability to analysis of its remarkable differentiation into spores and sterile stalk cells from a multicellular migrating pseudoplasmodium. The aggregation factor for amebae is cyclic AMP, which also serves as a recognition factor for bacterial food (secreted, e.g., by *Escherichia coli*) because it attracts the trophozoites. A defined medium for the most-studied form, *Dictyostelium discoideum,* common in forest litter, is available but yet unconfirmed. Reference 63 is one of many reviews and monographs dealing with this species. For descriptions of other genera consult recent papers in Reference 64.

ORDER 9. EUMYCETOZOIDA (MYCETAZOA, "TRUE" SLIME MOLDS)

Mycetazoa have a plasmodium and typical sporangia; a flagellate stage occurs in the life cycle. The organisms are free-living. Fruiting, i.e., sporulating bodies, of myxomycetes are often conspicuous to the naked eye on decaying wood in moist woodlands.

This order includes many genera. Defined media requiring heme have been devised for *Physarum polycephalum,* which therefore is attracting much attention. A comprehensive treatise with many illustrations in color is given in Reference 65.

ORDER 10. PLASMODIOPHORIDA

A large true plasmodium is developed inside host cells; the sporangia are not typical; the spores liberate a flagellate stage. The organisms are parasites in plants.

Plamodiophora brassicae is almost the only species studied; it causes the serious clubroot in cabbage and other crucifers, where intracellular invasion causes cellular hypertrophy. Other species infect sedges.

ORDER 11. LABYRINTHULIA

Groupings of spindle-shaped individuals that glide along filamentous tracks and form a slime net. Zoospores are reported but not yet confirmed. These marine organisms occur frequently on eelgrass (*Zostera*); they are a possible cause of a devastating eelgrass disease. The unique rapid gliding motility within anastomosing slimeways without obvious locomotory organelles has aroused keen interest. For a general review, see Reference 66.

Labyrinthula species are isolable on serum agar containing antibacterial antibiotics. Defined media are available for some isolates; some need sterols.

Labyrinthomyxa species multiply as amebae as well as spindle cells. One species, *Dermocystidium,* may be a serious oyster pathogen.[67]

CLASS 2. PIROPLASMEA

Small, pyriform, round, rod-shaped, or ameboid. Spores are absent. Piroplasmea have no flagella or cilia; locomotion is effected by body flection or gliding. Reproduction is asexual, either by binary fission or by schizogony. No pigment is formed from host cell hemoglobin. The organisms are heteroxenous (i.e., two or more hosts in the life cycle) and parasitic in vertebrate erythrocytes; ticks are known to be vectors. Arguments for including Piroplasmea in Sporozoa are advanced in Reference 68. However, electron microscopy reveals phagotrophy without the specialized mouth (conoid) structures characteristic of the group of Sporozoa containing the malaria parasite.

Babesia bigemina causes red-water fever in cattle; *B. canis* affects dogs; other *Babesia* species affect sheep, goats, pigs, and cats. Babesiasis of herbivores is common in the Near East, Africa, and Asia. Three cases of *Babesia* have been reported in splenectomized men. *Theileria parva* causes East Coast fever in cattle and goats in Africa; it may be highly pathogenic. *Theileria* has been cultivated in tissue culture. The organism may invade cells other than erythrocytes, notably lymphocytes. The piroplasms require special drugs; many drugs effective against Sporozoa and trypanosomatids are ineffective against these organisms.

CLASS 3. ACTINOPODEA

Spherical, typically floating forms, some attached secondarily. Pseudopodia are typically delicate and radiose, either axopodia or with filose or reticulate patterns. Actinopods are naked or have a test, which may be membranous, chitinoid, or of silica or strontium sulfate. Both asexual and sexual reproduction occur; the gametes are usually flagellated.

Subclass (1). Radiolaria

The central capsule is perforated by one, three, or more pores. Members of this group have a siliceous skeleton or spicules, and either filopodia, reticulopodia, or (occasionally) axopodia. They are generally radially symmetrical.

Radiolarians are found in marine habitats and are pelagic. Many have photosynthetic symbionts (zoochlorellae) and skeletons of jewel-like intricacy. Some deep-sea deposits are predominantly radiolarian ooze. No experimental studies have been made. For a sketch of the group, see Reference 69.

Subclass (2). Acantharia

The central capsule is a thin simple membrane without special pores; the skeleton is of celestite ($SrSO_4$), consisting of regularly arranged radial spines, with axopodia. Acantharia are marine organisms, considered by some,[69] as a suborder of radiolaria. No experimental studies have been made.

Subclass (3). Heliozoia

Without central capsule, usually naked; the skeleton, when present, is of discrete siliceous scales and spines. Heliozoa have axopodia or filopodia and occur mostly in fresh water.

Actinophrys sol and *Actinosphaerium* species are common in ponds, the latter are large. Some classical work on nuclear phenomena in *Actinophrys* has been done; recent work centers on microtubular contributions to axopodial organization and streaming movement. *Clathrulina,* with a polyhedral shell and stalk, is also common in eutrophic ponds. No stabilized laboratory cultures are available of any heliozoan.

Subclass (4). Proteomyxidia

Proteomyxidia have no test; some species have filopodia and reticulopodia. Flagellated swarmers and cysts are present in some species.

This group lumps together poorly known forms, many dwelling in soil and forest litter, and some plasmodial and spectacularly large. They are erratically obtainable (along with slime molds and acrasians) by use of intact bacterial or fungal cells as food ("bait") lawns on soft agar plates in moist chambers. No cultures are available.

SUBPHYLUM II. SPOROZOA

Spores are typically present; they are simple, without polar filaments and with one to many sporozoites. Sporozoa have a single type of nucleus. Cilia and flagella are absent, except for flagellated microgametes. Sexuality, when present, is syngamy. All Sporozoa are parasitic.

CLASS 1. TELOSPOREA

Spores are present; reproduction is sexual and asexual. Locomotion is effected by body flection or gliding; pseudopodia ordinarily absent, but if present, they are used for feeding, not locomotion. Flagellated microgametes are found in some groups.

Subclass (1). Gregarinia

Mature trophozoites are extracellular, large, and parasites of the digestive tracts and body cavities of invertebrates. They are common in insects and annelids. The much-studied genus *Selinidium* is especially common in marine polychaetes. Species of *Monocystis* are common in the seminal vesicles of earthworms.

The many gregarine genera are separated, among other characters, by differences in the organs of attachment at the anterior end, which penetrates the host cell. They are seldom pathogenic to the host. Sexual conjugation is conspicuous in some forms. Few biochemical studies have been made, but many life cycles have been worked out.

Subclass (2). Coccidia

Mature trophozoites are small and typically intracellular.

ORDER EUCOCCIDA

Schizogony is present; both asexual and sexual phases occur in the life cycle. The organisms are found in epithelial and blood cells of invertebrates and vertebrates.

Suborder (1). Adeleina

Males and females develop in association with each other (syzygy); the microgametocyte usually produces few microgametes; the sporozoites are enclosed in an envelope. The organisms may be either monoxenous (single host) or heteroxenous; they are generally non-pathogenic. Representative genera include *Adelina* in annelids and *Klossia helicina* in snail kidneys.

Suborder (2). Eimeriina

The macrogamete and microgametocyte develop independently. Syzygy is absent. The microgametocyte typically produces many microgametes. The zygote is non-motile; the oocyst does not enlarge during sporogony; the sporozoites are typically enclosed in a sporocyst. The organisms are monoxenous or heteroxenous.

Coccidia include parasites living in intestinal cells of all classes of vertebrates; *Toxoplasma,* an exception, invades other tissues. Some coccidia cause severe losses of poultry and cattle; synthesis of coccidiostats as feed additives has become a major industrial activity. Cats are so far the only hosts known to shed oocysts of *Toxoplasma gondii.* Toxoplasmosis is widespread in some human populations and may cause severe and sometimes fatal infection involving lungs, liver, and, most seriously, the brain. Cattle, sheep, and dogs in some areas are commonly infected. Antimalarial therapy is effective, especially pyrimethamine in combination with sulfonamides. The most important other genera are *Eimeria* and *Isospora. Besnoitia* causes wasting disease in cattle and horses; pseudocysts are located in the skin.

Much effort is being devoted to coccidia of domestic animals, aimed at reproducing the life cycle *in vitro;* oocysts, shed abundantly, serve as convenient, easily axenized inocula. Intracellular stages in various mammalian cell lines are readily obtained after germination of the oocysts and release of sporozoites with the aid of CO_2 + bile + trypsin.

Suborder (3). Haemosporina

The macrogamete and microgametocyte develop independently. Syzygy is absent. The microgametocyte produces a moderate number of microgametes. The zygote is motile in some forms; the oocyst enlarges during sporogony; the sporozoites are naked. Schizogony occurs in vertebrate, and sporogony in invertebrate hosts. The organisms are heteroxenous. Pigment is ordinarily formed from host cell hemoglobin.

Plasmodium. The malaria organism – cause of the most important infectious disease of man – is discussed in Reference 70. Until 1950, deaths due to malaria totaled at least 2.5 million a year, with half the world's population living at risk. The need to suppress the anopheline mosquito vector offers the weightiest argument for continued use of DDT; an adequate substitute for DDT is still lacking. The following are the main species: *P. vivax,* 48-hour cycle, governed by release of schizonts (asexual division forms) from erythrocytes; *P. malariae,* 72-hour cycle; *P. ovale* (uncommon), 48-hour cycle; *P. falciparum,* 48-hour cycle. The last-mentioned organism is the most dangerous species; it tends to be scarce in peripheral blood, and hence diagnosis may be missed. Unlike the other species, *P. falciparum* lacks an exoerythrocytic stage. The parasites concentrate in the capillaries and sinuses of internal organs; they may kill by causing embolism in the brain capillaries as infected erythrocytes may become sticky.

Chemotherapy. 4-Aminoquinoline schizontocides, e.g., chloroquine or amodiaquine, are employed for standard mass prophylaxis and treatment. Exoerythrocytic forms, i.e., those developing in the liver, require treatment with 8-aminoquinolines, e.g., primaquine. Combinations of pyrimethamine (Daraprim®) and sulfonamides (e.g., sulforthomidine or 4,4′-diaminodiphenylsulfone) are widely used for *P. falciparum* strains resistant to chloroquine and (fortunately seldom) quinine. Chloroquine-resistance is widespread in southeast Asia, Colombia, and Brazil. Repository drugs, i.e., poorly soluble drugs injected to form local deposits from which the active drug is slowly released, are widely used, e.g., salts of cycloguanil (similar to pyrimethamine) with emboic or pamoic acids (trivial names for certain high-molecular organic acids forming poorly soluble salts). The subject is extensively treated in Reference 71.

Malaria in Birds and Mice. Emergence of drug resistance has caused resumption of large-scale screening of antimalarial drugs by means of model infections in mice. The avian *P. gallinaceum* and *P. cathemerium* were extensively used in the World War II screening program; they have now been superseded by *P. berghei* of rodents, which is lethal to white mice and slightly less so to the golden hamster.

Plasmodium Species of Primates. The distinction between malaria of man and that of other primates is not absolute. *P. cynomolgi* of monkeys, closely similar to *P. vivax,* can infect man; the same is true of *P. knowlesi* of monkeys (24-hour cycle), which is lethal to the rhesus monkeys.

Biochemistry and Nutrition. Aside from the folic-reductase inhibitors, the mode of action of antimalarials is unknown, but the DNA-binding propensities of such drugs as chloroquine have elicited much speculation, as yet of little predictive value in designing practical antimalarials. Phagotrophic attack on the erythrocyte by *Plasmodium* has been demonstrated by electron microscopy. Nutritional studies have been mainly on duck malaria, *P. lophurae,* revealing complex requirements; CoA and *p*-aminobenzoate are essential.

***Plasmodium*-like Genera.** These differ from *Plasmodium* in that the only stage in the erythrocyte is the gametocyte; exoerythrocytic schizonts occur in various tissues. They occur in primates, birds, and reptiles. *Hepatocystis* in monkeys is transmitted by midges. *Haemoproteus* is common in birds; schizogony occurs chiefly in the lung. *Leucocytozoon* causes severe disease in domestic ducks; the vector is the blackfly (*Simulium*).

SUBPHYLUM III. CNIDOSPORA

Cnidospora have spores with one or more polar filaments and one or more sporoplasms. All are parasitic.

CLASS 1. MYXOSPORIDEA

Members of this class have spores of multicellular origin with one or more sporoplasms, and two or three (rarely one) valves.

ORDER MYXOSPORIDA

The spores of Myxosporida have one or two sporoplasms and one and six (typically two) polar capsules; each capsule has a coiled polar filament, which probably functions as an anchor; the spore membrane generally has two, and occasionally up to six, valves. Coelozoic or histozoic forms occur in cold-blooded vertebrates, almost exclusively fish. Many produce serious diseases, e.g., "twist" disease of salmon (*Myxosoma cerebralis*), where the organism attacks cartilage, including that of the skull. *Ceratomyxa* is found in the viscera of rainbow trout and in the gallbladder of marine fish.

CLASS 2. MICROSPORIDAE

Microsporidae have spores of unicellular origin with a single sporoplasm, a single valve, and one long, tubular polar filament through which sporoplasm emerges. They are cytozoic in invertebrates and lower (rarely higher) vertebrates. Among the members of this class are the species *Nosema bombycis,* cause of the fatal silkworm disease pébrine studied by Pasteur, and *N. apis,* which infects the honey-bee. *Glugea* forms dermal or visceral masses in fish. Both *Nosema* and *Glugea* contain species that parasitize trematode larvae in marine bivalves. *Plistophora myotrophica* infects skeletal muscles of the toad *Bufo bufo.* Colonies of anopheline mosquitoes are occasionally devastated by *Thelohania* or other microsporidia, suggesting cultivation of microsporidia for biological control of mosquitoes.

SUBPHYLUM IV. CILIOPHORA

Simple cilia or compound ciliary organelles occur in at least one stage of the life cycle; subpellicular infraciliature is universally present, even when the cilia are absent. Two types of nucleus exist, except in a few homokaryotic forms. Sexuality involves conjugation, autogamy, and cytogamy. Most species are free-living.

CLASS 1. CILIATA

Members of this class have the characters of the subphylum.[72]

Note: Designations of higher taxa for Sporozoa are scheduled for change in the next edition (date uncertain) of the classification proposed by the Society of Protozoologists; e.g., Sporozoa may be in the subphylum Apicomplexa, the Cnidospora are separated into two subphyla (Microspora and Myxospora), and the piroplasmas are back in the Apicomplexa (Dr. N. D. Levine, personal communication).

Subclass (1). Holotrichia (Holotrichs)

The somatic ciliature is often simple and uniform; the buccal ciliature, present in only two orders, is basically tetrahymenal and generally inconspicuous.

ORDER 1. GYMNOSTOMATIDA

Gymnostomatida have essentially no oral ciliature; the cytostome opens directly to the outside; the cytopharyngeal walls contain rods. Body morphology and ciliation are usually simple. Most members of this group are large. Common genera are listed below.

Chilodonella. Common in grazing on diatoms in streams. They have a conspicuous pharyngeal basket.

Coleps. Common in fresh-water aquaria and old infusions. The body is barrel-shaped, with conspicuous calcareous plates that make it resemble a hand grenade.

Didinium. A famous predator on *Paramecium,* much studied morphologically.

Dileptus. Large, with elephantine-like proboscis, and aggressively voracious, even attacking flatworms. This genus is fairly common.

Lacrymaria and Amphileptus. Has a long, swan-like, extensile proboscis, with the mouth at the base.

Nassula. A common organism, watermelon-shaped. One fresh-water species has a predilection for blue-green algae, resulting in brightly colored food vacuoles presumably due to retention of phycobilins.

Spathidium. Meat-cleaver-shaped; common among mosses.

Stephanopogon. Renowned for having only one type of nucleus; has not been studied in recent years.

Trachelocerca. Common in sand. It is like *Lacrymaria* in having a long, slender proboscis. The macronuclei do not divide, and hence are of cytological interest.

Prorodon and Holophrya. Fairly common, large fresh water-planktonic predators.

ORDER 2. TRICHOSTOMATIDA

The somatic cilia are typically uniform, but highly asymmetrical in some forms; a vestibular, but no buccal ciliature, is present in the oral region.

Balantidium coli. Common in the cecum and colon of domestic pigs and may cause dysentery in man; extraintestinal lesions are absent. It forms cysts; the sexual conjugation is known. Infection can be treated successfully with emetine, diiodohydroxyquin, or chloroquine.

Other Genera. *Colpoda* is extremely common in soil, forming very resistant cysts. *Dasytricha* and *Isotricha* are part of the rumen microflora of many ruminants. A monograph on rumen protozoa describing these and other ciliates is offered in Reference 73.

ORDER 3. CHONOTRICHIDA

The somatic ciliature is absent in mature individuals. Adults are vase-shaped, and attached to crustaceans by a non-contractile stalk. Reproduction takes place by budding. No laboratory studies have been made. The genus *Chilodochona* is common on crabs; *Spirochona* are found on fresh-water gammarids.

ORDER 4. APOSTOMATIDA

The somatic cilia of mature forms are spirally arranged, typically with a unique rosette near an inconspicuous cytostome. The life cycles are polymorphic, with marine crustaceans usually involved as hosts. *Anoplophrya* and other genera are common in the gut of earthworms. *Foettingeria actiniarum* undergoes its phoront stage (i.e., the encysted stage in which they are attached to the host and may undergo reorganization) in crustaceans, and its trophont (growth and feeding) stage in actiniid sea anemones, where it is common. *Gymnodiniodes* species are common on the gills of marine crustaceans; they have a complicated life cycle, correlated with that of the host.

ORDER 5. ASTOMATIDA

The somatic ciliature is typically uniform. The organisms are mouthless and often large. Some species have endoskeletons and holdfast organelles. Catenoid "colonies" are typical of some groups. Astomatida are mostly parasitic in oligochaetes. *Anoplophyra* species are common in the gut of earthworms.

ORDER 6. HYMENOSTOMATIDA

The somatic ciliature is typically uniform; the buccal cavity is ventral, with the ciliature fundamentally composed of one undulating membrane on the right and an adoral zone of three membranelles on the left. The organisms are often small.

Tetrahymena, a typical genus, is the subject of two monographs (References 74 and 75). Probably more biochemical work has been done on *T. pyriformis* than on all other ciliates put together, *Paramecium* excluded. Table 4 depicts a typical defined medium for *T. pyriformis*. The only nutritional peculiarity is the lipoic acid requirement.

The defined medium shown was devised for a micronucleate strain belonging to the mating type II of syngen one (*syngen* = a group of strains mating with one another). This is an indication that *Tetrahymena* represents the third type of eukaryote for which rigorously controlled biochemical genetics are presently feasible; the others are yeasts and heterothallic strains of *Chlamydomonas,* e.g., *C. reinhardi,* as noted previously. The cell membrane of *Tetrahymena* contains the triterpenoid tetrahymanol and only traces of sterols.

Tetrahymena includes species that change when they become cannibalistic, from a small-mouthed to a large-mouthed form, as in *T. vorax.* Some species require lipids (unsaturated fatty acids and sterols), e.g., *T. setifera.*[76]

Paramecium. Most recent work centers on *P. aurelia.*[77] Defined media are available. An outstanding nutritional peculiarity is its absolute requirement for plant sterols, notably stigmasterol; here *Paramecium* uniquely resembles the guinea pig. *Paramecium* has not yet been exploited as an assay organism for this

TABLE 4

DEFINED MEDIUM FOR *TETRAHYMENA PYRIFORMIS*

Ingredient	g/liter	Ingredient	g/liter	Ingredient	g/liter
K_3 citrate·H_2O	1.0	Ca pantothenate	0.03	L-Isoleucine	0.3
KH_2PO_4	0.1	Nicotinic acid	0.02	L-Leucine	0.5
$(NH_4)_2SO_4$	0.2	Thiamine HCl	0.005	L-Lysine HCl	0.3
$MgCO_3$	1.0	Na riboflavin PO_4·$2H_2O$	0.003	L-Methionine	0.25
$CaCO_3$	0.3	Pyridoxamine·2HCl	0.001	L-Phenylalanine	0.1
Fe [as Fe $(NH_4)_2(SO_4)_2$·$6H_2O$]	0.02	Pyridoxal HCl	0.001	L-Serine	0.15
Cu (as $CuSO_4$, anhydrous)	0.013	Folic acid	0.0005	L-Threonine	0.3
Mn (as $MnSO_4$·H_2O)	0.01	DL-Lipoic (thioctic) acid	0.0001	L-Tryptophan	0.12
Zn (as $ZnSO_4$·$7H_2O$)	0.01	Biotin	0.00001	L-Tyrosine	0.05
Mo [as $(NH_4)_6$ Mo_7 O_{24}·$4H_2O$]	0.005	Glucose	4.0	L-Valine	0.15
Co (as $CoSO_4$·$7H_2O$)	0.0001	L-Arginine HCl	0.4	Na_2 guanylate·H_2O (guanosine 2′, and 3′-monophosphates)	0.12
V (as NH_4VO_3)	0.0001	L-Glutamic acid HCl	2.5		
Uracil	0.06	L-Histidine HCl:H_2O	2.5		

pH = 5.4

Data taken from: Hutner, S. H., Baker, H., Frank, H., and Cox, D., in *Nutrition of Lower Organisms,* p. 89, R. N. Fienes, Ed. (1972). Reproduced by permission of the authors and the publishers, Pergamon Press Ltd., Oxford, England.

class of sterols, which presumably has an essential and wholly unknown function at least in metazoa and ciliates as well as higher plants. Endozoic bodies of *P. aurelia* and *P. caudatum* include the unique *kappa,* which controls secretion of a *Paramecium*-killing substance, and various rod-shaped bodies, which are almost certainly bacteria. *P. bursaria,* common in ponds, has symbiotic *Chlorella,* which have been cultivated axenically. Several other *Paramecium* species are common, but little experimental work has been done on them.[78]

Other Representative Hymenostome Genera. *Colpidium* is widespread in fresh water, also in the cavities of sea urchins; earlier it was confused with *Tetrahymena. Glaucoma* is much like *Tetrahymena; G. chattoni* is grown in defined media containing oleate. *Frontonia* is a large form common in ponds. *Pleuronema* has a conspicuous food-getting undulating membrane; it is common in fresh waters. *Cyclidium* has a similar undulating membrane; it is common in fresh water and soil, and often appears in infusions of grass or feces.

ORDER 7. THIGMOTRICHIDA

A tuft of thigmotactic somatic ciliature is typically present near the anterior end of the body; the buccal ciliature, if present, is located subequatorially on the ventral surface or at the posterior end. The organisms are usually parasitic in or on bivalve mollusks, usually in the mantle cavity. *Ancistrum mytili* is common in mussels and sometimes very abundant. *Kidderia mytili* is found on the muscles and foot of *Mytilus edulis,* the common mussel.

Subclass (2). Peritrichia (Peritrichs)

The somatic ciliature is essentially absent in mature forms; the oral ciliature is conspicuous, winding counterclockwise around the apical pole to the cytosome. Often the body is attached to the substrate by a contractile stalk or by a prominent adhesive basal disk. Colonial organization is common. The migratory larval form has an aborally located ciliary girdle.

ORDER PERITRICHIDA

These organisms are ubiquitous in fresh water and have the characters of the subclass. The "activated sludge" process of sewage disposal consists essentially of permitting mass growth of peritrichs, which are avid feeders on bacteria but require high O_2.

Vorticella is the most common genus; individuals exist as single organisms. In the colonial *Carchesium* the stalk is not contractile; in *Epistylis* individual stalks contract independently; in *Zoothamnium* the contractile stalks are joined so that the colony contracts as a whole. In *Telotrochidium* the free-swimming stage is prominent. *Cothurnia* is loricate, i.e., in a vase-like secreted container, with a stalk; it is common on crayfish. *Vaginicola* is loricate without a stalk; it is found in fresh-water and marine habitats. *Urceolaria* lacks a stalk; it is commensal on various invertebrates. *Trichodina* has a well-developed adhesive basal disk; it is commensal or parasitic on many aquatic animals and pathogenic to fish.

No peritrich has yet been grown in defined media but progress has been reported for a *Telotrichidium.* A brief review of peritrichs is given in Reference 79.

Subclass (3). Suctoria

The mature stage lacks an external ciliature of any kind. Typically the organisms are sessile, attached to the substrate by a non-contractile stalk. Ingestion is effected through few to many suctorial tentacles. The astomatous migratory larval stage, produced by budding, has some somatic cilia.

ORDER SUCTORIDA

The suctorida, which have the characters of the subclass, are common in fresh water and along coasts. They have attracted much attention because of their unique method of feeding and their complex life cycle, which, in *Tokophrya,* shows a sharply reduced life span as a consequence of overfeeding.[80] The common fresh-water *Discophrya piriformis,* like other suctorians maintained in culture, feeds on ciliates; the latter become immobilized on contact with the tentacles, which apparently secrete a toxin. Other common genera are *Acineta* in fresh, brackish, and salt water, *Podophrya,* and *Ephelota.*

Subclass (4). Spirotrichia (Spirotrichs)

The somatic ciliature is sparse in all but one order; cirri are the dominant feature of one order; the buccal ciliature is conspicuous, with an adoral zone, typically composed of many membranelles winding clockwise to the cytostome. Often the body is large.

ORDER 1. HETEROTRICHIDA (HETEROTRICHS)

The somatic cilia, when present, are usually uniform. The body is frequently large. Some species are pigmented. A few species are loricate, with migratory larval forms. Prominent genera are described below.

Bursaria. Common in fresh water; it may reach a size of 1 mm.

Metopus. Very common in stagnant ponds; one species is common in the digestive tracts of sea urchins.

Spirostomum. Up to 4 mm in size; very common in fresh water. It is immediately identifiable by its ribbon-like shape, rather blunt at each end, and its sinuous movement. Some experimental work has been done on its calcareous granules.

Blepharisma. Fairly common. It is easily identified by its meat-cleaver-like shape and its pink to magenta coloration, which is imparted by a photosensitizing hypericin-like pigment. The genus includes several species. Nuclear behavior has been extensively studied. Preliminary reports of axenic cultivation have not yet been followed up.

Condylostoma. Ellipsoid; common in fresh and salt water. No experimental work has been done.

Nytotherus. Common in the colon of amphibia and various invertebrates.

Stentor. Often used as object in the study of the control of cytoplasmic inheritance and intracellular organelles because it lends itself admirably to micrurgy.[81,82] Common species include the following: *S. coeruleus,* which grows up to 2 mm and contains the beautiful blue pigment stentorin (structure undetermined); *S. polymorphus,* which reaches 2 mm in size and has symbiotic *Chlorella* and *S. igneus,* which contains a rose-colored pigment.

Folliculina. Common in salt or fresh water. It has a spectacularly double-winged anterior end that bears membranelles emerging from a bent lorica.

Bursaria, Blepharisma, Frontonia, and Spirostomum. Usually available from biological study houses as mixed bacterized cultures.

ORDER 2. OLIGOTRICHIDA

A very small order. The somatic ciliature is either sparse or absent; the buccal membranelles are conspicuous, often extending around the apical end of the body. Typically these organisms are small and mostly marine. *Halteria grandinella* is very common in fresh water; it is immediately recognizable by its sudden erratic jumps. *Strombidium* species are common in fresh and brackish waters.

ORDER 3. TINTINNIDA

All members of this order are loricate but motile; the lorica exhibits a variety of shapes, sizes, and composition; the oral membranelles are conspicuous when extended from the lorica. Tintinnids are typically marine and pelagic. They are common in plankton hauls; some are common coastally. Artificial cultures are now maintained in several laboratories. *Tintinnopsis* is a common genus.

ORDER 4. ENTODINIOMORPHIDA

A simple somatic ciliature is absent; the oral membranelles are functional in feeding restricted to a small area; other membranellar tufts or zones are present in many species; the pellicle is firm, often drawn out posteriorly into spines. The organisms are parasitic in herbivores and commonly found in the rumen. These ciliates are inordinately complicated; Reference 73 should be consulted for details of distribution and physiology. All members of this order are strict anaerobes; considerable literature has accumulated on their fermentation patterns in temporary cultures.

ORDER 5. ODONTOSTOMATIDA

A small group, uncommon, occurring in organ-rich fresh waters. The somatic ciliature is usually sparse; the oral ciliature is reduced to eight membranelles, which are set in a row like teeth in a comb. The body is small, wedge-shaped, and laterally compressed; the pellicle sometimes has spines.

ORDER 6. HYPOTRICHIDA (HYPOTRICHS)

The cirri arranged in various patterns on the ventral body surface; the adoral zone of membranelles is prominent. The body is dorso-ventrally flattened. Hypotrichs comprise a huge group and are ubiquitous in fresh and brackish waters. No pure cultures are presently available, but some fresh-water species have been maintained for prolonged periods on live foods, including other ciliates. Consult Reference 83 for accounts of severe gout-like imbalances induced by feeding on nucleic acid-rich foods, such as other ciliates (whose macronuclei are predominantly RNA).

Common Genera. *Stylonychia* creeps, insect-like, on its stiff cirri, as do many other hypotrichs, e.g., *Oxytricha* and *Aspidisca*, the latter common in fresh waters and aquaria. *Euplotes* has attracted study because of the wave of reorganization undergone by the macronucleus before cell division, beginning as a band at each end.

REFERENCES

1. Honigberg, B. M., Balamuth, W., Bovee, E. C., Corliss, J. O., Gojdics, M., Hall, R. P., Kudo, R. R., Levine, R. D., Loeblich, A. R., Jr., Weiser, J., and Wenrich, D. H., *J. Protozool., 11,* 7 (1964).
2. Robinson, D. G., and Preston, R. D., *Br. Phycol. J., 6,* 113 (1971).
3. Schnepf, E., and Brown, R. M., Jr., in *Origin and Continuity of Cell Organelles,* pp. 299-342, J. Rewert and H. Ursprung, Eds. Springer Verlag, New York (1971).
4. Stransky, H., and Hager, A., *Arch. Mikrobiol., 73,* 315 (1970).
5. Goodwin, T. W. (Ed.), in *Aspects of Terpenoid Chemistry and Biochemistry,* pp. 315-356. Academic Press, New York (1971).
6. Boney, A. D., *A Biology of Marine Algae.* Hutchinson Educational, London, England (1966).
7. Baker, H., and Frank, O., *Clinical Vitaminology; Methods and Interpretation.* Interscience Publications, John Wiley and Sons, New York (1968).
8. Haines, T. H., *Prog. Chem. Fats Other Lipids, 11,* 297 (1971).
9. Haines, T. H., *Annu. Rev. Biochem., 27* (in press).
10. Shilo, M., in *Algal and Fungal Toxins,* Vol. 7, pp. 67-103, S. Kadis, A. Ciegler and S. J. Ajl, Eds. Academic Press, New York (1971).
11. Hibberd, D. J., *Br. Phycol. J., 6,* 207 (1971).
12. Paasche, E., *Annu. Rev. Microbiol., 22,* 71 (1968).
13. Isenberg, H. D., Moss, M. L., and Lavine, L. S., in *Current Practice in Orthopaedic Surgery,* pp. 202-237, J. P. Adams, Ed. C. V. Mosby, St. Louis, Missouri (1969).
14. Leadbeater, B. S. C., *Br. Phycol. J., 4,* 3 (1969).
15. Droop, M., *Symp. Soc. Gen. Microbiol., 13,* 171 (1963).
16. McLaughlin, J. J. M., and Zahl, P. A., in *Symbiosis,* Vol. 1, pp. 257-295, M. S. Henry, Ed. Academic Press, New York (1966).
17. Schantz, E., in *Algal and Fungal Toxins,* pp. 3-26, S. Kadis, A. Ciegler and S. J. Ajl, Eds. Academic Press, New York (1971).
18. Baslow, M. H., *Marine Pharmacology.* Williams and Wilkins, Baltimore, Maryland (1969).
19. Rae, P. M. M., *J. Cell Biol., 46,* 106 (1970).
20. Kubai, D. F., and Ris, H., *J. Cell. Biol., 40,* 508 (1969).
21. Harrington, G. W., Dunham, J., and Holz, G. G., Jr., *J. Protozool., 17,* 213 (1970).
22. Droop, M. R., *J. Mar. Biol. Assoc. U. K., 51,* 455 (1971).
23. Soyer, M.-O., *Z. Zellforsch., 104,* 29 (1970).
24. Wall, D., and Dale, B., *Micropaleontology, 16,* 47 (1970).
25. Keller, S. E., Hutner, S. H., and Keller, D. E., *J. Protozool., 15,* 792 (1968).

26. Schlenk, H., Sand, D. M., and Gellerman, J. L., *Biochim. Biophys. Acta, 187,* 201 (1969).
27. Kofoid, C. A., and Swezy, O., *The Free-Living Unarmored Dinoflagellates.* University of California Press, Berkeley, California (1921).
28. Leedale, G., *Euglenoid Flagellates.* Prentice-Hall, Englewood Cliffs, New Jersey (1967).
29. Buetow, D. E. (Ed.), *The Biology of Euglena.* Academic Press, New York (1968).
30. Hutner, S. H., Zahalsky, A. C., and Aaronson, S., *Methods Cell Physiol., 2,* 217 (1966).
31. Wolken, J. J., *Euglena,* 2nd ed. Meredith Press, New York (1967).
32. Schiff, J. A., Lyman, H., and Russel, G. K., *Methods Enzymol., 23A,* 143 (1971).
33. Levine, R. P., *Methods Enzymol., 23,* 119 (1971).
34. Sager, R., in *Autonomy and Biogenesis of Mitochondria and Chloroplasts,* pp. 250-259, N. K. Boardman, A. W. Linnane and R. M. Smillie, Eds. North-Holland Publications, Humanities Press, New York (1971).
35. Surzycki, S. J., Goodenough, U. W., Levine, R. P., and Armstrong, J. J., Control of Organelle Development, *Symp. Soc. Exp. Biol., 24,* 13 (1970).
36. Wessner, W., in *Photobiology of Microorganisms,* pp. 95-133, P. Halldal, Ed. Interscience Publications, John Wiley and Sons, New York (1970).
37. Starr, R. C., in *Changing Syntheses in Development,* pp. 59-100, M. N. Runner, Ed. Academic Press, New York (1971).
38. Honigberg, B. M., and Bennett, C. J., *J. Protozool., 18,* 687 (1971).
39. Gutteridge, W. E., and Macadam, R. F., *J. Protozool., 18,* 637 (1971).
40. Hutner, S. H., *Annu. Rev. Microbiol., 26,* (in press).
41. Garnham, P. C. C., *Bull. W.H.O., 44,* 477-490, 621-628 (1971).
42. Molyneux, D. H., *Parasitology, 59,* 737 (1969).
43. Mulligan, H. W. (Ed.), *The African Trypanosomiases.* Interscience Publications, John Wiley and Sons, New York (1970).
44. Hoare, C. A., *Trypanosomes of Mammals.* Blackwell Scientific Publications, F. A. Davis Co., Philadelphia, Pennsylvania (1972).
45. Adam, K. M. G., Paul, J., and Zaman, V., *Medical and Veterinary Protozoology: An Illustrated Guide.* Churchill Livingstone, Edinburgh, Scotland (1971).
46. Cancado, J. R. (Ed.), *Doenca de Chagas.* Universidade Federal de Minas Gerais, Belo Horizonte, Brazil (1968).
47. Stephens, L. E., *Pig Trypanosomiasis in Tropical Africa.* Commonwealth Agricultural Bureaux, Farnham Royal, Bucks, England (1966).
48. Shaw, J. J., *The Haemoflagellates of Sloths.* H. K. Lewis and Co., London, England (1969).
49. Honigberg, B. M., *J. Protozool., 10,* 20 (1963).
50. Stealey, J. R., and Watkins, W. M., *Biochem. J.* (in press).
51. Shorb, M., in *Biochemistry and Physiology of Protozoa,* Vol. 3, pp. 383-457, S. H. Hutner, Ed. Academic Press, New York (1964).
52. Manwell, R. D., *Introduction to Protozoology,* 2nd ed. Dover Publications, Inc., New York (1968).
53. Culbertson, C. G., *Annu. Rev. Microbiol., 25,* 231 (1971).
54. Yuyama, S., in *Developmental Aspects of the Cell Cycle,* pp. 41-56, I. L. Cameron, G. M. Padilla and A. M. Zimmerman, Eds. Academic Press, New York (1971).
55. Fulton, C., *Methods Cell Physiol., 4,* 341 (1971).
56. Neff, R. J., and Neff, R. H., Dormancy and Survival, *Symp. Soc. Exp. Biol., 23,* 51 (1969).
57. Griffiths, A. J., *Advan. Microb. Physiol., 4,* 106 (1970).
58. Chapman-Andresen, C., *Annu. Rev. Microbiol., 25,* 27 (1971).
59. Jeon, K. W. (Ed.), *The Biology of Amoeba.* Academic Press, New York (in press).
60. Spies, F., and Elbers, P. F., *J. Protozool., 19,* 102 (1972).
61. Leidy, J., Fresh-Water Rhizopods of North America, *U.S. Geological Surveys,* Vol. 12. U.S. Geological Society (1879).
62. Goke, G., *Meeresprotozoen (Foraminiferen, Radiolarien, Tintinninen).* Kosmol, Franckk'sche Verlagshandlung, Stuttgart, Germany (1963).
63. Newell, P. C., in *Essays in Biochemistry,* Vol. 7, pp. 87-126, P. N. Campbell and F. Dickens, Eds. Academic Press, New York (1971).
64. Olive, L. S., *Mycologia, 58,* 404 (1966); *J. Protozool., 13,* 164 (1967).
65. Martin, G. W., and Alexopolous, C. J., The *Myxomycetes.* University of Iowa Press, Iowa City, Iowa (1969).
66. Pokorny, K., *J. Protozool., 14,* 697 (1967).
67. Sprague, V., *Annu. Rev. Microbiol., 25,* 211 (1971).
68. Baker, J. R. (Ed.), in *Parasitic Protozoa,* pp. 8-126. Hutchinson University Library, London, England (1970).
69. Jepps, M. W., *The Protozoa, Sarcodina,* pp. 120-137. Oliver and Boyd, Edinburgh, Scotland (1956).
70. Garnham, P. C. C., *Malaria Parasites and Other Haemosporidia.* Blackwell Scientific Publications, F. A. Davis Co., Philadelphia, Pennsylvania (1967).
71. Peters, W., *Chemotherapy and Drug Resistance to Malaria.* Academic Press, New York (1970).
72. Corliss, J. O., *The Ciliated Protozoa.* Pergamon Press, Oxford, England (1961).

73. Hungate, R. E., *The Rumen and Its Microbes.* Academic Press, New York (1966).
74. Hill, D., *The Biochemistry and Physiology of Tetrahymena.* Academic Press, New York (1972).
75. Elliott, A. M. (Ed.), *Biology of Tetrahymena.* Dowden, Hutchinson and Ross, New York (in preparation).
76. Holz, G. G., Jr., in *Biochemistry and Physiology of Protozoa,* Vol. 3, pp. 199-242, S. H. Hutner, Ed. Academic Press, New York (1964).
77. Jurand, A., and Selman, G. G., *The Anatomy of Paramecium aurelia.* Macmillan Co., London, England, and New York (1969).
78. Van Wagtendonk, W. J. (Ed.), *Biology of Paramecium.* Elsevier, Amsterdam, The Netherlands (in press).
79. Finley, H. E., *J. Protozool., 16,* 1 (1969).
80. Rudzinska, M. A., *J. Protozool., 17,* 626 (1970); *Gerontologia, 6,* 206 (1962).
81. Tartar, V., *The Biology of Stentor.* Pergamon Press, Oxford, England (1961).
82. Margulis, L., *J. Protozool., 17,* 548 (1970).
83. Lilly, D. W., *Ann. N.Y. Acad. Sci., 53,* 910 (1953).

LITERATURE OF PROTOZOA

DR. S. H. HUTNER

Protozoology is served by several protozoological journals; these and the many monographs available seldom make it necessary to conduct prolonged searches to ascertain the status of any important topic. Protozoological journals are described below.

Journal of Protozoology. Published quarterly, the organ of the Society of Protozoologists; printed by Allen Press, Lawrence, Kansas. The annual supplement contains abstracts of papers presented before the British, Israeli, and Scandinavian affiliated societies and of the Groupement des Protistologues de Langue Francaise. A Brazilian affiliate of the Society of Protozoologists was organized in 1972. The journal is expanding its book review services, aiming at comprehensive coverage of protozoology.

Protistologica. Published quarterly, the organ of the aforementioned Groupement; published by the National Research Council of France (C.N.R.S., 15 Quai Anatole-France, Paris VIIe, France).

Acta Protozoologica. Published quarterly. Central office: Nencki Institute of Experimental Biology, Warszawa 22, Pasteura 3, Poland.

An international congress is held every four years; the third was held in Leningrad in 1969, the fourth is planned for France in 1973. The published proceedings of these congresses touch on virtually all developments.

Many papers dealing with protozoa are also found in journals of microbiology, e.g., *Journal of Bacteriology, Journal of General Microbiology*; in journals of parasitology, e.g., *Experimental Parasitology, Journal of Parasitology, Parasitology;* in journals of tropical medicine, e.g., *American Journal of Hygiene and Tropical Medicine, Transactions of the Royal Society of Tropical Medicine and Hygiene*; in journals of limnology and oceanography, e.g., *Limnology and Oceanography, Journal of the Marine Biological Association of the United Kingdom;* and in journals serving cell physiology and biochemistry.

The two most recent textbooks, both recommended, are the following:

Manwell, R. D., *Introduction to Protozoology,* 2nd ed. Dover Publications, Inc., New York (1968).
Grell, K. G., *Protozoologie,* 2nd ed. (in German). Springer Verlag, Berlin, Germany (1971).

A review of nutrition and metabolism, prepared by S. H. Hutner *et al.,* is published in *The Biology of Nutrition*, pp. 85-177, R. N. Fienes, Ed. Pergamon Press, Oxford, England (1972).

Protozoa are comprehensively illustrated in the following publication: Kudo, R. R., *Protozoology,* 5th ed. Charles C Thomas, Springfield, Illinois (1966).

Monographs covering protozoal immunology can be found in *Immunity to Parasitic Animals,* G. J. Jackson, *et al.,* Eds., published by Appleton-Century-Crofts, Inc., New York. They are listed here in order of appearance.

Weiser, J., Immunity of Insects to Protozoa, Vol. 1, pp. 129–147 (1969).
Lom, J., Cold-Blooded Vertebrate Immunity to Protozoa, Vol. 1, pp. 249–265 (1969).
McGhee, R. B., Avian Malaria, Vol. 2, pp. 331–369 (1970).
Cuckler, A. C., Coccidiosis and Histomoniasis in Avian Hosts, Vol. 2, pp. 371–397 (1970).
Balamuth, W., and Siddiqui, W. A., Amebae and Other Intestinal Protozoa, Vol. 2, pp. 439–468 (1970).
Honigberg, B. M., Trichomonads, Vol. 2, pp. 469–550 (1970).
Desowitz, R. S., African Trypanosomes, Vol. 2, pp. 551–596 (1970).
Goble, F. C., South American Trypanosomiasis, Vol. 2, pp. 597–689 (1970).
D'Alesandro, P. A., Nonpathogenic Trypanosomes of Rodents, Vol. 2, pp. 691–738 (1970).
Stauber, L. A., Leishmanias, Vol. 2, pp. 739–765 (1970).

Garnham, P. C. C., Primate Malaria, Vol. 2, pp. 767–791 (1970).
Zuckerman, A., Malaria of Lower Mammals, Vol. 2, pp. 794–829 (1970).
Ristic, M., Babesiosis and Theilerosis, Vol. 2., pp. 832–870 (1970).
Kozar, A., Toxoplasmosis and Coccidiosis in Mammalian Hosts, Vol. 2, pp. 871–912 (1970).

In addition to the literature mentioned above, two noteworthy specialized reviews and monographs are available:

Buetow, E. E., Preparation of Mitochondria from Protozoa and Algae, *Methods Cell Physiol., 4,* 84-115 (1970).
Marcial-Rojas, R. A. (Ed.), *Pathology of Protozoal and Helminthic Diseases.* Williams and Wilkins, Baltimore, Maryland (1971).

Monographs sponsored by the Society of Protozoologists, including one on coccidia (D. Hammond, Ed.), are almost ready to go to press or are already in press.

VIRUSES

INTRODUCTION TO THE SYSTEMATICS OF VIRUSES

DR. KARL MARAMOROSCH

According to the rules of the International Committee on Nomenclature of Viruses (ICNV), laid down in 1966 during the IX International Congress for Microbiology, the taxonomic system for viruses does not classify them by the hosts they infect, but by such criteria as chemistry (DNA or RNA viruses), symmetry (helical or cubical), and other characteristics of the virions. Consequently, viruses should no longer be grouped as viruses of bacteria, fungi, algae, mycoplasma, higher plants, invertebrate and vertebrate animals, but as members of a single "kingdom" of viruses (Virales). Lwoff and Tournier[3] have presented a partial classification of all viruses according to the unified system of classification (Table 1). It utilizes the nature of the genetic material, the symmetry of the capsid, the naked or enveloped nature of the nucleocapsid, the number of capsomeres for virions with cubical symmetry, and the diameter of the nucleocapsid for virions with helical symmetry.

The molecular weight of the nucleic acid, nature of the envelope, proportion of nucleotides, number of strands of nucleic acid, antigenicity of the viral proteins, and other characteristics are used for a more complete characterization of "families". Table 2 presents the characteristics of families according to Lwoff and Tournier.[3]

Viruses require ribosomes of host cells for their multiplication, and this requirement represents the ultimate degree of parasitism. The highly specific requirements for viral proliferation explain why many viruses have a fairly limited host range and why a few viruses appear to be limited to a single host. Years ago it was believed that certain viruses attack only vertebrates, invertebrates, higher plants, algae, fungi, or bacteria. While this is generally the rule, the exceptions are fairly numerous. The host range of viruses may include a number of species within a family, order or class. Certain viruses infect vertebrate and invertebrate animals (arbo viruses). There are no reported instances in which the same virus would infect a higher plant and a vertebrate animal, although there are a few viruses that alternate between invertebrate animals and higher plants. When viruses multiply in plant and animal hosts, the invertebrate animals act not only as vectors of plant-pathogenic viruses, but also as reservoirs. Among the invertebrate animals that are known to transmit plant-pathogenic viruses are insects, mites, and nematodes.[4] Certain plant viruses are transmitted by lower fungi, some are seed-borne, and several can be transmitted from plant to plant by parasitic higher plants.

Viruses seem to parasitize all forms of life. Viruses of ferns[5] and of mosses[1] have been reported, and in recent years viruses of mushrooms and of filamentous fungi have received increasing attention.[2] To date all viruses of lower fungi have been found to contain double-stranded RNA. Although there are no reports of viruses that affect protozoa, the eventual detection of such viruses can be expected.

Since this Handbook is not a treatise of systematic microbiology, but rather a handbook where people will be able to look up information about microorganisms, the viruses that are of primary interest to plant pathologists are described as "Viruses of Plants", even though some infect insect vectors and even cause diseases of these invertebrate animals. Viruses that primarily cause diseases of insects are described under the heading of "Viruses of Insects", viruses of bacteria and mycoplasma are listed as "Bacteriophages", and so on. This practical approach is based on the outmoded criterion of host affinity and disease, rather than on morphological and chemical characteristics of the virions.

REFERENCES

1. Blattný, C., Pilous, Z., and Osvald, V., *Ochr. Rostl., 22,* 136 (1949).
2. Hollings, M., and Stone, O. M., *Annu. Rev. Phytopathol., 9,* 93 (1971).
3. Lwoff, A., and Tournier, P., in *Comparative Virology,* p. 1, K. Maramorosch and E. Kurstak, Eds. Academic Press, New York (1971).
4. Maramorosch, K. (Ed.), *Viruses, Vectors, and Vegetation.* Interscience Publications, John Wiley and Sons, New York (1969).
5. Severin, H. H. P., and Tompkins, C. M., *Hilgardia, 20,* 81 (1950).

TABLE 1

CLASSIFICATION OF VIRUSES

Nucleic Acid	Capsid Symmetry	Naked (N) or Enveloped (E)	Helical Diameter or Number of Capsomeres	Taxonomic Designation
	H	N	50Å	Inoviridae
		E	?	Poxviridae
			12	Microviridae
			32	Parvoviridae
		N	42	Densoviridae
DNA	C		72	Papilloviridae
			252	Adenoviridae
			812	Iridoviridae
		E	162	Herpesviridae
	B	N Urovirales		Tailed bacteriophages
		N Rhabdovirales		
			90Å	Myxoviridae
			180Å	Paramyxoviridae
	H	E Sagovirales		Stomatoviridae
			?	Thylaxoviridae
RNA		N Gymnovirales	32	Napoviridae
			92	Reoviridae
	C		?	Blue tongue virus (sheep)
		E Togavirales	?	Encephaloviridae

Code

H = helical; C = cubic; B = binal

Taken from: Lwoff, A., and Tournier, P., The "LHT System" of 1969, in *Comparative Virology,* p. 6, K. Maramorosch and E. Kurstak, Eds. (1971). Reproduced by permission of Academic Press, New York.

TABLE 2

THE CHARACTERISTICS OF VARIOUS FAMILIES OF VIRUSES

				Virions with Cubic Symmetry			Virions with Helical Symmetry			
Viridae	Nucleic Acid	Symmetry	Naked (N) or Enveloped (E) Nucleocapsid	Number of Capsomers	Diameter (Å) of the Nucleocapsid	Diameter (Å) of the Envelope	Diameter and Length (Å) of the Nucleo-capsid	Dimensions (Å) of the Enveloped Virions	Mol Wt (x 10^6) of Nucleic Acid	Number of Nucleic Acid Strands
Ino–	D	H	N				5–6 x 760–850		1.7–3	1
Pox–	D	H?	E				?	2500 x 1600	160-240	2
								3000 x 2300		
Micro–	D	C	N	12	250				1.7	1
Parvo–	D	C	N	32	220				1.8	1
Denso–	D	C	N	42	200				160–240	1
Papilloma– (papova)	D	C	N	72	450–550				3–5	2
Adeno–	D	C	N	252	700				20–25	2
Irido–	D	C	N	812	1300				126	2
Herpes–	D	C	E	162	775	1500–2000			54–92	2
Uro–	D	BC	N							2
Rhabdo–	R	H	N				20 x 130			1
							10 x 1250			
Myxo–	R	H	E				90 x ?	1000	2–3	1
Paramyxo–	R	H	E				180 x ?	1200	7.5	1
Stomato– (rhabdo)	R	H	E				180 x ?	1750 x 680	6	1
Thylaxo–	R	H	E				?	10000	10	1
Napo–	R	C	N	32	220–270				1.1–2	1
Reo–	R	C	N	92	700				10	2
Cyano–	R	C	N	32 or 42	540					2
Encephalo–	R	C	E	?	?		600–800		2–3	1

Code

D = DNA; R = RNA; H = helical; C = cubic; B = binal

Taken from: Lwoff, A., and Tournier, P., Remarks on the Classification of Viruses, in *Comparative Virology*, pp. 1–42, K. Maramorosch and E. Kurstak, Eds. (1971). Reproduced by permission of Academic Press, New York.

VIRUSES OF PLANTS

DR. KARL MARAMOROSCH

Some plant pathologists and plant virologists consider the classification of viruses as a controversial subject, and there has been no agreement on how to classify them within the unified system of virus classification. There is considerable agreement, however, concerning certain groups of plant viruses.[13] A virus group has been defined as a collection of viruses or virus strains that share with a type member nearly all the main characteristics of the group. Groups of plant viruses have been proposed in a manner somewhat similar to that of "families" proposed for vertebrate viruses of the modern virus classification.[36] Latin names have been accepted for only a few groups, and categories in which only a single member was known have been listed as "monotypic" groups.

The following characteristics have been used in grouping plant viruses.

1. **Chemistry.** The nature of the nucleic acid: whether RNA or DNA; the nucleotide ratio (given in moles %); molecular weight; number of nucleic acid strands; whether ring-formed or not; and sedimentation coefficient of the nucleic acid. The characteristics of the protein, such as the number of different proteins, molecular weight of subunits, number of chemical subunits per particle, number of amino acid residues per subunit, enzymatic activities, percentage of particle weight, type and amount of other compounds, especially lipids, have also been used.

2. **Morphology.** Shape and symmetry of the virion, its size, and number and arrangement of subunits were among the main morphological features used in groupings.

3. **Physical Properties.** The sedimentation coefficient, weight (given in daltons), electrophoretic mobility, isoelectric point, thermal inactivation point in plant sap during 10 minutes, retention of infectivity in sap at 20°C, and the dilution end point in sap were compared.

4. **Virus–Host Interactions and Virus Vectors.** Criteria used to group plant viruses under this heading, pertaining to the behavior of viruses in hosts and the transmission of viruses by different vectors, have limited usefulness. These criteria were rejected by the International Committee on Virus Nomenclature (ICNV) because they contradicted the established rules of virus nomenclature. The behavior in hosts included the following: host range expressed by the number and type of plant species, vectors etc., as well as the expression of disease signs (called "symptoms" by plant virologists); affinity to certain tissues, such as mesophyll, phloem, or xylem; approximate concentration in crude extracts; types and location of inclusions; effectiveness of heat therapy; seed transmissibility; and geographic distribution. Interference between viruses ("cross protection") was also considered. Taxa of vectors, the number of different species known to transmit a given virus, presence or absence as well as length of the incubation ("latent") period, retention of the virus in vectors ("persistence"), multiplication in vectors, vertical ("transovarial") transmission, and vector stages (nymphal and/or imago) able to acquire and transmit a virus were compared.

Among the invertebrate vectors of plant viruses, arthropods, such as insects and mites, as well as nematodes have been known for a number of years.[24] More recently, fungus vectors have been established as carriers of at least six different plant-pathogenic viruses.[6] Lower fungi that transmit "soil-borne" viruses belong to the genera *Olpidium, Spongospora, Polymyxa,* and *Synchytrium.*

Typically rod-shaped plant viruses that are shorter than 600 nm are transmitted by vectors in the soil ("soil-borne"), and rod-shaped viruses that are longer than 600 nm are transmitted by aphids and eriophyid mites.

It is apparent that the different kinds of characteristics used in grouping of plant viruses are not of equal importance in defining taxonomic characteristics. The type of nucleic acid and the symmetry, which were given preference by those who classified vertebrate and invertebrate viruses, were given the same importance as virus–host interactions in plant virus characterization. This is an obvious weakness of the grouping, but it has been retained here because it serves a useful purpose. It might be mentioned here that natural or induced mutations of plant viruses are known, in which the ability to infect an insect vector was lost while other characteristics, such as gross chemical, morphological, and serological properties, remained unaltered. Mutations are not known to affect the symmetry of a virion nor to change the RNA to a DNA.

Mutants, as well as related strains, are known to differ in the sequence or kind of amino acids. Geographic distribution, also considered as a criterion in grouping plant viruses, seems to be of even less importance than virus–host interactions. Nevertheless, the groupings, based on all the above characteristics, have been found useful even though they are not based on sound taxonomic criteria. The following descriptions, with minor modifications, have been published.[35]

DNA PLANT VIRUSES

Cauliflower Mosaic Virus Group

The main characteristics are double-stranded DNA, a molecular weight of 5×10^6, and a guanidine cytosine ratio of 43%.[32] The virions are isometric, approximately 50 nm in diameter, 220 s, and lack accessory particles. The thermal inactivation point is between 75 and 80°C, and infectivity in crude sap is retained for a few days. The type member, cauliflower mosaic virus, is mechanically transmissible; it is carried in nature by aphid vectors, in which it persists a few hours. The host range is narrow, and the symptoms in plants are mosaic and mottling. Dahlia mosaic virus, in the same group, is serologically related by cauliflower mosaic virus.

RNA PLANT VIRUSES

Cowpea Mosaic Virus Group

Viruses in this group are spherical, with two different types of protein in their coat.[40] The virions possess three components, all isometric and about 30 nm in diameter, approximately 115, 95, and 55 s and containing 33, 22, and 0% RNA respectively; the two fastest sedimentation components are required for infectivity.[35] The RNA is single-stranded. The thermal inactivation point ranges from 60 to 80°C, and infectivity is retained in crude sap for one to a few weeks. The members of this group include squash mosaic virus, radish mosaic virus, broad bean mosaic virus, bean pod mottle virus, and red clover mottle virus. Their host range is fairly narrow, and the natural vectors are beetles. The different viruses in this group are serologically distantly related.

Alfalfa Mosaic Virus Group (Monotypic)

These organisms have at least five components of single-stranded RNA composition, four of bacilliform shape, and one spheroidal. They are inactivated in crude sap at 60 to 70°C and retain infectivity in extracts for a few days. The three heaviest components of the virus are required for infectivity.[3] The symptoms consist of mosaic, mottling, and ringspots. The host range is wide. Transmission is by aphids, since the virus growth is stylet-borne (persisting less than two hours); it is also transmitted mechanically.

Rhabdovirus Group

This group of single-stranded-RNA viruses resembles the vesicular-stomatitis virus group of animal viruses and might eventually be grouped together with the latter. Only one representative, potato yellow dwarf virus, has been well defined chemically. Morphologically all members of this group are bacilliform or bullet-shaped. Some members multiply in arthropod vectors as well as in plants; this is of special interest because the vertebrate-pathogenic viruses of the Rhabdo group often have alternating vertebrate and invertebrate hosts.

Certain generalizations can be made on the basis of limited data. Plant-pathogenic Rhabdo viruses that have alternate insect hosts are bullet-shaped or bacilliform, with an inner nucleoprotein core and an outer envelope. The nucleoproteins of these Rhabdoviruses do not have the typical helical symmetry of the rod-shaped plant viruses. The plant viruses listed below are tentatively grouped as Rhabdoviruses because the only common criterion established until now is the bullet shape of the virions: wheat striate mosaic virus, Russian wheat mosaic virus, eggplant mottle dwarf virus, maize mosaic virus, Gomphrena virus, rice transitory yellowing virus, sowthistle yellow vein virus, plantain virus, and lettuce necrotic yellows virus; the last-mentioned has been chemically studied in considerable detail. The vectors of the above possible members of the Rhabdo group are leafhoppers, aphids, and mites. No antigenic relationships have been found between plant-pathogenic and vertebrate-pathogenic members up to now.

Table 1 lists some of the characteristics of the plant-pathogenic viruses in this group. Apparently all Rhabdoviruses mature on pre-existing cell membranes, but vertebrate- and plant-pathogenic viruses seem to differ in their affinities to such membranes. Viruses of vertebrates utilize cytoplasmic membranes, whereas most, if not all, plant viruses mature on nuclear membranes;[37] this property might explain the complex structure of plant viruses. In width the viruses range from 65 to 90 nm, and in length from 120 to 500 nm. The envelope of the Rhabdoviruses encloses an internal helical nucleoprotein with transverse striations. There seems to be no difference in the arrangement of the nucleocapsid between Rhabdoviruses of vertebrates and those of invertebrates. An RNA-dependent RNA polymerase activity seems to be associated with particles of lettuce necrotic yellows virus. This parallels the findings of a polymerase associated with vesicular-stomatitis virus.[2,10] Potato yellow dwarf virus contains four major proteins in the relative ratio of 1:4:8:6, of molecular weights 22,000, 33,000, 56,000, and 78,000 respectively.[19] Potato yellow dwarf virus seems to bridge the taxonomic gap between the bacilliform viruses that replicate in plants and in invertebrates and the vertebrate Rhabdoviruses.

Tobacco Rattle Virus Group

These single-stranded-RNA viruses contain 5% nucleic acid. Two main components have been described, of which the larger is infective. The particles are straight, tubular, of helical symmetry, with a pitch of 2.5 nm. The larger particles are 180 to 210 nm long, 300 s, and the RNA has a molecular weight of approximately 2.4×10^6. The shorter nucleoprotein particles of various lengths contain the genetic information for coat protein. Thermal inactivation is between 70 and 80°C, and infectivity is retained in crude extracts for several months. Symptoms are mainly necrotic spots. The host range is wide. Transmission in nature is by nematodes of the genus *Trichodorus,* in which the viruses persist for several days or longer. Pea early browning virus, another member of this group, is serologically distantly related.

Tobacco Ringspot Virus Group (Nepovirus Group)

Single-stranded RNA has been demonstrated in at least three representatives of this group: tobacco ringspot, tomato ringspot, and arabis mosaic virus. There are three components, all isometric and 30 nm in diameter, approximately 125, 95, and 50 s. The two heavier components contain single-stranded RNA. Thermal inactivation is between 55 and 70°C, and longevity in sap is several days or weeks. Some representatives are seed-borne and mechanically transmissible. Natural vectors are nematodes of the genera *Xiphinema* and *Longidorus.* Although only few members of this group have been shown to be serologically related, the following seem to belong to the group: tobacco ringspot virus, arabis mosaic virus, strawberry latent ringspot virus, tomato ringspot virus, tomato black ring virus, raspberry ringspot virus, grapevine fanleaf virus, and cherry leaf roll virus.

The satellite of tobacco ringspot virus is composed of at least eleven components that differ in buoyant density. According to Schneider *et al.*,[30] all satellite particles contain numerous nucleic acid strands of uniform size, each strand weighing approximately 86,000 daltons. Each satellite particle of the lowest buoyant density has approximately twelve nucleic acid strands; each particle of higher density has one more nucleic acid strand than its nearest neighbor of lower density. Particles with the highest density contain approximately twenty-five nucleic acid strands per particle.

TABLE 1

RHABDO PLANT VIRUSES

Virus	Abbreviation	Host[a]	Reference[b]
Potato yellow dwarf	PYDV	Leafhopper	23
Sowthistle yellow vein	SYVV	Aphid	28
Gomphrena	GV		17
Lettuce necrotic yellows	LNYV	Aphid	12
Wheat striate mosaic	WSMV	Leafhopper	21
Melilotus latent	MLV		18
Plantain			16
Corn mosaic	CMV	Leafhopper	14
Rice transitory yellowing	RTYV		33
Northern cereal mosaic	NCMV		34
Broccoli necrotic yellows	BNYV		15
Eggplant mottled dwarf	EMDV		25
Russian winter wheat mosaic	RWWMV	Leafhopper	27
Clover enation	CEV		29
Cereal striate mosaic	CSMV	Plant hopper	7

[a] Not necessarily indicative of the host range; the viruses were originally isolated from these hosts.
[b] First morphological description to allow inclusion in the genus.

From: Hummeler, K., Bullet-Shaped Viruses, in *Comparative Virology,* K. Maramorosch and E. Kurstak, Eds. (1971). Reproduced by permission of Academic Press, New York.

Tobacco Mosaic Virus Group

The straight tubular particles, about 190 s, contain 5% single-stranded RNA of 2×10^6 molecular weight. The symmetry is helical, with a pitch of 2.3 nm. The particles that are infective are 300 nm in length. Some strains have a thermal inactivation point above 100°C, and infectivity in crude sap is retained for years. Host symptoms are mosaic and mottle or necrotic lesions. Transmission is by mechanical means; no efficient natural vectors are known. Members of this group are serologically related and include the following: tomato mosaic virus, ribgrass mosaic virus, cucumber green mottle mosaic virus, *Odontoglossum* ringspot virus, Sammon's *Opuntia* virus, and sunn hemp mosaic virus.

Turnip Yellow Mosaic Virus Group

Representatives of this single-stranded-RNA group contain 36% nucleic acid of 2×10^6 molecular weight. Some members have been shown to have isometric particles of about 110 s and 30 nm in diameter, with 180 subunits in pentameter-hexameter clusters. The molecular weight of the subunits is approximately 20,000, and there are accessory particles of 50 s that represented empty protein shells. The thermal inactivation point is between 70 and 90°C, and infectivity in crude sap is retained for several weeks. Symptoms caused by representatives of this group consist of mosaic and mottle, and the host range is narrow. The viruses have beetle vectors and are mechanically transmissible. Turnip yellow mosaic virus, cacao yellow mosaic virus, and wild cucumber mosaic virus are serologically related. A second group of related viruses (subgroup b) consists of the following: eggplant mosaic virus, *Dulcamara* mottle virus, *Belladonna* mottle virus, *Anonis* yellow mosaic virus, and Andean potato latent virus.

Tomato Bushy Stunt Virus Group

Virions contain 17% single-stranded RNA of 1.5 x 10^6 molecular weight. Particles are isometric, 30 nm in diameter and 140 s, with protein subunits in pairs. The inner and outer shells of these viruses result from different spatial arrangements of the same protein, whose molecular weight has been estimated at 42,000 by acrylamide gel electrophoresis and other methods.[25] Thermal inactivation is between 85 and 90°C; infectivity is retained in sap for several weeks. Host symptoms consist of mottling and distortion. Representatives of this group have a wide host range, are mechanically transmissible, and serologically related. Tomato bushy stunt is the best-characterized member. Other members are carnation ringspot virus (Italian), artichoke mottle crinkle virus, and petunia asteroid mosaic virus.

Tobacco Necrosis Virus Group

This group contains 20% single-stranded RNA of 1.5 x 10^6 molecular weight. The isometric particles have a diameter of 28 nm and 118 s, with protein subunits of 23,000 molecular weight. The "Urbana" isolate was estimated at 33,300,[22] and its RNA content was 18.7%. Thermal inactivation ranges from 65 to 95°C, and infectivity is retained for several months. The host range is wide; transmission is mechanical and by means of lower fungi of the genus *Olpidium.* The virions appear to be retained for several hours by fungal zoospores.

Pea Enation Mosaic Virus Group (Monotypic)

The only known representative of this group contains 29% single-stranded RNA of 1 x 10^6 molecular weight. The particles are isometric, 28 nm in diameter. Viral RNAs sediment at 34, 30, and 12 s and have molecular weights of 1.74 x 10^6, 1.44 x 10^6 and 0.28 x 10^6 daltons. The 1.44 x 10^6 dalton RNA component is infectious, whereas the largest and smallest are not and have no known function. There are two nucleoprotein components, 95 and 115 s, both infectious and containing 1.44 x 10^6 dalton RNA, but the largest and smallest RNA components are associated exclusively with the 115 s nucleoprotein.[11] The virus is mechanically transmissible to a narrow host range, causing mosaic, mottle, and enation symptoms. In natural aphid vectors the virus persists and possibly multiplies.

Brome Mosaic Virus Group (Bromovirus)

The well-defined, single-stranded-RNA virus of brome mosaic has three types of virions; all of these are required for infectivity.[20] The virions contain approximately 22% nucleic acid in either one RNA molecule of molecular weight 1.09 x 10^6 or 0.99 x 10^6 or in two RNA molecules of molecular weights 0.75 x 10^6 and 0.28 x 10^6. A mixture of the four RNA components is infectious; deletion of any but the smallest greatly reduces infectivity. The third-largest RNA component contains the protein coat of molecular weight 16,200. Cowpea chlorotic mottle has a molecular weight of 19,200, and broad bean mottle of 16,400.[1] Thermal inactivation is between 70 and 95°C; retention in crude sap is variable, and mechanical transmission to a restricted host range is reported. The members of this group, i.e., brome mosaic virus, cowpea chlorotic mottle virus, and broad bean mottle virus, have been serologically compared and found related.[31]

Cucumber Mosaic Virus Group (Cucumovirus)

The virions contain 18% single-stranded RNA of 1 x 10^6 molecular weight; the particles are isometric, 30 nm in diameter, and 98 s, with 180 protein subunits of molecular weight 32,000. Inactivation is between 60 and 70°C, and infectivity is retained in sap for a few days. Disease symptoms consist of a mosaic. The host range is wide. Transmission is by aphids as well as by mechanical means. Cucumber mosaic virus, tomato aspermy virus, and cucumber yellow mosaic virus are serologically distantly related. Peanut stunt virus may also belong to this group.

Potato Virus Y Group

The single-stranded-RNA virions are in the form of flexuous rods, of helical symmetry, with a pitch of 3.4 nm. Infective particles are 730 to 790 nm long. Purified tobacco etch virus, a member of this group, contains approximately 5% single-stranded RNA. The viral protein contains approximately 194 amino acids with a combined molecular weight of about 22,000.[8] Thermal inactivation is between 55 and 65°C, and infectivity is retained in sap for several days. The symptom is mosaic, and the host range is narrow. Transmission is by aphid vectors and by mechanical means. The viruses seem stylet-borne, since they are retained by aphids for less than two hours. The group contains numerous members: tobacco etch virus, bean yellow mosaic virus, common bean mosaic virus, cowpea mosaic (aphid-borne), beet mosaic, henbane mosaic, pea mosaic and soybean mosaic viruses, clover yellow vein virus, and others. Possible members also include celery, iris, lettuce, sugar cane, tulip, turnip, and watermelon mosaic viruses I and South African, cockfoot streak, malva vein clearing, narcissus yellow stripe, papaya ringspot, and plum pox viruses.

Carnation Latent Virus Group

Virions contain 6% single-stranded RNA in slightly flexuous rods with helical symmetry and a pitch of 3.4 nm. The length of infective virions ranges from 620 to 690 nm, and thermal inactivation ranges from 55 to 70°C. Infectivity is retained in sap for a few days, and symptoms are often nonapparent in a narrow host range. Aphid vectors of several species transmit the viruses in a stylet-borne manner; the viruses can be transmitted mechanically. A distant serological relationship has been demonstrated among members of the group. Carnation latent virus is the type member, and red clover vein mosaic virus is the best-determined. Other viruses are cactus virus 2, chrysanthemum virus B, passiflora latent virus, pea streak, potato M, and potato S viruses. Possible other members include poplar mosaic, hop latent, chicory blotch, and freesia mosaic viruses.

Potato Virus X Group

Flexuous rods with helical symmetry contain 6% single-stranded RNA and infective particles that are 480 to 580 nm long and sediment at 118 s. Inactivation of infectivity ranges from 65 to 75°C, and symptoms are mainly mosaic, mottle, or ringspot. Viruses are mechanically transmitted. They are serologically distantly related within the group. Members are potato virus X, hydrangea ringspot, white clover mosaic, and cactus virus X. Possible additional members include potato aucuba mosaic, papaya mosaic, cymbidium mosaic, artichoke curly dwarf, cassava common mosaic, and narcissus mosaic viruses.

Tomato Spotted Wilt Virus Group (Monotypic)

The single-stranded-RNA virions contain protein and lipid. The particles are isometric, 70 to 80 nm in diameter, and 560 s. Thermal inactivation is comparatively low: 42°C, and infectivity is retained in sap for a few hours. Symptoms are ringspot and necrotic spots. The host range, as well as the geographic distribution, is very wide. Transmission in nature is by thrips in which the virus persists.

Double-Stranded RNA-Viruses (Diploriboviridae)

Viruses in this group resemble the vertebrate-infecting reoviruses and can be divided into two subgroups: the reovirus group, which includes two plant viruses, and the orbivirus group,[4] which contains a number of viruses of filamentous fungi.

Reo group representatives contain 10 to 20% double-stranded RNA in several pieces, with a total molecular weight of 15×10^6, and a G + Ç ratio of 42 to 44%. The naked isometric capsid has icosahedral symmetry. Virus maturation occurs in the cytoplasm; virus particles may form crystalline arrays. The chief representatives include clover wound tumor virus (WTV) and rice yellow dwarf virus; both infect not only plants, but also invertebrate animals (leafhoppers) that act as their vectors in nature.

Orbi group representatives of lower fungi include *Penicillium stoloniferum* ATCC 14586 virus F ("fast") and S ("slow"). These are 34 nm in diameter; the RNA of F has 0.99 to 0.23 x 10^6 daltons plus single-stranded; the S has 1.01 to 0.95 x 10^6 daltons. Also in this group is a virus of *Aspergillus niger* IMI 146891, with a fast and slow component, both 40 nm in diameter. Possible members of the group include viruses recently found in *Penicillium brevi-compactum* NRRL 5260 and *Penicillium chrysogenum* ATCC 9480.[5,6,39]

Viroids

A new group of pathogenic agents, known to cause diseases of plants and perhaps also of animals, has recently been described; the name "viroid" was proposed for the causative agents. Viroids represent the smallest infectious agents known to date. In some respects they resemble nucleic acids isolated from virions, because they contain the genetic information necessary to induce the pathological condition of a specific disease as well as the information needed for replication. However, their size is much smaller than the size of the smallest nucleic acid of a known virus. It has been estimated that the molecular weight of the spindle tuber disease viroid is 50,000.[9] Phosphate buffer extracts from potato plants infected with the potato spindle tuber viroid contain infectious RNA, characterized by a sedimentation rate of approximately 10 s. The viroids are located in the nuclei of plant cells. They are sensitive to treatment with ribonuclease. The free RNA of the viroid appears to be resistant to exonucleases and could possibly be circular.

Exocortis disease of citrus and chrysanthemum stunt disease, previously considered to be caused by plant viruses, are now believed to be caused by viroids. There is also speculation that animal diseases, such as scrapie of sheep, are of similar etiology.

REFERENCES

1. Agrawal, H. O., and Tremaine, J. H., *Virology, 47,* 8 (1972).
2. Baltimore, D., Huang, A. S., and Stampfer, J., *Proc. Nat. Acad. Sci. U.S.A., 66,* 572 (1970).
3. Bol, J. F., van Vloten-Doting, L., and Jaspars, E. M. J., *Virology, 46,* 73 (1971).
4. Borden, E. C., Shope, R. E., and Murphy, F. A., *J. Gen. Virol., 13,* 261 (1971).
5. Bozart, R. F., Wood, H. A., and Mandelbrot, A., *Virology, 45,* 516 (1971).
6. Brakke, M. K., in *Viruses, Vectors, and Vegetation,* p. 527, K. Maramorosch, Ed. Interscience Publications, John Wiley and Sons, New York (1969).
7. Conti, M., *Phytopathol. Z., 66,* 275 (1969).
8. Damirdagh, I. S., and Shepherd, R. J., *Virology, 40,* 84 (1970).
9. Diener, T. O., in *Comparative Virology,* p. 433, K. Maramorosch and E. Kurstak, Eds. Academic Press, New York (1971).
10. Francki, R. I. B., and Randles, J. W., *Virology, 47,* 270 (1972).
11. Gonsalves, D., and Shepherd, R. J., *Virology, 48,* 709 (1972).
12. Harrison, B. D., and Crowley, N. C., *Virology, 26,* 297 (1965).
13. Harrison, B. D., Finch, J. T., Gibbs, A. J., Jollings, M., Shepherd, R. J., Valenta, V., and Wetter, C., *Virology, 45,* 356 (1971).
14. Herold, F., Bergold, G. H., and Weibel, J., *Virology, 12,* 335 (1960).
15. Hills, G. J., and Campbell, R. N., *J. Ultrastruct. Res., 24,* 134 (1968).
16. Hitchborn, J. H., Hills, G. J., and Hull, R., *Virology, 28,* 768 (1966).
17. Kitajima, E. W., and Costa, A. S., *Virology, 29,* 523 (1966).
18. Kitajima, E. W., Lauritis, J. A., and Swift, H., *J. Ultrastruct. Res., 29,* 141 (1969).
19. Knudson, D. L., and MacLeod, R., *Virology, 47,* 285 (1972).
20. Lane, L. C., and Kaesberg, P., *Nat. New Biol., 232,* 40 (1971).
21. Lee, P. E., *Virology, 23,* 145 (1964).
22. Lesnaw, J. A., and Reichmann, M. E., *Virology, 39,* 729 (1969).
23. MacLeod, R., Black, L. M., and Moyer, F. H., *Virology, 29,* 540 (1966).
24. Maramorosch, K. (Ed.), *Viruses, Vectors, and Vegetation.* Interscience Publications, John Wiley and Sons, New York (1969).

25. Martelli, C. P., *J. Gen. Virol., 5,* 319 (1969).
26. Michelin-Lausarot, P., Ambrosino, C., Steere, R. L., and Reichmann, M. E., *Virology, 41,* 160 (1970).
27. Razvjazkina, G. M., Poljakova, G. P., Stein-Margolina, V. A., and Cherny, N. E., in *Abstracts of the First International Congress on Plant Pathology, London,* p. 162 (1968).
28. Richardson, J., and Sylvester, E. S., *Virology, 35,* 347 (1968).
29. Rubio-Huertos, M., and Bos, L., *Neth. J. Plant Pathol., 75,* 329 (1969).
30. Schneider, I. R., Hull, R., and Markham, R., *Virology, 47,* 320 (1972).
31. Scott, H. A., and Slack, S. A., *Virology, 46,* 490 (1971).
32. Shepherd, R. J., Bruening, G. E., and Wakeman, R. J., *Virology, 41,* 339 (1970).
33. Shikata, E., and Chen, M.-J., *J. Virol., 3,* 271 (1969).
34. Shikata, E., and Lu, Y.-T., *Proc. Jap. Acad., 43,* 918 (1967).
35. van Kammen, A., and van Griensven, L. J. L. D., *Virology, 41,* 274 (1970).
36. Wildy, P., *Monogr. Virol., 5,* 1 (1971).
37. Wolanski, B. S., and Chambers, T. C., *Virology, 47,* 656 (1972).
38. Wood, H. A., and Bozarth, R. F., *Virology, 47,* 604 (1972).
39. Wood, H. A., Bozarth, R. F., and Mislivec, P. B., *Virology, 44,* 592 (1971).
40. Wu, G.-J., and Bruening, G., *Virology, 46,* 596 (1971).

VIRUSES OF INVERTEBRATES

DR. KARL MARAMOROSCH

Viruses that multiply in invertebrate animals are, by definition, viruses of invertebrates. Very little is known about viruses of nematodes, and most of the information available pertains to arthropod-infecting viruses. Since plant-pathogenic viruses that also multiply in arthropod vectors are dealt with under plant viruses, and since certain arthropod-borne (Arbo) viruses that have insect and vertebrate hosts are of primary interest to animal virologists, the discussion of viruses of invertebrates will be limited to viruses that usually cause fatal diseases in insects.

It has been found convenient to classify insect viruses on the basis of diseases induced in their hosts, the presence or absence of inclusion bodies and the shape of inclusions, and not on the basis of intrinsic properties of the virions. Although this approach does not provide a means of classification comparable to that used in other fields of modern virology, it will be followed here for practical purposes. Table 1 shows the groups of insect viruses classified according to the presence and morphology of inclusion bodies. Some characteristics of arthropod viruses that cause nuclear and cytoplasmic polyhedroses, granuloses, entomopoxviroses, and non-inclusion viroses are listed in Table 2 (after Reference 8).

Insect-pathogenic viruses have a wide range of sizes and shapes, as do plant-pathogenic and vertebrate-infecting viruses. The viruses that cause "inclusion body" diseases differ from other viruses because the virions are occluded in protein crystals, sometimes called "polyhedra" or "capsules", depending on their shape. Polyhedra-forming viruses can multiply in either the nuclei or the cytoplasm, and the diseases are therefore classified as nuclear or cytoplasmic polyhedroses. The capsule-shaped inclusions are present in the so-called granuloses diseases of insects and can be found in the nuclei as well as in the cytoplasm of insect cells.

Virions of nuclear polyhedroses are rod-shaped DNA viruses. Cytoplasmic polyhedroses are caused by icosahedral RNA viruses that multiply only in the cytoplasm. Polyhedra of both types of diseases contain from 100 to 1,000 virions. Inclusion bodies of granuloses diseases usually contain single rod-shaped DNA virions. No cell organelles or components other than virions have been found in inclusion bodies.

INSECT POX VIRUSES

A distinct group of DNA-containing insect-pathogenic viruses, morphologically and chemically resembling the pox viruses of vertebrates, is known as the insect pox viruses. These viruses are characterized in Table 3.[1] The main characteristic of the vertebrate pox viruses is the presence of 5 to 7.5% double-stranded DNA of approximately 160×10^6 molecular weight. The G + C ratio is 35 to 40%, and the virions are

TABLE 1

GROUPS OF ARTHROPOD VIRUSES

Inclusion Body	Shape of Inclusion	Type of Disease	Site of Viral Multiplication in Host Cell
Present	Polyhedron	Polyhedroses	Nuclear or cytoplasmic
	Ovoid	Granuloses	Nuclear or cytoplasmic
	Irregular	Entomopoxviroses	Cytoplasmic
Absent		Non-inclusion viroses	Cytoplasmic

Data taken from: Ignoffo, C. M., Viruses – Living Insecticides, *Curr. Top. Microbiol. Immunol., 42,* 134 (1968). Reproduced in modified form by permission of Springer-Verlag, New York.

TABLE 2

SOME CHARACTERISTICS OF ARTHROPOD VIRUSES

Virus Disease	Replication Site	Type of Nucleic Acid	Inclusion Body		Virion		
			Shape	Size, nm	Shape	Symmetry	Size, nm
Nuclear polyhedroses	Nucleus of mesodermal, endodermal and ectodermal cells	DNA	Polyhedron	200–2,000	Rod	Helical	400 x 80
Cytoplasmic polyhedroses	Cytoplasm of midgut epithelial cells	RNA	Polyhedron	500–2,500	Spherical	Cubical	30–80
Granuloses	Nucleus and cytoplasm of adipose, tracheal and epidermal cells	DNA	Ovoid	200 x 400	Rod	Helical	350 x 50
Entomopoxviroses	Cytoplasm of adipose cells, hemocytes	DNA	Irregular	2,000–20,000	Oval	Complex	400 x 250
Non-inclusion viroses	Cytoplasm of adipose cells	(?) DNA	None None	None None	Spherical Rod	Cubical Helical	35–150 200 x 10

Data taken from: Ignoffo, C. M., Viruses – Living Insecticides, *Curr. Top. Microbiol. Immunol., 42,* 135 (1968). Reproduced by permission of Springer-Verlag, New York.

brick-shaped or ovoid and belong to the largest viruses known. Particles are approximately 170–250 x 300–325 nm and approximately 5,000 s. They possess several enzymes, e.g., RNA polymerase, and probably NP antigen, in common to all members. The insect pox viruses are not as well characterized, but they seem to resemble closely the vertebrate virions. The current name "Entomopoxviruses" has been suggested for the members of this group that cause diseases of insects.

Iridescent Virus Group (Iridovirus)

This group comprises insect-pathogenic viruses that contain about 15% double-stranded DNA of approximately 130 x 10^6 molecular weight, and approximately 30% G + C. Particles are icosahedral, 130 nm in diameter, and approximately 2,200 s. Several proteins are contained in the complex particles, and the outer capsid contains about 1,500 capsomeres. The *Tipula* iridescent virus (Iridovirus-1) is the type species of this group. Other members are the *Chilo* iridescent virus and the *Serecesthis* iridescent virus, and possibly the *Aedes* and other iridescent viruses of mosquitoes.

Densonucleosis Virus Group

These small DNA viruses of *Galleria melonella* have particles of an average diameter of 20 nm, icosahedral symmetry, and roughly hexagonal outline. Full particles, according to Kurstak and Côté,[9] have a surface of regular shallow capsomeres; empty particles have capsomeres only around the periphery. Capsomeres measure 2.0 to 3.5 nm and have a 1.5-nm central hole; their total number has been estimated as 32.

Nuclear Polyhedrosis and Granulosis Virus Group (Baculovirus)

Viruses of this group contain double-stranded DNA of approximately 80 x 10^6 molecular weight and G + C content ranging from 35 to 60%. The virions are rod-shaped, 40–70 nm x 250–400 nm, with an outer and inner membrane and an inner electron-dense core. Virions may be occluded in crystalline protein inclusion bodies. Their bodies may be rounded and contain one or two particles (in some of the granuloses viruses), or they may be polyhedral and contain numerous particles (polyhedroses viruses). The virions are sensitive to ether and to heat. The diseases caused by these viruses affect Lepidoptera, Hymenoptera, Diptera and Neuroptera. The type species is the nuclear polyhedrosis virus of the silkworm, *Bombyx mori.* Other members include the nuclear polyhedrosis virus of *Porthetria dispar,* the granulosis virus of *Choristoneura fumiferana,* and the granulosis virus of *Plodia interpunctella.* Still other viruses that belong to this group have been isolated from over 100 species of insects.

Cytoplasmic Polyhedrosis Virus Group

This group is represented by the cytoplasmic polyhedrosis virus of the silkworm, *Bombyx mori.* It is characterized by 22% double-stranded RNA in several pieces, with a total molecular weight of 12.7 x 10^6. The virions are 60 nm in diameter, 440 s, icosahedral, and have large knobs. Particles are often occluded in large crystalline inclusions. In diseased Lepidoptera, virions develop only in the midgut cells. In Neuroptera, virions are first found in the midgut cells of larvae, but later they spread to other tissues. The polyhedra with virus particles are in the cell cytoplasm.

TABLE 3

MAIN CHARACTERISTICS OF POXVIRUSES OF INSECTS

Host	Size, nm	Shape	Internal Structure (Vertical Sections)	Nucleic Acid	References
Coleoptera					
Melolontha melolontha L. (Scarabaeidae)	450 x 250	Oval-shaped, mulberry-like surface	Unilaterally concave core that contains a filamentous rod	DNA	2, 11, 13
Demodena boranensis Bruch (Scarabaeidae)	420 x 230	Oval-shaped, mulberry-like surface	Unilaterally concave core that contains a filamentous rod		14
Othnonius batesi (Scarabaeidae)	470 x 265	Oval-shaped, mulberry-like surface	Unilaterally concave core that contains a filamentous rod		4
Figulus sublaevis (Lucanidae)	370 x 250	Oval-shaped, mulberry-like surface	Unilaterally concave core that contains a filamentous rod		12
Dermolepida albohirtum					4
Phyllopertha horticola L. (Rutelidae)	400 x 240	Oval-shaped, mulberry-like surface	Unilaterally concave core that contains a filamentous rod		15
Lepidoptera					
Operophtera brumata HB (Geometridae)	370 x 250	Oval-shaped			17
Acrobasia zelleri RAG. (Pyralididae)		Oval-shaped			17
Amsacta moorei (Arctiidae), original host, adapted on *Estigmene acrea* (Arctiidae)	350 x 250	Oval-shaped, mulberry-like surface	Rectangular core that contains rod-like structures	DNA	6
Oreopsyche angustella H.S. (Bsychidae)	360 x 260	Oval-shaped, mulberry-like surface	Rectangular or slightly biconcave core		10
Choristoneura biennis, adapted on *C. fumiferana*	400 x 300	Oval-shaped	Large dense core		3
Orthoptera					
Melanoplus sanguinipes F. (Acrididae)	320 x 250	Oval-shaped	Rectangular or dumbbell-shaped core		7

TABLE 3 (Continued)

MAIN CHARACTERISTICS OF POXVIRUSES OF INSECTS

Host	Size, nm	Shape	Internal Structure (Vertical Sections)	Nucleic Acid	References
Diptera					
Chironomus luridus (Chironomidae)	320 x 230 x 110	Cuboidal	Dumbbell-shaped core	DNA	5
Camptochironomus tentans (Chironomidae)	200–300 x 270–300 x 130–150	Cushion-shaped	Dumbbell-shaped core		16

Data from: Bergoin, M., and Dales, S., Comparative Observations on Poxviruses of Invertebrates and Vertebrates, in *Comparative Virology,* p. 169, K. Maramorosch and E. Kurstak, Eds. (1971). Reproduced by permission of Academic Press, New York.

REFERENCES

1. Bergoin, M., and Dales, S., in *Comparative Virology,* p. 169, K. Maramorosch and E. Kurstak, Eds. Academic Press, New York (1971).
2. Bergoin, M., Devauchelle, G., Duthoit, J.-L., and Vago, C., *C. R. Hebd. Séances Acad. Sci. Ser. D Sci. Nat. (Paris), 266,* 2126 (1968).
3. Bird, F. T., Sanders, C. J., and Burke, J. M., *J. Invertebr. Pathol., 18,* 159 (1971).
4. Goodwin, R. H., and Filshie, B. K., *J. Invertebr. Pathol., 13,* 317 (1969).
5. Götz, P., Huger, A. M., and Krieg, A., *Naturwissenschaften, 56,* 145 (1969).
6. Granados, R. R., and Roberts, D. W., *Virology, 40,* 230 (1970).
7. Henry, J. E., Nelson, B. P., and Jutila, J. W., *J. Virol., 3,* 605 (1969).
8. Ignoffo, C. M., *Curr. Top. Microbiol. Immunol., 42,* 129 (1968).
9. Kurstak, E., and Côté, J.-R., *C. R. Hebd. Séances Acad. Sci. Ser. D Sci. Nat. (Paris), 268,* 616 (1969).
10. Meynadier, G., Fosset, J., Vago, C., Duthoit, J.-L., and Bres, N., *Ann. Epiphyt. (Paris), 19,* 703 (1968).
11. Vago, C., *J. Insect Pathol., 5,* 275 (1963).
12. Vago, C., Amargier, A., Hurpin, B., Meynadier, G., and Duthoit, J.-L., *Entomophaga, 13,* 373 (1968).
13. Vago, C., and Croissant, O., *Entomophaga, 9,* 207 (1964).
14. Vago, C., Monsarrat, P., Duthoit, J.-L., Amargier, A., Meynadier, G., and Van Waerebeke, D., *C. R. Hebd. Séances Acad. Sci. Ser. D Sci. Nat. (Paris), 266,* 1621 (1968).
15. Vago, C., Robert, P., Amargier, A., and Duthoit, J.-L., *Mikroskopie, 25,* 378 (1969).
16. Weiser, J., *Acta Virol., 13,* 549 (1969).
17. Weiser, J., and Vago, C., *J. Invertebr. Pathol., 8,* 314 (1966).

VIRUSES OF VERTEBRATES

DR. BURTON I. WILNER

PRIMARY CLASSIFICATION OF VERTEBRATE VIRUSES

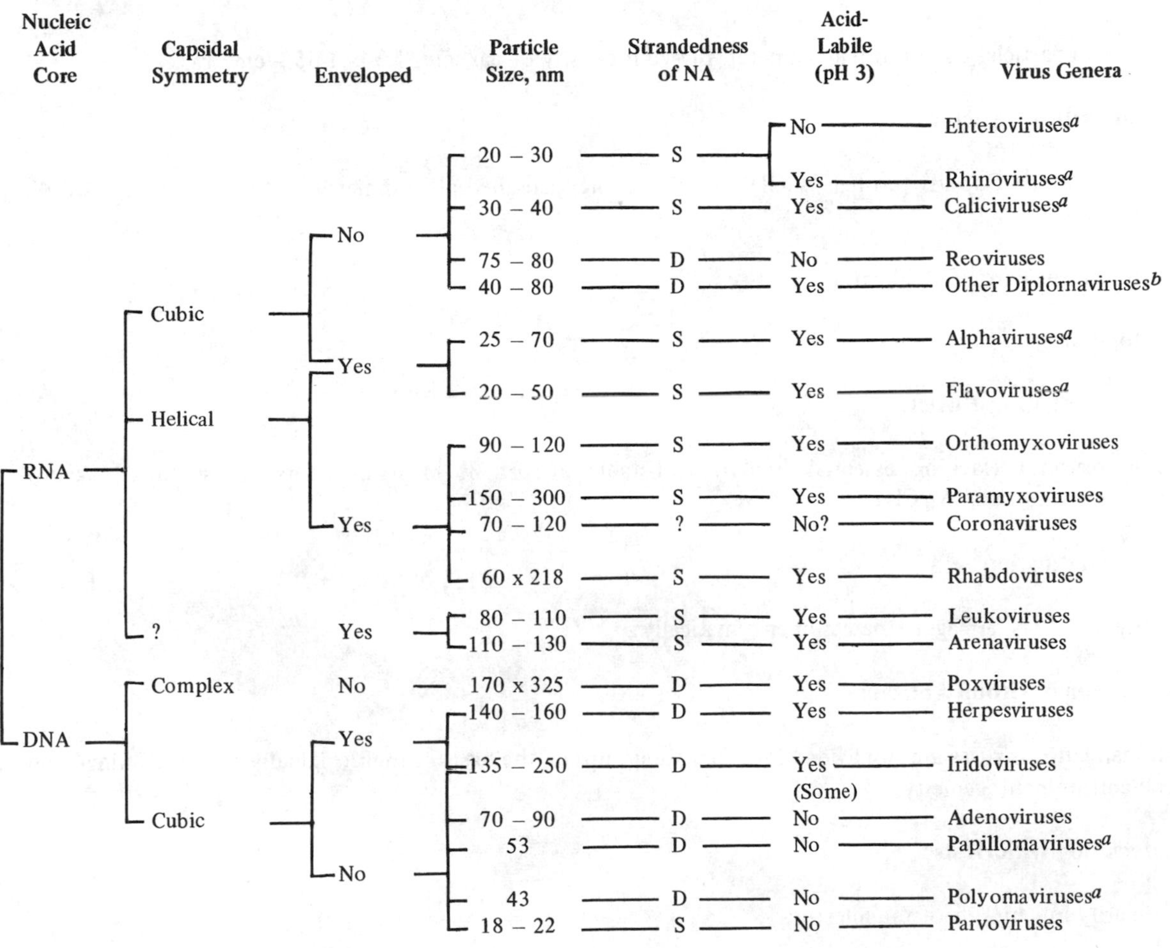

[a] The genera *Enterovirus, Rhinovirus*, and *Calicivirus* have been grouped as the family *Picornaviridae;* the genera *Papillomavirus* and *Polyomavirus* form the family Papovaviridae; a proposal has been made to include the genera *Alphavirus* and *Flavovirus* in a family to be called Togaviridae.

[b] Name suggested for viruses with double-stranded ribonucleic acid.

Code: NA = nucleic acid; S = single stranded; D = double stranded

Genus *Enterovirus* Family Picornaviridae

Members of this group may produce neural, gastrointestinal, respiratory, and exanthematous disease in humans and other animals. Infections with these agents are usually subclinical, however. Most of the viruses appear to be primarily species-host-specific.

GROUP CHARACTERISTICS

Nucleic Acid

Single-stranded RNA core; molecular weight of nucleic acid: 2.5 x 10^6 daltons; molar base composition: G = 24–28%, A = 25–29%, C = 22–26%, U = 23–27%.*

Virus Particle Form

Spherical particles 20–30 nm in diameter. Buoyant density of particles 1.34–1.35 g/cm^3 in CsCl.

Virus Particle Structure

No envelope. Capsid (protein shell) exhibits cubic, icosahedral symmetry with possibly 32 or 42 capsomeres.

Site of Virus Particle Replication

Cytoplasm.

Stability of Virus Particle

Ether-resistant (lacking essential lipids); acid-stable at pH 3. Most members are stabilized against heat-inactivation by $MgCl_2$.

Hemagglutination

Many members hemagglutinate non-enzymatically.

Possession of Group Antigens

Human enteroviruses are subdivided into four subgroups on the basis of their originally observed animal and cell culture pathogenicity.

Interaction with Hosts

Primarily inhabitants of the intestine.

AVIAN TYPES

Avian encephalomyelitis virus
Duck hepatitis virus

* G = guanine, A = adenine, C = cytosine, U = uracil.

BOVINE TYPES

Multiple isolates; serologic relationships among members have not yet been completely established; six serotypes do appear to be clearly defined.

1 LCR4[c]
2 BEV261[c]
3 BEV240A[c]
4 BEV180[c]
5 BEV243[c]
6 4S[c]

HUMAN TYPES

Poliovirus

Three serotypes:

1 Brunhilde
2 Lansing
3 Leon

Coxsackievirus A

Twenty-three primary serotypes, three subtypes (a, b, c), and one variant (v); virus names in parentheses are alternate designations for prototype strains.

1 T.T. (Tompkins)
2 Fl (Fleetwood)
3 J.Ol (Olson)
4 High Point (ALA-CDC)
5 G.S. (Swartz)
6 C.G. (Gdula)
7 W.P. (Parker), AB-IV (Russian)
8 C.D. (Donovan)
9 P.B. (Bozek)
10 M.I. (Kowalik)
11 Belgium 1
12 Texas 12
13 Flores
14 G-14
15 G-9
16 G-10
17 G-12
18 G-13
19 Dohi
20 IH Pool #35
20a Tulane #1623
20b Cecil P2647
20c Thai C-18
21 Kuykendall (= Coe virus)
22 Chulman
23 See echovirus 9
24 Joseph P3305
24v DN-19 (formerly echovirus 34)

Coxsackievirus B

Six serotypes; virus names in parentheses are alternate designations for prototype strains.

1 Conn-5
2 Ohio-1
3 Nancy
4 J.V.B (Benshoten)
5 Faulkner
6 Schmitt (1-15-21)

[c] Reference virus strains.

Echovirus

Thirty serotypes and two subtypes; virus names in parentheses are alternate designations for prototype strains.

1 Farouk
2 Cornelis
3 Morrisey
4 Pesascek
5 Noyce
6 D'Amori
6′ D-1 (Cox)
6″ Burgess
7 Wallace
8 Bryson (= echovirus 1)
9 Hill
10 See reoviruses
11 Gregory
12 Travis 2-85
13 Del Carmen
14 Tow
15 Ch 96-51
16 Harrington
17 CHHE-29
18 Metcalf
19 Burke
20 JV-1
21 Farina (E26D)
22 Harris
23 Williamson
24 DeCamp
25 JV-4
26 Coronel (11-3-6)
27 Bacon (1-36-4)
28 See *Rhinovirus*, human type 1A
29 JV-10
30 Bastianni
31 Caldwell
32 PR-10
33 Toluca-3
34 (Coxsackievirus A24 variant)

Other Human Enteroviruses

Fermon virus

MURINE TYPE

Mouse encephalomyelitis virus (*Poliovirus muris,* Theiler's virus), multiple serologically related, but not identical, strains.

FA strain   GDVII strain   TO (Falke) strain   TO (Yale) strain[c]

PORCINE TYPES

Eight serogroups and three possible subgroups (a, b, c); relative differences in the degree of cross-neutralization only permit designation of serogroups – rather than specific serotypes – among the seventy-two strains tested.

1 Teschen disease virus
Talfan disease virus
PS34 strain[d]
WR1 strain
Other strains

2 E4 strain
O3b strain[d]
T80 strain
Other strains

3 F34 strain
O2b strain[d]
PS14 strain
Other strains

4 F78 strain
PS36 strain[d]
PS38 strain
Other strains

[d]Proposed type strains.

5 F26 strain[d]
Other strains

6 F7 strain
PS37 strain[d]
T4 strain
Other strains

7 F43 strain[d]
WR2 strain
Other strains

8 PS26 strain
PS27 strain[d]
V13 strain
Other strains

a. PS32 strain
b. ECPO1 strain
c. PS30 strain

SIMIAN TYPES

Eighteen serotypes; more properly designated simian picornaviruses (rather than simian enteroviruses), since two members, SV28 and SA4, do not meet the criteria for enteroviruses; SV = simian virus, SA = simian agent.

1 SV2
2 SV6 (= SV29)
3 SV16
4 SV18
5 SV19
6 SV26 (= SV48)
7 SV28
8 SA4
9 SV35
10 SV42
11 SV43
12 SV44
13 SV45
14 SV46
15 SV47
16 SV49
17 SA5
18 A-13 (baboon)

OTHER ENTEROVIRUSES

Encephalomyocarditis (EMC) Virus

Affects cattle, human, monkeys, pigs, and rodents; multiple closely related, but not identical, strains.

AK strain
Columbia SK strain
Mengo strain
MM strain
SVW strain
Other strains

Nodamura Virus

Isolated from mosquitoes; arthropod-transmissible to susceptible vertebrates.

Genus *Rhinovirus*. Family Picornaviridae

The name of this genus is derived from the Greek *rhinos* for nose. Equine, human, and possibly bovine types produce respiratory disease in their respective hosts. Foot-and-mouth disease virus causes vesicular eruptions on the buccal mucosa and on the skin of the feet, on udders, and on other thin-skinned areas of susceptible animals.

GROUP CHARACTERISTICS

Nucleic Acid

Single-stranded RNA core; molecular weight of nucleic acid: 2.4–2.8 x 10^6 daltons; molar base composition: G = 20–24%, A = 26–34%, C = 20–28%, U = 22–26%.

Virus Particle Form

Spherical particles 20–30 nm in diameter. Buoyant density of particles 1.38–1.43 g/cm^3 in CsCl.

Virus Particle Structure

No envelope. Capsid (protein shell) exhibits cubic, icosahedral symmetry.

Site of Virus Particle Replication

Cytoplasm; low ceiling temperature of growth.

Stability of Virus Particle

Ether-resistant (lacking essential lipids). Acid-labile at pH 3. Some members are at least partially stabilized against heat-inactivation by $MgCl_2$.

Interaction with Hosts

Primarily inhabitants of the respiratory tract.

BOVINE TYPES

Three serologically related, but not identical, strains:

C-07 strain   Sd-1 strain   181/V strain

EQUINE TYPES

Two serotypes:

1. Equine respiratory virus
2. E26 virus

HUMAN TYPES

Eighty-nine serotypes and one subtype; virus names in parentheses are isolates found to be identical to prototype strains.

1A	Echovirus 28 (= JH-1 = GL 2060)	6	Thompson
1B	B632 (K779)	7	68-CV11
2	HGP	8	MRH-CV12
3	FEB	9	211-CV13
4	16/60	10	204-CV14
5	Norman	11	1-CV15

12 181-CV16
13 353 (5007-CV23, Chv/5/60)
14 1059
15 1734
16 11757
17 33342
18 5986-CV17
19 6072-CV18
20 15-CV19 (4462-63)
21 47-CV21
22 127-CV22 (C203F)
23 5124-CV24 (100319)
24 5146-CV25 (C147H)
25 5426-CV26 (K2218, 55216)
26 5660-CV27 (127-1)
27 5870-CV28
28 6101-CV29 (113E)
29 5582-CV30 (179E)
30 106F
31 140F
32 363
33 1200
34 137-3 (6692-CV42)
35 164A
36 342H
37 151-1 (1770-CV36)
38 CH79 (C201-3C)
39 209 (CH00052)
40 1794 (C148E)
41 56110 (C137F)
42 56882 (C248A)
43 58750 (E2 #133, 1936-CV43, WIS 258E, C04374)
44 71560
45 Baylor 1 (037211, E2 #46)
46 Baylor 2 (477-CV50, CH202)
47 Baylor 3 (1979M-CV46, CH310)
48 1505
49 8213
50 A2 #58
51 F01-4081 (19143, 605-CV45, 313G)
52 F01-3772 (16413, 515-CV34)
53 F01-3928 (C252B)
54 F01-3774 (2253-CV49)
55 WIS 315E (Baylor 4)
56 CH82 (6660-CV38, Baylor 5)
57 CH47
58 21-CV20
59 611-CV35 (1833-63)
60 2268-CV37
61 6669-CV39
62 1963M-CV40
63 6360-CV41
64 6258-CV44 (1647-63, Baylor 6)
65 425-CV47 (143-3, 4411-65)
66 1983-CV48
67 1857-CV51
68 F02-2317-Wood (SF23)
69 F02-2513-Mitchinson
70 F02-2547-Treganza
71 SF365
72 K2207 (4704-62, 410A)
73 107E
74 328A
75 328F
76 H00062 (SF123)
77 130-63
78 2030-65
79 101-1
80 277G
81 483F2
82 03647
83 Baylor 7 (191-1)
84 432D
85 50-525-CV54
86 121564-Johnson
87 F02-3607-Corn
88 CVD 01-0165-Dambrauskas
89 41467-Gallo

FOOT-AND-MOUTH DISEASE VIRUS

Seven primary serotypes; cattle, goats, pigs, and sheep are most commonly affected.

A	A_5 Westerwald[c]
O	O_1 Lombardy[c]
C	C-GC[c]
SAT-1	SAT-1/2 Rhodesia 5/66[c]
SAT-2	SAT-2/1 Rhodesia 1/48[c]
SAT-3	SAT-3/1 Rhodesia 7/34[c]
Asia-1	Asia-1 Pakistan 1/54[c]

Genus *Calicivirus*, Family Picornaviridae

GROUP CHARACTERISTICS

Nucleic Acid

Single-stranded RNA core; molecular weight of nucleic acid: approximately 2 x 10^6 daltons; molecular base composition: G = 21%, A = 29%, C = 25%, U = 25%.

Virus Particle Form

Spherical particles 30–40 nm in diameter. Buoyant density of particles 1.37–1.38 g/cm^3 in CsCl.

Virus Particle Structure

No envelope. Capsid (protein shell) exhibits cubic, icosahedral symmetry. Probably 32 capsomere.

Site of Virus Particle Replication

Cytoplasm.

Stability of Virus Particle

Ether-resistant (lacking essential lipids). Variable stability at pH 5; most are unstable at pH 3. Not stabilized against heat-inactivation by $MgCl_2$.

FELINE RESPIRATORY VIRUS

At least nine serotypes:

1 FRV-17[c]	4 CFI[c]	7 Bolin[c]
2 FCV[c]	5 F-20[c]	8 C-14[c]
3 FSV[c]	6 KCD[c]	9 F-17[c]

VESICULAR EXANTHEMA OF SWINE VIRUS

Thirteen serotypes; pigs and horses are affected.

1 1-34[c]	5 C52[c]	10 H54[c]
2 101-43[c]	6 D53[c]	11 I55[c]
3 A48[c]	7 E54[c]	12 J56[c]
4 B51[c]	8 F55[c]	13 K56[c]
	9 G55[c]	

Genus *Reovirus* and Other Diplornaviruses

The genus name is derived from the initials of "**Re**spiratory **o**rphan" viruses. The unofficial term "diplornaviruses" was coined to designate all viruses with double-stranded ribonucleic acid (diplo + RNA). Despite the widespread distribution of reoviruses and antibody to them, association with clinical disease is equivocal, at least in mammalian hosts. Infection with other diplornaviruses in their natural hosts varies from subclinical to fatal, depending on the particular virus. Several of these agents are biologically transmitted to vertebrates by insects: African horsesickness and bluetongue viruses by gnats and biting midges; Colorado tick fever virus, and possibly Corriparta and Eubenangee viruses and members of the Changuinola and Kemerovo subgroups, by ticks.

GROUP CHARACTERISTICS

Nucleic Acid

Double-stranded RNA core, with nucleic acid in several pieces; molecular weight of nucleic acid: approximately 15 x 10^6 daltons; molecular base composition: G = 21–22%, A = 28–29%, C = 21–22%, U= 28–29%.

Virus Particle Form

Hexagonal particles 75–80 nm in diameter for the reoviruses, 40–80 nm for the other diplornaviruses. Buoyant density of reovirus particles 1.31–1.38 g/cm^3 in CsCl.

Virus Particle Structure

No true envelope; occasional particles are observed enclosed in pseudomembranes of host origin; membranes are not required for virus infectivity. Capsid (protein shell) of the reoviruses is double-layered; that of other diplornaviruses, with few exceptions, is single-layered. There are 92 tubular capsomeres in the outer capsid layer of reoviruses, and 32 or 42 capsomeres in the capsid of most other diplornaviruses; in both cases the capsomeres appear to be arranged in cubic, icosahedral symmetry.

Sites of Replication and Maturation

Replication and maturation take place in the cell cytoplasm in close relation to the mitotic apparatus, particularly spindle tubes.

Stability of Virus Particles

Ether-resistant (lacking essential lipids). Reoviruses are highly stable at acid pH (3–5) and are stabilized against heat-inactivation with $MgCl_2$; most other diplornaviruses are not stable below pH 6 and are not stabilized against heat-inactivation with $MgCl_2$.

Hemagglutination

The primary serotypes of reoviruses agglutinate erythrocytes of human and bovine origin; the avian reoviruses and other diplornaviruses do not appear to hemagglutinate.

Possession of Group Antigens

The primary serotypes of reovirus share a common CF (complement-fixation) antigen.

Interaction with Hosts

Produce cytoplasmic inclusions and have a slow growth cycle.

PRIMARY REOVIRUS SEROTYPES

Cats, cattle, dogs, humans, monkeys, pigs, and other mammals are affected.

1 Lang[e] 2 Jones[e] 3 Abney[e]

AVIAN REOVIRUSES

Chickens and turkeys are affected.

Fahey-Crawley virus
Kawamura virus (5 serotypes)
Deshmukh virus (2 serotypes)
Turkey reovirus

OTHER DIPLORNAVIRUSES

Diplornaviruses isolated from arthropods are included with the vertebrate agents listed below, since they have been suspected of being transmissible to susceptible vertebrates.

African Horsesickness Virus

Nine serotypes; equines are affected.

1 A501[c]	4 Vryheid[c]	7 Karen[c]
2 OD[c]	5 VH[c]	8 18/60[c]
3 L[c]	6 114[c]	9 S2[c]

Bluetongue Virus

Sixteen serotypes; sheep, cattle, and other ruminants are affected.

1 Biggarsberg[c]	5 Mossop[c]	9 University Farm[c]	13 Westlands[c]
2 Vryheid[c]	6 Strathene[c]	10 Portugal[c]	14 Kolwani[c]
3 Cyprus[c]	7 Utrecht[c]	11 Nelspoort[c]	15 133/60[c]
4 Theiler[c]	8 Camp[c]	12 Byenespoort[c]	16 Pakistan[c]

Changuinola Virus Group

Two members:

Changuinola virus
Irituia virus

Colorado Tick Fever Virus

Affects humans.

Corriparta Virus

Isolated from ticks.

[e] Human prototype strains.

Cytoplasmic Polyhedrosis Virus

Isolated from various species of *Lepidoptera* and *Neuroptera*.

Epizoötic Hemorrhagic Disease of Deer Virus

Two serotypes; affects deer.

1 New Jersey

2 South Dakota

Epizoötic Mouse Diarrhea Virus

Affects mice.

Equine Encephalosis Virus

Affects equines.

Eubenangee Virus

Isolated from mosquitoes.

Infectious Pancreatic Necrosis Virus

Isolated from trout.

Kemerovo Virus Group

Six members:

Chenuda virus (ticks)
Kemerovo virus (birds, cattle, horses, humans)
Koliba virus (ticks)
Lipovnik virus (ticks)
Secaseni virus (ticks)
Tribec virus (ticks)

Nelson Bay Virus

Isolated from bats.

Sa11 (Simian Agent #11)

Affects African green monkeys.

"X" Virus

Isolated from bovine serum.

Genus *Alphavirus*, Family Togaviridae

Members of this genus were originally classified as the group A of arboviruses; hence the designation *Alphavirus*. All strains are biologically transmitted to vertebrates by mosquitoes. Infection of wild mammals, birds, and lower vertebrates apparently results in no injury; human and domestic-animal infections are likewise frequently mild and inapparent. Serious encephalitic disease is produced by some members, however, particularly in horses and occasionally in humans.

GROUP CHARACTERISTICS

Nucleic Acid

Single-stranded RNA core; molecular weight of nucleic acid: approximately 3 x 10^6 daltons; molar base composition: G = 26%, A = 29%, C = 25%, U = 20%.

Virus Particle Form

Spherical particles 25–70 nm in diameter. Buoyant density of particles 1.25 g/cm^3 in CsCl.

Virus Particle Structure

Envelope present. The ultrastructure of few members has been studied; possible helical symmetry.

Sites of Replication and Maturation

Replication in the cytoplasm. Maturation by budding from the host cell membrane.

Stability of Virus Particle

Ether-sensitive (possessing essential lipids). Acid-labile. Insensitive to trypsin (in contrast to flavoviruses).

Hemagglutination

All members hemagglutinate.

Possession of Group Antigens

All members are serologically related, as shown by hemagglutination-inhibition tests.

Interaction with Hosts

All members replicate in arthropod vectors.

ALPHAVIRUSES

1. Aura virus
2. Bebaru virus
3. Chikungunya virus
4. Eastern equine encephalitis virus
5. Getah virus
6. Highlands J virus
7. Mayaro virus
8. Middelburg virus
9. Mucambo virus
10. Ndumu virus
11. O'nyong-nyong virus
12. Paramaribo virus (= Semliki Forest virus)
13. Pixuna virus
14. Ross River virus
15. Sagiyama virus (= Getah virus)
16. Semliki Forest virus
17. Sindbis virus
18. Una virus
19. Uruma virus (= Mayaro virus)
20. Venezuelan equine encephalitis virus
21. Western equine encephalitis virus
22. Whataroa virus

Genus *Flavovirus*, Family Togaviridae

Members of this group originally formed the group B of arboviruses. The type species for the genus is yellow fever virus, hence the group name *Flavovirus* (*flavus* is Latin for yellow). Some members are biologically transmitted to vertebrates by mosquitoes, others by ticks. Arthropod-transmission of certain other members remains to be demonstrated. Epidemics of serious disease have been produced by some members, particularly yellow fever and dengue viruses.

GROUP CHARACTERISTICS

Nucleic Acid

Single-stranded RNA core; molecular weight of nucleic acid: approximately 3 x 10^6 daltons; molar base composition: G = 27%, A = 26%, C = 22%, U = 25%.

Virus Particle Form

Spherical particles 20–50 nm in diameter. Buoyant density of particles 1.25 g/cm^3 in CsCl.

Virus Particle Structure

Envelope present. Ultrastructure undetermined; both cubic and helical forms have been described.

Sites of Replication and Maturation

Replication takes place in the cytoplasm. Maturation by budding from the cell membrane.

Stability of Virus Particles

Ether-sensitive (possessing essential lipids). Acid-labile. Sensitive to trypsin (in contrast to alphaviruses).

Hemagglutination

All members hemagglutinate.

Possession of Group Antigens

All members are serologically related, as shown by hemagglutination-inhibition tests.

Interaction with Hosts

Replication in arthropods has not been confirmed for all members.

FLAVOVIRUSES

1. Apoi virus
2. Banzi virus
3. Bukalasa bat virus
4. Bussuquara virus
5. Cowbone Ridge virus
6. Dakar bat virus
7. Dengue virus, types 1 to 4 (type 5 = type 2; type 6 = type 1)
8. Edge Hill virus
9. Entebbe bat virus
10. Ilheus virus
11. Israel turkey meningoencephalitis virus

12. Japanese encephalitis virus
13. Kokobera virus
14. Kunjin virus
15. Kyasanur Forest virus
16. Langat virus
17. Louping ill virus
18. Modoc virus
19. Montana myotis leukoencephalitis virus
20. Murray Valley encephalitis virus
21. Negishi virus
22. Ntaya virus
23. Omsk hemorrhagic fever virus, types 1 and 2
24. Powassan virus
25. Rio Bravo virus (= U.S. bat salivary gland virus)
26. Spondweni virus
27. St. Louis encephalitis virus
28. Stratford virus
29. Tembusu virus
30. Tick-borne encephalitis virus
    a. Central European tick-borne encephalitis virus (subtype) (= diphasic meningoencephalitis virus)
    b. Far Eastern tick-borne encephalitis virus (subtype) (= Russian spring-summer encephalitis virus)
31. Uganda S virus
32. U.S. bat salivary gland virus
33. Usutu virus
34. Wesselsbron virus
35. West Nile virus
36. Yellow fever virus
37. Zika virus

Genus *Orthomyxovirus*

The term *Myxovirus* is derived from the Greek *myxa* for nasal mucus and was coined because of the unique reaction of influenza and related viruses with mucoproteins. *Ortho* (Greek for straight) indicates that these members conform in their behavior with mucoproteins, as well as in their structure, with the originally observed myxoviruses. These viruses produce respiratory disease in a variety of mammalian and avian species.

GROUP CHARACTERISTICS

Nucleic Acid

Single-stranded RNA core; nucleic acid apparently occurs in six separate pieces; molecular weight of nucleic acid: 2–4 x 10^6 daltons; molar base composition: G = 17–21%, A = 20–23%, C = 23–27%, U = 31–36%.

Virus Particle Form

Spherical or elongated particles 90–120 nm in diameter.

Virus Particle Structure

Helical nucleocapsid enclosed in an envelope studded with characteristic periodic projections. In most members the helix is 9–10 nm in diameter; in one member (influenza type C) it may be 18 nm in diameter.

Sites of Replication and Maturation

The nucleocapsid replicates in the nucleus. The virus particles mature at or near the cell surface and are released by budding from the cell membrane.

Stability of Virus Particles

Ether-sensitive (possessing essential lipids). Heat- and acid-labile. Sensitive to actinomycin D.

Hemagglutination

All members hemagglutinate enzymatically, i.e., viruses attach to mucoprotein receptors on the surface of red blood cells (RBC); hemagglutination is later followed by spontaneous elution due to destruction of the RBC receptors by neuraminidase enzyme in the virus.

Possession of Group Antigens

Three serotypes of orthomyxoviruses are distinguished on the basis of internal nucleoprotein complement-fixing (CF) antigens; families within types are differentiated by surface antigens, either CF or hemagglutinating. Antigenic alteration is common among subtypes, particularly of type A viruses.

TYPE A INFLUENZA VIRUSES

Avian Subtypes

A1	Fowl plage virus	Turkey/England/63[c]
A2	Virus N	Virus N[c]
A3	Avian influenza virus	Duck/England/56[c]
A4	Avian influenza virus	Duck/Czech/56[c]
A5	Avian influenza virus	Chicken/Scotland/59[c]
A6	Avian influenza virus	Turkey/Ontario/3724/63[c]
A7	Avian influenza virus	Duck/Ukraine/63[c]
A8	Avian influenza virus	Turkey/Ontario/6118/67[c]

Equine Subtypes

A1 A/Equi-1/Prague/56[c]
A2 A/Equi-2/Miami/63[c]

Human Subtypes

A0	A0/NWS/33	(Other related strains)
A1	A1/FM/1/47	(Other related strains)
A2	A2/Japan/305/57	(Other related strains)

Porcine Subtype

Swine influenza virus A/Swine/1976/31[c]

TYPE B INFLUENZA VIRUSES

Human Subtypes

B/Lee/40 (Other related strains)

TYPE C INFLUENZA VIRUSES

Human Subtype

C/Taylor/1233/47

Genus *Paramyxovirus*

Many, but not all, of these viruses exhibit the myxophilic two-phase interaction with mucoproteins; they differ structurally from the influenza orthomyxoviruses, however; hence the designation of *Paramyxovirus* (*para* is Greek for alongside). Members are primarily responsible for upper and lower respiratory disease in humans and other animals. Natural infection with some paramyxoviruses may result in neural or exanthematous disease (e.g., Newcastle disease virus and the serologically closely related measles, canine distemper, and rinderpest viruses).

GROUP CHARACTERISTICS

Nucleic Acid

Single-stranded RNA core; molecular weight of nucleic acid: 4–8 x 10^6 daltons; molar base composition: G = 24–25%, A = 20–26%, C = 24–27%, U = 22–31%.

Virus Particle Form

Spherical particles 150–300 nm in diameter.

Virus Particle Structure

Helical nucleocapsid enclosed in an envelope covered with characteristic periodic projections. The helix is 15–18 nm in diameter.

Sites of Replication and Maturation

Most members replicate in cytoplasm; some appear to develop in cell nucleoli (Newcastle disease, sendai, and measles viruses). Virus particles mature at or near the cell surface and are released by budding from the cell membrane.

Stability of Virus Particles

Ether-sensitive (possessing essential lipids). Heat- and acid-labile.

Hemagglutination

Most members hemagglutinate enzymatically (subgroup A), i.e., viruses attach to mucoprotein receptors on the surface of red blood cells (RBC); hemagglutination is later followed by spontaneous elution due to destruction of the RBC receptors by viral neuraminidase enzyme. Some paramyxoviruses do not contain neuraminidase (subgroup B); some do not even hemagglutinate (rinderpest and respiratory syncytial viruses).

Possession of Group Antigens

In contrast to orthomyxoviruses, paramyxoviruses are antigenically stable and are resistant to actinomycin D.

SUBGROUP A

1. Mumps virus (humans)
2. Newcastle disease virus (fowl)

3. Other avian paramyxoviruses
    Turkey/Canada/58 virus (turkeys)
    Yucaipa virus (chickens)
4. Parainfluenza-1 virus
    HA2 virus (humans)
    Sendai virus (mice)
5. Parainfluenza-2 virus
    CA (croup-associated) virus (humans)
6. Parainfluenza-3 virus
    HA1 virus (humans)
    SF4 virus (cattle)
    Other serologically identical strains isolated from horses and sheep
7. Parainfluenza-4 virus (humans)
    2 subtypes
8. Parainfluenza-5 virus (may be strains of parainfluenza-2 virus)
    AT-7 virus (humans)
    C958 virus (dogs)
    C-6A virus (dogs)
    DA virus (humans)
    SA virus (humans)
    SV5 virus (monkey kidney cell culture)
    WB virus (humans)

SUBGROUP B

1. Canine distemper virus (dogs, foxes, wolves)
2. Measles virus (humans)
3. Rinderpest virus (cattle)
4. Pneumonia virus of mice (mice)
5. Respiratory syncytial virus (humans)

Genus *Coronavirus*

The club-shaped projections on the surface of these virus particles resemble the corona of the sun; hence the name given to this genus. Avian and human types produce respiratory disease in their natural hosts; murine types cause inapparent infections, which, under conditions of stress, may result in hepatotropic or neurotropic disease. As indicated by its name, the porcine type produces gastroenteritis, with mortality approaching 100% in newborn pigs.

GROUP CHARACTERISTICS

Nucleic Acid

RNA core; strandedness unknown; molecular weight of nucleic acid and molar base composition not yet determined.

Virus Particle Form

More or less rounded particles 70–120 nm in diameter. Buoyant density of particles 1.15–1.16 g/cm^3 in CsCl.

Virus Particle Structure

Enveloped particles with characteristic petal-shaped projections. The nucleocapsid probably has a helical ultrastructure, with the helix 7–9 nm in diameter.

Sites of Replication and Maturation

Replication takes place in the cytoplasm. Maturation by budding into cytoplasmic vesicles.

Stability of Virus Particles

Ether-sensitive (possessing essential lipids). Murine types appear to be very acid-stable; the porcine type is reported to be acid-labile; others were not tested.

Hemagglutination

Avian types agglutinate erythrocytes after treatment with trypsin or ether; other members do not hemagglutinate.

Possession of Group Antigens

All murine types share common neutralizing and complement-fixing (CF) antigens. Serologic relationships have also been demonstrated between murine and some human types by CF tests.

Interaction with Hosts

Murine types and some avian strains induce syncytical formation on growth in cell culture. Some human members do not replicate in cell culture.

AVIAN TYPES

Avian Infectious Bronchitis Virus

At least two serotypes are claimed; affects chickens.

Massachusetts strain   Beaudette egg-adapted strain

HUMAN TYPES

Human Coronaviruses

At least four serotypes:

- B814 virus
- 229E virus
- LP virus (= 229E virus)
- OC38 virus
- OC43 virus

MURINE TYPES

Mouse Hepatitis Virus

At least three serotypes are claimed:

MHV-1 (Parkes) strain
MHV-2 (originally Craig) strain
MHV-3 (Nelson's PR-1) strain
JHM (Cheever) strain
H747 (Morris) strain
EHF-120 (Buescher) strain
Manaker strain

PORCINE TYPE

Transmissible Gastroenteritis of Swine Virus

Affects pigs.

Genus *Rhabdovirus*

The name selected for this genus is derived from the Greek *rhabdos* for rod because of the unique morphology of members of this group. Two important agents in this genus, rabies and vesicular stomatitis viruses, naturally infect a variety of species. Rabies virus produces encephalitis with occasional paralysis after a variable incubation period that extends to several months. Vesicular stomatitis is a relatively benign infection, but is important because the disease in cattle and pigs is clinically indistinguishable from vesicular exanthema and foot-and-mouth disease.

GROUP CHARACTERISTICS

Nucleic Acid

Single-stranded RNA core; molecular weight of nucleic acid: 3.5×10^6 daltons; molar base composition: G = 21 %, A = 29%, C = 21%, U = 29%.

Virus Particle Form

Bullet-shaped or bacilliform particles 60–70 x 130–220 nm. Buoyant density of particles 1.20 g/cm^3 in CsCl.

Virus Particle Structure

Helical nucleocapsid enclosed in an envelope with spikes 10 nm in length.

Site of Maturation

Viral particles mature by budding from cytoplasmic membranes.

Stability of Virus Particles

Ether-sensitive (possessing essential lipids). Acid-labile.

Hemagglutination

Some members agglutinate erythrocytes.

Interaction with Hosts

Some members replicate in arthropods as well as in vertebrates.

RHABDOVIRUSES

Bovine Ephemeral Fever Virus(?)

Affects cattle (= bovine epizoötic fever virus).

Hart Park Subgroup

1. Flanders virus (birds, mosquitoes)
2. Hart Park virus (mosquitoes)

Kern Canyon Virus

Isolated from bats.

M-1056 Virus

Isolated from vole-like rodents.

Mount Elgon Bat Virus

Isolated from bats.

Piry Subgroup

1. Chandipura virus (humans)
2. Piry virus (marsupials, rodents)

Rabies Subgroup

1. IbAn 27377 virus (shrews)
2. Lagos bat virus (bats)
3. Rabies virus (all mammals)

Salmonid Hematopoietic Necrosis Subgroup

1. Oregon Sockeye salmon disease virus (= Sacramento River Chinook salmon disease virus) (fish)
2. Infectious hematopoietic necrosis virus (fish)

Vervet Monkey Disease Agent(?)

Affects humans and monkeys (= Marburg virus).

Vesicular Stomatitis Subgroup

1. Indiana serotype

Subtype 1, classical strain (cattle, horses, humans, pigs, sandflies)
Subtype 2
Argentina virus (equines)
Cocal virus (mites)
Subtype 3, Brazil virus (equines)
2. New Jersey serotype (cattle, horses, humans, pigs)

Viral Hemorrhagic Septicemia Virus

Affects fish (= Egtved virus)

Genus *Leukovirus*

The name selected for this genus is derived from the leukemia/leukosis disease produced by many of the avian, feline, and murine types. All members cause neoplastic disease in at least their natural hosts.

GROUP CHARACTERISTICS

Nucleic Acid

Single-stranded RNA core; molecular weight of nucleic acid: approximately 10–12 x 10^6 daltons; molar base composition: G = 25–30%, A = 19–26%, C = 22–27%, U = 22–29%.

Virus Particle Form

Spherical particles approximately 100 nm in diameter.

Virus Particle Structure

Enveloped particles; some members have characteristic knob-like projections on the envelope surface. The symmetry of the nucleocapsid is uncertain, but possibly helical.

Site of Maturation

Viral particles mature by budding from cytoplasmic membranes.

Stability of Virus Particles

Ether-sensitive (possessing essential lipids). Acid-labile. Sensitive to actinomycin D.

Hemagglutination

None.

Possession of Group Antigens

Members of subgroups A and B share antigens.

Interaction with Hosts

Some members are oncogenic.

SUBGENUS A (AVIAN TUMOR VIRUSES)

Fowl Sarcoma Viruses

1. Bryan high-titer Rous sarcoma virus (BH-RSV)
2. Fujinami sarcoma virus (FSV)
3. Prague Rous sarcoma virus (PR-RSV)
4. Schmidt-Ruppin Rous sarcoma virus (SR-RSV)

Fowl Leukosis Viruses

1. Avian visceral leukosis viruses (RIF-1, RIF-2, RPL-12, RAV-49, RAV-50)
2. Avian myeloblastosis virus (AMV-1)
3. Avian osteopetrosis virus
4. Lymphoid leukosis-erythroblastosis virus (RAV-1, RAV-2)

SUBGENUS B (MURINE TUMOR VIRUSES)

Murine Leukemia Viruses

1. Breyere-Moloney virus
2. Buffett virus
3. Friend virus
4. Graffi (Mazurenko) virus
5. Gross virus
6. Kaplan virus
7. Manaker C-60 virus
8. Moloney virus
9. Rauscher virus
10. Rich virus
11. Other strains

Murine Sarcoma Viruses

1. Harvey sarcoma virus
2. Kirsten sarcoma virus
3. Moloney sarcoma virus

SUBGENUS C (FELINE TUMOR VIRUSES)

Feline Leukemia Virus

Multiple strains

Feline Sarcoma Virus

Multiple strains

SUBGENUS D

1. Mouse mammary tumor (Bittner) virus
2. Nodule-inducing virus

Genus *Arenavirus*

The name given to this group is derived from *arenosus,* Latin for sandy, because of the characteristic granules observed in the virus particles. Four members produce serious infections in humans; the others are apparently non-pathogenic for man, since no laboratory infections have occurred. The majority of the viruses are probably maintained in nature in a wild rodent-to-rodent cycle by means of persistent infection and chronic shedding of virus. Transmission by insects appears to be accidental.

GROUP CHARACTERISTICS

Nucleic Acid

RNA core, probably single-stranded; molecular weight and base composition of nucleic acid have not yet been determined.

Virus Particle Form

Round, oval, or irregularly shaped particles 60–280 nm in diameter in thin section, 90–220 nm in negative-contrast electron micrographs. Mean particle diameter is between 110 and 130 nm.

Virus Particle Structure

The particles consist of a well-defined unit membrane envelope with closely spaced projections and an unstructured interior that contains a variable number of electron-dense granules, 20–30 nm in diameter.

Sites of Replication and Maturation

Replication takes place within the cell cytoplasm. Particles are formed by budding, chiefly from the plasma membrane.

Stability of Virus Particles

Sensitive to lipid solvents. Acid-sensitive (below pH 5).

Hemagglutination

No hemagglutination has been reported.

Possession of Group Antigens

Members share a group-specific antigen that is demonstrable by immunofluorescence and, in some cases, by complement-fixation.

Interaction with Hosts

Replication in a variety of cell culture systems, often without visible cytopathic effects.

ARENAVIRUSES

1. Lassa virus (humans)
2. Lymphocytic choriomeningitis virus (dogs, humans, monkeys, pigs, rodents)

3. **Tacaribe Subgroup**
   1. Amapari virus (rodents)
   2. Junin (Argentinian hemorrhagic fever) virus (humans, rodents)
   3. Latino virus
   4. Machupo (Bolivian hemorrhagic fever) virus (humans, rodents)
   5. Parana virus (wild rodents)
   6. Pichinde virus (wild rodents)
   7. Pistillo virus
   8. Tacaribe virus (bats)
   9. Tamiami virus (wild cotton rats)

Genus *Poxvirus*

The poxviruses are divided into subgroups on the basis of morphology and antigenic relationships. Members of subgenus B differ in the outer structure of their particles from other poxviruses; they are also smaller and more ovoid. The close serologic relationship among members *within* subgenera, particularly in the case of subgenus A, is utilized in an immunological context: vaccinia virus, for example, is used to immunize humans against smallpox and monkeys against monkeypox infection.

GROUP CHARACTERISTICS

Nucleic Acid

Double-stranded DNA core; molecular weight of nucleic acid: 160 x 10^6 daltons; G + C (guanine + cytosine) mole fraction: 35–40%.

Virus Particle Form

Brick-shaped or ovoid particles, 170 x 300 nm to 250 x 325 nm in size. Buoyant density of particles 1.1–1.3 g/cm^3 in CsCl.

Virus Particle Structure

The viral particles are multilayered, but no true envelope is derived from host cell components. The ultrastructure is complex.

Sites of Replication and Maturation

Replication takes place wholly within the cell cytoplasm. Mature particles are released from microvilli at the cell surface.

Stability of Virus Particles

Variable sensitivity to ether and other lipid solvents. Acid-labile (pH 3).

Hemagglutination

Many members are shown to hemagglutinate, but possession of HA antigen may differ both among members and among the strains of the same poxvirus.

Possession of Group Antigens

All members exhibit non-genetic reactivation. All possess group-specific nucleoprotein (NP) antigen.

Interaction with Hosts

Production of basophilic or eosinophilic cytoplasmic inclusions. Many strains induce syncytia formation by cell fusion.

SUBGENUS A

1. Alastrim virus
2. Cowpox virus
3. Ectromelia (mousepox) virus
4. Monkeypox virus
5. Rabbitpox virus
6. Raccoonpox virus
7. Vaccinia virus
8. Variola (smallpox) virus

SUBGENUS B

1. Bovine papular stomatitis virus
2. Chamois contagious ecthyma virus
3. Orf virus (= contagious pustular dermatitis virus = contagious ecthyma virus = ovine ecthyma virus)
4. Pseudocowpox virus (= milkers's nodule virus = paravaccinia virus)
5. Ulcerative dermatosis virus

SUBGENUS C

1. Bovine lumpy skin disease virus
2. Goatpox virus
3. Sheeppox virus

SUBGENUS D

1. Canarypox virus
2. Fowlpox virus
3. Juncopox virus
4. Pigeonpox virus
5. Sparrowpox virus
6. Starlingpox virus
7. Turkeypox virus

Other Possible Members

1. Duckpox virus
2. Grousepox virus
3. Parakeetpox virus
4. Ptarmiganpox virus

SUBGENUS E

1. California myxoma virus
2. Hare fibroma virus

3. Myxoma virus
4. Rabbit fibroma virus
5. Squirrel fibroma virus

UNCLASSIFIED POXVIRUSES

1. Buffalopox virus
2. Camelpox virus
3. Horsepox virus
4. Molluscum contagiosum virus
5. Mouse papular virus
6. Rhinocerospox virus
7. Swinepox virus
8. Tana virus
9. Yaba monkey tumor virus
10. Yaba-like virus

Genus *Herpesvirus*

Microepidemiology of herpesvirus infections is characterized by a cell-to-cell virus spread. Some members of the genus are so strongly cell-associated that free virus is detected with difficulty, if at all; other herpesviruses are readily released from infected cells. Differentiation of subgroups on this basis is not practical, however, because of gradations in the degree of cell-association. Members of this group frequently produce latent infections in their natural hosts. Infections of heterologous species are generally serious and often fatal.

GROUP CHARACTERISTICS

Nucleic Acid

Double-stranded DNA core; molecular weight of nucleic acid: 54–92 x 10^6 daltons; G + C mole fraction: 57–74%.

Virus Particle Form

Particle size (complete virion with envelope) 140–160 nm in diameter. Buoyant density of particles 1.27–1.29 g cm^3 in CsCl.

Virus Particle Structure

Envelope present. Capsid (protein shell) exhibits cubic, icosahedral symmetry with 162 hollow, cylindrical capsomeres.

Sites of Replication and Maturation

The nucleocapsid is formed within the cell nucleus. The envelope is acquired as particles bud from the nuclear membrane into the cytoplasm.

Stability of Virus Particles

Ether-sensitive (envelope possesses essential lipids). Acid-labile.

Interaction with Hosts

Production of type A intranuclear inclusions in cell culture.

AMPHIBIAN AND REPTILIAN TYPES

1. Frog herpesviruses (FV = frog virus)
   FV4, FV5, FV6, FV7, FV8 (may be serologically identical)
   Lucké tumor virus
2. Iguana herpesvirus
3. Snake herpesvirus
4. *Xenopus* herpesvirus

AVIAN TYPES

1. Duck plague (duck enteritis) virus (wild and domestic ducks)
2. Herpesviruses of cormorants, owls, parrots, pigeons, and other birds
3. Infectious laryngotracheitis virus (chickens, pheasants)
4. Marek's disease virus (chickens)
5. Turkey herpesvirus, WTHV-1 strain (turkeys)

BOVINE TYPES

1. Bovine mammilitis virus (= Allerton virus)
2. Infectious bovine rhinotracheitis virus (= infectious pustular vulvovaginitis virus)
3. Malignant catarrhal fever virus

CANINE TYPE

Canine herpesvirus (multiple serologically identical isolates)

EQUINE TYPES

Five serotypes:

1. Rhinopneumonitis virus (abortion influenza virus; equine abortion virus)
2. LK virus
3. (Karpas)
4. (Hsiung)
5. (Bryans)

FELINE TYPE

Feline rhinotracheitis virus

HUMAN TYPES

1. Epstein-Barr virus; associated with infectious mononucleosis and Burkitt lymphoma
2. Herpes simplex virus (*Herpesvirus hominis*); two subtypes
3. Human cytomegalovirus
4. Varicella-zoster virus

LAPINE TYPES

1. Agent 923J
2. Cottontail rabbit herpesvirus
3. Virus III (*Herpesvirus cuniculi*)

OVINE TYPE

Sheep pulmonary adenomatosis (Jaagsiekte) virus

PISCINE TYPE

Fishpox virus

PORCINE TYPES

1. Pseudorabies virus (*Herpesvirus suis*; Aujeszky's disease virus)
2. Porcine cytomegalovirus (inclusion-body rhinitis virus)

RODENT TYPES

1. Ground squirrel cytomegalovirus
2. Guinea pig cytomegalovirus
3. Mouse cytomegalovirus
4. Mouse thymic virus (?)
5. Sand rat cytomegalovirus

SIMIAN TYPES

1. B virus (*Herpesvirus simiae*) (primarily in rhesus and cynomolgus monkeys)
2. *Herpesvirus saimiri* (isolated from squirrel monkeys)
3. Herpesvirus T (marmoset virus; *Herpesvirus tamarinus; Herpesvirus platyrrhinae*) (isolated from marmoset monkeys)
4. Liverpool vervet monkey virus (isolated from vervet monkeys)
5. Patas herpesvirus (isolated From patas monkeys)
6. SA6 (= Simian Agent #6) (simian cytomegalovirus) (isolated from vervet monkeys)
7. SA8 (= Simian Agent # 8) (isolated from vervet monkeys)

Genus *Iridovirus*

Extraction of insect members of this genus from infected tissues produces pellets with an iridescent appearance; hence the name given to the group. Differences in size, possession of a limiting membrane, and sensitivity to ether among the members suggest that this genus might properly be divided into subgroups or into separate genera.

GROUP CHARACTERISTICS
Nucleic Acid

Double-stranded DNA core; molecular weight of nucleic acid: approximately 130 x 10^6 daltons; G + C mole fraction: 29–32%.

Virus Particle Form

Hexagonal particles, some averaging 135 nm in diameter, others ranging from 175–215 nm and from 200–300nm.

Virus Particle Structure

Viral particles of some members (amphibian and porcine) possess envelopes; others (piscine member) are naked. The capsid (protein shell) of viral particles exhibits cubic, icosahedral symmetry, possibly with 812 to as many as 1,500 capsomeres.

Sites of Replication and Maturation

Replication takes place within the cell cytoplasm. Amphibian and porcine members mature by budding from the plasma membrane, from which they acquire a limiting membrane or envelope.

Stability of Virus Particles

Some members are sensitive to ether and other lipid solvents, others are not. Some are shown to be acid-labile.

IRIDOVIRUSES

1. African swine fever virus (pigs)
2. Amphibian cytoplasmic viruses (frogs and newts); multiple closely related, if not identical, strains (FV = fog virus; LT = Lucké-*Triturus* or newt virus)

FV1	FV10	FV13	FV16
FV2	FV11	FV14	FV17
FV3	FV12	FV15	FV18
FV9			FV19
LT1	LT2	LT3	L4B

   Tadpole edema virus
3. Gecko virus (lizards)
4. Lymphocystis virus (fish)

Genus *Adenovirus*

This group of viruses derives its name from their initial isolation from adenoid tissues. Adenoviruses, like herpesviruses, often produce latent infections in man. In their active form they may cause self-limiting respiratory and ocular disease. Injection of some members into laboratory animals may produce sarcomas. Adenoviruses are often associated with defective satellite viruses, which require them as helpers for replication.

GROUP CHARACTERISTICS

Nucleic Acid

Double-stranded DNA core; molecular weight of nucleic acid: 20-25 x 10^6 daltons; G + C mole fraction: 48–57%.

Virus Particle Form

Hexagonal particles 70–90 nm in diameter. Buoyant density of particle 1.34 g/cm^3 in RbCl.

Virus Particle Structure

No envelope. The capsid (protein shell) exhibits cubic, icosahedral symmetry with 252 hollow, cylindrical capsomeres.

Sites of Replication and Maturation

Nucleus.

Stability of Virus Particles

Ether-resistant (lacking essential lipids). Acid-stable.

Hemagglutination

Most members, except mouse types, agglutinate erythrocytes.

Possession of Group Antigens

All members, except avian types, share common CF antigen.

Interaction with Hosts

Production of characteristic cytopathic effects in cell culture and of type B intranuclear inclusions.

AVIAN TYPES

1. GAL_1 virus (**G**allus, **A**deno-**L**ike)
2. CELO virus (**C**hick-**E**mbryo-**L**ethal-**O**rphan)
3. Quail bronchitis virus; serologically related to CELO virus

Other strains have been isolated, but serologic relationship or possible identity with the above serotypes remains to be determined

BOVINE TYPES

At least three, and possibly four or more serotypes.

1 Bovine #10
2 Bovine #19
3 WBR1
THT/62 (serologic relationship or possible identity with the above serotypes remains to be confirmed)

CANINE TYPES

Infectious canine hepatitis (ICH) virus
A 26/61 virus (serologically related to ICH virus)

HUMAN TYPES

Thirty-three serotypes and one subtype; virus names in parentheses are alternate designations for prototype strains.

1 Adenoid 71
2 Adenoid 6
3 G.B.
4 RI-67
5 Adenoid 75
6 Tonsil 99
7 Gomen
7a S-1058
8 Trim
9 Hicks
10 J.J.
11 Slobitski
12 Huie
13 A.A.
14 DeWit
15 305(#955)
16 Ch. 79
17 Ch. 22
18 D.C.
19 AV-587 (3911)
20 AV-931 (D55-92)
21 AV-1645
22 AV-2711
23 AV-2732
24 AV-3153
25 BP-1
26 BP-2
27 BP-4
28 BP-5
29 BP-6
30 BP-7
31 1315/63
32 HH
33 DJ

MURINE TYPES

1. FL virus
2. E-20308 virus (serologic comparison with FL virus remains to be confirmed)

OPPOSSUM TYPE

IVIC-Di-7 virus

PORCINE TYPES

At least three serotypes:

1 25R
2 6618
3 A47

SIMIAN TYPES

Nineteen, and possibly twenty-four, serotypes (SV = simian virus, SA = simian agent, CV = chimpanzee virus); strain designations are given in parentheses.

1. SV1 (301)
2. SV11 (646776)
3. SV15 (AP4398)
4. SV17 (6630-1C)
5. SV20 (M-12)
6. SV23 (7495-2WK)
7. SV25 (8045-2WN)
8. SV30 (P-5)
9. SV31 (P-6)
10. SV32 (P-7)
11. SV33 (P-10)
12. SV34 (A-7644)
13. SV36 (P-9)
14. SV37 (E-4382, closely related to SV32)
15. SV38 (72707, related to SV33)
16. SA7
17. SA17
18. SA18
19. V340
20. C1 (Bertha)[f]

[f] Candidate simian adenovirus type.

Potential Candidate Types

Pan 5 (CV-23)
Pan 6 (CV-32)
Pan 7 (CV-33)
Pan 9 (CV-68)

Genus *Papillomavirus* Family Papovaviridae

The first part of the family name, papova-, is derived from three members that belong to the family: **Pa**pilloma-, **Po**lyoma-, and **Va**cuolating viruses.

GROUP CHARACTERISTICS

Nucleic Acid

Double-stranded circular DNA core; molecular weight of nucleic acid: 5×10^6 daltons; G + C mole fraction: 49%.

Virus Particle Form

Spherical particles 53 nm in diameter. Buoyant density of particles 1.34 g/cm^3 in CsCl.

Virus Particle Structure

No envelope. The capsid (protein shell) exhibits cubic symmetry, with 72 hollow, cylindrical capsomeres in a skew arrangement.

Site of Replication

Cell nucleus.

Stability of Virus Particles

Ether-resistant (lacking essential lipids). Acid- and heat-stable.

Hemagglutination

Several members hemagglutinate by reacting with neuraminidase-sensitive receptors.

Interaction with Hosts

Production of papillomata or warts in various animal hosts.

PAPILLOMAVIRUSES

1. Bovine papilloma virus
2. Canine oral papilloma virus
3. Deer fibroma virus
4. Human papilloma (wart) virus
5. Rabbit oral papilloma virus
6. Rabbit papilloma virus (= Shope papilloma virus)

Other possible members: viruses causing papillomata of goats, horses, monkeys, sheep, and other species.

Genus *Polyomavirus*, Family Papovaviridae

GROUP CHARACTERISTICS

Nucleic Acid

Double-stranded circular DNA core; molecular weight of nucleic acid: 3×10^6 daltons; G + C mole fraction: 41–49%.

Virus Particle Form

Spherical particles 43 nm in diameter. Buoyant density of particles 1.34 g/cm^3 in CsCl.

Virus Particle Structure

No envelope. The capsid (protein shell) exhibits cubic symmetry, with 72 hollow, cylindrical capsomeres in a skew arrangement.

Site of Replication

Cell nucleus.

Stability of Virus Particles

Ether-resistant (lacking essential lipids). Acid- and heat-stable.

Hemagglutination

Several members hemagglutinate by reacting with neuraminidase-sensitive receptors.

Interaction with Hosts

Production of latent and chronic infections in their natural hosts. Most members are oncogenic under certain conditions.

POLYOMAVIRUSES

1. K virus (mice)
2. Polyoma virus (mice)
3. Progressive multifocal leukoencephalopathy virus (humans)
4. Rabbit kidney vacuolating agent (rabbits)
5. Vacuolating agent (= SV40) (monkey kidney cell culture)

Genus *Parvovirus*

The name of the group is derived from the Latin *parvus* for small.

GROUP CHARACTERISTICS

Nucleic Acid

Single-stranded DNA core; single nucleic acid strands of subgroup B members are complementary and come together to form double strands *in vitro;* molecular weight of nucleic acid: 1.2–1.8 x 10^6 daltons; G + C mole fraction: 39%.

Virus Particle Form

Hexagonal particles 18–22 nm in diameter. Buoyant density of particles approximately 1.4 g/cm^3 in CsCl.

Virus Particle Structure

No envelope. The capsid (protein shell) exhibits cubic, icosahedral symmetry, with probably 32 capsomeres.

Sites of Replication and Maturation

Cell nucleus.

Stability of Virus Particles

Ether-resistant (lacking essential lipids). Acid- and heat-stable.

Hemagglutination

Some members of each subgroup agglutinate erythrocytes, others appear unable to hemagglutinate.

Interaction with Hosts

Members of subgroup B are defective and require other virus "helpers" to replicate.

SUBGENUS A

1. Hamster osteolytic viruses
   a. H-1 virus (human embryonic and tumor tissues)
      HT virus (human embryonic and tumor tissues)
   b. H-3 virus (human embryonic and tumor tissues)
      Hemorrhagic encephalopathy virus of rats (rats)
      Rat virus (= Kilham rat virus) (rats)
      X14 virus (rats)
   c. HB virus (human embryonic and tumor tissues)
2. Minute virus of mice (mice)
3. Porcine parvovirus (pigs)

59e/63
PRP
G10/1
Wavre

} (serologically closely related if not identical strains

SUBGENUS B

Adeno-associated (satellite) viruses; serotypes 1, 2, 3, and 4 (humans and monkeys)

SUBGENUS C

Probable or possible members

1. Feline panleukopenia (feline infectious enteritis, feline distemper) virus (=feline ataxia virus) (cats and other feline species)
2. Haden (hemadsorbing enteric) virus (cattle)
3. Mink enteritis virus (mink)
4. Minute virus of canines (dogs)

Unclassified Arboviruses

The arbovirus designation is based on epidemiologic rather than taxonomic considerations. Arboviruses are defined as viral agents that, in nature, infect hemophagous arthropods through ingested vertebrate blood, multiply in the arthropod tissues, and are transmitted by bite to susceptible vertebrates. Arboviruses are also considered to have other properties in common: ribonucleic acid core, sensitivity to lipid solvents, and pathogenicity for newborn mice by the intracerebral route. The arboviruses listed below are included here because they do not appear to belong to any of the recognized taxonomic groups. Some viruses for which no arthropod transmission has been confirmed are included because of serologic relationship to known arboviruses. In other cases, no serologic relationships exist, but arthropod transmission is suspected.

BUNYAMWERA SUPERGROUP

Bunyamwera Group

Batai virus
Bunyamwera virus
Cache Valley virus
Calovo virus (= Batai virus)
Chittoor virus (= Batai virus)
Germiston virus
Guaroa virus
Ilesha virus
Kairi virus
Lokern virus
Maguari virus
Olifantsvlei virus
Sororoca virus
Taiassui virus
Tensaw virus
Tucunduba virus
Ukauwa virus
Wyeomyia virus

Group C

Apeu virus
Caraparu virus
Gumbo Limbo virus
Itaqui virus
Madrid virus
Marituba virus
Murutucu virus
Nepuyo virus
Oriboca virus
Ossa virus
Restan virus

Bwamba Group

Bwamba virus
Pongola virus

California Encephalitis Group

Brooks virus
California encephalitis virus
Greeley virus
Hays-Vauxhall virus
Jamestown Canyon virus
Jerry Slough virus

Keystone virus
LaCrosse virus
Lumbo virus
Melao virus
Ottawa virus
San Angelo virus
Snowshoe hare virus
Tahyna virus
Trivittatus virus
Troica virus (= Tahyna virus)

Capim Group

Acara virus
Bushbush virus
Capim virus
Guajara virus
Moriche virus

Guama Group

Bertioga virus
Bimiti virus
Catu virus
Guama virus
Mahogany Hammock virus
Moju virus

Koongol Group

Koongol virus
Wongol virus

Patois Group

Mirim virus
Pahayokee virus
Patois virus
Shark River virus
Sontecomapan virus
Zegla virus

Simbu Group

Akabane virus
Buttonwillow virus
Ingwavuma virus
Manzanilla virus
Oropouche virus
Sango virus
Sathuperi virus
Simbu virus
Utinga virus

OTHER SMALL GROUPS

Anopheles A Group

Anopheles A virus
Lukuni virus

Anopheles B Group

Anopheles B virus
Boraceia virus

Bakau Group

Bakau virus
Ketapang virus

Ganjam Group

Dugbe virus
Ganjam virus

Hughes Group

Farallon virus
Hughes (Dry Tortugas) virus
Soldado virus

Kaisodi Group

Kaisodi virus
Lanjam virus

Nyando Group

Eratmapodites virus
Nyando virus

Phlebotomus Fever Group

Anhanga virus
Bujaru virus
Candiru virus
Chagres virus
I-47 virus
Icoaraci virus
Itaporanga virus
Karimabad virus
Naples virus
Punta Toro virus
Salehabad virus
Sicily virus

Quaranfil Group

Johnston virus
Quaranfil virus

Timbo Group

Chaco virus
Timbo virus

Turlock Group

Turlock virus
Umbre virus

UNGROUPED ARBOVIRUSES

Aruac virus
Congo virus (= Semunya virus)
Cotia virus
Crimean hemorrhagic fever virus
Embu virus
Ieri virus
Jurona virus
Kwatta virus
Lone Star virus
Mapputta virus
Marco virus
Mossuril virus
Nairobi sheep disease virus
Nariva virus
Nyamanini virus
Pacui virus
Potepli-63 virus
Rift Valley fever virus
Sawgrass virus
Semunya virus
Silverwater virus
Simian hemorrhagic fever virus
SM-214 virus
Tacaiuma virus
Tataguina virus
Tembe virus
Thogoto virus
Triniti virus
Uukuniemi virus
Wad Medani virus
Wanowrie virus
Witwatersrand virus

OTHER UNCLASSIFIED VERTEBRATE VIRUSES

Bovine Viral Diarrhea/Hog Cholera/Equine Arteritis Virus Group

GROUP CHARACTERISTICS

Nucleic Acid

RNA core.

Virus Particle Form

Spherical enveloped particles 46–58 nm in diameter; a larger size, 80–>100 nm, has also been reported. Buoyant density of particles 1.16–1.22 g/cm³ in CsCl.

Virus Particle Structure

The symmetry of the nucleocapsid is uncertain; both helical and cubic symmetry have been claimed. The envelope may be smooth or stubbled with tiny projections.

Site of Maturation

Particles may mature by budding from the cytoplasmic matrix into the cisternae of the endoplasmic reticulum.

Stability of Virus Particles

Ether-sensitive (possessing essential lipids).

Hemagglutination

None.

Possession of Group Antigens

Bovine and porcine members share neutralizing antigens; there is no apparent serologic relationship with the equine member.

Interaction with Hosts

Replication in cell culture without visible cytopathology.

MEMBERS

1. Bovine viral diarrhea (mucosal disease) virus	New York 1,[c] NADL[c]
2. Hog cholera (swine fever) virus	Ames[c]
3. Equine arteritis virus	Bucyrus[c]

Syncytial Viruses

GROUP CHARACTERISTICS

Nucleic Acid

Probably RNA.

Virus Particle Form

Enveloped particles 100–110 nm in diameter.

Virus Particle Structure

The nucleocapsid is of unknown symmetry. Some members have projections on the surface of the envelope.

Site of Maturation

Particles mature by budding from the cytoplasmic membrane.

Stability of Virus Particles

Ether-sensitive (possessing essential lipids). Acid-sensitive (pH 3). Heat-sensitive.

Hemagglutination

None.

Possession of Group Antigens

Isolates from different host species appear to be serologically distinct from one another.

Interaction with Hosts

All members induce formation of multinucleated giant cells (syncytia) on growth in cell culture. Some simian members possess RNA-dependent DNA polymerase enzyme.

MEMBERS

Bovine syncytial virus
Feline syncytial virus (multiple strains)
Hamster syncytial virus
Rabbit syncytial virus
Simian foamy virus (seven serotypes):

1 FV21 (rhesus, Formosan, and vervet monkeys)
2 FV34 (Formosan and vervet monkeys)
3 FV2014 (vervet monkeys)
4 1224 strain (squirrel monkeys)
5 1557 strain (Galagos monkeys)
6 CV-1 (Pan-1) (chimpanzees)
7 CV-11 (Pan-2) (chimpanzees)

Visna/Maedi Virus Group

GROUP CHARACTERISTICS

Nucleic Acid

Probably single-stranded RNA.

Virus Particle Form

Roughly spherical enveloped particles 80–110 nm in diameter. Buoyant density of particles 1.18–1.20 g/cm^3 in CsCl.

Virus Particle Structure

The nucleocapsid is of unknown symmetry. There are projections on the surface of the envelope.

Site of Maturation

Particles mature by budding from plasma membranes.

Stability of Virus Particles

Ether-sensitive (possessing essential lipids). Sensitive to actinomycin D.

Hemagglutination

None.

Possession of Group Antigens

Members are antigenically related, but not identical.

Interaction with Hosts

Members resemble the leukoviruses in the above properties, as well as in their dependence on cell DNA synthesis for replication and in their possession of RNA-dependent DNA polymerase enzyme. At least two members (visna and progressive pneumonia viruses) are also able to transform cells in culture, as do the tumor viruses. To date, however, members of this group have not been shown to be oncogenic.

MEMBERS

1. Maedi virus (sheep)
2. Progressive pneumonia virus (sheep)
3. Visna virus (sheep)

Equine Infectious Anemia Virus

CHARACTERISTICS

Nucleic Acid

RNA core.

Virus Particle Form

Mostly spherical enveloped particles ranging from 90 to 140 nm in diameter. Buoyant density of particles covers a broad range, with the peak at 1.15 g/cm^3 in CsCl.

Virus Particle Structure

The symmetry of the nucleocapsid is unknown. The envelope has projections.

Site of Maturation

Particles mature by budding from plasma membranes.

Stability of Virus Particles

Ether-sensitive (possessing essential lipids).

Hemagglutination

None.

Interaction with Hosts

Replication is dependent on cell DNA synthesis.

Rubella Virus

CHARACTERISTICS

Nucleic Acid

RNA core.

Virus Particle Form

Spherical enveloped particles 50–85 nm in diameter (55 nm average). Buoyant density of particles 1.22 g/cm^3 in CsCl.

Virus Particle Structure

The symmetry of the nucleocapsid is unknown. There are spicules on the envelope surface.

Site of Maturation

Particles mature by budding from marginal cytoplasmic membranes or into cytoplasmic vacuoles.

Stability of Virus Particles

Ether-sensitive (possessing essential lipids). Acid-sensitive. Heat-sensitive.

Hemagglutination

The virus will hemagglutinate non-enzymatically if inhibitors are removed from the culture medium in which the virus is grown.

Australia (AU) Antigen

An apparent virus particle associated with human serum hepatitis.

CHARACTERISTICS

Nucleic Acid

No nucleic acid associated with the particle has been detected and confirmed; a claim of RNA has been made, however.

Virus Particle Form

Spherical and tubular particles with an average diameter of 20 nm. Buoyant density of particles 1.21 g/cm^3 in CsCl.

Virus Particle Structure

Ultrastructure of the particle is unknown. Knob-like subunits are on the surface.

Stability of Virus Particles

Ether-resistant (lacking essential lipids). Heat-resistant.

Interaction with Hosts

Particles are observed in the blood of patients with viral hepatitis; they have not been cultivated in cell culture.

Scrapie Agent

CHARACTERISTICS

Nucleic Acid

Efforts to demonstrate nucleic acid have so far been unsuccessful.

Virus Particle Form

Infectious particles 16–26 nm (by filtration). Buoyant density of particles 1.32 g/cm^3 in CsCl.

Stability of Virus Particles

Extraordinary resistance to ether, heat, formalin, ultraviolet radiation, and ionizing radiation.

Interaction with Hosts

In-vitro replication has been demonstrated only in cultures of mouse brain cells. Visualization of particles with electron microscopy has been unsuccessful to date.

ISOLATES

1. Scrapie agent (sheep, goats)
2. Mink encephalopathy agent (mink) (may be related to the scrapie agent)

REFERENCES

1. Berge, T. O., and Stevens, D. A., *Catalogue of Viruses, Rickettsiae, Chlamydiae,* 4th ed. American Type Culture Collection, Rockville, Maryland (1971).
2. Western Hemisphere Committee on Animal Virus Characterization, Animal Reference Virus Recommendations, *Amer. J. Vet. Res., 31* 1915 (1970).
3. Wildy, P., Classification and Nomenclature of Viruses, *Monogr. Virol., 5* (1971).
4. Wilner, B. I., *A Classification of the Major Groups of Human and Other Animal Viruses,* 4th ed. Burgess Publishing Co., Minneapolis, Minnesota (1969).
5. Wilner, B. I., *A Classification of the Major Groups of Human and Other Animal Viruses,* 5th ed. (in preparation).

BACTERIOPHAGES
The Morphology of Bacteriophages

DR. H.–W. ACKERMANN

Bacterial viruses, morphologically, are fairly heterogeneous. Up to six morphological groups have been recognized.[1,10,36] Recently, some new types have been described: a phage with a lipid-containing envelope, a large cubic phage active on *Bdellovibrio,* and a rod-like phage of *Mycoplasma* (Figures 25 to 27).

All phages have nucleic acid and a protein shell; the lipid envelope is a unique feature. The nucleic acid is either double-stranded DNA, single-stranded DNA, or single-stranded RNA. The protein shell consists of regularly arranged subunits, which may or may not be assembled in capsomers and which form cubic or rod-shaped particles. The cubic particles may have tails for ejection of their nucleic acid. We therefore have three main groups: tailed, cubic, and filamentous phages. In tailed and cubic phages the nucleic acid is coiled inside the protein shell; in the filamentous fd-type phages it is distributed over the entire length of the particle and seems to be located in the middle of two rows of protein subunits,[16] unlike other viruses with helical symmetry. The above observations are summarized in Table 1.

Tailed phages form the largest group. They have isometric or elongated heads, and tails that are either long and contractile, long and non-contractile, or very short. They can be subdivided into eight morphological types.[1] The isometric heads seem to be octahedra.[1] The shape of elongated heads is more obscure; they may be bipyramidal hexagonal antiprisms[9] or prolate icosahedra.[7,27] The tails are basically hollow rods of helically arranged subunits, at least in long tails. Contractile tails have, in addition, a sheath that is also composed of helical rows of subunits. This sheath is always separated from the head by a free space, the neck. Furthermore, most phages possess facultative organites, such as collars, head or collar appendages (Figure 19), base plates, spikes, or tail fibers (for definitions, see Reference 36). Their role is not always understood. They may undergo functional changes and vary considerably in number and size, as is shown in Table 2.

Cubic and filamentous phages are much simpler in structure. However, it is noteworthy that RNA phages should be more properly called pentakis dodecahedra. The *Bdellovibrio* phage (Figure 26) is claimed to have a double capsid, but this feature needs confirmation. Nothing is known about the fine structure of the *Mycoplasma* phage (Figure 27).

Figures 1 to 29 show representatives of the various morphological groups or types. The considerable size variation within some types is illustrated by phages selected for their outstanding dimensions. The algal virus LPP-1 and a "killer particle" have been included for their particular interest or their characteristic morphology. Functional changes are illustrated only in T-even phages. No attempt was made to show the entire range of morphological features, because the facultative organites are difficult to resolve and descriptions are often poor (for further reading, see References 10 and 36). It is likely that new types will be discovered. For example, a wide series of interesting new phages of unknown hosts has been found in the intestinal tracts of animals.[31,32]

Table 3 identifies the morphological groups or types shown in Figures 1 to 29 and cites a representative example for each structure.

TABLE 1

MORPHOLOGIC CLASSIFICATION AND MAIN CHARACTERISTICS OF BACTERIOPHAGES

Main Group	Figure(s)	Nucleic Acid	Shape	Symmetry	Capsomers	Taxonomic Group: Reference 10	Taxonomic Group: Reference 36
Tailed	1-22	2-DNA	Complex	Binary	Variable	A, B, C	V, VI, III
Cubic	23	1-DNA	Icosahedron	Cubic	12	D	II
	24	1-RNA	Icosahedron	Cubic	32	E	II
	25[a]	2-DNA	Icosahedron	Cubic	?		
Filamentous	26[b]	1-DNA?	?	Cubic	?		
	27	DNA	Rod	Helical	None		
	28-29	1-DNA	Rod	Helical	None	F	I

[a]Lipid envelope.
[b]Double capsid(?).

TABLE 2

FACULTATIVE ORGANITES OF TAILED BACTERIOPHAGES

	Striations		Spikes		Tail Fibers			
	Number	References	Number	References	Number	References	Length, Å	References
Maximum	~100	29	12(?)	17,22,35,38	6	Frequent	~2,100	33
Minimum	2-4	25	3(?)	Frequent	1	Frequent	80	37

TABLE 3

SELECTED EXAMPLES OF BACTERIOPHAGES

Group or Type	Figure	Phage	Host	References
A1	1	M_S	*Clostridium*	6
	2	ΦIA	*Rhizobium*	21
	3	G[a]	*Bacillus*	2,18
	4	phi[b]	*Bacillus*	5,36
A2	5	T-even type[c]	*Aeromonas* Enterobacteria *Moraxella* *Rhizobium*	20, also*
	6	9266	Enterobacteria	30

TABLE 3 (Continued)

SELECTED EXAMPLES OF BACTERIOPHAGES

Group or Type	Figure	Phage	Host	References
	7	506	*Streptomyces*	19
B1	8	II	*Bacillus*	15
	9	P	*Corynebacterium*	26
	10	3ML	*Streptococcus*	11
	11	M_1	*Thermoactinomyces*	28
B2	12	L419	*Rhizobium*	21
	13	—[d]	*Staphylococcus*	†
B3	14	ΦCb13	*Caulobacter*	34
	15	7408b	Enterobacteria	30
	16	80	*Clostridium*	39
C1	17	P22	Enterobacteria	4,38,41,42
	18	LPP-1[e]	*Plectonema*	23,24,25
	19	ϕ29	*Bacillus*	3
C2	20	GA-1	*Bacillus*	8
	21	7	*Vibrio*	40
C3	22	F46/3039	Enterobacteria	12,13,14
D	23	ΦX type[f]	Enterobacteria	‡
E	24	RNA type	*Caulobacter* Enterobacteria *Pseudomonas*	‡
	25	PM2[g]	*Pseudomonas*	††
	26	HDC-1[h]	*Bdellovibrio*	††
	27	MVL1	*Mycoplasma*	††
F	28	fd type[i]	Enterobacteria	‡
	29	Pf	*Pseudomonas*	‡

* See Tailed Bacteriophages: Listing by Morphological Groups, Table 2, p. 586.
† See Tailed Bacteriophages: Listing by Morphological Groups, Table 4, p. 595.
See Bacterial Viruses Containing Single-Stranded Nucleic Acid, p. 617.
†† See Bacterial Viruses Containing Single-Stranded Nucleic Acid, Appendix, p. 626.

[a]Probably contracted.
[b]"Killer particle".
[c]Normal and contracted phage.
[d]"Idealized" phage.
[e]Thick capsid.
[f]Large capsomers.
[g]Lipid envelope.
[h]Double capsid(?).
[i]Excluding phage Xf.

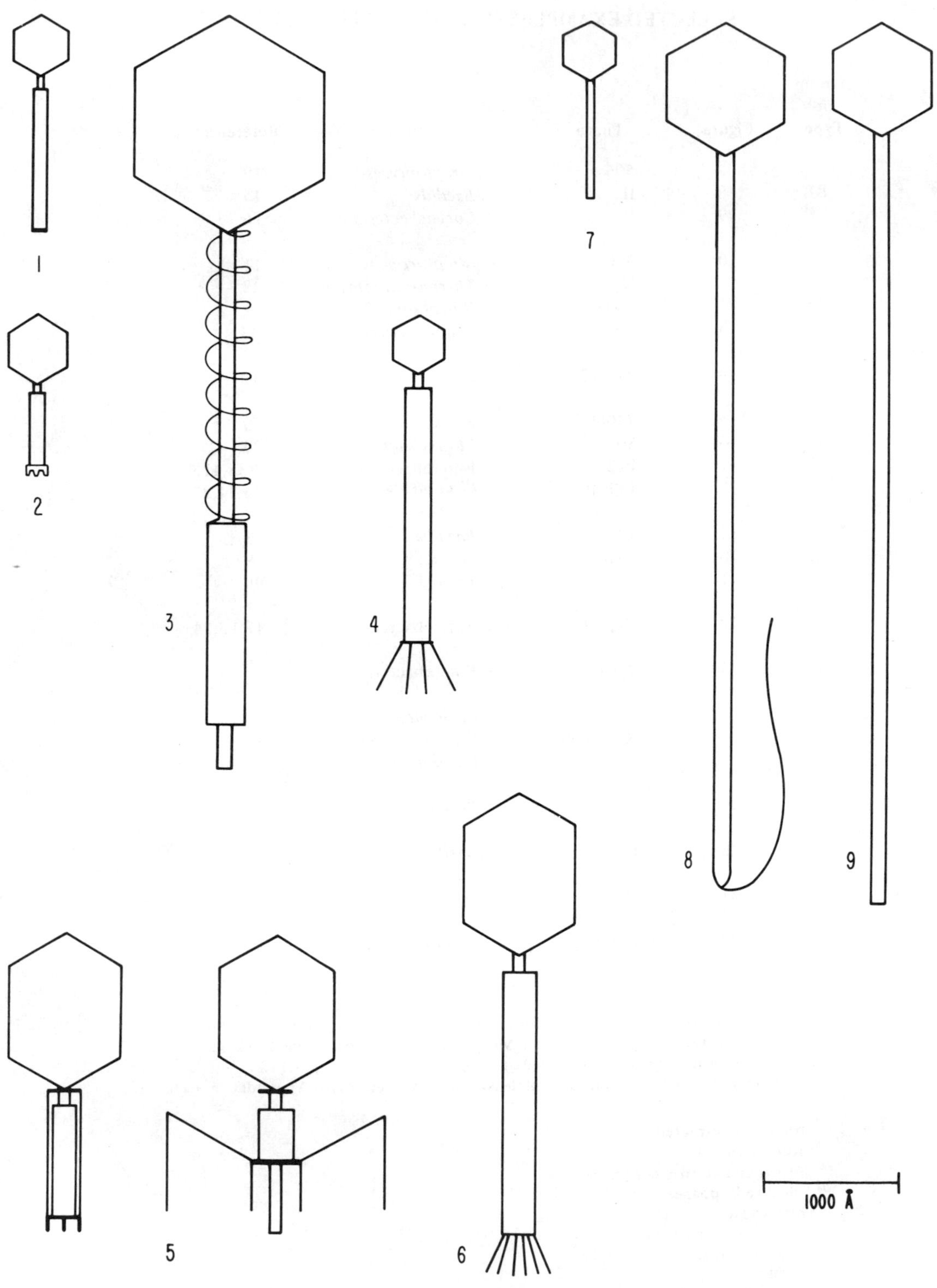

FIGURES 1-9.

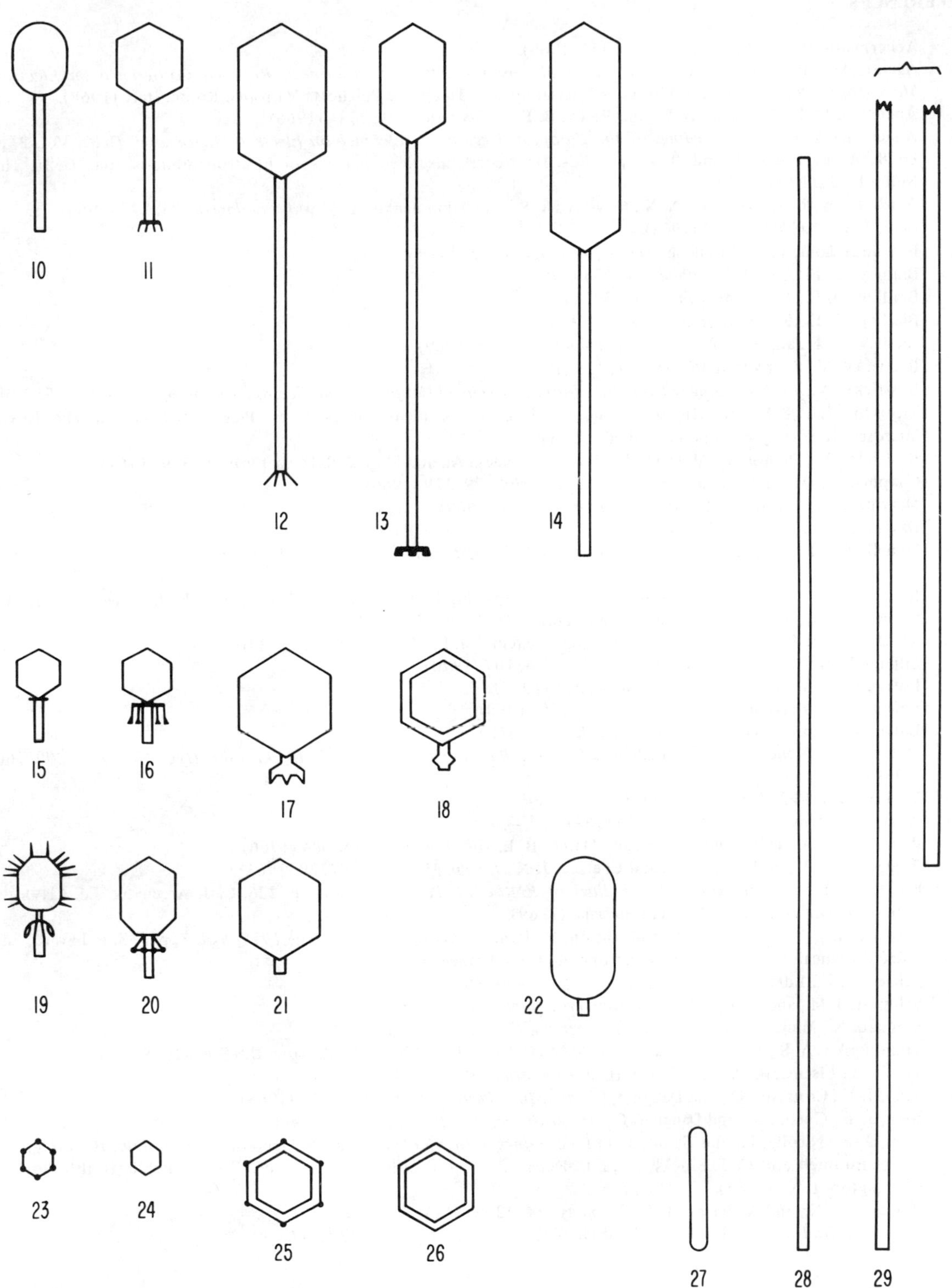

FIGURES 10-29.

REFERENCES

1. Ackermann, H.-W., *Pathol. Biol.*, *17*, 1003 (1969).
2. Ageno, M., and Donelli, G., in *Electron Microscopy 1968, 4th European Regional Conference on Electron Microscopy, Rome,* Vol. 2, p. 155, D. S. Bocciarelli, Ed. Tipografia Poliglotta Vaticana, Rome, Italy (1968).
3. Anderson, D. L., Hickman, D. D., and Reilly, B. E., *J. Bacteriol.*, *91*, 2081 (1966).
4. Anderson, T. F., in *Proceedings of the European Regional Conference on Electron Microscopy, Delft,* Vol. 2, p. 1008, A. L. Houwink and B. J. Spit, Eds. De Nederlandse Vereniging voor Electronenmicroscopie, Delft, The Netherlands (1960).
5. Azizbekyan, R. R., Belyaeva, N. N., Krivisky, A. S., and Tikhonenko, A. S., *Mikrobiologiya, 35*, 279 (1966).
6. Betz, J. V., *Virology, 36*, 9 (1968).
7. Boy de la Tour, E., and Kellenberger, E., *Virology, 27*, 222 (1965).
8. Bradley, D. E., *J. Gen. Microbiol.*, *41*, 233 (1965).
9. Bradley, D. E., *J. Gen. Microbiol.*, *38*, 395 (1965).
10. Bradley, D. E., *Bacteriol. Rev.*, *31*, 230 (1967).
11. Bradley, D. E., and Kay, D., *J. Gen. Microbiol.*, *23*, 553 (1960).
12. Bystrický, V., *J. Electron Microsc.*, *11*, 125 (1962).
13. Bystrický, V., in *Proceedings of the 3rd European Regional Conference on Electron Microscopy,* Vol. B, p. 557, M. Titlbach, Ed. Publishing House of the Czechoslovak Academy of Sciences, Prague (1964), and The Royal Microscopic Society of London (2nd ed., 1965).
14. Bystrický, V., Drahoš, V., Mulczyk, M., Przondo-Hessek, A., and Šlopek, S., *Acta Virol.*, *8*, 369 (1964).
15. Chapman, H. M., and Norris, J. R., *J. Appl. Bacteriol.*, *29*, 529 (1966).
16. Marvin, D. A., and Hohn, B., *Bacteriol. Rev.*, *33*, 172 (1969).
17. Davison, P. F., *Virology, 21*, 146 (1963).
18. Donelli, G., *Atti Accad. Naz. Lincei Rend. Cl. Sci. Fis. Mat. Natur. Ser. VII, 44*, 95 (1968).
19. Gesheva, R. L., *Mikrobiologiya, 39*, 134 (1970).
20. Kellenberger, E., Bolle, A., Boy de la Tour, E., Epstein, R. H., Franklin, N. C., Jerne, N. K., Reale-Scafati, A., Séchaud, J., Bendet, I., Goldstein, D., and Lauffer, M. A., *Virology, 26*, 419 (1965).
21. Khadgi-Murat, L. N., Smirnova, E. I., and Rautenstein, Ya. I., *Mikrobiologiya, 36*, 140 (1967).
22. Liljemark, W. F., and Anderson, D. L., *J. Virol.*, *6*, 107 (1970).
23. Luftig, R., and Haselkorn, R., *J. Virol.*, *1*, 344 (1967).
24. Luftig, R., and Haselkorn, R., *Virology, 34*, 675 (1968).
25. Luftig, R., and Haselkorn, R., *Virology, 34*, 664 (1968).
26. Mazáček, M., Renčová, J., and Hradečná, Z., *Zentralbl. Bakteriol. Parasitenk. Infektonskr. Hyg. Abt. Orig.*, *209*, 366 (1969).
27. Moody, M. F., *Virology, 26*, 567 (1965).
28. Patel, J. J., *Arch. Mikrobiol.*, *69*, 294 (1969).
29. Peshkoff, M. A., Tikhonenko, A. S., and Marek, B. I., *Mikrobiologiya, 35*, 684 (1966).
30. Prozesky, O. W., de Klerk, H. C., and Coetzee, J. N., *J. Gen. Microbiol.*, *41*, 29 (1965).
31. Richie, A. E., in *27th Annual Proceedings of EMSA, St. Paul, Minnesota,* p. 226, C. J. Arcenaux, Ed. Claytor's Publishing Division, Baton Rouge, Louisiana (1969).
32. Ritchie, A. E., Robinson, I. M., and Allison, M. J., in *Microscopie Electronique 1970,* Vol. 3, p. 333, P. Favard, Ed. Société Francaise de Microscopie Electronique, Paris, France (1970).
33. Schade, S. Z., Adler, J., and Ris, H., *J. Virol.*, *1*, 599 (1967).
34. Schmidt, J. M., and Stanier, R. Y., *J. Gen. Microbiol.*, *39*, 95 (1965).
35. Shimizu, N., Miura, K., and Aoki, H., *J. Biochem. (Tokyo)*, *68*, 265 (1970).
36. Tikhonenko, A. S., *Ultrastruktura virusov bakterii.* Izdadelstvo "Nauka", Moscow, U.S.S.R. (1968).
37. To, C. M., Eisenstark, A., and Töreci, H., *J. Ultrastruct. Res.*, *14*, 441 (1966).
38. Vieu, J.-F., Croissant, O., and Dauguet, C., *Ann. Inst. Pasteur (Paris)*, *109*, 160 (1965).
39. Vieu, J.-F., Guélin, A., and Dauguet, C., *Ann. Inst. Pasteur (Paris)*, *109*, 157 (1965).
40. Vieu, J.-F., Nicolle, P., and Gallut, J., in *Proceedings of the Cholera Research Symposium, Honolulu, Hawaii,* p. 34, O. A. Bushnell and C. S. Brookhyser, Eds. Department of Health, Education and Welfare, Public Health Service, Washington, D. C. (1965).
41. Yamamoto, N., and Anderson, T. F., *Virology, 14*, 430 (1961).
42. Young, B. G., Hartman, P. E., and Moudrianakis, E. N., *Virology, 28*, 249 (1966).

Tailed Bacteriophages: Listing by Morphological Groups

DR. H.-W. ACKERMANN

One of the largest viral groups if tailed phages; knowledge of it is rapidly expanding. Most phages belong to three basic morphological types – A1, B1, and C1.[1] There is increasing evidence that all isometric heads are octahedra, whereas the shape of elongated heads has not been determined or is disputed. As far as known, all tailed phages contain double-stranded deoxyribonucleic acid.

Much work remains to be done. It is probable that identical phages have been described under different names. An effort towards closer identification should be made, since measurements often reflect differences, due to the utilization of different electron microscopes, and incomplete descriptions are quite numerous. Thus, many micrographs had to be re-examined, especially for the presence of tail appendages. Re-examination of some phages can also give surprising results, as occurred for the *Bacillus* phages α and $\phi\mu$-4, which were found to have long, non-contractile tails.[2,189] Finally, phage nomenclature is in a very primitive state, and as a result, many different phages bear similar names.

The following tables should be helpful in the comparison and tracing of phages. Dealing with bacterial or viral taxonomy could not always be avoided.

LISTINGS OF PHAGES

1. Phages are arranged in the following order: gross morphology, host genus, and name.
2. Most morphological groups have their own tables, but three small groups are combined in Table 6.
3. Host genera are listed alphabetically, since there is no universally accepted system for bacteria. However, the following exceptions should be noted:

   (a)*Clostridium* includes *Plectridium* and *Welchia.*

   (b)The Enterobacteriaceae include the genus *Yersinia.* Members of this family are considered as belonging to the same "genus". Generic names such as *Escherichia* are, therefore, given only to provide greater specificity.

   (c)*Moraxella* includes *Bacterium anitratum* and *Herellea.*

   (d)*Streptomyces* includes hosts referred to by Russian authors as *Actinomyces.*
4. Phage names are listed in the following order: Latin alphabet, Greek alphabet, Arabic numerals, and Roman numerals. This also applies to complex names.
5. Should a series of morphologically identical phages be described in the same reference, the first phage mentioned should be considered as the eponymous phage (column 2). The remaining phages are listed in column 3 unless one phage of the series has been studied by several different workers; in that case the phage is listed separately.
6. Synonyms and mutants are indicated by italics. Synonyms are listed in column 2, and mutants in column 3.

DIMENSIONS

1. Data are listed regardless of the staining method, since the method was not always specified and uranyl acetate or formate was shown to give acceptable negative staining. However, when a phage was stained by several methods, data obtained after staining by phosphotungstic acid or its salts were preferred.
2. Assuming that most workers describe dimensions of isometric heads as diameters between opposite peaks, side-to-side distances or edge lengths of octahedra have been recalculated as diameters between opposite peaks.
3. In contractile phages, tail dimensions concern non-contracted tails only.

OMISSIONS

1. Shadowed phages.
2. Other phages on micrographs of very poor quality (some 15 papers); exceptions concern rare and interesting findings, such as a short-tailed staphylococcal phage.[197]
3. Data that cannot be assigned to individual phages; for example, summary descriptions in congress reports.
4. Phages without known hosts, such as phages observed in rumen content.
5. Bacteriocins and "killer particles"; however, inducible complete phages without known biological activity have been included in the tables.

NOTES

1. The list of references was closed on December 31, 1970. Some papers were not available. This is indicated by a numeral following the cited reference; for example, 223-49 shows that the paper cited in Reference 223 as Reference 49 is not available.
2. Algal viruses (cyanophages) possessing a tail have been included in Table 5.

CODE

NN	no name	<	less than
+	present	~	about
-	absent	*	indistinct "knob"
>	more than	blank space:	absence of data

TABLE 1

GROUP A1

(Long, Contractile Tail, Isometric Head)

Host	Phage	Identical Phages and Mutants	Dimensions, Å			Base Plate	Spikes	Fibers	References
			Head	Tail					
			Diameter	Length	Width				
Achromobacter[a]	NN		700	1700	110	+	+	–	123
	NN	NN (for 1)	850	1400	260	+	+	–	123
	NN	NN (for 9)	800	1200	180	+		–	123
Aeromonas	37	56RR2, 59.1	600–700	1140–1300		+			178
Agrobacterium	PIIBNV6		650	1300	200	+	+	–	26
Alcaligenes	A6		900	1100	160			+	143
	A11/A79		650	1300	190	–			143
Bacillus	AR1		850	1300	180	+	+		13
	AR2		950	1800	200	+	+		13
	AR3		900	1800	200	+	+		13
	AR9		1500	2700	290–370	+	–	+	18
	CP-51					+	+		245
	G		~1600	~4000		+	+		7,71
	GE1					+			31
	GF-2		1150	2800		+			30
	GT-1		800	2300		+			59
	GT-2		900–1000	2300		+			59
	GT-3		900	2400		+			59
	GT-4		870	2300		+			59
	GT-5		900	2300		+			59
	GT-6		870	2000		*			59
	GT-7		870	2600		*			59

TABLE 1 (Continued)

GROUP A1

(Long, Contractile Tail, Isometric Head)

Host	Phage	Identical Phages and Mutants	Dimensions, Å: Head Diameter	Tail Length	Tail Width	Base Plate	Spikes	Fibers	References
Bacillus (cont.)	GV-6		850–920			+	?	?	58
	Nº 1		810–1000	2000	170	+	+	+	222,224
	N17		810–850	2000	180	+	+		223–50
	N19		950	2600	150	+	+	+	224
	PBS1		1200	2000	220	+	–	+	75
	SP01		980	1550	175	+	+		228
	SP3		1150	2890	220	+	?	+	77
	SP8	*SP8*, SP82*	880–1000	1570–1650	170–200	+	+	–	64,228
	SP50		840–1000	1700–2300	180–250	+	+		76,83,182, 228
	SW		1010	1725	277	+	+		160
	Type V		700–800	1700–2000	100–200	+	+		216
	Type VI		700–900	900	130	+	+	+	216
	Vx		1140	1700	150	+	+		210
	ϕe		930	1520	160	+	+		228
	Φ25		750	1300		+	+		128
	2C		850–880	1250–1420	150	+	+		147,228
	III		905	2320	165	+	+		51
	NN		980	4250–4500		+		+	221
Caryophanon	$Cs1_{x13}b$		650–800	1200–1400					179
Chondrococcus	C2		600	1000	200	+			110
Clostridium	HM3		770	1000	150	?		+	166
	KT		850	~1000		+		+	98
	M_s		480	1117	130	+			19
	NN		600–650	1000–1100	200				184
	NN		1080	3540	320	+	–	–	239
	NN		540	1000					78

Enterobacteriaceae									
Salmonella	Beccles	Taunton	620	1550	130	–	–	+	2
Escherichia	E1					+		+	27
Shigella	H-Sh								67
Yersinia	Lucas 110	Lucas 404	700	1000		+			164
Yersinia	Lucas 303		600	1050					164
Escherichia	Mu1		610	1080	180	+	+	+	227
Salmonella	01, 7	9	600–760	900–1030	150	+	–	+	3,129
Salmonella	02	Dundee, 03	760	1050	150	+	–	+	2
Escherichia	PK	*PKvir*	480–520	1000–1040	150–170			+	102
Salmonella	P1		900–980	2200–2500	180–200	+		+	10,240
Salmonella	P2		680	1400	180	+		+	10
Salmonella	P3	P9a	550	1180	120			+	165
Salmonella	P10		550	950	120	+		+	165
Serratia	SMP		1350	2350	275	+			32
Serratia	SMP5	SMB3	570	1250					32
Serratia	SM2		1350	1700–1900		+	+		32,146
Salmonella	Vi I, *Vil*		900–950	960–1100	160–180	+	–	+	4,36,121
Polyvalence	Wϕ		650	1400					90
Serratia	κ		~740	1150	120–150	+			130
Polyvalence	ϕ2		660–720	1100	220	+		+	27,36
Proteus	13vir		600	1019	167	+	+		187
Proteus	57		619	882	178	+	+		187
Proteus	67b		589	981	174	+	+		187
Proteus	78		611	889	163	+	+		187
Proteus	107/69		660	1320					56
Escherichia	186		500	1250	~125				16
Erwinia	NN						+	+	237
Haemophilus	HP1		700	1200	200	+			93
Hydrogenomonas	H20		1000	1200	200	+		+	109
Lactobacillus	206	514	720	1380		+	+	–	112
	222a		690	1380	160	+	+	–	112
	300		820	1230	150	+	+	–	112
	315		720	1480	160	+	+	–	112
	356		820	1270	160	+	+	–	112
Leptospira	NN		450–500	1000–1200	100–120	?			194

TABLE 1 (Continued)

GROUP A1

(Long, Contractile Tail, Isometric Head)

Host	Phage	Identical Phages and Mutants	Dimensions, Å: Head	Dimensions, Å: Tail		Base Plate	Spikes	Fibers	References
			Diameter	Length	Width				
Moraxella	G4		840	1080	120	+	–	+	5
	HP2		750	1150		+	+		22
	HP3		750	1050		+	+		22
	HP4		750	1100		+	+		22
	NN								232
Myxococcus	Mx-1		750	1000		+	+		44
Neisseria	Group I[b]	Group II?[b]	700	1500	150				181
Pseudomonas	E79		760	1500	170	+		+	214
	PB-1, *PB1*		750–800	1400–1500		+		+	34,37
	Psa1		700–750	1200	200	+		+	162
	PX4		630			+		+	168
	PX7	CB3	520			+		+	168
	SLP1		900	950		+	+		31
	ΦW-14		850	1400	200	+	+	+	117
	φ-2	φ-LT?							55
	6C	41C, 41TD	650–660	1250	140	+		+	15
	7v					+			79
	12S		800	1000					28
Rhizobium	ΦIA		490–558	670	96–125	+	+		108
Saprospira	NN		1100–1200	1050	220	+	+		127
Staphylococcus	P2	P3, P4, P8, P9, P10	910	2400	190	+	+	–	223–49
	S3K	119, 130, 131, 200	750	2000	100	+	+		122
	Twort		750–1060	2000	170	+	+		122,234
	06	40, 58	800	1700		+			151
	1623		900	2250–2350	180	+	+	–	223–49

Streptococcus	RZh		900	2000–2100	200	+	+	–	223–49
Streptomyces	Type IV		500	720					191
Vibrio	Group II[b]		621–656	810	166	+	+		52,63
	SE	*kappa* type?	450–550	800–1000	150–200	+	+		218
	5		560–580	1040	165	+		+	238
	32f	111c, 191				+		+	82
Xanthomonas	XP5		600	>800	200	+	+		89

Total: >153 phages

[a]As defined by Prévot.[185]

[b]Number of phages not indicated.

TABLE 2

GROUP A2

(Long, Contractile Tail, Elongated Head)[a]

Host	Phage	Identical Phages and Mutants	Dimensions, Å Head Length	Head Width	Tail Length	Tail Width	Base Plate	Spikes	Fibers	References
Aeromonas	40RR2.8t		1000	900	950		+	+	+	178
Enterobacteriaceae										
Polyvalence	C4		1000	750	1000	200	+	+	+	36
Shigella	C16		1200	890	1200	180	+	+	+	2
Polyvalence	D_d- Vi, *DDVI*		900–1200	690–900	860–1350	180–200	+	+	+	149,153,223–49
Klebsiella	Kl 70/11, *K-3*		"as T4"		"as T4"		+	+	+	47,48,188
Escherichia	O111		937		750	266				24
Yersinia	PST		1250	950	1150	250	+	+	+	113
Serratia	SMB2	SMP2	"as T-even"		"as T-even"		+	+	+	32
Escherichia	T2	*T2r*$^+$	950–1450	650–900	950–1250	130–200	+	+	+	33,61,105, 223–49
Escherichia	T4		1000–1150	710–850	950	180	+	+	+	33,106
Escherichia	T6						+	+	+	11
Escherichia	α1		1350	810			+	+	+	27,33
Escherichia	3		1060	740	990	195	+	+	+	3
Polyvalence	3T$^+$		900	650	950	150				36,104
Proteus	9/0		1100–1200	800–820	1300–1400	180	+			223–49
Polyvalence	11F, *Ox-1*		900	650	920	150	+	+		36,104
Proteus	50		970	670	1252	159	+	+		187
Polyvalence	66F,*Ox-6*[b]		900	700	950	150	+	+	+	27,36,104
Salmonella	66t-,*φ66t-*[b]		1350	825			+	+	+	27,33
Proteus	5845		1022	763	1252	156	+	+	+	187
Proteus	8893		926	741	1296	178	+	+	+	187
Providencia	9266		1193	822	2152	248	–	+	+	187

Moraxella	HP1	1050	850	1150		+	+	?	22
	NN	"as T-even"		"as T-even"		+	+	+	232
Rhizobium	NN	1100	850	1000	175	+	+	+	158
Thermoactinomyces	108/106	1100	840	1400	170				223–49

Total: 27 phages

[a]Ratio of length to diameter = 1.2–1.4.
[b]Identical?

TABLE 3

GROUP B1

(Long, Non-Contractile Tail, Isometric Head)

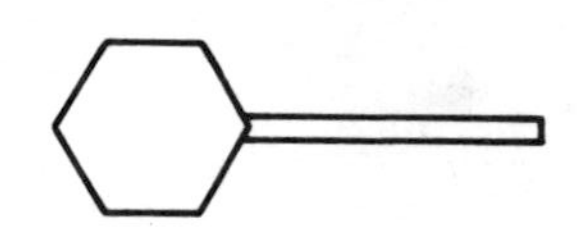

Host	Phage	Identical Phages and Mutants	Dimensions, Å			Base Plate	Spikes	Fibers	References
			Head	Tail					
			Diameter	Length	Width				
Achromobacter[a]	NN		500	2100	110				123
	NN	NN (for 2)	550	1500	85				123
	NN		550	1150	85				123
Agrobacterium	Lv-1	$La_6 k_1$, Lcg, Lc-58, Lr-4	630–710	2210	95	–		+	247
	PB-6	PA 6, PS 8	700–800	2000–3000			+		199
Alcaligenes	A5/A6		750	2400	90	–	–		143
	A5/415	A20/415	750	2400	100				143
	A64/A62		750	2500	100	–	–	–	143
	A74/A3		750	2500	120				143
	A86/A88		800	2500	170				143
	8764		600	1700	110	–	–	+	143
Arthrobacter	NN		630–750	2200		?			62
Asticcacaulis	ΦAc20		650–700	1500			–		208
Bacillus	A		640	2000	95	*			66
	B		640	2000	80	*			66
	C		590–620	2000	75	*			66
	CP-53					+			245
	D		580–620	2155	75	*			66
	D		520–525	1700–1720	80				223–49

Bacillus (cont.)	G		550	2260	76	*			159
	GT-8		600	1200		*			59
	GV-1		630	1700	90	*			58
	GV-2		630	1750	110	*			58
	GV-4		610	1800	110	*			58
	№ 2		780	4250	120			+	223–49
	N5		940–960	5300–5350	100			+	223–49
	N6P		560	2150				+	223–49
	PBLA		660	"long"		*			99
	P2	D	400–500	1400	40–60				216
	SPO2		495	1770	105	+	+		25
	SPP1		520	1400	65			+	195
	STI		550	1900–2000	70–90	+	+	–	223-49
	t		620	1700	70	+	+		8
	TP-1	*TP-1c*	650	2400	120				242
	TP-84		530	1300	30–50				204
	Tϕ3		570	1250	100		–	?	73
	VA-9		730–735	4000	90–95	+	+		223–49
	α		570	1215	90	–	–	–	2,53,54
	$\phi\mu$-4		550	2250	70	–	–	–	189
	ϕ105		495–520	1770–2250	105	+	+		21,25
	30	37	400–500	1300	40–60				216
	35	36,38	400–500	1500	40–60				216
	I		730	1925	75	–	–		51
	II		990	5450	135	–	–	+	51
	IV		710	1940	85	–	–		51
Brevibacterium	Ap85III		500	2030	150				167
	P465		500–600	2750	100				167
Caryophanon	$Csl_{x13}a$		650–800	3700–4500	80	+		+	179
	Ct_{kas}		770–790	4250–4350	120	+	+		180
Caulobacter	ϕ101		550	1170					72
	ϕ102		650	1560					72
	ϕ118		580	2000–2150					72
	ϕ151		600	1900–2100					72
Clostridium	CPT1		700–800	2000	80	–	–		141
	CPT4		640	1700	100				140
	HM7		750	2000	100		+		166
	M_L		830	3370	100	*			19
	NN		800	2400	100	+	+	–	92

TABLE 3 (Continued)

GROUP B1

(Long, Non-Contractile Tail, Isometric Head)

Host	Phage	Identical Phages and Mutants	Dimensions, Å: Head	Dimensions, Å: Tail		Base	Spikes	Fibers	References
			Diameter	Length	Width	Plate			
Escherichia	q	w		1300	70	*		+	29
Serratia	SMPA		650	2250					32
Salmonella	S1BL		500	1300	100	–	–		36
Escherichia	T1		620	1300	100	–			36
Escherichia	T3C[d]								74
Escherichia	T5		600–900	1400–2000	70–120	–		+	35,36,137, 223–49
Salmonella	Vi II		640–670	1310–1400	80–90	+	+	–	4,233
Escherichia	β4	γ2	600	1500		–		+	27
Escherichia	λ		540–650	1400–1530	70–170	–	–	+	76,105,107
Proteus	π2600		735	1600	100	+	+	–	220
Serratia	σ		650–700	1400–1450	100				130
Salmonella	φ1		750	1800	90	–	–		27,36
Escherichia	φ80	*φ80pt₁*	710	1800	60	–		+	212
Polyvalence	χ', φχ		750	2200	140	–	+	+	154,205
Salmonella	1	Jersey, 2, 3a, 3aI, 1010	680	1160	85	–	+	–	2
Escherichia	2		840	1760	80	+	–	–	3
Escherichia	4		830	1585	90	–	–	+	3
Escherichia	6		610	1335	75	+	+	–	3
Proteus	7/R49		total length	2310			+	–	116
Proteus	9		826	1837	89	–	–	+	187
Proteus	12		593	2163	115	–		+	187
Proteus	14		660	2045	115	–		+	187
Proteus	47		556	1481	85	–	–		187
Klebsiella	483	L1, 231, 490, 632	450–500	1100–1400	40–60				87,215
Aerobacter	886		470	1300	70				223–49
Proteus	2600/D52		655–675	1800	100	+	+	–	223–49
Escherichia	3000		550	1400				+	29
Proteus	7476/322		533	1600	77	–			187
Proteus	7478/325		604	1741	115	–			187
Proteus	7479		737	1852	85	–	–	+	187

Proteus	7480		634	2185	126	–			187
Providencia	9000/9402		682	2304	148			+	187
Providencia	9213/9211a		626	1696	137	+	+		187
Erysipelothrix	NN		460–520	>800	60				213
Haemophilus	N3		600	200		–			203
Lactobacillus	C5[e]		580	2100	120				202
	PL-1		630	2750	125				241
Clostridium (cont.)	NN		600–650	1000	50				184
	NN	NN (for 13)	500	1700–2200	100				95
	NN	NN (for 1)	700	1400	150	*			95
	NN[b]		500–600	1400	60–80	*			78
	NN[b]		500–600	2500	60–80				78
	NN[b]		900	2800	80				78
Corynebacterium	K		600–640	2750	90				96
	L		510–550	2350	95				96
	P		850	5700					148
	R29		640–650	3500–3700	110				223–49
	β	β^{vir}, B, Bh	520–590	2500–2700	90–120	–		+	96,144
	γ		550–560	2600	95			+	96
	γ19		600	2700					148
	δ		440–500	3300	75				96
	ρ		510–530	3250	75				96
Cytophaga	NCMB 384		600	860–1000		+			230
Dactylosporangium	A1		~750	2000	90		+		94,124
Enterobacteriaceae									
Proteus	Clichy 12			1600		+	+	–	233
Polyvalence	C1	F1	900	"long"		–		+	27
Polyvalence	DDUP		600–615	1400–1500	100				223–49
Shigella	F11		600	> 1500	80	–		–	47,48
Salmonella	P22-1		690	1800	100	*			243,246
Salmonella	P22-3		690						246
Salmonella	P22-12		800			–			246

TABLE 3 (Continued)

GROUP B1

(Long, Non-Contractile Tail, Isometric Head)

Host	Phage	Identical Phages and Mutants	Dimensions, Å			Base Plate	Spikes	Fibers	References
			Head	Tail					
			Diameter	Length	Width				
Lactobacillus (cont.)	Sa-S[e]		570	2300	120				202
	S-9[e]		590	2200	120				202
	S171[e]		630	2300	120				202
	3-793[e]		570	2160	120				202
	535/222a		500	1820	80	*			112
	II-5[e]		570	2150	120				202
	NN		550	2000	80	+			68
	NN		600	2150	110	+			68
Micrococcus	N1		580–640	2140–2300	90	*?		?	125,133
	N5		528	1500					125
	N6		647	2320					125
Mycobacterium	B-1		550	1400–1500	90	+	+	–	217
	B5		450	1200					103,114
	B7					*			103
	C2		600	2000	75	*		–	152
	GS-7								101
	MyF_3 P/59a		520	1230					155,156
	Phlei		700	1650–1700	100	+			50
	Polonus II		700	2500	105	+	+	–	223–49
	NN		total length	3000–3500		*			193
	NN		total length	1600					193
Nocardia	C		520	1920	100	–			42
	EC		520	1970	100	–			42
	R1		750	3300	100	+	–		196
Pseudomonas	B3		600	1630	80		+		214
	PP4		600	1950		+			34
	PS4		640	1700	120		?		146

Pseudomonas (cont.)	Pz	*Pc* (= *PC*)	600–650	1600–1650	100–110	*			34,36
	SD1		500	1880	62	–			209
	24		750	1500					28
	NN					*			79
Rhizobium	NN					*			158
Ristella	67.1	adl_2, 11, 67.3, 68.1	680	1500	120	–		+	186
Staphylococcus	CA-1	CA-2 to CA-5				*			132
	P42D		550	2300		*			28
	P52A		500	1500					28
	P71		600	1500					28
	UC-18		600	1650		*			23
	29		580	2250		*			122
	42D		560	2300		*			122
	52		500–550	1500–1800	110	*			28,36,122, 169
	52A		500	1500		*			122
	53		560	1700		*			122
	55		570	2250		*			122
	71		500–630	1500–1700	100				28,36,122
	72		550	1500					28
	75		500	2400		*			122
	77		550–600	2200–2500	100	*			36,122
	80		580	1750		*			122
	93		550	2300			–		151
	102	Z_1, 88A, 107, 129-16	550–600	1900–2000		*			85
	187		580–600	1700–1800	120	*			28,36,122
	581	b^{581}	550	2400		*			28
	ϕD		~500	2300			?		69,70
Streptococcus	P9	P3	584–625	3000	120	–			41
	28		550	1800–1900	100	+	+	–	223–49
	881	S2, 940, 1051, 1057				*			39
	NN		500	2000	110		+		84
	NN		564–614		75				200
Streptomyces	AP2863								138
	K		915	3000	150	+	+		60
	№ 121		580	1400–1600	80–90				192
	№ 1527		580	1300–1400	80–90				192
	R1		730	1735	120		+		60

TABLE 3 (Continued)

GROUP B1

(Long, Non-Contractile Tail, Isometric Head)

Host	Phage	Identical Phages and Mutants	Dimensions, Å			Base Plate	Spikes	Fibers	References
			Head	Tail					
			Diameter	Length	Width				
Streptomyces (cont.)	SAP1		590	1770	100				120
	SAP2		680–690	1680	100				120
	SAP3		690	1660	100				120
	Type II		610	1800		*			191
	Type V		500	1500		*			191
	3A	Nº 3	580	1300–1600	80–90				192,223
	506		430	900					88
	8238	AP-3					+		139
	NN		~630	~1480		+			190
	NN		~518	~1220		+			190
Thermoactinomyces	119		600	1800–1850	100	+	+	–	223–49
Thermomonospora	NN		680	1250	80				176
Vibrio	Group IV[b]		738–836	1528	107	*			52,63
	Vfi-6	Vfv-3				–	–		81
	V-45		~500	2400	70	+			43
Xanthomonas	Pg60	20, 22	~600	~1600					231

Total: >261 phages

[a] As defined by Prevot.[185]
[b] Number of phages not indicated.
[c] Identical?
[d] Claimed to be a mutant of phage T3.
[e] No biological activity.

TABLE 4

GROUP B2

(Long, Non-Contractile Tail, Elongated Head[a])

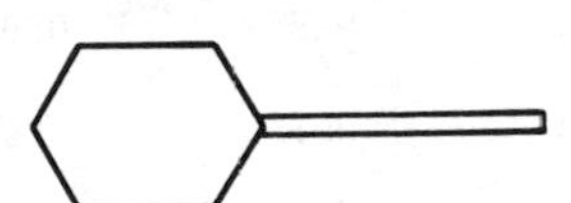

Host	Phage	Identical Phages and Mutants	Dimensions, Å				Base Plate	Spikes	Fibers	References
			Head		Tail					
			Length	Width	Length	Width				
Agrobacterium	PT11		1000	550	1250–1350	90	*			26
Bacillus	Type IV		700	400	1400	40–60				216
Caulobacter	76		885	600	2750–2800	100–110	–	–	+	223-49
Clostridium	F1	K, S2, 1	900	540–550	1230	60–100	–	–	–	19,20
	NN		850–950	460–500	2700	80	*			78
Enterobacteriaceae										
Serratia	SMB1		~ 650		1350		–			32
Escherichia	ZG/3A		750	600	1900	75			+	29
Escherichia	5		775	550	1585	80	+	–	–	3
Proteus	$\frac{24}{25}$14		715	604	1715	108	–			187
Mycobacterium	B1						–	+		103
	D29		700	400	1700–1900	70–90				80
	HP		850	450	1700	100	+	+		217
	Polonus I		1100	500	2200	150	*			223-49
	Roy		920–940	530–540	2500					142
Pseudomonas	PB-2		1000	700	1750		?			34
	PX2		700	500	1100		–			168
	PX5		800	600	1400		+			168

TABLE 4 (Continued)

GROUP B2

(Long, Non-Contractile Tail, Elongated Head[a])

Host	Phage	Identical Phages and Mutants	Dimensions, Å				Base Plate	Spikes	Fibers	References
			Head		Tail					
			Length	Width	Length	Width				
Rhizobium	F4/L425I	F5/L422	1130	652	2413	83	*			108
	F4/L425II		855	615	2002	74	*			108
	F5		1125	639	2270	76	*			108
	L419		1151	667	2160	80	*			108
	NN		1100	600	2500	100	*			158
	NN		750	600	1400		*			158
Staphylococcus	Z_4		900–1000		3000–3200		+	+		85
	3A		900–1000	400	3000–3200	75	+	+		85,122
	3B		800–920	420–600	3000	75–100	*			36,122
	3C		830	400	2810–4210	75	*			122,169
	6		850–920	400–550	3000	75–100	*			36,122
	16		800	400	3200					170
	21		800	450	3000					170
	42E		780–1000	490	3000–3200	75	+	+		85,122
	47		920	500	3000	75	*			122
	54		920	450	3000	75	*			122
	70		850–980	530–550	3000	75		+		28,122
	88		800	450	2650	60		+		170
	174		950	450	2850			+		170
	594n	b^{594n}	960	550	3000		+	+		28
	NN						+	+		70
Streptococcus	3ML		550	400	1000	90	–	–		36
Streptomyces	MSP2		620	520	1670		*			38
	MSP8		700	550	1500	100	*			115
	R2		1050	550	1700	95	+	+		60
	Type I		770	610	1800					191
	Type Ia		720	610	1380–1500		*			191

	Type III		1100	550	2500		*			191
	Type IV		830	810	~ 6000					191
	4		700	450	1500–1550	110		+	–	223-49
Thermoactinomyces	M_1	M_3	620	500	900	80	*		?	177
Xanthomonas	Xp12, *XP12*		760	550	1330	60	+			118,119
	φPS	φRS, φSD, φSL, φ56, φ112, 1	750	500	1300			+		231

Total: 62 phages

[a]Ratio of length to diameter = 1.2–2.2.

TABLE 5

GROUP C1

(Short Tail, Isometric Head)

Host	Phage	Identical Phages and Mutants	Dimensions, Å: Head Diameter	Dimensions, Å: Tail Length	Dimensions, Å: Tail Width	Base Plate	Spikes	Fibers	References
Acetobacter	AA-1		~ 650	185		+	+		30
Achromobacter[a]	NN		450	170					123
Agrobacterium	PR-1001	PR-590a. PS-192, PsR-1012	~ 600	"small"		+	+		198
	PIIBNV6-C		~ 550	120–140					26
Algae									
Plectonema	AT		"as LPP-1"						201
Plectonema	GIII, *LPP-1G*		560–700	150–370	150	?			135,171,201
Plectonema	LPP-1		700	100–200	150				134,135,136
Plectonema	LPP-2[b]		660	"short"					201
Chlorella	NN		410	240	100		+		226
Brucella	F_1		590–660	~ 200–230		?			49,173,175
	F_2	F_3 to F_{11}	590–660	230		?			49
	I	Im, 75, 84	500–551	280–310	71–86				131
	M	513, X	600	160					223-49
	Tb		500–600	~ 200–310	71–86				131,173,175
	10/1	FP_2, FO_1, F_1m, F_1U, F25, F25U, F44, F45, F48, P, 3, 6, 7, 12m, 24/II, 45/III, 212/XV	600	~ 200					173,175

	371/XXXIX		500–600	200–310	71–86				131,173,174, 175
	NN		550–650	140–150					36,40
	NN		~ 650	~ 160					150
Clostridium	HM2		450	300	55–80		+		166
	HMT		~ 450	~ 300			+		97
	80		400	300	65	?	+		236
	NN		500–600	150–200					45
	NN	NN (for 14)	400	300–400			+		95
	NN						+		172
Enterobacteriaceae									
Yersinia	EV		530	190		–	–	–	223-49
Yersinia	H		~ 500	~ 220					157
Yersinia	Kotljarova 903		519–553	385–422					12
Salmonella	MG40		~ 500	"short"		+	+		91
Escherichia	N4		700	"small"		+	+		206,207
Providencia	PL26	PL25, PL37	600	160		+	+		57
Proteus	Pm1		575	"very short"			+		220
Salmonella	P4	P9c	550	150	90	+	+		165
Salmonella	P22, PLT_{22}		530–800	75–175	70–75	+	+	?	10,235,243, 246
Salmonella	P22a1	P22-4, P22-7, P22-11	800	75		+	+		246
Serratia	Sa1		~ 500	"very short"			+		161
Escherichia	s_d, S_d		640–700	200		+	+	–	111,225
Serratia	SM4		570	"short"		+	+		32
Escherichia	T3		550	150	80	–	+	?	27,36
Escherichia	T7		700	150–200	150	–			65,135
Salmonella	Vi III	Vi IV, V, VI, VII	580	130	80	+	+	–	4
Yersinia	Y	PTB, R	700	150					113
Salmonella	ϵ_{15}		640–650	160		+	+	–	235
Salmonella	ϵ_{34}		600–605	175		+	+	?	235
Serratia	η		640–680	100–150		?	+		130
Proteus	π1					+	+		219
Escherichia	ϕ04-Cf		550	"short"					248
Escherichia	1		690	180	60	–	+	–	3
Salmonella	1, 37		590–605	185		+	+	?	235
Salmonella	1(40)		580–590	170		+	+	?	235
Salmonella	1, 42_2		605–615	180		+	+	?	235
Salmonella	3b	B.A.OR., Worksop	650	200	80	+	+	–	2
Salmonella	6, 14(18)		595–600	175		+	+	?	235
Escherichia	8		590	195	85	–	+	–	3

TABLE 5 (Continued)

GROUP C1

(Short Tail, Isometric Head)

Host	Phage	Identical Phages and Mutants	Dimensions, Å Head			Base Plate	Spikes	Fibers	References
			Head	Tail					
			Diameter	Length	Width				
Enterobacteriaceae *(cont.)*									
Proteus	12/57		589	185		+	+		187
Salmonella	14(6, 7)		600–610	180		+	+	?	235
Proteus	15	65	518	148		+	+		187
Salmonella	27		600–605	185		+	+	?	235
Proteus	34/13		630	167		+	+		187
Klebsiella	380		450–500			+	+		87
Proteus	7480b		367	330	63	–			187
Providencia	9211/9295		533	148		+	+		187
Providencia	9213/9211b		608	152		+	+	–	187
Providencia	9248		611	185		–			187
Proteus	10041/2815		544	182		+	+		187
Proteus	VI		570	180		+	+		223-49
Hydrogenomonas	NN	NN (for 4)	580	200					183
Moraxella	BP1		~ 500	~ 200					229
Pseudomonas	gh-1		500	"short"				+	126
	Pf		540	100					163
	PX1	PX12, PX14	500	170					168
	PX3	PX10	500	270					168
	Φ-MC		550	"short"		+			244
	12B		500–700			+	+		28
	14		400	130					28
	15		400	150					28
	20		500	300					28
	21		600	200					28

Rhizobium	F9		470–491	~ 100					108
	NN		500	125					158
Rhodopseudomonas	Rp1.Rp1		380–390	200	55–60				86
Sphaerophorus	277		500	"very short"					100
Staphylococcus	44AHJD		< 400			+			197
Streptomyces	ϕ17	ϕ*17-5*	565–625	105–140				+	14,145
Thermoactinomyces	114		560	170		–	+		223-49
Vibrio	Group I[c]		706–740	no tail?			+		52,63
	Group III[c]		611–644	178	174	*			52,63

Total: > 157 phages

a As defined by Prévot.[185]
b Several isolates (GM, SPI, WA).
c Number of phages not indicated.

TABLE 6

MISCELLANEOUS GROUPS

Host	Phage	Identical Phages and Mutants	Dimensions, Å: Head Length	Head Width	Tail Length	Tail Width	Base Plate	Spikes	Fibers	References
Group B3 (Long, Non-Contractile Tail, Elongated Head[a])										
Caulobacter	ΦCb13	ΦCb3, ΦCb6	1700	500	2000–2500	100	–	–		208
	ϕCbK		1950	640	2750		–			6
Group C2 (Short Tail, Elongated Head[b])										
Bacillus	GA-1		570	400	350	60	–	+	–	30
	GV-3		580	400	350	110	–	+	–	58
	GV-5		540	380	350	110	–	+		58
	Nf		430	330	~ 400	50	–	+		211
	ϕ29		415	315	325	60	–	+	–	9
Vibrio	7		740	660	90					238
Group C3 (Short Tail, Elongated Head[c])										
Escherichia	F46/3039		1000–1200	400–500	"very small"					46,47,48
	NN[d]		1400–1600	400–700	200	200				17

Total: 12 phages

TOTAL FOR ALL TABLES: > 672 phages

[a] Ratio of length to diameter = 3–3.4. [b] Ratio of length to diameter = 1.1–1.4. [c] Ratio of length to diameter = 2.4–2.7. [d] Shadowed.

REFERENCES

1. Ackermann, H.-W., *Pathol. Biol., 17,* 1003 (1969).
2. Ackermann, H.-W., Berthiaume, L., and Kasatiya, S. S., *Can. J. Microbiol., 18,* 77 (1972).
3. Ackermann, H.-W., and Berthiaume, L., *Can. J. Microbiol., 15,* 859 (1969).
4. Ackermann, H.-W., Berthiaume, L., and Kasatiya, S. S., *Can. J. Microbiol., 16,* 411 (1970).
5. Ackermann, H.-W., and Turcotte, J., *Rev. Can. Biol., 29,* 317 (1970).
6. Agabian-Keshishian, N., and Shapiro, L., *J. Virol., 5,* 795 (1970).
7. Ageno, M., and Donelli, G., in *Electron Microscopy 1968, 4th European Regional Conference on Electron Microscopy, Rome,* Vol. 2, p. 155, D. S. Bocciarelli, Ed. Tipografia Poliglotta Vaticana, Rome, Italy (1968).
8. Aiello, I., Frontali, C., and Tangucci, F., *Ann. Ist. Super. Sanita, 1,* 119 (1965).
9. Anderson, D. L., Hickman, D. D., and Reilly, B. E., *J. Bacteriol., 91,* 2081 (1966).
10. Anderson, T. F., in *Proceedings of the European Regional Conference on Electron Microscopy, Delft,* Vol. 2, p. 1008, A. L. Houwink and B. J. Spit, Eds. De Nederlandse Vereniging voor Electronenmicroscopie, Delft, The Netherlands (1960).
11. Anderson, T. F., and Stephens, R., *Virology, 23,* 113 (1964).
12. Arutyunov, Yu. I., *Zh. Mikrobiol. Epidemiol. Immunobiol.,* p. 106 (1970).
13. Azizbekyan, R. R., Belyaeva, N. N., Krivisky, A. S., and Tikhonenko, A. S., *Mikrobiologiya, 35,* 279 (1966).
14. Bacq, C.-M., and Horne, R. W., *J. Gen. Microbiol., 32,* 131 (1963).
15. Baigent, N. L., DeVay, J. E., and Starr, M. P., *N.Z.J. Sci., 6,* 75 (1963).
16. Baldwin, R. L., Barrand, P., Fritsch, A., Goldthwait, D. A., and Jacob, F., *J. Mol. Biol., 17,* 343 (1966).
17. Bartsch, G., *Jenaer Rundsch., 5,* 7 (1960).
18. Belyaeva, N. N., and Azizbekyan, R. R., *Mikrobiologiya, 36,* 1054 (1967)..
19. Betz, J. V., *Virology, 36,* 9 (1968).
20. Betz, J. V., and McClung, L. S., *Bacteriol. Proc.,* p. 119 (1964).
21. Birdsell, D. C., Hathaway, G. M., and Rutberg, L., *J. Virol., 4,* 264 (1969).
22. Blouse, L., and Twarog, R., *Can. J. Microbiol., 12,* 1023 (1966).
23. Blouse, L., and Twarog, R., *Proc. Soc. Exp. Biol. Med., 123,* 949 (1966).
24. Borisov, L. B., and Moguchy, A. M., *Zh. Mikrobiol. Epidemiol. Immunobiol.,* p. 77 (1962).
25. Boyce, L., Eiserling, F. A., and Romig, W. R., *Biochem. Biophys. Res. Commun., 24,* 398 (1969).
26. Boyd, R. J., Hildebrandt, A. C., and Allen, O. N., *Arch. Mikrobiol., 73,* 47 (1970).
27. Bradley, D. E., *J. Gen. Microbiol., 31,* 435 (1963).
28. Bradley, D. E., *J. Ultrastruct. Res., 8,* 552 (1963).
29. Bradley, D. E., *J. Gen. Microbiol., 35,* 471 (1964).
30. Bradley, D. E., *J. Gen. Microbiol., 41,* 233 (1965).
31. Bradley, D. E., *J. Roy. Microscop. Soc., 84,* 257 (1965).
32. Bradley, D. E., *J. Appl. Bacteriol., 28,* 271 (1965).
33. Bradley, D. E., *J. Gen. Microbiol., 38,* 395 (1965).
34. Bradley, D. E., *J. Gen. Microbiol., 45,* 83 (1966).
35. Bradley, D. E., *Bacteriol. Rev., 31,* 230 (1967).
36. Bradley, D. E., and Kay, D., *J. Gen. Microbiol., 23,* 553 (1960).
37. Bradley, D. E., and Robertson, D., *J. Gen. Virol., 3,* 247 (1968).
38. Bradley, S. G., and Ritzi, D., *J. Gen. Virol., 1,* 285 (1967).
39. Brailsford, M. D., and Hartman, P. A., *Can. J. Microbiol., 14,* 397 (1968).
40. Brinley-Morgan, W. J., Kay, D., and Bradley, D. E., *Nature (London), 188,* 74 (1960).
41. Brock, T. D., Johnson, R. M., and DeVille, W. B., *Virology, 25,* 439 (1965).
42. Brownell, G. H., Adams, N. J., and Bradley, S. G., *J. Gen. Microbiol., 47,* 247 (1967).
43. Bryner, J. H., Ritchie, A. E., Foley, J. W., and Berman, D. T., *J. Virol., 6,* 94 (1970).
44. Burchard, R. P., and Dworkin, M., *J. Bacteriol., 91,* 1305 (1966).
45. Bychkov, K. Y., *Zh. Mikrobiol. Epidemiol. Immunobiol.,* p. 39 (1964).
46. Bystrický, V., *J. Electron Microsc., 11,* 125 (1962).
47. Bystrický, V., in *Proceedings of the 3rd European Regional Conference on Electron Microscopy, Prague,* Vol. B, p. 557, M. Titlbach, Ed. Publishing House of the Czechoslovak Academy of Sciences, Prague (1964), and The Royal Microscopic Society of London (2nd ed., 1965).
48. Bystrický, V., Drahoš, V., Mulczyk, M., Przondo-Hessek, A., and Šlopek, S., *Acta Virol., 8,* 369 (1964).
49. Calderone, J. G., and Pickett, M. J., *J. Gen. Microbiol., 39,* 1 (1965).
50. Castelnuovo, G., Giuliani, H. I., Luchini di Giuliani, E., and Arancia, G., *J. Gen Virol., 4,* 253 (1969).
51. Chapman, H. M., and Norris, J. R., *J. Appl. Bacteriol., 29,* 529 (1966).
52. Chatterjee, S. N., Das, J., and Baruah, D., *Indian J. Med. Res., 53,* 934 (1965).
53. Chiozzotto, A., Coppo, A., Donini, P., and Graziosi, F., in *Proceedings of the European Regional Conference on Electron Microscopy, Delft,* Vol. 2, p. 1012, A. L. Houwink and B. J. Spit, Eds. De Nederlandse Vereniging voor Electronenmicroscopie, Delft, The Netherlands (1960).

54. Chiozzotto, A., Coppo, A., Donini, P., and Graziosi, F., *Sci. Rep. Ist. Super. Sanita, 1,* 112 (1961).
55. Chow, C. T., and Yamamoto, T., *Can. J. Microbiol., 15,* 1179 (1969).
56. Coetzee, J. N., de Klerk, H. C., and Smit, J. A., *J. Gen. Virol., 1,* 561 (1967).
57. Coetzee, J. N., Smit, J. A., and Prozesky, O. W., *J. Gen. Microbiol., 44,* 167 (1966).
58. Colasito, D. J., and Rogoff, M. H., *J. Gen. Virol., 5,* 267 (1969).
59. Colasito, D. J., and Rogoff, M. H., *J. Gen. Virol., 5,* 275 (1969).
60. Coyette, J., and Calberg-Bacq, C.-M., *J. Gen. Virol., 1,* 13 (1967).
61. Cummings, D. J., and Kozloff, L. M., *Biochim. Biophys. Acta, 44,* 445 (1960).
62. Daems, W. T., *Antonie van Leeuwenhoek J. Microbiol. Serol., 26,* 16 (1963).
63. Das, J., and Chatterjee, S. N., in *Electron Microscopy 1966, 6th International Congress for Electron Microscopy, Kyoto,* Vol. 2, p. 139, R. Uyeda, Ed. Maruzen Co. Ltd., Tokyo, Japan (1966).
64. Davison, P. F., *Virology, 21,* 146 (1963).
65. Davison, P. F., and Freifelder, D., *J. Mol. Biol., 5,* 635 (1962).
66. Dawson, I. M., Smillie, E., and Norris, J. R., *J. Gen. Microbiol., 28,* 517 (1962).
67. Dekegel, D., and Beumer-Jochmans, M. P., *J. Ultrastruct. Res., 16,* 278 (1966).
68. Dentan, E., Sozzi, T., and Bauer, H., *J. Microsc. (Paris), 9,* 567 (1970).
69. Dobardzic, R., Côté, J. R., Payment, P., and Sonea, S., *Rev. Can. Biol., 29,* 253 (1970).
70. Dobardzic, R., Sonea, S., and Côté, J. R., *Rev. Can. Biol., 29,* 1 (1970).
71. Donelli, G., *Atti Accad. Naz. Lincei Rend. Cl. Sci. Fis. Mat. Natur. Ser. VII, 44,* 95 (1968).
72. Driggers, L. J., and Schmidt, J. M., *J. Gen. Virol., 6,* 421 (1970).
73. Egbert, L. N., and Mitchell, H. K., *J. Virol., 1,* 610 (1967).
74. Eisenstark, A., Maaloe, O., and Birch-Andersen, A., *Virology, 15,* 56 (1961).
75. Eiserling, F. A., *J. Ultrastruct. Res., 17,* 342 (1967).
76. Eiserling, F. A., and Boy de la Tour, E., *Pathol. Microbiol., 28,* 175 (1965).
77. Eiserling, F. A., and Romig, W. R., *J. Ultrastruct. Res., 6,* 540 (1962).
78. Eklund, M. W., Poysky, F. T., and Boatman, E. S., *J. Virol., 3,* 270 (1969).
79. Feary, T. W., Fisher, E., and Fisher, T. N., *J. Bacteriol., 87,* 196 (1964).
80. Feltynowski, A., and Sikorska, E., *Med. Dosw. Mikrobiol., 17,* 153 (1965).
81. Firehammer, B. D., and Border, M., *Amer. J. Vet. Res., 29,* 2229 (1968).
82. Fletcher, R. D., *Amer. J. Vet. Res., 26,* 361 (1965).
83. Földes, J., and Trautner, J. A., *Z. Vererbungsl., 95,* 57 (1964).
84. Follett, E. A. C., *J. Gen. Virol., 1,* 281 (1967).
85. Frank, H., Lorbacher, H., and Blobel, H., *Z. Naturforsch. Teil B, 24,* 729 (1969).
86. Freund-Mölbert, E., Drews, G., Bosecker, K., and Schubel, B., *Arch. Mikrobiol., 64,* 1 (1968).
87. Gabrilovich, I. M., Lukelyeva, S. I., Polupanov, W. S., and Stefanov, S. B., *Arch. Immunol. Ther. Exp., 16,* 870 (1968).
88. Gesheva, R. L., *Mikrobiologiya, 39,* 134 (1970).
89. Ghei, O. K., Eisenstark, A., To, C. M., and Consigli, R. A., *J. Gen. Virol., 3,* 113 (1968).
90. Glover, S. W., and Kerszman, G., *Genet. Res., 9,* 135 (1967).
91. Grabnar, M., and Hartman, P. E., *Virology, 34,* 521 (1968).
92. Guélin, A., Beerens, H., and Petitprez, A., *Ann. Inst. Pasteur (Paris), 111,* 141 (1966).
93. Harm, W., and Rupert, C. S., *Z. Vererbungsl., 94,* 336 (1963).
94. Higgins, M. L., and Lechevalier, M. P., *J. Virol., 3,* 210 (1969).
95. Hirano, S., and Yonekura, Y., *Acta Med. Univ. Kagoshima, 9,* 41 (1967).
96. Holmes, R. K., and Barksdale, L., *J. Virol., 5,* 783 (1970).
97. Hongo, M., Murata, A., and Ogata, S., *Agr. Biol. Chem., 33,* 337 (1969).
98. Hongo, M., Murata, A., Ogata, S., Kono, K., and Kato, F., *Agr. Biol. Chem., 32,* 773 (1968).
99. Huang, W. M., and Marmur, J., *J. Virol., 5,* 237 (1970).
100. Huet, M., and Thouvenot, H., *Ann. Inst. Pasteur (Paris), 106,* 867 (1964).
101. Imaeda, T., and San Blas, F., *J. Gen. Virol., 5,* 493 (1969).
102. Jesaitis, M. A., and Hutton, J. J., *J. Exp. Med., 117,* 285 (1963).
103. Juhasz, S. E., and Bönicke, R., *Can. J. Microbiol., 11,* 235 (1965).
104. Kay, D., and Fildes, P., *J. Gen. Microbiol., 27,* 143 (1962).
105. Kellenberger, E., *Advan. Virus Res., 8,* 1 (1961).
106. Kellenberger, E., Bolle, A., Boy de la Tour, E., Epstein, R. H., Franklin, N. C., Jerne, N. K., Reale-Scafati, A., Séchaud, J., Bendet, I., Goldstein, D., and Lauffer, M. A., *Virology, 26,* 419 (1965).
107. Kemp, C. L., Howatson, A. F., and Siminovitch, L., *Virology, 36,* 490 (1968).
108. Khadgi-Murat, L. N., Smirnova, E. I., and Rautenstein, Ya. I., *Mikrobiologiya, 36,* 140 (1967).
109. Khavina, E. S., SaVelieva, N. D., and Rautenstein, Ya. I., *Mikrobiologiya, 37,* 471 (1968).
110. Kingsbury, D. T., and Ordal, E. J., *J. Bacteriol., 91,* 1327 (1966).
111. Kiselev, N. A., Tikhonenko, T. I., Kaftanova, A. S., and Kiselev, F. L., *Biokhimiya, 28,* 1065 (1963).
112. de Klerk, H. C., and Coetzee, J. N., *J. Gen Microbiol., 38,* 35 (1965).

113. Knapp, W., and Zwillenberg, L. O., *Arch. Gesamte Virusforsch., 14,* 563 (1964).
114. Kölbel, H., *Beitr. Klin. Erforsch. Tuberk. Lungenkr., 132,* 331 (1965).
115. Kolstad, R. A., and Bradley, S. G., *J. Bacteriol., 87,* 1157 (1964).
116. Krizsanovich, K., de Klerk, H. C., and Smit, J. A., *J. Gen. Virol., 4,* 437 (1969).
117. Kropinski, A. M. B., and Warren, R. A. J., *J. Gen. Virol., 6,* 85 (1970).
118. Kuo, T., Huang, T., and Teng, M., *J. Mol. Biol., 34,* 373 (1968).
119. Kuo, T. T., Huang, T. C., Wu, R. Y., and Chen, C. P., *Can. J. Microbiol., 14,* 1139 (1968).
120. Kuroda, S., and Bradley, S. G., *Advan. Frontiers Plant Sci., 19,* 29 (1969).
121. Kwiatkowski, B., and Taylor, A., *Acta Microbiol. Pol. Ser. A, 2,* 13 (1970).
122. Lapchine, L., and Enjalbert, L., *J. Microsc. (Paris), 4,* 33 (1965).
123. Lapchine, L., Goze, A., Moillo, A., and Enjalbert, L., *J. Microsc. (Paris), 8,* 503 (1969).
124. Lechevalier, H., Lechevalier, M. P., and Higgins, M. L., *Laval Med., 39,* 621 (1968).
125. Lee, C. S., and Davidson, N., *Virology, 40,* 102 (1970).
126. Lee, L. F., and Boezi, J. A., *J. Bacteriol., 92,* 1821 (1966).
127. Lewin, R. A., Crothers, D. M., Correll, D. L., and Reimann, B. E., *Can. J. Microbiol., 10,* 75 (1964).
128. Liljemark, W. F., and Anderson, D. L., *J. Virol., 6,* 107 (1970).
129. Lindberg, A. A., *J. Gen. Microbiol., 48,* 225 (1967).
130. von Lohr, R., *Z. Allg. Mikrobiol., 10,* 269 (1970).
131. Lomov, Yu. M., *Zh. Mikrobiol. Epidemiol. Immunobiol.,* p. 117 (1969).
132. Lorbacher, H., and Blobel, H., *Zentralbl. Bakteriol. Parasitenk. Infektionskr. Hyg. Abt. Orig., 210,* 371 (1969).
133. Lovett, P. S., and Shockman, G. D., *J. Virol., 6,* 125 (1970).
134. Luftig, R., and Haselkorn, R., *J. Virol., 1,* 344 (1967).
135. Luftig, R., and Haselkorn, R., *Virology, 34,* 675 (1968).
136. Luftig, R., and Haselkorn, R., *Virology, 34,* 664 (1968).
137. Lunt, M. R., and Kay, D., *J. Gen. Virol., 3,* 459 (1968).
138. Mach, F., in *Proceedings of the European Regional Conference on Electron Microscopy, Delft,* Vol. 2, p. 1016, A. L. Houwink and B. L. Spit, Eds. De Nederlandse Vereniging voor Electronenmicroscopie, Delft, The Netherlands (1960).
139. Mach, F., in *Electron Microscopy 1962, 5th International Congress for Electron Microscopy, Philadelphia, Pennsylvania,* Vol. 2, p. S-10, S. S. Breese, Ed. Academic Press, New York and London (1962).
140. Mahony, D. E., and Easterbrook, K. B., *Can. J. Microbiol., 16,* 983 (1970).
141. Mahony, D. E., and Kalz, G. G., *Can. J. Microbiol., 14,* 1085 (1968).
142. Mankiewicz, E., Liivak, M., and Dernuet, S., *J. Gen. Microbiol., 55,* 409 (1969).
143. Maré, I. J., de Klerk, H. C., and Prozesky, O. W., *J. Gen. Microbiol., 44,* 23 (1966).
144. Mathews, M. M., Miller, P. A., and Pappenheimer, A. M., *Virology, 29,* 402 (1966).
145. Mattern, I. E., and Bacq, C.-M., *Biochim. Biophys. Acta, 78,* 221 (1963).
146. Matthews, M. A., and Bradley, D. E., in *Proceedings of the 3rd European Regional Conference on Electron Microscopy, Prague,* Vol. B, p. 543, M. Titlbach, Ed. Publishing House of the Czechoslovak Academy of Sciences, Prague (1964), and The Royal Microscopic Society of London (2nd ed., 1965).
147. May, P., May, E., Granboulan, N., Branboulan, N., and Marmur, J., *Ann. Inst. Pasteur (Paris), 115,* 1029 (1968).
148. Mazáček, M., Renčová, J., and Hradečná, Z., *Zentralbl. Bakteriol. Parasitenk. Infektionskr. Hyg. Abt. Orig., 209,* 366 (1969).
149. Mazzarelli, M., Lysenko, A. M., Klimenko, S. M., Tikhonenko, T. I., Rudneva, I. A., and Kokurina, N. K., *Vop. Virusol., 12,* 471 (1967).
150. McDuff, C. R., Jones, L. M., and Wilson, J. B., *J. Bacteriol., 83,* 324 (1962).
151. Meekins, W. E., and Blouse, L. E., *Amer. J. Vet. Res., 30,* 917 (1969).
152. Menezes, J., and Pavilanis, V., *Experientia (Basel), 25,* 1112 (1969).
153. Mesyanzhinov, V. V., Mazzarelli, M., Bogomoleva, T. A., and Poglazov, B. F., *Biokhimiya, 34,* 768 (1969).
154. Meynell, E. W., *J. Gen. Microbiol., 25,* 253 (1961).
155. Mohelská, H., in *Proceedings of the 3rd European Regional Conference on Electron Microscopy, Prague,* Vol. B, p. 551, M. Titlbach, Ed. Publishing House of the Czechoslovak Academy of Sciences, Prague (1964), and The Royal Microscopic Society of London (2nd ed., 1965).
156. Mohelská, H., and Makovcová, A., *Z. Allg. Mikrobiol., 8,* 29 (1968).
157. Molnar, D. M., and Lawton, W. D., *J. Virol., 4,* 896 (1969).
158. Moskalenko, L. N., and Rautenstein, Ya. I., *Mikrobiologiya, 38,* 489 (1969).
159. Murphy, J. S., and Philipson, L., *J. Gen. Physiol., 45,* 155 (1969).
160. Naroditsky, B. S., Gofman, Yu. P., Scheludchenko, V. B., and Tikhonenko, T. I., *Vop. Virusol., 14,* 469 (1969).
161. Neri, M. G., Zampieri, A., and Dettori, R., *G. Microbiol., 14,* 115 (1966).
162. Neri, M. G., Zampieri, A., and Dettori, R., *G. Microbiol., 14,* 107 (1966).
163. Niblack, J. F., and Gunsalus, I. C., *Bacteriol. Proc.,* p. 115 (1965).
164. Nicolle, P., Mollaret, H., Hamon, Y., and Vieu, J.-F., *Ann. Inst. Pasteur (Paris), 112,* 86 (1967).
165. Nutter, R. L., Bullas, L. R., and Schultz, R. L., *J. Virol., 5,* 754 (1970).

166. Ogata, S., Nagao, N., Hidaka, Z., and Hongo, M., *Agr. Biol. Chem., 33,* 1541 (1969).
167. Oki, T., and Ogata, K., *Agr. Biol. Chem., 32,* 241 (1968).
168. Olsen, R. H., Metcalf, E. S., and Todd, J. K., *J. Virol., 2,* 357 (1968).
169. Ortel, S., *Zentralbl. Bakteriol. Parasitenk. Infektionskr. Hyg. Abt. Orig., 198,* 423 (1965).
170. Ortel, S., *Z. Allg. Mikrobiol., 9,* 565 (1969).
171. Padan, E., Shilo, M., and Kislev, N., *Virology, 32,* 234 (1967).
172. Paquette, G., *Ph.D. Thesis.* Université de Montréal, Canada (1970).
173. Parnas, J., Bradley, D. E., and Burdzy, K., *Zentralbl. Bakteriol. Parasitenk. Infektionskr. Hyg. Abt. Orig., 191,* 459 (1963).
174. Parnas, J., and Chiozzotto, A., *Zentralbl. Bakteriol. Parasitenk. Infektionskr. Hyg. Abt. Orig., 186,* 84 (1962).
175. Parnas, J., and Sarnecka-Szunke, B., *Arch. Exp. Veterinaermed., 19,* 497 (1965).
176. Patel, J. J., *Arch. Mikrobiol., 65,* 401 (1969).
177. Patel, J. J., *Arch. Mikrobiol., 69,* 294 (1969).
178. Paterson, W. D., Douglas, R. J., Grinyer, I., and McDermott, L. A., *J. Fish. Res. Board Can., 26,* 629 (1969).
179. Peshkoff, M. A., Stefanov, S. B., Shadrina, I. A., and Marek, B. I., *Mikrobiologiya, 36,* 136 (1967).
180. Peshkoff, M. A., Tikhonenko, A. S., and Marek, B. I., *Mikrobiologiya, 35,* 684 (1966).
181. Phelps, L. N., *J. Gen. Virol., 1,* 529 (1967).
182. Pohjanpelto, P., and Nyholm, M., *Arch. Gesamte Virusforsch., 17,* 481 (1965).
183. Pootjes, C. F., Mayhew, R. B., and Korant, B. D., *J. Bacteriol., 92,* 1787 (1966).
184. Prescott, L. M., and Altenbern, R. A., *J. Virol., 1,* 1085 (1967).
185. Prévot, A.-R., *Traité de systématique bactérienne,* Vol. 2. Dunod, Paris, France (1961).
186. Prévot, A.-R., Vieu, J.-F., Thouvenot, H., and Brault, H., *Bull. Acad. Nat. Med. (Paris), 154,* 681 (1970).
187. Prozesky, O. W., de Klerk, H. C., and Coetzee, J. N., *J. Gen. Microbiol., 41,* 29 (1965).
188. Przondo-Hessek, A., and Šlopek, S., *Arch. Immunol. Ther. Exp., 15,* 545 (1967).
189. Rabussay, D., Zillig, W., and Herrlich, P., *Virology, 41,* 91 (1970).
190. Rautenstein, Ya. I., Gesheva, R. L., and Egorova, S. A., *Mikrobiologiya, 39,* 685 (1970).
191. Rautenstein, Ya. I., Moskalenko, L. N., and Zhunaeva, V. V., *Mikrobiologiya, 39,* 358 (1970).
192. Rautenstein, Ya. I., Tikhonenko, A. S., and Retinskaya, V. L., *Mikrobiologiya, 31,* 49 (1962).
193. Rieber, M., and Imaeda, T., *J. Virol., 4,* 542 (1969).
194. Ritchie, A. E., and Ellinghausen, H. C., in *27th Annual Proceedings of EMSA, St. Paul, Minnesota,* p. 228, C. J. Arcenaux, Ed., Claytor's Publishing Division, Baton Rouge, Louisiana (1969).
195. Riva, S., Polsinelli, M., and Falaschi, A., *J. Mol. Biol., 35,* 347 (1968).
196. Riverin, M., Beaudoin, J., and Vézina, C., *J. Gen. Virol., 6,* 395 (1970).
197. Rosenblum, E. D., and Tyrone, S., *J. Bacteriol., 88,* 1737 (1964).
198. Roslycky, E. B., Allen, O. N., and McCoy, E., *Can. J. Microbiol., 9,* 199 (1963).
199. Rousseaux, J., Kurkdjian, A., and Beardsley, R. E., *Ann. Inst. Pasteur (Paris), 114,* 237 (1968).
200. Russell, H., Norcross, N. L., and Kahn, D. E., *J. Gen. Virol., 5,* 315 (1969).
201. Safferman, R. S., Morris, M.-E., Sherman, L. A., and Haselkorn, R., *Virology, 39,* 775 (1969).
202. Sakurai, T., Takahashi, T., and Arai, H., *Jap. J. Microbiol., 14,* 333 (1970).
203. Samuels, J., and Clarke, J. K., *J. Virol., 4,* 797 (1969).
204. Saunders, G. F., and Campbell, L. L., *J. Bacteriol., 91,* 340 (1966).
205. Schade, S. Z., Adler, J., and Ris, H., *J. Virol., 1,* 599 (1967).
206. Schito, G. C., Meloni, G. A., and Pesce, A., *Boll. Ist. Sieroter. Milan., 44,* 338 (1965).
207. Schito, G. C., Rialdi, R., and Pesce, A., *Biochim. Biophys. Acta, 129,* 482 (1966).
208. Schmidt, J. M., and Stanier, R. Y., *J. Gen. Microbiol., 39,* 95 (1965).
209. Shargool, P. D., and Townsend, E. E., *Can. J. Microbiol., 12,* 885 (1966).
210. Shimizu, N., Miura, K., and Aoki, H., *J. Biochem. (Tokyo), 68,* 265 (1970).
211. Shimizu, N., Miura, K., and Aoki, H., *J. Biochem. (Tokyo), 68,* 277 (1970).
212. Shinagawa, H., Hosaka, Y., Yamagishi, H., and Nishi, Y., *Biken J., 9,* 135 (1966).
213. Shved, A. D., Manjakov, V. F., Kishko, Ya. G., and Revenko, I. P., *Vop. Virusol., 13,* 735 (1968).
214. Slayter, H. S., Holloway, B. W., and Hall, C. E., *J. Ultrastruct. Res., 11,* 274 (1964).
215. Stefanov, S. B., and Gabrilovich, I. M., *Mikrobiologiya, 36,* 297 (1967).
216. Stefanov, S. B., Rautenstein, Ya. I., and Khachatrian, L. S., *Mikrobiologiya, 35,* 1064 (1966).
217. Takeya, K., and Amako, K., *Virology, 24,* 461 (1964).
218. Takeya, K., Zinnaka, Y., Shimodori, S., Nakayama, Y., Amako, Y., Amako, K., and Iida, K., in *Proceedings of the Cholera Research Symposium, Honolulu, Hawaii,* p. 24, O. A. Bushnell and C. S. Brookhyser, Eds. U.S. Department of Health, Education and Welfare, Public Health Service, Washington, D.C. (1965).
219. Taubeneck, U., *Biol. Zentralbl., 86* (Suppl.), 45 (1967).
220. Taubeneck, U., *Z. Allg. Mikrobiol., 8,* 281 (1968).
221. Taubeneck, U., *Z. Allg. Mikrobiol., 10,* 227 (1970).
222. Tikhonenko, A. S., *Dokl. Akad. Nauk SSSR, 138,* 1449 (1961).
223. Tikhonenko, A. S., *Ultrastruktura virusov bakterii.* Izdadelstvo "Nauka", Moscow, U.S.S.R. (1968).

224. Tikhonenko, A. S., and Belyaeva, N. N., *Mikrobiologiya, 36,* 475 (1967).
225. Tikhonenko, A. S., Velikodvorskaya, G. A., and Belykh, R. A., *Mikrobiologiya, 33,* 824 (1964).
226. Tikhonenko, A. S., and Zavarzina, N. B., *Mikrobiologiya, 35,* 848 (1966).
227. To, C. M., Eisenstark, A., and Töreci, H., *J. Ultrastruct. Res., 14,* 441 (1966).
228. Truffaut, N., Revet, B., and Soulie, M.-O., *Eur. J. Biochem., 15,* 391 (1970).
229. Twarog, R., and Blouse, L. E., *J. Virol., 2,* 716 (1968).
230. Valentine, A. F., Chen, P. K., Colwell, R. R., and Chapman, G. B., *J. Bacteriol., 91,* 819 (1966).
231. Vidaver, A. K., and Schuster, M. L., *J. Virol., 4,* 300 (1969).
232. Vieu, J.-F., Personal Communication.
233. Vieu, J.-F., and Croissant, O., *Arch. Roum. Pathol. Exp. Microbiol., 25,* 305 (1966).
234. Vieu, J.-F., Croissant, O., and Dauguet, C., *J. Microsc. (Paris), 3,* 403 (1964).
235. Vieu, J.-F., Croissant, O., and Dauguet, C., *Ann. Inst. Pasteur (Paris), 109,* 160 (1965).
236. Vieu, J.-F., Guélin, A., and Dauguet, C., *Ann. Inst. Pasteur (Paris), 109,* 157 (1965).
237. Vieu, J.-F., and Hoarau, M., Personal Communication.
238. Vieu, J.-F., Nicolle, P., and Gallut, J., in *Proceedings of the Cholera Research Symposium, Honolulu, Hawaii,* p. 34, O. A. Bushnell and C. S. Brookhyser, Eds. U.S. Department of Health, Education and Welfare, Public Health Service, Washington, D.C. (1965).
239. Vinet, G., Berthiaume, L., and Fredette, V., *Rev. Can. Biol., 27,* 73 (1968).
240. Walker, D. H., and Anderson, T. F., *J. Virol., 5,* 765 (1970).
241. Watanabe, K., Takesue, S., Jin-Nai, K., and Yoshikawa, T., *Appl. Microbiol., 20,* 409 (1970).
242. Welker, N. E., and Campbell, L. L., *J. Bacteriol., 89,* 175 (1965).
243. Yamamoto, N., and Anderson, T. F., *Virology, 14,* 430 (1961).
244. Yamamoto, T., and Chow, C. T., *Can. J. Microbiol., 14,* 667 (1968).
245. Yelton, D. B., and Thorne, C. B., *J. Bacteriol., 102,* 573 (1970).
246. Young, B. G., Hartman, P. E., and Moudrianakis, E. N., *Virology, 28,* 249 (1966).
247. Zimmerer, R. P., Hamilton, R. H., and Pootjes, C., *J. Bacteriol., 92,* 746 (1966).
248. Zubrzycki, L., Green, J., and Spaulding, E. H., *J. Gen. Microbiol., 45,* 113 (1966).

Tailed Bacteriophages: Molecular Weights, Lengths, and Percentages of Nucleic Acids, and Molecular Weights of Complete Phages

DR. H.-W. ACKERMANN

As far as determined, all tailed phages possess double-stranded DNA. Molecular weights and lengths of nucleic acid vary over wide ranges, as do weights of complete phages. There is a strong correlation between these data and the dimensions of phage capsids.

Molecular weights of nucleic acids have been quoted regardless of the method of determination; however, radiobiological data have been omitted as not reliable. Sedimentation constants have been omitted also, because they are employed for calculations of molecular weights; this could likewise apply to contour lengths, but quoting them was considered as interesting in itself. Molecular weights of complete phages are listed because comparison with the molecular weights of the nucleic acids is desirable.

Listing of phages follows the principles used in the preceding tables; however, the phages are grouped by host genera first and by morphology second. Mutants and "normal" phages are listed together, with a few exceptions, such as the density mutant λdg and phage λ. Mutants are indicated by italics. The table includes the phycovirus LPP-1 and some defective phages or "killer particles".

CODE

MW	molecular weight	~	about
NN	no name	?	not certain

Host	Morphological Group	Phage	Nucleic Acid: MW, 10^6 daltons	Nucleic Acid: Length, μ	Nucleic Acid: Weight, %	Phage MW, 10^6 daltons	References
Agrobacterium	B1	PS8	43				58
Bacillus		PBLB[a]	8.4				43
		PBSH[a]	9.3–12.4	4.7	16.5	68	41
		PBSX[a]	8.1–12	4.3			42,72,98
		PBS1	190	82			44,120
		SP01	~ 100				75
	A1	*SP82*	130			180	38
		SP50	94.3–108.6	49.1–49.7		~ 100	14,31,83,84
		SW	130	65	41.6		68,69
		Vx	120		~ 40		100
		ϕ25	100	50			60
		2C	97	50			65
		G			45–55	91	67
		PBLA	34–35				43
		SP02	25–26	13.2			15,88
		SPP1	25				87
		TP-1	12.1				118
	B1	TP-84	22.4–27	13.9	54	50.5	91,92
		Tϕ3	23.2–28.7	12.1			29
		a	34–40			68–70	6
		ϕμ-4	26	11.4			80
		ϕ105	24–26	11.6			13
	C2	Nf	11.4–12.5	6.4	~ 40		101
		ϕ29	10.9–11.3	5–5.8			3,4
	?	F	33				116
Brevibacterium	B1	Ap85III	20.1			59	73,74
		P465	2.9?			103?	73,74
	?	P4	11.4			149.4?	73,74
		P468II	7.9			30.1	73,74
Caulobacter	B3	ϕCbK			57		2

Clostridium	B2	F_1			55–60		11,12
Corynebacterium	B1	*βhv64tox*⁻	78.3				64
Enterobacteriaceae							
Escherichia		Mu1	27.8–28	14.5	~ 40		110,112
Salmonella		O1				131	76
Escherichia		Plkc[b]	58–60	29.6			1,45
Escherichia		P2	22	13.1–13.2	38	58	10,46,63
Serratia	A1	κ	117–119		34		79
Escherichia		13 vir	26.3				82
Escherichia		15[a]	27–33	11.9–12.8			32,55
Escherichia		186	19.2–20.5	10.1			114,115
Escherichia		D_d-*Vi*			48.5		66
Escherichia	A2	T2	103–134	49–56		215–220	23,26,49,90,104, 105,107,108
Escherichia		T4	112–140				5
Klebsiella		L1	22.1				34
Escherichia		T1	31–35	16			18,19,53
Escherichia		T5	76–84	38.3–38.8			1,21,89,106,108
Escherichia		λ	30.6–38	15.3–17.3	51.5–60	57–62	20,23,24,28,55, 62,102,117
Escherichia	B1	λ*dg*	32.6		50		47,104
Escherichia		φ80	26.5				102
Escherichia		φ80pt₁	30.3				102
Polyvalence		χ	41.6	21.7	41–46	90	93
Klebsiella		490	35.5				34
Escherichia		N4	40.5		47	83	94,95,96,97
Providencia		P125	24.4	12.6	48	54.3	77,78
Salmonella		P22	26–27	13.5–13.7			85,106
Escherichia		sd	66–88	36	43.5		48,109
Escherichia	C1	T3	23	11.6–14			9,53
Escherichia		T7	19–28.5	12–12.5	50	38	1,19,21,23,27,30, 33,57,86,104
Serratia		η	72–104		36–45		79
Klebsiella		380	29.6				34
Escherichia		N15	27–34				81,103
Escherichia	?	φN	60.5–81				7

Host	Morphological Group	Phage	Nucleic Acid MW, 10^6 daltons	Nucleic Acid Length, μ	Nucleic Acid Weight, %	Phage MW, 10^6 daltons	References
Micrococcus	B1	N1	30.8–33	14.1			54,119
		N5	23.2	10.6			54
		N6	30.5	13.9			54
Moraxella	C1	BP1	23				113
Mycobacterium	B1	B-1	104.2				111
Myxococcus	A1	MX-1			55		22
Plectonema[c]	C1	LPP-1	25–27	13.2	40	51.1	36,37,61
Pseudomonas	A1	PB-1		24–25			16,17
		Psa1	24.1				70
		ϕ-2	84.3		45		25
	B1	SD1	66.3		55	65.1–104.8	99
	C1	gh-1	22.6		36		56,57
		Pf	25				71
		ϕ-MC	46.4		27		25
	?	2	70.5–72.1				39, 40
Saprospira	A1	NN	90		45		59
Streptomyces	B2	MSP8	69.9–75.3		57	110	51,52
		Type II	219.4?	41.4–41.8		531.2?	50
	C1	ϕ17	12	25			8
Xanthomonas	A1	XP5	51.2				35

[a] Defective phage.

[b] Not studied by electron microscopy.

[c] Blue-green alga.

REFERENCES

1. Abelson, J., and Thomas, C. A., *J. Mol. Biol., 18,* 262 (1966).
2. Agabian-Keshishian, N., and Shapiro, L., *J. Virol., 5,* 795 (1970).
3. Anderson, D. L., Hickman, D. D., and Reilly, B. E., *J. Bacteriol., 91,* 2081 (1966).
4. Anderson, D. L., and Mosharrafa, E. T., *J. Virol., 2,* 1185 (1968).
5. Aten, J. B. T., and Cohen, J. A., *J. Mol. Biol., 12,* 537 (1965).
6. Aurisicchio, S., in *Procedures in Nucleic Acid Research,* p. 562, G. L. Cantoni and D. R. Davies, Eds. Harper and Row, New York and London (1966).
7. de Backer, R., and Bourgaux, P., *Ann. Inst. Pasteur (Paris), 113,* 825 (1967).
8. Bacq, C.-M., and Dierickx, L., *Biochim. Biophys. Acta, 61,* 19 (1962).
9. Bendet, I., Schachter, E., and Lauffer, M. A., *J. Mol. Biol., 5,* 76 (1962).
10. Bertani, L. E., and Bertani, G., *J. Gen. Virol., 6,* 201 (1970).
11. Betz, J. V., *Virology, 36,* 9 (1968).
12. Betz, J. V., and McClung, L. S., *Bacteriol. Proc.,* p. 119 (1964).
13. Birdsell, D. C., Hathaway, G. M., and Rutberg, L., *J. Virol., 4,* 264 (1969).
14. Biswal, N., Kleinschmidt, A. K., Spatz, H. C., and Trautner, T. A., *Mol. Gen. Genet., 100,* 39 (1967).
15. Boyce, L., Eiserling, F. A., and Romig, W. R., *Biochem. Biophys. Res. Commun., 34,* 398 (1969).
16. Bradley, D. E., *Bacteriol. Rev., 31,* 230 (1967).
17. Bradley, D. E., and Robertson, D., *J. Gen. Virol., 3,* 247 (1968).
18. Bresler, S. E., Kiselev, N. A., Manjakov, V. F., Mosevitsky, M. I., and Timkovsky, A. L., *Mol. Biol., 1,* 271 (1967).
19. Brody, E. N., Mackal, R. P., and Evans, E. A., *J. Virol., 1,* 76 (1967).
20. Buchwald, M., Steed-Glaister, P., and Siminovitch, L., *Virology, 42,* 375 (1970).
21. Bujard, H., *J. Mol. Biol., 49,* 125 (1969).
22. Burchard, R. P., and Dworkin, M., *J. Bacteriol., 91,* 1305 (1966).
23. Burgi, E., and Hershey, A. D., *Biophys. J., 3,* 309 (1963).
24. Caro, L. C., *Virology, 25,* 226 (1965).
25. Chow, C. T., and Yamamoto, T., *Can. J. Microbiol., 15,* 1179 (1969).
26. Cummings, D. J., and Kozloff, L. M., *Biochim. Biophys. Acta, 44,* 445 (1960).
27. Davison, P. F., and Freifelder, D., *J. Mol. Biol., 5,* 635 (1962).
28. Dyson, R. D., and van Holde, K. E., *Virology, 33,* 559 (1967).
29. Egbert, L. N., *J. Virol., 3,* 528 (1969).
30. Eigner, J., and Doty, P., *J. Mol. Biol., 12,* 549 (1965).
31. Földes, J., and Trautner, J. A., *Z. Vererbungsl., 95,* 57 (1964).
32. Frampton, E. W., and Mandel, M., *J. Virol., 5,* 8 (1970).
33. Freifelder, D., and Kleinschmidt, A. K., *J. Mol. Biol., 14,* 271 (1965).
34. Gabrilovich, I. M., Lukelyeva, S. I., Polupanov, W. S., and Stefanov, S. B., *Arch. Immunol. Ther. Exp., 16,* 870 (1968).
35. Ghei, O. K., Eisenstark, A., To, C. M., and Consigli, R. A., *J. Gen. Virol., 3,* 133 (1968).
36. Goldstein, D. A., and Bendet, I. J., *Virology, 32,* 614 (1967).
37. Goldstein, D. A., Bendet, I. J., Lauffer, M. A., and Smith, K. M., *Virology, 32,* 601 (1967).
38. Green, D. M., *J. Mol. Biol., 10,* 438 (1964).
39. Grogan, J. B., and Johnson, E. J., *Bacteriol. Proc.,* p. 120 (1964).
40. Grogan, J. B., and Johnson, E. J., *Virology, 24,* 235 (1964).
41. Haas, M., and Yoshikawa, H., *J. Virol., 3,* 233 (1969).
42. Hirokawa, H., and Kadlubar, F., *J. Virol., 3,* 205 (1969).
43. Huang, W. M., and Marmur, J., *J. Virol., 5,* 237 (1970).
44. Hunter, B. I., Yamagishi, H., and Takahashi, I., *J. Virol., 1,* 841 (1967).
45. Ikeda, H., and Tomizawa, J. I., *J. Mol. Biol., 14,* 85 (1965).
46. Inman, R. B., and Bertani, G., *J. Mol. Biol., 44,* 533 (1969).
47. Kaiser, A. D., and Hogness, D. S., *J. Mol. Biol., 2,* 392 (1960).
48. Kiselev, N. A., Tikhonenko, T. I., Kaftanova, A. S., and Kiselev, F. L., *Biokhimiya, 28,* 1065 (1963).
49. Kleinschmidt, A. K., Lang, D., Jacherts, D., and Zahn, R. K., *Biochim. Biophys. Acta, 61,* 857 (1962).
50. Kochkina, Z. M., *Dokl. Akad. Nauk SSSR, 185,* 697 (1969).
51. Kolstad, R. A., and Bradley, S. G., *J. Bacteriol., 87,* 1157 (1964).
52. Kolstad, R. A., and Bradley, S. G., *J. Bacteriol., 91,* 1372 (1966).
53. Lang, D., Bujard, H., Wolff, B., and Russell, D., *J. Mol. Biol., 23,* 163 (1967).
54. Lee, C. S., and Davidson, N., *Virology, 40,* 102 (1970).
55. Lee, C. S., Davis, R. W., and Davidson, N., *J. Mol. Biol., 48,* 1 (1970).
56. Lee, L. F., and Boezi, J. A., *J. Bacteriol., 92,* 1821 (1966).
57. Lee, L. F., and Boezi, J. A., *J. Virol., 1,* 1274 (1967).
58. Leff, J., and Beardsley, R. E., *C. R. Hebd. Séances Acad. Sci. Ser. D Sci. Natur. (Paris), 270,* 2505 (1970).
59. Lewin, R. A., Crothers, D. M., Correll, D. L., and Reimann, B. E., *Can. J. Microbiol., 10,* 75 (1964).

60. Liljemark, W. F., and Anderson, D. L., *J. Virol., 6,* 107 (1970).
61. Luftig, R., and Haselkorn, R., *J. Virol., 1,* 344 (1967).
62. MacHattie, L. A., and Thomas, C. A., *Science (Washington), 144,* 1142 (1964).
63. Mandel, M., *Mol. Gen. Genet., 99,* 88 (1967).
64. Matsuda, M., and Barksdale, L., *J. Bacteriol., 93,* 722 (1967).
65. May, P., May, E., Granboulan, E., Granboulan, N., and Marmur, J., *Ann. Inst. Pasteur (Paris), 115,* 1029 (1968).
66. Mazzarelli, M., Lysenko, A. M., Klimenko, S. M., Tikhonenko, T. I., Rudneva, I. A., and Kokurina, N. K., *Vop. Virusol., 12,* 471 (1967).
67. Murphy, J. S., and Philipson, L., *J. Gen. Physiol., 45,* 155 (1962).
68. Naroditsky, B. S., Tikhonenko, T. I., and Klimenko, S. M., *Vop. Med. Khim., 15,* 476 (1969).
69. Naroditsky, B. S., Ulanov, B. P., and Tikhonenko, T. I., *Biofizika, 15,* 187 (1970).
70. Neri, M. G., Zampieri, A., and Dettori, R., *G. Microbiol., 14,* 107 (1966).
71. Niblack, J. F., and Gunsalus, I. C., *Bacteriol. Proc.,* p. 115 (1965).
72. Okamoto, K., Mudd, J. A., Mangan, J., Huang, W. M., Subbaiah, T. V., and Marmur, J., *J. Mol. Biol., 34,* 413 (1968).
73. Oki, T., and Ogata, K., *Agr. Biol. Chem., 32,* 234 (1968).
74. Oki, T., and Ogata, K., *Agr. Biol. Chem., 32,* 241 (1968).
75. Okubo, S., Strauss, B., and Stodolsky, M., *Virology, 24,* 552 (1964).
76. Pawlitschek, W., Laue, F., and Völz, H., *Naturwissenschaften, 49,* 526 (1962).
77. Pitout, M. J., Conradie, J. D., and van Rensburg, A. J., *J. Gen. Virol., 4,* 577 (1969).
78. Pitout, M. J., and van Rensburg, A. J., *J. Gen. Virol., 4,* 615 (1969).
79. Pons, F. W., *Biochem. Z., 346,* 26 (1966).
80. Rabussay, D., Zillig, W., and Herrlich, P., *Virology, 41,* 91 (1970).
81. Ravin, V. K., and Shulga, M. G., *Virology, 40,* 800 (1970).
82. van Rensburg, A. J., *J. Gen. Virol., 5,* 437 (1969).
83. Reznikoff, W. S., *Diss. Abstr., 28,* 4426 (1967).
84. Reznikoff, W. S., and Thomas, C. A., *Virology, 37,* 309 (1969).
85. Rhoades, M., MacHattie, L. A., and Thomas, C. A., *J. Mol. Biol., 37,* 21 (1968).
86. Richardson, C. C., *J. Mol. Biol., 15,* 49 (1966).
87. Riva, S., Polsinelli, M., and Falaschi, A., *J. Mol. Biol., 35,* 347 (1968).
88. Romig, W. R., *Bacteriol. Rev., 32,* 349 (1968).
89. Rubenstein, I., *Virology, 36,* 356 (1968).
90. Rubenstein, I., Thomas, C. A., and Hershey, A. D., *Proc. Nat. Acad. Sci. U.S.A., 47,* 1113 (1961).
91. Saunders, G. F., and Campbell, L. L., *Biochemistry, 4,* 2836 (1965).
92. Saunders, G. F., and Campbell, L. L., *J. Bacteriol., 91,* 340 (1966).
93. Schade, S. Z., and Adler, J., *J. Virol., 1,* 591 (1967).
94. Schito, G. C., Meloni, G. A., and Pesce, A., *Boll. Ist. Sieroter. Milan., 44,* 338 (1965).
95. Schito, G. C., Molina, A. M., Pesce, A., and Romanzi, C. A., *Boll. Ist. Sieroter. Milan., 44,* 345 (1965).
96. Schito, G. C., Rialdi, R., and Pesce, A., *Biochim. Biophys. Acta, 129,* 482 (1966).
97. Schito, G. C., Rialdi, G., and Pesce, A., *Biochim. Biophys. Acta, 129,* 491 (1966).
98. Seaman, E., Tarmy, E., and Marmur, J., *Biochemistry, 3,* 607 (1964).
99. Shargool, P. D., and Townsend, E. D., *Can. J. Microbiol., 12,* 885 (1966).
100. Shimizu, N., Miura, K., and Aoki, H., *J. Biochem. (Tokyo), 68,* 265 (1970).
101. Shimizu, N., Miura, K., and Aoki, H., *J. Biochem. (Tokyo), 68,* 277 (1970).
102. Shinagawa, H., Hosaka, Y., Yamagishi, H., and Nishi, Y., *Biken J., 9,* 135 (1966).
103. Shul'ga, M. G., and Ravin, V. K., *Mol. Biol., 3,* 421 (1969).
104. Studier, F. W., *J. Mol. Biol., 11,* 373 (1965).
105. Taylor, N. W., Epstein, H. T., and Lauffer, M. A., *J. Amer. Chem. Soc., 77,* 1270 (1955).
106. Thomas, C. A., *J. Gen. Physiol., 49,* 143 (1966).
107. Thomas, C. A., and MacHattie, L. A., *Proc. Nat. Acad. Sci. U.S.A., 52,* 1297 (1964).
108. Thomas, C. A., and Pinkerton, T. C., *J. Mol. Biol., 5,* 356 (1962).
109. Tikhonenko, T. I., Velikodvorskaya, G. A., and Zemtsova, E. V., *Biokhimiya, 27,* 726 (1962).
110. To, C. M., Eisenstark, A., and Töreci, H., *J. Ultrastruct. Res., 14,* 441 (1966).
111. Tokunaga, T., and Nakamura, R. M., *J. Virol., 2,* 110 (1968).
112. Torti, F., Barksdale, C., and Abelson, J., *Virology, 41,* 567 (1970).
113. Twarog, R., and Blouse, L. E., *J. Virol., 2,* 716 (1968).
114. Wang, J. C., *J. Mol. Biol., 28,* 403 (1967).
115. Wang, J. C., and Schwartz, H., *Biopolymers, 5,* 953 (1967).
116. Watanabe, K., and Szybalski, W., *Jap. J. Microbiol., 11,* 153 (1967).
117. Weigle, J., Meselson, M., and Paigen, K., *J. Mol. Biol., 1,* 379 (1959).
118. Welker, N. E., and Campbell, L. L., *J. Bacteriol., 89,* 175 (1965).
119. Wetmur, J. G., Davidson, N., and Scaletti, J. V., *Biochem. Biophys. Res. Commun., 25,* 684 (1966).
120. Yamagishi, H., *J. Mol. Biol., 35,* 623 (1968).

Amino Acid Composition of Tailed Bacteriophages

DR. ALBERT GOZE

The data for the following table were compiled from numerous sources. Most of the authors of the cited references hydrolyzed the phage protein by using the standard method of constant-boiling hydrochloric acid. The amino acids were determined according to the method of Moore and Stein,[9] slightly modified by some authors. In previous reports the hydrolysates were analyzed by paper chromatography, but in more recent articles the Autoanalyzer was used. However, not all the amino acids were measured by the above method. Tryptophan and the cystine-cysteinecomplex were independently determined.

CODE

A = grams of amino acid per 100 grams of protein
B = moles per 100 moles of amino acids
C = number of amino acid residues for the minimum molecular weight (MMW) indicated
D = mole residues per 10^5 grams of protein

The values given represent the averages of several determinations made at various hydrolysis steps. In some instances the values were calculated by extrapolation of experimental data to zero time; in other instances the values correspond to a given time of hydrolysis. These corrections were necessary because some amino acids undergo appreciable degradation during the experiment and others are incompletely hydrolyzed. A dash (–) indicates that the measurement was not made.

Quite often the values given by the authors are expressed in different units. In most cases the units of the original article were conserved. However, for the sake of uniformity, three series of values, expressed by the author as grams of amino acid per 100 grams of virus, have been changed to grams of amino acid per 100 grams of protein.[7,14]

Hosts	Bacillus		Escherichia														
Phages	SW	TP-84	N4		s_d	T2	T2H	T2L Sheath				T2L Core		T4	T4D	T3	
References See Code	10 B	14 B	15 B	15 C	16 A	7[k] A	5 B	13 A	12 C	2 A	2 C	13 A	13 C	11 B	5 B	6 B	7[k] A
Alanine	9.9	7.34	8.63	12	10.96	7.64	11.1	6.53	50.7	7.6	46	5.01	343	9.40	11.8	9.7	9.36
Arginine	2.7	3.85	3.56	5	6.71	5.02	4.4	7.31	25.6	5.70	17	6.07	189	6.51	4.1	7.2	5.95
Aspartic acid	13.6	9.49	12.11	17	8.31	11.57	11.3	12.2	58.6	13.00	52	14.08	630	11.97	11.1	11.7	11.49
Half-cystine	Traces	0.24[h]	0[h]	0[h]	5.10[n]	–	–	0.371	2.27	0.23	1	0.105	5	–	–	–	–
Glutamic acid	11.5	14.54	10.70	15	11.18	11.79	10.2	9.76	42.2	9.94	36	11.38	429	11.97	10.4	10.5	11.28
Glycine	15.1	9.83[l]	8.94	12	8.31	9.39[d]	9.5	3.91	37.7	4.98	35	5.88	503	7.34	9.4	9.8	8.08[d]
Histidine	0.6	2.22	1.45	2	1.88	0.87	0.4	0.265	1.08	0.55	2	1.06	35	<2.6	1.0	1.7	1.7
Isoleucine	5.4	5.72	6.45	9	12.35[i]	6.55	7.0	6.32[e]	30.7[e]	7.5	31	6.54[e]	282[e]	3.90	6.5	4.9	4.68
Leucine	8.0	6.80	7.70	11	–	5.89	6.3	6.11[e]	29.6[e]	7.0	28	5.16[e]	222[e]	6.51	6.0	9.8	9.36
Lysine	5.4	5.07	5.71	8	3.81	6.33	6.0	4.21	18.1	6.21	23	4.81	183	8.46	5.4	6.3	5.95
Methionine	1.9	1.96	3.31	5	Traces	2.18	2.2	0.945	4.28	0.81	3	1.75	65	<1.3	2.1	1.7	1.91
Phenylalanine	3.8	2.85	3.50	5	5.69	5.46	4.5	4.77	17.6	5.00	16	5.45	180	4.16	4.3	3.5	3.4
Proline	3.8	6.47	3.52	5	5.48[m]	3.93	3.8	3.50	19.6	3.1	14	6.47	324	5.00	4.1	4.2	4.46
Serine	5.3	5.77[d]	5.29	8	1.73	5.24[d]	5.6	5.50[d]	35.1[d]	5.78	29	6.66[d]	373[d]	4.77	5.9	2.9	4.04[d]
Threonine	8.0	6.04[d]	8.47	13	7.17	5.90[d]	6.1	7.47[d]	40.6[d]	7.78	35	8.90[d]	429[d]	7.00	7.0	5.3	6.81[d]
Tryptophan	–	–	2.05[b]	2[b]	–	–	–	2.68[c]	7.75[c]	–	–	1.15[c]	30[c]	–	–	–	–
Tyrosine	–	3.16	3.65	5	3.27	6.33[d]	4.1	5.58	18.7[f]	6.35	19	4.31	129[g]	3.74	4.1	4.2	5.1[d]
Valine	8.9	8.22[e]	6.98	10	7.98	5.89	7.0	5.54[e]	30.9[e]	6.61	30	5.81[e]	285[e]	6.51	6.9	6.6	6.38
Amide NH_3							46.5	1.30	44.9[b]	1.94	(61)	1.82	557		46.6[a]		
MMW[o]			18,500						51,600		45,300		493,000				

Hosts	Escherichia												Klebsiella	Mycobacterium	Plectonema*	
Phages	T5	Lambda					λ Head		λ Tail		λ Vir		380	R1	LPP-1	
References See Code	5 B	17 B	17 C	4 B	3 B	3 C	3 B	3 C	3 B	3 C	17 B	17 C	1 B	1 B	8 A	8 D
Alanine	10.4	10.3	20	9.5	10	18	8.3	25	10.9	38	10.3	20	11.3	13.9	6.94	98
Arginine	4.4	5.6	11	5.3	4.9	9	5.9	18	3.5	12	5.4	11	9.2	3.5	6.24	40
Aspartic acid	11.5	8.9	18	11.0	9.9	18	11.0	33	9.8	34	9.4	19	12.4	10.9	11.21	97
Half-cystine	–	0	0	–	0.6[h]	1[h]	0.37[h]	1[h]	0.3[h]	1[h]	0	0	–	–	0.25	2
Glutamic acid	10.6	10.2	20	12.9	11.4	20	13.3	40	8.6	30	9.9	20	8.9	10.6	11.79	91
Glycine	8.9	8.5	17	7.3	8.6	15	6.7	20	9.5	33	8.5	17	9.4	12.3	4.62	81
Histidine	1.2	0.5	1	0.4	0.55	1	0.33	1	0.28	1	0.5	1	1.9	0.9	0.91	7
Isoleucine	4.9	3.1	6	3.3	3.3[e]	6[e]	3.7[e]	11[e]	2.8[e]	10[e]	3.1	6	4.0	4.8	5.95	53
Leucine	8.0	6.1	12	7.5	6.4	11	7.1	21	4.5	16	7.0	14	8.0	8.3	7.29	64
Lysine	6.3	5.5	11	4.5	5.6	10	5.5	17	5.3	18	5.3	11	8.1	2.6	3.54	28
Methionine	1.5	3.1	6	3.0	2.8[j]	5[j]	3.3[j]	10[j]	2.1[j]	7[j]	2.9	6	0.8	0.6	2.26	17
Phenylalanine	4.5	4.2	8	4.0	3.7	6	3.9	12	2.9	10	4.2	8	3.8	4.1	5.92	40
Proline	3.2	5.6	11	5.3	4.4	8	5.3	16	5.4	19	5.4	11	3.3	5.4	4.81	50
Serine	6.2	7.4	15	8.6	7.2[d]	13[d]	6.5[d]	20[d]	8.4[d]	29[d]	7.3	14	6.4	4.9	6.02	69
Threonine	6.4	8.8	18	6.5	8.4[d]	15[d]	6.3[d]	19[d]	11.6[d]	40[d]	8.9	17	6.5	7.4	8.28	82
Tryptophan	–	–	–	–	0.77[b]	1[b]	0.61[b]	2[b]	1.2[b]	4[b]	–	–	–	–	3.00[c]	16[c]
Tyrosine	4.3	4.1	8	4.0	3.4	6	3.8	12	2.0	7	3.8	8	3.8	2.7	2.50	15
Valine	7.6	7.4	15	7.0	8.0[e]	14[e]	8.1[e]	25[e]	10.9[e]	38[e]	7.5	15	7.7	7.1	6.07	61

Hosts	Escherichia												Kleb-siella	Mycobac-terium	Plectonema*	
Phages	T5	Lambda					λ Head		λ Tail		λ Vir		380	R1	LPP-1	
References	5	17	17	4	3	3	3	3	3	3	17	17	1	1	8	8
See Code	B	B	C	B	B	C	B	C	B	C	B	C	B	B	A	D
Amide NH_3	64.3[a]														2.40	141
MMW[o]		21,400			19,200		33,900		36,200		21,700					

*Blue-green alga.

[a] Mole % of the total dicarboxylic acid.

[b] Determined independently.

[c] Determined by UV-absorption spectrum.

[d] Extrapolated values corrected for losses during hydrolysis.

[e] Values calculated from the data of 72-hour hydrolysates to correct for incomplete hydrolysis.

[f] Calculated from the spectra, value is 17.5.

[g] Calculated from the spectra, value is 125.

[h] Total cystine plus cysteine was measured by oxidation to cysteic acid with performic acid.

[i] Leucine and isoleucine calculated together.

[j] Gave the same values determined either as methionine or as methionine sulfone.

[k] Basic amino acids are determined in a separate column.

[l] 22-hour value.

[m] Total proline plus hydroxyproline determined independently.

[n] Uncorrected for partial decomposition during hydrolysis.

[o] The protein analyzed to obtain this figure may be of more than one kind. In no case is the exact structure of the proteins involved known.

REFERENCES

1. Anisimova, N. I., cited by Gabrilovitch, N. in *Lysogeny.* Izdadelstvo "Bielorusj", Minsk, U.S.S.R. (1970).
2. Brenner, S., Streisinger, G., Horne, R. W., Champe, S. P., Barnett, L., Benzer, S., and Rees, M. W., *J. Mol. Biol., 1,* 281 (1959).
3. Buchwald, M., Steed-Glaister, P., and Siminovitch, L., *Virology, 42,* 375 (1970).
4. Dyson, R. D., and Van Holde, K. E., *Virology, 33,* 559 (1967).
5. Fitch, W. M., and Susman, M., *Virology, 26,* 754 (1965).
6. Fraser, D., and Jerrel, E. A., *J. Biol. Chem., 205,* 291 (1953).
7. Fraser, D., *J. Biol. Chem., 227,* 711 (1957).
8. Goldstein, D. A., Bendet, I. J., and Lauffer, M. A., *Virology, 32,* 601 (1967).
9. Moore, S., and Stein, W. H., *J. Biol. Chem., 192,* 663 (1951).
10. Naroditzky, B. S., Tikhonenko, T. I., and Klimenko, S. M., *Vop. Med. Khim., 15,* 476 (1969).
11. Polson, A., and Wyckoff, W. G., *Science (Washington), 108,* 501 (1948).
12. Sarkar, N., Sarkar, S., and Kozloff, L. M., *Biochemistry, 3,* 511 (1964).
13. Sarkar, N., Sarkar, S., and Kozloff, L. M., *Biochemistry, 3,* 517 (1964).
14. Saunders, G. F., and Campbell, L. L., *J. Bacteriol., 91,* 340 (1966).
15. Schito, G. C., Molina, A. M., and Pesce, A., *Biochem. Biophys. Res. Commun., 28,* 611 (1967).
16. Tikhonenko, T. I., Velikodvorskaya, G. A., and Zemtsova, E. V., *Biokhimiya, 27,* 726 (1962).
17. Villarejo, M., Hua, S., and Evans, E. A., *J. Virol., 1,* 928 (1967).

Bacterial Viruses Containing Single-Stranded Nucleic Acid

DR. R. L. WISEMAN, DR. J. M. BOWES AND DR. A. K. DUNKER

Three different classes of bacterial virus containing single-stranded nucleic acid have been identified: the RNA viruses (f2 type), the isometric DNA viruses (φX174 type), and the filamentous DNA viruses (fd type). Although nucleic acid isolated from the virion is a single-stranded polynucleotide chain, replication involves a double-stranded form. The following tables include only those viruses for which the size and type of nucleic acid are known.

Data on base composition of viral nucleic acids will be given in Volume II, now in preparation. Data about coat proteins of all small bacterial viruses are collected in Tables 1 and 2. All other data concerning small viruses are collected according to type of virus rather than type of data, since the different types of virus have little in common.

All of these viruses infect only Gram-negative bacteria. The adsorption site of the f2-type and fd-type viruses is the sex pilus, which is found only on bacteria harboring certain types of plasmid, such as the F-factor or I-factor. For these viruses, the host range depends on the presence or absence of the plasmid rather than on the strain of bacteria.

Although many different viruses of each type have been isolated, most viruses of a given type are very similar. The differences in measured parameters are probably more apparent than real. Therefore a line giving the physical properties of an "idealized virus" is included in Tables 3, 5, and 7.

All references are given at the end of this subchapter. Review articles for RNA viruses are References 31, 34, 47, 80, 88, and 97; for isometric DNA viruses, References 31, 64, 68, and 79; and for filamentous DNA viruses, References 31, 54, 64, and 68. A purification method applicable to all small viruses is described in Reference 96.

TABLE 1

AMINO ACID COMPOSITION OF COAT PROTEINS[a]

Amino Acid	RNA Virus[b] Coat Protein		RNA Virus[b] A-Protein[d]		Isometric DNA Virus[f] Capsid[g]		Isometric DNA Virus[f] Spikes	Filamentous DNA Virus fd B-Protein[h]		Filamentous DNA Virus Ifl B-Protein[i]	
	Mole %[c]	Number	Mole %[e]	Number	Mole %	Number	Mole %	Mole %	Number	Mole %	Number
Lys	4.6	6	4.6	16	4.6	20	5.9	9.9	5	8.9	4
His	0	0	1.4	5	3.0	13	1.3	0	0	0	0
Arg	3.1	4	7.2	25	6.9	31	3.4	0	0	2.0	1
Cys	1.6	2			0.9	4	1.0	0	0	0	0
Asp	2.3	3	8.6	30	10.5	47	12.7	6.3	3	6.1	3
Asn	8.5	11									
Met	1.6	1	1.5	5	2.3	10	2.4	1.8	1	1.5	1
Thr	7.0	9	6.2	22	7.7	34	8.6	5.9	3	7.4	4
Ser	10.1	13	8.8	30	7.0	31	7.4	7.9	4	8.5	4
Glu	3.9	5	8.5	30	9.5	42	11.6	6.3	3	6.1	3
Gln	4.6	6									
Pro	4.6	6	4.6	16	7.2	32	3.0	2.5	1	0	0
Gly	7.0	9	8.8	30	7.9	35	10.5	8.1	4	6.5	3
Ala	10.8	14	10.5	36	6.4	28	11.8	17.8	9	2.1	10
Val	10.8	14	8.1	28	4.2	19	6.7	8.0	4	1.2	6
Ile	6.2	8	4.3	15	5.2	23	4.6	8.0	4	2.1	1
Leu	5.4	8	6.0	32	8.0	35	6.4	4.6	2	7.9	4
Tyr	3.1	4	3.7	13	3.8	17	1.3	3.3	2	1.7	1
Phe	3.1	4	4.0	14	4.9	22	1.2	5.8	3	6.1	3
Trp	1.6	2						2.0	1	2.0	1
Total		129		347		443			49		49

[a] "Mole%" means the number of moles of the given amino acid per 100 moles of total amino acid. "Number" means the best value for an integral number of amino acids per protein molecule, based on the weight of the protein. For RNA virus coat protein and fd coat protein these values are based on sequence determination (see Table 2). A blank space indicates that the measurement was not taken; 0 means that the given amino acid was not detected.

[b] For R17.[81,91]

[c] Calculated in Reference 81 from sequence in Reference 90.

[d] Assumed values for cys and trp have not been included.

[e] Assuming a weight of 38,000 AMU.

[f] For ϕX174.[20] The compositions of ϕX174 and S13 total coat proteins are virtually identical.[63] Also see References 14 and 62.

[g] Without spikes.

[h] From Reference 1. Composition of the total proteins from fd, M13, fl, and ZJ/2 differs by only one or two amino acids. The virion contains, in addition to the B-protein (bulk coat protein), a small fraction of A-protein, a minor component required for adsorption.

[i] From Reference 92.

TABLE 2
AMINO ACID SEQUENCE OF COAT PROTEINS

RNA Virus

		1									10										20										30
fr[93-95]	H-	Ala-	Ser-	Asn-	Phe-	Glu-	Glu-	Phe-	Val-	Leu-	Val-	Asn-	Asp-	Gly-	Gly-	Thr-	Gly-	Asp-	Val-	Lys-	Val-	Ala-	Pro-	Ser-	Asn-	Phe-	Ala-	Asn-	Gly-	Val-	Ala-
f2[90]		–	–	–	–	Thr-	Gln	–	–	–	–	–	–	–	–	–	–	Asn	–	Thr	–	–	–	–	–	–	–	–	–	–	–
R17[91]		–	–	–	–	Thr-	Gln	–	–	–	–	–	–	–	–	–	–	Asn	–	Thr	–	–	–	–	–	–	–	–	–	–	–
R17′[91]		–	–	–	–	Thr-	Gln	–	–	–	–	–	–	–	–	–	–	Asn	–	Thr	–	–	–	–	–	–	–	–	–	–	–
MS2[89]		–	–	–	–	Thr-	Gln	–	–	–	–	–	–	–	–	–	–	Asn	–	Thr	–	–	–	–	–	–	–	–	–	–	–
Qβ[45]	H-	Ala-	Lys-	Leu-	Glu-	Thr-	Val-	Thr-	Leu-	Gly-	Asn-	Ile-	Gly-	Lys-	Asp-	Gly-	Lys-	Gln-	Thr-	Leu-	Val-	Leu-	Asp-	Pro-	Arg-	Gly-	Val-	Asn-	Pro-	Thr-	Asn-

Filamentous DNA Virus

											40										50										60
fd[1]	H-	Ala-	Glu-	Gly-	Asp-	Asp-	Pro-	Ala-	Lys-	Ala-	Ala-	Phe-	Asp-	Ser-	Leu-	Gln-	Ala-	Ser-	Ala-	Thr-	Glu-	Tyr-	Ile-	Gly-	Tyr-	Ala-	Trp-	Met-	Val-	Val-	Val-
fr		Glu-	Trp-	Ile-	Ser-	Ser-	Asn-	Ser-	Arg-	Ser-	Gln-	Ala-	Tyr-	Lys-	Val-	Thr-	Cys-	Ser-	Val-	Arg-	Gln-	Ser-	Ser-	Ala-	Asn-	Asn-	Arg-	Lys-	Tyr-	Thr-	Val-
f2		–	–	–	–	–	–	–	–	–	–	–	–	–	–	–	–	–	–	–	–	–	–	–	Gln	–	–	–	–	–	Ile
R17		–	–	–	–	–	–	–	–	–	–	–	–	–	–	–	–	–	–	–	–	–	–	–	Gln	–	–	–	–	–	Ile
R17′		–	–	–	–	–	–	–	–	–	–	–	–	–	–	–	–	–	–	–	–	–	–	–	Gln	–	–	–	–	–	Ile
MS2		–	–	–	–	–	–	–	–	–	–	–	–	–	–	–	–	–	–	–	–	–	–	–	Gln	–	–	–	–	–	Ile
Qβ		Gly-	Val-	Ala-	Ser-	Leu-	Ser-	Gln-	Ala-	Gly-	Ala-	Val-	Pro-	Ala-	Leu-	Glu-	Lys-	Arg-	Val-	Thr-	Val-	Ser-	Val-	Ser-	Gln-	Pro-	Arg-	Asn-	Arg-	Lys-	Asn

											70										80										90
fd		Ile-	Val-	Gly-	Ala-	Thr-	Ile-	Gly-	Ile-	Lys-	Leu-	Phe-	Lys-	Lys-	Phe-	Thr-	Ser-	Lys-	Ala-	Ser-	OH										
fr		Lys-	Val-	Glu-	Val-	Pro-	Lys-	Val-	Ala-	Thr-	Gln-	Val-	Gln-	Gly-	Gly-	Val-	Glu-	Leu-	Pro-	Val-	Ala-	Ala-	Trp-	Arg-	Ser-	Tyr-	Met-	Asn-	Leu-	Glu-	Leu
f2		–	–	–	–	–	–	–	–	–	–	Thr-	Val	–	–	–	–	–	–	–	–	–	–	–	–	–	Leu	–	–	–	–
R17		–	–	–	–	–	–	–	–	–	–	Thr-	Val	–	–	–	–	–	–	–	–	–	–	–	–	–	Leu	–	Met	–	–
R17′		–	–	–	–	–	–	–	–	–	–	Thr-	Val	–	–	–	–	–	–	–	–	–	–	–	–	–	Leu	–	Met	–	–
MS2		–	–	–	–	–	–	–	–	–	–	Thr-	Val	–	–	–	–	–	–	–	–	–	–	–	–	–	Leu	–	Met	–	–
Qβ		Tyr-	Lys-	Val-	Gln-	Val-	Lys-	Ile-	Gla-	Asn-	Pro-	Thr-	Ala-	Cys-	Thr-	Ala-	Asn-	Gly-	Ser-	Cys-	Asp-	Pro-	Ser-	Val-	Thr-	Arg-	Gln-	Ala-	Tyr-	Ala-	Asp

											100										110										120
fr		Thr-	Ile-	Pro-	Val-	Phe-	Ala-	Thr-	Asx-	Asp-	Asp-	Cys-	Ala-	Leu-	Ile-	Val-	Lys-	Ala-	Leu-	Gln-	Gly-	Thr-	Phe-	Lys-	Thr-	Gly-	Ile-	Ala-	Pro-	Asn-	Thr
f2		–	–	–	Ile	–	–	–	Asn-	Ser	–	–	Glu	–	–	–	–	–	Met	–	–	Leu-	Leu	–	Asp	–	Asn-	Pro-	Ile-	Pro-	Ser
R17		–	–	–	Ile	–	–	–	Asn-	Ser	–	–	Glu	–	–	–	–	–	Met	–	–	Leu-	Leu-	–	Asp	–	Asn-	Pro-	Ile-	Pro-	Ser
R17′		–	–	–	Ile	–	–	–	Asn-	Ser	–	–	Lys	–	–	–	–	–	Met	–	–	Leu-	Leu	–	Asp	–	Asn-	Pro-	Ile-	Pro-	Ser
MS2		–	–	–	Ile	–	–	–	Asn-	Ser	–	–	Lys	–	–	–	–	–	Met	–	–	Leu-	Leu	–	Asp	–	Asn-	Pro-	Ile-	Pro-	Ser
Qβ		Val-	Thr-	Phe-	Ser-	Phe-	Thr-	Gln-	Tyr-	Ser-	Thr-	Asp-	Glu-	Glu-	Arg-	Ala-	Phe-	Val-	Arg-	Thr-	Glu-	Leu-	Ala-	Ala-	Leu-	Leu-	Ala-	Ser-	Pro-	Leu-	Leu

									129				
fr		Ala-	Ile-	Ala-	Asn-	Ser-	Gly-	Ile-	Tyr-	OH			
f2		–	–	–	–	–	–	–	–				
R17		–	–	–	–	–	–	–	–				
R17′		–	–	–	–	–	–	–	–				
MS2		–	–	–	–	–	–	–	–				
Qβ		Ile-	Asp-	Ala-	Ile-	Asp-	Gln-	Leu-	Asn-	Pro-	Ala-	Tyr-	OH

TABLE 3

PHYSICAL PROPERTIES OF RNA VIRUS

Virus[a]	Diameter,[b] Å	Weight,[e] AMU x 10^{-6}	Nucleic Acid[h]		Buoyant Density,[j] g/cm³	$S_{20}w$,[l] Svedbergs	E_{260},[m] OD/mg/ml	E_{max}/E_{min}	References
			Weight, %	Weight,[e] AMU x 10^{-6}					
f2	210–230		31.5[i]						48
MS2	250–260	3.6, 3.9[f]		1.1	1.42,[k] 1.38–1.46	79, 81	7.8, 8.0	1.41,[n] 1.45	60, 83
R17	266[c]	3.6	30.5[i]	1.05		79–80	7.7	1.65[n,o]	21, 23, 27
M12	270	4.2[f]	31.7			79	8.2[n]		33
fr	220,[d] 263[c]	4.1[g], 4.4[f]		1.3[g]	1.46	79			52, 53, 99
β						75–90			59
FH5	250				1.44				36
f_{can1}						80			18
μ2	240		32–34			78			16
ZJ/1	255								6
Qβ	250	4.2, 4.3[f]			1.44	84	8.0	1.35[o]	60
7s	250								22
PP7	250								7
φCb23r	210–230								73
"Idealized"	260	3.6	31	1.1	1.44	80	8		

a All of the listed viruses have been isolated from *Escherichia coli* except for 7s and PP7 (*Pseudomonas aeruginosa*) and φCb23r (*Caulobacter*). The *Escherichia coli* viruses f2, MS2, R17, M12, fr, β, and FH5 form a serological group,[75] distinct from Qβ.[60]

b From electron microscopy, unless otherwise stated.

c From small-angle X-ray diffraction.

d For a dehydrated sphere weighing 4.1×10^6 AMU, a diameter of 210 Å was calculated.

e From light scattering, unless otherwise stated.

f From sedimentation-diffusion.

g From sedimentation-viscosity.

h From the nitrogen–phosphorus ratio, unless otherwise stated.

i From the weight of isolated RNA divided by the weight of the virion.

j From CsCl equilibrium gradients, unless otherwise stated.

k From sedimentation rates in D_2O/H_2O mixtures.

l Sedimentation coefficient in a solvent with the same density and viscosity as water at 20°C.

m Absorbance at 260 nm of a 1-mg/ml solution, with a 1-cm path length. Not corrected for light scattering, unless otherwise stated.

n Corrected for light scattering.

o Estimated from published spectra.

TABLE 4

GENE PRODUCTS OF RNA VIRUSES

Gene Designation[a]	Probable Gene Function	Weight of Gene Product,[b] AMU	Intracellular Proteins		Proteins per Virion[g]
			Weight%[e]	Number[f]	
A	Maturation factor	38,000	11–12	4	1
B	Coat (control of RNA synthesis)	13,729[c]	78–81	80	180
C	RNA-polymerase	60,000[d]	8–10	1	0

[a] These designations are for R17. The gene order is 5′-A-B-C-3′.[39]

[b] From SDS-gel electrophoresis on R17, unless otherwise stated.[13,40,44,57,81,82]

[c] From amino acid sequence.

[d] For MS2. Pierre Spahr found the same value for R17 polymerase (unpublished).

[e] For MS2, in bacteria treated with actinomycin D to inhibit host protein synthesis.[57] See Reference 80 for further discussion.

[f] Relative number, from the weight of the gene product and the weight% intracellular proteins, normalized to 1 for gene C.

[g] Idealized number, assuming that the capsid is constructed according to a T = 3 design[15,34,48,80,81] with one A-protein replacing a group of three B-proteins (A. K. Dunker, unpublished).

TABLE 5

PHYSICAL PROPERTIES OF ISOMETRIC DNA VIRUS

Virus[a]	Diameter,[b] Å	Weight, AMU x 10^{-6}	Nucleic Acid Weight %	Nucleic Acid Weight, AMU x 10^{-6}	Buoyant Density, g/cm^3	$S_{20,w}$,[g] Svedbergs	E_{260},[h] OD/mg/ml	E_{260}/E_{280}	References
φX174	225–275	6.2[c]	25.5	1.7[c]	1.40[e] 1.407[f]	114	8.1	1.53	17, 29, 50 71, 77, 78, 87
S13	250								50
φR	300								41, 42
ST1	260	6.8[d]				120.5[d]			5
α3	270								8
"Idealized"	275	6.4	25.5	1.6					10

[a] The host for these viruses is *Escherichia coli* C, except for ST1, which grows on *E. coli* K12.
[b] From electron microscopy.
[c] From light scattering.
[d] From sedimentation velocity relative to φX174.
[e] CsCl.
[f] RbCl.
[g] Sedimentation coefficient in a solvent with the same density and viscosity as water at 20°C.
[h] Absorbance at 260 nm of a 1-mg/ml solution, with a 1-cm path length.

TABLE 6

GENE PRODUCTS OF ISOMETRIC DNA VIRUS

Gene Designation[a]	Previous Gene Designations: φX174 (Sinsheimer)	φX174 (Hayashi)	S13 (Tessman)	Probable Gene Function[b]	Weight of Gene Product,[c] AMU	Intracellular Proteins, Molar Ratios[d]	Proteins per Virion[e]
A	VI	C	IV	RF-DNA synthesis			
B	IV	B	II	Spike(?)	5,000(?)	~0.1	60
C	VIII	H(?)	VI(?)	SS-DNA synthesis(?)			
D	V	D	VII	SS-DNA synthesis(?)	15,000	2	
E	I	G	V	Lysis	13,500	1.2	
F	VII	E	I	Capsid	48,000	1.0	60
G	III	F	IIIa	Spike	19,000	0.9	60
H	II	A	IIIb	Spike	40,000	0.2	12
I		I		Unknown			

[a] New gene designations are discussed in Reference 3. The genetic map is circular, and the order is A-B-C-D-E-F-G-H;[3] the location of gene I is not known.

[b] From References 3, 25, 37, 38, 76, 85, and 86.

[c] From SDS-gel electrophoresis. The correlation of intracellular proteins with gene products is not certain. See References 4, 10, 11, 26, 28, and 55 for further discussion.

[d] From Reference 4, normalized to 1 for gene F.

[e] Proposed in Reference 10. Also see References 28 and 55. The 5000-AMU protein in the virion may not be the gene B product.

TABLE 7

PHYSICAL PROPERTIES OF FILAMENTOUS VIRUS

Virus[a]	Length,[b] Å	Diameter,[b] Å	Weight, AMU x 10^{-6}	Nucleic acid: Weight, %	Nucleic acid: Weight AMU x 10^{-6}	Buoyant Density,[d] g/cm^3	$S_{20,w}$,[e] Svedbergs	E_{260},[f] OD/mg/ml	E_{max}/E_{min}[g]	References
fd	7,600–8,700	50, 56[c]	11.3	12.2	2.3	1.29	40	3.74	1.39	9,24,32,43, 51,52,53
f1	8,500	43–63	11.0	11.3		1.28	45	3.67	1.39	70,98
M13	8,500–9,000	60			2.0	1.29				56,64,69,72
ZJ/2	8,300	68								6
Ec9	6,000–9,000	50		14.8			39		1.31	12,19,56
AE2	8,000	50		14.0			43	3.80		61
HR	8,000	60				1.31				35
δA	8,300	60	17.6	10.0	1.7	1.30	46	2.60	1.38	58
Xf	9,770	84				1.27			1.23	46
"Idealized[h] Ff"	8,700	55	18.8	11	2	1.3	41	3.7	1.4	
If1	13,000	55[c]	24.0	11.9	3	1.30	45	3.19	1.29	56,92
If2	13,000									56
Pf1	14,000			12.0						84
"Idealized[h] If"	13,000	55	28.2	11	3	1.3	45	3.2	1.3	

[a] All viruses in the upper part of the table, except Xf, are known to be members of Ff species (they adsorb to F-pili). All viruses in the lower part, except Pfl, are known to be members of the If species (they adsorb to I-pili).

[b] From electron microscopy, unless otherwise stated.

[c] From X-ray diffraction (minimum lattice dimensions).

[d] In CsCl equilibrium gradients.

[e] Sedimentation coefficient in a solvent with the same density and viscosity as water at 20°C.

[f] Absorbance at 260 nm of a 1-mg/ml solution, with a 1-cm path length.

[g] E_{max} is between 260 nm and 270 nm; E_{min} is between 244 nm and 245 nm.

[h] Assuming six protein molecules weighing 5169 AMU every 16.1 Å along the virion.[54]

TABLE 8

GENE PRODUCTS OF FILAMENTOUS VIRUS

Gene Designation[a]	Probable Gene Function[b]	Weight of Gene Product,[c] AMU	Proteins per Virion
1	Unknown		
2	RF-DNA synthesis		
3	A-protein	70,000	4[e,f]
4	Unknown		
5	SS-DNA synthesis	8,000	
6	Unknown		
7	Unknown		
8	B-protein	5,169[d]	3,200[f]

[a] The genetic map is circular, and the order is 2-7-5-8-3-6-1-4.[49]

[b] From References 65-67, and 74.

[c] From SDS-gel electrophoresis, unless otherwise stated.[30]

[d] From amino acid sequence.[1]

[e] From References 2, 30, 66, and 92.

[f] Assuming that the virion weighs 18.8×10^6 AMU.

APPENDIX

Besides the above-mentioned cubic and filamentous phages, some cubic or rod-shaped phages with particular features have been described.

1. *Pseudomonas* phage PM2
   Icosahedron measuring ~ 600 Å in diameter, with a double-layered lipid-containing envelope ~ 70–100 Å thick,[100,104,106] molecular weight 58 (51–120) x 10^6 daltons; contains 10.5–16% lipids, of which 75–90% are phospholipids;[100,104] ~ 75% protein.[104] DNA: 10–14%,[100,104] double-stranded, circular,[100,102] molecular weight 6–6.9 x 10^6 daltons,[101,102] 3.02 μ in length.[102]
2. *Bdellovibrio* phage HDC-1
   Cubic, 600–700 Å in diameter; is claimed to have two coats and single-stranded DNA.[105]
3. *Mycoplasma* phage MV-L1
   Rod measuring 146 x 898 Å; contains DNA.[103]

REFERENCES

1. Asbeck, F., Beyreuther, K., Köhler, H., von Wettstein, G., and Braunitzer, G., *Hoppe-Seyler's Z. Physiol. Chem.*, *350*, 1047 (1969).
2. Beaudoin, J., *Ph.D. Thesis.* University of Wisconsin, Madison, Wisconsin (1970).
3. Benbow, R. M., Hutchison, C. A., Fabricant, J. D., and Sinsheimer, R. L., *J. Virol.*, *7*, 549 (1971).

4. Benbow, R. M., Mayol, R. F., Picchi, J. C., and Sinsheimer, R. L., *J. Virol., 10,* 99 (1972).
5. Bowes, J. M., *Ph.D. Thesis.* University of California, Davis, California (1971).
6. Bradley, D. E., *J. Gen. Microbiol., 35,* 471 (1964).
7. Bradley, D. E., *J. Gen. Microbiol., 45,* 83 (1966).
8. Bradley, D. E., Dewar, C. A., and Robertson, D., *J. Gen. Virol., 5,* 113 (1969).
9. Bujard, H., *J. Mol. Biol., 49,* 125 (1969).
10. Burgess, A. B., *Proc. Nat. Acad. Sci. U.S.A., 64,* 613 (1969).
11. Burgess, A. B., and Denhardt, D. T., *J. Mol. Biol., 44,* 377 (1969).
12. Calendi, E., Dettori, R., and Neri, M. G., *G. Microbiol., 14,* 227 (1966).
13. Capecchi, M. R., *J. Mol. Biol., 21,* 173 (1966).
14. Carusi, E. A., and Sinsheimer, R. L., *J. Mol. Biol., 7,* 388 (1963).
15. Caspar, D. L. D., and Klug, A., in *Cold Spring Harbor Symp. Quant. Biol., 27,* 1 (1962).
16. Ceppellini, M., Dettori, R., and Poole, F., *G. Microbiol., 11,* 9 (1963).
17. Daems, W. Th., Eigner, J., van der Sluys, I., and Cohen, J. A., *Biochim. Biophys. Acta, 55,* 801 (1962).
18. Davern, C. I., *Aust. J. Biol. Sci., 17,* 719 (1964).
19. Dettori, R., and Neri, M. G., *G. Microbiol., 13,* 111 (1965).
20. Edgell, M. H., Hutchison, C. A., and Sinsheimer, R. L., *J. Mol. Biol., 42,* 547 (1969).
21. Enger, M. D., Stubbs, E. A., Mitra, S., and Kaesberg, P., *Proc. Nat. Acad. Sci. U.S.A., 49,* 857 (1963).
22. Feary, T. W.,Fisher, E., and Fisher, T. N., *J. Bacteriol., 87,* 196 (1964).
23. Fischbach, F. A., Harrison, P. M., and Anderegg, J. W., *J. Mol. Biol., 13,* 638 (1965).
24. Frank, H., and Day, L. A., *Virology, 42,* 144 (1970).
25. Funk, F. D., and Sinsheimer, R. L., *J. Virol., 6,* 12 (1970).
26. Gelfand, D. H., and Hayashi, M., *J. Mol. Biol., 44,* 501 (1969).
27. Gesteland, R. F., and Boedtker, H., *J. Mol. Biol., 8,* 496 (1964).
28. Godson, G. N., *J. Mol. Biol., 57,* 541 (1971).
29. Hall, C. E., Maclean, E. C., and Tessman, I., *J. Mol. Biol., 1,* 192 (1959).
30. Henry, T. J., and Pratt, D., *Proc. Nat. Acad. Sci. U.S.A., 62,* 800 (1969).
31. Hoffmann-Berling, H., Kaerner, H. C., and Knippers, R., *Advan. Virus Res., 12,* 329 (1966).
32. Hoffmann-Berling, H., Marvin, D. A., and Dürwald, H., *Z. Naturforsch., 18b,* 876 (1963).
33. Hofschneider, P. H., *Z. Naturforsch., 18b,* 203 (1963).
34. Hohn, T., and Hohn, B., *Advan. Virus Res., 16,* 43 (1970).
35. Hsu, Y.-C., *Bacteriol. Rev., 32,* 387 (1968).
36. Huppert, J., Ryter, A., and Fouace, J., in *Viruses, Nucleic Acids and Cancer,* p. 68. Williams and Wilkins, Baltimore, Maryland (1963).
37. Jeng, Y., Gelfand, D., Hayashi, M., Shleser, R., and Tessman, E. S., *J. Mol. Biol., 49,* 521 (1970).
38. Jeng, Y. C., and Hayashi, M., *Virology, 40,* 406 (1970).
39. Jeppesen, P. G. N., Steitz, J. A., Gesteland, R. F., and Spahr, P. F., *Nature, 226,* 230 (1970).
40. Kamen, R., *Nature, 228,* 527 (1970).
41. Kay, D., *J. Gen. Microbiol., 27,* 201 (1962).
42. Kay, D., and Bradley, D. E., *J. Gen. Microbiol., 27,* 195 (1962).
43. Knippers, R., and Hoffmann-Berling, H., *J. Mol. Biol., 21,* 293 (1966).
44. Kondo, M., Gallerani, R., and Weissmann, C., *Nature, 228,* 525 (1970).
45. Konigsberg, W., Maita, T., Katz, J., and Weber, K., *Nature, 227,* 271 (1970).
46. Kuo, T.-T., Huang, T.-C., and Chow, T.-Y., *Virology, 39,* 548 (1969).
47. Lodish, H. F., *Progr. Biophys. Mol. Biol., 18,* 285 (1968).
48. Loeb, T., and Zinder, N. D., *Proc. Nat. Acad. Sci. U.S.A., 47,* 282 (1961).
49. Lyons, L. B., and Zinder, N. D., *Virology, 49,* 45 (1972).
50. Maclean, E. C., and Hall, C. E., *J. Mol. Biol., 4,* 173 (1962).
51. Marvin, D. A., *J. Mol. Biol., 15,* 8 (1966).
52. Marvin, D. A., and Hoffmann-Berling, H., *Nature, 197,* 517 (1963).
53. Marvin, D. A., and Hoffmann-Berling, H., *Z. Naturforsch., 18b,* 884 (1963).
54. Marvin, D. A., and Hohn, B., *Bacteriol. Rev., 33,* 172 (1969).
55. Mayol, R. F., and Sinsheimer, R. L., *J. Virol., 6,* 310 (1970).
56. Meynell, G. G., and Lawn, A. M., *Nature, 217,* 1184 (1968).
57. Nathans, D., Oeschger, M. P., Polmar, S. K., and Eggen, K., *J. Mol. Biol., 39,* 279 (1969).
58. Nishihara, T., and Watanabe, I., *Virus, 17,* 118 (1967).
59. Nonoyama, M., Yuki, A., and Ikeda, Y., *J. Gen. Appl. Microbiol., 9,* 299 (1963).
60. Overby, L. R., Barlow, G. H., Doi, R. H., Jacob, M., and Spiegelman, S., *J. Bacteriol., 91,* 442 (1966).
61. Panter, R. A., and Symons, R. H., *Aust. J. Biol. Sci., 19,* 565 (1966).
62. Poljak, R. J., *Virology, 35,* 185 (1968).
63. Poljak, R. J., and Suruda, A. J., *Virology, 39,* 145 (1969).
64. Pratt, D., *Annu. Rev. Genet., 3,* 343 (1969).

65. Pratt, D., and Erdahl, W. S., *J. Mol. Biol., 37,* 181 (1968).
66. Pratt, D., Tzagoloff, H., and Beaudoin, J., *Virology, 39,* 42 (1969).
67. Pratt, D., Tzagoloff, H., Erdahl, W. S., and Henry, T. J., in *The Molecular Biology of Viruses,* p. 219, J. S. Colter and W. Paranchych, Eds. Academic Press, New York (1967).
68. Ray, D. S., in *Molecular Basis of Virology,* p. 222, H. Fraenkel-Conrat, Ed. Reinhold Book Corporation, New York (1968).
69. Ray, D. S., Preuss, A., and Hofschneider, P. H., *J. Mol. Biol., 21,* 485 (1966).
70. Rossomando, E. F., and Zinder, N. D., *J. Mol. Biol., 36,* 387 (1968).
71. Rueckert, R. R., Zillig, W., and Huber, K., *Virology, 17,* 204 (1962).
72. Salivar, W. O., Tzagoloff, H., and Pratt, D., *Virology, 24,* 359 (1964).
73. Schmidt, J. M., and Stanier, R. Y., *J. Gen. Microbiol., 39,* 95 (1965).
74. Schwedes, A., *Ph.D. Thesis.* University of Heidelberg, Heidelberg, Germany (1969).
75. Scott, D. W., *Virology, 26,* 85 (1965).
76. Siegel, J. E. D., and Hayashi, M., *J. Virol., 4,* 400 (1969).
77. Sinsheimer, R. L., *J. Mol. Biol., 1,* 37 (1959).
78. Sinsheimer, R. L., *J. Mol. Biol., 1,* 43 (1959).
79. Sinsheimer, R. L., *Progr. Nucl. Acid Res. Mol. Biol., 8,* 115 (1968).
80. Stavis, R. L., and August, J. T., *Annu. Rev. Biochem., 39,* 527 (1970).
81. Steitz, J. A., *J. Mol. Biol., 33,* 937 (1968).
82. Strauss, E. G., and Kaesberg, P., *Virology, 42,* 437 (1970).
83. Strauss, J. H., and Sinsheimer, R. L., *J. Mol. Biol., 7,* 43 (1963).
84. Takeya, K., and Amako, K., *Virology, 28,* 163 (1966).
85. Tessman, E. S., *Virology, 25,* 303 (1965).
86. Tessman, E. S., in *The Molecular Biology of Viruses,* p. 193, J. S. Colter and W. Paranchych, Eds. Academic Press, New York (1967).
87. Tromans, W. J., and Horne, R. W., *Virology, 15,* 1 (1961).
88. Valentine, R. C., Ward, R., and Strand, M., *Advan. Virus Res., 15,* 1 (1969).
89. Wallis, M., and Naughton, M. A., in *Atlas of Protein Sequence and Structure,* p. 283, M. O. Dayhoff and R. V. Eck, Eds. National Biomedical Research Foundation, Silver Spring, Maryland (1967-1968).
90. Weber, K., *Biochemistry,6,* 3144 (1967).
91. Weber, K., and Konigsberg, W., *J. Biol. Chem., 242,* 3563 (1967).
92. Wiseman, R. L., Dunker, A. K., and Marvin, D. A., *Virology, 48,* 230 (1972).
93. Wittmann, H. G., and Wittmann-Liebold, B., *Cold Spring Harbor Symp. Quant. Biol., 31,* 163 (1966).
94. Wittmann-Liebold, B., *Z. Naturforsch., 21b,* 1249 (1966).
95. Wittmann-Liebold, B., and Wittmann, H. G., *Mol. Gen. Genet., 100,* 358 (1967).
96. Yamamoto, K. R., Alberts, B. M., Benzinger, R., Lawhorne, L., and Treiber, G., *Virology, 40,* 734 (1970).
97. Zinder, N. D., *Annu. Rev. Microbiol., 19,* 455 (1965).
98. Zinder, N. D., Valentine, R. C., Roger, M., and Stoeckenius, W., *Virology, 20,* 638 (1963).
99. Zipper, P., Kratky, O., Herrmann, R., and Hohn, T., *Eur. J. Biochem., 18,* 1 (1971).

Appendix

100. Espejo, R. T., and Canelo, E. S., *Virology, 34,* 738 (1968).
101. Espejo, R. T., and Canelo, E. S., *Virology, 37,* 495 (1969).
102. Espejo, R. T., Canelo, E. S., and Sinsheimer, R. L., *Proc. Nat. Acad. Sci. U.S.A., 63,* 1164 (1969).
103. Gourlay, R. N., Bruce, J., and Garwes, D. J., *Nature New Biol., 229,* 118 (1971).
104. Harrison, S. C., Caspar, D. L. D., Camerini-Otero, R. D., and Franklin, R. M., *Nature New Biol., 229,* 197 (1971).
105. Hashimoto, T., Diedrich, D. L., and Conti, S. F., *J. Virol., 5,* 97 (1970).
106. Silbert, J. A., Salditt, M., and Franklin, R. M., *Virology, 39,* 666 (1969).

Particulate Bacteriocins

DR. H.-W. ACKERMANN AND G. BROCHU

Particulate bacteriocins are phage-like particles or phage components of complex and definite structure that are bacteriocidal and non-infectious. They have been called "lethal phages"[24] and are characterized by high molecular weight and visibility in the electron microscope.[10] However, some "low-molecular-weight bacteriocins" have recently been studied by electron microscopy and were shown to be spherical particles of 80 to 240 Å in diameter.[4,15,19]

Except for the "killer-particles" found in *Bacillus,* there is no morphological difference between particulate bacteriocins and phages or phage tails without known biological activity. Both are produced spontaneously or upon induction and occur in a wide range of bacteria (see References 10, 20, and 59). There may even be serological relationships between bacteriocidal phage tails and temperate phages.[29,36,62] Particulate bacteriocins should therefore be called what they are: defective phages[20,59] with biological activity. The term "bacteriocin" should apply to agents of low molecular weight and without complex structure.

Moreover, there may be no distinction between particulate bacteriocins and phage "ghosts", or even infectious phages, provided that there is no suitable host. *Staphylococcus aureus,* for example, produces a temperate phage that may multiply in some strains and kill others.[13,14] Similar observations have been made on the temperate phage 2845 of *Bacterium anitratum,* which has an elongated head and a long, non-contractile tail.[2,12] In addition, some strains of *Bacterium anitratum* produce morphologically intact particles of similar shape, which apparently behave as bacteriocins. They have not been isolated and may be infectious phages.[12]

Particulate bacteriocins may be classified in three main groups:

1. Tailed particles that have full or empty heads and resemble usual phages. The particles so far observed belong to the basic morphological types A1 and B1.[1] They have isometric heads and long tails that are either contractile (A1) or not (B1).
2. "Killer particles" with characteristic small heads and long, contractile tails (see p.576.Morphology of Bacteriophages, Figure 4). Their heads contain bacterial DNA.[23,49]
3. Phage tails that are either contractile or not.

Empty phage heads found in *Streptococcus* have been associated with bacteriocidal activity, but this could result from low-molecular-weight enzymes attached to large particles.[40]

The following table comprises particulate bacteriocins whose bacteriocidal activity has been proven, or is at least probable, and which have been studied by electron microscopy. However, shadowed particles have been omitted. Since the "killer particles" of *Bacillus* or the particles released by *Escherichia coli*[15] might be identical,[20] they are listed together, as well as the contractile phage tails of *Proteus, Pseudomonas,* and *Vibrio.* These phage tails appear to have similar morphology, but biological differences have not been demonstrated or, as in some pyocins, are very slight.[29,37] Dimensions, if given, are therefore averages. Names are listed alphabetically, but it should be noted that many particles have not been named by the original investigators.

CODE

NN	no name	-	absent
+	present	blank space:	absence of data

Host	Name	Morphology	Dimensions, Å			Base Plate	Spikes	Fibers	References
			Head	Tail					
			Diameter	Length	Width				
Bacillus	GA-2, No. 1M, PBLB, PBSH, PBSX, phi, SPα, type 1, μ, NN[a]	Type A1	400	2030	180	+	+	+	6,7,8,10,16, 22,27,30,34,44, 49,51,52,54,60, 61
Chromobacterium	Type A[b]	Contractile tail	–	2185	160	+	–	+	3
	Type B[b]	Contractile tail	–	1440	165	+	–	+	3
Clostridium	NN	Contractile tail	–	1200–1500	120–150	+	–	+	33
Enterobacteriaceae									
Aerobacter	NN	Contractile tail[d]	–						10
Escherichia	Colicin D	Type B1	300						10
	Colicin H	Type A1	650	1800		+			11
	Colicin 15, phage 15,ϕ15A, Ψ	Type A1	610	1130	160	+	+	+	17,18,31,43,45, 46,47,50,63
Proteus	45, NN[a]	Contractile tail	–	1500	225	+	+	+	53,57,58,59
Yersinia	NN	Contractile tail	–	750–800			+		48,62
Lactobacillus	NN[c]	Type A1[e,f]				+		+	41
	NN[c]	Type B1				+		+ or –	41
	NN[c]	Type B1				–		+	41
Listeria	Monocin	Type A1[f]	1100	2400	200	+	+	+	11
Mycobacterium	NN	Type B1							32
Pseudomonas	A1$_{mc}$, C10, Götze, "pyocin", R, R$_{mc}$, R$_{sp}$, R2, R3, R4	Contractile tail	–	1110	150	+	+	+	5,9,10,21,25,26, 28,35,37,64
	28	Non-contractile tail	–	1000–1200 (up to 4000)	90	–	–	+	5,55,56
Streptococcus	NN	Isometric heads[e]	500	–	–	–	–	–	40
Vibrio	Vibriocin, NN	Contractile tail	–	1100	230				38,39,42

[a] Several descriptions. [b] One or both particles may be bacteriocidal. [c] Number of particles not indicated. [d] Presence of numerous empty and full elongated heads.
[e] Heads often empty. [f] Heads often absent.

REFERENCES

1. Ackermann, H.-W., *Pathol. Biol., 17,* 1003 (1969).
2. Ackermann, H.-W., and Brochu, G., *J. Microsc. (Paris),* (in press).
3. Ackermann, H.-W., and Gauvreau, L., *Zentralbl. Bakteriol. Parasitenk. Infektionskr. Hyg. Abt. Orig.,* (in press).
4. Amako, K., Tokiwa, H., and Takeya, K., *Jap. J. Microbiol., 14,* 505 (1970).
5. Amako, J., Yasunaka, K., and Takeya, K., *J. Gen. Microbiol., 62,* 107 (1970).
6. Azizbekyan, R. R., Belyaeva, N. N., Krivisky, A. S., and Tikhonenko, A. S., *Mikrobiologiya, 35,* 279 (1966).
7. Boyce, L., Eiserling, F. A., and Romig, W. R., *Biochem. Biophys. Res. Commun., 34,* 398 (1969).
8. Bradley, D. E., *J. Gen. Microbiol., 41,* 233 (1965).
9. Bradley, D. E., in *Electron Microscopy 1966, 6th International Congress for Electron Microscopy, Kyoto,* Vol. 2, p. 115, R. Uyeda, Ed. Maruzen Co. Ltd., Tokyo, Japan (1966).
10. Bradley, D. E., *Bacteriol. Rev., 31,* 230 (1967).
11. Bradley, D. E., and Dewar, C. A., *J. Gen. Microbiol., 45,* 399 (1966).
12. Brochu, G., *Ph.D. Thesis.* Université Laval, Québec, P. Q., Canada (in preparation).
13. Dobardzic, R., Payment, P., and Sonea, S., *Can. J. Microbiol., 17,* 847 (1971).
14. Dobardzic, R., Sonea, S., Côté, J. R., and Rohr, R., *Rev. Can. Biol., 30,* 79 (1971).
15. Durner, K., *Z. Allg. Mikrobiol., 10,* 93 (1970).
16. Eiserling, F. A., and Romig, W. R., *Bacteriol. Proc.,* p. 118 (1964).
17. Endo, H., Ayabe, K., Amako, K., and Takeya, K., *Virology, 25,* 469 (1965).
18. Frampton, E. W., and Brinkley, B. R., *J. Bacteriol., 90,* 446 (1965).
19. Gagliano, V. J., and Hinsdell, R. D., *J. Bacteriol., 104,* 117 (1970).
20. Garro, A. J., and Marmur, J., *J. Cell. Physiol., 76,* 253 (1970).
21. Garyaev, P. P., and Poglazov, B. F., *Biokhimiya, 33,* 585 (1969).
22. Haas, M., and Yoshikawa, H., *J. Virol., 3,* 233 (1969).
23. Haas, M., and Yoshikawa, H., *J. Virol., 3,* 248 (1969).
24. Hamon, Y., and Péron, Y., *Zentralbl. Bakteriol. Parasitenk. Infektionskr. Hyg. Abt. Orig., 206,* 439 (1968).
25. Higerd, T. B., Baechler, C. A., and Berk, R. S., *J. Bacteriol., 93,* 1976 (1967).
26. Higerd, T. B., Baechler, C. A., and Berk, R. S., *J. Bacteriol., 98,* 1378 (1969).
27. Hirokawa, H., and Kadlubar, F., *J. Virol., 3,* 205 (1969).
28. Homma, J. Y., Goto, S., and Shionoya, H., *Jap. J. Exp. Med., 37,* 373 (1967).
29. Homma, J. Y., and Shionoya, H., *Jap. J. Exp. Med., 37,* 395 (1967).
30. Huang, W. M., and Marmur, J., *J. Virol., 5,* 237 (1970).
31. Ikeda, H., Inuzuka, M., and Tomizawa, J., *J. Mol. Biol., 50,* 457 (1970).
32. Imaeda, T., and Rieber, M., *J. Bacteriol., 68,* 557 (1968).
33. Inoue, K., and Iida, H., *J. Virol., 2,* 537 (1968).
34. Ionesco, H., Ryter, A., and Schaeffer, P., *Ann. Inst. Pasteur (Paris), 107,* 764 (1964).
35. Ishii, S. I., Nishi, Y., and Egami, F., *J. Mol. Biol., 13,* 428 (1965).
36. Ito, S., and Kageyama, M., *J. Gen. Appl. Microbiol., 16,* 231 (1970).
37. Ito, S., Kageyama, M., and Egami, F., *J. Gen. Appl. Microbiol., 16,* 205 (1970).
38. Jayawardene, A., and Farkas-Himsley, H., *Nature (London), 219,* 79 (1968).
39. Jayawardene, A., and Farkas-Himsley, H., *Microbios, 1B,* 87 (1969).
40. Keogh, B. P., and Shimmin, P. D., *J. Dairy Res., 36,* 87 (1969).
41. de Klerk, H. C., and Hugo, N., *J. Gen. Virol., 8,* 231 (1970).
42. Lang, D., McDonald, T. O., and Gardner, E. W., *J. Bacteriol., 95,* 708 (1968).
43. Lee, C. S., Davis, R. W., and Davidson, N., *J. Mol. Biol., 48,* 1 (1970).
44. May, P., May, E., Granboulan, E., Granboulan, N., and Marmur, J., *Ann. Inst. Pasteur (Paris), 115,* 1029 (1968).
45. Medoff, G., and Swartz, M. N., *J. Gen. Virol., 4,* 15 (1969).
46. Mennigmann, H. D., *J. Gen. Microbiol., 41,* 151 (1965).
47. Mennigmann, H. D., *Zentralbl. Bakteriol. Parasitenk. Infektionskr. Hyg. Abt. Orig., 196,* 207 (1965).
48. Nicolle, P., Mollaret, H., Hamon, Y., and Vieu, J. F., *Ann. Inst. Pasteur (Paris), 112,* 86 (1967).
49. Okamoto, K., Mudd, J. A., Mangan, J., Huang, W. M., Subbaiah, T. V., and Marmur, J., *J. Mol. Biol., 34,* 413 (1968).
50. Sandoval, H. K., Reilly, H. C., and Tandler, B., *Nature (London), 205,* 522 (1965).
51. Seaman, E., Tarmy, E., and Marmur, J., *Biochemistry, 3,* 607 (1964).
52. Shimizu, N., Miura, K., and Aoki, H., *J. Biochem. (Tokyo), 68,* 265 (1970).
53. Smit, J. A., Hugo, N., and de Klerk, H. C., *J. Gen. Virol., 5,* 33 (1969).
54. Stickler, D. J., Tucker, R. G., and Kay, D., *Virology, 26,* 142 (1965).
55. Takeya, K., Minamishima, Y., and Amako, K., in *Electron Microscopy 1966, 6th International Congress for Electron Microscopy, Kyoto,* Vol. 2, p. 135, R. Uyeda, Ed. Maruzen Co. Ltd., Tokyo, Japan (1966).
56. Takeya, K., Minamishima, Y., Ohnishi, Y., and Amako, K., *J. Gen. Virol., 4,* 145 (1969).
57. Taubeneck, U., *Z. Naturforsch. Teil B, 18,* 989 (1963).
58. Taubeneck, U., *Biol. Zentralbl. (Suppl.), 86,* 45 (1967).

59. Taubeneck, U., *Z. Allg. Mikrobiol., 9,* 315 (1969).
60. Tikhonenko, A. S., *Ultrastruktura virusov bakterii,* ref. 50. Izdadelstvo "Nauka", Moscow, U.S.S.R. (1968).
61. Tikhonenko, A. S., and Bespalova, I. A., *Mikrobiologiya, 30,* 867 (1961).
62. Vieu, J. F., Croissant, O., and Hamon, Y., *C. R. Hebd. Séances Acad. Sci., Ser. D , Sci. Natur. (Paris), 264,* 181 (1967).
63. Yudelevich, A., and Gold, M., *J. Mol. Biol., 40,* 77 (1969).
64. Yui, C., *J. Biochem. (Tokyo), 69,* 101 (1971).

Phage Typing

DR. S. S. KASATIYA AND DR. P. NICOLLE

Isolation of a specific lytic agent was first reported by Twort.[1] Felix Hubert D'Hérelle coined the term "bacteriophagum",[2] and later "bacteriophage",[3] after having noticed that the agent was a living parasite of bacteria. He divided the bacterial species into two groups based on the lytic action of a bacteriophage.[4] As early as 1934, Marcuse[5] used five phages to divide *Salmonella typhi* into different phage types and reported that the strains isolated from the same patient at different time intervals belonged to the same phage type. Since then, many workers have isolated phages specific for different microbial species.

The microbial species are classified into various types according to their morphological, biological and biochemical characters. According to the host range of more or less specific phages, biotypes or serotypes of a given species may further be subdivided into reasonable phage types. This subdivision is called phage typing. In an efficient phage-typing scheme, the microbial strains isolated from patients, carriers, contaminated food and vectors during an outbreak all belong to the same phage type.

Phages may be virulent, temperate, or adapted. Phage-typing schemes employ either virulent or temperate (*Staphylococcus aureus*), adapted (*Salmonella typhi*), or a combination (*Salmonella paratyphi* B, *Salmonella typhimurium*) of these phages. Virulent phages are normally isolated from sewage, but they may be obtained from feces or carrier microbial strains. They always lyse the microorganism they infect and produce clear lytic plaques easy to observe. Temperate phages, though very host-specific, are isolated from lysogenic bacteria, which carry them in intracellular form as prophage. These lysogenic microorganisms are potential producers of phage or phages as a stable heritable character; spontaneous lysis results from the transition of prophage to vegetative state under certain circumstances, but the mechanism is not fully known. However, the lysogenic cells are immune to lytic infection by the same or related phage or phages. Adapted phages were described by Craigie and Yen,[6] who, on the basis of the constant mutation rate, justified the presence of mutants in a phage population. By selectively propagating any one of these mutants on certain strains of *Salmonella typhi,* they obtained different new phage preparations that showed high affinity to the new host and suggested that the behavior of the adapted phage is conditioned by the strain on which it had last been propagated. The newly adapted phage is immunologically indistinguishable from its unadapted parent but shows marked difference in host range. The authors proposed that these phage preparations, if used at the highest dilution giving confluent lysis with the homologous strain – known as routine test dilution (RTD) – and thus eliminating all but the dominant mutant, made it possible to identify *Salmonella typhi* strains similar to the ones on which they were last propagated.

For phage typing, the fresh culture to be examined is spread on the surface of a nutrient agar plate and the phages are spotted in standard amounts, usually those of the RTD.[7] The lytic reactions are read after six to eighteen hours incubation at 37°C. In order to avoid the risk of accidental contamination of the original phages with other bacteriophages, propagation and distribution of typing phages for *Salmonella typhi* and *Salmonella paratyphi* B is done by the International Centre for Phage Typing, Colindale, London, England.

THE IMPORTANCE OF PHAGE TYPING

Phage typing is a very useful epidemiological tool for tracing the origin of infection in epidemics by further subdividing the biotypes or serotypes of microorganisms. It establishes a relationship between the sick and the carriers in an epidemic or endemic focus due to the same phage type. It also indicates the distribution of various types of a pathogenic species throughout the world. It is very useful for investigating outbreaks of food poisoning by staphylococci, *Salmonella,* etc., and serves to establish a correlation between the strains isolated from the patients, contaminated food and the source of infection. Species-specific phages play an important role in taxonomy of bacterial species, such as phages of *Bacillus anthracis, Pasteurella pestis* and *Brucella,* which are employed in the identification of their specific hosts. Thus they may also be employed in the purification of cultures. Phage typing can also be used for the identification of structural or antigenic characters of bacteria, due to the specific affinity of some bacteriophages to certain bacterial structures, e.g., smooth and rough strains, Vi antigen of *Salmonella typhi,* or flagella of *Escherichia coli.* Phage typing is employed to characterize bacterial strains in various fundamental and applied research programs.

There is considerable literature on the isolation and epidemiological applications of phage typing of *Salmonella typhi, Salmonella paratyphi* B, *Salmonella typhimurium, Staphylococcus aureus,* and some other bacteria, but we have listed in the following table only the references that are of historical importance, that are landmarks in phage-typing schemes, or that are the latest publications covering the information on the phages or phage-typing schemes described earlier. The abbreviation (sp.) in the column giving the number of phage types indicates that the phages have been employed for species identification within a genus that shows that phage typing is feasible.

In addition, the most important phage-typing schemes, widely used throughout the world for their epidemiological value, have been reproduced in the appendix. In these tables the degrees of lysis are represented by various symbols; e.g., CL = confluent lysis, SCL = semiconfluent lysis, <SCL and <CL = intermediate degrees of lysis, OL = confluent "opaque" lysis, and ± to +++ = increasing numbers of discrete plaques. Different symbols are used for *Staphylococcus aureus* phage typing; e.g., 5 = ++ reaction in the same dilution as on the propagating strain, 4 = ++ reaction in a dilution 10 to 10^2 times more concentrated than that giving ++ on the propagating strain, 3 = ++ reaction in a dilution 10^3 to 10^4 times more concentrated than that giving ++ on the propagating strain, 2 = ++ reaction in a dilution 10^5 to 10^6 times more concentrated that that giving ++ on the propagating strain, and 1 = very weak lysis. High-titer phages may "inhibit" the growth of many of the strains when used undiluted, but produce no discrete plaques when diluted. In some cases this inhibition may simulate confluent lysis, but generally it appears as a thinning of the growth in the drop area. Such reactions are recorded as 0.

GUIDE TO LITERATURE ON PHAGE TYPING

Microorganisms	Number of Phages Isolated or Selected	Sources and Types of Phages	Number of Phage Types	Reference
Phage-Typing Schemes With International Agreement				
Salmonella				
paratyphi B	12	Feces; adapted	48	8
	10	Feces; adapted	10	9
	6	Temperate, adapted	9	10
typhi	72	Adapted (various authors)	72	11
	11	Adapted	11	6
	87	Adapted (various authors)	87	12
Staphylococcus				
aureus	27	Temperate, adapted	16	13
	18	Temperate, adapted	21	14
Phage-Typing Schemes Without International Agreement				
Aeromonas				
salmonicida	13	Sewage, water; temperate	13	15
Agrobacterium				
radiobacter	7	Sewage, soil; temperate	4	16
	19	Sewage, soil	4	17
Asticcacaulis	3	Sewage, pond water	2 (sp)	18
Bacillus				
pumilus	2	Not given	3	19
Brucella				
abortus }	3	Soil; adapted	3 (sp)	20
melitensis }	3	Temperate	4 (sp)	21
suis }	15	Manure; temperate	3 (sp)	22
	1	Soil	3 (sp)	23

Caulobacter				
bacteroides	9			
crescentus	6	Sewage, soil, pond water	6 (sp)	18
fusiformis	1			
vibrioides	4			
Corynebacterium				
anaerobic	6	Temperate	11	24
diphtheriae	19	Temperate, adapted	9	25
	3	Temperate	4	26
	24	Temperate	19	27
	8	Temperate, adapted	9	28
Escherichia				
coli				
026:B6		Feces; temperate	5	29
055:B5	28	Feces; temperate	9	29
0111:B4		Feces; temperate	11	29
0119:B14	10	Sewage	8	30
0124:K72 (B17)	12	Temperate	11	31
0127:B8	9	Sewage	9	32
bovine	24	Sewage, feces	57	33
urinary	13	Sewage	109	34
Klebsiella				
rhinoscleromatis	15	Sewage, stools	12	35
gastrointestinal	12	Feces	13	36
respiratory	15	Sewage, stools	15	37
Listeria				
monocytogenes	5	Temperate	8	38

GUIDE TO LITERATURE ON PHAGE TYPING (Continued)

Microorganisms	Number of Phages Isolated or Selected	Sources and Types of Phages	Number of Phage Types	Reference
Phage-Typing Schemes Without International Agreement (continued)				
Mycobacterium sp.	6	Soil, sewage	5 (sp)	39
	6	Soil, stools; adapted	6 (sp)	40
tuberculosis	8	Soil; adapted (various authors)	2	41
	5	Soil; adapted (various authors)	3	42
	4	Soil	4	43
	3	Soil; adapted (various authors)	2	44
fast-growing	7	Soil; temperate	7 (sp)	45
	4	Soil; manure	3 (sp)	46
saprophytes	14	(Various authors)	9 (sp)	47
slow-growing	17	(Various authors)	3 (sp)	48
Neisseria				
meningitidis	5	Body fluids, nasopharynx	18	49
Proteus				
hauseri	20	Sewage; temperate	10	50
	15	Sewage, water; temperate	14	51
	12	Sewage, water; temperate	10	52
mirabilis OXK *vulgaris* OX19 OX2 OXL	52	Sewage; temperate	5 (sp)	53

Pseudomonas				
aeruginosa	21	Water; temperate	6	54
	15	Sewage, water; temperate	unknown	55
	13	Temperate	64	56
	12	Sewage; temperate	12	57
	13	ATCC phage; temperate	11	58
Rhizobium	28	Sewage, soil	31	59
Ristella				
pseudo-insolita	9	Sewage; temperate	30	60
	6	Adapted	24	60
Salmonella				
adelaide	6	Temperate	6	61
blockley	3	Temperate	14	62
bovis-morbificans	10	Temperate	4	63
braenderup	3	Temperate	15	64
dublin	6	Temperate	11	65
enteritidis	6	Sewage, feces	8	66
gallinarum	2	Sewage, feces, dung water	2	67
heidelberg	8	Feces, intestinal contents	22	68
minnesota R forms	5	(Various authors)	7	69
oranienburg	14	Temperate	5	70
panama	8	Surface water; adapted	8	71
paratyphi A	4	Sewage; adapted	4	72
paratyphi B	6	Temperate, adapted; complementary to Felix and Callow, 1951	9	10

GUIDE TO LITERATURE ON PHAGE TYPING (Continued)

Microorganisms	Number of Phages Isolated or Selected	Sources and Types of Phages	Number of Phage Types	Reference
Phage-Typing Schemes Without International Agreement (continued)				
Salmonella (continued)				
pullorum	4	Sewage, feces, dung water	4	67
thompson	7	Temperate	11	73
typhimurium	31	Temperate, adapted of Callow (1959)	90	74
	29	Temperate, adapted	34	75
	12	Temperate, adapted	12	76
	12	Sewage, feces; temperate, adapted	24	77
	6	Canal water; temperate, adapted	9	78
	32	Surface water; adapted	32	79
	6	Sewage; temperate	11	80
	20	Canal water; adapted	26	81
Shigella				
flexneri				
different serotypes	12	Temperate	40	82
3a	14	Temperate	9	83
sonnei				
R forms, phase II	12	Sewage, feces, manure; temperate	68	84
	3	Temperate	3	85
	15	3 from feces, 12 from Hammarström (1949)	38	86
S forms			12	87
R forms, phases I and II	10	Temperate	20	87
Staphylococcus				
epidermidis	18	Temperate, adapted	7	88

Streptococcus				
bovis *durans* *faecalis zymogenes* var. liquefaciens *faecium*	30	Monkey feces, water	30	89
cremoris	10	Cheese; temperate	12	90
faecalis	3	Sewage, monkey feces, nasopharyngeal washing; temperate	4	91
hemolytic	4	Sludge, feces	8	92
lactis	12 24	Cheese; temperate (Various authors)	16 24	90 93
Vibrio				
cholerae	4	Stools, water; temperate	5	94
cholerae *El-Tor* NAG (non-agglutinable)	8	Stools, water; temperate	8	95
Xanthomonas				
malvacearum	14	Infected leaves	2	96
Yersinia				
enterocolitica	10 10	Temperate Temperate	10 5	97 98

APPENDIX

REACTIONS OF *SALMONELLA TYPHIMURIUM* TYPE STRAINS WITH ROUTINE TEST DILUTIONS OF THE NEW TYPING PHAGES

Type Strains		Phages in Routine Test Dilutions													
Old	**New**	**1**	**2**	**3**	**4**	**5**	**6**	**7**	**8**	**9**	**10**	**11**	**12**	**13**	**14**
1	1	**CL**	CL	CL	CL	CL	CL	CL	–	–	SCL	SCL	CL	CL	CL
1a	2	–	**CL**	CL	CL	CL	CL	–	–	±	CL	CL	CL	CL	CL
1a var. 1	3	–	+++	**CL**	CL	CL	CL	–	–	+	CL	SCL	–	–	CL
1b	4	–	++	–	**CL**	CL	CL	–	–	±	CL	SCL	+++	CL	+++
3	5	–	–	–	+++	**CL**	+++	–	–	–	–	++	–	–	+++
3a	6	–	++	–	++	±	**CL**	–	–	–	+++	–	–	–	–
2d	7	–	–	–	–	–	–	**CL**	±	–	–	–	–	–	–
4	8	–	–	–	–	–	–	–	**CL**	CL	CL	CL	–	–	–
2	9	–	–	–	–	–	–	–	CL	**CL**	CL	CL	CL	CL	–
	10	–	–	–	–	–	–	–	–	±	**CL**	SCL	CL	CL	+++
	11	–	–	–	–	–	–	–	–	–	–	**SCL**	SCL	CL	–
	12	–	–	–	–	–	–	–	–	–	–	–	**CL**	CL	–
	12a	–	–	–	–	–	–	–	–	–	–	–	**CL**	CL	–
2a	13	–	–	–	–	–	–	–	–	–	–	SCL	–	**CL**	OL
	14	–	–	–	–	–	–	–	–	–	–	SCL	++	SCL	**OL**
2c	15	–	–	–	–	–	–	–	–	–	–	–	+++	SCL	–
	15a	–	–	–	–	–	–	–	–	–	–	++	CL	CL	–
2b	16	–	–	–	–	–	–	–	–	–	–	+++	–	–	OL
	17	–	–	–	–	–	–	–	–	–	–	–	–	–	–
	18	–	–	–	–	–	–	–	–	–	–	–	–	–	–
	19	–	–	–	–	–	–	–	–	–	–	–	–	–	–
2d	20	–	–	–	–	–	–	–	–	–	–	–	–	–	–
	20a	–	–	–	–	–	–	–	–	–	–	–	–	–	–
Untypable	21	–	–	–	–	–	–	–	–	–	–	–	–	–	–
Untypable	22	–	–	–	–	–	–	–	–	–	–	–	–	–	–
Untypable	23	–	–	–	–	–	–	–	–	–	–	–	–	–	–
Untypable	24	–	–	+++	–	–	–	–	–	–	–	–	–	–	CL
Untypable	25	–	–	–	–	–	–	–	–	–	–	–	–	–	–
Untypable	26	–	–	–	–	–	–	–	–	–	–	–	–	–	–
Untypable	27	–	–	–	–	–	–	–	–	–	–	–	–	–	–
2c	28	–	–	–	–	–	–	–	+++	–	–	–	+++	SCL	–
Untypable	29	–	–	–	–	–	–	–	–	–	–	–	–	–	–
Untypable	30	–	–	–	–	–	–	–	SCL	–	–	±	–	–	–
Untypable	31	–	–	–	–	–	–	–	–	–	–	–	–	–	–

Old	**New**	**15**	**16**	**17**	**18**	**19**	**20**	**21**	**22**	**23**	**24**	**25**	**26**	**27**	**28**	**29**
1	1	CL	CL	CL	CL	CL	CL	CL	CL	SCL	CL	SCL	CL	CL	±	CL
1*a*	2	CL	CL	CL	–	CL	CL	CL	CL	CL	CL	CL	CL	CL	++	CL
1*a* var. 1	3	CL	CL	CL	–	CL	CL	SCL	SCL	CL	CL	+++	SCL	CL	SCL	CL
1*b*	4	CL	CL	++	–	CL	SCL	±	CL	CL	–	CL	+++	CL	+++	CL
3	5	+++	+++	–	–	++	±	–	SCL	–	–	–	–	–	–	–
3*a*	6	–	+++	+++	–	–	–	+++	+++	+++	–	–	–	SCL	++	CL
2*d*	7	–	–	–	CL	±	SCL	–	–	–	–	–	–	–	–	–
4	8	–	+++	–	–	–	++	±	CL	CL	–	±	++	–	±	CL
	9	–	SCL	–	–	–	++	–	CL	CL	–	±	+++	–	–	CL
	10	–	SCL	++	–	–	++	++	CL	CL	–	±	++	++	–	CL
2	11	–	–	OL	CL	SCL	OL	–	–	–	–	–	–	–	–	–
	12	–	–	+++	–	–	–	–	–	–	–	–	–	–	–	–
	12*a*	–	–	–	CL	–	–	+++	–	–	–	–	–	OL	–	–
2*a*	13	–	++	–	+++	–	–	+++	–	–	–	–	OL	OL	++	OL
	14	–	–	–	–	–	–	–	–	–	–	–	+++	+++	–	OL
2*c*	15	**SCL**	–	OL	OL	OL	OL	–	–	–	–	–	–	++	+	–
	15*a*	**OL**	–	OL	–	OL	OL	–	–	–	–	–	–	++	++	–
	16	–	**OL**	+++	+++	–	–	–	±	–	–	SCL	+++	OL	±	OL
	17	+++	–	**OL**	+++	+++	SCL	–	–	–	–	–	–	+	++	–
2*b*	18	SCL	–	OL	**CL**	SCL	SCL	–	–	–	–	–	–	++	++	–
	19	SCL	–	–	–	**SCL**	SCL	–	–	–	–	–	–	+++	+++	–
	20	–	–	–	–	–	**OL**	–	–	–	–	–	–	–	–	–
2*d*	20*a*	–	–	–	CL	±	**SCL**	–	–	–	–	–	–	–	–	–
Untypable	21	–	–	–	–	–	–	**OL**	++	–	–	–	–	–	–	–
Untypable	22	–	–	–	–	–	–	±	**OL**	–	–	–	–	–	–	–
Untypable	23	–	–	–	–	–	–	–	–	**OL**	CL	+++	+++	–	–	++
Untypable	24	–	–	–	–	–	–	–	–	–	**CL**	–	SCL	–	–	–
Untypable	25	–	++	–	–	–	–	–	–	±	+++	**OL**	OL	–	–	+++
Untypable	26	–	–	–	–	–	–	–	–	–	+++	+++	**OL**	–	–	–
Untypable	27	–	–	–	–	–	–	+++	–	–	–	–	–	**OL**	–	–
2*c*	28	SCL	–	SCL	SCL	SCL	SCL	–	–	–	–	–	–	±	**OL**	–
Untypable	29	–	–	–	–	–	–	–	–	–	–	–	–	–	–	**OL**
Untypable	30	–	–	–	–	–	–	–	–	–	–	–	–	–	–	–
Untypable	31	–	–	–	–	–	–	–	–	–	–	SCL	+	–	–	+++

Taken from Callow, B. R., *J. Hyg.*, *57*, 352 (1959). Reproduced by permission of Cambridge University Press, New York.

REACTIONS OF THE Vi-TYPE STRAINS OF *SALMONELLA TYPHI* WITH THE TYPING PHAGES IN ROUTINE TEST DILUTIONS[a]

Vi-Type Strains	Adapted Vi-Phage Preparations A	B1	B2	B3	C1	C2	C3	C4	C5	C6	C7	C8
A	CL	CL	CL	CL	CL	CL	CL	CL	CL	CL	CL	CL
B1	±	CL	-	+++	-	-	+	+++	-	+	-	-
B2	+++	-	CL	++	+++	-	-	-	-	-	-	-
B3	-	-	-	CL	-	-	-	-	-	-	-	-
C1	+	+++	++	+++	CL	CL	CL	CL	CL	CL	CL	CL
C2	-	-	-	-	-	CL	-	-	-	-	-	-
C3	-	-	-	-	+	CL	CL	+	-	+	±	+
C4	-	-	-	-	-	-	-	CL	-	-	-	-
C5	-	-	-	-	-	-	-	-	CL	-	-	-
C6	-	-	-	-	-	-	-	-	-	CL	-	-
C7	-	-	-	-	+	-	-	+++	+	+++	CL	+++
C8	-	-	-	-	-	-	-	-	-	-	-	CL
C9	-	-	-	±	+++	-	-	-	-	-	-	-
D1	-	±	-	-	±	CL	-	-	-	-	-	-
D2	-	-	-	-	-	-	-	-	-	-	-	-
D4	-	-	-	-	-	-	-	-	-	-	-	-
D5	-	±	-	-	±	CL	-	-	-	-	-	-
D6	-	-	-	-	-	SCL	+++	-	-	-	-	-
D7	-	-	-	-	-	-	-	-	-	-	-	-
D8	-	-	-	-	-	-	-	-	-	-	-	-
D9	-	-	-	-	-	-	-	-	-	-	-	-
D10	-	-	-	-	-	-	-	-	-	-	-	-
D11	-	-	-	-	-	-	-	-	-	-	-	-
E1	-	+	-	-	±	-	-	-	-	-	-	-
E2	-	-	-	-	-	-	-	-	-	-	-	-
E3	-	-	-	-	-	-	-	-	-	-	-	-
E4	-	-	-	-	-	-	-	-	-	-	-	-
E5	-	-	-	-	-	-	-	-	-	-	-	-
E6	-	-	-	-	-	-	-	-	-	-	-	-
E7	-	-	-	-	-	-	-	-	-	-	-	-
E8	-	-	-	-	-	-	-	-	-	-	-	-
E9	-	-	-	-	-	+	-	-	-	-	-	-
E10	-	-	-	-	-	-	-	-	-	-	-	-
F1	-	-	-	-	-	-	-	-	-	-	-	-
F2	-	-	-	-	-	-	-	-	-	-	-	-
F3	-	-	-	-	-	-	-	-	-	-	-	-
F4	-	-	-	-	-	-	-	-	-	-	-	-
F5	-	-	-	-	-	-	-	-	-	-	-	-
G	-	-	-	-	±	-	-	-	-	-	-	-
H	-	+	+	+	+	-	+	+	+	-	-	-
J1	-	-	-	-	-	-	-	-	-	-	-	-
J2	-	-	-	-	-	-	-	-	-	-	-	-
J3	-	-	-	-	-	-	-	-	-	-	-	-
K1	-	-	-	-	-	-	-	-	-	-	-	-
K2	-	-	-	-	-	-	-	-	-	-	-	-
L1	-	-	-	-	-	-	-	-	-	-	-	-
L2	-	-	-	-	-	-	-	-	-	-	-	-
M1	-	-	-	-	-	-	-	-	-	-	-	-
M2	-	-	-	-	-	-	-	-	-	-	-	-
M3	-	-	-	-	-	-	-	-	-	-	-	-
N	-	-	-	-	-	-	-	-	-	-	-	-
O	-	-	-	+++	-	-	-	+++	-	-	-	-
T	-	+++	-	-	-	-	+++	++	-	-	-	-
25	-	-	-	-	-	-	-	-	-	-	-	-
26	-	-	-	-	-	-	-	-	-	-	-	CL
27	-	-	-	-	-	-	-	-	-	-	-	-
28	-	-	-	-	-	-	-	-	-	-	-	-
29	-	+	+++	+++	±	CL	CL	+	-	-	-	+
32	-	±	±	±	+	±	±	±	±	±	±	±
34	-	-	-	-	-	-	-	-	-	-	-	-
35	-	+++	-	-	-	-	-	-	-	-	-	-
36	-	-	-	-	-	-	-	-	-	-	-	-
37	-	-	-	-	-	-	-	-	-	-	-	-
38	-	-	-	-	-	-	-	-	-	-	-	-
39	-	±	-	-	-	-	-	-	-	-	-	-
40	-	-	-	-	-	-	-	-	-	-	-	-
41	-	+++	++	+++	+++	+++	+++	+++	+++	++	++	+++
42	-	++	-	SCL	++	-	+++	CL	-	SCL	SCL	CL
43	-	-	-	++	-	-	-	+++	-	-	-	-
44	-	-	-	-	-	-	-	-	-	-	-	-
45	-	-	-	-	-	-	-	CL	-	-	-	-
46	-	-	-	-	-	-	-	-	-	-	-	-

[a] Homologous and group reactions are shown in bold type.

REACTIONS OF THE Vi-TYPE STRAINS OF *SALMONELLA TYPHI* WITH THE TYPING PHAGES IN ROUTINE TEST DILUTIONS[a] (Continued)

Vi-Type Strains	Adapted Vi-Phage Preparations											
	C9	D1	D2	D4	D5	D6	D7	D8	D9	D10	D11	E1
A	CL	CL	CL	CL	CL	CL	CL	CL	CL	CL	CL	CL
B1	+++	++	-	-	-	-	-	-	-	-	-	+
B2	+++	++	++	+++	+++	-	+++	++	+	+++	-	+++
B3	CL	+++	++	-	-	-	-	SCL	-	SCL	-	-
C1	**CL**	+++	-	+	++	++	+++	+	±	+	-	++
C2	+++	-	-	-	-	++	-	-	-	-	-	-
C3	+++	-	-	-	-	+++	-	-	-	-	-	-
C4	-	-	-	-	-	-	-	-	-	-	-	-
C5	-	-	-	-	-	-	-	-	-	-	-	-
C6	-	-	-	-	-	-	-	-	-	-	-	-
C7	±	-	-	-	-	-	-	-	-	-	-	-
C8	-	-	-	-	-	-	-	-	-	-	-	-
C9	**CL**	-	-	-	-	-	-	-	-	-	-	-
D1	++	**CL**	**CL**	**CL**	**CL**	**CL**	**CL**	**CL**	**CL**	**CL**	**CL**	-
D2	-	-	**CL**	±	±	-	-	-	-	-	-	-
D4	-	-	-	**CL**	-	-	++	+	+	+	-	-
D5	-	±	-	-	**CL**	**CL**	-	**CL**	**CL**	-	-	±
D6	-	+	-	-	-	**CL**	-	-	-	-	+++	-
D7	-	-	-	-	-	-	**CL**	-	-	**SCL**	-	-
D8	-	-	-	-	-	-	-	**CL**	-	**SCL**	-	-
D9	-	-	-	+++	-	-	**CL**	**SCL**	**CL**	**CL**	-	-
D10	-	-	-	-	-	-	-	-	-	**CL**	-	-
D11	-	-	-	-	-	-	-	-	-	-	**CL**	-
E1	±	-	-	-	-	-	-	-	-	-	-	**CL**
E2	-	-	-	-	-	-	-	-	-	-	-	-
E3	-	-	-	-	-	-	-	-	-	-	-	-
E4	-	-	-	-	-	-	-	-	-	-	-	-
E5	-	-	-	-	-	-	-	-	-	-	-	-
E6	-	-	-	-	-	-	-	-	-	-	-	-
E7	-	-	-	-	-	-	-	-	-	-	-	-
E8	-	-	-	-	-	-	-	-	-	-	-	-
E9	-	-	-	-	-	++	-	-	-	-	±	-
E10	-	-	-	-	-	-	-	-	-	-	-	-
F1	-	-	-	-	-	-	-	-	-	-	-	-
F2	-	-	-	-	-	-	-	-	-	-	-	-
F3	-	-	-	-	-	-	-	-	-	-	-	-
F4	-	-	-	-	-	-	-	-	-	-	-	-
F5	-	-	-	-	-	-	-	-	-	-	-	-
G	-	-	-	-	-	-	-	-	-	-	-	-
H	+	-	-	-	-	-	±	-	-	±	-	±
J1	-	-	-	-	-	-	-	-	-	-	-	-
J2	-	-	-	-	-	-	-	-	-	-	-	-
J3	-	-	-	-	-	-	-	-	-	-	-	-
K1	-	-	-	-	-	-	-	-	-	-	-	-
K2	-	-	-	-	-	-	-	-	-	-	-	-
L1	-	-	-	-	-	-	-	-	-	-	-	-
L2	-	-	-	-	-	-	-	-	-	-	-	-
M1	-	-	-	-	-	-	-	-	-	-	-	-
M2	-	-	-	-	-	-	-	-	-	-	-	-
M3	-	-	-	-	-	-	-	-	-	-	-	-
N	+++	-	-	-	-	-	-	+++	-	-	-	-
O	+++	-	-	-	-	-	-	-	-	-	-	-
T	-	+++	-	±	±	+++	±	-	-	-	-	-
25	-	-	-	-	-	-	-	-	-	-	-	-
26	-	-	-	-	-	-	-	-	-	-	-	-
27	-	-	-	-	-	-	-	-	-	-	-	-
28	-	-	-	-	-	-	-	-	-	-	-	-
29	+++	±	-	-	-	CL	-	-	-	-	CL	-
32	±	-	±	-	-	-	-	-	-	-	-	±
34	-	-	-	-	-	-	-	-	-	-	-	-
35	-	-	-	+	-	-	-	++	-	-	-	-
36	-	-	-	-	-	-	-	-	-	-	-	-
37	-	-	-	-	-	-	-	-	-	-	-	-
38	-	-	-	-	-	-	-	-	-	-	-	-
39	-	±	-	-	-	-	±	-	-	-	-	±
40	-	-	-	-	-	-	-	-	-	-	-	-
41	++	+++	+++	+++	+++	+++	+++	+++	+++	+	+	+++
42	CL	-	-	-	-	-	CL	-	-	-	CL	+++
43	+++	-	-	-	-	-	-	-	-	-	-	-
44	-	-	-	-	-	-	-	-	-	-	-	-
45	+++	-	-	-	-	-	-	-	-	-	-	-
46	-	-	-	-	-	-	-	-	-	-	-	-

a Homologous and group reactions are shown in bold type.

REACTIONS OF THE Vi-TYPE STRAINS OF *SALMONELLA TYPHI* WITH THE TYPING PHAGES IN ROUTINE TEST DILUTIONS[a] (Continued)

Vi-Type Strains	Adapted Vi-Phage Preparations											
	E2	E3	E4	E5	E6	E7	E8	E9	E10	F1	F2	F3
A	CL	CL	CL	CL	CL	CL	CL	CL	CL	CL	CL	CL
B1	-	-	-	+	++	+++	++	-	+++	+++	+++	-
B2	-	+++	CL	+++	+++	+++	+++	-	SCL	+++	-	+++
B3	-	++	+++	-	-	-	-	-	+++	-	-	-
C1	-	±	++	±	±	+	±	-	++	++	+++	+
C2	-	-	-	-	-	-	-	+	-	-	-	-
C3	-	-	-	-	-	+++	-	+++	-	-	+++	-
C4	-	-	-	-	-	-	-	-	-	-	-	-
C5	-	-	-	-	-	-	-	-	-	-	-	-
C6	-	-	-	-	-	-	-	-	-	-	-	-
C7	-	-	-	-	-	-	-	-	-	-	-	-
C8	-	-	-	-	-	-	±	-	-	-	-	-
C9	-	-	-	-	-	-	-	-	-	-	-	-
D1	-	SCL	CL	-	-	±	-	CL	±	-	±	-
D2	-	-	+	-	-	-	-	-	-	-	-	-
D4	-	-	+	-	-	-	-	-	-	-	-	-
D5	-	SCL	++	-	-	-	-	CL	±	-	-	-
D6	-	-	-	-	-	+++	-	+++	-	-	+++	-
D7	-	-	-	-	-	-	-	-	-	-	-	-
D8	-	-	-	-	-	-	-	-	-	-	-	-
D9	-	-	-	-	-	-	-	-	-	-	-	-
D10	-	-	-	-	-	-	-	-	-	-	-	-
D11	-	-	-	-	-	-	-	-	-	-	-	-
E1	**CL**	**CL**	**CL**	**CL**	**CL**	**CL**	**CL**	**CL**	**CL**	±	-	-
E2	**CL**	-	-	-	-	-	-	-	-	-	-	-
E3	-	**CL**	**SCL**	-	-	-	-	**SCL**	-	-	-	-
E4	-	-	**CL**	-	-	-	-	-	-	-	-	-
E5	-	-	-	**CL**	-	-	-	-	-	-	-	-
E6	-	-	-	-	**CL**	-	-	-	**++**	-	-	-
E7	-	-	-	-	-	**CL**	-	**CL**	**+++**	-	-	-
E8	-	-	-	-	-	-	**CL**	-	-	-	-	-
E9	-	-	-	-	-	-	-	**CL**	-	-	-	-
E10	-	-	-	-	-	-	-	-	**CL**	-	-	-
F1	-	-	-	-	-	-	-	-	-	**CL**	**CL**	**CL**
F2	-	-	-	-	-	-	-	-	-	**±**	**CL**	-
F3	-	-	-	-	-	-	-	-	-	-	-	**CL**
F4	-	-	-	-	-	-	-	-	-	**+++**	**++**	-
F5	-	-	-	-	-	-	-	-	-	-	**±**	**±**
G	-	-	-	-	-	±	-	-	-	-	-	-
H	-	-	+	±	±	+	-	-	+	±	+	+
J1	-	-	-	-	-	-	-	-	-	-	-	-
J2	-	-	-	-	-	-	-	-	-	-	-	-
J3	-	-	-	-	-	-	-	-	-	-	-	-
K1	-	-	-	-	-	-	-	-	-	-	-	-
K2	-	-	-	-	-	-	-	-	-	-	-	-
L1	-	-	-	-	-	-	-	-	-	-	-	-
L2	-	-	-	-	-	-	-	-	-	-	-	-
M1	-	-	-	-	-	-	-	-	-	-	-	-
M2	-	-	-	-	-	-	-	-	-	-	-	-
M3	-	-	-	-	-	-	-	-	-	-	-	-
N	-	-	-	-	-	-	-	-	-	-	-	-
O	-	-	-	-	-	+++	-	-	++	-	-	-
T	-	-	-	-	-	+++	-	+++	CL	-	+++	-
25	-	-	-	-	-	-	-	-	-	-	-	-
26	-	-	-	-	-	-	CL	-	-	-	-	-
27	-	-	-	-	-	-	-	-	-	-	-	-
28	-	-	-	-	-	-	-	-	SCL	-	-	-
29	-	-	-	-	-	CL	-	CL	++	+	CL	-
32	-	-	±	CL	-	±	-	-	±	-	±	±
34	-	-	-	-	-	-	-	-	-	-	-	-
35	-	-	-	-	-	-	-	-	-	-	++	-
36	-	-	-	-	-	-	-	-	-	-	-	-
37	-	-	-	-	-	-	-	-	-	-	-	-
38	-	-	-	-	-	-	-	-	-	-	-	-
39	-	-	-	-	-	-	-	-	±	-	±	±
40	-	-	-	-	-	-	-	-	-	-	-	-
41	-	+	+++	+	+	+++	+	+++	++	++	+++	+++
42	-	-	-	++	-	CL	+++	-	+++	+++	SCL	CL
43	-	-	-	-	-	-	-	-	-	-	-	-
44	-	-	-	-	-	-	-	-	-	-	-	-
45	-	-	-	-	-	+++	-	-	-	-	-	-
46	-	-	-	-	-	-	-	-	-	-	-	-

[a] Homologous and group reactions are shown in bold type.

REACTIONS OF THE Vi-TYPE STRAINS OF *SALMONELLA TYPHI* WITH THE TYPING PHAGES IN ROUTINE TEST DILUTIONS[a] (Continued)

Vi-Type Strains	Adapted Vi-Phage Preparations											
	F4	F5	G	H	J1	J2	J3	K1	K2	L1	L2	M1
A	CL	CL	CL	CL	CL	CL	CL	CL	CL	CL	CL	CL
B1	+++	+++	+++	+++	-	+	-	-	+++	±	-	+++
B2	+++	+++	+++	+++	SCL	-	SCL	++	-	+++	+++	+++
B3	-	-	-	-	SCL	+++	SCL	+++	±	SCL	SCL	CL
C1	-	+++	++	+	++	+++	+	-	++	++	-	++
C2	-	-	-	-	-	-	-	-	±	-	-	-
C3	-	-	-	-	±	±	-	-	++	±	-	±
C4	-	-	-	-	-	-	-	-	-	-	-	-
C5	-	-	-	-	-	-	-	-	-	-	-	-
C6	-	-	-	-	-	-	-	-	-	-	-	-
C7	-	-	-	-	-	++	+	-	±	-	-	++
C8	-	-	-	-	-	-	-	-	-	-	-	-
C9	-	-	-	-	±	-	-	-	-	-	-	±
D1	-	±	+	-	-	+++	-	-	SCL	-	-	-
D2	-	-	-	-	-	-	-	-	-	-	-	-
D4	-	-	-	-	-	-	-	-	±	-	-	-
D5	-	-	-	-	-	±	-	-	+++	-	-	-
D6	-	-	-	-	-	-	-	-	+++	-	-	+
D7	-	-	-	-	-	-	-	-	-	-	-	-
D8	-	-	-	-	-	-	-	-	-	-	-	-
D9	-	-	-	-	-	-	-	-	-	-	-	-
D10	-	-	-	-	-	-	-	-	-	-	-	-
D11	-	-	-	-	-	-	-	-	-	-	-	-
E1	+	±	-	-	-	-	-	-	-	-	-	-
E2	-	-	-	-	-	-	-	-	-	-	-	-
E3	-	-	-	-	-	-	-	-	-	-	-	-
E4	-	-	-	-	-	-	-	-	-	-	-	-
E5	-	-	-	-	-	-	-	-	-	-	-	-
E6	-	-	-	-	-	-	-	-	-	-	-	-
E7	-	-	-	-	-	-	-	-	±	-	-	-
E8	-	-	-	-	-	-	-	-	-	-	-	-
E9	-	-	-	-	-	-	-	-	-	-	-	-
E10	-	-	-	-	-	-	-	-	-	-	-	-
F1	**CL**	**CL**	-	-	-	-	-	-	-	-	-	-
F2	-	-	-	-	-	-	-	-	-	-	-	-
F3	-	-	-	-	-	-	-	-	-	-	-	-
F4	**CL**	**CL**	-	-	-	-	-	-	-	-	-	-
F5	-	**CL**	-	-	-	-	-	-	-	-	-	-
G	-	-	**CL**	-	-	-	-	-	-	-	-	-
H	±	++	+	CL	±	+	±	±	±	±	±	+
J1	-	-	-	-	**CL**	**CL**	**CL**	-	-	-	-	-
J2	-	-	-	-	-	**CL**	-	-	-	-	-	-
J3	-	-	-	-	-	-	**CL**	-	-	-	-	-
K1	-	-	-	-	-	-	-	**CL**	**CL**	-	-	-
K2	-	-	-	-	-	-	-	-	**CL**	-	-	-
L1	-	-	-	-	-	-	-	-	-	**CL**	**CL**	±
L2	-	-	-	-	-	-	-	-	-	**+++**	**CL**	-
M1	-	-	-	-	-	-	-	-	-	-	-	**CL**
M2	-	-	-	-	-	-	-	-	-	-	-	-
M3	-	-	-	-	-	-	-	-	-	-	-	-
N	**+++**	-	-	-	-	-	-	-	-	-	**+++**	-
O	-	-	-	-	-	-	-	-	++	-	±	++
T	-	CL	-	-	-	-	-	-	+++	-	-	-
25	-	-	-	-	-	-	-	-	-	-	-	-
26	-	-	-	-	-	-	-	-	-	-	-	-
27	-	-	-	-	-	-	-	-	-	-	-	-
28	-	-	-	-	-	-	-	-	-	-	-	-
29	-	-	+	±	±	-	-	-	CL	±	-	+++
32	-	±	±	±	-	±	-	-	-	-	-	-
34	-	-	-	-	-	-	-	-	-	-	-	-
35	-	-	-	-	-	-	-	-	-	-	-	-
36	-	-	-	-	-	-	-	-	-	-	-	-
37	-	-	-	-	-	-	-	-	-	-	-	-
38	-	-	-	-	-	-	-	-	-	-	-	-
39	-	±	-	-	-	±	-	-	-	-	-	-
40	-	-	-	-	-	-	-	-	-	-	-	-
41	+++	SCL	+++	+++	++	+++	+	+	++	+++	+	SCL
42	-	+++	+++	+++	-	CL	-	-	CL	+++	-	+++
43	SCL	-	-	-	-	-	-	-	-	-	-	+++
44	-	-	-	-	-	-	-	-	-	-	-	-
45	-	-	-	-	-	-	-	-	+++	±	-	±
46	-	-	-	-	-	-	-	-	-	-	-	-

[a] Homologous and group reactions are shown in bold type.

REACTIONS OF THE Vi-TYPE STRAINS OF *SALMONELLA TYPHI* WITH THE TYPING PHAGES IN ROUTINE TEST DILUTIONS[a] (Continued)

Vi-Type Strains	Adapted Vi-Phage Preparations											
	M2	M3	N	O	T	25	26	27	28	29	32	34
A	CL	CL	CL	CL	CL	CL	CL	CL	CL	CL	CL	CL
B1	+++	++	–	++	++	+++	+++	++	SCL	+++	+	+
B2	+++	++	CL	+	SCL	SCL	CL	SCL	CL	–	+++	–
B3	CL	SCL	CL	CL	CL	SCL	CL	SCL	CL	±	SCL	+++
C1	++	++	±	–	++	±	+++	±	SCL	+	–	–
C2	–	±	–	–	–	–	–	–	–	–	–	–
C3	±	++	–	–	±	–	±	±	+	±	–	–
C4	–	–	–	–	–	–	–	–	–	–	–	–
C5	–	–	–	–	–	–	–	–	–	–	–	–
C6	–	–	–	–	–	–	–	–	–	–	–	–
C7	++	–	–	–	–	–	±	–	–	–	–	–
C8	–	–	–	–	–	–	+++	–	–	–	–	–
C9	+	±	–	–	±	–	+	–	++	–	–	–
D1	–	CL	+	–	–	–	–	–	±	–	±	–
D2	–	–	–	–	–	–	–	–	–	–	–	–
D4	–	++	–	–	–	–	–	–	–	–	–	–
D5	–	+++	–	–	–	–	–	–	–	–	–	–
D6	+	+++	–	–	–	–	–	–	–	++	–	–
D7	–	++	–	–	–	–	±	–	–	–	–	–
D8	–	++	–	–	–	–	–	–	–	–	–	–
D9	–	++	–	–	–	–	–	–	–	–	–	–
D10	–	–	–	–	–	–	–	–	–	–	–	–
D11	–	–	–	–	–	–	–	–	–	–	–	–
E1	±	±	–	–	–	–	–	–	–	–	–	–
E2	–	–	–	–	–	–	–	–	–	–	–	–
E3	–	–	–	–	–	–	–	–	–	–	–	–
E4	–	–	–	–	–	–	–	–	–	–	–	–
E5	–	–	–	–	–	–	–	–	–	–	–	–
E6	–	–	–	–	–	–	–	–	–	–	–	–
E7	–	±	–	–	–	–	–	–	–	–	–	–
E8	–	–	–	–	–	–	±	–	–	–	–	–
E9	–	±	–	–	–	–	–	–	–	–	–	–
E10	–	–	–	–	–	–	–	–	±	–	–	–
F1	–	–	–	–	–	–	–	–	–	–	–	–
F2	–	–	–	–	–	–	–	–	–	–	–	–
F3	–	–	–	–	–	–	–	–	–	–	–	–
F4	–	–	–	–	–	–	–	–	–	–	–	–
F5	–	–	–	–	–	–	–	–	–	–	–	–
G	–	–	–	–	–	–	–	–	–	–	–	–
H	+	±	±	–	±	±	+	–	+++	±	–	–
J1	–	–	–	–	–	–	–	–	–	–	–	–
J2	–	–	–	–	–	–	–	–	–	–	–	–
J3	–	–	–	–	–	–	–	–	–	–	–	–
K1	–	–	–	–	–	–	–	–	–	–	–	–
K2	–	–	–	–	–	–	–	–	–	–	–	–
L1	–	–	–	–	–	–	–	–	±	–	–	–
L2	–	–	–	–	–	–	±	–	–	–	–	–
M1	**CL**	**CL**	–	–	–	–	–	–	–	–	–	–
M2	**CL**	–	–	–	–	–	–	–	–	–	–	–
M3	–	**CL**	–	–	–	–	–	–	–	–	–	–
N	–	–	**CL**	–	±	–	±	–	–	–	–	–
O	+++	++	–	**CL**	–	–	–	–	+++	++	–	–
T	–	+++	++	–	**CL**	–	–	–	CL	SCL	–	–
25	–	–	–	–	–	**CL**	–	–	–	–	–	–
26	–	–	–	–	–	–	**CL**	–	–	–	–	–
27	–	–	–	–	–	–	–	**CL**	–	–	–	–
28	–	–	–	–	–	–	–	–	**CL**	–	–	–
29	+++	CL	–	±	–	–	±	–	+++	**CL**	–	±
32	±	±	–	–	–	–	–	–	±	±	**CL**	
34	–	–	–	–	–		–	–	–	–	–	**CL**
35	±	–	–	–	–	–	–	–	–	–	–	–
36	–	–	–	–	–	–	–	–	–	–	–	–
37	–	–	–	–	–	–	–	–	–	–	–	–
38	–	–	–	–	–	–	–	–	–	–	–	–
39	–	–	–	–	–	–	±	–	±	–	–	–
40	–	–	–	–	–	–	–	–	–	–	–	–
41	SCL	SCL	++	+	+++	+	SCL	±	CL	++	+	+
42	SCL	–	–	+++	–	–	–	–	+++	+++	–	SCL
43	+++	–	–	–	–	–	–	–	–	–	–	–
44	–	–	–	–	–	–	–	–	–	–	–	–
45	+	++	–	–	SCL	–	–	–	+++	+++	–	–
46	–	–	+++	–	–	–	–	–	–	–	–	–

[a] Homologous and group reactions are shown in bold type.

REACTIONS OF THE Vi-TYPE STRAINS OF *SALMONELLA TYPHI* WITH THE TYPING PHAGES IN ROUTINE TEST DILUTIONS[a] (Continued)

Vi-Type Strains	Adapted Vi-Phage Preparations											
	35	36	37	38	39	40	41	42	43	44	45	46
A	CL	CL	CL	CL	CL	CL	CL	CL	CL	CL	CL	CL
B1	++	SCL	-	+++	+++	CL	+++	+++	CL	CL	+++	CL
B2	+	-	SCL	CL	CL	-	CL	++	CL	-	SCL	-
B3	+++	-	SCL	SCL	CL	+	CL	SCL	CL	++	-	CL
C1	-	+++	-	SCL	+++	CL	SCL	±	++	++	+	±
C2	-	±	-	-	-	+++	-	-	-	-	-	-
C3	-	SCL	-	++	±	CL	++	±	++	±	-	-
C4	-	-	-	-	-	-	-	-	-	-	-	-
C5	-	-	-	-	-	-	-	-	-	-	-	-
C6	-	-	-	-	-	-	-	-	-	-	-	-
C7	-	±	-	-	-	-	-	±	±	+	-	-
C8	-	-	-	-	-	-	-	-	-	-	-	-
C9	-	-	-	+	+	-	+	-	+	-	-	-
D1	-	±	-	±	±	±	±	-	-	-	-	++
D2	-	-	-	-	-	-	-	-	-	-	-	-
D4	-	-	-	-	-	-	-	-	-	-	-	-
D5	-	-	-	±	-	+	-	-	-	-	-	+
D6	-	SCL	-	-	-	CL	-	+++	-	-	-	±
D7	-	-	-	-	-	-	-	-	-	-	-	-
D8	-	-	-	-	-	-	-	-	-	-	-	-
D9	-	-	-	-	-	-	-	-	-	-	-	+
D10	-	-	-	-	-	-	-	-	-	-	-	-
D11	-	-	-	-	-	-	-	-	-	-	-	-
E1	-	-	-	±	±	±	±	-	±	-	-	SCL
E2	-	-	-	-	-	-	-	-	-	-	-	+++
E3	-	-	-	-	-	-	-	-	-	-	-	-
E4	-	-	-	-	-	-	-	-	-	-	-	-
E5	-	-	-	-	-	-	-	-	-	-	-	-
E6	-	-	-	-	-	-	-	-	-	-	-	-
E7	-	±	-	-	-	+	-	-	-	-	-	±
E8	-	-	-	-	-	-	-	-	-	-	-	-
E9	-	±	-	-	-	++	-	-	-	-	-	-
E10	-	-	-	-	-	-	-	-	-	-	-	-
F1	-	-	-	±	±	±	-	-	-	-	-	-
F2	-	-	-	-	-	-	-	-	-	-	-	-
F3	-	-	-	-	-	-	-	-	-	-	-	-
F4	-	-	-	-	-	-	-	-	-	-	-	-
F5	-	-	-	-	-	-	-	-	-	-	-	-
G	-	±	-	±	-	-	-	-	±	-	-	±
H	±	++	±	+	+++	+++	+++	±	++	±	±	CL
J1	-	-	-	-	-	-	-	-	-	-	-	-
J2	-	-	-	-	-	-	-	-	-	-	-	-
J3	-	-	-	-	-	-	-	-	-	-	-	-
K1	-	-	-	-	-	-	-	-	-	-	-	-
K2	-	-	-	-	-	-	-	-	-	-	-	-
L1	-	±	-	±	-	-	±	-	±	-	-	±
L2	-	-	-	±	-	-	±	-	-	-	-	±
M1	-	-	-	-	-	-	-	-	-	-	-	-
M2	-	-	-	-	-	-	-	-	-	-	-	-
M3	-	-	-	-	-	-	-	-	-	-	-	-
N	-	-	-	-	-	-	-	-	CL	-	-	±
O	-	CL	-	++	-	CL	±	SCL	CL	+++	-	CL
T	+++	SCL	±	++	CL	CL	++	+++	±	-	-	+
25	-	-	-	-	-	-	-	-	-	-	-	-
26	-	-	-	-	-	-	-	-	-	-	-	-
27	-	-	-	-	-	-	-	-	-	-	-	CL
28	-	-	-	-	SCL	±	-	-	-	-	-	-
29	++	CL	-	++	+++	CL	±	CL	+++	+++	+++	±
32	-	-	-	+	±	+	±	-	±	-	-	+
34	+++	-	-	CL	-	-	-	-	-	-	-	CL
35	**CL**	-	-	CL	-	-	-	-	+	-	-	CL
36	-	**CL**	-	-	-	CL	-	-	-	-	-	-
37	-	-	**CL**	-	-	-	-	-	-	-	-	±
38	-	-	-	**CL**	-	-	-	-	-	-	-	-
39	-	-	-	±	**CL**	-	±	-	±	±	-	-
40	-	±	-	-	-	**CL**	-	-	-	-	-	-
41	+	SCL	+	CL	+++	CL	CL	±	CL	+++	+++	CL
42	SCL	CL	-	+++	+++	CL	-	**CL**	SCL	SCL	-	CL
43	-	-	-	-	-	-	-	-	**CL**	-	-	CL
44	-	-	-	-	-	-	-	-	-	**CL**	-	-
45	+++	SCL	-	++	±	SCL	-	+++	-	±	**CL**	±
46	-	-	-	SCL	-	-	-	-	-	-	-	**CL**

[a] Homologous and group reactions are shown in bold type.

Taken from Bernstein, A., and Wilson, E. M. J., *J. Gen. Microbiol.*, *32*, 350-351 (1963). Reproduced by permission of the Society for General Microbiology, Reading, Berkshire, England, and Cambridge University Press, New York.

REACTIONS OF TYPES, SUBTYPES, AND VARIATIONS OF *SALMONELLA PARATYPHI* B WITH TYPING PHAGES IN ROUTINE TEST DILUTIONS[a]

Type Strains	Typing Phages										
	1	2	3a	3b	Jersey	Beccles	Taunton	B.A.O.R.	Dundee	Batter-sea	Work-sop	1010
1	CL	CL	+++	+∓	CL	+∓+	+∓+	–	+++	–	–	–
1 var. 1	CL	CL	+∓	+∓	CL	SCL	SCL	+∓	SCL	–	–	CL
1 var. 2	CL	CL	CL	CL	CL	CL	CL	OL	+++	SCL	++	CL
1 var. 3	CL	CL	CL	CL	CL	+∓	–	OL	SCL	+++	++	CL
1 var. 4	CL	CL	CL	CL	CL	+∓	+	OL	SCL	+++	++	–
1 var. 5	SCL	CL	SCL	–	SCL	–	–	±	–	OL	–	SCL
1 var. 6	+++ SCL	+++ SCL	–	–	CL	–	–	–	–	OL	–	SCL
1 var. 7	+++	+++	–	OL	–	+++ OL	SCL	–	SCL	–	++	CL
1 var. 8	OL	OL	+∓	–	OL	–	–	–	+++	–	–	CL
1 var. 9	+++	+++	–	OL	+∓	CL	CL	OL	SCL	–	+++	CL
1 var. 10	CL	CL	CL	–	CL	CL	OL	–	∓	+++	–	CL
1 var. 11	OL	CL	–	–	CL	–	–	OL	–	+++	–	CL
1 var. 12	CL	CL	CL	CL	CL	–	–	OL	–	OL	SCL	CL
2	–	CL	–	–	±	–	–	–	+++	–	–	–
2 var. 1	–	CL	–	–	–	SCL	CL	–	SCL	–	–	OL
3a	–	–	CL	CL	–	SCL	OL	OL	SCL	+∓	++	OL
3a var. 2	–	–	CL	CL	–	–	–	–	+++	–	++	SCL
3a var. 4	–	–	OL	+++	–	–	–	SCL	SCL	–	±	OL
3a var. 6	–	–	OL	OL	–	SCL	OL	OL	+++	–	++	–
3a var. 7	–	–	CL	CL	–	CL	SCL	OL	<SCL	CL	CL	CL
3aI	–	–	OL	–	–	–	–	–	+++	–	–	+++
3aI var. 1	–	–	OL	–	–	OL	OL	+++	SCL	–	–	OL
3aI var. 4	–	–	OL	–	–	+∓	+∓	–	+++	–	–	OL
3b	–	–	–	OL	–	SCL	OL	OL	SCL	–	++	OL
3b var. 1	–	–	–	+++	–	SCL	OL	OL	–	–	–	OL
3b var. 2	–	–	–	OL	–	–	–	–	+++	–	++	+++
3b var. 3	–	–	–	OL	–	–	–	OL	OL	–	++	SCL
3b var. 6	–	±	–	OL	OL	SCL	OL	OL	+++	–	++	CL
3b var. 7	–	++	–	SCL	–	–	–	–	+++	SCL	<CL	CL
Jersey	–	±	–	–	CL	SCL	OL	–	+++	–	–	CL
Jersey var. 1	–	–	–	–	OL	–	–	–	+++	–	–	CL
Jersey var. 2	–	–	–	++	SCL	SCL	SCL	–	++	–	–	–
Beccles	–	±	–	–	–	SCL	OL	–	+++	–	–	OL
Beccles var. 1	–	–	–	++∓	–	SCL	OL	+++	+++	–	–	+++
Beccles var. 2	–	–	–	–	–	CL	CL	–	OL	–	–	–
Beccles var. 3	–	–	–	–	–	SCL	SCL	–	–	–	–	OL
Beccles var. 4	–	–	–	–	–	SCL	OL	–	–	+++	–	OL
Beccles var. 5	–	–	–	–	–	SCL	OL	–	–	–	–	–
Beccles var. 6	–	±	–	–	–	SCL	SCL	–	<OL	CL	–	CL
Scarborough	–	+	–	–	–	+∓+	+++ SCL	–	SCL	–	–	OL
Taunton	–	–	–	–	–	–	SCL	–	SCL	–	–	OL
Taunton var. 1	–	–	–	–	–	–	SCL	–	SCL	–	–	–
B.A.O.R.	–	–	–	–	–	–	–	OL	–	–	–	OL
Dundee	–	–	–	–	–	–	–	–	SCL	–	–	SCL
Dundee var. 1	–	±	–	–	–	–	–	–	+++	–	–	–
Dundee var. 2	–	–	–	–	–	–	–	–	SCL	<SCL <OL	–	CL
Battersea	–	–	–	–	–	– ±	– ±	–	–	OL	–	±
Worksop	–	–	–	+	–	–	–	–	–	±	OL	+++

[a] Where a range of reactions may be found with a particular phage, the highest expected reaction is shown beneath the lowest, thus: + ∓.

Taken from Anderson, E. S., in *The World Problem of Salmonellosis,* pp. 94-95, E. Van Oye, Ed. (1964). Reproduced by permission of W. Junk, The Hague, Netherlands.

LYTIC SPECTRA OF *STAPHYLOCOCCUS AUREUS* PHAGES

Test Strain	Phages																									
	29	52	52A	79	80	3A	3B	3C	55	71	6	7	42E	47	53	54	75	77	42D	81	187	42B	47C	52B	69	73	78
29	5	0	0	0	0	.	.	.	.	.	.	.	.	.	.	.	.	.	.	.	.	.	.	.	3	.	.
52	0	5	4	0	4	.	.	.	.	.	.	.	.	.	.	.	.	.	.	.	.	.	0	3	.	.	.
52A/79	.	3	5	5	3	.	.	.	.	.	.	.	.	.	.	.	.	.	.	.	.	.		1	.	.	.
80	.	1	1	.	5	.	.	.	.	.	.	1	2	.	.	.	.	.	.	5	.	0	0	.	.	3	.
2009	3	5	.	.	.	.	.	.	.	.	.	.	.	.	.	.	.	.	.	.	.	0	0	4	3	.	.
3A	1	1	1	1	1	5	3	4	4	4	1	1	1	1	1	1	.	1	1	1	.	1	1	2	1	1	1
3B	.	.	.	.	.	3	5	5	5	5	.	.	.	.	.	.	1	.	.	.	.	.	1	.	.	.	.
71	1	.	.	.	.	.	.	5	5	5	.	.	.	.	.	.	1	.	.	.	.	.	.	.	.	.	.
8719	.	.	.	.	.	.	0	.	.	5	.	.	.	.	.	.	.	.	.	.	.	.	.	.	.	.	.
42C	2	0	0	0	0	3	3	4	0/2	4	.	0	3	2	3	3	2	0	.	3	.	2	.	.	.	.	.
42E	0	0	0	0	0	.	.	.	.	.	.	.	5	.	3	2	.	.	0/2	3	.	2	1	.	.	.	3
47	3	3	3	3	3	.	.	.	.	.	2	2	.	5	5	3	5	5	.	0	.	2	4	5	1	2	.
53	.	.	.	.	.	.	.	.	.	.	0	0	.	0	5	4	5	5	.	1	.	.	.	.	.	0	.
54	1	.	.	2	.	2	2	2	.	.	.	5	3	5	5	5	5	5	.	3	.	4	4	2	1	4	3
75	.	.	.	2	1	.	2	2	.	.	.	0	1	0	4	0	5	5	.	.	.	.	.	.	1	0	.
77	.	.	.	.	2	.	.	.	.	.	.	0	.	2	4	0	.	5	.	.	.	.	.	4	1	.	0

Notes

1. Phages propagated on strains not included in the test set are also tested on their propagating strains, on which the reaction is by definition "5".
2. The few minor differences between this table and the corresponding tables previously published derive partly from further experience with the phages and partly from the results of a comparative test of the phages currently in use in five national laboratories.
3. The notation 0/2 is used for reactions which are variable and which may appear as inhibition reactions on one occasion and as true lytic reactions on another.

Taken from Blair, J. E., and Williams, R. E. O., *Bull. World Health Organ.*, *24*, 778(1961). Reproduced by permission of the World Health Organization, Geneva, Switzerland.

REFERENCES

1. Twort, F. W., *Lancet, 2,* 1241 (1915).
2. D'Hérelle, F. H., *C.R. Acad. Sci. (Paris), 165,* 373 (1917).
3. D'Hérelle, F. H., *Le bactériophage, son rôle dans l'immunité.* Masson et Cie., Paris (1921).
4. D'Hérelle, F. H., *Le bactériophage et son comportement.* Masson et Cie., Paris (1926).
5. Marcuse, R., *J. Pathol. Bacteriol., 38,* 409 (1934).
6. Craigie, J., and Yen, C. H., *Can. J. Public Health, 29,* 448 (1938).
7. Adams, M. H., *Bacteriophages.* Interscience Publications, John Wiley and Sons, New York (1959).
8. Anderson, E. S., in *The World Problem of Salmonellosis,* p. 92, E. Van Oye, Ed. W. Junk, The Hague, Netherlands; Humanities Press, New York (1964).
9. Felix, A., and Callow, B. R., *Lancet, 2,* 10 (1951).
10. Scholtens, R. T., *Antonie van Leeuwenhoek J. Microbiol. Serol., 25,* 403 (1959).
11. Bernstein, A., and Wilson, M. J., *J. Gen. Microbiol., 32,* 349 (1963).
12. Nicolle, P., Vieu, J. F., Diverneau, G., Brault, J., and Klein, B., *Bull. Acad. Nat. Med. (Paris), 154,* 481 (1970).
13. Blair, J. E., and Williams, R. E. O., *Bull. World Health Organ., 24,* 771 (1961).
14. Wilson, G. S., and Atkinson, J. D., *Lancet, 1,* 647 (1945).
15. Popoff, M., and Vieu, J. F., *C.R. Acad. Sci. (Paris), 270,* 2219 (1970).
16. Conn, H. J., Bottcher, E. J., and Randall, C., *J. Bacteriol., 49,* 359 (1945).
17. Roslycky, E. B., Allen, O. N., and McCoy, E., *Can. J. Microbiol., 8,* 71 (1962).
18. Schmidt, J. M., and Stanier, R. Y., *J. Gen. Microbiol., 39,* 95 (1965).
19. Lovett, P. S., *Bacteriol. Proc.,* p. 207 (1971).
20. Drimmelen, G. C., *Bull. World Health Organ., 23,* 127 (1960).
21. Jacob, M. M., *Nature, 219,* 752 (1968).
22. Parnas, J., *Pathol. Microbiol. Suppl., 3,* 1 (1963).
23. Philippon, A., *Ann. Inst. Pasteur (Paris), 115,* 367 (1968).
24. Prévot, A. R., and Thouvenot, H., *Ann. Inst. Pasteur (Paris), 101,* 966 (1961).
25. Fahey, J. E., *Can. J. Public Health, 43,* 167 (1952).
26. Rische, H., and Endemann, D., *Arch. Roum. Pathol. Exp. Microbiol., 21,* 337 (1962).
27. Saragea, A., and Maximesco, P., *Bull. World Health Organ., 35,* 681 (1966).
28. Thibaut, J., and Fredericq, P., *C.R. Soc. Biol., 150,* 1039 (1956).
29. Nicolle, P., LeMinor, S., Hamon, Y., LeMinor, L., and Brault, G., *Rev. Hyg. Med. Soc., 8,* 523 (1960).
30. Kasatiya, S. S., *D. Sc. Thesis.* University of Paris, Paris, France (1963).
31. Deak, Z., *Acta Microbiol. Acad. Sci. Hung., 12,* 261 (1965).
32. Ackermann, H.-W., Nicolle, P., LeMinor, S., and LeMinor, L., *Ann. Inst. Pasteur (Paris), 103,* 523 (1963).
33. Smith, H. W., and Crabb, W. E., *J. Gen. Microbiol., 15,* 556 (1956).
34. Parisi, J. T., Russell, J. C., and Merlo, R. J., *Appl. Microbiol., 17,* 721 (1969).
35. Przondo-Hessek, A., Slopek, S., and Miodonska, J., *Arch. Immunol. Ther. Exp., 16,* 402 (1968).
36. Milch, H., and Deak, S., *Acta Microbiol., 11,* 250 (1965).
37. Slopek, S., Przondo-Hessek, A., Milch, H., and Deak, S., *Arch. Immunol. Ther. Exp., 15,* 589 (1967).
38. Sword, C. P., and Pickett, M. J., *J. Gen. Microbiol., 25,* 241 (1961).
39. Buraczewska, M., Manowska, W., and Rdultowska, H., *Med. Dosw. I. Mikrobiol., 18,* 255 (1966).
40. Redmond, W. B., Cater, J. C., and Ward, D. M., *Amer. Rev. Resp. Dis., 87,* 257 (1963).
41. Baess, I., *Acta Pathol. Microbiol. Scand., 76,* 464 (1969).
42. Bates, J., and Mitchison, D. A., *Amer. Rev. Resp. Dis., 100,* 189 (1969).
43. Froman, S., Will, D. W., and Bogen, E., *Amer. J. Public Health, 44,* 1326 (1954).
44. Tokunaga, T., Maruyama, Y., and Murohashi, T., *Amer. Rev. Resp. Dis., 97,* 469 (1968).
45. Juhasz, S. E., and Bönicke, R., *Can. J. Microbiol., 11,* 235 (1965).
46. Rodda, G. M. J., *Aust. J. Exp. Biol. Med. Sci., 42,* 457 (1964).
47. Tokunaga, T., and Murohashi, T., *Jap. J. Med. Sci. Biol., 16,* 21 (1963).
48. Murohashi, T., Tokunaga, T., Mizuguchi, Y., and Maruyama, Y., *Amer. Rev. Resp. Dis., 8,* 664 (1963).
49. Cary, S. G., and Hunter, D. H., *J. Virol., 1,* 538 (1967).
50. France, D. R., and Markham, N. R., *J. Clin. Pathol. (London), 21,* 97 (1968).
51. Pavlatou, M., Hassikou-Kaklamani, E., and Zantioti, M., *Ann. Inst. Pasteur (Paris), 108,* 402 (1965).
52. Vieu, J. F., *Zentralbl. Bakteriol. Parasitenk. Infektionskr. Hyg. Abt. Orig., 171,* 612 (1958).
53. Vieu, J. F., and Capponi, M., *Ann. Inst. Pasteur (Paris), 108,* 103 (1965).
54. Graber, C. D., Latta, R. L., Vogel, E. H., and Brame, R. E., J. A., *Ann. Inst. Pasteur (Paris), 37,* 54 (1962).
55. Lindberg, R. B., Latta, R. L., Brame, R. E., and Moncrief, J. A., *Bacteriol. Proc.,* p. 81 (1964).
56. Meitert, E., *Arch. Roum. Pathol. Exp. Microbiol., 24,* 439 (1965).
57. Pavlatou, M., and Klakamani, E., *Ann. Inst. Pasteur (Paris), 101,* 914 (1961).
58. Postic, B., and Finland, M., *J. Clin. Invest., 40,* 2064 (1961).
59. Staniewski, R., *Can. J. Microbiol., 16,* 1003 (1970).

60. Prévot, A. R., Vieu, J. F., Thouvenot, H., and Brault, G., *Bull. Acad. Nat. Med. Paris, 154,* 681 (1970).
61. Atkinson, N., and Klauss, C., *Aust. J. Exp. Biol. Med. Sci., 33,* 375 (1955).
62. Sechter, I., and Gerichter, C. B., *Ann. Inst. Pasteur (Paris), 116,* 190 (1969).
63. Atkinson, N., Geytenbeek, H., Swann, M. C., and Wollaston, J. M., *Aust. J. Exp. Biol., 30,* 333 (1952).
64. Sechter, I., and Gerichter, C. B., *Appl. Microbiol., 16,* 1708 (1968).
65. Smith, H. W., *J. Gen. Microbiol., 5,* 919 (1951).
66. Lilleengen, K., *Acta Pathol. Microbiol. Scand., 27,* 625 (1950).
67. Lilleengen, K., *Acta Pathol. Microbiol. Scand., 30,* 194 (1952).
68. Ibrahim, A. E., *Appl. Microbiol., 18,* 748 (1969).
69. Schmidt, G., and Lüderitz, O., *Zentralbl. Bakteriol. Parasitenk. Infektionskr. Hyg. Abt. Orig., 210,* 381 (1969).
70. Bordini, A., *C.R. Acad. Sci., 270,* 567 (1970).
71. Guinée, P. A. M., and Scholtens, R. T., *Antonie van Leeuwenhoek J. Microbiol. Serol., 33,* 25 (1967).
72. Banker, D. D., *Nature, 175,* 309 (1955).
73. Smith, H. W., *J. Gen. Microbiol., 5,* 472 (1951).
74. Anderson, E. S., in *Health Congress Papers,* p. 96. R. S. H. Health Congress, Torquay, Devonshire, England (1960).
75. Callow, B. R., *J. Hyg., 57,* 346 (1959).
76. Felix, A., *J. Hyg. J. Gen. Microbiol., 14,* 208 (1956).
77. Lilleengen, K., *Acta Pathol. Microbiol. Scand., Suppl. 77* (1948).
78. Popovici, M., Nestoresco, N., Szégli, L., Bercovivi, C., Iosub, C., and Besleaga, V., *Arch. Roum. Pathol. Exp. Microbiol., 21,* 359 (1962).
79. Scholtens, R. T., *Antonie van Leeuwenhoek J. Microbiol. Serol., 28,* 373 (1962).
80. Sechter, I., and Gerichter, C. B., *Ann. Inst. Pasteur (Paris), 113,* 399 (1967).
81. Wilson, V. R., Hermann, G. J., and Balows, A., *Appl. Microbiol., 21,* 774 (1971).
82. Slopek, S., and Mulczyk, M., *Arch. Immunol. Ther. Exp., 8,* 417 (1960).
83. Ogawa, T., Inagaki, Y., Takamatsu, M., Yamamoto, T., and Yoshikane, M., *Nagoya Med. J., 10,* 119 (1964).
84. Hammarström, E., *Acta Med. Scand., Suppl. 223* (1949).
85. Rische, H., *Arch. Immunol. Ther. Exp., 16,* 392 (1968).
86. Slopek, S., Krukowska, A., and Mulczyk, M., *Arch. Immunol. Ther. Exp., 16,* 519 (1968).
87. Gromkova, R., and Trifonova, A., *Zentralbl. Bakteriol. Parasitenk. Infektionskr. Hyg. Abt. Orig., 204,* 212 (1967).
88. Van Boven, C. P. A., *Antonie van Leeuwenhoek J. Microbiol. Serol., 35,* 232 (1969).
89. Baldovin, A. C., Balteanu, E., Mihalco, F., Beloiu, I., and Pleceas, P., *Arch. Roum. Pathol. Exp. Microbiol., 21,* 385 (1962).
90. Hunter, G. J. E., *J. Hyg., 44,* 264 (1946).
91. Ciuca, M., Baldovin, A. C., Mihalco, F., Beloiu, I., and Caffé, I., *Arch. Roum. Pathol. Exp. Microbiol., 18,* 519 (1959).
92. Evans, A. C., *U.S. Public Health Rep., 49,* 1386 (1934).
93. Wilkowske, H. H., Nelson, F. E., and Parmelee, C. E., *Iowa Agricultural Station Research Project No. 652* (1954).
94. Mukerjee, S., *Bull. World Health Organ., 28,* 337 (1963).
95. Nicolle, P., Gallut, J., Ducrest, P., and Quiniou, J., *Rev. Hyg. Med. Soc., 10,* 91 (1962).
96. Hayward, A. C., *J. Gen. Microbiol., 35,* 287 (1964).
97. Nicolle, P., Mollaret, H., Hamon, Y., and Vieu, J. F., *Ann. Inst. Pasteur (Paris), 112,* 86 (1967).
98. Niléhn, B., and Ericsson, H., *Acta Pathol. Microbiol. Scand., 75,* 177 (1969).

METHODOLOGY

STERILIZATION, DISINFECTION, AND ANTISEPSIS

DR. EUGENE R. L. GAUGHRAN AND DR. PAUL M. BORICK

STERILIZATION

Sterilization is the process by which living organisms are removed or killed to the extent that they are no longer detectable in standard culture media in which they have previously been found to proliferate. The methods most commonly employed and the conditions required to effect sterilization are listed below.

Heat, Dry

150–160°C (300–320°F) . . . . > 3 hours
160–170°C (320–340°F) . . . . 2–3 hours
170–180°C (340–355°F) . . . . 1–2 hours

Heat, Moist (Steam under Pressure)

121°C (250°F) . . . . 15–30 minutes
132°C (270°F) . . . . 3–5 minutes
140°C (284°F) . . . . 1–2 minutes

Heat, Boiling Liquids

Boiling point of liquids, e.g., cumene (isopropylbenzene):
152°C (306°F) . . . . Indirect, 1 hour
Direct, 2 minutes

Filtration

Type	Pressure
Nitrocellulose discs (Millipore)	Positive or negative
Asbestos–cellulose pads (Seitz)	Positive or negative
Unglazed procelain candles (Pasteur-Chamberland)	Negative
Diatomaceous earth (Berkfeld)	Negative
Sintered glass	Negative

Radiation

Source	Minimum Dose	Time
Cobalt 60	2.5 Mrad	Dependent on curies
Cesium 137	2.5 Mrad	Dependent on curies
Electron accelerator	2.5 Mrad	Instantaneous
Van de Graaff unit	2.5 Mrad	Instantaneous

Chemical, Gaseous

Gas Employed	Temperature	Time	Concentration	Relative Humidity
Ethylene oxide, 100%	> 25°C (77°F)	> 3 hours	400–1000 mg/liter	40–70%
Ethylene oxide diluted with dichlorodifluoromethane[a]	> 25°C (77°F)	> 3 hours	400–1000 mg/liter	40–70%
Ethylene oxide diluted with carbon dioxide[a]	> 25°C (77°F)	> 3 hours	400–1000 mg/liter	40–70%
Propylene oxide, 100%	> 25°C (77°F)	> 3 hours	800–2000 mg/liter	30–60%
Propylene oxide diluted with dichlorodifluoromethane	> 25°C (77°F)	> 3 hours	800–2000 mg/liter	30–60%
Propylene oxide diluted with carbon dioxide	> 25°C (77°F)	> 3 hours	800–2000 mg/liter	30–60%
Formaldehyde	> 18°C (65°F)	24 hours	5–15 mg/liter	Saturated
β-Propiolactone	> 18°C (65°F)	> 3 hours	5–10 mg/liter	Saturated

Chemical, Liquid

Solution for Sporicidal Action	Time
1% Formaldehyde (water or alcohol)	18–24 hours
8% Formaldehyde (70% ethanol or isopropanol)[b]	3–10 hours
2% Aqueous alkaline glutaraldehyde[b]	3–10 hours
1% Peracids	3–10 hours
2.5*N* Hydrochloric acid	1–3 hours
2.5*N* Sodium hydroxide	6–10 hours
1% Chlorine (10,000 ppm)	3–10 hours
1% Iodine	3–10 hours

STERILITY TESTING

Sterility is the freedom from living microorganisms achieved by the application of a process in which living organisms are removed or killed to the extent that they are no longer detectable in standard culture media in which they have previously been found to proliferate. Assurance of sterility depends upon the verification and reproducibility of the sterilization process. Nevertheless, biological tests are necessary to detect viable microorganisms. Suitable tests and media may be found in the following sources.

[a] Advantage: less flammable or explosive than 100% ETO.

[b] Approved by EPA for liquid chemical sterilization.

The Pharmacopeia of the United States of America, 18th revision. U.S.P., Bethesda, Maryland (1970).

National Formulary XIII and *First Supplement to National Formulary XIII.* American Pharmaceutical Association, Washington, D.C. (1970).

BBL Manual of Products and Laboratory Procedures, 5th ed. Bio Quest, Division of Becton, Dickinson and Company, Cockeysville, Maryland (1968).

Case Bacteriological Culture Media. Case Laboratories, Inc., Chicago, Illinois (1960).

Colab Product Manual. Consolidated Laboratories, Inc., Chicago Heights, Illinois (1971).

Difco Manual, 9th ed. Difco Laboratories, Inc., Detroit, Michigan (1953).

Difco Supplementary Literature. Difco Laboratories, Inc., Detroit, Michigan (1966).

Microbiological Culture Media and Associated Products. Pfizer Diagnostics, New York (1969).

The Oxoid Manual, 3rd ed. Oxoid Ltd., London, England (1971).

Guidelines for Sterility Tests on Products Sterilized by Steam under Pressure

Product	Minimal Testing Units	Medium	Temperature	Incubation Time, days
Liquid	10	Fluid thioglycollate	30–35°C	7
Solid	10	Fluid thioglycollate	30–35°C	10
Biological indicators	10	Optimal	Optimal	7
Simulated product	10	Optimal	Optimal	7

Guidelines for Sterility Tests on Products Sterilized by Filtration

Product	Minimal Testing Units	Medium	Temperature	Incubation Time, days[c]
Regular samples	30	Soybean casein digest liquid	20–25°C	14
Regular samples	30	Fluid thioglycollate	30–35°C	14

[c] When membrane filtration is employed for sterility testing, seven days incubation time may be adequate.

Guidelines for Sterility Tests on Products Sterilized by Other Methods

Product	Minimal Testing Units	Medium	Temperature	Incubation Time, days
Regular samples	20	Soybean casein digest liquid	20–25°C	14
Regular samples	20	Fluid thioglycollate	30–35°C	14
Biological indicators (inoculated carriers)	10	Optimal	Optimal	7
Products containing biological indicators[d] (inoculated simulated products)	10	Optimal	Optimal	7

Biological Indicators Recommended for Use in Various Sterilization Processes

Process	Microorganism[e]
Moist heat	*Bacillus stearothermophilus*
Dry heat	*Bacillus subtilis*[e]
Irradiation	*Bacillus pumilus*
Gas (ethylene oxide)	*Bacillus subtilis* var. *niger*
Liquids	*Bacillus subtilis* or *Clostridium sporogenes*

DISINFECTION AND ANTISEPSIS

Disinfection is the treatment of inanimate objects with a chemical in order to destroy all pathogenic microorganisms except spores. Some disinfectants fall short of this broad spectrum of activity; others go beyond it due to their ability to kill bacterial spores.

Antisepsis is the application of a chemical agent to living tissues in order to control growth of microorganisms either by killing them or preventing their growth.

Since many interrelated conditions (e.g., time, temperature, presence of organic matter, etc.) influence the action of disinfectants and antiseptics, the following tables attempt to indicate in general terms the antimicrobial activity usually attributed to these agents.

[d] When biological indicators (e.g., spore strips) and regular products are tested, half the minimal units of regular samples recommended above may be used.

[e] Since the above microorganisms are recommended as guidelines, organisms showing an order of resistance to the sterilization process comparable to or greater than that suggested may be used as biological indicators.

TABLE 1.

EFFECTS OF VARIOUS DISINFECTANTS ON MICROORGANISMS

+ = active; – = inactive

Class	Example	Concentration, %	Vegetative Bacteria	Tubercle Bacillus	Bacterial Spores	Higher Fungi	Virus
Alcohols, aliphatic	Ethyl	70	+	+	–	+	+
	Isopropyl	70–90	+	+	–	+	–
Aldehydes	Formaldehyde	1–8	+	+	+	+	+
	Glutaraldehyde	2	+	+	+	+	+
Bisguanidines	Chlorhexidine	0.1–1.0	+	–	–	+	–
β-Propiolactone	β-Propiolactone	1	+	+	+	+	+
Cresols	Coal tar disinfectants	1–5	+	–	–	+	–
Epoxides	Ethylene oxide	1	+	+	+	+	+
	Propylene oxide	1	+	+	+	+	+
Halogens	Organic and inorganic halogen-releasing compounds	0.005–0.02 available halogen	+	–	–	+	+
Organotin compounds	Tri-*n*-butylin salts	0.01–0.02	+	–	–	+	–
Peracids	Peracetic acid	1	+	+	+	+	+
Phenols	Phenol, phenyl phenol, and halogenated derivatives	0.5–5.0	+	+	–	+	+*
Surfactants							
Anionic	Alkyl aryl sulfonates†	0.1–0.25	+	–	–	–	–
Cationic	Quaternary ammonium salts‡	0.02	+	–	–	+	–
Amphoteric	Alkyl di(aminoethyl)glycine‡	1–5	+	–	–	+	–

* Phenol only. † Acid pH(<3). ‡ Alkaline pH.

TABLE 2.

EFFECTS OF VARIOUS ANTISEPTICS ON MICROORGANISMS

+ = active; – = inactive

Class	Example	Application	Concentration, %	Gram-Positive Bacteria	Gram-Negative Bacteria	Fungi
Acridines	Aminoacridine HCl	Wounds	0.1	+	–	+
Alcohols	Ethyl	Intact skin	70	+	+	+
	Isopropyl	Intact skin	70–90	+	+	+
Bisguanidines	Chlorhexidine	Intact skin	0.5	+	+	+
		Wounds	0.05	+	–	–
Bisphenols	Hexachlorophene	Intact skin	2–3	+	–	+
		Wounds	0.5	+	–	–
Carbanilides	Trichlorocarbanilide	Intact skin	2	+	–	+
Heavy silver salts	Silver nitrate	Burns	0.5	+	+	+
Iodine compounds	I_2-NaI or KI, aqueous	Wounds	2–5	+	+	+
	I_2-NaI or KI in 50% alcohol	Intact skin	2	+	+	+
	Iodophors	Intact skin and wounds	1	+	+	+
Mercurials, organic	Sodium (ethylmercurithio) salicylate	Intact skin: tincture	0.1	+	–	–
		Wounds: aqueous	0.1	+	–	–
Nitrofurans	Nitrofurazone	Wounds	0.2	+	+	–
Oxyquinolines	Chloroquinaldol	Wounds	3	+	–	+
Peroxides	Hydrogen peroxide	Wounds	1.5–3.0	+	–	–

TABLE 2 (Continued)

EFFECTS OF VARIOUS ANTISEPTICS ON MICROORGANISMS

Class	Example	Application	Concentration, %	Gram-Positive Bacteria	Gram-Negative Bacteria	Fungi
Pyrithiones	Zinc pyridinethione	Intact skin	0.1–0.5	+	+	+
Salicylanilides	Chloro- and bromosalicylanilides	Intact skin	2	+	–	+
Sulfa compounds	Mafenide HCl	Wounds	5	+	+	–
Surfactants						
Cationic	Quaternary ammonium salts	Intact skin	0.1–0.5	+	+	+
		Wounds	0.1–0.3	+	–	–
Amphoteric	Alkyl and acyl amino acids	Intact skin	1	+	+	–
Thiocarbamates	Tolnaftate	Wounds	1	–	–	+
Triphenylmethane dyes	Crystal violet	Wounds	0.5	+	–	+
Xylenols	2,4-Dichloro-*sym-meta*-xylenol	Intact skin	2	+	–	–

BIBLIOGRAPHY

1. Bernarde, M. A. (Ed.), *Disinfection.* Marcel Dekker, Inc., New York (1970).
2. Borick, P. M., Antimicrobial Agents as Liquid Chemosterilizers. *Biotechnol. Bioeng., 7,* 435 (1965).
3. Borick, P. M., Chemical Sterilizers (Chemosterilizers), *Advan. Appl. Microbiol., 10,* 291 (1968).
4. Borick, P. M., and Borick, J. A., in *Quality Control in the Pharmaceutical Industry,* pp. 1–38, M. S. Cooper, Ed. Academic Press, New York (1972).
5. Bruch, C. W., Gaseous Sterilization, *Annu. Rev. Microbiol., 15,* 245 (1961).
6. Corum, C. J. (Ed.), Federal Regulations and Practical Control Microbiology for Disinfectants, Drugs and Cosmetics, *Special Publication No. 4.* Society for Industrial Microbiology, Washington, D.C. (1969).
7. Davis, J. G., Chemical Sterilization, *Progr. Ind. Microbiol., 8,* 141 (1968).
8. Goldsmith, M., Ionizing Radiation and the Sterilization of Medical Products, *Proceedings of the First International Symposium.* Taylor and Francis Ltd., London, England (1964).
9. Hedgecock, L. W., *Antimicrobial Agents.* Lea and Febiger, Philadelphia, Pennsylvania (1967).
10. Horn, H., *Biologische Prüfung und Leistungskriterien der Sterilisation.* VEB Verlag Volk und Gesundheit, Berlin, Germany (1968).
11. Kereluk, K., and Lloyd, R. S., Ethylene Oxide Sterilization, A Current Review of Principles and Practices, *J. Hosp. Res., 7,* 7 (1969).
12. Lawrence, C. A., and Block, S. S. (Eds.), *Disinfection, Sterilization, and Preservation.* Lea and Febiger, Philadelphia, Pennsylvania (1968).
13. McCulloch, E. C., *Disinfection and Sterilization,* 2nd ed. Lea and Febiger, Philadelphia, Pennsylvania (1945).
14. *National Formulary XIII* and *First Supplement to National Formulary XIII.* American Pharmaceutical Association, Mack Publishing Co., Easton, Pennsylvania (1970).
15. Parisi, A. N., and Borick, P. M., Pharmaceutical Sterility Testing, *Contam. Control., 8,* 31 (1969).
16. Perkins, J. J., *Principles and Methods of Sterilization in Health Sciences,* 2nd ed. Charles C Thomas, Springfield, Illinois (1969).
17. Radiosterilization of Medical Products and Recommended Code of Practice, *Proceedings Series.* International Atomic Energy Agency, Vienna, Austria (1967).
18. Radiosterilization of Medical Products, Pharmaceutical and Bioproducts, *Technical Reports Series, No. 72.* International Atomic Energy Agency, Vienna, Austria (1967).
19. Recent Development in the Sterilization of Surgical Materials, *Symposium Held at the University of London, April 11–13, 1961.* The Pharmaceutical Society of Great Britain, Smith and Nephew, Ltd., London, England (1961).
20. Richards, J. W., *Introduction to Industrial Sterilization.* Academic Press, London, England (1968).
21. Rubbo, S. D., and Gardner, J. F., *A Review of Sterilization and Disinfection as Applied to Medical, Industrial and Laboratory Practice.* Lloyd-Duke Ltd., London, England (1965).
22. Stellmacher, W., Scholz, K., and Preissler, K., *Desinfektion.* Gustav Fischer, Jena, Germany (1970).
23. Sykes, G., *Disinfection and Sterilization,* 2nd ed. E. and F. N. Spon Ltd., London, England (1965).
24. *The Pharmacopeia of the United States of America,* 18th revision. U.S.P., Bethesda, Maryland (1970).
25. Wallhäusser, K. H., and Schmidt, H., *Sterilisation, Desinfektion, Konservierung, Chemotherapie; Verfahren, Wirkstoffe, Prüfungmethoden.* Georg Thieme, Stuttgart, Germany (1967).

ENUMERATION OF MICROORGANISMS

DR. JOHN J. GAVIN AND DR. DENNIS P. CUMMINGS

The following pages present general information regarding the methodology used to *estimate* the number of microorganisms in a given sample. The scope is limited to those methods used for the enumeration of cells; methods for the measurement of cellular mass are not considered in this section. Specific procedural details have not been included, because these will vary according to the interest of the investigator. For example, if one is interested in *Mycoplasma*, the number of cells can be estimated by direct observation of the sample or by turbidimetric measurement; however, prior to estimating their numbers it is necessary to concentrate these organisms rather than to dilute them, as is done with other types of cells. In applications that concern legal regulations, specific details of the procedures to be used are promulgated both in official and in non-official but expert manuals. Further, the investigator must decide the degree of importance that is attached to the enumeration, since no *absolute* value is obtainable. The value of the numbers is directly related to the skill and precision of the individual worker, his knowledge of his particular needs, and a not-too-rigid interpretation of his data. In many research situations it is advisable to determine cell mass rather than to enumerate the organisms.

The problem is most complex when working with mixed cell populations such as occur in samples of natural origin, foods, soils, raw materials used in pharmaceutical preparations, water, air, and other sources. No universal culture medium is available that will provide optimal growth conditions for all the microorganisms that could be present. Other factors are the specific interaction between various organisms in competition for nutrients and the elaboration of stimulatory and/or inhibitory metabolic by-products by certain species, which change the environmental conditions so that they favor the growth of some types of cells and suppress the development of others. Relative numbers are also of importance, because dilution to obtain a countable range could eliminate a large number of organisms if a single organism were predominant. Each dilution in a tenfold dilution series eliminates 90% of the organisms contained in the previous dilution. Even with pure cultures there is no assurance that observed counts correlate with actual numbers. Viable counts in any nutrient-agar medium are only accurate to the extent of dispersion of the microorganisms throughout the diluting medium. Single colonies will develop from a single cell; but single colonies may also develop from clumps of 10, 100 or 1000 cells, a factor that cannot be determined.

Physical methods, such as particle-counting and turbidimetric measurement, do not differentiate between viable and non-viable cells. In addition, absorbance measurements may be more nearly related to total cellular mass than to the number of cells.[a] Environmental changes that take place during the growth of various organisms in pure culture may lead to aggregation simulating lysis. Turbidity measurements, in such cases, would underestimate the actual count.

Mycelial cultures are difficult to evaluate by turbidimetric methods. The hyphae clump, and the culture becomes thick with aggregates. While it is possible to dilute such cultures to obtain absorbance values, the error is considerable. It is impossible to obtain more than a general impression of the quantity of these and other organisms that tend to grow in clusters by any enumeration method. Modifications of standard procedures can be applied in specific instances if enumeration is desired, but measurement of cellular mass would be more reliable.

Direct enumeration by microscopic examination may produce ambiguous results. The best precision is obtained when viral particles in pure culture are counted. The virions are easily recognizable in the electron microscope; however, as with the counting of other cells by this method, one cannot distinguish between viable (infectious) and non-viable (non-infectious) particles.

Counting of plaque or focal lesions approaches the ideal enumeration method for virions, because each plaque arises from a single viral particle. This is qualified by the fact that the absolute efficiency of infection (i.e., the total number of virions present in a sample as determined by electron-microscope count

[a] Koch, A. L., *Biochem. Biophys. Acta, 51,* 429 (1961).

relative to the total number of infectious particles) varies widely among different viruses and even for the same virus counted in different host-cell systems, due not only to the presence of non-infectious particles but also to the failure of potential infectious particles to initiate an infection in a specific cell under appropriate conditions for multiplication. Plaque overlapping is a further source of error.

Despite the shortcomings of enumerative methods, it is possible to obtain valuable information from such procedures if the investigator (a) has an understanding of the nature of the sample to be tested, (b) gives attention and care to the preparation of the sample, (c) utilizes the proper methodology in context with the nature of the sample and the type(s) of microorganism(s) to be enumerated, (d) includes calibration standards when required and/or possible, (e) provides sufficient replicates to obtain statistically precise, if not accurate in the absolute sense, results, and (f) realizes that numerical results are only indicative.

Standard references dealing, in part, with procedures for the enumeration of microorganisms include the following:

1. *Federal Register*
   Promulgates regulations enforced by the Food and Drug Administration, the Division of Biological Standards (NIH), and the Department of Agriculture.
2. *National Formulary XIII*
   American Pharmaceutical Association, Washington, D.C.
3. *NASA Standards for the Microbiological Examination of Space Hardware*
   NHB 5340.1A, October 1968. U.S. Government Printing Office, Washington, D.C.
4. *Official Methods of Analysis of the Association of Official Agricultural Chemists*
   Association of Official Agricultural Chemists, Washington, D.C.
5. *Recommended Methods for the Examination of Foods*
   American Public Health Association, Inc., New York.
6. *Standard Methods for the Examination of Water and Waste Water,* 12th edition
   American Public Health Association, Inc., New York.
7. *Standard Methods for the Examination of Dairy Products,* 12th edition
   American Public Health Association, Inc., New York.
8. *Standard Methods for the Examination of Sea Water and Shellfish,* 4th edition
   American Public Health Association, Inc., New York.
9. *The United States Pharmacopeia,* 18th revision
   U.S.P., Bethesda, Maryland.

An excellent general reference is the series *Methods in Microbiology* (J. R. Norris and D. W. Ribbons, Eds.), Academic Press, New York (1969). Volumes 1, 3A, and 3B contain information that is applicable to the enumeration of microorganisms. A further reference containing pertinent information is *Isolation Methods for Microbiologists* (D. A. Shopton and G. W. Gould, Eds.), Academic Press, London, England (1969).

MOST-PROBABLE-NUMBER (MPN) ESTIMATES

Theory

On the basis that microorganisms are uniformly distributed in liquid medium, it may be presumed that repeated samples of equal volumes from a single source will contain, *on the average,* the same number of microorganisms. This average is the most probable number (MPN). If a number of samples of varying volumes (for example, 5 samples of 10.0 ml each, 5 samples of 1.0 ml each, and 5 samples of 0.1 ml each) are inoculated into individual test tubes of nutrient medium and observed for growth after a suitable incubation period, it is possible to calculate the most probable number of organisms per unit volume, usually 100 ml, of the original samples.

The following tables indicate the estimated number of bacteria of the coliform group present in 100 ml of water according to various combinations of positive and negative results in the quantities used for the test. They are basically the tables originally computed by McCrady,[b] with certain amendments due to more precise calculations by Swaroop;[c] a few values have also been added to the tables from other sources, corresponding to further combinations of positive and negative results likely to occur in practice. Swaroop[d] has tabulated limits within which the real density of coliform organisms is likely to fall; his paper should be consulted by those anxious to know the precision of these estimates.

TABLE 1

ONE TUBE OF 50 ml AND FIVE TUBES OF 10 ml

50-ml Tubes Positive	10-ml Tubes Positive	MPN per 100 ml
0	0	0
0	1	1
0	2	2
0	3	4
0	4	5
0	5	7
1	0	2
1	1	3
1	2	6
1	3	9
1	4	16
1	5	18+

Data taken from: The Bacteriological Examination of Water Supplies, *Reports on Public Health and Medical Subjects, No. 71* (1957). Reproduced by permission of the Controller of Her Majesty's Stationery Office, London, England.

Basic Equipment Required

Sufficient glassware of adequate size to contain the selected volume(s) of sample plus medium; suitable environmental conditions for the desired application.

Sources of Error

Preparation of non-liquid samples, both in the extraction of microorganisms and in the even distribution of the material in the diluent used.

Applications

Because this method is amenable for the testing of a variety of samples, there are many applications for it. Although relatively inaccurate, it permits detection of very low concentrations of microorganisms.

TURBIDIMETRIC MEASUREMENT

Theory

The development of turbidity in suitable liquid nutrient medium inoculated with a given microorganism *is* a function of growth, reflecting increases in both mass and cell number. For enumeration purposes, changes

[b] McCrady, M. H., *Can. J. Public Health, 9,* 201 (1918).
[c] Swaroop, S., *Indian J. Med. Res., 26,* 353 (1938).
[d] Swaroop, S., *Indian J. Med. Res., 39,* 107 (1951).

Table 2

ONE TUBE OF 50 ml, FIVE TUBES OF 10 ml, AND FIVE TUBES OF 1 ml

50-ml Tubes Positive	10-ml Tubes Positive	1-ml Tubes Positive	MPN per 100 ml	50-ml Tubes Positive	10-ml Tubes Positive	1-ml Tubes Positive	MPN per 100 ml
0	0	0	0	1	2	1	7
0	0	1	1	1	2	2	10
0	0	2	2	1	2	3	12
0	1	0	1	1	3	0	8
0	1	1	2	1	3	1	11
0	1	2	3	1	3	2	14
0	2	0	2	1	3	3	18
0	2	1	3	1	3	4	20
0	2	2	4	1	4	0	13
0	3	0	3	1	4	1	17
0	3	1	5	1	4	2	20
0	4	0	5	1	4	3	30
1	0	0	1	1	4	4	35
1	0	1	3	1	4	5	40
1	0	2	4	1	5	0	25
1	0	3	6	1	5	1	35
1	1	0	3	1	5	2	50
1	1	1	5	1	5	3	90
1	1	2	7	1	5	4	160
1	1	3	9	1	5	5	180+
1	2	0	5				

Data taken from: The Bacteriological Examination of Water Supplies, *Reports on Public Health and Medical Subjects, No. 71* (1957). Reproduced by permission of the Controller of Her Majesty's Stationery Office, London, England.

TABLE 3

FIVE TUBES OF 10 ml, FIVE TUBES OF 1 ml, AND FIVE TUBES OF 0.1 ml

10-ml Tubes Positive	1-ml Tubes Positive	0.1-ml Tubes Positive	MPN per 100 ml	10-ml Tubes Positive	1-ml Tubes Positive	0.1-ml Tubes Positive	MPN per 100 ml
0	0	0	0	1	0	3	8
0	0	1	2	1	1	0	4
0	0	2	4	1	1	1	6
0	1	0	2	1	1	2	8
0	1	1	4	1	2	0	6
0	1	2	6	1	2	1	8
0	2	0	4	1	2	2	10
0	2	1	6	1	3	0	8
0	3	0	6	1	3	1	10
1	0	0	2	1	4	0	11
1	0	1	4	2	0	0	5
1	0	2	6	2	0	1	7

TABLE 4

THREE TUBES EACH INOCULATED WITH 10 ml, 1 ml, AND 0.1 ml OF SAMPLE

10-ml Tubes Positive	1-ml Tubes Positive	0.1-ml Tubes Positive	MPN per 100 ml	10-ml Tubes Positive	1-ml Tubes Positive	0.1-ml Tubes Positive	MPN per 100 ml
0	0	1	3	2	0	1	14
0	0	2	6	2	0	2	20
0	0	3	9	2	0	3	26
0	1	0	3	2	1	0	15
0	1	1	6	2	1	1	20
0	1	2	9	2	1	2	27
0	1	3	12	2	1	3	34
0	2	0	6	2	2	0	21
0	2	1	9	2	2	1	28
0	2	2	12	2	2	2	35
0	2	3	16	2	2	3	42
0	3	0	9	2	3	0	29
0	3	1	13	2	3	1	36
0	3	2	16	2	3	2	44
0	3	3	19	2	3	3	53
1	0	0	4	3	0	0	23
1	0	1	7	3	0	1	39
1	0	2	11	3	0	2	64
1	0	3	15	3	0	3	95
1	1	0	7	3	1	0	43
1	1	1	11	3	1	1	75
1	1	2	15	3	1	2	120
1	1	3	19	3	1	3	160
1	2	0	11	3	2	0	93
1	2	1	15	3	2	1	150
1	2	2	20	3	2	2	210
1	2	3	24	3	2	3	290
1	3	0	16	3	3	0	240
1	3	1	20	3	3	1	460
1	3	2	24	3	3	2	1,100
1	3	3	29	3	3	3	1,100+
2	0	0	9				

Data taken from: Jacobs, M. B., and Gerstein, M. J., *Handbook of Microbiology,* copyright 1960, D. Van Nostrand Company, Inc. Reproduced by permission of Van Nostrand Reinhold Company, New York.

TABLE 3 (Continued)

FIVE TUBES OF 10 ml, FIVE TUBES OF 1 ml, AND FIVE TUBES OF 0.1 ml

10-ml Tubes Positive	1-ml Tubes Positive	0.1-ml Tubes Positive	MPN per 100 ml
2	0	2	9
2	0	3	12
2	1	0	7
2	1	1	9
2	1	2	12
2	2	0	9
2	2	1	12
2	2	2	14
2	3	0	12
2	3	1	14
2	4	0	15
3	0	0	8
3	0	1	11
3	0	2	13
3	1	0	11
3	1	1	14
3	1	2	17
3	1	3	20
3	2	0	14
3	2	1	17
3	2	2	20
3	3	0	17
3	3	1	20
3	4	0	20
3	4	1	25
3	5	0	25
4	0	0	13
4	0	1	17
4	0	2	20
4	0	3	25
4	1	0	17
4	1	1	20
4	1	2	25
4	2	0	20
4	2	1	25
4	2	2	30
4	3	0	25
4	3	1	35
4	3	2	40

10-ml Tubes Positive	1-ml Tubes Positive	0.1-ml Tubes Positive	MPN per 100 ml
4	4	0	35
4	4	1	40
4	4	2	45
4	5	0	40
4	5	1	50
4	5	2	55
5	0	0	25
5	0	1	30
5	0	2	45
5	0	3	60
5	0	4	75
5	1	0	35
5	1	1	45
5	1	2	65
5	1	3	85
5	1	4	115
5	2	0	50
5	2	1	70
5	2	2	95
5	2	3	120
5	2	4	150
5	2	5	175
5	3	0	80
5	3	1	110
5	3	2	140
5	3	3	175
5	3	4	200
5	3	5	250
5	4	0	130
5	4	1	170
5	4	2	225
5	4	3	275
5	4	4	350
5	4	5	425
5	5	0	250
5	5	1	350
5	5	2	550
5	5	3	900
5	5	4	1600
5	5	5	1800+

Data taken from: The Bacteriological Examination of Water Supplies, *Reports on Public Health and Medical Subjects, No. 71* (1957). Reproduced by permission of the Controller of Her Majesty's Stationery Office, London, England.

in turbidity can be correlated with changes in cell numbers. Standard curves can be constructed, which – within an appropriate concentration range – may be used to estimate the microbial population from observed turbidity values.

Basic Equipment Required

A suitable instrument for the measurement of turbidity; numerous filter photometers, spectrophotometers, and direct-reading turbidimeters (nephelometers).

Optional Equipment and/or Accessories

Optically matched tubes; calibration standards; recording instruments; specialized equipment.

Sources of Error

INSTRUMENTAL

1. Non-linearity of response of the light-sensitive devices and associated measuring circuits.
2. Variation in the intensity of the light source.
3. Stray light.
4. Light scatter.
5. Temperature rise in the measuring photocell.
6. Dust, scratches, and imperfections in the optical system.
7. Clarity and color of the suspending medium.

MICROBIAL

1. Cell shape.
2. Cell aggregation (clumping).
3. Cell settlement with uneven distribution of particles.
4. Cell lysis.
5. Viable/non-viable cell ratio.

Applications

Studies of growth and metabolism of microorganisms; evaluation of stimulating and inhibiting effects of various compounds; osmotic effects; microbiological assay; control of commercial production of the products of microbial biosynthesis.

PARTICLE-COUNTING

Theory

Cells suspended in an electrolyte that passes through an electrical field of standard resistance within a small aperture will alter this resistance as they pass through the aperture. If a constant voltage is maintained, a cell passing through will cause a transient decrease in current as the resistance changes; if a constant current is maintained, a transient increase in voltage occurs. Such changes are amplified and recorded electronically, providing a count of the number of cells flowing through the opening. By relating the count obtained to either a metered volume or to a fixed time period for flow under a constant pressure head, an estimate of the total number of cells in a sample may be obtained.

Basic Equipment Required

Commercially available counting equipment.

Sources of Error

1. Foreign particles or bubbles.
2. Ratio of signal to electronic noise background.
3. Elevated background counting rates when the sensitivity of detection is increased for counting low concentrations of cells.
4. Partial blocking of the aperture.
5. High cell concentrations, which increase the probability of the coincident passage of two or more cells through the aperture.

Applications

All types of cell enumeration.

FOCAL-LESION DETERMINATION

Theory

Infection of a suitable host with a specific virus can produce a focal-lesion response that may be quantitated if – within an appropriate range of dilutions of virus – the average number of lesions is a linear function of the quantity of virus. The method is a true measure of the virus because the lesion is at the site of the activity of an individual viral particle. The relationship between the lesion count and virus dilution is consistent with a Poisson distribution.

Basic Equipment Required

Dependent upon the test system employed: focal lesions produced on the skin of whole animals, on the chorioallantoic membrane of chicken embryos, or in cell monolayers; focal lesions produced in agar plate cultures of bacterial viruses; focal lesions produced on the leaves of a plant rubbed with a mixture of virus and abrasive.

Sources of Error

1. Non-specific inhibition of the response by unknown factors.
2. Dilution errors.
3. Counting errors.
4. Overlapping of lesions.
5. Size of the lesions and the number countable on the surface area.
6. Insufficient replication to provide statistical significance.

Applications

The enumeration of plant, animal, and bacterial viruses.

QUANTAL MEASUREMENT

Theory

The concentration of viruses may be calculated by statistical methods if an all-or-none (mortality) response results from infection. By use of a dilution series, the endpoint of activity is considered to be the highest dilution of the virus at which there were 50% or more positive responses. The exact endpoint is determined, by interpolation from the cumulative frequencies of positive and negative response observed at various dilutions, to be that dilution at which there are 50% positive and 50% negative responses. The reciprocal of the dilution yields an estimate of the number of viral units per inoculum volume of the undiluted sample and is expressed in multiples of the 50% endpoint.

Basic Equipment Required

Dependent upon the test system employed: groups of animals, chick embryos, or tubes of cell cultures.

Sources of Error

1. Dilution errors.
2. Insufficient replication to provide statistical significance.

Applications

Titration of animal viruses.

VIABLE-CELL COUNT

Theory

A single microorganism will, if viable, give rise to a visible colony under appropriate conditions for growth. Thus, if a microbial population is dispersed in a suitable diluent and added to a medium that allows for the fixation of individual cells at single, discrete points, the number of cells in a sample can be calculated by counting the number of colonies that develop after incubation and multiplying the number so obtained by the dilution factor.

Basic Equipment Required

Petri dishes; glassware for dilutions; calibrated pipettes; suitable environmental conditions for the desired application (incubators).

Optional Equipment and/or Accessories

Commercial colony counter, either with magnification alone or with magnification combined with an electronic counting probe; membrane filters; wire or glass spreaders; mechanical mixers, dip slides for specific applications; tubes for the roll-tube method; Pasteur pipettes for the drop-count method.

Sources of Error

1. Factors that influence the development of inoculated cells into viable colonies.
2. Preparation of non-liquid samples, both in the extraction of microorganisms from the sample and in the even distribution of the material in the diluent used.
3. Pipetting in the preparation of the dilution series.
4. Cell aggregation (clumping).
5. Adhesion of cells to the spreaders.

Applications

All situations where an estimate of the viable number of microorganisms is required.

DIRECT MICROSCOPIC EXAMINATION

Theory

The number of cells in a given sample may be determined by the microscopic examination of a portion of the material, counting the number of cells observed. By use of an aliquot of known volume or by the inclusion of a counting standard, the number of cells in the original sample can be calculated. The precision and accuracy of this method are related to the number of cells per field and the number of fields counted.

Basic Equipment Required

Standard light microscope; some type of support for the sample, slides, counting chamber, and membrane filter; biological stains for certain applications; internal counting standard, such as latex particles or India ink, if required; electron microscope for virus determinations.

Optional Equipment and/or Accessories

Phase-contrast microscope; fluorescent microscope; commercial counting chambers; capillary tubes; slide cultures.

Sources of Error

1. The precision of the individual performing the count.
2. Differences in the thickness of the layer of the cell suspension.
3. Cell aggregation (clumps or chains).
4. Staining techniques, if used, may cause error in either direction: loss of cells while preparing slides, or counting precipitated stain as cells.
5. Surface/volume ratio of the container in which the cell suspension is prepared or stored.
6. Preparation of samples other than liquid.
7. Use of vital stains when the ratio of viable to dead cells is determined.

Applications

Milk and water examination; tissue cultures; clinical microbiology; virology.

MICROSCOPY
Filters

DR. FRED J. ROISEN AND DANIEL W. MCNEIL

Filters may conveniently be described as attenuators of light. Generally they are simple but precisely constructed devices and are most commonly employed in microscopy to 1) limit the wavelength distribution of a broadband spectral source or 2) to reduce the intensity of a radiation source without altering the spectral distribution.

Spectral distribution of white light is changed as follows:

Color passing through the filter	Colors eliminated by the filter
Red	Blue and green
Blue	Red and green
Green	Red and blue
Yellow (red plus green)	Blue
Magenta (red-blue)	Green
Cyan (blue-green)	Red
Gray	Equal portions of red, green, and blue

Reduction of intensity of radiation is as follows:

If the filter has an optical density of:	The following percentage of light is transmitted:
.10	80
.20	63
.30	50
.40	40
.50	32
.60	25
.70	20
.80	16
.90	13
1.00	10
2.00	1.0
3.00	0.10
4.00	0.010

Filters are especially useful in photomicrography to improve the contrast of lightly colored specimens and to adjust the spectral distribution (color temperature) to suit a particular color film. Filters may be classified as follows according to their applications. Filter types may be conveniently classified according to principle of operation. Virtually all filters employed today function by producing absorption, reflection, interference and/or dispersion effects in the transmitted beam. Each of these filter types will be described below.

Conversion filters for color films	Used to convert color temperature of various light sources to match color balance of color films.
Light-balancing filters	Used to produce subtle changes in color balance (to "cooler" or "warmer" appearance) with color films.
Color compensating (CC) filters	Used in image-forming systems to change overall color balance as well as to compensate for deficiencies in lighting.
Color printing filters	Used singly or in combination for color correction of enlarger light sources in color printing. Not suitable for image-forming systems because made of plastic.
Photometric filters	Designed primarily for visual use with a photometer to reduce color differences between illuminants operating at different color temperatures.
Safelight filters	Provide darkroom illumination compatible with the speed and spectral sensitivity characteristics of sensitized materials.
Variable contrast filters	Provide contrast control for variable-contrast printing papers. (Kodak Polycontrast; Du Pont Varigam®)

ABSORPTIVE FILTERS[1]

Filters based upon the principle of absorption were among the first types to be employed in microscopy and may easily be improvised by filling a flat cell with colored liquid. Due to their thickness such filters are unsatisfactory for use in image-forming optical systems and when used with the light microscope should be restricted to the illuminating system. Liquid filters are seldom used today but are still useful especially as heat-absorbing filters when high intensity light sources such as mercury and xenon arcs are needed. A typical filter of this type consists of a dilute water solution of copper or ferrous sulphate in a flat walled cell having a path length of about 2-5 cm.

More compact, efficient, and less messy than liquid filters are a variety of laminar optical materials available from various manufacturers. These are most commonly fabricated from gelatin, glass, plastic, or combinations of these materials.

Gelatin filters are available in a great variety of colors and densities from companies such as Eastman Kodak and Ilford, Ltd. They are among the thinnest filters available and are the filters chosen for optical systems requiring a precise and invariable optical pathlength. Gelatin filters are available in 2 to 5 in. squares and are relatively inexpensive. They will not, however, tolerate elevated temperatures and may be expected to fade with prolonged use. Heat and blocking filters will retard fading. Gelatin filters are frequently supplied cemented between glass plates to improve mechanical and environmental stability. Precise spectrophotometric data are available from the manufacturers.[2,3]

Glass filters are usually supplied by microscope manufacturers and have numerous advantages over other types such as gelatin or plastic. Glass is dimensionally more stable than other optical materials, is heat resistant, more easily cleaned and is much less subject to fading. Glass filters are for these reasons most satisfactory in illuminating systems for the light microscope. Because of the greater thickness of most colored glass filters, they may require adjustments in mechanical tube length when used between the objective and eyepiece in photomicrography.[4]

Plastic filters are most commonly dyed cellulose acetate. Surface quality and homogeneity are generally poor; hence plastic filters are suitable only for illuminating systems. Like gelatin, plastic is subject to fading and must be replaced periodically.

An important form of absorptive filter is the familiar heat-absorbing glass. This is usually in the form of a lightly tinted glass plate having the ability to pass visible wavelengths while blocking the entrance of infrared radiation into the microscope. It is very important to have such a filter in the illuminating system of the microscope to prevent specimen distortion, death of living specimens and operator fatigue. Most microscope manufacturers supply heat-absorbing filters built into their illuminators.[4,5]

Certain types of neutral density filters may also be considered as absorptive in action and are used to control the intensity of illumination. Gelatin filters, such as the Wratten® No. 96 series,[2] consist of fine carbon particles uniformly dispersed in a gelatin film. Since considerable light is scattered by the carbon, filters of this type are not satisfactory for use in image-forming systems.[2]

REFLECTIVE FILTERS

Filters of this type generally consist of a metal film which has been deposited on a rigid transparent substrate by vacuum evaporation.[6] Several metals are commonly employed in this application, including aluminum, chromium, and inconel. The latter is usually the metal of choice for producing a linear attenuation over the range of the ultraviolet, through the visible, and well into the infrared. Neutral density filters of this type are ideally suited for illumination control and since scattering of light is minimal may be readily employed in image-forming systems.

PLEOCHROIC FILTERS

Certain optically anisotropic materials such as the mineral tourmaline exhibit a property known as pleochroism. This is defined as variation in absorptivity dependent on the plane of vibration of the incident light.[6] When light passes through such a material, only the rays vibrating in one plane are passed,[7] the rays vibrating in the other plane being reflected and/or absorbed. This transmitted light is then said to be polarized and may be employed to learn much useful information about oriented materials such as crystals, fibers, plastics, etc. Polarized light may be produced by a variety of devices such as prisms of calcite (Nichol prisms), stacked glass plates, or most commonly by special filters known commercially as Polaroids.®

In microscopy polarizing filters are usually employed in pairs located near the aperture diaphragm (condenser) and above the objective in the body tube. Although polarizing filters may be used with any microscope, certain other accessories are necessary for quantitative work. These include a graduated, rotatable stage for the measurement of azimuth, various compensators for the determination of retardation, and strain-free objectives and condenser. Additional accessories are also necessary for many specialized types of observation and measurement.

INTERFERENCE FILTERS

Filters of this type depend for their action on the modification of illumination intensity by the superposition of multiple wavefronts.[7] An interference filter generally consists of a semitransparent reflective coating separated from a second such coating by a thin layer of transparent dielectric such as cryolite. The thickness of this layer determines the spectral properties of the filter. These coatings are usually applied to glass by vacuum evaporation and are often protected by cementing a covering glass to the coated surface. Interference filters are capable of isolating narrower portions of the spectrum than are other filter types; also they absorb less heat and do not fade.

A useful variation of the above filter type is the interference wedge. This filter differs in having the dielectric deposited as a layer of gradually increasing thickness along the length of the filter. When viewed by white transmitted light, Newton's color series is observed. Interference wedges may be employed as inexpensive substitutes for monochromators and are available in both calibrated and uncalibrated forms covering a range of from one to ten orders.

To achieve certain spectral characteristics, interference filters are often constructed by depositing thin layers of dielectric materials on a rigid substrate. These layers alternate in refractive index, and in certain instances as many as fifty layers may be deposited. Such filters are quite fragile and should be handled with care. Multilayer coatings of this type are very often efficient polarizers and should be checked for this property prior to use. When used with the polarizing microscope, a depolarizer such as ground glass may be employed to prevent aberrant results.

Since interference filters depend for their action on the multiple reflections of light within a thin transparent layer, they are especially sensitive to angular changes with respect to the incident beam. As the angle of incidence increases from the normal (90°), the path length through the filter increases and the transmitted frequency shifts toward the blue. It is especially important, therefore, in quantitative microscopy to orient interference filters normal to the incident beam.

Because of the very narrow passband obtainable from certain interference filters, they may be used as "yardsticks" in certain specialized applications. In determining refractive indexes by the immersion method, interference filters may be used to isolate the sodium D line for which refractive indexes are most commonly reported. Birefringence determinations by the Senarmont[5] method are greatly simplified with the use of an interference filter. Dry mass measurements with the interference microscope require a knowledge of the illuminating wavelength. Interference filters are especially useful in isolating the excitation wavelength in fluorescence microscopy and in analyzing the emitted wavelengths.

CARE AND CLEANING OF FILTERS

With the exception of elements permanently mounted in sealed optical systems, filters and other optical elements may occasionally require cleaning. This may consist of merely brushing away dust and lint with a soft camel's-hair brush. Fingerprints and other stains may be readily removed by moistening a piece of lens tissue with lens cleaning fluid (hexane or an 0.25% solution of sodium lauryl sulfate in water) and gently wiping the stain away. (When using this fluid or any other liquid on gelatin filters or cemented filters, care should be exercised to prevent the fluid from coming in contact with the edge of the filter. Gelatin filters are lacquered to protect them from moisture, etc., but the edges are unprotected.) Several different pieces of tissue should be used in rapid succession to prevent redeposition of the stain.

When not being used, filters should be kept in a storage case away from heat and moisture. In humid environments gelatin and certain cemented and interference filters should be kept in a desiccator.

TABLE 1

LIST OF MANUFACTURERS OF FILTERS

Source	Filters		
	Absorptive	Interference	Reflective
Bausch & Lomb Rochester, New York 14602	X	X	X
Carl Zeiss, Inc. 444 Fifth Avenue New York, New York 10018	X	X	X
Corning Optical Products Department Corning Glass Works Corning, New York 14830	X	X	
Eastman Kodak Company Rochester, New York 14650	X	X	X
Fish-Schurman Corporation 70 Portman Road New Rochelle, New York 10802	X	X	X
Ilford Incorporated W. 70 Century Road Box 288 Paramus, New Jersey 07652	X		
Spectrum Systems Division of Barnes Engineering Co. 211 Second Avenue Waltham, Massachusetts 02154	X	X	

TABLE 2

BRIEF DESCRIPTIONS OF KODAK FILTERS[2]

Kodak® Wratten® Filter		
Color	Number	Description and Uses
Colorless	0	Attenuation provided by gelatin of 0.1 mm thickness. Useful as a dummy filter in setting up optical systems requiring gelatin filters, when optical pathlength is critical.
Pink (Skylight Filter)	1A	Pale pink. Absorbs ultraviolet radiation. Reduces excess bluishness of outdoor scenes photographed in open shade under a clear, blue sky.
Yellows	2A	Pale yellow. Absorbs ultraviolet radiation below 405 nm. Used with black-and-white materials to reduce haze at high altitudes and as a barrier filter in fluorescence photography.
	2B	Pale yellow. Absorbs ultraviolet radiation below 390 nm. Slightly more effective than No. 2A in reducing haze at high altitudes and also when an excess of ultraviolet radiation is present. Used for fluorescence photography and for the optical system of color printers used for printing Eastman® color print film and Ektacolor® papers.
	2C	Pale yellow. Absorbs ultraviolet radiation below 385 nm. Slightly less effective than No. 2B.
	2E	Pale yellow. Absorbs ultraviolet radiation below 415 nm. Similar to a No. 2B but absorbs more violet. Recommended for filter packs when printing Ektachrome® paper.
	3	Light yellow. Provides partial correction for excess blue in black-and-white aerial photography and motion-picture photography.
	3N5	No. 3 + neutral density of 0.5. Permits use of larger lens aperture to reduce depth of field in motion-picture work.
	4	Yellow. Absorbs ultraviolet and corrects color response of panchromatic emulsions to match approximately the color-brightness response of the eye to outdoor scenes, including sky. Contrast control with Kodak® magenta contact screen (for photogravure).
	6*	Light Yellow. Provides partial correction of panchromatic materials to visual luminance values of outdoor scenes.
	8*	Yellow. Transmits less blue than No. 4. Widely used for correct rendition of sky, clouds, and foliage in black-and-white photography with Type B (high red-sensitive) panchromatic materials.
	8N5	No. 8 + neutral density of 0.5. Permits use of larger lens aperture to reduce depth of field in motion-picture work.
	9*	Deep yellow. Similar to No. 8 but tends to overcorrect sky rendition for more dramatic effect.
	11*	Yellowish-green. Corrects color response of Type B panchromatic emulsions to match color-brightness response of the eye to objects exposed to tungsten illumination. Reproduces greens slightly lighter in daylight.

TABLE 2 (Continued)

BRIEF DESCRIPTIONS OF KODAK FILTERS

Kodak® Wratten® Filter		
Color	**Number**	**Description and Uses**
Yellows (cont.)	12	Deep yellow. Minus-blue filter (see No. 32 for minus-green and No. 44A for black-and-white infrared films in infrared sensitive films for absorption of blue light.
	13*	Dark yellowish-green. Provides correction similar to No. 11, except that it is used for highly green-sensitive materials.
	15*	Deep yellow. Darkens sky in landscape photography more dramatically than No. 8 or No. 9. Useful for copying documents on yellowed paper. Used for black-and-white infrared photography and for special effects in infrared color photography. Used also for fluorescence photography.
	HF-3 HF-4 HF-5	Light yellow. Designed especially for color aerial photography to provide haze penetration.
	18A	Visibly opaque. Only transmits ultraviolet radiation between about 300 and 400 nm (e.g., 365 nm line of mercury spectrum) and infrared radiation. Used for ultraviolet reflection photography.
Oranges and reds	16	Yellow-orange. Permits greater overcorrection of sky than No. 15. Absorbs small amount of green.
	21	Orange. Contrast filter used for blue and blue-green absorption.
	22	Deep orange. Contrast filter with greater green absorption than No. 21. Used in photomicrography to increase contrast of blue preparations. Transmits only yellow radiation from mercury-vapor illumination.
	23A	Light red. Contrast filter with greater green absorption than No. 21 or 22. Suitable for two-color projection and special effects in black-and-white motion-picture work. Used for color separation work.
	24	Red. Formerly used for "two-color photography" (for daylight with green No. 57; for tungsten with green No. 40 or 60). Used also for white-flame-arc tricolor projection.
	25*	Red tricolor. Used for color separation work and tricolor printing, two-color general viewing, contrast effects in commercial and outdoor black-and-white photography and haze penetration in aerial work. Used also to remove blue in infrared photography.
	26	Red. For anaglyph viewing for a three-dimensional effect with a No. 55 (green).
	29*	Deep red tricolor. Used for color separation and tricolor printing work. Tricolor projection (tungsten) with No. 47 and 61.
Magentas and violets	30	Light magenta. Contrast filter for green absorption, especially in photomicrography. Contrast control with the Kodak® magenta contact screen (for photogravure).
	31	Magenta. Contrast filter for stronger green absorption than No. 30.
	32	Magenta. Minus-green (see No. 12 for minus-blue and No. 44A for minus-red).

TABLE 2 (Continued)

BRIEF DESCRIPTIONS OF KODAK FILTERS

Kodak® Wratten® Filter Color	Number	Description and Uses
Magentas and violets (cont.)	33	Magenta. Contrast filter for strongest green absorption. Used for photomechanical color masking.
	34	Deep violet. Contrast filter for green absorption. Offers slightly less blue and more red absorption than No. 32.
	34A	Violet. For minus-green and plus-blue separation.
	35	Purple. Contrast filter for total green absorption and partial blue and red absorption. Used in photomicrography.
	36	Dark violet. Contrast filter for total green absorption. More red and less blue absorption than No. 35.
Blues and blue-greens	38	Light blue. Contrast filters with some ultraviolet and some red absorption. Useful for correcting tendency of reds to reproduce too lightly in tungsten illumination.
	38A	Blue. Contrast filter for some ultraviolet and green absorption and much red absorption. In photomicrography, for increasing contrast in records of faintly yellow or orange preparations.
	39 (Glass)	Blue. Glass contrast filter for printing motion-picture duplicates.
	40	Light green. Formerly used for "two-color photography" (tungsten) with No. 24 (red).
	44	Light blue-green. Minus-red filter with much ultraviolet absorption.
	44A	Light blue-green. Minus-red (see No. 12 for minus-blue and No. 32 for minus-green).
	45	Blue-green. Contrast filter for ultraviolet and red absorption with some blue and green absorption. Used in photomicrography.
	45A	Blue-green. Similar to No. 45 with slightly less absorption of blue and green. Offers highest resolving power in visual microscopy.
	46	Blue. Used for three-color projection with No. 29 and 57.
	47*	Blue tricolor. Used for color separation work. For contrast effects in commercial and outdoor black-and-white photography. Tricolor projection (tungsten) with No. 29 and 61.
	47A	Light blue. Used for exciting fluorescein in medical applications of fluorescence photography.
	47B	Deep blue tricolor. Used for color separation and tricolor printing work.
	48	Deep blue. Provides some green absorption and strong absorption in the yellow, red and ultraviolet.
	48A	Deep blue. Similar to No. 48, except ultraviolet absorption and blue absorption to about 425 nm is less than No. 48.

TABLE 2 (Continued)

BRIEF DESCRIPTIONS OF KODAK FILTERS

Kodak® Wratten® Filter		
Color	Number	Description and Uses
Blues and blue-greens (cont.)	49*	Dark blue. Similar to No. 48A, except blue absorption is generally lower. Used for color separation work.
	49B	Dark blue. Similar to No. 48A and 49, except blue absorption is generally lower than either. Used for color separation work.
	50	Deep blue monochromat. Transmits mercury line at 436 nm and to a lesser extent, lines at 398, 405, and 408 nm.
Greens	52	Light green. Absorbs some blue and red.
	53	Green. Absorbs much blue and red and some green.
	54	Deep green. Contrast filter with total blue and red absorption and much green absorption.
	55	Green. For anaglyph viewing for a three-dimensional effect with No. 26 (red).
	56	Light green. Absorbs some blue and red. Useful for dropping out green-brown graph lines when copying charts.
	57	Green. Formerly used for "two-color photography" (daylight) with No. 24 (red). Often used with a No. 15 in tricolor printing from Eastman® color negative films.
	57A	Green. Absorbs some blue and much red.
	58*	Green tricolor. Used for color separation and tricolor printing work. Used to obtain contrast effects in commercial photography and micrography.
	59	Light green. Contrast filter similar to No. 57A, except blue absorption is slightly greater, while yellow, green, and red absorption is less.
	59A	Light green. Similar to No. 59 with less red and ultraviolet absorption.
	60	Green. Formerly used for "two-color photography" (tungsten) with No. 24 (red).
	61*	Deep green tricolor. Used with No. 29 and 47 for tricolor projection (tungsten) and for color separation and tricolor printing work.
	64	Light blue-green. Provides some red absorption.
	65	Blue-green. Similar to No. 64, except blue, red, and green absorption is greater.
	65A	Blue-green. Provides less blue and green absorption and greater red absorption than No. 65.
	66	Very light green. Contrast filter used in medical photography for emphasizing pink lesions in "orthochromatic" renderings with panchromatic materials. Also used in microscopy with pink- and red-stained preparations.
Narrow-band	70	Dark red. Narrow-band monochromat used for making separation positives from color negative films. Also used for three-color printing on color papers.
	72B	Dark orange-yellow monochromat.

TABLE 2 (Continued)

BRIEF DESCRIPTIONS OF KODAK FILTERS

Kodak® Wratten® Filter		
Color	Number	Description and Uses
Narrow-band (cont.)	73	Dark yellow-green monochromat.
	74	Dark green monochromat. Transmits ten percent of green radiation and virtually no yellow radiation from mercury-vapor illumination.
	75	Dark blue-green monochromat.
Photometric	78 78A 78B 78C	Bluish series of Kodak® Wratten® photometric filters.
	86 86A 86B 86C	Yellowish series of Kodak® Wratten® photometric filters.
	79	Light blue. Used in photographic sensitometry to correct 2360°K to 5500°K.
Conversion	80A 80B 80C	Blue series of conversion filters for color films.
	85 85N3 85N6 85N9 85B 85BN3 85BN6 85C	Amber series of conversion filters for color films.
Light balancing	81 81A 81B 81C 81D 81EF	Yellowish series of light-balancing filters.
	82 82A 82B 82C	Bluish series of light-balancing filters.
Miscellaneous	87 87A 87B 87C 88A	Visibly opaque series of filters used primarily in infrared photography to absorb unwanted visible light.
	89B	Visibly opaque. Used for infrared photography, especially aerial.
	90	Dark grayish-amber. Monochrome viewing filter. Used to visually approximate the relative tones of gray produced by different colors under daylight illumination in black-and-white prints.

TABLE 2 (Continued)

BRIEF DESCRIPTIONS OF KODAK FILTERS

Kodak® Wratten® Filter Color	Number	Description and Uses
Miscellaneous (cont.)	92	Red. Used with No. 93 and 94 for densitometric measurement of color films and papers.
	93	Green. Used with No. 92 and 94 for densitometric measurement of color films and papers.
	94	Blue. Used with No. 92 and 93 for densitometric measurement of color films and papers.
	96	Neutral. Neutral density of 1.0. For uniform light attenuation throughout the visible spectrum. Transmits infrared. Available in 12 additional densities with a transmittance range of 80% to 0.01%.
	97	Dichroic filter used to detect visually the red fluorescence of chlorophyll in green foliage.
	98	Blue. Equivalent to No. 47B plus No. 2B filter. Used for making separation positives from color negative films. Also used for three-color printing on color papers.
	99	Green. Equivalent to No. 61 plus No. 16 filter. Used for making separation positives from color negative films. Also, for three-color printing on color papers.
	102	Yellow-green. Used to convert the response characteristics of a barrier-layer photocell (as in a densitometer) to the luminosity response of the eye.
	106	Amber. Used to convert the response characteristics of an S-4 type photocell (as in a densitometer) to the luminosity response of the eye. Note: Due to the relatively high sensitivity of an S-4 type photocell in the red and near infrared, density readings above 2.0 should be interpreted with caution when measuring dye samples.

* The additional letter (or letter and number) designation for a number of Wratten® filters has been dropped from the official name of these filters. Designations such as K1, K2, A, G, etc., were assigned many years ago when the number of filters required for photographic work was quite small, and when a simple identification system was adequate. As the need for additional filters became apparent, Kodak® adopted its present numbering system for Wratten® filters, but continued to retain the original designation. The necessary introduction of filters having a number-plus-letter designation (such as the 3N5, 23A, 47B, and others) has caused some confusion and delay in filling customer orders because of the new use for these same letters. For this reason, Kodak® has dropped the alphameric designations and urges its customers to use only the current designations.

Because the published literature frequently contains references to the older designations only, both current and discontinued designations are shown at the right.

Current Designations	Discontinued Designations
No. 6	K1
No. 8	K2
No. 9	K3
No. 11	X1
No. 13	X2
No. 15	G
No. 25	A
No. 29	F
No. 47	C5
No. 49	C4
No. 58	B
No. 61	N

Taken from *Kodak Filters for Scientific and Industrial Uses* (B-3). Reproduced with the permission of the Eastman Kodak Company.

TABLE 3

EXAMPLE OF APPLICATION OF KODAK® LIGHT–BALANCING FILTERS

Light source	Film Corrected For 3200°K (Kodak® type B)	3400°K (Kodak® type A)	Daylight
Ribbon filament lamp, 3000°K	82A	82C	80A + 82A
Coil filament lamp up to 100W, 3100°K	82	82B	80A + 82
Same as 300–750 W, 3200°K	None	82A	80A
Photoflood, 3400°K	81A	None	80B

REFERENCES

1. Smith, Warren J., *Modern Optical Engineering,* pp. 160-177, McGraw-Hill, New York (1966).
2. *Kodak Filters for Scientific and Industrial Uses* (B-3), The Eastman Kodak Company, Department 454, Rochester, New York 14650.
3. Ilford Ltd., Ilford, Essex, England.
4. *Corning Color Filter Glasses,* Corning Glass Works, Corning, New York (1970).
5. Hartshorne, N. H., and Stuart, A., *Crystals and the Polarizing Microscope,* pp. 208-214, American Elsevier, New York (1970).
6. Jenkins, Francis A., and White, Harvey E., *Fundamentals of Optics,* pp. 211-221, 284-285, 492-496; Mc-Graw-Hill, New York (1957).
7. Martin, L. C., *The Theory of the Microscope*, pp. 372-374, American Elsevier, New York (1966).

Stains for Light Microscopy

DR. HUBERT A. LECHEVALIER AND DR. FRED J. ROISEN

General information about the chemistry of dyes used as biological stains, their nomenclature, and methods of testing can be found in Reference 1. The following is a selection of staining methods that may be of assistance to microbiologists.

STAINING SOLUTIONS

ACETO CARMINE

Heat an excess of powdered carmine in 45% acetic acid to boiling. Cool, then filter.[2]

ACETO CARMINE, BILLING'S MODIFICATION

Add 1 g of carmine to 100 ml of boiling 45% acetic acid. Boil for about 2 minutes, cool, then filter. To half of this mixture add a few drops of a solution of ferric hydroxide in 45% acetic acid until the liquid becomes bluish red, but without visible precipitate, then add the untreated portion of the aceto-carmine mixture.[2]

ALCOHOLIC SAFRANIN

Dissolve 0.25 g of safranin 0 in 10 ml of 95% ethyl alcohol. Mix with 100 ml of distilled water.[2]

AMMONIUM OXALATE CRYSTAL VIOLET

Dissolve 2 g of crystal violet in 20 ml of 95% ethyl alcohol. Mix with 80 ml of 1% (w/v) aqueous ammonium oxalate.[2]

AQUEOUS NIGROSIN

Add 6 to 8 g of nigrosin to 100 ml of distilled water and dissolve by placing the mixture for 30 minutes in a boiling water bath. Replace any water lost by evaporation, then add 0.5 ml of formalin. Filter twice through a double thickness of filter paper.[2]

AQUEOUS SAFRANIN

Prepare a 0.5% (w/v) solution of safranin in water.[2]

CARBOL ROSE BENGAL

Dissolve 1 g of rose Bengal in 100 ml of 5% (w/v) aqueous phenol. It is sometimes well to add 0.01 to 0.03 g of calcium chloride ($CaCl_2$) to the solution.[2]

COTTON BLUE

Prepare 0.1% (w/v) cotton blue solution in lactic acid (U.S.P. grade). This may be further diluted with water.[3]

CRYSTAL VIOLET IN DILUTE ALCOHOL

Dissolve 2 g of crystal violet in 20 ml of 95% ethyl alcohol. Mix with 80 ml of distilled water.[2]

LOEFFLER'S ALKALINE METHYLENE BLUE

Dissolve 0.3 g of methylene blue in 30 ml of 95% ethyl alcohol. Mix with 100 ml of 0.01% (w/v) aqueous potassium hydroxide (KOH) solution.[2]

MELZER'S SOLUTION

Dissolve 1.5 g of potassium iodide (KI) and 0.5 g of iodine in 20 ml of water, then add 22 g of chloral hydrate.[3]

METHYLENE BLUE IN DILUTE ALCOHOL

Dissolve 0.3 g of methylene blue in 30 ml of 95% ethyl alcohol. Mix with 100 ml of distilled water.[2]

SCHIFF REAGENT

Dissolve 0.5 g of basic fuchsin by pouring over it 100 ml of boiling distilled water. Cool to 50°C. Filter and add 10 ml of 1*N* hydrochloric acid and 0.5 g of anhydrous potassium metabisulfite to the filtrate. Allow the solution to stand in the dark overnight. The solution should become colorless or pale straw-colored. If not completely decolorized, add 0.25 to 0.50 g of charcoal, shake thoroughly, and filter immediately. This solution will keep for several weeks in a tightly stoppered bottle.

SUDAN BLACK B

Dissolve 0.5 g of Sudan black B in 100 ml of ethylene glycol.

ZIEHL'S CARBOLFUCHSIN

Dissolve 0.3 g of basic fuchsin in 10 ml of 95% ethyl alcohol. Mix with 100 ml of 5% aqueous phenol.[2]

General Stains for Bacteria

Loeffler's alkaline methylene blue, methylene blue in dilute alcohol, ammonium oxalate crystal violet, alcoholic safranin, and aqueous safranin solutions are all good general stains for bacterial preparations. Bacterial endospores will show as unstained elements

Negative Staining

India ink is a good, easy-to-use substance for negative staining in wet mounts. The ink, or a dilution thereof, is mixed with the microbial suspension, giving a black to sepia background against which the unstained microbial elements stand out clearly. Unfortunately, India ink is usually contaminated with bacteria and is, therefore, not very suitable for their study.

Aqueous nigrosin solution can be used as follows. With the help of a wire loop, stir a very small sample of a bacterial (or other) colony into a loopful of nigrosin on a coverslip. Place a second coverslip with a small drop of nigrosin (without bacteria) over the first one. The second slip should be turned through an angle of 45 degrees relative to the first one. Let the two drops flow together; then, gripping the corners of upper and lower slip, slide the two swiftly and gently apart (in a sideways movement). Let the two films so produced dry in air. Place the best one, face down, on a slide. Tack the corners down with wax.[5]

Gram Stains for Bacteria

HUCKER'S MODIFICATION[2]

Staining Schedule

1. Stain bacterial smears for 1 minute with ammonium oxalate crystal violet.
2. Wash in tap water.
3. Immerse for 1 minute in an iodine solution prepared by adding 1 g of iodine and 2 g of potassium iodide (KI) to 300 ml of distilled water.
4. Wash in tap water; blot dry.
5. Decolorize for 30 seconds, with gentle agitation, in 95% alcohol; blot dry.
6. Counterstain for 10 seconds with alcoholic safranin solution.
7. Wash in tap water.
8. Dry, then examine the smears.

Results

Gram-positive organisms: blue
Gram-negative organisms: red

Acid-Fast Stains for Bacteria[2]

Staining Schedule

1. Prepare smears and fix them on a flat surface over boiling water.

2. Stain for 3 to 5 minutes with Ziehl's carbolfuchsin, applying heat to permit gentle steaming.

3. Rinse in tap water.

4. Decolorize in 95% alcohol containing 3% by volume of concentrated hydrochloric acid (HCl) until only a suggestion of pink remains.

5. Wash in tap water.

6. Counterstain with saturated aqueous methylene blue or Loeffler's methylene blue.

7. Wash in tap water.

8. Dry, then examine the smears.

Results

Acid-fast bacteria: red
Other organisms: blue

Fat-Staining

Spread bacteria in a loop of distilled water on a grease-free coverslip. Let the film dry, then invert it over a drop of Sudan black B. Cytoplasm will be colorless and fat droplets will be dark.[5]

OBSERVATION MEDIA FOR FUNGI

Fungi can, of course, be observed in water mounts, but clearing can be achieved by using 5 to 20% potassium hydroxide (KOH) or lactophenol.

LACTOPHENOL OF AMANN[6]

Dissolve 1 g of crystalline phenol in a mixture of 1 g lactic acid, 2 g glycerol and 1 g distilled water. Keep in a dark bottle.

General Stains for Fungi

Fungi can be stained for general observation by some of the stains used for bacteria. Methylene blue and cotton blue solutions are especially useful. Concentrated cotton blue may be diluted in lactophenol.

Melzer's solution will give reddish to black reactions with various parts of mycelia. In general, it is a detector of starch, glycogen and related substances, giving the so-called amyloid (blue to black) reaction.

Periodic Acid-Schiff Stain for Fungi in Animal Tissues[4]

Staining Schedule

1. Fix the specimen to be examined.

2. Dehydrate, embed, and section it.

3. Deparaffinize the sections, then dehydrate them with absolute alcohol.

4. Wash in distilled water.

5. Immerse in 1% periodic acid for 5 minutes.

6. Wash in running tap water for 15 minutes.

7. Stain in Schiff reagent for 10 to 15 minutes.

8. Transfer directly to two changes of either of the following solutions for 5 minutes each:

(1)	10% potassium metabisulfite	5 ml
	1*N* hydrochloric acid	5 ml
	Distilled water	100 ml
(2)	Thionyl chloride	5 ml
	Distilled water	100 ml

9. Wash in running tap water for 10 minutes.

10. Counterstain with light green.

11. Dehydrate, clear, and mount.

Results

Fungal elements: red

Toluidine Blue-Safranin Stain for Fungi in Plant Tissues[7]

Staining Schedule

1. Flood sections of plant tissue with 1% (w/v) toluidine blue in a 1% (w/v) solution of borax in distilled water; place on a hot plate at 60 to 85°C until steam or a few bubbles appear.

2. After washing in tap water, flood with a 0.25*N* solution of HCl in 50% (v/v) ethanol.

3. Decolorize in a 0.2*N* solution of NaOH in 50% (v/v) ethanol.

4. After washing in tap water, counterstain with 1% (w/v) safranin in water.

5. After washing in water, the sections may be dehydrated in alcohol and mounted as desired.

Results

Fungal hyphae and spores: deep red
Plant tissues: blue or greenish blue

REFERENCES

1. Lillie, R. D. (Ed.), *H. J. Conn's Biological Stains, A Handbook on the Nature and Uses of the Dyes Employed in the Biological Laboratory,* 8th ed. Williams and Wilkins, Baltimore, Maryland (1969).
2. Conn, H. J., Darrow, M. A., and Emmel, V. M., *Staining Procedures Used by the Biological Stain Commission,* 2nd ed. Williams and Wilkins, Baltimore, Maryland (1960).
3. Kühner, R., and Romagnesi, H., *Flore Analytique des Champignons Supérieurs.* Masson, Paris, France (1953).
4. Kligman, A. M., and Mescon, H., *J. Bacteriology, 60,* 415 (1950).
5. Robinow, C., *Personal Communication.*
6. Segretain, G., Drouhet, E., and Mariat, F., *Diagnostic de laboratoire en mycologie médicale.* Editions de la Tourelle, St. Mandé, France (1958).
7. Pomerleau, R., *Can. J. Bot., 48,* 2043 (1970).

A Fixation and Embedding Procedure for Thin-Sectioning Bacteria

DR. M. L. HIGGINS

The basic procedure described here was selected and developed in collaboration with Dr. Bijan K. Ghosh, Rutgers University.

APPARATUS

1. Centrifuge, preferably refrigerated, such as International Model PR-2

2. Water bath, 45°C

3. Dissection microscope

4. Rotator, such as Scientific Industries Model 150V

5. Oven, variable in temperature from 37 to 60°C, such as the Fischer Isotemp Junior Model

REAGENTS

1. Glutaraldehyde solution

2. Buffered salt solution[1]

   0.05*M* phosphate buffer, pH 6.2; 0.08*M* KCl; 0.01*M* MgAc (0.05*M* cacodylate buffer can be used in place of 0.05*M* phosphate buffer).

3. Kellenberger base[2]

   Sodium barbital, 2.94 g; sodium acetate (hydrated), 1.94 g; sodium chloride, 3.40 g; distilled water to make 100 ml.

4. Kellenberger buffer[2]

   Kellenberger base, 5.0 ml; distilled water, 13.0 ml; 0.1*N* HCl, 7.0 ml; 1*M* $CaCl_2$, 0.25 ml; the pH must be between 6.0 and 6.2.

5. Kellenberger tryptone broth[2]

   Tryptone, 0.1 g; NaCl, 0.05 g; Kellenberger buffer, 10.0 ml. (One part of the tryptone solution is commonly diluted with nine or more parts of Kellenberger buffer to reduce the tryptone precipitate formed during OsO_4 fixation.)

6. Kellenberger osmium solution[2]

   OsO_4, 1.0 g; Kellenberger buffer, 100 ml.

7. Kellenberger fixative[2]

   Kellenberger osmium solution, 1.0 ml; Kellenberger tryptone broth (or diluted tryptone broth), 0.1 ml; combine just before use.

8. Kellenberger uranyl acetate solution[2]

   Uranyl acetate, 0.05 g; Kellenberger buffer, 10.0 ml; mix for twenty minutes with a magnetic stirrer, then filter through two thicknesses of Whatman No. 1 filter paper.

9. Agar, 2%

   Add 0.2 g of agar to 10.0 ml of boiling Kellenberger buffer.

10. Ethanol, absolute

11. Propylene oxide

12. Epon 812

    This embedding medium is compounded according to Luft's procedure[3] as modified by Ladd Research Industries, Inc., Burlington, Vermont, to account for batch variation in weight per epoxide equivalent in the epoxy resins. The ratio of Luft's mixture A to mixture B varies with the biological material used and with the cutting conditions; as a starting point, a ratio of three parts A to seven parts B, plus 0.14 ml of DMP-30 is suggested.

13. Uranyl acetate, alcoholic

Add 0.4 g of uranyl acetate to 10 ml of 50% ethanol. After mixing for twenty minutes with a magnetic stirrer, filter the solution through two thicknesses of Whatman No. 1 filter paper. The solution should be used within one hour of preparation.

14. CO_2-free water

Boil distilled water for twenty minutes with boiling stones.

15. Reynolds lead citrate[4]

Place 1.33 g of lead nitrate, 1.76 g of sodium citrate, and 30 ml of CO_2-free distilled water in a 50-ml volumetric flask. Shake the flask vigorously for about one minute, then allow it to stand for thirty minutes, shaking it intermittently (about once every minute). Add 8 ml of freshly prepared 1*N* sodium hydroxide (made with CO_2-free water) to the suspension, then bring the volume to 50 ml with CO_2-free water. After mixing by inversion, the solution should be clear. It can be stored in a disposable syringe for several months, provided that the tip is properly sealed.

PROCEDURE

Glutaraldehyde is added to a liquid culture containing about 2,000 to 4,000 μg of bacteria (dry weight) to a final concentration of 3% glutaraldehyde. A two-hour period of fixation at room temperature is usually sufficient for cells that were in the exponential phase of growth. After fixation, the cells are pelleted (about 1,000 x *g* for 15 minutes) and washed overnight in 25 ml of cold buffered salt solution (BSS).[1] Stationary-phase cells require longer periods of fixation; after the normal two-hour primary glutaraldehyde fixation, the cells are pelleted and fixed for an additional sixteen to twenty-four hours in 25 ml of BSS containing 3% glutaraldehyde. The cells are then again pelleted and washed overnight in 25 ml of cold BBS without glutaraldehyde.

After the cells from either fixation schedule have been washed overnight, they must receive three to four more washes (each thirty minutes) in 20 ml of cold BSS. On completion of the last BSS wash, the cells are suspended for fifteen minutes at room temperature in 5 ml of Kellenberger buffer (KB).[2] The cells are then transferred to plastic 12-ml (17 mm x 119 mm) conical centrifuge tubes. After removal of the KB, they are post-fixed overnight at room temperature in 2 to 3 ml of Kellenberger osmium fixative (KF).[2] The next morning the osmium solution is removed, and the cells are suspended in 5.0 ml of Kellenberger uranyl acetate[2] for two hours.

Prior to dehydration and infiltration, the cells are embedded in agar to avoid further rounds to centrifugation; this can also be done after glutaraldehyde fixation. While the cells are being pelleted in the uranyl acetate solution, a tube of molten 2% agar and pairs of prebalanced 50-ml swinging centrifuge buckets (each containing about 20 ml of water and adapted with rings that permit the buckets to take 12-ml conical tubes) are placed in a 45°C water bath. When the water in the buckets and the agar solution have both been equilibrated to 45°C, the uranyl acetate solution is decanted from the cell pellets. The conical tubes are then placed in the water bath for about one minute before adding the agar. Working quickly to avoid drying of the pellets, about 0.5 ml of agar is added to each conical tube. The cells are suspended in the agar by vortical agitation, and the tubes are placed in the centrifuge buckets. To concentrate the cells before the agar hardens, the buckets are rapidly transferred from the water bath to a refrigerated centrifuge precooled to 10°C, brought without delay to 1,000 x g, and centrifuged for fifteen minutes. The tubes are then placed for about fifteen to thirty minutes in an ice bath. The resulting hardened pellets are removed by inserting a pointed spatula between the pellet and the side of the tube, rotating the pellet once, inverting the tube, and rapping it once sharply on a solid surface. The agar-enrobed cells are placed under a dissection microscope, cut with a razor blade into cubes about 0.25 mm^2 in size, and transferred to capped 12 mm x 75 mm tubes of disposable polypropylene (Falcon Plastics, Oxnard, California) containing 50% ethanol.

The blocks are dehydrated and infiltrated by sequentially adding, in 3- to 5-ml amounts, the following: one change of cold 70% ethanol overnight; one change of 95% ethanol for ten minutes; four changes of 100% (absolute) ethanol for thirty minutes each (to insure proper dehydration, it is best to use a fresh bottle of absolute ethanol); four changes of 100% propylene oxide for fifteen minutes each, trying not to exceed one hour of total exposure; finally, one change of a solution containing 50% propylene oxide and 50% Epon 812 mixture, plus accelerator, for three to four hours. Except for the 70% ethanol wash, all operations are carried out at room temperature. The blocks can be left in 70% ethanol almost indefinitely without adverse effects.

After the final wash (50% propylene oxide and 50% Epon 812 mixture), the vials are capped and rotated for three to four hours (the time is not critical), then uncapped and rotated overnight; this step cannot be done in an atmosphere of high humidity. The next morning, blocks as close to 0.25 mm^2 as possible are selected with a dissection microscope and transferred to plastic capsules containing fresh Epon 812 mixture. Any bubbbles found in the Epon-containing capsules can be removed by placing the capped capsules in plastic 16 mm x 188 mm tubes and centrifuging them for one or two minutes at 120 x g. This can be repeated, if necessary, during the early phases of the curing process. The blocks are cured by heating at 37°C overnight, 48°C all day, then 60°C for at least two days.

After the blocks have been sectioned and the resulting ribbons picked up on appropriately coated grids, the sections are post-stained in alcoholic uranyl acetate for twenty minutes at 45°C. The grids are submerged in uranyl acetate stain contained in the wells of microculture slides. To prevent evaporation and stain precipitation, the well slides are placed in plastic petri dishes on top of several layers of filter paper saturated with 50% ethanol. The grids are washed free of stain by dipping them individually with forceps, thirty to fifty times each, in three changes of distilled water. The sections are then counterstained for ten minutes at room temperature with lead citrate, which is usually diluted; one part lead stain added to five parts 0.01*N* sodium hydroxide (prepared with CO_2-free water) is a good concentration to try initially. It is convenient to lead-stain in a plastic petri dish with a paraffin-lined bottom. Sodium hydroxide pellets are placed on one side of the dish in an effort to absorb carbon dioxide. With forceps, the grids are inserted into drops of lead stain applied to the paraffin surface with a Pasteur pipette. After staining, each of the grids is transferred in the petri dish to three drops of 0.02*N* sodium hydroxide for five minutes, and finally the grids are individually washed in streams of CO_2-free water.

REFERENCES

1. Higgins, M. L., and Shockman, G. D., *J. Bacteriol., 103,* 244 (1970).
2. Kellenberger, E., Ryter, A., and Sechaud, J., *J. Biophys. Biochem. Cytol., 4,* 671 (1958).
3. Luft, J. H., *J. Biophys. Biochem. Cytol., 9,* 409 (1961).
4. Reynolds, E. S., *J. Cell Biol., 17,* 208 (1963).

Melting and Boiling Points of Metals Used in Shadow-Casting

VIRGINIA L. THOMAS

Metal	Melting Point, °C	Boiling Point, °C
Aluminum	660	2057
Chromium	1890	2482
Germanium	938	2830
Gold	1063	2966
Gold 60:palladium 40	1410	
Iridium	2443	4500
Molybdenum	2610	5560
Palladium	1552	2900
Platinum	1769	4300
Platinum 80:iridium 20	1875	
Silver	961	2210
Tungsten	3380	5927

Characteristics of Some Electron Microscopes

VIRGINIA L. THOMAS

TABLE 1

TRANSMISSION ELECTRON MICROSCOPES

AEI Scientific Apparatus, Inc., England

Model	EM8
Point-to-point resolution	3 Å
Magnification range	2,500X to 400,000X; scan, 200X
Acceleration voltage	40, 60, 80, kv; 120 kv "overvolt" for stabilizing
Condenser lens	Double
Diffraction cameras	Selective area
Airlocks	Specimen chamber, camera chamber
Anticontamination	Standard equipment; liquid N_2
Photographic facilities	24 plates (3¼" x 3¼" or 2½" x 3½"), or 50 exposures on 70-mm film
Timer and exposure meter	Standard equipment
Vacuum system	Automatic; rotary pump, oil diffusion pump with cold trap, buffer tank, dry- nitrogen bleed valve

Model	EM801 Biological
Point-to-point resolution	5 Å
Magnification range	1,000X to 160,000X; scan, 200X
Acceleration voltage	40, 60, 80 kv; 120 kv "overvolt" for stabilizing
Condenser lens	Double

TABLE 1 (Continued)

TRANSMISSION ELECTRON MICROSCOPES

Diffraction cameras	Selective area
Airlocks	Specimen chamber, camera chamber
Anticontamination	Standard equipment; liquid N_2
Photographic facilities	24 plates (3¼" x 3¼" or 2½" x 3½"), or 50 exposures on 70-mm film
Timer and exposure meter	Standard equipment
Vacuum system	Automatic; rotary pump, oil diffusion pump with cold trap, buffer tank, dry-nitrogen bleed valve
Model	EM802 Metallurgical
Point-to-point resolution	5 Å with high-resolution goniometer stage
Magnification range	1,000X to 160,000X
Acceleration voltage	40, 60, 80 kv; 120 kv "overvolt" for stabilizing
Condenser lens	Double
Diffraction cameras	Selective area; four camera lengths
Airlocks	Specimen chamber, camera chamber
Anticontamination	Standard equipment; liquid N_2
Photographic facilities	24 plates (3¼" x 3¼" or 2½" x 3½"), or 50 exposures on 70-mm film
Timer and exposure meter	Standard equipment
Vacuum system	Automatic; rotary pump, oil diffusion pump with cold trap, buffer tank, dry-nitrogen bleed valve
Model	Corinth 275
Point-to-point resolution	10 Å
Magnification range	600X to 100,000X; scan, 250X
Acceleration voltage	60 kv
Condenser lens	Double
Diffraction cameras	Selective area
Airlocks	Specimen chamber
Anticontamination	Optional
Photographic facilities	50 exposures on 70-mm film
Timer and exposure meter	Standard equipment
Vacuum system	Automatic; oil diffusion pump, rotary pump

Akashi Seisakusho Ltd., Japan

Model	MAAK-1
Point-to-point resolution	3 Å
Magnification range	300X to 500,000X in 22 steps
Acceleration voltage	25, 50, 75, 100 kv
Condenser lens	Double
Diffraction cameras	Selective area; high resolution optional
Airlocks	Specimen chamber, camera chamber
Anticontamination	Standard equipment; liquid-N_2 traps on diffusion pumps
Photographic facilities	24 exposures, 82 mm x 119 mm
Timer and exposure meter	Automatic
Vacuum system	Automatic; two rotary pumps, three oil diffusion pumps

Hitachi, Japan[a]

Model	HU-12
Point-to-point resolution	3 Å

[a] American distributor: Perkin-Elmer

TABLE 1 (Continued)

TRANSMISSION ELECTRON MICROSCOPES

Magnification range	1,000X to 500,000X in steps; 500X to 500,000X continuous; scan, 250X
Acceleration voltage	25, 50, 75, 100, 125 kv
Condenser lens	Double
Diffraction cameras	Selective area; high resolution optional
Airlocks	Specimen chamber, camera chamber, gun chamber
Anticontamination	Standard equipment; cold traps on oil diffusion pumps
Photographic facilities	24 exposures, 3¼" x 4"
Timer and exposure meter	Standard equipment
Vacuum system	Automatic; two rotary pumps, two oil diffusion pumps, buffer tank
Model	HU-11E
Point-to-point resolution	3.5 Å
Magnification range	300X to 400,000X
Acceleration voltage	25, 50, 75, 100 kv
Condenser lens	Double
Diffraction cameras	Selective area; high resolution optional
Airlocks	Specimen chamber, camera chamber, gun chamber
Anticontamination	Standard equipment; liquid N_2
Photographic facilities	18 plates (3¼" x 4")
Timer and exposure meter	Standard equipment
Vacuum system	Automatic; two rotary pumps, oil diffusion pump
Model	HS-8
Point-to-point resolution	8 Å
Magnification range	1,000X to 100,000X
Acceleration voltage	25, 50 kv; 75 kv optional
Condenser lens	Double
Diffraction cameras	Selective area; high resolution optional
Airlocks	Specimen chamber
Anticontamination	Standard equipment; cold trap on oil diffusion pump
Photographic facilities	18 plates (3¼" x 4")
Timer and exposure meter	Standard equipment
Vacuum system	Push-button; two rotary pumps, oil diffusion pump

JEOLCO, Japan

Model	JEM-100B
Point-to-point resolution	3 Å
Magnification range	300X to 500,000X in steps; scan, 250X
Acceleration voltage	20, 40, 60, 80, 100 kv
Condenser lens	Double
Diffraction cameras	Selective area; high resolution optional
Airlocks	Specimen chamber, camera chamber, gun chamber
Anticontamination	Standard equipment; cold trap on oil diffusion pump
Photographic facilities	24 exposures, 2¼" x 3½" or 3¼" x 4" or 3¼" x 4¾"; 50 exposures on 70-mm film optional
Timer and exposure meter	Standard equipment
Vacuum system	Automatic; rotary pump, oil diffusion pump, oil ejector pump, buffer tank
Model	JEM-100U
Point-to-point resolution	4 Å
Magnification range	1,000X to 300,000X in steps; scan, 150X, 5,000X

TABLE 1 (Continued)

TRANSMISSION ELECTRON MICROSCOPES

Acceleration voltage	20, 60, 80, 100 kv
Condenser lens	Double
Diffraction cameras	Selective area; high resolution optional
Airlocks	Specimen chamber, camera chamber, gun chamber
Anticontamination	Standard equipment; cold trap on oil diffusion pump
Photographic facilities	24 exposures, 2½" x 3½" or 3¼" x 4" or 3¼" x 4¾"
Timer and exposure meter	Standard equipment
Vacuum system	Automatic; two rotary pumps, oil diffusion pump
Model	JEM-120
Point-to-point resolution	5 Å
Magnification range	600X to 250,000X
Acceleration voltage	25, 50, 80, 120 kv
Condenser lens	Double
Diffraction cameras	Selective area; high resolution optional
Airlocks	Specimen chamber, camera chamber, gun chamber
Anticontamination	Standard equipment
Photographic facilities	24 exposures, 3¼" x 4"
Timer and exposure meter	Standard equipment
Vacuum system	Automatic
Model	JEM-T8
Point-to-point resolution	7 Å
Magnification range	600X to 100,000X
Acceleration voltage	40, 60 kv
Condenser lens	Simplified double
Diffraction cameras	Selective area; high resolution optional
Airlocks	Specimen chamber, camera chamber
Anticontamination	Standard equipment; cold trap on oil diffusion pump
Photographic facilities	12 plates (2½" x 3½" or 3¼" x 4") or 24 exposures on 70-mm film
Timer and exposure meter	Standard equipment
Vacuum system	Two rotary pumps, oil diffusion pump
Model	JEM-50B
Point-to-point resolution	100 Å lattice
Magnification range	2,000X, 3,000X, 4,000X
Acceleration voltage	50 kv
Condenser lens	Anode chamber aperture only
Diffraction cameras	Diffraction attachment available
Airlocks	Specimen chamber
Anticontamination	Optional
Photographic facilities	36 exposures on 35-mm film
Timer and exposure meter	None
Vacuum system	Push-button; rotary pump, oil diffusion pump (air-cooled)
Model	JEM-30C
Point-to-point resolution	100 Å lattice
Magnification range	2,000X, 3,000X, 4,000X
Acceleration voltage	30 kv
Condenser lens	Anode chamber aperture only
Diffraction cameras	Diffraction attachment available
Airlocks	Specimen chamber
Anticontamination	Optional

TABLE 1 (Continued)

TRANSMISSION ELECTRON MICROSCOPES

Photographic facilities	36 exposures on 35-mm film
Vacuum system	Push-button; rotary pump, oil diffusion pump (air-cooled)

Philips, Holland

Model	EM300
Point-to-point resolution	3.5 Å
Magnification range	2,800X to 500,000X; low magnification, 220X to 3,900X
Acceleration voltage	20, 40, 60, 80, 100 kv
Condenser lens	Double
Diffraction cameras	Selective area; high resolution optional
Airlocks	Specimen chamber, camera chamber, gun chamber, projection chamber, column
Anticontamination	Standard equipment; liquid N_2
Photographic facilities	15 exposures, 2½" x 3½" or 3¼" x 3¼" or 3¼" x 4"; 40 exposures on 35-mm film or 50 exposures on 70-mm film optional
Timer and exposure meter	Standard equipment
Vacuum system	Automatic; rotary pump, mercury diffusion pump, oil diffusion pump, buffer tank

Model	EM201
Point-to-point resolution	6 Å
Magnification range	1,500X to 200,000X; scan, 200X
Acceleration voltage	40, 60, 80, 100 kv
Condenser	Single mini-lens
Diffraction cameras	Selective area; high resolution optional
Airlocks	Specimen chamber, camera chamber
Anticontamination	Optional
Photographic facilities	16 exposures, 3 ¼" x 4"; 40 exposures on 35-mm film or 50 exposures on 70-mm film
Timer and exposure meter	Standard equipment
Vacuum system	Automatic; rotary pump (two-stage), oil diffusion pump, buffer tank

Siemens, Germany

Model	Elmiskop 101
Point-to-point resolution	3.5 Å
Magnification range	285X to 32,000X in steps; 1,600X to 280,000X continuous
Acceleration voltage	40, 60, 80, 100 kv; 120 kv optional
Condenser lens	Double
Diffraction cameras	Selective area, high resolution, regular diffraction
Airlocks	Specimen chamber, camera chamber
Anticontamination	Standard equipment; liquid N_2
Photographic facilities	12 exposures, 6.5 cm x 9 cm, or 40 exposures on 70-mm film
Timer and exposure meter	Standard equipment
Vacuum system	Push-button rotary pump, mercury diffusion pump, oil diffusion pump, buffer tank

Model	Elmiskop 1A
Point-to-point resolution	3.5 Å
Magnification range	200X to 200,000X continuous
Acceleration voltage	40, 60, 80, 100 kv
Condenser lens	Double
Diffraction cameras	Selective area, high resolution, regular diffraction
Airlocks	Specimen chamber, camera chamber
Anticontamination	Standard equipment; liquid N_2
Photographic facilities	12 exposures, 6.5 cm x 9 cm; 40 exposures on 70-mm film optional

TABLE 1 (Continued)

TRANSMISSION ELECTRON MICROSCOPES

Timer and exposure meter	Optional
Vacuum system	Non-automatic rotary pump, mercury diffusion pump, oil diffusion pump, buffer tank
Model	Elmiskop 51
Point-to-point resolution	35 Å
Magnification range	1,250X, 2,500X, 5,000X, 12, 500X
Acceleration voltage	50 kv
Condenser lens	Anode chamber aperture only
Diffraction cameras	None
Airlocks	Main column only
Antidecontamination	None
Photographic facilities	15 plates (6.5 cm x 9 cm) or 30 exposures on 6.5 cm x 9 cm film
Timer and exposure meter	None
Vacuum system	Non-automatic rotary pump, oil diffusion pump (air-cooled)

Tesla, Czechoslovakia[b]

Model	Tesla
Point-to-point resolution	25 Å
Magnification range	1,000X to 30,000X in ten steps
Acceleration voltage	40, 60 kv
Condenser lens	Anode chamber aperture only
Diffraction cameras	Selective area
Airlocks	Specimen chamber
Anticontamination	None
Photographic facilities	9 plates (2" x 2") or 36 exposures on 35-mm film
Timer and exposure meter	None
Vacuum system	Non-automatic rotary pump, diffusion pump (air-cooled)

Carl Zeiss, Germany

Model	EM-10
Point-to-point resolution	3.5 Å
Magnification range	100X to 200,000X in twenty-five steps
Acceleration voltage	40, 60, 80, 100 kv
Condenser lens	Double
Diffraction cameras	Selective area, dark-field microscopy
Airlocks	Specimen chamber, camera chamber, column
Anticontamination	Standard equipment; < 0.1 Å per minute, liquid N_2
Photographic facilities	30 exposures, 3¼" x 4" plate or sheet film or 35-mm roll film or 70-mm roll film
Timer and exposure meter	Standard equipment; fully automatic
Vacuum system	Automatic; two rotary pumps, oil diffusion pump
Model	EM-9S-2
Point-to-point resolution	7 Å
Magnification	140X to 60,000X in six steps; 700X to 60,000X continuous
Acceleration voltage	60 kv
Condenser lens	Single
Diffraction cameras	Selective area, low angle
Airlocks	Specimen chamber

[b] American distributor: Ultrascan

TABLE 1 (Continued)

TRANSMISSION ELECTRON MICROSCOPES

Anticontamination	None
Photographic facilities	20 exposures, 2¾" x 2¾" plate or film; 60 exposures on 70-mm film optional
Timer and exposure meter	Standard equipment
Vacuum system	Automatic; rotary pump, oil diffusion pump

TABLE 2

ULTRAHIGH-VOLTAGE TRANSMISSION ELECTRON MICROSCOPES

AEI Scientific Apparatus, Inc., England

Model	EM7
Point-to-point resolution	5 Å
Magnification range	Normal, 1,000X to 160,000X; high, 16,000X to 1,600,000X; low, 63X to 1,000X
Acceleration voltage	100 kv to 1,200 kv
Condenser lens	Double
Diffraction cameras	Selective area, high resolution
Airlocks	Specimen chamber, camera chamber
Anticontamination	Standard equipment; liquid N_2
Photographic facilities	48 exposures, 3¼" x 4"
Timer and exposure meter	Automatic
Vacuum system	Automatic

Hitachi, Japan

Model	HU200F
Point-to-point resolution	4.2 Å lattice
Magnification range	1,000X to 200,000X
Acceleration voltage	50, 100, 125, 150, 175, 200 kv
Condenser lens	Double
Diffraction cameras	Selective area, high resolution
Airlocks	Specimen chamber, camera chamber
Anticontamination	Standard equipment; cold trap on oil diffusion pump, liquid N_2
Photographic facilities	18 exposures, 3¼" x 4"
Timer and exposure meter	Automatic
Vacuum system	Automatic; three rotary pumps, oil diffusion pump

JEOLCO, Japan

Model	JEM-200A
Point-to-point resolution	5 Å
Magnification range	1,000X to 200,000X in steps
Acceleration voltage	50, 100, 150, 200 kv
Condenser lens	Double
Diffraction cameras	Selective area; high resolution optional
Airlocks	Specimen chamber, camera chamber, gun chamber
Anticontamination	Standard equipment; cold traps on oil diffusion pumps
Photographic facilities	24 exposures, 2½" x 3½" or 3¼" x 4", or 24 exposures on 70-mm film
Timer and exposure meter	Automatic
Vacuum system	Automatic; three rotary pumps, two oil diffusion pumps, buffer tank

TABLE 3

SCANNING ELECTRON MICROSCOPES

Advanced Metals Research Corporation, U.S.A.

Model	900
Resolution (guaranteed)	100 Å
Magnification range	5X to 200,000X
Acceleration voltage	1 kv to 30 kv in 1-kv steps; 50 kv available
Imaging modes	Secondary electrons, backscattered electrons, specimen current, induced current, cathodoluminescence, transmission electrons
Anticontamination	Gun chamber airlock, specimen chamber airlock, column airlock, cryogenic baffle
Display facilities	Two oscilloscopes: 800 lines resolution for display, 2,500 lines resolution for recording
Vacuum system	Automatic; oil diffusion pump with cryogenic baffle, rotary pump, buffer tank
Specimen stage	Eucentric goniometer, ±¾″ X-Y motion, 2″ Z motion, −5° to +65° tilt, ±180° rotation
Other features	Specimen chamber, 10″ x 10″ x 14″

Model	1000
Resolution (guaranteed)	150 Å
Magnification range	5X to 100,000X
Acceleration voltage	1, 5, 10, 20, 30 kv
Imaging modes	Secondary electrons, backscattered electrons, specimen current, induced current, cathodoluminescence, transmission electrons
Anticontamination	Specimen chamber airlock, column airlock, cryogenic baffle
Display facilities	Two oscilloscopes: 800 lines resolution for display, 2,500 lines resolution for recording
Vacuum system	Automatic; oil diffusion pump with cryogenic baffle, rotary pump, buffer tank
Specimen stage	Goniometer, ±½″ X-Y motion, 1″ Z motion, ±90° tilt, 360° rotation
Other features	Specimen chamber, 10″ diameter x 7″ high

Applied Research Laboratories, Inc., U.S.A.

Model	ARL-SEM
Resolution (guaranteed)	200 Å
Magnification range	4X to 50,000X
Acceleration voltage	0 to 30 kv; 50 kv available
Imaging modes	Secondary electrons, sample current, backscattered electron, X-rays
Anticontamination	9½″ cold dome for liquid N_2 or water, gun chamber airlock, specimen chamber airlock, column airlock
Display facilities	Two oscilloscopes for simultaneous display: P-7 phosphor for viewing, P-11 phosphor for photography; oscilloscope camera with Polaroid back
Vacuum system	Oil diffusion pump, rotary pump, extra rotary pump for sample and filament change
Specimen stage	±½″ X-Y motion, 5½″ Z motion, −12° to +78° tilt, 360° rotation; maximum specimen size, 3″ diameter x 2″ high
Other features	Light microscope, 11X to 560X; dispersive X-ray spectrometer available

Cambridge Scientific Instruments Ltd., England[a]

Model	Stereoscan S4
Resolution (guaranteed)	150 Å in secondary emission
Magnification range	10X to 200,000X
Acceleration voltage	1 to 30 kv
Imaging modes	Secondary electrons, backscattered electrons, cathodoluminescence, specimen current, transmission electrons; X-rays optional

[a] American distributor: Kent-Cambridge.

TABLE 3 (Continued)

SCANNING ELECTRON MICROSCOPES

Anticontamination	Two liquid N_2 baffles, two water-cooled baffles and alumina trap, dry N_2 backfilling, gun chamber airlock, specimen chamber airlock
Display facilities	Two display oscilloscopes: 600 lines resolution for viewing, 1,000 lines resolution for recording; wave form monitor oscilloscope; 12″ TV monitor display optional
Vacuum system	Automatic and manual; two oil diffusion pumps, one rotary pump with two buffer tanks
Specimen stage	Series 100: 16 mm X motion, 13 mm Y motion, 10 mm Z motion, 0 to 90° tilt, 360° rotation; Series 200: goniometer, 25 mm X motion, 50 mm Y motion, 12 mm Z motion, –5° to +45° tilt, 360° rotation, ±10° stereotilt
Other features	Series 100: specimen size up to 12 mm diameter; Series 200: specimen size 50 mm x 50 mm x 30 mm

Model	Stereoscan 600
Resolution (guaranteed)	250 Å in secondary emission
Magnification range	20X to 50,000X
Acceleration voltage	1.5, 7.5, 15, 25 kv
Imaging modes	Secondary electrons, backscattered electrons; specimen current and X-rays optional
Anticontamination	Water-cooled baffle; alumina trap and dry N_2 backfilling optional, specimen chamber airlock
Display facilities	Two oscilloscopes: 95 mm x 124 mm, 600 lines resolution for viewing, 1,000 lines resolution for recording
Vacuum system	Automatic; oil diffusion pump, rotary pump
Specimen stage	13 mm X-Y motion, 20 mm Z motion, 0 to 90° tilt, 360° rotation; electrical image shift (20 μm) provided
Other features	Specimen size up to 2″ diameter

Coates and Welter Instrument Company, U.S.A.

Model	CWIKSCAN 100
Resolution (guaranteed)	250 Å
Magnification range	3X to 70,000X
Acceleration voltage	0.1 kv to 20 kv
Imaging modes	Secondary electrons, backscattered electrons
Anticontamination	Specimen chamber airlock, ion pumps
Display facilities	12″ TV, 15 pictures per second, 1,155 lines resolution; photographic camera optional
Vacuum system	Ion pumps, rotary pump
Specimen stage	0.8″ X-Y motion, 45° tilt; choice of tilt or rotation
Other features	Field emission gun; specimen size up to 3″ diameter

ETEC Corporation, U.S.A.

Model	Autoscan
Resolution (guaranteed)	200 Å
Magnification range	7X to 240,000X
Acceleration voltage	2.5 kv to 30 kv
Imaging modes	Secondary electrons, backscattered electrons, cathodoluminescence, voltage contrast
Anticontamination	Specimen stage airlock, water-cooled baffle; liquid N_2 traps available
Display facilities	Two oscilloscopes: 7¼″ x 7¼″, 1,000 lines resolution for viewing, 9 cm x 9 cm, 2,500 lines resolution for photography; TV display, 2,500 lines resolution, available
Vacuum system	Automatic; oil diffusion pump, rotary pump
Specimen stage	Eucentric goniometer, ±0.75″ X-Y motion, 1″ Z motion, 360° rotation; basic goniometer, ±0.5″ X-Y motion, 1″ Z motion, 360° rotation
Other features	Specimen size up to 3″ diameter x 2″ thick

TABLE 3 (Continued)

SCANNING ELECTRON MICROSCOPES

Hitachi, Japan[b]

Model	HSM-2
Resolution (guaranteed)	150 Å
Magnification range	12X to 200,000X
Acceleration voltage	3 kv to 40 kv
Imaging modes	Secondary electrons, absorption electrons
Anticontamination	Gun chamber airlock, specimen airlock, column airlock, liquid N_2 traps
Display facilities	Two oscilloscopes: 120 mm x 90 mm screen, 250 to 1,000 scanning lines, one slow phosphor, one fast phosphor
Vacuum system	Automatic; oil diffusion pump, two rotary pumps, buffer tank
Specimen stage	Goniometer (motor-driven), ±9 mm X-Y motion, 10 mm to 65 mm Z motion, –5° to +50° tilt, ±180° rotation
Other features	Specimen size up to 38 mm diameter x 12 mm high
Model	SSM-2
Resolution (guaranteed)	250 Å
Magnification range	20X to 20,000X
Acceleration voltage	4, 10, 20 kv
Imaging modes	Secondary electrons, backscattered electrons
Anticontamination	Liquid N_2 cold trap
Display facilities	Two oscilloscopes, three visual display speeds; TV scan system and voltage contrast imaging optional
Vacuum system	Oil diffusion pump, rotary pump
Specimen stage	Goniometer, ±7 mm X-Y motion, –35° to +60° tilt, 360° rotation
Other features	Specimen size up to 15 mm diameter x 15 mm high

JEOLCO, Japan

Model	JSM-50A
Resolution (guaranteed)	100 Å
Magnification range	20X to 140,000X
Acceleration voltage	0 to 50 kv
Imaging modes	Secondary electrons, backscattered electrons, absorption electrons, cathodoluminescence, transmission electrons, Y modulation
Anticontamination	Specimen chamber airlock, gun chamber airlock, water-cooled baffle; liquid N_2 traps optional
Display facilities	Two oscilloscopes: 600 lines resolution for viewing, 800 lines resolution for recording
Vacuum system	Automatic; oil diffusion pump, two rotary pumps
Specimen stage	Goniometer, 0 to 15 mm X motion, 0 to 25 mm Y motion, 13 mm to 37 mm Z motion, 0 to 45° tilt, ±180° rotation
Other features	Specimen chamber, 10 mm diameter x 10 mm high or 33 mm diameter x 25 mm high
Model	JSM-Sl
Resolution (guaranteed)	250 Å
Magnification range	22X to 140,000X
Acceleration voltage	4 kv to 10 kv
Imaging modes	Secondary electrons

[b] American distributor: Perkin-Elmer.

TABLE 3 (Continued)

SCANNING ELECTRON MICROSCOPES

Anticontamination	Liquid N_2 cold finger
Display facilities	Two oscilloscopes: 600 lines resolution for viewing, 800 lines resolution for recording
Vacuum system	Oil diffusion pump, two rotary pumps
Specimen stage	Linear, ±5 mm X-Y motion, –5° to +45° tilt, 360° rotation
Other features	Specimen size up to 1″ diameter

Materials Analysis Company, U.S.A.

Model	SXII
Resolution (guaranteed)	150 Å
Magnification range	20X to 100,000X
Acceleration voltage	1.0 kv to 30 kv
Imaging modes	Secondary electrons, backscattered electrons, specimen current; Y-deflector modulation and X-ray optional
Anticontamination	Chevron water baffle, roughing and isolation valves
Display facilities	Two oscilloscopes: high speed for viewing, high resolution for recording
Vacuum system	Automatic; oil diffusion pump, rotary pump, built-in leak detector system
Specimen stage	Goniometer, ±½″ X-Y motion, 0 to 1.5″ Z motion, –5° to +90° tilt, 360° rotation
Other features	Specimen chamber, 9″ diameter x 6″ high

Ultrascan, U.S.A.

Model	SM-3
Resolution (guaranteed)	150 Å; 50 Å with LaB_6 gun
Magnification range	10X to 100,000X; 140,000X with change in working distance
Acceleration voltage	1 kv to 30 kv; 50 kv optional
Imaging modes	Secondary electrons, backscattered electrons, cathodoluminescence, transmission electrons, charge collection current, specimen absorption current; X-ray analysis and Auger analysis optional
Anticontamination	Liquid N_2 baffle; automatic valves isolate column, specimen chamber, pumping system
Display facilities	Two oscilloscopes: 10″ diameter, 1,000 lines resolution for viewing, 7″ diameter, 1,000 lines (4,000 lines optional) resolution for recording; second viewing oscilloscope or TV display optional
Vacuum system	Automatic; ion pump (10^{-9} torr) or oil diffusion pump (10^{-6} torr), rotary pump
Specimen stage	Goniometer, motorized 5-axis stage controlled from console, 25 mm X, Y, and Z motion, –3° to +50° tilt, 360° rotation
Other features	Specimen chamber, 11.0″ diameter x 8.0″ high

Photomicroscopy

DR. FRED J. ROISEN

Photomicroscopy should be employed for either of two purposes: to accurately record micrographic images and to increase resolution. The success with which these goals are attained will depend upon proper selection of a number of factors: a) the optical system, b) selection of correct light sources, c) type of film, d) development of the film, e) use of proper filters.

The following tables will aid in the selection of the proper combination of these factors.

TABLE 1

TRANSMITTED LIGHT OPTICAL SYSTEMS

Type	Use
Bright field	Stained specimens
Oblique light	Provides information concerning three-dimensional aspects of a specimen, i.e., indentation, surface imperfection, etc. This information is based on an asymmetrical illumination of the specimen.
Dark field	Produces a bright image on a black background. Employed to best advantage when specimen is extremely small and colorless and occupies only a portion of the field. Optimum results obtained when object has approximately the same refractive index as background. The specimen details are rendered visible by the rays being either diffracted, diffused, or refracted by the object.
Phase contrast	Contrast is generated by optical means. Phase differences of the light waves passing through the object are observable in the image field as differences in light intensity. Particularly valuable for studying living tissue cultures. The disadvantage is that out of focus objects above and below the plane of focus generate "edge effects" (halos).
Interference microscopy	Light passing through the specimen is made to interfere with light which has not passed through the specimen. The use of a compensator in the reference beam allows this system to provide measurements of specimen thickness and refractive index.
Polarized microscopy	Employed in the study of birefringent objects (i.e. membranes and fibrous elements). Since birefringent objects appear brightest over background illumination at extinction, microphotography occurs under light-limited conditions and fast films, and/or long exposures are necessitated. The disadvantage is that it cannot be used through plastic.
Differential interference (Nomarski) microscopy	The image formed is based on the rate of change of refractive index. This extremely sensitive system can be employed with favorable results for living unstained objects having low refractive differences from their background. The thin plane of focus provides an ideal optical system for photomicrographic sectioning. Since this method utilizes polarized light, it is especially adapted for the study of fibrous arrays. The two major disadvantages are that it allows only a minimal working distance and specimen cannot be viewed through plastics.
Fluorescence microscopy	The specimen behaves as a self-luminous object. The image is seen as a result of molecules absorbing the transmitted light and then re-emitting light of lower energy and longer wavelength. Since only a fraction of the incident light is re-emitted, the conditions of fluorescence microscopy generally are light limited.

FILMS

The resolution obtained with photomicrography can be superior to visual observation for two reasons: (a) the image can be formed with shorter (nonvisible) wavelengths, and (b) new thin emulsions with low light scattering coefficients offer a greater acuity than the human eye. In general there is a direct correlation between film speed and grain size; therefore, to insure maximum detail, it is a good practice to utilize films with ASA values close to the minimum acceptable value for the given conditions. The sharpest image can be obtained with an ultrathin film of extremely fine-grain size. One way of increasing resolution is to use (where possible) long exposures which allow the use of slow, fine-grain films. The period of exposure can be regulated either by using neutral density filters or by adjusting the intensity of illumination; neutral density filters are the preferred technique for color photomicrography (see section on filters).

Tables 3 and 4 contain information that will aid in the proper selection of black-and-white and color 35-mm films.

PROCESSING BLACK-AND-WHITE FILMS

After exposure it is best to process films as rapidly as possible; if necessary, exposed films may be stored for several months at 20°C without serious loss of detail. The selection of an appropriate developer is as important as the selection of the film since developers affect clarity, contrast, granularity, speed, and tonal rendition.

Most films listed in Table 3 were found to perform close to their published ASA values when developed in either FrX-22, Kodak D-19®, Kodak D-76®, or Kodak Microdol-X®. Development was performed at 68°F (20°C) for the time indicated in the instruction sheets supplied by the respective manufacturers. It is possible to enhance film speed without significant loss of resolution by employing one of several commercially available fine-grain developers.

a) *Acufine® Film Developer.*[1] Increases both resolution and film speed (Table 3) while simultaneously producing uniform grain size and good tonal range. Acufine® was judged to yield a high degree of reproducibility. (Average increase in ASA: 2.5 X)

b) *Diafine®.*[1] A two-stage fine-grain developer allows a wide latitude of exposure, "one developing time for all films", increased contrast, and maximum resolution. The average increase in film speed (ASA) as a result of Diafine® development was 3.5 X (See Table 3). Of all the tested developers, Diafine® gave the best reproducibility.

c) *Isodol® Film Developer.*[2] Used without dilution, produces moderately fine-grain, contrasty negatives. By varying length of development time and temperature, it is possible to extend given ASA values by a factor of 2 X.

d) *Microphen®.*[3] A fine-grain developer affording moderate increases in ASA (5/3 X) with no apparent loss in image quality.

In all of the preceding tables, the recommended data are based on approximations and therefore should be used as an aid in selecting likely combinations for good photomicrography. Best results will be obtained only after careful trial exposures have been made under a variety of conditions.

[1] Acufine Inc., 439 E. Illinois St., Chicago, Illinois 60611
[2] GAF Corporation, 140 West 51 Street, New York, New York 10020
[3] Ilford Inc., W. 70 Century Road, Paramus, New Jersey 07652

TABLE 2

ARTIFICIAL LIGHT SOURCES FOR PHOTOMICROGRAPHY

Type (Rated Voltage-Rated Output)	Color Temperature, °K, at Rated Voltage	Comments
Carbon arc lamp	3,800-5,500	Provides high levels of illumination suited for light-limited conditions (i.e., polarized light microscopy). Color temperature varies with type of carbon electrode and the number of hours used.
High pressure mercury vapor lamps, i.e., Osram® HBO 75, HBO 100 w/2, HBO 200 w/4	Do not emit continuous spectra; thus no color temperature can be assigned.	Bluish-white light of high intensity poorly suited for color photomicrography. Fine monochromatic light source for black-and-white photography. Ideal sources for fluorescence microscopy. Relative luminous intensity of cited examples: 1, 1, 4, respectively.
Incandescent lamps	2,500-2,800	Coiled tungsten filaments. Color temperature close to artificial light values of color films. Widespread distribution. Provide an inexpensive acceptable illumination for bright-field, oblique-light, and some phase-contrast microscopy.
Low voltage microscope lamps		
6 v 15 w	2,850	Suited for bright-field, oblique, dark-field and phase-contrast photomicrography when only moderate illumination is required. Color temperature falls with use.
12 v 60 w	3,050	Same as 6 v 15 w. In addition, high enough level of illumination for polarized light and interference microscopy.
12 v 100 w	3,100	Recommended for most applications in photomicrography (exceptions are differential interference cinematography and fluorescence microscopy). Good for color photography since it requires virtually no filtration for use with artificial light films of type A.
Projection lamps	2,900-3,200	Can be adapted for operation as short-term high-intensity light sources. Since they are available for operation on 110 v and 220 v lines, they do not require expensive regulation transformers. These lamps provide a moderately good light source for bright-field, oblique-light, dark-field, phase-contrast, and polarized light microscopy. Because of their direct connection to available current, they are subject to continual line fluctuations.

TABLE 2 (Continued)

ARTIFICIAL LIGHT SOURCES FOR PHOTOMICROGRAPHY

Type (Rated Voltage-Rated Output)	Color Temperature, °K, at Rated Voltage	Comments
Photo lamps		
at maximum voltage at moderate voltage	3,400 3,200	Provide short-term adequate light source for color photomicrography under most optical conditions except for interference and fluorescence microscopy.
Quartz-iodine lamp		
12 v 150 w (halogen) (also available as 24 v 250 w bulb)	3,400	Rich blue-white light of reasonably high intensity. Satisfies most of the requirements for polarized light, interference, and differential interference color photomicrography. Good general purpose light source. Color temperature remains constant for life of bulb. Not suitable for fluorescence microscopy.
Continuous light xenon lamp	6,000	The recommended high intensity light source. Its spectral output is equivalent to daylight; it is therefore ideally suited to color photomicrography with all optical systems including cinematography (except for fluorescence microscopy). Color daylight film can be used with little or no filtration. Color temperature constant.
Electronic flash xenon-flash	4,000-5,500	Good for black-and-white and color photography. Ideal for laboratory where vibration is a problem. Suitable for rapidly moving objects. Do not use type L color films. Needs little or no color correction for daylight films.

TABLE 3

CHARACTERISTICS OF SOME COMMON BLACK-AND-WHITE FILM (35 mm)

Film	ASA STD	ASA Acufine®	ASA Diafine®	Resolving Graininess	Color Power	Sensitivity	Comments
Kodak Panatomic X®	40	100	200	Extremely fine	Very high	Panchromatic	Excellent definition, optimally suitable for high magnification enlargements
Kodak Plus-X Pan®	125	320	640	Extremely fine	High	Panchromatic	Medium speed, excellent acutance
Kodak Tri-X Pan®	400	1,200	2,400	Fine	Medium	Panchromatic	Fine granularity, useful with high shutter speeds and great field depth
Kodak High Contrast Pan®	6	20	40	Extremely fine	Very high (200 lines/mm)	Panchromatic	For moderate contrast specimens, for high magnification enlargement
Kodak High Contrast Copy Film®	64	175	350	Extremely fine	Extremely high (200-250 lines/mm)	Panchromatic	Extremely high acutance (sharpness), good for low contrast objects
GAF® Professional 2681	250	500	–	Fine	High	Panchromatic	Polyester base which offers resistance to negative injury
GAF 125 Ansco (Versapan)®	125	250	500	Extremely fine	High	Panchromatic	High acutance, long tone scale
GAF 500 Ansco (Super Hypan)®	500	1,000	1,000	Moderate	Medium	Panchromatic	
Ilford® Pan-F	50	N.A.[a]	N.A.[a]	Extremely fine	Very high	Panchromatic	Excellent definition, optimally suitable for high magnification enlargements
Ilford® FP-4	125	200 (Microphen®)	400	Fine	High	Panchromatic	High acutance, optimally suited for enlargements, great exposure latitude

[a] N.A. = not available.

TABLE 3 (Continued)

CHARACTERISTICS OF SOME COMMON BLACK-AND-WHITE FILM (35 mm)

Film	ASA STD	ASA Acufine®	ASA Diafine®	Resolving Graininess	Color Power	Sensitivity	Comments
Ilford® HP-4	400	650 (Microphen®)	800	Fine	Medium	Panchromatic	Fine granularity, best used with high shutter speeds
Fuji Film Neopan SSS®	200	600 (If developed with Pandol 400-1,600)	1,200	Fine	Medium	Panchromatic	High speed, wide exposure latitude, great field depth

TABLE 4

CHARACTERISTICS OF SOME COMMON COLOR FILM (35-mm)

Film	Color Temperature, °K	ASA Daylight	Resolving Power	Graininess	Color Balance	ASA and Correction Filter		Comments
						(Photolamp) A, 3400°K	(Tungsten) B, 3200°K	
Kodak® Photo-micrography Color Film So-456	3,200	16	Very high	Extremely fine	Tungsten	5 (#80B)	4 (#80A)	High contrast, transparency, high definition, true color
Kodachrome II®	5,500	25	80-120 lines/mm	Very fine	Daylight blue-flash	8 (#80B)	6 (#80A)	Reversal transparency, very bright colors
Kodachrome II A®	3,400	25 (#85)	80-120 lines/mm	Very fine	Photoflood	40	32 (#82A)	Reversal transparency, very bright colors
Kodachrome X®	5,500	64	60-100 lines/mm	Fine	Daylight blue flash	20 (#80B)	16 (#80A)	Reversal transparency, dark colors
Ektachrome X®	5,500	64	50-100 lines/mm	Fine	Daylight blue flash	20 (#80B)	16 (#80A)	Reversal transparency, very bright colors
Kodacolor X®	5,500	80	50-100 lines/mm	Fine	Daylight	25 (#80B)	20 (#80A)	Negatives
High Speed Ektachrome (Daylight Type EH)®	5,500	160	40-80 lines/mm	Moderate	Daylight or blueflash	50 (#80B)	40 (#80A)	Transparency, very bright colors
High Speed Ektachrome Type B (EHB)®	3,200	80 (#85B)	50-100 lines/mm	Fine	3200K or tungsten	100 (#81A)	125	Transparency, very bright colors
Agfachrome 50L®	3,100	32 (#85B)	77-100 lines/mm	Very fine	Artificial light	40 (#81A)	50	Transparency, for long exposure (greater than 2 sec) bright sharp colors
Agfachrome 50S®	5,500	50	77-100 lines/mm	Very fine	Daylight electronic flash, blue flash bulbs	16 (#80B)	12 (#80A)	Transparency, for short exposure (less than 2 sec) bright sharp colors

TABLE 4 (Continued)

CHARACTERISTICS OF SOME COMMON COLOR FILM (35 mm)

Film	Color Temperature, °K	ASA Daylight	Resolving Power	Graininess	Color Balance	ASA and Correction Filter: (Photolamp) A, 3400°K	ASA and Correction Filter: (Tungsten) B, 3200°K	Comments
Agfachrome CT-18®	5,500	50	80 lines/mm	Fine	Daylight	–	–	Transparency, very bright
Agfacolor CNS®	5,500	80	45-110 lines/mm	Fine	Daylight electronic flash, blue flash bulbs	–	–	Negatives, prints
GAF 64 Anscochrome®	5,500	64	50-100 lines/mm	Medium	Daylight	32 (#80B)	16 (#80A)	Transparency, pastel colors
GAF T/100 Anscochrome®	3,200	64 (#85B)	50-100 lines/mm	Fine	Tungsten	80 (#81A)	100	Transparency, warm colors
GAF 200 Anscochrome®	5,500	200	40-80 lines/mm	Moderately coarse	Daylight	100 (80B)	50 (#80A)	Transparency, high shutter speeds, small lens opening, pastel coloration
GAF 500 Anscochrome®	5,500	500	N.A.	Coarse	Daylight	250 (#80B)	125 (#80A)	Transparency, low-level illumination, small lens opening, muted colors
Fujichrome R100®	5,500	100	N.A.	Very fine	Daylight	–	25 (#80A)	Transparency
Fujicolor N100®	5,500	100	N.A.	Very fine	Daylight	–	25 (#80A)	Prints, wide exposure range, enlargements

FILTERS

Filters have a wide application in photomicroscopy. They are used to enhance image formation, balance color temperature of illuminators, control exposures, and to process color films. Many of these uses have been discussed in the section on filters in this volume. Only a few specific comments are included here.

Filters soften similar hues and darken complementary ones. Therefore, in order to develop maximum contrast between a color specimen and its background, a filter of complementary color should be used. For examples see Table 5.

Specimens with moderate to heavy staining in one hue commonly suffer from lack of intrinsic detail. This difficulty can frequently be overcome by photographing the specimen through a filter whose hue and intensity are similar to that of the specimen.

For color photomicrography, conversion filters can be used to match the light source with the photographic emulsion. As has been noted in Table 2, the common light sources used in photomicroscopy emit light of differing color temperatures. Kodak® Wratten® filters can be used as light-balancing filters (Table 6).

Illustration: If Kodachrome X® film (color temperature 5,500°K) is to be used with a 12 v 100 w illuminator (emitting light with a color temperature of 3,100°K), then it is necessary to use two filters: 80A and 82, or 5,500°K = 3,100°K + 2,300°K + 100°K.

Many photomicroscope manufacturers provide suggested starting point combinations for the routine operation of their equipment with representative films (Table 7).

TABLE 5

FILTERS FOR INCREASED SPECIMEN CONTRAST OVER BACKGROUND

Specimen Color	Filter Color
red	green
orange	blue
yellow	violet
green	red
blue	orange
violet	yellow

TABLE 6

USE OF LIGHT-BALANCING FILTERS

Blue Series for Increases		Yellow Series for Decreases	
Filter Number	Approximate °K	Filter Number	Approximate °K
82	+100	81	−100
82A	+200	81A	−200
82B	+300	81B	−300
82C	+400	81C	−400
80A	+2,300	81D	−500
80B	+2,100	81E F	−600
		85A	−2,300
		85B	−2,100

TABLE 7

SUGGESTED STARTING POINT COMBINATIONS FOR PHOTOMICROSCOPY EQUIPMENT[a]

Type of Film	Speed		6 Volt 15 Watt	12 Volt 60 Watt	12 Volt 100 Watt	Xenon XBO 150	Electronic Flash (Strobe)	CSI 250 Watt
	Daylight	Artificial Light						
35-mm Film								
Kodachrome X®	19 DIN 64 ASA		80A + 82B	80A + 82B	80A + 82B	CC05Y	CC05Y	80D
Kodachrome II®	15 DIN 25 ASA		80A	80A	80A	CC10Y	CC10Y	80D
Kodachrome II A®		17 DIN 40 ASA	80 D	80 D	82A	not recommended	not recommended	
Ektachrome X®	19 DIN 64 ASA		80A + 82A +BG20[b]	80A + 82A +BG20	80A +BG20	CC15Y	CC10Y	80D
Kodak High Speed Ektachrome Daylight®	23 DIN 160 ASA		80A + 80D	80A	80A	CC10Y	CC10Y	
Kodak High Speed Ektachrome Artificial Light®		22 DIN 125ASA	80D	82A	82A	not recommended	not recommended	
Anscochrome 200®	24 DIN 200 ASA		80A + 80D +BG20	80A +BG20	80A +BG20	CC15Y	CC15Y	
Agfachrome CT 18®	18 DIN 50 ASA		80A + 80D +BG20	80A +BG20	80A +BG20	none	none	
Sheet Film								
Polaroid®[c] Polar Color 58	19 DIN 75 ASA		Tiffen 820 B +BG20	Tiffen 840B (for red and orange specimen, Tiffen 820 B and BG20)	82C +BG20	CC10Y +BG20	none	82B +BG20

TABLE 7 (Continued)

SUGGESTED STARTING POINT COMBINATIONS FOR PHOTOMICROSCOPY EQUIPMENT

	Speed							
Type of Film	Daylight	Aritifcial Light	6 Volt 15 Watt	12 Volt 60 Watt	12 Volt 100 Watt	Xenon XBO 150	Electronic Flash (Strobe)	CSI 250 Watt
Ektachrome Daylight®	18 DIN 50 ASA		not recommended	not recommended	not recommended	CC10Y +BG20	CC10Y +BG20	
Ektachrome Type B®		15 DIN 25 ASA	CC40B +BG20	CC20B +BG20	CC10B +BG20	not recommended	not recommended	81EF+ CC30R

[a] Applicable to Zeiss Photomicroscope I.

[b] BG 20 didymium Filter (thickness: 1 mm)

[c] Develop 60 seconds for red and orange specimens; develop 75 seconds for blue specimens.

Reproduced by permission of Carl Zeiss, Inc., New York.

The following selected references provide valuable information for anyone seriously considering photomicroscopy.

1. Allen, R. M., *Photomicrography,* 2nd ed. D. Van Nostrand, New York (1958).
2. Barron, A. L. E., *Using the Microscope.* Chapman and Hall, Ltd., London (1965).
3. Eastman Kodak Co., *Black and White Films in Rolls,* 3rd ed., Information Book AF-13. Rochester, New York (1967).
4. Eastman Kodak Co., *Color Films,* Color Data Book E-77, 5th ed. Rochester, New York (1968).
5. Eastman Kodak Co., *Photography Through the Microscope,* 5th ed., Technical Bulletin E-2. Rochester, New York (1970).
6. Eastman Kodak Co., *Wratten Filters.* Scientific and Technical Data Book B-3, 22nd ed. Rochester, New York (1968).
7. Lawson, D. F., *The Technique of Photomicrography.* G. Newnes, London (1960).
8. Schenk, R., and Kistler, G., *An Introduction to the Principles of the Microscope and Its Application to the Practice of Photomicrography* (Translation by F. Bradley). Chapman and Hall, Ltd., London (1962).
9. Szabó, D., *Medical Colour Photomicrography.* (Translation by P. Palotay). Akademiai Kiado, Budapest (1967).
10. Always refer to the instructional leaflets supplied by the respective manufacturer. They provide practical information on proven and tried methods.

PRESERVATION OF MICROORGANISMS

S. P. LAPAGE AND K. F. REDWAY

INTRODUCTION

Preservation may be by subculture (including in living animals and cell lines), by reduced metabolism, or by suspended animation (or nearly so). Emphasis in recent years has been on preservation by drying, freeze-drying (lyophilization), and freezing, including storage in liquid nitrogen. General accounts[1,2] give references to earlier work, and practical techniques[3] are described.

METHODS

Serial Subculture

This is the least satisfactory method, due to the inherent difficulties of mislabeling, contamination, loss of cultures, and changes in the cultures. However, it is simple and may be necessary for organisms that do not withstand freezing or drying. The metabolic rate of the organisms may be reduced by storage on minimal media, e.g., Dorset's egg or meat extract agar[4] for bacteria. Containers should be airtight, and a suitable storage temperature should be chosen. Storage under mineral oil is widely used for fungi[5] and, although messy, prolongs survival and prevents dehydration.

Storage at Low Temperatures[a]

Storage at 4°C prolongs survival of many organisms, but is not successful for all. Storage at temperatures from 0 to -30°C is not recommended, because eutectic mixtures of water and electrolytes (common in this range of temperatures, although they may also occur at lower temperatures) may damage the stored microorganisms.[6,7] Storage at -20°C has proved successful for some fungi.[8]

Storage in Distilled Water

Successful storage of *Pseudomonas solanacearum* in distilled water at room temperature for ten years or more has been reported.[9] Coliform organisms have not survived under similar conditions.[10] The method does not appear to have been widely used.

Drying

Sterile Soil or Sand. These have been used for the preservation of spore-bearing organisms.[5,11,12] Addition of the organisms to sterile soil and drying at room temperature may be suitable. Kieselguhr and silica gel[13–16] have also been used.

Paper Disks. These have been used for the short-term storage of bacteria.[17]

Gelatin disks. These are prepared by mixing bacteria with molten gelatin, allowing the mixtures to set in drops and then drying them in a desiccator.[18] The method is simple and satisfactory for many microorganisms.

Predried plugs. Substances such as peptone, starch or dextran, either alone or in mixtures,[19] have been used in this form. The plugs are freeze-dried from solutions of the substances, the organisms are added in a small volume of liquid, and the plugs are then subjected to a vacuum. The suspensions are said to dry

[a] Also see Freezing.

immediately and do not freeze;[19] thus the method may prove successful for organisms that do not withstand freeze-drying.

L-Drying.[20,21] Drying from the liquid state[22] is a related method, in which microorganisms immersed in a water bath are dried by a vacuum pump. The temperature of the suspensions in the bath is controlled so that organisms do not freeze. A desiccant is included in the system, to remove water withdrawn from the suspensions.

Freezing[6]

The process of injury to cells during freezing is not fully understood.[23] More resistant microorganisms, e.g., most bacteria, yeasts, and many fungi, are readily frozen. More sensitive organisms may require protective agents to prevent intracellular freezing, which leads to death.[24] Commonly used protective agents are glycerol[25] and dimethyl sulfoxide (DMSO).[6,23,26] Protective agents have been reviewed.[27]

The principles of freezing are as follows: the rate of cooling should be slow (1°C per minute) to -20°C, then as fast as possible; electrolytes should be kept to a minimum; and the rate of rewarming should be as rapid as possible.[3] The frozen cultures may be stored at -30°C or below; -70°C is commonly used. Storage has proved successful at -196°C (or thereabouts) in liquid-nitrogen systems (see Separate Taxa). Storage at -196°C is said to reduce the metabolism to virtually nil[28] and is probably preferable to storage at temperatures that are not as low.

Freeze-Drying[29]

This widely used method is that of choice for microorganisms that will withstand the process. The ampoules produced are easily transported, and special storage apparatus, apart from lower temperature, is not required. Viability may be longer if ampoules are stored at 4°C, -30°C, or even lower, in preference to room temperature.[30]

The machinery and techniques[3,21] used for freeze-drying differ, but the process essentially consists of the removal of water by sublimation from suspensions frozen either by freezing agents or by application of a vacuum, as in centrifugal freeze-drying. Many media have been used for the suspension of organisms during freeze-drying (see Separate Taxa). The medium should contain the following:[31] a substance capable of maintaining the residual moisture content at the optimal level, e.g., carbohydrates such as glucose, sucrose or sorbitol (not all carbohydrates are protective, and some are deleterious[32,33]); a compound that contains amino groups, which neutralize carbonyl groups said to be toxic; minimal electrolytes; and a substance that forms a 'cake' when dried, to serve as mechanical protection of the dried organisms, to aid the sublimation of liquid, and to prevent dispersion of the organisms when the ampoule is opened.

The residual water content in the freeze-dried preparations may play a part in their longevity.[34-41] Most freeze-drying processes involve a primary stage with a relatively low vacuum and high humidity, and a secondary stage with high vacuum and a desiccant to reduce humidity; nonetheless, death can still occur during the secondary stage.[42]

Sordelli's Method. This is a simple method of drying. An inner tube containing a small volume of the microorganisms emulsified in horse serum is placed in an outer tube containing P_2O_5 and a vacuum is applied for a few minutes; the outer tube is then sealed under the vacuum. The method has been described, and survival of many microorganisms for years has been obtained.[43-45]

Recovery

The percentage of recovery of dried microorganisms may be affected by the conditions of recovery; e.g., for some bacteria the temperature of recovery,[46-48] the recovery medium,[48-50] and even the volume of the recovery medium added to the dried preparation may be of importance.[48] The rate of warming of frozen microorganisms can also affect the percentage of recovery.[23]

Viability

Both freezing and freeze-drying may cause some loss in viability, often extensive, with losses being 100- or 1000-fold. This is not important if the aim of preservation is to maintain viable cultures, provided that the initial suspension had adequate numbers to sustain the loss. If methods of preservation are designed to produce products with a given number of viable organisms, e.g., vaccines, different techniques may be required and greater accuracy may be needed. The expected viabilities of a wide range of microorganisms have been given,[51,52] and the survival of a wide range of bacteria has been investigated.[21]

Changes after Freezing, Drying and Freeze-Drying

Little extensive work has been done on this subject, although it is known that damage to particular enzymes[53–56] and the DNA[57–60] can occur. Changes in antigenic structure have been reported in some cases, but not in others.[61] In some cases pathogenicity may be altered by preservation and storage,[62] in others it may not be affected.[63]

In practice, after revival from the dried or frozen state, it is advisable to subculture the strain several times to enable it to attain all of the characteristics it had before processing. Studies have shown no detectable changes in some cases.[64,65]

SEPARATE TAXA

Algae (Microalgae Only)

Microalgae are usually maintained by periodic transfer of illuminated cultures. When the temperature is lowered (e.g., to 10–15°C) and the light intensity decreased, some species need only annual transfer; others require subculture more frequently.[66] Storage in dry sand was successful for *Nostoc*,[67] and some algae can be maintained in soil.[68] *Tolpothrix tenuis* and *Calothrix brevissima* have been preserved for at least two years by both freeze-drying and drying on a special volcanic earth.[69] Some algae can be freeze-dried,[70–73] and members of the *Chlorophyta, Chrysophyta* and *Cyanophyta* have survived four to ten months by this method.[74] *Euglena gracilis* and other unicellular green algae have been preserved in liquid nitrogen for up to three years by the use of cryoprotective agents.[75,76]

Bacteria[21,77–79 b]

Most bacteria can be readily maintained by periodic transfer on a variety of media[80] and by storage under oil.[81,82] Longevity can be increased by storage at low temperatures. However, may species can be successfully preserved by freeze-drying[21,77,83,84] or in liquid nitrogen.[52,85] Liquid nitrogen may be required for storage of organisms that do not readily withstand freeze-drying, e.g., autotrophic bacteria. Some of these organisms can be preserved by L-drying (q.v.) or predried peptone plugs (q.v.).

Widely used media for suspension are 7.5% glucose horse serum,[21] 7.5% glucose in a mixture of 1/3 nutrient broth and 2/3 horse serum (*Mist desiccans*),[86] and skim milk.[87,88] Other media are reviewed in References 77 and 78. Special techniques may be useful for particular species, e.g., sterile soil for *Bacillus* species,[11] or porcelain beads.[89]

Gelatin disks (q.v.) or Sordelli's method (q.v.) are useful methods for a laboratory that lacks specialized equipment. Paper disks (q.v.) may be valuable for easy distribution and short-term storage, but their value in long-term storage is untested.

Many species have survived in the freeze-dried state for up to twenty years or more.[21,51,90–92]

[b] See also mycoplasmas and spirochaetes.

Bacteriophages

Bacteriophages can often be stored in broth at 4°C,[93-95] which was found to be more satisfactory than freeze-drying for long-term storage.[95] Drying and freeze-drying have been used successfully for some phages,[93-100] as has freezing at various temperatures.[94] However, storage in liquid nitrogen appears the most satisfactory method for long-term preservation.[96,101-103]

Cell Lines

Primary cells, diploid cells, and continuous cell lines can be successfully preserved in liquid nitrogen[104] with the addition of DMSO or glycerol. Most cell lines can be readily stored at -70°C.

Fungi[5,105 c]

Many fungi and yeasts withstand freeze-drying.[106-114] Suitable suspending media are serum,[65,115] sugars, and skim milk. Simple desiccation in a suitable medium can also be used for some species.[44,116]

Fungi can be readily preserved in liquid nitrogen,[64,117-126] which in conjunction with protective agents has proved successful with groups of fungi that do not withstand freeze-drying, e.g., *Entomophorales*, aquatic phycomycetes, strictly mycelial forms, and species with extra large or delicate spores.[105]

Storage of fungi under mineral oil is a common technique.[5,127,128] Periodic transfer is also used on various media.[105,129] Special methods, such as preservation in sterile soil,[5,130,131] anhydrous silica gel,[13-15] and allowing agar slopes to dry up,[5] have been successfully employed. Storage at -20°C or thereabouts has also proved successful.[8,132-135]

Freeze-dried fungi and yeasts have survived many years.[109,110,113] No long-term studies of survival in liquid nitrogen are yet available, but strains of *Blastomyces, Cercospora, Histoplasma, Madurella* and *Paracoccidioides* have survived up to four years,[136] and urediospores of *Puccinia graminis tritici* have survived for five years.[122]

Mycoplasmas

Many species withstand freeze-drying and storage for years,[137-140] although initial survival may be poor. Many protective compounds used for the freeze-drying of bacteria and viruses have no effect on the survival of mycoplasmas.[139] Preservation by freezing at various temperatures and storage in liquid nitrogen have also proved successful with species difficult to freeze-dry, and the methods have been compared.[139,140] Otherwise, mycoplasmas are fully maintained by subculture and preserved on blocks of agar at approximately -30°C, though the more delicate species (e.g., *M. pneumoniae*) survive better at -70°C.

Protozoa

Preservation of protozoa has been carried out by subculture,[141] often because there is no alternative method. The temperature of storage may be reduced, e.g., to 10–15°C.

Freeze-drying has been used for a few protozoa: amoebae,[142,143] *Stentor* and *Frontonia* species,[142] slime molds,[143] and *Mastigina* species.[144] Successful preservation of *Strigomonas oncopelti* by drying on starch-peptone plugs has been reported,[145] and also survival of this organism for up to five years after L-drying (q.v.) from a 50% glucose solution.[146,147]

However, storage at low temperatures, normally using a cryoprotective agent such as glycerol or DMSO, is a more common method. Species of the following genera have been preserved in this way: *Crithidia,*[148] *Entamoeba,*[149] *Leishmania,*[150,151] *Paramecium,*[152] *Plasmodium*[150,153-156] *Tetrahymena,*[152] *Toxoplasma,*[157] *Trichomonas,*[150,158-161] and *Trypanosoma.*[148,150,162-164] The cultures were usually stored at approximately -79°C with the use of solid carbon dioxide.

[c] Also see Yeasts.

Storage in liquid nitrogen has been successfully employed for the preservation of coccidia,[165,166] species of *Crithidia,*[167] *Entamoeba,*[167,168] *Paramecium,*[152] *Tetrahymena,*[169,170] *Toxoplasma,*[171] *Trichomonas,*[167,168] and *Trypanosoma.*[167,168]

Survival times at low temperatures vary considerably with the species and storage temperature, but some protozoa remain viable for months[159] and even years.[167]

Spirochaetes

These organisms have often proved difficult to preserve, except by subculture, which in the case of some *Treponema* species requires living animals. Drying in peptone–starch plugs and also L-drying (q.v.) have been used for the successful preservation of *Leptospira*[147,172-174] and the Reiter treponeme.[147,175] Some species of *Leptospira, Borrelia* and *Treponema* have been freeze-dried.[176,177] *Borrelia anserina* has been successfully frozen,[178] as has *Treponema pallidum*[179] and various species of *Leptospira.*[180,181]

Viruses[94,182-184,d]

In general, plant viruses are easier to preserve than animal viruses. Long-term storage of some plant viruses can be achieved by drying infected plant tissue[183] or by freeze-drying.[183,185,186] Some labile plant viruses have been preserved frozen at various temperatures[183,187] and by storage in liquid nitrogen.[183,188]

Animal viruses are often preserved for short periods of time by refrigeration (approximately 4°C) in normal saline, or for transport in glycerol solutions.[184] Freeze-drying[184,189-194] with various suspending media has been used for the preservation of some animal viruses: measles virus,[190,195-197] influenza virus,[39,41,198,199] polio virus,[190,200,201] vaccinia virus,[40,202-207] rubella virus,[208] foot-and-mouth disease virus,[209-212] various herpes viruses,[213-215] and the scrapie agent.[216] Many other viruses are held in lyophilized form by the American Type Culture Collection.[217] Freezing at various temperatures with protective substances can also be used for preserving animal viruses.[184,189-195,218 221] The animal viruses vary greatly in their resistance to freezing, freeze-drying and storage.[192,194] Some viruses need continuous subculture in living cells, whereas among those capable of being preserved some may survive only a matter of days, others for years.[51]

Rickettsiaceae and Chlamydiaceae, though probably more related to bacteria than viruses, are handled by viral techniques, and most species can be satisfactorily freeze-dried.[217]

Yeasts[44,108,222]

Methods used for fungi are satisfactory for yeasts. The mineral oil technique[223] has been used. Most yeasts freeze-dry readily[224-231] and have been dried on cellulose tufts[232] or preserved in liquid nitrogen.[233] Freeze-dried yeasts survive well,[222,234,235] but attention may have to be paid to the reconstitution medium if the original fermentative powers are to be retained.[236]

REFERENCES

1. Martin, S. M., Conservation of Microorganisms, *Annu. Rev. Microbiol., 18,* 1 (1964).
2. Clark, W. A., and Loegering, W. Q., Functions and Maintenance of a Type-Culture Collection. *Annu. Rev. Phytopathol., 5,* 319 (1967).
3. Muggleton, P. W., The Preservation of Cultures. *Prog. Ind. Microbiol., 4,* 191 (1963).
4. Kauffmann, F., in *The Bacteriology of Enterobacteriaceae,* p. 368.Munksgaard, Copenhagen, Denmark (1966).
5. Fennell, D. I., Conservation of Fungous Cultures. *Bot. Rev., 26,* 79 (1960).
6. Meryman, H. T., Review of Biological Freezing, in *Cryobiology,* p. 1, H. T. Meryman, Ed. Academic Press, London, England (1966).

[d] Also see Bacteriophages.

7. Speck, M. L., and Cowman, R. A., Preservation of Lactic Streptococci at Low Temperatures, in *Proceedings of the First International Conference on Culture Collections,* p. 241, H. Iizuka, and T. Hasegawa, Eds. University of Tokyo Press, Tokyo, Japan (1970).
8. Carmichael, J. W., Viability of Mold Cultures Stored at –20°C, *Mycologia, 54,* 432 (1962).
9. Berger, L. R., Proposed Basis for the Storage of Viable Bacterial Clones in Distilled Water, in *Proceedings of the First International Conference on Culture Collections,* p. 265, H. Iizuka, and T. Hasegawa, Eds. University of Tokyo Press, Tokyo, Japan (1970).
10. Strange, R. E., Dark, F. A., and Ness, A. G., The Survival of Stationary Phase *Aerobacter Aerogenes* Stored in Aqueous Suspension, *J. Gen. Microbiol., 25,* 61 (1961).
11. Gordon, R. E., and Rynearson, T. K., Maintenance of Strains of *Bacillus* Species, in *Culture Collections: Perspectives and Problems,* p. 118, S. M., Martin, Ed. University of Toronto Press, Toronto, Ont., Canada (1963).
12. Raper, K. B., General Methods for Preserving Cultures, in *Culture Collections: Perspectives and Problems,* p. 81, S. M. Martin, Ed. University of Toronto Press, Toronto, Ont., Canada (1963).
13. Perkins, D. D., Preservation of *Neurospora* Stock Cultures with Anhydrous Silica Gel, *Can. J. Microbiol., 8,* 591 (1962).
14. Ogata, W. N., Preservation of *Neurospora* Stock Cultures with Anhydrous Silica Gel, *Neurospora News Lett., 1,* 13 (1962).
15. Barratt, R. W., General Discussion, in *Culture Collections: Perspectives and Problems,* p. 99, S. M., Martin, Ed. University of Toronto Press, Toronto, Ont., Canada (1963).
16. Grivell, A. R., and Jackson, J. F., Microbial Culture Preservation with Silica Gel, *J. Gen. Microbiol., 58,* 423 (1969).
17. Coe, A. W., and Clark, S. P., Short-Term Preservation of Cultures for Transmission by Post, *Mon. Bull. Min. Health Public Health Lab. Serv., 25,* 97 (1966).
18. Lord Stamp, The Preservation of Bacteria by Drying, *J. Gen. Microbiol., 1,* 251 (1947).
19. Annear, D. I., The Preservation of Bacteria by Drying in Peptone Plugs, *J. Hyg., 54,* 487 (1956).
20. Shewan, J. M., Personal Communication (1969).
21. Lapage, S. P., Shelton, J. E., Mitchell, T. G., and MacKenzie, A. R., Culture Collections and the Preservation of Bacteria, in *Methods in Microbiology,* Vol. 3A, p. 135, J. R. Norris, and D. W. Ribbons, Eds. Academic Press, London, England (1970).
22. Annear, D. I., Observations on Drying Bacteria from the Frozen and from the Liquid State, *Aust. J. Exp. Biol. Med. Sci., 36,* 211 (1958).
23. Mazur, P., Physical and Chemical Basis of Injury in Single-Celled Microorganisms Subjected to Freezing and Thawing, in *Cryobiology,* p. 213, H. T. Meryman, Ed. Academic Press, London, England (1966).
24. Nei, T., Araki, T., and Matsusaka, T., Freezing Injury of Aerated and Non-aerated Cultures of *Escherichia coli,* in *Freezing and Drying of Microorganisms,* p. 3, T. Nei, Ed. University of Tokyo Press, Tokyo, Japan (1969).
25. Polge, C., Smith, A. U., and Parkes, A. S., Revival of Spermatozoa after Vitrification and Dehydration at Low Temperatures, *Nature, (london), 164,* 666 (1949).
    Smith, A. U., and Parkes, A. S., Revival of Spermatozoa after Vitrification and Dehydration at Low Temperatures, *Nature, (London), 164,* 666 (1949).
26. Greiff, D., and Myers, M., Effect of Dimethyl Sulphoxide on the Cryo-Tolerance of Mitochondria, *Nature (London), 190,* 1202 (1961).
27. Vos, O., and Kaalen, M. C. A. C., Prevention of Freezing Damage to Proliferating Cells in Tissue Culture – A Quantitative Study of a Number of Agents, *Cryobiology, 1,* 249 (1965).
28. Meryman, H. T., Mechanics of Freezing in Living Cells and Tissues, *Science, 124,* 515 (1956).
29. Meryman, H. T. (Ed.), Freeze-Drying, in *Cryobiology,* p. 609, Academic Press, London, England (1966).
30. Greiff, D., and Rightsel, W. A., An Accelerated Storage Test for Predicting the Stability of Suspensions of Measles Virus Dried by Sublimation *in vacuo, J. Immunol., 94,* 395 (1965).
31. Greaves, R. I. N., Fundamental Aspects of Freeze-Drying Bacteria and Living Cells, in *Aspects Théoriques et Industriels de la Lyophilisation,* p. 407, L. Rey, Ed. Hermann, Paris, France (1964).
32. Morichi, T., Nature and Action of Protective Solutes in Freeze-Drying of Bacteria, in *Proceedings of the First International Conference on Culture Collection,* p. 351, H. Iizuka, and T. Hasegawa, Eds. University of Tokyo Press, Tokyo, Japan (1970).
33. Marshall, B. J., and Scott, W. J., The Effects of Some Solutes on Preservation of Dried Bacteria during Storage *in vacuo,* in *Proceedings of the First International Conference on Culture Collections,* p. 363, H. Iizuka, and T. Hasegawa, Eds. University of Tokyo Press, Tokyo, Japan (1970).
34. Scott, W. J., The Effects of Residual Water on the Survival of Dried Bacteria during Storage, *J. Gen. Microbiol., 19,* 624 (1958).
35. Scott, W. J., A Mechanism Causing Death During Storage of Dried Microorganisms, in *Recent Research in Freezing and Drying,* p. 188, A. S. Parkes, and A. U. Smith, Eds. Blackwell, Oxford, England (1960).
36. Rey, L. (Ed.), L'humidité résiduelle des produits lyophilisés. Nature-Origine et méthodes d'étude, in *Aspects Théoriques et Industriels de la Lyophilisation,* p. 199, Hermann, Paris, France (1964).
37. Nei, T., Araki, T., and Souza, H., Studies of the Effect of Drying Conditions on Residual Moisture Content and Cell Viability in the Freeze-Drying of Microorganisms, *Cryobiology, 2,* 68 (1965).
38. Nei, T., Souza, H., and Araki, T., Effect of Residual Moisture Content on the Survival of Freeze-Dried Bacteria during Storage under Various Conditions, *Cryobiology, 2,* 276 (1966).

39. Greiff, D., and Rightsel, W. A., Stability of Suspensions of Influenza Virus Dried to Different Contents of Residual Moisture by Sublimation *in vacuo, Appl. Microbiol., 16,* 835 (1968).
40. Suzuki, M., Relation between Reduction of Vaccinia Virus Titer and Residual Moisture Content, in *Freezing and Drying of Microorganisms,* p. 111, T. Nei, Ed. University of Tokyo Press, Tokyo, Japan (1969).
41. Greiff, D., Stabilities of Suspensions of Influenza Virus Dried by Sublimation of Ice *in vacuo* to Different Contents of Residual Moisture and Sealed under Different Gases, *Appl. Microbiol., 20,* 935 (1970).
42. Takano, M., and Terui, G., Correlation of Dehydration with Death of Microbial Cells in the Secondary Stage of Freeze-Drying, in *Freezing and Drying of Microorganisms,* p. 131, T. Nei, Ed. University of Tokyo Press, Tokyo, Japan (1969).
43. Rhodes, M., Viability of Dried Bacterial Cultures, *J. Gen. Microbiol., 4,* 450 (1950).
44. Rhodes, M., Preservation of Yeasts and Fungi by Desiccation, *Trans. Brit. Mycol. Soc., 33,* 35 (1950).
45. Soriano, S., Sordelli's Method for Preservation of Microbial Cultures by Desiccation in Vacuum, in *Proceedings of the First International Conference on Culture Collections,* p. 269, H. Iizuka and T. Hasegawa, Eds. University of Tokyo Press, Tokyo, Japan (1970).
46. Speck, M. L., and Myers, R. P., The Viability of Dried Skim-milk Cultures of *Lactobacillus bulgaricus* as Affected by the Temperature or Reconstitution, *J. Bacteriol., 52,* 657 (1946).
47. Wasserman, A. E., and Hopkins, W. J., Studies in the Recovery of Viable Cells of Freeze-Dried *Serratia marcescens, Appl. Microbiol., 5,* 295 (1957).
48. Leach, R. H., and Scott, W. J., The Influence of Rehydration on the Viability of Dried Microorganisms, *J. Gen. Microbiol., 21,* 295 (1959).
49. Baird-Parker, A. C., and Davenport, E., The Effect of Recovery Medium on the Isolation of *Staphylococcus aureus* after Heat Treatment and after the Storage of Frozen or Dried Cells, *J. Appl. Bacteriol., 28,* 390 (1965).
50. Weiler, W. A., and Hartsell, S. E., Diluent Composition and the Recovery of *Escherichia coli, Appl. Microbiol., 18,* 956 (1969).
51. Rhoades, H. E., Effects of 20 Years' Storage on Lyophilized Cultures of Bacteria, Molds, Viruses, and Yeasts, *Amer. J. Vet. Res., 31,* 1867 (1970).
52. Clark, W. A., The American Type Culture Collection: Experiences in Freezing and Freeze-Drying Microorganisms, Viruses, and Cell Lines, in *Proceedings of the First International Conference on Culture Collections,* p. 309, H. Iizuka, and T. Hasegawa, Eds. University of Tokyo Press, Tokyo, Japan (1970).
53. MacLeod, R. A., Smith, L. D. H., and Gelinas, R., Metabolic Injury to Bacteria. 1. Effect of Freezing and Storage on the Requirements of *Aerobacter aerogenes* and *Escherichia coli* for Growth, *Can. J. Microbiol., 12,* 61 (1966).
54. Moss, C. W., and Speck, M. L., Identification of Nutritional Components in Trypticase Responsible for Recovery of *Escherichia coli* Injured by Freezing, *J. Bacteriol., 91,* 1098 (1966).
55. Speck, M. L., and Cowman, R. A., Metabolic Injury to Bacteria Resulting from Freezing, in *Freezing and Drying of Microorganisms,* p. 39, T. Nei, Ed. University of Tokyo Press, Tokyo, Japan (1969).
56. Vaschenko, L. N., The Effect of Nutritional Conditions on the Growth of Bacterial Cells Intact and Injured in the Process of Lyophilization, *Zh. Mikrobiol. Epidemiol. Immunobiol., 46,* 107 (1969).
57. Webb, S. J., Mutation of Bacterial Cells by Controlled Desiccation, *Nature (London), 213,* 1137 (1967).
58. Webb, S. J., Some Effects of Dehydration on the Genetics of Microorganisms, in *Freezing and Drying of Microorganisms,* p. 153, T. Nei, Ed. University of Tokyo Press, Tokyo, Japan (1969).
59. Servín-Massieu, M., Sanchez-Torres, L. E., and Pallares, F., Gene Unstabilization and Mutation by Freeze-Drying of Bacteria, *Bacteriological Proc.,* p. 62 (1968).
60. Servín-Massieu, M., and Cruz-Camarillo, R., Variants of *Serratia marcescens* Induced by Freeze-Drying, *Appl. Microbiol., 18,* 689 (1969).
61. Jennens, M. G., The Effect of Desiccation on Antigenic Structure, *J. Gen. Microbiol., 10,* 127 (1954).
62. Priestley, F. W., Freeze-Drying of the Organism of Contagious Bovine Pleuropneumonia, *J. Comp. Pathol. and Ther., 62,* 125 (1952).
63. Sorrells, K. M., Speck, M. L., and Warren, J. A., Pathogenicity of *Salmonella gallinarum* after Metabolic Injury by Freezing, *Appl. Microbiol., 19,* 39 (1970).
64. Hwang, S. W., Longevity of Fungal Cultures in Liquid Nitrogen Refrigeration, in *Proceedings of the First International Conference on Culture Collections,* p. 251, H. Iizuka, and T. Hasegawa, Eds. University of Tokyo Press, Tokyo, Japan (1970).
65. Mehrotra, B. S., Tandon, G. D., Maurya, J. N., Chopra, B. K., and Prasad, R., Preservation of Industrial Cultures by Lyophilization, in *Proceedings of the First International Conference on Culture Collections,* p. 319, H. Iizuka, and T. Hasegawa, Eds. University of Tokyo Press, Tokyo, Japan (1970).
66. Starr, R. C., Culture Collections of Algae, in *Culture Collections: Perspectives and Problems,* p. 136, S. M. Martin, Ed. University of Toronto Press, Toronto, Ont., Canada (1963).
67. Venkataraman, G. S., A Method of Preserving Blue-Green Algae for Seeding Purposes, *J. Gen. Appl. Microbiol., 7,* 96 (1961).
68. Dietz, A., General Discussion, in *Culture Collections: Perspectives and Problems,* p. 116, S. M. Martin, Ed. University of Toronto Press, Toronto, Ont., Canada (1963).

69. Watanabe, A., Some Devices for Preserving Blue-Green Algae in Viable State, *J. Gen. Appl. Microbiol., 5,* 153 (1959).
70. Holm-Hansen, O., Effect of Varying Residual Moisture Content on the Viability of Lyophilized Algae, *Nature (London), 198,* 1014 (1963).
71. George, E. A., Discussion I, in *Culture Collections: Perspectives and Problems,* p. 139, S. M. Martin, Ed. University of Toronto Press, Toronto, Ont., Canada (1963).
72. Holm-Hansen, O., Viability of Lyophilized Algae, *Can. J. Bot., 42,* 127 (1964).
73. Holm-Hansen, O., Factors Affecting the Viability of Lyophilized Algae, *Cryobiology, 4,* 17 (1967).
74. Daily, W. A., and McGuire, J. M., Preservation of Some Algal Cultures by Lyophilization, *Butler Univ. Bot. Stud., 11,* 139 (1953–54).
75. Hwang, S. W., and Horneland, W., Survival of Algal Cultures after Freezing by Controlled and Uncontrolled Cooling, *Cryobiology, 1,* 305 (1965).
76. Hwang, S. W., Problems in Superlow Temperature Preservation of Microorganisms, in *Freezing and Drying of Microorganisms,* p. 169, T. Nei, Ed. University of Tokyo Press, Tokyo, Japan (1969).
77. Heckly, R. J., Preservation of Bacteria by Lyophilization, in *Advances in Applied Microbiology,* Vol. 3, p. 1, W. W. Umbreit, Ed. Academic Press, New York (1961).
78. Fry, R. M., Freezing and Drying of Bacteria, in *Cryobiology,* p. 665, H. T. Meryman, Ed. Academic Press, London, England (1966).
79. Sturdza, S. A., Dessiccation directe et lyophilisation en bactériologie (revue générale), *Arch. Roum. Pathol. Exp. Microbiol., 30,* 25 (1971).
80. Lapage, S. P., Shelton, J. E., and Mitchell, T. G., Media for the Maintenance and Preservation of Bacteria, in *Methods in Microbiology,* Vol. 3A, p. 1, J. R. Norris, and D. W. Ribbons, Eds. Academic Press, London, England (1970).
81. Hartsell, S. E., The Preservation of Bacterial Cultures under Paraffin Oil, *Appl. Microbiol., 1,* 36 (1953).
82. Hartsell, S. E., Maintenance of Cultures under Paraffin Oil, *Appl. Microbiol., 4,* 350 (1956).
83. Floodgate, G. D., and Hayes, P. R., The Preservation of Marine Bacteria, *J. Appl. Bacteriol., 24,* 87 (1961).
84. Lelliott, R. A., The Preservation of Plant Pathogenic Bacteria, *J. Appl. Bacteriol., 28,* 181 (1965).
85. Jarvis, J. D., Wynne, C. D., and Telfer, E. R., Storage of Bacteria in Liquid Nitrogen, *J. Med. Lab. Technol., 24,* 312 (1967).
86. Fry, R. M., and Greaves, R. I. N., The Survival of Bacteria during and after Drying, *J. Hyg., 49,* 220 (1951).
87. Hornibrook, J. W., A Useful Menstruum for Drying Organisms and Viruses, *J. Lab. Clin. Med., 35,* 788 (1950).
88. Heckly, R. J., Anderson, A. W., and Rockenmacher, M., Lyophilization of *Pasteurella pestis, Appl. Microbiol., 6,* 255 (1958).
89. Hunt, G. A., Gourevitch, A., and Lein, J., Preservation of Cultures by Drying on Porcelain Beads, *Journal of Bacteriology, 76,* 453 (1958).
90. Steel, K. J., and Ross, H. E., Survival of Freeze-Dried Bacterial Cultures, *J. Appl. Bacteriol., 26,* 370 (1963).
91. Harrison, A. P., and Pelczar, M. J., Damage and Survival of Bacteria during Freeze-Drying and during Storage over a Ten-Year Period, *J. Gen. Microbiol., 30,* 395 (1963).
92. Aktan, M., A Laboratory Investigation on the Percentage of Viable Cells in Cultures of Bacteria 12 Years after Lyophilisation, *Inform. Bull. Int. Center Inform. Distrib. Type Cult., 2* (4), 29 (1967–68).
93. Wahl, R., Discussion II, in *Culture Collections: Perspectives and Problems,* p. 159, S. M. Martin, Ed. University of Toronto Press, Toronto, Ont., Canada (1963).
94. Williams, R. E. O., and Asheshov, E. A., Preservation of Viruses and Bacteriophage, in *Culture Collections: Perspectives and Problems,* p. 147, S. M. Martin, Ed. University of Toronto Press, Toronto, Ont., Canada (1963).
95. Clark, W. A., Comparison of Several Methods for Preserving Bacteriophages, *Appl. Microbiol., 10,* 466 (1962).
96. Clark, W. A., and Geary, D., The Collection of Bacteriophages at the American Type Culture Collection, in *Freezing and Drying of Micro-organisms,* p. 179, T. Nei, Ed. University of Tokyo Press, Tokyo, Japan (1969).
97. Annear, D. I., The Preservation of Bacteriophage by Drying, *J. Appl. Bacteriol., 20,* 21 (1957).
98. Zierdt, C. H., Preservation of Staphylococcal Bacteriophage by Means of Lyophilization, *Amer. J. Clin. Pathol., 31,* 326 (1959).
99. Prouty, C. C., Storage of the Bacteriophage of the Lactic Acid Streptococci in the Desiccated State with Observations on Longevity, *Appl. Microbiol., 1,* 250 (1953).
100. Davies, J. D., and Kelly, M. J., The Preservation of Bacteriophage H1 of *Corynebacterium ulcerans* U 103 by Freeze-Drying, *J. Hyg., 67,* 573 (1969).
101. Clark, W. A., Horneland, W., and Klein, A. G., Attempts to Freeze Some Bacteriophages to Ultralow Temperatures. *Appl. Microbiol., 10,* 463 (1962).
102. Meyle, J. S., and Kempf, J. E., Preservation of T_2 Bacteriophage with Liquid Nitrogen, *Appl. Microbiol., 12,* 400 (1964).
103. Clark, W. A., and Klein, A., The Stability of Bacteriophages in Long-Term Storage at Liquid Nitrogen Temperatures, *Cryobiology, 3,* 68 (1966).
104. Stevenson, R. E., Preservation of Cultured Cell Lines at Low Temperatures, in *Aspects Théoriques et Industriels de la lyophilization,* p. 279, L. Rey, Ed. Hermann, Paris, France (1964).

105. Onions, A. H. S., Preservation of Fungi, in *Methods in Microbiology,* Vol. 4, p. 113, C. Booth, Ed. Academic Press, London, England (1971).
106. Raper, K. B., and Alexander, D. F., Preservation of Molds by the Lyophil Process, *Mycologia, 37,* 499 (1945).
107. Sharp, E. L., and Smith, F. G., Preservation of *Puccinia* urediospores by Lyophilization, *Phytopathology, 42,* 263 (1952).
108. Haynes, W. C., Wickerham, L. J., and Hesseltine, C. W., Maintenance of Cultures of Industrially Important Microorganisms, *Appl. Microbiol., 3,* 361 (1955).
109. Mehrotra, B. S., and Hesseltine, C. W., Further Evaluation of the Lyophil Process for the Preservation of Aspergilli and Penicillia, *Appl. Microbiol., 6,* 179 (1958).
110. Hesseltine, C. W., Bradle, B. J., and Benjamin, C. R., Further Investigations on the Preservation of Molds, *Mycologia, 52,* 762 (1960).
111. Staffeldt, E. E., Observations on Lyophil Preservation and Storage of *Pythium* species, *Phytopathology, 51,* 259 (1961).
112. Boyd, I., and Bullock, K., Freeze-Dried Preparations of *Penicillium spinulosum, J. Pharm. Pharmacol., 18,* Suppl. 28S (1966).
113. Ellis, J. J., and Roberson, J. A., Viability of Fungus Cultures Preserved by Lyophilization, *Mycologia, 60,* 399 (1968).
114. Price, L. K., Simple Freeze-Dry Method for Fungal Cultures, *Lab. Pract., 19,* 388 (1970).
115. Fennell, D. I., Raper, K. B., and Flickinger, M. H., Further Investigations on the Preservation of Mold Cultures, *Mycologia, 42,* 135 (1950).
116. Goldie-Smith, E. K., Maintenance of Stock Cultures of Aquatic Fungi, *Journal of the Elisha Mitchell Sci. Soc., 72,* 158 (1956).
117. Hwang, S. W., Effects of Ultralow Temperatures on the Viability of Selected Fungus Strains, *Mycologia, 52,* 527 (1960).
118. Hwang, S. W., Long-Term Preservation of Fungus Cultures with Liquid Nitrogen Refrigeration, *Appl. Microbiol., 14,* 784 (1966).
119. Hwang, S. W., Investigation of Ultralow Temperature for Fungal Cultures. I. An Evaluation of Liquid-Nitrogen Storage for Preservation of Selected Fungal Cultures, *Mycologia, 60,* 613 (1968).
120. Hwang, S. W., and Howells, A., Investigation of Ultralow Temperature for Fungal Cultures. II. Cryoprotection Afforded by Glycerol and Dimethyl Sulfoxide to 8 Selected Fungal Cultures, *Mycologia, 60,* 622 (1968).
121. Davis, E. E., Preservation of Myxomycetes, *Mycologia, 57,* 986 (1965).
122. Loegering, W. Q., McKinney, H. H., Harmon, D. L., and Clark, W. A., A Long-Term Experiment for Preservation of Urediospores of *Puccinia graminis tritici* in Liquid Nitrogen, *Plant Dis. Rep., 45,* 384 (1961).
123. Loegering, W. Q., Harmon, D. L., and Clark, W. A., Storage of Urediospores of *Puccinia graminis tritici* in Liquid Nitrogen, *Plant Dis. Rep., 50,* 502 (1966).
124. Wellman, A. M., and Walden, D. B., Qualitative and Quantitative Estimates of Viability for Some Fungi after Periods of Storage in Liquid Nitrogen, *Can. J. Microbiol., 10,* 585 (1964).
125. Wellman, A. M., Growth of Some Fungi before and after Cryogenic Storage, in *Proceedings of the First International Conference on Culture Collections,* p. 255, H. Iizuka and T. Hasegawa, Eds. University of Tokyo Press, Tokyo, Japan (1970).
126. Bugbee, W. M., and Kernkamp, M. F., Storage of Pycniospores of *Puccinia graminis secalis* in Liquid Nitrogen, *Plant Dis. Rep., 50,* 576 (1966).
127. Dade, H. A., in *Herb. I.M.I. Handbook,* p. 40. Commonwealth Mycological Institute, Kew, Surrey, England (1960).
128. Little G. N., and Gordon, M. A., Survival of Fungus Cultures Maintained under Mineral Oil for Twelve Years, *Mycologia, 59,* 733 (1967).
129. Booth, C. (Ed.), Fungal Culture Media, in *Methods in Microbiology,* Vol. 4, p. 49. Academic Press, London, England (1971).
130. Bakerspigel, A., Soil as a Storage Medium for Fungi, *Mycologia, 45,* 596 (1953).
131. Bakerspigel, A., A Further Report on the Soil Storage of Fungi, *Mycologia, 46,* 680 (1954).
132. Meyer, E., The Preservation of Dermatophytes at Subfreezing Temperatures, *Mycologia, 47,* 664 (1955).
133. Kramer, C. L., and Mix, A. J., Deep-Freeze Storage of Fungus Cultures, *Trans. Kans. Acad. Sci., 60,* 58 (1957).
134. O'Brien, M. J., and Webb, R. E., Preservation of Conidia of *Albugo occidentalis* and *Peronospora effusa,* Obligate Parasites of Spinach, *Plant Dis. Rep., 42,* 1312 (1958).
135. Wester, R. E., Drechsler, C., and Jorgensen, H., Effect of Freezing on Viability of the Lima Bean Downy Mildew Fungus (*Phytophthora phaseoli* Thaxt), *Plant Dis. Rep., 42,* 413 (1958).
136. Hwang, S. W., Longevity of Human Pathogenic Fungi Preserved at –165°C, *Bacteriol. Proc.,* p. 61 (1966).
137. Kelton, W. H., Storage of *Mycoplasma* Strains, *J. Bacteriol., 87,* 588 (1964).
138. Tully, J. G., and Ruchman, I., Recovery, Identification, and Neurotoxicity of Sabin's Type A and C Mouse Mycoplasmas (PPLO) from Lyophilized Cultures, *Proc. Soc. Exp. Biol. and Med., 115,* 554 (1964).
139. Addey, J. P., Taylor-Robinson, D., and Dimic, M., Viability of Mycoplasmas after Storage in Frozen or Lyophilised states. *J. Med. Microbiol., 3,* 137 (1970).

140. Norman, M. C., Franck, E. B., and Choate, R. V., Preservation of *Mycoplasma* Strains by Freezing in Liquid Nitrogen and by Lyophilization with Sucrose, *Appl. Microbiol., 20,* 69 (1970).
141. Taylor, A. E. R., and Baker, J. R., Cultivation of Protozoa, in *The Cultivation of Parasites in vitro,* p. 3. Blackwell, Oxford, England (1968).
142. Hjelm, K. K., and Moller, K. M., A Freeze-Drying Procedure for Protozoa, *C. R. Trav. Lab. Carlsberg, 33,* 301 (1963).
143. Raper, K. B., General Discussion, in *Culture Collections: Perspectives and Problems,* p. 145, S. M. Martin, Ed. University of Toronto Press, Toronto, Ont., Canada (1963).
144. Wickerham, L. J., and Page, F. C., Cultivation and Lyophilization of *Mastigina* sp., *J. Protozool., 17,* 518 (1970).
145. Annear, D. I., Preservation of *Strigomonas oncopelti* in the Dried State, *Nature (London), 178,* 413 (1956).
146. Annear, D. I., Recovery of *Strigomonas oncopelti* after Drying from the Liquid State, *Australian Journal of Experimental Biology and Medical Science, 39,* 295 (1961).
147. Annear, D. I., Recoveries of *Strigomonas oncopelti,* leptospirae and Reiter's treponeme from desiccates after storage for 5 years. *Aust. J. Exp. Biol. Med. Sci., 43,* 683 (1965).
148. O'Connell, K. M., Hutner, S. H., Fromentin, H., Frank, O., and Baker, H., Cryoprotectants for *Crithidia fasciculata* Stored at −20°C, with Notes on *Trypanosoma gambiense* and *T. conorhini, J. Protozool., 15,* 719 (1968).
149. Fulton, J. D., and Smith, A. U., Preservation of *Entamoeba histolytica* at −79°C in the Presence of Glycerol, *Ann. Trop. Med. Parasitol., 47,* 240 (1953).
150. Weinman, D., and McAllister, J., Prolonged Storage of Human Pathogenic Protozoa with Conservation of Virulence: Observations on the Storage of Helminths and Leptospiras, *Amer. J. Hyg., 45,* 102 (1947).
151. Most, H., Alger, N., and Yoeli, M., Preservation of *Leishmania donovani* by Low-Temperature Freezing, *Nature (London), 201,* 735 (1964).
152. Wang, G. T., and Marquardt, W. C., Survival of *Tetrahymena pyriformis* and *Paramecium aurelia* Following Freezing, *J. Protozool., 13,* 123 (1966).
153. Manwell, R. D., The Low-Temperature Freezing of Malaria Parasites, *Amer. J. Trop. Med., 23,* 123 (1943).
154. Manwell, R. D., and Edgett, R., The Relative Importance of Certain Factors in the Low-Temperature Preservation of Malaria Parasites, *Amer. J. Trop. Med., 23,* 551 (1943).
155. Wolfson, F., Effect of Preservation by Freezing upon the Virulence of *Plasmodium* for Ducks, *Amer. J. Hyg., 42,* 155 (1945).
156. Saunders, G. M., and Scott, V., Preservation of *Plasmodium vivax* by Freezing, *Science, 106,* 300 (1947).
157. Eyles, D. E., Coleman, N., and Cavanaugh, D. J., Preservation of *Toxoplasma gondii* by Freezing, *J. Parasitol., 42,* 408 (1956).
158. McEntegart, M. G., The Maintenance of Stock Strains of Trichomonads by Freezing, *J. Hyg., 52,* 545 (1954).
159. McEntegart, M. G., Prolonged Survival of *Trichomonas vaginalis* at −79°C, *Nature (London), 183,* 270 (1959).
160. Levine, N. D., Mizell, M., and Houlahan, D. A., Factors Affecting the Protective Action of Glycerol on *Trichomonas foetus* at Freezing Temperatures, *Exp. Parasitol., 7,* 236 (1958).
161. Levine, N. D., Andersen, F. L., Losch, M. B., Notzold, R. A., and Mehra, K. N., Survival of *Tritrichomonas foetus* Stored at −28 and −95°C after Freezing in the Presence of Glycerol, *J. Protozool., 9,* 347 (1962).
162. Polge, C., and Soltys, M. A., Preservation of Trypanosomes in the Frozen State, *Trans. Roy. Soc. Trop. Med. Hyg., 51,* 519 (1957).
163. Cunningham, M. P., and Harley, J. M. B., Preservation of Living Metacylic Forms of the. *Trypanosoma brucei* Subgroup. *Nature (London), 194,* 1186 (1962).
164. Polge, C., Freezing and Freeze-Drying in Parasitology, in *Aspects Thèoriques et Industriels de la Lyophilisation,* p. 417, L. Rey, Ed. Hermann, Paris, France (1964).
165. Doran, D. J., and Vetterling, J. M., Preservation of Coccidial Sporozoites by Freezing, *Nature (London), 217,* 1262 (1968).
166. Norton, C. C., and Joyner, L. P., The Freeze Preservation of Coccidia, *Res. Vet. Sci., 9,* 598 (1968).
167. Diamond, L. S., Freeze-Preservation of Protozoa, *Cryobiology, 1,* 95 (1964).
168. Diamond, L. S., Meryman, H. T., and Kafig, E., Preservation of Parasitic Protozoa in Liquid Nitrogen, in *Culture Collections: Perspectives and Problems,* p. 189, S. M. Martin, Ed. University of Toronto Press, Toronto, Ont., Canada (1963).
169. Hwang, S. W., Davis, E. E., Alexander, M. T., Freezing and Viability of *Tetrahymena pyriformis* in Dimethylsulfoxide, *Science, 144,* 64 (1964).
170. Simon, E. M., and Hwang, S. W., *Tetrahymena:* Effect of Freezing and Subsequent Thawing on Breeding Performance, *Science, 155,* 694 (1967).
171. Paine, G. D., and Meyer, R. C., *Toxoplasma gondii* Propagation in Cell Cultures and Preservation at Liquid Nitrogen Temperatures, *Cryobiology, 5,* 270 (1969).
172. Annear, D. I., Preservation of Leptospirae by Drying, *J. Pathol. Bacteriol., 72,* 322 (1956).
173. Annear, D. I., Observations on the Preservation by Drying of Leptospirae and Some Other Bacteria, *Aust. J. Exp. Biol. Med. Sci., 36,* 1 (1958).
174. Annear, D. I., The Preservation of Leptospires by Drying from the Liquid State, *J. Gen. Microbiol., 27,* 341 (1962).
175. Annear, D. I., Preservation of the Reiter Treponeme by Drying from the Liquid State, *J. Bacteriol., 83,* 932 (1962).

176. Otsuka, S., and Manako, K., Studies on the Preservation of Leptospirae by Freeze-Drying, *Jap. J. Microbiol., 5,* 141 (1961).

177. Hanson, A. W., and Cannefax, G. R., Recovery of *Treponema* and *Borrelia* after Lyophilization, *J. Bacteriol., 88,* 811 (1964).

178. Hart, L., Freeze Preservation of *Borrelia anserina, Aust. Vet. J., 46,* 455 (1970).

179. Hardy, P. H., and Nell, E. E., Maintenance of Virulence and Motility of *Treponema pallidum* in the Frozen State, *Bacteriol. Proc.,* p. 109 (1971).

180. Torney, H. L., and Bordt, D. E., Viability Quantitation of Leptospires after Rapid and Controlled Rate Freezing, *Cryobiology, 5,* 352 (1969).

181. Stalheim, O. H. V., Viable, Avirulent *Leptospira interrogans* Serotype *pomona* Vaccine: Preservation in Liquid Nitrogen, *Appl. Microbiol., 22,* 726 (1971).

182. Harris, R. J. C. (Ed.), The Preservation of Viruses, in *Biological Applications of Freezing and Drying,* p. 201. Academic Press, New York (1954).

183. McKinney, H. H., and Silber, G., Methods of Preservation and Storage of Plant Viruses, in *Methods in Virology,* Vol. 4, p. 491, K. Maramorosch and H. Koprowski, Eds. Academic Press, New York (1968).

184. Ward, T. G., Methods of Storage and Preservation of Animal Viruses, in *Methods in Virology,* Vol. 4, p. 481, K. Maramorosch and H. Koprowski, Eds. Academic Press, New York (1968).

185. Hollings, M., and Lelliott, R. A., Preservation of Some Plant Viruses by Freeze-Drying, *Plant Pathol., 9,* 63 (1960).

186. Worley, J. F., and Schneider, I. R., Long-Term Storage of Purified Southern Bean Mosaic Virus Freeze-Dried in the Presence of Lactose, *Phytopathology, 56,* 1327 (1966).

187. Best, R. J., On Maintaining the Infectivity of a Labile Plant Virus by Storage at −69°, *Virology, 14,* 440 (1961).

188. McKinney, H. H., Greeley, L. W., and Clark, W. A., Preservation of Plant Viruses in Liquid Nitrogen, *Plant Dis. Rep., 45,* 755 (1961).

189. Greiff, D., The Effect of Freezing, Low Temperature Storage and Drying by Vacuum Sublimation on the Activities of Viruses and Cellular Particulates, in *Recent Research in Freezing and Drying,* p. 167, A. S. Parkes and A. U. Smith, Eds. Blackwell, Oxford, England (1960).

190. Greiff, D., and Rightsel, W., The Effects on the Activities of Viruses of Rates and Temperatures of Freezing, Cyclic Freezing and Thawing, Storage at Low Temperatures, or Drying by Sublimation *in vacuo,* in *Aspects Théoriques et Industriels de la Lyophilisation,* p. 369, L. Rey, Ed. Hermann, Paris, France (1964).

191. Greiff, D., and Rightsel, W., Freezing and Freeze-Drying of Viruses, in *Cryobiology,* p. 697, H. T. Meryman, Ed. Academic Press, London, England (1966).

192. Greiff, D., and Rightsel, W., Recent Research in Freezing, Drying and Kinetics of Thermal Degradation of Viruses, in *Lyophilisation Recherches et Applications Novelles,* p. 103, L. Rey, Ed. Hermann, Paris, France (1966).

193. Greiff, D., and Rightsel, W. A., Stabilities of Suspensions of Viruses after Freezing or Drying by Vacuum Sublimation and Storage, *Cryobiology, 3,* 432 (1967).

194. Rightsel, W. A., and Greiff, D., Freezing and Freeze-Drying of Viruses, *Cryobiology, 3,* 423 (1967).

195. Greiff, D., Rightsel, W. A., and Schuler, E. E., Effects of Freezing, Storage at Low Temperatures, and Drying by Sublimation *in vacuo* on the Activities of Measles Virus, *Nature (London), 202,* 624 (1964).

196. Damjanović, V., and Klašnja, A., Freeze-Drying of the Attenuated Measles Virus (Belgrade's Strain), *Mikrobiologija (Beograd), 5,* 87 (1968).

197. Mareš, I., Kittnar, E., Srbová, H., and Časny, J., The Lyophilization of Measles Virus, *J. Hyg. Epidemiol. Microbiol. Immunol., 13,* 279 (1969).

198. Annear, D. I., and Beswick, T. S. L., A Note on the Preservation of Influenza Virus, *J. Hyg., 54,* 509 (1956).

199. Beardmore, W. B., Clark, T. D., and Jones, K. V., Preservation of Influenza Virus Infectivity by Lyophilization, *Appl. Microbiol., 16,* 362 (1968).

200. Tyrrell, D. A. J., and Ridgwell, B., Freeze-Drying of Certain Viruses, *Nature (London), 206,* 115 (1965).

201. Portocală, R., Samuel, I., and Popescu, M., Effect of Lyophilization on Picornaviruses, *Arch. Gesamte Virusforsch., 28,* 97 (1969).

202. Collier, L. H., The Preservation of Vaccinia Virus by Freeze-Drying, in *Freezing and Drying,* p. 133, R. J. C. Harris, Ed. Institute of Biology, London, England (1951).

203. Collier, L. H., The Development of a Stable Smallpox Vaccine, *J. Hyg., 53,* 76 (1955).

204. Slonim, D., and Kittnar, E., Remarks on Lyophilization and Stability of Lyophilized Vaccinia Virus, *Česk. Epidemiol. Mikrobiol. Immunol., 16,* 325 (1967).

205. Suzuki, M., Studies on Freeze-Drying of Vaccinia Virus: Effect of Suspending Media on Infectivity Titers, *Jap. J. Vet. Res., 16,* 87 (1968).

206. Suzuki, M., Suspending Media for the Preservation of Dried Vaccinia Virus, in *Proceedings of the First International Conference on Culture Collections,* p. 287, H. Iizuka and T. Hasegawa, Eds. University of Tokyo Press, Tokyo, Japan (1970).

207. Suzuki, M., Effect of Suspending Media on Freeze-Drying and Preservation of Vaccinia Virus, *J. Hyg., 68,* 29 (1970).

208. Hekker, A. C., Smith, L., and Huisman, P., Stabilizer for Lyophilization of Rubella Virus, *Arch. Gesamte Virusforsch., 29,* 257 (1970).

209. Verge, J., Goret, P., and Merteax, C., Nouvelle note sur la conservation de quelques ultravirus par la dessication, *Ann. Inst. Pasteur, 72,* 499 (1946).
210. Fellowes, O. N., Freeze-Drying of Foot-and-Mouth Disease Virus and Storage Stability of the Infectivity of Dried Virus at 4°C, *Appl. Microbiol., 13,* 496 (1965).
211. Ramyar, H., and Traub, E., Lyophilizing Foot-and-Mouth Disease Virus at Low Drying Temperature, *Amer. J. Vet. Res., 28,* 1605 (1967).
212. Fellowes, O. N., Comparison of Cryobiological and Freeze-Drying Characteristics of Foot-and-Mouth Disease Virus and of Vesicular Stomatitis Virus, *Cryobiology, 4,* 223 (1968).
213. Dundarov, S., Trendafilova, P., and Andonov, P., Lyophilization of Herpes Simplex Virus, *Ann. Res. Inst. Epidemiol. Microbiol. (Sofia), 14,* 167 (1969).
214. Calneck, B. W., Hitchner, S. B., and Adldinger, H. K., Lyophilization of Cell-Free Marek's Disease Herpes virus and a Herpes Virus from Turkeys, *Appl. Microbiol., 20,* 723 (1970).
215. Jung, M., and Krech, U., Lyophilisation of Human Cytomegaloviruses, *Pathol. Microbiol., 37,* 47 (1971).
216. Pattison, I. H., Jones, K. M., and Kimberlin, R. H., Observations on a Freeze-Dried Preparation Containing the Scrapies Agent, *Res. Vet. Sci., 10,* 214 (1969).
217. American Type Culture Collection, *Catalogue of Viruses Rickettsiae Chylamydiae,* 4th ed. (1971).
218. Olitsky, P. K., Casals, J., Walker, D. L., Ginsberg, H. S., and Horsfall, F. L., Preservation of Viruses in a Mechanical Refrigerator at −25°C, *J. Lab. Clin. Med., 34,* 1023 (1949).
219. Allen, E. G., Kaneda, B., Girardi, A. J., Scott, T. F. M., and Sigel, M. M., Preservation of Viruses of the Psittacosis-Lymphogranuloma Venereum Group and Herpes Simplex under Various Conditions of Storage, *J. Bacteriol., 63,* 369 (1952).
220. Melnick, J. L., Preservation of Viruses by Freezing, *Fed. Proc., Suppl. 15,* S-280 (1965).
221. Dobrowolska, H., and Kańtoch, M., Survival of Poliomyelitis Virus in a Frozen State, *Med. Dósw. Mikrobiol., 21,* 305 (1969).
222. Beech, F. W., and Davenport, R. R., Isolation, Purification and Maintenance of Yeasts, in *Methods in Microbiology,* Vol. 4, p. 153, C. Booth, Ed. Academic Press, London, England (1971).
223. Henry, B. S., The Viability of Yeast Cultures Preserved under Mineral Oil, *J. Bacteriol., 54,* 264 (1947).
224. Wickerham, L. J., and Andreasen, A. A., The Lyophil Process: Its Use in the Preservation of Yeasts, *Wallerstein Lab. Commun., 5,* 165 (1942).
225. Guibert, L., and Bréchot, P., La lyophilisation des levures, *Ann. Inst. Pasteur, 88,* 750 (1955).
226. Kirsop, B., Maintenance of Yeasts by Freeze-Drying, *J. Inst. Brew., 61,* 466 (1955).
227. Atkin, L., Moses, W., and Gray, P. P., The Preservation of Yeast Cultures by Lyophilization, *J. Bacteriol., 57,* 575 (1949).
228. Bréchot, P., Guibert, L., and Croson, M., La lyophilisation des levures, *Ann. Inst. Pasteur, 95,* 62 (1958).
229. Brady, B. L., Some Observations on the Freeze-Drying of Yeasts, in *Recent Research in Freezing and Drying,* p. 243, A. S. Parkes and A. U. Smith, Eds. Blackwell, Oxford, England (1960).
230. Wynants, J., Preservation of Yeast Cultures by Lyophilization, *J. Inst. Brew., 68,* 350 (1962).
231. Rose, D., Some Factors Influencing the Survival of Freeze-Dried Yeast Cultures, *J. Appl. Bacteriol., 33,* 228 (1970).
232. Annear, D. I., Preservation of Yeasts by Drying, *Aust. J. Exp. Biol. Med. Sci., 41,* 575 (1963).
233. Tsuji, K., Liquid Nitrogen Preservation of *Saccharomyces carlsbergensis* and Its Use in a Rapid Biological Assay of Vitamin B_6 (Pyridoxine), *Appl. Microbiol., 14,* 456 (1966).
234. Wickerham, L. J., and Flickinger, M. H., Viability of Yeasts Preserved Two Years by the Lyophil Process, *Brew. Dig., 21,* 55, 65 (1946).
235. Burns, M. E., Survival of Lyophilized Yeasts, *Sabouraudia, 1,* 203 (1962).
236. Scheda, R., and Yarrow, D., The Instability of Physiological Properties Used as Criteria in the Taxonomy of Yeasts, *Arch. Mikrobiol., 55,* 209 (1966).

ENRICHMENT CULTURE

DR. SHELDON AARONSON

Enrichment (elective) culture is the creation of a special environment *in vitro* for the growth and multiplication of microorganisms. This special environment permits the selection of the desired microorganism or microbial group from a mixture of microorganisms with which it is likely to be found in its natural environment. The special environment may function in one of the following ways: (1) by permitting the desired microorganism to outgrow all the others in the medium and thus make it the most common and therefore easy to isolate; (2) by inhibiting the growth of all microorganisms except the desired species or group; (3) by permitting the desired microorganism to multiply or perhaps metabolize in such a conspicuous way that it can be distinguished from other microorganisms.

The enrichment culture was first exploited in the late nineteenth century by several of the most famous microbiologists of that century, including M. W. Beijerinck and S. Winogradsky. Through their brilliant use of this technique and through the work of their twentieth-century colleagues we have learned a great deal about the role of microorganisms in the natural cycles: carbon, iron, nitrogen, oxygen, and sulfur; this technique has also been exploited to isolate microorganisms capable of specific metabolic reactions for the study of metabolic pathways or for the experimental or industrial conversion of one molecule to another, more desirable molecule.

Enrichment culture may be performed in liquid media by successive incubation in the same selective medium until the population is predominantly the desired organism, or by successive incubation in increasing concentrations of the selective chemical or physical environment. Another method is that of continuous-liquid culture, in which a selective medium enhances the growth of a favored microorganism while washing away its products and the less-favored species (see Specific References 1,2, and 3 for details). Enrichment cultures may also be performed on media solidified with agar (0.5–2.0%), gelatin (1–2%), or silica gel (see Specific Reference 4 for details). Solid media have the advantage of permitting the selection of the desired microorganism in an isolated and probably purified state. It is common to use liquid media for enrichment and solid media for the isolation of the desired microorganism.

The medium to be used for a specific organism must be tailored to that organism. However, all media and enrichments should include the following:

1. Energy. Energy may be supplied solely as organic molecules for heterotrophs (see below), or as reduced inorganic molecules or ions (H_2, CO, H_2S, S, NH_4+, NO_2-, N, Fe++, Mn++, etc.) for chemoautotrophs, or as artificial or sunlight for photosynthetics. Photosynthetic bacteria do better with light rich in the red part of the spectrum (tungsten lamp); on the other hand, light rich in the blue part of the spectrum tends to select blue-green algae (fluorescent lamp). Most photosynthetic eukaryotes do best with light relatively rich in the red end of the spectrum.

2. Source of Carbon. Autotrophic organisms may be satisfied by bubbling carbon dioxide through their medium or by supplying carbonate. Most microorganisms are heterotrophic, and their carbon and energy requirements may be met by any one or by a combination of the organic supplements listed in Table 1 or the compounds listed in Table 2. Many heterotrophs, as well as some algae, require exogenous organic nutrients they cannot synthesize; these may include one or more of the compounds listed in Table 3. A good rule to remember here is that the more organic material the medium contains, the more likely you are to select for heterotrophic "weeds", which quickly utilize the available organic compounds and swamp the autotrophs or more fastidious heterotrophs.

3. Source of Trace and Major Elements. All microorganisms need a source of sulfur (usually sulfate), nitrogen (usually nitrate, ammonium), and phosphorus (phosphate), as well as the trace elements. Examples of trace element solutions for marine and fresh-water microorganisms are shown in Table 4.

4. pH, Temperature, and Oxygen Tension. These must be varied to suit the needs and limits of the desired microorganism. Marine microorganisms tend to prefer a somewhat alkaline pH and lower temperatures (10–20°C); fresh-water microorganisms show wider environmental preferences, depending on their natural source.

Specific groups of microorganisms may be excluded from growth in appropriate media by the use of the antimicrobial agents shown in Table 5. The inhibition of fast-growing microorganisms allows slower growing or more fastidious microorganisms to appear under circumstances where they might normally be overgrown or suppressed.

One of the simplest demonstrations of an enrichment culture is the Winogradsky column (Figure 1), which permits the enrichment for different photosynthetic microorganisms on the basis of their need for a reducing (anaerobic) environment. The bottom of a transparent tube (glass or plastic) that contains wet bits of paper mixed with several grams of calcium sulfate and calcium carbonate as well as some mud is gradually filled with fine mud, preferably marine, from which pebbles, twigs, leaves, and other debris have been removed. Avoid air bubbles. The column of mud is allowed to settle until there is a clear liquid zone of approximately ¼″ on top. Excess water is poured off. The tube is tightly covered with a transparent plastic cover (e.g., Saran) to minimize evaporation, and then it is incubated for two to four weeks in sunlight or tungsten bulb light at room temperature. Do not allow the column of mud to dry or to become too warm. Add distilled water as needed. After a time, areas of the mud surface will become rust-colored, red, or dark green, indicating the appearance of specific types of photosynthetic bacteria. The liquid area on top of the mud column will appear pigmented and contain a variety of aerobic algae. Some of the algae may be isolated on the selected agar media in Table 6, some photosynthetic bacteria may be isolated in the anaerobic media shown in Table 7.

Procedures for the isolation of a variety of other special microbial groups are outlined in the tables below. Table 8 describes media for sulfur bacteria. Table 9 describes media for the isolation of bacteria and/or fungi that hydrolyze a variety of organic substrates.

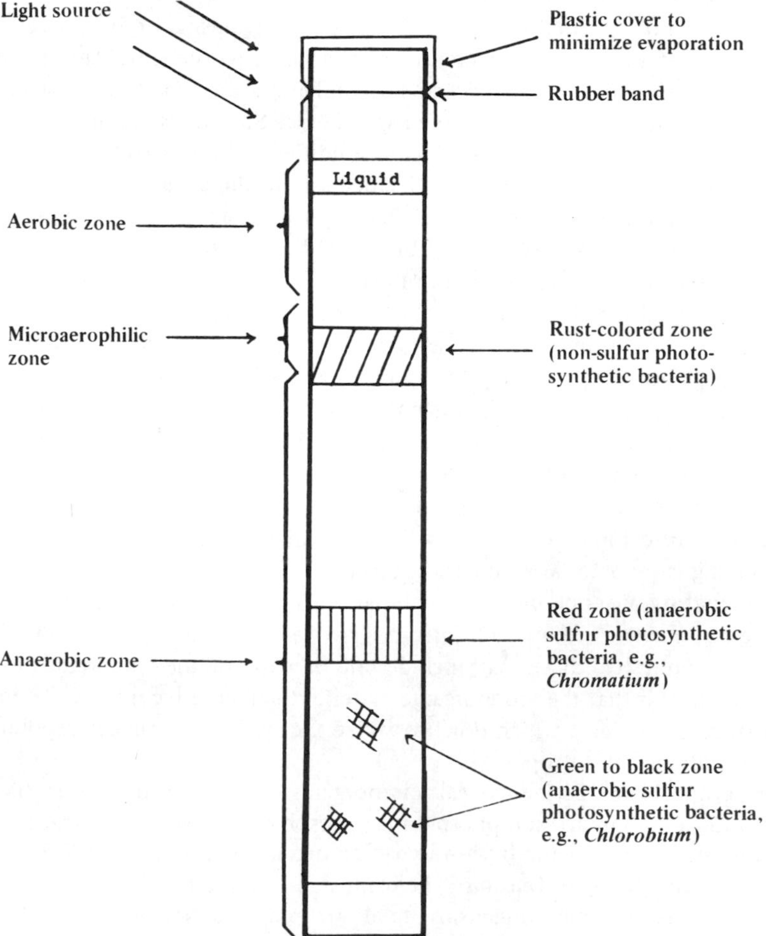

FIGURE 1. Winogradsky Column.

TABLE 1

ORGANIC SUPPLEMENTS

Supplement	Concentration, g/liter	Major Constituents: Amino Acids	Nucleic Acid Derivatives	Water-Soluble Vitamins	Lipids: Fatty Acids	Lipids: Steroids	Minerals
Soybean lecithin	0.1				+	+	
TEM[a]	0.1				+	+	
Tween 40	0.1				+	+	
Tween 60	0.1				+	+	
Tween 80	0.1				+	+	
Gelatin hydrolysate[b]	5.0	+					
Trypticase[c]	5.0	+		+			
Hycase[d]	5.0	+		+			
Protease peptone	5.0	+		+			
Liver extract	1.0	+	+	+			
Soil extract[e]	5.0 ml	+		+			

[a] TEM = diacetyl tartaric ester of tallow glycerides.
[b] Except L-methionine and L-tryptophan.
[c] Enzymatic digest of casein.
[d] Acid digest of casein.
[e] See Specific Reference 4 for preparation.

TABLE 2

SOME CARBON AND ENERGY SOURCES[a,b]

Carbohydrates, Sugars, and Sugar Alcohols

Arabinose
Cellulose
Dextrin
Fructose
Galactose
Glucose
Glycerol, dulcitol, sorbitol
Glycogen
Inulin
Lactose
Maltose
Mannitol, adonitol
Melibiose
Raffinose
Rhamnose
Ribose
Starch
Trehalose
Xylose

TCA Cycle Acids

Acetic acid
Aconitic acid
Citric acid
Fumaric acid
α-Ketoglutaric acid
Malic acid
Oxaloacetic acid
Succinic acid

Compounds Leading to the TCA Cycle

AMINO ACIDS
Alanine
Arginine
Aspartic acid, asparagine
Glutamic acid, glutamine
Glycine
Serine

FATTY ACIDS
Butyric acid
Glyoxylic acid
Lactic acid
Propionic acid
Pyruvic acid

[a] Use at 1 to 10 g per liter.
[b] Any one of these molecules provided with required trace or major elements may be used to isolate microorganisms capable of using the chosen molecule as the sole source of carbon and energy. Other organic molecules, such as hydrocarbons, aromatic compounds, etc., may also be used.

TABLE 3

EXOGENOUS ORGANIC NUTRIENTS AND THEIR RECOMMENDED AMOUNTS[a]

Organic Nutrient	Amount, mg/liter	Present in Simple Vitamin Mixture	Organic Nutrient	Amount, mg/liter	Present in Simple Vitamin Mixture
Water-Soluble Vitamins					
p-Aminobenzoic acid	0.1	+	Nicotinic acid	1.0	+
Betaine	1.0	–	Putrescine 2HCl	0.2	–
Biotin	0.01	+	Pyridoxal ethylacetal	0.1	+
Calcium pantothenate	1.0	+	Pyridoxamine 2HCl	0.2	+
DL-Carnitine HCl	0.4	–	Pyridoxine	0.2	–
Choline H_2·citrate	16.	+	Riboflavine	0.1	+
Cystamine 2HCl	0.1	–	Sodium riboflavine $PO_4 \cdot 2H_2O$	0.1	+
Ferulic acid	0.1	–	Spermidine $PO_4 \cdot 3H_2O$	0.1	–
Folic acid	0.5	+	Thiamine NO_3	0.4	+
p-Hydroxybenzoic acid	0.1	–	DL-Thioctic acid	0.05	+
Inositol	10.0	+	Vitamin B_{12}	0.004	+
Purines and Pyrimidines					
Adenine	5.0		Sodium guanylate H_2O	20.0	
Adenosine	10.0		Sodium inosinate	40.0	
Adenylic acid (2′, 3′, or 5′)	20.0		Thymidine	10.0	
Cytidylic acid	20.0		Thymine	5.0	
Deoxyguanosine	10.0		Uracil	5.0	
Guanine	4.0		Uridine	10.0	
Guanosine H_2O	10.0		Xanthine	5.0	
Hypoxanthine	4.0		Xanthosine	10.0	
Orotic acid	4.0				
Amino Acids					
DL-Alanine	0.4		DL-Methionine	0.06	
L-Arginine (free base)	0.3		DL-Phenylalanine	0.04	
DL-Aspartic acid	0.5		L-Proline	0.04	
L-Glutamic acid	1.0		DL-Serine	0.1	
Glycine	0.5		DL-Threonine	0.1	
L-Histidine	0.2		DL-Tryptophan	0.05	
DL-Isoleucine	0.05		L-Tyrosine	0.04	
DL-Leucine	0.05		DL-Valine	0.05	
DL-Lysine HCl	0.45				

[a] Any one of these molecules provided with required trace or major elements may be used to isolate microorganisms capable of using the chosen molecule as the sole source of carbon and energy. Other organic molecules, such as hydrocarbons, aromatic compounds, etc., may also be used.

TABLE 4

MAJOR AND MINOR ELEMENTS FOR SELECTED MICROORGANISMS*

	Fresh-Water			Marine		
	1	2	3	4	5	6
Compound for Major Elements, g/liter						
NH_4Cl	0.5		0.5			
$(NH_4)_2SO_4$		0.5				
KCl		0.03		0.3		
NaCl				24.0	23.0	25.0
$MgCl_2 \cdot 6H_2O$				3.0	11.0	
$MgSO_4 \cdot 7H_2O$	1.0	0.3	2.0	0.3		5.0
Na_2SO_4					4.0	
$MgCO_3$	0.4					
$NaHCO_3$					2.0	
KNO_3			0.1	0.1		
$Ca(NO_3)_2 \cdot 4H_2O$					0.1	
$CaCO_3$	0.05		0.2	0.2		0.25
K_2HPO_4					0.02	0.1
KH_2PO_4	0.3	*a*	*a*	0.02		
Trace Elements, mg/liter						
Fe	2.0	1.0	4.0	3.0	0.5	2.0
Zn	1.0	0.01	10.0	4.0	0.3	20.0
Mn	0.5	0.09	8.0	10.0	0.1	20.0
Cu	0.08	0.0003	1.0	0.003	0.1	
Co	0.1	0.003	1.0	0.03	*b*	*b*
B	0.1		4.0	2.0	0.1	
Mo	0.05	.002	0.8	0.5	0.1	
V	0.01					
Ca		20.0	10.0			
Sr				6.5		
Al				0.25		
Rb				0.1		
Li				0.05		
I				0.025		
Br				32.5		
pH	5.0	5.5	7.0	7.5	8.0	8.0–8.2

Code: 1. *Ochromonas danica*[5]; 2. *Phacus pyrum*[6]; 3. *Micrococcus sodonensis*[7]; 4. *Gyrodinium* sp.[8]; 5. *Stichococcus* sp.[8]; 6. *Labyrinthula* sp.[8]

a Supplied as Na_2 glycerophosphate $5H_2O$, 5.0 g.

b Supplied as vitamin B_{12}, 0.004 mg.

TABLE 5

ANTIBIOTICS INHIBITING SPECIFIC MICROBIAL GROUPS

Antibiotic	Concentration to be Used, mg/liter
Antiprokaryotic	
Bacitracin	50 mg
Chloramphenicol	0.5–75 mg
D-Cycloserine	1–400
Penicillin G	0.2–100
Streptomycin	0.1–20
Tetracyclines	0.1–50
Antieukaryotic	
Amphotericin B	0.1–20
Candicidin	0.1–20
Cycloheximide (actidione)	0.2–100
Filipin	0.1–20
Griseofulvin	1–20
Nystatin (mycostatin)	0.1–20

TABLE 6

SOME MEDIA FOR THE SELECTION OF SPECIFIC ALGAL GROUPS[a]

Additions, g/liter[b]	Blue-Green Algae	Blue-Green Algae	Blue-Green Algae	Acid-Living Phytoflagellates[c]	Desmids	Diatoms[d]
$Ca(NO_3)_2 \cdot 4H_2O$	0.04		0.025			0.1
Citric acid	0.003					
Ferric ammonium citrate	0.003	0.5	1.0			
Ferric chloride $6H_2O$			0.001	0.001		
K_2HPO_4	0.01	0.1	1.0	0.2	0.01	0.02
KNO_3		5.0	0.05		0.1	
$MgSO_4 \cdot 7H_2O$	0.025	0.05	0.25	0.1	0.01	
Na_2CO_3	0.02		1.5[e]			
$NaSiO_3 \cdot 9H_2O$	0.025					0.05
NH_4NO_3			1.0	1.0		
Soil extract					50.0 ml	
Tryptone						1.0
B_{12}						1.0 μg
Fe, as $Fe(NH_4)_2(SO_4)_2 \cdot 6H_2O$						0.5 mg
Zn, as $ZnSO_4 \cdot 7H_2O$						0.3 mg
B, as H_3BO_3						0.1 mg
Co, as $CoSO_4 \cdot 7H_2O$						0.1 mg
Cu, as $CuSO_4 \cdot 5H_2O$						0.1 mg
Mn, as $MnSO_4 \cdot H_2O$						0.1 mg
Mo, as $(NH_4)_6Mo_7O_{24} \cdot 4H_2O$						0.1 mg
Agar	15	15	1–15		7.5	10
pH	8–9.5	6–8	6–8		5–8	–
Reference	9	10	11	12	10	13

[a] Use fresh- or salt-water mud or water as the inoculum, at room temperature or at the temperature of the source of inoculum.
[b] Unless indicated as mg, ml, or μg.
[c] $CaCO_3$ may be added to make the medium alkaline.
[d] Use filtered sea water.
[e] Add aseptically.

TABLE 7

MEDIA FOR THE ENRICHMENT AND ISOLATION OF SOME PHOTOSYNTHETIC BACTERIA[14a]

Additions, g/liter medium	Anaerobic Sulfide Oxidizers: *Chlorobium limicola*	*Chlorobium thiosulfatophilum*	*Chromatium*	*Thiopedia*	*Rhodopseudomonas Rhodospirillum*
$(NH_4)_2SO_4$	1	1	1	1	
$NaHCO_3$[b]	2	2	2	2	2
$NaS \cdot 9H_2O$[b]	1	0.2	0.2	0.2	
$Na_2S_2O_3 \cdot 5H_2O$[b]		1	1	1	
Sodium malate			1	1	1
Yeast extract (or hydrolyzate)					1
Ethanol (or other alcohol)					20 ml
pH	7–7.2[c]	7–7.2[c]	8–8.4[c]	8–8.4[c]	7–7.5
Color of growth	green	green	red	red	rust

[a] Fill the container (a 50-ml bottle or smaller tube) with medium to the top. Add agar (15 g per liter) for solid media. All the above media contain the following ingredients (in g/liter, unless otherwise indicated): KH_2PO_4, 1; NH_4Cl, 1; $MgCl_2 \cdot 6H_2O$, 0.5; NaCl (for marine forms only), 10; vitamin B_{12}, 1.0 μg; 1.0 ml of each of the following trace metals: $FeCl_3 \cdot 6H_2O$, 1.5 mg; $MnCl_2 \cdot 6H_2O$, 5 mg; $CuCl_2 \cdot 2H_2O$, 1 mg; $NiCl_4 \cdot 6H_2O$, 2 mg $Na_2MOO_4 \cdot Co(NO_3)_2 \cdot 6H_2$ 0, 0.05; $CuSO \cdot 5H_2O$, 0.005.[b]

[b] Sterilize by filtration and add separately just before use to yield the above concentrations.

[c] Adjust the pH with H_3PO_4.

TABLE 8

MEDIA FOR THE ENRICHMENT AND ISOLATION OF SOME SULFUR BACTERIA[14]

	Sulfur (Thiosulfate) Oxidizers				
	Thiobacillus (Thiobacterium)				
Additions, g/liter	*thiooxidans*	*thioparus*	*denitrificans*[a]	*ferrooxidans*	Sulfate Reducers
Sulfur	10	10			
(or $Na_2S_2O_3 \cdot 5H_2O$)[b]	5	5	5		
$(NH_4)_2SO_4$	2–4	2–4			0.15
KH_2PO_4	2–4	2–4	2[c]	0.05	0.5
$CaCl_2$	0.25	0.25			
$MgSO_4 \cdot 7H_2O$	0.5	0.5		0.5	2
$FeSO_4 \cdot 7H_2O$	0.01	0.01	0.01[c]	1[c]	0.5
NH_4Cl			0.5		
$MgCl_2 \cdot 6H_2O$			0.5		
KNO_3			2		
$NaHCO_3$			1[c]		
KCl				0.05	
$Ca(NO_3)_2$				0.01	
$CaSO_4$					1
Yeast extract					1
Sodium lactate					3.5
Ascorbic acid					1
Thioglycollic acid					1
Agar (or silica gel)	30[d]	10–20	15	30[d]	15
pH	[5	[7.5	[7	3.5	7–7.5

[a] Anaerobically at 30° C.
[b] Sterilize by steaming if sulfur is the substrate.
[c] Sterilize separately and then add.
[d] Sterilize agar separately as a 15% solution and then add to medium.

TABLE 9

ENRICHMENT AND ISOLATION FOR BACTERIA AND/OR FUNGI THAT HYDROLYZE ORGANIC SUBSTRATES WITH VISIBLE CLEARING OR PIT FORMATION AROUND THE COLONIES ON SOLID MEDIUM

Addition	Grams per liter[a]									
Sodium alginate	25									
Agar		15								
Cellulose powder			10							
Chitin				1						
Soluble starch					5					
Skim milk (casein)						20				
Lipid (olive oil, tributyrin, etc.)							50			
Chondroitin sulfate								0.4[b]		
Sodium hyaluronidate									0.4[b]	
Pectin										5
$CaCl_2 \cdot 2H_2O$										0.2
K_2HPO_4	0.3		0.5							
KNO_3		1								
$FeCl_3 \cdot 6H_2O$										0.01
NaCl	30									
$MgSO_4 \cdot 7H_2O$			0.5							
NH_4Cl			0.5							
Trypticase	2.5									
Soil extract (see Specific Reference 4)			0.5 ml							
Yeast extract				1						1
Brain-heart infusion broth								1 liter	1 liter	
Bovine albumin, fraction V								10	10	
Agar	15	15	15	15			15	10	10	15
Nutrient agar					15	15				
pH	7		7	7.5–7.8	7	7	7	6.8	6.8	7
Source of inoculum	soil, salt water	soil, salt water	soil	soil, salt water	soil	soil	soil	humans, animals		plants, soil
Reference	15	16	4	17	4	4	18	19	19	20
Special notes	1	2	3	3,4	3	3	3	3,6	3,6	3,7

Instructions:

1. Preparation of the agar top layer: sodium alginate, NaCl, K_2HPO_4. Preparation of the bottom layer: K_2HPO_4, trypticase, agar.
2. Look for pits around the colonies.
3. Look for a clear halo.
4. Seal the plates to prevent evaporation and incubate them for 4 to 40 days in the dark at room temperature.
5. After incubation, flood the plates with Gram's iodine (iodine, 1 g; KI,2 g; water to 100 ml).
6. After incubation, flood the plates with 2*N* acetic acid for 10 minutes. Non-degraded substrate precipitates with the albumin, leaving a clear zone around colonies that produce enzyme.
7. After incubation, flood the plates with 1% aqueous solution of hexadecyltrimethylammonium bromide (J. T. Baker Chemical Co.); areas of pectolysis usually appear within 15 minutes.

[a] Unless otherwise indicated.
[b] Sterilize by filtration and add aseptically.

REFERENCES

General References

1. Aaronson, S., *Experimental Microbial Ecology.* Academic Press, New York (1970).
2. Alexander, M., *Microbial Ecology.* John Wiley and Sons, Inc., New York (1971).
3. Schlegel, H. G. (Ed.), *Anreicherungskultur und Mutantenauslese.* G. Fischer Verlag, Stuttgart, Germany (1965).

Specific References

1. Jannasch, H. W., *Arch. Mikrobiol., 45,* 323 (1963).
2. Jannasch, H. W., *Arch. Mikrobiol., 59,* 165 (1967).
3. Jannasch, H. W., *Limnol. Oceanogr., 12,* 264 (1967).
4. Aaronson, S., *Experimental Microbial Ecology.* Academic Press, New York, (1970).
5. Aaronson, S., and Scher, S., *J. Protozool., 7,* 156 (1960).
6. Provasoli, L., and Pintner, I. J., *The Ecology of Algae,* p. 84, C. A. Tryon, and R. T., Hartmen, Eds. University of Pittsburgh Press, Pittsburgh, Pennsylvania (1960).
7. Aaronson, S., *J. Bacteriol., 69,* 67 (1955).
8. Provasoli, L., McLaughlin, and Droop, M. R., *Arch. Mikrobiol., 25,* 392 (1957).
9. Gerloff, G. C., Fitzgerald, G. P., and Skoog, F., *Amer. J. Bot., 37,* 216 (1950).
10. Starr, R. C., *Amer. J. Bot., 51,* 1013 (1964).
11. Kratz, W. A., and Myers, J., *Amer. J. Bot., 42,* 282 (1955).
12. Pringsheim, E. G., in *Manual of Phycology,* p. 347, G. M. Smith, Ed. Chronica Botanica Co., Waltham, Massachusetts (1951).
13. Lewin, J. C., and Lewin, R. A., *Can. J. Microbiol., 6,* 127 (1960).
14. Postgate, J. R., *Lab. Pract., 15,* 1239 (1966).
15. Yaphe, W., *Nature, 196,* 1120 (1962).
16. Stanier, R. Y., *J. Bacteriol., 42,* 527 (1941).
17. Lear, D. W., Jr., in *Symposium on Marine Microbiology,* p. 594, C. H. Oppenheimer, Ed. Charles C Thomas, Springfield, Illinois (1963).
18. Rhodes, M. E., *J. Gen. Microbiol., 21,* 221 (1959).
19. Smith, R. F., and Willett, N. P., *Appl. Microbiol., 16,* 1434 (1968).
20. Hankin, L., Zucker, M., and Sands, D. C., *Appl. Microbiol., 22,* 205 (1971).

GENERAL REFERENCE DATA

GLOSSARY

DR. HUBERT LECHEVALIER

Abiotic substances Basic inorganic and organic substances, such as water, carbon dioxide, oxygen, calcium, etc.

Acaryotic Without a nucleus; a stage in the life cycle of a cell when nuclear material is not stainable.

Accession number The number given to a culture when it is deposited in a collection.

Acervulus (*pl.* -li) A cushion-like mass of conidia-producing hyphae. It has no peridium or covering of fungal nature.

Acetabuliform Saucer-shaped.

Achromogenic Producing no pigment.

Aciculiform Needle-shaped.

Acid-fast (1) Not decolorized by acid treatment. (2) Not decolorized after staining with carbol fuchsin.

Acrogenous Growing at the apex.

Acronym A word made up of the first letters of several words.

Acropetal A type of growth in which the apical part is the youngest.

Acropleurogenous Borne at the tip and on the side.

Adansonian A type of classification in which equal weight is given to each character of an organism.

Adelphotaxy A mutual attraction leading to the accumulation of certain sporangiospores at the mouth of the sporangium that contained them.

Adiaspiromycosis Haplomycosis; pulmonary mycotic disease of rodents and carnivores caused by *Emmonsia parva* or *E. crescens.*

Adnate Of gills or tubes of basidiomycetes, broadly attached to the stipe.

Adoral zone The portion of the protozoan surface that includes the peristome and its adornments.

Aeciospore A one-celled spore produced in spring by macrocyclic rust fungi. Its germination provides the mycelium that produces the uredial stage.

Aecium (*pl.* -ia) — A cup-shaped structure in which the aeciospores of rust fungi are produced in spring.

Aerobic — Growing in the presence of air.

Aerogenic — Producing gas.

Aerotolerant — Normally anaerobic, but able to grow somewhat in the presence of air.

Aethalium (*pl.* -ia) — A sessile fruiting body formed by a mass of plasmodium; for example, the fructification of *Fuligo septica.*

Agglutination — The clumping of cells distributed in a fluid. Agglutination is caused by specific substances called agglutinins.

Aggressin — A substance produced by a bacterium and contributing to its virulence by interfering with the host defense mechanism.

Agnotobiotic culture — A culture in which the nutrient source is unknown biological material. In slime mold cultivation, this usually means unknown microbial flora.

Akinete — A non-motile resting spore. The spore is formed within a vegetative cell by thickening of the cell wall and deposition of food reserves.

Aleuriospore — A terminal or lateral chlamydospore.

Allantoid — Sausage-shaped.

Alleles — Genes located at corresponding sites of homologous chromosomes. Allelic characters, controlled by alleles, are mutually exclusive.

Allelochemic — Pertaining to chemicals produced by one species and having an effect on another species.

Allelopathy — Suppression of the growth of a plant by chemicals produced by another plant.

Allochronic — Occurring at different periods of time.

Allopatric — Located in different geographical regions.

Amber — The name of the nonsense codon UAG. The presence of this codon indicates the termination of a growing polypeptide chain.

Amerospore — A spore formed of one cell.

Amino acyl adenylate — An activated compound involved in the formation of a covalent bond between an amino acid to its adaptor (soluble RNA).

Amino acyl synthetase — An enzyme catalyzing the process involved in hooking an amino acid to its adaptor (soluble RNA).

Amitosis — Nuclear division by constriction.

Amphibolic — A reversible pathway that may be catabolic or anabolic.

Amphimixis — The sexual union of cells and nuclei.

Ampulliform — Flask-shaped.

Amygdaliform — Almond-shaped.

Amyloid — Staining grayish to blackish-violet in Melzer's reagent.

Anabolism — The metabolic building of complex chemicals from simpler materials.

Anaerobic — Growing in the absence of air.

Anaerogenic — Does not produce gas.

Analogous — Similar, but not coming from a common ancestor.

Analogue — In chemistry, a chemical closely related to another one.

Anamnestic response — The secondary immune response.

Anaphylaxis — A violent reaction in a subject following injection of an antigen to which the subject is sensitized.

Anaplerotic — Replenishing, refilling.

Anastomosis (*pl.* -ses) — Crosswise fusion, as between two hyphae or ridges or gills.

Androgynous — With the antheridium and the oögonium on the same hypha.

Anemophilous — Liking wind; being carried by wind.

Angiocarpous — Closed at least until the spores are mature (as in fruit bodies of Gasteromycetes and Tuberales).

Angstrom — $\frac{1}{10}$ millimicron (or 10^{-8} cm).

Angustate — Narrow.

Anisogamy — The copulation of unequal gametes.

Anisotropic — Having different properties along different axes.

Annulus (*pl.* -li) — The ring-like structure left on the stipes of mature agarics as a remnant of the partial veil.

Anoxybiotic	Not able to utilize oxygen for growth.
Antheridium (*pl.* -dia)	The male gametangium.
Antherozoid	A motile male gamete.
Antibiotic	A chemical substance that is produced by a microorganism and inhibits the growth of other microbes at low concentrations (of the order of 1 μg/ml).
Antibody	A modified form of globulin that is produced by reticuloendothelial cells of vertebrates and can react with an antigen.
Antigen	A substance containing reactive groups by which it can react specifically with an antibody.
Antiseptic	An agent preventing or arresting microbial growth in living tissues.
Apiculus (*pl.* -li)	A short, sharp projection.
Aplanogamete	A non-flagellated gamete.
Aplanogamy	The union of non-flagellated gametes.
Aplanospore	A non-motile spore.
Apogamy	Apomictic development of diploid sex cells produced in the absence of meiosis.
Apomixis	The vegetative development of sexual cells into new individuals without being fertilized.
Apophysis (*pl.* -ses)	A swelling at the tip of a sporangiophore, below the sporangium; a swollen filament.
Apothecium (*pl.* -ia)	An ascocarp in which the hymenium is exposed.
Appressed	Closely flattened down.
Appressorium (*pl.* -ia)	The swelled part of a germ tube or hypha used for attachment to the host to be infected.
Arachnoid	Cobweb-like.
Arboreal	Pertaining to trees; growing on or in trees.
Arbuscular	Shrub-like.
Arcuate	Bow-shaped.
Arenaceous	Having to do with sand.

Armilla (*pl.* -ae) — A frill-like annulus.

Arthrospores — The conidia formed by fragmentation of a hypha.

Articulate — Jointed.

Ascocarp — Fruit body containing asci.

Ascogonium (*pl.* -ia) — A cell that, after fertilization, produces dicaryotic ascogenous hyphae.

Ascomycetes — Fungi known to form asci. Examples: cup fungi (*Peziza*), many yeasts, most of the fungi forming lichens.

Ascus (*pl.* -ci) — In the fungi, a sac-like cell in which sexual spores are formed. Most ascomycetes form eight spores per ascus.

Aseptate — Without cross walls.

Asteroid — Star-like.

Atomate — Having a powdered surface.

Aureous — Golden.

Auriculate — Ear-shaped.

Autogamy — (1) The fusion of two nuclei, both produced within the same female organ. (2) The process of internal self-fertilization in *Paramecium*, whereby a heterozygous animal gives rise to completely homozygous individuals.

Automixis — The copulation of two closely related cells, such as two daughter cells of the same haploid mother cell.

Autospore — An aplanospore with the same shape as a vegetative cell of a given species. Many Chlorococcales form autospores.

Autotrophic — Capable of utilizing inorganic substances as food, with carbon dioxide as the source of carbon.

Auxanography — The detection of growth inhibition or growth promotion by diffusion in a gel.

Auxospore — In diatoms, spores formed by cells whose size approach the minimum for a given species.

Auxotrophic — Biochemically deficient strains of microorganisms that require the product of the blocked pathway for growth.

Avellaneous — Hazel-colored.

Axenic culture — A pure culture. Used most often to refer to pure cultures of organisms usually difficult to grow in the absence of bacteria.

Axile — In or of the axis.

Axoneme — The inner fiber of flagella or cilia.

Axostyle — An elongated structure running along the body of some flagellates and spirochaetes.

Azygospore — A zygospore formed by parthenogenesis.

Bacciform — Shaped like a berry.

Bacilliform — Rod-shaped.

Bacteremia — The presence of bacteria in the blood.

Bactericidal — Able to kill bacteria.

Bacteriophages — Viruses attacking bacteria, often called "phages" for short.

Bacteriostatic — Preventing the growth of bacteria without killing them.

Ballistospore — A spore projected when mature.

Base-analogue — A purine or pyrimidine not naturally found in nucleic acids, but capable of being incorporated into it *in vivo* by virtue of its resemblance to one of the naturally occurring bases.

Basidiocarp — A fruit body containing basidia.

Basidiomycetes — Fungi forming their sexual spores on basidia. Examples: toadstools *(Amanita, Boletus)*, common cultivated mushrooms *(Agaricus)*, rust fungi, and smuts.

Basidium (*pl.* -dia) — The organ holding the sexual spores of basidiomycetes at the end of sterigmata.

Basifugal — Acropetal.

Basionym — A synonym resulting from a different combination. Example: *Waksmania rosea=Microbispora rosea.*

Basipetal — A type of growth in which the apical part is the oldest.

Basonym — Basionym.

Becke line — The bright line that appears at the boundary of two regions of different refractive indices.

Benthic — Benthonic.

Benthonic Occurring at the bottom of a body of water.

Benthos (1) The deep bottom of the ocean. (2) Benthonic organisms.

Bibulous Capable of absorbing water.

Bifid Split in the middle.

Bilateral In Agaricaceae, the hyphae of gill trama diverging towards the hymenium when viewed from gill base to gill edge.

Binomen The generic name + the specific name.

Binomial The use of two words (generic + specific) to name species.

Binominal Binomial.

Biogeocoenosis Ecosystem.

Biomass The weight of microbial growth.

Biome A large land-biotic community unit, such as the tundra.

Biophage An organism that consumes other living organisms.

Biota The biological population.

Biotic community The living part of an ecosystem.

Blastospore A spore formed by budding.

Blepharoplast The name for a centriole in flagellates.

Botryose Clustered like grapes.

Botuliform Allantoid.

Brachysporous With short spores.

Branching (1) True branching: branching that involves the cellular living material, as in fungi. (2) False branching: branching that involves only a non-living sheath, as in some blue-green algae.

Budding The production of a daughter cell by the development of a protrusion of the mother cell, as in yeasts.

Bulbil A small sclerotium.

Bullate Umbonate.

Bulliform Bubble-shaped.

Bursiform — Bag-like.

Caducous — Falling off early (or easily).

Caespitose — Forming tufts.

Calcarate — Having a spur.

Calcareous — Chalky; limy.

Callose — With a hard surface.

Calorie — The amount of energy necessary to raise the temperature of 1 ml of water 1°C.

Calvous — Bald; bare.

Calyciform — Cup-shaped.

Calyptrate — With a cap.

Campanulate — Bell-shaped.

Capillitium *(pl.* -ia) — A mass of non-living tubes, fibers or filaments found among the spores of Myxomycetes and Gasteromycetes. In Myxomycetes, the capillitium branches out from the columella.

Capneic — Able to grow in the presence of CO_2.

Capsid — The structure formed of the protein subunits (structure units) of a virion.

Capsomere — A group of viral protein subunits (structure units). Capsomeres are arranged symmetrically; they are present in virions with icosahedral symmetry, and absent in virions with helical symmetry. Capsomers can be seen with the electron microscope.

Capsule — A cellular envelope located outside the cell wall.

Carboxyphilic — Liking carbon dioxide.

Carpogonium (*pl.* -ia) — The female sex organ, mainly of red algae.

Carpophore — The fruit body of higher fungi; the stem of the same.

Carpospore — Spores formed by the division of the zygote; mainly used in the case of red algae.

Caryogamy — The fusion of two nuclei following plasmogamy.

Caryosome — Karyosome.

Castaneous	Chestnut-colored.
Catabolism	The metabolic activities involved in the degradation of complex chemicals into simpler products.
Catalyst	A substance capable of speeding up a chemical reaction without being used in the reaction.
Catenate	In chain; forming a chain.
Catenulate	Catenate.
Caudate	With a tail.
Cell wall	The cellular envelope located just outside the plasma (or cytoplasmic) membrane; its chemical composition differs from one group of organisms to another.
Central capsule	A porous membrane that separates the ectoplasm from the endoplasm in radiolarians.
Centriole	The basal body of a cilium; in various groups of organisms called basal granule, basal corpuscle, kinetosome, and blepharoplast.
Centromere	A specialized portion of a chromosome to which the spindle attaches during mitosis.
Centrum (*pl.* -ra)	Thc totality of structures within the perithecial or cleistothecial wall.
Cephalophore	A fructification in which the conidia are held in an apical spherical mucilaginous mass.
Cephalosporium (*pl.* -ia)	A cephalophore; a genus of imperfect fungi forming cephalophores.
Cerebriform	Brain-like; convoluted.
CF	Complement fixation.
Chain termination	The phenomenon whereby the synthesis of a growing polypeptide chain ceases because of the presence in the messenger RNA of a codon that is not translated (nonsense codon).
Character	In taxonomy, one of the attributes or features that make up and distinguish a taxon or an individual.
Cheilocyst	A cystidium located at the edge of the gill of an agaric.
Cheilocystidium (*pl.* -ia)	Cheilocyst.
Chemoautotrophy	The energy provided by dark chemical reactions involving inorganic substances. Carbon dioxide is the source of carbon.

Chemolithotrophy	(1) Chemoautotrophy. (2) Mineral nutritional requirement.
Chemoorganotrophy	(1) Heterotrophy. (2) Organic nutritional requirement.
Chemotaxis	The phenomenon of having living cells moving toward or away from a chemical substance.
Chemotrophy	The energy provided by dark chemical reactions.
Chiastobasidium	A clavate holobasidium in which nuclear spindles are placed crosswise.
Chlamydospore	A thick-walled, asexual resting spore of a fungus. It can be intercalary or terminal.
Chloroplast	A plastid containing photosynthetic pigments.
Chondroid	Having the consistency of cartilage.
Chromatid	One of the two strands of a duplicating chromosome.
Chromatophore	An organelle of procaryotic organisms that contains photosynthetic pigments.
Chromogenic	Color-producing.
Chryseous	Aureous.
Cilium (*pl.* -ia)	A short, vibrating cellular filament.
Cinereous	Ash-gray.
Circadian rhythm	The ability to repeat something at about 24-hour intervals.
Circinate	Coiled into a ring.
Circumscription	The definition of the limits of a taxon.
Cirrate	Curled.
Cirrus (*pl.* -ri)	An organelle formed by the fusion of a number of cilia.
Cisterna (*pl.* -ae)	The space between two unit membranes; for example, the perinuclear cisterna is the space between the two membranes enveloping the nuclear material.
Cistron	The smallest number of nucleotides that must be present together on the same chromosome in order to permit the synthesis of a given enzyme.
Cladistic	Indicating the degree of phyletic relatedness.

Clamp or clamp connection — A small semicircular protuberance of the mycelium of certain basidiomycetes.

Class — A taxonomic rank between division and order. The following endings are used for names of classes:

higher plants	-opsida
algae	-phyceae
fungi	-mycetes
bacteria	-no set ending

Classification — The orderly arrangement of units into groups of larger units.

Clathrate — Latticed.

Clavate — Club-shaped.

Cleistothecium (*pl.* -ia) — A closed ascocarp from which spores are released by rupture of the ascocarp wall.

Climate regime — The physical factors affecting living organisms in an ecosystem.

Clone — The asexually produced progeny of an individual.

Cloning — A procedure designed to give rise to recognizably distinct clones.

Clypeate — Shield-shaped.

Coccoid — Spheroid.

Coccolith — The platelets found in the shell of some protozoa; often found as fossils in chalk.

Cochleate — Conchate.

Codon — Three neighboring nucleotides (code triplets) in nucleic acid that specify the incorporation of a particular amino acid into a nascent protein; for example, three adenine residues in a row (AAA) or two adenine and a guanine residue (AAG) are codons for lysine.

Coelozoic — Living in one of the cavities of a host.

Coenocytic — Multinucleate. Examples of coenocytic structures are the thallus of most phycomycetes and the plasmodia of myxomycetes and plasmodiophorales.

Coenzyme I — DPN.

Colony — The growth arising on or in a solid medium from a small inoculum, often a single propagule.

Columella (*pl.* -ae) — A sterile central axis within a sporangium.

Commensal — Living in an indifferent association with another organism.

Competence — Ability of a bacterial cell to bind irreversibly DNA of high molecular weight, thereby becoming resistant to deoxyribonuclease. It is the first important step of transformation.

Complete medium — A complex medium that satisfies the nutritional requirements of as many of the usually encountered auxotrophic mutants as possible, usually containing such things as amino acids, water-soluble vitamins, nucleic acid bases, etc.

Concanavalin A — The phytohemagglutinin of the Jack bean.

Conchate — Shaped like a shell.

Concolor — The same color.

Conditional lethal mutation — A mutant whose viability depends on growth conditions.

Confluent — Running together; getting fused.

Conidium (*pl.* -ia) — Any asexual spore, but not a sporangiospore or an intercalary chlamydospore.

Conjugation — (1) In fungi, isogamic copulation. (2) In protozoa, the temporary union of two individuals of the same species, during which some nuclear material is exchanged. (3) In bacteria, the transfer of genetic material from one cell to another by direct contact.

Conspecific — Belonging to the same species.

Context — The inner "tissue" of the fruit body of higher fungi.

Contractile vacuole — A pulsating vacuole found in some protozoa, believed to help maintain proper osmotic pressure.

Conversion (viral) — The acquisition of new traits by a cell upon infection by a virus.

Coprophagic — Eating dung.

Coprophytic — Dung-inhabiting.

Cordate — Heart-shaped.

Coremium (*pl.* -ia) — A group of hyphae closely packed together, forming an upright column and producing spores at its apex.

Corepressor — A metabolite that combines with a repressor to inhibit the enzyme or enzymes involved in their biosynthesis.

Cornute — Horn-shaped.

Corona The whorls of apical hairs.

Cortex The tough outer layer of a fruit body.

Cortina A partial veil persisting to cover mature gills in some agarics. It is of a fibrillous or cobweb-like nature rather than membranous.

Cotype Any specimen that is part of an author's material used in the description of a new taxon if no single specimen has been described as the holotype.

Crateriform Goblet-shaped.

Cribrose Sieve-like.

Crista (*pl.* -ae) A ridge; a crest; a fold of the inner membrane of a mitochondrion.

Crossing over The exchange of genetic material between homologous chromosomes.

Crozier A hook on ascogenous hyphae that develops into an ascus.

Cruciform Cross-shaped.

Crustose Forming a crust.

Cryophilic Liking cold.

Cryptogam A plant producing spores, but no flower (and no seed). The cryptogams include the algae, the fungi, the mosses, and the ferns.

Culmination The terminal phase of the formation of a sorocarp by cellular slime molds.

Cyaneous Dark blue.

Cyathyform Shaped like a goblet with flaring margin.

Cyst (1) A resting form surrounded by a resistant wall. (2) A zoöspore after it has lost its flagellum and has formed a heavy wall.

Cystidium (*pl.* -ia) (1) A large cell located among basidia. (2) Variously shaped sterile cells found on the gills, pileus, and stipe of a fungus.

Cystites Coccoid cells formed by arthrobacters, several times larger than the majority.

Cystosorus A group of united cysts or resting spores found in Plasmodiophorales and Chytridiales.

Cystospore An encysted zoöspore, as found in the Chytridiales.

Cytogamy (1) In general, the union of cells in the sexual process. (2) In protozoa, a type of conjugation with no nuclear exchange between individuals, with each individual undergoing autogamy after nuclear division.

Cytokinesis The division of the extranuclear portion of the protoplast.

Cytopathic Cytopathologic.

Cytopathology Pathology at the cellular level.

Cytoproct Cytopyge.

Cytopyge Cell anus.

Cytostome Cell mouth.

Daedaloid Shaped like a maze.

Dalton The weight of one atom of hydrogen; one dalton = 1.66 x 10^{-24} g.

Damping-off The rotting of seedlings or cuttings at soil level.

Datum (*pl.* -a) A fact.

Decurrent Of gills or tubes, descending along the stipe.

Defective virus A virus able to reproduce in a suitable host only with the help of another virus.

Degenerative codons More than one codon coding for the same amino acid.

Dehiscence The opening of a closed fruit body upon ripening.

Deliquescing Melting away.

Dendrogram A tree-like diagram.

Dendroid Tree-like.

Denitrification The microbial breakdown of nitrates and nitrites accompanied by the evolution of nitrogen.

Dentate Toothed.

Desiccate To remove moisture.

Desquamate Not scaly.

Detritivore An organism ingesting organic detritus.

Deuteromycetes Fungi Imperfecti.

Dextrorse Clockwise.

Diapedesis The passage of blood cells through the walls of intact vessels. The migrating cells pass between the cells forming the wall of the blood vessels.

Diastole The period of filling of a contractile vacuole.

Diauxie The development of an organism in a medium containing more than one substrate, with two (successive) log phases separated by a lag phase.

Dicaryotic Having two nuclei of different types.

Diclinous With the antheridium and the oögonium on different hyphae.

Dictyospore A muriform spore.

Didymous With two cells.

Dimidiate With a semicircular outline.

Dimorphism The ability to exist under two forms; mainly used to refer to the ability of some fungi to exist either under a mycelial or under a yeast form.

Diplanetism Having two motile stages.

Diploid Having $2n$ number of chromosomes; having a pair of each chromosome.

Disinfectant An agent that kills growing microbes, but not necessarily resting spores.

Division A taxonomic rank between kingdom and class. The following endings are used for names of divisions:

plants	-phyta
fungi	-mycota

DNA polymerase The enzyme catalyzing the condensation of deoxyribonucleoside triphosphates into DNA. The enzyme recognizes only the sugar phosphate portions of the nucleotide precursors. Preformed DNA acts as a template to insure proper base sequence.

Dolipore The barrel-shaped septal pore of a basidiomycete hypha.

Eburneous — Ivory-colored.

Echinidium (*pl.* -ia) — Cells with numerous hair-like projections found in fungi; brush-like cells.

Echinulate — With spines.

Ecology — The study of the relationships between organisms and their environment.

Ecosystem — Ecological system; all the organisms in a given space and the physical environment with which they react.

Ecotone — The transition zone between two or more diverse biotic communities.

Ecotype — A locally adapted population of a euryecious species.

Ectoplasm — A layer of hyaline protoplasm surrounding the granular endoplasm of amebas and myxomycetes.

Effused — Spread out over the substratum.

Effused-reflexed — Effused and turned back at the margin.

Egg — (1) A large non-motile female gamete. (2) The young sporocarp of some higher fungi before the rupture of the volva.

Elater — A free thread of the capillitium of some myxomycetes.

Emarginate — Of lamellae, notched near the stem.

Emendation — Revision of the definition of a taxon.

Endergonic — Requiring energy.

Endobiotic — Living inside another living organism.

Endogenous — Formed within.

Endomixis — Self-fertilization.

Endophyte — An organism living inside another one.

Endoplasm — A mass of granular protoplasm surrounded by a layer of hyaline ectoplasm. These are observed in amebas and myxomycetes and can be converted one into the other.

Endoplasmic reticulum — The system of intercommunicating tubules extending throughout the cytoplasm of eucaryotic cells. These are formed of unit membranes.

Endosome Caryosome.

Endospore (1) In bacteria, the heat-resistant, dipicolinic acid-containing spore of bacilli and clostridia. (2) Any spore formed endogenously.

Enzyme A catalytic protein that mediates biochemical reactions.

Epibiotic Living on the surface of another organism.

Epiphyte An organism growing on the outside of another one.

Episome A non-chromosomal replicon that may be, at times, incorporated into the bacterial chromosome.

Epitheca Outer (larger) half of a frustule.

Epithet The specific name; for example, *fradiae* is the epithet in *Streptomyces fradiae.*

Epixylous Growing on wood.

Erumpent Breaking through.

Erysipelas An infectious disease causing redness and swelling of skin and subcutaneous tissue.

Eucarpic Fructifications being formed on only a part of the thallus.

Eucaryotic With a true nucleus; that is, one in which the nuclear material is surrounded by a membrane.

Euphotic (1) Well lit. (2) Pertaining to well-lit places where photosynthesis can take place.

Eury- Wide, as in euryecious – having a wide selection of habitat.

Eutrophication (1) The addition of nutrients. (2) Organic pollution resulting from man's activity.

Evaginate Without a sheath.

Evanescent Disappearing soon after formation.

Exciple Excipulum.

Excipulum (*pl.* -la) The main structural part of an apothecium.

Exergonic Producing energy.

Exine The outer coat of a bacterial cyst.

Exobiology The study of extraterrestrial life.

Exomixis Cross-fertilization.

Exospore (1) A spore formed at the end of a sporulating filament. (2) One of the layers of a spore envelope.

Exosporium (*pl.* -ia) One of the envelopes of a spore, usually the outermost.

Exsiccata Sets of dried and labeled botanical specimens for sale or exchange.

F Fertility factor; an episome that determines sex in bacteria.

Family A taxonomic rank between order and genus. The following endings are used for names of families:

plants and bacteria	-aceae
animals	-idae

Feedback inhibition The inhibition of enzymatic activity in a metabolic pathway by the end product of the pathway.

Ferredoxin An electronegative electron carrier that occurs in anaerobes and is known to be an iron protein of low molecular weight.

Filipod A threadlike appendage.

Fimbria (*pl.* -ae) Fine hair-like appendage found on the surface of a bacterial cell.

Fission, binary The process of division typical of bacteria; the division of a cell by the formation of a median wall without the formation of mitotic or meiotic nuclear figures.

Flabelliform Fan-shaped.

Flagellin The protein of which a flagellum is composed.

Flagellum (*pl.* -la) A whip-like cellular organelle.

Flavous Golden-yellow.

Fluorescence The property of some substances of producing light when acted upon by radiant energy of wavelengths shorter than that of the light produced.

Fluorochrome A fluorescing chemical.

Foramen An opening.

Form species, - genus, - family Taxa based on morphological properties. The organisms grouped in these taxa are imperfect.

Forma specialis (*pl. -ae -les*) — A taxonomic rank for a parasitic species adapted to a specific host, whose name it takes.

Forward mutation — A mutation from the wild-type allele to a mutant allele, resulting in a selective advantage under given environmental conditions.

Fragmentation — The division of a hypha into rods or cocci.

Frame-shift mutation — An addition or deletion of a number of contiguous nucleotide pairs, not in multiples of three, in a gene. Since the codon consists of three contiguous pairs, the reading frame of translation of the corresponding RNA is put out of normal register.

Free — Of gills or pores of fungi, not attached to the stipe.

Frustule — The cell wall of diatoms.

Fruticose — Shrub-like.

Fuligineous — Sooty.

Fulvous — Reddish-brown.

Funiform — Rope-like.

Furcate — Forked.

Fuscous — Dusky; dull gray.

Fusiform — Spindle-shaped.

Fusoid — Fusiform.

GC content — The ratio of the molar concentration of guanine + cytosine to that of the total base content in DNA.

Galeriform — Cap-shaped.

Gametangium (*pl.* -ia) — A structure containing gametes.

Gamete — A reproductive sex-cell. The union of two gametes of opposite sex results in nuclear fusion and the formation of a zygote.

Gametophyte — The haploid plant bearing sexual organs.

Gangliospore — A spore formed by the transformation of the swollen tip of a hypha.

Gemma (*pl.* -ae) — A chlamydospore of a phycomycete.

Gemmation — The formation of gemmules.

Gemmiform — Bud-shaped.

Gemmule — A bud from which certain protozoa develop.

Gene — A specific area of a chromosome that determines a specific trait.

Genetic markers — The genes that differentiate the parents in a cross, allowing the inheritance of particular chromosomal regions from one or the other parent by the progeny to be experimentally determined.

Genome — (1) The entire set of genetic informational material. (2) In haploid microbes, the DNA representing the total set of chromosomal genes in an organism.

Genotype — (1) The sum of the heritable determinants of an individual. (2) The type species of a genus.

Genus (*pl.* -nera) — A taxon that groups related species.

Geobiocoenosis — Ecosystem.

Germ pore — The area in a spore wall through which a germ tube may pass.

Ghost — In virology, the virion empty of nucleic acid.

Gill — Lamella.

Glaucous — Sea-green.

Gleba (*pl.* -ae) — The sporing portion of an angiocarpous fruit body.

Gnotobiotic — Known biological makeup.

Golgi apparatus — A cellular system of canals and sacs somewhat similar to the endoplasmic reticulum. It is believed to concentrate and distribute cellular chemicals rather than be the seat of their synthesis.

Gonangium (*pl.* -ia) — A closed structure in which reproductive elements are formed.

Gonidium (*pl.* -ia) — The algal part of a lichen thallus; a motile element formed by a non-motile bacterium or alga.

Gram stain — See page 683. Gram-positive organisms are blue, Gram-negative organisms are red.

Grana — Lamellar organelles that contain photosynthetic pigments and are photochemically active.

Granulomatosis infanti-*septica* — An intrauterine infection of the human newborn, characterized by numerous foci of necrosis and high mortality; caused by *Listeria monocytogenes.*

Granulose A type of viral disease of insects, in which single virions are surrounded by a protein coat.

Grex The pseudoplasmodium of Acrasiales.

Gymnocarpous Having an exposed hymenium while the spores mature, as in the fruit bodies of agarics.

H-antigen A flagellar antigen.

Halophilic Salt-loving.

Haploid Having *n* numbers of chromosomes; that is, one chromosome of each type.

Hapten An antigen that can only react with an antibody. A hapten becomes an immunogen when coupled to a suitable carrier.

Haustorium (*pl.* -ia) A special hyphal branch of parasitic fungi that penetrates the cell of the host in order to absorb food.

Heliotropic Turning towards the sun.

Helper virus A virus permitting the development of a defective virus.

Helveous Pale yellow.

Hemagglutination The agglutination of red blood cells.

Heterocaryosis The presence of nuclei of more than one genetic constitution in the same cytoplasmic unit.

Heterocyst In blue-green algae, a cell of uncertain function with a thick wall and transparent contents.

Heterofermentation A fermentation with multiple end products.

Heterokontous Having flagella of different lengths.

Heteromerous In lichens, having a thallus in which separate layers of fungal hyphae and algal cells can be seen.

Heterophyllous Having lamellae of different form and/or length.

Heteropolymer A macromolecule composed of different chemical moieties.

Heterothallism A condition in which sexual conjugation is possible only through the interaction of genetically different thalli, which are usually morphologically identical.

Heterotrophic Requiring complex organic substances as food.

Hfr	High-frequency recombinant; i.e., a strain of enterobacteria in which the F episome is incorporated in the bacterial chromosome. Such strains recombine very frequently.
High-energy bond	A chemical bond that releases, upon hydrolysis, at least 5000 calories per mole.
Hilum (*pl.* -la)	A marking, especially on a spore, at the point of attachment to the sporophore.
Hirsute	Covered with hair.
Hispid	Covered with stiff hairs.
Holdfast	A structure made to anchor an organism to a substratum.
Holobasidium (*pl.* -ia)	A non-septate basidium.
Holocarpic	Of a fungus, the whole thallus becoming the fruiting body.
Holological	Having to do with the whole.
Holophytic	Obtaining food like a plant (by uptake of solutes).
Holotype	A specimen designated by the author of a taxonomic description as the nomenclatural type.
Holozoic	Obtaining food like an animal (by ingestion of solid particles).
Homeostasis	The tendency to resist change.
Homeotype	A specimen that was compared with the nomenclatural type and was found to belong to the same species.
Homobasidium (*pl.* -ia)	A typical club-shaped, non-septate basidium.
Homofermentation	Fermentation yielding a single end product.
Homoiomerous	In lichens, having a thallus in which algal cells and hyphae are not stratified.
Homologous	Similar in structure and in origin, but not necessarily in function.
Homologous chromosomes	Morphologically identical chromosomes that pair during meiosis.
Homonym	The name of a taxon that duplicates a previously published taxonomic name.
Homopolymer	A macromolecule composed of the same repeating chemical entities.
Homosporous	Having only one kind of asexual spores.

Homothallism — A condition in which sexual conjugation occurs within the same thallus.

Hormogonium (*pl.* -ia) — In blue-green algae, a short section of a trichome that is more motile than the rest of the trichome. Hormogonia slip away from parent filaments to develop into new trichomes.

Hugh and Leifson Test — A test used to determine whether carbohydrates are broken down by anaerobic fermentation or by oxidation.

Humus — An ill-defined yellow-brown to black colloidal substance resulting from the decomposition of organic matter on the surface of the earth.

Hyaline — Colorless.

Hybrid — (1) The result of cross-mating between two species. (2) The result of cross-mating between two individuals differing in one or more characters. (3) In etymology, a word made up of parts of words from different languages.

Hydnoid — Having a toothed hymenophore.

Hygrophanous — Somewhat translucent.

Hymeniform — Resembling the hymenium; composed of a palisade of cells.

Hymenium (*pl.* -ia) — In fungi, the sporing layer of a fructification. The cells of the hymenium are usually arranged as a palisade.

Hymenolichens — Basidiolichens.

Hymenophore — The portion of a carpophore bearing the hymenium.

Hypha (*pl.* -ae) — One of the filaments forming the thallus (mycelium) of many fungi and actinomycetes.

Hyphomycetes — Moniliales + Mycelia Sterilia.

Hypnospore — A thick-walled resting spore.

Hypogenous — Produced beneath.

Hyponym — The name of a taxon that has been poorly described.

Hypothallus (*pl.* -li) — In myxomycetes, the material not used in the formation of a sporangium.

Hypotheca — The inner (smaller) half of a frustule.

Hypothecium (*pl.* -ia) — A layer of hyphae immediately beneath the hymenium of an apothecium.

Icosahedron A three-dimensional body with twenty identical sides.

Identification The assignment of unknown entities into a classification–nomenclatural system.

Imbricate Overlapping like shingles.

Immunogen An antigen that can induce the formation of an antibody.

Imperfect Pertaining to stages of growth with no sexual structures.

Imperfect fungi Fungi without a known sexual cycle.

Incertae sedis Of uncertain taxonomic position.

Inducer A chemical compound whose presence causes the production of the enzymes necessary for its utilization.

Induction The mechanism by which a prophage is induced to proceed with a lytic cycle. UV is an inducing agent.

Indusium (*pl.* -ia) A cover; a tunic.

Infraciliature In protozoa, the system of granules and fibrils underlying the cilia.

Infundibuliform Funnel-shaped.

Inoculum (*pl.* -la) A living microbial cell (or cells) that is transferred to a substratum, where it will presumably proliferate.

Inoperculate Without a lid.

Intercalary Not apical; between two similar structures.

Intine The inner coat of a bacterial cyst.

Invaginated Enclosed in a sheath.

Inversely bilateral In Agaricaceae, the hyphae of gill trama diverging from the hymenium towards the plane of symmetry of the gill when viewed from gill base to gill edge.

Irpiciform Having flattened teeth.

Isidium (*pl.* -ia) The minute growth of the upper cortex of some lichens.

Isoantigens Antigens produced by some individuals of one species and not by others; for example, blood group substances.

Isogametangia Male and female (+ and −) gametangia that are of equal size.

Isogamete A gamete similar to that of the opposite sex.

Isogamy The copulation between morphologically identical gametes.

Isolate Strain.

Isomers Different compounds with the same molecular formula.

Isotropic Having the same properties along all axes.

Isotype A specimen that is part of the material collected with the holotype.

***Kappa* particles** Particles found in the cytoplasm of certain paramecia.

Karyogamy Caryogamy.

Karyosome A large, deeply staining body found in the nuclei of some flagellates, protozoa and plasmodiophorales. It elongates and divides by constriction during nuclear division.

Keratinophilic Liking keratin (hairs, hoofs, horns, nails, feathers).

Key In taxonomy a succinct presentation of the properties of taxa to help in their identification.

Kinetoplast A nucleic acid-containing organelle associated with the blepharoplast of some flagellates.

Kinetosome The name of a centriole in ciliates.

Laccate Varnished.

Lacuna (*pl.* -ae) Depression.

Lageniform Flask-shaped.

Lamella (*pl.* -ae) The ridge to blade-like projection bearing the hymenium of agarics.

Lanceolate Lance-shaped.

Leaky mutant A strain not completely lacking the function that distinguishes it from the wild type; for example, a leaky auxotroph does not fail completely to grow on a medium lacking the nutrient required for optimal growth.

Lectotype A substitute for a missing or not designated holotype selected from the original material investigated by the author of a taxonomic name.

Lentic	Having to do with still waters.
Leucoplast	A colorless plastid.
L-form	A strain of bacterium that has lost the ability to form a cell wall.
Ligand	A general term to refer to either a hapten or an antigen.
Lingulate	Tongue-shaped.
Lithotroph	An organism deriving its carbon primarily from carbon dioxide.
Loculus (*pl.* -li)	A small chamber.
Lomasomes	Systems of unit membrane tubules and vesicles, forming bodies located between the cell wall and the plasma membrane of certain fungi and algae.
Lophotrichiate	Having a tuft (or tufts) of flagella.
Lorica (*pl.* -ae)	The shell of some protozoa.
Lotic	Having to do with running waters.
Lumen	A central cavity.
Luminescence	The production of light at low temperature. Fluorescence and phosphorescence are examples of luminescence.
Lymphocyte	A type of white blood cell, formed in the reticular tissue of lymph nodes.
Lyophile	A freeze-dried material.
Lysis	Dissolution.
Lysogenic bacteria	Bacteria containing a prophage.
Lysogenic viruses	Viruses whose genome may become a prophage.
Lysosome	A membrane-bound cellular particle rich in lytic enzymes.
Lysozyme	An enzyme attacking the murein of the cell wall of bacteria.
Macroconidium	The larger conidium of a fungus that forms conidia of two different sizes. Macroconidia usually have more diagnostic value than microconidia.
Macroconsumer	Phagotroph.
Macrocyclic rusts (or long-cycled rusts)	Rusts producing at least one type of binucleate spore in addition to teleutospores.

Macrocyst (1) In cellular slime molds, an encysted aggregate, the whole mass of cysts being surrounded by a common envelope. (2) In myxomycetes, the spherules composing the sclerotium.

Macronucleus The larger (and vegetative) of the two types of nuclei of ciliates.

Map unit An expression of the frequency of recombination between two genes. One map unit = 1% frequency of recombination.

Mastigont The complex of organelles underlying a flagellum.

Matrix (1) A substance located between cells. (2) An arrangement of mathematical data.

Medium (*pl.* -ia) A nutritive substrate used to grow microorganisms.

Megalospherite One of the forms under which a foraminiferan exists. It is small in size, uninucleate, and has a large proloculum.

Meiosis A process of nuclear division found in eucaryotic organisms that permits the passage from a diploid to a haploid state. The essential characteristics of meiosis are the pairing of chromosomes, crossing over, and the reduction in chromosome number.

Meiospore A spore in whose formation meiosis is invloved; for example, ascospores, basidiospores, and sporangiospores of myxomycetes.

Melzer's reagent Potassium iodide, 1.5 g; iodine, 0.5 g; water, 20 ml; chloral hydrate, 22 g.

Membrane (or unit membrane) The term used to refer to the basic structure of all cell membranes. A unit membrane is about 75 Å thick and consists of two osmophilic layers separated by a less dense one; each layer is roughly 25 Å thick.

Meristem-arthrospore Meristospore.

Meristospores Spores formed in chains by the successive elongation and division of the tip of the conidiophore.

Merological Having to do with a part.

Meromyxis The process by which only a part of the genetic information of a cell is transferred to another.

Merosporangium (*pl.* -ia) A sporangium containing a single row of spores.

Merozoite In the life cycle of sporozoa, a young cell resulting from the multiple division of a mature vegetative cell.

Merozygote The zygote formed during partial fecundation or meromyxis.

Mesophilic Growing at temperatures in the range of 20 to 37°C.

Mesosome An organelle resulting from invagination of the plasma membrane within the cytoplasm of bacteria.

Messenger RNA (mRNA) RNA that has a base composition complementary to that of a portion of a chromosome and serving as a template for protein synthesis.

Metabolic block An interruption of a biosynthetic pathway as a result of a mutation, causing lack or inactivity of an enzyme.

Metabolism The chemical processes of living organisms.

Metachromatic granules A deposit of polyphosphates found in bacterial cells.

Metula (*pl.* -ae) A branch of a sporophore holding phialides.

Microconidium The smaller conidium of a fungus that forms conidia of two different sizes.

Microconsumer Saprotroph.

Microcyst In cellular slime molds, a single ameba after it has formed a cyst.

Microcyclic rusts (or short-cycled rusts) Rust fungi that produce only one type of binucleate spore, the teleutospore.

Micron (μ) 10^{-4} cm.

Micronucleus The smaller (and generative) of the two types of nuclei in ciliates.

Microspherite One of the forms under which a foraminiferan exists. It is large in size, multinucleate, and has a small proloculum.

Micropyle A small opening.

Minimal medium A usually chemically defined medium that is nutritionally the simplest on which the wild-type strain of a microorganism can grow; auxotrophic mutants grow on a minimal medium only when it is supplemented with one or more nutrients.

Missence mutation The alteration of a codon, changing its meaning from one amino acid to another.

Mitochondrion (*pl.* -ia) A cellular organelle composed of an outer membrane and an inner membrane folded into cristae. It is the seat of activity of enzymes of the citric acid cycle and those involved in energy transfer.

Mitosis A process of nuclear division found in eucaryotic organisms that does not involve any change in chromosome number. The essential characteristic of mitosis, after the formation of the chromosomes, is their exact duplication, with each new nucleus getting a copy of the original.

Mixotrophy — The simultaneous utilization of autotrophic and heterotrophic metabolic pathways.

Mold — A vernacular, ill-defined term used to refer to any filamentous fungus that has a cottony appearance. Examples: *Mucor, Rhizopus, Aspergillus, Penicillium.*

Monera — The kingdom of procaryotic organisms with unicellular or simple colonial organization, comprised of bacteria and blue-green algae.

Monocentric — Having only one center of growth and development and one organ of reproduction per thallus.

Monocytosis — An increase in the proportion of monocytes in the blood.

Monogenic — Controlled by one gene.

Monomer — A basic unit of a polymer.

Monomorphic — Having only one characteristic morphology (in contrast to pleomorphic).

Monoplanetism — Having only one motile phase.

Monothetic — Having a unique set of features. Dichotomous keys lead to monothetic units.

Morphogenesis — (1) The evolution and development of form. (2) The process undergone by an individual in order to attain the morphology typical of the species.

Morphopoiesis — The process by which a morphologically characterizable biological entity is assembled from a pool of subunits.

Mucedinaceae — Moniliaceae.

Mucoid forms — Bacteria producing moist, glistening colonies that have a tendency to run together. The bacterial cells under this form are often capsulated.

Murein — The mucopeptidic macromolecule that forms the sacculus of bacteria.

Muriform — Shaped like a wall; with a brick-like pattern.

Murine — Having to do with mice; mouse gray.

Mushrooms — Fungi that form fleshy fruiting bodies.

Mutagen	An agent that causes mutations or increases their frequency.
Mutant	An individual or strain that differs from the wild type by one or more mutations.
Mutation	A change in the DNA base sequence that is not recombinational.
Mutator gene	A gene that leads to greatly increased frequency of mutation of some or all genes of the strain bearing it.
Muton	The smallest unit of the chromosome that, if changed, can cause a mutation. A change in a pair of DNA nucleotides could produce a mutation.
Mycelium (*pl.* -lia)	A mass of hyphae.
Mycetocyte	A special microorganism-containing cell found in insects carrying hereditary microorganisms.
Mycetoma	A tumor-like mass with fistules from which pus-containing granules escape. Mycetomas are infections caused by fungi or actinomycetes. They evolve slowly and are the result of an infected wound.
Mycetome	A group of mycetocytes.
Mycetozoa	Myxomycetes.
Mycoplasmas	A group of pleomorphic bacteria that lack a cell wall.
Myoneme	The contractile filament found in the cytoplasm of some protozoa.
Myxameba (*pl.* -bae)	(1) A zoöspore after becoming ameboid. (2) The amebas of the myxothallophytes.
Myxogastres	Myxomycetes.
Navicular	Shaped like a boat.
Nematocyst	Stinging organelle.
Neotype	A substitute for a missing or not designated holotype.
Net plasmodium	The pseudoplasmodium formed by Labyrinthulales, which has a net-like appearance; also called filoplasmodium.
Neuraminidase	An enzyme found in viruses with envelopes. It attacks specific receptor sites on the membrane of susceptible cells and it cleaves N-acetylneuraminic acid of red blood cells.

Neutral spore — In red algae, the spore formed by division and redivision of a vegetative cell. The spore is usually liberated as a naked ameboid cell, which later secretes a wall.

Nitrification — The microbial oxidation of ammonia to nitrate.

Nitrogen fixation — The biological utilization of atmospheric nitrogen in the synthesis of organic compounds.

Nodoc — The triplet base sequence of a molecule of transfer RNA that is complementary to that of the codon in the molecule of messenger RNA.

Nomenclatural type — The specimen or strain to which the name of the taxon is permanently attached.

Nomenclature — The naming of units delineated by classification.

Nomen confusum — The name applied to a taxon that, upon further study, was found to be composed of two or more taxa.

Nomen conservandum — A name that has officially been conserved against another.

Nomen dubium — A name of uncertain application.

Nomen nudum — A name proposed for a taxon that was not adequately described.

Nomen oblitum — A name that has not been used for a long time (50 years).

Nomen rejiciendum — A name officially rejected for use.

Non-cellular — In fungi, having no isodiametric cells.

Non-selected markers — Genetic markers that are not selected for by cultivation on a particular medium. Thus the progeny resulting from cross-mating of parents with different growth requirements are free to vary in respect to such markers.

Nonsense codon — One coding for nothing; it may indicate termination of a polypeptide chain.

Nonsense mutation — The alteration of a codon, changing it from one coding for an amino acid to a nonsense codon.

Nosology — The science and classification of diseases.

Nuclear cap — A body, often crescent-shaped, that is located next to the nucleus of some zoöspores or gametes; it is formed of a mass of ribosomes.

Nucleocapsid — The structure formed of the capsid and the enclosed nucleic acid.

Nucleoid — The procaryotic "nucleus".

O-antigen A somatic antigen.

Ocellus (*pl.* -li) Eyespot.

Ochre The name of the nonsense codon UAA. The presence of this codon indicates the termination of a growing polypeptide chain.

Oidium (*pl.* -ia) (1) One of the conidia borne in short chains, as in the genus *Oidium.* (2) A cylindrical arthrospore.

Oligogenic Controlled by a few genes.

Oligotrophic Receiving few nutrients.

Oncogenic Capable of stimulating the production of a tumor.

Ontogeny The course of development of an individual organism.

Oöcyst The cyst containing the zygote (the same as an oögonium).

Oögamy The fertilization of a large non-motile egg by a small motile male gamete.

Oögonium (*pl.* -ia) A female gametangium that contains one or more oöspheres.

Oökinete In protozoa, a motile zygote.

Oösphere A naked, non-motile female gamete that is larger than its male counterpart.

Oöspore The resting spore resulting from the fecundation of a non-motile gamete called the oösphere.

Operator The part of a chromosome that is capable of interacting with a repressor and controlling the function of an operon.

Operculate Having an operculum.

Operculum A lid that permits the discharge of spores from a sporangium or from an ascus.

Operon A group of genes under the control of an operator and a repressor; it is a DNA chromosomal region, which is transcribed into one single molecule of messenger RNA.

Opsonin A specific antibody that promotes the phagocytic destruction of bacteria.

Organelle A subcellular unit, such as a mitochondrion or a flagellum.

Organotroph An organism that derives its carbon primarily from organic compounds.

Orthogenesis The progressive evolution of organism in successive generations along predestined lines.

Osmotroph Saprotroph.

Ostiole An opening through which spores escape a fruiting body.

Oxybiontic Aerobic.

Papilla (*pl.* -ae) Any small, nipple-shaped elevation.

Paramylum A storage material of algae, probably a glucan.

Paraphysis (*pl.* -ses) A sterile filament of the hymenium.

Parasexual cycle A mechanism through which recombination of hereditary properties occurs without meiosis. In filamentous fungi this process consists of the union of nuclei in a heterocaryotic mycelium. Segregation and recombination occur by crossing over at mitosis and are followed by haploidization.

Paratype A specimen or strain cited in an original description that is not the holotype or an isotype.

Parenthesome The septal cap of the dolipore.

Parietal Attached to the wall.

Parthenogenesis Reproduction by the development of a gamete that has not been fertilized by another gamete.

Pasteur effect The inhibition of fermentation reactions by oxygen.

Pasteurization The heating of a solution for a few minutes at less than 100°C in order to reduce the microbial population.

Pathogen A disease-producing organism.

Peduncle A stalk.

Pelagic Floating freely.

Peplomers Protein subunits of the viral peplos.

Peplos Envelope.

Perfect fungi Fungi with a known sexual cycle.

Peridiole A small, spheroid disseminative structure that contains the gleba of some gasteromycetes.

Peridium (*pl.* -ia) — The limiting wall of a sporangium or of a fruit body.

Periphysis (*pl.* -ses) — A hair-like element found lining the ostiolar canal of fruiting bodies.

Periplasm — (1) In procaryotic organisms, the space between the plasma membrane and the cell wall. (2) In eucaryotic organisms, the peripheral layer of the cytoplasm.

Peristome — The area surrounding the mouth of the cytostome.

Perithecium (*pl.* -ia) — An ascocarp that is almost closed (the opening being a pore or a slit), in which a hymenium is formed.

Peritrichous — Having flagella (or other filamentous appendages) all over the surface.

Petite mutant — A mutant of yeast that lacks functional mitochondria and therefore grows more slowly than the normal type under aerobic conditions.

Phage — Bacteriophage.

Phage type — Subdivision, usually a subspecies, based on susceptibility to phages.

Phagocytosis — The intracytoplasmic engulfment of particles by a cell (also see *pinocytosis*).

Phagotroph — A heterotrophic organism that ingests other organisms or particulate organic matter.

Phagotrophic — Engulfing and digesting solid particles of food.

Phenetic — Based on overall similarity.

Phenon — A group of strains that have high mutual similarity.

Phenotype — The sum of the apparent characteristics of an individual.

Pheromone — A chemical that serves to give messages between members of a species.

Phialide — A hyphal structure with one or more openings from which conidia are formed basipetally. The phialide does not increase in length during conidial production.

Phialopore — A small opening through which some colonial microorganisms may turn themselves inside out, as in the case of *Volvox*.

Phialospore — A spore produced by a phialide.

Phobotaxis — A random reaction to avoid a source of undesirable stimulus.

Phosphorescence — Fluorescence that continues after the exciting radiation has stopped.

Photoautotrophic — Able to grow in a completely mineral medium, using light as a source of energy.

Photoauxotrophic — Able to grow in mineral medium, using light as a source of energy, but requiring certain organic growth factors.

Photolithotrophy — Energy provided by a photochemical reaction, with CO_2 used as the carbon source.

Photoorganotrophy — Energy provided by a photochemical reaction, with utilization of organic carbon sources.

Phototrophy — Energy provided by photochemical reactions.

Phragmoplast — A barrel-shaped structure found between the two nuclei during the telophase.

Phragmospore — A septate spore.

Phycomycetes — A heterogeneous group of fungi that have few septa when they form a mycelium. Sexual reproduction, when known, may be by the conjugation of swarm cells, or by formation of oöspores or of zygospores.

Phyletic — Based on natural evolutionary relationships.

Phylogenetic — Phyletic.

Phylogeny — The history of the evolution of an organism.

Phylum (*pl.* -la) — An important group of organisms.

Pileate — Having a pileus.

Pileus (*pl.* -ei) — The top part of a basidiocarp, also called cap, that bears the hymenium on its surface.

Pilus (*pl.* -li) — (1) A flagellum-like appendage used in bacterial conjugation. (2) A fimbria.

Pinocytosis — The absorption of liquid droplets by cells by invagination of the cellular membrane, with resulting formation of fluid-filled vacuoles.

Pinosome — A vacuole formed by pinocytosis.

Piriform — Pyriform.

Plage — A smooth area on a spore close to the hilum.

Plankton — The aggregate of pelagic organisms of small size.

Planogamete — A motile gamete.

Plaque — The clear area produced by a bacteriophage in bacterial culture on a solid medium.

Plasmalemma — Plasma membrane.

Plasma membrane — A membrane surrounding the cytoplasm. The electron microscope revealed that it is a unit membrane.

Plasmagel — The stiff layer of cytoplasm located between the plasma membrane and the plasmasol in ameboid organisms.

Plasmasol — The inner and more fluid part of the cytoplasm of ameboid organisms.

Plasmid — Non-nuclear, self-replicating unit that is not an essential part of a cell and carries genetic information.

Plasmodiocarp — A type of sessile fructification of myxomycetes that retains the shape of the larger veins of the plasmodium; for example, the fructification of *Cienkowskia reticulata.*

Plasmodium (*pl.* -dia) — A multinucleate mass of protoplasm that has no firm wall and is capable of motion, such as that formed by myxomycetes.

Plasmogamy — The cytoplasmic union of two cells. If these two cells are gametes, nuclear fusion or karyogamy follows.

Plasmolemma — Plasma membrane.

Plasmotomy — Cytoplasmic fission of a coenocytic individual not linked to nuclear divisions.

Plastid — A region of the plant protoplast that has become structurally specialized mainly in connection with carbohydrate synthesis. The most obvious examples of plastids are the chloroplasts of chlorophyll-containing plants.

Pleochroism — The change in color that occurs during the rotation of an object under polarized light. The change in color is due to unequal absorption of polarized light, depending on the path of travel in the object.

Pleomorphism — The condition of producing more than one type of sporulation in the course of the life cycle, such as in rust fungi. Bacteria that do not produce cells that are all alike are also called pleomorphic; for example, the Corynebacteriaceae.

Plethysmothallus In algae, a dwarf filamentous diploid thallus that bears sporangia.

Pleurocystidia The cystidia found on the faces (sides) of gills or tubes.

Pleuropodial Having a lateral stipe.

Plexus A network.

Ploidy The numerical chromosomal status of a cell.

Plurilocular Having several small chambers.

Podetium A stalked structure of the thallus of some lichens that bears apically reproductive organs.

Polar mutation A mutation that reduces or abolishes the function not only of the gene in which it occurs but also of the genes in the same operon on the side away from the operator.

Polycentric Having more than one center growth and development and more than one reproductive organ per thallus.

Polyene A substance with several conjugated carbon-to-carbon double bonds.

Polygenic Controlled by several genes.

Polyhedrose A type of viral disease of insects, characterized by the formation of polyhedral proteinaceous crystals in which virions are embedded.

Polymorphic (1) Having many shapes. (2) Capable of existing under several forms.

Polyribosome A group of ribosomes held together by a molecule of messenger RNA. Polyribosomes are the site of polypeptide synthesis.

Polysome Polyribosome.

Polythetic Having many shared features (also see *monothetic*).

Porospore A spore formed through small terminal or lateral pores on the wall of the conidiophore.

PPLO Pleuropneumonia-like organisms, mycoplasmas.

Primary mycelium (1) In fungi, the haploid mycelium. (2) In actinomycetes, the substrate mycelium.

Probasidium (*pl.* -ia) The spherical body from which the basidium of some heterobasidiomycetes develops.

Procaryotic — Without a true nucleus; that is, one in which no membrane surrounds the nuclear material.

Producer organism — In ecology, an autotrophic organism.

Proloculum (*pl.* -la) — The first chamber formed in the test of a foraminiferan.

Promitosis — A type of mitosis that takes place in karyosome-containing nuclei, in which the nuclear membrane is retained during the whole process.

Promycelium (*pl.* -ia) — The basidium of rust fungi and smuts.

Propagule — A part of an organism capable of growth; for example, one bacterial cell, a piece of mycelium, a spore, etc.

Propagulum (*pl.* -la) — A viable shoot capable of reproducing a plant. Brown algae form propagula that float away from the mother thallus to settle upon a favorable substratum and develop into a new thallus.

Prophage — The genome of a temperate phage when inserted in the chromosome of a lysogenic bacterium.

Prosorus (*pl.* -ri) — A cell that eventually divides to yield the sorus of some Chytridiales.

Prostheca (*pl.* -ae) — Semirigid bacterial appendages bound by the cell wall.

Protein subunits — In virions, the polypeptidic part of the viral molecule forming a structural entity. A virion has a number of identical protein subunits symmetrically arranged. *Synonym:* structure unit.

Protobasidium (*pl.* -ia) — A septate basidium.

Protocaryotic — Procaryotic.

Protoctista — The kingdom of lower eucaryotic organisms that lack differentiation into tissues; it is comprised of algae other than the blue-green, of fungi, and of protozoa.

Protologue — All the information given when the name of an organism is first published.

Proton — The positive core of an ordinary hydrogen atom; it is equivalent to the hydrogen ion in mass and equivalent but opposite to the electron in charge.

Protoperithecium (*pl.* -ia) — The haploid structure that matures into the perithecium after diploidization.

Protoplast — (1) The contents of a cell. (2) A bacterium that lacks a cell-wall.

Prototrophic — Able to synthesize growth factors from simpler substances.

Protype — Neotype.

Provirus — The state of a virus when its genome is incorporated into that of its host.

Pseudocapillitium (*pl.* -ia) — A mass of threads found in aethalia, originating not from the columella but from the partly developed sporangial wall.

Pseudoplasmodium (*pl.* -ia) — A slimy structure formed by the massing of separate cells and capable of moving as a unit, such as those formed by the Acrasiales.

Pseudopodium (*pl.* -ia) — A temporary extension of the cytoplasm of ameboid organisms.

Pseudoseptum (*pl.* -ta) — A thickened ring in a hypha that looks like a septum.

Pseudospore — A spore-like structure that returns to active life without casting off a wall of any kind, such as those formed in the genera *Sappinia* and *Guttulinopsis* (Acrasiales).

Psychrophilic — Liking cold; used for bacteria with growth temperature optima below 20°C.

Pubescent — Covered with downy hairs.

Pulvinate — Cushion-shaped.

Punky — Soft and tough; fibrous.

Pycnidium (*pl.* -ia) — A perithecium-like structure in which asexual spores are formed.

Pycnium (*pl.* -ia) — Spermogonium.

Pyreniform — Nut-shaped.

Pyrenoid — A body associated with chloroplasts, probably playing a role in starch formation.

Pyriform — Pear-shaped.

Quantasomes — The elementary photosynthetic particles associated with thylakoids.

R — Of bacterial growth, rough, appearing hard and dry.

Raceme — A fructification in which flowers or spores are borne on short peduncles that branch from a common axis.

Radiciform — Root-shaped.

Raphe — A furrow or a ridge that marks the line of union of symmetrical parts.

Ravelase — An enzyme that transforms double-stranded DNA into the single-stranded form.

Recombination — (1) The redistribution of hereditary properties; for example, daughter cells have a different gene combination than the parent cell. (2) The rearrangement of genetic material after transfer from one individual to another or from one chromosome to another.

Recon — The smallest unit interchangeable by genetic recombination; it may be one nucleotide.

Reflexed — Turned up.

Regulator gene — A gene whose product, in conjunction with a small-molecular-weight compound (corepressor or inducer), influences the activity of one or more structural genes.

Reniform — Kidney-shaped.

Replica plating — The transfer of colonies growing on a solid medium to fresh media without changing their relative position.

Replicon — An independent self-replicating genetic unit of bacteria. The genetic equipment of bacteria consists of a number of replicons; the largest of these is called the bacterial chromosome.

Repressor — A protein produced under the direction of a regulatory gene.

Resupinate — Lying flat on the substratum.

Reverse mutation — A mutation from a mutant allele to the wild-type allele; this usually involves a return of the nucleotide sequence of the DNA to that in the wild-type gene.

Rhapidosome — A rod-shaped particle composed of protein and ribonucleic acid, found in the cells of some bacteria.

Rhizine — One of the cords of hyphae that anchor foliose lichen thalli to the substratum.

Rhizoid — A root-like structure.

Rhizomorph — A hard, dense strand made of hyphae. The outer hyphae form a protective sheath for the softer core; often, a continuous central lumen exists. Such structures permit the spread of the fungus and the passage of food from one part of the fungal mass to another.

Rhizomycelium — A mycelium limited to short, thin branches that resemble a root structure.

Rhizoplast (or rhizoblast) — The thread joining the blepharoplast to the nucleus.

Rhizopodium (*pl.* -ia) — A pseudopodium with anastomosing branches.

Rhizosphere — The portion of the soil that is intimately associated with roots.

Ribosome — A particulate ribonucleoprotein with a molecular weight of 2.5 to 4.5 million and containing about 35% protein and 65% RNA; ribosomes range in diameter from 100 to 200 Å.

Ribosomal RNA (rRNA) — The RNA found in ribosomes; it accounts for about 80% of the RNA present in a bacterial cell.

Rimose — With cracks.

Ring — Annulus.

RNA — Ribonucleic acid.

RNA polymerase — An enzyme that catalyzes the condensation of ribonucleoside triphosphate into RNA. The enzyme recognizes only the sugar phosphate portions of the nucleotide precursors. DNA serves as a template to insure proper base sequence.

Rough forms — Of bacteria, non-capsulated and producing dull to granular colonies.

Rufous — Reddish.

S — (1) Of bacterial growth, smooth, with a glistening, moist, soft appearance; (2) Svedberg, the unit of sedimentation.

Sacculus (*pl.* -li) — The rigid part of the cell wall of a bacterium; it is formed of a macromolecule of murein.

Sagittate — Arrow-shaped.

Salsuginous — Associated with brackish waters.

Saltation — Mutation.

Saprobic — Rich in organic matter, usually from decaying organisms.

Saprophage — An organism that feeds on dead organic matter.

Saprophytic — Using dead organic matter as food.

Saprotroph — A heterotrophic organism that breaks down complex organic compounds and absorbs some of the decomposition products.

Sarciniform — Packet-shaped.

Sarcoma A tumor consisting of tissue that is composed of closely packed cells embedded in a fibrillar or homogeneous substance, mainly cells derived from connective tissue.

Saxicolous Living among rocks.

Schizogony In protozoa, the multiple division of a mature vegetative cell or schizont.

Schizont In protozoa, a mature vegetative cell undergoing schizogony.

Scleroid Having a hard texture.

Sclerotium (*pl.* -ia) A compact knot of mycelium, whose inner part is usually rich in reserve food material and whose outer part forms a hard protective layer.

Scyphiform Cup-shaped.

Seceding Of lamellae, at first adnate, later separating from the stem.

Secondary mycelium (1) In fungi, the dicaryotic mycelium. (2) In actinomycetes, the aerial mycelium.

Sector In microbial culture on a solid medium, a part of the colony that differs in appearance from the rest. This part is usually pie-shaped and is the expression of a mutation or other phenotypic change.

Segmentation The fragmentation of aerial hyphae.

Segregation The redistribution of chromosomes during the division of nuclei. This process leads to new interchromosomal combinations.

Selected markers Genetic markers that are selected by cultivation on a particular medium; for example, a gene that confers resistance to an antibiotic is selected by cultivation on a medium containing this antibiotic.

Seminicolous Growing on seeds.

Septum (*pl.* -ta) A wall across the hyphae of fungi. Some septa have holes that permit the flow of cytoplasm across them.

Seriate Arranged in a series; in a row.

Serological type A subdivision, usually a subspecies, based on antigenic differences.

Sessile Without a stem.

Seta (*pl.* -ae) A bristle.

Sex pilus (1) A flagellum-like appendage used in bacterial conjugation. (2) A fimbria. (See Pilus.)

Sexual cycle A mechanism of recombination of hereditary properties, involving the union of two gametes that have *n* chromosomes. The 2*n* zygote formed eventually produces, by meiosis, gametes of *n* chromosome number. In other terms, sexual reproduction involves plasmogamy, karyogamy, and meiosis.

Sinistrorse Counterclockwise.

Slime flux A slimy material exuded by plants, on (or in) which fungi or bacteria are growing.

Slug A terrestrial gastropod whose name is used to refer to the pseudoplasmodia of Acrasiales.

Smooth forms Of bacteria, often capsulated and producing smooth, glistening colonies.

Soluble RNA (sRNA) Transfer RNA (tRNA).

Somatic Having to do with the non-reproductive part of an organism.

Soralium (*pl.* -ia) A powdery clump of soredia that is visible with the naked eye.

Sordid Dingy.

Soredium (*pl.* -ia) A small cluster of lichen gonidia surrounded by a few fungal hyphae.

Sorocarp The fruiting body of Acrasiales; it usually consists of a stalk (sorophore) and a head (sorus).

Sorogen The rising cell mass of cellular slime molds in the process of forming the sorocarp.

Sorophore (1) The stalk of the sorocarp. (2) The stalk of a sorus.

Sorus (*pl.* -ri) (1) The fruiting structure of certain fungi; for example, the spore mass of rust fungi and smuts, or that of cellular slime molds. (2) In the Chytridiales, a group of sporangia.

Species A group of individuals that seem closely related.

Specific refractive increment The amount by which the refractive index of a diluted substance increases for every 1% rise in concentration (weight/volume).

Spermatangium (*pl.* -ia) Antheridium; mainly used in the case of red algae.

Spermatium (*pl.* -ia) A non-motile male gamete that unites with a receptive hypha or with a trichogyne.

Spermogonium (*pl.* -ia)	A pycnidial structure of rust fungi that produces male spermatia and female receptive hyphae.
Sphaerocyst	A spherical cell found in the pseudotissues of *Lactarius* and *Russula*.
Spheroplast	A bacterium with a defective cell wall. When the cell wall is completely lacking, the bacterial cell is called a protoplast.
Spherule	A large, thin-walled, spherical structure formed by some zoöpathogenic fungi.
Spiculospore	A spore formed at the tip of a pointed structure.
Spinulose	Covered with spinules (little spines).
Spontaneous mutant	A mutant that arises as a result of an event unknown to or uncontrolled by the experimenter.
Sporangiole	A small sporangium without columella, containing few spores.
Sporangium	An enclosed organ within which asexual spores are formed.
Spore	(1) In general, the name for a reproductive structure of cryptogams. (2) In protozoa, the resistant form of a sporozoan developing from the zygote.
Spore print	The spore deposit obtained by placing the pileus of a fungus, hymenium downwards, on a flat surface, such as paper.
Sporidesmins	Mycotoxins; sulfur-containing peptides produced by *Pithomyces chartarum*.
Sporidium (*pl.* -ia)	The basidiospore of a smut or a rust fungus.
Sporocarp	A structure that holds spore-producing hyphae or cells.
Sporocladium (*pl.* -ia)	A sporogenous branch that holds the phialides of *Kickxellaceae*.
Sporocyte	A cell that is enclosed within a sporangium and divides to yield spores.
Sporozoite	In protozoa, a spore formed by the division of the oöcyst; it is the final stage in the sexual cycle of Sporozoa.
Squamose	Covered with scales.
Squamulose	(1) Covered with minute scales. (2) In lichens, a type of thallus intermediate between foliose and crustose.
Squarrose	Covered with scales curving away from the surface.
Staling	The loss of vitality of old cultures.

Statospore A spore formed by some Chrysophyta; usually spherical, with a silicified wall that has one opening closed by a plug.

Staurospore A star-shaped spore.

Steno- Narrow, as in stenoecious – having a narrow selection of habitat.

Stephanocyst A bicelled structure that is composed of a cup-like basal cell subtended by a clamp connection and a terminal globose cell. Such structures are found in the members of the genus *Hyphoderma.*

Stereoisomers Molecules that have identical structural formulae but different spatial arrangements of some of their groups.

Sterigma (*pl.* -ata) A sporophore, especially one that holds basidiospores on basidia.

Sterilization The killing of all living cells.

Stigma (*pl.* -ata) A pigmented body found in flagellates, associated with plastids; it is also called eyespot and is believed to be a light-sensitive organelle.

Stipe A stalk.

Stipitate Having a stipe, a stem.

Stolon A horizontal hypha that produces vertical structures from place to place; a runner.

Strigose Having coarse hairs.

Stroma (*pl.* -ata) (1) In fungi, a mass of vegetative hyphae in (or on) which reproductive bodies are formed. (2) In chloroplasts, the site of carbon dioxide fixation.

Structural gene A gene that controls the specificity of a protein; it determines the amino acid sequence of enzyme protein.

Structure unit In virions, a protein subunit.

Substrate (1) A nutrient used by microorganisms. (2) A substance transformed by an enzyme.

Suppressor gene A gene that can reverse the phenotypic expression of mutations caused by changes in other genes.

Suppressor mutation A mutation occurring in a mutant, causing partial or complete reversal to the wild-type phenotype though taking place at a distant unlinked locus of the DNA.

Suspensor A hyphal structure that supports a gametangium.

Swarmer (or swarm cell) — In fungi, a motile flagellated cell that often acts as a gamete.

Symbiosis — The beneficial association of organisms.

Sympatric — Occurring in the same area.

Synapsis — The intimate association of homologous chromosomes during the first prophase of meiosis.

Synergism — The effect produced by a mixture of agents that is more than the sum of the effects of each agent.

Syngamy — Fertilization.

Synnema (*pl.* -ata) — Coremium.

Syntype — Two or more specimens designated by the author of a description as nomenclatural types.

Synzoöspore — A large multiflagellate spore, as formed by *Vaucheria*.

Systematics — Taxonomy.

Systole — A contraction, as that of contractile vacuoles.

Tautomerism — A form of isomerism in which there is ready interconversion between two isomers.

Tautonymy — The use of the same word as a generic and as a specific term.

Taxon (*pl.* -xa) — A taxonomic group, such as a species, a genus, an order, etc.

Taxonomy — The science of classification and associated nomenclature.

TCA — (1) Tricarboxylic acid. (2) Trichloroacetic acid.

Teleutospores — The basidia-producing spores of rust fungi (Uridinales) and smuts (Ustilaginales).

Teliospore — Teleutospore.

Telium (*pl.* -ia) — The sorus that contains the teleutospores or winter-resting spores of rust fungi.

Temperate bacteriophage — A bacteriophage capable of establishing a lysogenic relationship with the bacterium it infects.

Teratology — The science of abnormalities.

Test — A shell, as that of a foraminiferan.

Thallus The vegetative body of a thallophyte; it cannot be differentiated into roots, stems and leaves.

Theca (*pl.* -ae) An envelope.

Thermocline An area in a water body where temperature changes rapidly with depth.

Thermophilic Liking high temperatures (40°C to 70°C).

Thylakoids Stacks of membranous structures that comprise the grana of chloroplasts.

Tinea A dermatomycosis.

Tinsel-type flagellum A flagellum with lateral branches or "hairs".

Tomentose Woolly.

Topotype The material collected from the same location as a type specimen or a type culture.

Trama (*pl.* -ae) The hyphal pseudotissue that supports the hymenium.

Transaminase A transferase enzyme that transfers an amino group from one chemical compound to another.

Transcription The transfer of genetic information from DNA to RNA.

Transducer A device that receives power from one system and supplies it to another system. The output and input power may be of the same or of different types.

Transduction The incorporation in the bacterial chromosome of informational bacterial DNA carried by a phage.

Transfer RNA (tRNA) Small RNA molecules capable of linking, with the help of a ribosome, an activated amino acid to messenger RNA.

Transformation (1) In bacteria, the transfer of genetic information by free extracellular DNA. (2) In higher organisms, the change in a cell that makes malignancy likely.

Translation The process by which messenger RNA directs the sequence of amino acids in a polypeptide.

Transliterate To transcribe from one alphabet to another; for example, from Greek to Latin (see Greek, Russian, and German alphabets, page 869).

Translocation The rearrangement of chromosome segments in new sequences as a result of breakage and rejoining.

Trichocyst — An organelle capable of sudden discharge, found in some protozoa.

Trichogyne — A receptive hypha of the female organ of certain fungi and algae.

Trichome — A single row of cells, as found in some filamentous blue-green algae.

Trinomen — The combination of the generic, specific and subspecific names.

Trophocyst — A hyphal vesicle from which a sporangiophore grows.

Trophozoite — The vegetative stage of the life cycle in protozoa; a vegetative cell of a protozoan.

Tunicate — Covered with membrane.

Tyndallization — A form of sterilization that consists of repeated pasteurization.

Type method — A system of nomenclature in which a taxonomic name is permanently attached to a specimen, a culture or a taxon.

Type species — A species that is characteristic of a genus; the oldest validly published species of a genus. A given genus that does not include its type species cannot be conceived; for example, the type species of the genus *Bacillus* is *B. subtilis*, and one cannot change the definition of the genus *Bacillus* so that *B. subtilis* is excluded from it.

Typology — The study of types.

Ubiquinones (co-enzymes Q) — A group of soluble benzoquinones that occur in the majority of aerobic organisms; they are part of the respiratory electron carrier chain.

Umbel — A fructification in which spores or flowers are borne on stalks that are about equal and that originate from a common center.

Umbilicus — In lichens, the hyphal peduncle that anchors the thallus of certain foliose lichens.

Umbonate — Having an umbo or boss.

Unit membrane — See Membrane

Uredium (*pl.* -ia) — A fructification in which, in summer, the uredospores of macrocyclic rusts are formed.

Uredospore — The one-celled spores of rust fungus, formed in late spring and during the summer; it germinates to produce the dicaryotic mycelium, which bears more uredia.

Utriculiform	Bladder-shaped.
Vaccine	Any material used for preventive inoculation.
Vacuole	A cytoplasmic inclusion, bound by a membrane or not, that contains a liquid or a gas.
Vacuome	A network of vacuoles.
Variant	Mutant.
Variety	A subspecies.
Vegetative	Not forming spores.
Veil	(1) Partial veil: the layer of pseudotissue that joins the rim of the pileus to the stipe of developing fruit bodies of agarics. (2) Universal veil: the envelope that surrounds immature fruit bodies of certain agarics and gasteromycetes.
Verrucose	Covered with warts.
Verticillate	Arranged around a point on an axis.
Vi antigen	A surface antigen that masks O-antigens.
Villose	Covered with villi.
Villus (*pl.* -li)	Long, hair-like appendage.
Virion	(1) A virus particle. (2) The mature virus.
Virosis (*pl.* -ses)	A disease caused by a virus.
Virulent bacteriophage	A bacteriophage incapable of establishing a lysogenic relationship with the bacterium it infects; infection, therefore, leads to lysis.
Virus	An organism, usually submicroscopic, that contains only RNA or DNA and requires for reproduction cells of organisms that contain both types of nucleic acids.
Volva (*pl.* -ae)	A cup-like structure at the base of the stipe of some agarics and gasteromycetes; it is a remainder of the universal veil.
Volutin	The linear polymer of orthophosphate found in metachromatic granules.
Whiplash-type flagellum	A flagellum without lateral branches.

Wild type — Usually the starting strain of an organism, isolated from nature, from which mutant individuals or strains are subsequently derived.

Xerophilic — Liking dry places.

Xylophilous — Liking wood; growing on wood.

Zöögamete — A flagellated gamete.

Zoöspore — A flagellated spore.

Zygote — The result of the fusion of two gametes.

Zymogram — The banding of a mixture of enzymes by electrophoresis.

SANITARY BACTERIOLOGY

DR. M.S. FINSTEIN

TABLE 1

PRINCIPAL BACTERIAL INDICATORS OF FECAL POLLUTION OF WATER

Coliform Bacteria	Fecal Coliform Bacteria	Fecal Streptococci	Clostridia
Characteristics			
Gram-negative rods, facultative, ferment lactose at 35°C	Gram-negative rods, facultative, ferment lactose at 44.5°C	Gram-positive cococci, facultative, grow in the presence of sodium azide	Gram-positive endospore-forming rods, anaerobic, yield stormy fermentation of milk
Component Species			
Escherichia coli, E. freundii *Enterobacter aerogenes* *(Aerobacter aerogenes)* *Klebsiella pneumoniae*	Primarily *Escherichia coli*	*Streptococcus faecalis, S. bovis, S. equinus, (S. faecium?)*	Primarily *Clostridium perfringens*
Abundance in Human Feces[a]			
Escherichia coli: $10^{8.4}$ ($10^{4.3-9.3}$) *Enterobacter aerogenes* + *Klebsiella pneumoniae:* $10^{7.6}$ ($10^{4.0-9.0}$)	*Escherichia coli:* $10^{8.4}$ ($10^{4.3-9.3}$)	*S. faecalis* + enterococci: $10^{7.4}$ ($10^{3.3-9.3}$)	$10^{7.2}$ ($10^{3.7-9.0}$)
Occurrence in Human Feces[b]			
Escherichia coli: 99% *Enterobacter aerogenes* + *Klebsiella pneumoniae:* 62%	*Escherichia coli:* 99%	100%	30%
Specificity for Feces of Warm-Blooded Animals			
Escherichia coli: relatively specific; sparse in non-polluted environments	*Escherichia coli:* relatively specific; sparse in non-polluted environments	Not highly specific; not uncommon in non-polluted environments	Uncertain
Enterobacter aerogenes: not highly specific, not uncommon in non-polluted environments			

TABLE 1 (Continued)

PRINCIPAL BACTERIAL INDICATORS OF FECAL POLLUTION OF WATER

Coliform Bacteria	Fecal Coliform Bacteria	Fecal Streptococci	Clostridia
Principal Advantages			
Most conservative indicator	Specificity; little aftergrowth[c]	Little or no aftergrowth; it is possible to distinguish between human and animal fecal pollution on the basis of the *Streptococcus* species in the water and on the basis of the fecal-coliform: fecal-streptococcus ratio	May indicate pollution remote in time
Principal Disadvantages			
Non-specificity; tendency to aftergrowth, particularly *Enterobacter aerogenes*	Compared to the coliform group, may not provide adequate safety margin for judging water potability	Rapid die-off outside the intestinal tract	Spores may persist long after danger from pollution is past

[a]Bacteria per gram wet feces; subjects free of intestinal-disease symptoms.
[b]Percentage of positive specimens.
[c]Proliferation after chlorination of sewage effluent or upon dilution with clean water.

Data compiled from: *Standard Methods for the Examination of Water and Wastewater*, 13th ed., American Public Health Association (1971); *Sanitary Significance of Fecal Coliforms in the Environment (Publication No. WP-20-3)*, U.S. Department of the Interior (1966); Ketyi, I., and Barna, K., *Acta Microbiol. Acad. Sci. Hung.*, *11*, 173(1964); Zubrzycki, L., and Spaulding, E., *J. Bacteriol.*, *83*, 968(1962); others.

TABLE 2

SUMMARY OF THE EVOLUTION OF U.S. PUBLIC HEALTH SERVICE BACTERIOLOGICAL STANDARDS FOR DRINKING WATER[a]

1914[b]	1925[c]	1943[d]	1946[e]	1962[f]
Standard Portion				
10 ml	10 ml	10 ml or 100 ml	10 ml or 100 ml	10 ml or 100 ml
Standard Sample				
5 portions	5 portions	5 portions of 10 ml or of 100 ml	5 portions of 10 ml or of 100 ml	5 portions of 10 ml or of 100 ml
Sampling Procedure				
Not specified	Number and frequency of samples should be sufficient to indicate water quality (to be regulated by a certifying authority, e.g., the State Health Department)	Samples to be taken from representative points throughout distribution system frequently enough to adequately determine water quality (to be regulated by a certifying authority); minimum number of samples per month is specified, based on the size of the population served	Samples to be taken from representative points throughout distribution system frequently enough to adequately determine water quality (to be regulated by a certifying authority); minimum number of samples per month is specified, based on the size of the population served	As for 1943 and 1946, except that sampling sites and frequency are to be regulated jointly by a certifying authority and reporting agency
Approximate Maximum Mean Coliform Density Permitted per 100 ml[g]				
2.2 (MPN)	1 (MPN)	1 (MPN)	1 (MPN)	1 (MPN or membrane filter count)

TABLE 2 (Continued)

SUMMARY OF THE EVOLUTION OF U.S. PUBLIC HEALTH SERVICE BACTERIOLOGICAL STANDARDS FOR DRINKING WATER[a]

1914[b]	1925[c]	1943[d]	1946[e]	1962[f]
Variability Permitted Among Samples				
None	Occasional excessive coliform density, restricted according to the number of samples taken	Occasional excessive coliform density, restricted according to the number of samples taken and on a monthly basis; excessive density is not permitted in consecutive samples	Occasional excessive coliform density, restricted according to the number of samples taken and on a monthly basis; excessive density is not permitted in consecutive samples	As for 1943 and 1946, except that the language is slightly more precise; the permissible variability is stipulated also in terms of membrane-filter count
Remarks				
Standard includes the plate count (nutrient agar, 37°C, 24 hours) which should not exceed 100 colonies per ml	Plate count deleted from Standard	Samples are to be freed of disinfecting agent within 20 minutes of collection When excessive coliform density is noted, daily samples are to be taken until satisfactory results are obtained	Samples taken in response to excessive coliform density shall be considered special and shall not help satisfy the minimal sample number requirement Supply shall not be prohibited because of excessive coliform density if immediate effective remedial action is taken	As for 1946, except that, when excessive density occurs, inoculation of decimal-dilution series is required to determine the definitive MPN The membrane-filter procedure is accepted

[a]Bacteriological quality is only one of the parameters upon which the standards are based.
[b]*Public Health Reports,* Vol. 29.
[c]*Public Health Reports,* Vol. 40.
[d]*Public Health Reports,* Vol. 58.
[e]*Public Health Reports,* Vol. 61.
[f]*Public Health Service Publication No. 956.*
[g]The bacteriological group used throughout the years is essentially the same, although the terminology and definition change somewhat; MPN = most probable number.

TABLE 3

SUMMARY OF BACTERIOLOGICAL STANDARDS FOR DRINKING WATER IN COUNTRIES OTHER THAN THE UNITED STATES[a]

Coliform Bacteria	Others
World Health Organization International Standards[b]	
Treated Water	
<1 coliform per 100 ml in 90% of samples examined throughout one year, no sample >10, nor consecutive samples >8-10	
Untreated Water	
<10 coliforms per 100 ml in 90% of samples examined throughout one year, no sample >20, nor consecutive samples >15; not more than 40% of coliforms detected shall be members of the fecal coliform group; if >20 coliforms per 100 ml are present, treatment of the water supply should be considered	
European Standards[c]	
One or more coliforms per 100 ml permitted in 5% of the samples examined, provided that positive results are not obtained in two or more consecutive samples and that at least 100 samples, regularly distributed over the year, are examined	
Belgium (1967)	
No coliforms per 20 ml, no *E. coli* per 100 ml	No *streptococcus faecalis* per 100 ml
Bulgaria (1955)	
Towns >50,000 population: no coliforms per 100 ml	Aerobic plate count ≤100 per ml
Towns ≤50,000 population: no coliforms per 30 ml	(37°C, 24 hours)
Canada (1956)	
≤2.2 coliforms per 100 ml	
England[d]	
Chlorinated Piped Supply	
No coliforms per 100 ml, no *E. coli* I per 100 ml	
Non-chlorinated Piped Supply	
Excellent: no coliforms per 100 ml, no *E. coli* I per 100 ml	
Satisfactory: 1–3 coliforms per 100 ml, no *E. coli* I per 100 ml	

TABLE 3 (Continued)

SUMMARY OF BACTERIOLOGICAL STANDARDS FOR DRINKING WATER IN COUNTRIES OTHER THAN THE UNITED STATES[a]

Coliform Bacteria	Others
Suspicious: 4–10 coliforms per 100 ml, no *E. coli* I per 100 ml	
England (continued)	
Non-chlorinated Piped Supply (cont.)	
Unsatisfactory: >10 coliforms per 100 ml, 0 or more *E. coli* I per 100 ml	
France (1962)	
Treated Water No *E. coli* per 100 ml	Treated Water No fecal streptococci per 50 ml, small quantities of sulfite-reducing clostridia
Untreated Water No *E. coli* per 100 ml	Untreated Water No fecal streptococci per 50 ml, no sulfite-reducing clostridia per 20 ml
German Democratic Republic (1953)	
Untreated Well Water No coliforms per 100 ml	Aerobic plate count <100 per ml
Mexico (1955)	
<2 coliforms per 100 ml	Aerobic plate count <200 per ml (37°C, 24 hours); bacteria that liquefy gelatin or are chromogenic or fetid in gelatin (20°C, 48 hours) shall be absent from 1 ml
South Africa[e]	
<10 coliforms per 100 ml, no *E. coli* I per 100 ml	Aerobic plate count <100 per ml
Spain (1969)	
Satisfactory: no coliforms per 100 ml, *E. coli* absent	Satisfactory: aerobic plate count <50–65 per ml (nutrient agar, 37°C, 48 hours); fecal streptococci, sulfite-reducing clostridia, and the bacteriophages specific to *E. coli* and *Shigella* shall be absent
Tolerable: 1 or 2 coliforms per 100 ml, *E. coli* absent	Tolerable: aerobic plate count ≤100 per ml (nutrient agar, 37°C, 48 hours); fecal streptococci and sulfite-reducing clostridia <1 or 2 per 100 ml; the bacteriophages specific to *E. coli* and *Shigella* shall be absent

TABLE 3 (Continued)

SUMMARY OF BACTERIOLOGICAL STANDARDS FOR DRINKING WATER IN COUNTRIES OTHER THAN THE UNITED STATES[a]

Coliform Bacteria	Others
Sweden (1970)	
Suitable: <1 coliform per 100 ml	Suitable: aerobic plate count <100 per ml (22°C, 48 hours)
Conditionally suitable: 1–9 coliforms per 100 ml	Conditionally suitable: aerobic plate count ⩾100 per ml (22°C, 48 hours)
Unsuitable: >10 coliforms per 100 ml	Unsuitable: not to be judged unsuitable solely on the basis of aerobic plate count
Yugoslavia (1961)	
Treated Water No coliforms per 100 ml	Treated Water Aerobic plate count <10 per ml, no *Streptococcus faecalis* or *Proteus* per 100 ml
Untreated Water May contain >10 coliforms per 100 ml, provided they are not of fecal origin (*E. coli*)	Untreated Water Aerobic plate count ⩽300 per ml, no *Streptococcus faecalis* or *Welchia perfringens* per 100 ml

[a] Unless noted otherwise, data are taken from *International Digest of Health Legislation*, World Health Organization, Geneva, Switzerland. The year (shown in parentheses) indicates when the data were recorded in the *Digest*. Bacteriological quality is only one of the parameters upon which the standards are based. No distinction is made between treated and untreated supplies unless specifically stated.

[b] *International Standards for Drinking Water,* 2nd ed. World Health Organization, Geneva, Switzerland (1963).

[c] *European Standards for Drinking Water,* 2nd ed. World Health Organization, Geneva, Switzerland (1970).

[d] *The Bacteriological Examination of Water Supplies*. Reports on Public Health and Medical Subjects, No. 71, Her Majesty's Stationery Office, London, England (1957).

[e] *Specification for Water for Domestic Supplies*. South African Bureau of Standards, Pretoria, South Africa (1951).

TABLE 4

SUMMARY OF RECOMMENDATIONS FOR BACTERIOLOGICAL QUALITY OF WATER INTENDED FOR VARIOUS USES IN THE UNITED STATES[a]

Bacterial Group	Recommended Criteria (Limits per 100 ml Water)
General Recreation[b]	
Fecal coliform	Average not to exceed 2,000; maximum 4,000 (except in specified outfall mixing zones)
Secondary-Contact Recreation[c]	
Fecal coliform	Should not exceed log mean of 1,000, nor exceed 2,000 in more than 10% of the samples
Primary-Contact Recreation[d]	
Fecal coliform	Sample five or more times per month during the recreation season; should not exceed log mean of 200, nor exceed 400 in more than 10% of the samples
Public Surface Water Supplies[e]	
Coliform	Permissible: 10,000; desirable: <100[f]
Fecal coliform	Permissible: 2,000; desirable: <20[f]
Farmstead (Individual) Water Supplies	
Coliform	No more than 1; if the water is routinely disinfected, up to 100 can be tolerated
Crop Irrigation	
Coliform	Monthly arithmetic average not to exceed 5,000; limit of single sample 20,000[g]
Fecal coliform	Monthly arithmetic average not to exceed 1,000; limit of single sample 4,000[g]
Shellfish-Growing Areas[h]	
Coliform	Approved area: median MPN should not exceed 70, and not more than 10% of the samples should ordinarily exceed 230
Coliform	Conditionally approved area: refers to an area potentially subject to pollution; conditional approval is based on meeting requirements as for an approved area and on assurances that contributory-waste treatment facilities are performing satisfactorily

TABLE 4 (Continued)

SUMMARY OF RECOMMENDATIONS FOR BACTERIOLOGICAL QUALITY OF WATER INTENDED FOR VARIOUS USES IN THE UNITED STATES[a]

Bacterial Group	Recommended Criteria (Limits per 100 ml Water)
Shellfish-Growing Areas (continued)	
Coliform	Restricted area: refers to an area from which shellfish may be marketed only after purification by effective procedures (relaying or controlled purification); median MPN should not exceed 700, and not more than 10% of the samples should exceed 2,300
Coliform	Prohibited area: refers to an area from which shellfish may be marketed by special permit only, after purification by effective procedures (relaying is preferred over controlled purification); water is of poorer quality than indicated for restricted areas

[a] Unless noted otherwise, data are taken from *Report of the Committee on Water Quality Criteria,* Federal Water Pollution Control Administration, U.S. Department of the Interior (1968). Bacteriological quality is only one of the parameters upon which the recommendations are based. The same criteria are recommended for fresh, estuarine, and marine recreational waters.
[b] Refers to waters not officially designated for recreational use; the criteria, however, provide for relatively safe secondary-contact recreation (boating, fishing, and limited contact incident to shoreline activity).
[c] Refers to waters officially designated for recreational use involving the activities named above.
[d] Refers to waters officially designated for recreational use involving swimming, surfing, water-skiing, etc.
[e] To be treated to meet the limits of U.S. Public Health Service standards for drinking water.
[f] Monthly arithmetic averages are based on an adequate number of samples. Coliform limit may be relaxed if fecal coliform concentration does not exceed the specified limit.
[g] Suggested as guidelines only, especially for use on crops intended for direct human or animal consumption; average limits are based on at least two consecutive samples per month during the irrigation season.
[h] Data taken from *National Shellfish Sanitation Program Manual of Operations,* Part I, Sanitation of Shellfish Growing Areas (Public Health Service Publication No. 33). U.S. Department of Health, Education and Welfare, Washington, D. C. (1965).

TABLE 5

SUMMARY OF BACTERIOLOGICAL STANDARDS OF WATER FOR VARIOUS INTENDED USES IN COUNTRIES OTHER THAN THE UNITED STATES[a]

Political Unit	Coliform Bacteria	Others (Aerobic Plate Count)
Swimming Pools		
Alberta, Canada (1959)	Not more than 15% of any series of samples, nor more than two consecutive samples of any series collected when the pool is in use, shall show coliforms in any of five 10-ml portions	≤200 per ml (nutrient agar, 37°C, 24 hours)
British Columbia, Canada (1969)	Not more than 15% of any series of samples, nor more than two consecutive samples in any series of samples taken at least one week apart, may show the presence of coliform organisms in any of five 10-ml portions	
Costa Rica (1962)	A weekly sample taken from each swimming pool and comprising five portions of 10 ml shall be tested, and at least 80% of five or more consecutive samples shall be free of coliforms in all five 10-ml portions of each sample	≤200 per ml in 80% of five or more consecutive samples
Hong Kong (1970)	*E. coli* absent in 100 ml	
South Africa (1968)	Fecal *E. coli* absent in 100 ml	
Spain (1961)	No fecal coliforms in two of every five samples of 5 ml taken on the same day when the pool is in use	≤100 per ml (37°C, 24 hours) under normal use, or ≤200 per ml when baths are used to capacity
Sweden (1970)	Suitable: <10 fecal coliforms per 100 ml	Suitable: <1,000 per ml (22°C, 48 hours); <100 per ml (35°C, 24 hours)
	Conditionally suitable: 10–100 fecal coliforms per 100 ml	Conditionally suitable: ≥ 1,000 per ml (22°C, 48 hours); ≥ 100 per ml (35°C, 24 hours)
	Unsuitable: >100 fecal coliforms per 100 ml	Unsuitable: not to be regarded as unsuitable solely on the basis of high count
Swimming Areas		
Canton of Vaud, Switzerland (1964)	No coliforms in a volume equal to or less than 100 ml	≤500 per ml (20–22°C, 5 days)
Hungary (1967)	≤10,000 coliforms per 100 ml	≤20,000 per ml
New Zealand (1966)	Not to consistently exceed 1,000 coliforms per 100 ml	
Paraguay (1962)	Not more than 15% of samples per season shall reveal the presence of the coliform group (confirmed test) in any of five 10-ml samples	Not more than 15% of samples per season >200 colonies per ml (35°C, 24 hours)

TABLE 5 (Continued)

SUMMARY OF BACTERIOLOGICAL STANDARDS OF WATER FOR VARIOUS INTENDED USES IN COUNTRIES OTHER THAN THE UNITED STATES[a]

Political Unit	Coliform Bacteria	Others (Aerobic Plate Count)
Swimming Areas (continued)		
Roumania (1967)	10,000 coli bacilli per 100 ml	
Sweden (1970)	Suitable: <1,000 coliforms per 100 ml, <100 fecal coliforms per 100 ml	
	Conditionally suitable: >1,000 coliforms per 100 ml, 100-1,000 fecal coliforms per 100 ml	
	Unsuitable: ⩾1,000 coliforms per 100 ml, >1,000 fecal coliforms per 100 ml	
Watering Domestic Animals		
Sweden (1970)	Suitable: <50 coliforms per 100 ml, <2 fecal coliforms per 100 ml	Suitable: low, <100 per ml (22°C, 48 hours); medium 100–1,000 per ml (22°C, 48 hours)
	Conditionally suitable: 50–500 coliforms per 100 ml, 2-9 fecal coliforms per 100 ml	Conditionally suitable: >1,000 per ml (22°C, 48 hours)
	Unsuitable: >500 coliforms per 100 ml, ⩾10 fecal coliforms per 100 ml	Unsuitable: not to be regarded as unsuitable solely on the basis of high count
Food Industry		
Sweden (1970)	Suitable: <1 coliform per 100 ml	Suitable: <100 per ml (22°C, 48 hours)
	Conditionally suitable: 1–9 coliforms per 100 ml	Conditionally suitable: ⩾100 per ml (22°C, 48 hours)
	Unsuitable: ⩾10 coliforms per 100 ml	Unsuitable: not to be regarded as unsuitable solely on the basis of high count
Shellfish-Growing Areas		
New Zealand (1966)	Coliforms not to consistently exceed 50 per 100 ml	
Oyster Beds		
Chile (1951)	Approved areas: coliform monthly average <70 per 100 ml	
	Provisionally approved areas: coliform monthly average 70–700 per 100 ml	

TABLE 5 (Continued)

SUMMARY OF BACTERIOLOGICAL STANDARDS OF WATER FOR VARIOUS INTENDED USES IN COUNTRIES OTHER THAN THE UNITED STATES[a]

Political Unit	Coliform Bacteria	Others (Aerobic Plate Count)
Oyster Beds (continued)		
Portugal (1968)	Average of two monthly samples ≤50 coliforms per 100 ml, or ≤100 coliforms per 100 ml in either sample	
Drinking Water Source to Be Treated by Simple Chlorination		
Chile (1954)	Monthly average ≤50 coliforms per 100 ml	
Drinking Water Source to Be Treated by Complete Chlorination		
Chile (1954)	Monthly average ≤5,000 coliforms per 100 ml; at least 80% of the samples ≤5,000 coliforms per 100 ml	
Effluent Discharged to Receiving Water Used as Drinking Water Source or for Foodstuff Production		
Yugoslavia (1966)	Effluent must not increase coliform count of receiving water to >6,000 per 100 ml	
Effluent Discharged to Receiving Water Used for Fish Culture, Bathing, or Watering Animals		
Yugoslavia (1966)	Effluent must not increase coliform count of receiving water to >3,000 per 100 ml	

[a] Data taken from *International Digest of Health Legislation,* World Health Organization, Geneva, Switzerland. The year (shown in parentheses) indicates when the data were recorded in the *Digest*. Bacteriological quality is only one of the parameters upon which the standards are based.

SAFETY RULES FOR INFECTIOUS DISEASE LABORATORIES[a]

GENERAL RULES

A. Only authorized employees, students, and visitors should be allowed to enter infectious disease laboratories or utility rooms and attics serving these laboratories.

B. Food, candy, gum, or beverages for human consumption should not be taken into infectious disease laboratories.

C. Smoking should not be permitted in any area in which work on infectious or toxic substances is in progress. Employees who have been working with infectious materials should thoroughly wash and disinfect their hands before smoking.

D. Library books and journals should not be taken into rooms where work with infectious agents is in progress.

E. An effort should be made to keep all other surplus materials and equipment out of these rooms.

F. Drinking fountains should be the sole source of water for drinking by human occupants.

G. According to the level of risk, the wearing of laboratory or protective clothing may be required for persons entering infectious disease laboratories. Likewise, showers with a germicidal soap may be required before exit.

H. Contaminated laboratory clothing should not be worn in clean areas or outside the building.

DISINFECTION AND STERILIZATION

A. All infectious or toxic materials, equipment, or apparatus should be autoclaved or otherwise sterilized before being washed or disposed of. Each individual working with infectious material should be responsible for its sterilization before disposal.

B. Infectious or toxic materials should not be placed in autoclaves overnight in anticipation of autoclaving the next day.

C. To minimize hazard to firemen or disaster crews, at the close of each workday, all infectious or toxic material should be (1) placed in the refrigerator, (2) placed in the incubator, or (3) autoclaved or otherwise sterilized before the building is closed.

D. Autoclaves should be checked for operating efficiency by the frequent use of Diack, or equivalent, controls.

E. All laboratory rooms containing infectious or toxic substances should designate separate areas or containers labeled: INFECTIOUS – TO BE AUTOCLAVED or NOT INFECTIOUS – TO BE CLEANED. All infectious disease work areas, including cabinetry, should be prominently marked with the Biohazards Warning Symbol.

[a] Taken from: Hellman, A. (Ed.), *Biohazard Control and Containment in Oncogenic Virus Research.* U.S. Government Printing Office, Washington, D.C. (1969). Reproduced by permission of the U.S. Department of Health, Education and Welfare, National Institutes of Health, U.S. Public Health Service.

F. Floors, laboratory benches, and other surfaces in buildings in which infectious substances are handled should be disinfected with a suitable germicide as often as deemed necessary by the supervisors. After completion of operations involving planting, pipetting, centrifuging, and similar procedures with infectious agents, the surroundings should be disinfected.

G. Floor drains throughout the building should be flooded with water or disinfectant at least once each week in order to fill traps and prevent backing up of sewer gases. (New construction plans should omit floor drains wherever possible.)

H. Floors should be swept with push brooms only. The use of a floorsweeping compound is recommended because of its effectiveness in lowering the number of airborne organisms. Water used to mop floors should contain a disinfectant. (Elimination of sweeping through use of vacuum cleaners or wet mopping only is highly desirable, if exhaust is vented through absolute filters.)

I. Stock solutions of suitable disinfectants should be maintained in each laboratory.

J. All laboratories should be sprayed with insecticides as often as necessary to control flies and other insects. Vermin proofing of all exterior building openings should be given close attention.

K. No infectious substances should be allowed to enter the building drainage system without prior sterilization.

L. Mechanical garbage disposal units should not be installed for use in disposing of contaminating wastes. These units release considerable amounts of aerosol.

SAFETY CABINETS AND SIMILAR DEVICES

A. A ventilated safety cabinet should be used for all procedures with infectious substances, such as opening of test tubes, flasks, and bottles; using pipettes; making dilutions; inoculating; necropsying animals; grinding; blending; opening lyophile tubes; operating a sonic vibrator; operating a standard table model centrifuge, etc.

B. A safety box or safety shaker tray should be used to house or safeguard all containers of infectious substances on shaking machines.

C. A safety centrifuge cabinet or safety centrifuge cup should be used to house or safeguard all centrifuging of infectious substances. When centrifuging is done in a ventilated cabinet, the glove panel should be in place with the glove ports covered. A centrifuge in operation creates reverse air currents that may cause escape of agent from an open cabinet.

D. A respirator or gasmask should be worn when changing a glove or gloves attached to a cabinet if an infectious aerosol may possibly be present in the cabinet.

PIPETTES

A. No infectious or toxic materials should be pipetted by mouth.

B. No infectious mixtures should be prepared by bubbling expiratory air through a liquid with a pipette.

C. No infectious material should be blown out of pipettes.

D. Pipettes used for the pipetting of infectious or toxic materials should be plugged with cotton.

E. Contaminated pipettes should be placed horizontally in a pan containing enough suitable disinfectant to allow complete immersion. They should not be placed vertically in a cylinder. The pan and pipettes should be autoclaved as a unit and replaced by a clean pan with fresh disinfectant.

F. Infectious material should not be mixed by pipetting.

SYRINGES

A. Only syringes of the Luer-Lok type should be used with infectious materials.

B. Use an alcohol-soaked pledget around the stopper and needle when removing a syringe and needle from a rubber-stoppered vaccine bottle.

C. Expel excess fluid and bubbles from a syringe vertically into a cotton pledget soaked with disinfectant, or into a small bottle of cotton.

D. Before and after injection of an animal, swab the site of injection with a disinfectant.

GENERAL PRECAUTIONS AND RECOMMENDATIONS

A. Before centrifuging, inspect tubes for cracks, inspect the inside of the trunnion cup for rough walls caused by erosion or adhering matter, and carefully remove bits of glass from the rubber cushion. A germicidal solution added between the tube and trunnion cup not only disinfects the outer surface of both of these, but also provides an excellent cushion against shocks that might otherwise break the tube.

B. Avoid decanting centrifuge tubes. If you must do so, afterwards wipe off the outer rim with a disinfectant; otherwise, the infectious fluid will spin off as an aerosol. Avoid filling the tube to the point that the rim ever becomes wet with culture.

C. Water baths and Warburg baths used to inactivate, incubate, or test infectious substances should contain a disinfectant. For cold water baths, 70% propylene glycol is recommended.

D. When the building vacuum line is used, suitable traps or filters should be interposed to insure that pathogens do not enter the fixed system.

E. Deepfreeze and dry ice chests and refrigerators should be checked and cleaned out periodically to remove any broken ampules, tubes, etc., containing infectious material. Use rubber gloves and respiratory protection during this cleaning. All infectious or toxic material stored in refrigerators or deepfreezes should be properly labeled.

F. Insure that all virulent fluid cultures or viable powdered infectious materials in glass vessels are transported, incubated, and stored in easily handled, nonbreakable, leakproof containers that are large enough to contain all the fluid or powder in case of leakage or breakage of the glass vessel.

G. All inoculated Petri plates or other inoculated solid media should be transported and incubated in leakproof pans or other leakproof containers.

H. Care must be exercised in the use of membrane filters to obtain sterile filtrates of infectious materials. Because of the fragility of the membrane and other factors, such filtrates cannot be handled as noninfectious until culture or other tests have proved their sterility.

I. Develop the habit of keeping your hands away from your mouth, nose, eyes, and face. This habit may prevent self-inoculation.

J. No person should work alone on an extremely hazardous operation.

K. Broth cultures should be shaken in a manner that avoids wetting the plug or cap.

L. Diagnostic serum specimens carrying a risk of serum hepatitis should be handled with rubber gloves.

ANIMALS

Animal Cages

All animal cages should be marked to indicate the following information:

A. Uninoculated animals.

B. Animals inoculated with noninfectious material.

C. Animals inoculated with infectious substances.

Cages used for infected animals should be cared for in the following manner:

A. Careful handling procedures should be employed to minimize the dissemination of dust from cage refuse and animals.

B. Cages should be sterilized by autoclaving. Refuse, bowls, and watering devices will remain in the cage during sterilization.

C. All watering devices should be of the nondrip type.

D. Each cage should be examined each morning and at each feeding time so that dead animals can be removed.

Handling Infected Animals

A. *Special attention* should be given to the humane treatment of all laboratory animals in accordance with the Principles of Laboratory Animal Care as promulgated by the National Society for Medical Research.

B. Monkeys should be tuberculin-tested and examined for herpetic lesions.

C. Persons regularly handling monkeys should receive periodic chest X-ray examination and other appropriate tuberculosis detection procedures.

D. When animals are to be injected with pathogenic material, the animal caretaker should wear protective gloves and the laboratory workers should wear surgeon's gloves. Every effort should be made to restrain the animal to avoid accidents that may result in disseminating infectious material.

E. Heavy gloves should be worn when feeding, watering, or removing infected animals. Under no circumstances will the bare hands be placed in the cage to move any object.

F. Animals in cages with shavings should be transferred to clean cages once each week unless otherwise directed by the supervisor. If cages have false screen platforms, the catch pan should be replaced before it becomes full.

G. Infected animals to be transferred between buildings should be placed in aerosol-proof containers.

Animal Rooms

A. Doors to animal rooms should be kept closed at all times except for necessary entrance and exit.

B. Unauthorized persons should not be permitted entry to animal rooms.

C. A container of disinfectant should be kept in each animal room for disinfecting gloves, boots, and general decontamination. Floors, walls, and cage racks should be washed with disinfectant frequently.

D. Floor drains in animal rooms should be flooded with water or disinfectant periodically to prevent backing up of sewer gases. (Drains should be avoided where possible.)

E. Shavings or other refuse on floors should not be washed down the floor drain, if these are present.

F. An effective poison should be maintained in animal rooms to kill escaped rodents.

G. Special care should be taken to prevent live animals, especially mice, from finding their way into disposable trash.

Necropsy of Infected Animals

A. Necropsy of infected animals should be carried out in ventilated safety cabinets.

B. Rubber gloves should be worn when performing necropsies.

C. Surgeon's gowns should be worn over laboratory clothing during necropsies.

D. Fur of the animal should be wet with a suitable disinfectant.

E. Animals should be pinned down or fastened on wood or metal in a metal tray.

F. Upon completion of necropsy, all potentially contaminated material should be placed in suitable disinfectant or left in the necropsy tray. The entire tray should be autoclaved at the conclusion of the operation.

G. The inside of the ventilated cabinet and other potentially contaminated surfaces should be disinfected with a suitable germicide.

H. Grossly contaminated rubber gloves should be cleaned in disinfectant before removal from the hands, preparatory to sterilization.

I. Dead animals should be placed in proper leakproof containers and thoroughly autoclaved before being placed outside for removal and incineration.

REGULATIONS CONCERNING THE SHIPMENT OF PATHOGENS

Three Federal statutes – The Animal Quarantine Act of 1903, the Plant Quarantine Act of 1912, and the Federal Plant Pest Act of 1957 – prohibit the importation and movement of pests, pathogens, vectors, and other articles that might harbor these organisms unless authorized by the U.S. Department of Agriculture. These acts and related State laws, cooperatively enforced, are intended to protect American agriculture by preventing the unauthorized movement and establishment of dangerous organisms.

PLANT PESTS, PATHOGENS, AND VECTORS

Plant pests, pathogens, and vectors cannot be imported into or through the United States unless accompanied by a permit from USDA's Plant Quarantine Division.

Plant pests subject to Federal Domestic quarantines or Federal–State cooperative programs, and parasites associated with these pests, cannot be moved intrastate and interstate without a permit. These permits should be requested from USDA's Plant Pest Control Division. Permits for the movement of all other plant pests, pathogens, or vectors are required and should be requested from the Plant Quarantine Division.

ANIMAL PESTS, PATHOGENS, AND VECTORS

Pests, pathogens, and vectors of livestock and poultry diseases cannot be imported into the United States except under a permit issued by USDA's Animal Health Division.

Permits should be requested from the Animal Health Division for the intrastate and interstate movement of animal pests and vectors which are enzootic pathogens of high virulence or for any organism for which a national animal diseases eradication or control program exists. Newly isolated agents may be placed in this category until their significance and distribution are known. Diagnostic specimens from livestock or poultry should have the concurrence of the receiving State livestock sanitary officials before being moved.

When animal pathogens and vectors are indigenous to all states, a specific permit usually is not required. This is also true of diagnostic specimens from livestock or poultry having diseases readily found throughout the country and of everyday disease significance.

APPLICATION FORMS AND PERMITS

Type of Organism	Contact
Plant pests coming under Federal domestic quarantines or cooperative Federal–State programs	Director, Plant Pest Control Division U.S. Department of Agriculture Federal Center Building Hyattsville, Md 20782
All plant pests being imported and those plant pests being moved interstate that are not under Federal domestic quarantine	Director, Plant Quarantine Division U.S. Department of Agriculture Federal Center Building Hyattsville, Md 20782
Importation or movement of animal pathogens or their vectors	Director, Animal Health Division U.S. Department of Agriculture Federal Center Building Hyattsville, Md 20782

Note: The Public Health Service, U.S. Department of Health, Education and Welfare, has regulations governing the shipment of pathogens and vectors of diseases of man. Requests for information relating to these regulations should be addressed as follows:

Importation, and distribution after importation, of pathogens, vectors, or potential vectors	Chief, Foreign Quarantine Program National Communicable Disease Center United States Public Health Service Atlanta, Ga 30333 or Public Health Service Quarantine Stations at United States ports of entry

INTERNATIONAL ASSOCIATION OF MICROBIOLOGICAL SOCIETIES

The International Association of Microbiological Societies (IAMS) is the Division of Microbiology of the International Union of Biological Sciences. The Secretary General of the IAMS is Dr. N. E. Gibbons, IAMS, P. O. Box 3123, Station C, Ottawa, Ontario, Canada, K1Y 4J4.

MEMBER SOCIETIES AS OF DECEMBER 1970

Argentina	Asociacion Argentina de Microbiologia
Australia	Australian Society for Microbiology
Austria	Österreichische Gesellschaft für Mikrobiologie und Hygiene
Belgium	Société Belge de Microbiologie
Brazil	Sociedad Brasileira de Microbiologia
Bulgaria	Microbiologicna Sekzia, Sajuz na Naucnite Rabotnizi v Bulgaria (Section of Microbiology, Union of Scientific Workers in Bulgaria)
Canada	Canadian Society of Microbiologists (Société canadienne des Microbiologistes)
Chile	Asociacion Chilena de Microbiologia
China (Republic of)	The Chinese Society of Microbiology
Costa Rica	Asociacion Costarricense de Microbiologia
Czechoslovakia	Československá společnost mikrobiologická (Czechoslovak Society for Microbiology)
Denmark	Danmarks Mikrobiologiske Selskab
Ecuador	Sociedad Ecuatoriana de Microbiologia
Finland	Societas Biochemica, Biophysica et Microbiologica Fenniae
France	Société francaise de Microbiologie
Germany (Democratic Republic)	Sektion Mikrobiologie, Biologische Gesellschaft

Germany (Federal Republic)	Deutsche Gesellschaft für Hygiene und Mikrobiologie
Great Britain	Society for General Microbiology
Greece	Greek Society of Microbiology
Hungary	Hungarian Society of Microbiology
India	Association of Microbiologists of India
Iran	Andjoman Microbiologie Iran (Iranian Society of Microbiology)
Israel	The Microbiological Society of Israel
Italy	Società Italiana di Microbiologia
Japan	Japan National Committee of Microbiology
Korea	The Korean Society for Microbiology
Mexico	La Asociacion Mexicana de Microbiologia
Morocco	La Société des Sciences Naturelles et Physiques du Maroc
Netherlands	Nederlands Vereniging voor Microbiologie (Netherlands Society for Microbiology)
New Zealand	New Zealand Microbiological Society
Norway	Norsk Forening for Mikrobiologi (Norwegian Society for Microbiology)
Poland	Polskie Towarzystwo Mikrobiologow
Roumania	Société Roumaine de Microbiologie
South Africa	South African Society for Plant Pathology and Microbiology
Spain	Sociedad Española de Microbiología
Sweden	Svenska Föreningen för Mikrobiologi (Swedish Microbiological Society)
Switzerland	Schweizerische Mikrobiologische Gesellschaft (Swiss Society for Microbiology)
Thailand	The Microbiological Society of Thailand
Turkey	Turk Mikrobiyoloji Cemiyeti

United Arab Republic	Society of Applied Microbiology
United States of America	American Society for Microbiology
U.S.S.R.	I. I. Mechnikov's All-Union Scientific Medical Society of Epidemiologists, Microbiologists and Infectionists
	All-Union Society of General Microbiology
Uruguay	Sociedad Uruguaya de Microbiologia
Venezuela	Sociedad Venezolana de Microbiologia
South Vietnam	Institut Pasteur du Viet-Nam
Yugoslavia	Yugoslav Society for Microbiology

RULES FOR NOMENCLATURE

DR. HUBERT LECHEVALIER

Taxonomy, also called *systematics,* is the science of characterizing and arranging organisms in an orderly fashion. One can recognize three aspects to this science: (1) *classification,* which is the orderly arrangement of units into groups; (2) *nomenclature,* which consists in giving names to the units and groups; and (3) *identification,* which consists in assigning organisms to the proper units while giving them proper names.[1]

Taxonomy is an enterprise of man and is filled with human imperfections. International societies of zoology, botany, microbiology and virology publish, from time to time, international codes of nomenclature, which aim at standardizing nomenclature in order to facilitate exchange of information among men. The following is a summary of the International Code of Nomenclature of Bacteria, published in 1966.[3] Microbiologists interested in the nomenclature of algae and fungi are reminded to refer to the botanical code,[2] those interested in protozoa to the zoological code,[4] and virologists to a monograph edited by Wildy in 1971.[5]

The basic aims of a code of nomenclature are to assure that the same taxon is always called by the same name and that this name is not used for other taxa. In practice, the same name may be used for an animal and for a plant (if they are not algae, fungi, or protozoa), and different investigators call the same taxon by different names (personal idiosyncrasies). The codes of nomenclature aim at the fixity of names and at the rejection of names that may cause confusion.

Scientific names of all taxa are either taken from Latin or are latinized if taken from other tongues. A taxon is a word used to refer to any taxonomic group. The following major categories of taxa are recognized:

Category	Name Ending
Individual	
Species	
Series	
Section	
Genus	
Tribe	-eae
Family	-aceae
Order	-ales
Class	
Division	

The *correct* name of a taxon is the earliest one published in accordance with the rules of the Code. Exceptions to this rule of priority are made by placing the name of a taxon on a list of *nomina conservanda.*

A *nomenclatural type* is the constituent element of a taxon to which the name of the taxon is permanently attached (for example, the type species of a genus).

The name of a taxon has no claim to recognition unless it is *effectively* and *validly published.* An *effective publication* is one that has been printed and has been sold and/or distributed to the general public or to bacteriological institutions. Publication in a patent does not constitute effective publication. A *valid publication* of a taxon is one that has been effectively published in conjunction with a description of the said taxon. The rank of the taxon should be stated and the name of the next higher taxon indicated. For example, if one describes a new genus, one should indicate the family to which it belongs. The description of a new taxon should make it clear in which way it differs from related taxa of the same rank. A name proposed without a description (or reference to an adequate one) is a *nomen nudum.* To be validly published, the name of a species must be published as a binary combination and its description should be based on the properties of strains growing in pure culture. This rule does not apply to organisms that cannot be grown in pure culture, such as *Mycobacterium leprae,* for example.

In order to prevent confusion, the name(s) of the author(s) who first validly published the name of a taxon is cited after the taxonomic name. When a taxonomic name is altered in rank, the original author is cited in parentheses, followed by the name(s) of the author(s) who effected the alteration.

The *legitimate* (published according to the rules) name of a taxon cannot be rejected because it is disagreeable or because another name would be more appropriate, but an *illegitimate* name must be rejected. A common cause of illegitimacy is that a proposed name is a homonym of a previously published name of a taxon of bacteria, fungi, algae, viruses, or protozoa.

A name that is a persistent source of error (*nomen ambiguum*), one that is applied to a taxon of dubious integrity, such as an impure culture (*nomen confusum*), and a name of uncertain application (*nomen dubium*) should be rejected and put on the list of *nomina rejicienda.* A *nomen oblitum,* a forgotten name, one that has not been used within the past fifty years, is rejected by zoologists but is, unfortunately, accepted by bacteriologists.

Tautonymy, the use of the same word for the generic and the specific names, is not permitted in bacterial nomenclature.

Provisions for changes in rules, for exceptions to rules, and for their interpretation have been made through the establishment of an International Committee on Nomenclature of Bacteria and its Judicial Commission. These bodies are organized by the International Association of Microbiological Societies.

REFERENCES

1. Cowan, S. T., *A Dictionary of Microbial Taxonomic Usage.* Oliver and Boyd, Edinburgh, Scotland (1968).
2. *International Code of Botanical Nomenclature Adopted by the llth International Botanical Congress, Seattle.* International Association for Plant Taxonomy, Utrecht, The Netherlands (1969).
3. International Code of Nomenclature of Bacteria, *Int. J. Syst. Bacteriol., 16,* 459 (1966).
4. *International Code of Zoological Nomenclature.* International Commission on Zoological Nomenclature, London, England (published in installments, 196-).
5. Wildy, P. (Ed.), Classification and Nomenclature of Viruses, *Monographs in Virology,* No. 5. S. Karger, Basel, Switzerland (1971).

IMPORTANT CULTURE COLLECTIONS

PROFESSOR V. B. D. SKERMAN

Since 1967 the World Federation of Culture Collections has been engaged in the preparation of a *World Directory of Collections of Cultures of Microorganisms*.*

The World Federation of Culture Collections has recognized the Department of Microbiology of the University of Queensland as a center for data on organisms maintained in collections and enquiries relating to the location of cultures in collections may be directed to this department. The *World Directory* contains most of the relevant information. However, with the collaboration of the collections, the Centre is establishing an IBM card reference system that will be updated as frequently as new information relating to the collections becomes available, and it is hoped that we can periodically produce a statement on changes in the collection.

Parallel to this, the Centre is also developing a system for data input and retrieval intended to receive detailed data on individual cultures and make them available on demand probably at the cost of the computation costs involved. A further statement on this development will appear at the proper time in the *International Journal of Systematic Bacteriology*.

The following is a list of some of the collections that are listed in the *World Directory*. Together with the address of each collection, we have indicated the number that the collection has been assigned in the *World Directory* and in parentheses an indication of the nature of the microorganisms which are held in the collections. The abbreviations are as follows:

A – algae
B – bacteria
F – fungi
P – protozoa
T – tissue cultures
V – viruses
Y – yeasts

The collections listed here have been selected from the *World Directory* largely on the basis of geographical location rather than specific importance. It should be noted that the countries of the British Commonwealth have produced catalogs of their cultures, copies of which are available from Her Majesty's Stationery Office, Kingsway, London.

Where an asterisk appears against the name of the country, there is no recognized national culture collection.

Country	Number of Known Collections	Address	Assigned Number
Algeria	1	The Director Institut Pasteur d'Algerie Rue Docteur Laveran Alger Algeria	144 (B, V)
Argentina	10	Prof. R. E. Halbinger Director Coleccion Catedra Microbiologia Agricola Facultade Agronomia y Veterinaria Universidad de Buenos Aires Avda San Martin 4453 Buenos Aires Argentina	307 (A, B, F, Y)

*Martin, S. M., and Skerman, V. B. D., Eds. Interscience Publications, John Wiley and Sons, New York (1972).

Country	Number of Known Collections	Address	Assigned Number
Australia* The Australian Culture Collection Committee produces a catalog which is available from Her Majesty's Stationery Office.	34	Prof. V. B. D. Skerman Director Department of Microbiology University of Queensland St. Lucia Queensland 4067 Australia	13 (A, B, F, P, Y) Catalog available
Belgium	2	Prof. Dr. J. de Ley Director Laboratorium voor Microbiologie Faculte Wetenschappen Rijksuniversiteit Casinoplein, 21 Gent Belgium	296 (B)
Brazil*	4	Dr. R. N. Neder Director Instituto Zimotecnico Culture Collection Instituto Zimotecnico Caixo Postal 56 Piracicaba Sao Paulo Brazil.	294 (B, F, Y)
		Prof. I. Suassuna Director Laboratorio de Enterobacterias Departmento de Microbiologia Medica Instituto de Microbiologia Univ. Federal de Rio de Janeiro Avenida Pasteur 250 Rio de Janeiro Estado de Guanabara Brazil	313 (B)
Bulgaria	1	Dr. M. Zheleva Director Bulgarian Type Culture Collection Institute for State Control of Medical Preparations Ministry of Health Vladimir Zaimov No. 26 Sofia Bulgaria	66 (B, F, Y) Catalog available
Burma	1	Dr. Ko Ko Gyi Director, Burma Pharmaceutical Industry Type Culture Collection BPI Road Gyogon Rangoon Burma	292 (B, F, V, Y)

Country	Number of Known Collections	Address	Assigned Number
Canada The Canadian Culture Collection Committee produces a catalog which is available from Her Majesty's Stationery Office.	35	Dr. J. K. Shields Director Reference Collection of Microorganisms Inhabiting Wood Forest Products Laboratory Montreal Road Ottawa 7 Ontario Canada	38 (F, Y) Catalog available
		Dr. J. W. Carmichael Director Mold Herbarium and Culture Collection University of Alberta Edmonton Alberta Canada	73 (A, F, Y) Catalog available
		Dr. R. H. Allen Director Canadian Communicable Disease Centre Department of National Health and Welfare Tunney's Pasture Ottawa 3 Ontario Canada	156 (B, V) Catalog available
Czechoslovakia	12	Prof. T. Martinec Czechoslovak Collection of Microorganisms J. E. Purkyne University of Brno Tr. Obrancu Miru 10 Brno Czechoslovakia	65 (B, F) Catalog available
		Dr. E. Minarik Director Yeast Collection Research Institute for Viticulture and Enology Czechoslovak Collection of Microorganisms Matuskova 21 Bratislava Czechoslovakia	28 (Y) Catalog available
		Dr. J. Sourek Director Czechoslovak National Collection of Type Cultures Institute of Epidemiology and Microbiology Srobarova 48 Prague 10 Czechoslovakia	130 (B, F, P, V, Y) Catalog available

Country	Number of Known Collections	Address	Assigned Number
Czechoslovakia (*cont.*)		Dr. A. Kockova-Kratochvilova Director Slovenska Akademie Vied Chemicky Ustav Bratislava Dubravska cesta Czechoslovakia	333 (Y)
Denmark	3	Dr. F. Orskov Director WHO International Escherichia Centre Statens Serum Institute 80 Amager Boulevard Copenhagen S, DK 2300 Denmark	158 (B)
		Prof. G. J. Bonde Bacillus Collection Institute of Hygiene University of Aarhus Universitetsparken DK-8000 Aarhus Jylland Denmark	267 (B) Catalog available
Ecuador*	1	Dr. D. Uriguen B. Director Instituto Nacional de Higiene 'Leopoldo Izquieta Perez' Direccion Nacional de Salud P. O. Box 3961 Guayaquil, Guayas Ecuador	302 (B)
Finland	1	Dr. H. G. Gyllenberg Director Department of Microbiology University of Helsinki Helsinki 71 Finland	223 (B, F, Y)
Eire*	8	Prof. M. J. Geoghegan The Director Department of Industrial Microbiology University College Ardmore, Stillorgan Road Dublin 4 Ireland	227 (B, F)
France	6	Prof. R. Buttiaux Director Centre de Collection de Type Microbiens Institut Pasteur de Lille 20, Boulevard Louis XIV Lille France	135 (B, V, Y)

Country	Number of Known Collections	Address	Assigned Number
France (*cont.*)		Dr. M. Carraz Institut Pasteur Lyon Rue Pasteur Lyon 69 France	185 (B)
		Dr. P. Thibault Director Collection de l'Institut Pasteur Institut Pasteur 25, Rue du Dr. Roux Paris 15 France	174 (B) Catalog available
Germany, East	5	Dr. P. Hubsch Director Botanisches Institut Mykologie Weimar Friedrich-Schiller-Universitaet Jena Frh. v. Stein-Allee 2 Weimar, Deutsche Demokratische Republik	216 (F, Y) Catalog available
		Dr. H. Prauser Director Kulturensammlungen des Zentral Institutes für Mikrobiologie und Experimentelle Therapie Jena Deutsche Akademie der Wissenschaften zu Berlin Beuthenbergstrasse 11 Jena, 69 Deutsche Demokratische Republik	217 (A, B, F, P, V, Y)
		Dr. W. Luthardt Director 'Forstschutz Eberswalde' German Academy of Agricultural Sciences Alfred-Moller-Str., Institut für Forstwissenschaften DDR-13 Eberswalde German Democratic Republic	314 (B, F, Y) Catalog available
Germany, West	11	Dr. S. Hofmann Director Robert Koch Institute Federal Health Office Nordufer 20 1 Berlin 65 West Berlin Federal Republic of Germany	162 (B, F, V, P, Y)
		Dr. W. Koch Director Sammlung von Algenkulturen Pflanzenphysiologisches Institut der Universitaet Göttingen Nikolausbergerweg 18 34 Göttingen Bundesrepublik Deutschland	192 (A, B, P) Catalog available

Country	Number of Known Collections	Address	Assigned Number
Germany, West (*cont.*)		Dr. I. Benda Director Bayerische Landesanstalt für Wein-, Obst-, und Gartenbau Residenzpl. 3 84 Wurzburg Bundesrepublik Deutschland	264 (F, Y) Catalog available
		Prof. Dr. N. Pfennig Director Institut für Mikrobiologie D-34 Göttingen Griesebachstrasse 8 West Germany	274 (B) Catalog available
Great Britain The United Kingdom Culture Collection Committee produces a catalog which is available from Her Majesty's Stationery Office.	31	Dr. E. I. Garvie Director National Collection of Dairy Organisms National Institute for Research in Dairying University of Reading Shinfield Reading, Berkshire England	118 (B, V) Catalog available
		Dr. E. A. George Director Culture Centre of Algae and Protozoa 36 Storey's Way Cambridge, CB3 0DT England	140 (A, B, F, P) Catalog available
		Dr. S. P. Lapage Director National Collection of Type Cultures Central Public Health Laboratory Colindale Avenue London, N.W. 9 England	154 (B) Catalog available
		Dr. A. H. S. Onions Director The Commonwealth Mycological Institute Ministry of Technology Ferry Lane Kew, Surrey United Kingdom	214 (B, F, Y) Catalog available
		Dr. J. M. Shewan, Director, National Collection of Marine Bacteria, National Collection of Industrial Bacteria, Ministry of Technology, P. O. Box 31, 135 Abbey Road, Aberdeen, AB9 8DG Scotland, United Kingdom	Catalog available 238 (B, V, Y) 239 (B, V)

Country	Number of Known Collections	Address	Assigned Number
Great Britain (*cont.*)		Dr. A. H. Cook Director British National Collection of Yeast Cultures Brewing Industry Research Foundation Nutfield Surrey Great Britain	169 (Y) Catalog available
		Dr. R. A. Lelliott Director National Collection of Plant Pathogenic Bacteria Plant Pathology Laboratory Ministry of Agriculture, Fisheries and Food Hatching Green Harpenden, Hertfordshire England	126 (B, V)
Greece	2	Dr. J. Papavassiliou Director Department of Microbiology National University of Athens Goudi-Ampelokipi Athens (609) Greece	281 (B, Y)
Hungary	6	Dr. B. Lanyi Director Hungarian National Collection of Medical Bacteria National Institute of Public Health Gyali UT 2-6 Budapest IX Hungary	258 (B) Catalog available
		Dr. J. Domok Director Hungarian National Collection of Animal Viruses Department of Virology National Institute of Public Health Gyali Budapest Hungary	259 (V)
India	12	Prof. P. Nandi Director Department of Microbiology Bose Institute 93/1 Acharya Prafulla Chandra Road Calcutta-9 West Bengal India	119 (B, F, V, Y) Catalog available

Country	Number of Known Collections	Address	Assigned Number
India (*cont.*)		Dr. B. K. Bakshi Director National Type Culture Collection Forest Research Institute and Colleges Forest Pathology Branch New Forest Dehra Dun Uttar Pradesh India	132 (F) Catalog available
		Dr. V. S. Krishnamachar Director National Collection of Industrial Micro-organisms National Chemical Laboratory Council of Scientific and Industrial Research Poona-8 Maharashtra India	3 (A, B, F, P, T, Y) Catalog available
Indonesia	1	Dr. S. Saono Director Culture Collection of the Treub Laboratory Treub Laboratory of the National Biological Institute The Botanical Garden Bogor Java Indonesia	298 (B, F, Y) Catalog available
Iran*	3	Dr. M. Kaveh Director State Razi Institute Razi Culture Collection Ministry of Agriculture P. O. Box 656 Tehran State Razi Institute-Hessarak Karadj Tehran Iran	35 (B, F, Y, P, V)
		Prof. A. Shimi Director Veterinary Faculty's Culture Collection Department of Microbiology Tehran Veterinary College Eisenhower Avenue Tehran Iran	183 (B, V, Y)
Israel	3	Dr. C. B. Gerichter Director Government Central Laboratories Jaffa Road Machne Jehuda Jerusalem, P.O.B. 6115 Jerusalem Israel	295 (B, V)

Country	Number of Known Collections	Address	Assigned Number
Italy	14	Prof. G. Florenzano Director Centro di Studio dei Microorganismi Autotrofi Istituto di Microbiologia Agraria e Tecnica, Universita di Firenze Piazzale della Cascine 27 Firenze Italy	147 (A, B, F, Y)
		Prof. A. Sanna Director Istituto di Microbiologia Universita di Parma Ospedale Maggiore Parma Italy	20 (B, F, Y, P, V)
		Dr. F. Fatichenti Director Yeast Collection Istituto di Microbiologia Agraria e Tecnica Enrico de Nicola 07100 Sassari Sardegna Italy	201 (Y)
		Prof. T. Castelli Director Collezione dei Lieviti Vinari Istituto de Microbiologia Agraria e Tecnica, University of Perugia Bg. XX Giugno 06100 Perugia Italy	180 (B, Y) Catalog available
		Dr. G. Giammanco Director C.E.I.M. Istituto di Igiene dell' Universita Via del Vespro 113 90127 Palermo Italy	213 (B)
Jamaica	1	Prof. L. S. Grant Director Department of Microbiology University of the West Indies Mona Kingston 7 Jamaica	221 (B, V)
Japan The Japanese Federation of Culture Collections of Microorganisms produces a catalog of cultures edited by and available from Prof. Hiroshi Iizuka.	13	Prof. H. Iizuka Director Institute of Applied Microbiology The University of Tokyo Bunkyo-ku Tokyo Japan	190 (A, B, F, Y) Catalog available

Country	Number of Known Collections	Address	Assigned Number
Lebanon	1	Dr. G. A. Garabedian Director Department of Bacteriology and Virology American University of Beirut Beirut Lebanon	60 (B, V)
Malaysia	2	Dr. Ting Wen-Poh Director Department of Agriculture Swettenham Road Kuala Lumpur Malaysia	278 (B, F, V, A)
Netherlands	3	Dr. J. A. von Arx Director Centraalbureau voor Schimmelcultures Oosterstraat 1 Baarn Netherlands	133 (B, F, Y) Catalog available
		Dr. H. Dikken Director WHO/FAO Leptospirosis Reference Laboratory Institute of Tropical Hygiene Department of the Royal Tropical Institute Mauritskade 57 Amsterdam Netherlands	196 (B)
New Zealand	5	Dr. D. W. Dye Director Plant Diseases Division, Department of Scientific and Industrial Research Private Bag Auckland New Zealand	52 (B, F, V, Y) Catalog available
Nigeria*	1	Prof. S. H. Z. Naqvi Director Fungus Culture Collection University of Lagos, University Road Yaba, Lagos Nigeria	47 (F, Y)
Norway*	3	Dr. K. Eimhjellen Director Norges Tekniske Hogskoles Collection Technical University of Norway Department of Biochemistry Trondheim Nth Norway	40 (B, F, Y)

Country	Number of Known Collections	Address	Assigned Number
Pakistan	1	Dr. M. I. Huq Director Bacteriology Section Pak-SEATO Cholera Research Laboratory Mahakhali Dacca-12 East Pakistan Pakistan	116 (B, V)
Philippines	4	Dr. J. S. Sumpaico Director Philippine Type Culture Collection Bureau of Research and Laboratories Department of Health P. O. Box 911 Manila Philippines	46 (B, V)
Poland	7	Dr. T. Szulga Director Central Centre of Microorganisms Collections Ul. Chalubinskiego 4 Wroclaw Poland	106 (B)
Portugal	4	Prof. J. Pinto-Lopes Director Colleccao de Culturas de Fungos Micologia-Faculdade de Ciencias Lisboa Portugal	26 (F, Y) Catalog available
		Dr. N. Van Uden Director Laboratorio de Microbiologia Centro de Biologia Instituto Gulbenkian de Ciencia Rua da Quinta Grande Oeiras Portugal	43 (Y)
Puerto Rico	3	Dr. J. E. Pérez Director Agricultural Experimental Station Department of Plant Pathology and Botany P. O. Box H Rio Piedras Puerto Rico 00928	309 (B)
Rhodesia*	2	Officer-in-charge, Microbiology Laboratory Grasslands Rhizobium Collection Grasslands Research Station P.B. 701 Marandellas Rhodesia	34 (B)

Country	Number of Known Collections	Address	Assigned Number
Roumania	5	Dr. A. Sasarman Director Colectia Nationala Institutul Cantacuzino Dr. I. Cantacuzino Institute 103, Splaiul Independentei Bucuresti 35 Roumania	233 (B) Catalog available
		Dr. I. Lazar Director Collection of Plant Pathogenic Bacteria and Saprophytic Fungi Tr. Savulescu Institute of Biology 296, Splaiul Independentei Bucharest Roumania	234 (B, F, V) Catalog available
Senegal	1	Dr. Y. Robin Director Centre Regional O.M.S. de Reference pour les Arboviruses Institut Pasteur de Dakar 36, Avenue Pasteur Dakar Republique du Senegal	81 (V) Catalog available
Sudan	1	Dr. A. M. Abbas Director Sudan National Collection Sudan Medical Research Institute P.O. Box 287 Khartoum Sudan	293 (B, P)
Switzerland	3	Dr. V. H. Bonifas National Collection for Switzerland Institut de Microbiologie de l'Université de Lausanne 19 Av. César Roux Lausanne Switzerland	
Tchad	1	Dr. A. Provost Director Microbiologia animale, Laboratorie de Farcha Institut d'Elevage et de Médecine Vétérinaire des Pays Tropicaux B.P. 433 Fort-Lamy Tchad	10 (B, V)
Thailand*	3	Dr. P. Tuchinda Director Virus Research Institute Department of Medical Sciences Bamrungmuang Road, Yod-se Bangkok Thailand	82 (V) Catalog available

Country	Number of Known Collections	Address	Assigned Number
Turkey*	2	Prof. E. T. Cetin Director Department of Microbiology, Tropical Diseases and Parasitology Medical Faculty Istanbul University Istanbul Turkey	101 (B, F, V, Y)
U.A.R.*	2	Dr. M. A. El-Fadl Director Agricultural Microbiology Division (A.M.D.), Ministry of Agriculture Soils Department Building University St. Orman, Giza United Arab Republic	284 (A, B, F, Y)
Uganda*	1	Dr. R. J. Onyango Director Trypanosomes – Human and Animal East African Trypanosomiasis Research Organization P. O. Box 96 Tororo Eastern Region Uganda	289 (P)
U.S.A.	32	Dr. W. A. Clark Director American Type Culture Collection 12301 Parklawn Drive Rockville, Maryland 20852 U.S.A.	1 (A, B, F, P, T, V, Y) Catalog available
		Dr. R. C. Starr Culture Collection of Algae Department of Botany University of Indiana Bloomington, Indiana 47401 U.S.A.	78 (A) Catalog available
		Dr. C. W. Hesseltine Director ARS Culture Collection Northern Regional Research Laboratory 1815 North University Street Peoria, Illinois 61604 U.S.A.	97 (A, B, F, P, Y) Catalog available
		Dr. W. N. Ogata Director Fungal Genetics Stock Center Humboldt State College Arcata, California 95521 U.S.A.	115 (F) Catalog available

Country	Number of Known Collections	Address	Assigned Number
U.S.A. (*cont.*)		Dr. M. P. Starr International Collection of Phytopatho-genic Bacteria Department of Bacteriology University of California Davis, California 95616 U.S.A.	74 (B)
U.S.S.R.	2	Dr. V. Kuznetsov Collection of Cultures of of the U.S.S.R. Research Institute for Antibiotics Nagatinskaya 3A Moscow, M-105 U.S.S.R.	337 (B, Y) Catalog available
		Prof. V. Kudriavzev U.S.S.R. All-Union Collection of Microorganisms Institute of Microbiology USSR Academy of Sciences Profsojuznaja 7 Moscow, B-133 U.S.S.R.	342 (B, F, Y) Catalog available
Venezuela*	3	Dr. A. Divo Director Catedra de Microbiologia Facultad de Medicina Universidad de Carabobo Valencia Estado Carabobo **Venezuela**	303 (B, F, V, Y)

LITERATURE GUIDE FOR MICROBIOLOGY

EMMA C. GERGELY

PERIODICALS

A microbiologist approaches periodicals from a number of aspects during his productive years.

1. He may need to locate primary-source material on a particular subject or by a particular author. **Abstract and Index Services** often provide the route of access.

2. He may be expected to use standard abbreviations for periodical titles in the bibliography of a paper he is submitting for publication. Conversely, he may find it necessary to decipher an abbreviation of a periodical title cited in a pertinent article. Numerous **Lists of Periodical Abbreviations** are available.

3. He may want to determine what periodicals have been published or are currently being published on a given subject. **Subject Lists of Periodicals** are designed for this purpose.

4. He may want to locate a particular journal in a near-by library and obtain it on interlibrary loan or in photocopy form. This process can be initiated through **Union Lists of Periodicals.**

These approaches are outlined below. The outline is not meant to be comprehensive but to provide a glimpse of some tools that will be helpful to the microbiologist by providing the necessary information. In several cases the tool encompasses more than one approach.

Abstract and Index Services

Bibliography of Agriculture, 1942–

Data provided by the National Agricultural Library, U.S. Department of Agriculture. Publisher: CCM Information Corporation, 909 Third Avenue, New York, New York 10022, U.S.A.

Prior to 1970, this bibliography was published and distributed by the U.S. Government Printing Office, Washington, D.C. In 1970, distribution and publication were assumed by the CCM Information Corporation, New York. Data continue to be provided by the National Agricultural Library.

Books, journal articles, pamphlets, government documents, and special reports from sources all over the world, pertaining to agriculture and its allied sciences and received in the National Agricultural Library, are indexed in the *Bibliography.* Soil microbiology, microorganisms pathogenic to plants and animals, microscopy, antibiotics relating to veterinary medicine, and other areas of interest are extensively covered.

Each issue of this computerized index consists of five sections: the main entries, a check list of new government publications, a list of books recently acquired by the library, a subject index, and an author index. Each item, with full bibliographic data and with an assigned 6-digit identification number, is indexed under one of eight broad subject categories in the main entries. Approximately six subject headings are derived for each entry, generally from key words of the title, and are cited in the subject index.

The *Bibliography* can be used both for retrospective searching and as a current-awareness tool.

Biological Abstracts, 1926–

Preceded by *Abstracts of Bacteriology,* volumes 1–9 (1917–1925) and *Botanical Abstracts,* volumes 1–15 (1918–1925), which combined in 1926 to form *Biological Abstracts.* Publisher: BioSciences Information Service of Biological Abstracts, 2100 Arch Street, Philadelphia, Pennsylvania 19103, U.S.A.

This semimonthly abstract journal covers the biological literature of many countries. Each issue is divided into 84 sections, under which pertinent references can be found. Of special interest to the microbiologist would be the sections on the following subjects: enzymes; microorganisms, general; general and systematic bacteriology; morphology and cytology of bacteria; physiology and biochemistry of bacteria; genetics of bacteria and viruses; bacteriological apparatus and methods; tissue culture, apparatus and media, virology; general immunology; medical microbiology; clinical microbiological methods; public health; chemotherapy; food and industrial microbiology; soil microbiology. Over 140,000 papers were abstracted in 1970.

Each issue contains an author index (covering every author), cross index (a subject coordinator), biosystematic index (a taxonomic guide), and B.A.S.I.C. (a subject index), which are cumulated yearly.

B.A.S.I.C. is assembled by computer and is based on the significant terms in the author's title and on key words in the abstract and body of the paper. These search terms, permuted and arranged alphabetically by the computer, are preceded and followed by additional subject words. Absence of cross references can present a serious problem to the uninitiated. Unless the user tries all variants of a particular term, he may very well miss some significant papers, because the index provides no assistance in linking terms that have identical meanings. For example, there is no link between the entries under DNA and deoxyribonucleic acid, under Poly-A and polyadenylic acid, and under Poly-U and polyuridylic acid.

Bioresearch Index, a monthly companion to *Biological Abstracts,* began publication in 1965 and indexes literature not found in *Biological Abstracts,* such as symposia, reviews, letters, notes, bibliographies, preliminary reports, semi-popular journals, trade journals, annual institutional reports, or selected government reports.

Bulletin de l'Institut Pasteur, abstract–index–review periodical, 1903–1970, review periodical, 1971–

Publisher: Masson & Cie, éditeurs, 120 Bd St-Germain, Paris 6, France.

Published under the direction of the Institut Pasteur, each volume between 1903 and 1970 contains analytical reviews on particular microbiological topics, book reviews, and abstracts of periodical articles, listed under various classifications. Abstracts were discontinued in 1968 and the papers were merely indexed by classification. At the end of 1970, the *Bulletin* discontinued its indexing service. West European, Russian, and American literature of medical microbiology was emphasized between 1903 and 1970.

Starting with 1971, the *Bulletin* confines itself to presenting analytical reviews on particular topics only. Consequently, it resembles *Bacteriological Reviews.* Review journals are extremely helpful to a microbiologist who is embarking on a new phase of research, because a review article will summarize the work conducted in a special field and the bibliography at the end of the review refers him to the original sources of the information. A thorough review can considerably reduce the time that a microbiologist would ordinarily spend poring over abstract and index journals.

Subject and author indexes appear in each volume.

Bulletin Signalètique, 1940–

340: Microbiologie, Virologie, Immunologie. Publisher: Centre de Documentation du C.N.R.S., 15 quai Anatole France, 75 Paris VIIe, France.

Published by the Centre du Documentation du Centre National de la Recherche Scientifique, this is only one of 36 sections of the *Bulletin Signalétique.* Three other sections of peripheral interest to microbiologists would be 320: Biochimie, Chimie Analytique Biologique, Biophysique, Génie Biologique et Médical; 370: Biologie et Physiologie Végétale; 761: Microscopie Electronique, Diffraction Electronique.

This French abstract journal is strongly recommended because of its wide coverage and its precise divisions of the field. In addition, each issue contains a detailed subject index, divided into three parts, and an author index, both of which are cumulated yearly.

The comprehensive subject index recognizes aliases, such as *Actinomyces rimosus: Streptomyces rimosus* and *Actinomyces acnes: Corynebacterium acnes,* by cross references; it also has an easy-to-use format. In these respects it is superior to the subject index of *Biological Abstracts.*

Chemical Abstracts, 1907–

Publisher: Chemical Abstracts Service, Marketing Department, University Post Office, Columbus, Ohio 43210, U.S.A.

This excellent and comprehensive semimonthly abstract journal covers the world's periodical and patent literature of chemistry and allied fields. Within the scope of the journal are papers dealing with pigment formation by microbes, natural products of microbial metabolism, biochemistry of microbes, chemistry of antibiotics, nutrient requirements of microorganisms, enzymes, chemical composition of cell walls and protoplasm, photosynthesis, bactericidal action, bactericides, industrial microbiology, fermentation, sanitation and sewage disposal, soil microbiology, physiology of microorganisms, and genetics of microbes from a chemical viewpoint. Specific names of microbes, such as *Bacillus, Actinomyces, Micromonospora, Clostridium,* or *Neisseria,* are used in the subject index.

Cumulative indexes arranged by author, subject, patent, and formula are issued yearly. Decennial indexes for the years 1907–1916, 1917–1926, 1927–1936, 1937–1946, and 1947–1956 were published; the sixth cumulative index covers 1957–1961. The seventh cumulative index, covering 1962–1966, was recently published.

Chemical Abstracts functions effectively not only as an alerting service for current research but also as a retrospective searching tool because of its long history of publication and its excellent indexes.

Current Contents, 1958–

Life Sciences. Publisher: Institute for Scientific Information, 325 Chestnut Street, Philadelphia, Pennsylvania 19106, U.S.A.

This current-awareness journal reproduces on a weekly basis the tables of contents of periodicals in the life sciences, often in advance of their distribution. Over 600 titles are covered yearly. It is not designed to be used as a retrospective searching tool.

Each issue contains, in addition to the tables of contents of periodicals, an author index with addresses.

Current List of Medical Literature, 1941–1959

Formerly published by the National Library of Medicine. Superseded by *Index Medicus.*

International in scope, this publication scanned the medical periodicals. Biographical references, scientific reports to Congress, and, starting in 1957, government documents dealing with medicine were also treated.

Journal titles are arranged alphabetically, and the articles are listed and numbered according to their appearance in the journal itself. Subject and author indexes in each volume direct the reader to the numbered article.

Dissertation Abstracts International, 1970–

Section B, The Sciences and Engineering. Publisher: University Microfilms, A Xerox Company, Ann Arbor, Michigan 48106, U.S.A. Formerly published as *Dissertation Abstracts,* volumes 1–30, No. 11 (1938–1970).

Beginning with issue 1 of volume 27, *Dissertation Abstracts International* is divided into two sections: Humanities (A), and Sciences and Engineering (B). Abstracts of doctoral dissertations from more than 250 institutions in the United States and Canada appear on a monthly basis. Each issue is broadly divided by subjects, such as bacteriology, biochemistry, biology, biology–genetics, health sciences: immunology, chemistry: pharmaceutical, etc. Pertinent abstracts appear under each subject heading. Each dissertation is listed under one subject only, even though it may overlap into another area. Author indexes in the monthly issues are cumulated yearly.

Copies of dissertations can be purchased in either microfilm or Xerox form from University Microfilms, Inc., Ann Arbor, Michigan. University Microfilms also offers to the individual a computerized information retrieval service called DATRIX (Direct Access to Reference Information Service, a Xerox service), which permits retrieval of references to dissertations through key words from the files of *Dissertation Abstracts.* A basic fee of $5 is charged for a DATRIX inquiry. This covers the first ten references in the bibliography; additional references are priced at 10 cents each.

Exerpta Medica, 1947–

Publisher: Excerpta Medica, Herengracht 119–123, Amsterdam, Holland.

Each of the 41 sections of this publication represents a broad subject area in the medical sciences. The following sections would be of significance to the microbiologist: Microbiology; Public Health, Social Medicine and Hygiene; Immunology, Serology and Transplantation; Biochemistry; Virology.

International in scope, each section contains monthly author and subject indexes, which are cumulated either semiannually or annually.

Genetics Abstracts, 1968–

Publisher: Information Retrieval Ltd., Chansitor House, 38 Chancery Lane, London WC 2, England.

Appearing monthly, this secondary journal surveys the periodical literature in the field of genetics as follows: molecular genetics, structure and function of chromosomes; mapping, structure and action of genes; complex loci; mutagenesis; radiation genetics; recombination; taxonomy; immunogenetics; genetic resistance; behavioral genetics; statistical genetics; evolution; ecological genetics; human genetics; medical genetics; animal genetics and breeding; plant genetics and breeding; algal genetics; fungal genetics; bacterial genetics; viral genetics. The monthly author indexes are cumulated at the completion of the volume. The key word subject and taxonomic index is published quarterly and cumulated every five years.

Until a substantial backfile can be accumulated, this journal has limited value for retrospective searching. Its most important function, at present, is as a current-awareness alerting service.

Index Medicus, 1960–

Publisher: Superintendent of Documents, U.S. Government Printing Office, Washington, D.C. 20402, U.S.A.

Periodical literature of microbiology with a medical slant is presented here. Proceedings of congresses and symposia are excluded unless they are published in periodicals.

Each monthly issue, produced by the National Library of Medicine, contains four sections: authors and subjects of periodical articles, and authors and subjects of the Bibliography of Medical Reviews. At the end of the year the volume is cumulated and titled *Cumulated Index Medicus.* Each entry includes the name(s) of the author(s), the title of the paper in English, name of the periodical, volume, page numbers, and year of publication. A key to the journal title abbreviations appears in the January issue of each volume.

Medical subject headings with form, geographic and topical subheadings as well as "see references" are published in issue 1, part 2, of each volume. Since there are no "see references" in the subject indexes of subsequent issues, the user should consult the medical subject headings list before initiating a literature search by subject. For example, the list of medical subject headings indicates that articles on deoxyribonucleic acid are all grouped under DNA in the subject index rather than under deoxyribonucleic acid. In addition to broad subject headings, such as bacteriological techniques, bacterial proteins, bacteriolysis, bacteriophage, etc., there are specific subject headings, such as *Mycobacterium tuberculosis, Clostridium, Neisseria, Bacillus,* and *Azotobacter.* Often the subject headings overlap.

International Center for Information on Antibiotics Information Bulletin, 1964–

In collaboration with WHO. Publisher: International Center of Information on Antibiotics, c/o L. Delcambe, Bd de la Constitution 32, B. 4000 Liège, Belgium.

The center serves as a repository for antibiotics and for cultures producing them. It maintains contact with laboratories that have culture collections and makes compilations of antibiotic-producing microorganisms. Bulletins are published irregularly. The first bulletin appeared in 1964.

Bulletins 2 and 3 contain lists of antibiotics and related substances, giving their origin, lot and purity. Bulletin 4 lists culture collections, including codes and addresses. Bulletins 4, 5, 6, 7, and 8 contain alphabetical indexes of antibiotic, antiviral and antitumor substances; if the product is unnamed, a numerical code designation or the name of the first discoverer is used for identification. An alphabetical index listing the names of the producing microorganisms is also published in each of these bulletins.

International Journal of Systematic Bacteriology, 1966–

Preceded by *International Bulletin of Bacteriological Nomenclature,* volumes 1–15 (1951–1965). Publisher: International Association for Microbiology, c/o American Society for Microbiology, 4715 Cordell Avenue, Bethesda, Maryland 20014.

Indispensable for anyone interested in bacterial systematics, this journal is published for the International Association of Microbiological Societies by the American Society for Microbiology. In addition to research papers dealing with bacterial systematics, it publishes the official business of the International Committee on Nomenclature of Bacteria and its Judicial Commission, such as amendments to the *International Code of Nomenclature of Bacteria.* Information from the World Federation of Culture Collections is also disseminated through this journal.

Jahresbericht über die Fortschritte in der Lehre von den pathogenen Mikroorganismen, umfassend Bakterien, Pilze und Protozoen,

Volumes 1–27. P. von Baumgarten and R. Tangl, Editors. Cumulative author and subject index: volume 1 (1885)–10 (1894).

Found only in the larger research libraries, the *Jahresbericht* enables the microbiologist to identify and trace the older German, French, American, Russian, and English literature of pathogenic microbiology.

Each volume is divided into two sections: Section 1, books, and Section 2, original papers. Subdivision A of Section 2 deals with specific morphological groups of microorganisms, such as cocci, bacilli, spirilla, pleomorphic bacteria, Actinomyces, Botryomyces, Hyphomyces and Blastomyces, protozoa, and viruses. Subdivision B deals with general microbiology, and subdivision C with general methods, disinfecting practices and techniques.

Pertinent references are cited under the various categories, alphabetically arranged by the authors' names. In addition, papers that, according to the referees, have some degree of importance are abstracted and signed by a referee.

Microbiology Abstracts, 1965–

Publisher: CCM Information Corporation, 909 Third Avenue, New York, New York 10022, U.S.A.

Starting with volume 2 (1966), the publication was divided into two sections: Section A, Industrial Microbiology, and Section B, General Microbiology and Bacteriology. Pertinent microbiological literature from 1,723 journals forms the basis for the two sections.

Section A covers yearly approximately 8,000 periodical articles and patents related to industrial microbiology, such as products of microorganisms, fermentation, food microbiology, deterioration of fruit and vegetables, plant diseases, plant protection, forestry, soil microbiology, protection of materials, hydrocarbons, antimicrobial agents, sterilization and preservation, therapy, veterinary medicine, antibiotics, vaccines, pollution and environmental hygiene, and methodology. Each issue contains an author index as well as a patentee and assignee index.

Section B covers yearly approximately 8,500 periodical articles on the various aspects of general microbiology, such as biochemistry, morphology, spores, effects of physical and chemical factors, resistance, diseases of man, experimental diseases of animals, epidemiology, ecology, aquatic microbiology, immunology, vaccines, toxins, identification and taxonomy, methodology, media and cultures, genetics, rickettsia, and scrapie agents. Each issue contains an author index.

Book notices and notification of proceedings appear at the end of each issue. Cumulative yearly author and subject indexes are published for each section.

Full bibliographic data, including an abstract in most cases, are given for each reference. Patent references, both U.S. and foreign, cite patent number, date of application, date of patent, and equivalence to any other patent.

Its relatively short history of publication limits the value of this periodical as a retrospective searching tool. It is useful, however, for alerting the microbiologist to recently published material.

Nucleic Acids Abstracts, 1971–

Publisher: Information Retrieval Ltd., 38 Chancery Lane, London WC 2, England.

Commencing publication in January 1971, this monthly periodical monitors 2,750 primary journals and abstracts 400 to 450 pertinent articles in each of its issues. Within its scope are purines, pyrimidines, nucleosides, nucleotides, oligonucleotides, transfer RNA, protein biosynthesis, RNA, DNA, immunology, nucleoproteins, and enzymes. Each issue contains an author index. A comprehensive author and subject index is compiled when the annual volume is completed.

Pandex Current Index to Scientific and Technical Literature, 1969–

Publisher: CCM Information Corporation, 909 Third Avenue, New York, New York 10022, U.S.A.

Published biweekly in computer print-out format, Pandex scanned 2,400 scientific, medical and technological journals, 35,000 U.S. Government technical reports, and 6,000 books in 1969. Each paper book title, or technical report appears on the average under six different subject headings. An author index appears in each issue as well. Quarterly and annual cumulative indexes are published in microform (microfiche and microfilm).

Journal titles are abbreviated according to the *ASTM CODEN for Periodical Titles.* Each volume has a complete listing of the abbreviations used in the journal titles that are cited in the index.

Pandex Weekly Magnetic Tape Service provides the bibliographic data in either 7- or 9-track magnetic tape for use on the subscriber's own IBM 360 computer. The service includes programs for print-out, retrospective search, and SDI (Selective Dissemination of Information) as part of the subscription. Pandex also provides individualized search programs on a weekly, biweekly, or monthly basis. Inquiries concerning this personalized service should be directed to the publisher.

Quarterly Cumulative Index Medicus, 1927–1956

American Medical Association, Chicago, volumes 1–60. Formed by the merger of *Index Medicus* and *Quarterly Cumulative Index to Current Medical Literature.* Superseded in 1960 by *Index Medicus.*

Each volume is divided into two sections; one lists books in alphabetical order of the authors' names, the other indexes periodical literature in dictionary fashion by author and subject.

Only those aspects of microbiology that have medical significance are covered. The subject heading Bacteria is subdivided by the names of specific organisms as well as by general topics, such as morphology, virulence, viability, or culture. Other pertinent subject headings used are Fungi, Microscopy, Immunity, Microorganisms, and Viruses.

Science Citation Index, 1961–

Publisher: Institute for Scientific Information, 325 Chestnut Street, Philadelphia, Pennsylvania 19106, U.S.A.

Published by the Institute for Scientific Information, the computer-produced index approaches the literature from a specific key article upon which later research has been based. The individual initiates the search with a recent pertinent reference that he has previously obtained from a book or subject index. This reference is located in the Citation Index, which in turn refers him to the Source Index for related papers. In 1969, 2,180 source journals were examined.

Independence of technical nomenclature characterizes this unique index.

Virology Abstracts, 1967–

Publisher: CCM Information Corporation, 909 Third Avenue, New York, New York 10022, U.S.A.

This secondary periodical publishes monthly abstracts of articles in primary journals on virology, especially basic virology, biochemistry, biophysics, cell cultures, electron microscopy, genetics, immunology and epidemiology, identification, interferon and interferon-like substances, morphology, viral resistance, viruses and bacteria, viruses of man and animals, viruses of plants, and viruses of insects.

An author index is printed in each issue; these are cumulated yearly. The annual subject indexes are based on key words in the titles.

Handicapped by a small backfile, *Virology Abstracts* has at present limited use as a retrospective searching tool. Its main value, for the time being, is in keeping the scientist aware of current research in virology.

Zentralblatt für Bakteriologie, Parasitenkunde, Infektionskrankheiten und Hygiene, 1902–

Volumes 1–16 (1887–1894); volumes 17–30 (1895–1901), Abteilung, 1, Medizinisch-hygienische Bakteriologie, Virusforschung und tierische Parasitologie. Replaced by two publications in 1902: Abteilung I, Originale (volume 31), consisting of original papers, and *Referate* (volume 31), consisting of abstracts of papers arranged by classification. Publisher: Gustav Fischer Verlag, Stafflenbergstrasse 36, Schliessfach 431, 7 Stuttgart 5, West Germany.

In addition to coverage of the different families and species of microorganisms, the journal presents abstracts of papers on disinfection, chemotherapy, immunity, blood, and environmental hygiene. The section on environmental hygiene, a recent innovation, presents papers on the microbiology of air, soil, water and food in relation to health.

Author and subject indexes are issued upon completion of the volume only. Unfortunately a list of periodicals abstracted is not published.

This journal is useful as a tool for retrospective searches of West European medical microbiology literature.

Abbreviations of Periodical Titles

ACCESS, 1969

Key to the Source Literature of the Chemical Sciences. Copyright 1969, American Chemical Society, Columbus, Ohio. Publisher: Chemical Abstracts Service, Marketing Department, Ohio State University, Columbus, Ohio 43210, U.S.A.

The abbreviations are based on the *American Standard for Periodical Title Abbreviations: Z 39.5,* 1963, and the *ASTM CODEN for Periodical Titles,* 1966.

Additional details on *ACCESS* will be found under the heading **Union Lists.**

American National Standard for the Abbreviation of Titles of Periodicals, Z 39.5, 1969

Publisher: American National Standards Institute, 1430 Broadway, New York, New York 10018, U.S.A.

This is a revision of the *American Standard for Periodical Title Abbreviations Z 39.5,* 1963. It is compatible with the abbreviations that appear in *British Standard 4148, 1969, Recommendations for the Abbreviations of Titles of Periodicals* (British Standards Institution Documentation Standard Committee, Subcommittee on Abbreviations and Codes in Documentation).

The *Standard* outlines the rules for abbreviations. A separate word abbreviation list is used in connection with the *Standard;* it is titled *Revised and Enlarged Word-Abbreviation List for American National Standard for Periodical Title Abbreviations, Z 39.5, 1963* and was prepared by the National Clearinghouse for Periodical Title Word Abbreviations, c/o Chemical Abstracts Service, Columbus, Ohio (December 21, 1966).

BIOSIS, 1970

List of Serials with Title Abbreviations. Publisher: BioSciences Information Service of Biological Abstracts, 2100 Arch Street, Philadelphia, Pennsylvania 19103, U.S.A.

The list contains more than 7,663 titles of serials issued in 98 different countries and territories, arranged alphabetically by title. It includes serials covered not only by *Biological Abstracts* but also by *Bioresearch Index.*

Two types of abbreviations are shown for the majority of titles. Each title has been assigned an abbreviation based on the *American National Standards Institute's Standard for Periodical Title Abbreviations Z 39.5.* An ASTM CODEN precedes the serial title, if one had been devised by October 20, 1970.

The list is updated and revised as necessary.

CODEN for Periodical Titles, 2nd edition, 1966

Two volumes plus supplement (1968). Prepared and maintained for the ASTM Special Committee on Numerical Reference Data by the Wyandotte–ASTM Punched Card Project (ASTM Data Series DS 23A). L. E. Kuentzel, Editor. Publisher: American Society for Testing and Materials, 1916 Race Street, Philadelphia, Pennsylvania 19103, U.S.A.

Volume 1:
Periodical titles, alphabetical by CODEN; non-periodical titles, numero-alphabetical by title; non-periodical titles, alphabetical by title; deleted CODEN.
Volume 2:
Periodical titles, alphabetical by title.
Supplement:
Same as volumes 1 and 2 combined.

This directory provides unique 5-letter codes, or CODEN, for the titles of periodicals, so that concise and accurate abbreviations are available for handling references by machine. The first four letters of each CODEN have some mnemonic relation to the title being coded; the fifth letter represents a "grid" pattern. CODEN serve as concise links between information stored in computer memories and the material as published in the primary source. They save time and space in publishing long lists of references or bibliographies, such as *Chemical Titles,* that are machine-generated publications.

Index Medicus

Publisher: Superintendent of Documents, U.S. Government Printing Office, Washington, D.C. 20402, U.S.A.

A key to the abbreviations of journals that are indexed by *Index Medicus* appears each year in the January issue (No. 1, part 2) of *Index Medicus.* The abbreviations were devised by the National Library of Medicine.

World List of Scientific Periodicals Published in the Years 1900–1960, 4th edition, 1963

Three volumes. Publisher: Butterworths, London, England.

Abbreviations for periodicals appear in accordance with the *British Standard for the Abbreviations of Titles of Periodicals.*

Additional information on the *World List* will be found under the heading **Union Lists.**

Subject Lists of Periodicals

Periodicals Relevant to Microbiology and Immunology, 1968

A World List. G. Tunevall, Editor. Wiley-Interscience, New York, 1969.

Sponsored by the Permanent Committee for Microbiological and Immunological Documentation of the International Association of Microbiological Societies, this comprehensive volume lists periodicals alphabetically by title and country of origin. Data include title and abbreviation of the title, address of the publisher, special field and nature of the periodical, languages used, whether or not summaries of the articles are published, frequency of appearance, rates of subscription, volumes issued, and bibliographical periodicals that summarize or index the articles published in the periodical.

The Standard Periodical Directory, 3rd edition, 1970

Leon Garry, Editor. Copyright 1969, Oxbridge Publishing Co., Inc., New York.

More than 50,000 American and Canadian periodicals are listed under 224 subject classifications. In this particular volume, periodicals are rather broadly defined to include magazines, journals, directories, serials, transactions and proceedings of professional societies, bulletins, and yearbooks, as well as museum, religious, ethnic, social, and literary publications. Surprisingly, cover-to-cover translations of Russian journals and some British periodicals are found in the volume. This is simply because they are issued by publishers based in the United States.

The volume is composed of three sections. Cross Index: a listing of subjects and key-word index to related subjects. Periodical Listing by Subject Classification: this is the principal section of the volume and is arranged alphabetically by classification. Index: a list of periodicals, arranged alphabetically by titles.

Each entry for a periodical supplies title, address, name(s) of the editor(s), publisher and advertising director, scope, year founded, frequency, subscription rate, price of a single copy, distribution limitations, if any, circulation size, advertising rate, and name and address of the auditor.

Ulrich's International Periodical Directory, 13th edition, 1969–1970

A classified guide to a selected list of current periodicals, foreign and domestic. Two volumes plus supplements. Copyright 1969, R. R. Bowker Co., New York.

More than 40,000 periodicals in print are listed alphabetically under 223 main subject headings and subheadings. If a periodical covers more than one subject, cross references direct the reader to the subject under which the full citation is found. Each entry for a periodical includes the name of the publication, transliterated title, subtitle, title change, sponsoring organization, language used in the paper, first year of publication, subscription price, name of the editor, name and address of the publisher, and special features. Periodicals on microbiology are cited in volume 1, pages 205–207, under the heading Biological Sciences – Microbiology.

Annual and irregular serials are covered in a companion volume titled *Irregular Serials and Annuals: An International Directory,* R. R. Bowker Co., New York, 1967.

Union Lists of Periodicals

Union Lists are compilations of periodicals that indicate libraries holding each title. They can be international, national, regional, or even local in scope. For example, often a university library system comprised of many specialized libraries issues a union list of the periodicals held by all its branches and thereby provides ready access to its various collections.

ACCESS, 1969

Key to the Source Literature of the Chemical Sciences. Copyright 1969, American Chemical Society, Columbus, Ohio. Supersedes List of *Periodicals Abstracted by Chemical Abstracts* (1908–1960) and *Chemical Abstracts – List of Periodicals* (1961–1967). Supplements continued as *Chemical Abstracts Service Source Index Quarterly,* starting in 1970. Publisher: Chemical Abstracts Service, Marketing Department, Ohio State University, Columbus, Ohio 43210, U.S.A.

The catalog serves as an international union list, as a list of periodical abbreviations, and as a file of bibliographic data on primary-source publications.

The 1969 edition, reflecting the serial and non-serial publications abstracted by *Chemical Abstracts* since 1907, has 30,798 entries, of which 10,399 are currently published serials. In addition to the listings of serials, there are 2,784 entries from scientific and technical meetings and 1,813 entries from monographs, identifying individual contributors. The edition also has 9,333 cross references, many of which refer from a variant form of the title to the form selected as the main entry. Libraries maintaining files of particular titles are also identified.

Parts of the introduction to *ACCESS* are quoted below:[a]

ENTRY ARRANGEMENTS

Entries are arranged alphabetically, letter by letter, according to the abbreviated form of the title. Titles as they appear on the publications have been abbreviated according to the *American Standard for Periodical Title Abbreviations Z 39.5* (1963). These abbreviations are indicated by printing the part of the title that forms the abbreviation in bold-face type.

SERIALS

Entries for serials contain the following data when applicable.

1. Complete title as it appears on the serial, with the standard title abbreviation indicated in bold-face type.
2. ASTM CODEN with computer-calculated check character.
3. English translation of complete title if title is in a language other than English, French, German, Latin, or Spanish.
4. Reference to a former title, if there was one. If an entry for the former title is present, the reference is displayed as a title abbreviation. If the former title has no entry in this edition, then it appears spelled out in full.
5. Languages in which the papers, summaries, and tables of contents are printed.
6. Publication history, showing the beginning volume and issue numbers and date. Any suspensions or disruptions in the publication history are also given. Entries for serials no longer published include the volume number, issue number, and the date of the last issue published.
7. For current serial entries, the frequency of issue, the number of volumes per year (if other than one), and a current volume number/year correlation are given.

[a] From *ACCESS,* 1969 edition. Reproduced by permission of the copyright owners, American Chemical Society, Columbus, Ohio.

8. The current price per year has been included for each current periodical or serial if this data were available from examination of the issues. Prices are given in U.S. dollars or in the monetary units of the countries in which the periodicals and serials are published.

9. The address of the publisher or other source from which the serial may be ordered is given for current serials.

10. The main entry for the serial, prepared according to American Library Association cataloging rules.

11. Reference to successor title or titles, if the entry is for a serial that has undergone a title change. This reference will appear as a title abbreviation, followed by "which see" if an entry for the successor title is present. If not, the successor title will appear in full.

12. If publication of the serial has been discontinued, this will be noted.

13. Library-holdings data are given for each serial if such data were available to CAS. Holdings are shown by inclusive years. Libraries are identified by their National Union Catalog symbols. Symbols for U.S. libraries appear first, followed by symbols for foreign libraries, filed alphabetically by country.

Examples

GU; IC 1907+; MiD 1910–1942, 1948*–1958*.

Explanations

A symbol standing alone signifies that a library has a full set of the serial, i.e., GU, The University of Georgia Library, has all the issues. IC 1907+ means that the file at the Chicago Public Library is complete from 1907. The data following MiD, the symbol for the Detroit Public Library, means that the library has a complete file for the years 1910–1942, nothing for the years 1943–1947, and its file for 1948–1958 is incomplete. Incomplete files are shown by asterisks appended to the beginning and ending years. A directory of participating libraries in which the National Union Catalog symbols are expanded to full names and addresses for the libraries is to be found on pages 11-22.

New Serial Titles, 1966

A Union List of Serials Commencing Publication after December 31, 1949. Volume 1–2, 3 (changes in titles). R. R. Bowker Co., New York.

The cumulated volumes noted above serve as a supplement to the 3rd edition of the *Union List of Serials* and were prepared under the sponsorship of the Joint Committee on the Union List of Serials. The basic volumes are updated by monthly issues called *New Serial Titles,* which are cumulated annually and over 5- or 10-year periods. *New Serial Titles* is compiled and edited by the Serial Record Division of the Library of Congress and is distributed by the Card Division.

Entries follow the form used in *Union List of Serials.* Libraries holding a particular title are noted by code letters along with their initial volume.

Union Catalog of Medical Periodicals, 1967–1968

Two volumes. Publisher: Medical Library Center of New York, 17 East 102nd Street, New York, New York 10029, U.S.A.

A regional union list of medical and paramedical periodicals that gives the holdings of 68 libraries in the New York metropolitan area.

Union List of Serials in Libraries of the United States and Canada, 3rd edition, 1965

Five volumes. Edna Brown Titus, Editor. H. W. Wilson Co., New York. Sponsored by the Joint Committee on the Union List of Serials with the cooperation of the Library of Congress and funded by a grant from the Council on Library Resources, Inc.

Serials published throughout the world form this list. There are a total number of 226,987 entries, of which 70,538 are cross references. A typical entry includes: title of serial, beginning date, cessation date if applicable, history of title changes if applicable, codes for libraries holding the title, and the beginning volume in their collection.

Intended primarily to be a national finding list of serials in the United States and Canada, the list has peripheral use in verifying serial titles.

World List of Scientific Periodicals Published in the Years 1900–1960, 4th edition, 1963

Three volumes. Peter Brown and George B. Stratton, Editors. Butterworths, London, England. Updated by annual supplements of the *British Union Catalogue of Periodicals.*

Intended primarily as a national finding list of periodicals, this file is arranged alphabetically by the important words of the titles. It also presents abbreviations for each entry in accordance with the *British Standard for the Abbreviations of Titles of Periodicals.*

Only the holdings of British libraries are given; however, the periodical titles themselves are those published throughout the world.

GOVERNMENT PUBLICATIONS

Government documents, often a vastly underrated source of scientific information by the inexperienced, encompass the immense quantity of literature published in all countries by or on behalf of some governing body, either central, local, departmental, or agency. The literature can be subdivided into three broad categories: material published by international organizations set up through the cooperation of several national governments; material published by national governments, their departments and agencies; material issued by subordinate administrative groups such as regional or local governmental divisions.

World Health Organization[b]

WHO, an international organization, issues many publications of prime interest to microbiologists. It is one of the specialized agencies connected with the United Nations. Through this organization, which came into being in 1948, the public-health and medical professions of more than 120 countries exchange their knowledge and experience and collaborate in an effort to achieve the highest possible level of health throughout the world. WHO is not concerned with problems that individual countries or territories can solve with their own resources. It deals with problems that can be satisfactorily solved only through the cooperation of all or several countries – for example, the eradication or control of malaria, schistosomiasis, smallpox, and other communicable diseases, as well as of some cardiovascular diseases and cancer. Progress towards better health throughout the world also demands international cooperation in many other activities: setting up international standards for biological substances, for pesticides and for pesticide spraying equipment; compiling an international pharmacopoeia; drawing up and administering the International Health Regulations; revising the international lists of diseases and causes of death; assembling and disseminating epidemiological information; recommending non-proprietary names for drugs; promoting the exchange of scientific knowledge. In many parts of the world there is need for improvement in maternal and child health, nutrition, nursing, mental health, dental health, social and occupational health, environmental health, public-health administration, professional education and training, and health education of the public. A large share of the organization's resources is devoted to giving assistance and advice in these fields and to making available – often through publications – the latest information on these subjects. Since 1958, an extensive international program of collaborative research and research coordination has added substantially to knowledge in many fields of medicine and public health. This program is constantly developing, and its many facets are reflected in WHO publications.

[b] From *WHO Chronicle,* *24*(11), 1970. Reproduced by permission of the World Health Organization, Geneva, Switzerland.

WHO Chronicle

Published by WHO for the medical and public-health professions, this publication provides a monthly record of the principal health activities undertaken in various countries with WHO assistance. It also contains summaries and detailed accounts of the other publications issued by this organization.

WHO Chronicle is published in Chinese, English, French, Russian, and Spanish.

Bulletin of the World Health Organization

This bulletin contains technical articles in English or French, contributed by physicians and scientists engaged in public-health work. It appears monthly; each issue has about 120 pages.

World Health

An illustrated magazine for the general public, this publication aims at giving an idea of WHO activities throughout the world and at showing some of the more striking aspects of public-health work. *World Health* is published every month in English, French, Portuguese, Russian, and Spanish.

Other WHO Publications

The *Technical Report Series,* the *Monograph Series,* the *World Health Statistics Report,* the *International Digest of Health Legislation,* etc. are described in a catalog that can be supplied free of charge on request.

Sources of WHO Publications

Algeria:	Société Nationale d'Edition et de Diffusion, 3 Bd Zirout Youcef, Algiers
Argentina:	Editorial Sudamericana S.A., Humberto 1° 545, Buenos Aires
Australia:	Hunter Publications, 23 McKillop Street, Melbourne C 1
	United Nations Association of Australia, Victorian Division, 364 Lonsdale Street, Melbourne, Victoria 3000
Austria:	Gerold & Co., Graben 31, Vienna 1
Belgium:	Office international de Librairie, 30 avenue Marnix, Brussels
Burma:	*see* India, WHO Regional Office
Cambodia:	The WHO Representative, P. O. Box 111, Phnom-Penh
Canada:	Information Canada, Ottawa
Ceylon:	*see* India, WHO Regional Office
China:	The WHO Representative, 5 Chungshan Road South, Taipei, Taiwan
	The World Book Co. Ltd, 99 Chungking South Road, Section 1, Taipei, Taiwan
Colombia:	Distrilibros Ltd, Pio Alfonso Garcia, Carrera 4a, Nos 36–119, Cartagena

Congo:	Librairie congolaise, 12 avenue des Aviateurs, Kinshasa
Costa Rica:	Imprenta y Libreria Trejos S.A., Apartado 1313, San Jose
Cyprus:	MAM, P. O. Box 1674, Nicosia
Denmark:	Ejnar Munksgaard Ltd, Norregarde 6, Copenhagen
Ecuador:	Libreria Cientifica S.A., P. O. Box 362, Luque 223 Guayaquil
Fiji:	The WHO Representative, P. O. Box 113, Suva
Finland:	Akateeminen Kirjakauppa, Keskuskatu 2, Helsinki 10
France:	Librairie Arnette, 3 rue Casimir-Delavigne, Paris 6^e
Germany:	Govi Verlag GmbH, Beethovenplatz 1–3, Frankfurt am Main 6
	W. E. Saarbach, Postfach 1510, Follerstrasse 2, 5 Cologne 1
	Alex. Horn, Spiegelgasse 9, 62 Wiesbaden
Greece:	G. C. Eleftheroukadis S.A., Librairie internationale, rue Nikis 4, Athens (T 126)
Haiti:	Max Bouchereau, Librairie "A la Caravelle", Boite postale 111-B, Port-au-Prince
Hungary:	Kultura, P. O. Box 149, Budapest 62
	Akadémiai Könyvesbolt, Váci utca 22, Budapest V
Iceland:	Snaebjörn Jonsson & Co., P. O. Box 1131, Hafnarstraeti 9, Reykjavik
India:	WHO Regional Office for South-East Asia, World Health House, Indraprastha Estate Ring Road, New Delhi 1
	Oxford Book & Stationery Co., Scindia House, New Delhi; 17 Park Street, Calcutta (subagent)
Indonesia:	*see* India, WHO Regional Office
Iran:	Mesrob Grigorian, Naderi Avenue, Arbab-Guiv Building, Teheran
Ireland:	The Stationery Office, Dublin
Israel:	Heiliger & Co., 3 Nathan Strauss Street, Jerusalem
Italy:	Edizioni Minerva Medica, Corso Bramante 83–85, Turin; Via Lamarmora 3, Milan
Japan:	Maruzen Co. Ltd, P. O. Box 5050, Tokyo International, 100-31 Japan
Kenya:	The Caxton Press Ltd, Gathani House, Huddersfield Road, P. O. Box 1742, Nairobi

Korea:	The WHO Representative, Central P. O. Box 540, Seoul
Laos:	The WHO Representative, P. O. Box 343, Vientiane
Lebanon:	Librairie au Papyrus, Immeuble Abdel Baki, rue Cinéma Colisée, Hamra, Beirut
Luxembourg:	Librairie Trausch-Schummer, place du Théatre, Luxembourg
Malaysia:	The WHO Representative, P. O. Box 2550, Kuala Lumpur
	Jubilee (Book) Store Ltd, 97 Jalan Tuanku Abdul Raman, P. O. Box 629, Kuala Lumpur
Mexico:	La Prensa Médica Mexicana, Ediciones Cientificas, Paseo de las Facultades 26, Mexico City 20, D. F.
Mongolia:	*see* India, WHO Regional Office
Morocco:	Editions La Porte, 281 avenue Mohammed V, Rabat
Nepal:	*see* India, WHO Regional Office
Netherlands:	N. V. Martinus Nijhoff's Boekhandel en Uitgevers Maatschappij, Lange Voorhout 9, The Hague
New Zealand:	Government Printing Office or Government Bookshops at:
	Rutland Street, P. O. Box 5344, Auckland; 130 Oxford Terrace, P. O. Box 1721, Christchurch; Alma Street, P. O. Box 857, Hamilton; Princes Street, P. O. Box 1104, Dunedin; Mulgrave Street, Private Bag, Wellington
	R. Hill & Son Ltd, Ideal House, Gilles Avenue and Eden Street, Newmarket, Auckland S.E. 1
Nigeria:	University Bookshop Nigeria Ltd, University of Ibadan, Ibadan
Norway:	Johan Grundt Tanum Forlag, Karl Johansgt. 41, Oslo
Pakistan:	Mirza Book Agency, 65 Shahrah Quaid-E. Azam, P. O. Box 729, Lahore 3
	Shilpa Niketan, 29 D.I.T: Super Market, Mymensingh Road, P. O. Box 415, Dacca 2
Paraguay:	Agencia de Librerias Nizza S.A., Estrella No. 721, Asunción
Peru:	Distribuidora Inca S.A., Apartado 3115, Emilio Althaus 470, Lima
Philippines:	WHO Regional Office for the Western Pacific P. O. Box 2932, Manila
Poland:	Skladnica Ksiegarska, ul. Mazowieka 9, Warsaw (except periodicals)
	BKWZ Ruch, ul. Wronia 23, Warsaw (periodicals only)
Portugal:	Livraria Rodrigues, 186 Rua Aurea, Lisbon

South Africa: Van Schaik's Bookstore (Pty) Ltd, P. O. Box 724, Pretoria

Spain: Commercial Atheneum S.A., Consejo de Ciento 130–136, Barcelona 15; General Moscardó 29, Madrid 20

Libreria Diaz de Santos, Lagasca 95, Madrid 6

Sweden: Aktiebolaget C. E. Fritzes Kungl. Hovbokhandel, Fredsgatan 2, Stockholm 16

Switzerland: Medizinischer Verlag Hans Huber, Marktgasse 9, Bern

Tanzania: *see* Kenya

Thailand: *see* India, WHO Regional Office

Tunisia: Société Tunisienne de Diffusion, 5 avenue de Carthage, Tunis

Turkey: Librairie Hachette, 469 avenue de l'Indépendance, Istanbul

Uganda: *see* Kenya

United Arab Republic: Al Ahram Bookshop, 10 Avenue el Horreya, Alexandria

United Kingdom: H. M. Stationery Office: 49 High Holborn, London WC 1; 13a Castle Street, Edinburgh 2; 109 St. Mary Street, Cardiff CFI, IJW; 7–11 Linenhall Street, Belfast BT 2, 8 AY; Brazennose Street, Manchester 2; 258–259 Broad Street, Birmingham 1; 50 Fairfax Street, Bristol 1

Postal orders: P. O. Box 569, London SE 1

U.S.A.: The American Public Health Association, Inc., 1740 Broadway, New York, New York 10019

USSR: For readers in the USSR requiring Russian editions: Komsomolskij prospekt 18, Medicinskaja Kniga, Moscow

For readers outside the USSR requiring Russian editions: Kuzneckij most 18, Mezdunarodnaja Kniga, Moscow G-200

Venezuela: The University Society Venezolana C.A., Apartado 50785, Caracas

Libreria del Este, Avenida Francisco de Miranda 52, Edificio Galipan, Caracas

Vietnam: The WHO Representative, P. O. Box 242, Saigon

Yugoslavia: Jugoslovenska Knjiga, Terazije, 27/II, Belgrade

Orders may also be addressed to: World Health Organization, Distribution and Sales, Geneva, Switzerland, but must be paid for in British, American, or Swiss currency.

Government Reports[c]

Some countries have made their atomic-energy reports available as follows:

Australia: Australian Atomic Research Establishment Library, Private Mail Bag, Sutherland, New South Wales

Belgium: Centre d'Etude de l'Energie Nucléaire (CEN), Studiecentrum voor Kernenergie (SCK), Technical Information Department, Mol-Donk

CERN: CERN European Organization for Nuclear Research, Geneva 23 (free of charge)

EURATOM: Office central de vente des publications des Communautés Européennes, 2 place de Metz, Luxembourg

France: CEA reports 1–2199: Service Central de Documentation du C.E.A., Centre d'Etudes Nucléaires de Saclay, Boite Postale n° 2, 91 Gif-sur-Yvette

CEA reports 2200 and above: La Documentation Francaise, 31 quai Voltaire, 75 Paris (7e)

SCPRI reports: Ministère de la Santé Publique, Service Central de Protection contre les Rayonnements Ionisants, Boite Postale n° 78, Le Vesinet; Telex 25667

Germany: Zentralbibliothek der Kernforschungsanlage Juelich, Juelich

KFK reports: Gesellschaft für Kernforschung mbH, 75 Karlsruhe, Postfach 947

HMI reports: Hahn-Meitner Institut für Kernforschung Berlin, Bibliothek, Glienickerstrasse 100, 1 Berlin 39 (free of charge)

IKF and AED reports: Gmelin Institut, 40–42 Varrentrappstrasse, Frankfurt am Main

DESY reports: Deutsches Elektronen-Synchrotron (Desy), Dokumentation, Notkestieg 1, 2000 Hamburg 52

Israel: Israel Atomic Energy Commission, Soreq Nuclear Research Centre, Yavne (free of charge)

Italy: Comitato Nazionale per l'Energia Nucleare, Divisione Affari Internazionali e Studi Economici, Ufficio Pubblicazioni, Via Belisario 15, Rome

CISE reports: CISE Press Office, Casella Postale 3986, Milan

Japan: Institute for Nuclear Study, University of Tokyo, Tanashi-machi, Kitatama-gun, Tokyo (free of charge)

JAERI reports: Staff Supply Cooperative of Japan Atomic Energy Research Institute (JAERI), Tokai-mura, Nakagun, Ibaraki-ken

IPPJ reports: Research Information Center, Institute of Plasma Physics, Nagoya University, Nagoya

[c] From *Chemical Abstracts, 74,* xii–xiii (1971). Reproduced by permission of the publishers.

Netherlands:	RCN reports: Reactor Centrum Nederland, 112 Scheveningseweg, s-Gravenhage
New Zealand:	N. Z. Institute of Nuclear Sciences, D.S.I.R., Private Bag, Lower Hutt
Pakistan:	Atomic Energy Center, P. O. Box 658, Lahore
Poland:	Nuclear Energy Information Center, Palace of Culture and Science, Warsaw (free of charge)
Spain:	JEN reports: Servicio de Documentation, Junta de Energie Nuclear, Ciudad Universitaria, Madrid 3 (free of charge)
Switzerland:	EIR reports: Eidgenössisches Institut für Reaktorforschung, c/o Bibliothek, 5303 Würenlingen
Turkey:	TAEC, Cekmece Nuclear Research Center, P. O. Box 1, Airport, Istanbul (free of charge)

Abbreviations of Availability Sources

CAN	AECL reports: Scientific Document Distribution Office, Atomic Energy of Canada Ltd, Chalk River, Ontario, Canada
	NRC reports: National Research Council of Canada, Ottawa 2, Ontario, Canada
DENM	Reports of the Atomic Energy Commission Research Establishment, Riso: on exchange from Danish Atomic Energy Commission Library, Riso, Roskilde; also for sale by Jul. Gjellerup, 87 Solvgade, Copenhagen K, Denmark
	Danatom reports: on loan from Research Establishment, Riso, Roskilde; also from Danatom, Aurehojvej 2, Hellerup, Denmark
Dep	AEC reports: copy on deposit in libraries both inside and outside the United States
	Non-U.S. reports: copy on deposit in libraries in the United States
DTIE	U.S. Atomic Energy Commission, Division of Technical Information Extension, P. O. Box 62, Oak Ridge, Tennessee 37830, U.S.A.
	Conference papers: available on loan only
Gmelin	AED reports: Atomic Energy Documentation Center, Gmelin Institut, 40–42 Varrentrappstrasse, Frankfurt am Main, Germany
GPO	Superintendent of Documents, Government Printing Office, Washington, D.C. 20402, U.S.A.
HMSO	Her Majesty's Stationery Office, London, England
IAEA	International Atomic Energy Agency, Vienna, Austria, and its sales agencies
JCL	Special Libraries Association Translation Center, John Crerar Library, 35 West 33rd Street, Chicago, Illinois 60616, U.S.A.

NAS	National Academy of Sciences, National Research Council, Washington, D.C. 20418, U.S.A.
NORW	Reports from Forsvarets Forskningsinstitutt: Norwegian Defence Research Establishment Library, P. O. Box 25, Kjeller, Norway
	KR reports (on exchange basis) and reports of the Norwegian Reactor School: Institute for Atomenergi Library, P. O. Box 40, Kjeller, Norway
NRC	National Research Council of Canada, Ottawa, Ontario, Canada
NTIS	National Technical Information Service, 5285 Port Royal Road, Springfield, Virginia 22151, U.S.A.*
SWED	AE reports: Aktiebolaget Atomenergi, Library, Box 43041, Stockholm 43, Sweden
	Microcopies of AE reports: International Documentation Center, Turnba, Sweden
	VDDIT reports: Aktiebolaget Atomenergi, Information Office, Scientific and Technical Information, Box 43041, Stockholm 43, Sweden
	Other series: not generally available to the public; organizations in the atomic-energy field may obtain copies from Aktiebolaget Atomenergi, Box 43041, Stockholm 43, Sweden
UK	Her Majesty's Stationery Office (HMSO), London, England. Copies not for sale by HMSO may be purchased in microcopy form from Micro Methods Limited, East Ardsley, Wakefield, Yorks, England
UKAEA	United Kingdom Atomic Energy Authority, 11 Charles II Street, London, SW 1, England
UN	Sales Section, Room 1059, United Nations Headquarters, New York, New York 10017, U.S.A.
USGS	U.S. Geological Survey, Washington, D.C. 20242, U.S.A. (free of charge)
ZLDI	Zentralstelle für Luftfahrtdokumentation und Information, Maria Theresiastrasse 21, Munich 27, Germany

United States of America

U.S. Documents are distributed by a number of offices, each of which announces the documents for which it is responsible through index and abstract publications. The principal distributors are listed below:

Superintendent of Documents
U.S. Government Printing Office
Washington, D.C. 20402

* NTIS has established uniform costs for all reports sold by it. Reports issued by organizations outside the United States will be sold by NTIS only to purchasers within the United States. Purchasers outside the United States may acquire such reports from the non-U.S. source.

National Technical Information Service
U.S. Department of Commerce
Springfield, Virginia 22151

Commissioner of Patents
Box 9
Patent Office
Washington, D.C. 20231

To some extent, government bodies also distribute the documents they originate.

In addition, a number of libraries throughout the United States have been designated as either full or partial depositories for government documents. Local public libraries can direct the user to the nearest depository, where these document collections are readily accessible.

Monthly Catalog of United States Government Publications, 1895–

Publisher: Superintendent of Documents, U.S. Government Printing Office, Washington, D.C. 20402, U.S.A.

Government publications printed by the Government Printing Office are announced monthly. Each issue contains previews, listing important titles for prepublication orders, a catalog of publications arranged under the originating agencies, and an index. The catalog indicates where a copy of the document can be obtained, its price (some are free), and whether it is a depository item. The index is an alphabetical file of titles or subjects based on titles. If no specific subject is assigned to bulletins, circulars or reports, the entry is under the originating agency only. The indexes, which have limited value because the subjects depend on titles, are cumulated at the end of the year.

Public Health Service, National Communicable Disease Center, Food and Drug Administration, National Science Foundation, and Department of Agriculture are a few of the agencies that publish documents of significance to microbiologists. The *Monthly Catalog* is an authoritative alerting service for new government publications.

Nuclear Science Abstracts, 1948–

Published by the U.S. Atomic Energy Commission, Division of Technical Information, the semimonthly issues cover the international nuclear science literature: government, industrial, and research reports; books; patents; conference papers; periodical articles. Arranged broadly by subject, each issue contains a section on the life sciences, with subdivisions such as biochemistry, physiology and molecular biology, genetics, and radiation effects in microorganisms.

The four indexes (subject, personal author, corporate author, and report number) in each issue are cumulated quarterly, semiannually, and annually. Five-year cumulations of the author and subject indexes were compiled between 1948 and 1966.

U.S. Atomic Energy Commission reports can be purchased from the National Technical Information Service, U.S. Department of Commerce, Springfield, Virginia 22151. Price and necessary bibliographic information are all supplied in *Nuclear Science Abstracts.* Libraries having USAEC and foreign reports on deposit are also listed in each issue.

Subscription to *Nuclear Science Abstracts* can be obtained from the Superintendent of Documents, U.S. Government Printing Office, Washington, D.C. 20402, U.S.A.

Official Gazette of the United States Patent Office, 1872–

Government Printing Office, Washington, D.C. The patents granted by the United States government are arranged by class and subclass number in each issue. A brief description is supplied for each patent. The List of Patentees, List of Design Patentees, and Index to Subjects of Invention in each issue are cumulated yearly as an *Index of Patents.*

Several publications should be used in connection with the *Official Gazette:*

1. *Manual of Classification*

   Contains the entire classification schedule, which lists the number and descriptive title of each class and subclass.

2. *Index to Classification*

   The pertinent class and subclass are shown for the alphabetically arranged subject headings.

3. *Classification Bulletins*

   Supplements to the *Manual of Classification.*

U.S. Government Research and Development Reports, 1965–

Volume 40. Formerly *Bibliography of Scientific and Industrial Reports,* volumes 1 (1946)–20 (1953), and *U.S. Government Research Reports,* volumes 21 (1954)–39 (1964). Publisher: National Technical Information Service, Operations Division, Springfield, Virginia 22151.

This semimonthly computer-produced abstract journal announces reports of U.S. government-sponsored research and development, U.S. government-sponsored translations, and some foreign reports, which are written in English and are non-confidential. The reports cited in this publication can be purchased from the National Technical Information Service in either paper copy or microfiche form.

Each issue is divided by subject into 22 "Fields" and each "Field" is divided into subcategories called "Groups". The Microbiology Group is found in Field #6: Biological and Medical Sciences.

Quarterly and annual cumulative indexes of subject, personal author, corporate author, contract/grant number, and accession/report number facilitate retrospective searches.

Patents

A patent is a contract between an individual or company and a government, giving to the inventor or assignee a monopoly on his invention for a specified time; after the expiration date, the inventor makes his invention freely available to the public. The specified time varies with the government. Each country announces in some manner the patents it has granted. All United States patents are announced in the *Official Gazette of the U.S. Patent Office.* In addition, many United States and foreign patents are abstracted in *Chemical Abstracts.*

A patent can also be considered a government document and a primary source of information. It is often of particular interest to the industrial microbiologist.

Procurement of Patents[d]

Copies of patents can be obtained as follows:

Australia:	Commissioner of Patents, Patent Office, Canberra, A.C.T.
Austria:	Österreichisches Patentamt, Druckschriftenverschleiss, Postfach 95, A-1014 Vienna

[d] From *Chemical Abstracts, 74,* xii (1971). Reproduced by permission of the publishers.

Belgium: Service de la Proprieté Industrielle et Commerciale, 24–26, rue DeMot, B-1040 Brussels

Britain: Comptroller-General, The Patent Office, Sales Branch, Block C, Station Square House, St. Mary Cray, Orpington, Kent, BR 53rd

Canada: Commissioner of Patents, Ottawa

Czechoslovakia: Dr. Karel Neumann, Advokatni poradna c. 10, Zitna 25, Prague I

Denmark: Direktoratet for Patent- og Varemaerkevaesenet, Nyropsgade 45, Copenhagen V

Finland: Patentti-ja rekisterihallitus, Bulevardi 21, Helsinki 18

France: Service d'Edition et de Vente Des Publications Officielles, 39 rue de la Convention, Paris 15

Germany: Deutsches Patentamt, Dienststelle Berlin, Gitschinerstrasse 97–103, 1000 Berlin 61

Germany (East): Zeitungsvertrieb Gebrüder Petermann, Kurfürstenstrasse 111, 1000 Berlin West 30

Hungary: Licencia Hungarian Company for the Commercial Exploitation of Inventions, P. O. Box 207, Budapest 5

India: Patent Numbers 1–85,000: Controller of Patents and Designs, 214 Lower Circular Road, Calcutta 17

Patent numbers 85,001 and up: Officer-in-Charge, Government of India Book Depot, 8 Hastings Street, Calcutta 1

Israel: Commissioner of Patents, Designs and Trade Marks, P. O Box 767, Jerusalem

Italy: Libreria dello Stato, Piazza G. Verdi 10, 00100 Rome

Japan: Hatsumei-Kyokai (Invention Association), 17 Shiba, Nishikubo-Akefune Cho, Minato-ku, Tokyo

Information on patent searches, translations and publications: Patent Data Center, Inc., P. O. Box 180, Shiba, Tokyo, 105-91 Japan

Netherlands: The Patent Office (Octrooiraad), Willem Witsenplein 6, The Hague

Norway: Styret for det industrielle rettsvern, P. O. 8160, Oslo-Dep., Oslo 1

Poland: Foreign Trade Enterprise POLSERVICE, Warszawa-Poznanska 15

Romania: Camera de Comert a Republicii Socialiste Romania, Biroul de Brevete si Inventii pentru Strainatate, Bucharest, B-dul N, Balcescu 22

South Africa: The Patent Office, P. O. Box 429, Pretoria

Spain: Photocopies are available from Spanish patent agents who are members of the Colegio Oficial de Agentes de la Propiedad Industrial; requests for names and addresses of Spanish patent agents should be made to Colegio Oficial de Agentes de la Propiedad Industrial, Montera 13, Madrid 14

Sweden: Patentverkets bibliotek, Box 5055, S-102, 42 Stockholm 5

Switzerland: Eidgenössisches Amt für geistiges Eigentum, 3003 Bern

United States: Commissioner of Patents, Washington, D.C. 20231

USSR: The abstracts are obtained from *Otkrytiya, Izobreteniya, Promyshlennye Obraztsy, Tovarnye Znaki,* in which abstracts only are published

Photocopies of most foreign patents are available from the U.S. Patent Office, Washington, D.C. 20231.

COLLEGES AND UNIVERSITIES OFFERING DEGREES IN MICROBIOLOGY [a]

Institution	Location	Department	Head of Department	Degrees Offered
Alabama				
Auburn University	Auburn	Microbiology, School of Agriculture	J. A. Lyle	MS,[b] PhD[b]
Tuskegee Institute	Tuskegee	Microbiology, School of Veterinary Medicine	Bernard B. Watson	MS[b]
University of Alabama	Birmingham	Microbiology, College of Medicine and Dentistry	J. C. Bennett	MS, PhD
	Tuscaloosa (University)	Microbiology, College of Arts and Sciences	Margaret Green	BS, MS, PhD[b]
Arizona				
Arizona State University	Tempe	Botany and Microbiology	James E. Canright	BS, MS, PhD
University of Arizona	Tucson	Microbiology and Medical Technology	Wayburn S. Jeter	BS, MS, PhD
Arkansas				
University of Arkansas	Fayetteville	Botany and Bacteriology	Lowell F. Bailey	BA, BS, MA, MS, PhD
	Little Rock	Microbiology, School of Medicine	Carl E. Duffy	MS, PhD
California				
California State College	Fullerton	Biological Sciences	Donald B. Bright	BA,[b] MA[b]
	Long Beach	Microbiology	Frank E. Swatek	BS, MS
	Los Angeles	Microbiology	Joseph T. Seto	BS, BS, MS
California State Polytechnic College	Pomona	Biological Sciences	Fred Shafia	BS, MS[b]
Chico State College	Chico	Biological Sciences	William L. Stephens	BA
Fresno State College	Fresno	Biology	K. M. Standing	AB, BS, MA
Immaculate Heart College	Los Angeles	Biology	H. E. Wachowski	BA, MA
Loma Linda University	Loma Linda	Microbiology	C. E. Winter	MS, PhD
Loyola University	Los Angeles	Biology	Carl G. Kadner	BS[b]
Mt. St. Mary's College	Los Angeles	Biology	S. Schwanzara	AB,[b] BS
Sacramento State College	Sacramento	Biological Sciences	Marlin L. Bolar	AB,[b] MS[b]
San Diego State College	San Diego	Microbiology	W. L. Baxter	AB, BS, MS

Institution	Location	Department	Head of Department	Degrees Offered
San Francisco State College	San Francisco	Microbiology	W. G. Wu	BA,[b] MA[b]
San Jose State College	San Jose	Biological Sciences	H. R. Patterson	BA, BS, MA
Stanford University	Stanford	Medical Microbiology, School of Medicine	Sidney Raffel	BS, MS, PhD
University of California	Berkeley	Bacteriology and Immunology	Leon Wolfsy	BA, MA, PhD
		Medical Microbiology and Immunology, School of Public Health	J. L. Hardy	BA, MA, MPH, PhD
	Davis	Bacteriology	H. J. Phaff	AB, MA, PhD
	Irvine	Molecular Biology and Biochemistry	Robert C. Warner	BS,[b] MA,[b] PhD[b]
	Los Angeles	Bacteriology	S. C. Rittenberg	AB, MA, PhD
		Medical Microbiology and Immunology	J. H. Fahey	MS, PhD
	Riverside	Biology	F. C. Vasek	BA,[b] MA,[b] PhD[b]
	San Diego (La Jolla)	Marine Biology Curriculum, Scripps Institution of Oceanography	A. A. Benson	MS, PhD
	San Francisco	Microbiology, Medical Center	Ernest Jawetz	PhD
	Santa Barbara	Biological Sciences	Henry A. Harbury	BA,[b] MA,[b] PhD[b]
University of Southern California	Los Angeles	Biological Sciences	Bernard C. Abbott	BS,[b] MS,[c] PhD[c]
		Microbiology, School of Medicine	Irving Gordon	PhD
Colorado				
Colorado State University	Fort Collins	Microbiology	J. E. Ogg	BS, MS, PhD
University of Colorado	Boulder	Biology	H. M. Smith	BA, MA, PhD
	Denver	Microbiology, School of Medicine	L. M. Kozloff	PhD
University of Denver	Denver	Biological Sciences	Ronald R. Cowden	BA, BS, MS
Connecticut				
University of Connecticut	Storrs	Animal Diseases	R. W. Leader	MS, PhD
		Biological Sciences Group, Microbiology	Phillip I. Marcus	BS, MS, PhD
Yale University	New Haven	Microbiology, School of Medicine	E. A. Adelberg	PhD
Delaware				
University of Delaware	Newark	Biological Sciences	W. S. Vincent	BA,[b] MS,[b] PhD[b]
District of Columbia				
American University	Washington, D.C.	Biology	R. R. Anderson	BS,[b] MS[b]

Institution	Location	Department	Head of Department	Degrees Offered
Catholic University of America	Washington, D.C.	Biology	E. R. Kennedy	MS, PhD
Georgetown University	Washington, D.C.	Biology	George B. Chapman	MS,[b] PhD[b]
		Microbiology, School of Medicine and Dentistry	Arthur K. Saz	MS, PhD
George Washington University	Washington, D.C.	Microbiology, School of Medicine	Robert C. Parlett	MS, PhD
Florida				
Florida Atlantic University	Boca Raton	Biological Sciences	H. A. Hoffmann	BS,[b] MS[b]
Florida State University	Tallahassee	Biological Sciences, Division of Bacteriology	Mary Noka Hood	BS, MS, PhD
University of Florida	Gainesville	Microbiology, College of Arts and Sciences	P. H. Smith	BS, MS, PhD
		Microbiology, College of Medicine	P. A. Small, Jr.	MS, PhD
University of Miami	Miami	School of Marine and Atmospheric Science, Division of Fish and Applied Estuary Ecology	J. L. Runnels	MS,[d] PhD[d]
		Microbiology, School of Medicine	Bennett Sallman	MS, PhD
University of West Florida	Pensacola	Biological Sciences	Thomas S. Hopkins	BS, MS
Georgia				
Emory University	Atlanta	Microbiology	Morris Tager	MS, PhD
Medical College of Georgia	Augusta	Microbiology, School of Medicine and Dentistry	Robert B. Dienst	MS, PhD
University of Georgia	Athens	Dairy Science	H. B. Henderson	BS, MS, PhD[e]
		Medical Microbiology, School of Veterinary Medicine	J. B. Gratzek	MS, PhD
		Microbiology	W. J. Payne	BS, MS, PhD
Hawaii				
University of Hawaii	Honolulu	Microbiology	A. A. Benedict	BA, MS, PhD
Idaho				
Idaho State University	Pocatello	Microbiology and Biochemistry	R. W. McCune	BS, MS

Institution	Location	Department	Head of Department	Degrees Offered
University of Idaho	Moscow	Bacteriology	V. A. Cherrington	BS, MS, PhD
Illinois				
Illinois Institute of Technology	Chicago	Biology	T. Hayashi	BS, MS, PhD
Loyola University	Chicago (Maywood)	Microbiology, Stritch School of Medicine	H. J. Blumenthal	MS, PhD
Northwestern University	Chicago	Microbiology, School of Medicine	Guy P. Youmans	PhD
	Evanston	Biological Sciences	John I. Hubbard	MS,[b] PhD[b]
Quincy College	Quincy	Biological Sciences	Frances M. Cardillo O.S.F.	BS[b]
Roosevelt University	Chicago	Biology	Gerald Seaman	BS, MS
Southern Illinois University	Carbondale	Microbiology	Maurice Ogur	BA, BS, MA, PhD
University of Chicago	Chicago	Microbiology	Bernard S. Strauss	MS, PhD
University of Illinois	Chicago	Microbiology, Medical Center	Sheldon Dray	MS, PhD
	Urbana	Microbiology	R. D. DeMoss	BS, MS, PhD
		Veterinary Pathology and Hygiene, College of Veterinary Medicine	L. E. Hanson	MS, PhD
Indiana				
De Pauw University	Greencastle	Botany and Bacteriology	H. R. Youse	AB
Indiana University	Bloomington	Microbiology	Dean Fraser	AB, MA, PhD
	Indianapolis	Microbiology, School of Medicine	E. W. Shrigley	MS, PhD
Purdue University	Lafayette	Animal Sciences	W. R. Woods	MS, PhD
		Biological Sciences	Henry Koffler	BS, MS, PhD
		Veterinary Microbiology, Pathology and Public Health	R. M. Claflin	MS, PhD
University of Notre Dame	Notre Dame	Microbiology	Morris Pollard	MS, PhD
Iowa				
Iowa State University	Ames	Bacteriology	W. R. Lockhart	BS, MS, PhD
		Dairy Microbiology	G. W. Reinbold	MS, PhD
		Food Technology	H. W. Walker	MS, PhD
		Veterinary Microbiology and Preventive Medicine	R. A. Packer	MS, PhD

Institution	Location	Department	Head of Department	Degrees Offered
University of Iowa	Iowa City	Microbiology	J. R. Porter	BA, BS, MS, PhD
Kansas				
Kansas State College	Pittsburg	Biology	R. W. Kelting	BS,[b] MS
Kansas State University	Manhattan	Division of Biology	L. E. Roth	BS, MS, PhD
University of Kansas	Kansas City	Microbiology, School of Medicine	A. A. Werder	MA, PhD
	Lawrence	Microbiology	David Paretsky	BA, MA, PhD
Wichita State University	Wichita	Biology	A. Saracheck	BS, MS
Kentucky				
University of Kentucky	Lexington	Microbiology, School of Biological Sciences	S. F. Conti	BS, MS, PhD
University of Louisville	Louisville	Microbiology	R. D. Higginbotham	MS, PhD
Louisiana				
Louisiana Polytechnic Institute	Ruston	Botany and Bacteriology, School of Agriculture and Forestry	Dallas D. Lutes	BS, MS
Louisiana State University	Baton Rouge	Microbiology	M. D. Socolofsky	BS, MS, PhD
	New Orleans	Microbiology, School of Medicine	Calderon Howe	MS, PhD
		Tropical Medicine, School of Medicine and Medical Parasitology	J. C. Swartzwelder	MS,[f] PhD[f]
Loyola University of the South	New Orleans	Biological Sciences	John H. Mullahy, S. J.	BS,[b] MS[b]
McNeese State College	Lake Charles	Microbiology	Victor Monsour	BS, MS
Northwestern State College	Natchitoches	Microbiology	Paul Donaldson	BS, MS
Southern University Agricultural and Mechanical College	Baton Rouge	Biology and Bacteriology	James B. Bryant, Jr.	BS
Tulane University	New Orleans	Microbiology and Immunology, School of Medicine	M. F. Shaffer	MS, PhD, DS[b]
University of Southwestern Louisiana	Lafayette	Microbiology	William L. Flannery	BS, MS, PhD
Maine				
University of Maine	Orono	Microbiology	Darrell Pratt	BS, MS, PhD[b]

Institution	Location	Department	Head of Department	Degrees Offered
Maryland				
Goucher College	Baltimore	Biological Sciences	Ann M. Lacy	AB[b]
Johns Hopkins University	Baltimore	Biology	Saul Roseman	PhD[b]
		Microbiology, School of Medicine	Daniel Nathan	MA, PhD
University of Maryland	Baltimore	Microbiology, School of Dentistry	Donald E. Shay	MS, PhD
		Microbiology, School of Medicine	Charles L. Wisseman, Jr.	MS, PhD
	College Park	Microbiology	B. G. Young	BS, MS, PhD
Massachusetts				
Boston University	Boston	Biological Sciences	George P. Fulton	MS,[b] PhD[b]
		Microbiology, School of Medicine	E. E. Baker	MA, PhD
Brandeis University	Waltham	Biology	H. T. Epstein	BA, MA, PhD
Harvard University	Boston	Microbiology and Molecular Genetics, School of Medicine	Jonathan Beckwith	PhD
		Microbiology, School of Public Health	Roger L. Nichols	MPH, MS, DPH, DSc
	Cambridge	Biology	John R. Raper	PhD
Massachusetts Institute of Technology	Cambridge	Biology	Boris Magasanik	PhD[b]
		Nutrition and Food Sciences	Nevin S. Scrimshaw	MS, MS,[g] DSc,[g] PhD[g]
Northeastern University	Boston	Biology	F. D. Crisley	MS,[b] PhD[b]
Smith College	Northampton	Biological Sciences	George de Villa Franca	AB, MA, PhD[b]
Southeastern Massachusetts University	North Dartmouth	Biology	John J. Reardon	BS,[b] MS[b]
Tufts University School of Medicine	Boston	Molecular Biology and Microbiology	Moselio Schaechter	PhD[j]
University of Massachusetts	Amherst	Food Science and Technology	F. J. Francis	BS,[g] MS,[g] PhD[g]
		Microbiology	C. D. Cox	BS,[e] MS,[e] PhD[e]
Wellesley College	Wellesley	Biological Sciences	Delaphine G. R. Wyckoff	BA,[b] MA[b]
Michigan				
Central Michigan University	Mt. Pleasant	Biology	LaVerne L. Curry	BA,[b] BS,[b] MS[b]
Hope College	Holland	Biology	Norman J. Norton	BS[b]
Michigan State University	East Lansing	Microbiology and Public Health	Philipp Gerhardt	BS, MS, PhD

Institution	Location	Department	Head of Department	Degrees Offered
University of Michigan	Ann Arbor	Biology	F. H. Test, R. J. Lowry	BS,[b] MS[b]
		Botany	C. B. Beck	BS, MS, DPH,
		Epidemiology, School of Public Health	Fred M. Davenport	MPH, MS, DPH PhD
		Microbiology	Frederick C. Neidhardt	BS, MS, PhD
Wayne State University	Detroit	Biology	D. L. DeGiusti	BS, MS, PhD
		Microbiology, School of Medicine	L. M. Weiner	MS, PhD
Western Michigan University	Kalamazoo	Biology	Clarence J. Goodnight	MS[b]
Minnesota				
University of Minnesota	Minneapolis	Microbiology	Dennis W. Watson	BA, BS, MS, PhD
	St. Paul	Veterinary Microbiology and Public Health	B. S. Pomeroy	MS, PhD
Mississippi				
Mississippi State College for Women	Columbus	Biological Sciences	Harry L. Sherman	BS, MS
Mississippi State University	State College	Microbiology	R. G. Tischer	BS, MS, PhD
University of Mississippi	Jackson	Microbiology, School of Medicine	C. C. Randall	MS, PhD
	University	Biology	Lyman A. Magee	BA,[b] BS,[b] MA,[b] MS,[b] PhD
University of Southern Mississippi	Hattiesburg	Microbiology	P. K. Stocks	BS, MS, PhD
Missouri				
St. Louis University	St. Louis	Microbiology, School of Medicine	Morton M. Weber	MS, PhD
University of Missouri	Columbia	Botany	Billy Cumbie	AB, MA, PhD
		Food Science and Nutrition	Dee M. Graham	MS, PhD
		Microbiology	Frank Engley, Jr.	MS, PhD
		Veterinary Microbiology	Raymond W. Loan	MS, PhD
	Kansas City	Biology	W. C. Bell	BS, MS
Washington University	St. Louis	Microbiology, School of Medicine	H. N. Eisen	PhD
Montana				
Montana State University	Bozeman	Botany and Microbiology	William G. Walter	BS, MS, PhD
University of Montana	Missoula	Microbiology	M. Nakamura	AB, MS, PhD

Institution	Location	Department	Head of Department	Degrees Offered
Nebraska				
Creighton University	Omaha	Medical Microbiology, School of Medicine	F. M. Ferraro	MS, PhD
University of Nebraska	Lincoln	Microbiology, College of Arts and Sciences	T. L. Thompson	BS, MS, PhD
	Omaha	Medical Microbiology, College of Medicine	H. W. McFadden, Jr.	MS, PhD
New Hampshire				
Dartmouth	Hanover	Microbiology, School of Medicine	Clarke T. Gray	PhD
University of New Hampshire	Durham	Microbiology	T. G. Metcalf	BA, MS, PhD
New Jersey				
New Jersey College of Medicine and Dentistry	Newark	Microbiology, New Jersey Medical School	B. A. Briody	MS, PhD
Rutgers, The State University	Camden	Biology, College of Southern Jersey.	Gerard Weissman	BA[b]
	Newark	Zoology and Physiology, Newark College of Arts and Sciences	Daniel C. Wilhoft	BA[b]
	New Brunswick	Bacteriology, College of Arts and Sciences	W. W. Umbreit	BA, BS, MS, PhD
		Bacteriology, Douglass College	H. Christine Reilly	AB
		Biochemistry and Microbiology, College of Agriculture	J. D. Macmillan	MS, PhD
		Institute of Microbiology	J. O. Lampen	MS, PhD
		Microbiology, School of Medicine	R. W. Schlesinger	MS, PhD
New Mexico				
New Mexico Highlands University	Las Vegas	Biology	R. G. Lindeborg	BS,[b] MS[b]
New Mexico State University	Las Cruces	Biology	W. A. Dick-Peddie	BS,[b] MS,[b] PhD[b]
University of New Mexico	Albuquerque	Biology	Loren D. Potter	BS,[b] MS,[b] PhD[b]
		Microbiology, School of Medicine	Leroy C. McLaren	PhD[b]
New York				
Adelphi University	Garden City, L. I.	Biology	A. B. Burdick	MS[b]

Institution	Location	Department	Head of Department	Degrees Offered
Albany Medical College, Union University	Albany	Microbiology	S. V. Covert	MS, PhD
Albert Einstein College of Medicine, Yeshiva University	Bronx	Microbiology and Immunology, Sue Golding Graduate Division	E. J. Hehre	PhD
Brooklyn College	Brooklyn	Biology	D. D. Hurst	BA,[b] BS,[b] MA,[b] PhD[b]
Columbia University	New York	Biological Sciences Microbiology	Cyrus Levinthal Harry M. Rose	BA,[b] MA, PhD[b] MA, PhD
Cornell University	Ithaca	Microbiology Section, Division of Biological Sciences	A. J. Gibson	BS, MS, PhD
Cornell University Graduate School of Medical Sciences	New York	Biology Unit, Sloan Kettering Division	Dorris J. Hutchison	PhD
Cornell University Medical College	New York	Microbiology	W. F. Scherer	MS, PhD
Hamilton College	Clinton	Biology	N. J. Gerold	BA
New York Medical College	New York	Microbiology, Graduate School of Basic Medical Sciences	Sidney Shulman	MS, PhD
New York State Veterinary College, Cornell University	Ithaca	Veterinary Microbiology	D. W. Bruner	MS, PhD
New York University School of Medicine	New York	Microbiology	Milton R. J. Salton	MS, PhD
Rensselaer Polytechnic Institute	Troy	Biology	W. H. Johnson	MS,[b] PhD[b]
Rockefeller University	New York	School of Graduate Studies	Maclyn McCarty	PhD
St. Bonaventure University	St. Bonaventure	Biology	Alfred Finocchio	BS, MS,[b] PhD[b]
St. John's University	Jamaica	Biology	Alfred V. Liberti	MS, PhD
State University College	Geneseo	Biology	Edward Ritter	BA,[b] BS,[b] MA[b]
State University of New York	Buffalo	Microbiology	F. Milgrom	MA, PhD
State University of New York Downstate Medical Center	Brooklyn	Microbiology and Immunology	Stephen L. Morse	PhD

Institution	Location	Department	Head of Department	Degrees Offered
State University of New York Upstate Medical Center	Syracuse	Microbiology	G. G. Holz, Jr.	MS, PhD
Syracuse University	Syracuse	Bacteriology and Botany	D. G. Lundgren	BS, MS, PhD
University of Rochester	Rochester	Biology, College of Arts and Sciences	Thomas T. Bannister	BA,[b] MS,[b] PhD[b]
		Microbiology, School of Medicine and Dentistry	F. E. Young	MS, PhD
Vassar College	Poughkeepsie	Biology	E. Tokay	BA,[b] MA,[b] MS[b]
Wagner College	Staten Island	Bacteriology and Public Health	Edythe Kershaw	BS, MS
North Carolina				
Bowman Gray School of Medicine, Wake Forest University	Winston-Salem	Microbiology	Quentin N. Myrvik	MS, PhD
Duke University	Durham	Microbiology and Immunology	Wolfgang Joklik	MA, PhD
East Carolina University	Greenville	Biology	G. J. Davis	BA,[b] BS,[b] MA[b]
North Carolina State University	Raleigh	Microbiology	James B. Evans	BS,[b] MS, PhD
University of North Carolina	Chapel Hill	Bacteriology and Immunology	G. P. Manire	BS,[b] MS, PhD
	Greensboro	Biology	Bruce Eberhart	AB,[b] MA[b]
North Dakota				
North Dakota State University	Fargo	Bacteriology	K. J. McMahon	BS, MS
University of North Dakota	Grand Forks	Microbiology, School of Medicine	R. G. Fischer	MS, PhD
Ohio				
Case Western Reserve University	Cleveland	Microbiology, School of Medicine	L. O. Krampitz	PhD
Medical College of Ohio	Toledo	Microbiology	Earl H. Freimer	MA, PhD
Miami University	Oxford	Microbiology	C. K. Williamson	AB, MS, PhD[h]
Ohio State University	Columbus	Microbiology, College of Medicine	H. G. Cramblett	MS, PhD

Institution	Location	Department	Head of Department	Degrees Offered
		Microbial and Cellular Biology	P. R. Dugan	BS, MS, PhD
Ohio University	Athens	Zoology and Microbiology	W. J. Peterson	BA, BS, MS, PhD
Ohio Wesleyan	Delaware	Botany and Bacteriology	A. A. Ichida	BA
University of Cincinnati	Cincinnati	Microbiology, College of Medicine	H. C. Lichstein	MS, PhD
University of Dayton	Dayton	Biology	G. B. Noland	BS,[b] MS,[b] PhD
Oklahoma				
Oklahoma State University	Stillwater	Microbiology	Lynn L. Gee	BA, BS, MS, PhD
University of Oklahoma	Norman	Botany and Microbiology	Howard W. Larsh	BS, MS, PhD
	Oklahoma City	Microbiology and Immunology, School of Medicine	L. Vernon Scott	MS, PhD
Oregon				
Oregon State University	Corvallis	Microbiology	P. R. Elliker	BS, MS, PhD
University of Oregon	Portland	Microbiology, School of Medicine	A. W. Frisch	MS, PhD
Pennsylvania				
Bryn Mawr College	Bryn Mawr	Biology	Robert L. Conner	AB,[b] MA,[b] PhD[b]
Bucknell University	Lewisburg	Biology	Jack Harclerode	BS,[b] MS[b]
Carnegie-Mellon University	Pittsburgh	Biological Sciences	Martha P. Eggers	BS,[b] MS
Duquesne University	Pittsburgh	Biological Sciences	Julius S. Greenstein	BS, MS
Hahnemann Medical College	Philadelphia	Microbiology	Amedeo Bondi	MS, PhD
Jefferson Medical College	Philadelphia	Microbiology	R. W. Schaedler	MS, PhD
Lehigh University	Bethlehem	Biology	Saul B. Barber	AB,[b] MS,[b] PhD[b]
Medical College of Pennsylvania	Philadelphia	Microbiology	Kurt Paucker	PhD
Penn State University	Hershey	Microbiology, College of Medicine	Fred Rapp	MS, PhD
	University Park	Microbiology	E. H. Cota-Robles	BS, MS, PhD

Institution	Location	Department	Head of Department	Degrees Offered
Philadelphia College of Pharmaceutical Sciences	Philadelphia	Bacteriology	Bernard Witlin	BS, MS,[i] PhD[i]
Temple University	Philadelphia	Biology	S. T. Takats	AB, AM, PhD
		Microbiology, School of Medicine	E. H. Spaulding	MS, PhD
University of Pennsylvania	Philadelphia	Biology	L. D. Peachey	AB, PhD
		Microbiology, School of Dental Medicine	N. B. Williams	PhD
		Microbiology, School of Medicine	H. S. Ginsberg	PhD
		Microbiology, School of Veterinary Medicine	W. C. Wilcox	PhD
University of Pittsburgh	Pittsburgh	Biophysics and Microbiology Faculty of Arts and Sciences	M. A. Lauffer	BS, MS, PhD
		Epidemiology and Microbiology, Graduate School of Public Health	Monto Ho	MS, DPH, DSc, PhD
		Microbiology, School of Dental Medicine	David Platt	MS, PhD
		Microbiology, School of Medicine	J. S. Youngner	MS, PhD
Rhode Island				
Brown University	Providence	Microbiology and Molecular Biology Section	S. Lederberg	MS,[b] PhD[b]
University of Rhode Island	Kingston	Bacteriology and Biophysics	Norris P. Wood	BS, MS, PhD
South Carolina				
Clemson University	Clemson	Microbiology	M. J. B. Paynter	BS, MS, PhD[b]
Medical College of South Carolina	Charleston	Microbiology	Ben H. Boltjes	MS, PhD
University of South Carolina	Columbia	Biology	B. Theodore Cole	MS,[b] PhD[b]
South Dakota				
South Dakota State University	Brookings	Bacteriology	Robert M. Pengra	BS, MS
University of South Dakota	Vermillion	Microbiology, School of Medicine	Paul F. Smith	MA, PhD

Institution	Location	Department	Head of Department	Degrees Offered
Tennessee				
East Tennessee State University	Johnson City	Health Science, Division of Microbiology	William L. Gaby	BS, MS
Memphis State University	Memphis	Biology	Carl Dee Brown	BS,[b] MS,[b] PhD[b]
University of Tennessee	Knoxville	Microbiology	Arthur Brown	BS, MS, PhD
	Memphis	Microbiology, Medical Units	B. A. Freeman	MS, PhD
Vanderbilt University	Nashville	Microbiology, School of Medicine	A. S. Kaplan	PhD
		Molecular Biology	Oscar Touster	MA,[c] PhD[c]
Texas				
Baylor College of Medicine	Houston	Microbiology	Vernon Knight	MS, PhD
		Virology and Epidemiology	Joseph L. Melnick	PhD
Baylor University	Dallas	Microbiology, Graduate Division, College of Dentistry	Eugene R. Zimmermann	MS, PhD
North Texas State University	Denton	Biology	J. K. G. Silvey	BS,[b] MS,[b] PhD[b]
Texas Agricultural and Mechanical University	College Station	Biology	J. van Oberbeek	BS, MS, PhD
		Veterinary Microbiology	L. C. Grumbles	MS, PhD
Texas Christian University	Fort Worth	Biology	J. Durward Smith	BS, MS
Texas Technical University	Lubbock	Biology	R. C. Jackson	BS, MS
Texas Woman's University	Denton	Biology	K. A. Fry	AB,[b] BS,[b] MS,[b] PhD[b]
University of Houston	Houston	Biology	G. D. Aumann	BA,[b] BS,[b] MS,[b] PhD[b]
University of Texas	Arlington	Biology	William R. Meacham	BS,[b] MS[b]
	Austin	Microbiology	L. Joe Berry	BA, MA, PhD
	Dallas	Division of Biology	Royston C. Clowes	MS,[j] PhD[j]
		Microbiology, School of Medicine	S. F. Sulkin	MA, PhD
	Galveston	Microbiology, Medical Branch	W. F. Verwey	MA, PhD
	Houston	Biology, Graduate School of Biomedical Sciences	Felix L. Haas	MA, PhD
		Pathobiology and Comparative Medicine, School of Public Health	Robert H. Kokernot	MS, PhD

Institution	Location	Department	Head of Department	Degrees Offered
Utah				
Brigham Young University	Provo	Microbiology	Don H. Larsen	BS, MS, PhD
University of Utah	Salt Lake City	Microbiology, College of Medicine	L. P. Gebhardt	BS, MA, MS, PhD
Utah State University	Logan	Bacteriology and Public Health	R. S. Spendlove	BS, MS, PhD
Weber State College	Ogden	Microbiology	Sheldon P. Hayes	BS
Vermont				
University of Vermont	Burlington	Medical Microbiology, College of Medicine	W. R. Stinebring	MS, PhD
		Microbiology and Biochemistry	D. B. Johnstone	MS, PhD
Virginia				
College of William and Mary	Williamsburg	Biology	M. A. Byrd	MA[b]
University of Virginia	Charlottesville	Microbiology	R. R. Wagner	MS, PhD
Virginia Commonwealth University	Richmond	Microbiology; Health Science Division	S. Gaylen Bradley	MS, PhD
Virginia Polytechnic Institute	Blacksburg	Biology	R. A. Paterson	BS,[b] MS, PhD
Virginia State College	Petersburg	Biology	B. F. Woodson	BS, MS
Washington				
Central Washington State College	Ellensburg	Biological Sciences	P. C. Dumas	BA,[b] MS[b]
Seattle Pacific College	Seattle	Biology	Charles Shockey	BS[b]
University of Washington	Seattle	Microbiology	J. C. Sherris	BS, MS, PhD
Washington State University	Pullman	Bacteriology and Public Health	H. M. Nakata	BS, MS, PhD
West Virginia				
West Virginia University	Morgantown	Microbiology, Medical Center	J. M. Slack	MS, PhD
		Plant Pathology, Bacteriology	H. L. Barnett	MS, PhD
Wisconsin				
Medical College of Wisconsin	Milwaukee	Microbiology	Sidney E. Grossberg	MS, PhD

Institution	Location	Head of Department	Department	Degrees Offered
University of Wisconsin	Madison	Bacteriology Medical Microbiology, School of Medicine	J. B. Wilson D. L. Walker	BA, BS, MS, PhD BS, MS, PhD
Wisconsin State University	Oshkosh	Biology	M. A. Rouf	BS, MS
Wyoming				
University of Wyoming	Laramie	Division of Microbiology and Veterinary Medicine	J. O. Tucker	BS, MS

[a] Compiled by the American Society for Microbiology, Washington, D. C., from responses to mail questionnaires.

[b] Degree in Biological Sciences, Biology, Medical Sciences or Botany, with emphasis in Microbiology.

[c] Degree in Cellular and Molecular Biology, with emphasis in Microbiology.

[d] Degree in Fishery Sciences, with emphasis in Microbiology.

[e] Degree in Dairy Science—Food Science, with emphasis in Microbiology.

[f] Degree in Medical Parasitology.

[g] Degree in Food Science or Biochemical Engineering, with emphasis in Microbiology.

[h] Degree in cooperation with the Ohio State University.

[i] Degree in Pharmaceutical Science with Microbiology major.

[j] Degree in Molecular Biology, with emphasis in Microbiology.

Data from *ASM News, 38,* 196–201 (1972). Reproduced by permission of the American Society for Microbiology, Washington, D. C.

FOREIGN ALPHABETS

Greek

Letter		Name	Transliteration
Α	α	alpha	a
Β	β	beta	b
Γ	γ	gamma	g
Δ	δ	delta	d
Ε	ϵ	epsilon	e
Ζ	ζ	zeta	z
Η	η	eta	e (or ē)
Θ	θ	theta	th
Ι	ι	iota	i
Κ	κ	kappa	k
Λ	λ	lambda	l
Μ	μ	mu	m
Ν	ν	nu	n
Ξ	ξ	xi	x
Ο	ο	omicron	o
Π	π	pi	p
Ρ	ρ	rho	r
Σ	σ, ς[a]	sigma	s
Τ	τ	tau	t
Υ	υ	upsilon	y
Φ	φ	phi	ph
Χ	χ	chi	ch, kh
Ψ	ψ	psi	ps
Ω	ω	omega	o (or ō)

Russian

Letter		Transliteration
А	а	a
Б	б	b
В	в	v
Г	г	g
Д	д	d
Е	е	e, ye
Ж	ж	zh
З	з	z
И	и	i
І[b]	і[b]	
Й	й	ĭ, i
К	к	k
Л	л	l
М	м	m
Н	н	n
О	о	o
П	п	p
Р	р	r
С	с	s
Т	т	t
У	у	u
Ф	ф	f
Х	х	kh, x
Ц	ц	ts, c
Ч	ч	ch, č
Ш	ш	sh, š
Щ	щ	shch, šč
Ъ[c]	ъ[c]	
Ы	ы	i
Ь	ь	'
Ѣ[d]	ѣ[d]	
Э	э	e
Ю	ю	yu, ju
Я	я	ya, ja
Ѳ[e]	ѳ[e]	
Ѵ[b]	ѵ[b]	

German

Letter		Transliteration
𝔄	𝔞	a
𝔄̈	𝔞̈	ae
𝔅	𝔟	b
ℭ	𝔠	c
𝔇	𝔡	d
𝔈	𝔢	e
𝔉	𝔣	f
𝔊	𝔤	g
ℌ	𝔥	h
ℑ	𝔦	i
ℑ	𝔧	j
𝔎	𝔨	k
𝔏	𝔩	l
𝔐	𝔪	m
𝔑	𝔫	n
𝔒	𝔬	o
𝔒̈	𝔬̈	oe
𝔓	𝔭	p
𝔔	𝔮	q
ℜ	𝔯	r
𝔖	ſ 𝔰[a]	s
𝔗	𝔱	t
𝔘	𝔲	u
𝔘̈	𝔲̈	ue
𝔙	𝔳	v
𝔚	𝔴	w
𝔛	𝔵	x
𝔜	𝔶	y
ℨ	𝔷	z

[a] At end of word.
[b] Abolished in 1918 in favor of И.
[c] In middle of words as sign of division; often replaced by apostrophe (mute).
[d] Abolished in 1918 in favor of е.
[e] Abolished in 1918 in favor of ф.

AMERICAN STANDARD ABBREVIATIONS

Abbreviations are shortened forms of common and scientific expressions, units of weight and measurement, or names of chemical compounds, which may be occasionally used in tabulation or text. In tabular matter they should only be used where space must be conserved; in text they should be avoided as far as possible. Short words should be spelled out.

BASIC RULES

1. Abbreviations should not be used where their meaning may be misunderstood.

2. The same abbreviations are used for singular and plural forms.

3. Periods should be omitted unless the abbreviation can be confused with a common English word.

4. Abbreviations for units of weight and measurements should be used only after numerical values.

5. Abbreviations in the form of exponents ("2" for square and "3" for cube) should never be used in text, because they are easily mistaken for footnote references. For U.S. customary units, the abbreviations "sq" and "cu" are usually preferred.

6. The letters of such abbreviations as ACTH should not be spaced (not A C T H).

ABBREVIATIONS

$[\alpha]_D^{25}$	specific optical rotation at 25°C for D (Na) line
A	adenine
Å	Angstrom unit
a	acid are
abs	absolute
abt	about
Ac	acetyl
ac	alternating current (as noun)
a-c	alternating-current (as adjective)
acet	acetone
Acetyl-CoA	acetyl coenzyme A
ACTH	adrenocorticotropic hormone
ADP	adenosine 5′-diphosphate (pyro)
Ala	alanine
alc	alcohol
alk	alkali
alt	altitude
am.	amplitude
amal	amalgam amalgamated
amor	amorphous
amorph	amorphous
AMP	adenosine 5′-phosphate
amp	ampere

anhydr	anhydrous
antilog	antilogarithm
ap	apothecaries'
approx	approximately
aq	aqua aqueous water
aq reg	aqua regia
Arg	arginine
Asp-NH_2	aspartic acid
asym	asymmetrical
atm, atmos	atmosphere atmospheric
at. no.	atomic number
ATP	adenosine 5′-triphosphate (pyro)
at. wt.	atomic weight
aux	auxiliary
av	average avoirdupois
avdp	avoirdupois
avg	average
avoir	avoirdupois
az	azimuth
bar.	barometer
Bé	Baumé
BG	Birmingham gauge
bl	blue
blk	black
B.O.D.	biochemical oxygen demand
bp	boiling point
br	brown
Btu	British thermal unit
Bu	butyl
BWG	Birmingham wire gauge
bz	benzene
C	centigrade cytosine
c	candle carat cent centi- (10^{-2}) cold concentration cycles per second
ca	candle centare
ca	*circa* (about or approximately)
cal	calorie
cc	cubic centimeter
CDP	cytidine 5′-diphosphate (pyro)
cemf	counter-electromotive force
cent.	centigrade
cf	*confer* (compare)
cfm	cubic feet per minute
cfs	cubic feet per second
cg	centigram
cgs	centimeter-gram-second system of units
cgse	cgs electrostatic system of units

cgsm	cgs electromagnetic system of units
chem	chemical
chl	chloroform
c-hr	candle-hour
cif	cost, insurance, and freight
cir	circular
circum	circumference
cir mil	circular mil
cl	centiliter
cm	centimeter
cm^2	square centimeter
cm^3	cubic centimeter
c m	circular mil
CM-cellulose	O-(carboxymethyl)-cellulose
CMP	cytidine 5′-phosphate
cn	cosine of the amplitude
CoA	coenzyme A
CoASAc	acetyl coenzyme A
CoASH	coenzyme A
coef	coefficient
Col	colicinogenic factor
colog	cologarithm
colorl	colorless
comm'l	commercial
conc	concentrate concentrated
cond	condensing conductivity
const	constant
cor(r)	corrected
cos	cosine
$\cos^{-1}$	anticosine arc or angle whose cosine is . . . inverse cosine
cosec	cosecant
cosh	hyperbolic cosine
$\cosh^{-1}$	inverse hyperbolic cosine
cot	cotangent
$\cot^{-1}$	arc or angle whose cotangent is . . .
coth	hyperbolic cotangent
$\coth^{-1}$	inverse hyperbolic cotangent
covers	coversed sine
CP	chemically pure
cp	candle power center of pressure centipoise circular pitch
cry.	crystal crystalline
cryst	crystal crystalline
csc	cosecant
$\csc^{-1}$	arc or angle whose cosecant is . . .
csch	hyperbolic cosecant
csch^{-1}	inverse hyperbolic cosecant

c to c	center to center
CTP	cytidine 5′-triphosphate (pyro)
Ctu	centigrade thermal unit
cu	cubic
cu cm	cubic centimeter
cwt	hundredweight
cyl	cylinder
Cys-SH	cysteine
Cys-SO_3H	cysteic acid
d	day deci- (10^{-1}) decomposes density derivative
d	dextrorotary
db	decibel
dc	direct current (as noun)
d-c	direct-current (as adjective)
DDT	1,1,1-trichloro-2,2-bis(*p*-chlorphenyl)ethane
DEAE-cellulose	O-(diethylaminoethyl)cellulose
dec	decomposes
def	definition
deg	degree
deliq	deliquescent
den.	density
dens	density
deRib	deoxyribose
DFP	diisopropyl phosphorofluoridate
dg	decigram
diam	diameter
dil	dilute
dissd	dissolved
dk	dark deka- (10)
dkg	dekagram
dkl	dekaliter
dkm	dekameter
dl	deciliter
dm	decimeter
DMSO	dimethylsulfoxide
dn	*delta* amplitude
DNA	deoxyribonucleic acid
DNP	2,4-dinitrophenyl
DOPA	3,4-dihydroxyphenylalanine
doz	dozen
dp	diametral pitch degree of polymerization double pole
DPN	diphosphopyridine nucleotide (see NAD)
DPNH, $DPNH_2$	reduced diphosphopyridine nucleotide (see NADPH, $NADPH_2$)
DPT	diphosphothiamine (thiamine-pyrophosphate, cocarboxylase)
dr	dram
dwt	pennyweight

$E^{1\%}_{1\,cm}$	absorbancy of 1% (1 g/100 ml) solution in a cell that has an absorption path of 1 cm
e or e^{θ}	electron
EC	enzyme commission number
EDTA	ethylenediaminetetraacetate
eff	efficiency
efflor	efflorescent
e.g.	*exempli gratia* (for example)
ehp	efficient horsepower
EL	elastic limit
el	elevation
elec	electric
em	electromagnetic unit of a quantity of electricity
emf	electromotive force
eq	equation
equiv	equivalent
es	electrostatic unit of a quantity of electricity
Et	ethyl
et al.	*et alii* (and others)
etc.	*et cetera* (and so forth)
eth	ether
et seq.	*et sequentes* (and the following)
evap	evaporation
ex	excess
exp	explodes exponential function
exsec	exterior secant
ext	external
F	Fahrenheit
f	farad from
FAD	flavin-adenine dinucleotide
$FADH_2$	reduced flavin-adenine dinucleotide
fahr	Fahrenheit
fath	fathom
FDNB	1-fluoro-2,4-dinitrobenzene
feath	feathery
ff	following
fl	fluid
fl dr	fluid dram
fl oz	fluid ounce
fluores	fluorescent
FMN	riboflavin 5′-phosphate
fnp	fusion point
fob	free on board
fp	freezing point
fpm	feet per minute
fps	feet per second foot-pound-second system of units
fpse	fps electrostatic system of units

fpsm	fps electromagnetic system of units
Fru	fructose
FS	factor of safety
ft	foot
ft-c	foot-candle
ft-L	foot-lambert
fur.	furlong
G	gravitation constant guanine
g	gram
gal	gallon
Gal	galactose
GC	gas chromatography
gcd	greatest common divisor
GDP	guanosine 5′-diphosphate (pyro)
gel	gelatinous
gi	gill
glac	glacial
Glc	glucose
GLC	gas–liquid chromatography
glit	glittering
Glu	glutamic acid
Glu-NH_2	glutamine
Gly	glycine
glyc	glycerine
gm	gram
GMP	guanosine 5′-phosphate
gpm	gallons per minute
gps	gallons per second
gr	grain gray
grn	green
GSH	glutathione
GSSG	oxidized glutathione
GTP	guanosine 5′-triphosphate (pyro)
gyr	gyration
h	hecto- (10^2) henry hot hour
ha	hectare
Hb	hemoglobin
HbCO	carboxyhemoglobin
HbO_2	oxyhemoglobin
hex.	hexagonal
hg	hectogram
His	histidine
hl	hectoliter
hm	hectometer
HMDS	hexamethyldisilazane
HOAc	acetic acid
hor	horizontal

horiz	horizontal
HP or hp	horsepower
h-p	high-pressure (as adjective)
hr	hour
hyg	hygroscopic
i	insoluble
ibid.	*ibidem* (in the same place)
ID	inside diameter
IDP	inosine 5′-diphosphate (pyro)
i.e.	*id est* (that is)
ign	ignites
ihp	indicated horsepower
Ile	isoleucine
im	intramuscular
IMP	inosine 5′-phosphate
in.	inch indigo
inc	inclusive
ins	insoluble
insol	insoluble
Int	international
int	internal
IP	isoelectric point
Ip	intraperitoneal
i-p	intermediate-pressure (as adjective)
ips	inches per second
IR	infrared
iso	isotropic
isom	isometric
isoth	isothermal
ITP	inosine 5′-triphosphate (pyro)
IV	intravenous
j	joule
K	ionization constant Kelvin
k	kilo- (10^3)
kc	kilocycles per second
kcal	kilocalorie
KDO	2-keto-3-deoxyoctonic acid
kg	kilogram
kg-cal	kilogram-calorie
kg-m	kilogram-meter
kg/m^3	kilograms per cubic meter
kgps	kilograms per second
kip	thousand pounds
kl	kiloliter
km	kilometer
kmps	kilometers per second
kv	kilovolt
kva	kilovolt-ampere
kw	kilowatt

L	lambert
l	liter long
l	levorotary
lat	latitude
lb	pound
lcm	least common multiple
LD	lethal dose
LD_{50}	dose lethal to 50% of animals tested
leaf.	leaflets
Leu	leucine
lgr	ligroin
lim	limit
lin	linear
liq	liquid
lm	lumen
ln	natural logarithm
lng	long
log	common logarithm
$\log_{10}$	common logarithm to base 10
$\log_e$	logarithm to base *e* (natural logarithm)
long.	longitude
l-p	low-pressure (as adjective)
lpw	lumens per watt
lt	light
lust.	lustrous
Lys	lysine
M	thousand molar
m	meter milli- (10^{-3}) minute
m-	*meta-*
ma	milliampere
Man	mannose
math	mathematical mathematics
max	maximum
Mb	myoglobin
MbCO	carboxymyoglobin
MbO_2	oxymyoglobin
mcg, μg	microgram
M_D	molecular rotation $\frac{[\alpha]_D \times \text{mol wt}}{100}$
Me	methyl
m/e	mass-to-charge ratio of molecular ion
med	medium
mep	mean effective horsepower
mEq	milliequivalent
Met	methionine
met.	metallic
meth	methyl
meth al	methyl alcohol

MetHb	methemoglobin
MetMb	metmyoglobin
mf	microfarad
mg	milligram
mgd	million gallons per day
mh	millihenry
mhcp	mean horizontal candle power
mi	mile
mic	microscopic
min	mineral minim (drop) minimum minute
mL	millilambert
ml	milliliter
MLD	minimum lethal dose
mm	millimeter
mmf	magnetomotive force
mμ (mmu)	millimicro- (10^{-9}) millimicron
mol	molecule
mol wt	molecular weight
monocl	monoclinic
mp	melting point
mph	miles per hour
mphps	miles per hour per second
MS	mass spectroscopy
mscp	mean spherical candle power
MSH	melanocyte-stimulating hormone
μ (mu)	micro- (10^{-6}) micron
μm	micrometer micron
$\mu\mu$ (mu mu)	micromicro- (10^{-12}) micromicron
$\mu\mu$f (mu mu f)	micromicrofarad
mv	millivolt
N	normal (equivalents per liter) number (in mathematical tables) numeric
n	normal
n	refractive index
NAD	nicotinamide-adenine dinucleotide (cozymase, coenzyme I, diphosphopyridine nucleotide)
NADH, $NADH_2$	reduced nicotinamide-adenine dinucleotide
NADP	nicotinamide-adenine dinucleotide phosphate (coenzyme II, triphospho-pyridine nucleotide)
NADPH, $NADPH_2$	reduced nicotinamide-adenine dinucleotide phosphate
need.	needles
nm	nanometer
NMN	nicotinamide mononucleotide
NMR	nuclear magnetic resonance

O-	attachment to oxygen
o-	*ortho-*
obs	observed
octahdr	octahedral
OD	optical density outside diameter
ohm-cm	ohm-centimeter
or.	orange
ORD	optical rotary dispersion
oz	ounce
oz ap	ounce, apothecaries'
oz av	ounce, avoirdupois
oz avdp	ounce, avoirdupois
oz fl	ounce, fluid
oz-ft	ounce-foot
oz-in.	ounce inch
oz t	ounce, troy
p-	*para-*
pa	pale
PABA	*para*-aminobenzoic acid
PAL	pyridoxal phosphate
PAMP	pyridoxamine phosphate
PCMB	*para*-chloromercuribenzoate
pct	percent
PEG	polyethylene glycol
PEP	phosphoenol pyruvate
perp	perpendicular
pf	power factor
pH	log of reciprocal of H_2 ion concentration
Phe	phenylalanine
P_i	inorganic orthophosphate
pK	log of 1/K
pl	plates
PNA	pentose nucleic acid
powd	powder
PP_i	inorganic pyrophosphate
PPLO	pleuropneumonia-like organism
ppm	parts per million
p'p't'd	precipitated
pr	prisms
precip	precipitated
Pro	proline
psf	pounds per square foot
psi	pounds per square inch
psia	pounds per square inch, absolute
p sol	partly soluble
pt	pint point
purp	purple
pyr	pyridine
Q	quantity

q	quintal (hundredweight)
qt	quart
q.v.	*quod vide* (which see)
R	radioactive mineral Rankine
rac	racemic
rad	radius
reg	regular
rev	revolution
R_f	movement of solute movement of solvent front
rhbdr	rhombohedral
rhomb	orthorhombic rhombic
Rib	ribose
rms	root of mean square
RNA	ribonucleic acid
rpm	revolutions per minute
rps	revolutions per second
RT	retention time
Rul	ribulose
S	Svedberg unit sedimentation coefficient
$S°_{20,w}$	sedimentation coefficient at 20°C in water corrected to infinite dilution of solute
S	attachment to sulfur
s	scruple second soluble stere
s ap	scruple, apothecaries'
sat or sat'd	saturated
sc	scales subcutaneous
scp	spherical candle power
sec	secant second secondary
sec^{-1}	arc or angle whose secant is . . .
sech	hyperbolic secant
segm	segment
Ser	serine
sh	short
sin	sine
sin^{-1}	arc or angle whose sine is . . .
sinh	hyperbolic sine
sl	slightly
sm	small
sn	sine of the amplitude
sol	soluble solution
soln	solution
sp.	species specific
specif	specification

sp gr	specific gravity
sp heat	specific heat
spp.	species (plural)
sq	square
std	standard
STP	standard temperature and pressure
subl	sublimes
sym	symmetrical
T	thymine
t	ton, metric troy
tab. or tabl	tablets
tan	tangent
tan^{-1}	arc or angle whose tangent is . . .
tanh	hyperbolic tangent
TB	tuberculosis
TCA	tricarboxylic acid trichloroacetic acid
TEAE-cellulose	*O*-(triethylaminoethyl)-cellulose
tech	technical
temp	temperature
tetr	tetragonal
tetrag	tetragonal
Thr	threonine
Thx	thyroxine
TLC	thin-layer chromatography
TMCS	trimethylchlorosilane
TMS	trimethylsilyl
tn	ton
TPN	triphosphopyridine nucleotide (see NADP)
TPNH, $TPNH_2$	reduced triphosphopyridine nucleotide
tr	transition
tricl	triclinic
trig.	trigonal
trim.	trimetric
Tris	tris(hydroxymethyl)amino-methane (2-amino-2-hydroxy-methylpropane-1,3-diol)
Trp	tryptophan
TS or ts	tensile strength
TTC	triphenyl tetrazoleum chloride
turp	turpentine
Tyr	tyrosine
U	uracil
UDP	uridine diphosphate (pyro)
UDPG	uridine diphosphate glucose
ult	ultimate
UMP	uridine monophosphate
uncor(r)	uncorrected
uns	unsymmetrical
UTP	uridine triphosphate (pyro)
UV	ultraviolet

v	very volt volume
v	*vide* (see)
va	volt-ampere
Val	valine
var	reactive volt-ampere
vel or veloc	velocity
vers	versed sine
vert	vertical
visc	viscous
vol	volume
volat	volatilizes
v/v	% volume in volume (number of ml of a constituent in 100 ml of solution)
w	water watt weight
wh	white
whr	watt-hour
wk(s)	week(s)
wpc	watts per candle
wt	weight
w/v	% weight in volume (number of g of a constituent in 100 ml of solution)
w/w	% weight in weight (number of g of a constituent in 100 g of solution)
Xul	xylulose
Xyl	xylose
yd	yard
yel	yellow
yr	year

TEMPERATURE CONVERSION

To convert degrees Celsius to degrees Fahrenheit or degrees Fahrenheit to degrees Celsius, locate the temperature to be converted in one of the bold-face columns. If the number represents degrees Fahrenheit, its equivalent in degrees Celsius is in the column at the left. If the number represents degrees Celsius, its equivalent in degrees Fahrenheit is in the column at the right.

To Convert			To Convert			To Convert		
To °C	←°F or °C→	To °F	To °C	←°F or °C→	To °F	To °C	←°F or °C→	To °F
−273.15	**−459.67**	—	−245.56	**−410**	—	−217.78	**−360**	—
−272.78	**−459**	—	−245	**−409**	—	−217.22	**−359**	—
−272.22	**−458**	—	−244.44	**−408**	—	−216.67	**−358**	—
−271.67	**−457**	—	−243.89	**−407**	—	−216.11	**−357**	—
−271.11	**−456**	—	−243.33	**−406**	—	−215.56	**−356**	—
−270.56	**−455**	—	−242.78	**−405**	—	−215	**−355**	—
−270	**−454**	—	−242.22	**−404**	—	−214.44	**−354**	—
−269.44	**−453**	—	−241.67	**−403**	—	−213.89	**−353**	—
−268.89	**−452**	—	−241.11	**−402**	—	−213.33	**−352**	—
−268.33	**−451**	—	−240.56	**−401**	—	−212.78	**−351**	—
−267.78	**−450**	—	−240	**−400**	—	−212.22	**−350**	—
−267.22	**−449**	—	−239.44	**−399**	—	−211.67	**−349**	—
−266.67	**−448**	—	−238.89	**−398**	—	−211.11	**−348**	—
−266.11	**−447**	—	−238.33	**−397**	—	−210.56	**−347**	—
−265.56	**−446**	—	−237.78	**−396**	—	−210	**−346**	—
−265	**−445**	—	−237.22	**−395**	—	−209.44	**−345**	—
−264.44	**−444**	—	−236.67	**−394**	—	−208.89	**−344**	—
−263.89	**−443**	—	−236.11	**−393**	—	−208.33	**−343**	—
−263.33	**−442**	—	−235.56	**−392**	—	−207.78	**−342**	—
−262.78	**−441**	—	−235	**−391**	—	−207.22	**−341**	—
−262.22	**−440**	—	−234.44	**−390**	—	−206.67	**−340**	—
−261.67	**−439**	—	−233.89	**−389**	—	−206.11	**−339**	—
−261.11	**−438**	—	−233.33	**−388**	—	−205.56	**−338**	—
−260.56	**−437**	—	−232.78	**−387**	—	−205	**−337**	—
−260	**−436**	—	−232.22	**−386**	—	−204.44	**−336**	—
−259.44	**−435**	—	−231.67	**−385**	—	−203.89	**−335**	—
−258.89	**−434**	—	−231.11	**−384**	—	−203.33	**−334**	—
−258.33	**−433**	—	−230.56	**−383**	—	−202.78	**−333**	—
−257.78	**−432**	—	−230	**−382**	—	−202.22	**−332**	—
−257.22	**−431**	—	−229.44	**−381**	—	−201.67	**−331**	—
−256.67	**−430**	—	−228.89	**−380**	—	−201.11	**−330**	—
−256.11	**−429**	—	−228.33	**−379**	—	−200.56	**−329**	—
−255.56	**−428**	—	−227.78	**−378**	—	−200	**−328**	—
−255	**−427**	—	−227.22	**−377**	—	−199.44	**−327**	—
−254.44	**−426**	—	−226.67	**−376**	—	−198.89	**−326**	—
−253.89	**−425**	—	−226.11	**−375**	—	−198.33	**−325**	—
−253.33	**−424**	—	−225.56	**−374**	—	−197.78	**−324**	—
−252.78	**−423**	—	−225	**−373**	—	−197.22	**−323**	—
−252.22	**−422**	—	−224.44	**−372**	—	−196.67	**−322**	—
−251.67	**−421**	—	−223.89	**−371**	—	−196.11	**−321**	—
−251.11	**−420**	—	−223.33	**−370**	—	−195.56	**−320**	—
−250.56	**−419**	—	−222.78	**−369**	—	−195	**−319**	—
−250	**−418**	—	−222.22	**−368**	—	−194.44	**−318**	—
−249.44	**−417**	—	−221.67	**−367**	—	−193.89	**−317**	—
−248.89	**−416**	—	−221.11	**−366**	—	−193.33	**−316**	—
−248.33	**−415**	—	−220.56	**−365**	—	−192.78	**−315**	—
−247.78	**−414**	—	−220	**−364**	—	−192.22	**−314**	—
−247.22	**−413**	—	−219.44	**−363**	—	−191.67	**−313**	—
−246.67	**−412**	—	−218.89	**−362**	—	−191.11	**−312**	—
−246.11	**−411**	—	−218.33	**−361**	—	−190.56	**−311**	—

To Convert		
To °C	←°F or °C→	To °F
−190	**−310**	—
−189.44	**−309**	—
−188.89	**−308**	—
−188.33	**−307**	—
−187.78	**−306**	—
−187.22	**−305**	—
−186.67	**−304**	—
−186.11	**−303**	—
−185.56	**−302**	—
−185	**−301**	—
−184.44	**−300**	—
−183.89	**−299**	—
−183.33	**−298**	—
−182.78	**−297**	—
−182.22	**−296**	—
−181.67	**−295**	—
−181.11	**−294**	—
−180.56	**−293**	—
−180	**−292**	—
−179.44	**−291**	—
−178.89	**−290**	—
−178.33	**−289**	—
−177.78	**−288**	—
−177.22	**−287**	—
−176.67	**−286**	—
−176.11	**−285**	—
−175.56	**−284**	—
−175	**−283**	—
−174.44	**−282**	—
−173.89	**−281**	—
−173.33	**−280**	—
−172.78	**−279**	—
−172.22	**−278**	—
−171.67	**−277**	—
−171.11	**−276**	—
−170.56	**−275**	—
−170	**−274**	—
—	**−273.15**	−459.67
−169.44	**−273**	−459.4
−168.89	**−272**	−457.6
−168.33	**−271**	−455.8
−167.78	**−270**	−454
−167.22	**−269**	−452.2
−166.67	**−268**	−450.4
−166.11	**−267**	−448.6
−165.56	**−266**	−446.8
−165	**−265**	−445
−164.44	**−264**	−443.2
−163.89	**−263**	−441.4
−163.33	**−262**	−439.6
−162.78	**−261**	−437.8
−162.22	**−260**	−436
−161.67	**−259**	−434.2
−161.11	**−258**	−432.4
−160.56	**−257**	−430.6
−160	**−256**	−428.8
−159.44	**−255**	−427
−158.89	**−254**	−425.2
−158.33	**−253**	−423.4
−157.78	**−252**	−421.6
−157.22	**−251**	−419.8

To Convert		
To °C	←°F or °C→	To °F
−156.67	**−250**	−418
−156.11	**−249**	−416.2
−155.56	**−248**	−414.4
−155	**−247**	−412.6
−154.44	**−246**	−410.8
−153.89	**−245**	−409
−153.33	**−244**	−407.2
−152.78	**−243**	−405.4
−152.22	**−242**	−403.6
−151.67	**−241**	−401.8
−151.11	**−240**	−400
−150.56	**−239**	−398.2
−150	**−238**	−396.4
−149.44	**−237**	−394.6
−148.89	**−236**	−392.8
−148.33	**−235**	−391
−147.78	**−234**	−389.2
−147.22	**−233**	−387.4
−146.67	**−232**	−385.6
−146.11	**−231**	−383.8
−145.56	**−230**	−382
−145	**−229**	−380.2
−144.44	**−228**	−378.4
−143.89	**−227**	−376.6
−143.33	**−226**	−374.8
−142.78	**−225**	−373
−142.22	**−224**	−371.2
−141.67	**−223**	−369.4
−141.11	**−222**	−367.6
−140.56	**−221**	−365.8
−140	**−220**	−364
−139.44	**−219**	−362.2
−138.89	**−218**	−360.4
−138.33	**−217**	−358.6
−137.78	**−216**	−356.8
−137.22	**−215**	−355
−136.67	**−214**	−353.2
−136.11	**−213**	−351.4
−135.56	**−212**	−349.6
−135	**−211**	−347.8
−134.44	**−210**	−346
−133.89	**−209**	−344.2
−133.33	**−208**	−342.4
−132.78	**−207**	−340.6
−132.22	**−206**	−338.8
−131.67	**−205**	−337
−131.11	**−204**	−335.2
−130.56	**−203**	−333.4
−130	**−202**	−331.6
−129.44	**−201**	−329.8
−128.89	**−200**	−328
−128.33	**−199**	−326.2
−127.78	**−198**	−324.4
−127.22	**−197**	−322.6
−126.67	**−196**	−320.8
−126.11	**−195**	−319
−125.56	**−194**	−317.2
−125	**−193**	−315.4
−124.44	**−192**	−313.6
−123.89	**−191**	−311.8

To Convert		
To °C	←°F or °C→	To °F
−123.33	**−190**	−310
−122.78	**−189**	−308.2
−122.22	**−188**	−306.4
−121.67	**−187**	−304.6
−121.11	**−186**	−302.8
−120.56	**−185**	−301
−120	**−184**	−299.2
−119.44	**−183**	−297.4
−118.89	**−182**	−295.6
−118.33	**−181**	−293.8
−117.78	**−180**	−292
−117.22	**−179**	−290.2
−116.67	**−178**	−288.4
−116.11	**−177**	−286.6
−115.56	**−176**	−284.8
−115	**−175**	−283
−114.44	**−174**	−281.2
−113.89	**−173**	−279.4
−113.33	**−172**	−277.6
−112.78	**−171**	−275.8
−112.22	**−170**	−274
−111.67	**−169**	−272.2
−111.11	**−168**	−270.4
−110.56	**−167**	−268.6
−110	**−166**	−266.8
−109.44	**−165**	−265
−108.89	**−164**	−263.2
−108.33	**−163**	−261.4
−107.78	**−162**	−259.6
−107.22	**−161**	−257.8
−106.67	**−160**	−256
−106.11	**−159**	−254.2
−105.56	**−158**	−252.4
−105	**−157**	−250.6
−104.44	**−156**	−248.8
−103.89	**−155**	−247
−103.33	**−154**	−245.2
−102.78	**−153**	−243.4
−102.22	**−152**	−241.6
−101.67	**−151**	−239.8
−101.11	**−150**	−238
−100.56	**−149**	−236.2
−100	**−148**	−234.4
−99.44	**−147**	−232.6
−98.89	**−146**	−230.8
−98.33	**−145**	−229
−97.78	**−144**	−227.2
−97.22	**−143**	−225.4
−96.67	**−142**	−223.6
−96.11	**−141**	−221.8
−95.56	**−140**	−220
−95	**−139**	−218.2
−94.44	**−138**	−216.4
−93.89	**−137**	−214.6
−93.33	**−136**	−212.8
−92.78	**−135**	−211
−92.22	**−134**	−209.2
−91.67	**−133**	−207.4
−91.11	**−132**	−205.6
−90.56	**−131**	−203.8

To Convert			To Convert			To Convert		
To °C	←°F or °C→	To °F	To °C	←°F or °C→	To °F	To °C	←°F or °C→	To °F
−90	**−130**	−202	−56.67	**−70**	−94	−23.33	**−10**	14
−89.44	**−129**	−200.2	−56.11	**−69**	−92.2	−22.78	**−9**	15.8
−88.89	**−128**	−198.4	−55.56	**−68**	−90.4	−22.22	**−8**	17.6
−88.33	**−127**	−196.6	−55	**−67**	−88.6	−21.67	**−7**	19.4
−87.78	**−126**	−194.8	−54.44	**−66**	−86.8	−21.11	**−6**	21.2
−87.22	**−125**	−193	−53.89	**−65**	−85	−20.56	**−5**	23
−86.67	**−124**	−191.2	−53.33	**−64**	−83.2	−20	**−4**	24.8
−86.11	**−123**	−189.4	−52.78	**−63**	−81.4	−19.44	**−3**	26.6
−85.56	**−122**	−187.6	−52.22	**−62**	−79.6	−18.89	**−2**	28.4
−85	**−121**	−185.8	−51.67	**−61**	−77.8	−18.33	**−1**	30.2
−84.44	**−120**	−184	−51.11	**−60**	−76	−17.78	**0**	32
−83.89	**−119**	−182.2	−50.56	**−59**	−74.2	−17.22	**1**	33.8
−83.33	**−118**	−180.4	−50	**−58**	−72.4	−16.67	**2**	35.6
−82.78	**−117**	−178.6	−49.44	**−57**	−70.6	−16.11	**3**	37.4
−82.22	**−116**	−176.8	−48.89	**−56**	−68.8	−15.56	**4**	39.2
−81.67	**−115**	−175	−48.33	**−55**	−67	−15	**5**	41
−81.11	**−114**	−173.2	−47.78	**−54**	−65.2	−14.44	**6**	42.8
−80.56	**−113**	−171.4	−47.22	**−53**	−63.4	−13.89	**7**	44.6
−80	**−112**	−169.6	−46.67	**−52**	−61.6	−13.33	**8**	46.4
−79.44	**−111**	−167.8	−46.11	**−51**	−59.8	−12.78	**9**	48.2
−78.89	**−110**	−166	−45.56	**−50**	−58	−12.22	**10**	50
−78.33	**−109**	−164.2	−45	**−49**	−56.2	−11.67	**11**	51.8
−77.78	**−108**	−162.4	−44.44	**−48**	−54.4	−11.11	**12**	53.6
−77.22	**−107**	−160.6	−43.89	**−47**	−52.6	−10.56	**13**	55.4
−76.67	**−106**	−158.8	−43.33	**−46**	−50.8	−10	**14**	57.2
−76.11	**−105**	−157	−42.78	**−45**	−49	−9.44	**15**	59
−75.56	**−104**	−155.2	−42.22	**−44**	−47.2	−8.89	**16**	60.8
−75	**−103**	−153.4	−41.67	**−43**	−45.4	−8.33	**17**	62.6
−74.44	**−102**	−151.6	−41.11	**−42**	−43.6	−7.78	**18**	64.4
−73.89	**−101**	−149.8	−40.56	**−41**	−41.8	−7.22	**19**	66.2
−73.33	**−100**	−148	−40	**−40**	−40	−6.67	**20**	68
−72.78	**−99**	−146.2	−39.44	**−39**	−38.2	−6.11	**21**	69.8
−72.22	**−98**	−144.4	−38.89	**−38**	−36.4	−5.56	**22**	71.6
−71.67	**−97**	−142.6	−38.33	**−37**	−34.6	−5	**23**	73.4
−71.11	**−96**	−140.8	−37.78	**−36**	−32.8	−4.44	**24**	75.2
−70.56	**−95**	−139	−37.22	**−35**	−31	−3.89	**25**	77
−70	**−94**	−137.2	−36.67	**−34**	−29.2	−3.33	**26**	78.8
−69.44	**−93**	−135.4	−36.11	**−33**	−27.4	−2.78	**27**	80.6
−68.89	**−92**	−133.6	−35.56	**−32**	−25.6	−2.22	**28**	82.4
−68.33	**−91**	−131.8	−35	**−31**	−23.8	−1.67	**29**	84.2
−67.78	**−90**	−130	−34.44	**−30**	−22	−1.11	**30**	86
−67.22	**−89**	−128.2	−33.89	**−29**	−20.2	−0.56	**31**	87.8
−66.67	**−88**	−126.4	−33.33	**−28**	−18.4	0	**32**	89.6
−66.11	**−87**	−124.6	−32.78	**−27**	−16.6	.56	**33**	91.4
−65.56	**−86**	−122.8	−32.22	**−26**	−14.8	1.11	**34**	93.2
−65	**−85**	−121	−31.67	**−25**	−13	1.67	**35**	95
−64.44	**−84**	−119.2	−31.11	**−24**	−11.2	2.22	**36**	96.8
−63.89	**−83**	−117.4	−30.56	**−23**	−9.4	2.78	**37**	98.6
−63.33	**−82**	−115.6	−30	**−22**	−7.6	3.33	**38**	100.4
−62.78	**−81**	−113.8	−29.44	**−21**	−5.8	3.89	**39**	102.2
−62.22	**−80**	−112	−28.89	**−20**	−4	4.44	**40**	104
−61.67	**−79**	−110.2	−28.33	**−19**	−2.2	5	**41**	105.8
−61.11	**−78**	−108.4	−27.78	**−18**	−0.4	5.56	**42**	107.6
−60.56	**−77**	−106.6	−27.22	**−17**	1.4	6.11	**43**	109.4
−60	**−76**	−104.8	−26.67	**−16**	3.2	6.67	**44**	111.2
−59.44	**−75**	−103	−26.11	**−15**	5	7.22	**45**	113
−58.89	**−74**	−101.2	−25.56	**−14**	6.8	7.78	**46**	114.8
−58.33	**−73**	−99.4	−25	**−13**	8.6	8.33	**47**	116.6
−57.78	**−72**	−97.6	−24.44	**−12**	10.4	8.89	**48**	118.4
−57.22	**−71**	−95.8	−23.89	**−11**	12.2	9.44	**49**	120.2

To Convert			To Convert			To Convert		
To °C	←°F or °C→	To °F	To °C	←°F or °C→	To °F	To °C	←°F or °C→	To °F
10	**50**	122	43.33	**110**	230	76.67	**170**	338
10.56	**51**	123.8	43.89	**111**	231.8	77.22	**171**	339.8
11.11	**52**	125.6	44.44	**112**	233.6	77.78	**172**	341.6
11.67	**53**	127.4	45	**113**	235.4	78.33	**173**	343.4
12.22	**54**	129.2	45.56	**114**	237.2	78.89	**174**	345.2
12.78	**55**	131	46.11	**115**	239	79.44	**175**	347
13.33	**56**	132.8	46.67	**116**	240.8	80	**176**	348.8
13.89	**57**	134.6	47.22	**117**	242.6	80.56	**177**	350.6
14.44	**58**	136.4	47.78	**118**	244.4	81.11	**178**	352.4
15	**59**	138.2	48.33	**119**	246.2	81.67	**179**	354.2
15.56	**60**	140	48.89	**120**	248	82.22	**180**	356
16.11	**61**	141.8	49.44	**121**	249.8	82.78	**181**	357.8
16.67	**62**	143.6	50	**122**	251.6	83.33	**182**	359.6
17.22	**63**	145.4	50.56	**123**	253.4	83.89	**183**	361.4
17.78	**64**	147.2	51.11	**124**	255.2	84.44	**184**	363.2
18.33	**65**	149	51.67	**125**	257	85	**185**	365
18.89	**66**	150.8	52.22	**126**	258.8	85.56	**186**	366.8
19.44	**67**	152.6	52.78	**127**	260.6	86.11	**187**	368.6
20	**68**	154.4	53.33	**128**	262.4	86.67	**188**	370.4
20.56	**69**	156.2	53.89	**129**	264.2	87.22	**189**	372.2
21.11	**70**	158	54.44	**130**	266	87.78	**190**	374
21.67	**71**	159.8	55	**131**	267.8	88.33	**191**	375.8
22.22	**72**	161.6	55.56	**132**	269.6	88.89	**192**	377.6
22.78	**73**	163.4	56.11	**133**	271.4	89.44	**193**	379.4
23.33	**74**	165.2	56.67	**134**	273.2	90	**194**	381.2
23.89	**75**	167	57.22	**135**	275	90.56	**195**	383
24.44	**76**	168.8	57.78	**136**	276.8	91.11	**196**	384.8
25	**77**	170.6	58.33	**137**	278.6	91.67	**197**	386.6
25.56	**78**	172.4	58.89	**138**	280.4	92.22	**198**	388.4
26.11	**79**	174.2	59.44	**139**	282.2	92.78	**199**	390.2
26.67	**80**	176	60	**140**	284	93.33	**200**	392
27.22	**81**	177.8	60.56	**141**	285.8	93.89	**201**	393.8
27.78	**82**	179.6	61.11	**142**	287.6	94.44	**202**	395.6
28.33	**83**	181.4	61.67	**143**	289.4	95	**203**	397.4
28.89	**84**	183.2	62.22	**144**	291.2	95.56	**204**	399.2
29.44	**85**	185	62.78	**145**	293	96.11	**205**	401
30	**86**	186.8	63.33	**146**	294.8	96.67	**206**	402.8
30.56	**87**	188.6	63.89	**147**	296.6	97.22	**207**	404.6
31.11	**88**	190.4	64.44	**148**	298.4	97.78	**208**	406.4
31.67	**89**	192.2	65	**149**	300.2	98.33	**209**	408.2
32.22	**90**	194	65.56	**150**	302	98.89	**210**	410
32.78	**91**	195.8	66.11	**151**	303.8	99.44	**211**	411.8
33.33	**92**	197.6	66.67	**152**	305.6	100	**212**	413.6
33.89	**93**	199.4	67.22	**153**	307.4	100.56	**213**	415.4
34.44	**94**	201.2	67.78	**154**	309.2	101.11	**214**	417.2
35	**95**	203	68.33	**155**	311	101.67	**215**	419
35.56	**96**	204.8	68.89	**156**	312.8	102.22	**216**	420.8
36.11	**97**	206.6	69.44	**157**	314.6	102.78	**217**	422.6
36.67	**98**	208.4	70	**158**	316.4	103.33	**218**	424.4
37.22	**99**	210.2	70.56	**159**	318.2	103.89	**219**	426.2
37.78	**100**	212	71.11	**160**	320	104.44	**220**	428
38.33	**101**	213.8	71.67	**161**	321.8	105	**221**	429.8
38.89	**102**	215.6	72.22	**162**	323.6	105.56	**222**	431.6
39.44	**103**	217.4	72.78	**163**	325.4	106.11	**223**	433.4
40	**104**	219.2	73.33	**164**	327.2	106.67	**224**	435.2
40.56	**105**	221	73.89	**165**	329	107.22	**225**	437
41.11	**106**	222.8	74.44	**166**	330.8	107.78	**226**	438.8
41.67	**107**	224.6	75	**167**	332.6	108.33	**227**	440.6
42.22	**108**	226.4	75.56	**168**	334.4	108.89	**228**	442.4
42.78	**109**	228.2	76.11	**169**	336.2	109.44	**229**	444.2

To Convert			To Convert			To Convert		
To °C	←°F or °C→	To °F	To °C	←°F or °C→	To °F	To °C	←°F or °C→	To °F
110	**230**	446	143.33	**290**	554	176.67	**350**	662
110.56	**231**	447.8	143.89	**291**	555.8	177.22	**351**	663.8
111.11	**232**	449.6	144.44	**292**	557.6	177.78	**352**	665.6
111.67	**233**	451.4	145	**293**	559.4	178.33	**353**	667.4
112.22	**234**	453.2	145.56	**294**	561.2	178.89	**354**	669.2
112.78	**235**	455	146.11	**295**	563	179.44	**355**	671
113.33	**236**	456.8	146.67	**296**	564.8	180	**356**	672.8
113.89	**237**	458.6	147.22	**297**	566.6	180.56	**357**	674.6
114.44	**238**	460.4	147.78	**298**	568.4	181.11	**358**	676.4
115	**239**	462.2	148.33	**299**	570.2	181.67	**359**	678.2
115.56	**240**	464	148.89	**300**	572	182.22	**360**	680
116.11	**241**	465.8	149.44	**301**	573.8	182.78	**361**	681.8
116.67	**242**	467.6	150	**302**	575.6	183.33	**362**	683.6
117.22	**243**	469.4	150.56	**303**	577.4	183.89	**363**	685.4
117.78	**244**	471.2	151.11	**304**	579.2	184.44	**364**	687.2
118.33	**245**	473	151.67	**305**	581	185	**365**	689
118.89	**246**	474.8	152.22	**306**	582.8	185.56	**366**	690.8
119.44	**247**	476.6	152.78	**307**	584.6	186.11	**367**	692.6
120	**248**	478.4	153.33	**308**	586.4	186.67	**368**	694.4
120.56	**249**	480.2	153.89	**309**	588.2	187.22	**369**	696.2
121.11	**250**	482	154.44	**310**	590	187.78	**370**	698
121.67	**251**	483.8	155	**311**	591.8	188.33	**371**	699.8
122.22	**252**	485.6	155.56	**312**	593.6	188.89	**372**	701.6
122.78	**253**	487.4	156.11	**313**	595.4	189.44	**373**	703.4
123.33	**254**	489.2	156.67	**314**	597.2	190	**374**	705.2
123.89	**255**	491	157.22	**315**	599	190.56	**375**	707
124.44	**256**	492.8	157.78	**316**	600.8	191.11	**376**	708.8
125	**257**	494.6	158.33	**317**	602.6	191.67	**377**	710.6
125.56	**258**	496.4	158.89	**318**	604.4	192.22	**378**	712.4
126.11	**259**	498.2	159.44	**319**	606.2	192.78	**379**	714.2
126.67	**260**	500	160	**320**	608	193.33	**380**	716
127.22	**261**	501.8	160.56	**321**	609.8	193.89	**381**	717.8
127.78	**262**	503.6	161.11	**322**	611.6	194.44	**382**	719.6
128.33	**263**	505.4	161.67	**323**	613.4	195	**383**	721.4
128.89	**264**	507.2	162.22	**324**	615.2	195.56	**384**	723.2
129.44	**265**	509	162.78	**325**	617	196.11	**385**	725
130	**266**	510.8	163.33	**326**	618.8	196.67	**386**	726.8
130.56	**267**	512.6	163.89	**327**	620.6	197.22	**387**	728.6
131.11	**268**	514.4	164.44	**328**	622.4	197.78	**388**	730.4
131.67	**269**	516.2	165	**329**	624.2	198.33	**389**	732.2
132.22	**270**	518	165.56	**330**	626	198.89	**390**	734
132.78	**271**	519.8	166.11	**331**	627.8	199.44	**391**	735.8
133.33	**272**	521.6	166.67	**332**	629.6	200	**392**	737.6
133.89	**273**	523.4	167.22	**333**	631.4	200.56	**393**	739.4
134.44	**274**	525.2	167.78	**334**	633.2	201.11	**394**	741.2
135	**275**	527	168.33	**335**	635	201.67	**395**	743
135.56	**276**	528.8	168.89	**336**	636.8	202.22	**396**	744.8
136.11	**277**	530.6	169.44	**337**	638.6	202.78	**397**	746.6
136.67	**278**	532.4	170	**338**	640.4	203.33	**398**	748.4
137.22	**279**	534.2	170.56	**339**	642.2	203.89	**399**	750.2
137.78	**280**	536	171.11	**340**	644	204.44	**400**	752
138.33	**281**	537.8	171.67	**341**	645.8	205	**401**	753.8
138.89	**282**	539.6	172.22	**342**	647.6	205.56	**402**	755.6
139.44	**283**	541.4	172.78	**343**	649.4	206.11	**403**	757.4
140	**284**	543.2	173.33	**344**	651.2	206.67	**404**	759.2
140.56	**285**	545	173.89	**345**	653	207.22	**405**	761
141.11	**286**	546.8	174.44	**346**	654.8	207.78	**406**	762.8
141.67	**287**	548.6	175	**347**	656.6	208.33	**407**	764.6
142.22	**288**	550.4	175.56	**348**	658.4	208.89	**408**	766.4
142.78	**289**	552.2	176.11	**349**	660.2	209.44	**409**	768.2

To Convert			To Convert			To Convert		
To °C	←°F or °C→	To °F	To °C	←°F or °C→	To °F	To °C	←°F or °C→	To °F
210	**410**	770	243.33	**470**	878	276.67	**530**	986
210.56	**411**	771.8	243.89	**471**	879.8	277.22	**531**	987.8
211.11	**412**	773.6	244.44	**472**	881.6	277.78	**532**	989.6
211.67	**413**	775.4	245	**473**	883.4	278.33	**533**	991.4
212.22	**414**	777.2	245.56	**474**	885.2	278.89	**534**	993.2
212.78	**415**	779	246.11	**475**	887	279.44	**535**	995
213.33	**416**	780.8	246.67	**476**	888.8	280	**536**	996.8
213.89	**417**	782.6	247.22	**477**	890.6	280.56	**537**	998.6
214.44	**418**	784.4	247.78	**478**	892.4	281.11	**538**	1000.4
215	**419**	786.2	248.33	**479**	894.2	281.67	**539**	1002.2
215.56	**420**	788	248.89	**480**	896	282.22	**540**	1004
216.11	**421**	789.8	249.44	**481**	897.8	282.78	**541**	1005.8
216.67	**422**	791.6	250	**482**	899.6	283.33	**542**	1007.6
217.22	**423**	793.4	250.56	**483**	901.4	283.89	**543**	1009.4
217.78	**424**	795.2	251.11	**484**	903.2	284.44	**544**	1011.2
218.33	**425**	797	251.67	**485**	905	285	**545**	1013
218.89	**426**	798.8	252.22	**486**	906.8	285.56	**546**	1014.8
219.44	**427**	800.6	252.78	**487**	908.6	286.11	**547**	1016.6
220	**428**	802.4	253.33	**488**	910.4	286.67	**548**	1018.4
220.56	**429**	804.2	253.89	**489**	912.2	287.22	**549**	1020.2
221.11	**430**	806	254.44	**490**	914	287.78	**550**	1022
221.67	**431**	807.8	255	**491**	915.8	288.33	**551**	1023.8
222.22	**432**	809.6	255.56	**492**	917.6	288.89	**552**	1025.6
222.78	**433**	811.4	256.11	**493**	919.4	289.44	**553**	1027.4
223.33	**434**	813.2	256.67	**494**	921.2	290	**554**	1029.2
223.89	**435**	815	257.22	**495**	923	290.56	**555**	1031
224.44	**436**	816.8	257.78	**496**	924.8	291.11	**556**	1032.8
225	**437**	818.6	258.33	**497**	926.6	291.67	**557**	1034.6
225.56	**438**	820.4	258.89	**498**	928.4	292.22	**558**	1036.4
226.11	**439**	822.2	259.44	**499**	930.2	292.78	**559**	1038.2
226.67	**440**	824	260	**500**	932	293.33	**560**	1040
227.22	**441**	825.8	260.56	**501**	933.8	293.89	**561**	1041.8
227.78	**442**	827.6	261.11	**502**	935.6	294.44	**562**	1043.6
228.33	**443**	829.4	261.67	**503**	937.4	295	**563**	1045.4
228.89	**444**	831.2	262.22	**504**	939.2	295.56	**564**	1047.2
229.44	**445**	833	262.78	**505**	941	296.11	**565**	1049
230	**446**	834.8	263.33	**506**	942.8	296.67	**566**	1050.8
230.56	**447**	836.6	263.89	**507**	944.6	297.22	**567**	1052.6
231.11	**448**	838.4	264.44	**508**	946.4	297.78	**568**	1054.4
231.67	**449**	840.2	265	**509**	948.2	298.33	**569**	1056.2
232.22	**450**	842	265.56	**510**	950	298.89	**570**	1058
232.78	**451**	843.8	266.11	**511**	951.8	299.44	**571**	1059.8
233.33	**452**	845.6	266.67	**512**	953.6	300	**572**	1061.6
233.89	**453**	847.4	267.22	**513**	955.4	300.56	**573**	1063.4
234.44	**454**	849.2	267.78	**514**	957.2	301.11	**574**	1065.2
235	**455**	851	268.33	**515**	959	301.67	**575**	1067
235.56	**456**	852.8	268.89	**516**	960.8	302.22	**576**	1068.8
236.11	**457**	854.6	269.44	**517**	962.6	302.78	**577**	1070.6
236.67	**458**	856.4	270	**518**	964.4	303.33	**578**	1072.4
237.22	**459**	858.2	270.56	**519**	966.2	303.89	**579**	1074.2
237.78	**460**	860	271.11	**520**	968	304.44	**580**	1076
238.33	**461**	861.8	271.67	**521**	969.8	305	**581**	1077.8
238.89	**462**	863.6	272.22	**522**	971.6	305.56	**582**	1079.6
239.44	**463**	865.4	272.78	**523**	973.4	306.11	**583**	1081.4
240	**464**	867.2	273.33	**524**	975.2	306.67	**584**	1083.2
240.56	**465**	869	273.89	**525**	977	307.22	**585**	1085
241.11	**466**	870.8	274.44	**526**	978.8	307.78	**586**	1086.8
241.67	**467**	872.6	275	**527**	980.6	308.33	**587**	1088.6
242.22	**468**	874.4	275.56	**528**	982.4	308.89	**588**	1090.4
242.78	**469**	876.2	276.11	**529**	984.2	309.44	**589**	1092.2

To Convert			To Convert			To Convert		
To °C	←°F or °C→	To °F	To °C	←°F or °C→	To °F	To °C	←°F or °C→	To °F
310	**590**	1094	343.33	**650**	1202	376.67	**710**	1310
310.56	**591**	1095.8	343.89	**651**	1203.8	377.22	**711**	1311.8
311.11	**592**	1097.6	344.44	**652**	1205.6	377.78	**712**	1313.6
311.67	**593**	1099.4	345	**653**	1207.4	378.33	**713**	1315.4
312.22	**594**	1101.2	345.56	**654**	1209.2	378.89	**714**	1317.2
312.78	**595**	1103	346.11	**655**	1211	379.44	**715**	1319
313.33	**596**	1104.8	346.67	**656**	1212.8	380	**716**	1320.8
313.89	**597**	1106.6	347.22	**657**	1214.6	380.56	**717**	1322.6
314.44	**598**	1108.4	347.78	**658**	1216.4	381.11	**718**	1324.4
315	**599**	1110.2	348.33	**659**	1218.2	381.67	**719**	1326.2
315.56	**600**	1112	348.89	**660**	1220	382.22	**720**	1328
316.11	**601**	1113.8	349.44	**661**	1221.8	382.78	**721**	1329.8
316.67	**602**	1115.6	350	**662**	1223.6	383.33	**722**	1331.6
317.22	**603**	1117.4	350.56	**663**	1225.4	383.89	**723**	1333.4
317.78	**604**	1119.2	351.11	**664**	1227.2	384.44	**724**	1335.2
318.33	**605**	1121	351.67	**665**	1229	385	**725**	1337
318.89	**606**	1122.8	352.22	**666**	1230.8	385.56	**726**	1338.8
319.44	**607**	1124.6	352.78	**667**	1232.6	386.11	**727**	1340.6
320	**608**	1126.4	353.33	**668**	1234.4	386.67	**728**	1342.4
320.56	**609**	1128.2	353.89	**669**	1236.2	387.22	**729**	1344.2
321.11	**610**	1130	354.44	**670**	1238	387.78	**730**	1346
321.67	**611**	1131.8	355	**671**	1239.8	388.33	**731**	1347.8
322.22	**612**	1133.6	355.56	**672**	1241.6	388.89	**732**	1349.6
322.78	**613**	1135.4	356.11	**673**	1243.4	389.44	**733**	1351.4
323.33	**614**	1137.2	356.67	**674**	1245.2	390	**734**	1353.2
323.89	**615**	1139	357.22	**675**	1247	390.56	**735**	1355
324.44	**616**	1140.8	357.78	**676**	1248.8	391.11	**736**	1356.8
325	**617**	1142.6	358.33	**677**	1250.6	391.67	**737**	1358.6
325.56	**618**	1144.4	358.89	**678**	1252.4	392.22	**738**	1360.4
326.11	**619**	1146.2	359.44	**679**	1254.2	392.78	**739**	1362.2
326.67	**620**	1148	360	**680**	1256	393.33	**740**	1364
327.22	**621**	1149.8	360.56	**681**	1257.8	393.89	**741**	1365.8
327.78	**622**	1151.6	361.11	**682**	1259.6	394.44	**742**	1367.6
328.33	**623**	1153.4	361.67	**683**	1261.4	395	**743**	1369.4
328.89	**624**	1155.2	362.22	**684**	1263.2	395.56	**744**	1371.2
329.44	**625**	1157	362.78	**685**	1265	396.11	**745**	1373
330	**626**	1158.8	363.33	**686**	1266.8	396.67	**746**	1374.8
330.56	**627**	1160.6	363.89	**687**	1268.6	397.22	**747**	1376.6
331.11	**628**	1162.4	364.44	**688**	1270.4	397.78	**748**	1378.4
331.67	**629**	1164.2	365	**689**	1272.2	398.33	**749**	1380.2
332.22	**630**	1166	365.56	**690**	1274	398.89	**750**	1382
332.78	**631**	1167.8	366.11	**691**	1275.8	399.44	**751**	1383.8
333.33	**632**	1169.6	366.67	**692**	1277.6	400	**752**	1385.6
333.89	**633**	1171.4	367.22	**693**	1279.4	400.56	**753**	1387.4
334.44	**634**	1173.2	367.78	**694**	1281.2	401.11	**754**	1389.2
335	**635**	1175	368.33	**695**	1283	401.67	**755**	1391
335.56	**636**	1176.8	368.89	**696**	1284.8	402.22	**756**	1392.8
336.11	**637**	1178.6	369.44	**697**	1286.6	402.78	**757**	1394.6
336.67	**638**	1180.4	370	**698**	1288.4	403.33	**758**	1396.4
337.22	**639**	1182.2	370.56	**699**	1290.2	403.89	**759**	1398.2
337.78	**640**	1184	371.11	**700**	1292	404.44	**760**	1400
338.33	**641**	1185.8	371.67	**701**	1293.8	405	**761**	1401.8
338.89	**642**	1187.6	372.22	**702**	1295.6	405.56	**762**	1403.6
339.44	**643**	1189.4	372.78	**703**	1297.4	406.11	**763**	1405.4
340	**644**	1191.2	373.33	**704**	1299.2	406.67	**764**	1407.2
340.56	**645**	1193	373.89	**705**	1301	407.22	**765**	1409
341.11	**646**	1194.8	374.44	**706**	1302.8	407.78	**766**	1410.8
341.67	**647**	1196.6	375	**707**	1304.6	408.33	**767**	1412.6
342.22	**648**	1198.4	375.56	**708**	1306.4	408.89	**768**	1414.4
342.78	**649**	1200.2	376.11	**709**	1308.2	409.44	**769**	1416.2

To Convert			To Convert			To Convert		
To °C	←°F or °C→	To °F	To °C	←°F or °C→	To °F	To °C	←°F or °C→	To °F
410	**770**	1418	443.33	**830**	1526	476.67	**890**	1634
410.56	**771**	1419.8	443.89	**831**	1527.8	477.22	**891**	1635.8
411.11	**772**	1421.6	444.44	**832**	1529.6	477.78	**892**	1637.6
411.67	**773**	1423.4	445	**833**	1531.4	478.33	**893**	1639.4
412.22	**774**	1425.2	445.56	**834**	1533.2	478.89	**894**	1641.2
412.78	**775**	1427	446.11	**835**	1535	479.44	**895**	1643
413.33	**776**	1428.8	446.67	**836**	1536.8	480	**896**	1644.8
413.89	**777**	1430.6	447.22	**837**	1538.6	480.56	**897**	1646.6
414.44	**778**	1432.4	447.78	**838**	1540.4	481.11	**898**	1648.4
415	**779**	1434.2	448.33	**839**	1542.2	481.67	**899**	1650.2
415.56	**780**	1436	448.89	**840**	1544	482.22	**900**	1652
416.11	**781**	1437.8	449.44	**841**	1545.8	482.78	**901**	1653.8
416.67	**782**	1439.6	450	**842**	1547.6	483.33	**902**	1655.6
417.22	**783**	1441.4	450.56	**843**	1549.4	483.89	**903**	1657.4
417.78	**784**	1443.2	451.11	**844**	1551.2	484.44	**904**	1659.2
418.33	**785**	1445	451.67	**845**	1553	485	**905**	1661
418.89	**786**	1446.8	452.22	**846**	1554.8	485.56	**906**	1662.8
419.44	**787**	1448.6	452.78	**847**	1556.6	486.11	**907**	1664.6
420	**788**	1450.4	453.33	**848**	1558.4	486.67	**908**	1666.4
420.56	**789**	1452.2	453.89	**849**	1560.2	487.22	**909**	1668.2
421.11	**790**	1454	454.44	**850**	1562	487.78	**910**	1670
421.67	**791**	1455.8	455	**851**	1563.8	488.33	**911**	1671.8
422.22	**792**	1457.6	455.56	**852**	1565.6	488.89	**912**	1673.6
422.78	**793**	1459.4	456.11	**853**	1567.4	489.44	**913**	1675.4
423.33	**794**	1461.2	456.67	**854**	1569.2	490	**914**	1677.2
423.89	**795**	1463	457.22	**855**	1571	490.56	**915**	1679
424.44	**796**	1464.8	457.78	**856**	1572.8	491.11	**916**	1680.8
425	**797**	1466.6	458.33	**857**	1574.6	491.67	**917**	1682.6
425.56	**798**	1468.4	458.89	**858**	1576.4	492.22	**918**	1684.4
426.11	**799**	1470.2	459.44	**859**	1578.2	492.78	**919**	1686.2
426.67	**800**	1472	460	**860**	1580	493.33	**920**	1688
427.22	**801**	1473.8	460.56	**861**	1581.8	493.89	**921**	1689.8
427.78	**802**	1475.6	461.11	**862**	1583.6	494.44	**922**	1691.6
428.33	**803**	1477.4	461.67	**863**	1585.4	495	**923**	1693.4
428.89	**804**	1479.2	462.22	**864**	1587.2	495.56	**924**	1695.2
429.44	**805**	1481	462.78	**865**	1589	496.11	**925**	1697
430	**806**	1482.8	463.33	**866**	1590.8	496.67	**926**	1698.8
430.56	**807**	1484.6	463.89	**867**	1592.6	497.22	**927**	1700.6
431.11	**808**	1486.4	464.44	**868**	1594.4	497.78	**928**	1702.4
431.67	**809**	1488.2	465	**869**	1596.2	498.33	**929**	1704.2
432.22	**810**	1490	465.56	**870**	1598	498.89	**930**	1706
432.78	**811**	1491.8	466.11	**871**	1599.8	499.44	**931**	1707.8
433.33	**812**	1493.6	466.67	**872**	1601.6	500	**932**	1709.6
433.89	**813**	1495.4	467.22	**873**	1603.4	500.56	**933**	1711.4
434.44	**814**	1497.2	467.78	**874**	1605.2	501.11	**934**	1713.2
435	**815**	1499	468.33	**875**	1607	501.67	**935**	1715
435.56	**816**	1500.8	468.89	**876**	1608.8	502.22	**936**	1716.8
436.11	**817**	1502.6	469.44	**877**	1610.6	502.78	**937**	1718.6
436.67	**818**	1504.4	470	**878**	1612.4	503.33	**938**	1720.4
437.22	**819**	1506.2	470.56	**879**	1614.2	503.89	**939**	1722.2
437.78	**820**	1508	471.11	**880**	1616	504.44	**940**	1724
438.33	**821**	1509.8	471.67	**881**	1617.8	505	**941**	1725.8
438.89	**822**	1511.6	472.22	**882**	1619.6	505.56	**942**	1727.6
439.44	**823**	1513.4	472.78	**883**	1621.4	506.11	**943**	1729.4
440	**824**	1515.2	473.33	**884**	1623.2	506.67	**944**	1731.2
440.56	**825**	1517	473.89	**885**	1625	507.22	**945**	1733
441.11	**826**	1518.8	474.44	**886**	1626.8	507.78	**946**	1734.8
441.67	**827**	1520.6	475	**887**	1628.6	508.33	**947**	1736.6
442.22	**828**	1522.4	475.56	**888**	1630.4	508.89	**948**	1738.4
442.78	**829**	1524.2	476.11	**889**	1632.2	509.44	**949**	1740.2

To Convert			To Convert			To Convert		
To °C	←°F or °C→	To °F	To °C	←°F or °C→	To °F	To °C	←°F or °C→	To °F
510	**950**	1742	543.33	**1010**	1850	576.67	**1070**	1958
510.56	**951**	1743.8	543.89	**1011**	1851.8	577.22	**1071**	1959.8
511.11	**952**	1745.6	544.44	**1012**	1853.6	577.78	**1072**	1961.6
511.67	**953**	1747.4	545	**1013**	1855.4	578.33	**1073**	1963.4
512.22	**954**	1749.2	545.56	**1014**	1857.2	578.89	**1074**	1965.2
512.78	**955**	1751	546.11	**1015**	1859	579.44	**1075**	1967
513.33	**956**	1752.8	546.67	**1016**	1860.8	580	**1076**	1968.8
513.89	**957**	1754.6	547.22	**1017**	1862.6	580.56	**1077**	1970.6
514.44	**958**	1756.4	547.78	**1018**	1864.4	581.11	**1078**	1972.4
515	**959**	1758.2	548.33	**1019**	1866.2	581.67	**1079**	1974.2
515.56	**960**	1760	548.89	**1020**	1868	582.22	**1080**	1976
516.11	**961**	1761.8	549.44	**1021**	1869.8	582.78	**1081**	1977.8
516.67	**962**	1763.6	550	**1022**	1871.6	583.33	**1082**	1979.6
517.22	**963**	1765.4	550.56	**1023**	1873.4	583.89	**1083**	1981.4
517.78	**964**	1767.2	551.11	**1024**	1875.2	584.44	**1084**	1983.2
518.33	**965**	1769	551.67	**1025**	1877	585	**1085**	1985
518.89	**966**	1770.8	552.22	**1026**	1878.8	585.56	**1086**	1986.8
519.44	**967**	1772.6	552.78	**1027**	1880.6	586.11	**1087**	1988.6
520	**968**	1774.4	553.33	**1028**	1882.4	586.67	**1088**	1990.4
520.56	**969**	1776.2	553.89	**1029**	1884.2	587.22	**1089**	1992.2
521.11	**970**	1778	554.44	**1030**	1886	587.78	**1090**	1994
521.67	**971**	1779.8	555	**1031**	1887.8	588.33	**1091**	1995.8
522.22	**972**	1781.6	555.56	**1032**	1889.6	588.89	**1092**	1997.6
522.78	**973**	1783.4	556.11	**1033**	1891.4	589.44	**1093**	1999.4
523.33	**974**	1785.2	556.67	**1034**	1893.2	590	**1094**	2001.2
523.89	**975**	1787	557.22	**1035**	1895	590.56	**1095**	2003
524.44	**976**	1788.8	557.78	**1036**	1896.8	591.11	**1096**	2004.8
525	**977**	1790.6	558.33	**1037**	1898.6	591.67	**1097**	2006.6
525.56	**978**	1792.4	558.89	**1038**	1900.4	592.22	**1098**	2008.4
526.11	**979**	1794.2	559.44	**1039**	1902.2	592.78	**1099**	2010.2
526.67	**980**	1796	560	**1040**	1904	593.33	**1100**	2012
527.22	**981**	1797.8	560.56	**1041**	1905.8	593.89	**1101**	2013.8
527.78	**982**	1799.6	561.11	**1042**	1907.6	594.44	**1102**	2015.6
528.33	**983**	1801.4	561.67	**1043**	1909.4	595	**1103**	2017.4
528.89	**984**	1803.2	562.22	**1044**	1911.2	595.56	**1104**	2019.2
529.44	**985**	1805	562.78	**1045**	1913	596.11	**1105**	2021
530	**986**	1806.8	563.33	**1046**	1914.8	596.67	**1106**	2022.8
530.56	**987**	1808.6	563.89	**1047**	1916.6	597.22	**1107**	2024.6
531.11	**988**	1810.4	564.44	**1048**	1918.4	597.78	**1108**	2026.4
531.67	**989**	1812.2	565	**1049**	1920.2	598.33	**1109**	2028.2
532.22	**990**	1814	565.56	**1050**	1922	598.89	**1110**	2030
532.78	**991**	1815.8	566.11	**1051**	1923.8	599.44	**1111**	2031.8
533.33	**992**	1817.6	566.67	**1052**	1925.6	600	**1112**	2033.6
533.89	**993**	1819.4	567.22	**1053**	1927.4	600.56	**1113**	2035.4
534.44	**994**	1821.2	567.78	**1054**	1929.2	601.11	**1114**	2037.2
535	**995**	1823	568.33	**1055**	1931	601.67	**1115**	2039
535.56	**996**	1824.8	568.89	**1056**	1932.8	602.22	**1116**	2040.8
536.11	**997**	1826.6	569.44	**1057**	1934.6	602.78	**1117**	2042.6
536.67	**998**	1828.4	570	**1058**	1936.4	603.33	**1118**	2044.4
537.22	**999**	1830.2	570.56	**1059**	1938.2	603.89	**1119**	2046.2
537.78	**1000**	1832	571.11	**1060**	1940	604.44	**1120**	2048
538.33	**1001**	1833.8	571.67	**1061**	1941.8	605	**1121**	2049.8
538.89	**1002**	1835.6	572.22	**1062**	1943.6	605.56	**1122**	2051.6
539.44	**1003**	1837.4	572.78	**1063**	1945.4	606.11	**1123**	2053.4
540	**1004**	1839.2	573.33	**1064**	1947.2	606.67	**1124**	2055.2
540.56	**1005**	1841	573.89	**1065**	1949	607.22	**1125**	2057
541.11	**1006**	1842.8	574.44	**1066**	1950.8	607.78	**1126**	2058.8
541.67	**1007**	1844.6	575	**1067**	1952.6	608.33	**1127**	2060.6
542.22	**1008**	1846.4	575.56	**1068**	1954.4	608.89	**1128**	2062.4
542.78	**1009**	1848.2	576.11	**1069**	1956.2	609.44	**1129**	2064.2

WEIGHTS AND MEASURES[a]

THE METRIC SYSTEM

Prefixes Used in the Metric System

The prefixes listed in the table below are used in conjunction with the standard metric units: *meter* for length, *gram* for weight or mass, and *liter* for capacity.

Prefix	Meaning	Numerical Equivalents
pico-	one trillionth	$\frac{1}{1{,}000{,}000{,}000{,}000} = 10^{-12}$
nano-	one billionth	$\frac{1}{1{,}000{,}000{,}000} = 10^{-9}$
micro-	one millionth	$\frac{1}{1{,}000{,}000} = 10^{-6}$
milli-	one thousandth	$\frac{1}{1{,}000} = 10^{-3}$
centi-	one hundredth	$\frac{1}{100} = 10^{-2}$
deci-	one tenth	$\frac{1}{10} = 10^{-1}$
unit	one	1 = 1
deka-	ten	10 = 10
hecto-	one hundred	100 = 10^2
kilo-	one thousand	1,000 = 10^3
mega-	one million	1,000,000 = 10^6
giga-	one billion	1,000,000,000 = 10^9
tera-	one trillion	1,000,000,000,000 = 10^{12}

[a] Data taken from: *Merck Index,* 8th ed., pp. 1267–1273, Merck and Co., Inc., Rahway, New Jersey, 1968. By permission of the copyright owners.

Units of Length

The unit of length is the *meter* (m), which is defined as the length of the International Prototype Meter bar.

Length	Abbreviation	Equivalent in Meters	U.S. Equivalent
1 kilometer	km	10^3	0.6214 mile
1 meter	m	1	39.37 inches, or 3.2808 feet, or 1.09361 yards
1 centimeter	cm	10^{-2}	0.3937 inch
1 millimeter	mm	10^{-3}	0.03937 inch
1 micron	μ or μm	10^{-6}	
1 millimicron	mμ or nm	10^{-9}	
1 angstrom	Å or A	10^{-10}	

Units of Area

Area	Abbreviation	U.S. Equivalent
1 square meter	m^2	1,550 square inches, or 10.764 square feet, or 1.196 square yards
1 square centimeter	cm^2	0.155 square inch

Units of Volume

Volume	Abbreviation	U.S. Equivalent
1 cubic meter	m^3	35.3145 cubic feet, or 1.3079 cubic yards
1 cubic decimeter	dm^3	61.0234 cubic inches
1 cubic centimeter	cm^3	0.06102 cubic inch

Units of Weight

The unit of weight is the *gram* (g), which is defined as $\frac{1}{1,000}$ of the mass of the International Prototype Kilogram.

Weight	Abbreviation	Equivalent in Grams	U.S. Equivalent
1 kilogram	kg	10^3	2.2046 avoirdupois pounds
1 gram	g	1	0.03527 avoirdupois ounce
1 decigram	dg	10^{-1}	
1 centigram	cg	10^{-2}	
1 milligram	mg	10^{-3}	
1 microgram, or	μg, or	10^{-6}	
1 gamma	γ		

Units of Capacity

The unit of capacity is the *liter* (l), which is defined as the volume of one kilogram of water at 4°C and standard atmospheric pressure. One liter is equal to 1000.028 cubic centimeters.

Capacity	Abbreviation	Equivalent in Liters	U.S. Equivalent
1 liter	l	1	33.8148 fluid ounces, or 1.0567 quarts, or 0.2642 gallon
1 deciliter	dl	10^{-1}	
1 centiliter	cl	10^{-2}	
1 milliliter	ml	10^{-3}	0.0338 fluid ounce
1 microliter	μl	10^{-6}	

THE U.S. SYSTEMS

Units of Length

Length	Abbreviation	U.S. Equivalent	Metric Equivalent
1 nautical mile		6,076.1 feet	
1 statute mile	mi	5,280 feet	1.6093 kilometers
1 yard	yd	3 feet	91.440 centimeters, or 0.9144 meter
1 foot	ft	12 inches	30.480 centimeters, or 0.3048 meter
1 inch	in.		2.540 centimeters

Units of Area

Area	Abbreviation	Metric Equivalent
1 square yard	sq yd	0.8361 square meter
1 square foot	sq ft	929.0341 square centimeters, or 0.0929 square meter
1 square inch	sq in.	6.4516 square centimeters

Units of Volume

Volume	Abbreviation	Metric Equivalent
1 cubic yard	cu yd	0.7646 cubic meter
1 cubic foot	cu ft	28.317 cubic decimeters (or liters), or 0.02832 cubic meter
1 cubic inch	cu in.	16.3872 cubic centimeters (or milliliters)

Units of Weight

Weight, Avoirdupois	Abbreviation	U.S. Equivalent	Metric Equivalent
1 pound	lb	16 ounces	453.592 grams
1 ounce	oz		28.350 grams

Units of Capacity

Capacity, Apothecaries' Fluid Measure	Abbreviation	U.S. Equivalent	Metric Equivalent
1 gallon	gal	4 quarts	3.785 liters
1 quart	qt	32 fluid ounces, or 2 pints	946.322 milliliters
1 pint	pt	16 fluid ounces	473.166 milliliters
1 fluid ounce	fl oz		29.573 milliliters

COMPARISON OF UNITED STATES, BRITISH IMPERIAL, AND METRIC SYSTEMS OF LIQUID MEASURE

Liquid Measure	Equivalent
1 U.S. gallon	0.8327 Imperial gallon, or 3.785 liters, or 128 fluid ounces
1 Imperial gallon	1.2009 U.S. gallons, or 4.5460 liters, or 160 fluid ounces
1 U.S. pint	0.8327 Imperial pint, or 473.166 milliliters
1 Imperial pint	1.2009 U.S. pints, or 568.25 milliliters

VOLUME-TO-WEIGHT RELATIONSHIPS FOR WATER

Volume	Weight
1 U.S. fluid ounce	
at 4°C	1.0432 avoirdupois ounces
at 25°C	1.0390 avoirdupois ounces
1 U.S. gallon	
at 20°C	8.33 avoirdupois pounds
1 Imperial gallon	
at 60°F	10 avoirdupois pounds

Weight	Volume
1 avoirdupois ounce	0.9586 U.S. fluid ounce at 4°C
1 avoirdupois pound	0.1200 U.S. gallon at 20°C

LOGARITHMS

To find the common logarithm of a number, proceed as described below.

For a number of four digits, take out the tabular mantissa on a line with the first three digits of the number and under its fourth digit.

For a number of less than four digits, supply zeros to make a four-digit number, then take the value of the mantissa from the table as before. For example: log 2 = log 2.000 = 0.30103.

For a number of more than four digits, interpolation must be used. There are several precautions that must be observed when interpolating:

1. Linear interpolation, as described below, may only be used to add one extra digit to the argument (i.e., in the present case, for a four-digit argument).

2. Even though the mantissas given in the table are accurate to five decimal places, interpolated values are accurate only to the same number of places as in the argument, i.e., four places.

3. Because of the rapidly changing values in this region of the table, interpolation is not accurate if the first two digits of the argument are 11 or 12. If interpolation is required in this section of a four-place table, the value should be read instead from a suitable five-place table without interpolation.

If the above precautions cannot be observed, then higher order interpolation should be used.

Where applicable, linear interpolation is carried out as follows.

Take the tabular value of the mantissa for the first four digits; find the difference between the mantissa and the next greater tabular mantissa, then multiply the difference so found by the remaining figures of the number as a decimal and add the product to the mantissa of the first four digits. For example, to find log 46.762:

$$\log 46.76 = 1.66987;$$

tabular difference between this mantissa and that for 4677 is .00010

$$\begin{aligned} \therefore \log 46.762 &= 1.66987 + .2 \times .00010 \\ &= 1.66987 + .00002 \\ &= 1.66989 \end{aligned}$$

In the four-place logarithm table, a column of proportional parts is given at the end of each line. The number in the column under the fourth digit of the argument is the amount that must be added to the last place of any mantissa in that line to interpolate for the fourth digit. For example, to find log 33.74

$$\begin{aligned} \log 33.7 &= 1.5276 \\ \text{proportional part for } 4 &= \quad\quad 5 \\ \therefore \log 33.74 &= 1.5281 \end{aligned}$$

Since these numbers are averages for the entire line, they may be off by 1 in the last place.

To find the number corresponding to a given logarithm,[b] use the following procedure.[c]

[a] This description and examples refer specifically to five-place logarithms. For four-place mantissas there will be minor differences from this description.

[b] This number is called the antilogarithm, and is denoted by $\log^{-1}$. Since the logarithm function is the inverse of the exponential function, $\log_a^{-1} x = a^x$, any procedure or table that calculates antilogarithms may also be used to calculate exponentials, and vice-versa. In particular, tables of e^x may be used to compute antilogarithms to the base e.

[c] The procedure given refers to the five-place logarithm table. As before, any significant deviation for other tables will be noted.

If the mantissa is found exactly in the table, join the figure at the top, directly above the given mantissa, to the three figures on the line at the left, then place the decimal point according to the characteristic of the logarithm. For example,

$$\log^{-1} 3{,}3996{,}7 = \text{antilog } 3.39967 = 2510.$$

If the mantissa is not found exactly in the table, it is necessary to interpolate. For example, to find antilog 3.40028, we find in the table.

antilog 3.40019 = 2513.	
antilog 3.40037 = 2514.	
tabular difference	18

The required difference is 9, so we must add $\frac{9}{18} = .5$.

$$\therefore \text{antilog } 3.40028 = 2513.5$$

The same precautions must be observed for interpolation in finding antilogarithms as in finding logarithms.

Tables of natural logarithms are used in the same way as tables of common logarithms, except that they contain both the characteristics and the mantissas of the logarithms.

Examples of the use of logarithms in computation follow. Almost all computation with logarithms is done with common logarithms, since the computation of the characteristic is simpler, and since only the significant digits of the argument need be given in the table, without regard for the decimal point location. These examples all use the table of five-place common logarithms.

1. 52600 x 0.00381 x 2.74 = 549.11

log 52600	=	4.72099
log 0.00381	=	$\bar{3}$.58092[d]
log 2.74	=	0.43775
adding	=	2.73966
antilog	=	549.11

The sum is the logarithm of the product; the mantissa of the logarithm is 73966. On looking up this mantissa in the logarithm tables, we see that it corresponds to the digits 54911. The characteristic is 2, hence there are three figures before the decimal point. The number corresponding to the logarithm, called the antilogarithm, is 549.11.

2. 0.00123 ÷ 52.7 = 0.000 023 34

log 0.00123	=	$\bar{3}$.08991	or	log 0.00123	=	7.08991 – 10
log 52.7	=	1.72181		log 52.7	=	1.72181
subtracting	=	$\bar{5}$.336810				5.36810 – 10
antilog	=	0.000 023 34				

The characteristic $\bar{5}$(5.–10) shows four zeros after the decimal point before the first significant figure.

[d] For numbers less than one, the characteristic is negative, whereas the mantissa is positive.

3. $\frac{273 \times 780}{292 \times 760} \times 15 \times 0.09 = 1.2954$

log 273	= 2.43616	log 292	= 2.46538
log 780	= 2.89209	log 760	= 2.88081
log 15	= 1.17609	log denominator	= 5.34619
log 0.09	= $\bar{2}$.95424		
log numerator	= 5.45858		

log numerator	= 5.45858
log denominator	= 5.34619
subtracting	= 0.11237
antilog	= 1.2954

Division may be accomplished by multiplying by the reciprocal of a number, and the above may thus be considerably simplified. The logarithm of the reciprocal of a number, called the cologarithm, is readily obtained from the table by subtracting the logarithm of the number from zero. This may be readily read off from the table of mantissas. Change the sign of the characteristic algebraically, adding to it − 1, then mentally subtract each figure of the mantissa from 9, proceeding from left to right, subtracting the last figure from 10. The example then is:

log 273	= 2.43616
log 780	= 2.89209
log 15	= 1.17609
log 0.09	= $\bar{2}$.95424
colog 292	= $\bar{3}$.53462
colog 760	= $\bar{3}$.11919
	0.11239

4. $(0.00098)^4 = 9.224 \times 10^{-13}$

log 0.00098	= $\bar{4}$.99123	or	log 0.00098	= 6.99123 − 10
	4			4
	3.96492 (a)			27.96492 − 40
$\bar{4} \times 4$	= $\bar{16}$. (b)			or 7.96492 − 20
log $(0.00098)^4$	= $\bar{13}$.96492 (c)			or $\bar{13}$.96492
antilog	= 9.224×10^{-13}		antilog	= 9.224×10^{-13}

In the above it will be noted that the mantissa is always positive; hence the multiplication of the mantissa is shown at (a), (b) shows the multiplication of the characteristic, and (c) is the algebraic sum.

5. $\sqrt[5]{492} = 3.4546$

log 492 = 2.69197

Dividing the logarithm by 5 gives as the logarithm of the root 0.53839, the antilogarithm of this is 3.4546, both characteristic and mantissa being positive. When the characteristic is negative and not evenly divisible by the root to be taken, a modification of the logarithm is necessary, as the following example shows.

6. $\sqrt[3]{0.000372} = 0.07192$

$$\begin{aligned} \log 3.72 \times 10^{-4} &= 4.57054 \quad \text{(a)} \\ &= 26.57054 - 30 \ \text{(b)} \end{aligned}$$

Dividing (b) by 3 gives 8.85685 – 10, which may be written $\bar{2}.85685$ and is the logarithm of the root sought; the antilogarithm of the latter is 0.07191.

7. $(0.000\ 372)^{1.2} = 0.000\ 076\ 672$

$$\begin{aligned} \log 0.000\ 372 &= \bar{4}.57054 \\ &\text{or } 6.57054 - 10 \\ &\qquad\qquad 1.2 \\ &\underline{7.884648 - 12} \\ \text{antilog} &= 0.000\ 076\ 672 \end{aligned}$$

8. $(0.000372)^{-1.32} = 33642$

$$\begin{aligned} \text{colog } 0.000372 &= 3.42946 \\ &\quad \underline{1.32} \\ &\quad 4.52689 \\ \text{antilog} &= 33642 \end{aligned}$$

FOUR-PLACE MANTISSAS FOR COMMON LOGARITHMS

N	0	1	2	3	4	5	6	7	8	9	Proportional Parts								
											1	2	3	4	5	6	7	8	9
10	0000	0043	0086	0128	0170	0212	0253	0294	0334	0374	*4	8	12	17	21	25	29	33	37
11	0414	0453	0492	0531	0569	0607	0645	0682	0719	0755	4	8	11	15	19	23	26	30	34
12	0792	0828	0864	0899	0934	0969	1004	1038	1072	1106	3	7	10	14	17	21	24	28	31
13	1139	1173	1206	1239	1271	1303	1335	1367	1399	1430	3	6	10	13	16	19	23	26	29
14	1461	1492	1523	1553	1584	1614	1644	1673	1703	1732	3	6	9	12	15	18	21	24	27
15	1761	1790	1818	1847	1875	1903	1931	1959	1987	2014	*3	6	8	11	14	17	20	22	25
16	2041	2068	2095	2122	2148	2175	2201	2227	2253	2279	3	5	8	11	13	16	18	21	24
17	2304	2330	2355	2380	2405	2430	2455	2480	2504	2529	2	5	7	10	12	15	17	20	22
18	2553	2577	2601	2625	2648	2672	2695	2718	2742	2765	2	5	7	9	12	14	16	19	21
19	2788	2810	2833	2856	2878	2900	2923	2945	2967	2989	2	4	7	9	11	13	16	18	20
20	3010	3032	3054	3075	3096	3118	3139	3160	3181	3201	2	4	6	8	11	13	15	17	19
21	3222	3243	3263	3284	3304	3324	3345	3365	3385	3404	2	4	6	8	10	12	14	16	18
22	3424	3444	3464	3483	3502	3522	3541	3560	3579	3598	2	4	6	8	10	12	14	15	17
23	3617	3636	3655	3674	3692	3711	3729	3747	3766	3784	2	4	6	7	9	11	13	15	17
24	3802	3820	3838	3856	3874	3892	3909	3927	3945	3962	2	4	5	7	9	11	12	14	16
25	3979	3997	4014	4031	4048	4065	4082	4099	4116	4133	2	3	5	7	9	10	12	14	15
26	4150	4166	4183	4200	4216	4232	4249	4265	4281	4298	2	3	5	7	8	10	11	13	15
27	4314	4330	4346	4362	4378	4393	4409	4425	4440	4456	2	3	5	6	8	9	11	13	14
28	4472	4487	4502	4518	4533	4548	4564	4579	4594	4609	2	3	5	6	8	9	11	12	14
29	4624	4639	4654	4669	4683	4698	4713	4728	4742	4757	1	3	4	6	7	9	10	12	13
30	4771	4786	4800	4814	4829	4843	4857	4871	4886	4900	1	3	4	6	7	9	10	11	13
31	4914	4928	4942	4955	4969	4983	4997	5011	5024	5038	1	3	4	6	7	8	10	11	12
32	5051	5065	5079	5092	5105	5119	5132	5145	5159	5172	1	3	4	5	7	8	9	11	12
33	5185	5198	5211	5224	5237	5250	5263	5276	5289	5302	1	3	4	5	6	8	9	10	12
34	5315	5328	5340	5353	5366	5378	5391	5403	5416	5428	1	3	4	5	6	8	9	10	11
35	5441	5453	5465	5478	5490	5502	5514	5527	5539	5551	1	2	4	5	6	7	9	10	11
36	5563	5575	5587	5599	5611	5623	5635	5647	5658	5670	1	2	4	5	6	7	8	10	11
37	5682	5694	5705	5717	5729	5740	5752	5763	5775	5786	1	2	3	5	6	7	8	9	10
38	5798	5809	5821	5832	5843	5855	5866	5877	5888	5899	1	2	3	5	6	7	8	9	10
39	5911	5922	5933	5944	5955	5966	5977	5988	5999	6010	1	2	3	4	5	7	8	9	10
40	6021	6031	6042	6053	6064	6075	6085	6096	6107	6117	1	2	3	4	5	6	8	9	10
41	6128	6138	6149	6160	6170	6180	6191	6201	6212	6222	1	2	3	4	5	6	7	8	9
42	6232	6243	6253	6263	6274	6284	6294	6304	6314	6325	1	2	3	4	5	6	7	8	9
43	6335	6345	6355	6365	6375	6385	6395	6405	6415	6425	1	2	3	4	5	6	7	8	9
44	6435	6444	6454	6464	6474	6484	6493	6503	6513	6522	1	2	3	4	5	6	7	8	9
45	6532	6542	6551	6561	6571	6580	6590	6599	6609	6618	1	2	3	4	5	6	7	8	9
46	6628	6637	6646	6656	6665	6675	6684	6693	6702	6712	1	2	3	4	5	6	7	7	8
47	6721	6730	6739	6749	6758	6767	6776	6785	6794	6803	1	2	3	4	5	5	6	7	8
48	6812	6821	6830	6839	6848	6857	6866	6875	6884	6893	1	2	3	4	4	5	6	7	8
49	6902	6911	6920	6928	6937	6946	6955	6964	6972	6981	1	2	3	4	4	5	6	7	8
N	**0**	**1**	**2**	**3**	**4**	**5**	**6**	**7**	**8**	**9**	**1**	**2**	**3**	**4**	**5**	**6**	**7**	**8**	**9**

* Interpolation in this section of the table is inaccurate.

N	0	1	2	3	4	5	6	7	8	9	Proportional Parts 1	2	3	4	5	6	7	8	9
50	6990	6998	7007	7016	7024	7033	7042	7050	7059	7067	1	2	3	3	4	5	6	7	8
51	7076	7084	7093	7101	7110	7118	7126	7135	7143	7152	1	2	3	3	4	5	6	7	8
52	7160	7168	7177	7185	7193	7202	7210	7218	7226	7235	1	2	2	3	4	5	6	7	7
53	7243	7251	7259	7267	7275	7284	7292	7300	7308	7316	1	2	2	3	4	5	6	6	7
54	7324	7332	7340	7348	7356	7364	7372	7380	7388	7396	1	2	2	3	4	5	6	6	7
55	7404	7412	7419	7427	7435	7443	7451	7459	7466	7474	1	2	2	3	4	5	5	6	7
56	7482	7490	7497	7505	7513	7520	7528	7536	7543	7551	1	2	2	3	4	5	5	6	7
57	7559	7566	7574	7582	7589	7597	7604	7612	7619	7627	1	2	2	3	4	5	5	6	7
58	7634	7642	7649	7657	7664	7672	7679	7686	7694	7701	1	1	2	3	4	4	5	6	7
59	7709	7716	7723	7731	7738	7745	7752	7760	7767	7774	1	1	2	3	4	4	5	6	7
60	7782	7789	7796	7803	7810	7818	7825	7832	7839	7846	1	1	2	3	4	4	5	6	6
61	7853	7860	7868	7875	7882	7889	7896	7903	7910	7917	1	1	2	3	4	4	5	6	6
62	7924	7931	7938	7945	7952	7959	7966	7973	7980	7987	1	1	2	3	3	4	5	6	6
63	7993	8000	8007	8014	8021	8028	8035	8041	8048	8055	1	1	2	3	3	4	5	5	6
64	8062	8069	8075	8082	8089	8096	8102	8109	8116	8122	1	1	2	3	3	4	5	5	6
65	8129	8136	8142	8149	8156	8162	8169	8176	8182	8189	1	1	2	3	3	4	5	5	6
66	8195	8202	8209	8215	8222	8228	8235	8241	8248	8254	1	1	2	3	3	4	5	5	6
67	8261	8267	8274	8280	8287	8293	8299	8306	8312	8319	1	1	2	3	3	4	5	5	6
68	8325	8331	8338	8344	8351	8357	8363	8370	8376	8382	1	1	2	3	3	4	4	5	6
69	8388	8395	8401	8407	8414	8420	8426	8432	8439	8445	1	1	2	2	3	4	4	5	6
70	8451	8457	8463	8470	8476	8482	8488	8494	8500	8506	1	1	2	2	3	4	4	5	6
71	8513	8519	8525	8531	8537	8543	8549	8555	8561	8567	1	1	2	2	3	4	4	5	5
72	8573	8579	8585	8591	8597	8603	8609	8615	8621	8627	1	1	2	2	3	4	4	5	5
73	8633	8639	8645	8651	8657	8663	8669	8675	8681	8686	1	1	2	2	3	4	4	5	5
74	8692	8698	8704	8710	8716	8722	8727	8733	8739	8745	1	1	2	2	3	4	4	5	5
75	8751	8756	8762	8768	8774	8779	8785	8791	8797	8802	1	1	2	2	3	3	4	5	5
76	8808	8814	8820	8825	8831	8837	8842	8848	8854	8859	1	1	2	2	3	3	4	5	5
77	8865	8871	8876	8882	8887	8893	8899	8904	8910	8915	1	1	2	2	3	3	4	4	5
78	8921	8927	8932	8938	8943	8949	8954	8960	8965	8971	1	1	2	2	3	3	4	4	5
79	8976	8982	8987	8993	8998	9004	9009	9015	9020	9025	1	1	2	2	3	3	4	4	5
80	9031	9036	9042	9047	9053	9058	9063	9069	9074	9079	1	1	2	2	3	3	4	4	5
81	9085	9090	9096	9101	9106	9112	9117	9122	9128	9133	1	1	2	2	3	3	4	4	5
82	9138	9143	9149	9154	9159	9165	9170	9175	9180	9186	1	1	2	2	3	3	4	4	5
83	9191	9196	9201	9206	9212	9217	9222	9227	9232	9238	1	1	2	2	3	3	4	4	5
84	9243	9248	9253	9258	9263	9269	9274	9279	9284	9289	1	1	2	2	3	3	4	4	5
85	9294	9299	9304	9309	9315	9320	9325	9330	9335	9340	1	1	2	2	3	3	4	4	5
86	9345	9350	9355	9360	9365	9370	9375	9380	9385	9390	1	1	2	2	3	3	4	4	5
87	9395	9400	9405	9410	9415	9420	9425	9430	9435	9440	0	1	1	2	2	3	3	4	4
88	9445	9450	9455	9460	9465	9469	9474	9479	9484	9489	0	1	1	2	2	3	3	4	4
89	9494	9499	9504	9509	9513	9518	9523	9528	9533	9538	0	1	1	2	2	3	3	4	4
90	9542	9547	9552	9557	9562	9566	9571	9576	9581	9586	0	1	1	2	2	3	3	4	4
91	9590	9595	9600	9605	9609	9614	9619	9624	9628	9633	0	1	1	2	2	3	3	4	4
92	9638	9643	9647	9652	9657	9661	9666	9671	9675	9680	0	1	1	2	2	3	3	4	4
93	9685	9689	9694	9699	9703	9708	9713	9717	9722	9727	0	1	1	2	2	3	3	4	4
94	9731	9736	9741	9745	9750	9754	9759	9763	9768	9773	0	1	1	2	2	3	3	4	4
95	9777	9782	9786	9791	9795	9800	9805	9809	9814	9818	0	1	1	2	2	3	3	4	4
96	9823	9827	9832	9836	9841	9845	9850	9854	9859	9863	0	1	1	2	2	3	3	4	4
97	9868	9872	9877	9881	9886	9890	9894	9899	9903	9908	0	1	1	2	2	3	3	4	4
98	9912	9917	9921	9926	9930	9934	9939	9943	9948	9952	0	1	1	2	2	3	3	4	4
99	9956	9961	9965	9969	9974	9978	9983	9987	9991	9996	0	1	1	2	2	3	3	3	4
N	0	1	2	3	4	5	6	7	8	9	1	2	3	4	5	6	7	8	9

INDEX

A

B

C

D

E

F

G

H

N

O

P

R

S

T

U

V

W

X, Y, Z